AF182887

Handbook of Toxic Fungal Metabolites

Handbook of Toxic Fungal Metabolites

RICHARD J. COLE

United States Department of Agriculture
National Peanut Research Laboratory
Dawson, Georgia

RICHARD H. COX

National Institute of Environmental Health Sciences
Research Triangle Park, North Carolina

1981

ACADEMIC PRESS

A Subsidiary of Harcourt Brace Jovanovich, Publishers

New York London Toronto Sydney San Francisco

COPYRIGHT © 1981, BY ACADEMIC PRESS, INC.
ALL RIGHTS RESERVED.
NO PART OF THIS PUBLICATION MAY BE REPRODUCED OR TRANSMITTED IN ANY FORM OR BY ANY MEANS, ELECTRONIC OR MECHANICAL, INCLUDING PHOTOCOPY, RECORDING, OR ANY INFORMATION STORAGE AND RETRIEVAL SYSTEM, WITHOUT PERMISSION IN WRITING FROM THE PUBLISHER.

ACADEMIC PRESS, INC.
111 Fifth Avenue, New York, New York 10003

United Kingdom Edition published by
ACADEMIC PRESS, INC. (LONDON) LTD.
24/28 Oval Road, London NW1 7DX

Library of Congress Cataloging in Publication Data

Cole, Richard J.
Handbook of toxic fungal metabolites.

Includes bibliographies and index.
1. Mycotoxins--Handbooks, manuals, etc.
2. Mycotoxins--Spectra--Handbooks, manuals, etc.
3. Microbial metabolites--Toxicology--Handbooks, manuals, etc. 4. Food contamination--Handbooks, manuals, etc. 5. Feeds--Contamination--Handbooks, manuals, etc.
I. Cox, Richard H., Date. II. Title.
QP632.M9C65 615.9'5292 81-4082
ISBN 0-12-179760-0 AACR2

PRINTED IN THE UNITED STATES OF AMERICA

81 82 83 84 9 8 7 6 5 4 3 2 1

Contents

Preface

The science of mycotoxicology had its advent with the discovery of aflatoxin in the early 1960s in England. Since that time, interest and research on the aflatoxins and mycotoxins in general have steadily intensified. The result has been an ever-increasing awareness of the potential danger of these secondary metabolites to human and animal health.

The evaluation of the role and extent of mycotoxin contamination in food and feed requires the isolation of the toxic principle(s) from a suspect commodity or from cultures of fungi isolated from the suspected material. This handbook has been compiled with the aid of numerous individuals for the purpose of facilitating the identification of known or related mycotoxins. The book provides investigators in the field with a comprehensive accumulation of chemical, physical, spectral, and biological data on toxic fungal metabolites and related chemicals that would otherwise be widely scattered throughout the literature. Since mycotoxins represent a wide diversity of chemical species, the book will be particularly useful to other scientists interested in some aspect of a particular chemical species or related species other than its toxic nature.

The presentation, where possible, of actual copies of UV, IR, ^{1}H NMR, ^{13}C NMR, and mass spectra greatly facilitates the spectral identification of known mycotoxins or related metabolites by both chemists and researchers not knowledgeable in the interpretation of spectral data.

The handbook has been divided into twenty-one sections. Members were placed into sections on the basis of chemical relationships. The last four groups are the exception. These could not be placed into any group based on chemical considerations and, therefore, three groups were devised according to the genus most likely to produce them: *Aspergillus* toxins, *Penicillium* toxins, and *Fusarium* toxins. The final section contains metabolites that could not be classified under the aforementioned categories.

The handbook is oriented primarily toward fungal metabolites that elicit a toxic response in vertebrate animals; however, it does contain metabolites that show little or no known acute toxicity. The latter are included because of their chemical or close biosynthetic relationship to a toxin or group of toxins.

Richard J. Cole
*Richard H. Cox**

*Present address: Philip Morris U. S. A. Research Center, P. O. Box 26583, Richmond, Virginia 23261

Acknowledgments

Special thanks are due to Mr. Joe W. Dorner, USDA, Mr. C. Pape, University of Georgia, and Mrs. Betty Shope for their valuable assistance in the production of this book. We also wish to thank the following investigators who provided editorial assistance, authentic samples, and/or spectra for this book.

Dr. Matazo Abe
Tokyo University of Education
Tokyo, Japan

Dr. W. Acklin
Laboratorium für Organische Chemie
Eidgenössische Technische Hochschule Zurich
Zurich, Switzerland

Dr. D. C. Aldridge
Imperial Chemical Industries Limited
Alderley Park, Macclesfield
Cheshire, England

Dr. Duilio Arigoni
Laboratory for Organic Chemistry
Zurich, Switzerland

Dr. J.-C. Bouhet
Service de Biochimie
Centre D'Études Nucléaires de Saclay
Gif-sur-Yvette, France

Dr. H. P. Broquist
Vanderbilt University
Nashville, Tennessee

Dr. George Büchi
Massachusetts Institute of Technology
Cambridge, Massachusetts

Dr. Tom Burka
Vanderbilt University
Nashville, Tennessee

Dr. H. Burkhardt
Western Regional Research Center
USDA, SEA, AR
Berkeley, California

Dr. Vincent J. Carroll
Pfizer Chemicals Division
Brooklyn, New York

Professor Roy W. Curtis
Purdue University
West Lafayette, Indiana

Dr. Horace G. Cutler
Richard B. Russell Research Center
USDA, SEA, AR
Athens, Georgia

Dr. Makoto Enomoto
Sagamihara Kyodo Hospital
Sagamihara, Japan

Dr. Pierre Fromageot
Service de Biochimie
Centre D'Études Nucléaires de Saclay
Gif-sur-Yvette, France

Dr. Rex Gallagher
Department of Science and Industrial Research
Applied Biochemistry Division
Palmerston North, New Zealand

Dr. A. O. Geiszler
Abbott Laboratory
North Chicago, Illinois

Dr. W. O. Godtfredsen
Leo Pharmaceutical Products
Ballerup, Denmark

Professor Thomas Harris
Vanderbilt University
Nashville, Tennessee

Dr. Donald Harvan
National Institute of Environmental Health Science
Research Triangle Park, North Carolina

Dr. A. Wallace Hayes
Director, Toxicology Research
Rome and Haas Company
Spring House, Pennsylvania

Dr. J. G. Heathcote
University of Salford
Salford, England

Dr. Robert A. Hill
University Chemical Laboratory
Lensfield Road
Cambridge, England

Dr. Dennis Hsieh
University of California
Davis, California

Dr. K. Ishii
Science University of Tokyo
Tokyo, Japan

Dr. Alexander Keyl
Richard B. Russell Research Center
Athens, Georgia

Dr. David G. I. Kingston
Virginia Polytechnic Institute and State University
Blacksburg, Virginia

Dr. Hiroshi Kurata
National Institute of Hygienic Sciences
Tokyo, Japan

Dr. Stephen H. Larsen
Lilly Research Laboratories
Indianapolis, Indiana

Dr. Louise S. Lee
Southern Regional Research Center
USDA, SEA, AR
New Orleans, Louisiana

Dr. E. B. Lillehoj
Southern Regional Research Center
USDA, SEA, AR
New Orleans, Louisiana

Dr. J. C. MacDonald
National Research Council of Canada
Saskatoon, Saskatchewan, Canada

Dr. A. G. McInnes
Atlantic Regional Laboratory
National Research Council of Canada
Halifax, Nova Scotia, Canada

Dr. Merle Sid Masri
Western Regional Research Center
USDA, SEA, AR
Berkeley, California

Professor C. J. Mirocha
University of Minnesota
St. Paul, Minnesota

Dr. Nobuichi Morooka
Kagawa University
Kagawa, Japan

Professor Dr. E. Mutschler
de Johann Wolfgang Goethe-Universität
Frankfurt, Germany

Dr. S. Natori
National Institute of Hygienic Sciences
Tokyo, Japan

Dr. Stanley Nesheim
Department of Health, Education, and Welfare
Food and Drug Administration
Washington, D.C.

Dr. S. V. Pathre
University of Minnesota
St. Paul, Minnesota

Dr. A. E. Pohland
Department of Health, Education, and Welfare
Food and Drug Administration
Washington, D.C.

Mme. J. Polonsky
Institut de Chimie des Substances Naturelles
Gif-sur-Yvette, France

Dr. John Richard
National Animal Disease Laboratory
Ames, Iowa

Dr. John W. Ronaldson
Ruakura Agriculture Research Center
Hamilton, New Zealand

Dr. Harry Schroeder
USDA, SEA, AR
Texas A&M University
College Station, Texas

Dr. Peter M. Scott
Health Protection Branch
Health and Welfare Canada
Ottawa, Ontario, Canada

Dr. William G. Sorenson
The University of Oklahoma
Norman, Oklahoma

Dr. Pieter S. Steyn
National Chemical Research Laboratory
CSIRO
Pretoria, South Africa

Dr. G. M. Strunz
Canadian Forestry Service
New Brunswick, Canada

Professor Dr. Ch. Tamm
Institut für Organische Chemie der Universität Basel
Basel, Switzerland

Dr. A. Taylor
Atlantic Regional Laboratory
National Research Council of Canada
Halifax, Nova Scotia, Canada

Dr. M. Tenabe
Stanford Research Institute
Menlo Park, California

Professor Y. Ueno
Science University of Tokyo
Tokyo, Japan

Dr. A. C. Waiss, Jr.
Western Regional Research Laboratory
USDA, SEA, AR
Albany, California

Dr. Steve Weinreb
Fordham University
New York, New York

Dr. John M. Wells
Southeastern Fruit & Tree Nut Research Laboratory
USDA, SEA, AR
Byron, Georgia

Dr. Benjamin J. Wilson
Vanderbilt University
Nashville, Tennessee

Dr. David Wilson
University of Georgia College of Agriculture
Coastal Plain Station
Tifton, Georgia

Professor Yuzuru Yamamoto
Kanazawa University
Kanazawa, Japan

Professor Dr. M. Yamazaki
Chiba University
Chiba, Japan

Spectral Format

Samples and/or copies of the various spectra were solicited from various individuals and we are grateful to those who responded to our request. Copies of spectra received from individuals were traced and photographed, otherwise spectra were photographed directly from an original black-ink recording on blank chart paper.

UV spectra of samples were taken with a Beckman model DB-G* recording spectrophotometer in methanol solution unless otherwise indicated. The recorder speed was 2.54 cm per minute. Spectra were calibrated with a holmium oxide standard.

IR spectra of samples were obtained with a Perkin-Elmer model 257 recording spectrophotometer equipped with a 3× beam condenser and baseline attenuator. Unless otherwise indicated, samples for analysis were prepared as a mull or as a thin film on a KBr window.

Mass spectra were obtained from individuals or institutions when possible. Mass spectra of collected samples were obtained on an A.E.I. MS-902 double focus instrument. Electron-impact spectra were run at 70 eV with an accelerating voltage of 8000 volts and a source temperature between 175° and 200°C. Chemical-ionization spectra were obtained on the same model instrument fitted with an S.R.I.C. chemical-ionization source. Positive ion spectra were recorded using isobutane as the ionizing reagent gas at $\simeq$ 1 torr source pressure and source temperature at 175°–200°C.

Criteria for purity of samples were based on TLC analyses and correlation of the spectral data with those reported in the literature.

The appropriate reference scale with TMS at 0 ppm was added at the bottom of each NMR spectrum. For most of the ^{1}H spectra, a standard 10 ppm display

*Mention of a trademark, proprietary product, or vendor does not constitute a guarantee or warranty of the product by the U.S. Department of Agriculture or the National Institute of Environmental Health Sciences and does not imply its approval to the exclusion of other products or vendors that may also be suitable.

is used. Many of the compounds contain peaks outside this range, and these peaks are recorded with an offset baseline above the normal 10 ppm spectrum at the left side of the chart. For spectra for which there is no scale for this portion of the spectrum, one can determine the offset by observing where the spectrum is located with respect to the 10 ppm scale. For example, a recording between 8 and 10 ppm corresponds to a sweep between 10 and 12 ppm and a recording between 7 and 10 ppm corresponds to a sweep between 10 and 13 ppm, etc.

An attempt has been made to assign as many of the ^1H spectra as possible using known trends and comparison among similar compounds. Assignments of specific protons are presented above each spectrum assigned.

The ^{13}C spectra presented are those obtained with proton noise decoupling using the Fourier transform technique. Chemical shifts are reported relative to internal TMS using the convention that downfield shifts are assigned positive ppm values. For D_2O soluble samples, dioxane was used as the internal reference, and the chemical shifts were converted to the TMS scale using the conversion factor of 67.4 ppm. In a few cases for which long-term accumulation was required, no reference was added and the chemical shifts were referenced to the solvent peaks and later converted to the TMS scale using 77.0 ppm for $CDCl_3$ and 40.4 ppm for DMSO. Sufficient numbers of pulses were accumulated to provide adequate signal-to-noise ratios.

The assignment of chemical shifts of each ^{13}C spectrum is presented above each spectrum. Assignments are based on single frequency off-resonance decoupling results, comparison with related compounds, and data reported in the literature. The results of the single frequency off-resonance decoupling spectra are reported along with the chemical shifts as s, d, t, q for quaternary, methine, methylene, or methyl carbons, respectively. In a few cases, no sample was available and spectra were traced; therefore, no single frequency off-resonance decoupling results are available. These spectra were not assigned unless a sufficient number of spectra of related compounds were available.

The ^1H spectra in this collection were obtained on either Varian HA-100 or XL-100-12 spectrometers or a JEOL PFT-100 spectrometer using 5 mm sample tubes. The XL-100-12 spectrometer was equipped with the Varian 620-L disk data system, and the PFT-100 spectrometer with the JEOL EC-100 disk data system. TMS was used as an internal lock on the HA-100 instrument, whereas the deuterium resonance from the deuterated solvent provided the lock signal on the XL-100-12 and PFT-100 instruments. For samples run in the Fourier transform mode, standard operating conditions were a 90° pulse with a repetition time of 4.0 seconds using 8K data points.

^{13}C spectra were obtained with PFT-100 and XL-100-12 instruments using 10, 5, and 1.7 mm sample tubes, depending on the amount of sample available. The deuterium resonance from the deuterated solvent provided the lock signal and TMS was used as an internal reference in most cases. All samples were run in the Fourier transform mode and standard operating conditions were a pulse angle of 30° with a repetition of 3.0 seconds using 8K data points. An exponential weighting factor of -1 was applied to the free induction decay before Fourier transformation. Sweep widths of either 5000 Hz or 6250 Hz were used, depending on the type of carbonyl carbons present in the compound. Peak positions and calculated chemical shifts were obtained from computer printouts and the chemical shifts are accurate to ± 0.1 ppm.

Most samples were run in $CDCl_3$ and/or DMSO. As a general rule, the solubility of the compound in question was first checked in $CDCl_3$. When the compound was not sufficiently soluble in $CDCl_3$ alone, a few drops of DMSO was added. At this point, if the compound was not sufficiently soluble, the $CDCl_3$ was removed and the compound was dissolved in DMSO. In a few cases, the solubility of the compound in either D_2O or acetone was known and spectra were obtained in these solvents. For most compounds, the ^{13}C spectrum was obtained first and the ^{1}H spectrum was obtained on the same sample. Despite our best efforts, many of the ^{1}H spectra obtained in DMSO show a peak for H_2O absorbed from the atmosphere during the interval between sample preparation and determination of the spectrum.

The Aflatoxins

1

The aflatoxin group, notably aflatoxin B_1, has had the most profound impact on the development of the science of mycotoxicology. Their discovery in 1960 as the cause of "Turkey X" disease and subsequent implication in other mycotoxicoses has had the effect of stimulating an interest in and an awareness of other possible mycotoxin problems.

The four naturally occurring aflatoxins, B_1, B_2, G_1, and G_2, are acutely toxic and carcinogenic metabolites produced exclusively by *Aspergillus flavus* and the closely related fungus *A. parasiticus*. Other members of the group are derived from these four as metabolic products of microbial or animal systems (such as M_1, M_2, P_1, Q_1, and aflatoxicol) or produced spontaneously in response to the chemical environment (such as B_{2a}, G_{2a}, and D_1).

The aflatoxins are highly fluorescent, highly oxygenated, heterocyclic compounds characterized by dihydrodifurano or tetrahydrodifurano moieties fused to a substituted coumarin moiety. Aflatoxin B_1 is the most prevalent naturally occurring, acutely toxic, and carcinogenic member of the group.

Although the toxicology of the aflatoxins varies considerably among species and with regard to age, sex, and nutrition, the primary organ affected is the liver. Gross clinical signs are growth retardation and weight loss due to reduced feed intake and efficiency, followed by severe tenesmus a few days before death. Postmortem examination may show fatty infiltration of the liver, liver fibrosis, ascites, visceral edema, bile duct proliferation, and hepatic carcinoma, depending on species involved. The ability of aflatoxin B_1 to induce liver carcinoma varies considerably with species. Trout and ducklings are very susceptible, but most animal species appear less prone to aflatoxin-induced hepatic carcinoma.

More subtle effects of aflatoxin ingestion are breakdown in the immune response and synergism or antagonism with various vitamins.

Several excellent reviews and at least one book which discuss various aspects of the aflatoxins are available.[2,10,11,24,50]

The importance of the aflatoxins in mycotoxicoses of animals and man cannot be overemphasized or underestimated.

Aflatoxin group	Molecular weight	Molecular formula
Aflatoxin P_1	298.0477	$C_{16}H_{10}O_6$
Aflatoxin D_1	286.0841	$C_{16}H_{14}O_5$
Parasiticol (aflatoxin B_3)	302.0790	$C_{16}H_{14}O_6$
Aflatoxin B_1	312.0633	$C_{17}H_{12}O_6$
Aflatoxin Q_1	328.0582	$C_{17}H_{12}O_7$
Aflatoxin G_1	328.0582	$C_{17}H_{12}O_7$
Aflatoxin M_1	328.0582	$C_{17}H_{12}O_7$
Aflatoxin B_2	314.0790	$C_{17}H_{14}O_6$
Aflatoxicol A	314.0790	$C_{17}H_{14}O_6$
Aflatoxicol B	314.0790	$C_{17}H_{14}O_6$
Aflatoxin G_2	330.0739	$C_{17}H_{14}O_7$
Aflatoxin M_2	330.0739	$C_{17}H_{14}O_7$
Aflatoxin B_{2a}	330.0739	$C_{17}H_{14}O_7$
Aflatoxin G_{2a}	346.0688	$C_{17}H_{14}O_8$
Aflatoxicol *O*-ethyl ether A	342.1103	$C_{19}H_{18}O_6$
Aflatoxicol *O*-ethyl ether B	342.1103	$C_{19}H_{18}O_6$

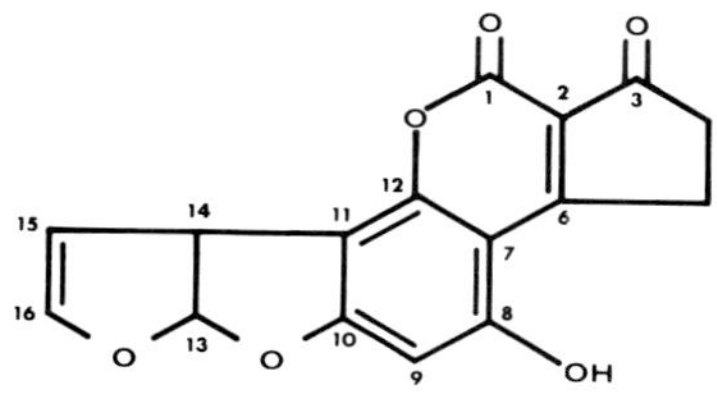

Common name[20]	Aflatoxin P_1
Molecular weight	298.0477
Molecular formula	$C_{16}H_{10}O_6$
General characteristics[6,20]	Pale yellow needles from methanol–benzene–hexane, mp >320°C $[\alpha]_D^{20}$ $-574°$ (C = 0.08 in methanol)
UV data[6]	λ_{max}^{EtOH} nm (ϵ): 226(20,400), 267(11,200), 342(14,900), 362(15,400), and 425(2,500)
Source[6]	Aflatoxin P_1 is the principal urinary metabolite of aflatoxin B_1 in rhesus monkeys. It occurred in the urine in unconjugated (3%), sulfate (10%), and glucuronide (50%) forms.
Toxicity data[6]	In a mouse bioassay using IP injection, aflatoxin P_1 showed considerably less toxicity than aflatoxin B_1. At a dosage of 100 mg/kg, no mortalities were observed; at 150 mg/kg there were 2 mortalities in 15 animals; and at 200 mg/kg, no mortalities occurred. Aflatoxin B_1 in the same assay had an LD_{50} of 9.5 mg/kg.

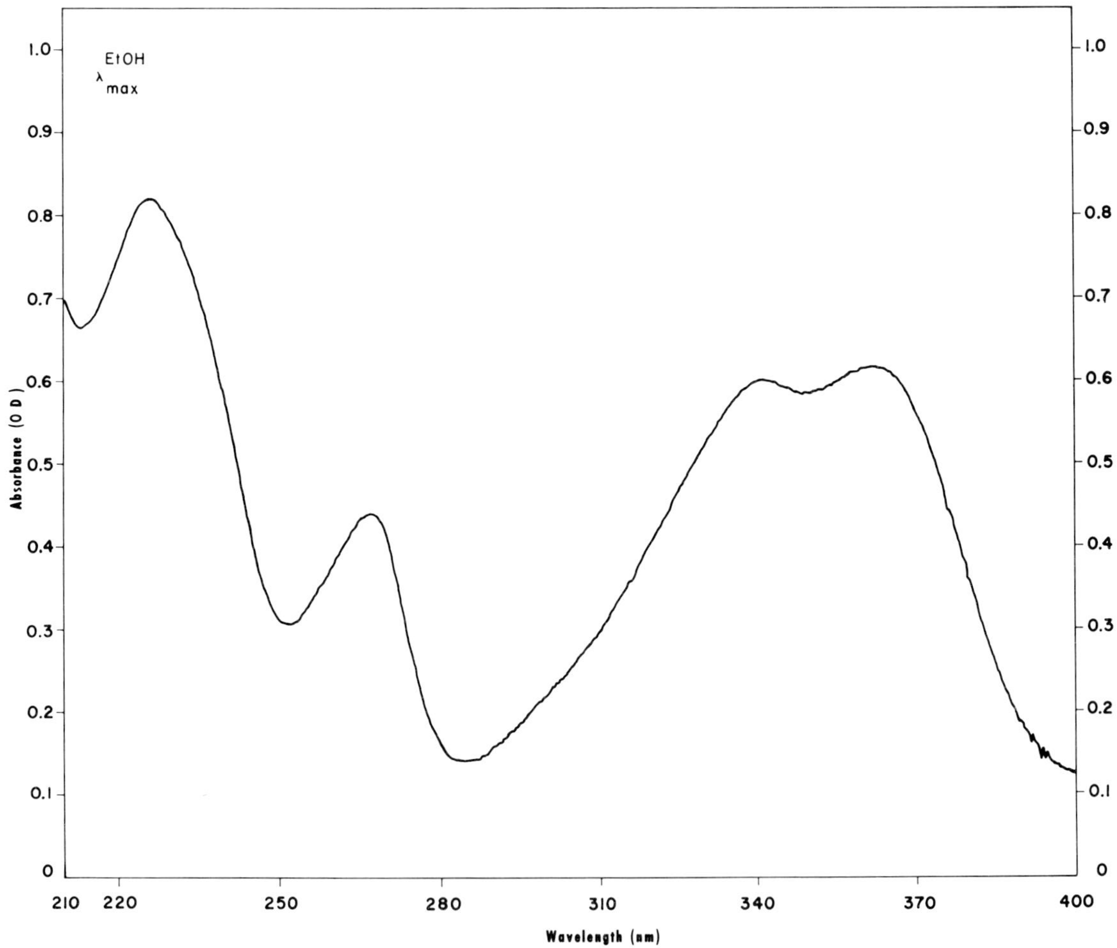

EtOH
λ_{max}
Absorbance (O D)
Wavelength (nm)
210
220
250
280
310
340
370
400
0
0.1
0.2
0.3
0.4
0.5
0.6
0.7
0.8
0.9
1.0

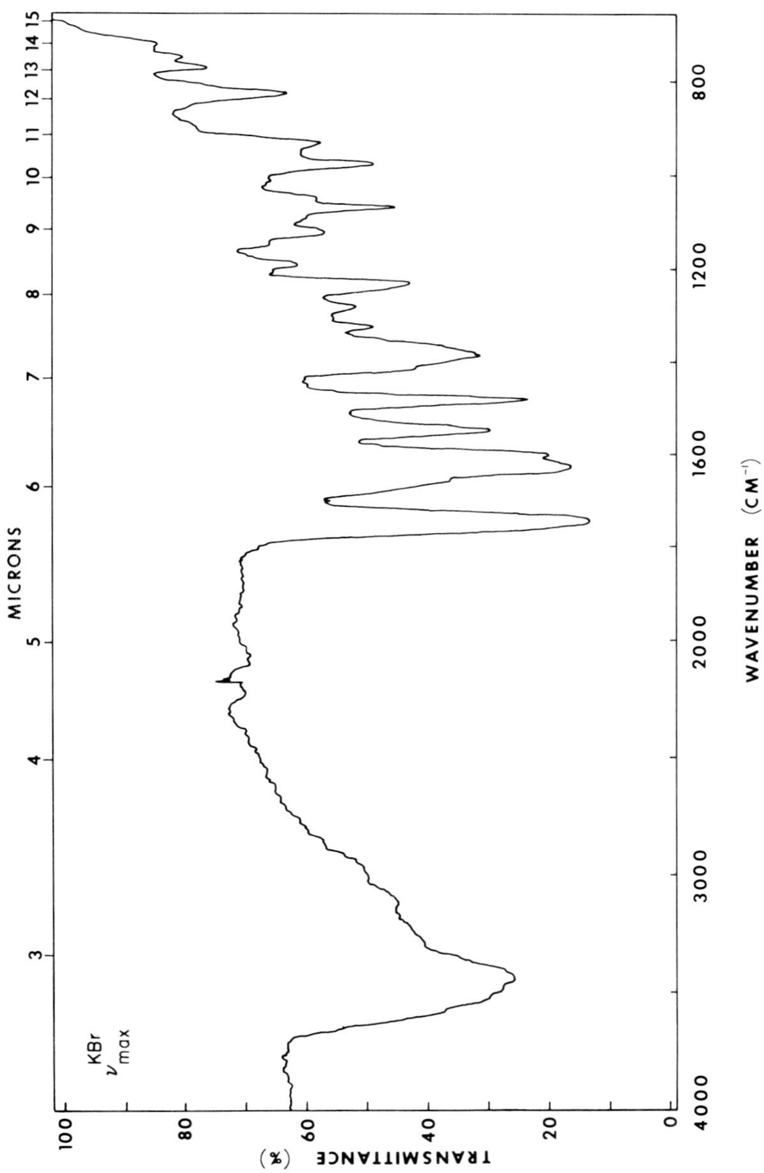
MICRONS
3
4
5
6
7
8
9
10
11
12
13
14
15
TRANSMITTANCE (%)
100
80
60
40
20
0
KBr
ν_{max}
WAVENUMBER (CM^{-1})
4000
3000
2000
1600
1200
800

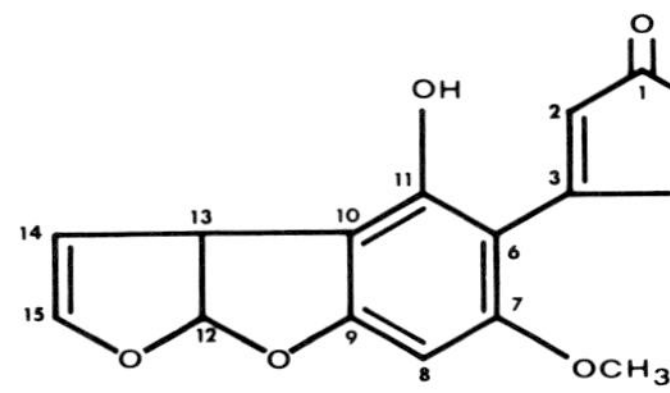

Common name	Aflatoxin D_1
Molecular weight	286.0841
Molecular formula	$C_{16}H_{14}O_5$
General characteristics[52]	Sublimed without melting at 220°–290°C; in sealed tube crystals decomposed without melting at 255°–258°C. Forms acetate derivative, mp 155°–157°C
UV data[36,52]	λ_{max}^{MeOH} nm (ϵ): 227(20,792) and 324(17,074) [corrected from 227(15,920) and 324(12,440)]
Source[36,52]	A major product formed from reacting aflatoxin B_1 with ammonium hydroxide at 100°C under pressure.
Toxicity data[35]	Specific toxicity data not known. It has been reported that the nonfluorescent compounds formed in alkaline solutions are strongly toxic (mixture includes aflatoxin D_1).
TLC data[52]	*Adsorbent:* silica gel *Solvent:* chloroform–acetone, 95:5 v/v R_f: 0.31 *Detection:* turns gray-brown after spraying with 10% $FeCl_3$; orange-yellow with 2,4-dinitrophenylhydrazine *Acetate* R_f: 0.78
^{13}*C NMR data*	C1, 208.6 s; C2, 131.5 d; C3, 170.9 s; C4, 31.9 t; C5, 34.2 t; C6, 106.1 s*; C7, 159.5 s; C8, 86.6 d; C9, 158.6 s; C10, 106.7 s*; C11, 151.3 s; C12, 111.5 d; C13, 47.6 d; C14, 103.1 d; C15, 144.1 d; C16, 55.8 q
^{1}*H NMR data*	H2, 6.24; H4, 2.40; H5, 3.10; H8, 6.25; H12, 6.76; H13, 4.70; H14, 5.40; H15, 6.63; H16, 3.75

* Assignments may be reversed.

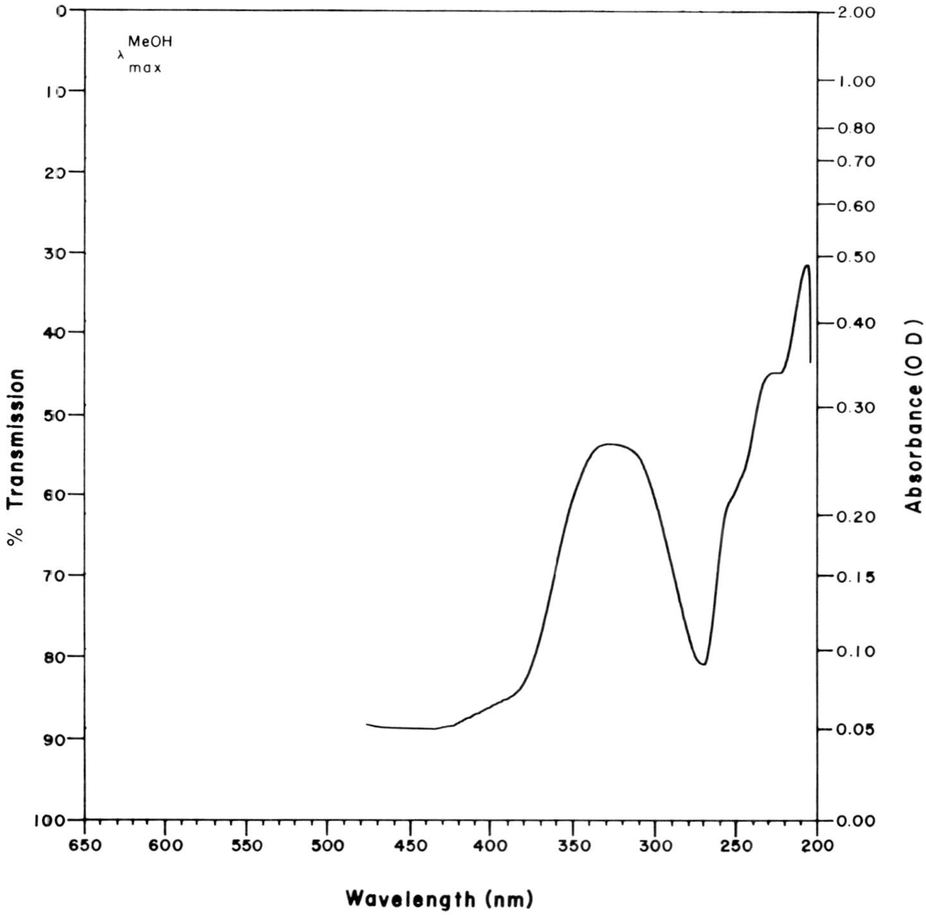

λ_{max}^{MeOH}
% Transmission
0
10
20
30
40
50
60
70
80
90
100
Absorbance (O D)
2.00
1.00
0.80
0.70
0.60
0.50
0.40
0.30
0.20
0.15
0.10
0.05
0.00
650
600
550
500
450
400
350
300
250
200
Wavelength (nm)

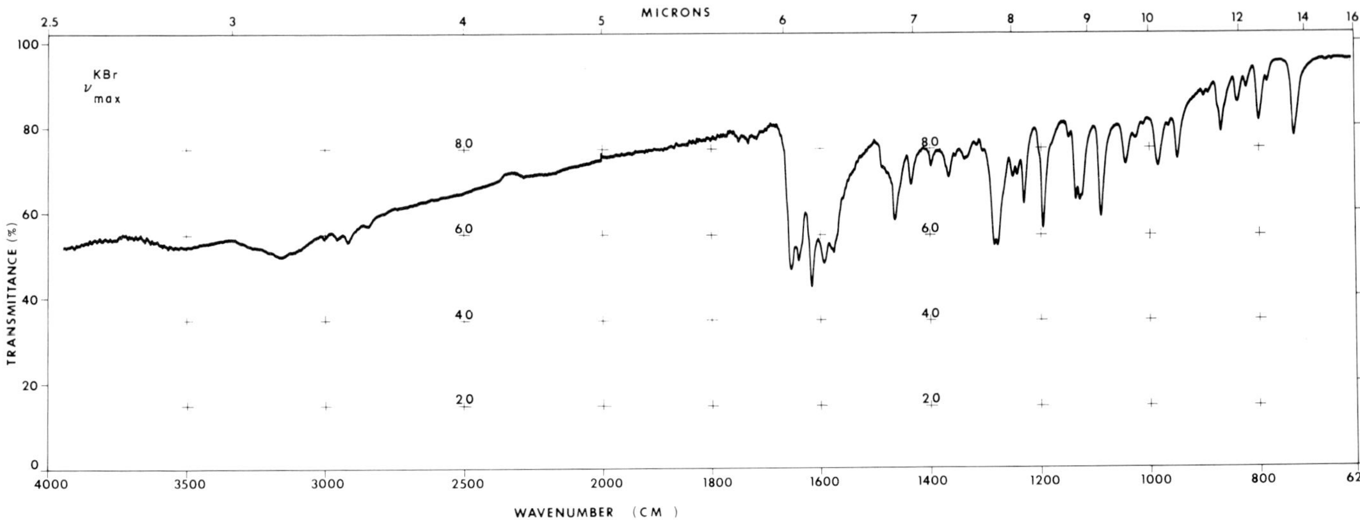
MICRONS
TRANSMITTANCE (%)
KBr
ν max
WAVENUMBER (CM)

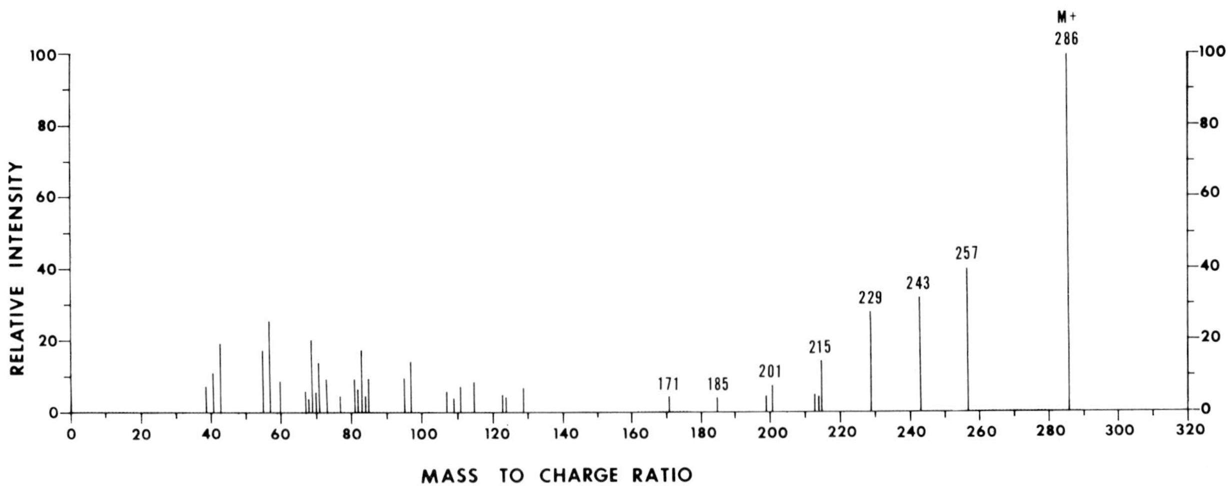
M+
286
257
243
229
215
201
185
171
RELATIVE INTENSITY
MASS TO CHARGE RATIO

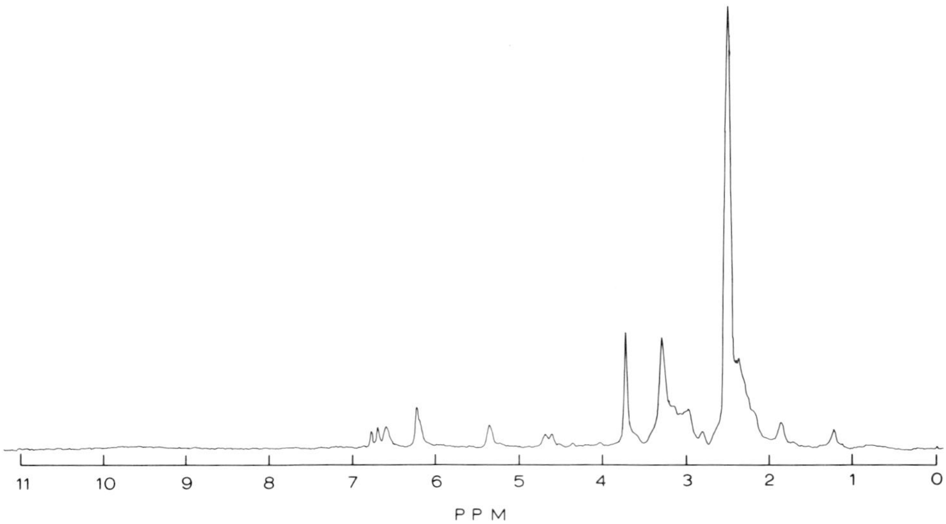

The 100 MHz ^1H NMR spectrum of aflatoxin D_1 in DMSO solution.

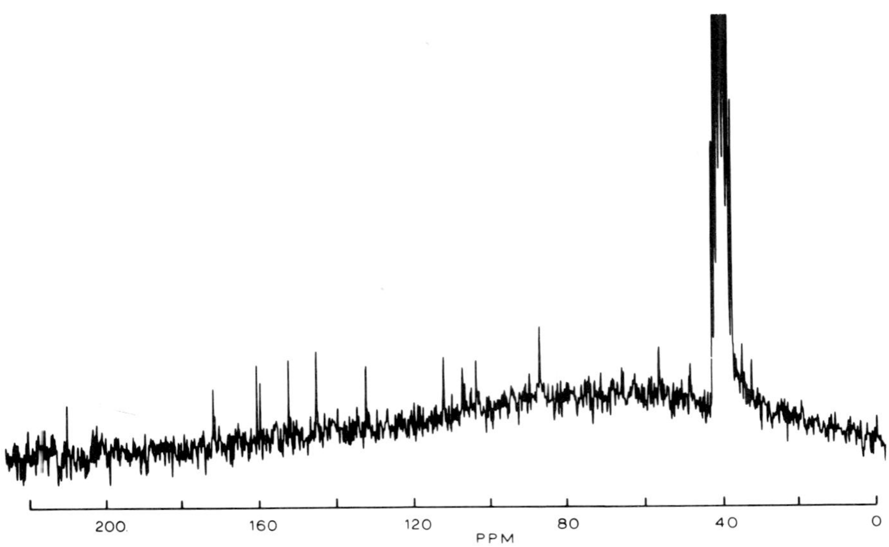

The 25 MHz ^{13}C NMR spectrum of aflatoxin D_1 in DMSO solution.

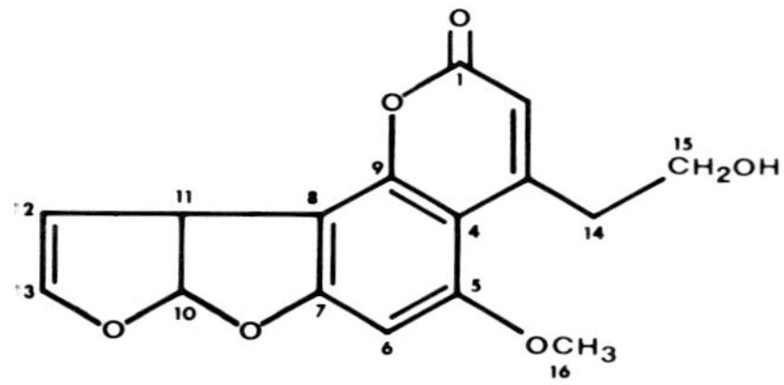

Common name	Parasiticol (Aflatoxin B_3)
Molecular weight	302.0790
Molecular formula	$C_{16}H_{14}O_6$
General characteristics[18,54]	Crystals from chloroform, mp 217°C
UV data[18,29,54]	λ_{max}^{MeOH} nm (ϵ): 229(sh)(10,000), 253(7,300), 262(7,550), and 326(9,350) λ_{max}^{MeOH} nm (ϵ): 225(sh)(12,600), 253(6,800), 262(7,400), and 325(9,700)
Fungal source[29,54]	*Aspergillus flavus* Link *A. parasiticus* Speare Parasiticol appears to be more prominent in older cultures of *A. flavus* and *A. parasiticus*. It is suggested to be the first step in the biological degradation of aflatoxin G_1 and is reported as an intermediate in the biodegradation of aflatoxin G_1 by *Rhizopus* spp.
Toxicity data[54]	Parasiticol had the same acute toxicity to ducklings as aflatoxin B_1, but it had little tendency to cause biliary hyperplasia. It was only 1/100 as toxic as B_1 in chick embryo studies.
^{13}C NMR data	C1, 158.6 s*; C2, 110.8 d; C3, 154.7 s; C4, 103.5 s; C5, 158.8 s*; C6, 91.0 d; C7, 160.6 s; C8, 106.8 s; C9, 150.7 s; C10, 112.6 d; C11, 47.2 d; C12, 102.0 d; C13, 145.1 d; C14, 41.1 t; C15, 59.9 t; C16, 56.4 q
^{1}H NMR data	H2, 5.94; H6, 6.62; H10, 6.85 (J = 7.0); H11, 4.72 (J = 7.0, 3.0); H12, 5.36 (J = 3.0, 3.0); H13, 6.66; H14, 3.00; H15, 3.61; H16, 3.85; OH, 4.59

* Assignments may be reversed.

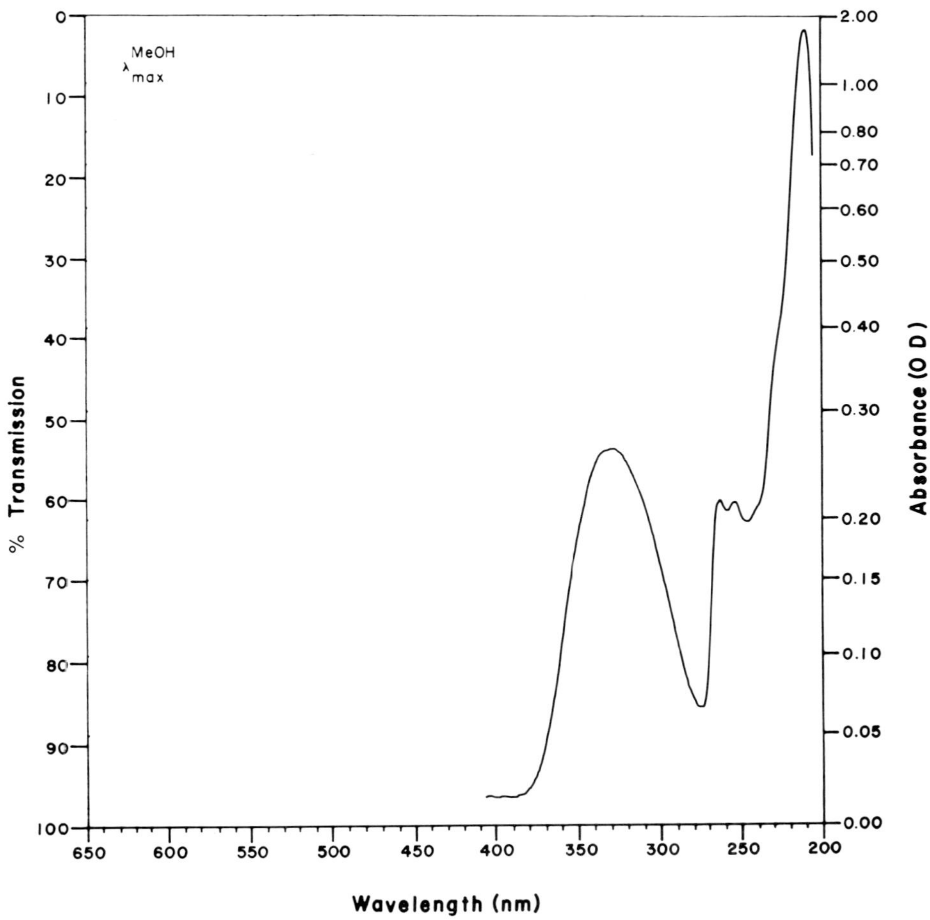
MeOH
λ max
% Transmission
Absorbance (O D)
Wavelength (nm)
0
10
20
30
40
50
60
70
80
90
100
2.00
1.00
0.80
0.70
0.60
0.50
0.40
0.30
0.20
0.15
0.10
0.05
0.00
650
600
550
500
450
400
350
300
250
200

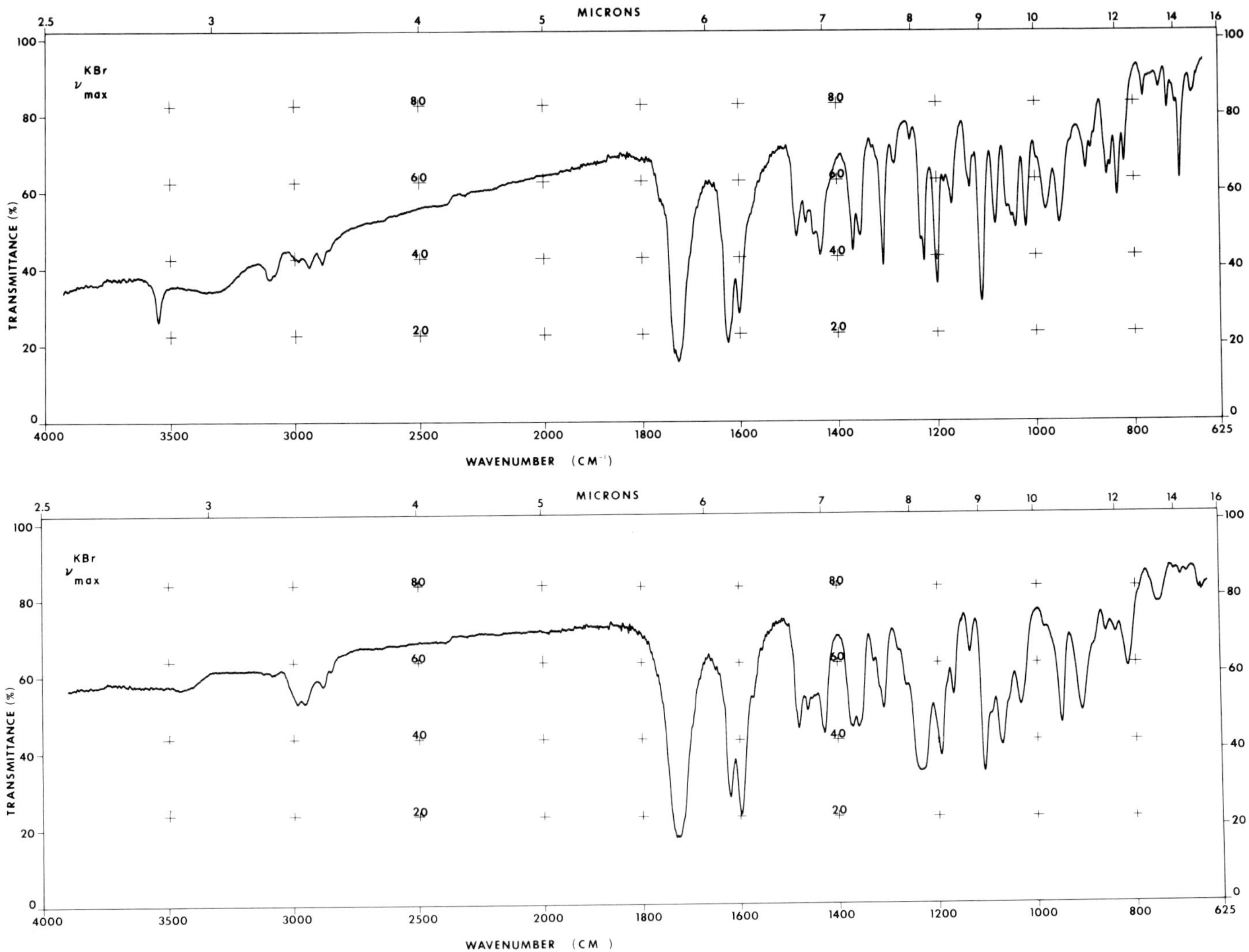

Upper spectrum: parasiticol; lower spectrum, parasiticol acetate.

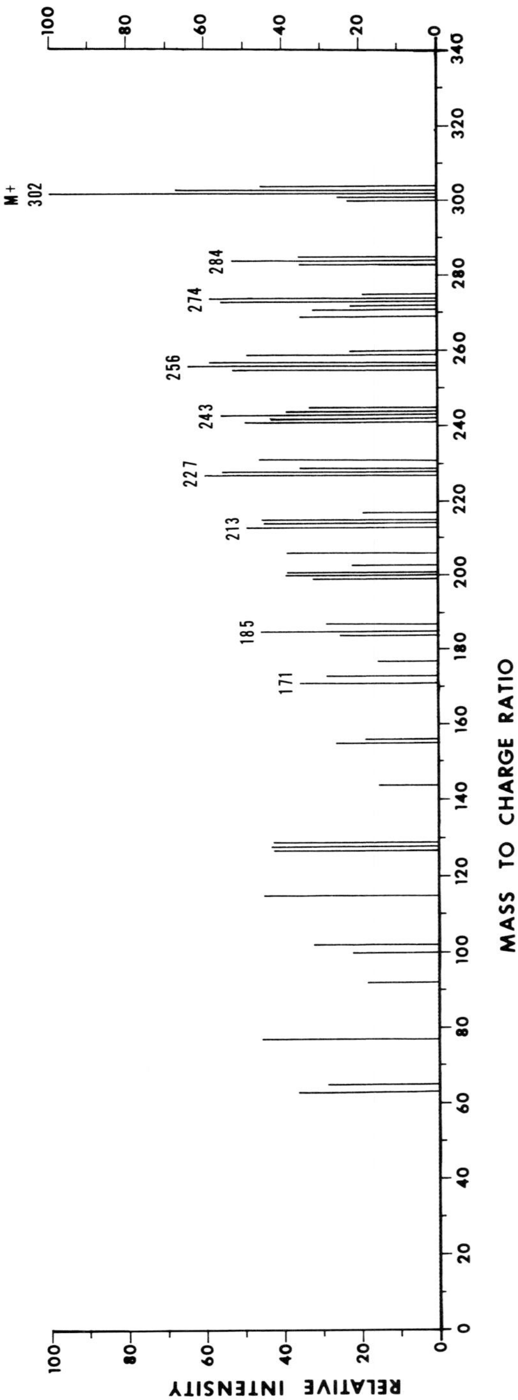
M+
302
284
274
256
243
227
213
185
171
MASS TO CHARGE RATIO
RELATIVE INTENSITY

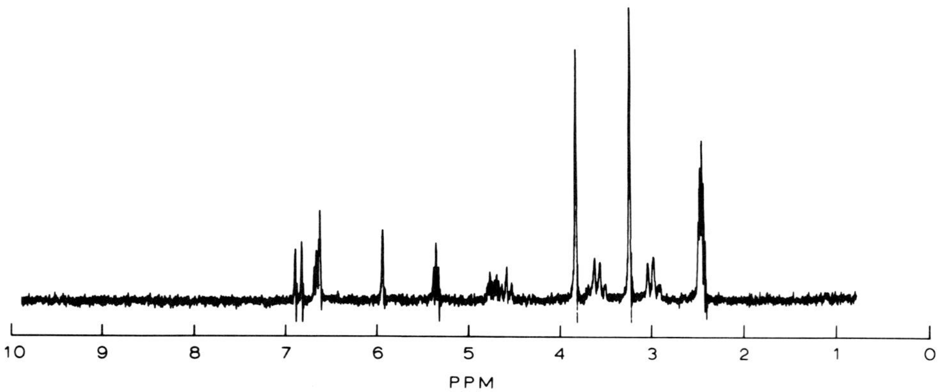

The 100 MHz ^{1}H NMR spectrum of parasiticol in DMSO solution.

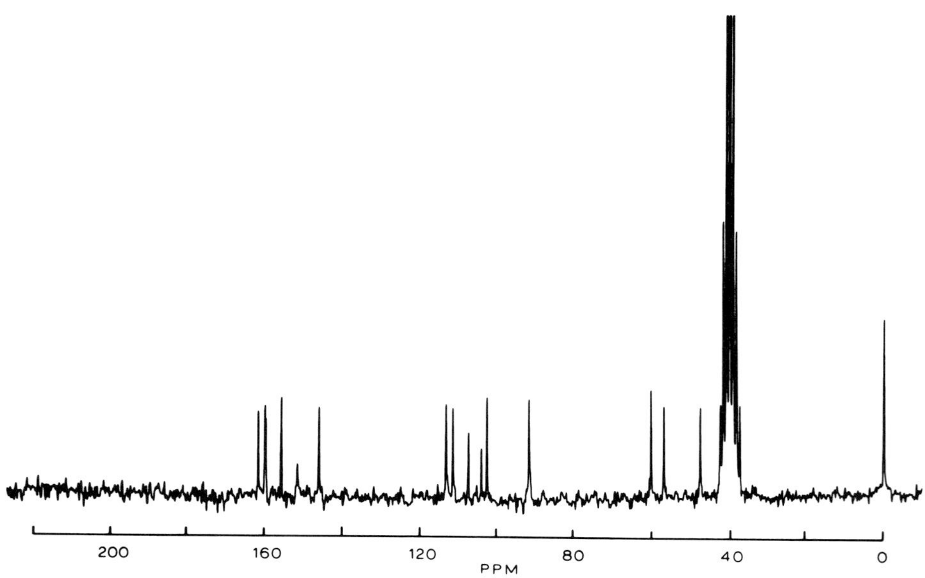

The 25 MHz ^{13}C NMR spectrum of parasiticol in DMSO solution.

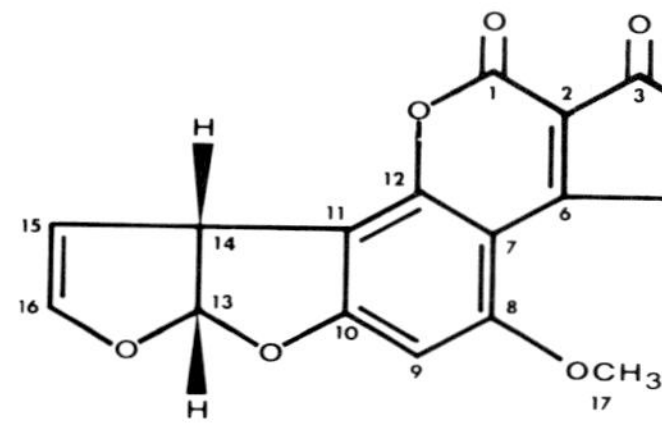

Common name Aflatoxin B_1

Molecular weight 312.0633

Molecular formula $C_{17}H_{12}O_6$

General characteristics[30,31] Crystals from chloroform, mp 268°–269°C (dec.) $[\alpha]_D$ −480° (C = 0.1 in dimethylformamide)

UV data[3] λ_{max}^{EtOH} nm (ϵ): 223(25,600), 265(13,400), and 362(21,800)
Fluorescence emission, 425 nm

Fungal source[21,48,58] *Aspergillus flavus* Link
A. parasiticus Speare

Toxicity data[8,49,61] The effects of aflatoxin B_1 vary with species, age, sex, and general nutrition. The primary organ affected is the liver with some changes occurring in other organs. The induction of liver lesions, liver carcinoma, and bile duct proliferation by aflatoxin B_1 has been demonstrated in several species of animals. Some species, such as the trout, duckling, and pig, are very susceptible to aflatoxicosis, whereas others, such as sheep and cattle, are more resistant. In Fischer rats and rainbow trout, it is the most potent hepatocarcinogen known.

The LD_{50} in 1-day-old rats dosed orally was 0.56 mg/kg; in weanling rats, 5.5 mg/kg (male), 7.4 mg/kg (female); in 100 gm male rats, 7.2 mg/kg; in 150 gm female rats, 17.9 mg/kg. The LD_{50} in Macaque monkeys dosed orally was 7.8 mg/kg.

TLC data[28] *Adsorbent:* silica gel G-HR
Solvent: chloroform–acetone, 93:7 v/v
R_f: 0.33
Detection: bright blue fluorescence under UV light

	NOTE: See Heathcote and Dutton[29] (pp. 83–86) for discussion on TLC of aflatoxins.
Other significant references	Isolation[42]; structure[4,59]; synthesis[5]; etiology[1]
^{13}C NMR data	C1, 154.7 s; C2, 117.0 s; C3, 200.6 s; C4, 35.0 t; C5, 29.0 t; C6, 176.5 s; C7, 103.7 s; C8, 161.0 s; C9, 90.6 d; C10, 165.3 s; C11, 107.5 s; C12, 152.5 s; C13, 113.2 s; C14, 47.8 d; C15, 102.3 d; C16, 144.8 d; C17, 56.4 q
^{1}H NMR data	H4, 2.56; H5, 3.34; H9, 6.38; H13, 6.75 (J = 7.0); H14, 4.72 (J = 7.0, 3.0); H15, 5.42 (J = 3.0, 3.0); H16, 6.40; H17, 3.93

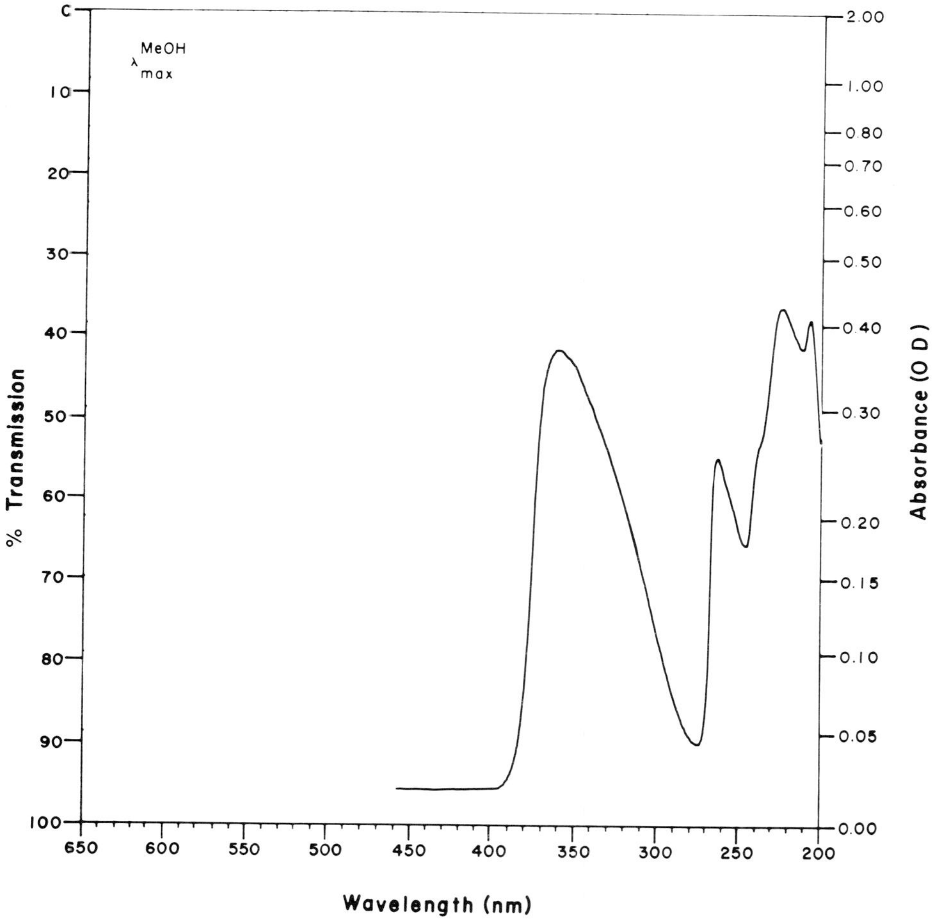
C
10
20
30
40
50
60
70
80
90
100
% Transmission
MeOH
λ max
2.00
1.00
0.80
0.70
0.60
0.50
0.40
0.30
0.20
0.15
0.10
0.05
0.00
Absorbance (O D)
650
600
550
500
450
400
350
300
250
200
Wavelength (nm)

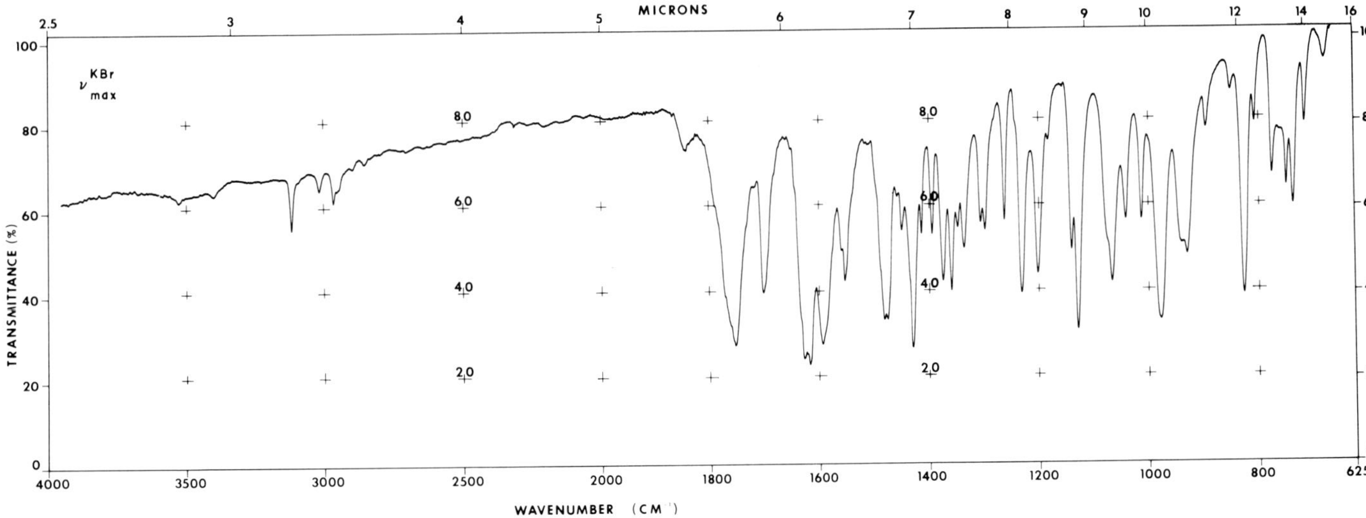
MICRONS
TRANSMITTANCE (%)
ν_{max}^{KBr}
WAVENUMBER (CM⁻¹)

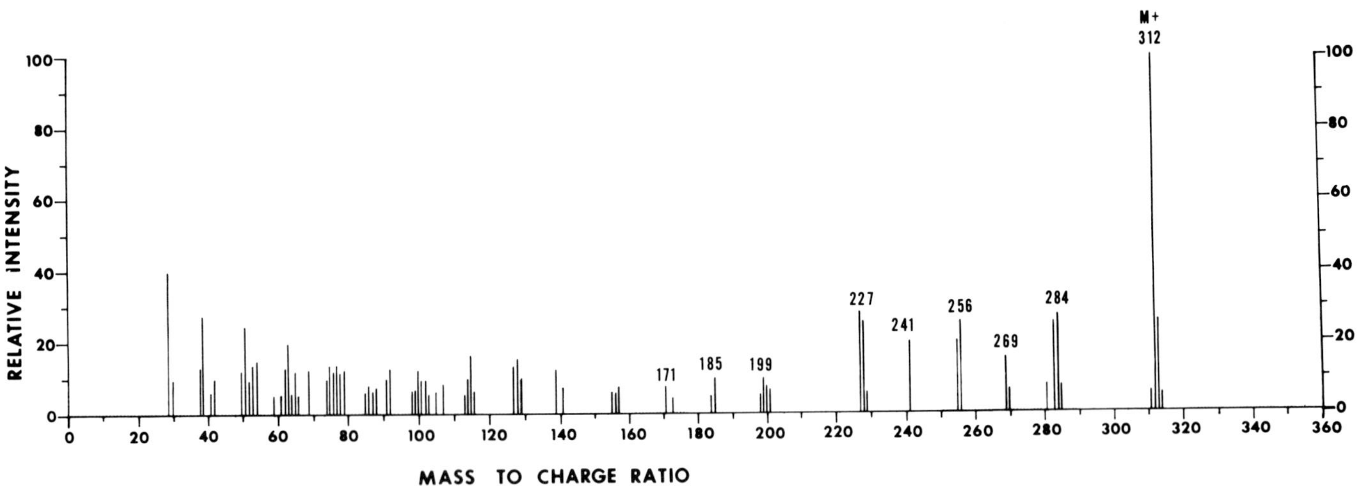
M+
312
284
269
256
241
227
199
185
171
RELATIVE INTENSITY
MASS TO CHARGE RATIO

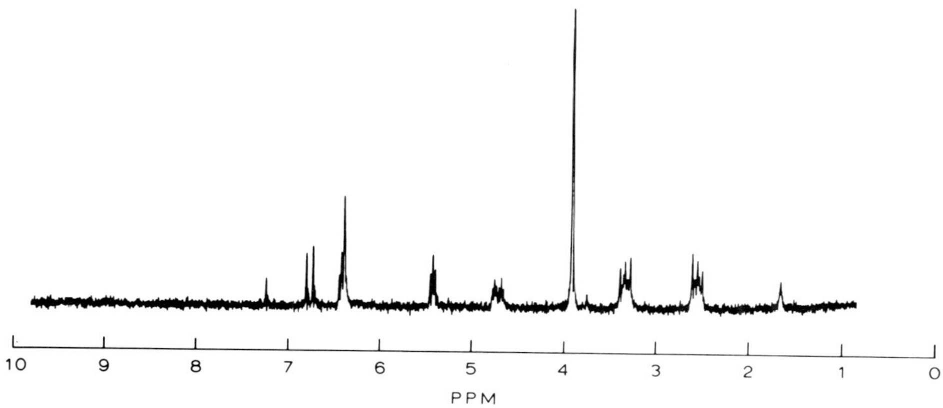

The 100 MHz ^1H NMR spectrum of aflatoxin B_1 in $CDCl_3$ solution.

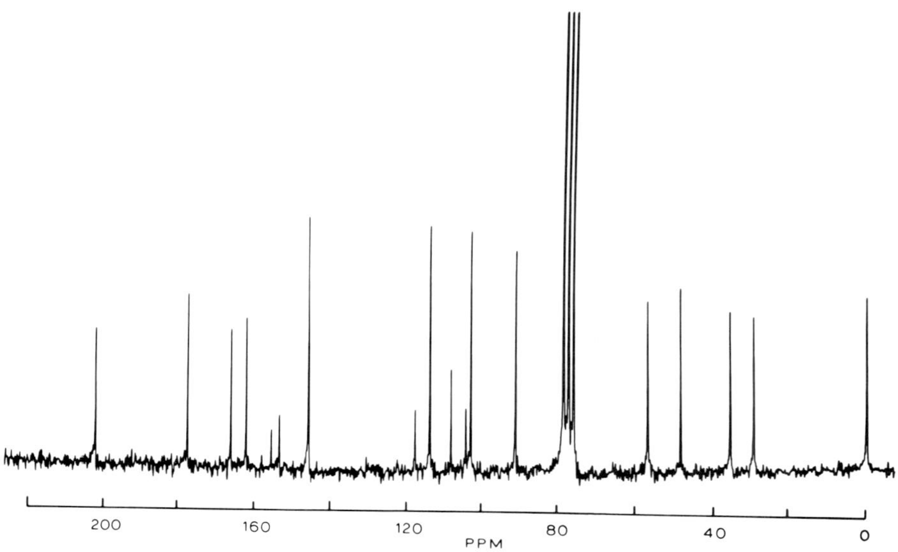

The 25 MHz ^{13}C NMR spectrum of aflatoxin B_1 in $CDCl_3$ solution.

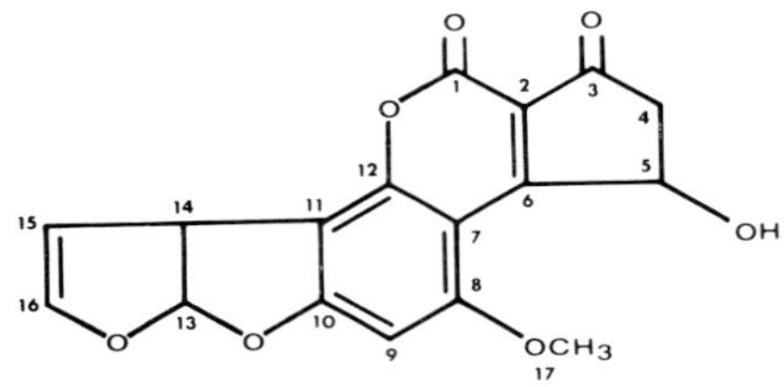

Common name[39]	Aflatoxin Q_1
Molecular weight	328.0582
Molecula formula	$C_{17}H_{12}O_7$
General characteristics[32,40]	Colorless needles from methanol–chloroform or hot acetonitrile, mp 295°C (dec.) Crystals from chloroform–hexane, mp 266°C (dec.)
UV data[7,40]	λ_{max}^{MeOH} nm (ϵ): 366(17,500), 267(11,450), and 223(19,030) λ_{max}^{EtOH} nm (ϵ): 365(18,800), 266(11,700), 242(sh)(10,000), and 224(20,500)
Source[7,39,40]	Aflatoxin Q_1 is a major metabolite of aflatoxin B_1 metabolism in monkey, rat, and human liver preparations (*in vitro*).
Toxicity data[33]	Aflatoxin Q_1 was approximately 18 times less toxic than aflatoxin B_1 by the air cell route in the chicken embryo test. No mutagenic activity was detected by the bacterial mutagenesis test, using *Salmonella typhimurium* TA 1538 with or without microsomal activities.
TLC data[7,32]	A *Adsorbent*: Adsorbosil-l silica gel *Solvent:* chloroform–acetone–*n*-hexane, 88:15:20 v/v/v R_f: 0.23 *Detection:* intense yellow-green fluorescence under UV light B *Adsorbent:* silica gel *Solvent:* chloroform–methanol, 20:1 v/v R_f: 0.25 *Detection:* yellow fluorescence in 350 nm UV light

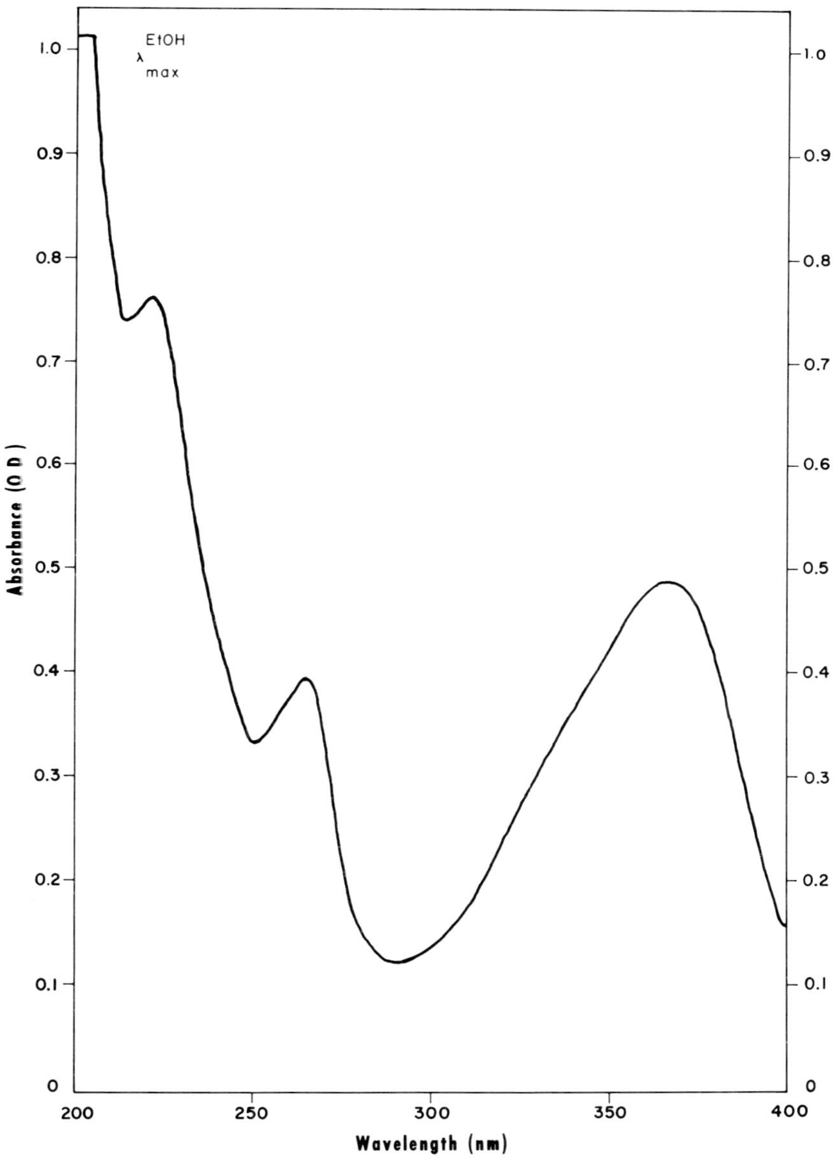

EtOH
λ max
Absorbance (O.D.)
Wavelength (nm)
0
0.1
0.2
0.3
0.4
0.5
0.6
0.7
0.8
0.9
1.0
200
250
300
350
400

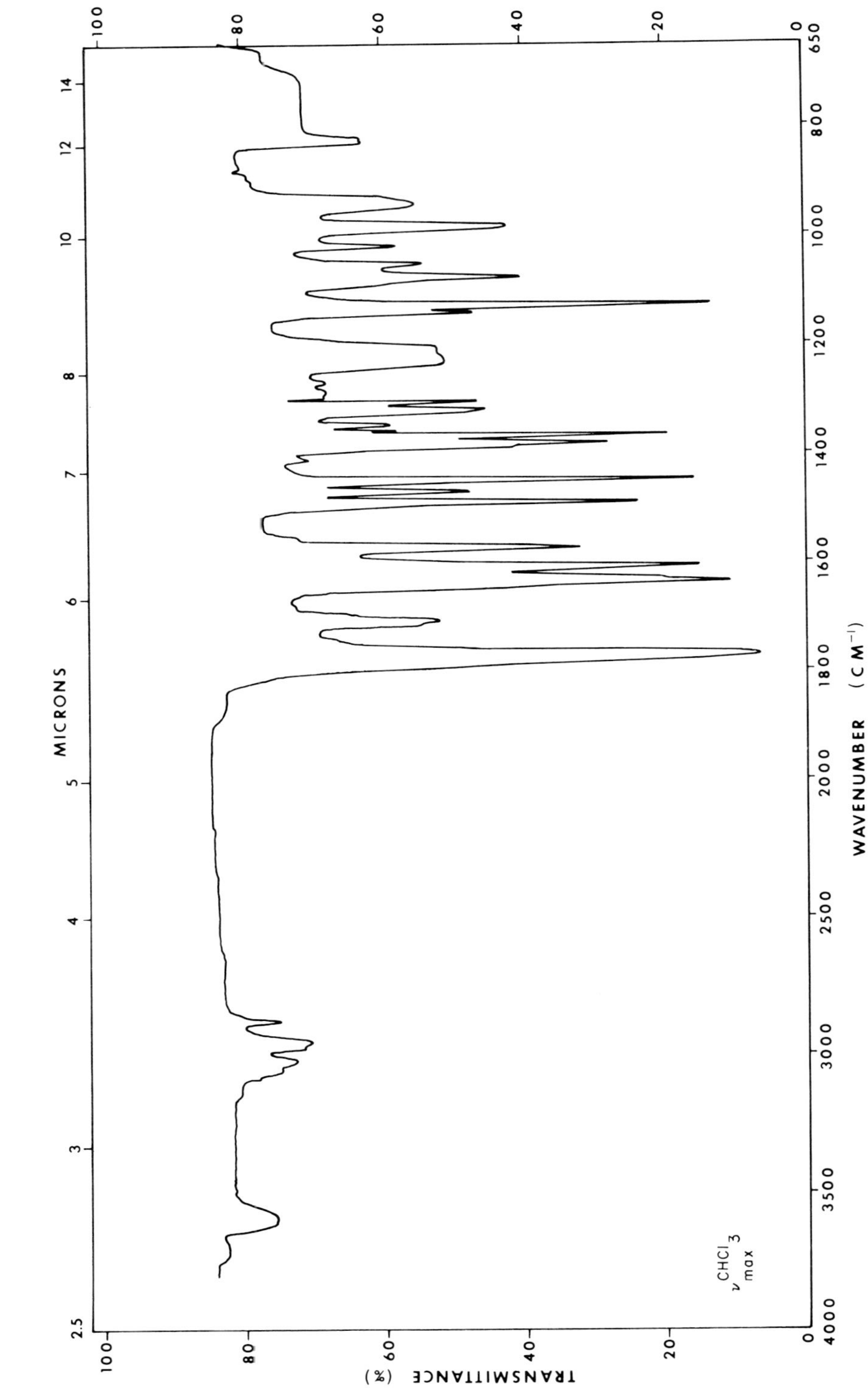
MICRONS
WAVENUMBER (CM^{-1})
TRANSMITTANCE (%)
$\nu_{max}^{CHCl_3}$

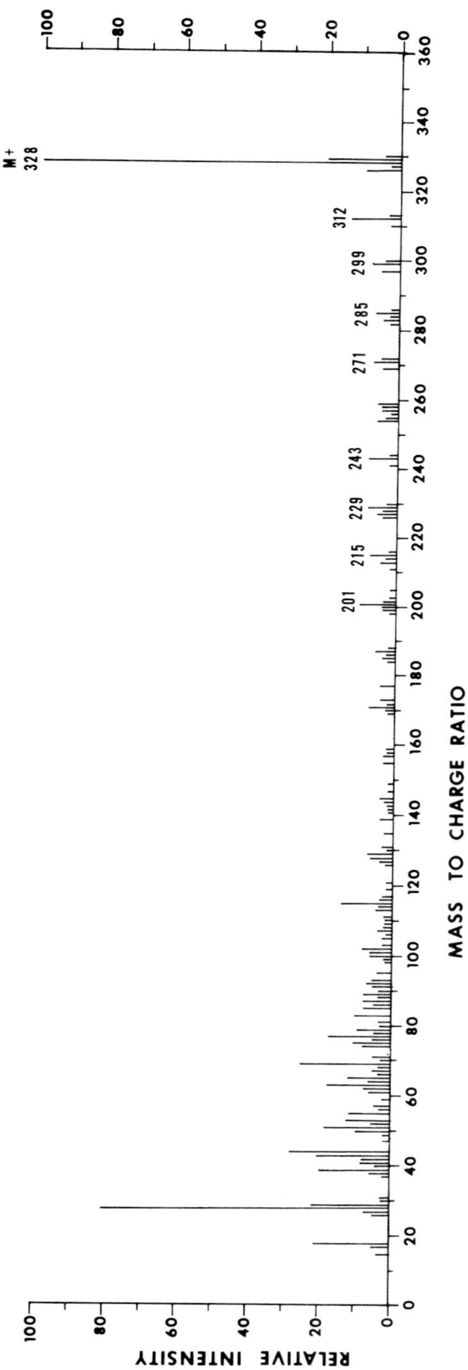
M+
328
312
299
285
271
243
229
215
201
MASS TO CHARGE RATIO
RELATIVE INTENSITY
0
20
40
60
80
100
120
140
160
180
200
220
240
260
280
300
320
340
360

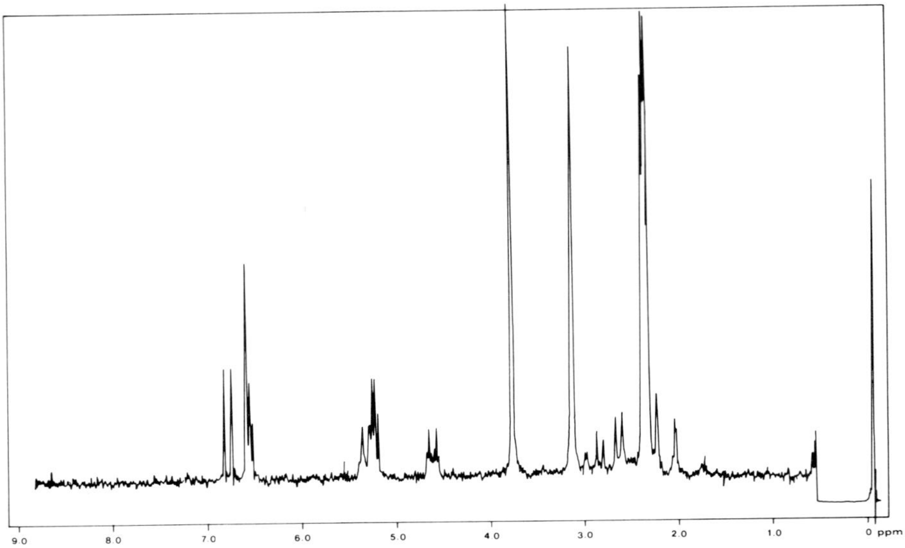

The 100 MHz ^{1}H spectrum of aflatoxin Q in $CDCl_3$.

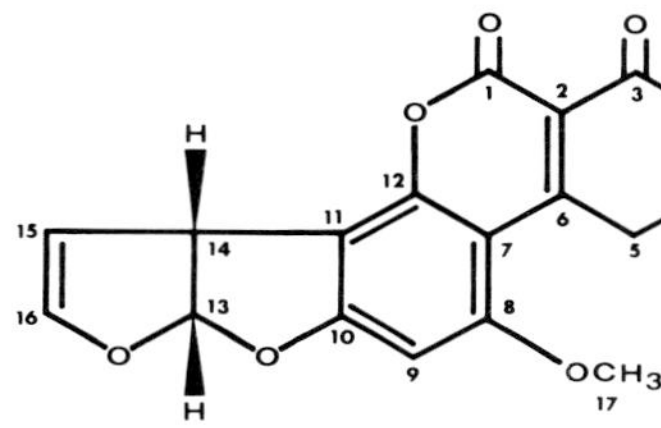

Common name	Aflatoxin G_1
Molecular weight	328.0582
Molecular formula	$C_{17}H_{12}O_7$
General characteristics[4,57]	Crystals from chloroform–methanol, mp 244°–246°C (dec.) Crystals from acetone, mp 257°–259°C $[\alpha]_D$ −556° (C = 1 in chloroform)
UV data[4]	λ_{max}^{EtOH} nm (ϵ): 243(11,500), 257(9,900), 264 (10,000), and 362(16,100) Fluorescence emission, 450 nm
Fungal source[13]	*Aspergillus flavus* Link *A. parasiticus* Speare
Toxicity data[9,15,19]	The toxicity of aflatoxin G_1 was similar to aflatoxin B_1; acute toxicity was less than B_1 but greater than B_2. The LD_{50} in ducklings was 39.2 μg/duckling using dimethylformamide as solvent; 45.7 μg/duckling (0.785 mg/kg) in another trial using dimethylformamide. The LD_{50} in the rat was twice that of aflatoxin B_1. The lesions induced by G_1 in ducklings are the same as from B_1; in the rat, the zone in the affected liver lobule was the same as in B_1 but the consistent pattern seen with B_1 was absent.
TLC data[18]	*Adsorbent:* silica gel G-HR *Solvent:* chloroform–acetone, 93:7 v/v R_f: 0.29 *Detection:* blue-green fluorescence under UV light
Other significant references	X-Ray crystallography[15]
^{13}C NMR data	C1, 154.8 s; C2, 113.2 s; C3, 159.9 s; C4, 64.3 t; C5, 28.8 t; C6, 161.1 s; C7, 106.9 s; C8, 161.0 s; C9,

91.0 d; C10, 164.6 s; C11, 107.5 s; C12, 151.7 s; C13, 113.2 d; C14, 47.7 d; C15, 102.3 d; C16, 144.8 d; C17, 56.4 q

^{1}H NMR data H4, 4.41; H5, 3.46; H9, 6.43; H13, 6.80 (J = 7.1); H14, 4.75 (J = 7.1, 2.5); H15, 5.45 (J = 2.5, 2.5); H16, 6.47 (J = 2.5); H17, 3.95

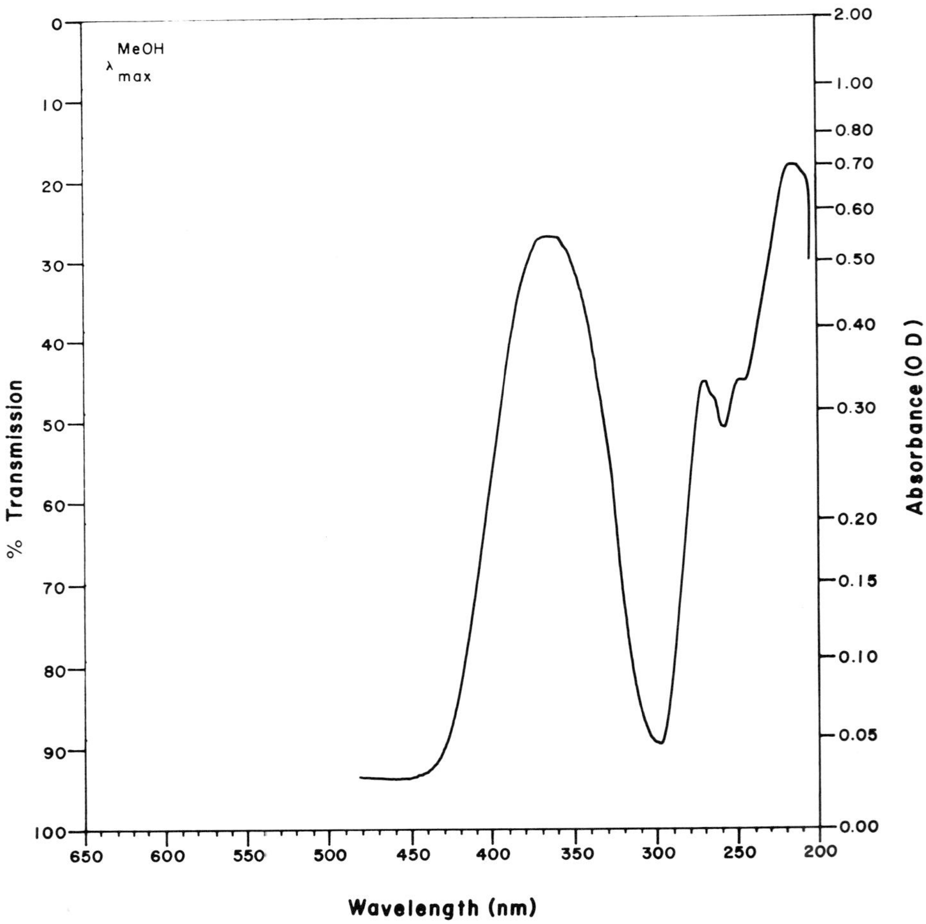

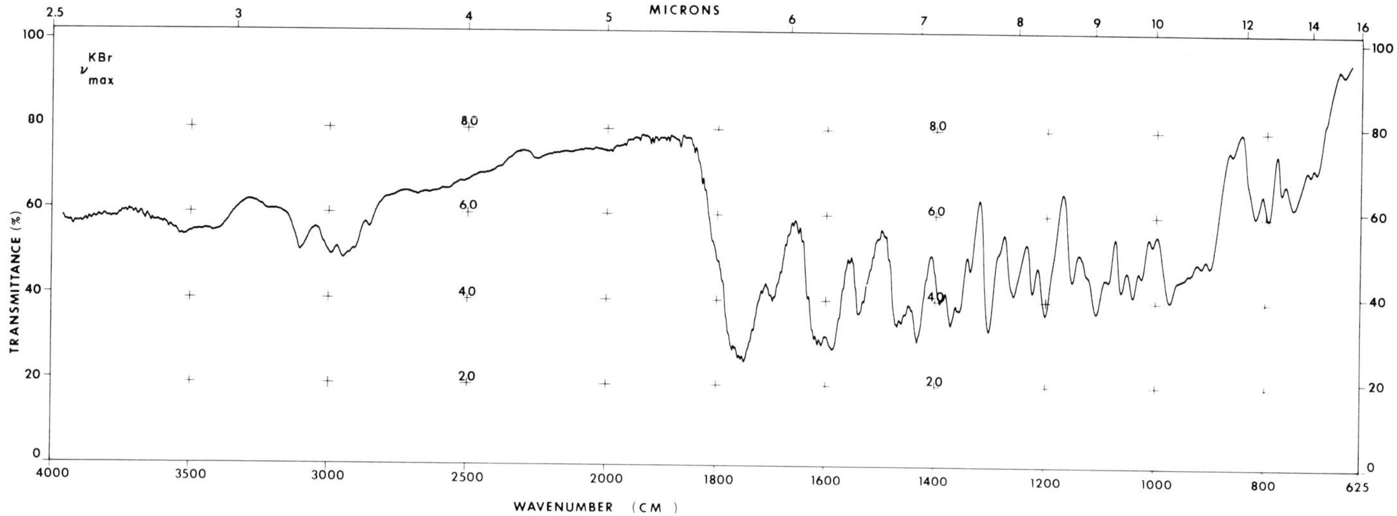
MICRONS
KBr
ν_{max}
TRANSMITTANCE (%)
WAVENUMBER (CM)

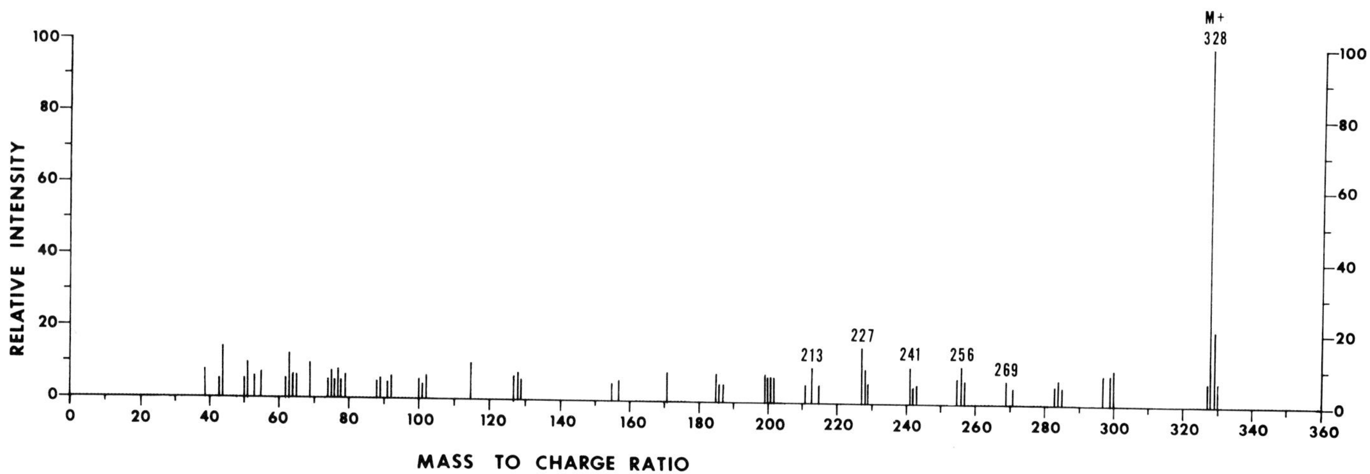
M+
328
213
227
241
256
269
RELATIVE INTENSITY
MASS TO CHARGE RATIO

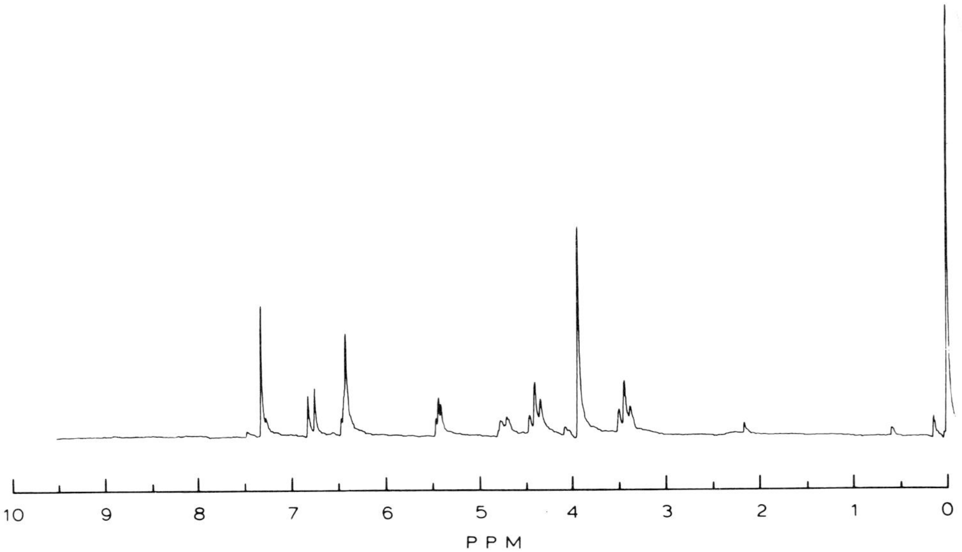

The 100 MHz ^{1}H NMR spectrum of aflatoxin G_1 in $CDCl_3$ solution.

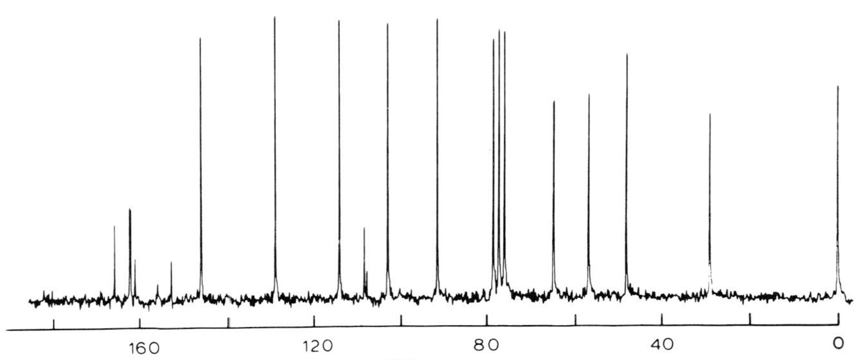

The 25 MHz ^{13}C NMR spectrum of aflatoxin G_1 in $CDCl_3$ solution.

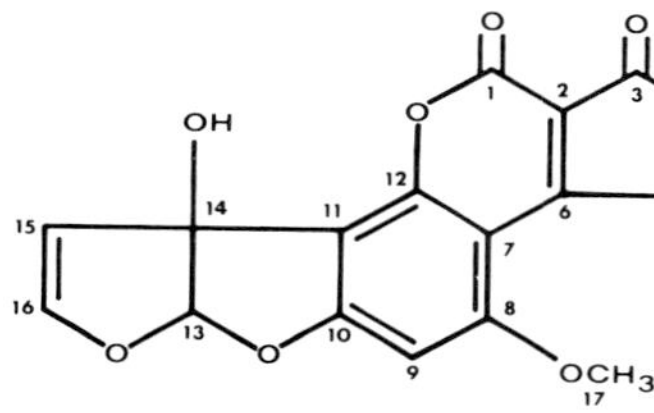

Common name	Aflatoxin M_1
Molecular weight	328.0582
Molecular formula	$C_{17}H_{12}O_7$
General characteristics[31]	Crystals from methanol, mp 299°C (dec.) $[\alpha]_D$ −280° (C = 0.1 in dimethylformamide)
UV data[31,55]	λ_{max}^{EtOH} nm (ϵ): 226(23,100), 265(11,600), and 357(19,000)
Fungal source[55]	*Aspergillus flavus* Link (NRRL 3251) *A. parasiticus* Speare
Other source[1]	Aflatoxin M_1 is excreted in the milk and urine of animals ingesting aflatoxin B_1. The conversion of B_1 to M_1 apparently occurs in the liver.
Toxicity data[12,31,46,51,62]	The LD_{50} value for aflatoxin M_1 in day old ducklings was 16.6 μg/day old duckling compared to 12 μg/duckling for aflatoxin B_1 determined simultaneously. Ducklings dosed with aflatoxin M_1 showed liver lesions indistinguishable from those induced by aflatoxin B_1; however, M_1 also induced a renal tubular necrosis not seen with B_1. In trout and rats, aflatoxin M_1 was slightly less effective than B_1 in inducing hepatic carcinoma.
TLC data[31,55,56]	A *Adsorbent:* Kieselgel G *Solvent:* chloroform–methanol, 97.3 v/v R_f: 0.34 *Detection:* blue fluorescence under long wave UV light

NOTE: For detailed discussion of TLC of aflatoxins M_1 and M_2 see Büchi *et al.*[6]

B

Adsorbent: Adsorbosil-1

Solvent: isopropyl alcohol–acetone–chloroform, 5:10:85 v/v/v

R_f: 0.46

Detection: blue fluorescence under long wave UV light

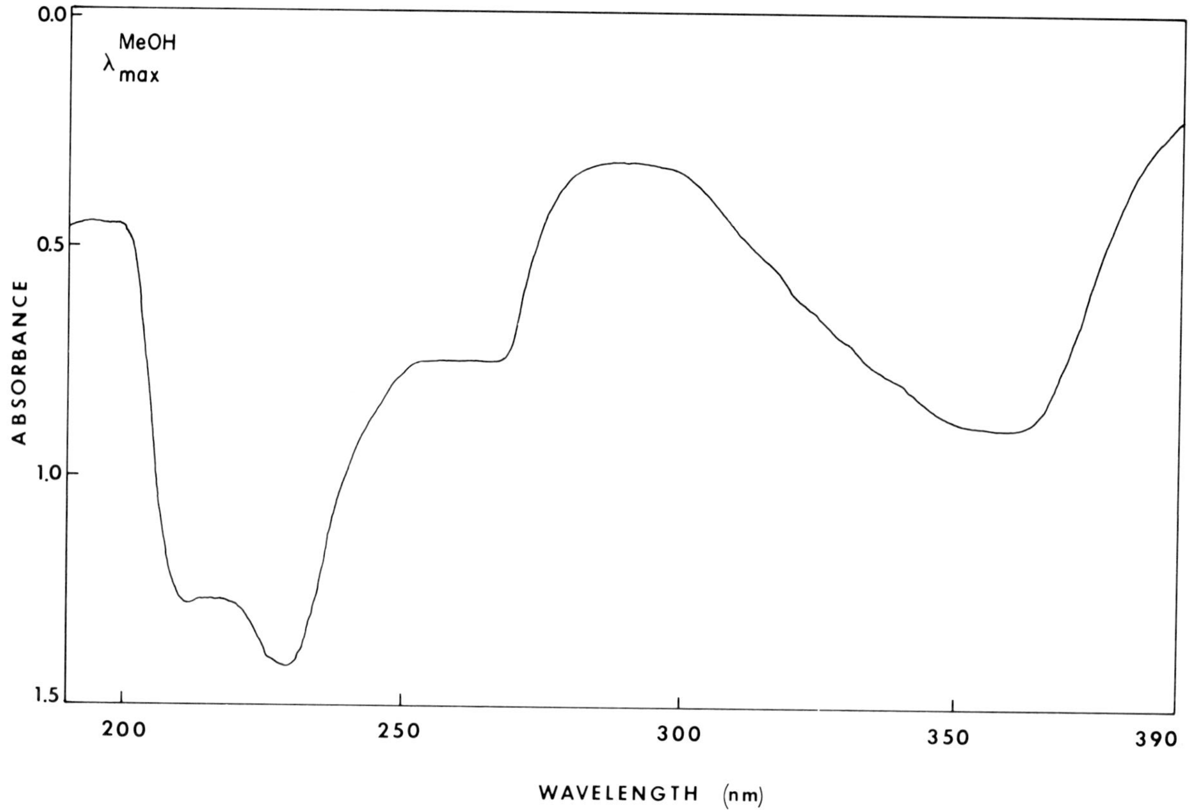

λ_{max}^{MeOH}
ABSORBANCE
0.0
0.5
1.0
1.5
200
250
300
350
390
WAVELENGTH (nm)

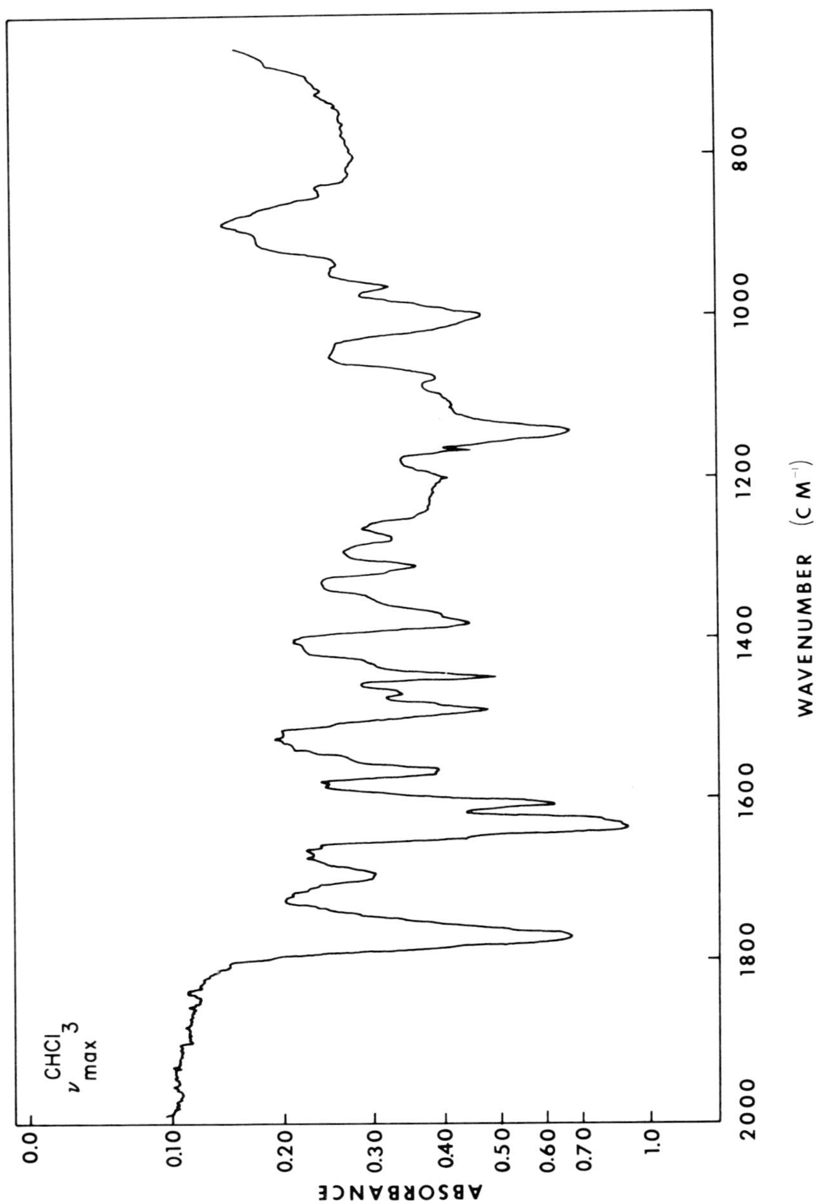

$\nu_{max}^{CHCl_3}$
ABSORBANCE
0.0
0.10
0.20
0.30
0.40
0.50
0.60
0.70
1.0
2000
1800
1600
1400
1200
1000
800
WAVENUMBER (CM⁻¹)

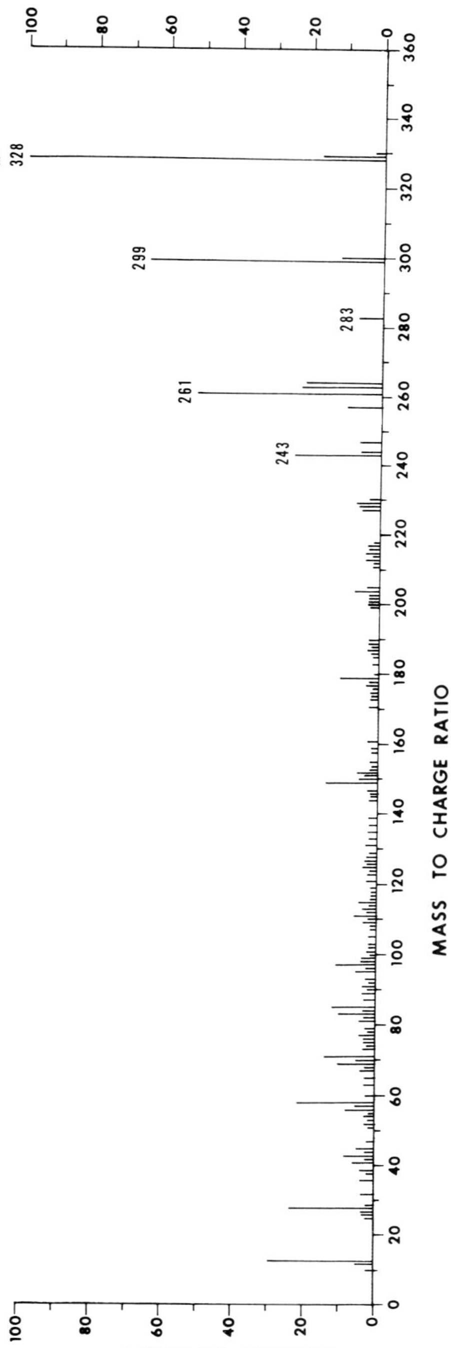

RELATIVE INTENSITY
0
20
40
60
80
100
MASS TO CHARGE RATIO
0
20
40
60
80
100
120
140
160
180
200
220
240
260
280
300
320
340
360
243
261
283
299
M+
328

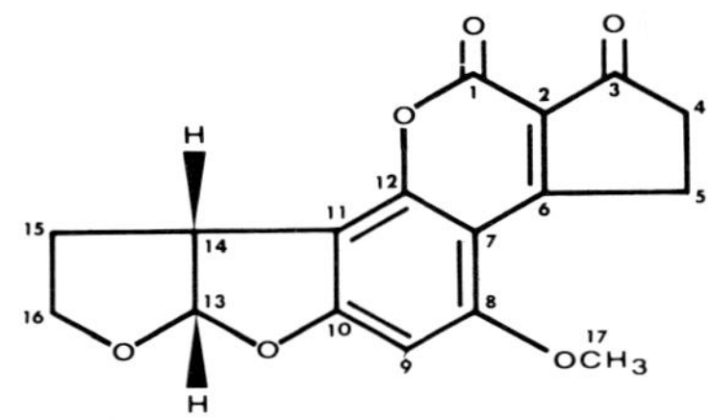

Common name	Aflatoxin B_2
Molecular weight	314.0790
Molecular formula	$C_{17}H_{14}O_6$
General characteristics[14,57,59]	Crystals from chloroform–pentane, mp 287°–289°C (dec.) $[\alpha]_D$ −430° (C = 1 in chloroform) Crystals from chloroform, mp 305°C (dec.) $[\alpha]_D$ −490° (C = *1.0* in chloroform)
UV data[57]	λ_{max} nm (ϵ): 220(20,500), 265(12,700), and 363(24,000) Fluorescence emission, 425 nm
Fungal source[14,57,59]	*Aspergillus flavus* Link *A. parasiticus* Speare
Toxicity data[14]	The toxic properties of aflatoxin B_2 were similar to those for aflatoxin B_1; however, potency was markedly reduced compared to aflatoxin B_1. A dosage in ducklings of 50 μg of aflatoxin B_2 was required to produce the same degree of bile duct proliferation produced by 3.9 μg of aflatoxin B_1.
TLC data[14]	*Adsorbent:* silica gel G-HR *Solvent:* chloroform–acetone, 93:7 v/v R_f: 0.31 *Detection:* bright blue fluorescence under UV light
Other significant references	X-Ray crystallography[60]
^{13}*C NMR data*	C1, 155.8 s; C2, 115.6 s; C3, 200.0 s; C4, 34.6 t; C5, 28.5 t; C6, 176.6 s; C7, 105.5 s; C8, 160.8 s; C9, 90.0 d; C10, 165.8 s; C11, 105.9 s; C12, 152.0 s; C13, 113.3 d; C14, 42.8 d; C15, 30.6 t; C16, 66.9 t; C17, 56.6 q

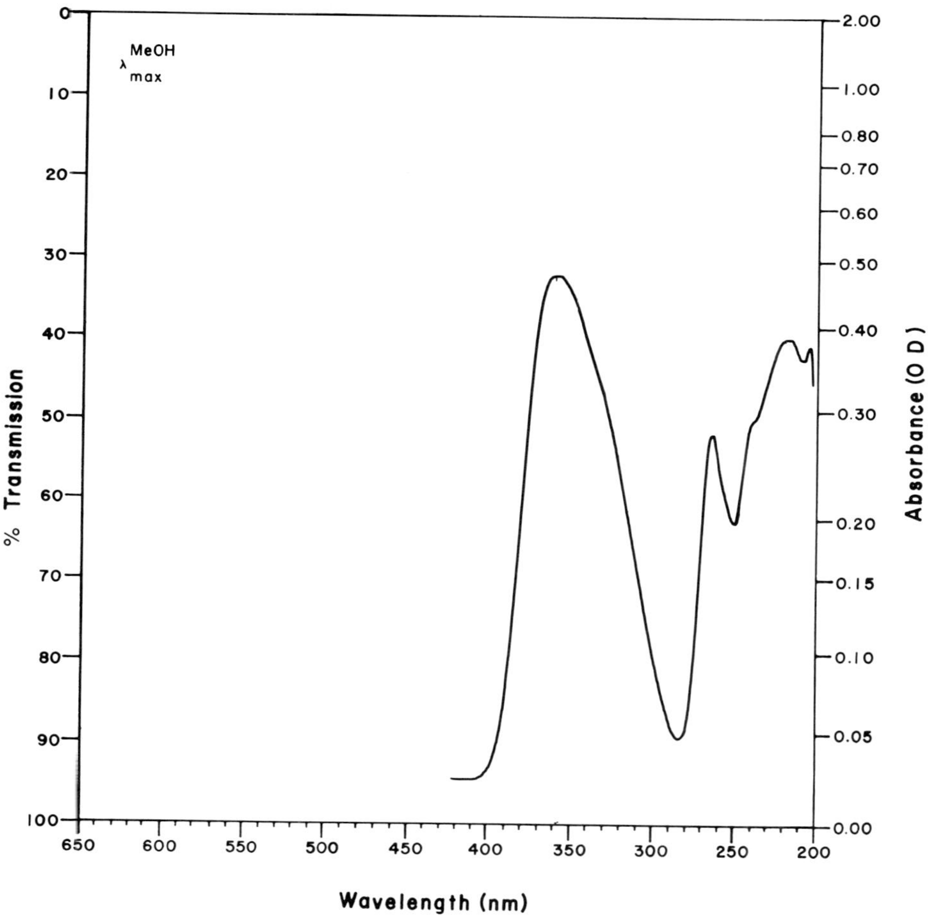

λ MeOH max
% Transmission
Absorbance (O D)
Wavelength (nm)
0
10
20
30
40
50
60
70
80
90
100
2.00
1.00
0.80
0.70
0.60
0.50
0.40
0.30
0.20
0.15
0.10
0.05
0.00
650
600
550
500
450
400
350
300
250
200

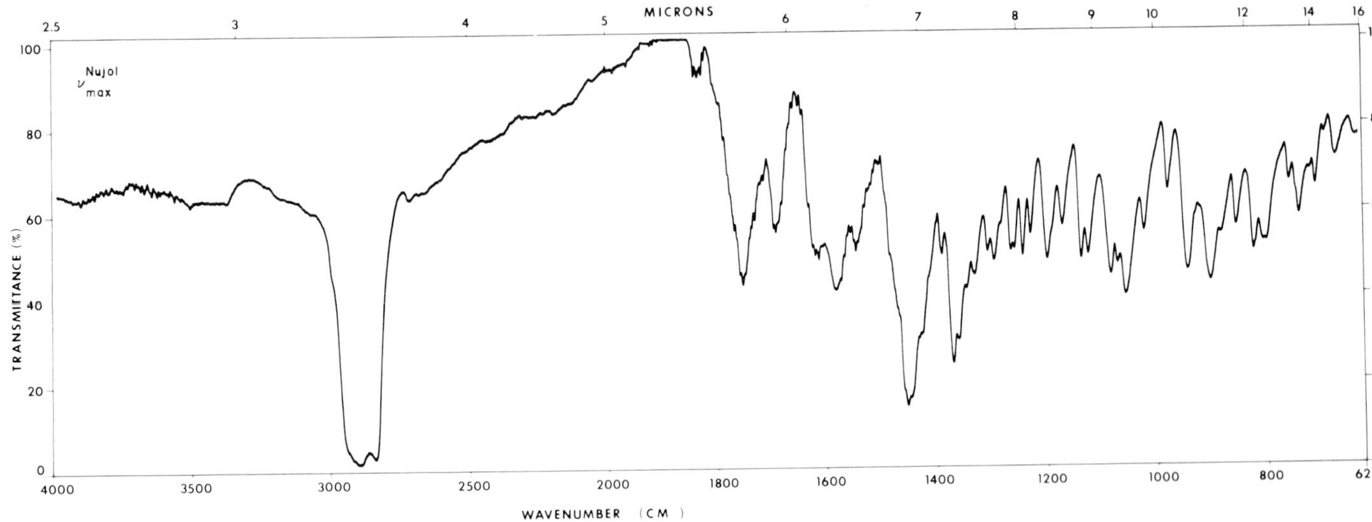
MICRONS
2.5
3
4
5
6
7
8
9
10
12
14
16
Nujol
ν_{max}
TRANSMITTANCE (%)
WAVENUMBER (CM)
4000
3500
3000
2500
2000
1800
1600
1400
1200
1000
800
625

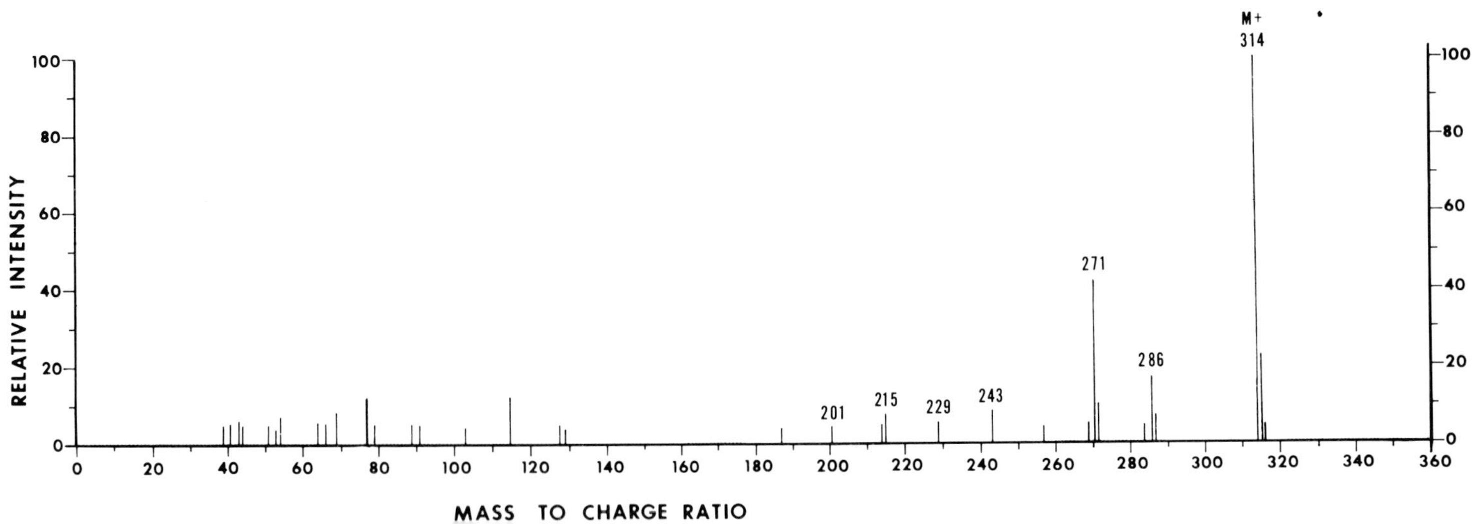
RELATIVE INTENSITY
M+
314
286
271
243
229
215
201
MASS TO CHARGE RATIO

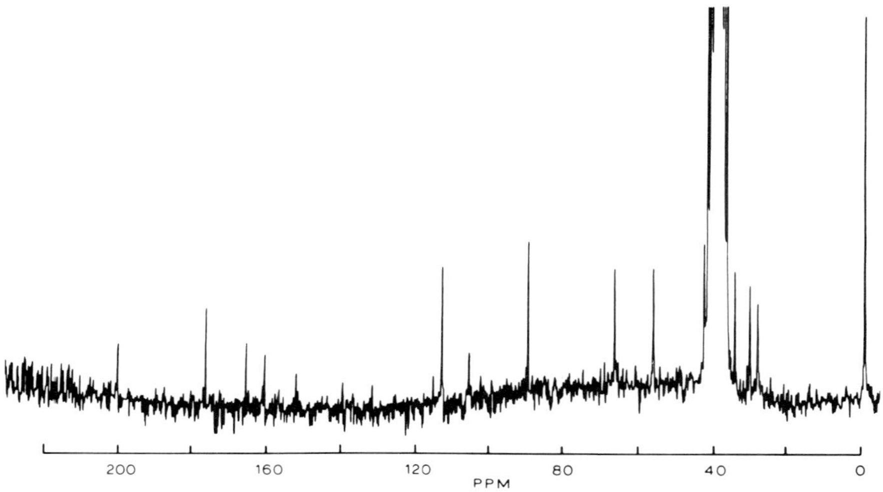

The 25 MHz ^{13}C NMR spectrum of aflatoxin B_2 in DMSO solution.

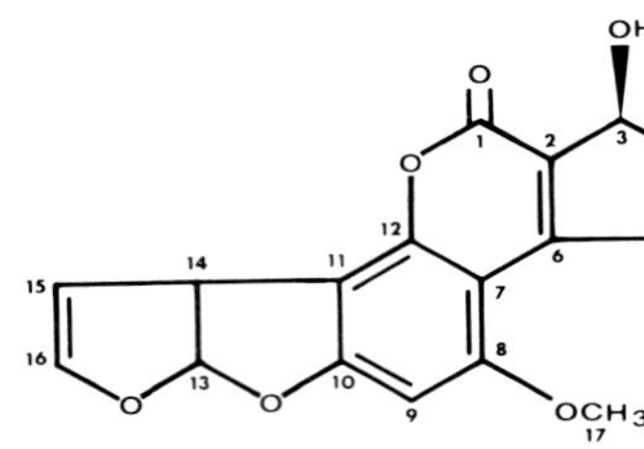

Common name	Aflatoxicol A (Ro)
Molecular weight	314.0790
Molecular formula	$C_{17}H_{14}O_6$
General characteristics[17]	Crystals from benzene–*n*-hexane, mp 224°–226°C
UV data[23,44]	λ_{max}^{EtOH} nm (ϵ): 332(14,100), 261(10,800), and 254(6,790) Fluorescence emission, 425 nm
Source[16,17,47,56]	Aflatoxicol A (Ro) is produced from the *in vitro* incubation of aflatoxin B_1 with submitochondrial liver fractions from several animal species. It is also produced from biological reduction of aflatoxin B_1 by *Tetrahymena pyriformis, Dactylium dendroides, Rhizopus* spp., and other microorganisms. It is reported to occur with a diastereoisomer (aflatoxicol B).
Toxicity data[22]	Reportedly 18 times less toxic than aflatoxin B_1 in the duckling biliary hyperplasia assay.
TLC data[17]	*Adsorbent:* silica gel G-HR *Solvent:* chloroform–acetone, 93:7 v/v R_f: 0.30 *Detection:* blue fluorescence under UV light

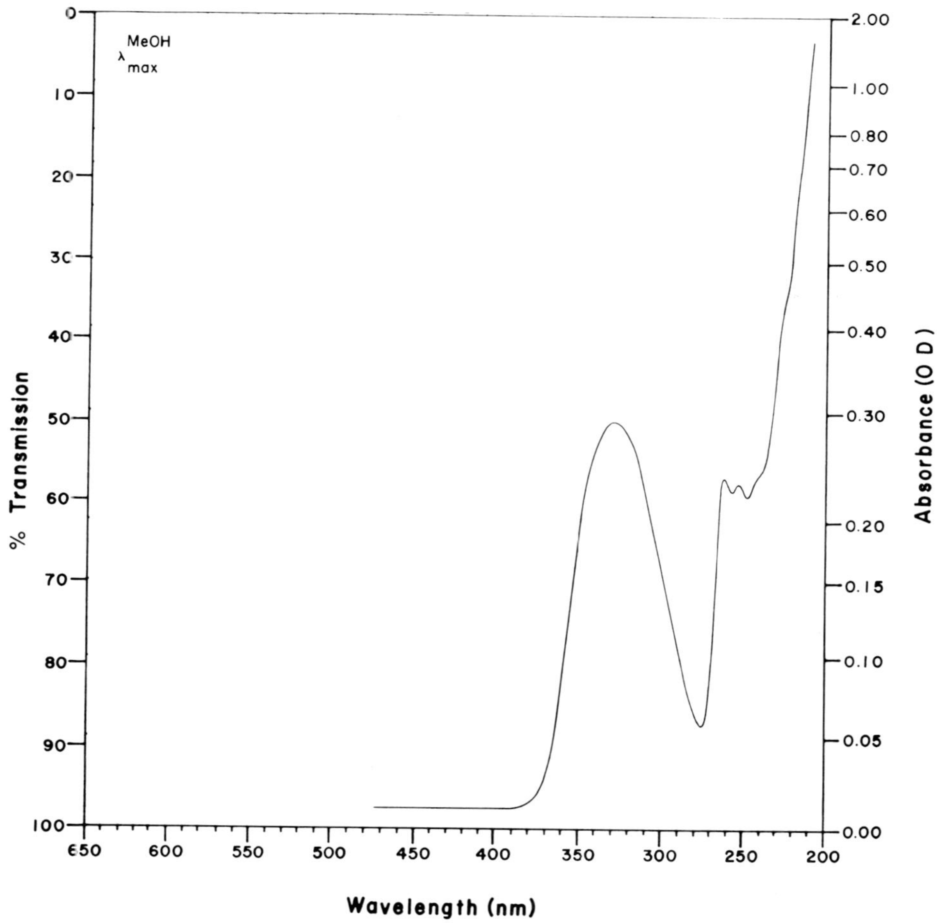

λ MeOH max
% Transmission
0
10
20
30
40
50
60
70
80
90
100
Absorbance (O D)
2.00
1.00
0.80
0.70
0.60
0.50
0.40
0.30
0.20
0.15
0.10
0.05
0.00
650
600
550
500
450
400
350
300
250
200
Wavelength (nm)

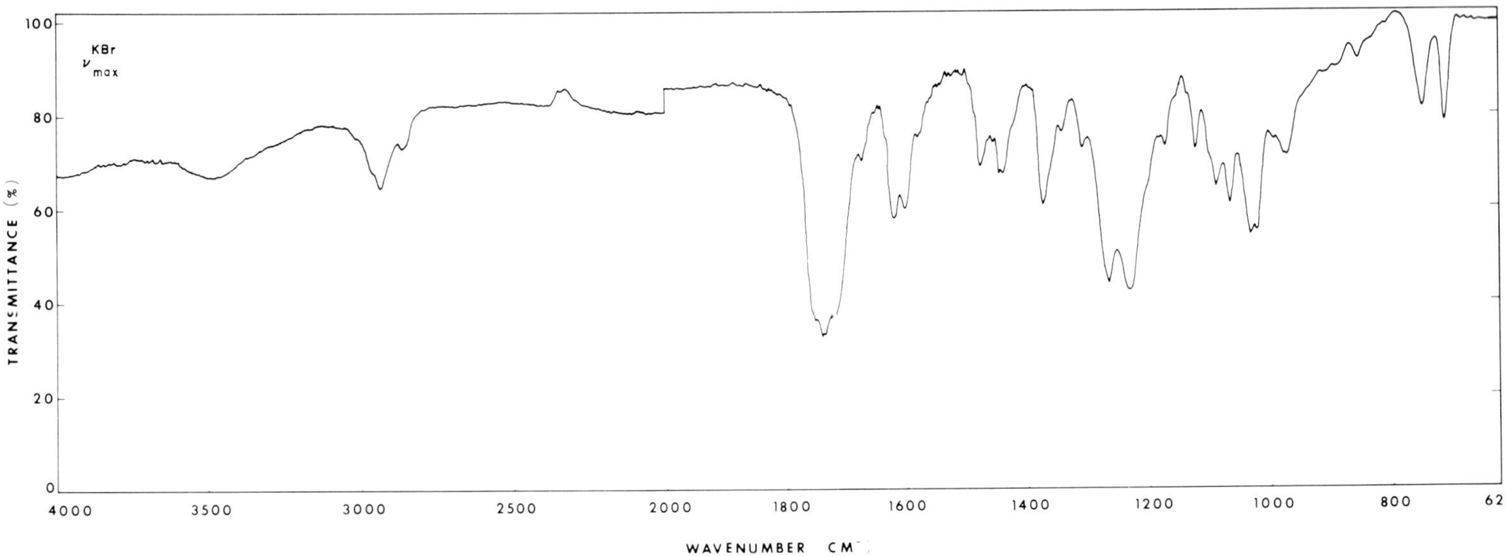

ν_{max}^{KBr}
TRANSMITTANCE (%)
WAVENUMBER CM

1-P

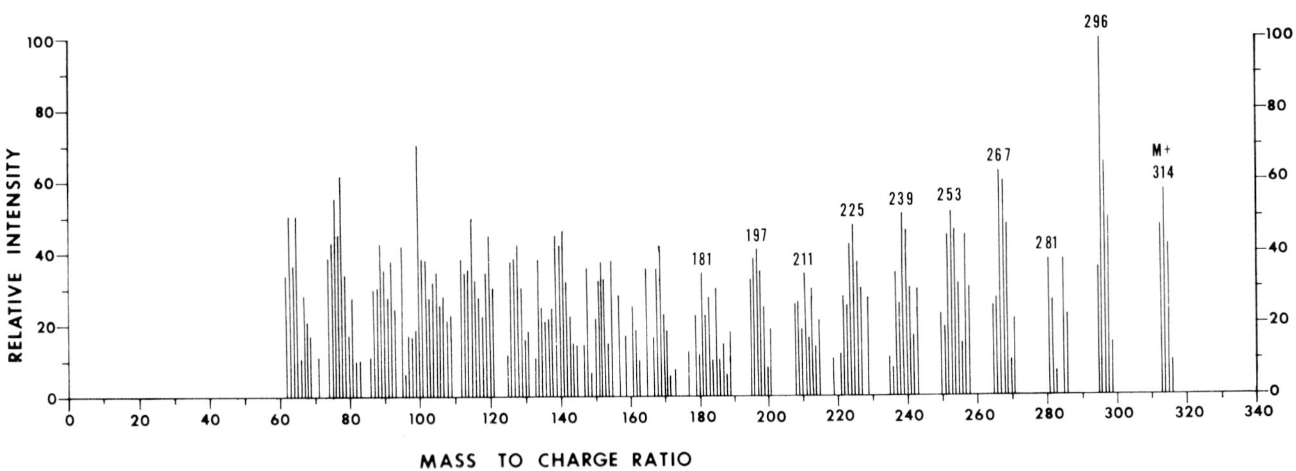

RELATIVE INTENSITY
MASS TO CHARGE RATIO
M+ 314
296
281
267
253
239
225
211
197
181

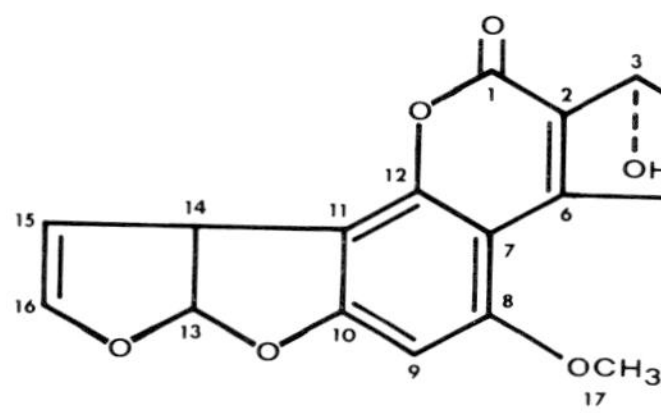

Common name	Aflatoxicol B
Molecular weight	314.0790
Molecular formula	$C_{17}H_{14}O_6$
General characteristics[17]	Crystals from benzene–*n*-hexane, decomposed over broad range, starting at 233°C
UV data[23,44]	λ_{max}^{EtOH} nm (ϵ): 325(14,100), 261(10,800), and 254(6,790)
Source[16,17,22,56]	Aflatoxicol B is obtained from biological reduction of aflatoxin B_1 by *Rhizopus* spp., *Dactylium dendroides, Tetrahymena pyriformis,* and other microorganisms.
Toxicity data	Biological activity unknown
TLC data[17]	*Adsorbent:* silica gel G-HR *Solvent:* chloroform–acetone, 93:7 v/v R_f: 0.26 *Detection:* blue fluorescence under UV light

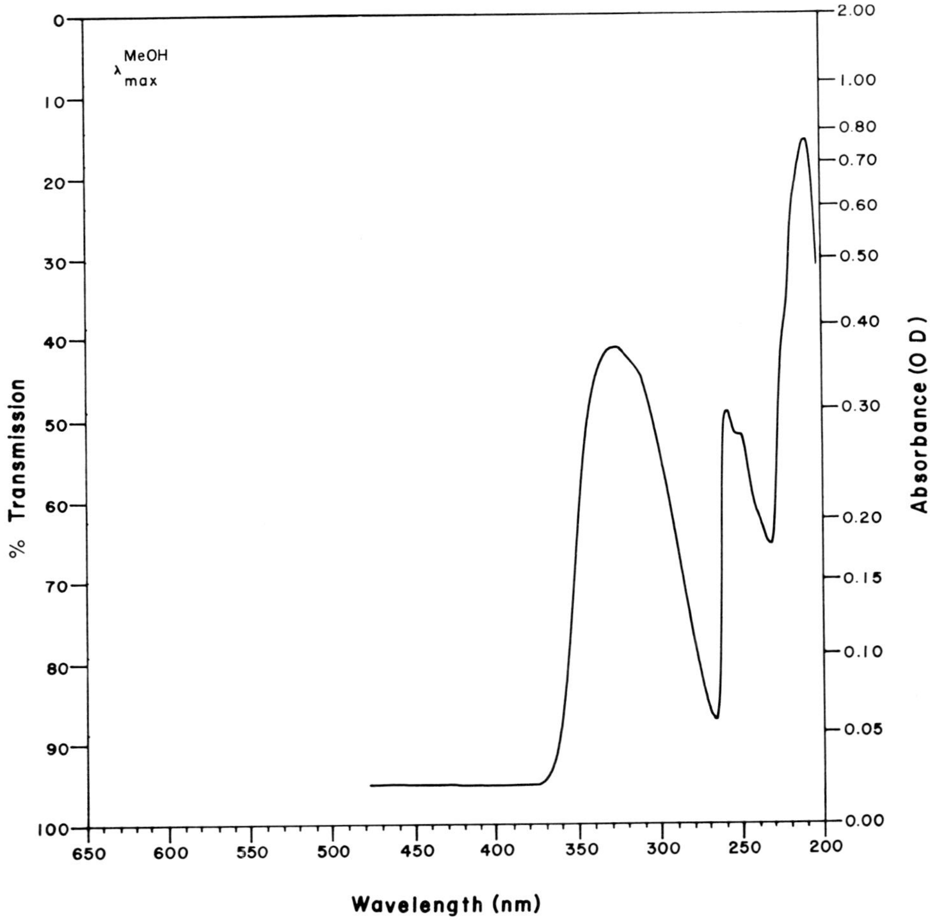
λ_{max}^{MeOH}
% Transmission
0
10
20
30
40
50
60
70
80
90
100
Absorbance (O D)
2.00
1.00
0.80
0.70
0.60
0.50
0.40
0.30
0.20
0.15
0.10
0.05
0.00
650
600
550
500
450
400
350
300
250
200
Wavelength (nm)

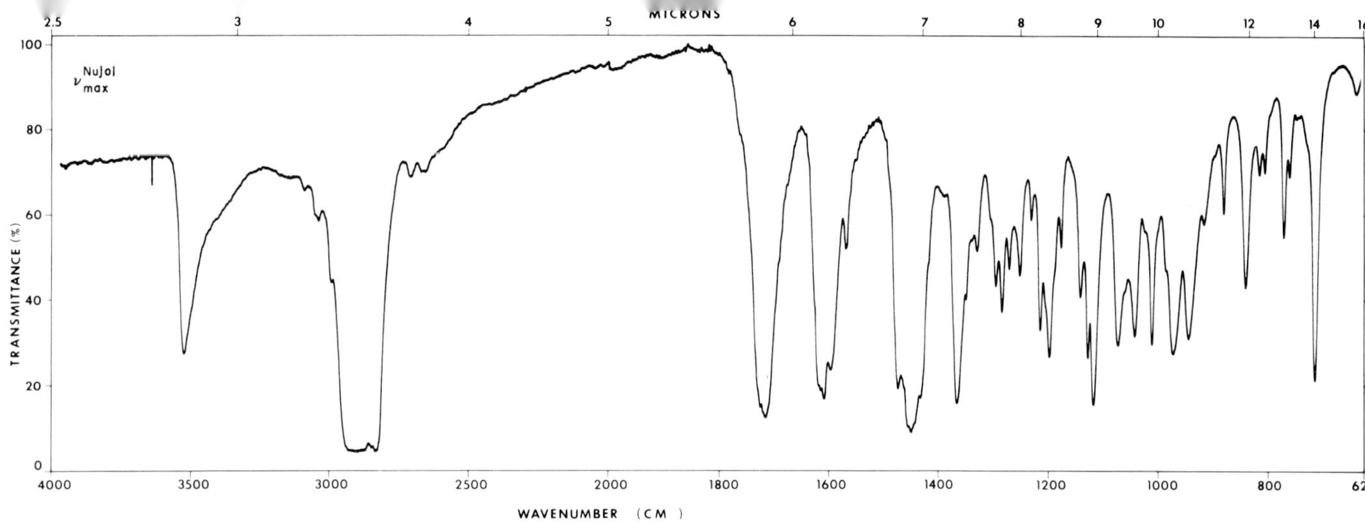
MICRONS
TRANSMITTANCE (%)
ν_{max}^{Nujol}
WAVENUMBER (CM)

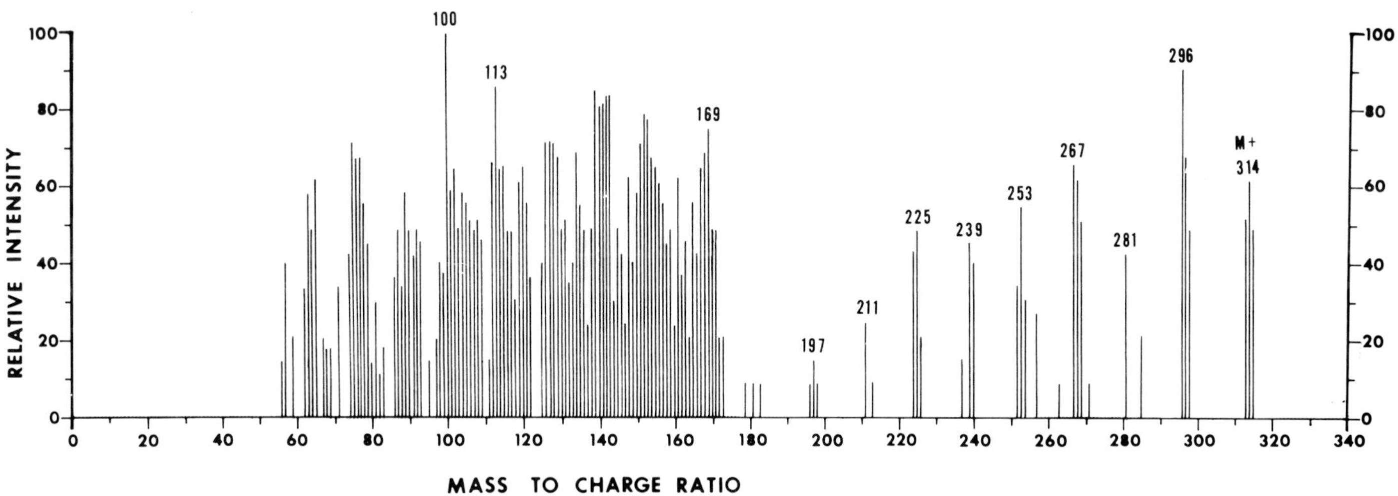
RELATIVE INTENSITY
100
113
169
197
211
225
239
253
267
281
296
M+
314
MASS TO CHARGE RATIO

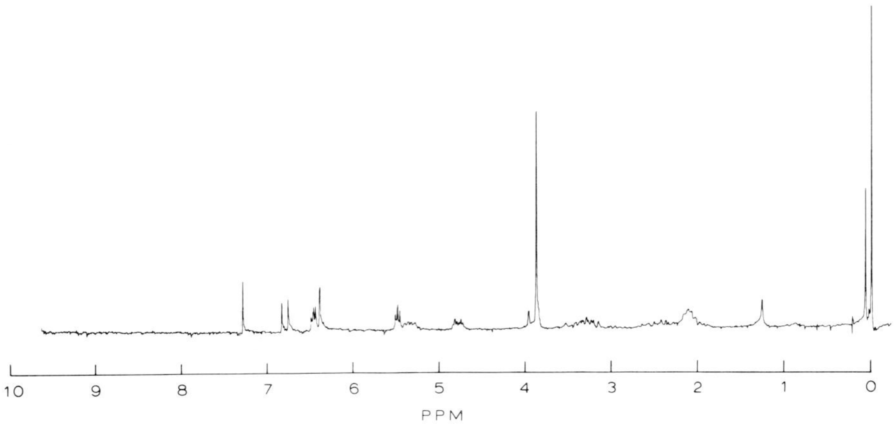

The 100 MHz ^{1}H NMR spectrum of aflatoxicol B in $CDCl_3$ solution.

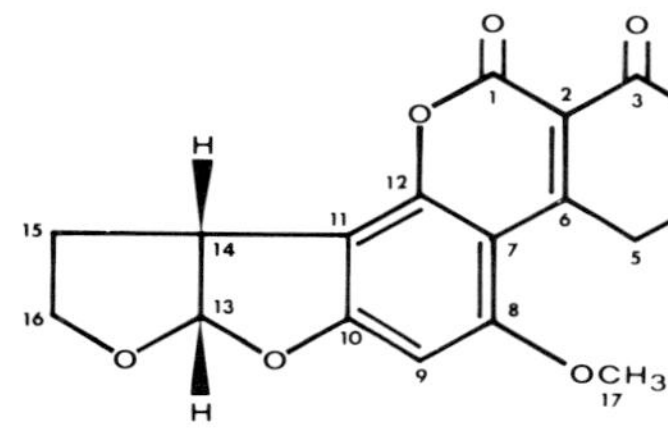

Common name	Aflatoxin G_2
Molecular weight	330.0739
Molecular formula	$C_{17}H_{14}O_7$
General characteristics[57]	Crystals from ethyl acetate, mp 230°C $[\alpha]_D$ −454° (C = 1.0 chloroform)
UV data[57]	λ_{max} nm (ϵ): 217(28,000), 245(12,900), 265(11,200), and 365(19,300) Fluorescence emission, 450 nm
Fungal source[57]	*Aspergillus flavus* Link *A. parasiticus* Speare
Toxicity data[37]	Aflatoxin G_2 is the least acutely toxic of the 4 major naturally occurring aflatoxins. The LD_{50} in ducklings was 172.5 μg/duckling using dimethylformamide as solvent in a 6-day test.
TLC data[18]	*Adsorbent:* silica gel G-HR *Solvent:* chloroform–acetone, 93:7 v/v R_f: 0.27 *Detection:* blue-green fluorescence under UV light
^{13}C *NMR data*	C1, 154.9 s; C2, 113.7 s; C3, 160.0 s; C4, 64.3 t; C5, 28.9 t; C6, 161.3 s; C7, 102.1 s; C8, 161.0 s; C9, 90.3 d; C10, 166.1 s; C11, 106.6 s; C12, 152.2 s; C13, 113.7 d; C14, 43.9 d; C15, 31.4 t; C16, 67.7 t; C17, 56.4 q
^{1}H *NMR data*	H4, 3.46; H5, 2.30; H9, 6.34; H13, 6.47; H14, 3.60; H15, 4.16; H16, 4.42; H17, 3.94

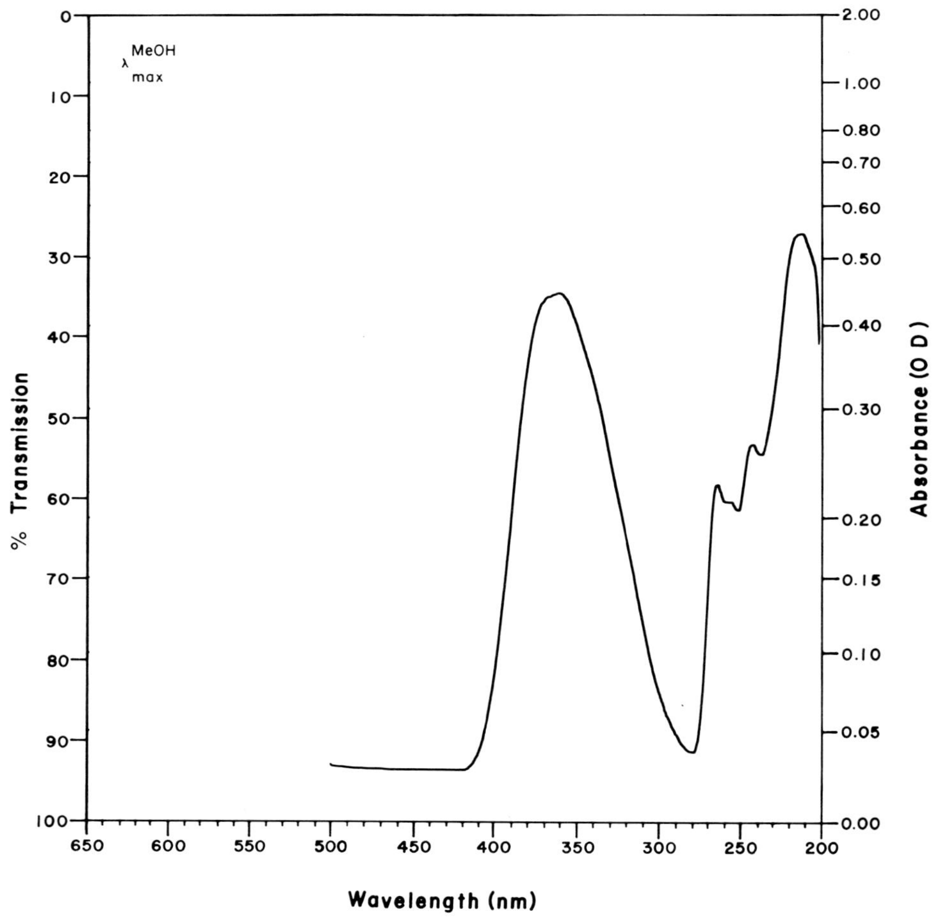

λ_{max}^{MeOH}
% Transmission
0
10
20
30
40
50
60
70
80
90
100
Absorbance (O D)
2.00
1.00
0.80
0.70
0.60
0.50
0.40
0.30
0.20
0.15
0.10
0.05
0.00
650
600
550
500
450
400
350
300
250
200
Wavelength (nm)

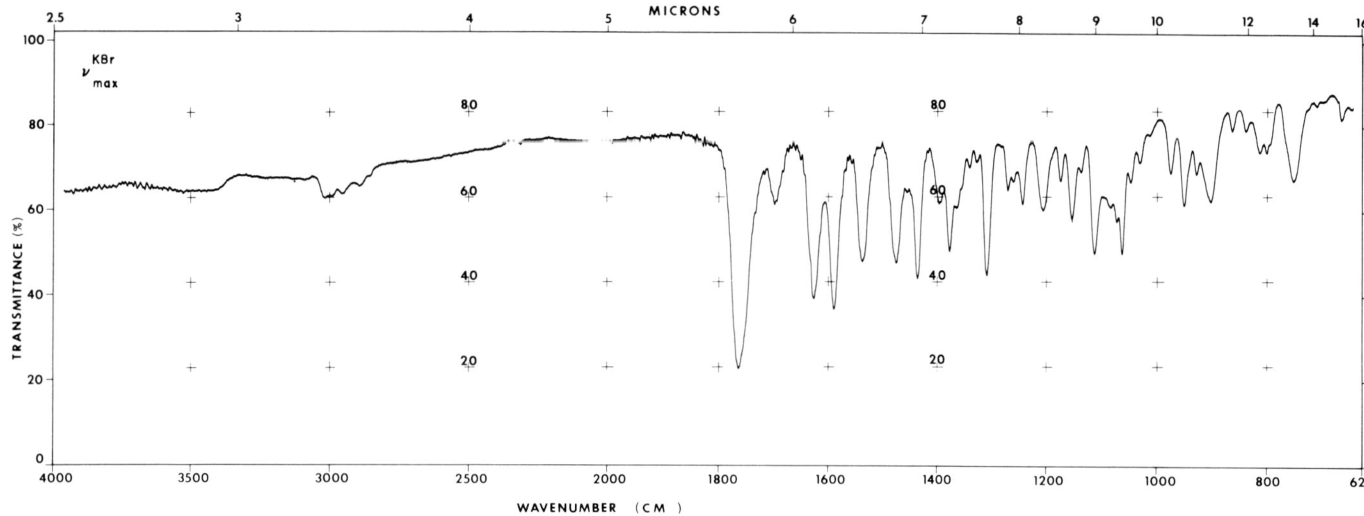
MICRONS
KBr
ν_{max}
TRANSMITTANCE (%)
WAVENUMBER (CM)

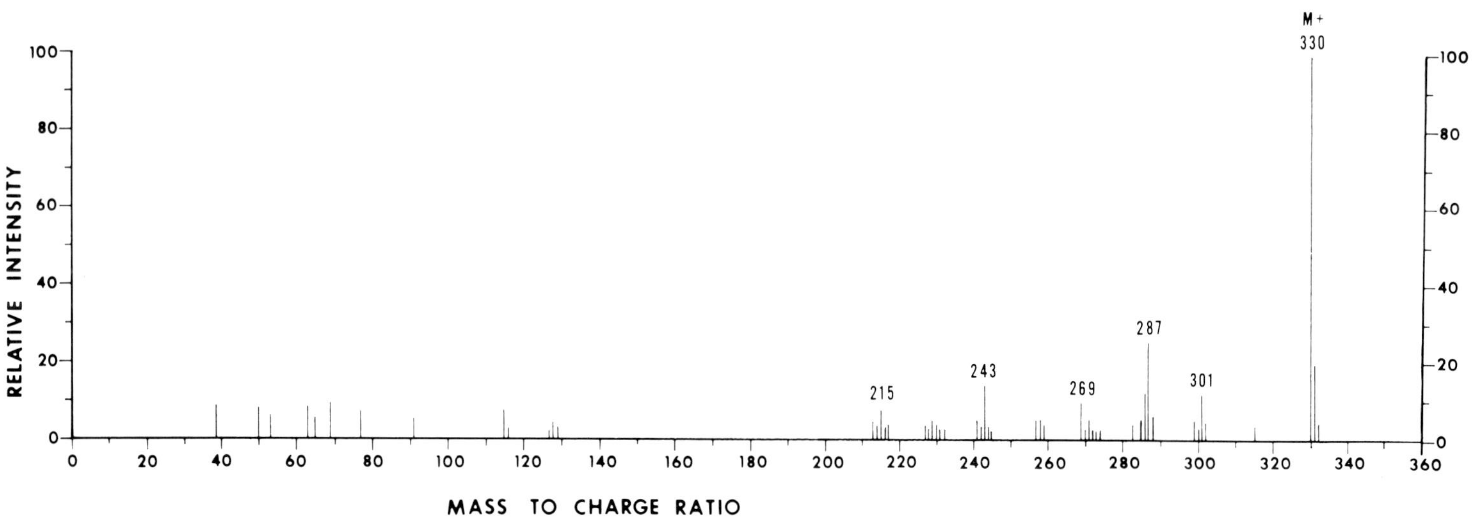
M+
330
301
287
269
243
215
RELATIVE INTENSITY
MASS TO CHARGE RATIO

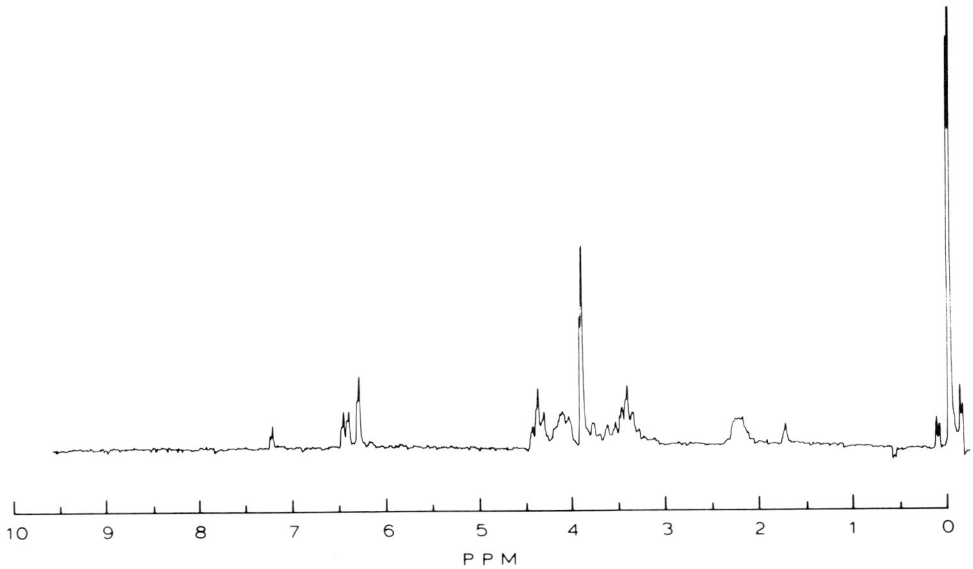

The 100 MHz ^1H NMR spectrum of aflatoxin G_2 in $CDCl_3$ solution.

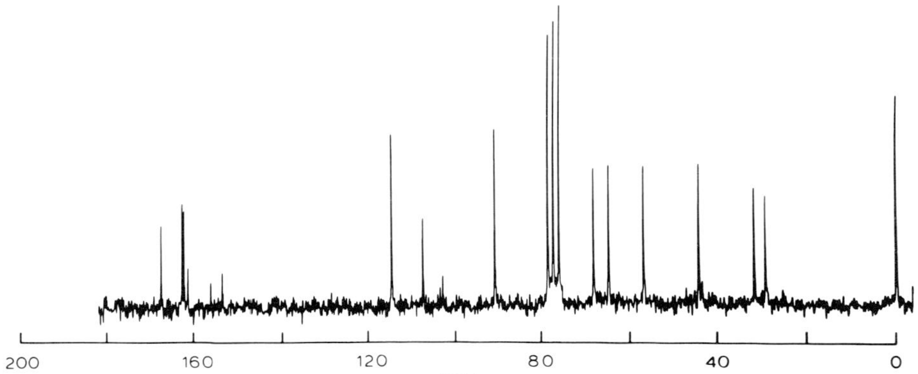

The 25 MHz ^{13}C NMR spectrum of aflatoxin G_2 in $CDCl_3$ solution.

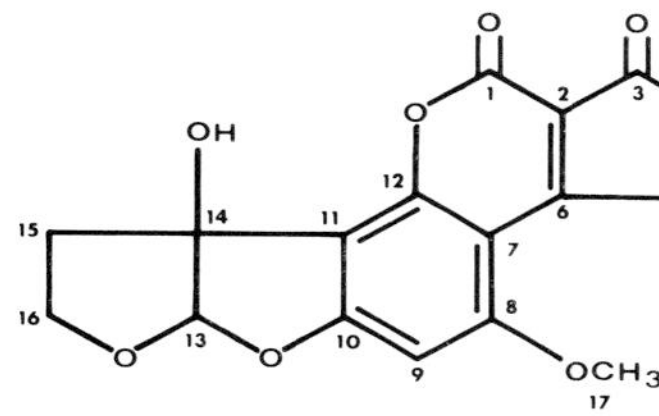

Common name	Aflatoxin M_2
Molecular weight	330.0739
Molecular formula	$C_{17}H_{14}O_7$
General characteristics[31]	Crystals from methanol–chloroform, mp 293°C (dec.)
UV data[31]	λ_{max}^{EtOH} nm (ϵ): 221(20,000), 264(10,900), and 357(21,000)
IR data[31]	$\nu_{max}^{CHCl_3}$ cm^{-1}: 3350, 1760, and 1690
Fungal source[31,55]	*Aspergillus flavus* Link *A. parasiticus* Speare
Other source[55]	Aflatoxin M_2 is excreted in the urine and milk of animals ingesting mixed aflatoxins.
Toxicity data[46]	The LD_{50} value for aflatoxin M_2 was 62 μg/day old duckling compared to 12 μg for aflatoxin B_1 in the same assay done simultaneously.
TLC data[55]	*Adsorbent:* Adsorbosil-1 silica gel *Solvent*: isopropyl alcohol–acetone–chloroform, 5:10:85 v/v/v R_f: 0.42 *Detection:* blue fluorescence under UV light *NOTE:* For detailed discussion of TLC of aflatoxin M_1 and M_2, see Asao *et al.*[3]

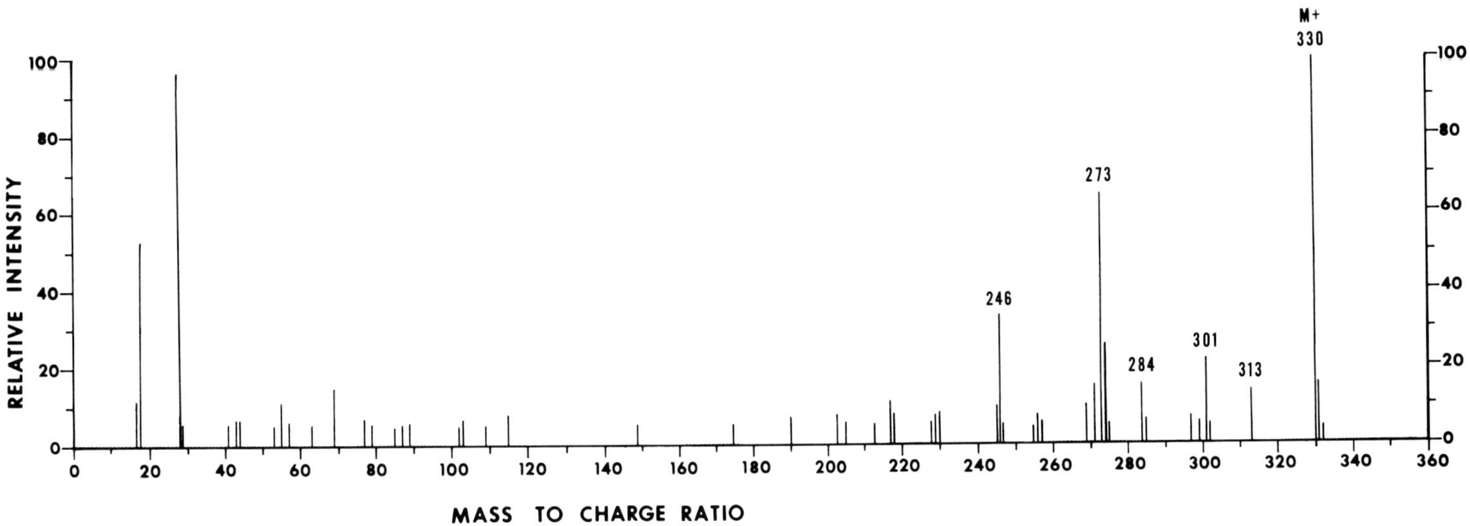
M+
330
313
301
284
273
246
RELATIVE INTENSITY
MASS TO CHARGE RATIO

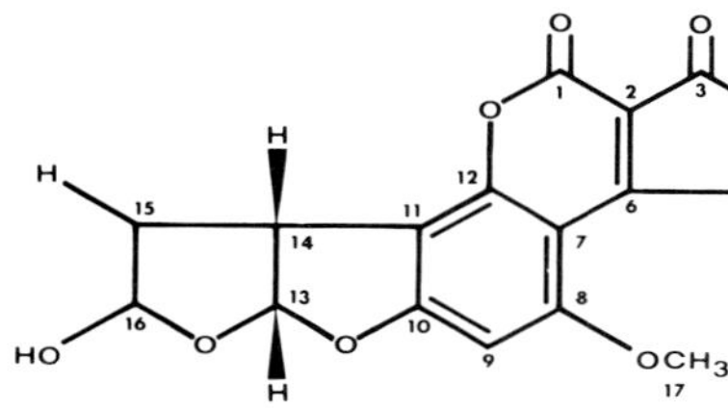

Common name	Aflatoxin B_{2a}
Molecular weight	330.0739
Molecular formula	$C_{17}H_{14}O_7$
General characteristics[27,30]	mp 240°C(dec.) *NOTE:* For isolation and characteristics of aflatoxin M_{2a} and GM_{2a}, see Holzapfel *et al.*[31]
UV data[27]	λ_{max}^{MeOH} nm (ϵ): 228(17,600), 256(10,300), and 363(20,400)
Fungal source[25,26,45]	*Aspergillus flavus* Link *A. parasiticus* Speare
Other source[43]	Biotransformation product of aflatoxin B_1 by the liver of some animals; a detoxification mechanism.
Toxicity data[38]	Aflatoxin B_{2a} is more than 200 times less toxic than aflatoxin B_1 as measured by the initiation of bile duct proliferation in the standard duckling assay. No acute toxicity was noted in Khaki Cambell ducklings (day-old) at levels up to 1200 μg/duckling.
TLC data[25]	*Adsorbent:* Kieselgel G *Solvent:* chloroform–methanol, 98:2 v/v R_f: 0.13 *Detection:* blue fluorescence under UV light
^{13}*C NMR data*	C1, 153.9 s; C2, 117.4 s; C3, 200.0 s; C4, 34.6 t; C5, 28.5 t; C6, 176.7 s; C7, 106.9 s; C8, 160.8 s; C9, 90.5 d; C10, 165.3 s; C11, 108.6 s; C12, 152.0 s; C13, 113.6 d; C14, 41.3 d; C15, 37.2 t; C16, 99.7 d; C17, 56.6 q

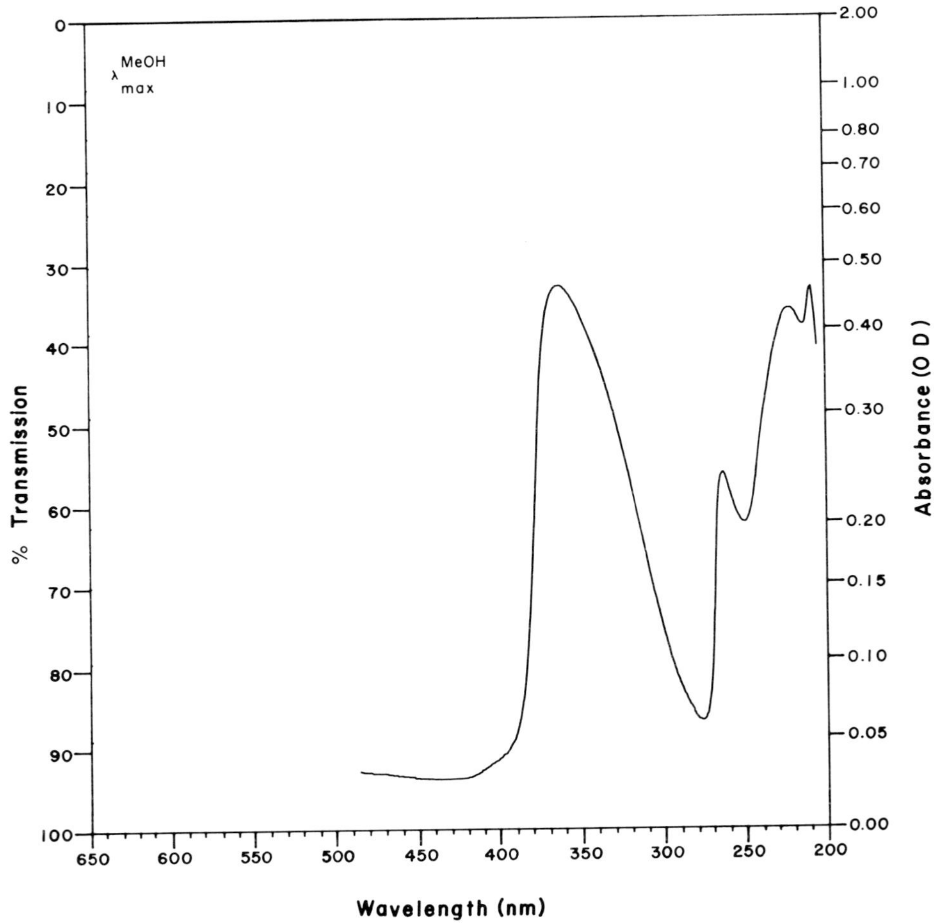
λ_{max}^{MeOH}
% Transmission
Absorbance (O D)
Wavelength (nm)
0
10
20
30
40
50
60
70
80
90
100
2.00
1.00
0.80
0.70
0.60
0.50
0.40
0.30
0.20
0.15
0.10
0.05
0.00
650
600
550
500
450
400
350
300
250
200

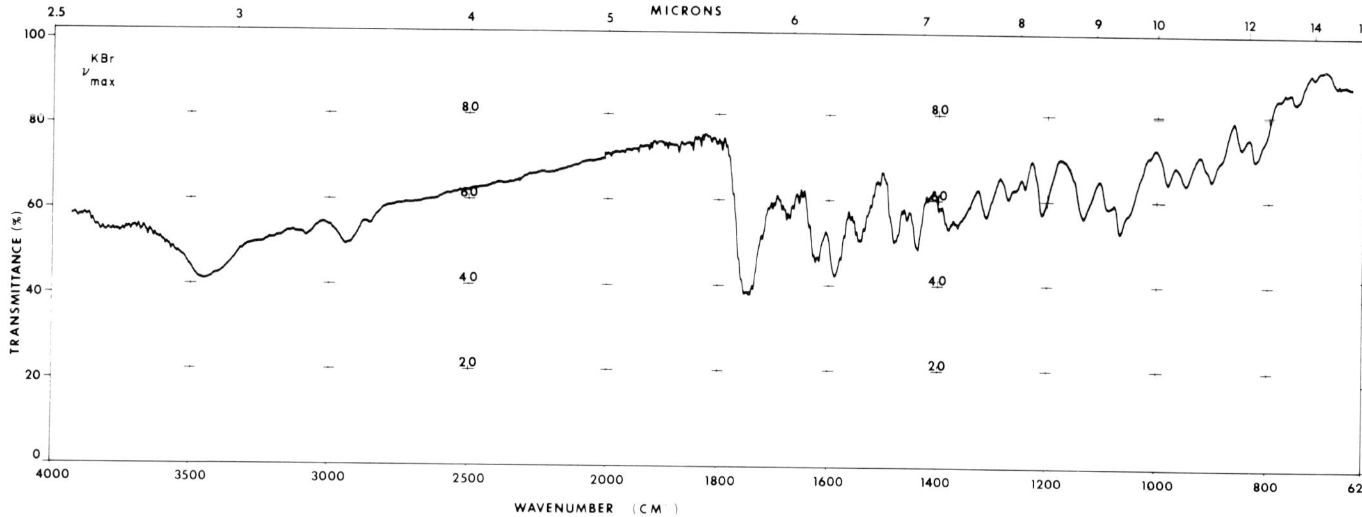
MICRONS
KBr
ν_{max}
TRANSMITTANCE (%)
WAVENUMBER (CM^{-1})

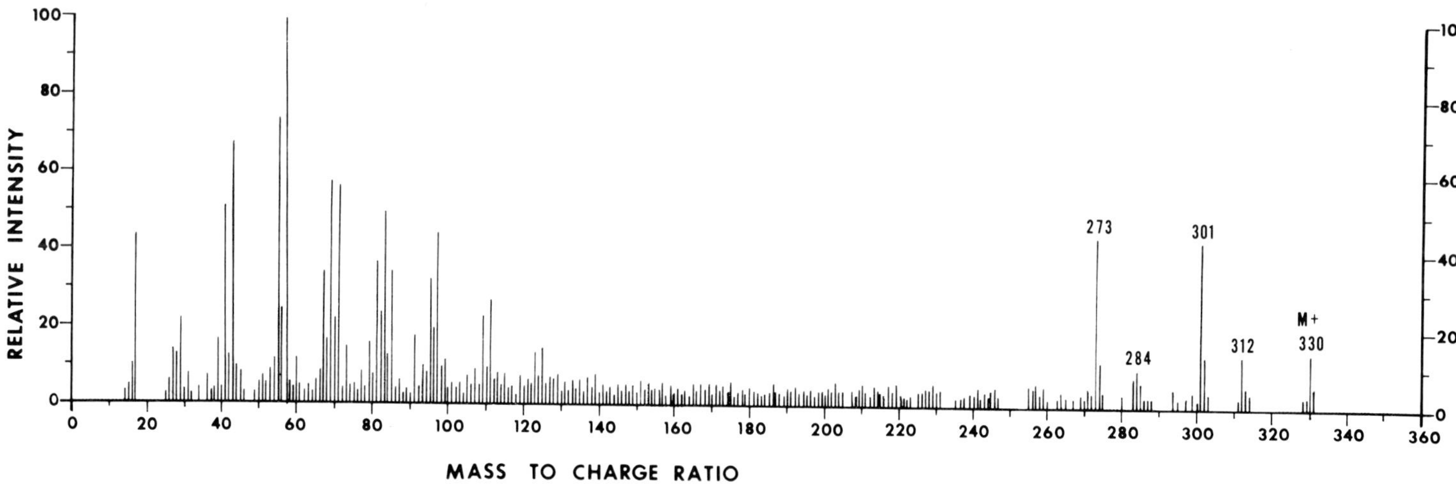
RELATIVE INTENSITY
273
284
301
312
M+
330
MASS TO CHARGE RATIO

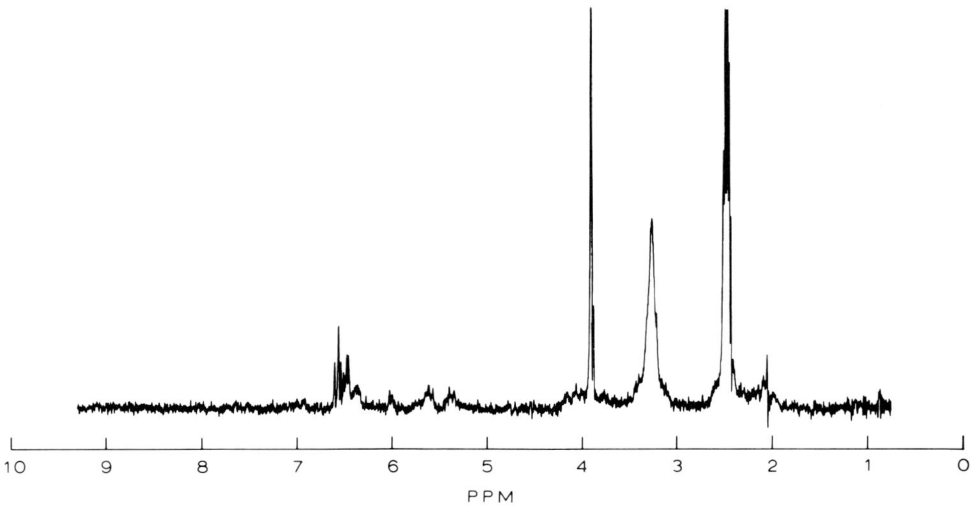

The 100 MHz ^{1}H NMR spectrum of aflatoxin B_{2a} in DMSO solution.

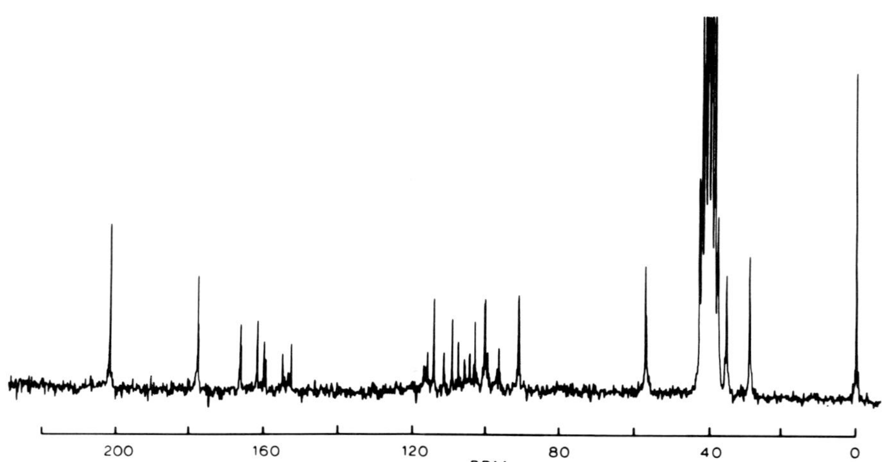

The 25 MHz ^{13}C NMR spectrum of aflatoxin B_{2a} in DMSO solution.

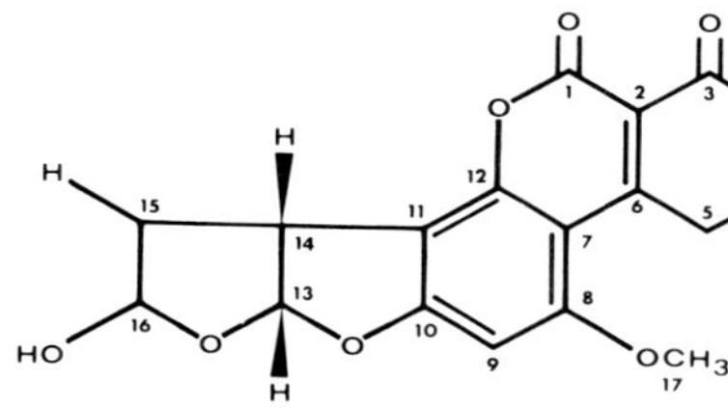

Common name	Aflatoxin G_{2a}
Molecular weight	346.0688
Molecular formula	$C_{17}H_{14}O_8$
General characteristics[25,26]	mp 190°C (dec.)
UV data[25,26]	λ_{max}^{MeOH} nm (ϵ): 223(18,600), 242(10,100), 262(8,700), and 365(18,000)
Fungal source[25]	*Aspergillus flavus* Link *A. parasiticus* Speare
Other source[43]	Also formed in livers of certain animals ingesting aflatoxin G_1; presumably a detoxification mechanism.
Toxicity data[26]	There were no significant differences in growth and no characteristic liver lesions in day-old Khaki Cambell ducklings dosed up to 1600 μg/duckling (LD_{50} of aflatoxin B_1 in same assay was 18.2 μg/duckling).
TLC data[26,27]	*Adsorbent:* Kieselgel G *Solvent:* chloroform–methanol, 98:2 v/v R_f: 0.10 *Detection:* blue fluorescence under UV light
^{13}C NMR data	C1, 153.9 s; C2, 111.0 s; C3, 159.6 s; C4, 64.1 t; C5, 28.4 t; C6, 161.9 s; C7, 105.3 s; C8, 161.6 s; C9, 91.1 d; C10, 164.9 s; C11, 108.6 s; C12, 151.6 s; C13, 113.7 d; C14, 41.9 d; C15, 41.9 t; C16, 91.1 d; C17, 56.7 q

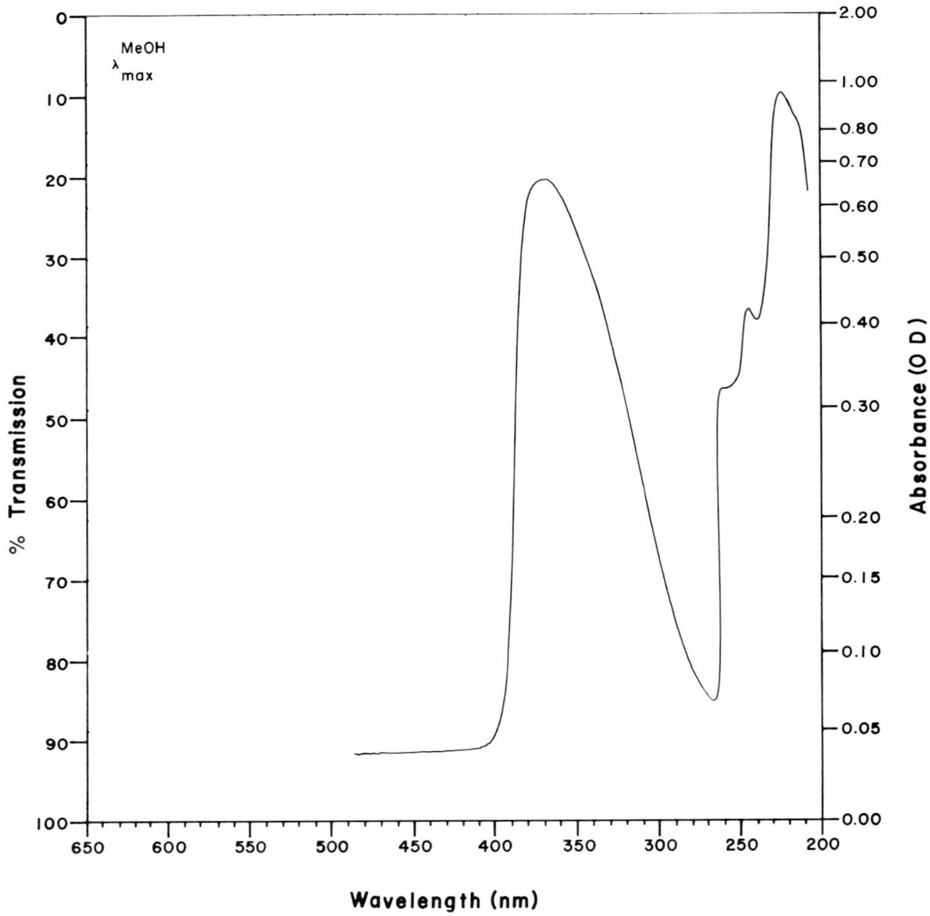
MeOH
λ max
% Transmission
Absorbance (O D)
Wavelength (nm)
0
10
20
30
40
50
60
70
80
90
100
2.00
1.00
0.80
0.70
0.60
0.50
0.40
0.30
0.20
0.15
0.10
0.05
0.00
650
600
550
500
450
400
350
300
250
200

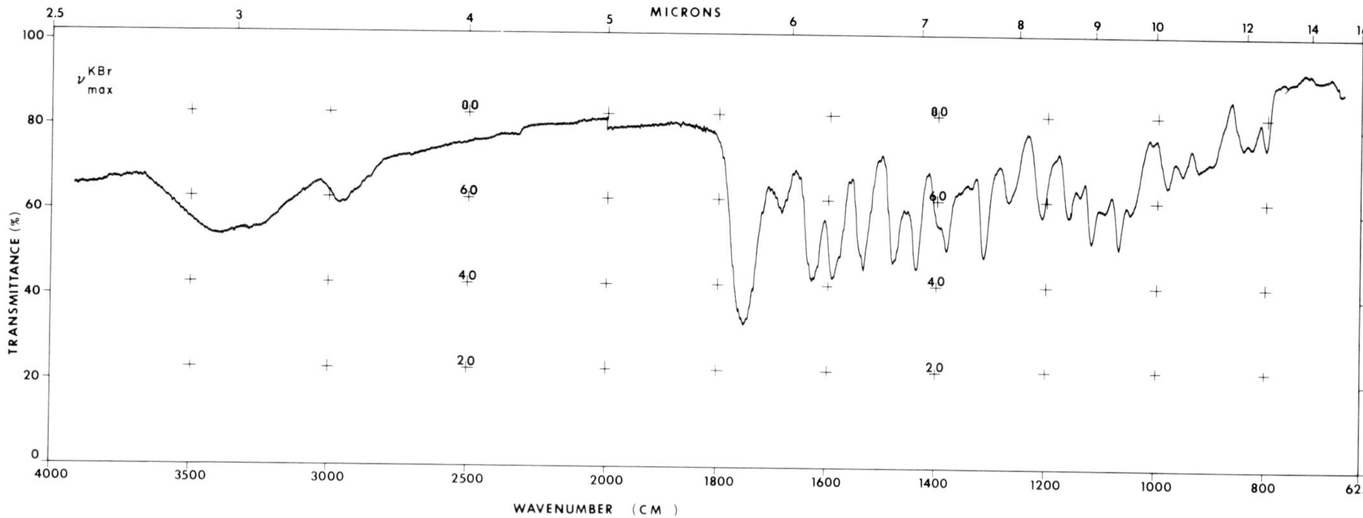
MICRONS
TRANSMITTANCE (%)
ν_{max}^{KBr}
WAVENUMBER (CM)

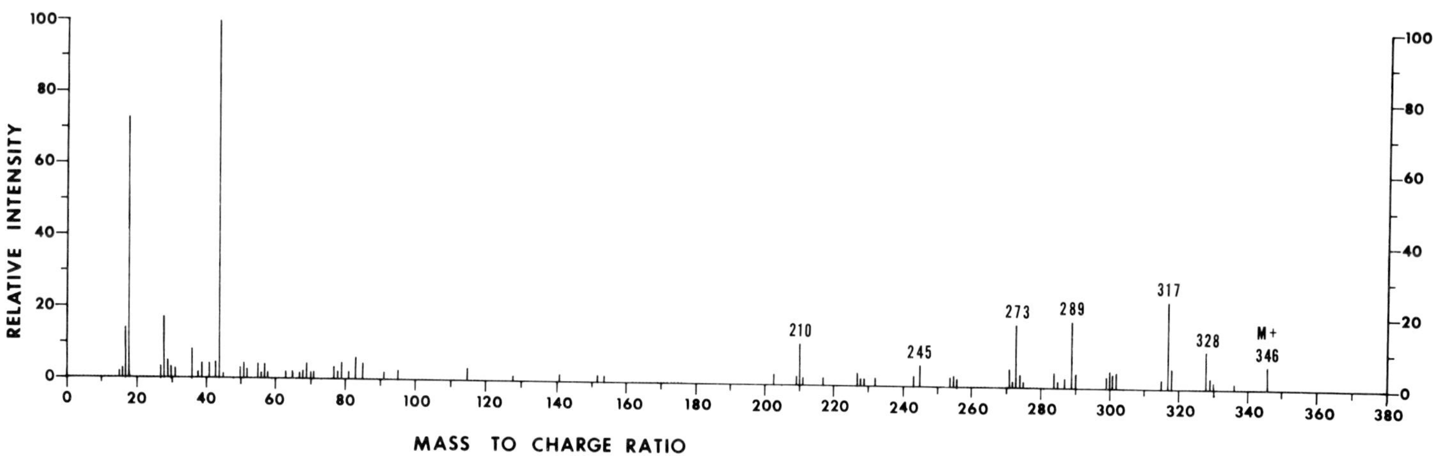
RELATIVE INTENSITY
MASS TO CHARGE RATIO
210
245
273
289
317
328
M+
346

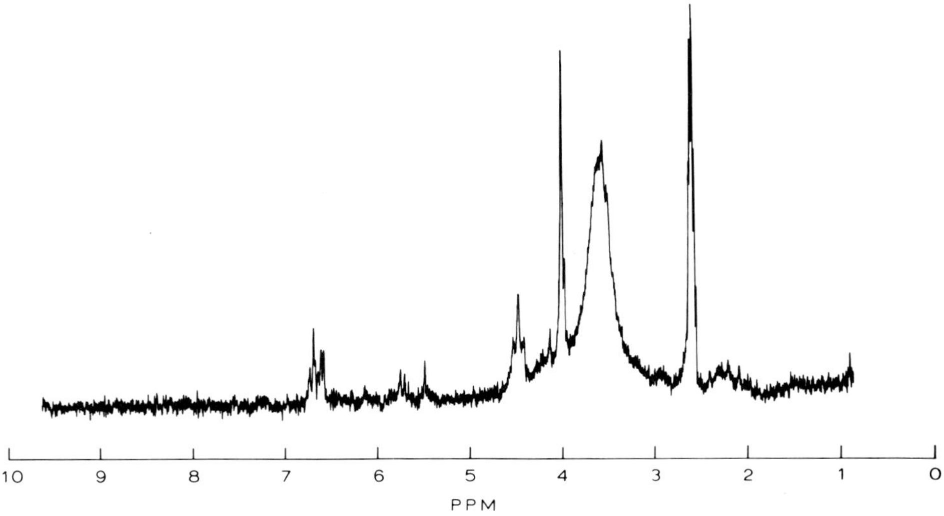

The 100 MHz ^{1}H NMR spectrum of aflatoxin G_{2a} in DMSO solution.

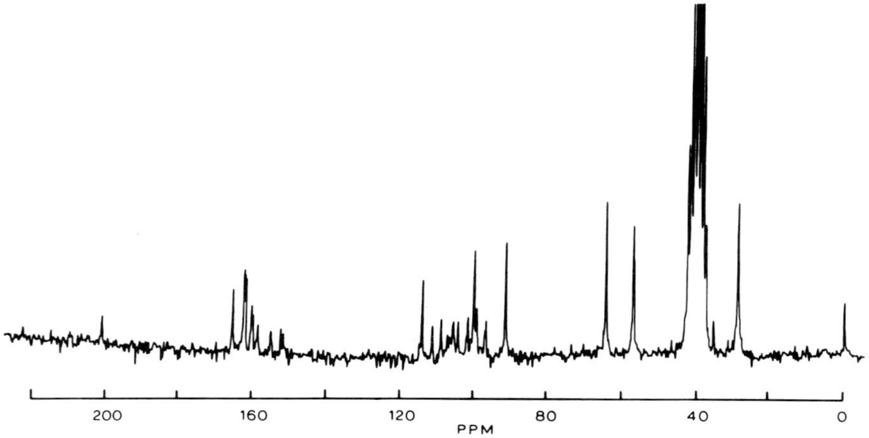

The 25 MHz ^{13}C NMR spectrum of aflatoxin G_{2a} in DMSO solution.

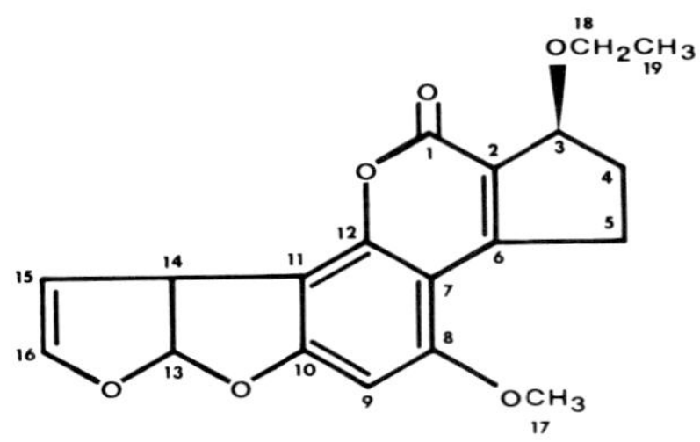

Common name	Aflatoxicol *O*-ethyl ether A
Molecular weight	342.1103
Molecular formula	$C_{19}H_{18}O_6$
General characteristics[17]	Colorless crystals from ethanol solution, mp 198°–200°C
UV data[17]	λ_{max}^{EtOH} nm (ϵ): 332(14,200), 261(9,660), and 255(8,830)
Source[17]	Produced spontaneously from aflatoxicol A during silica gel column chromatography using chloroform (ethanol preservative) as the eluting solvent.
Toxicity data	Toxicity unknown
TLC data[17]	*Adsorbent:* silica gel G-HR *Solvent:* chloroform–acetone, 93:7 v/v R_f: 0.81 *Detection:* blue fluorescence under UV light

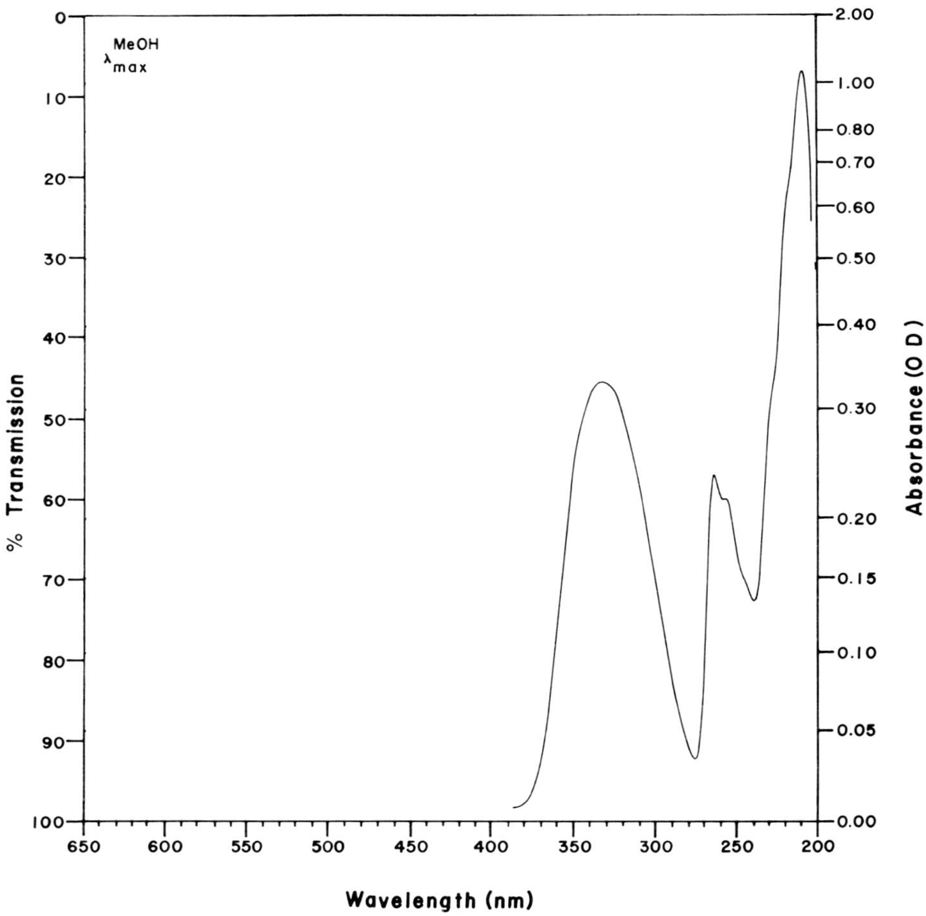

λ_{max}^{MeOH}
% Transmission
Absorbance (O D)
Wavelength (nm)
0
10
20
30
40
50
60
70
80
90
100
2.00
1.00
0.80
0.70
0.60
0.50
0.40
0.30
0.20
0.15
0.10
0.05
0.00
650
600
550
500
450
400
350
300
250
200

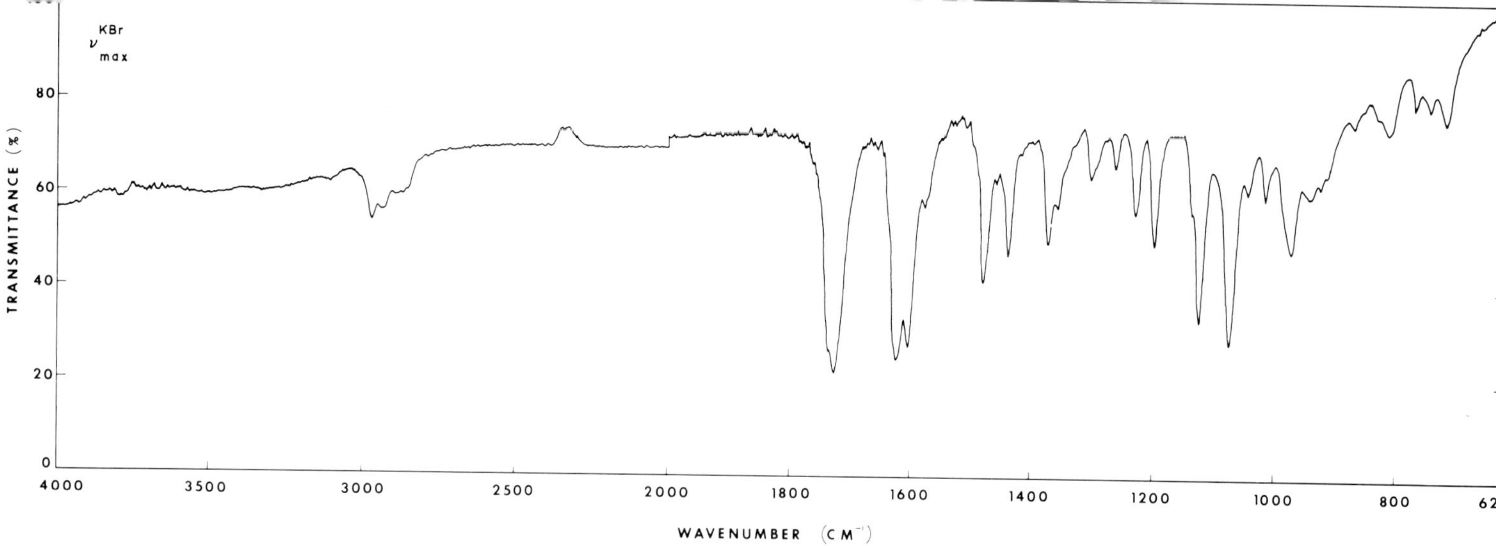

ν_{max}^{KBr}
TRANSMITTANCE (%)
80
60
40
20
0
4000
3500
3000
2500
2000
1800
1600
1400
1200
1000
800
625
WAVENUMBER (CM⁻¹)

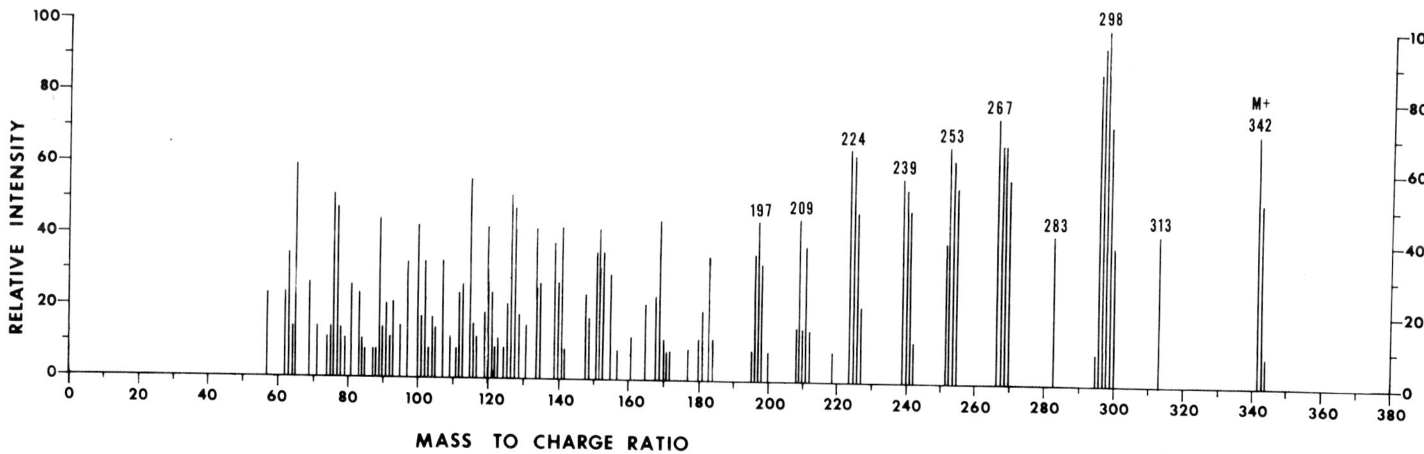

RELATIVE INTENSITY
100
80
60
40
20
0
197
209
224
239
253
267
283
298
313
M+
342
0
20
40
60
80
100
120
140
160
180
200
220
240
260
280
300
320
340
360
380
MASS TO CHARGE RATIO

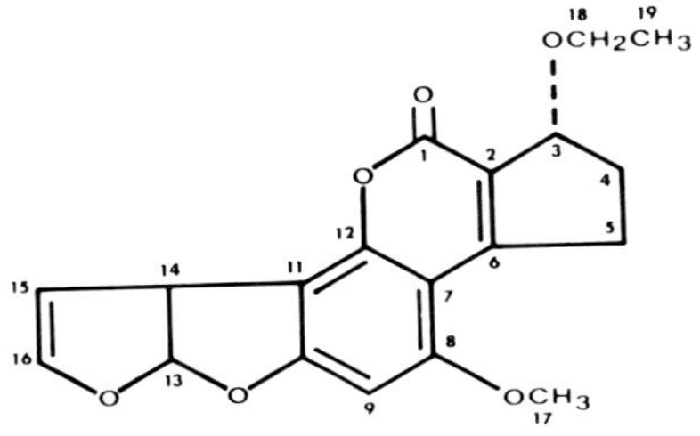

Common name	Aflatoxicol *O*-ethyl ether B
Molecular weight	342.1103
Molecular formula	$C_{19}H_{18}O_6$
General characteristics[17]	Colorless crystals from ethanol solution, mp 194°–196°C
UV data[17]	λ_{max}^{EtOH} nm (ϵ): 331(15,750), 261(12,280), and 255(11,170)
Source[17]	Produced spontaneously from aflatoxicol B during silica gel column chromatography using chloroform (ethanol preservative) as the eluting solvent.
Toxicity data	Toxicity unknown
TLC data[17]	*Adsorbent:* silica gel G-HR *Solvent:* chloroform–acetone, 93:7 v/v R_f: 0.73 *Detection:* blue fluorescence under UV light

Aflatoxicol *O*-ethyl ether B

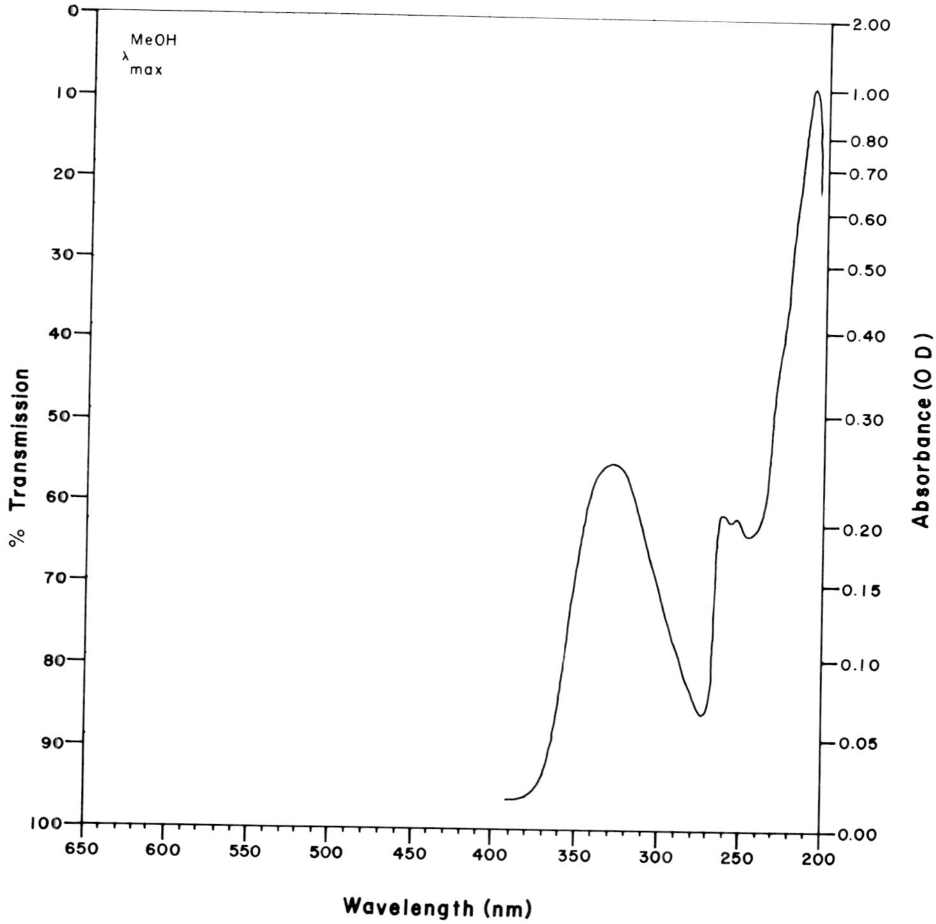

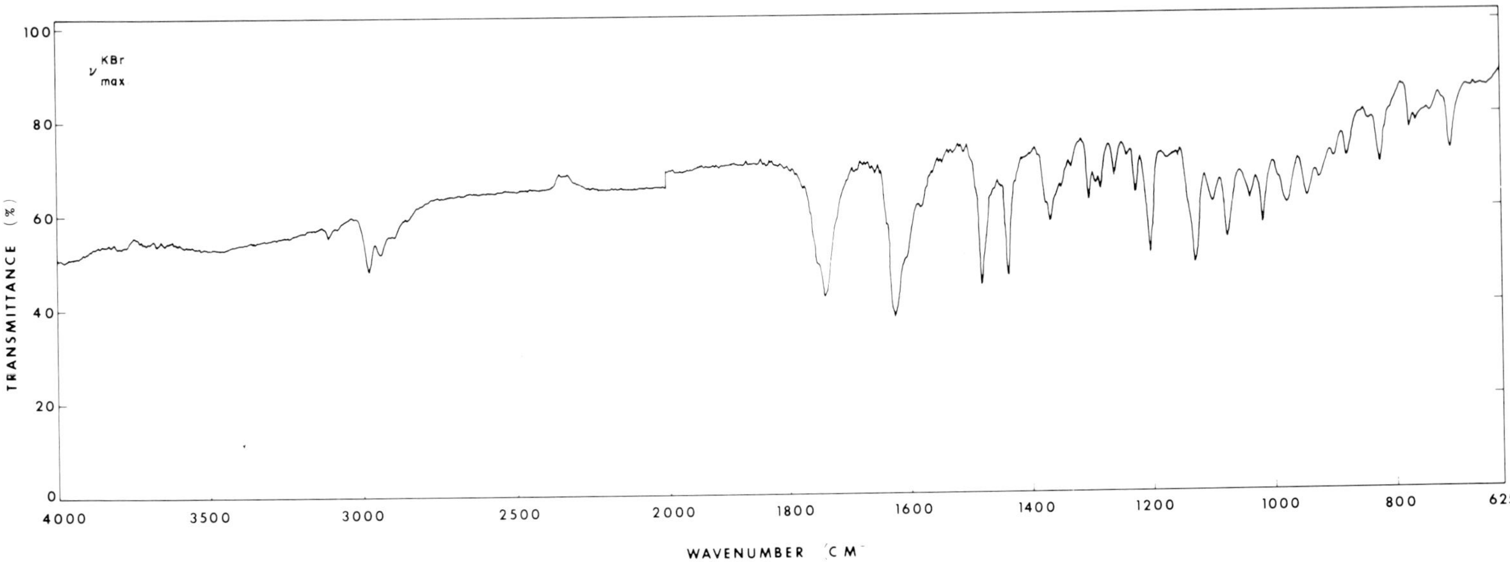

ν_{max}^{KBr}
TRANSMITTANCE (%)
100
80
60
40
20
0
4000
3500
3000
2500
2000
1800
1600
1400
1200
1000
800
625
WAVENUMBER (CM⁻

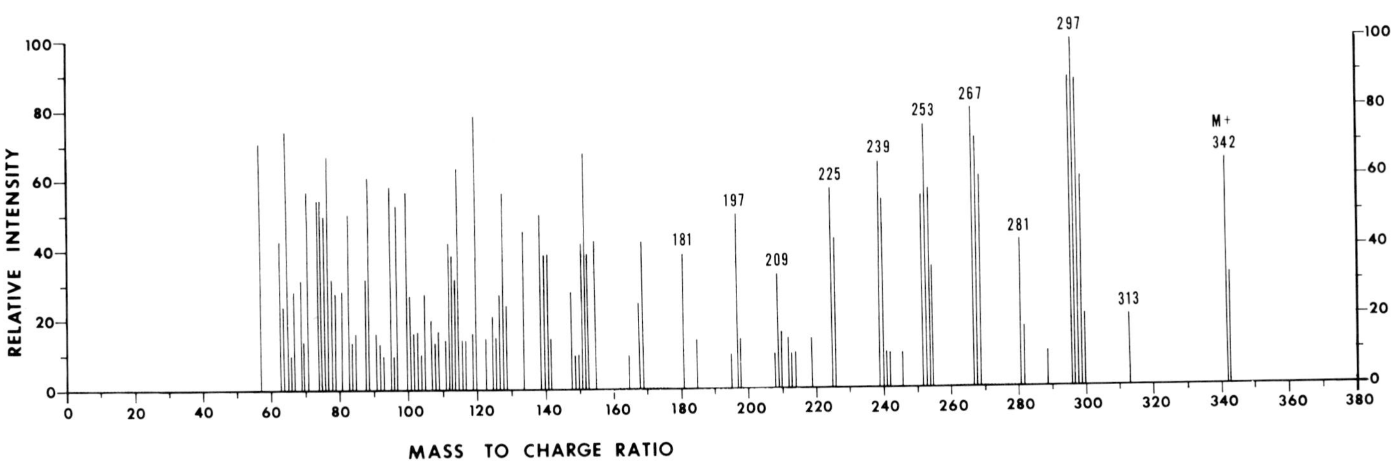

RELATIVE INTENSITY
MASS TO CHARGE RATIO
M+
342
313
297
281
267
253
239
225
209
197
181

REFERENCES

1. Allcroft, R., and Carnaghan, R. B. A., *Chem. Ind. (London)* p. 50 (1963).
2. Allcroft, R., *in* "Aflatoxin" (L. A. Goldblatt, ed.), p. 237. Academic Press, New York, 1969.
3. Asao, T., Büchi, G., Abdel-Kader, M. M., Chang, S. B., Wick, E. L., and Wogan, G. N., *J. Am. Chem. Soc.* **85,** 1706 (1963).
4. Asao, T., Büchi, G., Abdel-Kader, M. M., Chang, S. B., Wick, E. L., and Wogan, G. N., *J. Am. Chem. Soc.* **87,** 882 (1965).
5. Büchi, G., Foulkes, D. M., Jurono, M., and Mitchess, G. F., *J. Am. Chem. Soc.* **88,** 4534 (1966).
6. Büchi, G., Spitzner, D., Paglialunga, S., and Wogan, G. N., *Life Sci.* **13,** 1145 (1973).
7. Büchi, G., Muller, P. M., Roebuck, B. D., and Wogan, G. N., *Res. Commun. Chem. Pathol. Pharmacol.* **8,** 585 (1974).
8. Butler, W. H., *Br. J. Cancer* **18,** 756 (1964).
9. Butler, W. H., *J. Pathol.* **102,** 209 (1970).
10. Butler, W. H., *in* "Mycotoxins in Human Health" (I. F. H. Purchase, ed.), p. 141. Macmillan, New York, 1971.
11. Butler, W. H., *in* "Mycotoxins" (I. F. H. Purchase, ed.), p. 1. Am. Elsevier, New York, 1974.
12. Canton, J. H., Kroes, R., van Logten, M. J., van Schothorst, M., Stavenuiter, J. F. C., and VerHülsdonk, C. A. H., *Food Cosmet. Toxicol.* **13,** 441 (1975).
13. Carnaghan, R. B. A., Hartley, R. D., and O'Kelly, J., *Nature (London)* **200,** 1101 (1963).
14. Chang, S. B., Abdel-Kader, M. M., Wick, E. L., and Wogan, G. N., *Science* **142,** 1191 (1963).
15. Cheung, K. K., and Sim, G. A., *Nature (London)* **201,** 1185 (1964).
16. Ciegler, A., Lillehoj, E. B., Peterson, R. E., and Hall, H. H., *Appl. Microbiol.* **14,** 934 (1966).
17. Cole, R. J., Kirksey, J. W., and Blankenship, B. R., *J. Agric. Food Chem.* **20,** 1100 (1972).
18. Cole, R. J., and Kirksey, J. W., *J. Agric. Food Chem.* **19,** 222 (1971).
19. Cox, R. H., and Cole, R. J., *J. Org. Chem.* **42,** 112 (1977).
20. Dalezios, J., Wogan, G. N., and Weinreb, S. M., *Science* **171,** 584 (1971).
21. DeIongh, H., Beerthuis, R. K., Vles, R. D., Barrett, C. B., and Ord, W. O., *Biochim. Biophys. Acta* **65,** 548 (1962).
22. Detroy, R. W., and Hesseltine, C. W., *Nature (London)* **219,** 967 (1968).
23. Detroy, R. W., and Hesseltine, C. W., *Can. J. Biochem.* **48,** 830 (1970).
24. Detroy, R. W., Lillehoj, E. B., and Ciegler, A., *in* "Microbial Toxins" (S. Kadis, A. Ciegler, and S. J. Ajls, eds.), Vol. 6, pp. 3–178. Academic Press, New York, 1971.
25. Dutton, M. F., and Heathcote, J. G., *Biochem. J.* **101,** 21p (1966).
26. Dutton, M. F., and Heathcote, J. G., *Chem. Ind. (London)* p. 418 (1968).
27. Dutton, M. F., and Heathcote, J. G., *J. S. Afr. Chem Inst.* **22,** S107 (1969).
28. Goldblatt, L. A., ed., "Aflatoxin." Academic Press, New York, 1969.
29. Heathcote, J. G., and Dutton, M. F., *Tetrahedron* **25,** 1497 (1969).
30. Heathcote, J. G., and Hibbert, J. R., *Biochem. Soc. Trans.* **2,** 301 (1974).
31. Holzapfel, C. W., Steyn, P. S., and Purchase, I. F. H., *Tetrahedron Lett.* p. 2799 (1966).
32. Hsieh, D. P. H., Dalezios, J. I., Krieger, R. I., Masri, M. S., and Hadden, W. F., *J. Agric. Food Chem.* **22,** 515 (1974).
33. Hseih, D. P. H., Salhab, A. S., Wong, J. J., and Yang, S. L., *Toxicol. Appl. Pharmacol.* **30,** 237 (1974).
34. Hsieh, D. P. H., Seiber, J. N., Reece, C. A., Fitzell, D. L., Yang, S. L., Dalezios, J. I., LaMar, G. N., Budd, D. L., and Motell, E., *Tetrahedron* **31,** 661 (1975).
35. Kiermeier, F., and Ruffer, L., *Z. Lebensm.-Unters.-Forsch.* **155,** 129 (1974).
36. Lee, L. S., Stanley, J. B., Cucullu, A. F., Pons, W. A., Jr., and Goldblatt, L. A., *J. Assoc. Off. Anal. Chem.* **57,** 626 (1974).

37. Lijinsky, W., and Butler, W. H., *Proc. Soc. Exp. Biol. Med.* **123,** 151 (1966).
38. Lillehoj, E. B., and Ciegler, A., *Appl. Microbiol.* **17,** 516 (1969).
39. Masri, M. S., Booth, A. N., and Hsieh, D. P. H., *Life Sci.* **15,** 203 (1974).
40. Masri, M. S., Haddon, W. F., Lundin, R. E., and Hsieh, D. P. H., *J. Agric. Food Chem.* **22,** 512 (1974).
41. Nesbitt, B. F., O'Kelly, J., Sargeant, K., and Sheridan, A., *Nature (London)* **195,** 1962 (1962).
42. Pachler, K. G. R., Steyn, P. S., Vleggaar, R., Wessels, P. L., and Scott, D. B., *J. Chem. Soc., Perkin Trans. 1* p. 1182 (1976).
43. Patterson, D. S. P., and Roberts, B. A., *Food Cosmet. Toxicol.* **8,** 527 (1970).
44. Pawlowski, N. E., Schoenhard, G. L., Lee, D. J., Libbey, L. M., Loveland, P. M., and Sinnhuber, R. O., *J. Agric. Food Chem.* **25,** 437 (1977).
45. Pohland, A. E., Cushmac, M. E., and Andrellos, P. J., *J. Assoc. Off. Anal. Chem.* **51,** 907 (1968).
46. Purchase, I. F. H., *Food Cosmet. Toxicol.* **5,** 339 (1967).
47. Robertson, J. A., Tennisson, D. J., and Boudreaux, G. J., *J. Agric. Food Chem.* **18,** 1090 (1970).
48. Sargeant, K., Carnaghan, R. B. A., and Allcroft, R., *Chem. Ind. (London)* p. 53 (1963).
49. Shank, R. C., Johnsen, D. O., Tantichardenyos, P., Wooding, W. L., and Bourgeois, C. H., *Toxicol. Appl. Pharmacol.* **20,** 227 (1971).
50. Shank, R. C., *Adv. Chem. Ser.* **149,** 51 (1974).
51. Sinnhuber, R. O., Lee, D. J., Wales, J. H., Landers, M. K., and Keyl, A. C., *Fed. Proc., Fed. Am. Soc. Exp. Biol.* **29,** 568 (1970).
52. Stanley, J. B., Lee, L. S., Cucullu, A. F., and deGruy, I. V., *J. Agric. Food Chem.* **23,** 447 (1975).
53. Steyn, P. S., Vleggaar, R., Wessels, P. L., and Scott, D. B., *J. Chem. Soc., Chem. Commun.* p. 193 (1975).
54. Stubblefield, R. D., Shotwell, O. L., Shannon, G. M., Weisleder, D., and Rohwedder, W. K., *J. Agric. Food Chem.* **18,** 391 (1970).
55. Stubblefield, R. D., Shotwell, O. L., and Shannon, G. M., *J. Assoc. Off. Anal. Chem.* **55,** 762 (1972).
56. Tennisson, D. J., and Robertson, J. A., *Appl. Microbiol.* **15,** 1099 (1967).
57. van der Merwe, K. J., and Fourie, L., *Chem. Ind. (London)* p. 1660 (1963).
58. van der Zijden, A. S. M., Blanche Koelensmid, W. A. A., Boldingh, J., Barrett, C. B., Ord, W. O., and Philp, J., *Nature (London)* **195,** 1060 (1962).
59. van Dorp, D. A., van der Zijden, A. S. M., Beerthuis, R. K., Sparreboom, S., Ord, W. O., DeIongh, H., and Kenning, R., *Recl. Trav. Chim. Pays-Bas* **82,** 587 (1963).
60. van Soest, T. C., and Peerdeman, A. F., *Proc. Ned. Akad. Wet., Ser. B* **67,** 469 (1964).
61. Wogan, G. N., *Bacteriol. Rev.* **30,** 460 (1965).
62. Wogan, G. N., and Paglialunga, S., *Food Cosmet. Toxicol.* **12,** 381 (1974).

Sterigmatocystins

2

The sterigmatocystins are a group of closely related fungal metabolites produced by *Aspergillus* spp. and *Bipolaris* spp. Chemically, they are characterized by a xanthone moiety fused to a dihydrodifurano or tetrahydrodifurano moiety. The chemical differences among the various sterigmatocystins are the presence or absence of unsaturation at positions 2 and 3 of the difurano ring system, the substitution pattern on positions 6, 7, and 10 of the xanthone system, and/or the substituent on position 3 of the difurano system.

Sterigmatocystin, the most economically important member of the group, is acutely toxic and carcinogenic.[19] With the exception of aspertoxin, little is known about the toxicity of other members of the group.[23,24,30] Indications are that other members such as *O*-methylsterigmatocystin show little or no acute toxicity to vertebrate animals.

Biochemically, sterigmatocystin and probably other members of the group are biosynthetic precursors to the aflatoxins. Previous studies have indicated that sterigmatocystin and aflatoxin induce similar but not identical toxicological effects.[19]

Sterigmatocystin group	Molecular weight	Molecular formula
Sterigmatocystin	324.0633	$C_{18}H_{12}O_6$
Dihydrosterigmatocystin	326.0790	$C_{18}H_{14}O_6$
O-Methylsterigmatocystin	338.0790	$C_{19}H_{14}O_6$
Dihydro-*O*-methylsterigmatocystin	340.0946	$C_{19}H_{16}O_6$
Aspertoxin	354.0739	$C_{19}H_{14}O_7$
5-Methoxysterigmatocystin	354.0739	$C_{19}H_{14}O_7$
Dihydrodemethylsterigmatocystin	312.0633	$C_{17}H_{12}O_6$
Dimethoxysterigmatocystin	384.0844	$C_{20}H_{16}O_8$

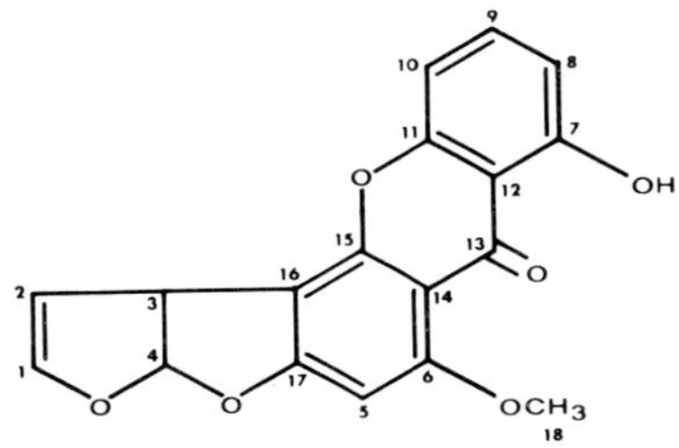

Common name	Sterigmatocystin (3*a*,12*c*-Dihydro-8-hydroxy-6-methoxyfuro[3′,2′:4,5]furo[3,2-*c*]xanthen-7-one)
Molecular weight	324.0633
Molecular formula	$C_{18}H_{12}O_6$
General characteristics[2,3,8,27]	Pale yellow crystals, mp 246°C (dec.) (of a sublimed sample) Sterigmatocystin is relatively insoluble in most solvents tested; solubility was best in chloroform (7138 mg/100 ml) and pyridine (1815 mg/100 ml) $[\alpha]_D^{20}$ −398° (C = 1.0 in chloroform)
UV data[14]	λ_{max}^{EtOH} nm (ϵ): 208(19,000), 235(24,500), 249(27,500), and 329(13,100)
Fungal source[5,14,15,20,21,25]	*Aspergillus versicolor* (Vuill.) Tiraboschi *A. nidulans* (Eidam) Wint. *Bipolaris sorokiniana* Saccardo in Shorok (Shoem) *A. aurantio-brunneus* (Atkins, Hindson, and Russell) *A. quadrilineatus* (Thom & Raper) *A. ustus* Bainier (Thom & Church) *A. variecolor* Also an intermediate in biosynthesis of aflatoxins by *A. parasiticus* and A. flavus
Toxicity data[19]	The LD_{50} values for sterigmatocystin in albino rats was 166 mg/kg (male rats, *per os,* dimethylformamide solvent); 60 mg/kg (male rats, IP, dimethylformamide solvent); 120 mg/kg (female rats, *per os,* wheat germ oil carrier); 65 mg/kg (male rats, IP, wheat germ oil carrier). The LD_{50} in male monkeys was 32 mg/kg (dosed IP, with DMSO as carrier). *Acute toxicity:* causes liver and kidney damage and renal necrosis in rats; it is cirrhogenic, causes hepatic damage in primates. In monkeys, oral

administration (20 mg/kg) every 14 days for several months caused chronic hepatitis and hyperplasia. It has been hypothesized that sterigmatocystin may be implicated in the etiology of chronic liver disease in man in Africa.

TLC data[14]

Adsorbent: silica gel
Solvent: chloroform–methanol, 98:2 v/v
R_f: 0.5
Detection: orange-red fluorescent spot under UV light; if sprayed with acetic acid, it exhibits a light yellow fluorescence.

GLC data[16]

Solid support: Chromosorb W
Liquid phase: 1.5% SE-30
Column conditions: 1.5 meter glass, 220°C
Relative retention time (relative to cholestane): 1.21

Other significant references

X-Ray crystallography[10,29]; isolation[8]; ^{13}C NMR and biosynthesis[7,17,18,26,28]; production methods[20]; TLC method in grain[1]

^{13}C NMR data

C1, 145.1 d; C2, 105.7 d; C3, 47.9 d; C4, 113.1 d; C5, 90.4 d; C6, 163.0 s; C7, 154.7 s; C8, 106.4 d; C9, 135.4 d; C10, 111.0 d; C11, 162.1 s; C12, 108.8 s; C13, 180.9 s; C14, 105.7 s; C15, 153.7 s; C16, 106.4 s; C17, 164.3 s; C18, 56.6 q

^{1}H NMR data

H1, 6.48 (J = 2.0, 2.5); H2, 5.42 (J = 2.5); H3, 4.75 (J = 2.0, 7.25); H4, 6.79 (J = 7.25); H5, 6.35; H8, 6.78 (J = 8.25, 1.0); H9, 7.45 (J = 8.25); H10, 6.70 (J = 8.25, 1.0); H18, 3.96

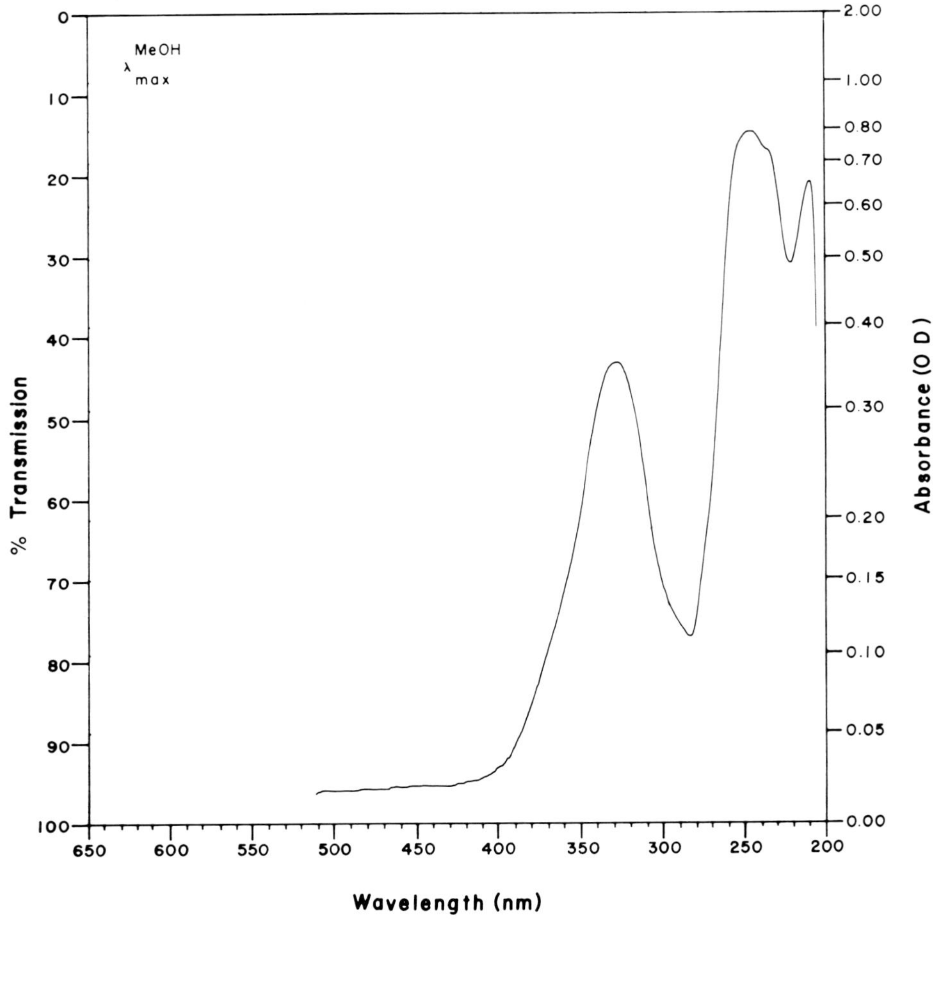

MeOH
λ max
% Transmission
0
10
20
30
40
50
60
70
80
90
100
Absorbance (O D)
2.00
1.00
0.80
0.70
0.60
0.50
0.40
0.30
0.20
0.15
0.10
0.05
0.00
650
600
550
500
450
400
350
300
250
200
Wavelength (nm)

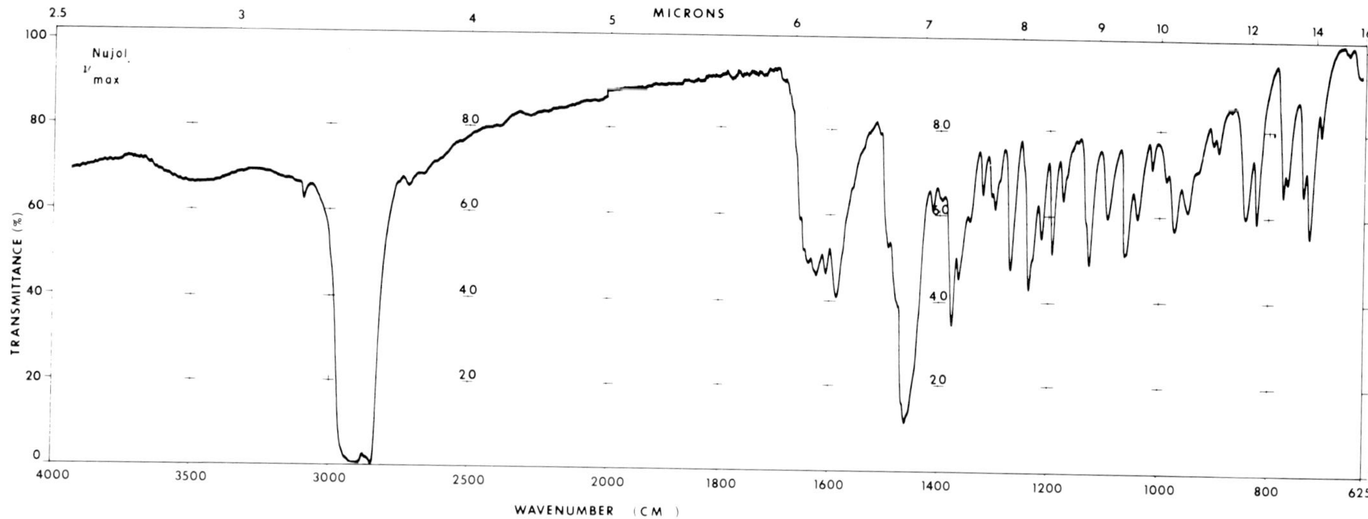
MICRONS
Nujol
ν_{max}
TRANSMITTANCE (%)
WAVENUMBER (CM)
2.5
3
4
5
6
7
8
9
10
12
14
16
4000
3500
3000
2500
2000
1800
1600
1400
1200
1000
800
625
0
20
40
60
80
100

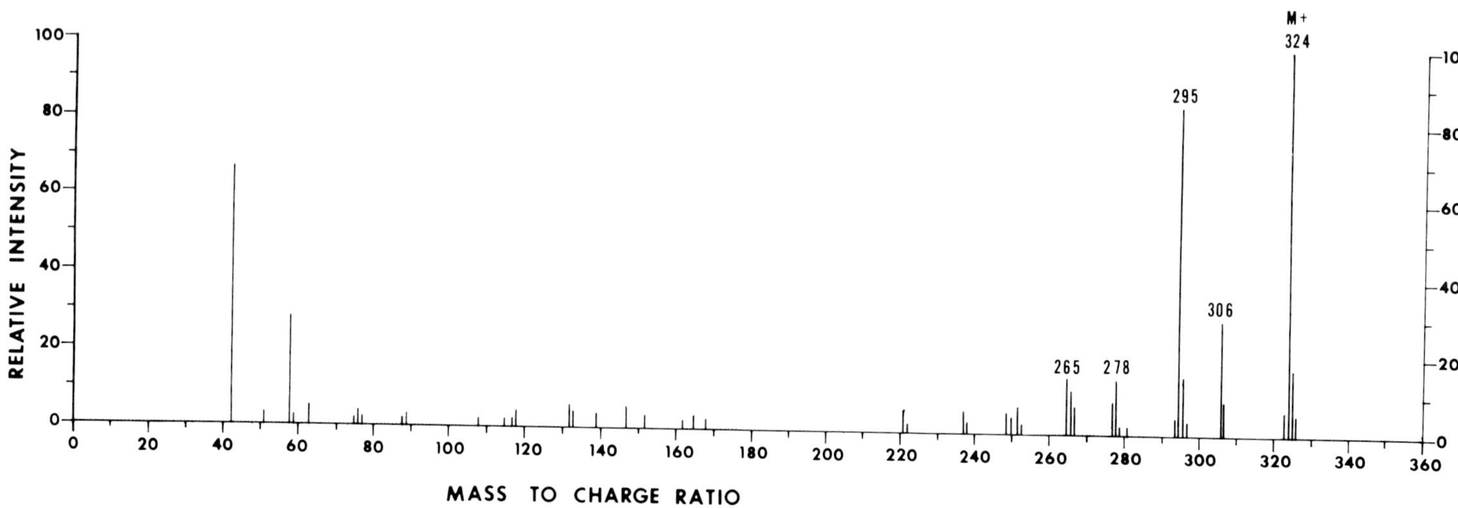
RELATIVE INTENSITY
MASS TO CHARGE RATIO
M+
324
295
306
265
278
0
20
40
60
80
100
120
140
160
180
200
220
240
260
280
300
320
340
360

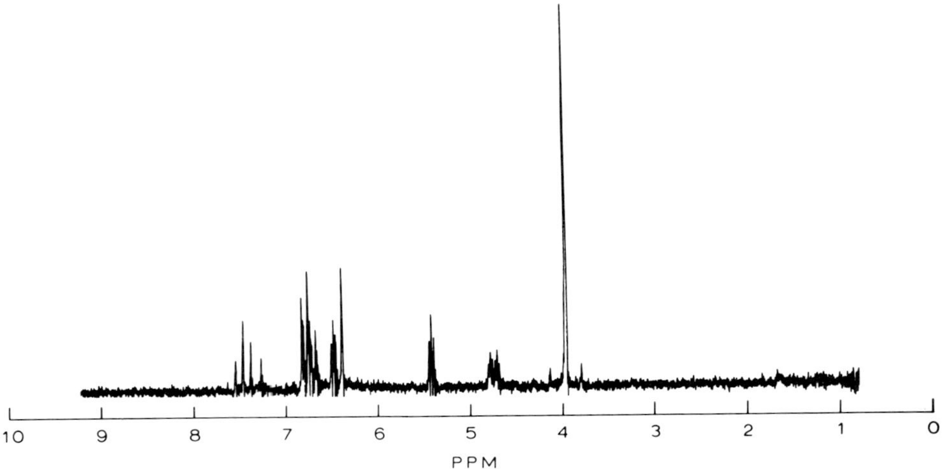

The 100 MHz ^{1}H NMR spectrum of sterigmatocystin in $CDCl_3$ solution.

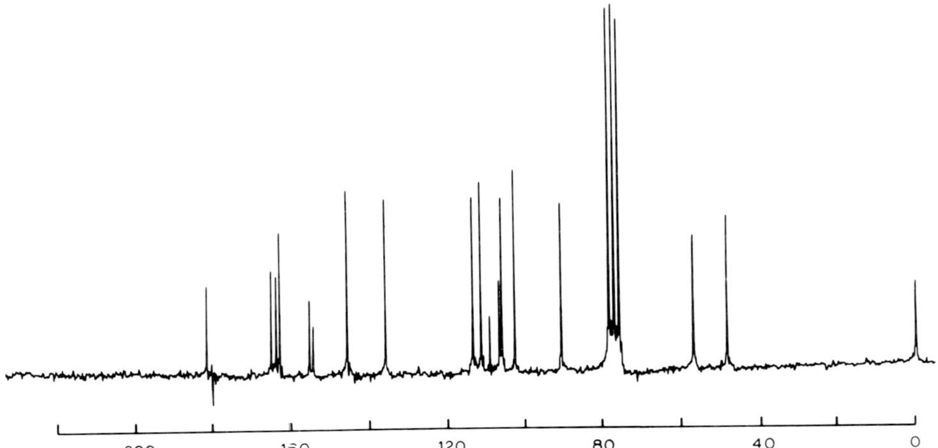

The 25 MHz ^{13}C NMR spectrum of sterigmatocystin in $CDCl_3$ solution.

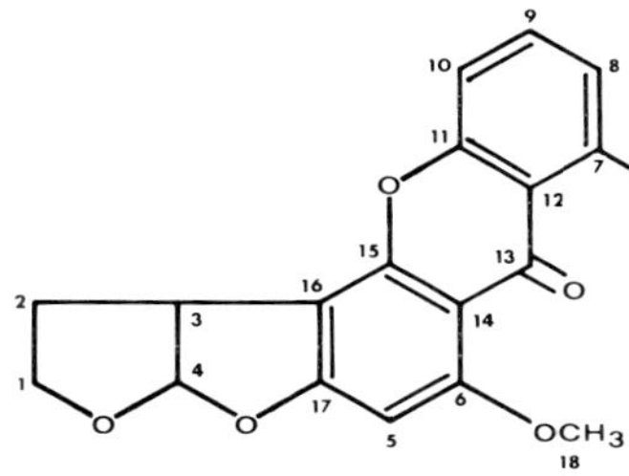

Common name	Dihydrosterigmatocystin (1,2-Dihydro-6-methoxy-7-hydroxydifuroxanthone)
Molecular weight	326.0790
Molecular formula	$C_{18}H_{14}O_6$
General characteristics[8,12]	Pale yellow needles, mp 230°C (dec.) $[\alpha]_D^{20}$ −311.7° (*C* = 0.85 in chloroform) Yellow plates from ethanol (sublimed sample), mp 229°–230°C
UV data[8,12]	λ_{max}^{EtOH} nm (ϵ): 233(27,600), 247(32,200), and 325(16,600) λ_{max} nm (ϵ): 208(20,400), 232(26,300), 247(30,900), and 325(16,300)
Fungal source[12]	*Aspergillus versicolor* (Vuill.) Tiraboschi
Toxicity data[9]	Dihydrosterigmatocystin markedly inhibited mitosis of primary kidney epithelial cells of *Cercopithecus aethiops* but had negligible effects on nucleolar morphology. It also markedly inhibited the incorporation of [^{3}H]thymidine and [^{3}H]uridine, which indicates impaired DNA and RNA synthesis.
TLC data[16]	*Adsorbent:* silica gel G-HR *Solvent:* chloroform–acetone, 93:7 v/v R_f: 0.83* *Detection:* red-orange fluorescent spot under UV light
Other significant references	^{13}C NMR[7]
^{13}C NMR data	C1, 67.6 t; C2, 31.4 t; C3, 44.2 d; C4, 113.1 d; C5, 89.6 d; C6, 163.1 s; C7, 154.6 s; C8, 105.5 d; C9, 135.2 d; C10, 110.7 d; C11, 161.9 s; C12, 108.7 s;

* Sterigmatocystin R_f 0.86 in same system

C13, 180.8 s; C14, 105.1 s; C15, 154.6 s; C16, 106.7 s; C17, 165.7 s; C18, 56.6 q

^1H NMR data H1, 4.10; H2, 2.23; H3, 3.62 (J = 5.5); H4, 6.41 (J = 5.5); H5, 6.25; H8, 6.72 (J = 8.0, 1.0); H9, 7.39 (J = 8.0); H10, 6.64 (J = 8.0, 1.0); H18, 3.90; 7-OH, 13.26

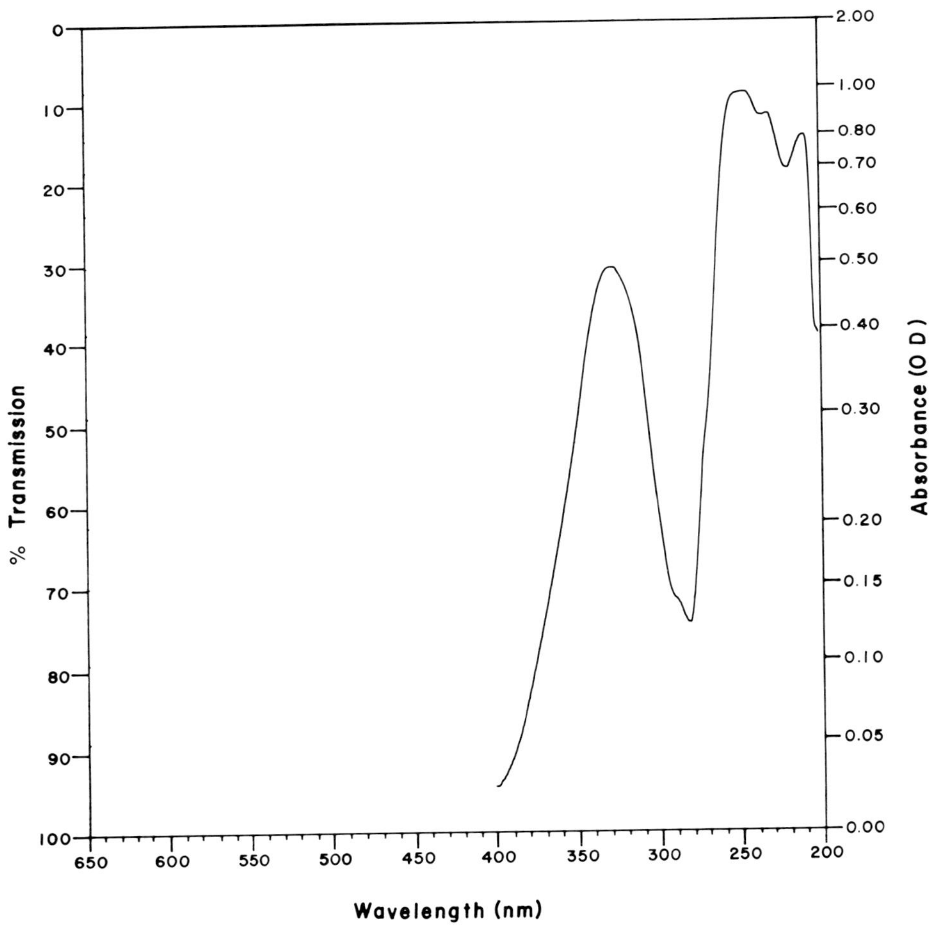

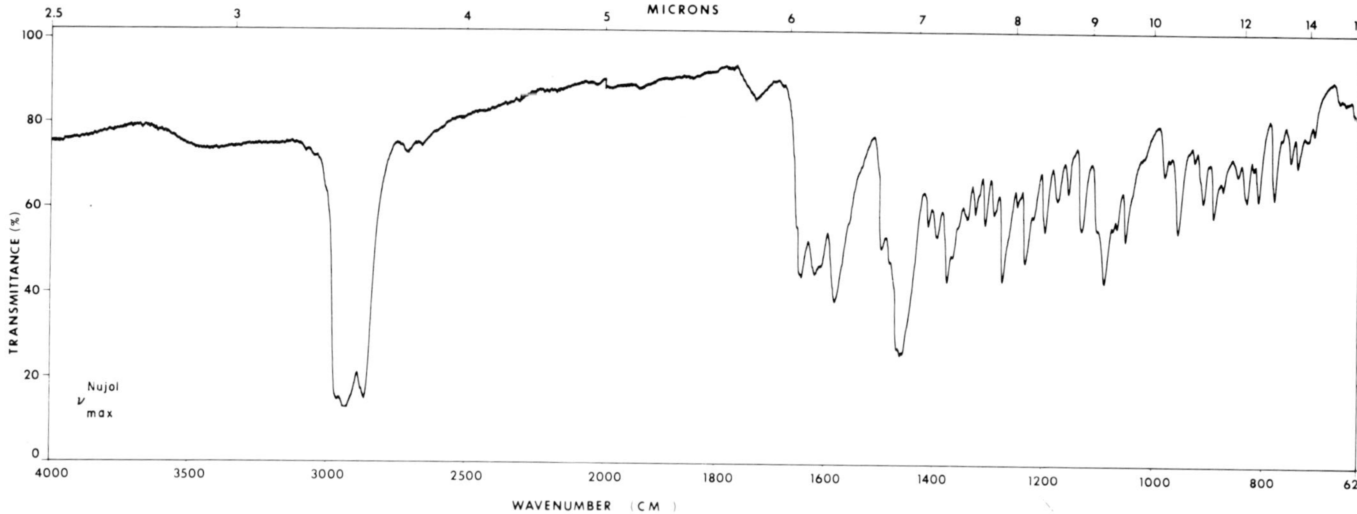
MICRONS
TRANSMITTANCE (%)
Nujol
ν max
WAVENUMBER (CM)

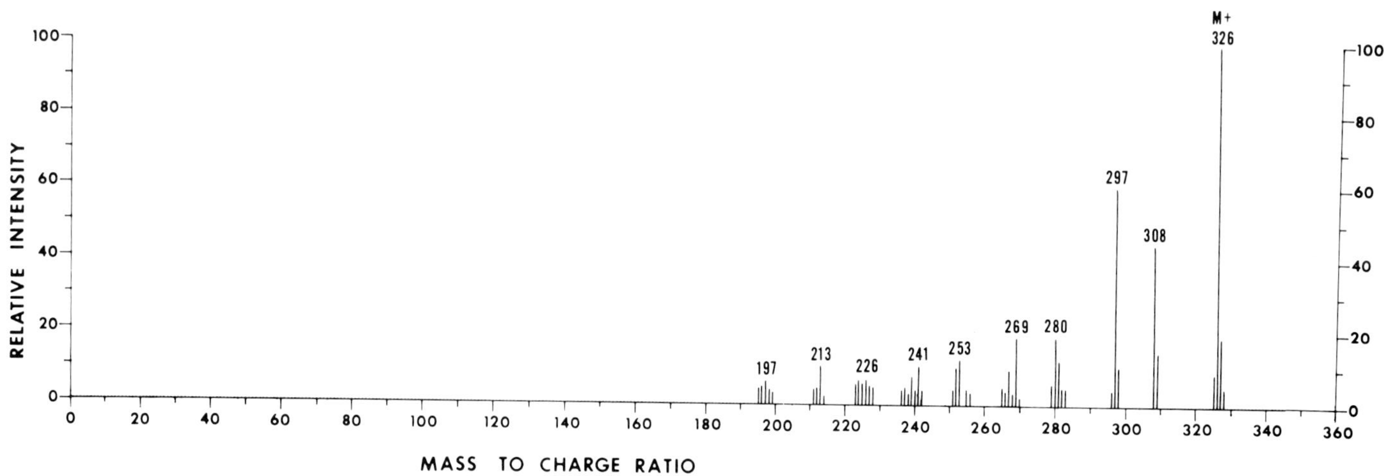
M+
326
308
297
280
269
253
241
226
213
197
RELATIVE INTENSITY
MASS TO CHARGE RATIO

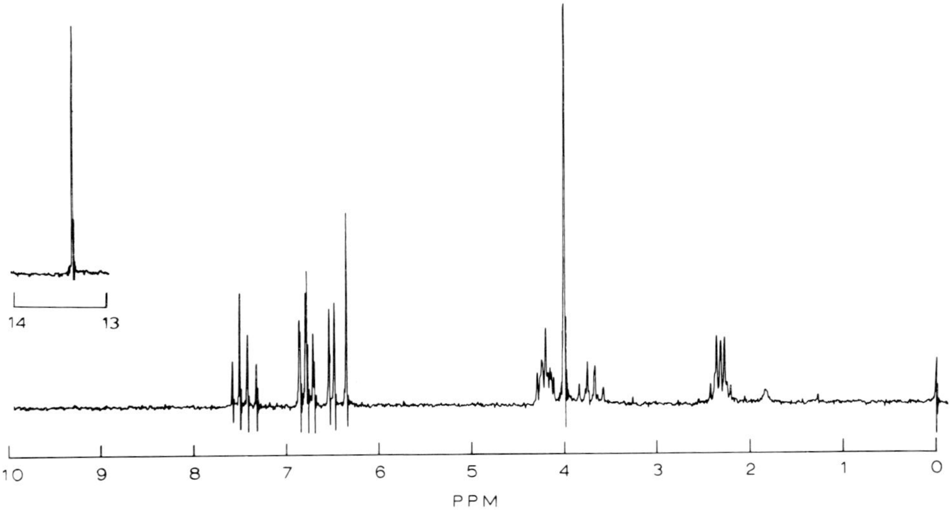

The 100 MHz ^{1}H NMR spectrum of dihydrosterigmatocystin in $CDCl_3$ solution.

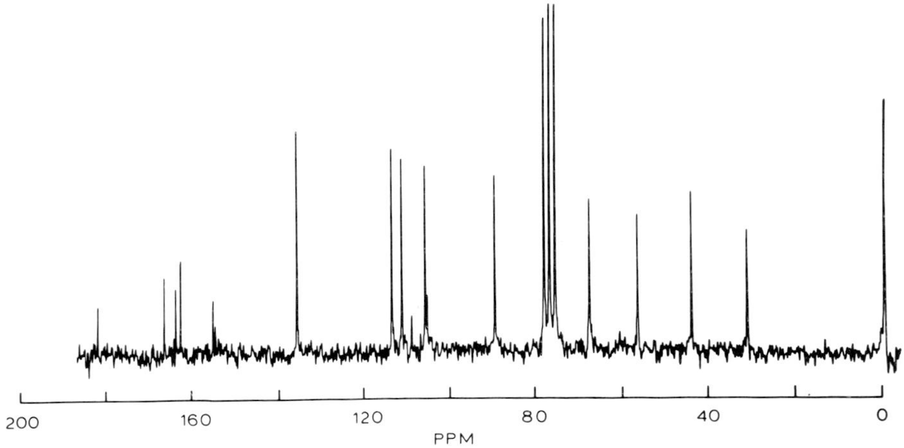

The 25 MHz ^{13}C NMR spectrum of dihydrosterigmatocystin in $CDCl_3$ solution.

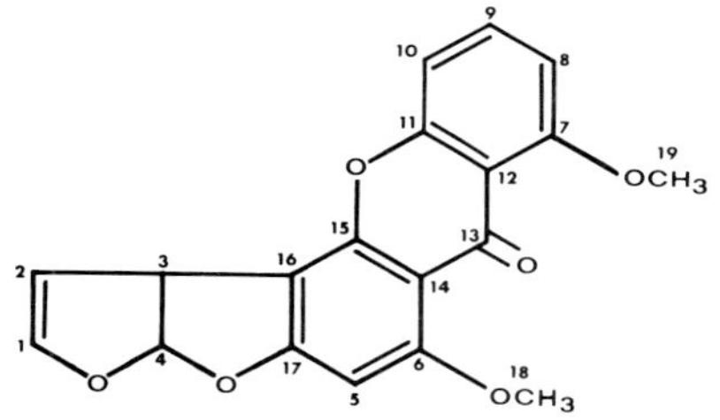

Common name	*O*-Methylsterigmatocystin (6,7-Dimethoxydifuroxanthone)
Molecular weight	338.0790
Molecular formula	$C_{19}H_{14}O_6$
General characteristics[4,10,22]	mp 274°C (colorless prisms from ethanol, synthetic compound) mp 265°C (dec.) from methanol and chloroform–heptane mp 265°–267°C, faintly yellow, slender rods from methanol
UV data[4,8]	λ_{max}^{MeOH} nm (ϵ): 236(40,700) and 310(16,500)
Fungal source[4,6]	*Aspergillus flavus* Link
Toxicity data	Toxicity unknown
TLC data[6]	*Adsorbent:* silica gel G-HR *Solvent:* chloroform–acetone, 93:7 v/v R_f: 0.38 *Detection:* blue fluorescent spot under UV light; turns to yellow fluorescent spot after spraying with 50% ethanolic H_2SO_4
Other significant references	^{13}C NMR[7]; synthesis[22]
^{13}C NMR data	C1, 145.0 d; C2, 102.6 d; C3, 48.1 d; C4, 112.9 d; C5, 90.3 d; C6, 162.8 s; C7, 156.5 s; C8, 106.2 d; C9, 133.4 d; C10, 108.9 d; C11, 160.4 s; C12, 106.2 s; C13, 174.6 s; C14, 105.6 s; C15, 152.9 s; C16, 106.2 s; C17, 162.7 s; C18, 56.3 q; C19, 56.3 q
^1H NMR data	H1, 6.48 (J = 2.5, 2.0); H2, 5.43 (J = 2.5); H3, 4.76 (J = 7.1, 2.0); H4, 6.78 (J = 7.1); H5, 6.36; H7, 6.92 (J = 8.3, 1.0); H8, 7.48 (J = 8.3); H9, 6.74 (J = 8.3, 1.0); H18, 3.95; H19, 3.91

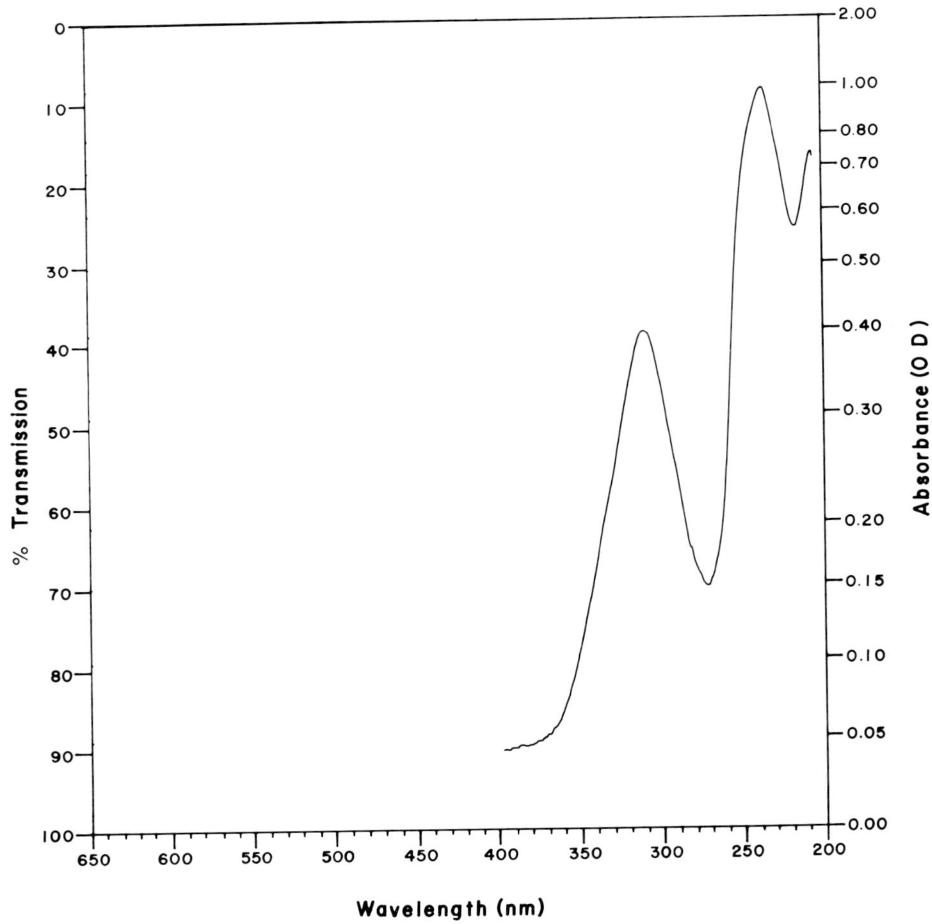
% Transmission
0
10
20
30
40
50
60
70
80
90
100
Absorbance (O D)
2.00
1.00
0.80
0.70
0.60
0.50
0.40
0.30
0.20
0.15
0.10
0.05
0.00
650
600
550
500
450
400
350
300
250
200
Wavelength (nm)

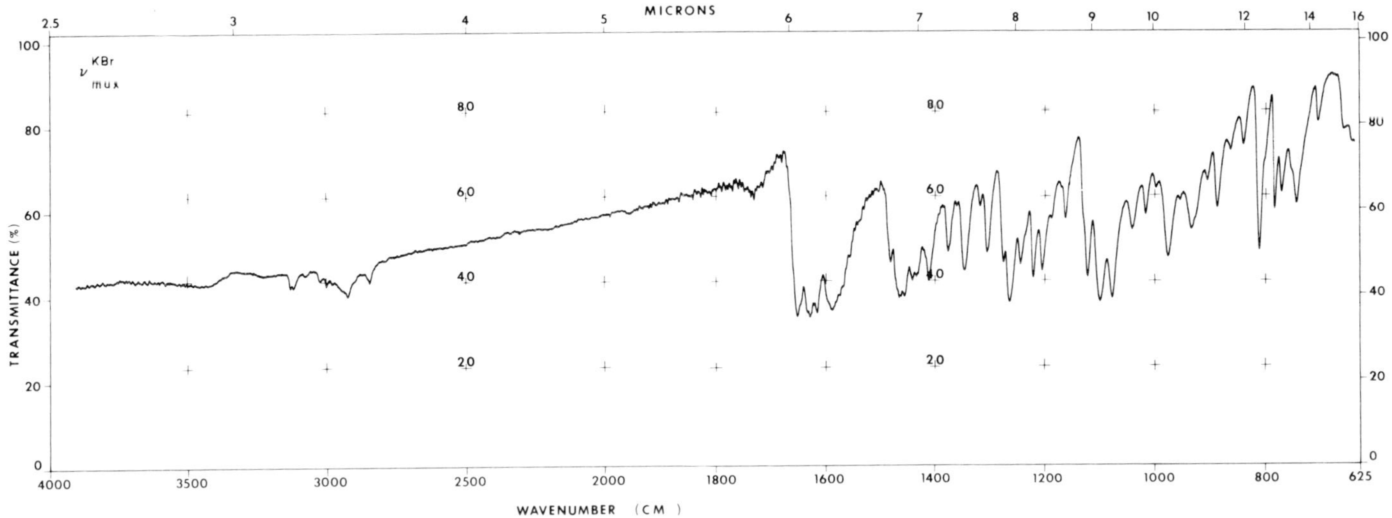

MICRONS
ν_{max}^{KBr}
TRANSMITTANCE (%)
WAVENUMBER (CM)

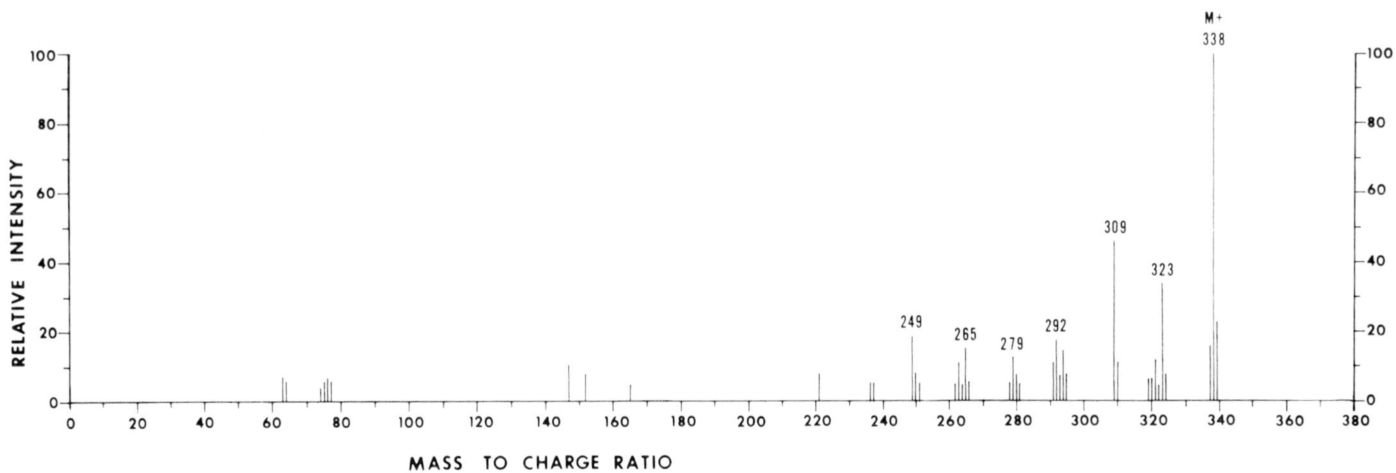

M+
338
323
309
292
279
265
249
RELATIVE INTENSITY
MASS TO CHARGE RATIO

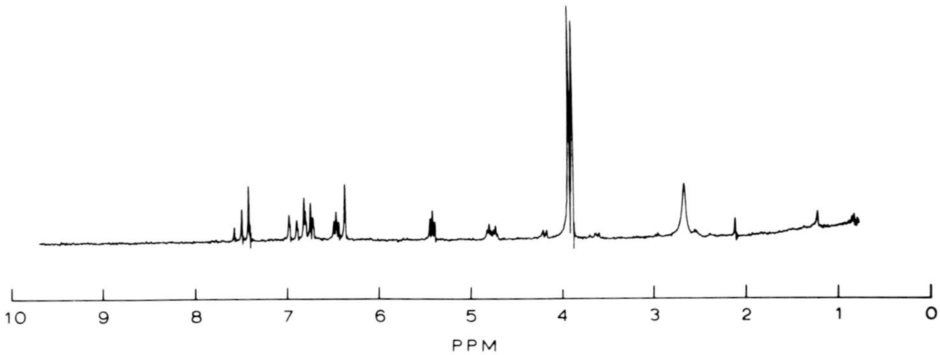

The 100 MHz ^{1}H NMR spectrum of *O*-methylsterigmatocystin in $CDCl_3$/DMSO solution.

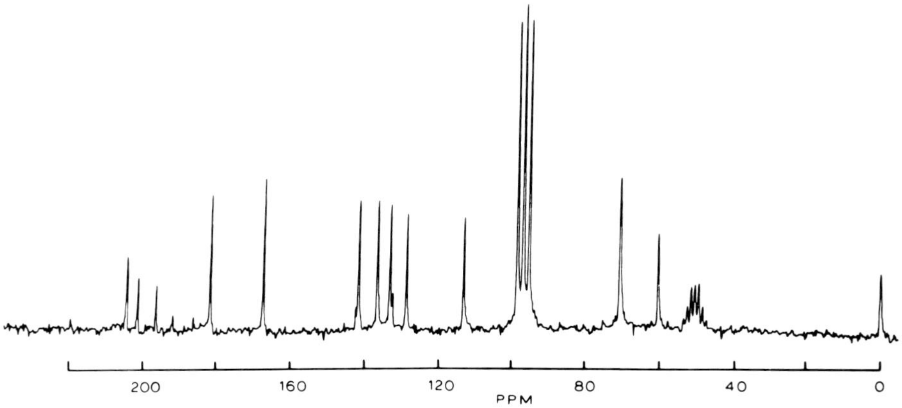

The 25 MHz ^{13}C NMR spectrum of *O*-methylsterigmatocystin in $CDCl_3$/DMSO solution.

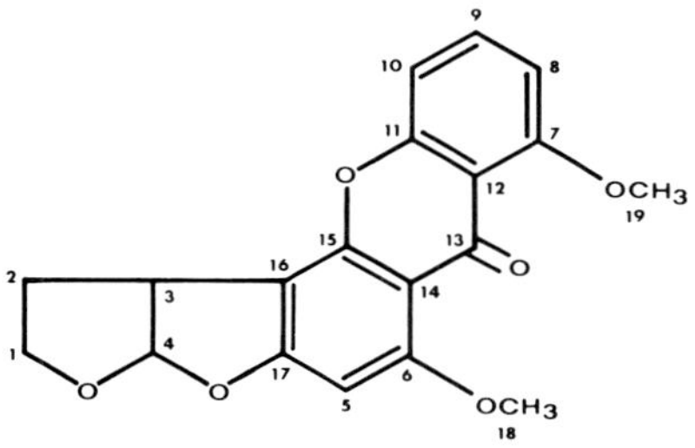

Common name	Dihydro-*O*-methylsterigmatocystin (1,2-Dihydro-6-methoxy-7-hydroxydifuroxanthone)
Molecular weight	340.0946
Molecular formula	$C_{19}H_{16}O_6$
General characteristics[6,8]	Colorless rods from methanol, mp 282°–283°C Colorless needles from chloroform–methanol, mp 281°–282°C
UV data[6,8]	λ_{max}^{MeOH} nm: 238 and 313 (ϵ_{max} not reported) λ_{max} nm (ϵ): 203(26,300), 237(38,900), and 311(17,300)
Fungal source[6,8]	*Aspergillus flavus*
Toxicity data[9]	The toxicity of dihydro-*O*-methylsterigmatocystin, as evaluated in primary cell culture (primary kidney epithelial cells of *Cercopithecus aethiops*) showed negligible effects on mitosis and nucleolar morphology; also negligible effects on the incorporation of [^{3}H]thymidine and [^{3}H]uridine, indicating no inhibition on DNA and RNA synthesis.
TLC data[6]	*Adsorbent:* silica gel G-HR *Solvent:* chloroform–acetone, 93:7 v/v R_f: 0.32 *Detection:* blue fluorescent spot under UV light; changing to yellow fluorescent after spraying with 50% ethanolic H_2SO_4
Other significant references	^{13}C NMR[7]
^{13}C NMR data	C1, 67.5 t; C2, 31.4 t; C3, 44.3 d; C4, 112.8 d; C5, 89.5 d; C6, 162.7 s; C7, 156.3 s; C8, 106.0 d; C9,

133.1 d; C10, 108.7 d; C11, 160.3 s; C12, 106.0 s; C13, 156.3 s; C14, 104.2 s; C15, 153.2 s; C16, 106.0 s; C17, 164.1 s; C18, 56.3 q; C19, 56.3 q

^{1}H NMR data H1, 4.20; H2, 2.34; H3, 3.72; H4, 6.48 (J = 6.0); H5, 6.34; H8, 6.97 (J = 8.0, 1.0); H9, 7.52 (J = 8.0); H10, 6.78 (J = 8.0, 1.0); H18, 4.00; H19, 3.96

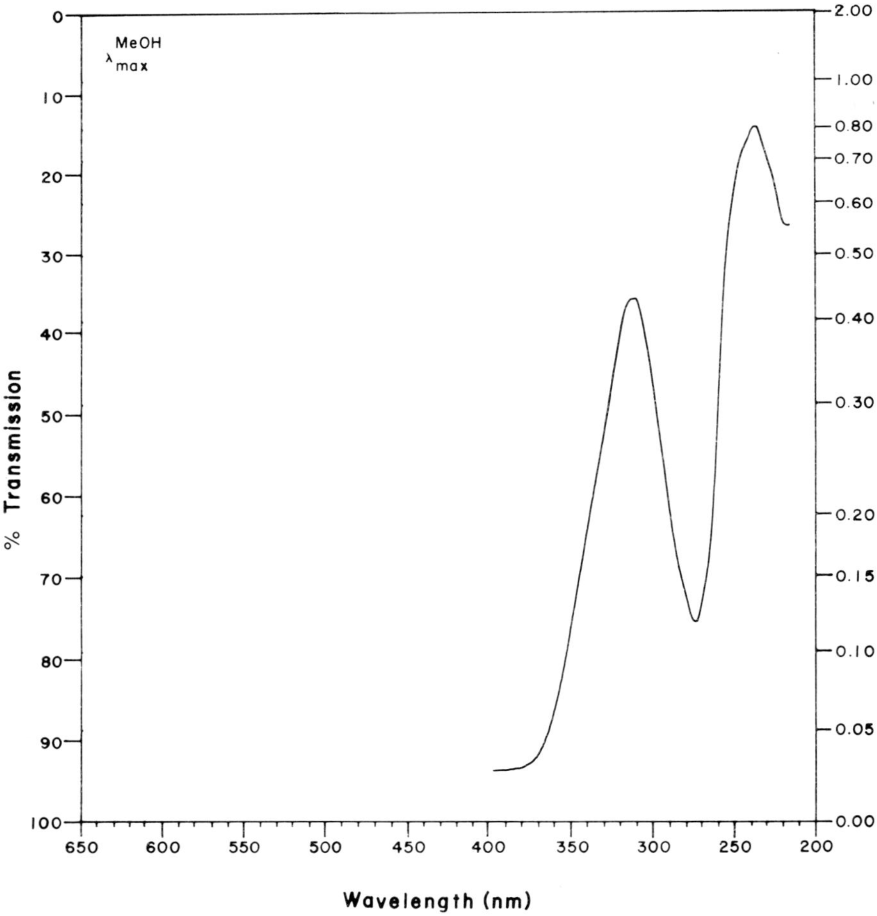

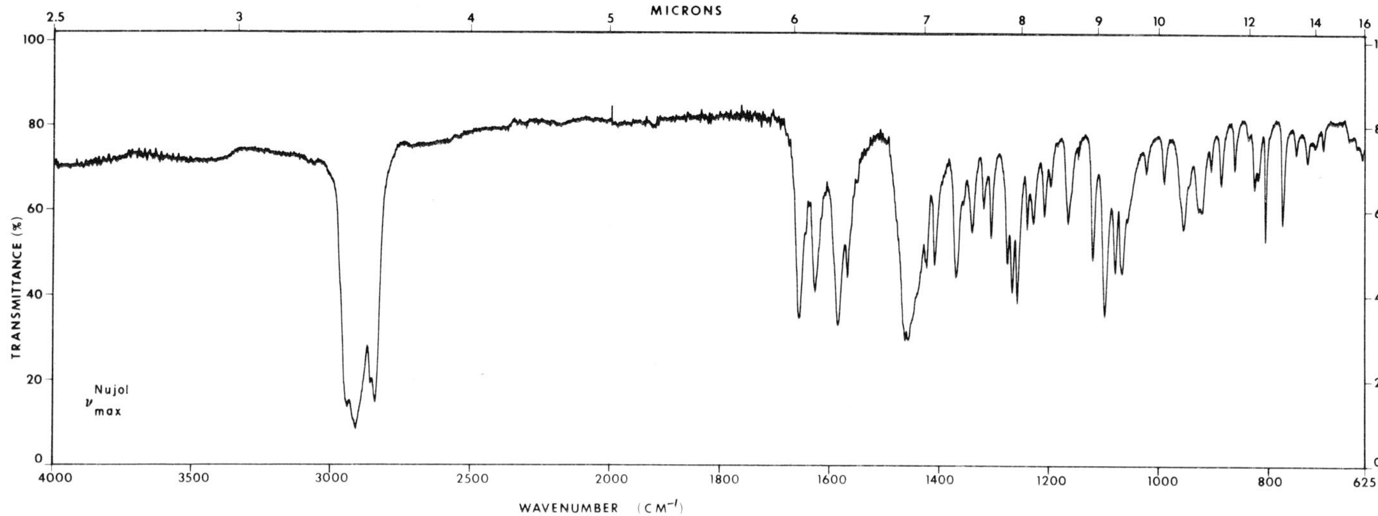
MICRONS
TRANSMITTANCE (%)
Nujol
ν max
WAVENUMBER (CM⁻¹)

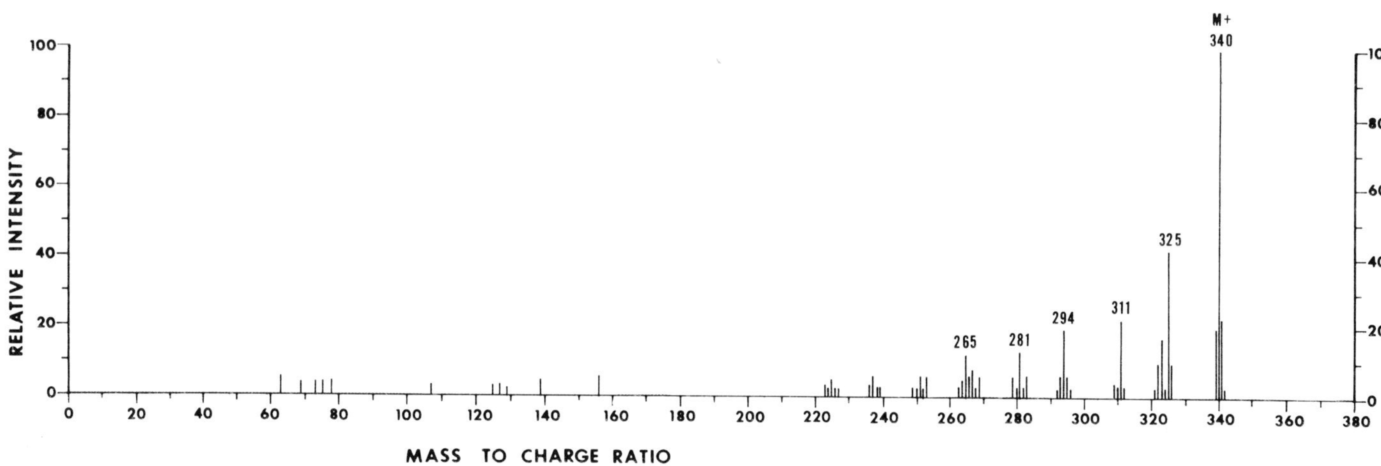
M+
340
325
311
294
281
265
RELATIVE INTENSITY
MASS TO CHARGE RATIO

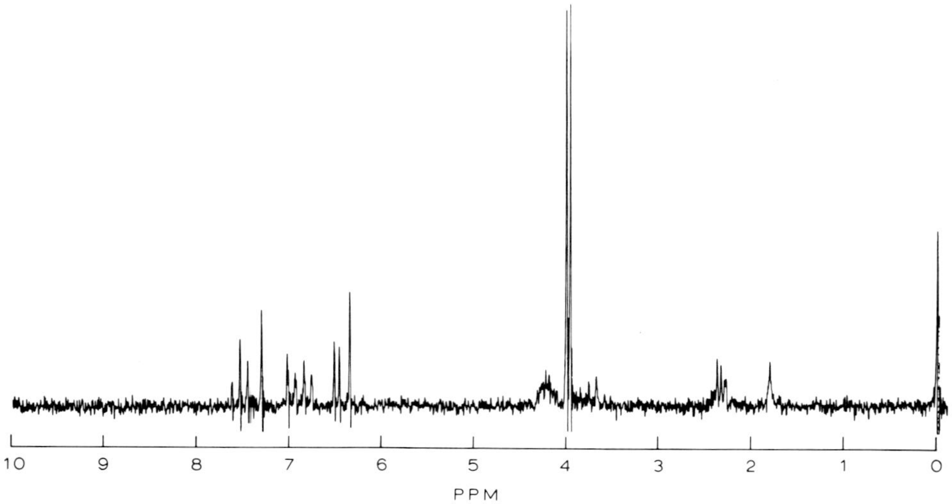

The 100 MHz ^{1}H NMR spectrum of dihydro-*O*-methylsterigmatocystin in $CDCl_3$ solution.

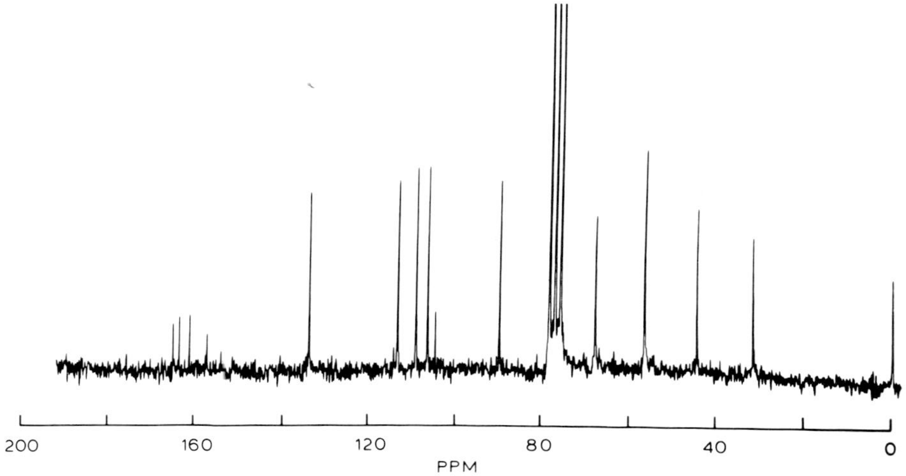

The 25 MHz ^{13}C NMR spectrum of dihydro-*O*-methylsterigmatocystin in $CDCl_3$ solution.

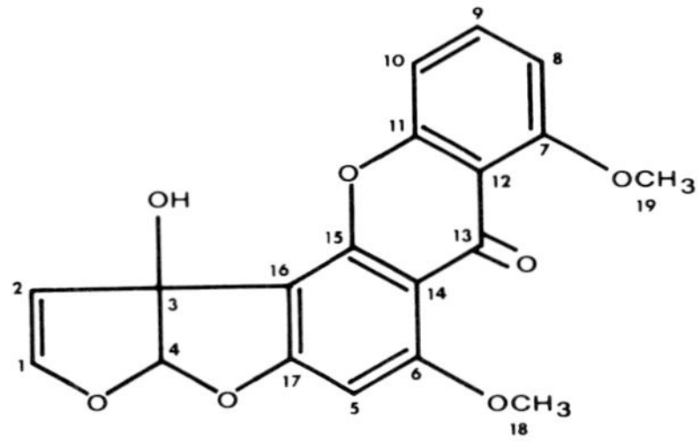

Common name	Aspertoxin (3-Hydroxy-6,7-dimethoxydifuroxanthone)
Molecular weight	354.0739
Molecular formula	$C_{19}H_{14}O_7$
General characteristics[23,24,30]	mp 240°–280°C (dec.) from dioxane/H_2O mp 325°–327°C, from dimethylformamide $[\alpha]_D^{27}$ −140° (C = 0.015, in dimethylformamide) Aspertoxin is insoluble in most organic solvents.
UV data[23,24,30]	λ_{max}^{MeOH} nm (ϵ): 241(33,900) and 310(12,100)
Fungal source	*Aspergillus flavus* Link
Toxicity data[30]	In developing chicken embryo, injections of 2.0 μg/egg (yolk or air sac) killed 100% of the embryos; while 0.7 μg/egg killed 50% of the embryos. Microscopically, the embryos showed beak malformations, generalized edema, loss of muscle tone, and hemorrhage from the umbilical vessels.
TLC data[23]	*Adsorbent:* silica gel *Solvent:* chloroform–acetic acid, 9:1 v/v R_f: 0.55–0.60 *Detection:* blue fluorescent spot under UV light
^1H NMR data[30]	H1, 6.72; H2, 5.76; H4, 6.38; H5, 6.56; H8, 7.05; H9, 7.61; H10, 6.95; H18, 3.93; H19, 3.93

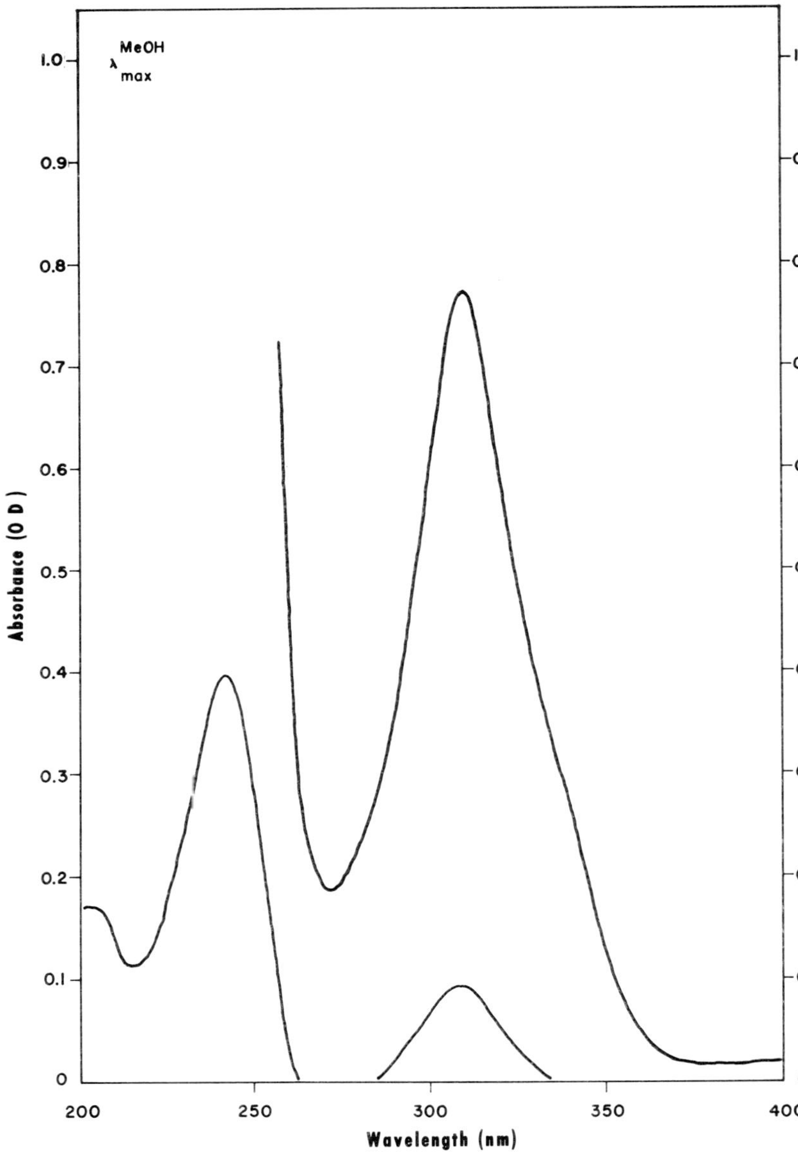

λ_{max}^{MeOH}
Absorbance (O D)
1.0
0.9
0.8
0.7
0.6
0.5
0.4
0.3
0.2
0.1
0
200
250
300
350
400
Wavelength (nm)

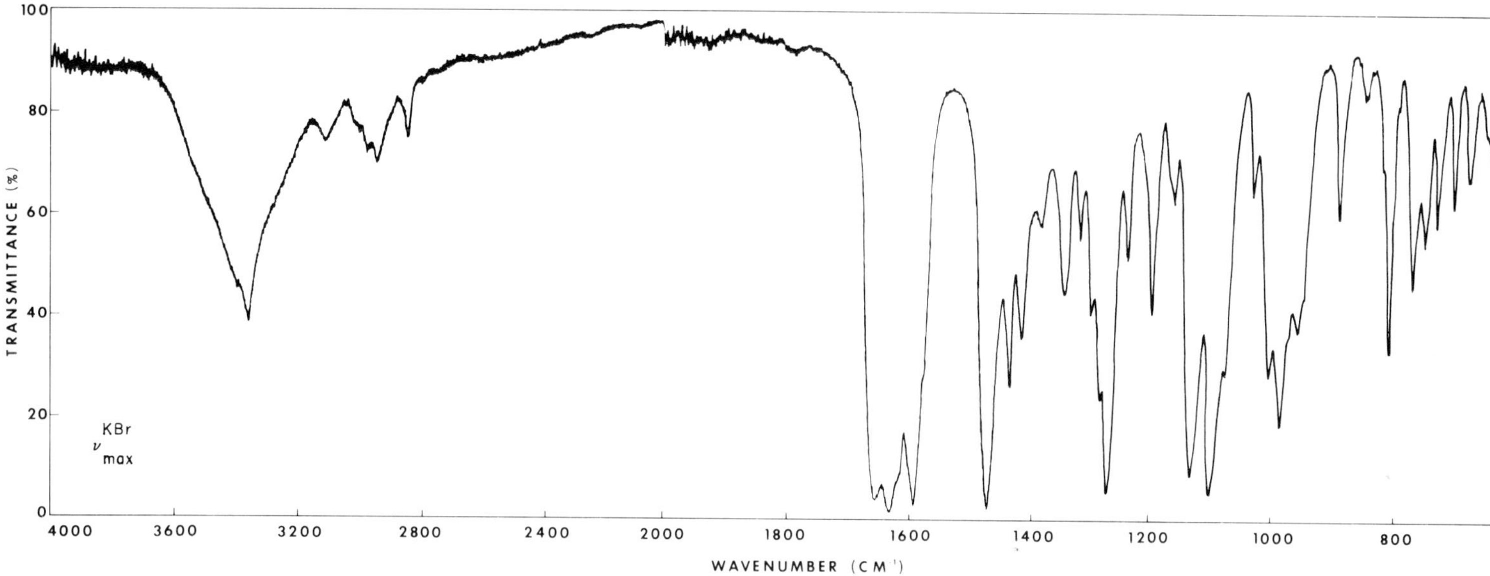

TRANSMITTANCE (%)
KBr
ν max
WAVENUMBER (CM⁻¹)

ASPERTOXIN

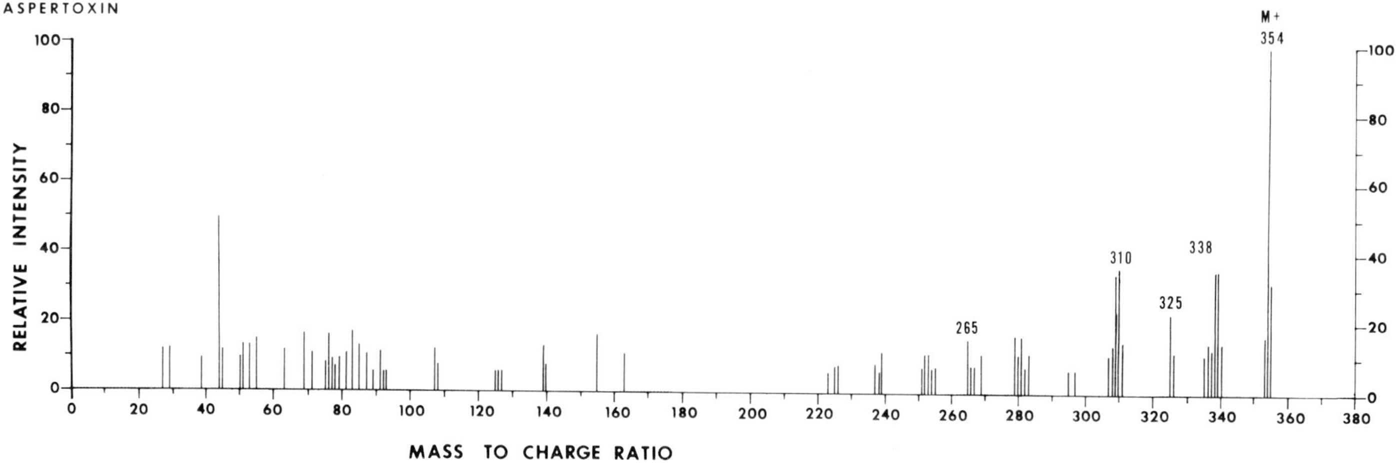

RELATIVE INTENSITY
M+
354
338
325
310
265
MASS TO CHARGE RATIO

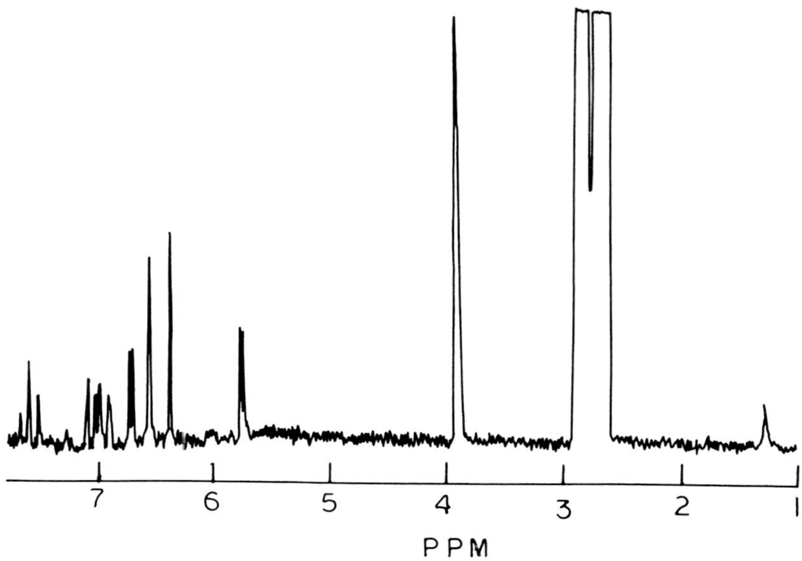

The 60 MHz ^{1}H NMR spectrum of aspertoxin in DMF-d_6.

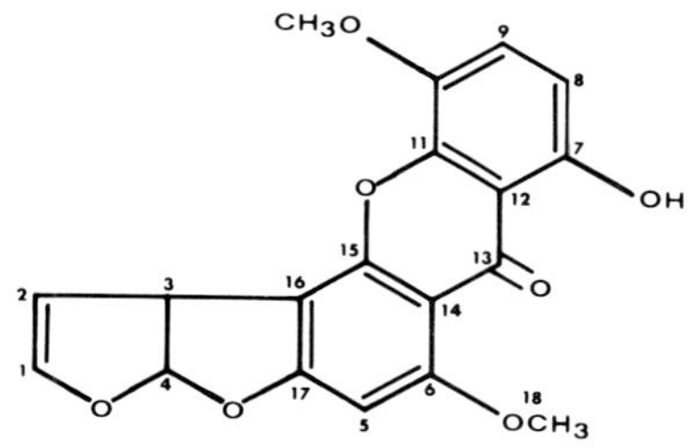

Common name	5-Methoxysterigmatocystin (7-Hydroxy-6,10-dimethoxydifuroxanthone)
Molecular weight	354.0739
Molecular formula	$C_{19}H_{14}O_7$
General characteristics[3,13]	Pale yellow needles from sublimed sample, mp 223°C (dec.) $[\alpha]_D^{20}$ −360° (C = 0.238 in chloroform)
UV data[3,13]	λ_{max}^{EtOH} nm (ϵ): 232(24,100), 248(26,800), 279(11,200), and 331(12,100)
¹H NMR data[3,13]	Similar to sterigmatocystin except for two coupled aromatic protons at 6.71 and 7.18 ppm
IR data[3,13]	ν_{max}^{KBr} cm^{-1}: 3447, 3383, 3121, 3096, 3016, 2971, 2927, 2853, 1662, 1634, 1618, 1595, 1559, 1498, 1462, 1444, 1419, 1398, 1360, 1345, 1339, 1302, 1285, 1262, 1248, 1233, 1223, 1199, 1187, 1168, 1148, 1140, 1099, 1078, 1060, 1042, 1017, 979, 960, 930, 881, 853, 825, 788, 767, 744, 737, and 724
Fungal source[3,13]	*Aspergillus versicolor* (Vuill.) Tiraboschi

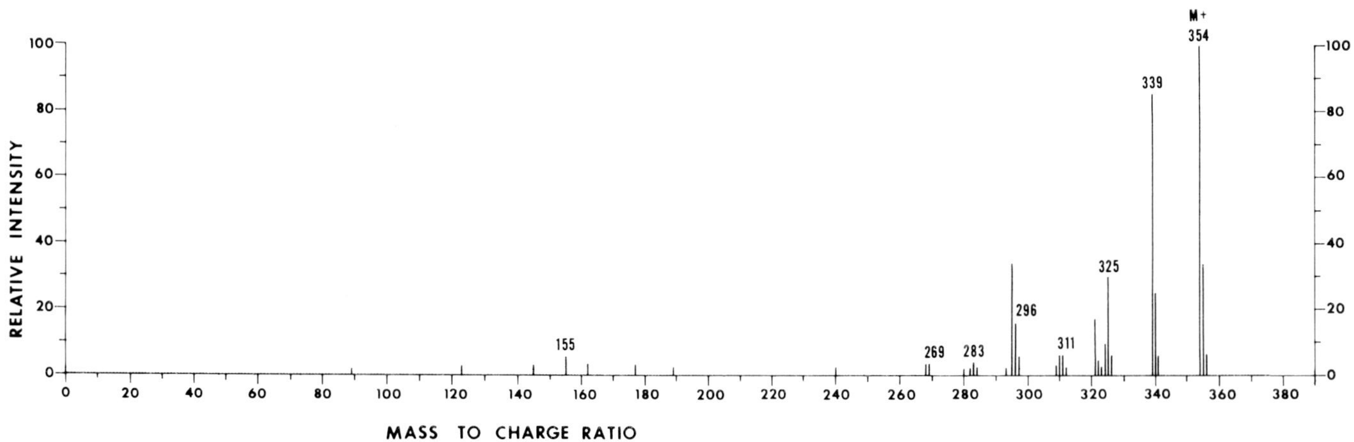

M+
354
339
325
311
296
283
269
155
RELATIVE INTENSITY
MASS TO CHARGE RATIO

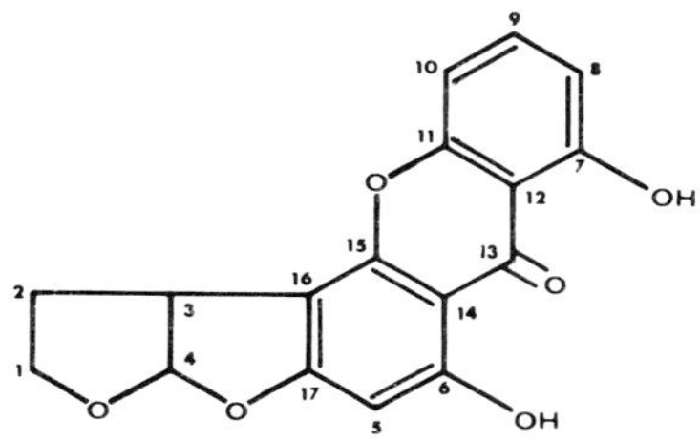

Common name	Dihydrodemethylsterigmatocystin (1,2-Dihydro-6,7-hydroxydifuroxanthone)
Molecular weight	312.0633
Molecular formula	$C_{17}H_{12}O_6$
General characteristics[12]	Pale yellow needles from acetone, mp 202°C (dec.) $[\alpha]_D^{25}$ −376.6° (C = 0.90 in chloroform) Green ferric reaction in ethanol
UV data[12]	λ_{max}^{EtOH} nm (ϵ): 233(26,800), 250(33,100) 259(30,300), and 335(19,400)
IR data[12]	ν_{max} cm^{-1}: 1659 and 1625
Fungal source[12]	*Aspergillus versicolor* (Vuill.) Tiraboschi

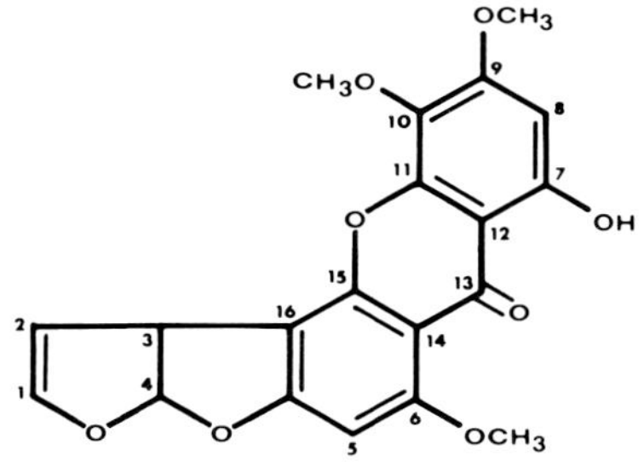

Common name	Dimethoxysterigmatocystin (6,9,10,-Trimethoxy-7-hydroxydifuroxanthone)
Molecular weight	384.0844
Molecular formula	$C_{20}H_{16}O_8$
General characteristics[11]	Pale yellow needles from acetone, mp 253°–254°C; dihydro derivative, mp 241°–242°C; acetate derivative, mp 188°–189°C (X-ray analysis); monomethyl ether derivative, mp 257°–258°C
UV data[11]	λ_{max}^{EtOH} nm (ϵ): 233(27,200), 284(34,000), 275(sh)(7,700), and 330(19,200)
IR data[11]	ν_{max}^{KBr} cm^{-1}: 3125, 1655, 1635, and 1610
^1H NMR data[11]	($CDCl_3$) H1, 6.64 (J = 2, 2.5); H2, 5.41 (J = 2, 2.5); H3, 4.80 (J = 2, 2.5, 7); H4, 6.75 (J = 7); H5,* 6.28; H8,* 6.32; 7OH, 13.20; OCH_3, 3.91, 3.94, 3.98 ppm

* Assignments uncertain.

REFERENCES

1. Athnasios, A. K., and Kuhn, G. O., *J. Assoc. Off. Anal. Chem.* **60,** 104 (1977).
2. Birkinshaw, J. H., and Hammady, I. M. M., *Biochem. J.* **65,** 162 (1957).
3. Bullock, E., Roberts, J. C., and Underwood, J. G., *J. Chem. Soc.* p. 4179 (1962).
4. Burkhardt, H. J., and Forgacs, J., *Tetrahedron* **24,** 717 (1968).
5. Chexal, K. K., Holker, J. S. E., Simpson, T. J., and Young, K., *J. Chem. Soc., Perkin Trans. I* p. 543 (1975).
6. Cole, R. J., Kirksey, J. W., and Schroeder, H. W., *Tetrahedron Lett.* p. 3109 (1970).
7. Cox, R. H., and Cole, R. J., *J. Org. Chem.* **42,** 112 (1977).
8. Davies, J. E., Kirkaldy, D., and Roberts, J. C., *J. Chem. Soc.* p. 2169 (1960).
9. Engelbrecht, J. C., and Altenkirk, B., *J. Natl. Cancer Inst.* **48,** 1647 (1972).
10. Fukuyama, K., Hamada, K., Tsukihara, T., Katsube, Y., Hamasaki, T., and Hatsuda, Y., *Bull. Chem. Soc. Jpn.* **49,** 1153 (1976).
11. Hamasaki, T., Nakagomi, T., Hatsuda, Y., Fukuyama, K., and Katsube, Y., *Tetrahedron Lett.* p. 2765 (1977).
12. Hatsuda, Y., Hamasaki, T., Ishida, M., Matsui, K., and Hara, S., *Agric. Biol. Chem.* **36,** 521 (1972).
13. Holker, J. S. E., and Kagol, S. A., *Chem. Commun.* p. 1574 (1968).
14. Holsapfel, C. W., Purchase, I. F. H., Steyn, P. S., and Gouws, L., *S. Afr. Med. J.* **40,** 1100 (1966).
15. Hsieh, D. P. H., Lin, M. T., and Yao, R. C., *Biochem. Biophys. Res. Commun.* **52,** 992 (1973).
16. Manabe, M., Minamisawa, M., and Matsuura, S., *J. Agric. Chem. Soc. Jpn.* **45,** 565 (1971).
17. Pachler, K. G. R., Steyn, P. S., Vleggaar, R., Wessels, P. L., and Scott, D. B., *J. Chem. Soc., Perkin Trans. I* p. 1182 (1976).
18. Pachler, K. G. R., Steyn, P. S., Vleggaar, R., Wessels, P. L., *J. Chem. Soc., Chem. Commun.* p. 355 (1975).
19. Purchase, I. F. H., and van der Watt, J. J., *in* "Mycotoxins in Human Health" (I. F. H. Purchase, ed.), p. 209. Macmillan, New York, 1971.
20. Rabie, C. J., Lubben, A., and Steyn, M., *Appl. Environ. Microbiol.* **32,** 206 (1976).
21. Rabie, C. J., Steyn, M., and van Schalkwyk, G. C., *Appl. Environ. Microbiol.* **33,** 1023 (1977).
22. Rance, M. J., and Roberts, J. C., *Tetrahedron Lett.* p. 2799 (1970).
23. Rodricks, J. V., Henery-Logan, K. R., Campbell, A. D., Stoloff, L., and Verrett, M. J., *Nature (London)* **217,** 688 (1968).
24. Rodricks, J. V., Lustig, E., Campbell, A. D., Stoloff, L., and Henery-Logan, K. R., *Tetrahedron Lett.* p. 2975 (1968).
25. Schroeder, H. W., and Kelton, W. H., *Appl. Microbiol.* **30,** 589 (1975).
26. Seto, H., Cary, L. W., and Tanabe, M., *Tetrahedron Lett.* p. 4491 (1974).
27. Steyn, M., *J. Assoc. Off. Anal. Chem.* **57,** 225 (1974).
28. Tanabe, M., Hamasaki, T., Seto, H., and Johnson, L., *Chem. Commun.* p. 1539 (1970).
29. Tanaka, N., Katsube, Y., Hatsuda, Y., Hamasaki, T., and Ishida, M., *Bull. Chem. Soc. Jpn.* **43,** 3635 (1970).
30. Waiss, A. C., Jr., Wiley, M., Black, D. R., and Lundin, R. E., *Tetrahedron Lett.* p. 3207 (1968).

3

Versicolorin Group

The versicolorins are a group of fungal metabolites that show little or no acute toxicity to vertebrate animals. The significance of this group of metabolites, as it relates to the science of mycotoxicology, is based on their chemical structure (presence of the dihydrodifurano and tetrahydrodifurano moieties identical to that in the aflatoxins and sterigmatocystins) and/or the demonstration that representative members of the group are biosynthetic precursors of the aflatoxins and sterigmatocystins. In addition to the dihydrodifurano and tetrahydrodifurano moieties, they contain anthraquinone moieties that differ from each other by the number and position of substituents on the anthraquinone ring.

Also included in the versicolorin group are versiconal hemiacetal acetate, versiconol acetate, versiconol, averufin, nidurufin, dimethylnidurufin, aversin, norsolorinic acid, and *O*-methylaversin, all biosynthetic precursors of the aflatoxins.

Versicolorin group	Molecular weight	Molecular formula
Versicolorin A	338.0426	$C_{18}H_{10}O_7$
Versicolorin B	340.0582	$C_{18}H_{12}O_7$
Versicolorin C	340.0582	$C_{18}H_{12}O_7$
Averufin	368.0895	$C_{20}H_{16}O_7$
Norsolorinic acid	370.1052	$C_{20}H_{18}O_7$
Versiconal hemiacetal acetate	400.0793	$C_{20}H_{16}O_9$
Versiconol acetate	402.0945	$C_{20}H_{18}O_9$
Versiconol	360.0844	$C_{18}H_{16}O_8$
Nidurufin	384.0844	$C_{20}H_{16}O_8$
Dimethylnidurufin	412.1157	$C_{22}H_{20}O_8$
Aversin	368.0896	$C_{20}H_{16}O_7$
O-Methylaversin	382.1052	$C_{21}H_{18}O_7$

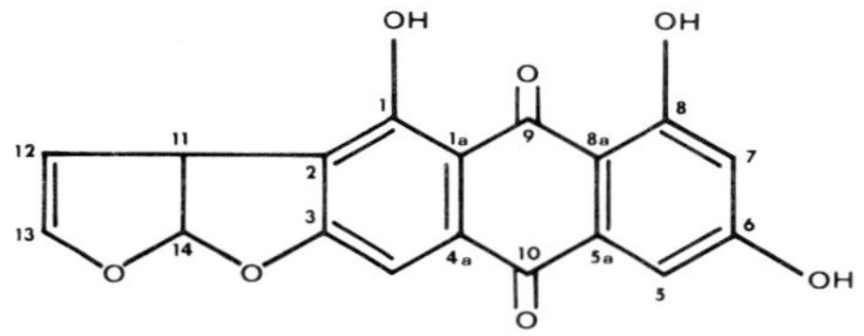

Common name	Versicolorin A (2,3-Bisfurano-1,6,8-trihydroxyanthraquinone)
Molecular weight	338.0426
Molecular formula	$C_{18}H_{10}O_7$
General characteristics[12,22]	Orange-yellow needles from acetone, mp 289°C (dec.) Fine yellow needles from hexane, mp 287°–288°C (dec.) $[\alpha]_D^{18}$ −354° (C = 0.75 in dioxane) Trimethyl ether, mp 241°C Versicolorin A was soluble in acetone, dioxane, ethyl acetate, ether and ethanol; sparingly soluble in chloroform and benzene; insoluble in water and aqueous bicarbonate. Versicolorin A is a biosynthetic precursor of the aflatoxins.
UV data[12,22]	λ_{max}^{EtOH} nm (ϵ): 222(31,488), 254(15,227), 265(17,756), 290(26,547), 321(12,118), and 453(8,166) λ_{max}^{EtOH} nm (ϵ): 222(28,100), 255(13,500), 267(18,200), 290(25,100), 326(6,700), and 450(7,000)
Fungal source[14,15,21,22]	*Aspergillus versicolor* *A. parasiticus* (mutant strain 1-11-105 wk-1)
Toxicity data	Toxicity unknown
TLC data[22]	*Adsorbent:* adsorbosil-l silica gel *Solvent:* benzene–acetic acid, 95:5 v/v R_f: 0.32 *Detection:* yellow-orange spot in visible light

^{13}C NMR data[11]

C1, 158.9 s; C1a, 109.4 s; C2, 121.1 s; C3, 164.7 s; C4, 102.4 d; C4a, 135.8 s; C5, 109.4 d; C5a, 135.3 s; C6, 165.8 s; C7, 108.5 d; C8, 165.8 s; C8a, 109.0 s; C9, 189.5 s; C10, 181.2 s; C11, 47.9 d; C12, 101.9 d; C13, 146.3 d; C14, 113.6 d

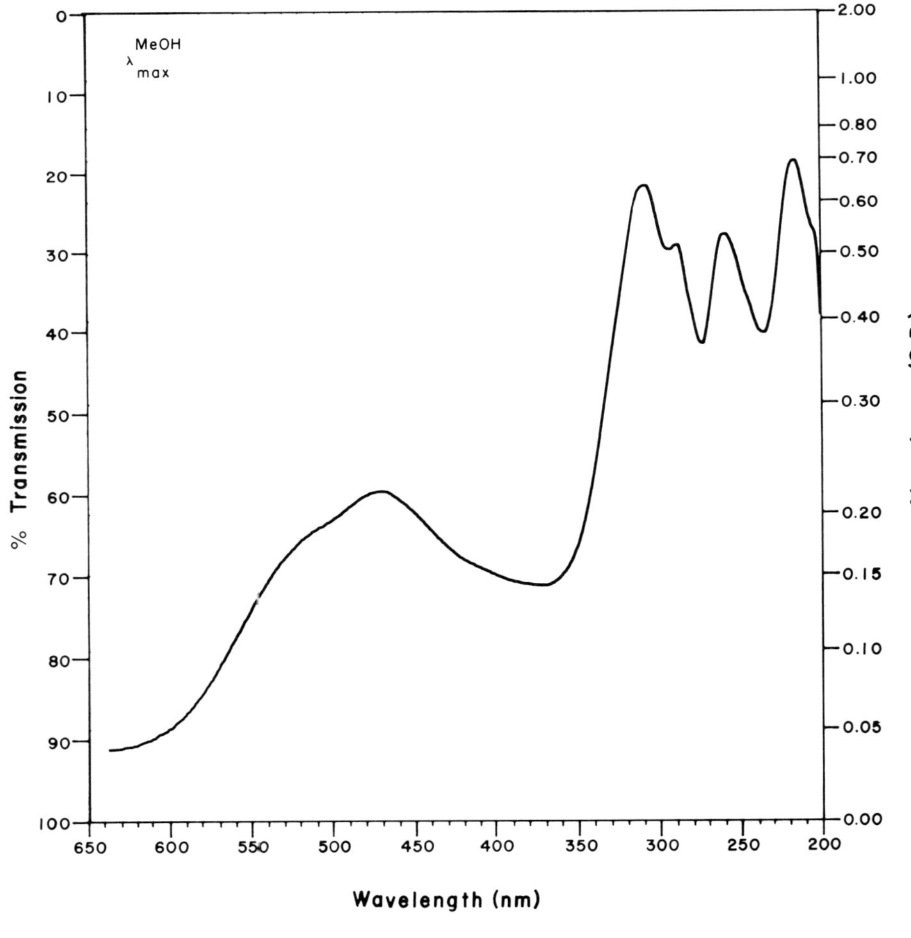

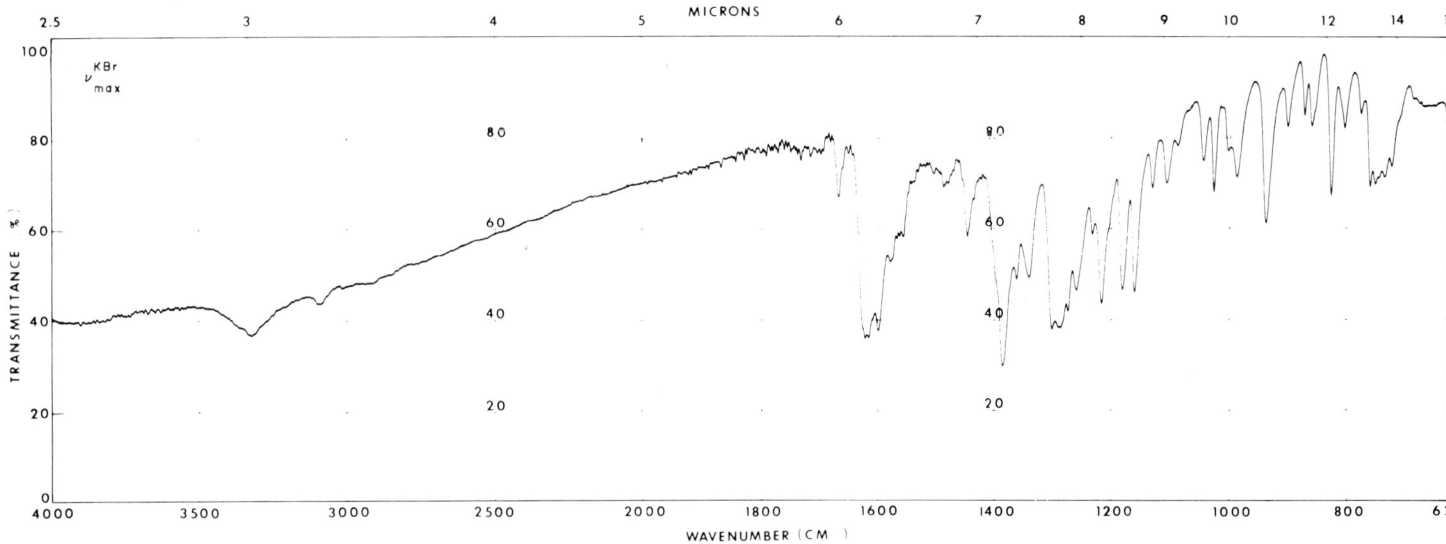
MICRONS
TRANSMITTANCE (%)
ν_{max}^{KBr}
WAVENUMBER (CM)

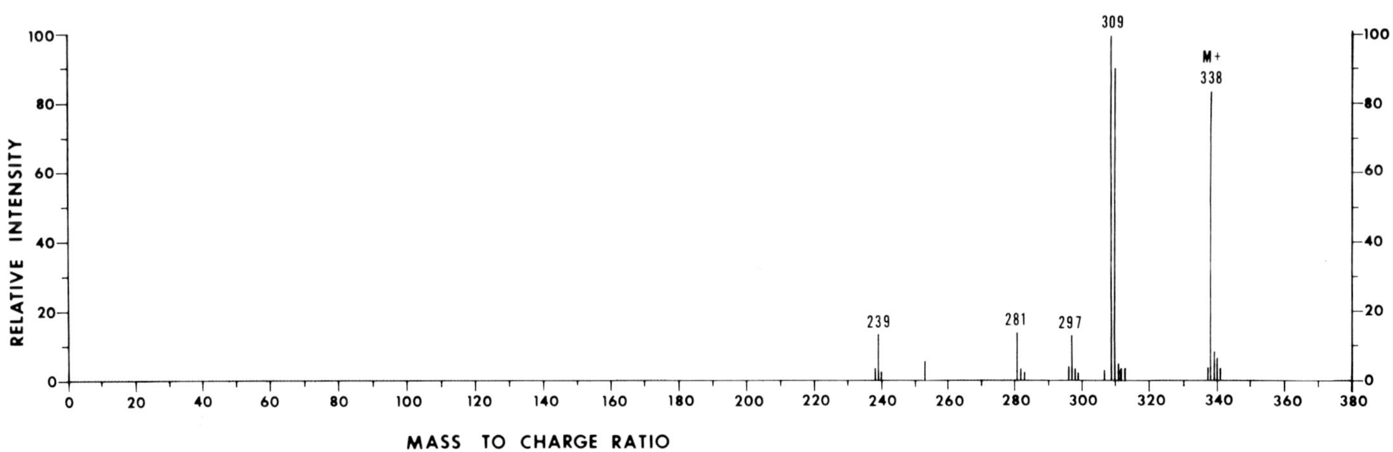
RELATIVE INTENSITY
MASS TO CHARGE RATIO
239
281
297
309
M+
338

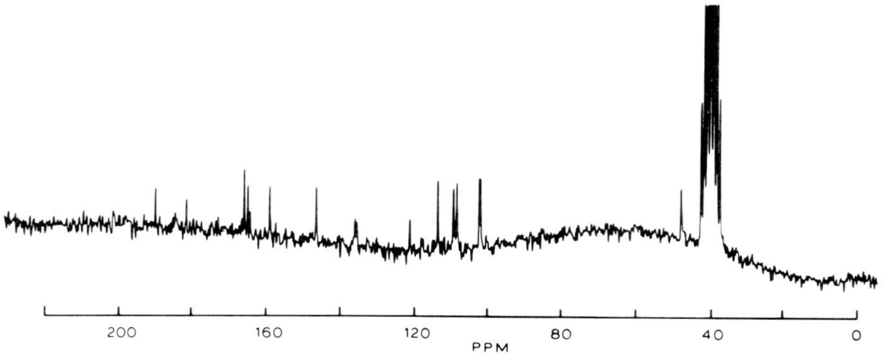

The 25 MHz ^{13}C NMR spectrum of versicolorin A in DMSO solution.

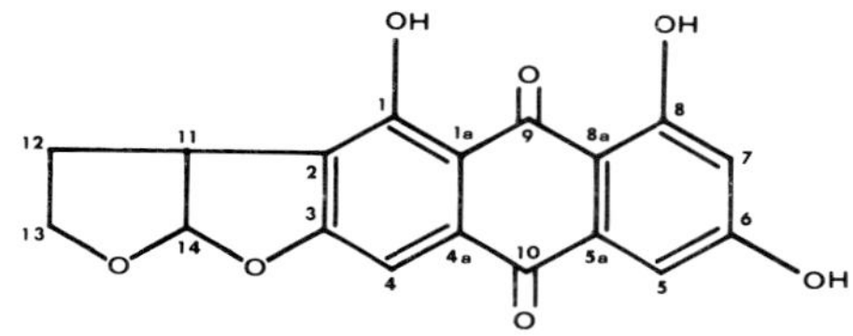

Common name	Versicolorin B (2,3,-Bistetrahydrofurano-1,6,8-trihydroxyanthraquinone)
Molecular weight	340.0582
Molecular formula	$C_{18}H_{12}O_7$
General characteristics[12]	Fine yellow needles from acetone, mp 298°C (dec.) Trimethyl ether, mp 216°C $[\alpha]_D^{25}$ −223° (C = 0.42 in dioxane) Versicolorin B was soluble in acetone, dioxane, ethyl acetate, ether and ethanol; sparingly soluble in chloroform and benzene; insoluble in water and aqueous sodium bicarbonate.
UV data[12]	λ_{max}^{EtOH} nm (ϵ): 223(23,900), 255(13,500), 266(19,500), 291(24,000), 324(12,800), and 450(8,700)
Fungal source[12,14,15]	*Aspergillus versicolor* (Vuill.) Tiraboschi
Toxicity data	Toxicity unknown
TLC data[23]	*Adsorbent:* silica gel *Solvent:* benzene–acetic acid, 95:5 v/v R_f: 0.23 *Detection:* yellow spot in visible light
Other significant references	Biosynthesis[3]

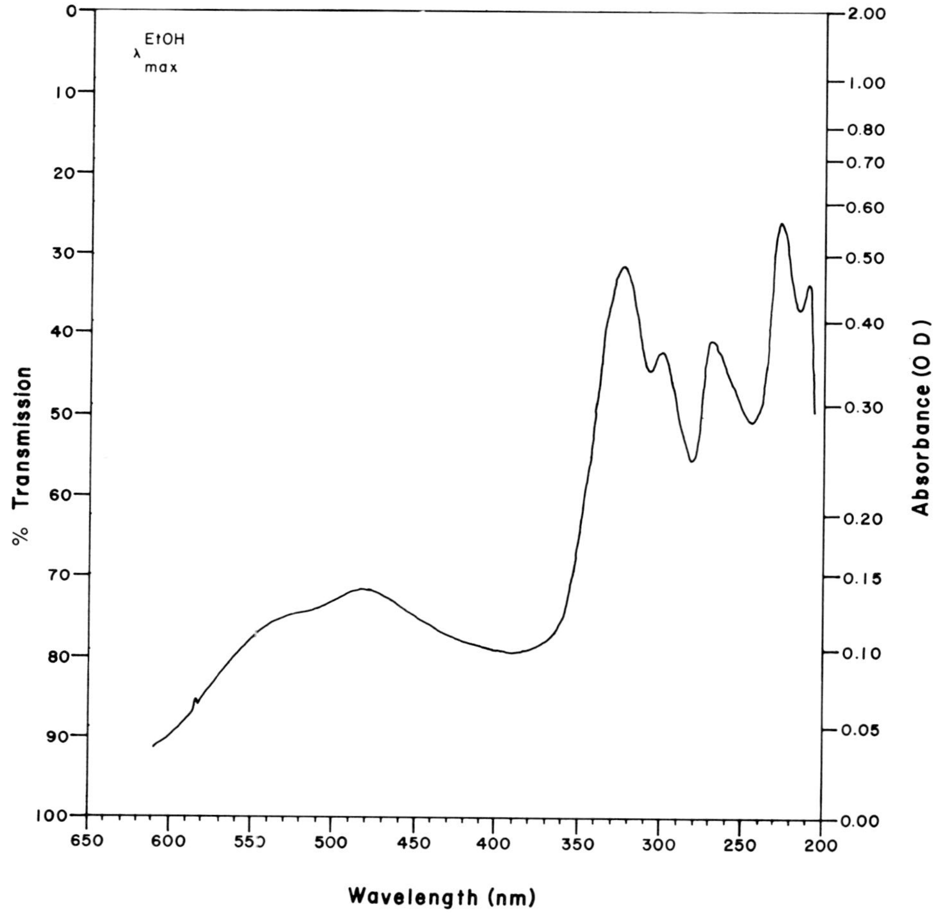

λ EtOH max
% Transmission
Absorbance (O D)
Wavelength (nm)
0
10
20
30
40
50
60
70
80
90
100
2.00
1.00
0.80
0.70
0.60
0.50
0.40
0.30
0.20
0.15
0.10
0.05
0.00
650
600
550
500
450
400
350
300
250
200

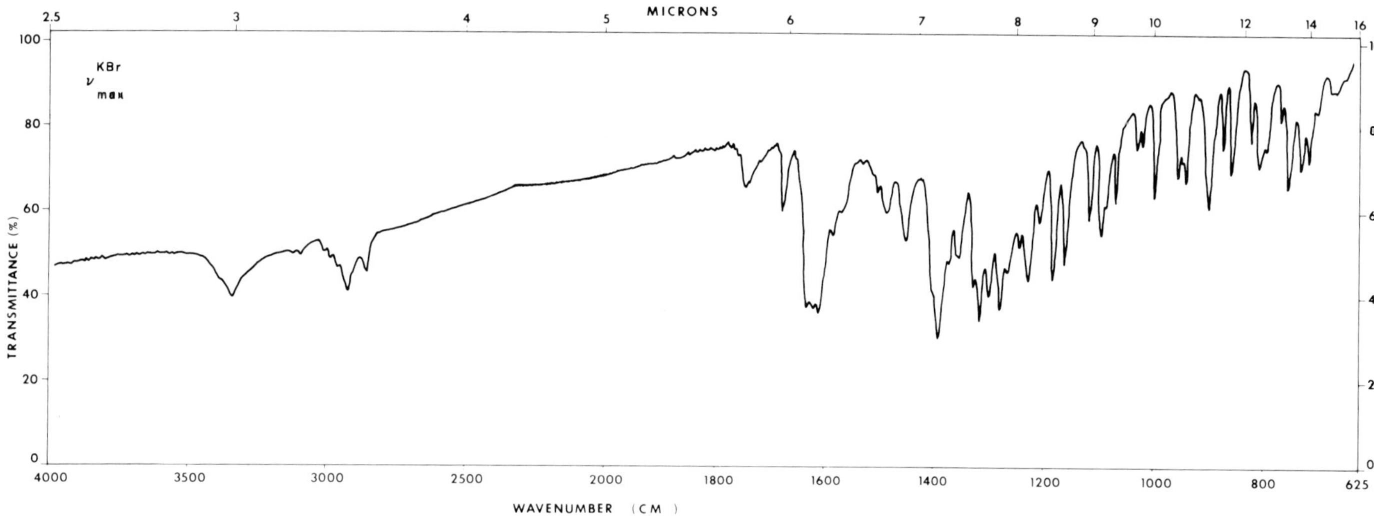
MICRONS
KBr
ν max
TRANSMITTANCE (%)
WAVENUMBER (CM)

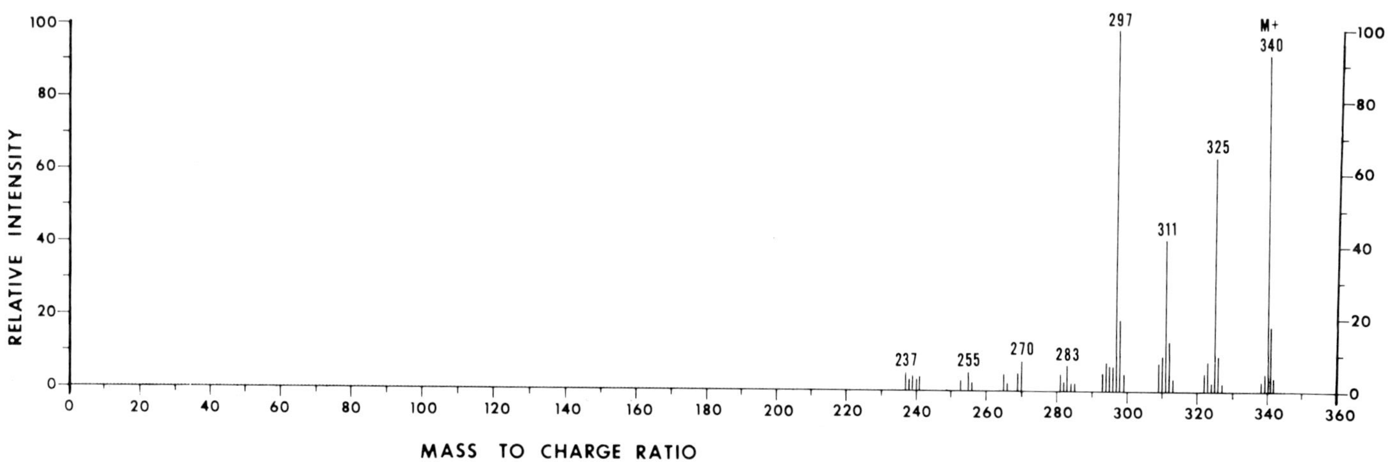
RELATIVE INTENSITY
MASS TO CHARGE RATIO
M+
340
325
311
297
283
270
255
237

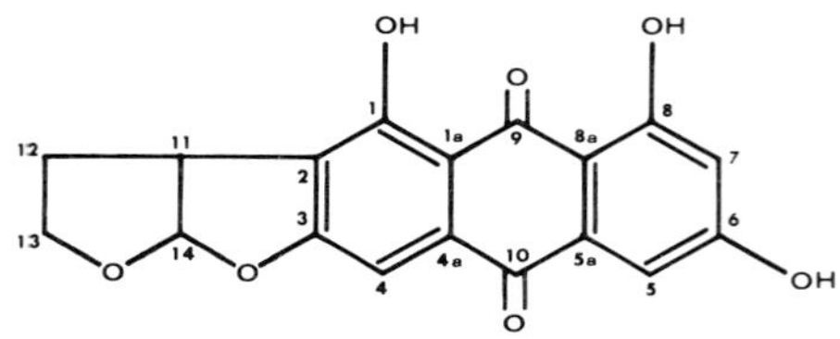

Common name	Versicolorin C (2,3-Bistetrahydrofurano-1,6,8-trihydroxyanthraquinone)
Molecular weight	340.0582
Molecular formula	$C_{18}H_{12}O_7$
General characteristics[12,14,23]	Orange-red needles from acetone, mp 310°C $[\alpha]_D^{25}$ O (*C* = 0.44 in dioxane) Trimethyl ether, mp 216°C $[\alpha]_D$ 0° Versicolorin C was soluble in acetone, dioxane, ethyl acetate, ether, and ethanol; sparingly soluble in chloroform and benzene; insoluble in water and aqueous bicarbonate. Versicolorin C is a biosynthetic precursor of the aflatoxins.
UV data[12]	λ_{max}^{EtOH} nm (ϵ): 223(28,800), 255(15,800), 267(20,400), 292(28,800), 326(10,000), and 450(10,700)
Fungal source[12,14,15]	*Aspergillus versicolor* (Vuill.) Tiraboschi
Toxicity data	Toxicity unknown
TLC data[23]	*Adsorbent:* silica gel *Solvent:* benzene–acetic acid, 95:5 v/v R_f: 0.23 *Detection:* yellow spot in visible light
Other significant references	X-ray structure[10]
^{13}C NMR data	C1, 158.6 s; C2, 119.7 s; C3, 165.1 s; C4, 100.9 d; C5, 108.6 d; C6, 164.8 s; C7, 107.6 d; C8, 163.8 s; C9, 188.4 s; C10, 180.1 s; C11, 43.3 t; C12, 30.0 t; C13, 66.9 d; C14, 112.9 d; C1a, 110.4 s; C4a, 134.2 s; C5a, 134.8 s; C8a, 107.9 s

Versicolorin C

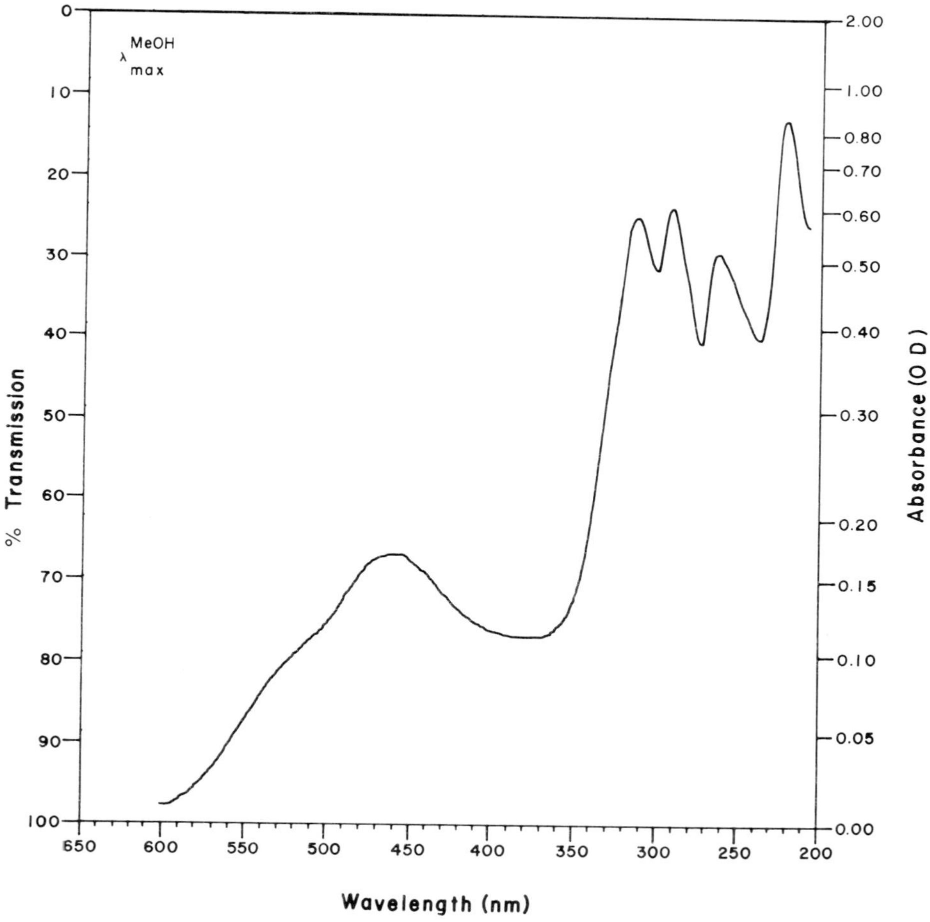

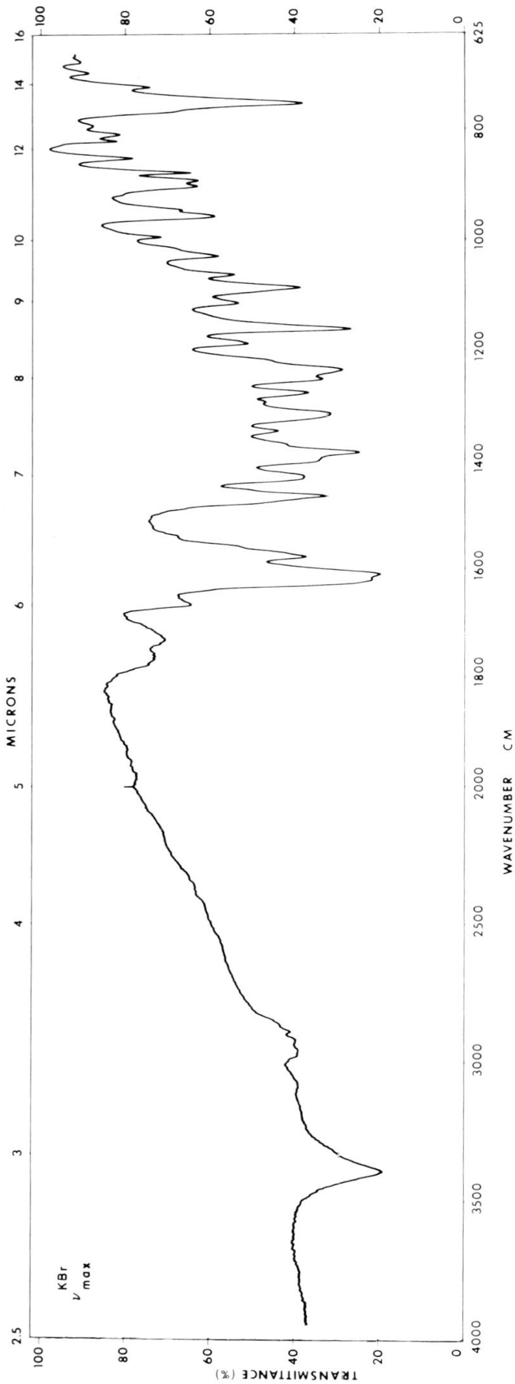
MICRONS
WAVENUMBER CM
TRANSMITTANCE (%)
KBr
ν_{max}

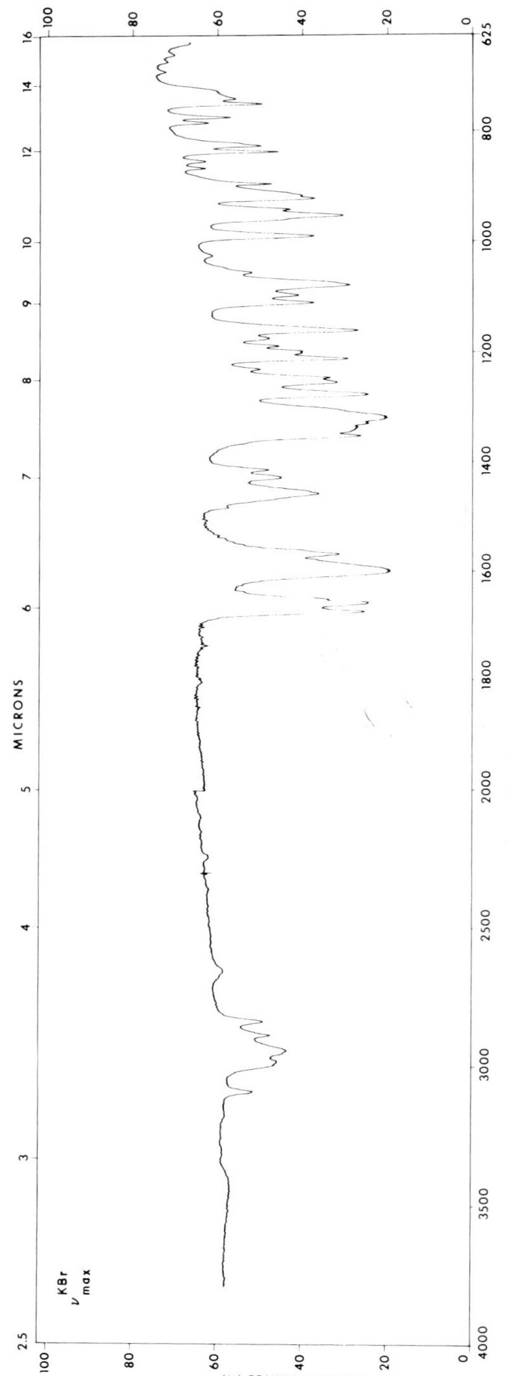
MICRONS
WAVENUMBER (CM)
TRANSMITTANCE (%)
KBr
ν_{max}

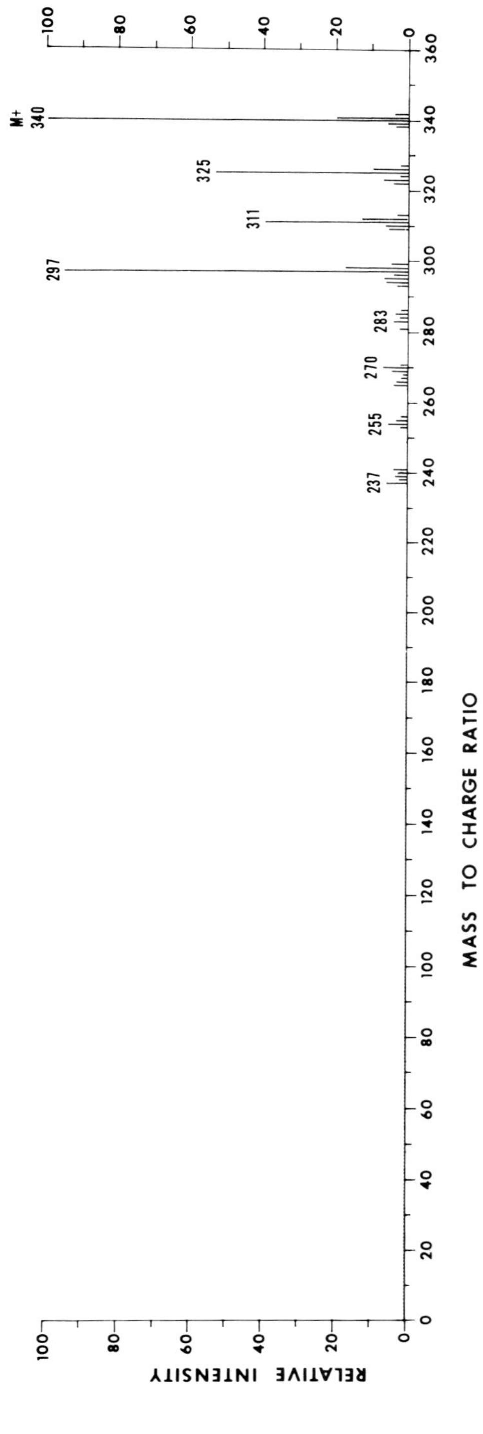

RELATIVE INTENSITY
MASS TO CHARGE RATIO
M+
340
325
311
297
283
270
255
237
0
20
40
60
80
100
120
140
160
180
200
220
240
260
280
300
320
340
360

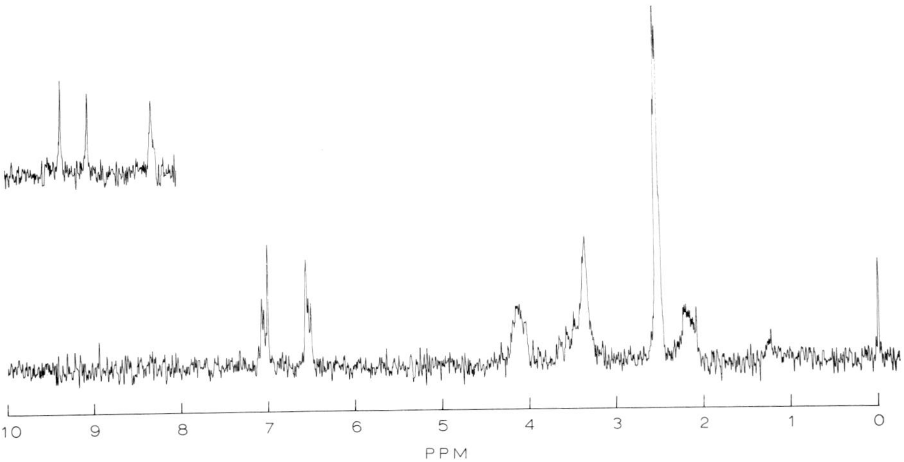

The 100 MHz ^{1}H NMR spectrum of versicolorin C in DMSO solution.

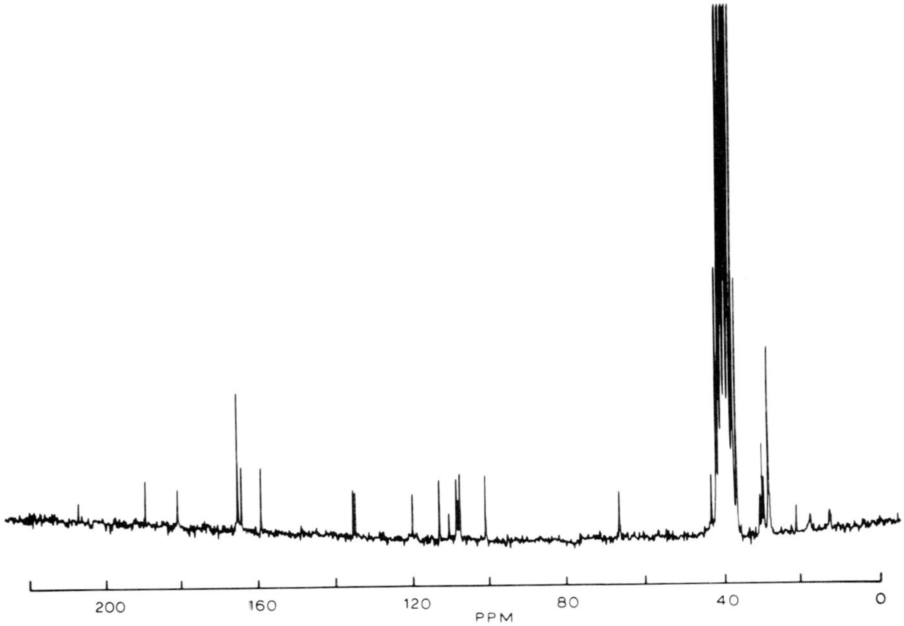

The 25 MHz ^{13}C NMR spectrum of versicolorin C in DMSO solution.

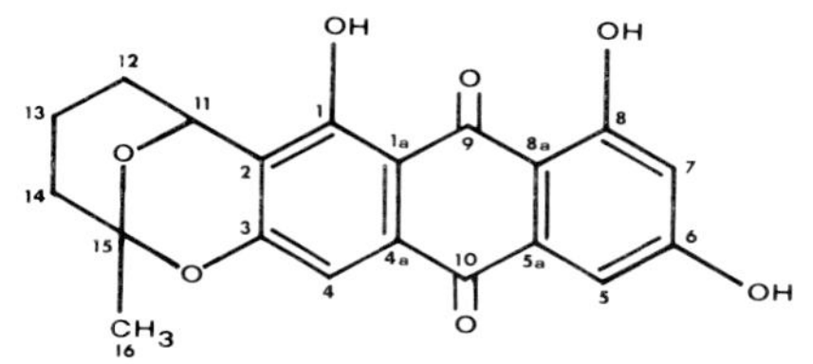

Common name	Averufin
Molecular weight	368.0895
Molecular formula	$C_{20}H_{16}O_7$
General characteristics[6,8,23,24]	Bright orange-red laths from acetone, mp 280°–282° (dec.) $[\alpha]_D^{23}$ = >1° (*C* = 0.30 in ethanol) Crystals from acetone, mp 283°–289°(dec.) Tri-*O*-acetyl derivative, yellow needles from ethanol, mp 297°–311°C $[\alpha]_D^{22}$ −15° (5 g/liter in chloroform) Averufin is unstable toward alkali; stable toward acid. It is a biosynthetic precursor of aflatoxin B_1.
UV data[6]	λ_{max}^{EtOH} nm (ϵ): 223(33,000), 256(sh)(16,500), 265(18,500), 286(sh)(24,800), 294(30,800), 319(12,500), and 454(10,500) Tri-*O*-acetyl derivative λ_{max}^{EtOH} nm (ϵ): 244(18,300), 248(18,300), 281(45,000), 282(45,000), 335(4,900), and 370(4,250)
Fungal source[6,7,24]	*Aspergillus versicolor* (Vuill.) Tiraboschi *A. parasiticus* (mutant, ATCC 15517) *A. ustus*
TLC data[6]	*Adsorbent:* silical gel (Eastman 6060) *Solvent:* chloroform–acetone–acetic acid, 97:2:1 v/v/v R_f: 0.5 *Detection:* red spot in visible light
Other significant references	Revised structure[16,25,26]; biosynthesis[3,8,23]; X-ray structure[19]
^{13}C *NMR data*	C1, 158.0 s; C2, 115.8 s; C3, 159.7 s; C4, 107.0 d;

C5, 108.8 d; C6, 165.1 s; C7, 107.9 d; C8, 164.1 s; C9, 188.6 s; C10, 180.5 s; C11, 65.9 d; C12, 26.8 t; C13, 15.3 t; C14, 35.1 t; C15, 101.1 s; C16, 27.3 q; C1a, 108.3 s; C4a, 132.9 s; C5a, 134.7 s; C8a, 108.5 s

^1H NMR data H4, 6.98; H5, 7.08 (J = 2.5); H7, 6.56 (J = 2.5); H11, 5.26; H16, 1.58; OH's, 11.30, 12.04, 12.42

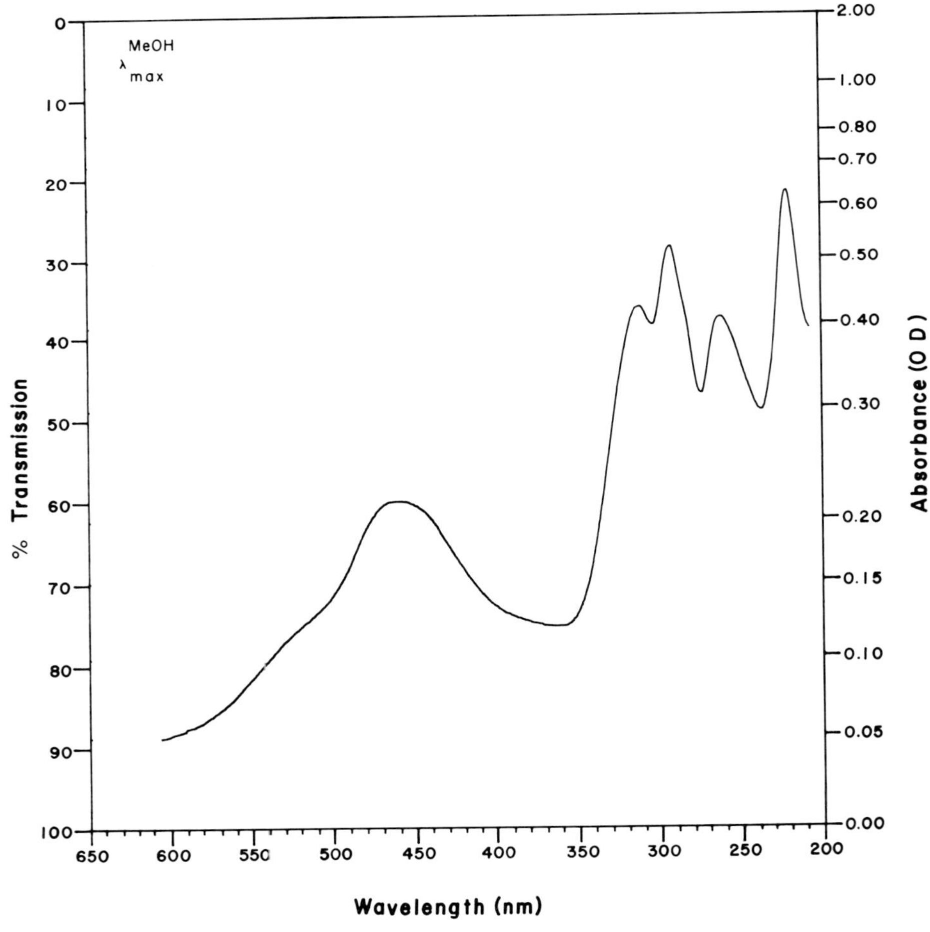

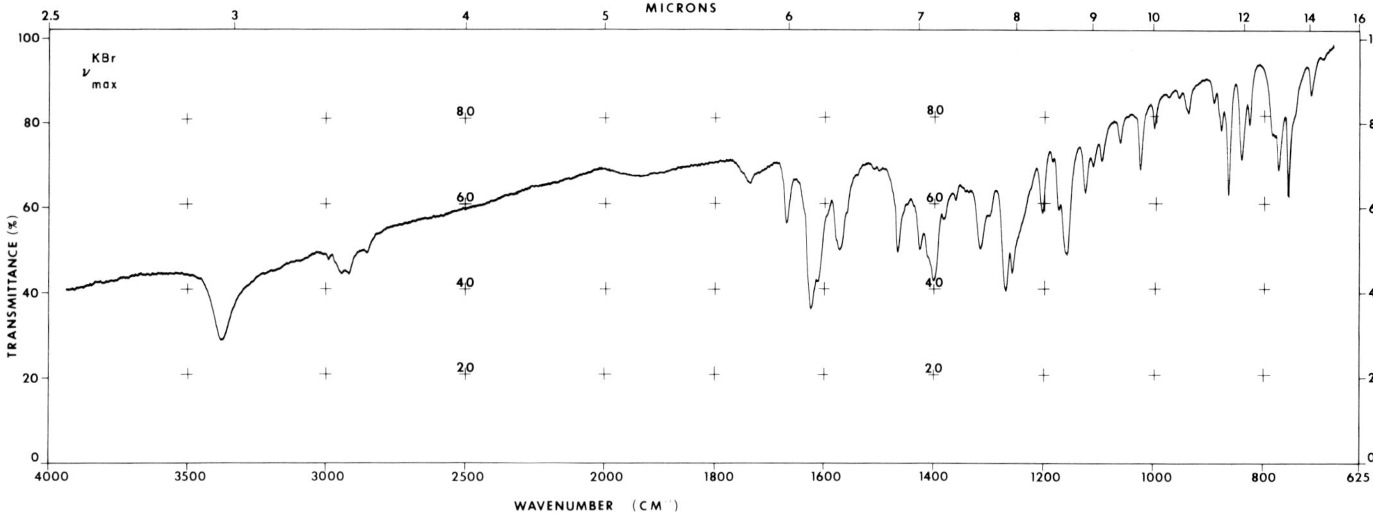
MICRONS
KBr
ν_{max}
TRANSMITTANCE (%)
WAVENUMBER (CM⁻¹)

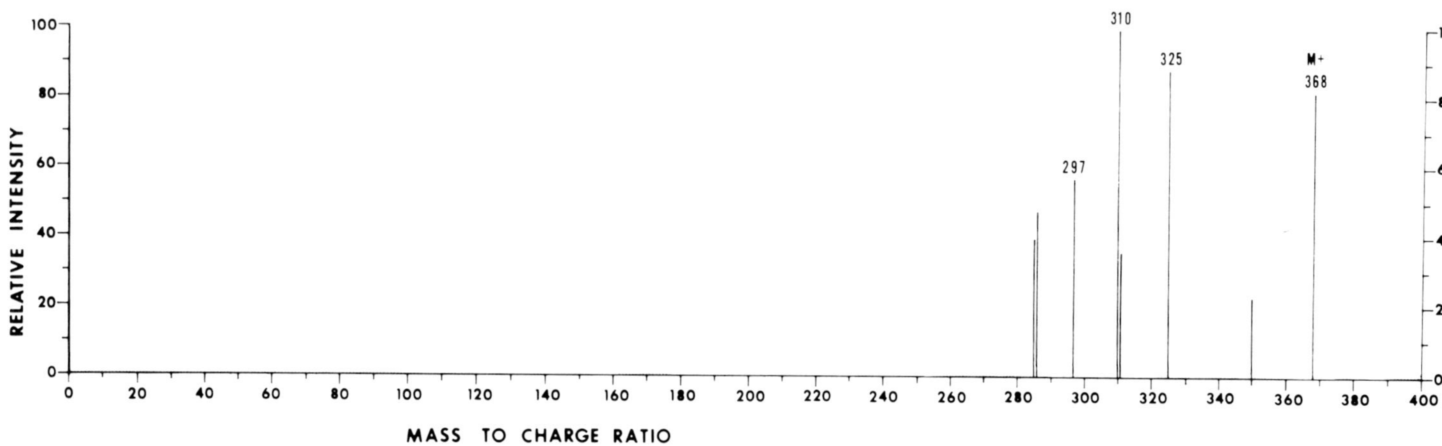
310
325
M+
368
297
RELATIVE INTENSITY
MASS TO CHARGE RATIO

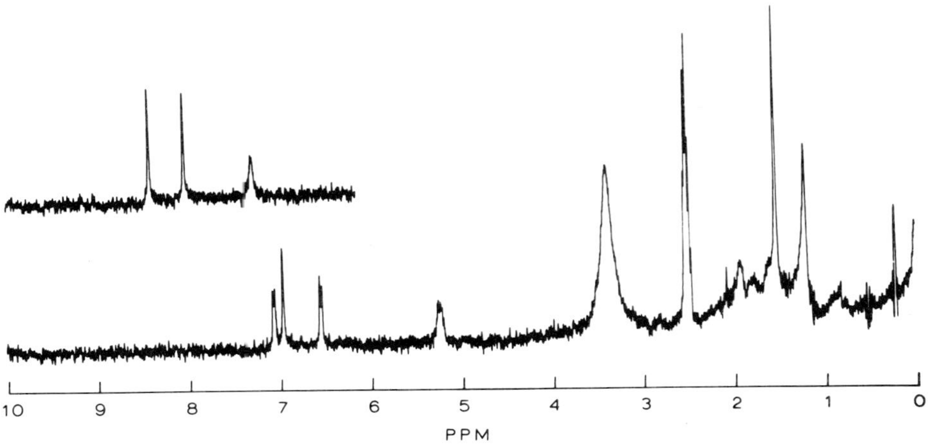

The 100 MHz ^1H NMR spectrum of averufin in DMSO solution.

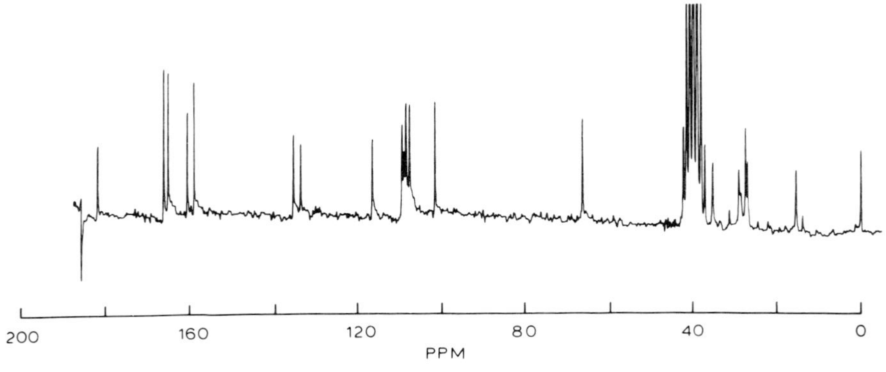

The 25 MHz ^{13}C NMR spectrum of averufin in DMSO solution.

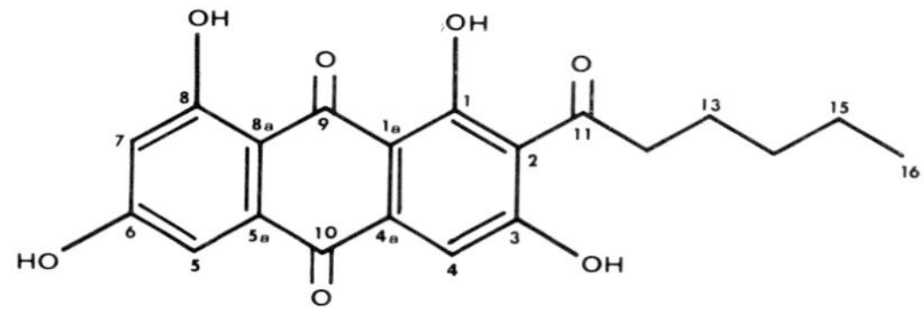

Common name — Norsolorinic acid (2-*n*-Hexanoyl-1,3,6,8-tetrahydroxyanthraquinone)

Molecular weight — 370.1052

Molecular formula — $C_{20}H_{18}O_7$

General characteristics[13,22] — Red prisms from acetone, mp 256°–257°C (dec.); forms tetramethyl ether, mp 131°–132°C.
Norsolorinic acid was sparingly soluble in acetone, ethanol, and methanol; it was insoluble in water, hexane, ether and in sodium bicarbonate; it dissolved in sodium carbonate and in sodium hydroxide to give a purple color.
Norsolorinic acid is a precursor in aflatoxin biosynthesis.

UV data[13,22] — λ_{max}^{EtOH} nm (ϵ): 235(24,500), 269(16,900), 284(18,600), 297(19,900), 314(22,900), and 465(7,760)
λ_{max}^{EtOH} nm (ϵ): 234(23,667), 265(16,650), 283(17,352), 297(sh)(19,872), 313(23,763), and 465(7,336)

Fungal source[1,13,22] — *Aspergillus versicolor* (Vuill.) Tiraboschi
A. parasiticus (mutant strain, NRRL-A-27,996)
The lichen, *Sorlorina crocea* (L.) Ach.

Toxicity data — Toxicity unknown

TLC data[18] — *Adsorbent:* adsorbosil-1
Solvent: A: chloroform–acetone–*n*-hexane, 85:15:20 v/v/v
B: ethyl acetate–2-propranol–water, 10:2:1 v/v/v
R_f: A: 0.69
B: 1.0
Detection: orange-red spot in normal light

^{13}C NMR data	C1, 161.0 s; C2, 121.1 s; C3, 163.0 s; C4, 108.7 d; C5, 108.0 d; C6, 165.0 s; C7, 107.8 d; C8, 164.1 s; C9, 188.0 s; C10, 181.1 s; C11, 202.6 s; C12, 43. 5 t; C13, 22.6 t; C14, 30.6 t; C15, 21.8 t; C16, 13.7 q; C1a; 108.9 s; C4a, 134.7 s; C5a, 134.5 s; C8a, 108.0 s
^{1}H NMR data	H4, 7.22; H5, 7.13 (J = 2.5); H7, 6.61 (J = 2.5); H12, 2.81; H13, 1.60; H14, 1.30; H15, 1.30; H16, 0.88

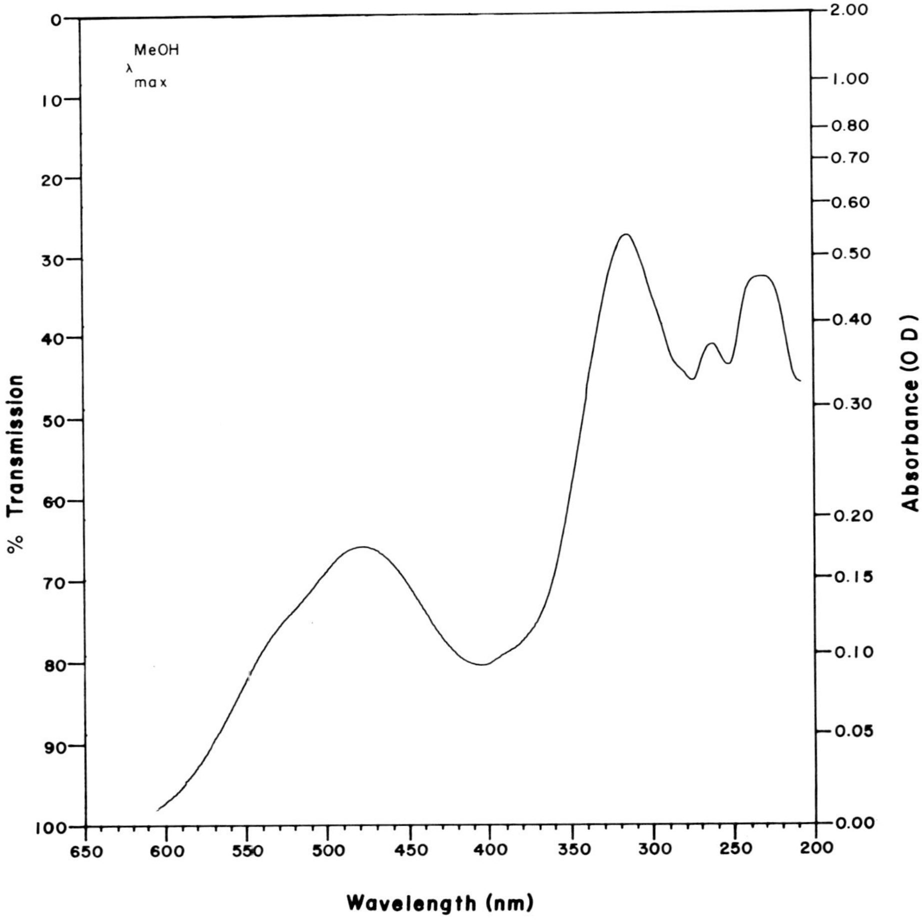

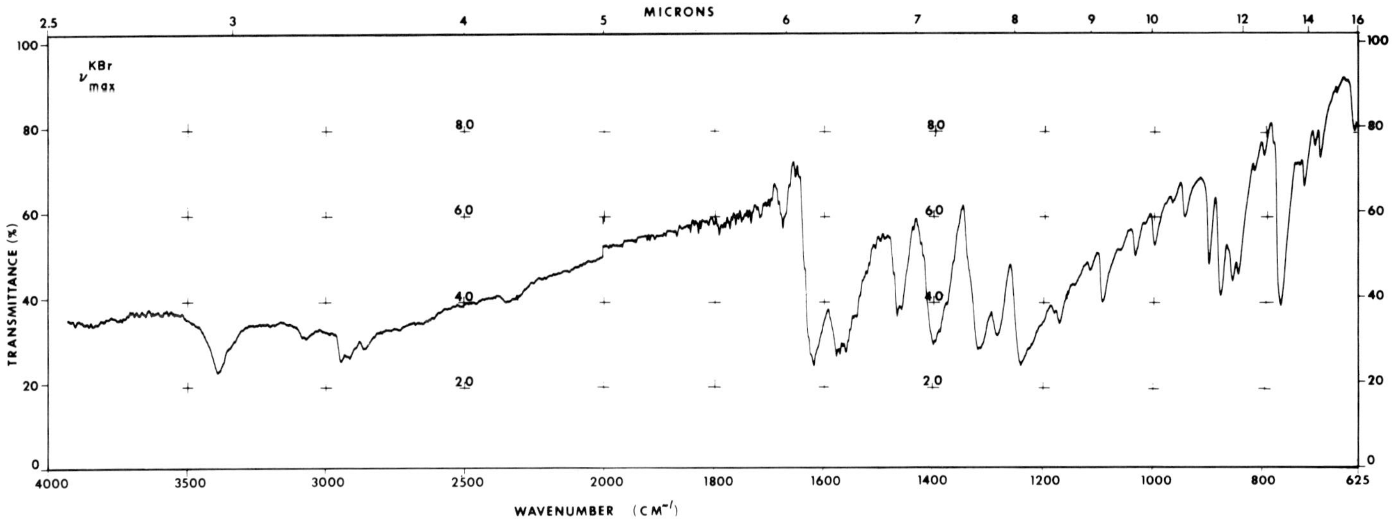
MICRONS
TRANSMITTANCE (%)
ν_{max}^{KBr}
WAVENUMBER (CM⁻¹)

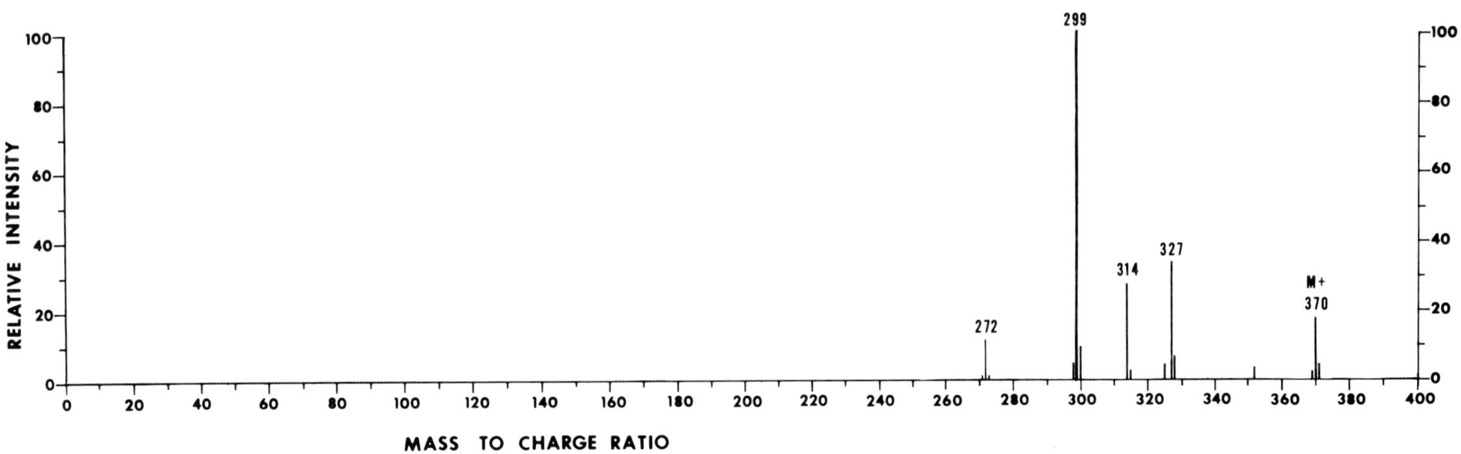
RELATIVE INTENSITY
MASS TO CHARGE RATIO
272
299
314
327
M+
370

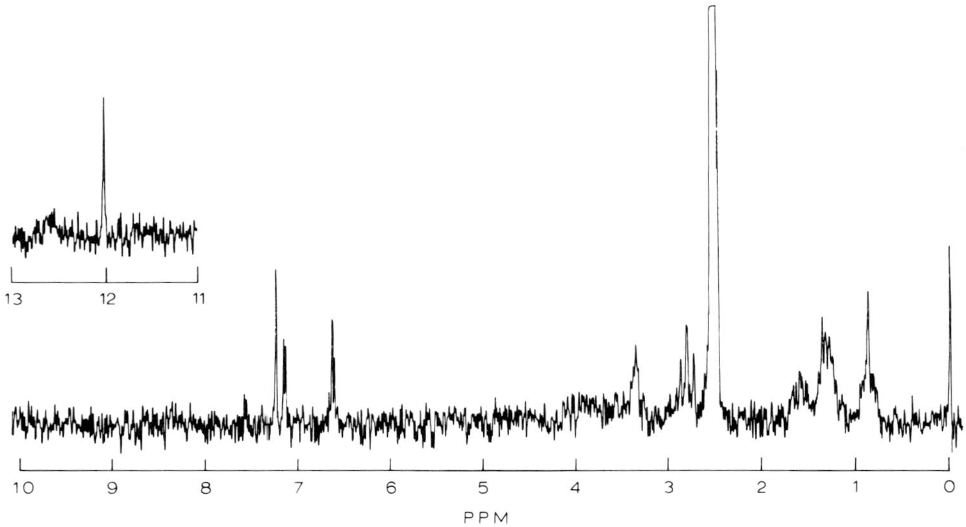

The 100 MHz ^{1}H NMR spectrum of norsolorinic acid in DMSO solution.

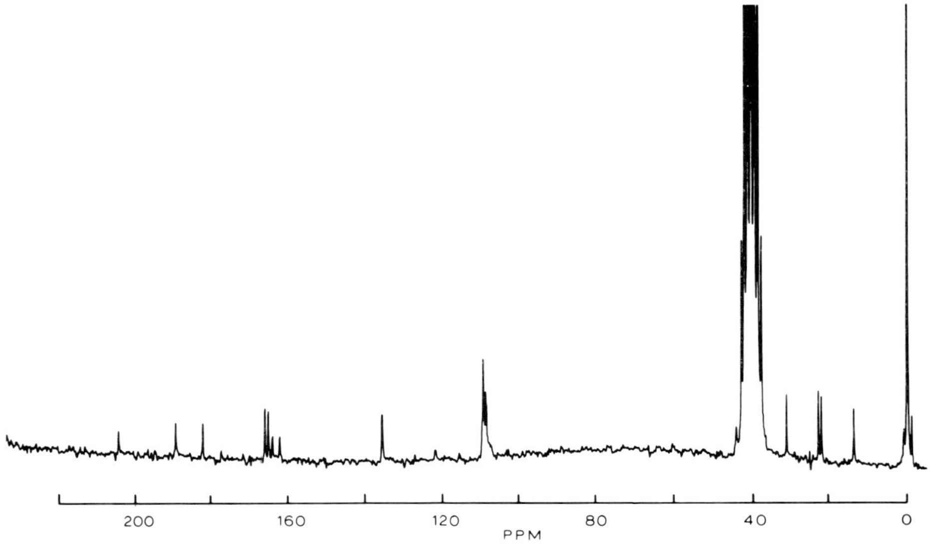

The 25 MHz ^{13}C NMR spectrum of norsolorinic acid in DMSO solution.

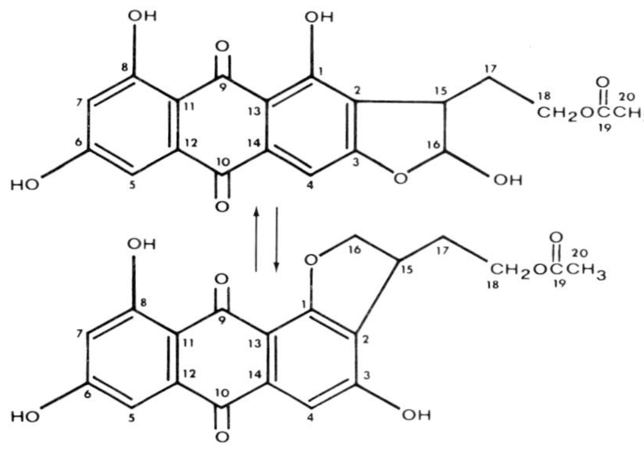

Common name
Versiconal hemiacetal acetate (2,3-Tetrafurano-1,6,8-trihydroxy-15-ethyl acetate)

Molecular weight
400.0793
Electron-impact mass spectral analysis shows 382.0688 due to $M^+ - H_2O$ ($C_{20}H_{14}O_8$); chemical ionization shows distinct *m/e* 401 due to $M^+ + H^+$.

Molecular formula
$C_{20}H_{16}O_9$

General characteristics[5,9,27,28,29]
Orange needles from chloroform–acetone, mp 216°–220°C; from acetone, mp 234°–236°C. In polar solvents, such as dimethyl sulfoxide, versiconal hemiacetal acetate exists as an equilibrium mixture of isomers; in acetone solution, the angular hemiacetal form is absent. Versiconal hemiacetal acetate accumulates in cultures of *A. flavus* and *A. parasiticus* treated with the insecticide dichlorvos, an inhibitor of aflatoxin biosynthesis. It is an intermediate in the biosynthesis of the aflatoxins.

UV data[27,28,29]
λ_{max}^{EtOH} nm (ϵ): 225(23,800), 267(14,000), 298(23,000), 323(11,300), and 480(7,260)
λ_{max}^{EtOH} nm (ϵ): 224(23,800), 225(sh)(13,000), 265(14,000), 295(23,000), 323(11,300), and 480(7,250)
λ_{max}^{MeOH} nm (ϵ): 223(25,700), 265(14,400), 291(2,400), 315(10,200), and 453(7,000)

Fungal source[27,29]	*Aspergillus flavus* *A. parasiticus*
Toxicity data	Toxicity unknown
TLC data[28]	*Adsorbent:* silicar TLC-7G silica gel *Solvent:* A: toluene–ethyl acetate, 27:12 v/v B: chloroform–acetone, 85:15 v/v R_f: A: 0.32 B: 0.33 *Detection:* orange-red spot in normal light
^{13}C NMR data	C1, 159.1 s; C2, 120.5 s; C3, 164.2 s; C4, 103.3 d; C5, 108.8 d; C6, 165.1 s; C7, 107.0 d; C8, 164.2 s; C9, 188.9 s; C10, 180.9 s; C11, 134.7 s; C12, 107.9 s; C13, 110.0 s; C14, 134.7 s; C15, 43.6 d; C16, 113.0 d; C17, 28.9 t; C18, 61.7 t; C19, 170.1 s; C20, 20.5 q

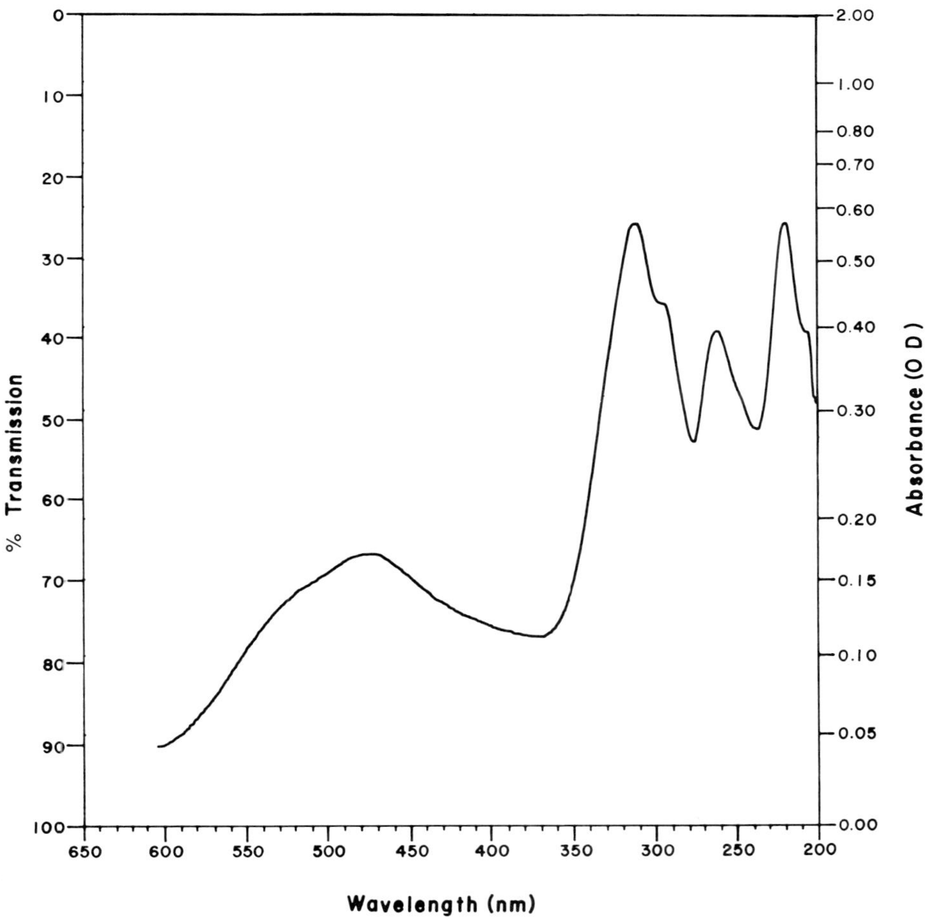
% Transmission
Absorbance (O D)
Wavelength (nm)
0
10
20
30
40
50
60
70
80
90
100
2.00
1.00
0.80
0.70
0.60
0.50
0.40
0.30
0.20
0.15
0.10
0.05
0.00
650
600
550
500
450
400
350
300
250
200

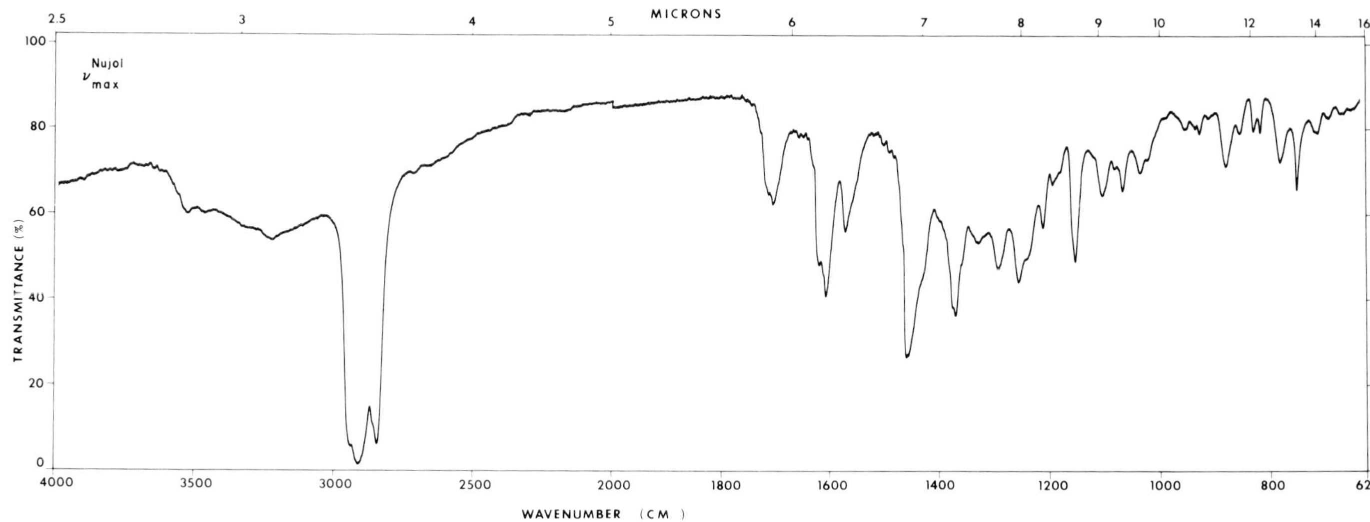
MICRONS
Nujol
ν_{max}
TRANSMITTANCE (%)
WAVENUMBER (CM)

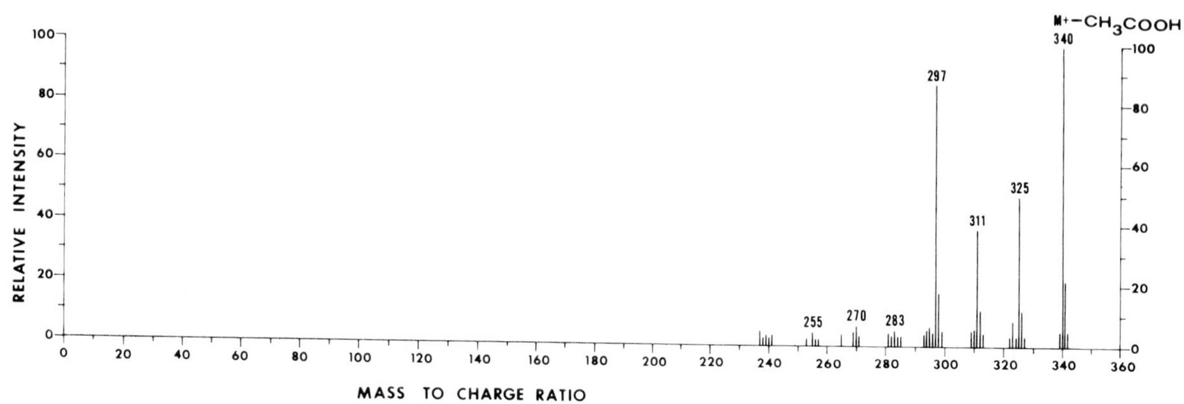
M+−CH3COOH
340
297
325
311
255
270
283
RELATIVE INTENSITY
MASS TO CHARGE RATIO

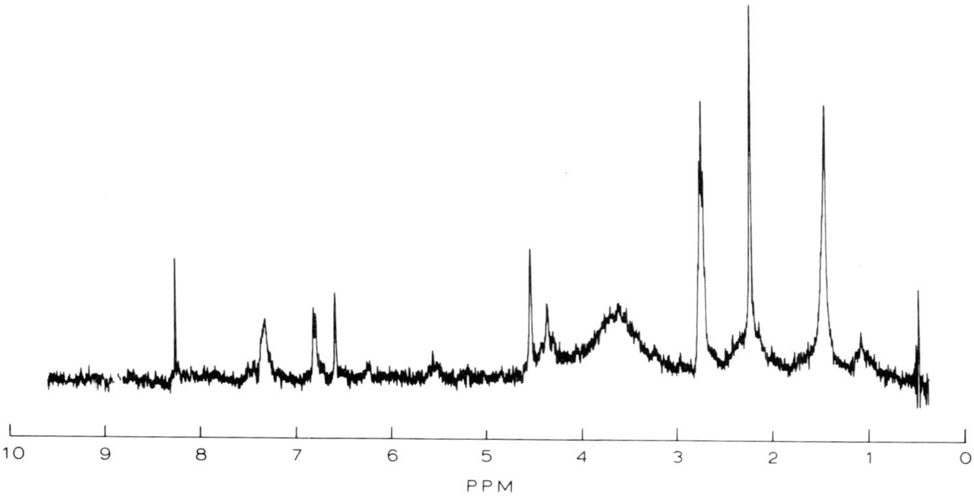

The 100 MHz ^{1}H NMR spectrum of versiconal hemiacetal acetate in DMSO solution.

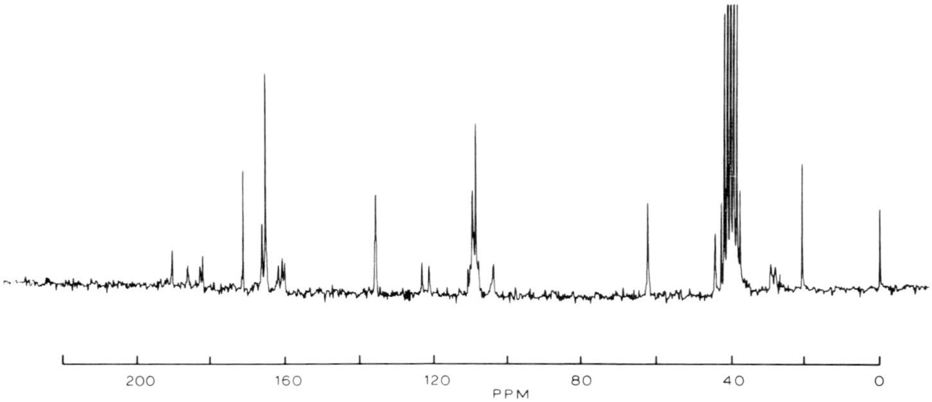

The 25 MHz ^{13}C NMR spectrum of versiconal hemiacetal acetate in DMSO solution.

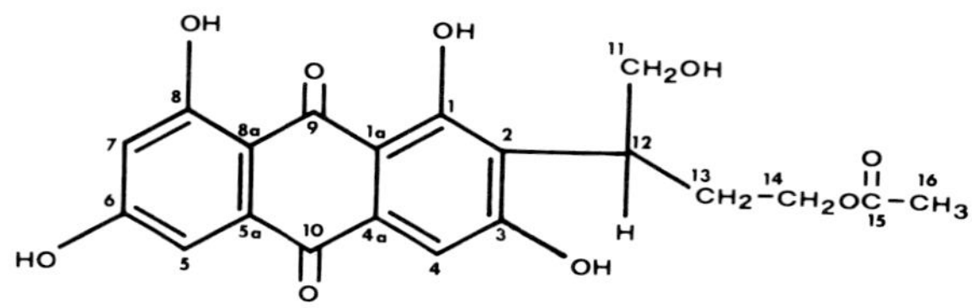

Common name	Versiconol acetate
Molecular weight	402.0945
Molecular formula	$C_{20}H_{18}O_9$
General characteristics[28]	Isolated as a glass
UV data[28]	λ_{max}^{MeOH} nm (ϵ): 225(30,900), 265(sh)(15,800), 294(25,700), 315(12,800), and 453(8,500)
IR data[28]	ν_{max}^{KBr} cm^{-1}: 1700(sh), 1620(broad), and 1600
Fungal source[28]	A blocked mutant of *Aspergillus parasiticus*
Toxicity data	Toxicity unknown
^{13}C NMR data[28]	(DMSO, $CDCl_3$, 1:1) C1, 163.0; C1a, 108.1; C2, 121.4; C3, 163.3; C4, 108.6; C4a, 132.2; C5, 108.6; C5a, 134.6; C6, 164.8; C7, 107.9; C8, 164.2; C8a, 108.6; C9, 188.8; C10, 181.0; C11, 62.9; C12, 35.0; C13, 27.8; C14, 62.9; C15, 169.8; C16, 20.8
^1H NMR data[28]	H4, 7.37; H5, 7.23 (J = 2.3); H7, 6.61 (J = 2.3); H11, 3.83; H12, NR; H13, 2.17; H14, 4.03; H16, 2.00; 1-OH, 13.02; 8-OH, 12.31

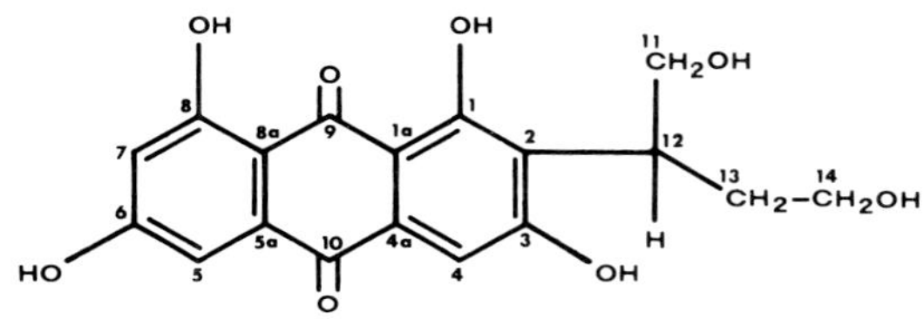

Common name	Versiconol
Molecular weight	360.0844 Electron-impact mass spectral analysis shows 342.0739 due to $M^+ - H_2O$
Molecular formula	$C_{18}H_{16}O_8$
General characteristics[14,28]	Crystals from acetone, mp 257°–259°C Orange-red needles, mp 265°C (dec.) $[\alpha]_D^{25}$ − 35.8° (C = 0.35 in dioxane)
UV data[14,28]	λ_{max}^{MeOH} nm (ϵ): 224(30,100), 266(14,100), 294(23,900), 315(12,300), and 455(7,400) λ_{max}^{EtOH} nm (ϵ): 224(44,600), 255(22,300), 265(22,300), 295(33,800), 322(13,800), and 460(10,400)
Fungal source[28]	A blocked mutant of *Aspergillus parasiticus*
^{13}C NMR data[28]	(DMSO, $CDCL_3$, 1:1) C1, 162.9; C1a; 108.1; C2, 122.9; C3, 163.1; C4, 108.9; C4a, 132.1; C5, 108.6; C5a, 134.7; C6, 164.7; C7, 107.9; C8, 164.2; C8a, 108.6; C9, 188.8; C10, 181.0; C11, 63.2; C12, 34.9; C13, 32.5; C14, 60.1
^1H NMR data[28]	(DMSO) H4, 7.18; H5, 7.23 (J = 2.3); H7, 6.54 (J = 2.3); H11, 3.0–4.0; H12, 3.0–4.0; H13, 1.95; H14, 3.0–4.0; 1-OH, 12.79; 8-OH, 12.15

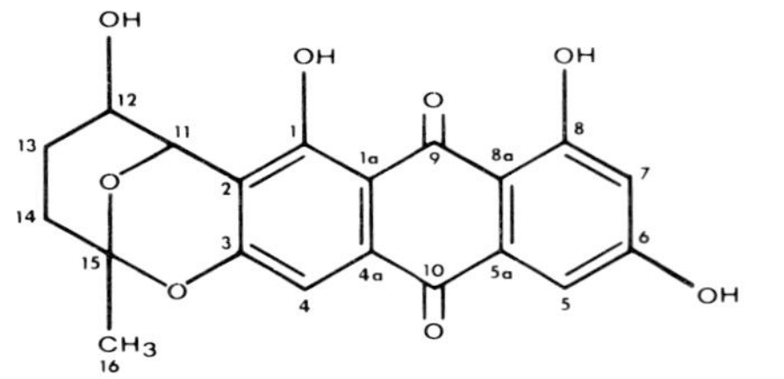

Common name	Nidurufin
Molecular weight	384.0844
Molecular formula	$C_{20}H_{16}O_8$
General characteristics[2]	Crystals from chloroform–methanol, mp 188°C
UV data[2]	λ_{max}^{EtOH} nm (ϵ): 223(33,500), 253(15,300), 264(19,000), 291(30,400), 319(10,700), and 450(10,000)
CD data[2]	$\Delta\epsilon$ nm: 230(−10.4), 240(0), 256(sh)(1.9), 260(2.84), 270(0), 280(0.95), 288(0), 296(−7.6), 308(0), 325(2.8), 348(0), 365(−0.76), 380(−0.57), 400(0), 444(4.17), and 480(0.76)
IR data[2]	ν_{max} cm^{-1}: 1673 and 1725
Fungal source[2]	*Aspergillus nidulans*
Toxicity data	Toxicity unknown

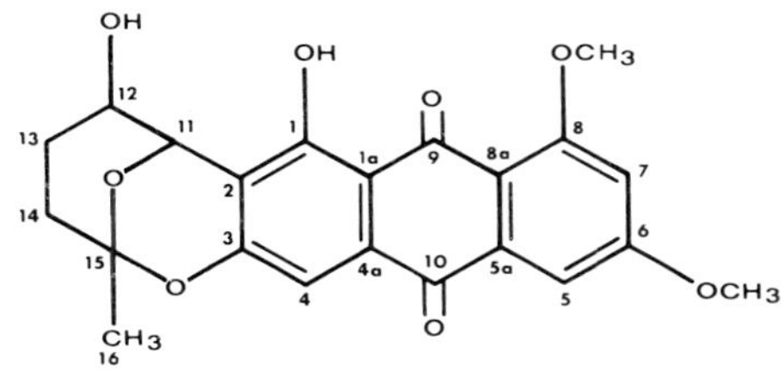

Common name	Dimethylnidurufin
Molecular weight	412.1157
Molecular formula	$C_{22}H_{20}O_8$
General characteristics[20]	Crystals from acetone–hexane, mp 211°–213°C $[\alpha]_D^{25}$ −77° (C = 0.15 in chloroform)
UV data[20]	λ_{max}^{EtOH} nm (ϵ): 224(48,200), 251(19,000), 288(30,900), 314(8,540), and 444(8,790)
IR data[20]	ν_{max}^{KBr} cm^{-1}: 3500, 3420, 2940, 1680, 1625, 1600, 1560, 1490, 1460, 1400, 1330, 1300, 1250, 1220, 1170, 1068, 1050, 1000, 970, 890, and 850
Fungal source[20]	*Aspergillus versicolor*
Toxicity data	Toxicity unknown
^{1}H NMR data[20]	H4, 7.28; H5, 7.46 (J = 2.5); H7, 6.86 (J = 2.5); H11, 5.30 (J = 2.0); H12, 4.16 m; H13,14, 1.6–2.5 m; H16, 1.64; 12-OH, 1.60; OCH_3, 4.00, 4.16

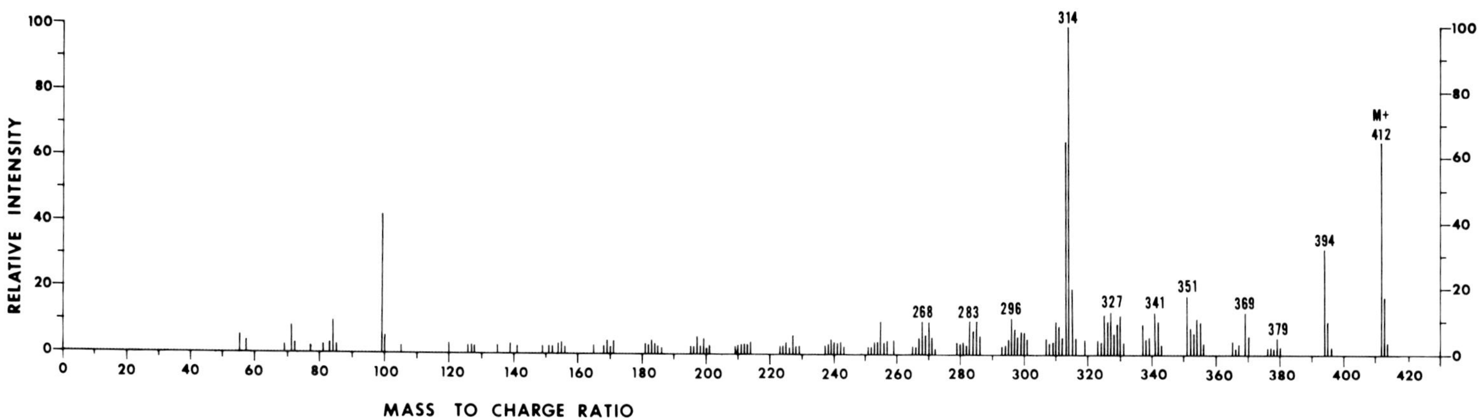

RELATIVE INTENSITY
MASS TO CHARGE RATIO
314
M+
412
394
268
283
296
327
341
351
369
379

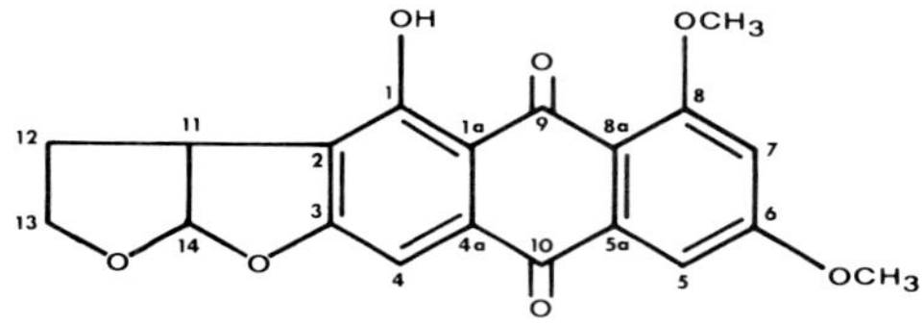

Common name	Aversin
Molecular weight	368.0896
Molecular formula	$C_{20}H_{16}O_7$
General characteristics[4]	Slender golden needles from acetone solution, mp 217°C $[\alpha]_D^{20}$ −222° (C = 0.248 in chloroform)
UV data[4]	λ_{max} nm (ϵ): 224(36,700), 251(13,400), 285(33,600), 313(8,900), 363(4,960), and 440(7,830) λ_{max}^{NaOH} nm (ϵ): 240(21,300), 263(34,800), 307(13,800), 355(5,040), 498(6,890), and 501(6,890)
IR data[4]	ν_{max}^{KBr} cm^{-1}: 3470, 3090, 2970, 2935, 2880, 2840, 1725, 1665, 1628, 1595, 1558, 1479, 1436, 1423, 1375, 1354, 1329, 1298, 1243, 1210, 1174, 1160, 1118, 1104, 1084, 1076, 1044, 990, 953, 911, 864, 821, 798, 771, 758, and 747
Fungal source[4]	*Aspergillus versicolor* (Vuill.) Tiraboschi
Toxicity data	Toxicity unknown

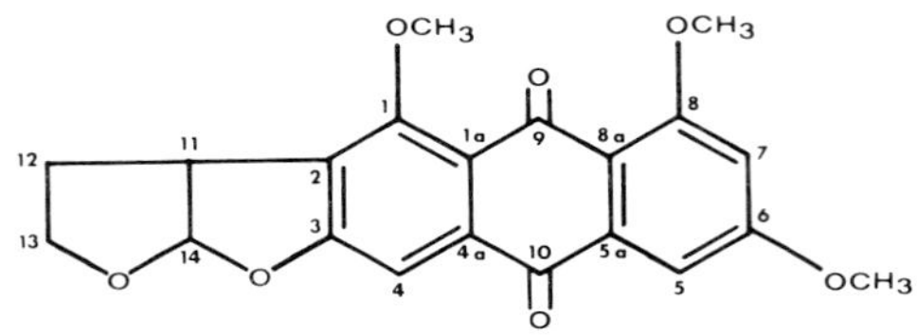

Common name	*O*-Methylaversin (Tri-*O*-methylversicolorin B)
Molecular weight	382.1052
Molecular formula	$C_{21}H_{18}O_7$
General characteristics[4,17]	Fine golden-yellow needles from chloroform–methanol, mp 216°–217°C $[\alpha]_D^{25}$ −127° (C = 0.1804 in chloroform) Golden-yellow crystals from methanol, mp 212°–213°C Sublimed sample, mp 229°–230°C
UV data[17]	λ_{max}^{EtOH} nm (ϵ): 222(33,800), 285(39,200), 348(4,700), and 407(4,200)
IR data[4]	ν_{max}^{KBr} cm^{-1}: 3460, 3100, 3089, 3053, 3019, 3004, 2986, 2947, 2937, 2880, 2845, 2664, 1673, 1667, 1602, 1572, 1510, 1463, 1440, 1431, 1419, 1357, 1349, 1332, 1323, 1307, 1281, 1255, 1238, 1209, 1201, 1188, 1178, 1163, 1114, 1103, 1085, 1060, 1035, 990, 958, 942, 925, 914, 897, 871, 850, 841, 836, 787, 778, and 753
Fungal source	*Aspergillus versicolor* (Vuill.) Tiraboschi
Toxicity data	Toxicity unknown
Other significant references	Synthesis[17]

REFERENCES

1. Anderson, H. A., Thomson, R. H., and Wells, J. W., *J. Chem. Soc. C* p. 1927 (1966).
2. Aucamp, P. J., and Holzapfel, C. W., *J. S. Afr. Chem. Inst.* **23,** 40 (1970).
3. Berger, Y., and Jadot, J., *Bull. Soc. Chim. Belg.* **85,** 271 (1976).
4. Bullock, E., Kirkaldy, D., Roberts, J. C., and Underwood, J. G., *J. Chem. Soc.* p. 829 (1963).
5. Cox, R. H., Churchill, F., Cole, R. J., and Dorner, J. W., *J. Am. Chem. Soc.* **99,** 3159 (1977).
6. Donkersloot, J. A., and Mateles, R. I., *Biochim. Biophys. Acta* **86,** 418 (1964).
7. Donkersloot, J. A., Mateles, R. I., and Yang, S. S., *Biochem. Biophys. Res. Commun.,* **47,** 1051 (1972).
8. Fitzell, D. L., Hsieh, D. P. H., Yao, R. C., and Lamar, G. N., *J. Agric. Food Chem.* **23,** 442 (1975).
9. Fitzell, D. L., Singh, R., and Hsieh, D. P. H., *J. Agric. Food Chem.* **25,** 1193 (1977).
10. Fukuyama, K., Tsukihara, T., Katsube, Y., Hamasaki, T., and Hatsuda, Y., *Bull. Chem. Soc. Jpn.* **48,** 2648 (1975).
11. Gorst-Allman, G. P., Steyn, P. S., Wessels, P. L., and Scott, D. B., *J. Chem. Soc., Perkin Trans. 1* p. 961 (1978).
12. Hamasaki, T., Hatsuda, Y., Terashima, N., and Renbutsu, M., *Agric. Biol. Chem.* **31,** 11 (1967).
13. Hamasaki, T., Renbutsu, M., and Hatsuda, Y., *Agric. Biol. Chem.* **31,** 1513 (1967).
14. Hatsuda, Y., Hamasaki, T., Ishida, M., and Yoshikawa, S., *Agric. Biol. Chem.* **33,** 131 (1969).
15. Hatsuda, Y., Juyama, S., and Terashima, N., *J. Agric. Chem. Soc. Jpn.* **29,** 11 (1955).
16. Holker, J. S. E., Kagal, S. A., Mulheirn, L. J., and White, P. M., *Chem. Commun.* p. 911 (1966).
17. Holmwood, G. M., and Roberts, J. C., *J. Chem. Soc. C* p. 3899 (1971).
18. Hsieh, D. P. H., Lin, M. T., Yao, R. C., and Singh, R., *J. Agric. Food Chem.* **24,** 1170 (1976).
19. Katsube, Y., Tsukihara, T., Tanaka, N., Ando, K., Hamasaki, T., and Hatsuda, Y., *Bull. Chem. Soc. Jpn.* **45,** 2091 (1972).
20. Kingston, D. G. I., Chen, P. N., and Vercellotti, J. R., *Phytochemistry* **15,** 1037 (1976).
21. Lee, L. S., Bennett, J. W., Goldblatt, L. A., and Lundin, R. E., *J. Am. Oil Chem. Soc.* **48,** 93 (1971).
22. Lee, L. S., Bennett, J. W., Cucullu, A. F., and Stanley, J. B., *J. Agric. Food Chem.* **23,** 1132 (1975).
23. Lin, M. T., and Hsieh, D. P. H., *J. Am. Chem. Soc.* **95,** 1668 (1973).
24. Pusey, D. F. G., and Roberts, J. C., *J. Chem. Soc.* p. 3542 (1963).
25. Roffey, P., and Sargent, M. V., *Chem. Commun.* p. 913 (1966).
26. Roffey, P., and Sargent, M. V., *J. Chem. Soc.* p. 2328 (1967).
27. Schroeder, H. W., Cole, R. J., Grigsby, R. D., and Hein, H., Jr., *Appl. Microbiol.* **27,** 394 (1974).
28. Steyn, P. S., Vleggaar, R., Wessels, L., Cole, R. J., and Scott, D., *J. Chem. Soc., Perkin Trans. 1* p. 451 (1979).
29. Yao, R. C., and Hsieh, P. H., *Appl. Microbiol.* **28,** 52 (1974).

4

Ochratoxins

In contrast to the discovery of the aflatoxins, which was prompted by a naturally occurring toxigenic syndrome of some magnitude, the ochratoxins were discovered by South African scientists during laboratory screening for toxigenic fungi.[18] These scientists frequently isolated strains of *Aspergillus ochraceus* that were toxic to laboratory animals.[18] They subsequently isolated a chlorine-containing metabolite designated ochratoxin A as the major toxic principle.[26] The ochratoxin group consists of ochratoxin A, its methyl and ethyl esters, ochratoxin B (dechloro-ochratoxin A), its methyl and ethyl esters, and 4-hydroxyochratoxin A. Ochratoxin A and its methyl and ethyl esters are the toxic members of the group while the other members show little or no acute toxicity. The primary target organs of ochratoxin A are the liver and kidney.

Chemically, the ochratoxins are composed of a 3,4-dihydro-3-methyliso-coumarin moiety linked via the 7-carboxy group to L-β-phenylalanine by an amide bond.

In addition to being produced by strains of *A. ochraceus*, the ochratoxins are also produced by certain strains of *Penicillium viridicatum* Westling.

Also included in this group, because of the obvious chemical relationship, are mellein and 4-hydroxymellein.

Several reviews covering various aspects of the ochratoxins have been published.[8,13,16,23]

Ochratoxin group	Molecular weight	Molecular formula
Mellein	178.0629	$C_{10}H_{10}O_3$
4-Hydroxymellein	194.0579	$C_{10}H_{10}O_4$
Ochratoxin A	403.0822	$C_{20}H_{18}O_6NCl$
Ochratoxin B	369.1212	$C_{20}H_{19}O_6N$
4-Hydroxyochratoxin A	419.0771	$C_{20}H_{18}O_7NCl$
Ochratoxin C	431.1135	$C_{22}H_{22}O_6NCl$

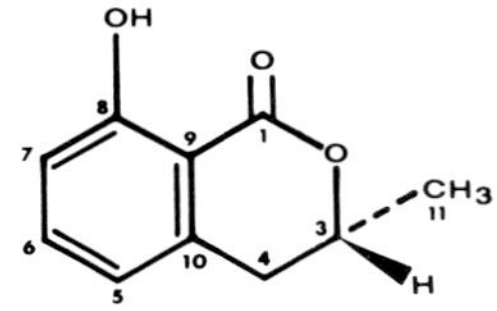

Common name	Mellein (Ochracin) [(−)3,4-Dihydro-8-hydroxy-3-methylisocoumarin]
Molecular weight	178.0629
Molecular formula	$C_{10}H_{10}O_3$
General characteristics[3]	Crystals (subl.), mp 54°–55°C
UV data[2,3]	λ_{max}^{EtOH} nm (ϵ): 212(20,000), 246(6,500), and 314(4,100)
Fungal source[14,28]	*Aspergillus melleus* Yukawa *A. ochraceus* Wilh.
Toxicity data	Toxicity unknown
TLC data	*Adsorbent:* silica gel G-HR *Solvent:* toluene–ethyl acetate–formic acid, 5:4:1 v/v/v/ R_f: 0.82 *Detection:* blue fluorescent spot under UV light
Other significant references	Synthesis[2]
^{13}C NMR data	C1, 169.8 s; C3, 76.1 d; C4, 34.7 t; C5, 117.8 d; C6, 136.0 d; C7, 116.2 d; C8, 162.1 s; C9, 108.2 s; C10, 139.3 s; C11, 20.8 q
^{1}H NMR data	H3, 4.76 (J = 7.0, 7.0); H4, 2.95 (J = 7.0); H5, 6.72 (J = 7.5); H6, 7.44 (J = 7.5, 8.0); H7, 6.91 (J = 8.0); H11, 1.55 (J = 7.0); 8-OH, 11.07

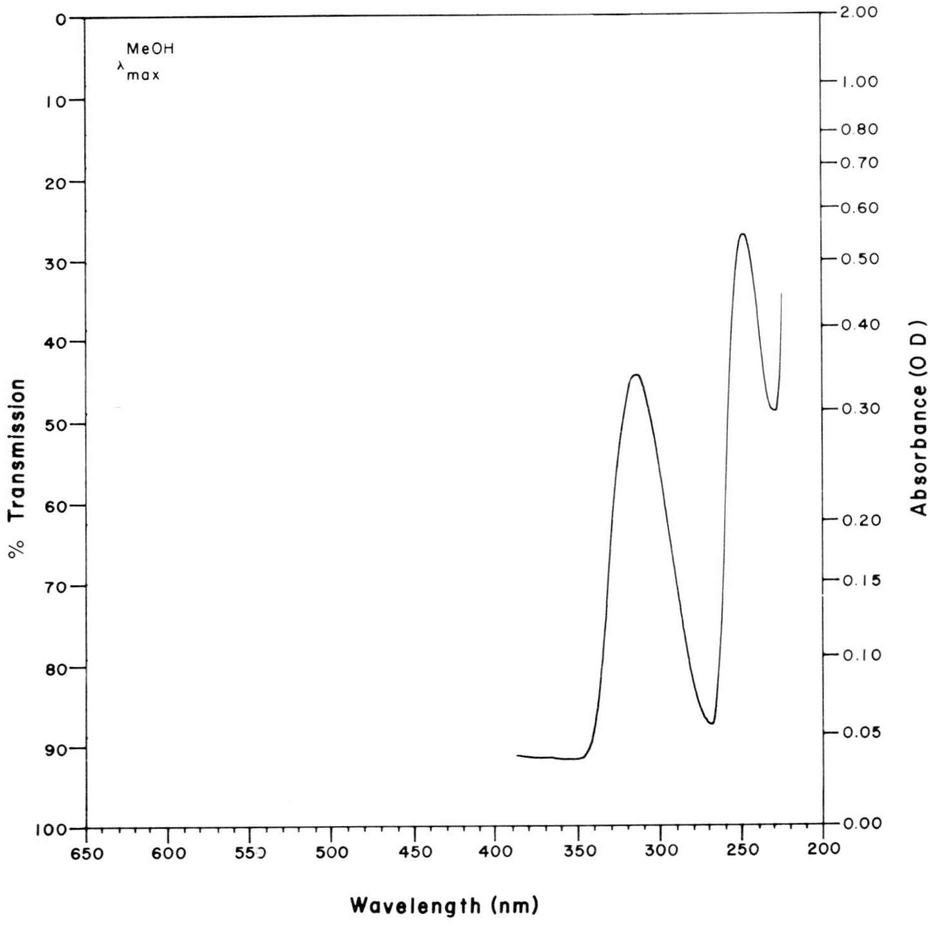

MeOH
λmax
% Transmission
Absorbance (O D)
Wavelength (nm)
0
10
20
30
40
50
60
70
80
90
100
2.00
1.00
0.80
0.70
0.60
0.50
0.40
0.30
0.20
0.15
0.10
0.05
0.00
650
600
550
500
450
400
350
300
250
200

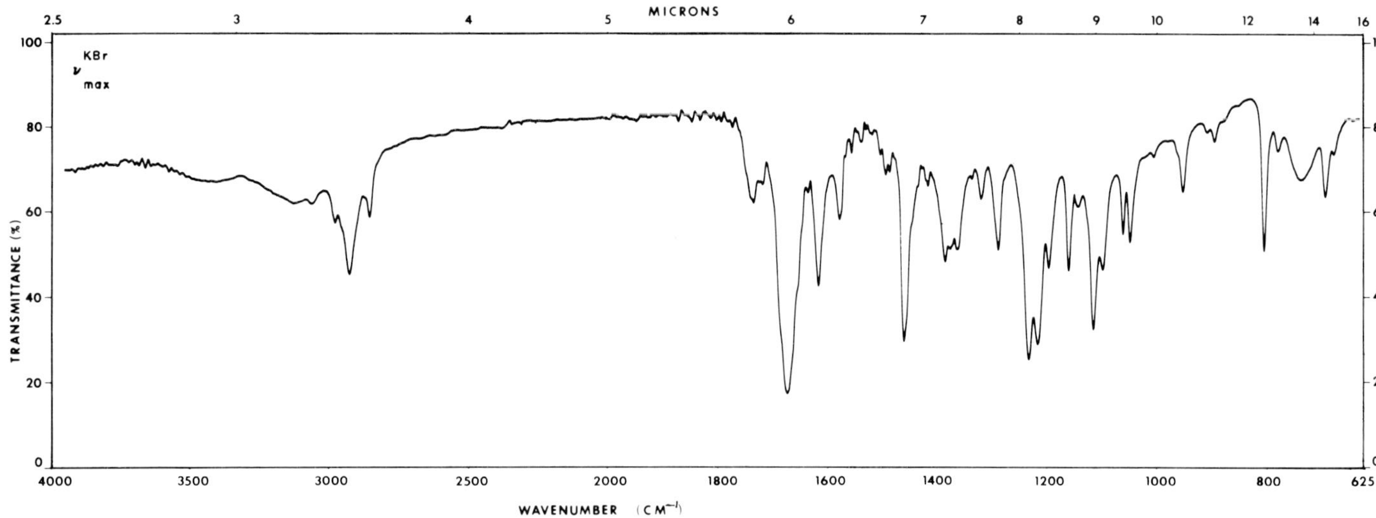
MICRONS
2.5
3
4
5
6
7
8
9
10
12
14
16
KBr
ν_{max}
TRANSMITTANCE (%)
100
80
60
40
20
0
4000
3500
3000
2500
2000
1800
1600
1400
1200
1000
800
625
WAVENUMBER (CM^{-1})

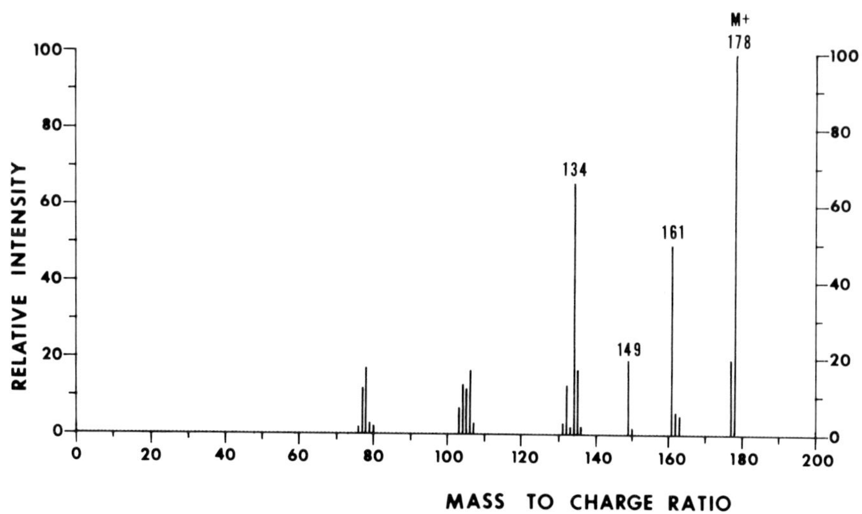
M+
178
134
161
149
RELATIVE INTENSITY
100
80
60
40
20
0
20
40
60
80
100
120
140
160
180
200
MASS TO CHARGE RATIO

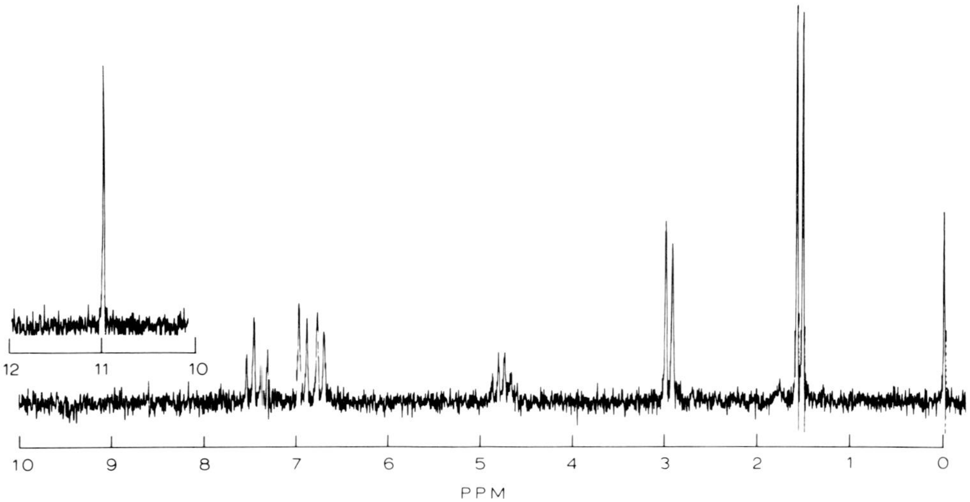

The 100 MHz ^{1}H NMR spectrum of mellein in $CDCl_3$ solution.

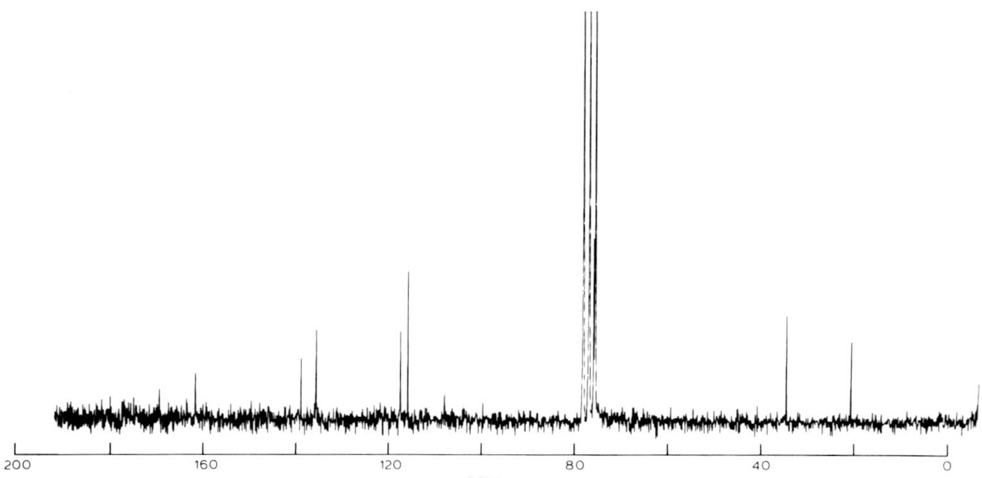

The 25 MHz ^{13}C NMR spectrum of mellein in $CDCl_3$ solution.

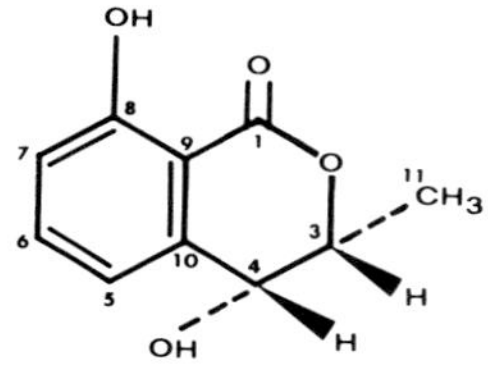

Common name	4-Hydroxymellein (3-Methyl-4,8-dihydroxy-3,4-dihydroisocoumarin)
Molecular weight	194.0579
Molecular formula	$C_{10}H_{10}O_4$
General characteristics[6,17]	Crystals from chloroform–hexane, mp 131°–132°C $[\alpha]_D^{25}$ −40° (*C* = 1.0 in chloroform)
UV data[6,17]	λ_{max}^{MeOH} nm (ϵ): 247(5,300) and 315(4,200)
Fungal source[1,4,6,17]	*Aspergillus ochraceus* Wilh. (NRRL 3174) *A. oniki* *Lasiodiplodia theobromae* (*cis*-4-hydroxymellein) *Apiospora camptospora*
Toxicity data	Toxicity unknown
TLC data[6]	*Adsorbent:* silica gel G-HR *Solvent:* toluene–ethyl acetate–formic acid, 5:4:1 v/v/v R_f: 0.74 *Detection:* blue fluorescent spot under UV light
^{13}*C NMR data*	C1, 168.1 s; C3, 79.7 d; C4, 69.1 d; C5, 117.7 d; C6, 136.4 d; C7, 115.9 d; C8, 161.7 s; C9, 106.5 s; C10, 140.9 s; C11, 29.6 q
^{1}*H NMR data*	H3, 4.63; H4, 4.63; H5, 7.04 (*J* = 7.5); H6, 7.57; H7, 7.02 (*J* = 8.2); H11, 1.52 (*J* = 6.5); 4-OH, 2.16; 8-OH, 11.01

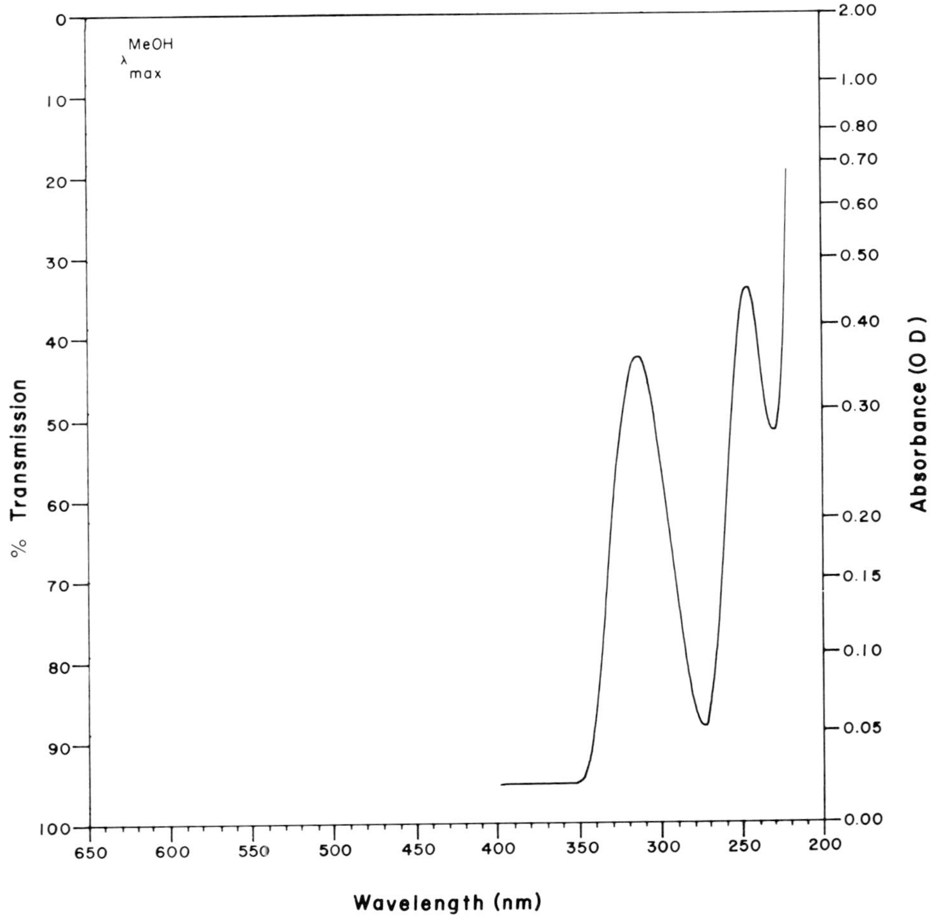

MeOH
λ max
0
10
20
30
40
50
60
70
80
90
100
% Transmission
2.00
1.00
0.80
0.70
0.60
0.50
0.40
0.30
0.20
0.15
0.10
0.05
0.00
Absorbance (O D)
650
600
550
500
450
400
350
300
250
200
Wavelength (nm)

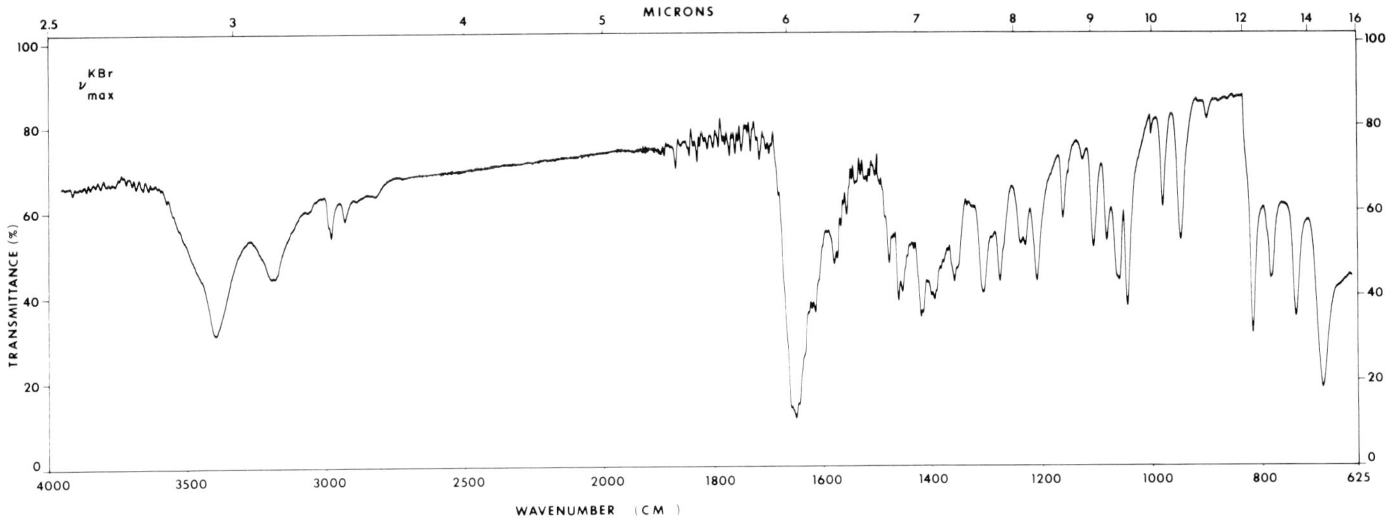

MICRONS
ν_{max}^{KBr}
TRANSMITTANCE (%)
WAVENUMBER (CM)

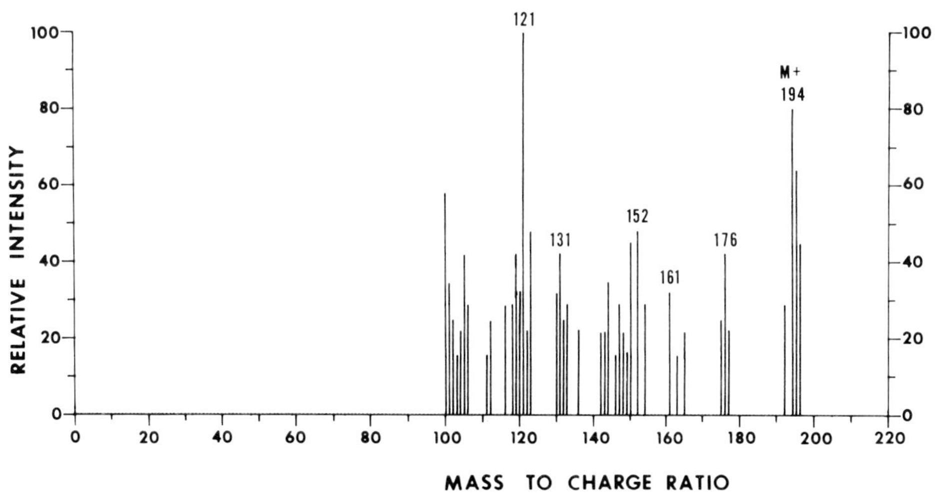

121
131
152
161
176
M+
194
RELATIVE INTENSITY
MASS TO CHARGE RATIO

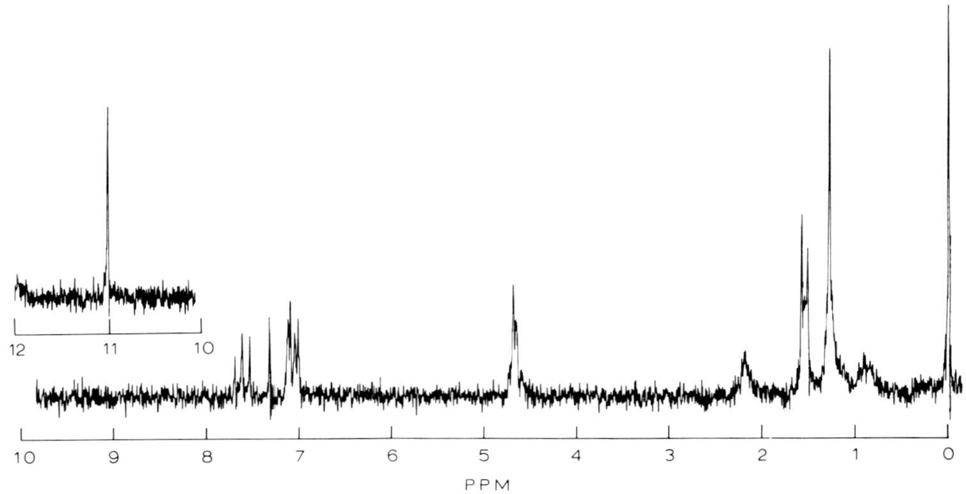

The 100 MHz ^1H NMR spectrum of 4-hydroxymellein in $CDCl_3$ solution.

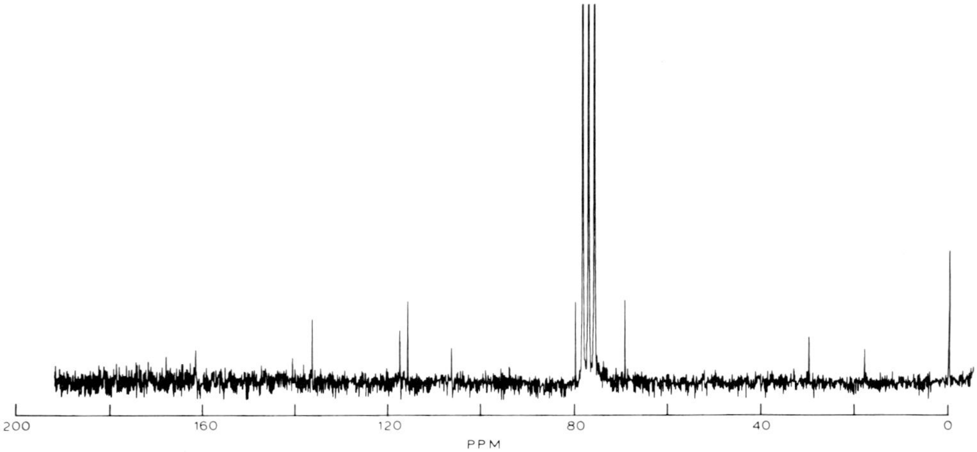

The 25 MHz ^{13}C NMR spectrum of 4-hydroxymellein in $CDCl_3$ solution.

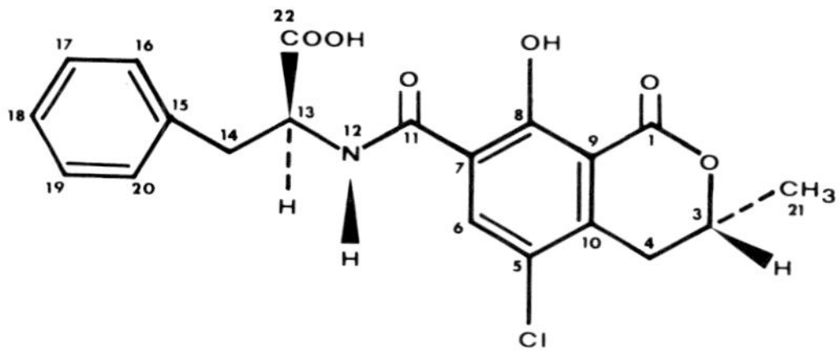

Common name	Ochratoxin A (7-Carboxy-5-chloro-8-hydroxy-3,4-dihydro-3*R*-methylisocoumarin-7-L-β-phenylalanine)
Molecular weight	403.0822
Molecular formula	$C_{20}H_{18}O_6NCl$
General characteristics[21,25]	Crystals from benzene, mp 94°–96°C Crystals from xylene, mp 169°C $[\alpha]_D$ −118° (C = 1.1 in chloroform)
UV data	λ_{max}^{MeOH} nm (ε): 215(36,800) and 333(6,400)
Fungal source[26,27]	*Aspergillus ochraceus* Wilh. (NRRL 3174) *A. sulphureus* (NRRL 4077) *A. melleus* (NRRL 3519; 3520) *Penicillium viridicatum* (ATCC 18411)
Toxicity data[7,9,11,12,20,24]	LD_{50} in weanling rats dosed orally was 22 mg/kg, LD_{50} in trout IP, 3.0 mg/kg. Ochratoxin A caused acute hepatic injury, fatty infiltration, hyaline degeneration and focal necrosis in livers of weanling rats and ducklings given a single oral dose of ochratoxin A. The effects were more marked in rats. Beagle dogs dosed with ochratoxin A *per os* developed anorexia, weight loss, emesis, tenesmus, passage of blood-stained mucus from rectum, pyrexia, dehydration, prostration and renal damage. Ochratoxin A is teratogenic in mice and acts as an inhibitor of mitochondrial transport systems.
TLC data[19]	*Adsorbent:* silica gel *Solvent:* benzene–acetic acid, 3:1 v/v R_f: 0.5 *Detection:* green fluorescent spot under UV light
Other significant references	Synthesis[20]; biosynthesis[22]

${}^{13}C$ *NMR data*	C1, 169.1 s; C3, 76.2 d; C4, 32.4 t; C5, 122.3 s; C6, 136.8 d; C7, 120.9 s; C8, 163.5 s; C9, 112.1 s; C10, 142.6 s; C11, 159.1 s; C13, 54.3 d; C14, 37.4 t; C15, 137.7 s; C16,20, 129.1 d; C17,19, 130.0 d; C18, 127.4 d; C21, 20.8 q; C22, 173.1 s
${}^{1}H$ *NMR data*	H3, 4.76; H4, 2.97; H6, 8.07; 12-NH, 8.58 (J = 7.3); H13, 4.76; H14, 3.18; H16–20, 7.25; H21, 1.46 (J = 6.1)

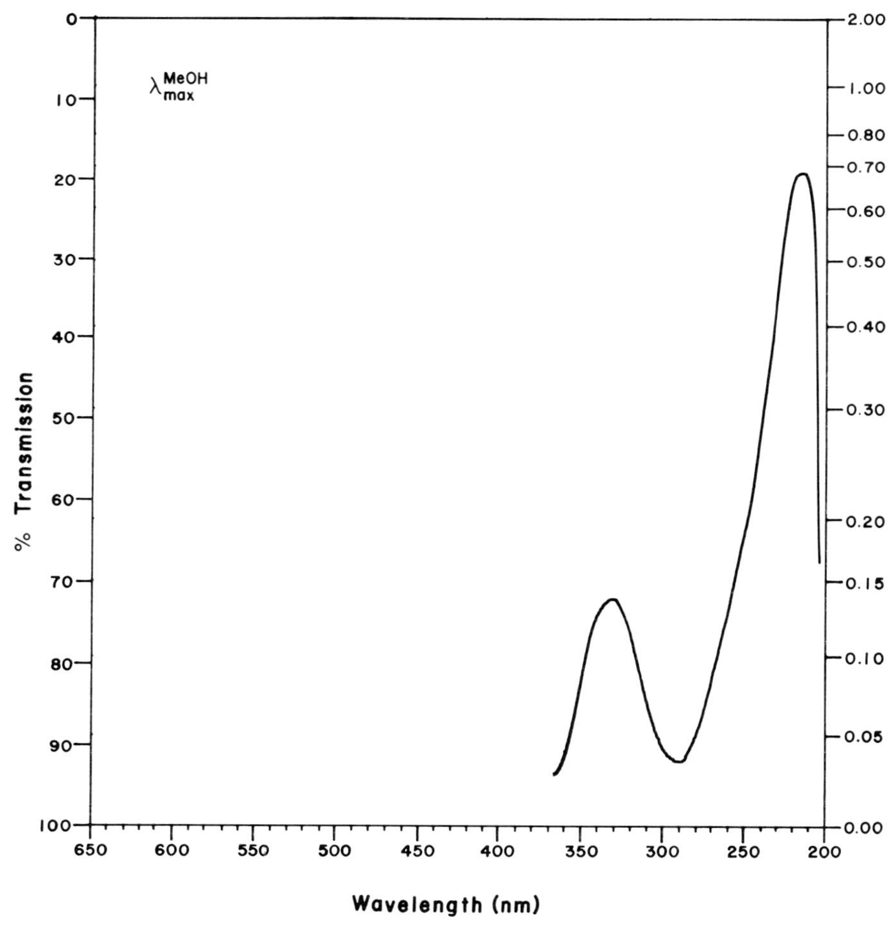

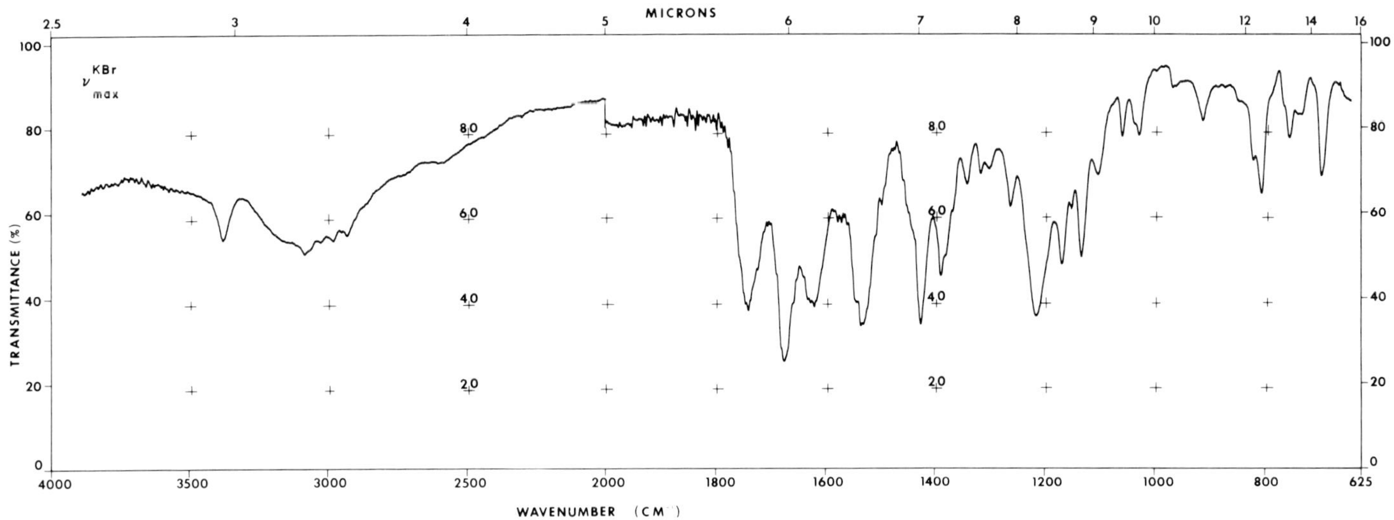
MICRONS
ν_{max}^{KBr}
TRANSMITTANCE (%)
WAVENUMBER (CM⁻¹)

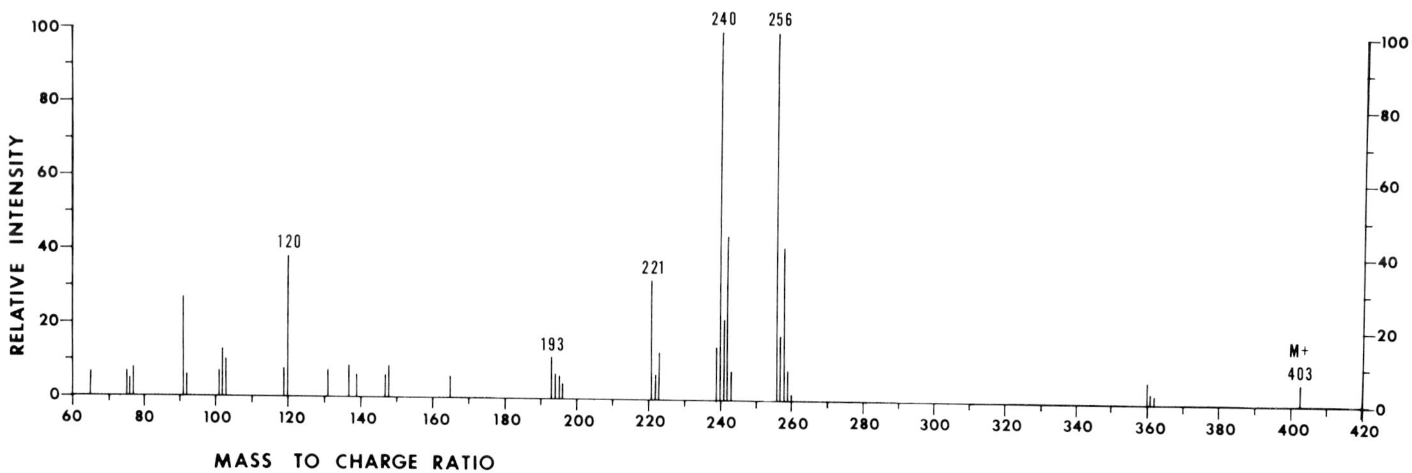
120
193
221
240
256
M+
403
RELATIVE INTENSITY
MASS TO CHARGE RATIO

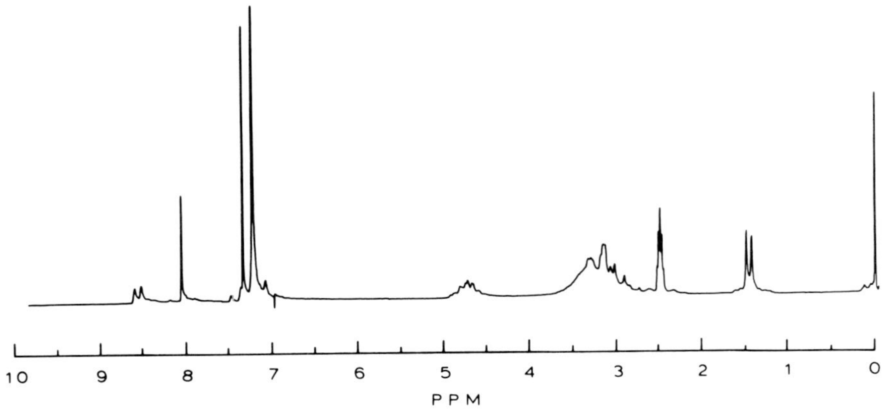

The 100 MHz ^{1}H NMR spectrum of ochratoxin A in DMSO solution.

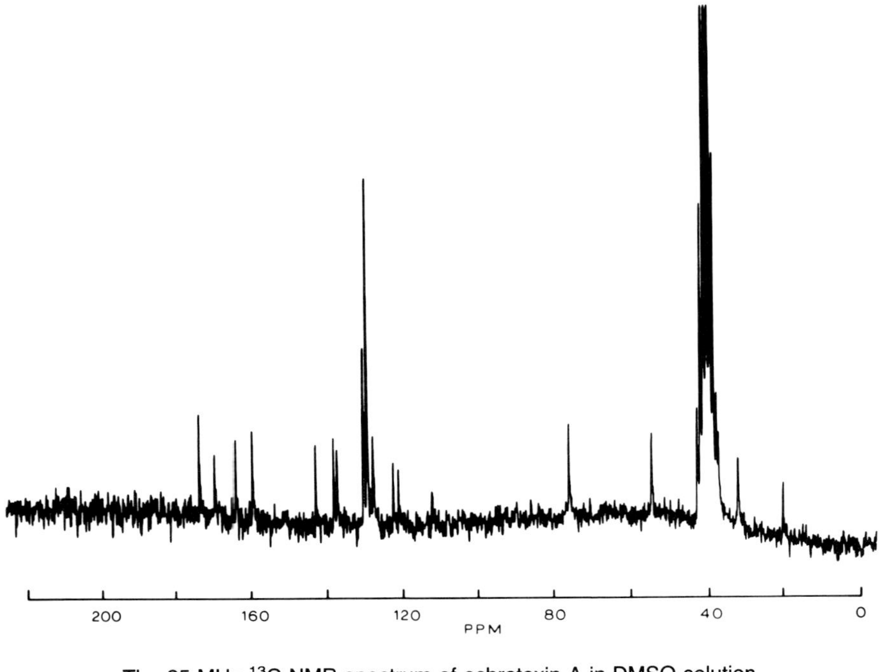

The 25 MHz ^{13}C NMR spectrum of ochratoxin A in DMSO solution.

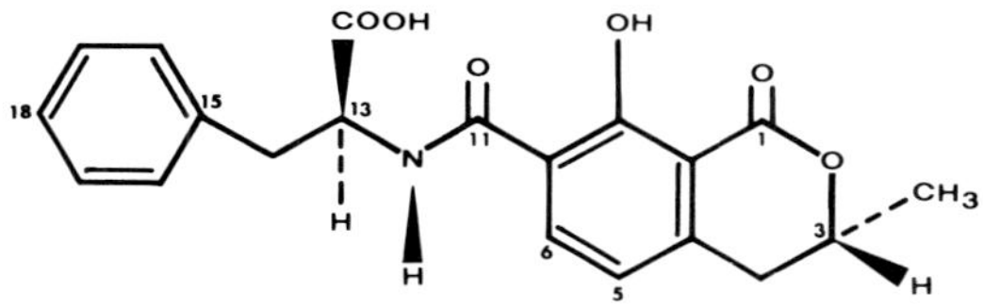

Common name	Ochratoxin B (7-Carboxy-8-hydroxy-3,4-dihydro-3*R*-methylisocoumarin-7-L-β-phenylalanine)
Molecular weight	369.1212
Molecular formula	$C_{20}H_{19}O_6N$
General characteristics[20,25]	Crystals from acidic methanol, mp 221°C $[\alpha]_D$ −35° (C = 0.15 in ethanol); −56° (C = 0.29 in methanol)
UV data[26]	λ_{max}^{EtOH} nm (ϵ): 218(37,200) an 318(6,900)
Fungal source	*Aspergillus ochraceus* Wilh. (NRRL 3174)
Toxicity data[5,15,21]	Ochratoxin B is considerably less toxic than ochratoxin A or C. The oral LD_{50} in day-old chicks was 54 mg/kg while ochratoxin A was 3.3–3.9 mg/kg in the same assay.
TLC data[19]	*Adsorbent:* silical gel *Solvent:* benzene–acetic acid, 4:1 v/v R_f: 0.35 *Detection:* blue fluorescent spot under UV light
^{13}C NMR data	C1, 168.7 s; C3, 75.1 d; C4, 33.1 t; C5, 117.3 d; C6, 135.5 d; C7, 117.9 s; C8, 162.0 s; C9, 107.7 s; C10, 142. 7 s; C11, 158.9 t; C13, 52.8 d; C14, 36.3 t; C15, 137.0 s; C16,20, 128.2 d; C17,19, 127.0 d; C18, 125.5 d; C21, 19.3 q; C22, 171.6 s
^1H NMR data	H3, 4.90; H4, 3.10; H5, 8.21 (J = 6.8); H6, 7.02 (J = 6.0); 12-NH, 8.64 (J = 6.1); H13, 4.90; H14, 3.10; H16–20, 7.28; H21, 1.45 (J = 5.4)

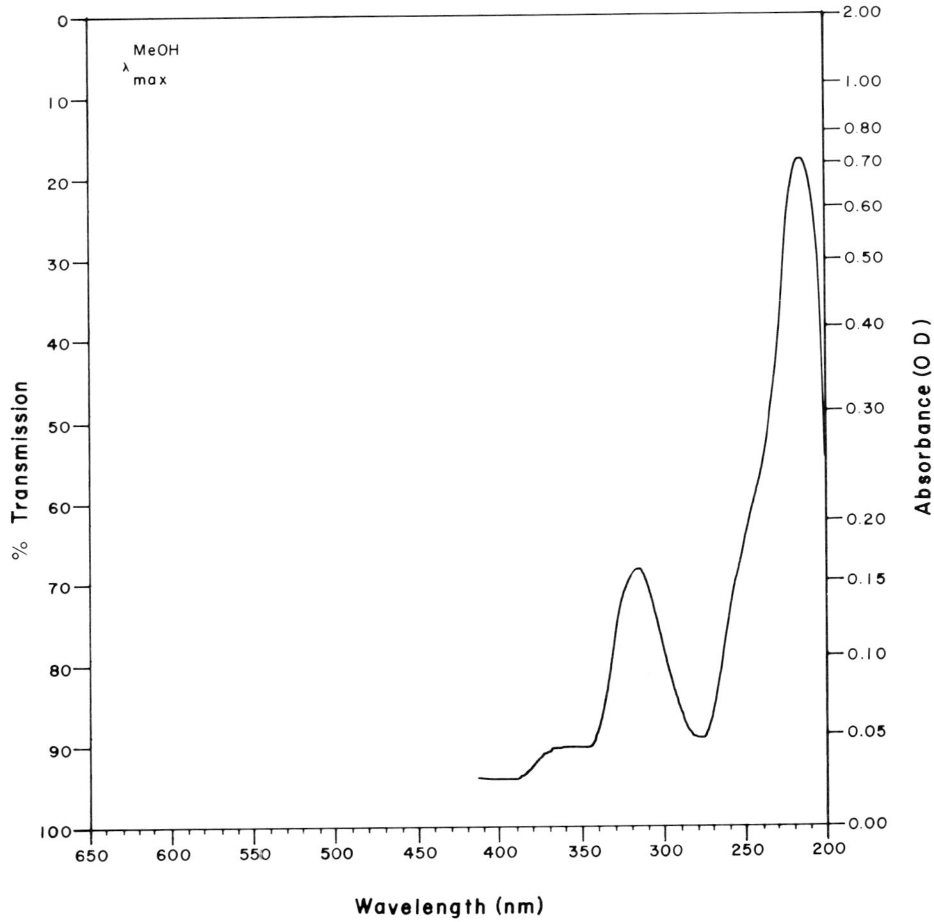

MeOH
λ max
0
10
20
30
40
50
60
70
80
90
100
% Transmission
2.00
1.00
0.80
0.70
0.60
0.50
0.40
0.30
0.20
0.15
0.10
0.05
0.00
Absorbance (O D)
650
600
550
500
450
400
350
300
250
200
Wavelength (nm)

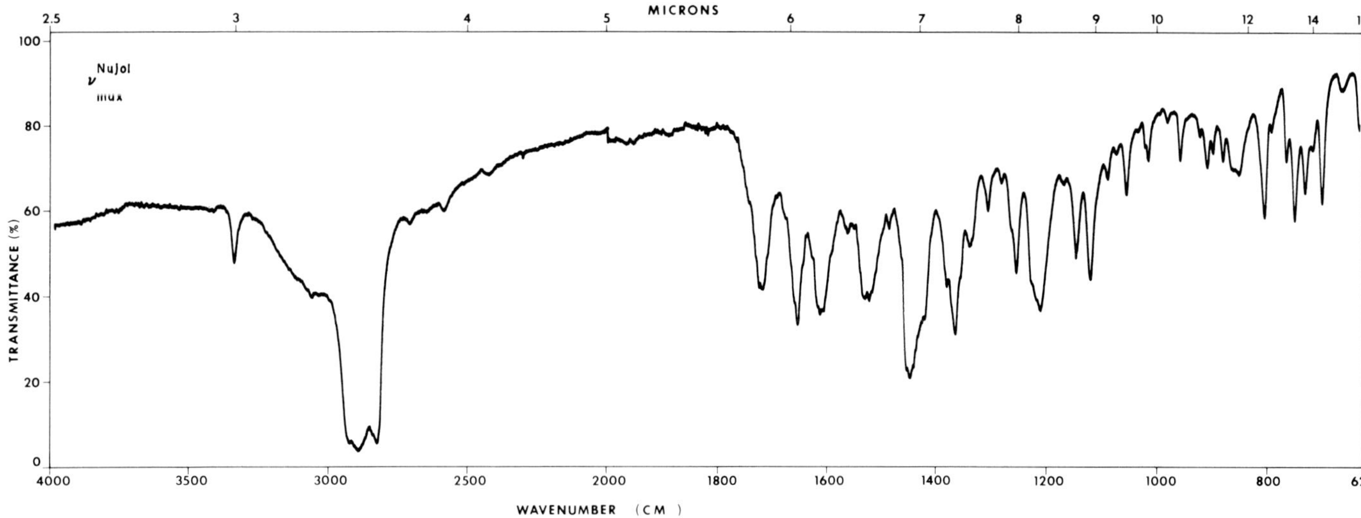
MICRONS
2.5
3
4
5
6
7
8
9
10
12
14
16
ν_{max}^{Nujol}
TRANSMITTANCE (%)
100
80
60
40
20
0
4000
3500
3000
2500
2000
1800
1600
1400
1200
1000
800
625
WAVENUMBER (CM)

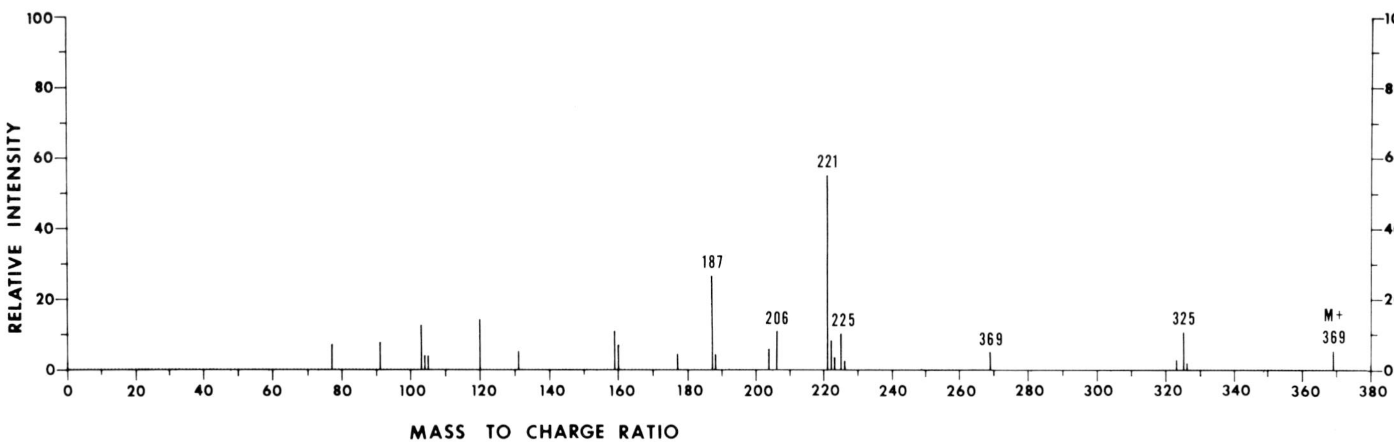
RELATIVE INTENSITY
100
80
60
40
20
0
187
206
221
225
369
325
M+
369
0
20
40
60
80
100
120
140
160
180
200
220
240
260
280
300
320
340
360
380
MASS TO CHARGE RATIO

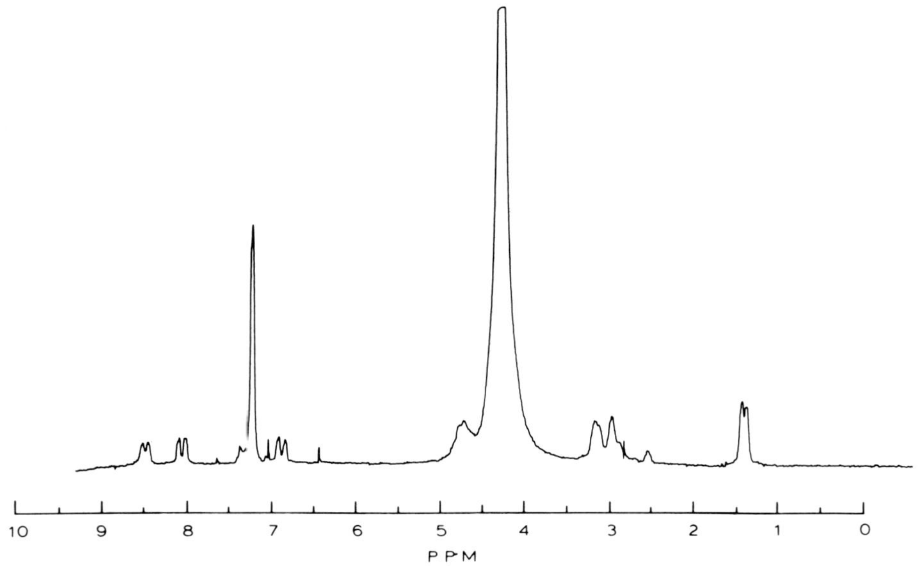

The 100 MHz ¹H NMR spectrum of ochratoxin B in $CDCl_3$/DMSO solution.

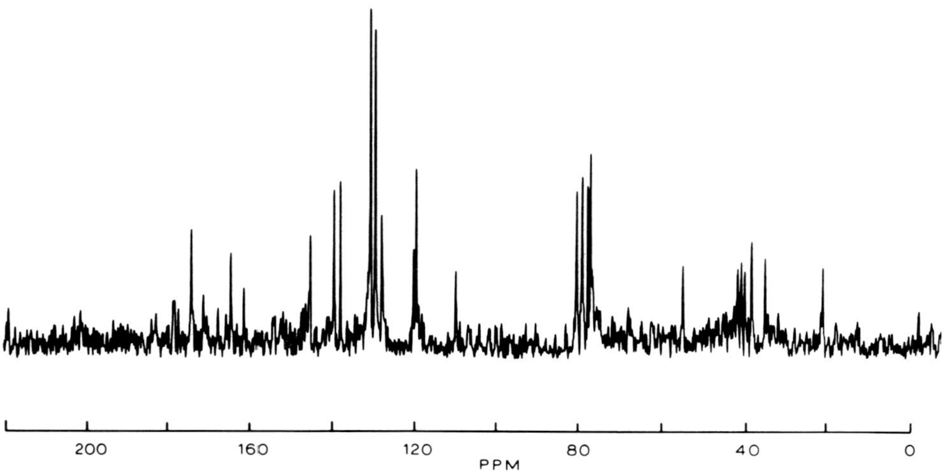

The 25 MHz ^{13}C NMR spectrum of ochratoxin B in $CDCl_3$/DMSO solution.

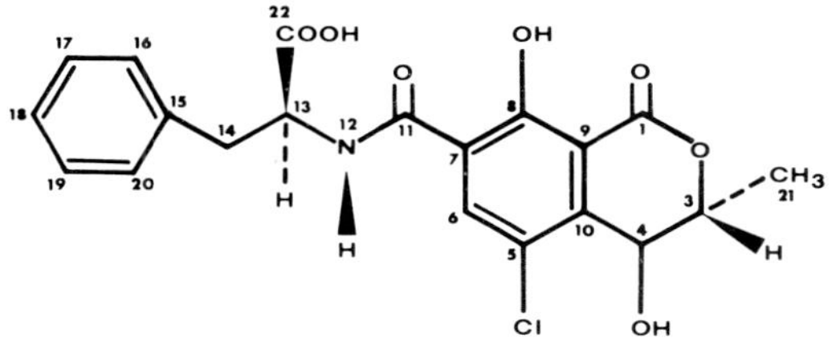

Common name	4-Hydroxyochratoxin A
Molecular weight	419.0771
Molecular formula	$C_{20}H_{18}O_7NCl$
General characteristics[10]	Colorless crystals from benzene, mp 216°–218°C
UV data[10]	λ_{max}^{EtOH} nm (ϵ): 213(32,500) and 334(6,400)
IR data[10]	$\nu_{max}^{CHCl_3}$ cm^{-1}: 1723, 2500, 3000, 1655, 1535, 1678, and 3380
Fungal source[10]	*Penicillium viridicatum* Westling (ATCC 18411)
Toxicity data[10]	4-hydroxyochratoxin A was excreted in the urine of male Wistar rats dosed with ochratoxin A (IP). 4-hydroxyochratoxin A had no effect when dosed to rats at a level of 40 mg/kg; ochratoxin A at this level caused 100% mortality. Therefore, 4-hydroxyochratoxin A may be a detoxification product in animals dosed with ochratoxin A.
TLC data[10]	*Adsorbent:* silica gel *Solvent:* benzene–acetic acid, 4:1 v/v R_f: 0.25 *Detection:* green fluorescent spot under UV light
^{1}*H NMR data*[10]	H6, 8.7; H2, 1.68(J = 7); H3, 4.8 (J = 7,2); and H4, 5.11 (J = 2)

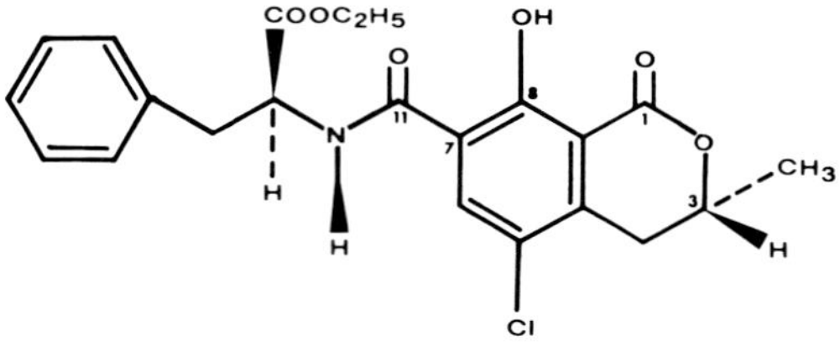

Common name	Ochratoxin C (ethyl ester of ochratoxin A)
Molecular weight	431.1135
Molecular formula	$C_{22}H_{22}O_6NCl$
General characteristics[21]	Amorphous compound $[\alpha]_D$ −100° (C = 1.2) Ochratoxin A methyl ester $[\alpha]_D$ −78° (C = 0.027)
UV data[21]	λ_{max} nm (ϵ): 214(30,000), 333(7,000), and 378(2,050)
IR data[21]	ν_{max} cm^{-1}: 1730 and 1680
Fungal source[25]	*Aspergillus ochraceus* Wilh.
Toxicity data[10,20,21,25]	Initially, ochratoxin C was considered to be relatively nontoxic; however, it was later reported the toxicity of *A. ochraceus* was due to the chlorine-containing metabolites only, including ochratoxin C and the methyl ester of ochratoxin A. Toxicity of ochratoxin A and methyl and ethyl esters to day-old Peking ducklings was 135–170 μg per animal.
TLC data[21]	*Adsorbent:* silica gel *Solvent:* benzene–acetic acid, 25:1 v/v R_f: 0.55 *Detection:* light green fluorescent spot under UV light

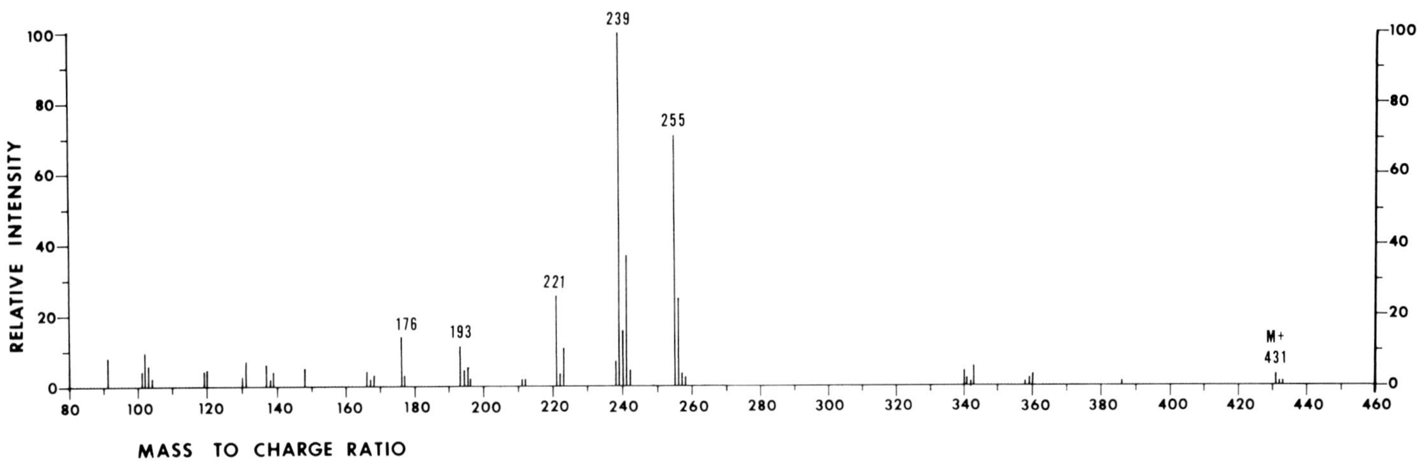

Ochratoxin C

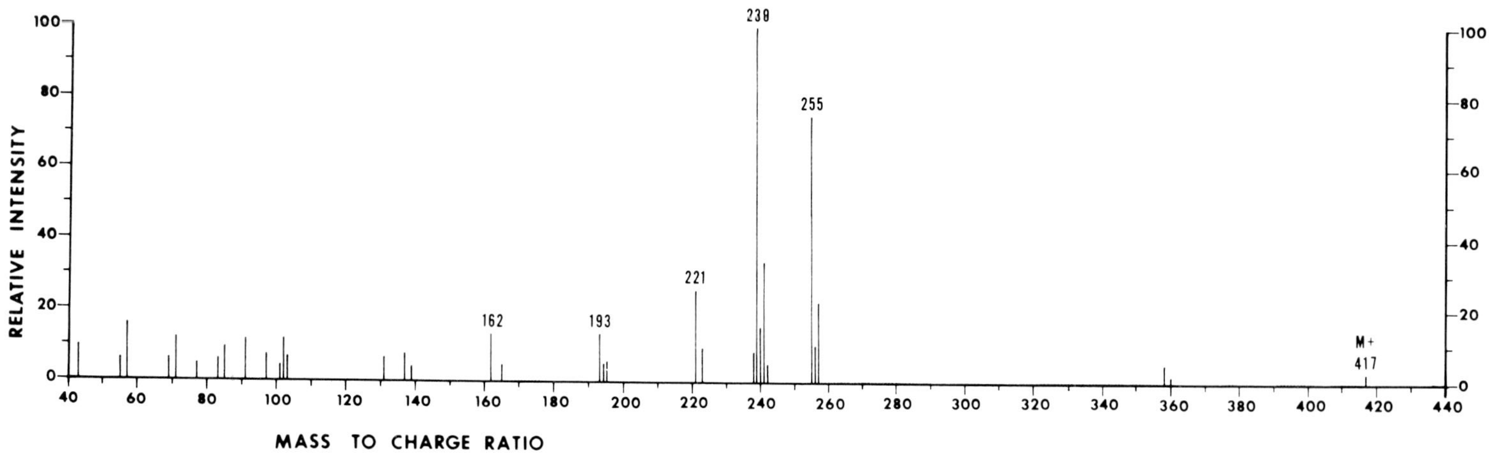

Ochratoxin A methyl ester

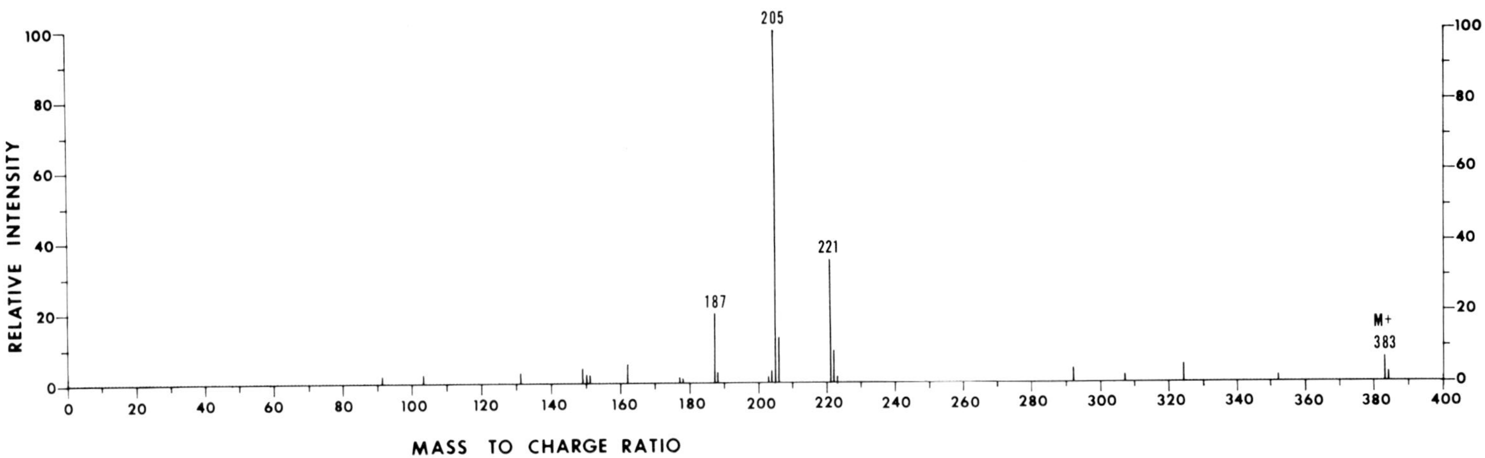

Ochratoxin B methyl ester

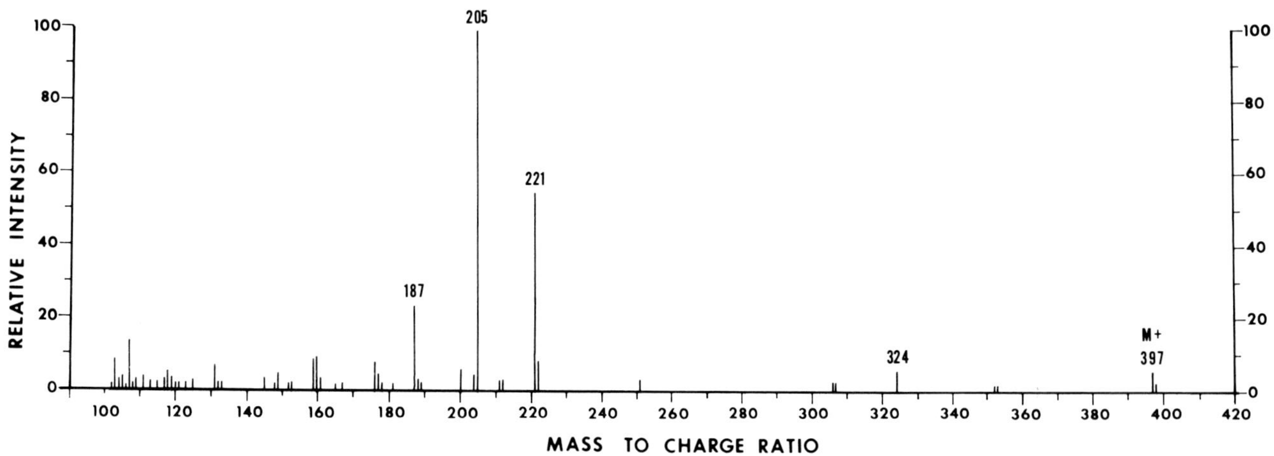

Ochratoxin B ethyl ester

REFERENCES

1. Aldridge, D. C., Galt, S., Giles, D., and Turner, W. B., *J. Chem. Soc. C* p. 1624 (1971).
2. Blair, J., and Newbold, G. T., *Chem. Ind. (London)* p. 93 (1955).
3. Blair, J., and Newbold, G. T., *Chem. Ind. (London)* p. 2871 (1955).
4. Burrows, B. F., unpublished.
5. Chu, F. S., and Chang, C. C., *J. Assoc. Off. Anal. Chem.* **54,** 1032 (1971).
6. Cole, R. J., Moore, J. H., Davis, N. D., Kirksey, J. W., and Diener, U. L., *J. Agric. Food Chem.* **19,** 909 (1971).
7. Doster, R. C., Sinnhuber, R. O., and Pawlowski, N. E., *Food Cosmet. Toxicol.* **12,** 499 (1975).
8. Harwig, J., *in* "Mycotoxins" (I. F. H. Purchase, ed.), p. 345. Elsevier, Am. New York, 1974.
9. Hayes, A. W., Hood, R. D., and Lee, H. L., Teratology **9,** 93 (1973).
10. Hutchison, R. D., Steyn, P. S., and Thompson, D. L., *Tetrahedron Lett.* p. 4033 (1971).
11. Meisner, H., and Chan, S., *Biochemistry* **13,** 2795 (1974).
12. Munro, I. C., Scott, P. M., Moodie, C. A., and Willes, R. F., *J. Am. Vet. Med. Assoc.* **163,** 1269 (1973).
13. Nesheim, S., *Adv. Chem. Ser.* **149,** 276 (1974).
14. Nishikawa, E., *J. Agric. Chem. Soc. Jpn.* **9,** 772 (1933).
15. Peckham, J. C., Doupnick, B., and Jones, O. H., *Appl. Microbiol.* **21,** 492 (1971).
16. Pitout, M. J., *in* "Mycotoxins in Human Health" (I. F. H. Purchase, ed.), p. 53. Macmillan, New York, 1971.
17. Sasaki, M., Kaneko, Y., Oshita, K., Takamatsu, H., Asao, Y., and Yokatsuka, T., *Agric. Biol. Chem.* **34,** 1296 (1970).
18. Scott, De B., *Mycopathol. Mycol. Appl.* **25,** 213 (1965).
19. Steyn, P. S., and van der Merwe, K. J., *Nature (London)* **211,** 418 (1966).
20. Steyn, P. S., and Holzapfel, C. W., *Tetrahedron* **23,** 4449 (1967).
21. Steyn, P. S., and Holzapfel, C. W., *J. S. Afr. Chem. Inst.* **20,** 186 (1967).
22. Steyn, P. S., and Holzapfel, C. W., *Phytochemistry* **9,** 1977 (1970).
23. Steyn, P. S., *in* "Mycrobial Toxins" (A. Ciegler, S. Kadis, and S. J. Ajl, eds.), Vol. 6, p. 179. Academic Press, New York, 1971.
24. Szczech, G. M., *Vet. Pathol.* **12,** 66 (1975).
25. van der Merwe, K. J., Steyn, P. S., and Fourie, L., *J. Chem. Soc.* p. 7083 (1965).
26. van der Merwe, K. J., Steyn, P. S., Fourie, L., Scott, De B., and Theron, J. J., *Nature (London)* **205,** 1112 (1965).
27. van Walbech, W., Scott, P. M., and Tahtcher, F. S., *Can. J. Microbiol.* **14,** 131 (1968).
28. Yabuta, T., and Sumiki, Y., *J. Agric. Chem. Soc. Jpn.* **9,** 1264 (1933).

The Trichothecenes

5

The trichothecenes comprise a group of sesquiterpenes produced by various species of imperfect fungi and characterized by the 12,13-epoxy-trichothec-9-ene ring system. The group can be divided into four subgroups based on chemical differences. The first two groups differ mainly by the presence or absence of a carbonyl function at C-8. Examples of trichothecenes that do not contain a carbonyl function at C-8 are T-2 toxin, HT-2 toxin, monoacetoxyscirpenol, diacetoxyscirpenol, trichodermin, and scirpentriol; those that do contain a carbonyl at C-8 are nivalenol, fusarenon-X, deoxynivalenol, deoxynivalenol monoacetate, and trichothecin. The last two groups, the roridins and verrucarins, contain a macrocyclic ester bridge between carbons 4 and 5. The roridins are macrocyclic diesters of verrucarol, whereas the verrucarins are macrocyclic triesters of verrucarol.

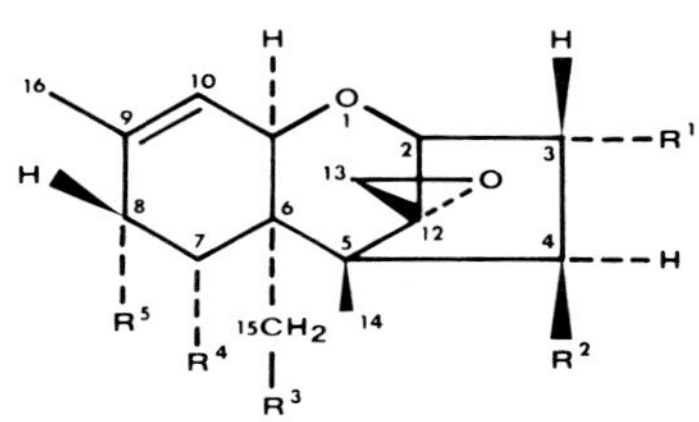

Basic Δ^9-trichothecene

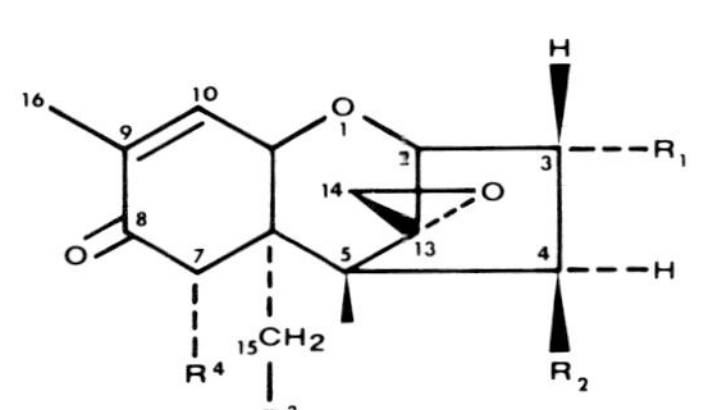

8-Ketotrichothecenes

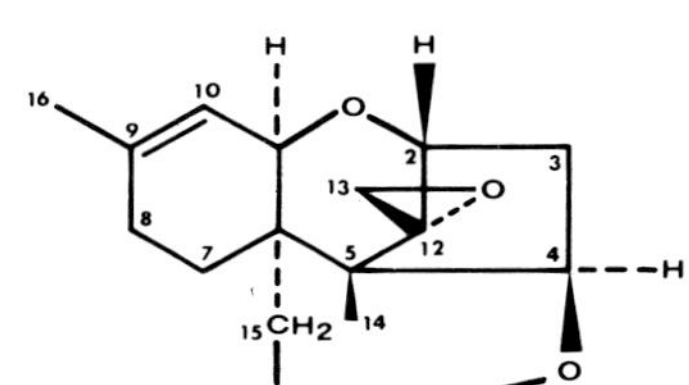

Basic verrucarin–roridin nucleus

The trichothecenes as a group show a wide range of biological activity such as antibacterial, antiviral, antifungal, and cytostatic activity; some are phytotoxic and all show some degree of animal toxicity, including insecticidal activity.[6,16,36,38,49,66,76] There exists a broad range of toxicity in vertebrate animals and the acute toxicities vary considerably. General gross clinical signs reported for trichothecenes are vomiting, diarrhea, anorexia, ataxia, hematuria, leukocytosis soon followed by severe leukopenia, inflammation of the gastrointestinal tract, degeneration of nerve cells in the central nervous system, degeneration and hemorrhaging of cardiac muscle, and lesions of lymph nodes, testes, and thymus.[4,24,93] Topical application or contact with skin causes dermal necrosis.[86] Tissue damage may be extensive, extending into the subcutaneous regions. Mortalities have been reported from IV, IP, oral, and topical administration of the trichothecenes. The trichothecenes are cytotoxic to mammalian cells in culture. Biochemically, they are potent inhibitors of protein and DNA synthesis in whole-cell and cell-free systems.[83,92,95] Some trichothecenes inhibit the initiation step of protein synthesis on polyribosomes and others inhibit the elongation and/or termination step in protein synthesis.

The trichothecenes have been strongly implicated in natural intoxications of animals and man. These mycotoxicoses are moldy corn toxicoses of cattle, swine, and poultry in the United States; Akakabi-Byo disease in Japan; Stachybotryo-toxicosis in the U.S.S.R; alimentary toxic aleukia; and dendrodochiotoxicosis.[7,33,94] Thus, they are potentially one of the most important mycotoxin groups to animals and man.

Several reviews concerning the various aspects of the trichothecene metabolites have been reported.[5,10,74,81]

Trichothecene Group	Molecular weight	Molecular formula
12,13-Epoxytrichothec-9-enes		
Trichodermol	250.1563	$C_{15}H_{22}O_3$
Verrucarol	266.1517	$C_{15}H_{22}O_4$
Scirpentriol	282.1461	$C_{15}H_{22}O_5$
T-2 tetraol	298.1410	$C_{15}H_{22}O_6$
Trichodermin	292.1668	$C_{17}H_{24}O_4$
Monoacetoxyscirpenol	324.1566	$C_{17}H_{24}O_6$
Diacetoxyscirpenol	366.1671	$C_{19}H_{26}O_7$
Neosolaniol	382.1620	$C_{19}H_{26}O_8$
Neosolaniol monoacetate	424.1725	$C_{21}H_{28}O_9$
HT-2 toxin	424.2096	$C_{22}H_{32}O_8$
T-2 toxin	466.2193	$C_{24}H_{34}O_9$
4,15-Diacetylverrucarol	350.1722	$C_{19}H_{26}O_6$
7α-Hydroxydiacetoxyscirpenol	382.1620	$C_{19}H_{26}O_8$

(Continued)

Trichothecene Group	Molecular weight	Molecular formula
7α,8α-Dihydroxydiacetoxy-scirpenol	398.1569	$C_{19}H_{26}O_9$
Calonectrin	350.1722	$C_{19}H_{26}O_6$
15-Deacetylcalonectrin	308.1617	$C_{17}H_{24}O_5$
Acetyl T-2 toxin	508.2298	$C_{26}H_{36}O_{10}$
Crotocin	332.1617	$C_{19}H_{24}O_5$
Crotocol	264.1361	$C_{15}H_{20}O_4$
Trichothecene	234.1614	$C_{15}H_{22}O_2$
4β,8α-Diacetoxy-12,13-epoxy-trichothec-9-ene-3α,15-diol	382.1467	$C_{19}H_{26}O_8$
4-Acetoxyscirpenediol	324.1572	$C_{17}H_{24}O_6$
4β,8α,15-Triacetoxy-12,13-epoxytrichothec-9-ene-3α,72-diol	440.1681	$C_{21}H_{28}O_{10}$
Triacetoxyscirpenol	408.1776	$C_{21}H_{28}O_8$
8-Ketotrichothecenes		
Deoxynivalenol	296.1254	$C_{15}H_{20}O_6$
Nivalenol	312.1203	$C_{15}H_{20}O_7$
Deoxynivalenol monoacetate	338.1359	$C_{17}H_{22}O_7$
Fusarenon-X	354.1308	$C_{17}H_{22}O_8$
Trichothecin	332.1617	$C_{19}H_{24}O_5$
Nivalenol diacetate	396.1419	$C_{19}H_{24}O_9$
Trichothecolone	264.1356	$C_{15}H_{20}O_4$
Trichodermone	248.1407	$C_{15}H_{20}O_3$
Macrocyclic diesters of verrucarol		
Satratoxin G	544.1935	$C_{28}H_{32}O_{11}$
Satratoxin H	528.2349	$C_{29}H_{36}O_9$
Roridin A	532.2661	$C_{29}H_{40}O_9$
Roridin D	530.2515	$C_{29}H_{38}O_9$
Roridin E	514.2449	$C_{29}H_{38}O_8$
Roridin H	512.2409	$C_{29}H_{36}O_8$
Vertisporin	544.2307	$C_{29}H_{36}O_{10}$
Isororidin E	514.2556	$C_{29}H_{38}O_8$
7β,8β-Epoxyisororidin E	528.2348	$C_{29}H_{36}O_9$
7β,8β-Epoxyroridin H	526.2193	$C_{29}H_{34}O_9$
7β,8β,2′,3′-Diepoxyroridin H	542.2142	$C_{29}H_{34}O_{10}$
Baccharin	562.2413	$C_{20}H_{38}O_{11}$
Macrocyclic triesters of verrucarol		
Verrucarin A	502.2201	$C_{27}H_{34}O_9$
Verrucarin B	500.2045	$C_{27}H_{32}O_9$
2′-Dehydroverrucarin A	500.2045	$C_{27}H_{32}O_9$
Verrucarin J	484.2114	$C_{27}H_{32}O_8$
Verrucarin K	486.2254	$C_{27}H_{34}O_8$

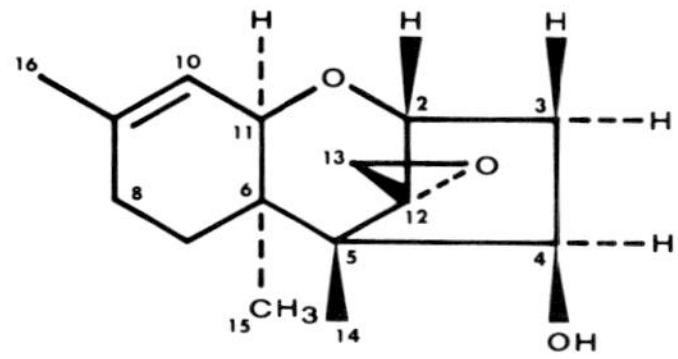

Common name[48]	Trichodermol (Roridin C)
Molecular weight	250.1563
Molecular formula	$C_{15}H_{22}O_3$
General characteristics[1,39]	Needle crystals from light petroleum, mp 116°–119°C Crystals from ether–*n*-hexane, mp 117.5°–118°C $[\alpha]_D^{20}$ −33.5° (*C* = 1.0 in chloroform)
UV data[1,39]	End absorption
Fungal source[41,48]	*Myrothecium roridum* Tode ex Fr.
TLC data[4]	A *Adsorbent*: silica gel G *Solvent*: chloroform–methanol, 98:2 v/v R_f: 0.36 *Detection*: H_2SO_4 spraying and heating at 110°C for 5 min B *Adsorbent*: alumina *Solvent*: chloroform–methanol, 98:2 v/v R_f: 0.86 *Detection*: H_2SO_4 spraying and heating at 110°C for 5 min
^{13}*C NMR data*[71a]	C2, 78.8; C3, 40.2; C4, 74.0; C5, 49.2; C6, 39.8; C7, 24.5; C8, 28.0; C9, 140.1; C10, 118.8; C11, 70.4; C12, 65.8; C13, 47.6; C14, 6.2; C15, 15.8; C16, 23.2
^{1}*H NMR data*[47b]	H2, 3.51; H3, 1.7-2.5; H4, 4.3; H7, 1.7-2.5; H8, 1.7-2.5; H10, 5.41; H11, 3.81; H13, 2.95; H14, 0.80; H15, 0.85; H16, 1.70

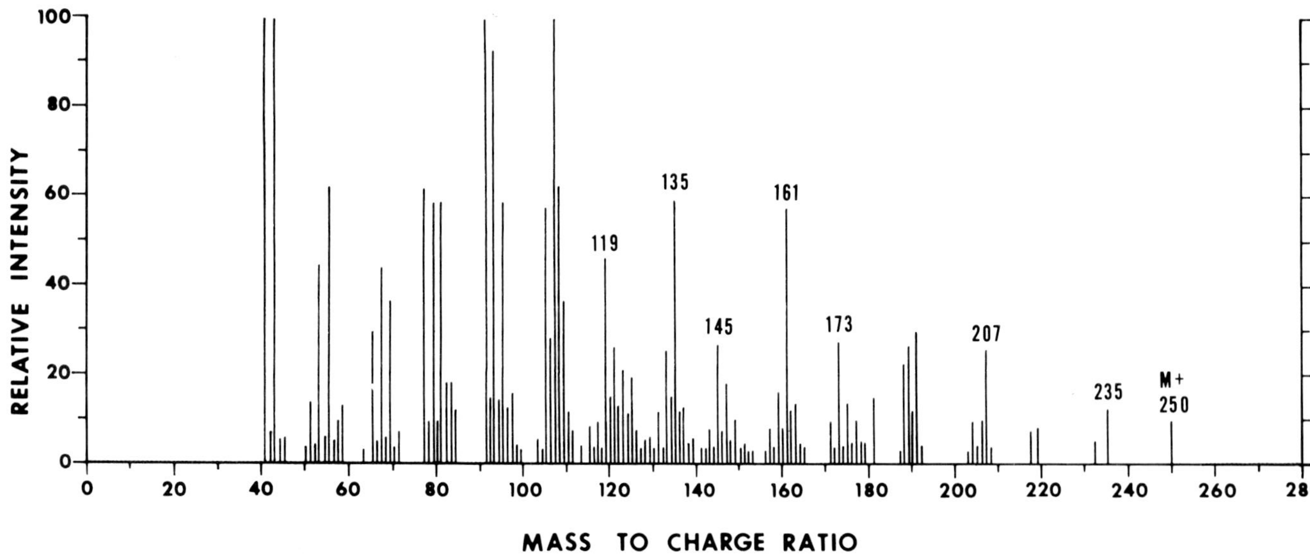

RELATIVE INTENSITY
MASS TO CHARGE RATIO
119
135
145
161
173
207
235
M+
250

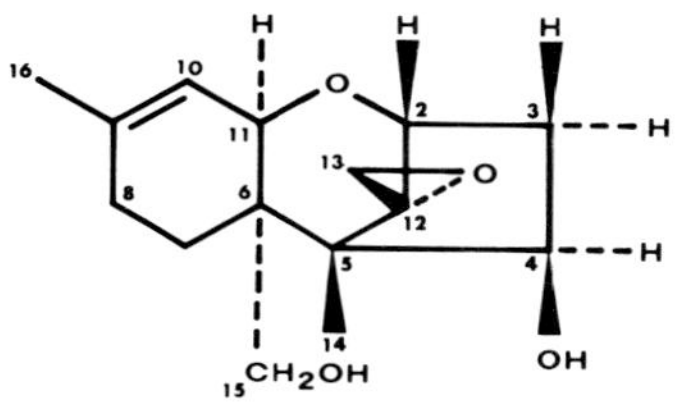

Common name	Verrucarol
Molecular weight	266.1517
Molecular formula	$C_{15}H_{22}O_4$
General characteristics[42,45]	Crystals from ether–methylene chloride, mp 158°–159°C Fine needles from acetone–ether, mp 155°–156°C $[\alpha]_D^{22}$ −39° (*C* = 1.069 in chloroform) Di-*O*-acetylverrucarol Colorless needles from acetone–ether–petroleum ether, mp 148°–150°C $[\alpha]_D^{25}$ −17° (*C* = 1.22 in chloroform) Di-*O*-benzoylverrucarol Crystals from ether–petroleum ether, mp 151°–152°C $[\alpha]_D^{24}$ −64° (*C* = 1.130 in acetone) Dihydroverrucarol Crystals from acetone–ether–petroleum ether, mp 149°–151°C $[\alpha]_D^{25}$ −6° (*C* = 1.101 in chloroform)
UV data[42]	λ_{max}^{EtOH} nm (ϵ): 195(7,900)
IR data[42]	$\nu_{max}^{CH_2Cl_2}$ cm^{-1}: 3610, 3570, 1675, 1380, 1335, 1079, 1045, 965, and 820
Fungal source	Formed as one of the products of alkaline hydrolysis of verrucarin A.
TLC data[4]	A *Adsorbent*: silica gel G *Solvent*: chloroform–methanol, 98:2 v/v R_f: 0.06 *Detection*: H_2SO_4 spraying and heating at 110°C for 5 min

B
Adsorbent: alumina
Solvent: chloroform–methanol, 98:2 v/v
R_f: 0.34
Detection: H_2SO_4 spraying and heating at 110°C for 5 min

Other significant references — Biosynthesis[3,56]

^{13}C NMR data[14] — C2, 78.5; C3, 39.5; C4, 74.0; C5, 48.7; C6, 43.7; C7, 20.9; C8, 28.0; C9, 140.4; C10, 118.7; C11, 66.4; C12, 65.6; C13, 47.4; C14, 6.8; C15, 62.1; C16, 22.9

^{1}H NMR data[14] — H2, 3.70; H3, 1.7-2.5; H4, 4.7; H7, 1.7-2.5; H8, 1.7-2.5; H10, 5.45; H11, 3.80; H13, 2.95; H14, 0.92; H15, 3.70; H16, 1.72

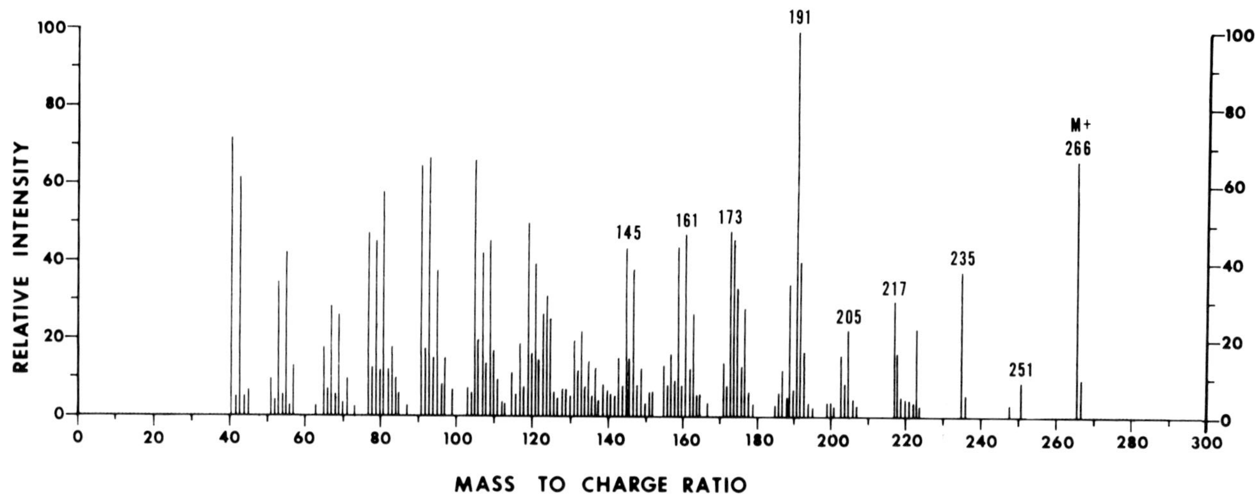
RELATIVE INTENSITY
MASS TO CHARGE RATIO
191
173
161
145
205
217
235
251
M+
266

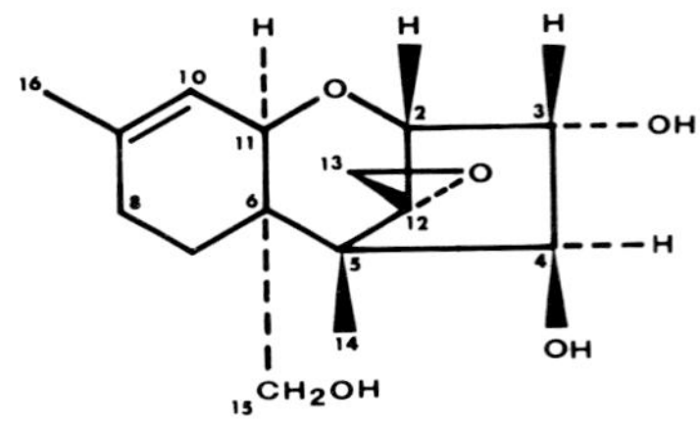

Common name	Scirpenetriol (3α, 4β, 15-Trihydroxy-12,13-epoxy-trichothec-9-ene)
Molecular weight	282.1461
Molecular formula	$C_{15}H_{22}O_5$
General characteristics[16]	Prisms from ethyl acetate, mp 193°C $[\alpha]_D^{23}$ 0° (C = 1.27 in acetone)
UV data	End absorption
IR data[16]	ν_{max}^{Nujol} cm^{-1}: 3480, 3455, 3405, 1676, and 830
Fungal source[16,68]	*Fusarium roseum* Gibbosum *F. equiseti*
Toxicity data[41]	LD_{50} in rats dosed IP was 0.81 mg/kg (single dose).
^1H NMR data[70]	H2, 3.78; H3, 4.08; H4, 4.12; H7, NR; H8, NR; H10, NR; H11, NR; H13a, 2.70; H13b, 3.05; H14, 0.90; H15, 3.59; H16, 1.72

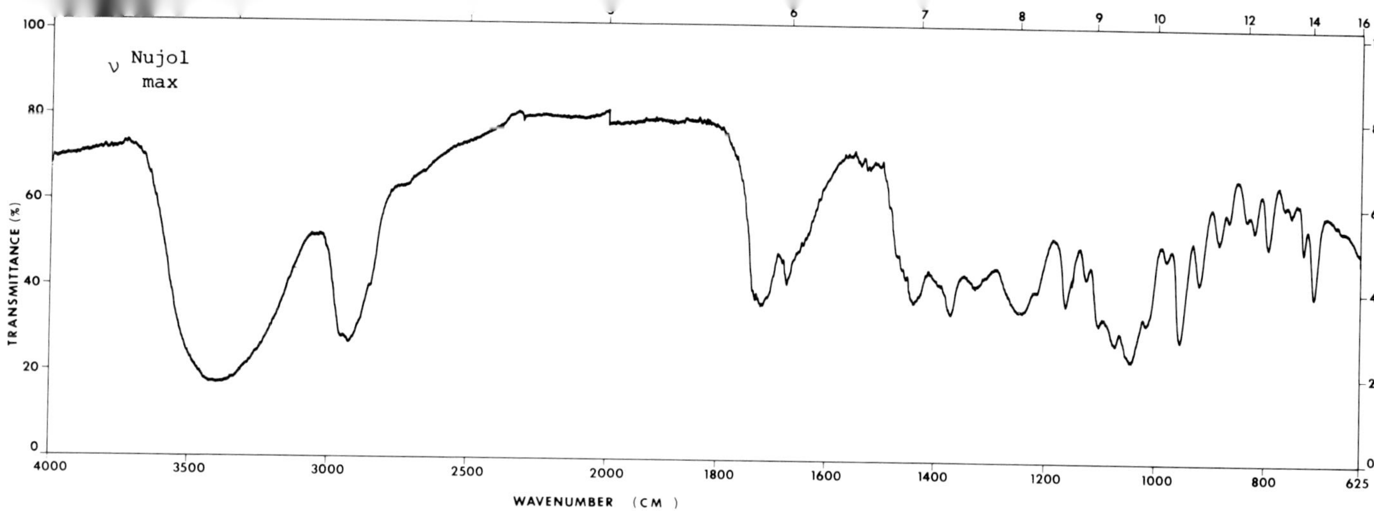
ν Nujol max
TRANSMITTANCE (%)
WAVENUMBER (CM)
0
20
40
60
80
100
4000
3500
3000
2500
2000
1800
1600
1400
1200
1000
800
625
6
7
8
9
10
12
14
16

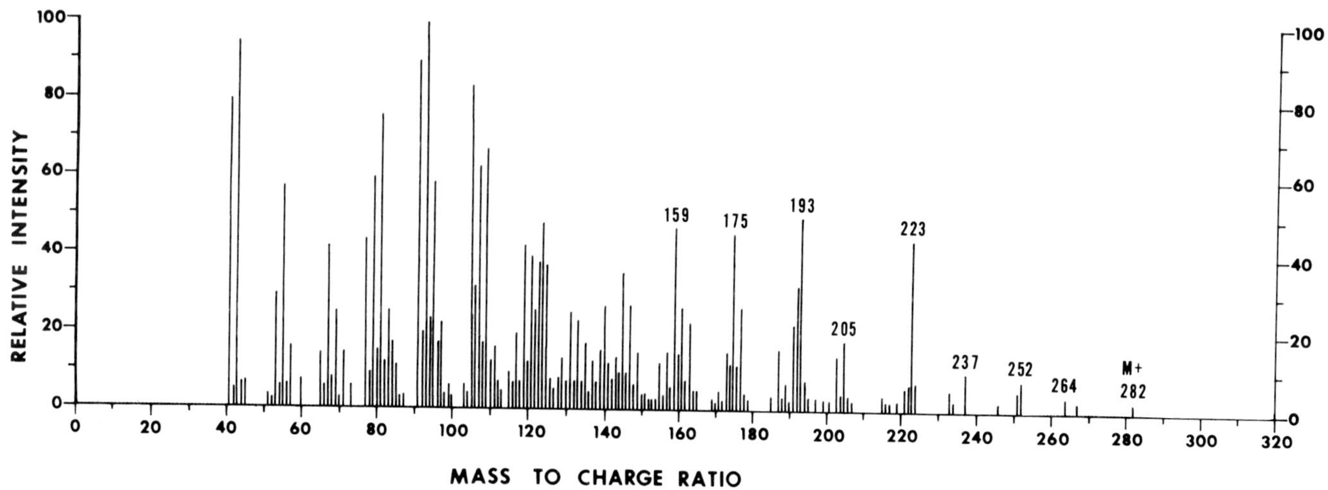
RELATIVE INTENSITY
MASS TO CHARGE RATIO
159
175
193
205
223
237
252
264
M+
282
0
20
40
60
80
100
120
140
160
180
200
220
240
260
280
300
320

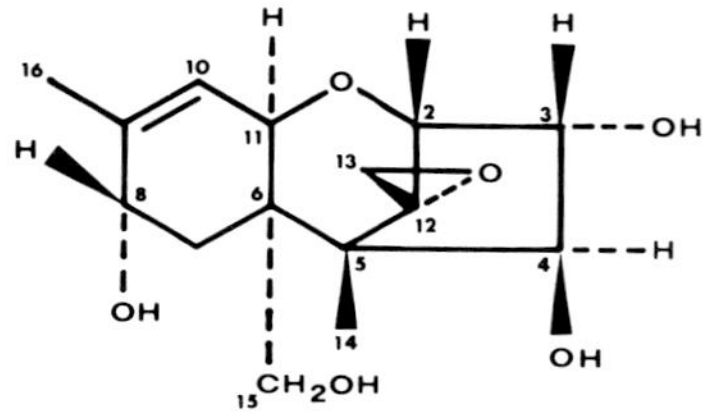

Common name	T-2 tetraol
Molecular weight	298.1410
Molecular formula	$C_{15}H_{22}O_6$
UV data[76]	End absorption
IR data[76]	ν_{max}^{KBr} cm^{-1}: 3400, 2930, 1625, 1450, and 1380
Fungal source[76]	*Fusarium poae* *F. sporotrichioides*
TLC data[7]	*Adsorbent*: silica gel G or H *Solvent:* ethanol–ethyl acetate–acetone, 1:4:4 v/v/v R_f: 0.52 *Detection*: H_2SO_4 spraying and heating at 150°C for 5 min
[13]*C NMR data*[25]	C2, 78.5; C3, 79.1; C4, 80.0; C5, 48.0; C6, 45.1; C7, 28.5; C8, 64.8; C9, 138.1; C10, 121.6; C11, 67.6; C12, 64.3; C13, 45.1; C14, 6.3; C15, 61.0; C16, 19.9

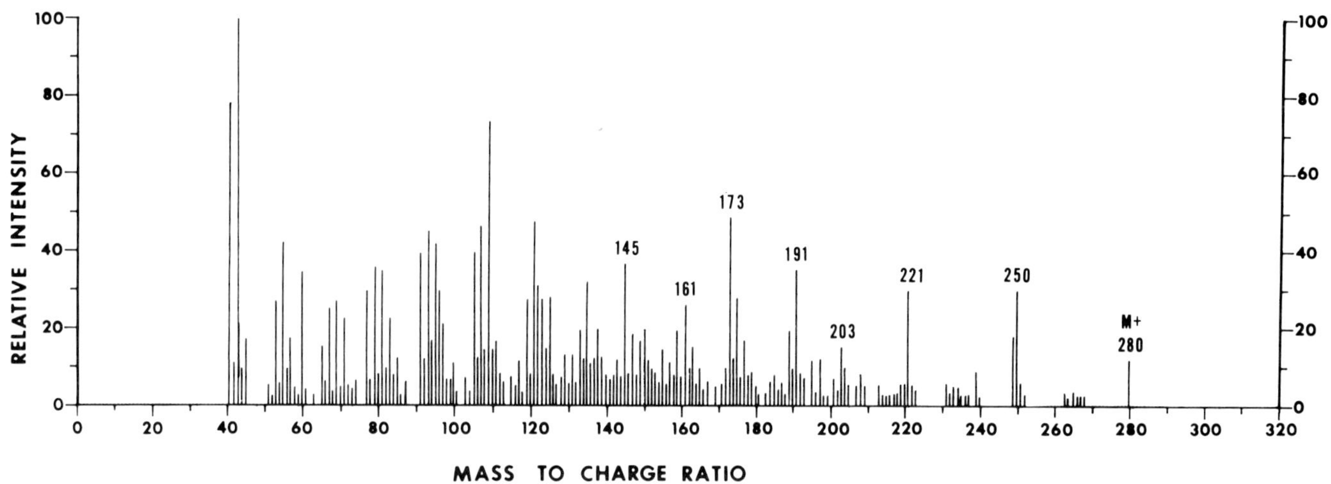
RELATIVE INTENSITY
MASS TO CHARGE RATIO
145
161
173
191
203
221
250
M+
280

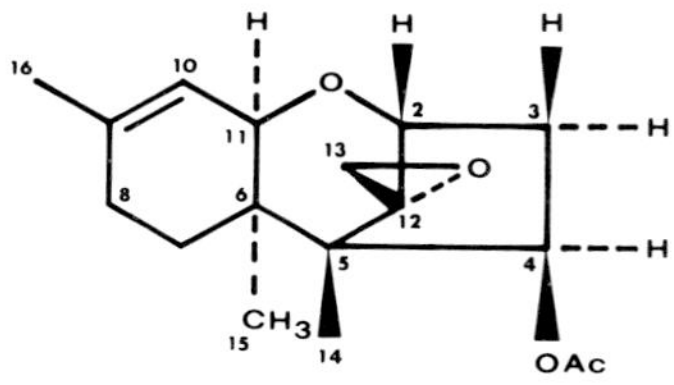

Common name	Trichodermin (4β-Acetoxy-12,13-epoxytrichothec-9-ene)
Molecular weight	292.1668
Molecular formula	$C_{17}H_{24}O_4$
General characteristics[5,18,38]	Synthetic ± trichodermin, mp 58°–60°C; natural trichodermin crystallized from pentane, mp 45°–46°C $[\alpha]_D^{20}$ −10.2° (*C* = 1.0 in chloroform) Crystals from pentane, mp 45°–46°C Insoluble in water; soluble in all common organic solvents.
UV data[38]	λ_{max} nm (ε): 205(2,400)
IR data[38]	ν_{max}^{KBr} cm^{-1}: 1730, 1682, 1225, 1245, and 1085
Fungal source[38,39]	*Trichoderma lignorum* *T. viride*
Toxicity data[5,59,96]	A potent inhibitor of protein synthesis in mammalian cells. Trichodermin inhibits the elongation and/or termination steps in protein synthesis. *In vitro,* it is a potent inhibitor of the peptidyl transferase activity required for elongation and/or termination. LD_{50} in mice dosed SC was 500–1000 mg/kg. Trichodermin is also a potent antifungal antibiotic.
TLC data[4]	A *Adsorbent:* silica gel G *Solvent:* chloroform–methanol, 98:2 v/v R_f: 0.67 *Detection:* H_2SO_4 spraying and heating at 110°C for 5 min B *Adsorbent*: silica gel G *Solvent*: benzene–tetrahydrofuran, 85:15 v/v

R_f: 0.51
Detection: H_2SO_4 spraying and heating at 110°C for 5 min

Other significant references — Synthesis[18,19]

^{13}C NMR data — C2, 79.1 d; C3, 36.7 t; C4, 75.0 d; C5, 48.9 s; C6, 40.4 s; C7, 24.5 t; C8, 28.0 t; C9, 140.1 s; C10, 118.6 d; C11, 70.5 d; C12, 65.5 s; C13, 47.8 t; C14, 5.8 q; C15, 16.0 q; C16, 23.3 q; $\underline{CH_3}C = O$, 21.1 q; $CH_3\underline{C} = O$, 170.8 s

^{1}H NMR data — H2, 3.80; H3, 1.9-2.5; H4, 5.60; H7, 1.9-2.5; H8, 1.9-2.5; H10, 5.4; H11, 3.7; H13a, 2.80; H13b, 3.10; H14, 0.95; H15, 0.75; H16, 1.70; $\underline{CH_3}C = O$, 2.05

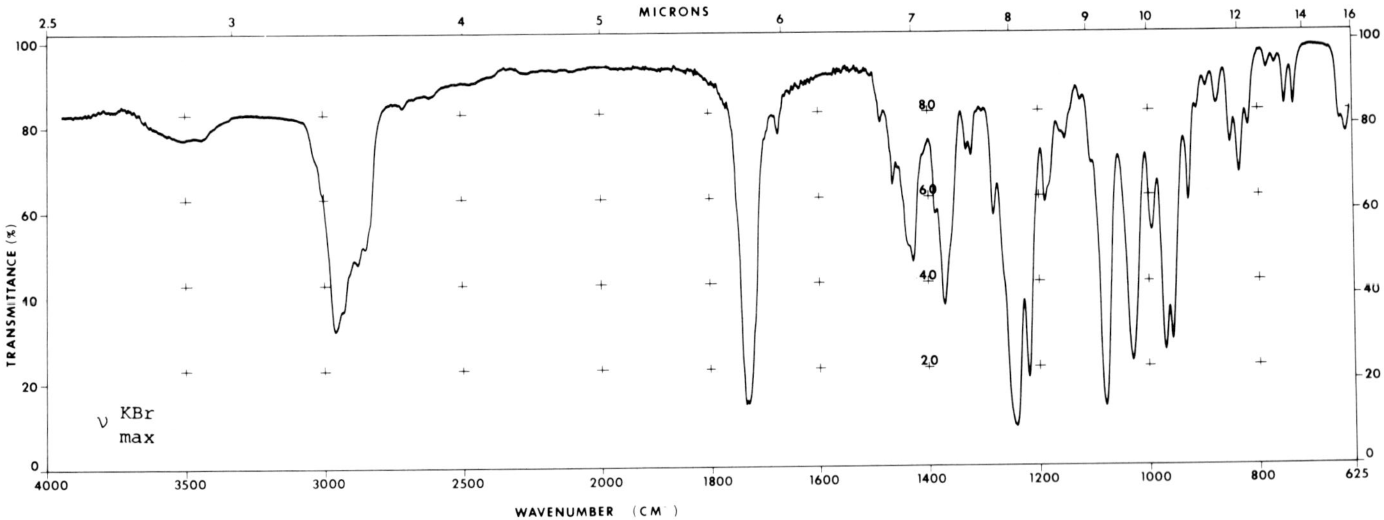
MICRONS
TRANSMITTANCE (%)
WAVENUMBER (CM⁻¹)
ν KBr max

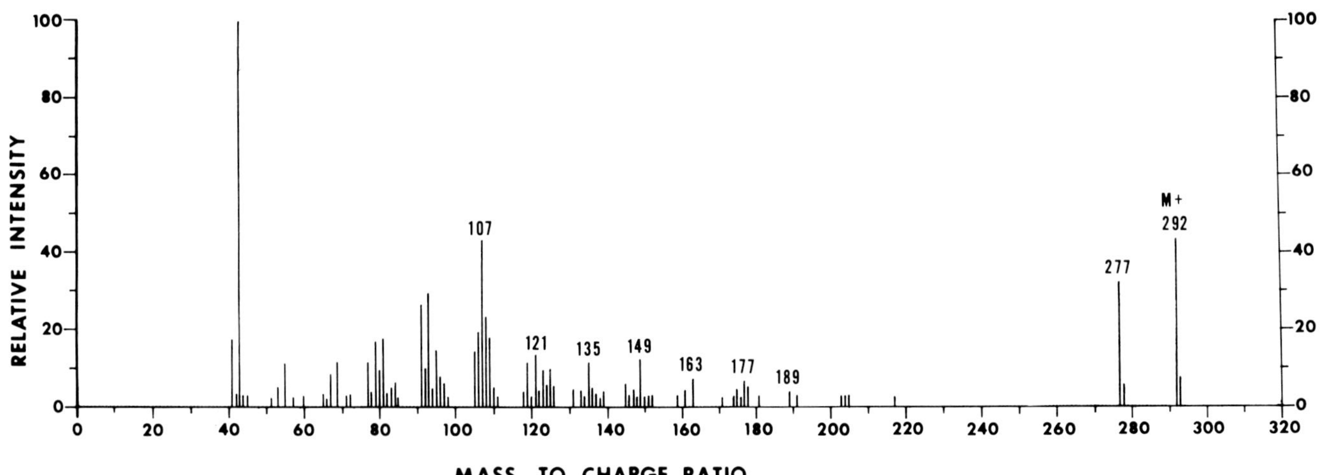
107
121
135
149
163
177
189
277
M+
292
RELATIVE INTENSITY
MASS TO CHARGE RATIO

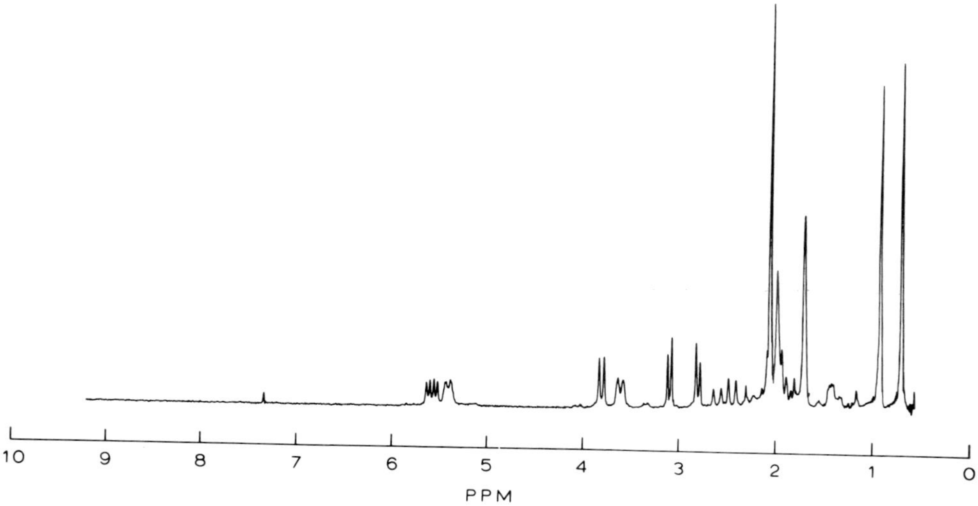

The 100 MHz ^{1}H NMR spectrum of trichodermin in $CDCl_3$ solution.

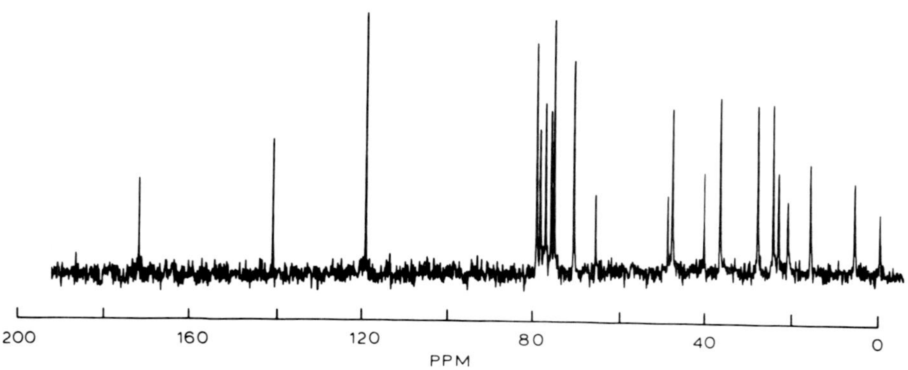

The 25 MHz ^{13}C NMR spectrum of trichodermin in $CDCl_3$ solution.

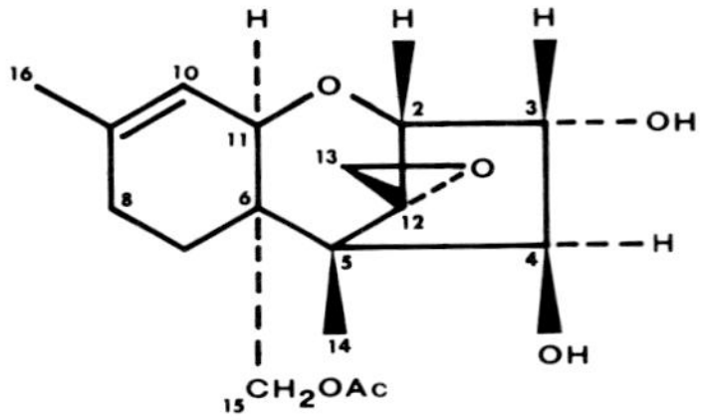

Common name: Monoacetoxyscirpenol (15-Acetoxy-3α, 4β-dihydroxy-12,13-epoxytrichothec-9-ene)

Molecular weight: 324.1566

Molecular formula: $C_{17}H_{24}O_6$

General characteristics[68]: Crystals from isooctane–ethyl acetate, mp 172°–173°C

UV data: End absorption

Fungal source[68,78]: *Fusarium roseum* Gibbosum
F. sulphureum

Toxicity data[68]: Bilateral inflammation of the beak area of birds ingesting the toxin, also gastrointestinal hemorrhaging. Application to shaved skin of rat resulted in hyperkeratosis and petechial hemorrhaging. The LD_{50} in 20-day-old white female weanling rats dosed SC was 0.752 mg/kg. A single 20 kg female pig injected IV with 20 mg monoacetoxyscirpenol (1 mg/kg) exhibited emesis within 0.5 hr, lethargy, a staggered gait, and death within 13 hr.

TLC data[68]:
Adsorbent: silica gel G
Solvent: chloroform–methanol, 90:10 v/v
R_f: 0.37
Detection: H_2SO_4 or *p*-anisaldehyde spraying and heating at 110° for 10 min. Purple to brown with H_2SO_4; purple with *p*-anisaldehyde.

GLC data[68]:
Liquid phase: 3% OV-1 on Gas Chrom Q (100-120 mesh)
Retention time: trifluoroacetate, 5–6 min; TMS, 9–10 min
Derivative for GLC: trimethylsilyl ether or trifluoroacetate

^1H NMR data[70]: H2, 3.63; H3, 4.22; H4, 4.31; H7, NR; H8, NR; H10, NR; H11, NR; H13a, 2.79; H13b, 3.04; H14, 0.83; H15, 4.05; H16, 1.73; CH_3CO, 2.05

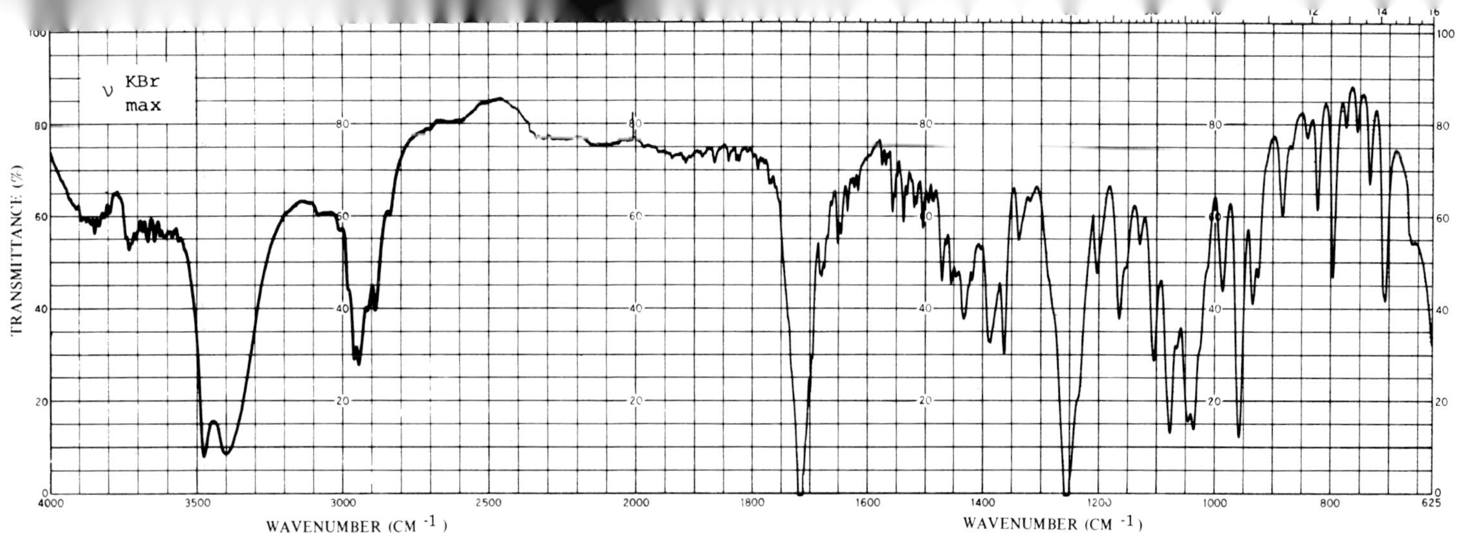

ν KBr max
TRANSMITTANCE (%)
WAVENUMBER (CM -1)
WAVENUMBER (CM -1)

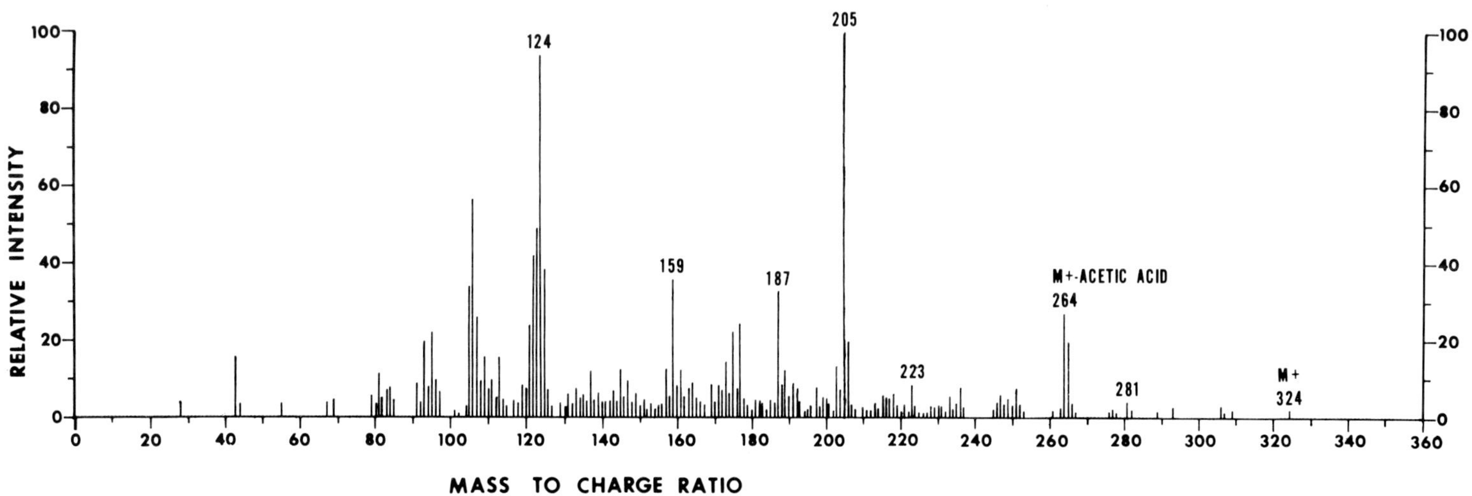

124
205
159
187
223
M+-ACETIC ACID
264
281
M+
324
RELATIVE INTENSITY
MASS TO CHARGE RATIO

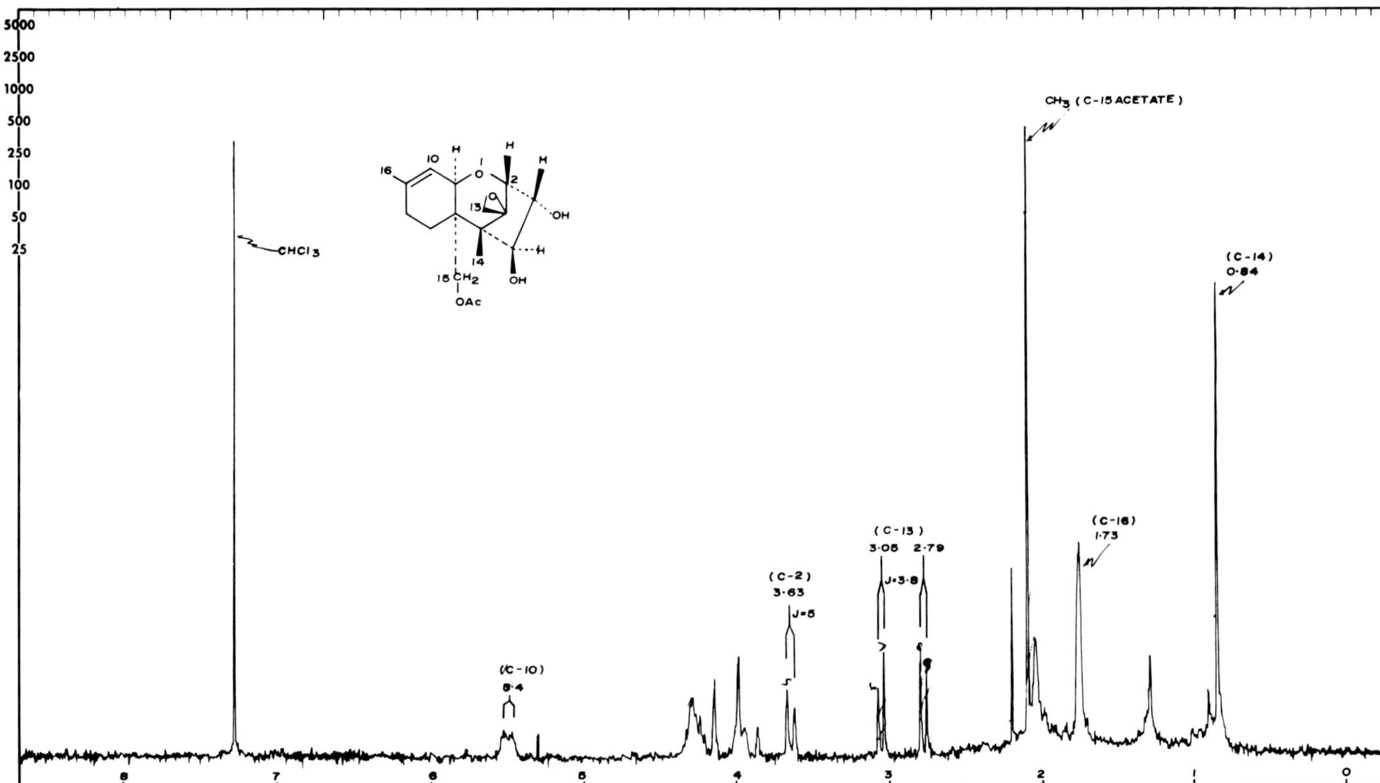
5000
2500
1000
500
250
100
50
25
CHCl3
16
10
H
I
H
2
H
13
O
OH
14
H
15 CH2
OAc
OH
CH3 (C-15 ACETATE)
(C-14)
0·84
(C-16)
1·73
(C-13)
3·05 2·79
J=3·8
(C-2)
3·63
J=5
(C-10)
5·4
8
7
6
5
4
3
2
1
0

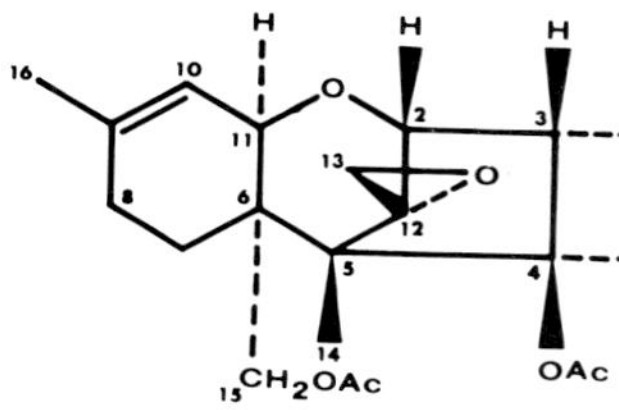

Common name	Diacetoxyscirpenol (3-Hydroxy-4,15-diacetoxy-12,13-epoxytrichothec-9-ene)
Molecular weight	366.1671
Molecular formula	$C_{19}H_{26}O_7$
General characteristics[22,29]	Crystals from ether, mp 162°–164°C Benzoate: prisms from ethanol, mp 178°C $[\alpha]_D^{24}$ −27° (C = 1.28)
UV data	End absorption
Fungal source[16,29,36,73,78]	*Fusarium tricinctum* *F. equiseti* *F. sambucinum* *F. lateritium* *F. graminearum* *F. semitectum* *F. sulphureum* *F. diversisporum* *F. scirpi* *Gibberella intricans*
Toxicity data[4,16,83,91]	The LD_{50} in mice dosed IP was 23.0 mg/kg; LD_{50} in chicken embryo test was 0.09 μg per egg. ID_{50} of protein synthesis in rabbit reticulocytes (whole cell) was 0.03 μg/ml; cell-free was 5.0 μg/ml; tumor cells was 0.30; rat liver (cell-free) was >50 μg/ml. Dermal toxicity to rabbit, mouse, and guinea pig. Diacetoxyscirpenol is also highly phytotoxic.
TLC data[4,92]	A *Adsorbent*: Kieselgel G *Solvent*: chloroform–methanol, 97:3 v/v R_f: 0.55 *Detection*: H_2SO_4 spraying and heating at 110°C for 5 min

B
Adsorbent: alumina
Solvent: chloroform–methanol, 98:2 v/v
R_f: 0.86
Detection: H_2SO_4 spraying and heating at 110°C for 5 min

GLC data[92]

Support: Shimalite W
Liquid phase: OV-17 (1.5%)
Retention time: 5.95
Relative retention time: (relative to Nivalenol) 1.67

Other significant references

Chemistry[21]

^{13}C NMR data

C2, 79.0 d; C3, 76.3 d; C4, 83.0 d; C5, 49.1 s; C6, 44.2 s; C7, 21.1 t; C8, 27.9 t; C9, 139.4 s; C10, 119.3 d; C11, 67.4 d; C12, 64.8 s; C13, 46.6 t; C14, 6.8 q; C15, 63.4 t; C16, 23.2 q; $\underline{CH_3}C = O$, 21.1 q, 21.1 q; $CH_3\underline{C} = O$, 170.3 s, 170.7 s

^1H NMR data

H2, 3.68; H3, 4.23; 3-OH, 3.68; H4, 4.38; H7, 2.19; H8, 2.19; H10, 5.86 (J = 4.6); H11, 5.65 (J = 4.6); H13a, 3.34 (J = 4.0); H13b, 3.57 (J = 4.0); H14, 0.91; H15a, 4.10 (J = 14.0); H15b, 4.42 (J = 14.0); H16, 1.95; $\underline{CH_3}C = O$, 2.28, 2.35

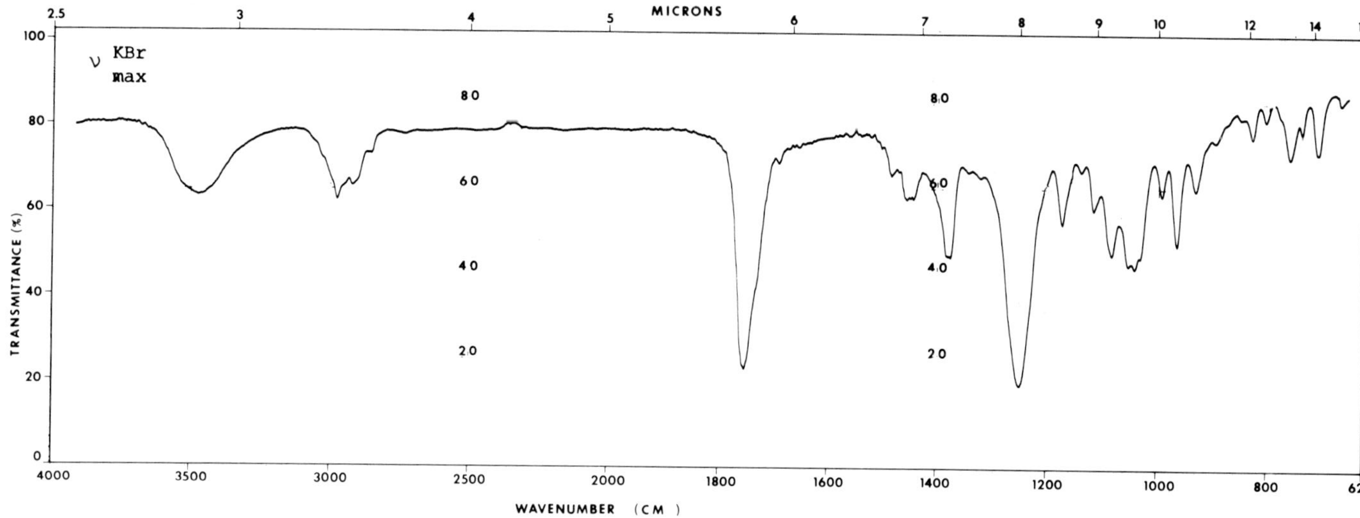
MICRONS
ν KBr max
TRANSMITTANCE (%)
WAVENUMBER (CM)

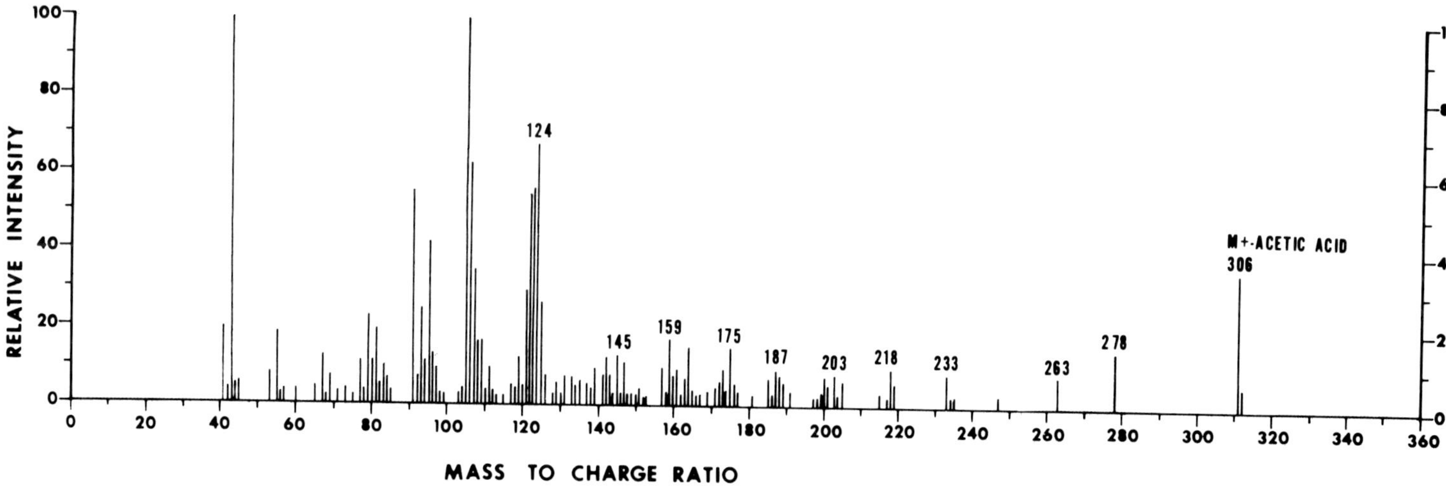
RELATIVE INTENSITY
124
145
159
175
187
203
218
233
263
278
M+-ACETIC ACID
306
MASS TO CHARGE RATIO

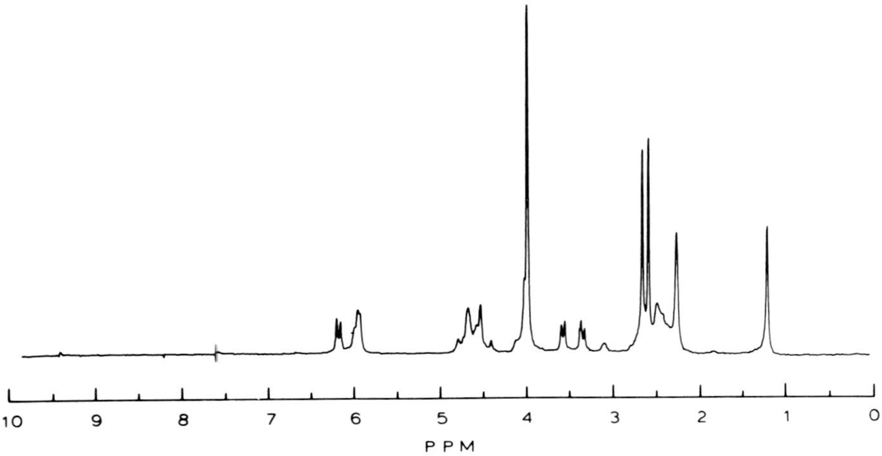

The 100 MHz ^1H NMR spectrum of diacetoxyscirpenol in DMSO solution.

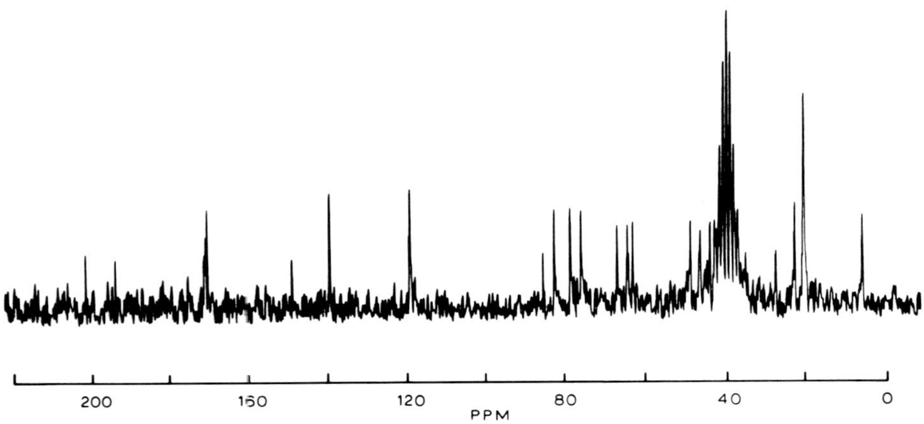

The 25 MHz ^{13}C NMR spectrum of diacetoxyscirpenol in DMSO solution.

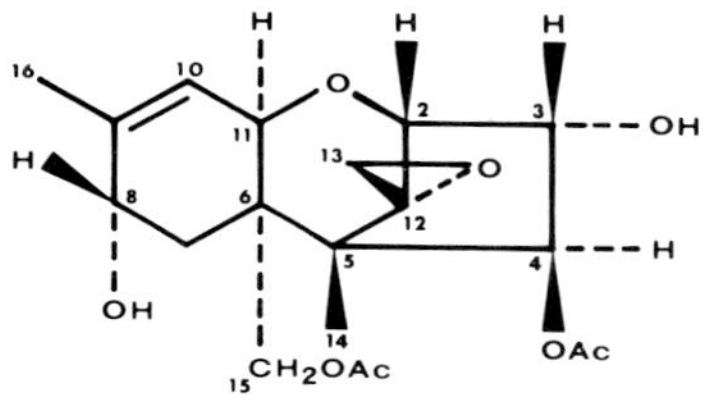

Common name	Neosolaniol (Solaniol) (4β,15-Diacetoxy-3α,8α-dihydroxy-12,13-epoxytrichothec-9-ene)
Molecular weight	382.1620
Molecular formula	$C_{19}H_{26}O_8$
General characteristics[49]	Crystals from ethyl acetate–*n*-hexane, mp 171°–172°C
UV data	End absorption
Fungal source[11,76,90,91]	*Fusarium culmorum* *F. solani* *F. poae* *F. sporotrichioides* *F. lateritium* *F. equiseti* *F. averaceum*
Toxicity data[93,95]	LD_{50} to mice dosed IP was 14.5 mg/kg; LD_{50} in chicken embryo test was 5.0 μg per egg. ID_{50} of protein synthesis in rabbit reticulocytes (whole cell) was 0.25 μg/ml; rat liver (cell-free) was 20 μg/ml. Clinical signs in animals receiving a fatal dose were marked cellular degeneration and karyorrhexis in actively dividing cells of thymus, lymph nodes, spleen, bone marrow, intestine, and testes reflecting the so-called ''radiomimetic'' property. Minimum dose for skin-irritant toxicity to rabbits was 1.0 μg.
TLC data[92]	*Adsorbent*: Kieselgel G *Solvent*: ethyl acetate–*n*-hexane, 3:1 v/v R_f: 0.15 *Detection*: H_2SO_4 spraying and heating at 110°C for 5 min

GLC data[83] (as TMS derivative)
Solid support: Shimalite W
Liquid phase: OV-17 (1.5%)
Retention time: 7.50
Relative retention time: (relative to Nivalenol) 2.11

^{1}*H NMR data*
H2, 3.69 (J = 3.8); H3, 4.2; H4, 5.27; H7, 2.18; H8, 4.2; H10, 5.68; (J = 6.0); H11, 4.2; H13a, 2.79 (J = 4.0); H13b, 3.04 (J = 4.0); H14, 0.86; H15a, 4.1-4.3; H15b, 4.1-4.3; H16, 1.87; $\underline{CH_3}C{=}O$, 2.04, 2.15

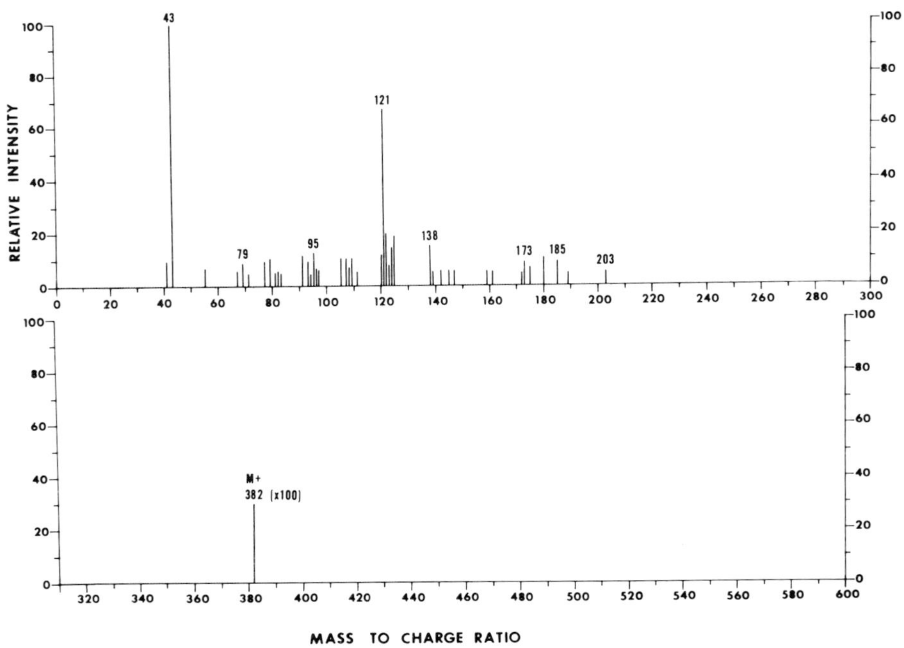

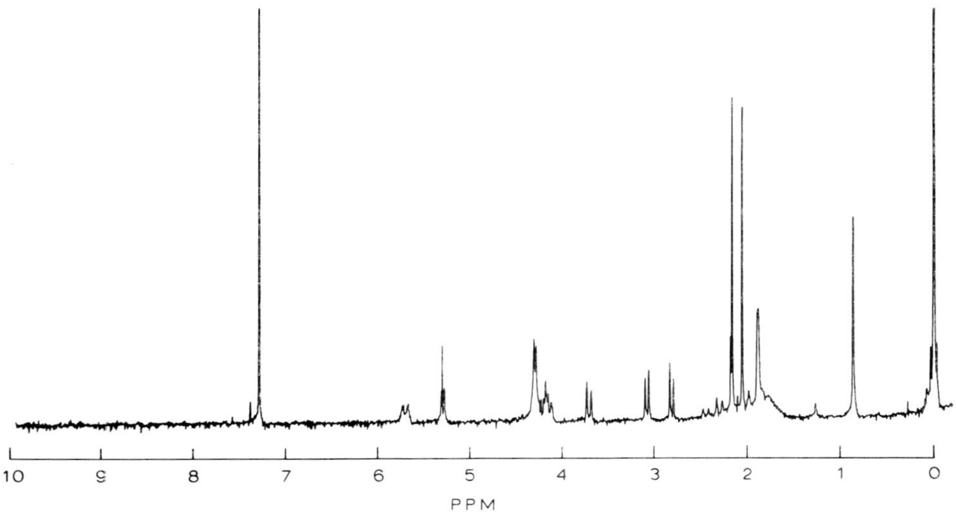

The 100 MHz ¹H NMR spectrum of neosolaniol in $CDCl_3$ solution.

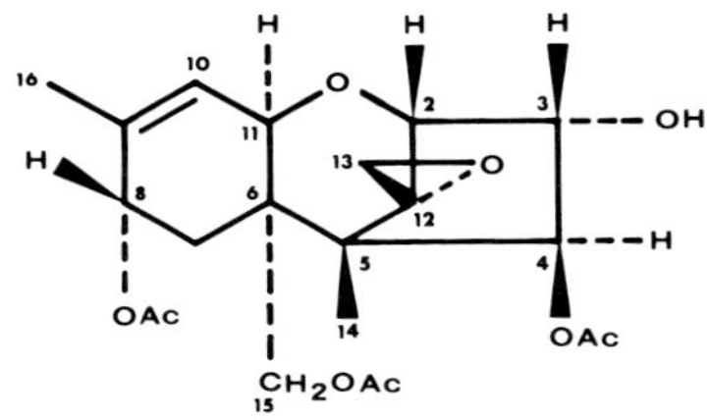

Common name	Neosolaniol monoacetate (3-Hydroxy-4β,8α,15-triacetoxy-12,13-epoxytrichothec-9-ene)
Molecular weight	424.1725
Molecular formula	$C_{21}H_{28}O_9$
General characteristics[54,59]	White needles from benzene–heptane, 1:1 v/v, mp 190°–190.5°C
UV data	End absorption
Fungal source	*Fusarium* spp.
Toxicity data[54,59]	LD_{50} in day-old cockerels dosed via crop intubation was 0.789 mg/kg. LD_{50} of T-2 toxin (dosed simultaneously) was 1.84 mg/kg. Also shows inhibition in wheat coleoptile assay down to 10^{-6} *M*. It showed a linear, rather than a curvilinear, response.
^{13}*C NMR data*	C2, 78.7d; C3, 78.3 d; C4, 84.5 d; C5, 47.1 s; C6, 43.0 s; C7, 27.4 t; C8, 68.5 d; C9, 136.1 s; C10, 123.8 d; C11, 67.4 d; C12, 64.3 s; C13, 48.6 t; C14, 7.0 q; C15, 64.4 t; C16, 20.3 q; $\underline{CH_3}C = O$, 21.0, 21.0, 21.0; $CH_3\underline{C} = O$, 170.0, 170.6, 172.5
^{1}*H NMR data*	H2, 3.69 (J = 5.0); H3, 4.18 (J = 5.0, 3.0); 3-OH, 3.36; H4, 5.30; H7a, 2.10 (J = 15.0); H7b, 2.36 (J = 15.0, 5.5); H8, 5.29 (J = 5.5); H10, 5.81 (J = 6.0); H11, 4.31 (J = 6.0); H13a, 2.81 (J = 4.0); H13b, 3.06 (J = 4.0); H14, 0.91; H15a, 4.08 (J = 13.4); H15b, 4.30 (J = 13.4); H16, 1.76; $\underline{CH_3}C = O$, 2.05, 2.05, 2.14

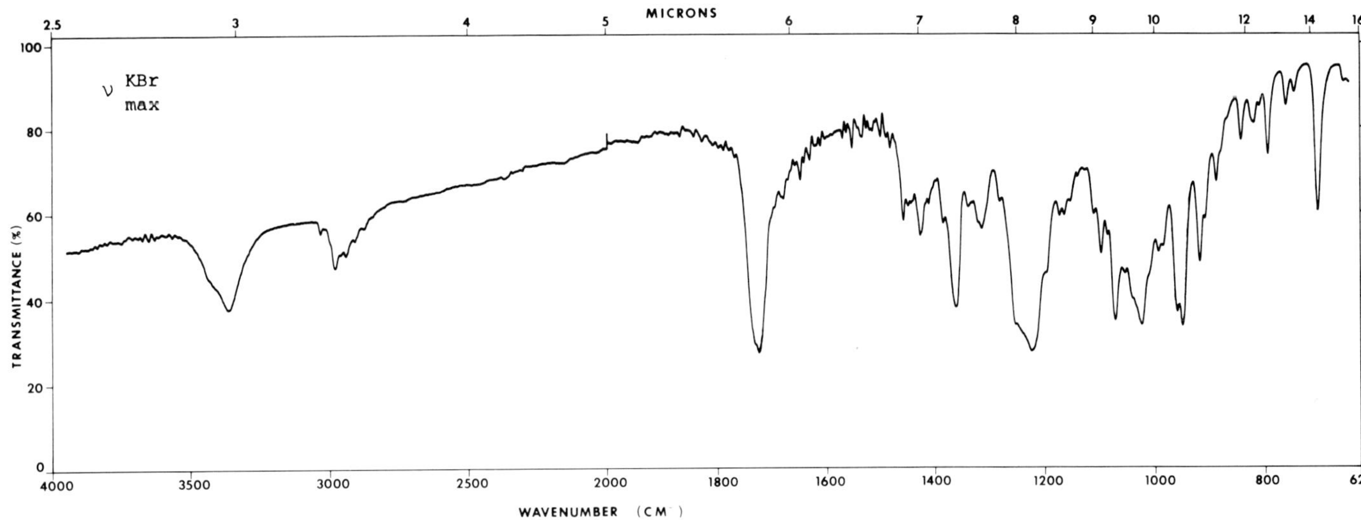
MICRONS
ν KBr max
TRANSMITTANCE (%)
WAVENUMBER (CM⁻¹)

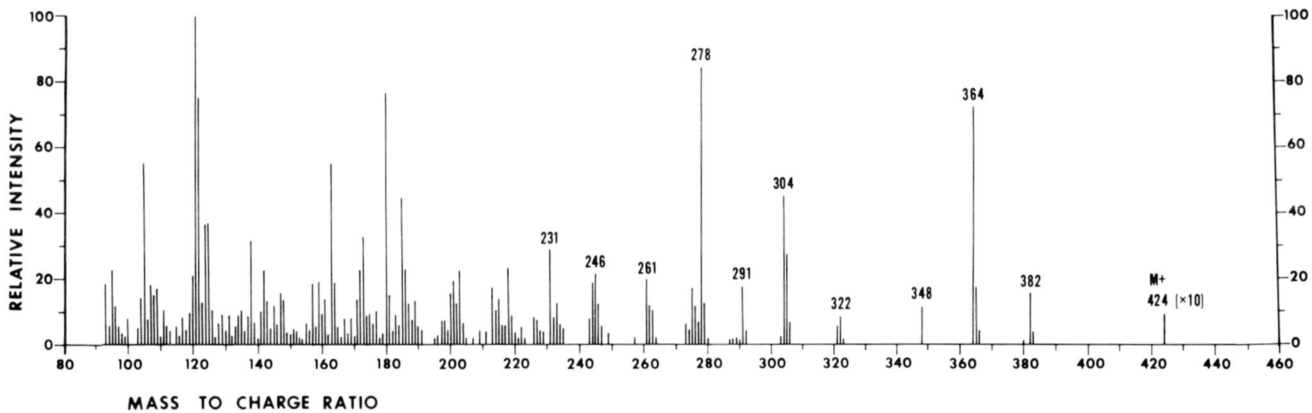
RELATIVE INTENSITY
231
246
261
278
291
304
322
348
364
382
M+
424 (×10)
MASS TO CHARGE RATIO

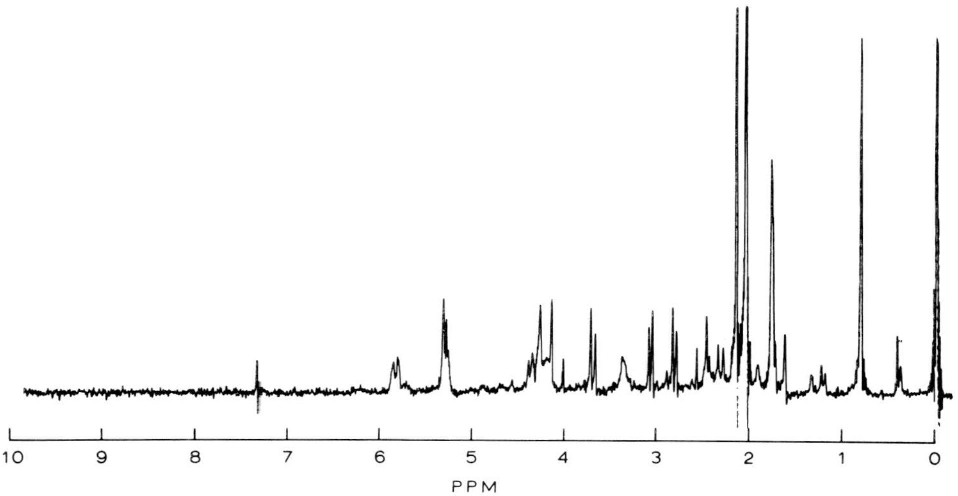

The 100 MHz ^{1}H NMR spectrum of neosolaniol monoacetate in $CDCl_3$ solution

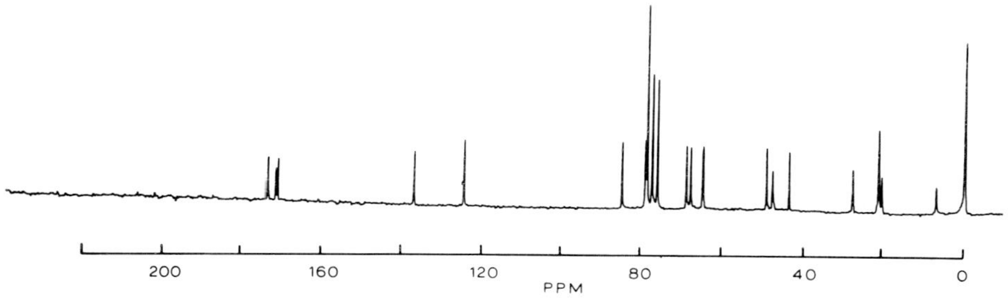

The 25 MHz ^{13}C NMR spectrum of neosolaniol monoacetate in $CDCl_3$ solution.

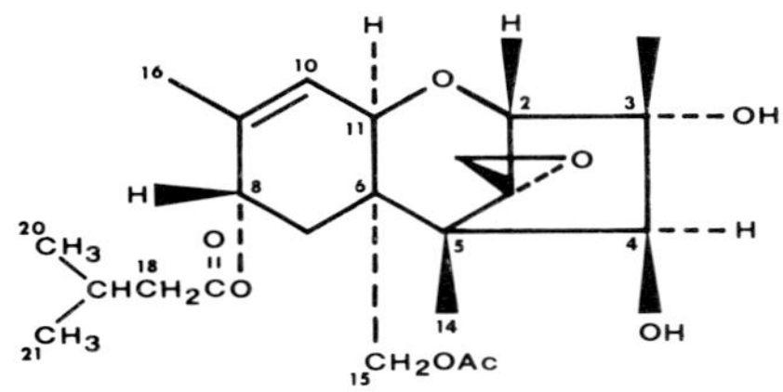

Common name[9]	HT-2 (3α,4β-Hydroxy-15-acetoxy-8α-[3-methyl-butyryloxy]-12,13-epoxytrichothec-9-ene)
Molecular weight	424.2096
Molecular formula	$C_{22}H_{32}O_8$
General characteristics	Pale yellow oil
UV data	End absorption
IR data	$\nu_{max}^{CHCl_3}$ cm^{-1}: 3400, 2950, 1720, 1635, and 1240
Fungal source[9,76]	*Fusarium culnorum* *F. poae* *F. sporotrichioides*
Toxicity data[4,5,94]	LD_{50} dosed IP to mice was 9.0 mg/kg; LD_{50} in chicken embryo assay was 0.5 μg per egg. ID_{50} of protein synthesis in rabbit reticulocytes (whole cell) was 0.03 μg/ml. It inhibits the initiation step in protein synthesis. It causes dermal necrosis similar to that of T-2.
TLC data[4]	A *Adsorbent*: silica gel G *Solvent*: chloroform–methanol, 98:2 v/v R_f: 0.09 *Detection*: H_2SO_4 spraying and heating at 110°C for 5 min B *Adsorbent*: alumina *Solvent*: chloroform–methanol, 98:2 v/v R_f: 0.32 *Detection*: H_2SO_4 spraying and heating at 110°C for 5 min
^{13}C NMR data	C2, 79.9 d; C3, 80.9 d; C4, 81.2 d; C5, 47.0 s; C6, 44.1 s; C7, 28.1 t; C8, 69.1 d; C9, 138.6 s; C10,

135.6 d; C11, 67.6 d; C12, 67.6 s; C13, 46.8 t; C14, 8.2 q; C15, 65.6 t; C16, 22.2 q; C1′, 172.5 s; C2′, 43.3 t; C3′, 26.5 d; C4′, 23.4 q; $\underline{CH_3}C=O$, 21.2 q; $CH_3\underline{C}=O$, 170.5 s

^{1}H NMR data H2, 3.66 (J = 5.0); H3, 4.26; H4, 4.42; H7a, 2.10; H7b, 2.20; H8, 4.26; H10, 5.78 (J = 6.0); H11, 4.32 (J = 6.0); H13a, 2.80 (J = 3.0); H13b, 3.06 (J = 3.0); H14, 0.80; H15a, 4.04 (J = 13.0); H15b, 4.34 (J = 13.0); H16, 1.76; H2′, 2.10; H3′, 2.10; H4′, 0.98 (J = 6.0); $\underline{CH_3}C=O$, 1.76

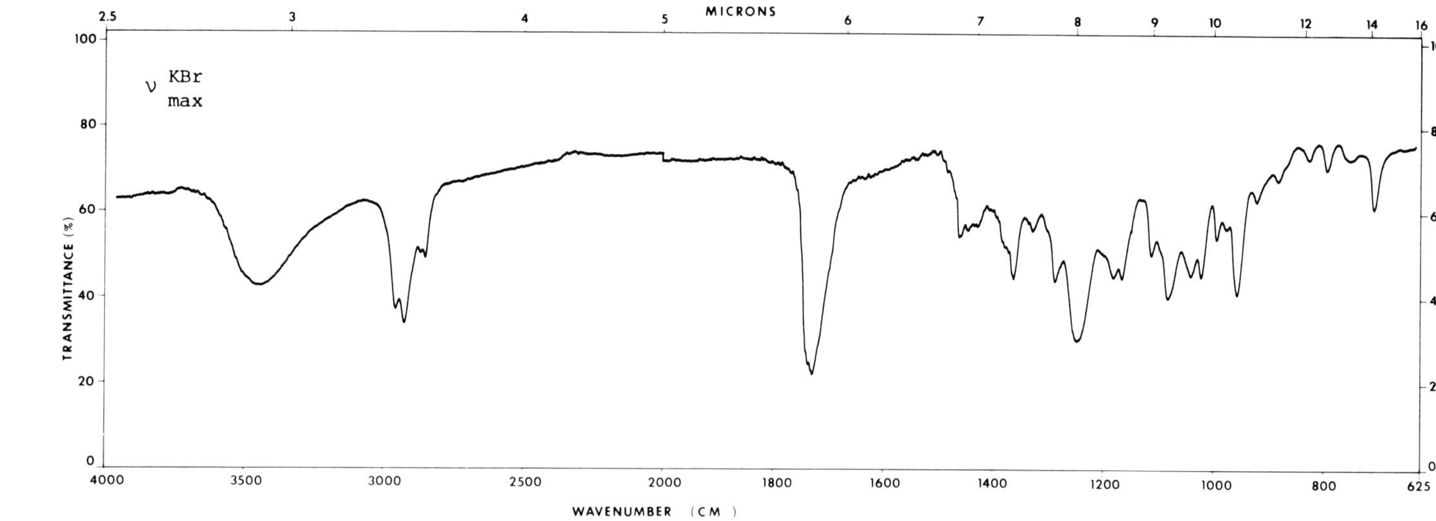
MICRONS
TRANSMITTANCE (%)
ν KBr max
WAVENUMBER (CM)

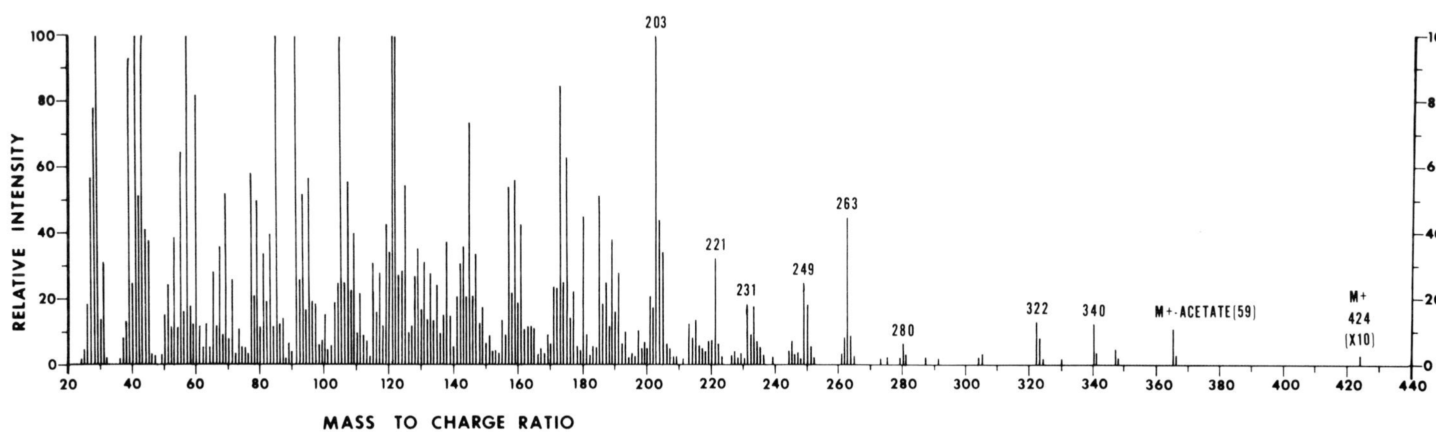
RELATIVE INTENSITY
203
221
231
249
263
280
322
340
M+-ACETATE(59)
M+
424
(X10)
MASS TO CHARGE RATIO

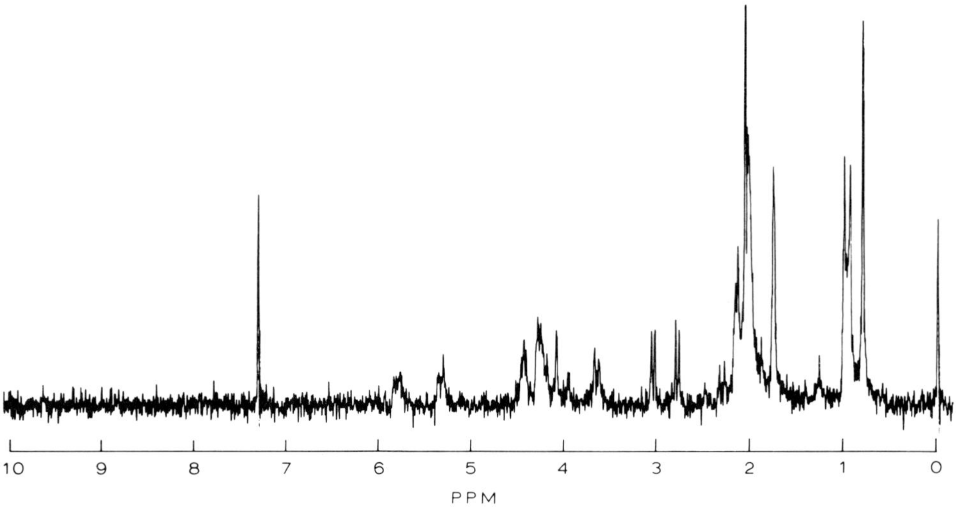

The 100 MHz ^{1}H NMR spectrum of HT-2 toxin in $CDCl_3$ solution.

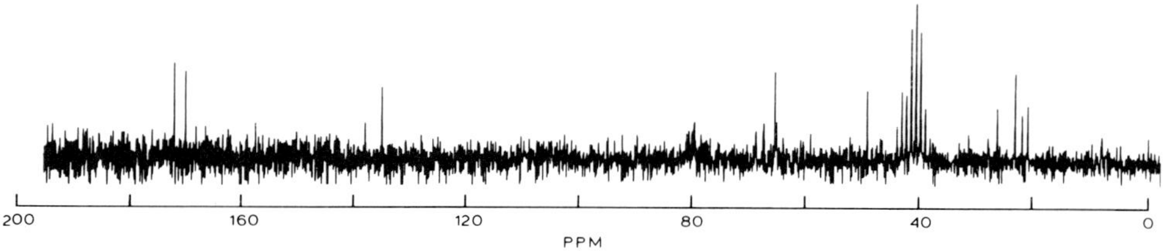

The 25 MHz ^{13}C NMR spectrum of HT-2 toxin in DMSO solution.

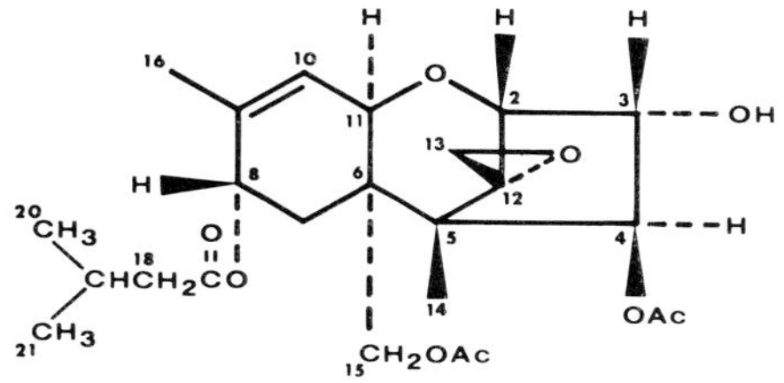

Common name	T-2 toxin (3-Hydroxy-4,15-diacetoxy-8-[3-methyl-butyryloxy]-12,13-epoxytrichothec-9-ene)
Molecular weight	466.2193
Molecular formula	$C_{24}H_{34}O_9$
General characteristics[7]	White needles from benzene–Skellysolve B, mp 151°–152°C $[\alpha]_D^{26}$ +15° (*C* = 2.58 in ethanol) Acetate derivative: amorphous solid from ether–pentane $[\alpha]_D^{30}$ +27° (*C* = 0.9 in 95% ethanol)
UV data	End absorption
Fungal source[8,9,18,71,76]	*Fusarium tricinctum* *F. culmorum (F. roseum)* *F. solani* *F. poae* *F. sporotrichioides* *Trichoderma lignorum*
Toxicity data[10,17,59,72,74,83,93]	Acute LD_{50} in rats and swine dosed orally was 4 mg/kg; LD_{50} in mice dosed IP was 3.04 mg/kg; LD_{50} in chicks dosed orally was 1.84 mg/kg; ID_{50} of rabbit reticulocytes (whole cell) was 0.03 g/ml; ID_{50} of rabbit reticulocytes (cell-free was 0.15 μg/ml. T-2 toxin inhibits the initiation step of protein synthesis on polyribosomes. It has emetic properties when administered orally, IP or IV. Minimum SC dosage to induce vomiting in ducklings was 0.1 mg/kg; in cat, 0.1–0.2 mg/kg. Major gross clinical signs in cats from T-2 toxin were emesis, vomiting, diarrhea, anorexia, ataxia of the hind legs, discharge from the eyes, and ejection of hemorrhagic fluid. Consecutive administration at sublethal dosages caused a marked decrease in

white cells. Necropsy showed extensive cellular damage in the bone marrow, intestine, spleen, and lymph nodes. Also evident were meningeal hemorrhage of the brain, bleeding in the lungs, and vacuolic degeneration of the renal tubes. T-2 causes dermal necrosis when applied to the surface of the skin. It causes primary oral lesions in animals ingesting it.

TLC data[4]

A
Adsorbent: silica gel G
Solvent: chloroform–methanol, 98:2 v/v
R_f: 0.40
Detection: H_2SO_4 spraying and heating at 110°C for 5 min
B
Adsorbent: alumina
Solvent: chloroform–methanol, 98:2 v/v
R_f: 0.88
Detection: H_2SO_4 spraying and heating at 110°C for 5 min

GLC data[83] (as TMS derivatives)

Support: Shimalite W
Liquid phase: OV-17 (1.5%)
Retention time: 8.55
Relative retention time: (relative to Nivalenol) 2.40

Other significant references

GLC[50]

^{13}C NMR data

C2, 78.5 d; C3, 76.8 d; C4, 84.2 d; C5, 48.3 s; C6, 42.9 s; C7, 27.8 t; C8, 67.8 d; C9, 135.8 s; C10, 123.4 d; C11, 67.2 d; C12, 64.4 s; C13, 47.1 t; C14, 6.8 q; C15, 64.1 t; C16, 20.2 q; C1′, 172.0 s; C2′, 43.4 t; C3′, 25.7 d; C4′, 22.4 q; $CH_3\underline{C}=O$, 169.5 s, 172.0 s; $\underline{C}H_3C=O$, 20.2, 21.0

^{1}H NMR data

H2, 3.69 (J = 5.0); H3, 4.18 (J = 5.0, 3.0); H4, 5.32 (J = 3.0); H7a, 2.33 (J = 15.0, 6.0); H7b, 2.49 (J = 15.0); H8, 5.28 (J = 6.0); H10, 5.81 (J = 5.4); H11, 4.35 (J = 5.4); H13a, 2.80 (J = 3.9); H13b, 3.06 (J = 3.9); H14, 0.81; H15a, 4.08; H15b, 4.25; H16,1.75; H2′, 1.93; H4′, 0.96; $CH_3C=O$, 2.04, 2.14

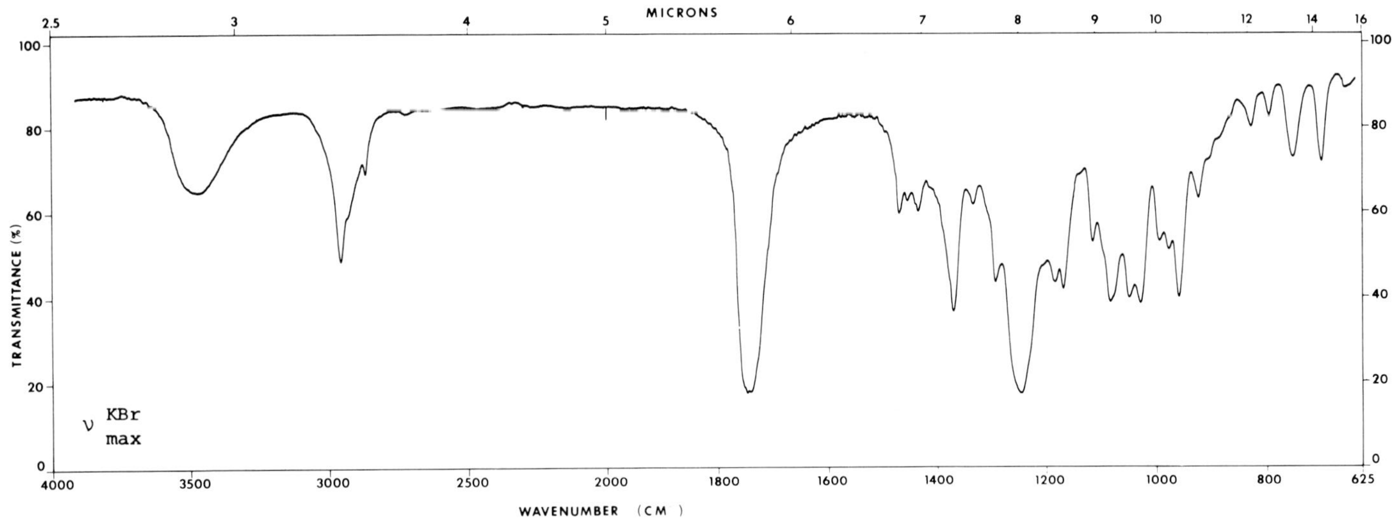
MICRONS
TRANSMITTANCE (%)
WAVENUMBER (CM)
ν KBr max

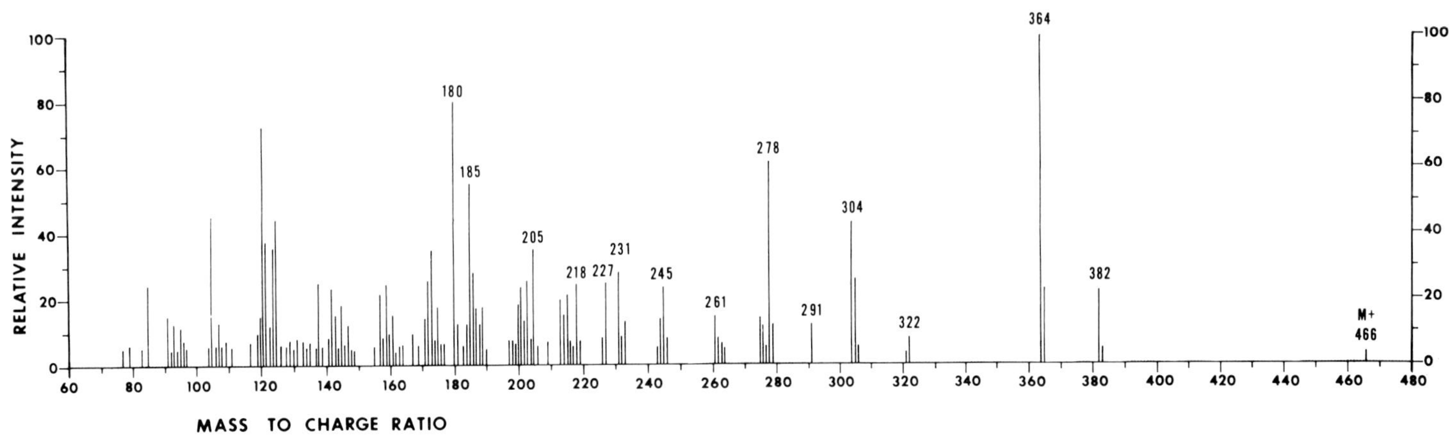
RELATIVE INTENSITY
MASS TO CHARGE RATIO
180
185
205
218
227
231
245
261
278
291
304
322
364
382
M+
466

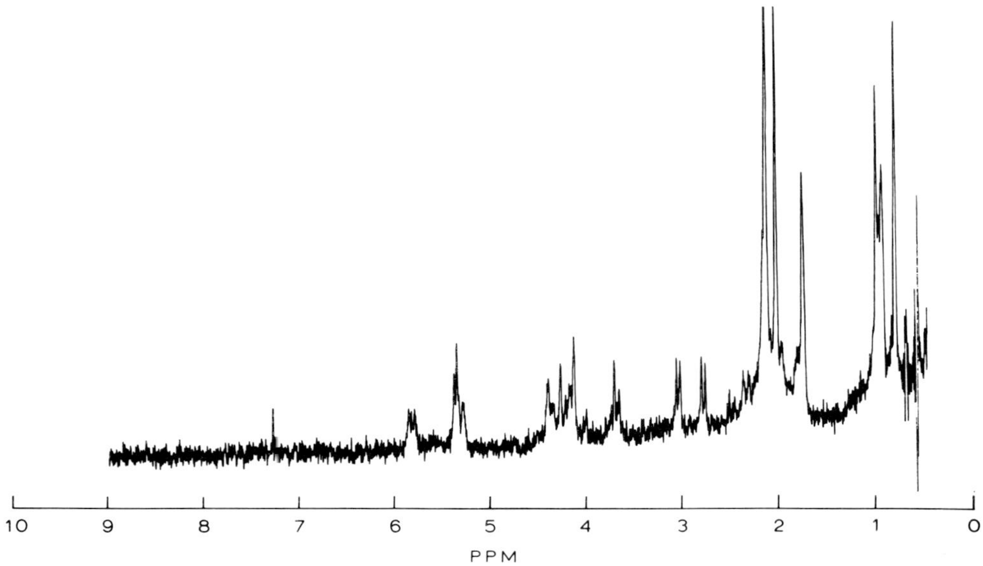

The 100 MHz ^1H NMR spectrum of T-2 toxin in $CDCl_3$ solution.

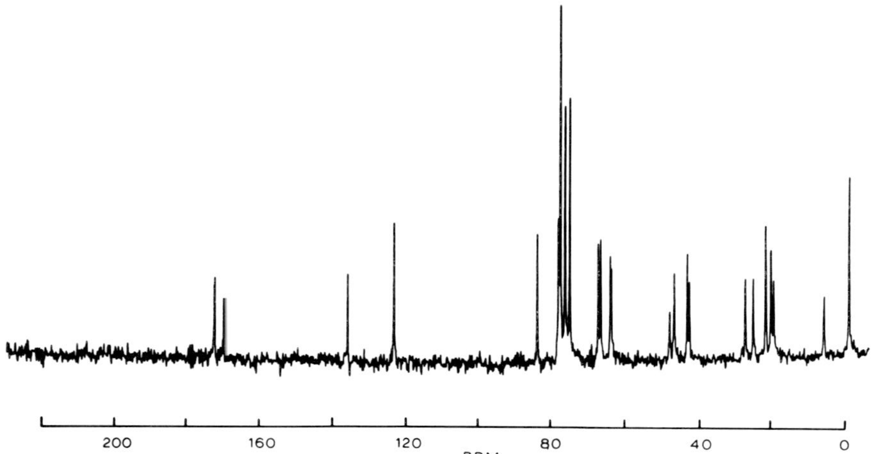

The 25 MHz ^{13}C NMR spectrum of T-2 toxin in $CDCl_3$ solution.

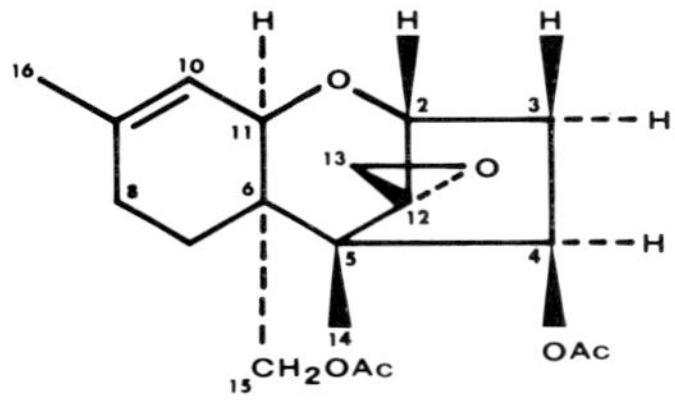

Common name	4,15-Diacetylverrucarol
Molecular weight	350.1722
Molecular formula	$C_{19}H_{26}O_6$
General characteristics[69]	White prisms from benzene–methanol, mp 147°–148°C $[\alpha]_D^{25}$ −14° (*C* = 1.01 in methanol) Readily soluble in most organic solvents such as methanol, acetone, chloroform, ether, and benzene; sparingly soluble in hexane and petroleum ether
UV data[69]	End absorption
IR data[69]	Spectrum presented in Okuchi *et al.*[69]
Fungal source[69]	*Myrothecium* spp.
Toxicity data[69]	Antifungal activity against *Trichophyton asteroides* (ED_{50} 0.1 μg/ml) and *T. interdigitale* (ED_{50} 10 μg/ml)
^{13}C NMR data[65]	C2, 79.0; C3, 36.7; C4, 75.4; C5, 48.9; C6, 43.4; C7, 21.4; C8, 28.1; C9, 138.7; C10, 119.7; C11, 66.8; C12, 65.2; C13, 47.3; C14, 6.8; C15, 63.7; C16, 23.0; CH_3CO, 20.2 (2), 169.6, 169.8
^1H NMR data[65]	H2, 3.80; H3, NR; H4, 3.24; H7, NR; H8, NR; H10, 5.39; H11, 3.75; H13a, 2.79; H13b, 3.09; H14, 0.79; H15a, 4.05; H15b, 4.15; H16, 1.70

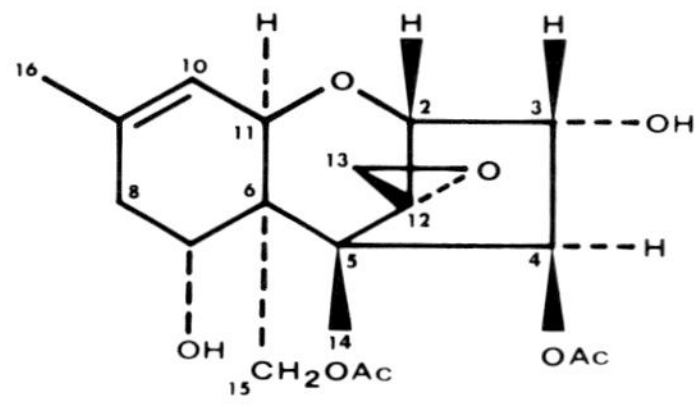

Common name	7α-Hydroxydiacetoxyscirpenol
Molecular weight	382.1620
Molecular formula	$C_{19}H_{26}O_8$
General characteristics[52]	Needles from chloroform–hexane, mp 201°–203°C Tetraacetoxy derivative Needles from benzene–*n*-hexane, mp 147°–149°C
UV data[52]	End absorption
IR data[52]	ν_{max}^{KBr} cm^{-1}: 3500, 2950, 1720, 1380, 1280, and 1060
Fungal source[52,91]	*Fusarium lateritium* *Fusarium* spp.
Toxicity data[91]	The LD_{50} to mice dosed IP was about 3.5 mg/kg, ID_{50} of protein synthesis in rabbit reticulocytes was 0.3 μg/ml.
TLC data	*Adsorbent:* silica gel *Solvent*: acetone–hexane, 1:1 v/v R_f: 0.5 *Detection*: spraying with 20% H_2SO_4 and heating at 110°C
^{1}*H NMR data*[52]	H2, 3.70; H3, 4.25; H4, 5.20; H7, 4.62; H8a, 2.00; H8b, 2.43; H10, 5.50; H11, 4.25; H13a, 3.07; H13b, 3.17; H14, 1.13; H15a, 4.24; H15b, 4.42; H16, 1.75; CH_3CO, 2.07, 2.15

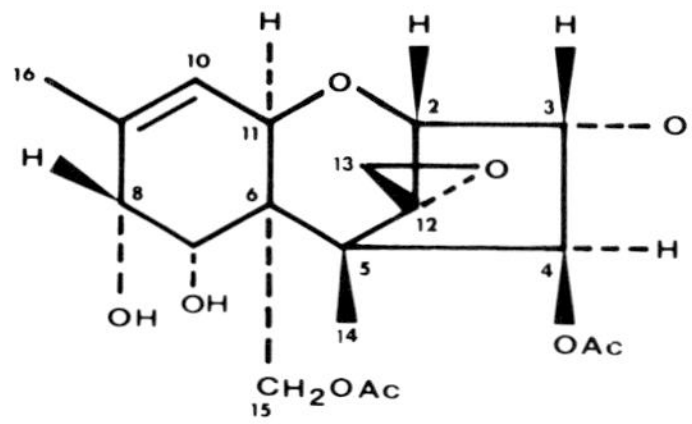

Common name	7α,8α-Dihydroxydiacetoxyscirpenol
Molecular weight	398.1569
Molecular formula	$C_{19}H_{26}O_9$
General characteristics[52]	Needles from ethanol–ether–*n*-hexane, mp 167°–169°C
UV data[52]	End absorption
IR data[52]	ν_{max}^{KBr} cm^{-1}: 3500, 2950, 1750, 1390, 1260, and 1050
Toxicity data[91]	The LD_{50} to mice dosed IP was about 6.0 mg/kg; ID_{50} of protein synthesis in rabbit reticulocytes was 0.6 μg/ml.
Fungal source[52,91]	*Fusarium lateritium* *Fusarium* spp.
TLC data	*Adsorbent*: silica gel *Solvent*: acetone–*n*-hexane, 1:1 v/v R_f: 0.38 *Detection*: spraying with 20% H_2SO_4 and heating at 110°C
^{1}*H NMR data*[52]	H2, 3.70; H3, 4.20; H4, 5.45; H7, 4.01; H8, 4.50; H10, 5.65; H11, 4.50; H13a, 3.05; H13b, 3.15; H14, 1.10; H15a, 4.29; H15b, 4.50; H16, 1.85; CH_3CO, 2.02, 2.13

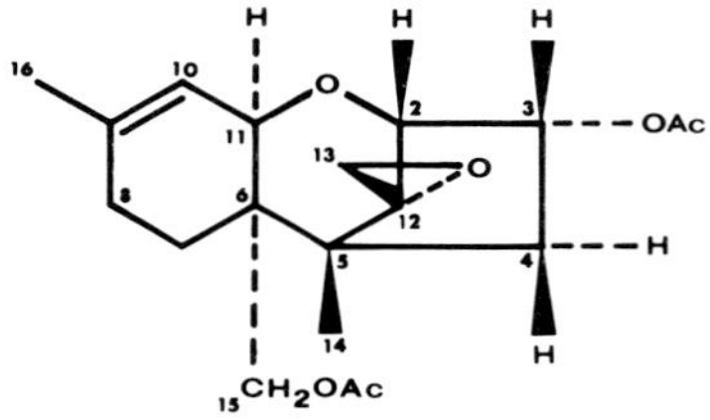

Common name	Calonectrin (3α,15-Diacetoxy-12,13-epoxytrichothec-9-ene)
Molecular weight	350.1722
Molecular formula	$C_{19}H_{26}O_6$
General characteristics[35]	Prisms from ether–petroleum ether, mp 83°–85°C $[\alpha]_D^{27}$ −14.6° (*C* = 1.0 in chloroform)
IR data[35]	ν_{max}^{KBr} cm^{-1}: 1745, 1725, 1240, and 965
Fungal source[35]	*Calonectria nivalis* Schaffnit
^{13}C *NMR data*[48]	C2, 78.1; C3, 71.3; C4, 39.4; C5, 45.4; C6, 43.0; C7, 21.0; C8, 28.2; C9, 140.3; C10, 119.0; C11, 68.2; C12, 65.0; C13, 48.5; C14, 12.2; C15, 63.7; C16, 23.2
^1H *NMR data*[35]	H2, 3.75; H3, 5.28; H4, NR; H7, NR; H8, NR; H10, 5.43; H11, 4.00 H13a, 2.86; H13b, 3.10; H14, 0.83; H15a, 3.82; H15b, 4.12; H16, 1.73; CH_3CO, 2.03, 2.11

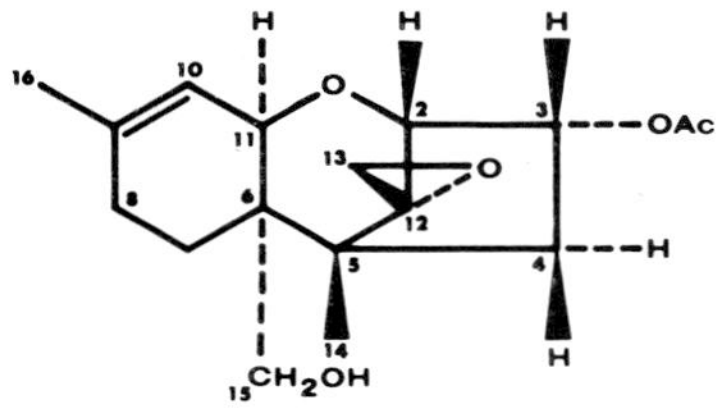

Common name	15-Deacetylcalonectrin (3α-Acetoxy-15-hydroxy-12,13 epoxytrichothec-9-en)
Molecular weight	308.1617
Molecular formula	$C_{17}H_{24}O_5$
General characteristics[35]	Prisms from ether–light petroleum, mp 184°–186°C $[\alpha]_D^{27}$ +11.2° (*C* = 1.0 in chloroform)
IR data[35]	ν_{max} cm^{-1}: 3550, 1740, 1235, and 960
Fungal source[35]	*Calonectria nivalis* Schaffnit
^1H NMR data[35]	H2, 3.73; H3, 5.18; H7, NR*; H8, NR; H10, 5.47; H11, 3.98; H13a, 2.86; H13b, 3.10; H14, 0.92; H15a, 3.42; H15b, 3.70; H16, 1.73; CH_3CO, 2.09

*NR indicates data not reported.

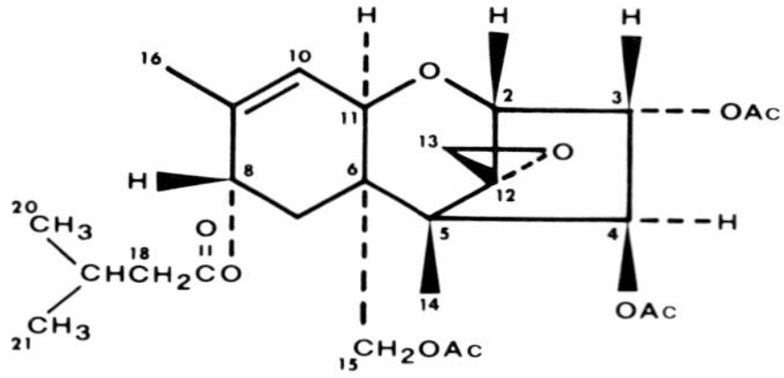

Common name	Acetyl T-2 toxin (3,4,15-triacetoxy-8-[3-methylbutryloxy]/12,13-epoxytrichothec-9-ene)
Molecular weight	508.2298
Molecular formula	$C_{26}H_{36}O_{10}$
General characteristics[57]	Oil
UV data[57]	End absorption
Fungal source[57]	*Fusarium poae* (NRRL 3287)
Toxicity data[57]	Acetyl T-2 toxin was much less toxic than T-2 toxin in pigeons. Causes emesis at higher dosages (18.2 mg/kg oral). No deaths occurred at this dosage level.
TLC data[57]	*Adsorbent*: Brinkman F_{254} silica gel *Solvent*: ethyl acetate–ethanol, 6:1 v/v R_f: 0.74 *Detection*: a quenching spot under UV light
^{13}C NMR data[25]	C2, 76.2; C3, 78.7; C4, 78.2; C5, 47.7; C6, 42.0; C7, 26.7; C8, 67.2; C9, 135.0; C10, 124.0; C11, 66.3; C12, 63.6; C13, 41.4; C14, 6.1; C15, 63.6; C16, 19.4; C17, 172.2; C18, 42.4; C19, 25.3; C20, 21.7; C21, 21.7; CH_3CO, 20.3 (3), 170.6; 170.1 (2)
^1H NMR data	H2, 3.78; H3, 5.07; H4, 5.85; H7, 1.85-2.1; H8, 5.20; H10, 5.65; H11, 4.14; H13a, 2.75; H13b, 2.95; H14, 0.65; H15a, 4.00; H15b, 4.27; H16, 1.66; H18, 1.93; H20, 0.96; H21, 0.96

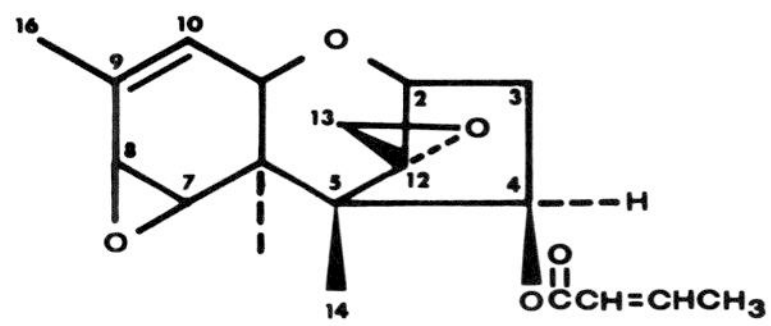

Common name[37]	Crotocin (Antibiotic T)
Molecular weight	332.1617
Molecular formula	$C_{19}H_{24}O_5$
General characteristics[2,46,47]	Colorless prisms from methanol, mp 126°–128°C Needles from ether–benzene, mp 116°–118°C $[\alpha]_D^{25}$ +7.17° (C = 3.98 in chloroform)
UV data[2,46,47]	λ_{max} nm (ϵ): 211(21,800) and 209
IR data[2,46,47]	ν_{max} cm^{-1}: 1710 and 1640
Fungal source[2,46,47]	*Cephalosporium crotocinigenum* *Trichothecium roseum*
Toxicity data[2,7,46,47]	LD_{50} of crotocin in mice dosed IV was 700 mg/kg; fungistatic; negligible antitumor activity.
TLC data[4]	*Adsorbent*: silica gel *Solvent:* ethanol–ethyl acetate–acetone, 1:4:4 v/v/v R_f: 0.67 *Detection*: 50% H_2SO_4 spraying and heating at 110°C for 5 min
^{13}*C NMR data*[25]	C2, 74.3; C3, 36.2; C4, 78.5; C5, 47.7; C6, 41.3; C7, 58.1; C8, 50.4; C9, 137.2; C10, 122.9; C11, 69.4; C12, 65.9; C13, 46.8; C14, 6.2; C15, 15.9; C16, 21.0; C17, 166.4; C18, 120.2; C19, 146.5; C20, 14.8

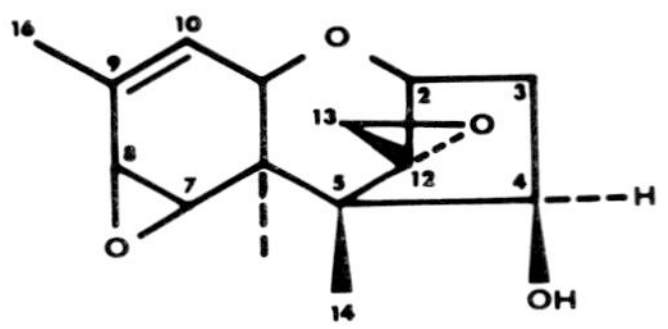

Common name	Crotocol
Molecular weight	264.1361
Molecular formula	$C_{15}H_{20}O_4$
General characteristics[47]	Crystals, mp 154°C $[\alpha]_D^{20}$ −6.4° (C = 2.01 in chloroform) Acetate derivative: mp 128°C
UV data[47]	λ_{max}^{EtOH} nm (ϵ): 210(4,830)
Source[47]	Formed upon mild alkaline hydrolysis of crotocin
TLC data[4]	*Adsorbent*: silica gel *Solvent*: ethanol–ethyl acetate–acetone, 1:4:4 v/v/v R_f: 0.50 *Detection*: 50% H_2SO_4 spraying and heating at 110°C for 5 min

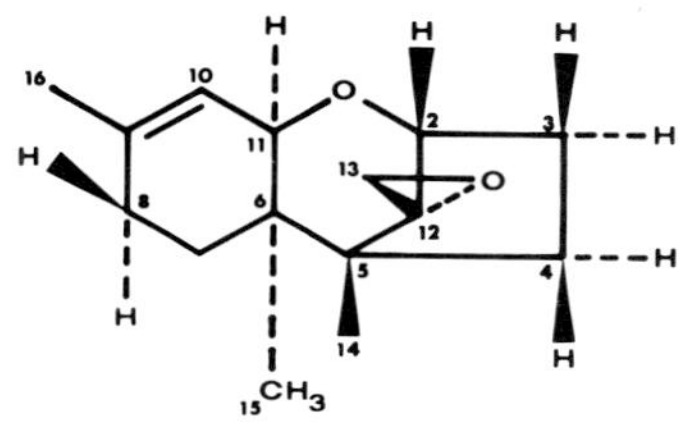

Common name	Trichothecene (12,13-Epoxytrichothec-9-ene)
Molecular weight	234.1614
Molecular formula	$C_{15}H_{22}O_2$
Fungal source[34,61]	*Trichothecium roseum*
Other significant references	Synthesis[34,61]

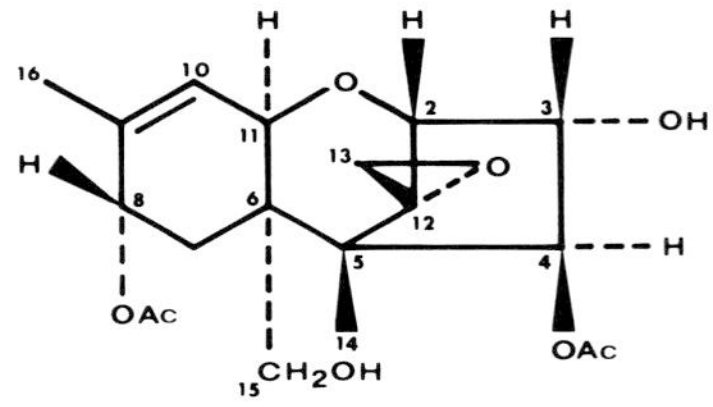

Common name	4β,8α-Diacetoxy-3α,15-dihydroxy-12,13-epoxy-trichothec-9-en
Molecular weight	382.1467
Molecular formula	$C_{19}H_{26}O_8$
UV data[51]	End absorption
IR data[51]	ν_{max}^{KBr} cm^{-1}: 3450, 2940, 1720, and 1230
Fungal source[51]	*Fusarium tricinctum* (Corda) Saccardo *F. equiseti*
Toxicity data[51]	LD_{50} in rats dosed IP was 1.2 mg/kg
TLC data[51]	*Adsorbent:* silica gel *Solvent:* ethyl acetate–petrol, 3:1 v/v R_f: 0.4 *Detection:* skyblue fluorescence in UV after spraying with *p*-anisaldehyde and heating
^1H NMR data[51]	H2, 3.66; H3, 4.24; H4, 5.40; H7a, 1.96; H7b, 2.34; H8, 5.36; H10, 5.82; H11, 4.27; H13a, 2.80; H13b, 3.04; H14, 0.84; H15a, 3.60; H15b, 3.91; H16, 1.75; CH_3CO, 2.09, 2.14

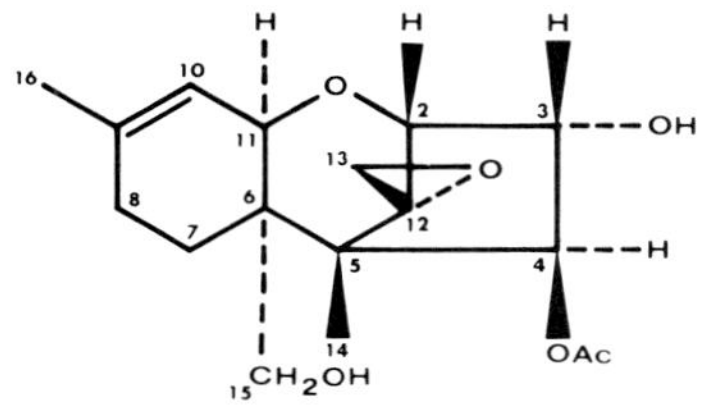

Common name	4-Acetoxyscirpenediol
Molecular weight	324.1572
Molecular formula	$C_{17}H_{24}O_6$
General characteristics[53]	Amorphous solid from benzene–hexane, mp 100°–110°C; mass spectrum showed major fragments at *m/e* 306, 278, 234, 219, and 43
UV data[53]	End absorption
IR data[53]	ν_{max} cm^{-1}: 3450, 2950, 1720, 1435, 1375, 1240, 1165, 1110, 1080, 1050, and 960
Fungal source[53,78]	*Fusarium roseum* Gibbosum *F. sulphureum*
Toxicity data	Caused dermal necrosis in white rats

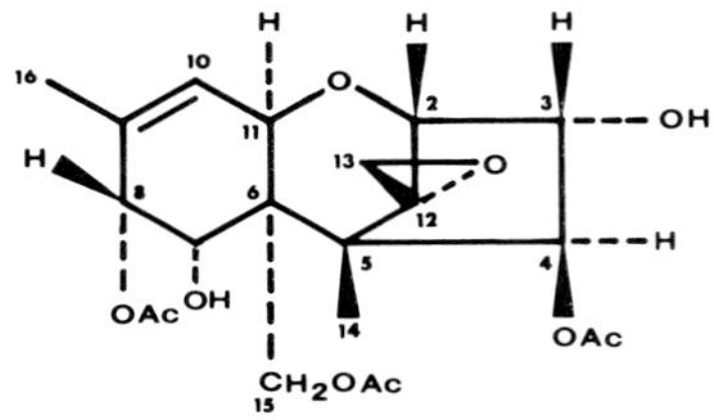

Common name	4β,8α,15-Triacetoxy-3α,7α-dihydroxy-12,13-epoxytrichothec-9-en
Molecular weight	440.1681
Molecular formula	$C_{21}H_{28}O_{10}$
General characteristics[40]	Crystals, mp 185°–188°C
UV data[40]	End absorption
Fungal source[40]	*Fusarium equiseti*
Toxicity data[40]	LD_{50} in rats dosed IP was 1.2 mg/kg

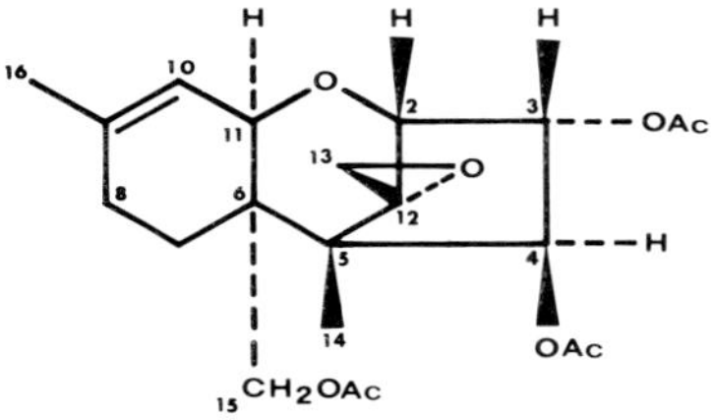

Common name	Triacetoxyscirpenol
Molecular weight	408.1776
Molecular formula	$C_{21}H_{28}O_8$
General characteristics	Colorless crystals from benzene–*n*-hexane, mp 124°–126°C
UV data	End absorption
Fungal source[78]	*Fusarium sulphureum*

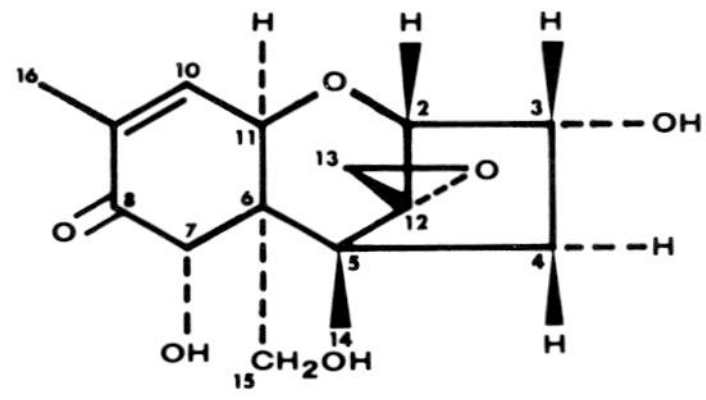

Common name[66,96–99]	Deoxynivalenol (Rd toxin) (Vomitoxin) (3α, 7α,15-Trihydroxy-12,13-epoxytrichothec-9-en-8-one)
Molecular weight	296.1254
Molecular formula	$C_{15}H_{20}O_6$
General characteristics[101]	Needles from ethyl acetate–petroleum ether, mp 151°–153°C Semicarbazone derivative, mp 199°–201°C Triacetate: colorless needles from ethyl acetate–petroleum ether, mp 155°–157°C $[\alpha]_D^{25}$ +6.35° (C = 0.07 in ethanol)
UV data[101]	λ_{max}^{EtOH} nm (ϵ): 218(4,500) λ_{max}^{EtOH} nm (ϵ): semicarbazone derivative: 268(18,200)
IR data	ν_{max}^{KBr} cm^{-1}: 3470, 3430, 3350, and 1680
Fungal source[101]	*Fusarium roseum*
Toxicity data[60,66,98,104]	LD_{50} in male mice (ddY strain) dosed IP was 70 mg/kg; females, 76.7 mg/kg. Also inhibited the multiplication of culture cells of *Tetrahymena pyriformis* at same level as fusarenon-X.
TLC data[101,102]	A *Adsorbent*: silica gel G *Solvent*: chloroform–methanol, 97:3 v/v R_f: 0.09 *Detection*: H_2SO_4 spraying and heating at 110°C for 5 min B *Adsorbent*: silica gel G *Solvent*: ethyl acetate–toluene, 3:1 v/v R_f: 0.21 *Detection*: H_2SO_4 spraying and heating at 110°C for 5 min
^1H NMR data	H2, 3.66; H3, 4.63; H4, 2.20; H7, 4.95; H10, 6.79; H11, 4.50; H13a, 3.14; H13b, 3.25; H14, 1.15; H15a, 3.66; H15b, 3.98; H16, 1.92

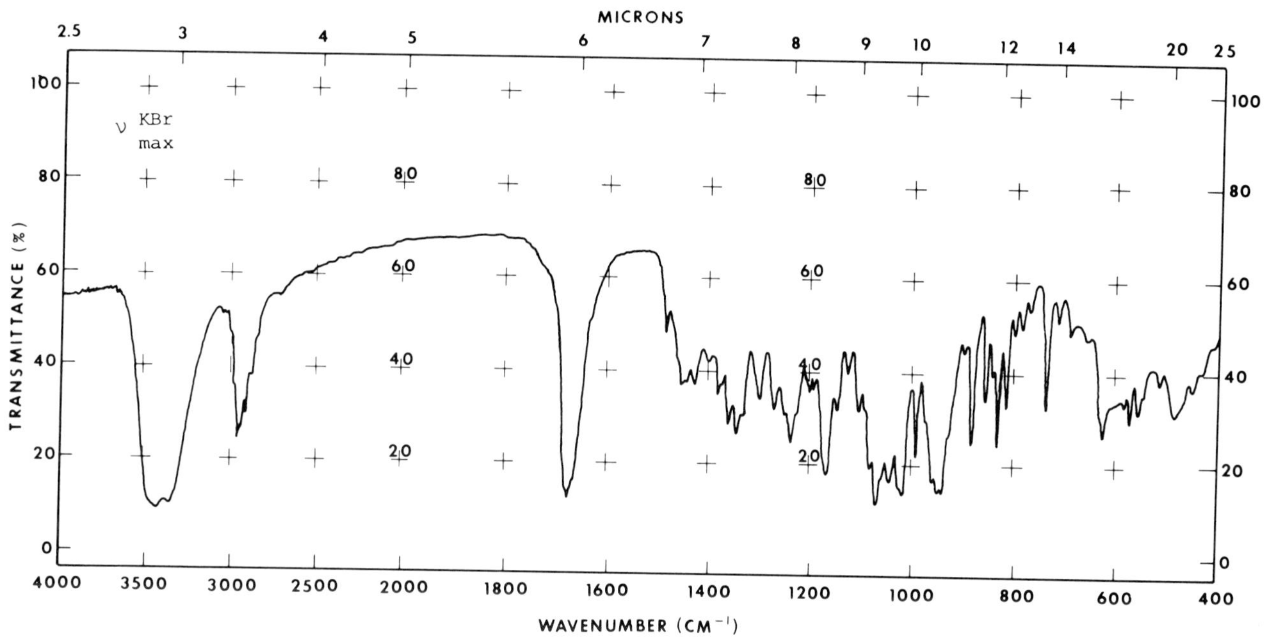
MICRONS
TRANSMITTANCE (%)
WAVENUMBER (CM⁻¹)
ν KBr max

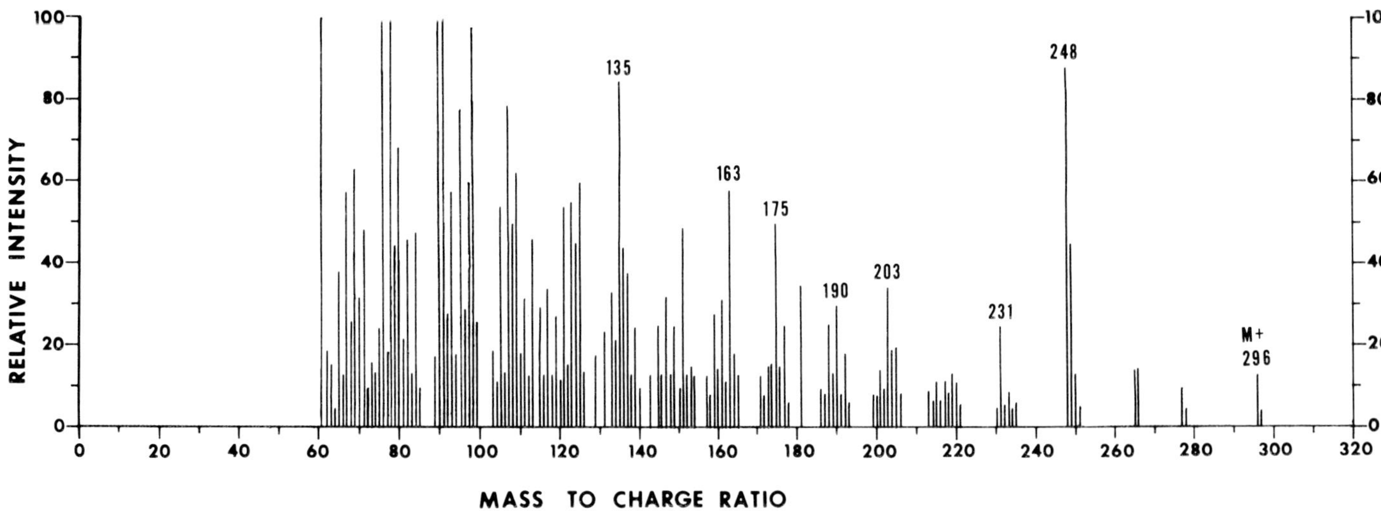
RELATIVE INTENSITY
MASS TO CHARGE RATIO
135
163
175
190
203
231
248
M+
296

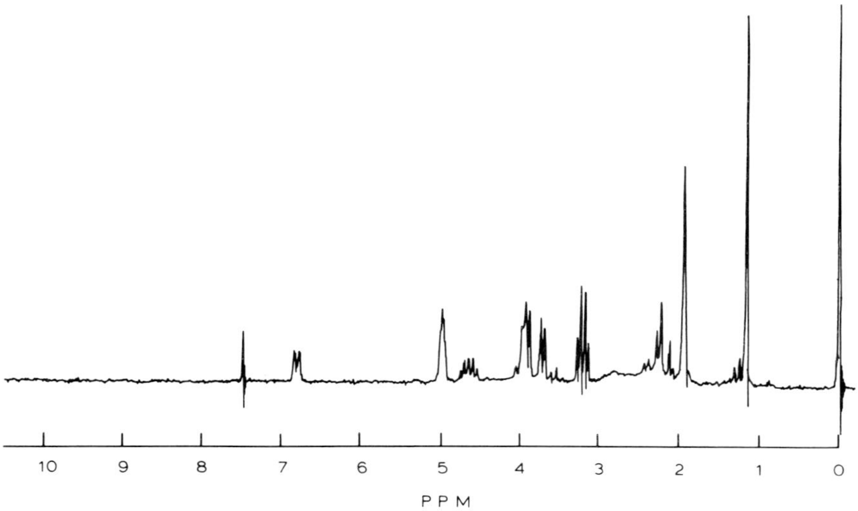

The 100 MHz ^{1}H NMR spectrum of deoxynivalenol in $CDCl_3$ solution.

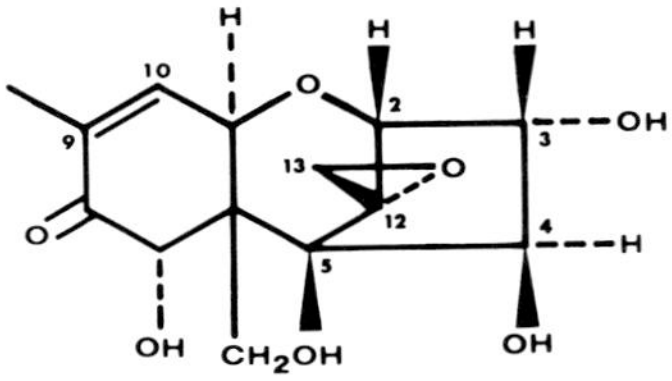

Common name[66,85] — Nivalenol (3α,4β,7α,15-Tetrahydroxy-12,13-epoxytrichothec-9-en-8-one)

Molecular weight — 312.1203

Molecular formula — $C_{15}H_{20}O_7$

General characteristics[85] — Crystals from methanol, mp 80°–90°C; dried in presence of P_2O_5 in reduced pressure, mp 222°–223°C (dec.)
$[\alpha]_D^{24}$ +21.54° (C = 1.3 in ethanol)
Tetraacetate: mp 168°–170°C

UV data[33,80] — λ_{max}^{MeOH} nm (ϵ): 218(7,500)
λ_{max}^{MeOH} nm (ϵ): tetraacetate: 227(7,900)

Fungal source[1] — *Fusarium nivale*

Toxicity data[16,17,20,42,85,86,89,92,94,100] — LD_{50} in mice dosed IP was 50 μg/10 gm (injures proliferating cells). Minimum SC dosage to induce vomiting in ducklings was 1.0 mg/kg. 1–10 μg/ml (Ehrlich ascites tumor cells) inhibited uptake of [^{14}C]leucine and [^{14}C]thymidine into protein and DNA. However, uptake of [^{14}C]uracil into RNA was not affected. Inhibition of protein synthesis occurs before inhibition of DNA synthesis. It inhibits the initiation step of protein synthesis on polyribosomes. Dermal toxicity in rabbit and guinea pig.

TLC data[4,85,92] —
A
Adsorbent: Kieselgel G
Solvent: chloroform–methanol, 5:1 v/v
R_f: 0.45
Detection: brown color with H_2SO_4 spraying and heating at 110°C for 5 min
B
Adsorbent: silica gel G
Solvent: toluene–ethyl acetate, 1:3 v/v

	R_f: 0.09 *Detection*: brown color with H_2SO_4 spraying and heating at 110°C for 5 min
GLC data[83] (as TMS derivative)	*Support*: Shimalite W *Liquid phase*: OV-17 (1.5%) *Retention time*: 3.55
^1H NMR data[33]	H2, 3.50; H3, NR; H4, 4.37; H7, 4.70; H10, 6.58; H11, 4.57; H13a, 2.87; H13b, 2.97; H14, 0.95; H15, 3.64; H16, 1.77

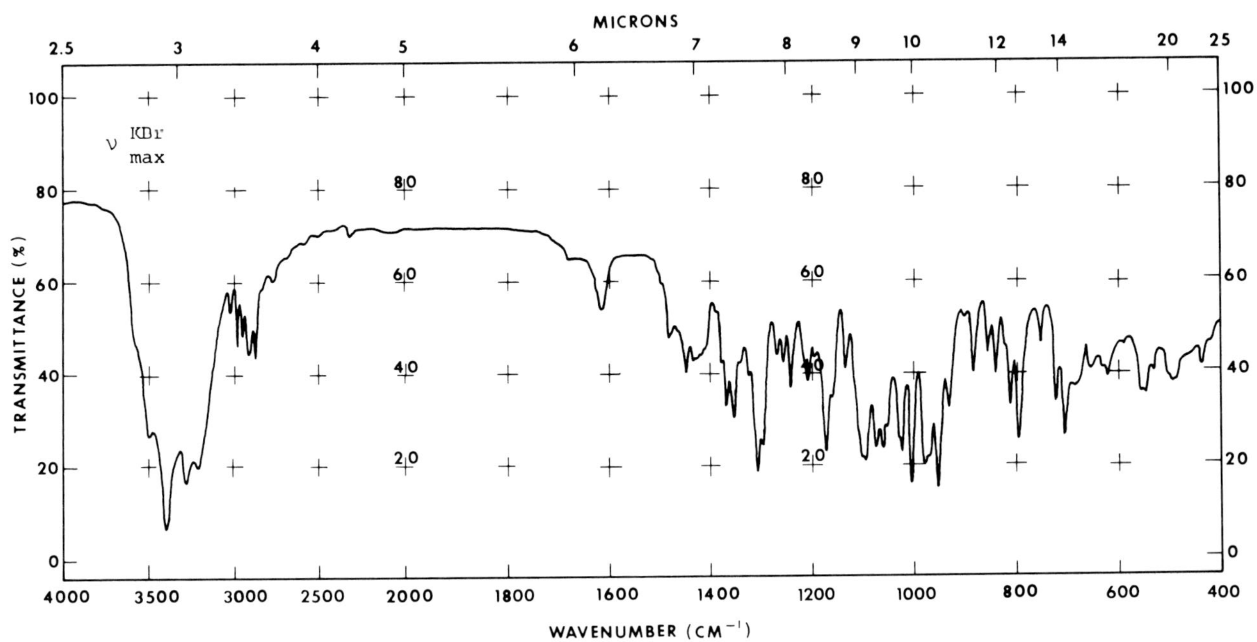
MICRONS
2.5 3 4 5 6 7 8 9 10 12 14 20 25
TRANSMITTANCE (%)
100 80 60 40 20 0
ν KBr max
WAVENUMBER (CM⁻¹)
4000 3500 3000 2500 2000 1800 1600 1400 1200 1000 800 600 400

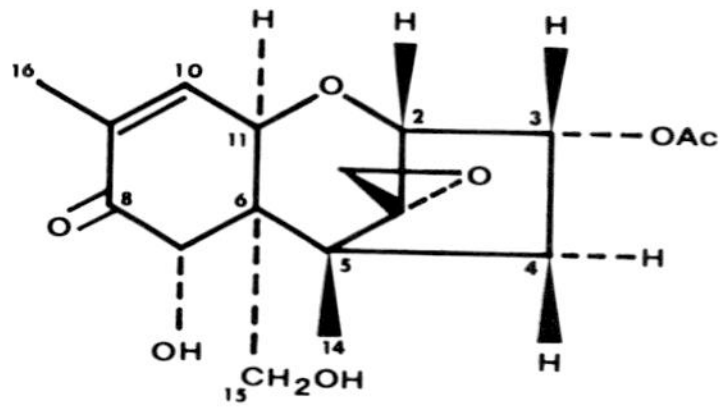

Common name	Deoxynivalenol monoacetate (3*a*-Acetoxy-7*a*,15-dihydroxy-12,13-epoxytrichothec-9-en-8-one)
Molecular weight	338.1359
Molecular formula	$C_{17}H_{22}O_7$
General characteristics[11,101]	Crystallized from ethyl ether–*n*-pentane, mp 185.5°–186°C $[\alpha]_D^{20}$ +430° (C = 0.28 in methanol)
UV data[101]	λ_{max}^{EtOH} nm (ϵ): 219(5,900)
IR data	ν_{max}^{KBr} cm^{-1}: 3480, 3400, 1740, and 1680
Fungal source[11,101]	*Fusarium roseum* *F. culmorum*
Toxicity data[101]	LD_{50} in ddS strain of male mice dosed IP was 49.4 mg/kg; females, 49.9 mg/kg.
TLC data	A *Adsorbent*: silica gel G *Solvent*: chloroform–methanol, 97:3 v/v R_f: 0.40 *Detection*: H_2SO_4 spraying and heating at 110°C for 5 min B *Adsorbent*: silica gel G *Solvent*: ethyl acetate–toluene, 3:1 v/v R_f: 0.43 *Detection*: H_2SO_4 spraying and heating at 110°C for 5 min
^{1}H NMR data	H2, 3.90 (J = 4.5); H3, 5.24; H4, 2.25; H7, 4.84; H10, 6.60 (J = 5.5, 1.0); H11, 4.70; H13a, 3.10 (J = 4.0); H13b, 3.18 (J = 4.0); H14, 1.16; H15a, 3.76 (J = 12.0); H15b, 3.91 (J = 12.0); H16, 1.90; CH_3CO, 2.14

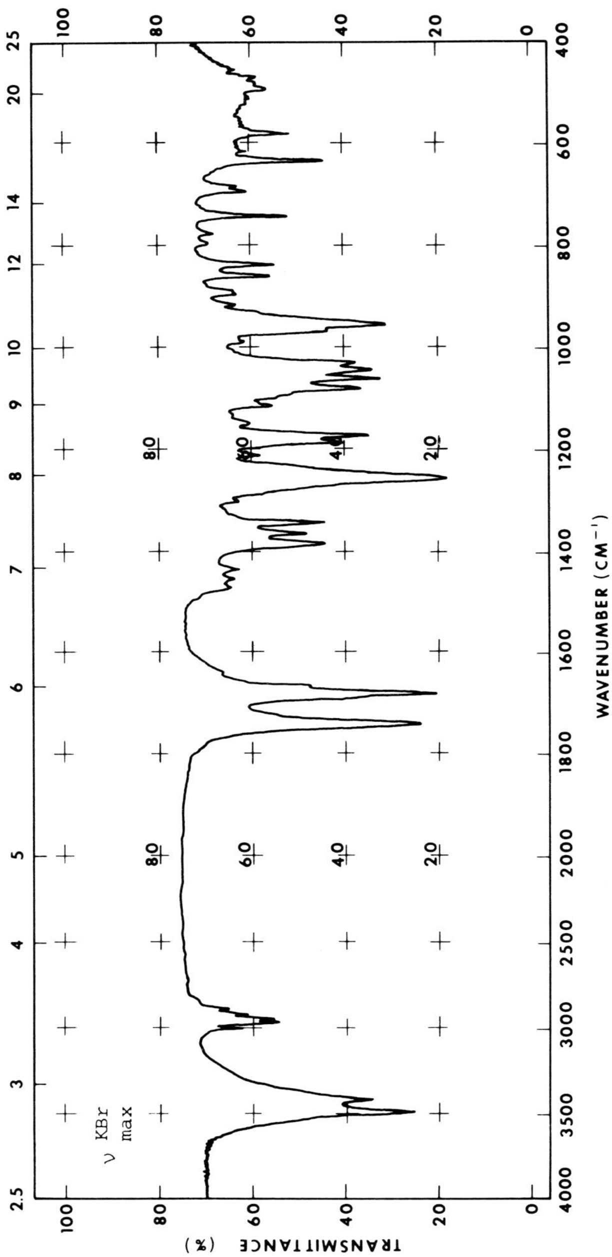
MICRONS
WAVENUMBER (CM⁻¹)
TRANSMITTANCE (%)
ν KBr max

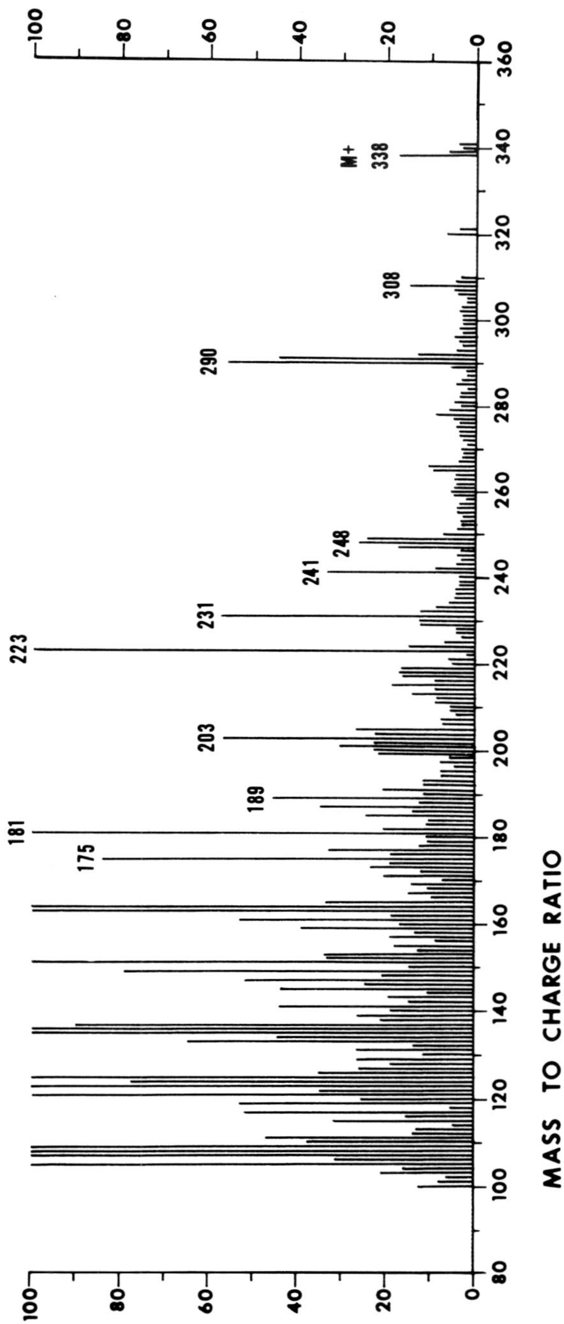
RELATIVE INTENSITY
MASS TO CHARGE RATIO
0
20
40
60
80
100
80
100
120
140
160
180
200
220
240
260
280
300
320
340
360
175
181
189
203
223
231
241
248
290
308
M+
338

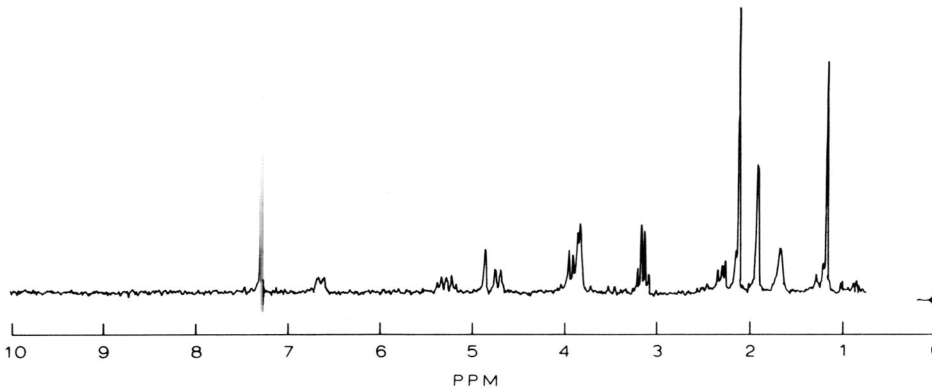

The 100 MHz ^1H NMR spectrum of deoxynivalenol monoacetate in $CDCl_3$ solution.

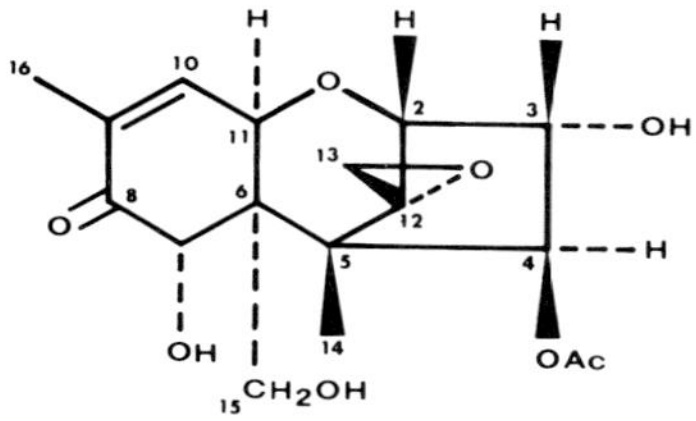

Common name[4] — Fusarenon-X (Nivalenol monoacetate; Fusarenon) (3α,7α,15-Trihydroxy-4β-acetoxy-12,13-epoxytrichothec-9-en-8-one)

Molecular weight[60] — 354.1308

Molecular formula — $C_{17}H_{22}O_8$

General characteristics[60,80,83,97] — Hexagonal bipyramid crystals from dichloromethane–*n*-pentane, mp 91°–92°C
Crystals from dichloroethane, mp 181°–182°C
$[\alpha]_D^{25}$ +58° (*C* = 1.0 in methanol)

UV data — λ_{max}^{MeOH} nm (ϵ): 220(6,500)

Fungal source[80,82] — *Fusarium nivale*
F. graminearum
F. equiseti
F. solani
F. lateritium
F. avenaceum

Toxicity data[33,90,94] — Application of 0.2–1.0 μg to dehaired back skin of rabbit, guinea pig, and mouse caused cytotoxic effects on epidermis, dermis, and hair follicles—no change in the subcutaneous tissues or muscle fibers. LD_{50} in male mice (ddY strain) dosed IP (0.9% saline) was ≃3.3 mg/kg; nonlethal doses produced mucoidal secretion around the eye 1 week after treatment. Emetic, minimum SC dose to induce vomiting in ducklings was 0.4–0.5 mg/kg; in cats, 0.3–0.5 mg/kg; in pigeons, 0.5–1.0 mg/kg (IV and PO)

TLC data[4,60,92] — A
Adsorbent: silica gel G
Solvent: chloroform–methanol, 97:3 v/v
R_f: 0.19

Detection: purple to yellow-green with H_2SO_4 spraying and heating at 110°C for 5 min

B

Adsorbent: silica gel G

Solvent: toluene–ethyl acetate, 1:3 v/v

R_f: 0.36

Detection: purple to yellow-green with H_2SO_4 spraying and heating at 110°C for 5 min

GLC data[83] (as TMS derivative)

Support: Shimalite W

Liquid phase: OV-17 (1.5%)

Retention time: 3.95

Relative retention time (relative to Nivalenol): 1.11

^{13}*C NMR data*

C2, 77.5 d; C3, 80.0 d; C4, 83.5 d; C5, 49.4 s; C6, 52.9 s; C7, 75.8 d; C8, 199.6 s; C9, 135.9 s; C10, 138.3 d; C11, 69.1 d; C12, 64.7 s; C13, 46.3 t; C14, 7.8 q; C15, 61.6 t; C16, 15.2 q; $\underline{CH_3}C=O$, 21.0 q; $CH_3\underline{C}=O$, 172.9 s

^{1}*H NMR data*

H2, 3.70; H3, 4.05; H4, 5.60; H7, 4.80; H10, 6.80; H11, 4.30; H13a, 3.05; H13b, 3.10; H14, 1.10; H15a, 3.80; H15b, 3.90; H16, 1.90; $\underline{CH_3}C=O$, 2.20

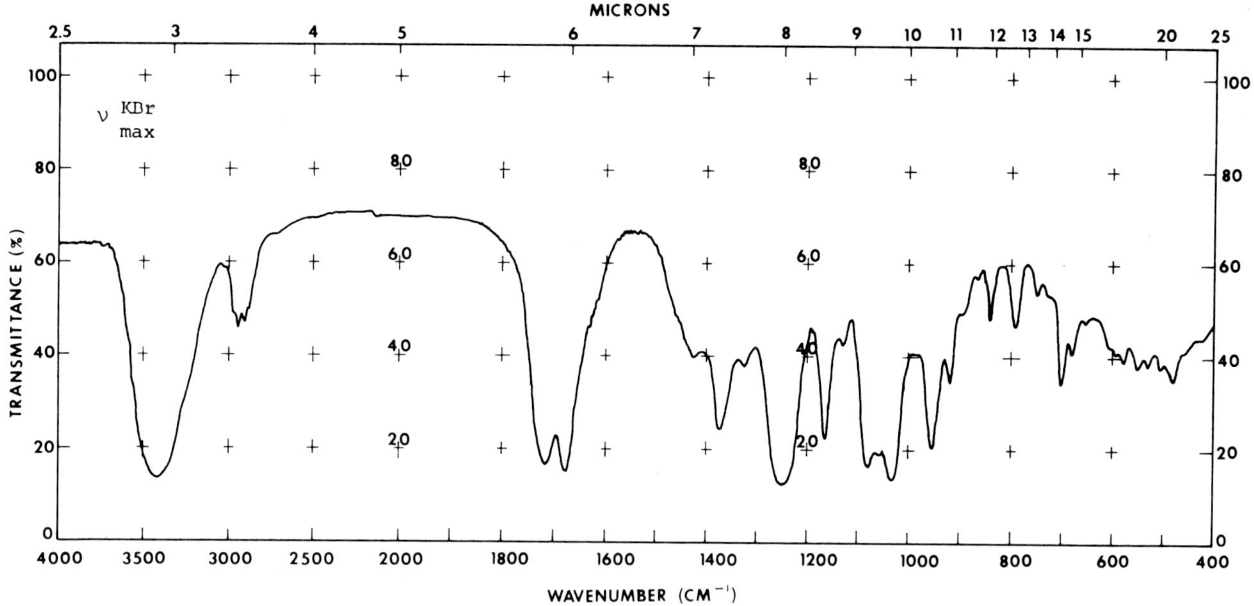
MICRONS
ν_{max}^{KBr}
TRANSMITTANCE (%)
WAVENUMBER (CM⁻¹)

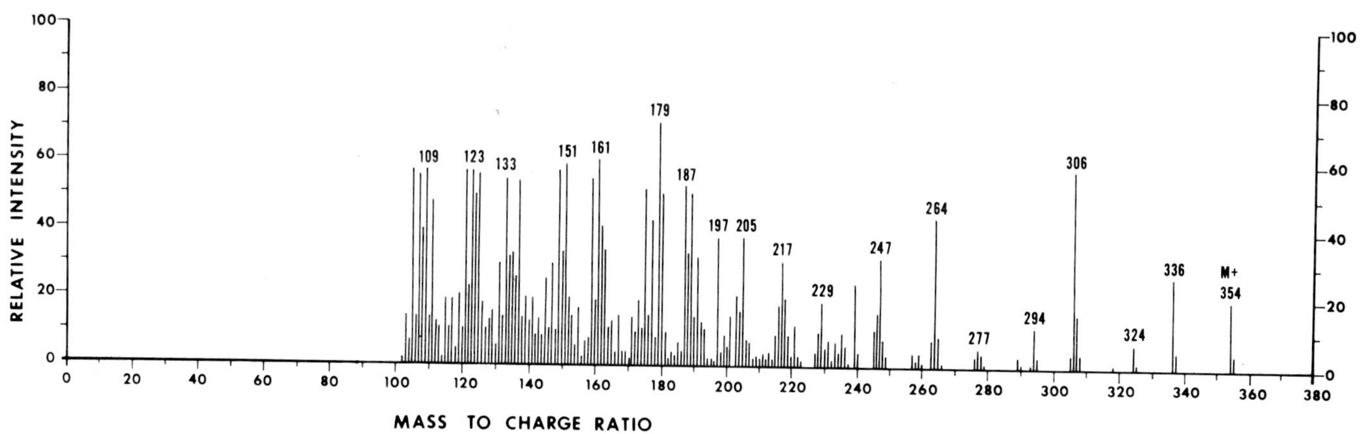
RELATIVE INTENSITY
MASS TO CHARGE RATIO
109
123
133
151
161
179
187
197
205
217
229
247
264
277
294
306
324
336
M+
354

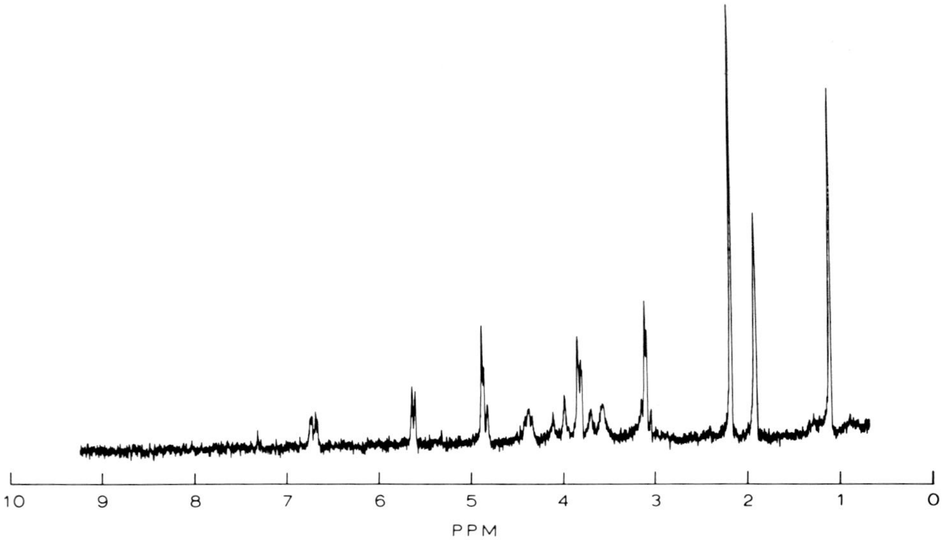

The 100 MHz ^1H NMR spectrum of fusarenon-X in $CDCl_3$ solution.

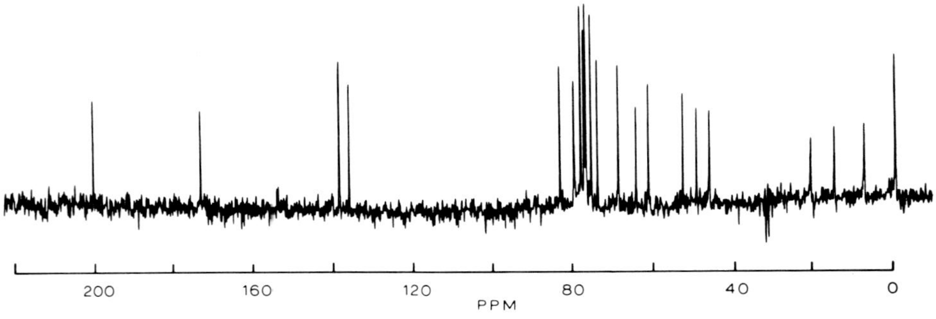

The 25 MHz ^{13}C NMR spectrum of fusarenon-X in $CDCl_3$ solution.

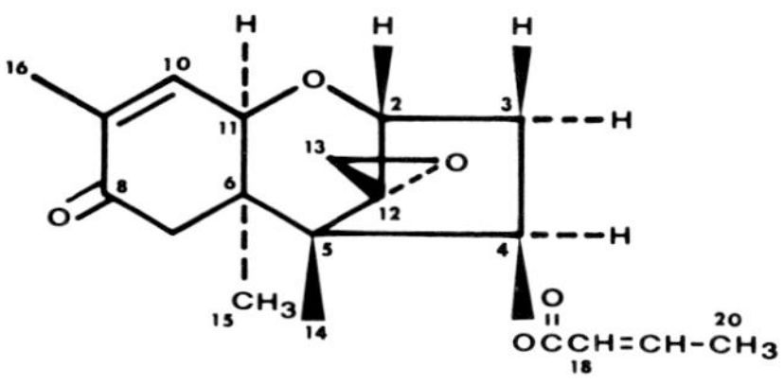

Common name[32]	Trichothecin (4β-Butyryloxy-12,13-epoxytrichothec-9-en-8-one)
Molecular weight	332.1617
Molecular formula	$C_{19}H_{24}O_5$
General characteristics[32]	mp 118°C $[\alpha]_D^{10}$ +44° (C = 1.0)
UV data[32]	λ_{max}^{MeOH} nm (ϵ): 215(19,000)
Fungal source[32]	*Trichothecium roseum* Link
Toxicity data[30,93]	LD_{50} in mice dosed IV was ~300 mg/kg; ID_{50} of protein synthesis in rabbit reticulocytes (whole cell) was 0.15 μg/ml. Inhibits the elongation or termination steps in protein synthesis on polyribosomes.
TLC data[4]	A *Adsorbent*: silica gel G *Solvent*: chloroform–methanol, 98:2 v/v R_f: 0.66 *Detection*: H_2SO_4 spraying and heating at 110°C for 5 min B *Adsorbent*: silica gel G *Solvent*: benzene–tetrahydrofuran, 85:15 v/v R_f: 0.53 *Detection*: H_2SO_4 spraying and heating at 110°C for 5 min
GLC data[83] (as TMS derivatives)	*Support:* Shimalite W *Liquid phase*: OV-17 (1.5%) *Retention time*: 8.55 *Relative retention time* (relative to Nivalenol): 2.40

Other significant references	Biosynthesis[23,55]
^{13}C NMR data	C2, 73.3 d; C3, 36.9 t; C4, 79.5 d; C5, 49.0 s; C6, 43.7 s; C7, 42.1 t; C8, 198.5 s; C9, 138.1 s; C10, 137.0 d; C11, 70.1 d; C12, 65.4 s; C13, 47.4 t; C14, 5.7 q; C15, 15.4 q; C16, 18.5 q; C1′, 166.1 s; C2′, 120.3 d; C3′, 145.9 d; C4′, 15.4 q
^{1}H NMR data	H2, 3.95; H3, 2.0-2.5; H4, 5.70; H7, 2.0-2.5; H10, 6.50; H11, 3.95; H13a, 2.80; H13b, 3.10; H14, 1.05; H15, 0.80; H16, 1.80; H2′, 5.80; H3′, 6.30; H4′, 2.10

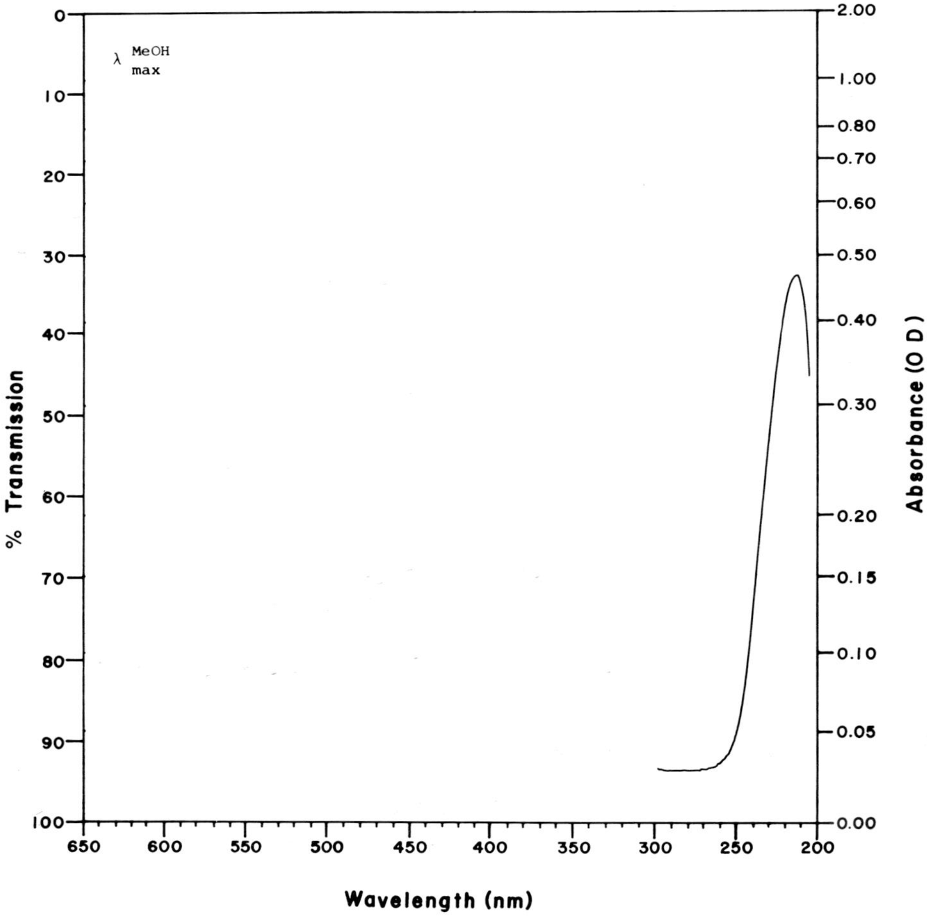

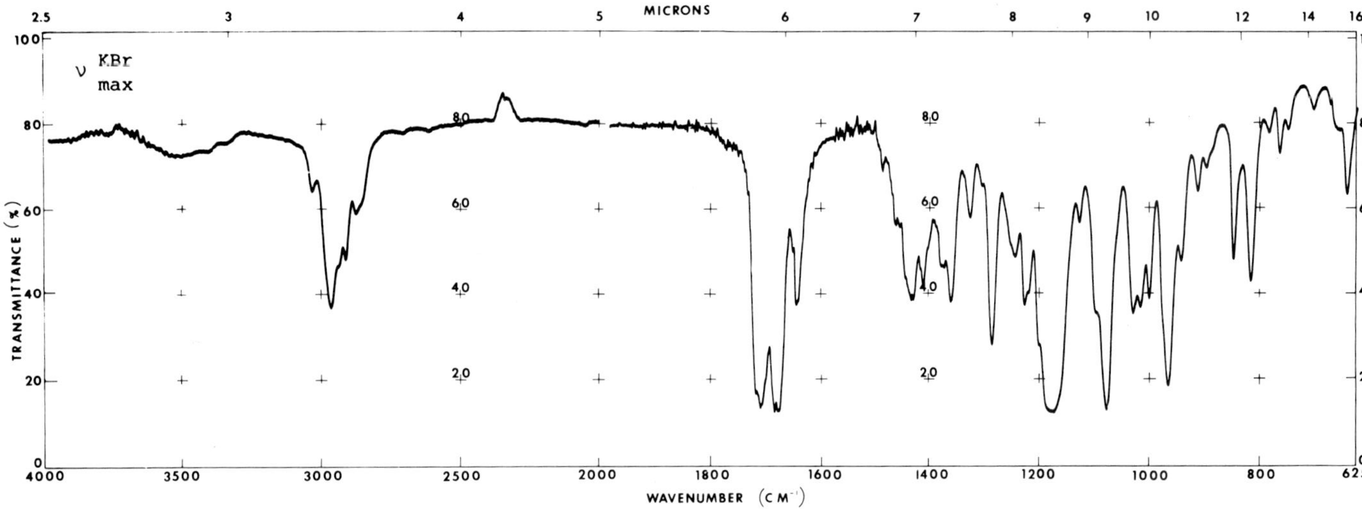
MICRONS
ν_{max}^{KBr}
TRANSMITTANCE (%)
WAVENUMBER (CM⁻¹)

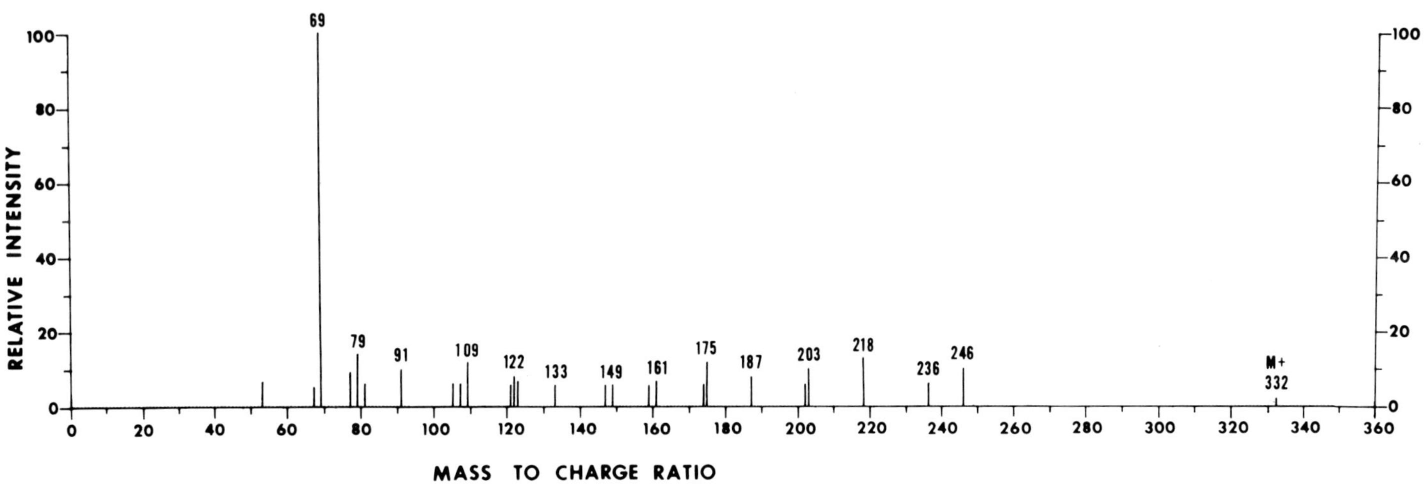
69
79
91
109
122
133
149
161
175
187
203
218
236
246
M+
332
RELATIVE INTENSITY
MASS TO CHARGE RATIO

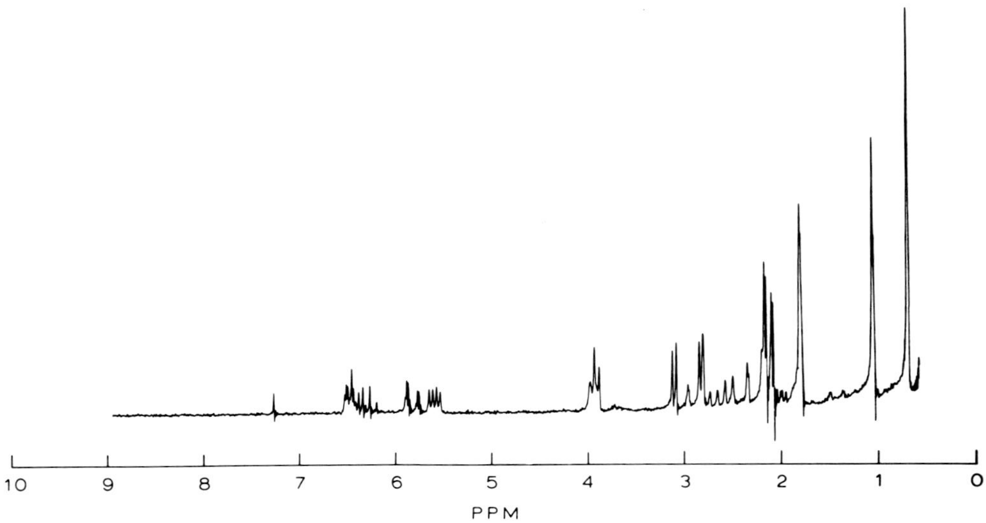

The 100 MHz ^{1}H NMR spectrum of trichothecin in $CDCl_3$ solution.

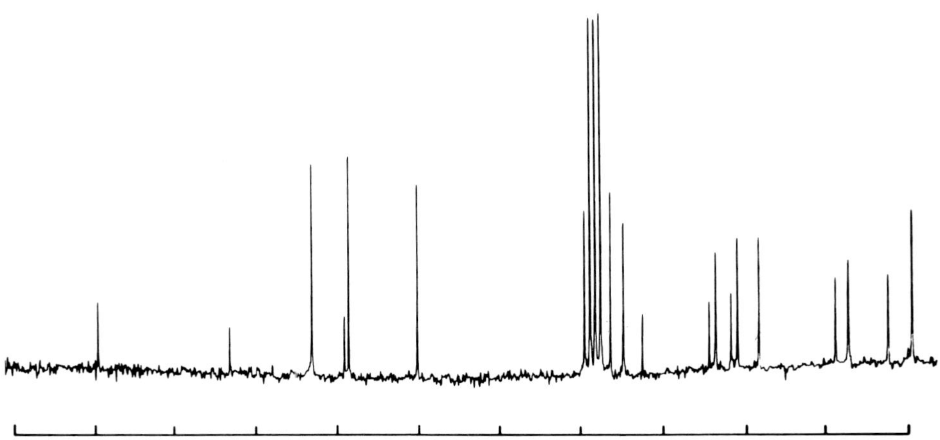

The 25 MHz ^{13}C NMR spectrum of trichothecin in $CDCl_3$ solution.

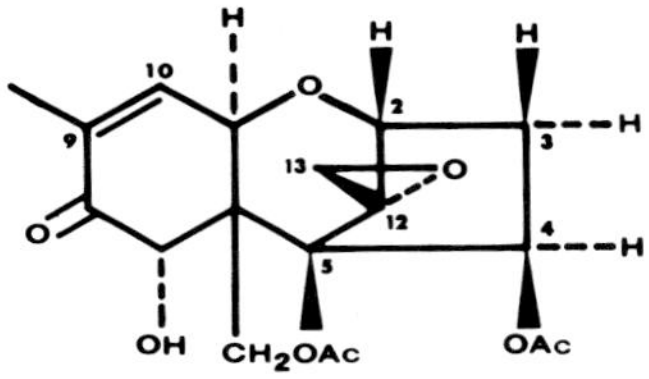

Common name	Nivalenol diacetate (3α,7α-Dihydroxy-4β,15-diacetoxyscirp-9-en-8-one)
Molecular weight	396.1419
Molecular formula	$C_{19}H_{24}O_9$
General characteristics[33,84]	Crystals from acetone–hexane, mp 135°–136°C $[\alpha]_D^{26}$ +64.3° (C = 1.0 in ethanol)
UV data[33,84]	λ_{max}^{MeOH} nm (ϵ): 220(6,200)
IR data[33,84]	ν_{max}^{KBr} cm^{-1}: 3420-3250, 2980-2920, 1735, 1680, 1235, and 1040
Fungal source[33,84,87]	*Fusarium nivale* *F. equiseti*
Toxicity data[33,84]	LD_{50} in male mice (ddY strain) dosed IP was 3.5 mg/kg.
Other significant references	^{1}H NMR[87]
^{1}H NMR data[33]	H2, 3.90; H3, 4.30; H4, 5.30; H7, 4.95; H10, 6.70; H11, 4.80; H13, 3.10; H14, 1.10; H15, 4.30; H16, 1.90; CH_3CO, 1.95, 2.20

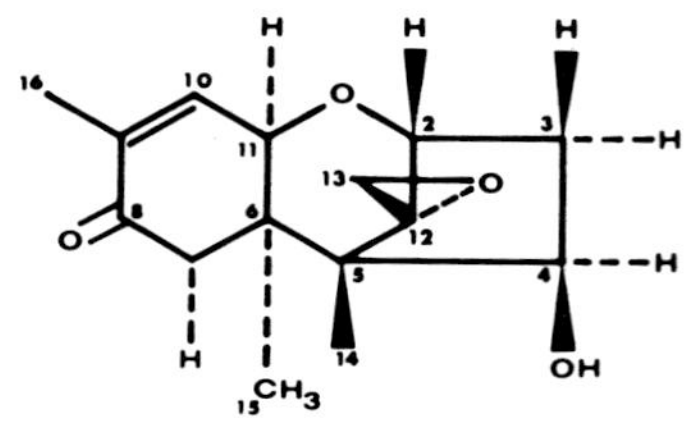

Common name	Trichothecolone
Molecular weight	264.1356
Molecular formula	$C_{15}H_{20}O_4$
General characteristics[31]	Colorless needles from benzene–light petroleum, mp 183°–184°C $[\alpha]_D^{19.5}$ +22.5° (C = 1.0 in chloroform) $[\alpha]_D^{8}$ +17.0° (C = 1.0 in ethanol) Acetyl derivative: prisms from methanol, mp 148°–149°C 2,4-Dinitrophenylhydrazone: deep orange rods from chloroform–ethanol, mp 261°–262°C
UV data[31]	λ_{max}^{MeOH} nm (ϵ): 226(8,000) λ_{max}^{MeOH} nm (ϵ): acetate derivative: 227(8,000)
IR data[31]	ν_{max}^{KBr} cm^{-1}: 3560 and 1680
Fungal source[31]	Hydrolysis product of trichothecin *Fusarium roseum* Link *Trichothecium roseum*
Other significant references	Chemistry[28]
^{13}C NMR data[48]	C2, 70.2; C3, 40.2; C4, 79.6; C5, 49.5; C6, 43.6; C7, 42.3; C8, 199.4; C9, 138.5; C10, 137.7; C11, 73.1; C12, 66.0; C13, 47.3; C14, 6.2; C15, 15.6; C16, 18.5
^{1}H NMR data[47]	H2, 3.90; H3a, 1.93; H3b, 2.43; H4, 4.35; H7, NR; H10, 6.49; H11, 3.94; H13a, 2.81; H13b, 3.12; H14, NR; H15, 1.00; H16, 1.80

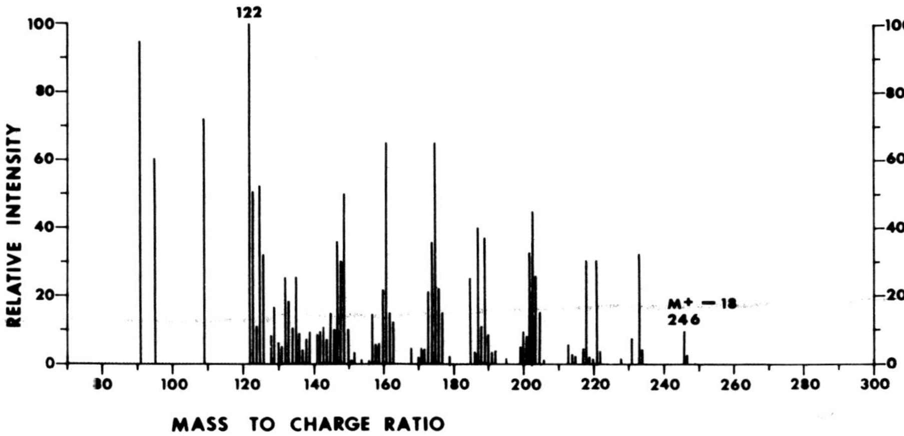
122
M+ − 18
246
RELATIVE INTENSITY
MASS TO CHARGE RATIO
100
80
60
40
20
0
80
100
120
140
160
180
200
220
240
260
280
300

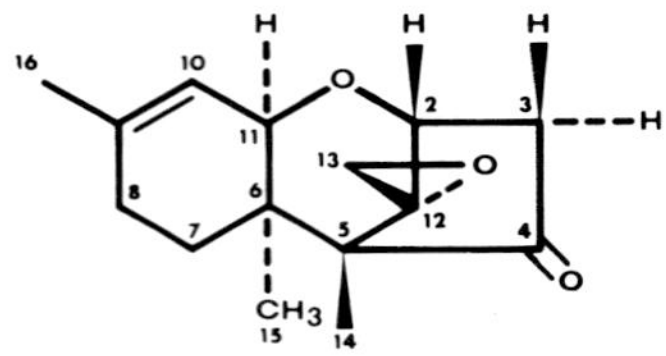

Common name	Trichodermone
Molecular weight	248.1407
Molecular formula	$C_{15}H_{20}O_3$
General characteristics[39]	Crystals from ether–hexane, mp 154°–155°C $[\alpha]_D^{20}$ −42.3° (C = 1.0 in chloroform)
UV data[39]	λ_{max}^{EtOH} nm (ϵ): 205(2,800)
IR data[39]	ν_{max}^{KBr} cm^{-1}: 1735
Source[39]	Oxidation of trichodermin
^{13}C NMR data[48]	C2, 76.2; C3, 42.0; C4, 214.9; C5, 55.3; C6, 42.7; C7, 23.9; C8, 27.9; C9, 140.9; C10, 118.7; C11, 71.4; C12, 65.3; C13, 50.1; C14, 5.4; C15, 15.4; C16, 23.4
^1H NMR data	H2, 4.22; H3, 2.80; H7, 2.01; H8, 2.01; H10, 5.41; H11, 4.38; H13, 3.15; H14, 0.84; H15, 0.39; H16, 1.70

Common name	Satratoxin G*
Molecular weight	544.1935
Molecular formula	$C_{28}H_{32}O_{11}$
General characteristics[26,27]	mp 167°–170°C
UV data[26,27]	λ_{max} nm (ϵ): 256(6,500)
Fungal source[26,27]	*Stachybotrys atra (S. alternans)*

* No structure known for satratoxin G.

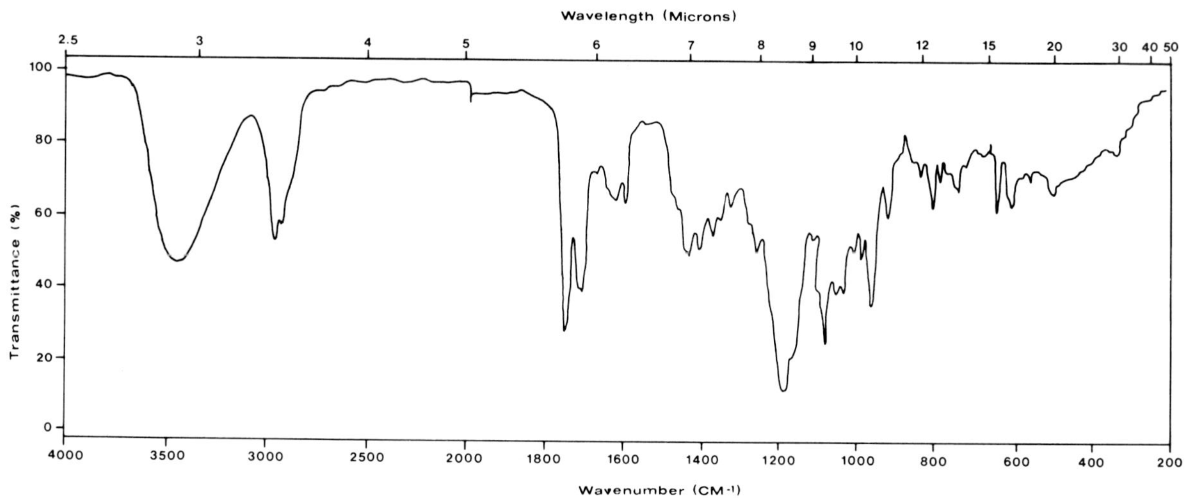
Wavelength (Microns)
Transmittance (%)
Wavenumber (CM-1)

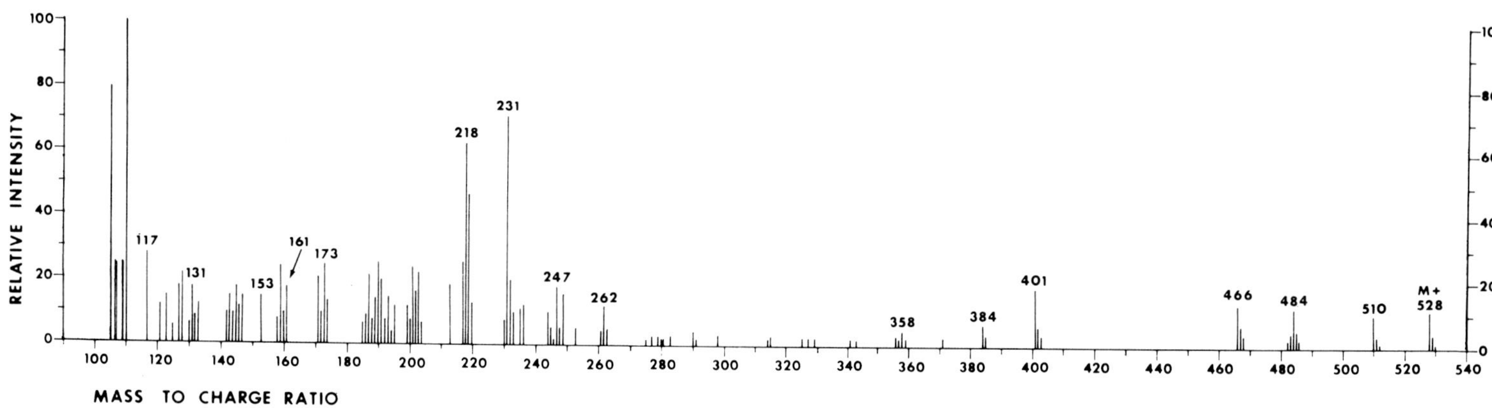
RELATIVE INTENSITY
MASS TO CHARGE RATIO
117
131
153
161
173
218
231
247
262
358
384
401
466
484
510
M+
528

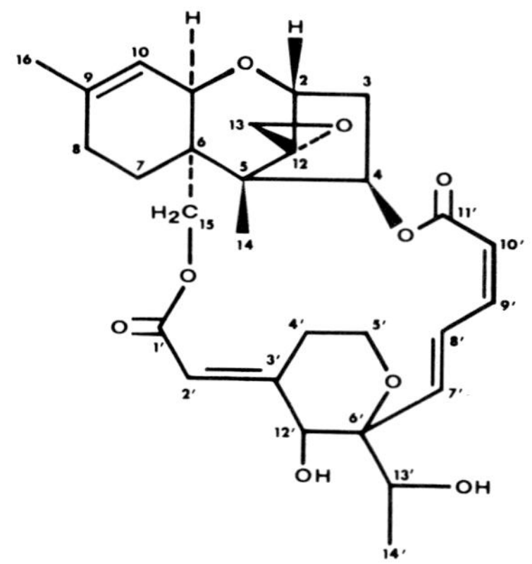

Common name	Satratoxin H
Molecular weight	528.2349
Molecular formula	$C_{29}H_{36}O_9$
General characteristics[26]	mp 162°–166°C
UV data[26]	λ_{max}^{MeOH} nm (ϵ): 225(14,700) and 255(10,400)
Fungal source[26,27]	*Stachybotrys atra*
Toxicity data[27]	Toxic to brine shrimp and positive in rabbit skin tests. Details not reported.
TLC data[26]	*Adsorbent*: silica gel *Solvent*: isopropanol–chloroform, 2:98 v/v R_f: not reported *Detection*: not reported
^{13}C NMR data[27]	C2, 79.1; C3, 34.4; C4, 74.2; C5, 49.0; C6, 43.4; C7, 20.4; C8, 27.6; C9, 140.2; C10, 119.0; C11, 68.2; C12, 65.4; C13, 48.0; C14, 7.6; C15, 64.2; C16, 23.3; C1′, 166.2; C2′, 119.0; C3′, 155.1; C4′, 25.3; C5′, 60.4; C6′, 81.4; C7′, 134.2; C8′, 132.2; C9′, 143.0; C10′, 120.4; C11′, 167.0; C12′, 73.7; C13′, 69.7; C14′, 15.7
^{1}H NMR data[27]	H2, 3.90; H3a, 2.20; H3b, 2.45; H4, 5.90; H7, 1.90; H8, 2.10; H10, 5.46; H11, 3.62; H13a, 2.81; H13b, 3.12; H14, 0.83; H15a, 3.88; H15b, 4.56; H16, 1.74; H2′, 5.85; H4′a, 2.60; H4′b, 3.74; H5′, 3.90; H7′, 6.09; H8′, 7.36; H9′, 6.63; H10′, 5.91; H12′, 3.97; H13′, 4.38; H14′, 1.16

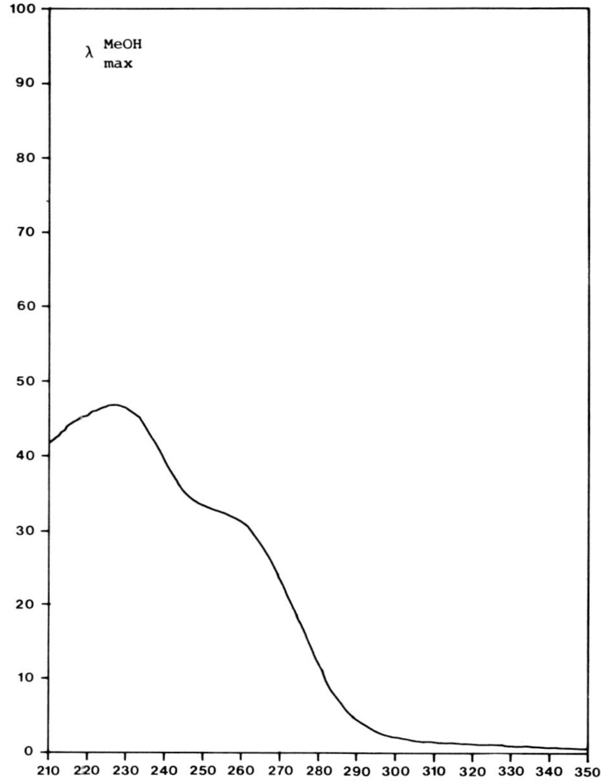

λ MeOH max
100
90
80
70
60
50
40
30
20
10
0
210 220 230 240 250 260 270 280 290 300 310 320 330 340 350

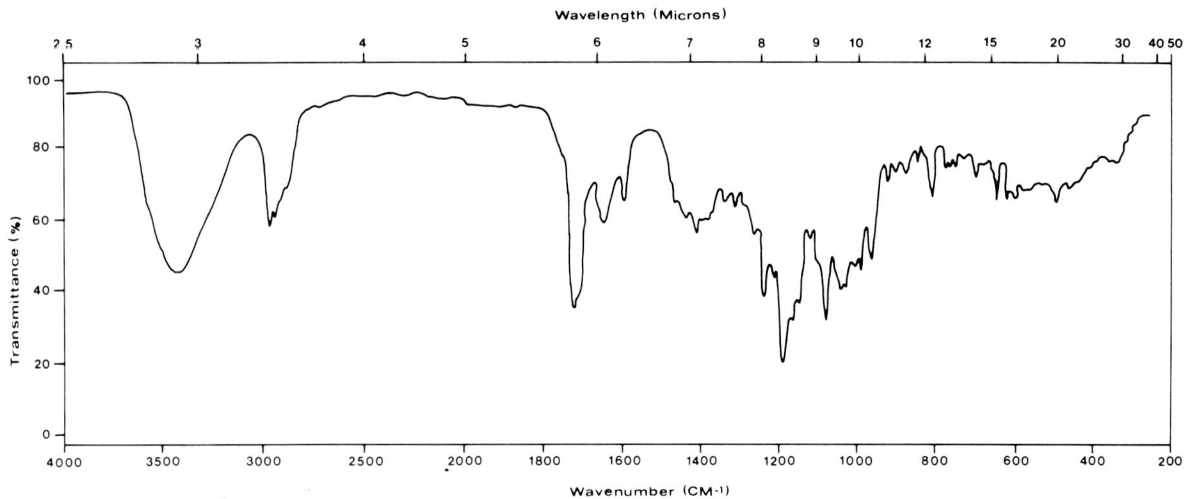
Wavelength (Microns)
Transmittance (%)
Wavenumber (CM-1)

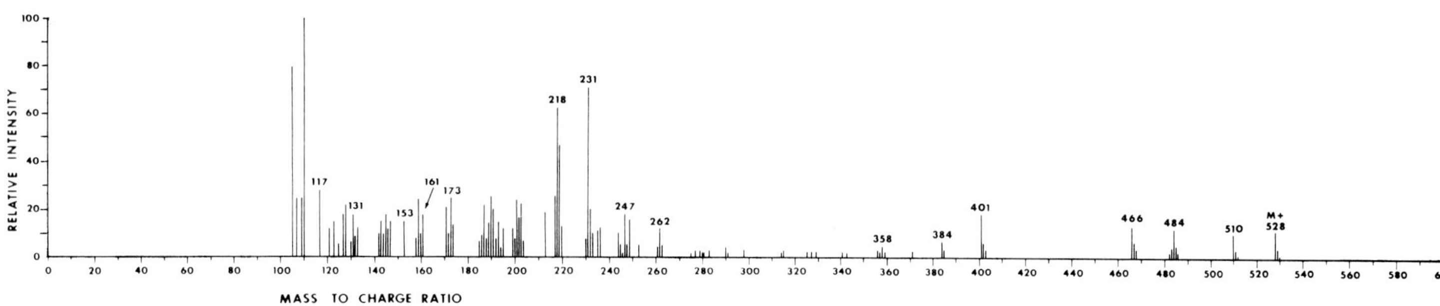
RELATIVE INTENSITY
MASS TO CHARGE RATIO
117
131
153
161
173
218
231
247
262
358
384
401
466
484
510
M+
528

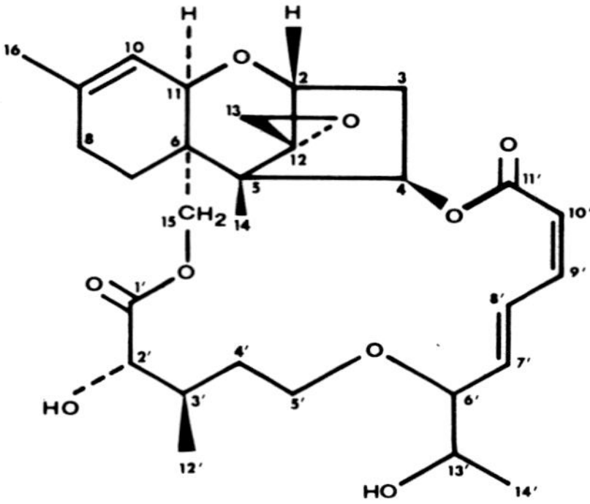

Common name[67] Roridin A

Molecular weight 532.2661

Molecular formula $C_{29}H_{40}O_9$

General characteristics[12,13,48]
Crystals from ether, mp 198°–204°C
$[\alpha]_D^{22}$ +130° (*C* = 1.36 in chloroform)
$[\alpha]_D^{22}$ + 192° (*C* = 1.38 in dioxane)
Di-*O*-acetylroridin A: needles from ether–petroleum ether, mp 165°–166°C
$[\alpha]_D^{22}$ + 168.5° (*C* = 1.108 in chloroform)

UV data[12,13] λ_{max}^{EtOH} nm (ϵ): 263(18,600)

IR data[12,13] ν_{max}^{KBr} cm^{-1}: 3546, 3472, 1742, 1709, 1704, 1701, 1637, 1631, and 1597

Fungal source[12,48,67]
Myrothecium verrucaria
M. roridum

Toxicity data[4,7] The LD_{50} value of roridin A in mice dosed IV was 1 mg/kg. In mouse tumor cells, it caused 50% inhibition of cell growth at 1 ng/ml.

TLC data[4]
A
Adsorbent: silica gel G
Solvent: chloroform–methanol, 98:2 v/v
R_f: 0.70
Detection: H_2SO_4 spraying and heating at 110°C for 5 min
B
Adsorbent: alumina

	Solvent: chloroform–methanol, 98:2 v/v R_f: 0.18 *Detection*: H_2O_4 spraying and heating at 110°C for 5 min
Other significant references	Circular dichroism[77]
^{13}C NMR data[14]	C2, 78.8; C3, 34.6; C4, 74.2; C5, 49.1; C6, 43.6; C7, 20.0; C8, 27.5; C9, 140.4; C10, 118.2; C11, 66.9; C12, 64.9; C13, 47.4; C14, 7.2; C15, 64.2; C16, 22.9; C1′, 174.5; C2′, 75.3; C3′, 36.7; C4′, 33.0; C5′, 69.5; C6′, 83.7; C7′, 139.0; C8′, 126.0; C9′, 143.6; C10′, 117.2; C11′, 166.3; C12′, 14.4; C13′, 70.4; C14′, 18.0
^1H NMR data[13]	H4, 5.85; H10, 5.44; H13a, 2.96; H14, 0.80; H15a, 4.44; H16, 1.74; H2′, 4.09; H7′, 6.00; H8′, 7.68; H9′, 6.66; H10′, 5.78; H12′, 1.08; H14′, 1.17

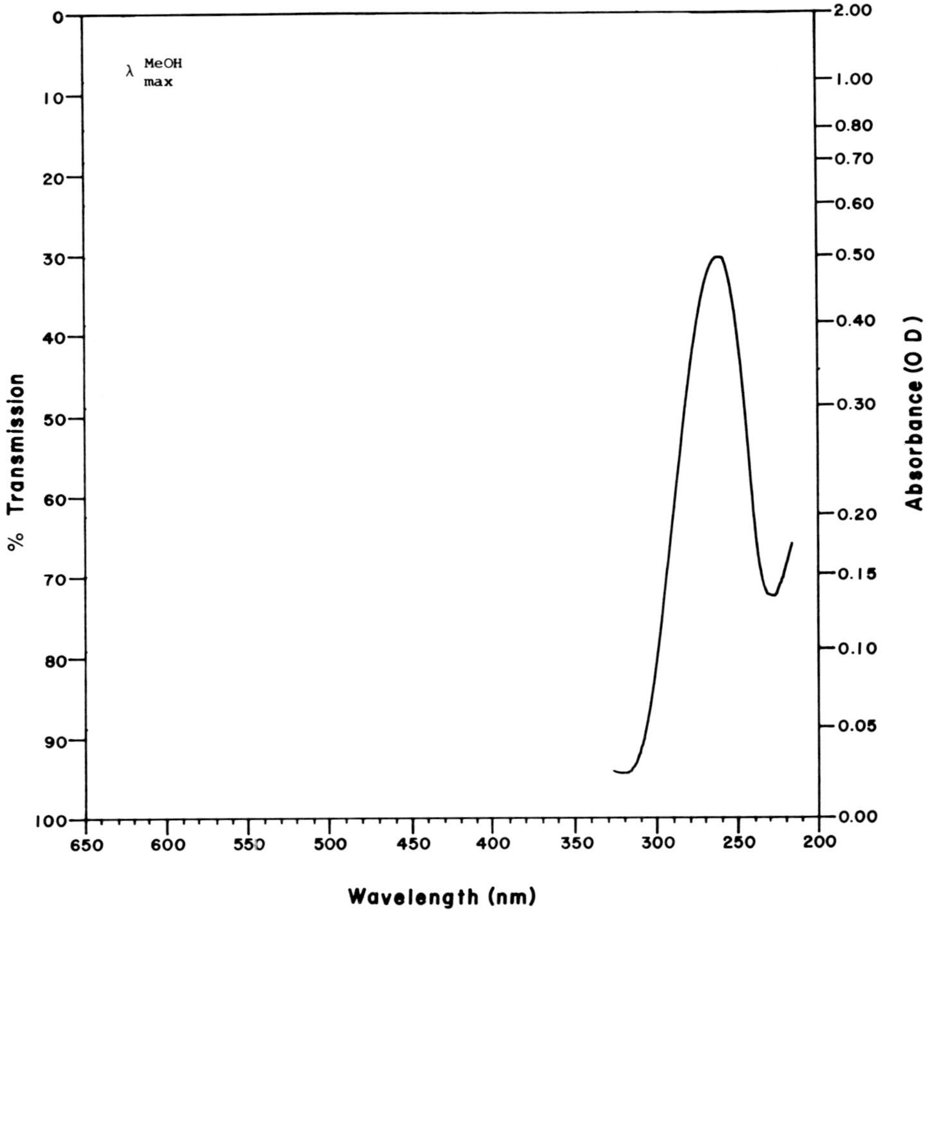

λ MeOH max
% Transmission
0
10
20
30
40
50
60
70
80
90
100
Absorbance (O D)
2.00
1.00
0.80
0.70
0.60
0.50
0.40
0.30
0.20
0.15
0.10
0.05
0.00
650
600
550
500
450
400
350
300
250
200
Wavelength (nm)

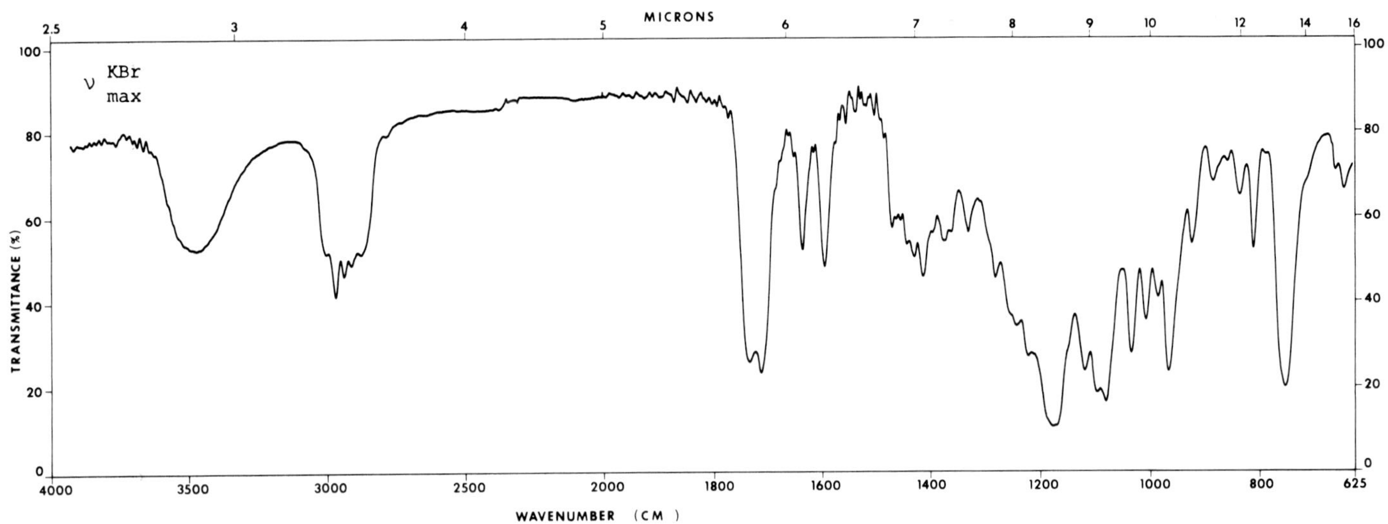

MICRONS
WAVENUMBER (CM⁻¹)
TRANSMITTANCE (%)
ν KBr max

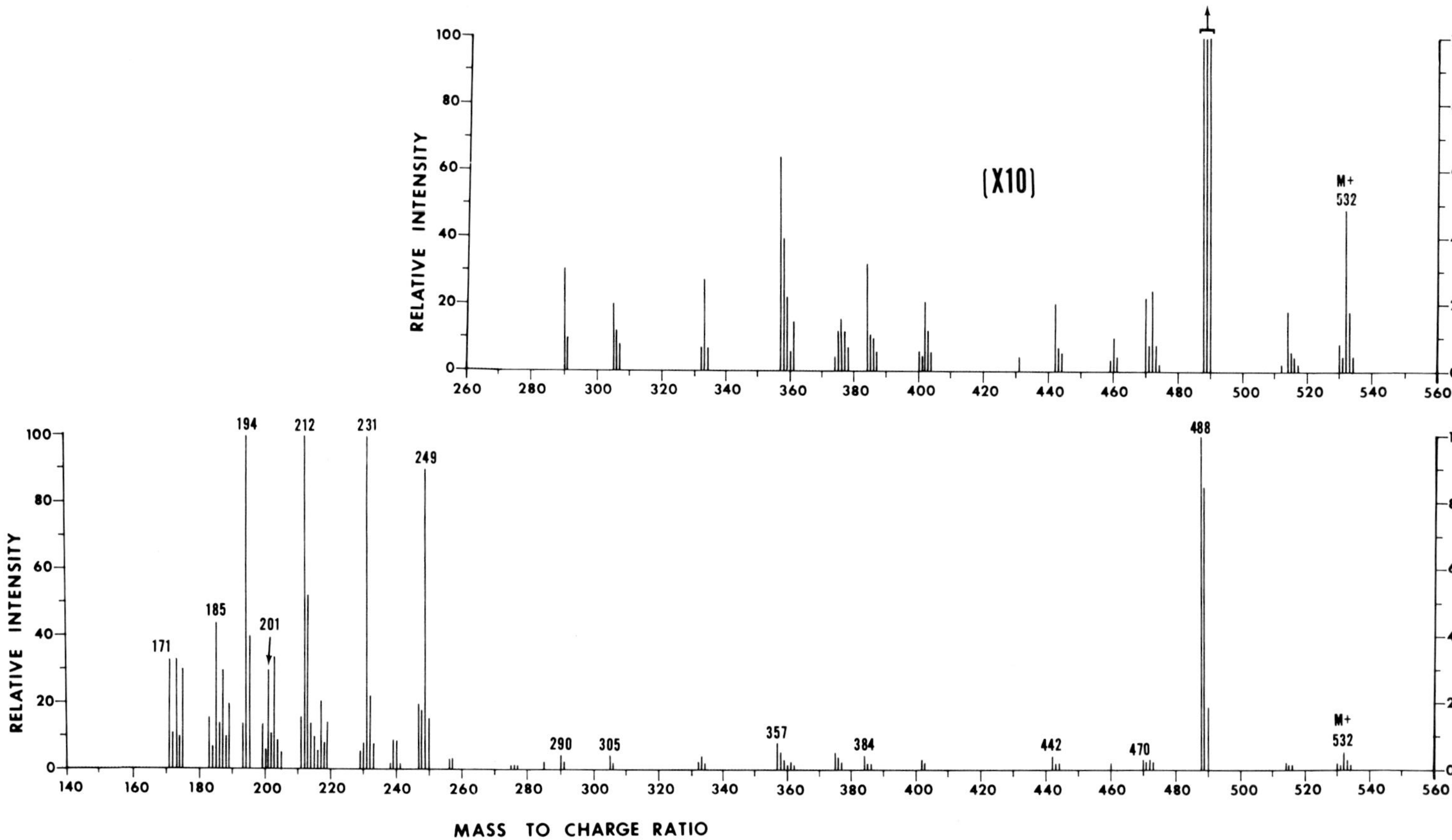

RELATIVE INTENSITY
(X10)
M+
532
171
185
194
201
212
231
249
290
305
357
384
442
470
488
M+
532
MASS TO CHARGE RATIO

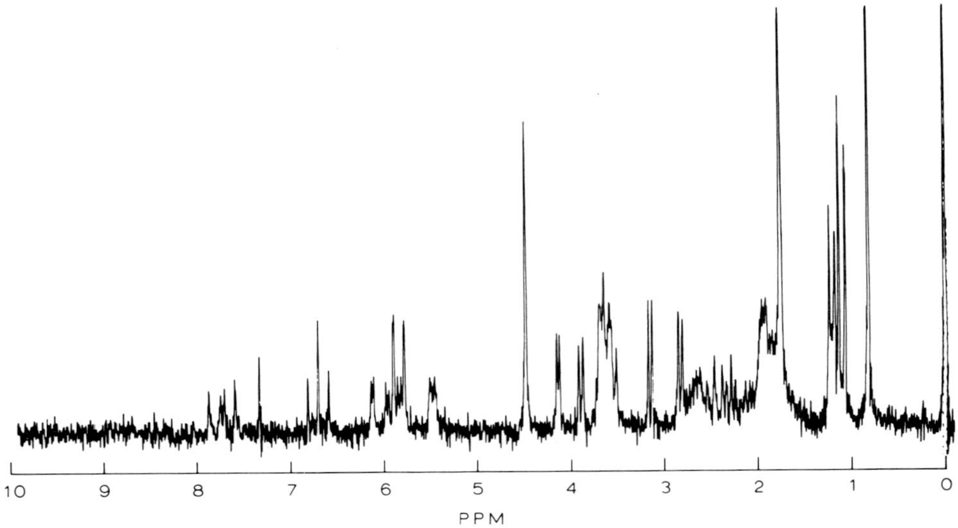

The 100 MHz ^{1}H NMR spectrum of roridin A in $CDCl_3$ solution.

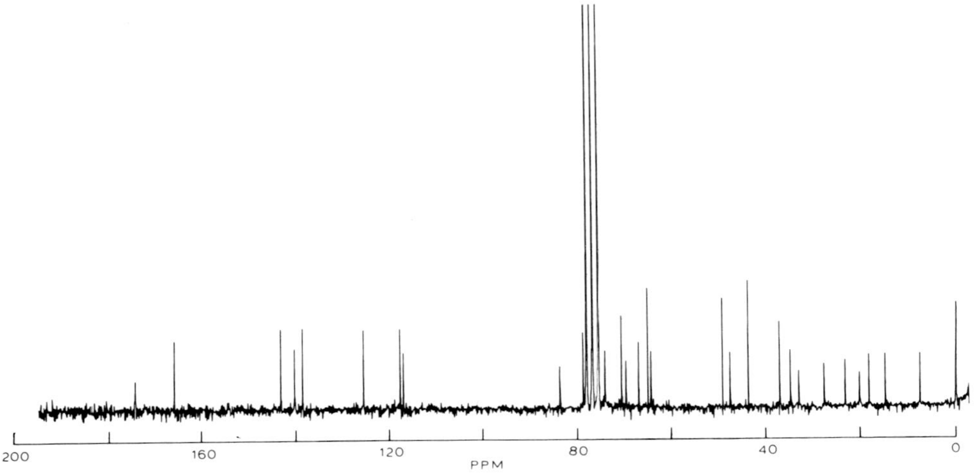

The 25 MHz ^{13}C NMR spectrum of roridin A in $CDCl_3$ solution.

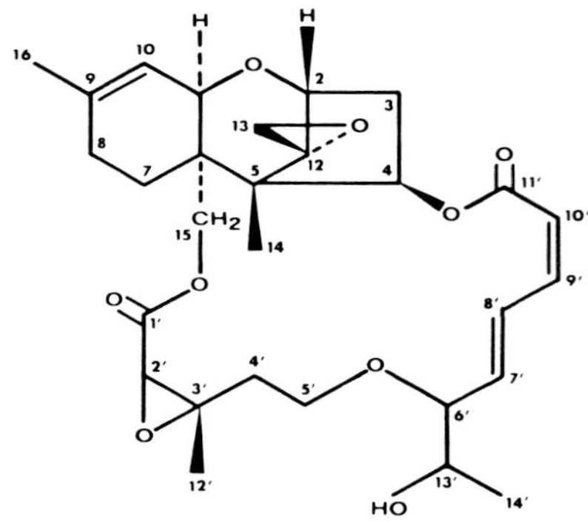

Common name	Roridin D
Molecular weight	530.2515
Molecular formula	$C_{29}H_{38}O_9$
General characteristics[12]	Needles from acetone–ether, mp 232°–235°C $[\alpha]_D^{23}$ +29° (*C* = 2.71 in chloroform)
UV data[12]	λ_{max}^{EtOH} nm (ϵ): 260(21,400)
IR data[12]	ν_{max}^{KBr} cm^{-1}: 3540-3440, 2970, 1747, 1711, 1641, 1601, 1420, 1190, 1105, 1083, 968, 818, 753, and 660
Fungal source[12]	*Myrothecium verrucaria* (Albertini and Schweinitz) Ditmar *ex* Fries *M. roridum* Tode *ex* Fr.
Toxicity data[12]	Antibiotic activity
TLC data[12]	*Absorbent*: Kieselgel G *Solvent*: ethyl ether (two consecutive runs) R_f: 0.18 *Detection*: iodine vapors
Other significant references	Circular dichroism[77]
^{13}C NMR data[14]	C2, 78.8; C3, 34.9; C4, 74.3; C5, 49.0; C6, 43.1; C7, 20.4; C8, 27.4; C9, 140.1; C10, 118.4; C11, 66.9; C12, 65.1; C13, 47.4; C14, 6.8; C15, 64.3; C16, 22.9; C1′, 167.8; C2′, 57.9; C3′, 62.9; C4′, 39.4; C5′, 67.3; C6′, 85.3; C7′, 138.1; C8′, 126.2; C9′, 142.9; C10′, 117.8; C11′, 166.1; C12′, 17.2; C13′, 70.5; C14′, 17.9

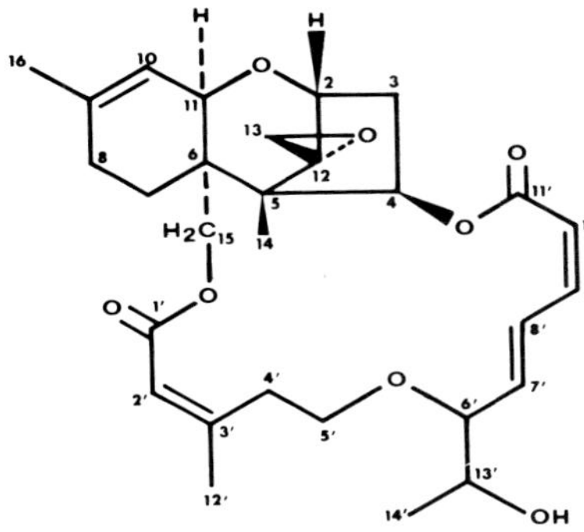

Common name	Roridin E
Molecular weight	514.2449
Molecular formula	$C_{29}H_{38}O_8$
General characteristics[12]	Colored needles from ether–pentane, mp 177°–178°C or 211°–212°C Needles from ether, 220°–221°C $[\alpha]_D^{23}$ −27° (C = 1.3 in chloroform) $[\alpha]_D^{25}$ −21° (C = 1.852 in chloroform)
UV data[12]	λ_{max}^{EtOH} nm (ϵ): 195(15,800), 223(25,100), and 263(19,900)
IR data[12]	$\nu_{max}^{CH_2Cl_2}$ cm^{-1}: 3570, 3050, 2975, 1712, 1647, 1603, 1365, 1220, 1180, 1148, 1142, 1096, 1090, 1080, 966, and 814
Fungal source[12]	*Myrothecium verrucaria* (Albertini and Schweinitz) Ditmar *ex* Fries. *M. roridum* Tode *ex* Fr.
Toxicity data	Antibiotic activity
TLC data[12]	*Adsorbent*: Kieselgel G *Solvent*: ethyl ether (two consecutive runs) R_f: 0.24 *Detection*: iodine vapors
Other significant references	Circular dichroism[77]
^{13}*C NMR data*[62]	C2, 79.3; C3, 35.8; C4, 74.2; C5, 48.4; C6, 42.8; C7, 21.6; C8, 27.7; C9, 140.0; C10, 117.8; C11, 67.2; C12, 65.6; C13, 48.1; C14, 6.7; C15, 63.7; C16,

	23.2; C1′, 165.8; C2′, 119.0; C3′, 159.0; C4′, 41.3; C5′, 69.8; C6′, 83.8; C7′, 138.1; C8′, 126.6; C9′, 143.7; C10′, 117.2; C11′, 166.4; C12′, 20.2; C13′, 70.5; C14′, 18.3
^{1}H NMR data[62]	H2, 3.82; H3a, 2.04; H3b, 2.53; H4, 6.20; H10, 5.47; H11, 3.89; H13a, 2.81; H13b, 3.12; H14, 0.79; H15a, 3.93; H15b, 4.32; H16, 1.71; H2′, 5.95; H6′, 3.70; H7′, 5.89; H8′, 7.51; H9′, 6.56; H10′, 5.73; H12′, 2.25; H13′, 3.70; H14′, 1.19

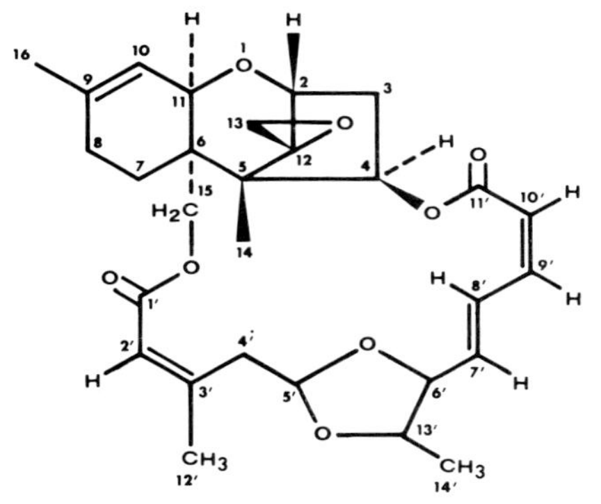

Common name	Roridin H (Verrucarin H)
Molecular weight	512.2409
Molecular formula	$C_{29}H_{36}O_8$
General characteristics[12,88]	Crystals from acetone–ether, mp >325°C (dec.) $[\alpha]_D^{23}$ +31° (C = 1.16 in chloroform) $[\alpha]_D^{23}$ +40.3° (C = 1.065 in benzene) $[\alpha]_D^{23}$ +31.5° (C = 0.855 in dioxane)
UV data[12,88]	λ_{max}^{EtOH} nm (ϵ): 195(15,800), 224(24,500), and 260(18,200)
IR data[12,88]	$\nu_{max}^{CH_2Cl_2}$ cm^{-1}: 2970, 2910, 1710, 1645, 1600, 1380, 1355, 1220, 1175, 1115, 1101, 1087, 1071, 1033, 1005, 992, 971, and 832
Fungal source[12,88]	*Myrothecium verrucaria*
Toxicity data[12,88]	Antibiotic activity
TLC data[12,88]	*Absorbent*: Kieselgel G *Solvent*: ethyl ether (two consecutive runs) R_f: 0.51 *Detection*: iodine vapors
Other significant references	Circular dichroism[77]
^{13}C NMR data[62]	C2, 79.0; C3, 34.8; C4, 74.0; C5, 48.9; C6, 43.2; C7, 20.5; C8, 27.6; C9, 139.9; C10, 118.6; C11, 67.6; C12, 65.3; C13, 47.3; C14, 7.0; C15, 63.0; C16, 22.9; C1′, 166.0; C2′, 119.0; C3′, 154.4; C4′, 47.7; C5′, 100.8; C6′, 81.9; C7′, 134.6; C8′, 126.2; C9′,

142.5; C10′, 118.9; C11′, 166.0; C12′, 18.2; C13′, 76.8; C14′, 16.3

^{1}H NMR data[62] H2, 3.80; H4, 5.90; H10, 5.42; H11, 3.64; H13a, 2.96; H14, 0.85; H15a, 4.15; H16, 1.69; H2′, 5.67; H4′a, 2.74; H5′, 5.58; H6′, 4.03; H7′, 5.90; H8′, 7.68; H9′, 6.55; H10′, 5.79; H12′, 2.27; H13′, 3.65; H14′, 1.32

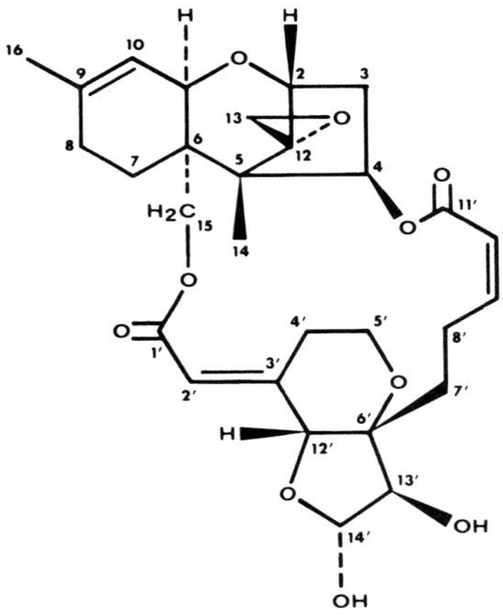

Common name	Vertisporin
Molecular weight	544.2307
Molecular formula	$C_{29}H_{36}O_{10}$
General characteristics[65]	A colorless amorphous powder, mp 176°–183°C $[\alpha]_D^{26}$ +62.5° Diacetate: mp 145°–155°C
UV data[65]	λ_{max}^{EtOH} nm (ϵ): 216(19,500)
IR data[65]	$\nu_{max}^{CHCl_3}$ cm^{-1}: 1723 and 171
Fungal source[65]	*Verticimonosporium diffractum*
Toxicity data[65]	Cytotoxic and antibiotic activity: ED_{50} to HeLa cells was 0.001 mg/ml; limited antifungal activity against *Trichophyton asteroides*.
^{13}C NMR data[65]	C2, 79.0; C3, 35.0; C4, 74.0; C5, 49.7; C6, 43.2; C7, 20.7; C8, 27.7; C9, 138.4; C10, 119.9; C11, 67.6; C12, 65.3; C13, 47.2; C14, 8.0; C15, 64.7; C16, 23.0; C1′, 165.7; C2′, 118.5; C3′, 152.6; C4′, 22.8; C5′, 64.9; C6′, 86.7; C7′, 23.6; C8′, 24.4; C9′, 149.4; C10′, 121.1; C11′, 166.0; C12′, 86.0; C13′, 75.7; C14′, 97.5
^{1}H NMR data[65]	H2, 3.82; H4, 5.80; H10, 5.40; H11, 3.57; H13a, 2.77; H13b, 3.09; H14, 0.80; H15a, 3.97; H15b, 4.24; H16, 1.70; H2′, 5.81; H9′, 6.41; H10′, 5.82; H12′, 4.19; H13′, 5.10; H14′, 6.61

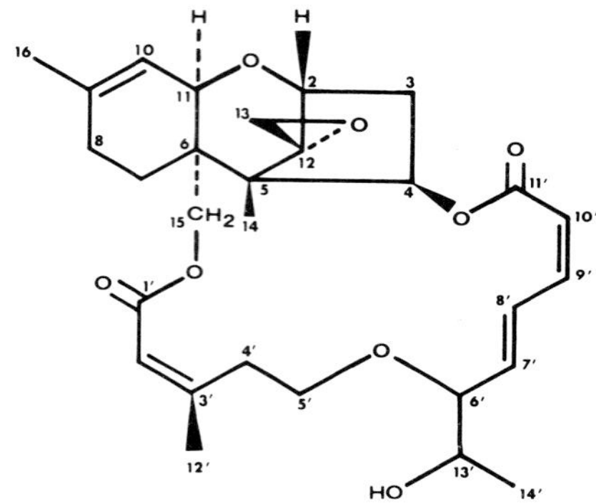

Common name	Isororidin E
Molecular weight	514.2556
Molecular formula	$C_{29}H_{38}O_8$
General characteristics[62,63]	Colorless prisms from ethyl acetate, mp 200°–202°C $[\alpha]_D$ −65.1° (in chloroform)
UV data[62,63]	λ_{max}^{EtOH} nm (ϵ): 223(24,000) and 262(16,000)
IR data[62,63]	$\nu_{max}^{CHCl_3}$ cm^{-1}: 3757, 1713, 1644, and 1598
Fungal source[62,63]	*Cylindrocarpon* spp.
^{13}C NMR data[62]	C2, 79.2; C3, 36.6; C4, 75.3; C5, 48.5; C6, 42.6; C7, 22.7; C8, 27.7; C9, 140.1; C10, 118.9; C11, 66.7; C12, 65.7; C13, 47.7; C14, 6.4; C15, 64.5; C16, 23.2; C1′, 166.3; C2′, 119.5; C3′, 158.0; C4′, 40.0; C5′, 67.0; C6′, 83.2; C7′, 135.3; C8′, 131.0; C9′, 142.0; C10′, 117.1; C11′, 166.3; C12′, 19.8; C13′, 67.6; C14′, 18.5
^1H NMR data[62]	H2, 3.84; H3a, 2.03; H3b, 2.58; H4, 6.35; H10, 5.50; H11, 4.09; H13a, 2.83; H13b, 3.15; H14, 0.80; H15a, 4.06; H15b, 4.16; H16, 1.71; H2′, 5.83; H6′, 3.74; H7′, 5.71; H8′, 7.55; H9′, 6.60; H10′, 5.82; H12′, 2.22; H13′, 3.70; H14′, 1.17

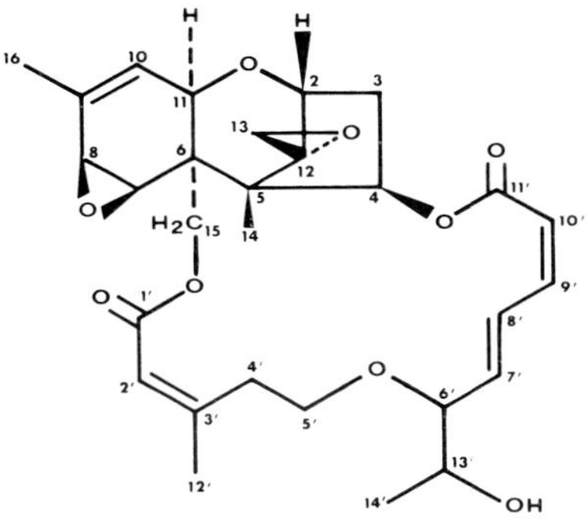

Common name	7β,8β-Epoxyisororidin E
Molecular weight	528.2348
Molecular formula	$C_{29}H_{36}O_9$
General characteristics[62,63]	Colorless prisms from acetone, mp 216°–219°C $[\alpha]_D^{24}$ −69.9° (in chloroform)
UV data[62,63]	λ_{max}^{EtOH} nm (ε): 221(26,300) and 262(16,700)
IR data[62,63]	$\nu_{max}^{CHCl_3}$ cm^{-1}: 3567, 1712, 1644, and 1598
Fungal source[62,63]	*Cylindrocarpon* spp.
^{13}C NMR data[62]	C2, 79.2; C3, 36.6; C4, 74.6; C5, 47.0; C6, 44.0; C7, 50.8; C8, 56.1; C9, 138.2; C10, 123.0; C11, 66.7; C12, 65.7; C13, 48.3; C14, 7.2; C15, 62.9; C16, 21.9; C1′, 165.9; C2′, 119.1; C3′, 159.0; C4′, 40.1; C5′, 67.0; C6′, 83.3; C7′, 135.5; C8′, 130.5; C9′, 142.4; C10′, 116.7; C11′, 116.2; C12′, 20.0; C13′, 69.7; C14′, 18.5
^1H NMR data[62]	H2, 3.90; H3a, 2.06; H3b, 2.55; H4, 6.30; H7, 3.34; H8, 3.17; H10, 5.75; H11, 4.21; H13a, 2.99; H13b, 3.21; H14, 1.02; H15a, 3.79; H15b, 4.07; H16, 2.00; H2′, 5.80; H6′, 3.75; H7′, 5.77; H8′, 7.54; H9′, 6.62; H10′, 5.83; H12′, 2.24; H13′, 3.70; H14′, 1.16

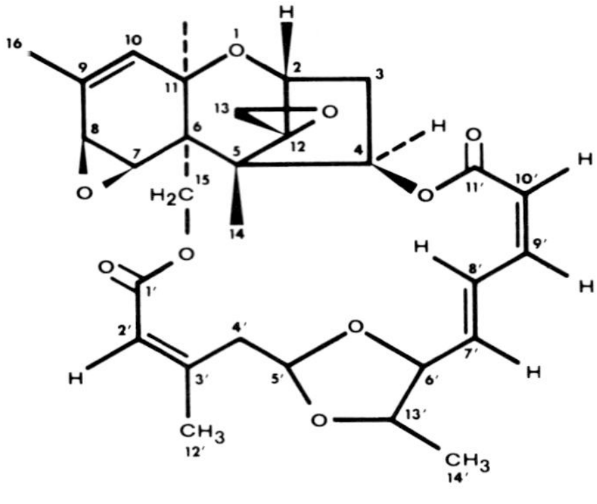

Common name	7β,8β-Epoxyroridin H
Molecular weight	526.2193
Molecular formula	$C_{29}H_{34}O_9$
General characteristics[62,63]	Colorless amorphous powder
IR data[62,63]	$\nu_{max}^{CHCl_3}$ cm^{-1}: 1717, 1646, and 1601
Fungal source[62,63]	*Cylindrocarpon* spp.
^{13}C NMR data[62]	C2, 79.3; C3, 35.0; C4, 73.6; C5, 48.2; C6, 44.2; C7, 50.9; C8, 57.2; C9, 138.0; C10, 123.0; C11, 67.9; C12, 65.5; C13, 47.9; C14, 8.2; C15, 65.8; C16, 22.0; C1′, 165.9; C2′, 118.7; C3′, 155.6; C4′, 47.7; C5′, 101.0; C6′, 82.0; C7′, 135.4; C8′, 126.2; C9′, 143.4; C10′, 118.4; C11′, 165.9; C12′, 18.4; C13′, 77.1; C14′, 16.5
^{1}H NMR data[62]	H2, 3.91; H3a, 2.15; H3b, 2.40; H4, 5.87; H7, 3.75; H8, 3.17; H10, 5.69; H11, 4.08; H13a, 2.98; H13b, 3.18; H14, 1.12; H15a, 3.41; H15b, 4.43; H16, 1.98; H2′, 5.68; H4′a, 2.32; H4′b, 2.41; H5′, 5.54; H6′, 4.06; H7′, 5.95; H8′, 7.76; H9′, 6.58; H10′, 5.79; H12′, 2.28; H14′, 1.34

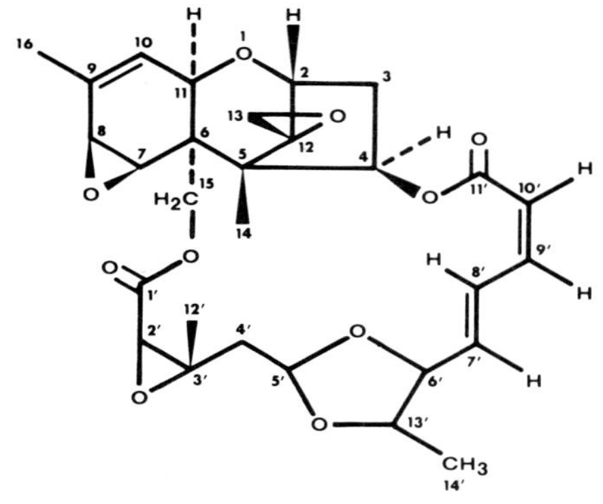

Common name	7β,8β,2′,3′-Diepoxyisororidin H
Molecular weight	542.2142
Molecular formula	$C_{29}H_{34}O_{10}$
General characteristics[62,63]	Colorless prisms from ethyl acetate, mp 291°–293°C (dec.)
IR data[62,63]	$\nu_{max}^{CHCl_3}$ cm^{-1}: 1755, 1710, 1645, and 1601
Fungal source[62,63]	*Cylindrocarpon* spp.
^{13}*C NMR data*[62]	C2, 79.2; C3, 34.8; C4, 73.9; C5, 48.5; C6, 44.7; C7, 50.9; C8, 56.3; C9, 138.6; C10, 122.4; C11, 67.3; C12, 65.3; C13, 48.0; C14, 8.5; C15, 65.6; C16, 22.0; C1′, 167.2; C2′, 59.2; C3′, 60.8; C4′, 44.0; C5′, 101.1; C6′, 82.7; C7′, 134.6; C8′, 126.8; C9′, 142.7; C10′, 119.0; C11′, 166.9; C12′, 17.1; C13′, 76.3; C14′, 15.8
^{1}*H NMR data*[62]	H2, 3.90; H3a, 2.26; H3b, 2.45; H4, 5.89; H7, 3.62; H8, 3.18; H10, 5.67; H11, 3.75; H13a, 2.97; H13b, 3.18; H14, 1.12; H15a, 4.44; H16, 2.01; H2′, 3.29; H4′a, 1.57; H4′b, 2.32; H5′, 5.37; H6′, 4.17; H7′, 5.98; H8′, 7.60; H9′, 6.62; H10′, 5.88; H12′, 1.60; H13′, 3.70; H14′, 1.33

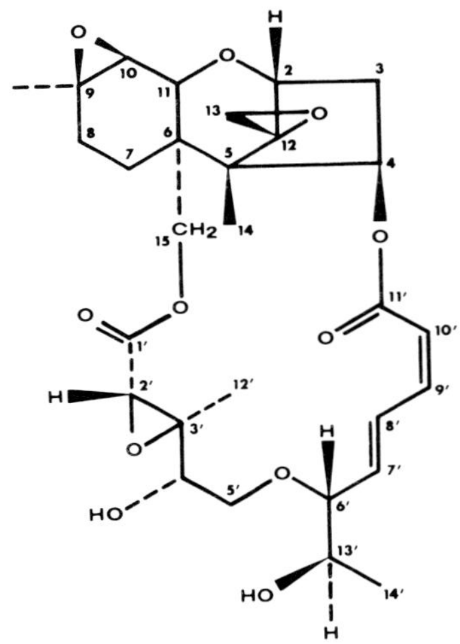

Common name	Baccharin
Molecular weight	562.2413
Molecular formula	$C_{29}H_{38}O_{11}$
General characteristics[58]	Crystals from dichloromethane–methanol, mp 200°–230°C $[\alpha]_D^{24}$ +41.5° (C = 2.2 in chloroform) Diacetate: mp 254°–256°C
UV data[58]	λ_{max}^{EtOH} nm (ϵ): 259(18,700) λ_{max}^{EtOH} nm (ϵ): diacetate: 262(21,000)
Toxicity data[58]	Significant inhibitory activity *in vitro* against cells derived from human carcinoma of the nasopharynx and *in vivo* against P-388 leukemia in mice.
Fungal source[58]	*Baccharis megapotamica* Spreng
^{1}H NMR data[58]	H3a, 2.48; H4, 5.80; H10, 3.11; H13a, 2.75; H13b, 3.16; H14, 0.75; H15a, 4.24; H15b, 4.42; H16, 1.37; H2′, 3.37; H7′, 5.98; H8′, 7.48; H9′, 6.60; H10′, 5.82; H12′, 1.65; H14′, 1.20

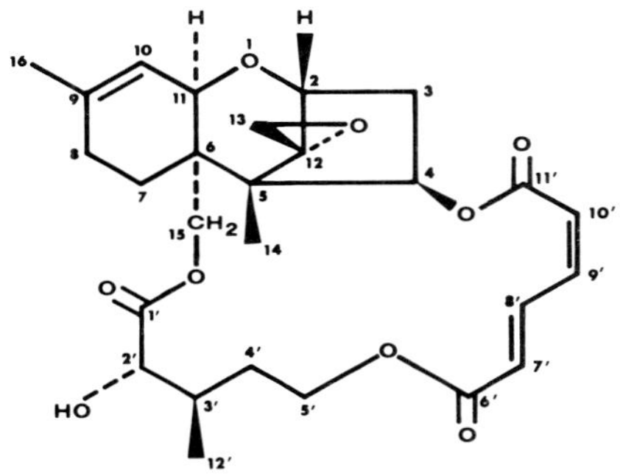

Common name	Verrucarin A
Molecular weight	502.2201
Molecular formula	$C_{27}H_{34}O_9$
General characteristics[12,48]	mp >360° (dec.) $[\alpha]_D^{22}$ +206° (C = 1.012 in dioxane) +260° (in chloroform)
UV data[12,48]	λ_{max}^{EtOH} nm (ϵ): 260(17,700)
Fungal source[12,48]	*Myrothecium verrucaria* *M. roridum*
Toxicity data[4]	Rats given verrucarin A developed creatinurea; *in vitro* it inhibited ATP-creatine phosphotransferase. It caused severe edema in body cavities. Major clinical signs in several different experimental animals were diarrhea, hematuria, vomiting, anorexia, loss of weight, ataxia, and thirst. At relatively low dosages, it caused leukocytosis followed by severe leukopenia. The LD_{50} values in mice were 1.5 mg/kg (IV) and 0.5 mg/kg (IP). It caused dermal toxicity: necrosis of epidermis with suppuration and extensive necrosis of dermal tissue with damage extending into subcuticulis. ID_{50} of protein synthesis in rabbit reticulocytes was 10–15 μg/ml.
TLC data[4]	A *Adsorbent:* silica gel G *Solvent:* chloroform–methanol, 98:2 v/v R_f: 0.70 *Detection:* H_2SO_4 spraying and heating at 110°C for 5 min

B
Adsorbent: alumina
Solvent: chloroform–methanol, 98:2 v/v
R_f: 0.28
Detection: H_2SO_4 spraying and heating at 110°C for 5 min

Other significant references	Circular dichroism[77]; stereochemistry[64]
^{13}*C NMR data*[14]	C2, 78.6; C3, 34.6; C4, 75.3; C5, 49.2; C6, 43.9; C7, 19.7; C8, 27.2; C9, 140.7; C10, 117.7; C11, 66.6; C12, 64.9; C13, 47.4; C14, 7.0; C15, 63.1; C16, 22.6; C1′, 174.3; C2′, 73.8; C3′, 32.9; C4′, 31.9; C5′, 60.8; C6′, 165.8; C7′, 127.2; C8′, 138.6; C9′, 138.6; C10′, 125.5; C11′, 165.1; C12′, 9.8
^{1}*H NMR data*	H4, 5.83; H10, 5.46; H13a, 2.97; H14, 0.87; H16, 1.79; H2′, 4.20; H7′, 6.06; H8′, 8.08; H9′, 6.70; H10′, 6.17; H12′, 0.89

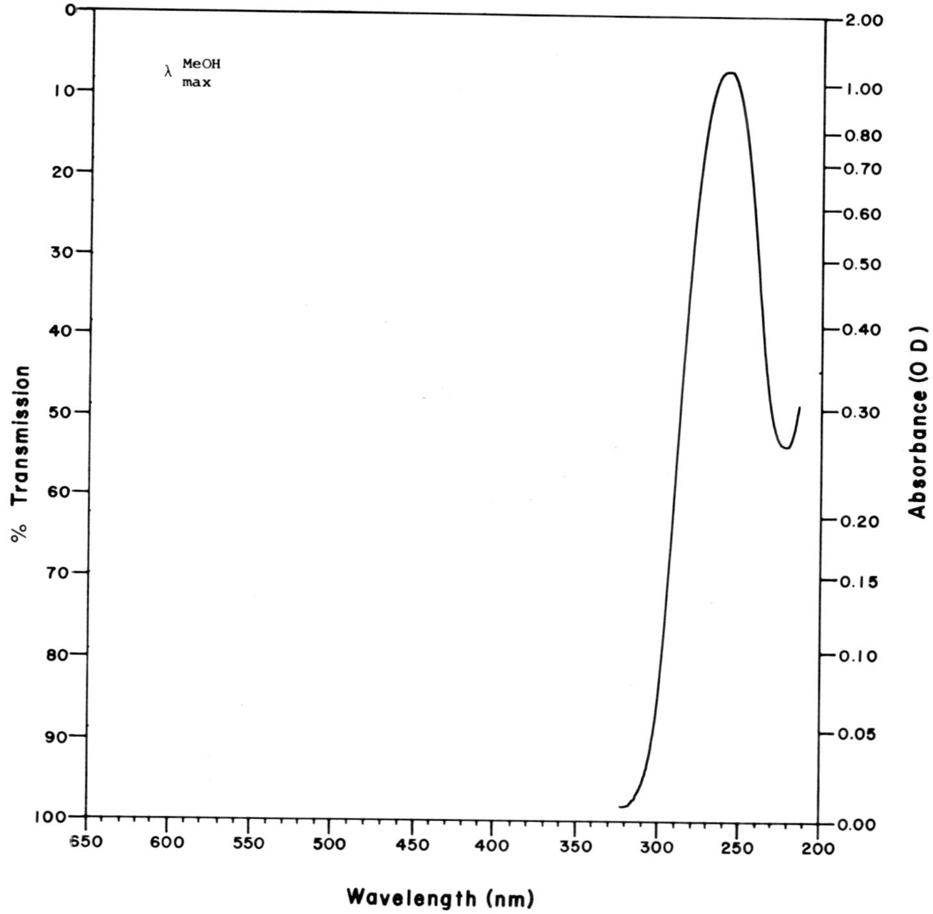
λ MeOH max
% Transmission
Absorbance (O D)
Wavelength (nm)
0
10
20
30
40
50
60
70
80
90
100
2.00
1.00
0.80
0.70
0.60
0.50
0.40
0.30
0.20
0.15
0.10
0.05
0.00
650
600
550
500
450
400
350
300
250
200

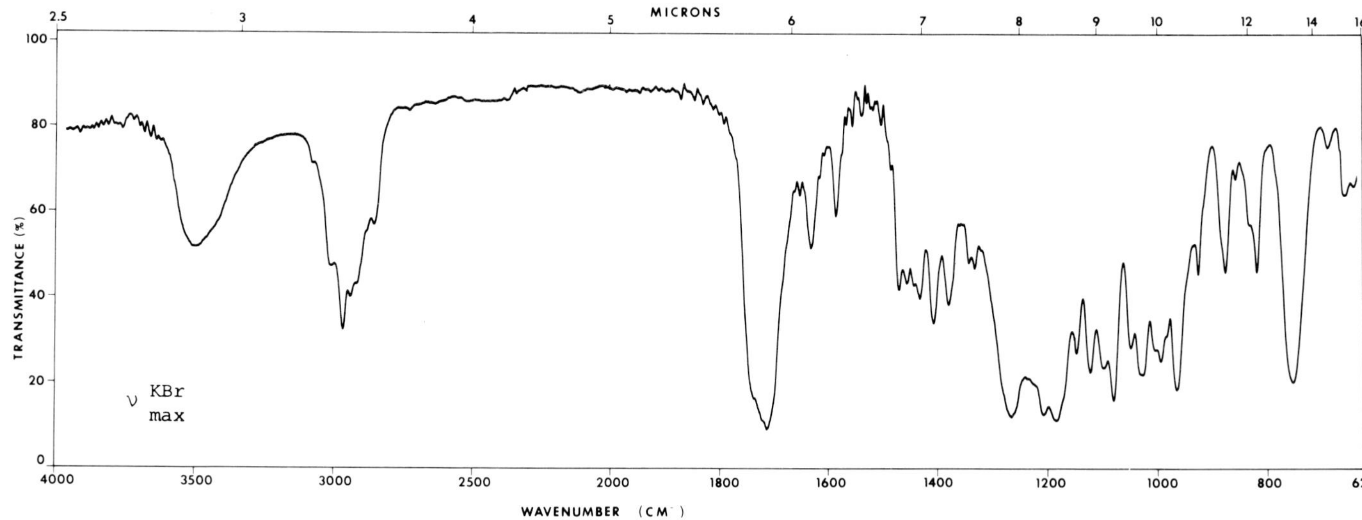
MICRONS
2.5
3
4
5
6
7
8
9
10
12
14
16
TRANSMITTANCE (%)
100
80
60
40
20
0
WAVENUMBER (CM⁻¹)
4000
3500
3000
2500
2000
1800
1600
1400
1200
1000
800
625
ν KBr max

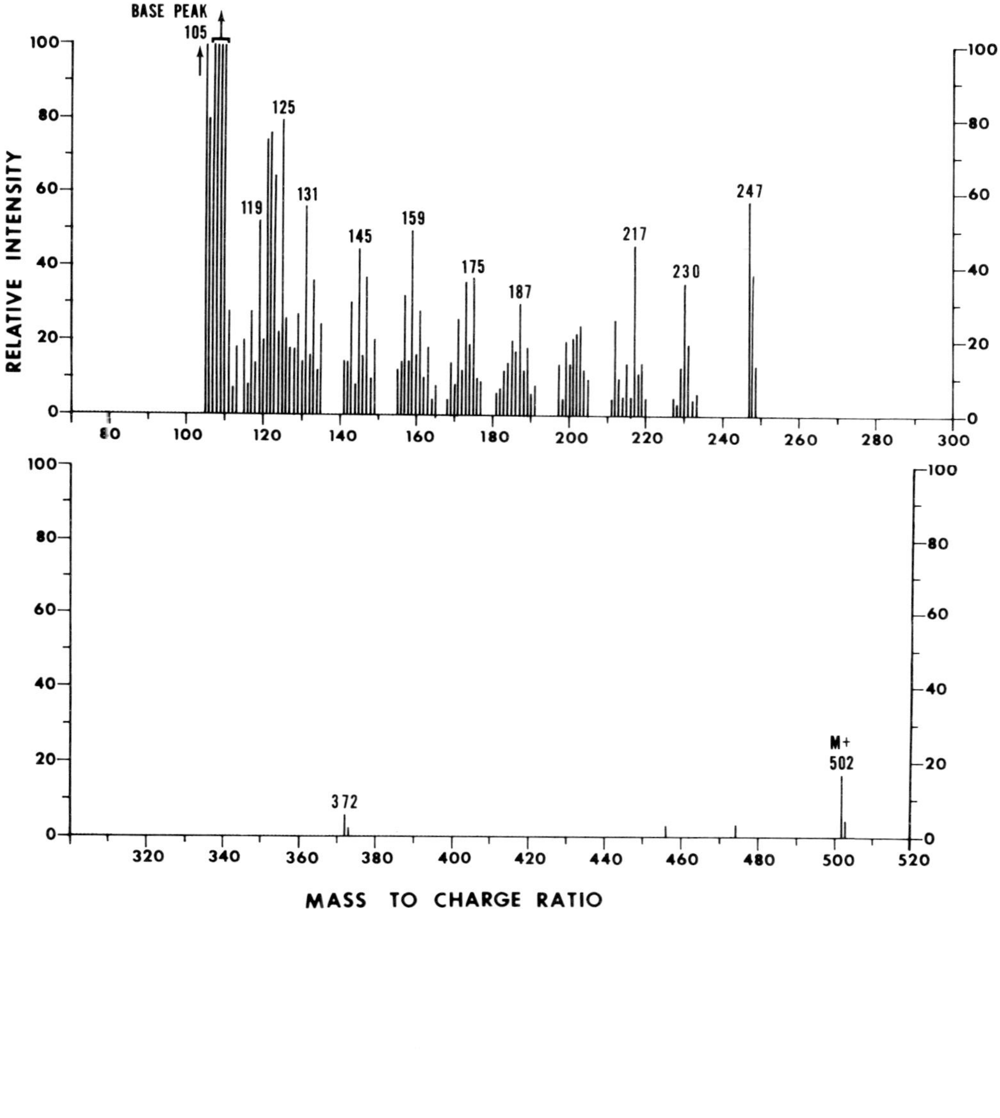
BASE PEAK
105
119
125
131
145
159
175
187
217
230
247
372
M+
502
RELATIVE INTENSITY
MASS TO CHARGE RATIO

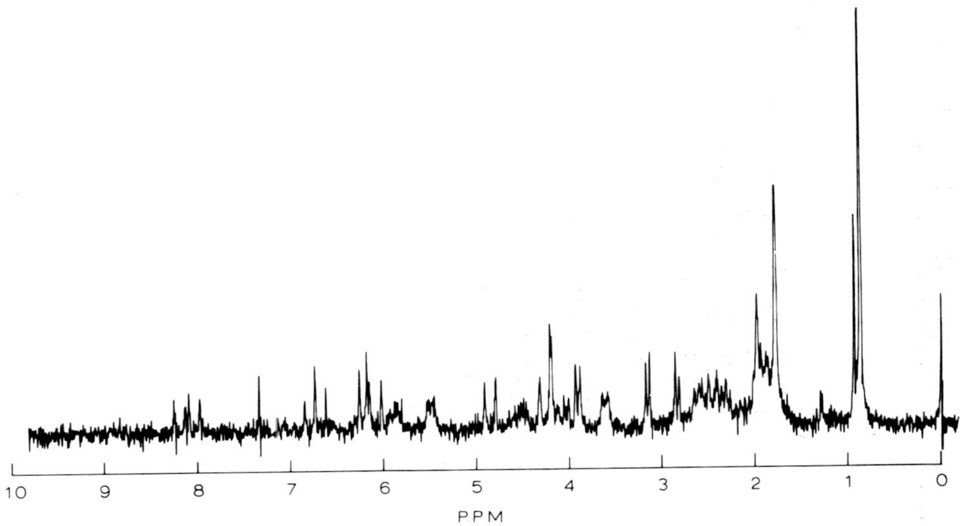

The 100 MHz ^1H NMR spectrum of verrucarin A in $CDCl_3$ solution.

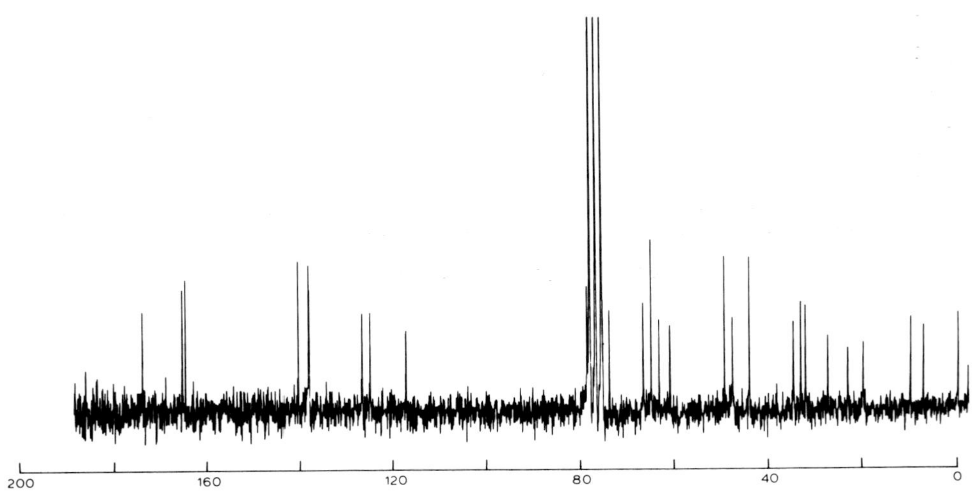

The 25 MHz ^{13}C NMR spectrum of verrucarin A in $CDCl_3$ solution.

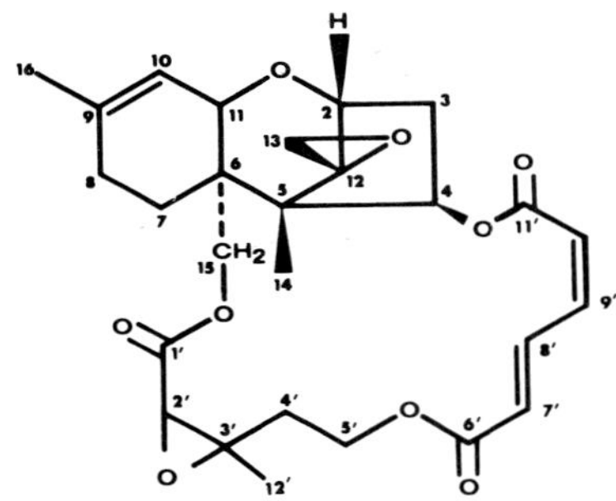

Common name	Verrucarin B
Molecular weight	500.2045
Molecular formula	$C_{27}H_{32}O_9$
General characteristics[48]	Crystals, mp >330°C $[\alpha]_D$ +94° (in chloroform) +101° (in dioxane) +147° (in benzene)
UV data[48]	λ_{max}^{EtOH} nm (ϵ): 258.5(23,400)
Fungal source[43,44,48]	*Myrothecium verrucaria* (Albertini and Schweinitz) Ditmar *ex* Fries *M. roridum*
Toxicity data[14,78]	Typical trichothecene type activity: dermal toxicity, antifungal, cytostatic, and acute toxicity to vertebrate animals. The LD_{50} in mice dosed IV was 7.0 mg/kg. The ED_{50} for cytostatic activity *in vitro* against Ehrlich ascites tumor cells was 0.003 μg/ml.
TLC data[12]	*Absorbent*: Kieselgel G *Solvent*: ethyl ether (two consecutive runs) R_f: 0.37 *Detection*: iodine vapors
Other significant references	Circular dichroism[77]
^{13}C *NMR data*[14]	C2, 78.7; C3, 34.8; C4, 75.4; C5, 49.0; C6, 43.6; C7, 19.8; C8, 27.4; C9, 140.4; C10, 118.0; C11, 66.9; C12, 64.9; C13, 47.5; C14, 7.3; C15, 63.6; C16, 22.8; C1′, 167.4; C2′, 58.0; C3′, 61.1; C4′, 36.9;

C5′, 60.4; C6′, 165.8; C7′, 127.2; C8′, 138.0; C9′, 138.0; C10′, 125.6; C11′, 164.8; C12′, 15.8

^{1}H NMR data H4, 5.90; H10, 5.47; H13a, 3.00; H14, 0.88; H16, 1.74; H2′, 3.41; H7′, 6.10; H8′, 7.98; H9′, 6.69; H10′, 6.19; H12′, 1.56

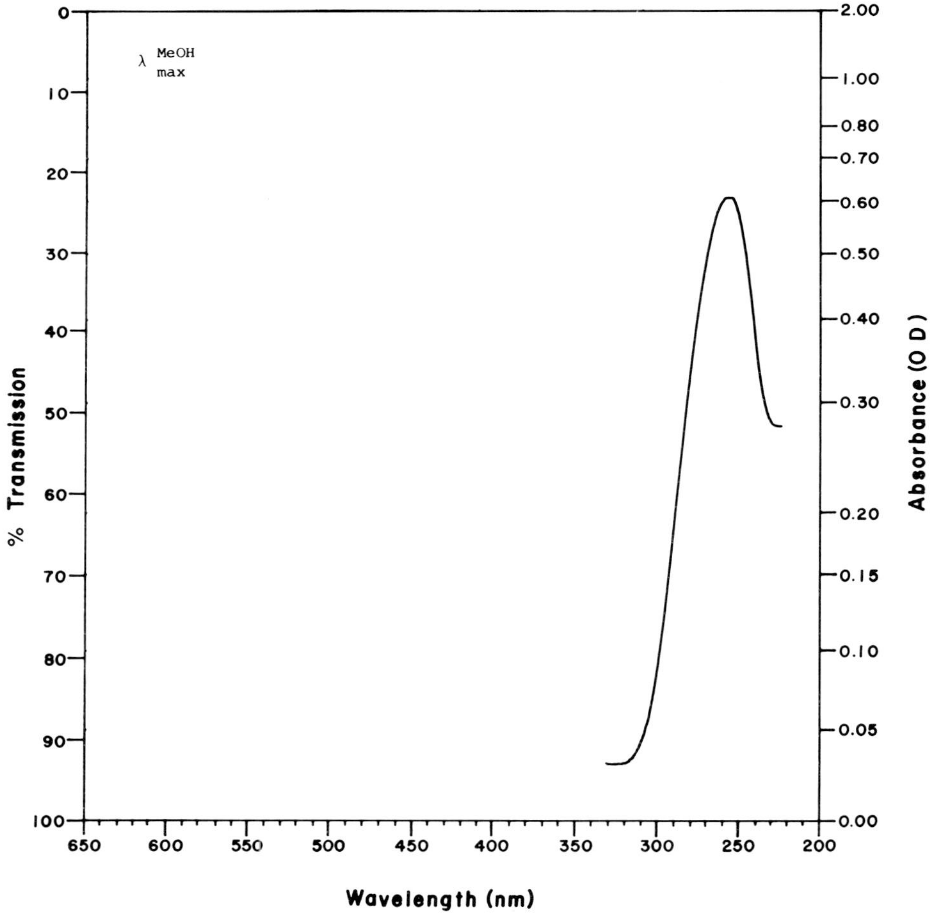

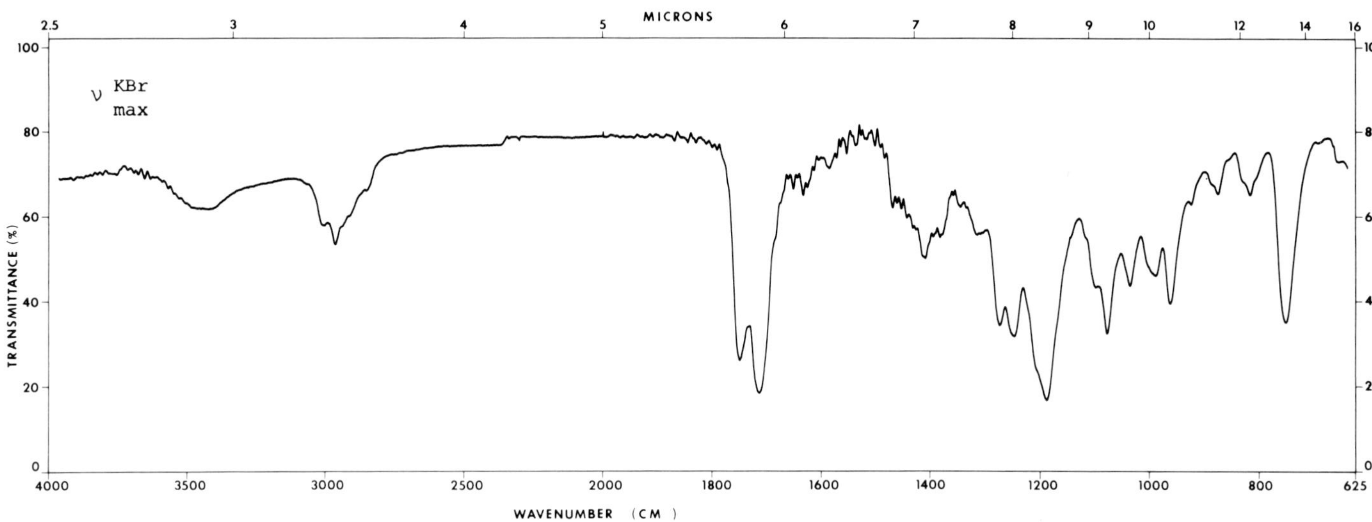
MICRONS
2.5
3
4
5
6
7
8
9
10
12
14
16
100
80
60
40
20
0
TRANSMITTANCE (%)
ν KBr max
4000
3500
3000
2500
2000
1800
1600
1400
1200
1000
800
625
WAVENUMBER (CM)

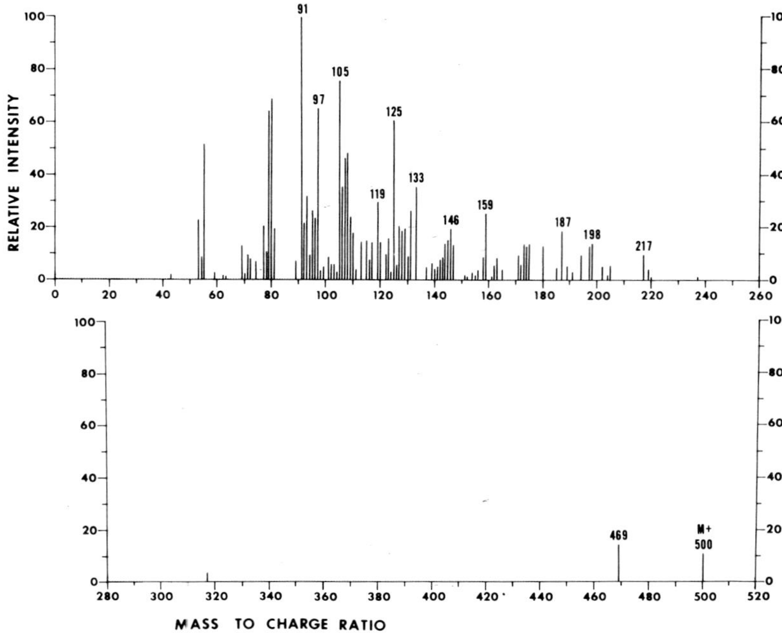
RELATIVE INTENSITY
91
105
97
125
133
119
146
159
187
198
217
469
M+
500
MASS TO CHARGE RATIO

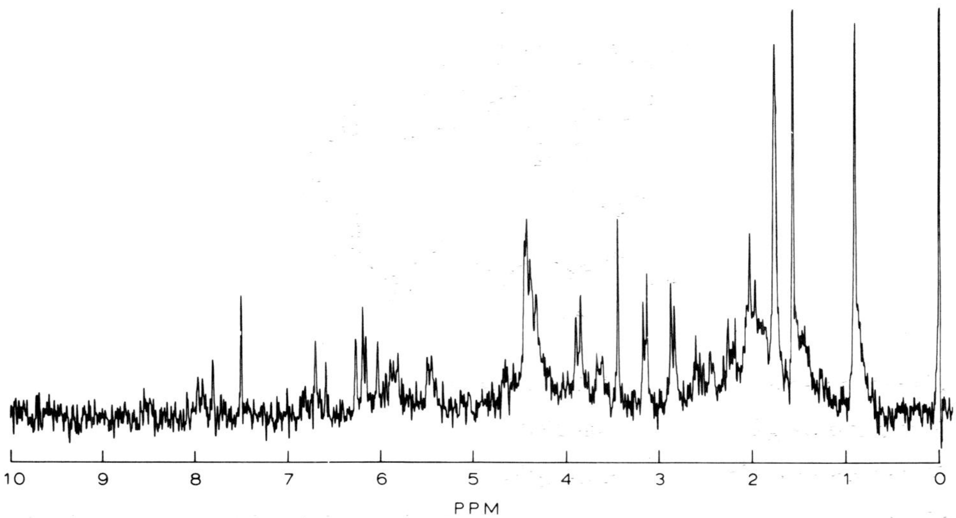

The 100 MHz ¹H NMR spectrum of verrucarin B in $CDCl_3$/DMSO solution.

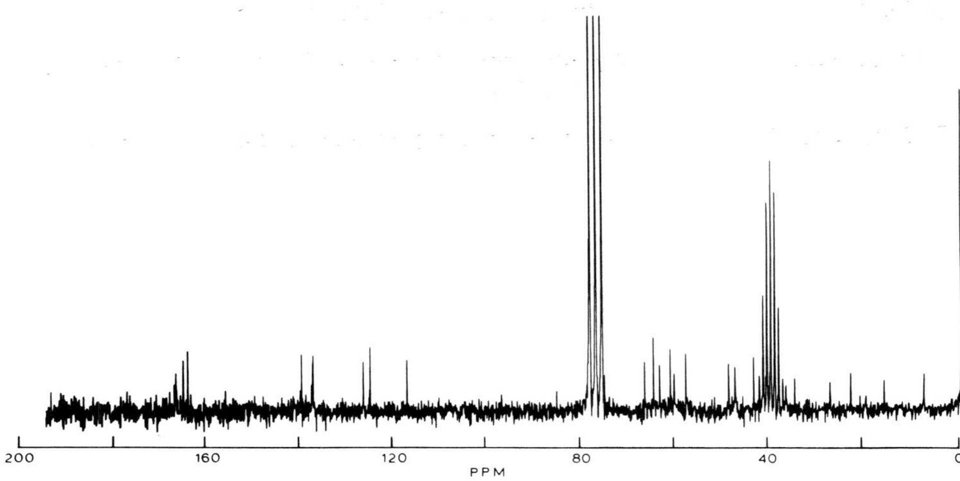

The 25 MHz ¹³C NMR spectrum of verrucarin B in $CDCl_3$/DMSO solution.

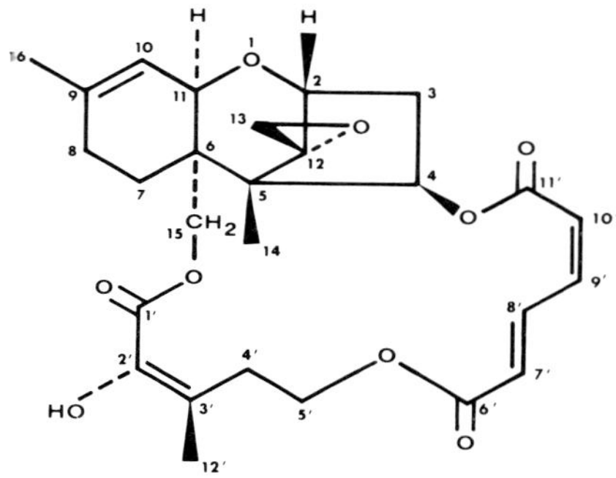

Common name	2′-Dehydroverrucarin A
Molecular weight	500.2045
Molecular formula	$C_{27}H_{32}O_9$
General characteristics[105]	Needles from acetone–ethyl ether, mp 233°–240°C $[\alpha]_D^{25}$ +118° (in chloroform)
UV data[105]	λ_{max}^{EtOH} nm (ϵ): 262(23,400)
IR data[105]	ν_{max}^{KBr} cm^{-1}: 1710, 1725, 1590, and 1630
Fungal source[105]	*Myrothecium roridum* Tode *ex* Fr., Gattungstyp bei Fries
Toxicity data[105]	Antibiotic activity similar to that of verrucarin A

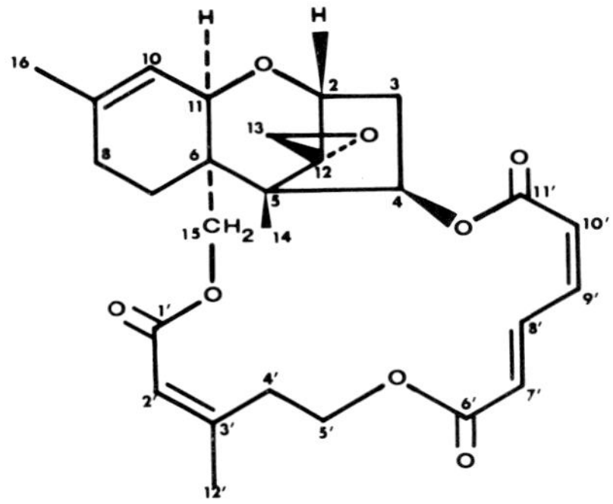

Common name	Verrucarin J
Molecular weight	484.2114
Molecular formula	$C_{27}H_{32}O_8$
General characteristics[12]	Colored needles from chloroform–ether; acetone–ether, mp >315°C (dec.) $[\alpha]_D^{22}$ +22° (*C* = 1.011 in chloroform) $[\alpha]_D^{23}$ +41° (*C* = 0.784 in benzene)
UV data[12]	λ_{max}^{EtOH} nm (ϵ): 196(15,500), 219(19,900), and 262(14,500)
IR data[12]	$\nu_{max}^{CH_2Cl_2}$ cm^{-1}: 2810, 1710, 1650, 1630, 1588, 1352, 1221, 1180, 1147, 1070-1088, 1042, 995, 968, 877, and 820
Fungal source[12]	*Myrothecium verrucaria* (Albertini and Schweinitz) Ditmar *ex* Fries *M. roridum* Tode *ex* Fr.
Toxicity data[12]	Antibiotic activity similar to that of verrucarin A
TLC data[12]	*Adsorbent*: Kieselgel G *Solvent*: ethyl ether (two consecutive runs) R_f: 0.42 *Detection*: iodine vapors
Other significant references	Circular dichroism[77]
^{1}*H NMR data*[27a]	H4, 5.90; H10, 5.47; H13a, 2.98; H14, 0.83; H15a, 4.30; H16, 1.72; H2′, 3.50; H4a′, 2.50; H5′, 3.82; H7′, 6.05; H8′, 8.12; H9′, 6.60; H10′, 6.10; H12′, 2.28

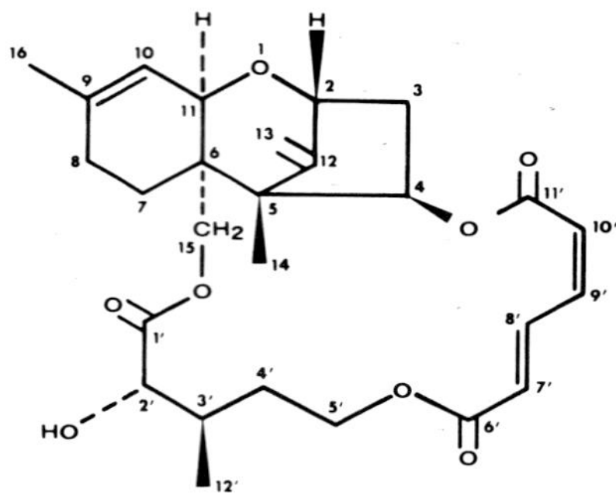

Common name	Verrucarin K
Molecular weight	486.2254
Molecular formula	$C_{27}H_{34}O_8$
General characteristics[15]	Crystals from dichloromethane–ether, mp >320°C (dec.) $[\alpha]_D^{23}$ +218° (C = 0.58 in chloroform) Monoacetyl derivative: colorless needles from acetone–ether–petroleum ether, mp 199°–202°C $[\alpha]_D^{23}$ +143° (C = 0.83 in chloroform)
UV data[15]	λ_{max}^{EtOH} nm (ϵ): 259(15,500)
IR data[15]	ν_{max} cm^{-1}: 1710, 3550, 1630, and 1585
Fungal source[15]	*Myrothecium verrucaria* (Albertini and Schweinitz) Ditmar *ex* Fries
^{13}C NMR data[15]	C2, 78.5; C3, 35.9; C4, 75.5; C5, 52.0; C6, 44.2; C7, 18.7; C8, 27.4; C9, 140.5; C10, 118.2; C11, 66.5; C12, 151.6; C13, 106.3; C14, 12.1; C15, 63.5; C16, 23.2; C1′, 174.5; C2′, 74.0; C3′, 61.1; C4′, 32.1; C5′, 61.1; C6′, 165.9; C7′, 127.4; C8′, 138.7; C9′, 138.7; C10′, 125.6; C11′, 165.3; C12′, 10.0
^{1}H NMR data[15]	H4, 5.81; H10, 5.39; H13a, 4.71; H13b, 5.18; H14, 1.09; H15, 3.66; H16, 1.70; H7′, 6.05; H8′, 8.05; H9′, 6.67; H10′, 6.08; H12′, 0.89

REFERENCES

1. Achilladelis, B. A., Adams, P. M., and Hanson, J. R., *J. Chem. Soc., Perkin Trans. 1* p. 1425 (1972).
2. Achilladelis, B. A., and Hanson, J. R., *Phytochemistry* **8,** 765 (1969).
3. Achini, R., Muller, B., and Tamm, C., *J. Chem. Soc. D* p. 404 (1971).
4. Bamburg, J. R., *Clin. Toxicol.* **5,** 495 (1972).
5. Bamburg, J. R., *Adv. Chem. Ser.* **149,** 144 (1976).
6. Bamburg, J. R., Marasas, W. F., Riggs, N. V., Smalley, E. B., and Strong, F. M., *Biotechnol. Bioeng.* **10,** 445 (1968).
7. Bamburg, J. R., Riggs, N. V., and Strong, F. M., *Tetrahedron* **24,** 3329 (1968).
8. Bamburg, J. R., Smalley, E. B., Riggs, N. V., and Strong, F. M., *Abstr. Pap. 152nd Annu. Meet. Am. Chem. Soc.* No. A-93, Div. Agric. Food Chem. (1966).
9. Bamburg, J. R., and Strong, F. M., *Phytochemistry* **8,** 2405 (1969).
10. Bamburg, J. R., and Strong, F. M., *in* "Microbial Toxins" (S. Kadis, A. Ciegler, and S. J. Ajl, eds.), Vol. 7, p. 207. Academic Press, New York, 1971.
11. Blight, M. M., and Grove, J. F., *J. Chem. Soc., Perkin Trans. 1* p. 1691 (1974).
12. Böhner, B., Fetz, E., Härri, E., Sigg, H. P., Stoll, C., and Tamm, C., *Helv. Chim. Acta* **48,** 1079 (1965).
13. Böhner, B., and Tamm, C., *Helv. Chim. Acta* **49,** 2526 (1966).
14. Breitenstein, W., and Tamm, C., *Helv. Chim. Acta* **58,** 1172 (1975).
15. Breitenstein, W., and Tamm, C., *Helv. Chim. Acta* **60,** 1522 (1977).
16. Brian, P. W., Dawkins, A. W., Grove, J. F., Hemming, H. G., Lowe, D., and Norris, G. L. F., *J. Exp. Bot.* **12,** 1(1961).
17. Chung, C. W., Trucksess, M. W., Giles, A. L., Jr., and Friedman, L., *J. Am. Off. Anal. Chem.* **57,** 1121 (1974).
18. Colvin, E. W., Malchenko, S., Raphael, R. A., and Roberts, J. S., *J. Chem. Soc., Perkin Trans. 1* p. 1989 (1973).
19. Colvin, E. W., Raphael, R. A., and Roberts, J. S., *J. Chem. Soc. D* p. 858 (1971).
20. Cundliffe, E., Cannon, M., and Davies, J., *Proc. Natl. Acad. Sci. U.S.A.* **71,** 30 (1974).
21. Dawkins, A. W., *J. Chem. Soc. C* p. 116 (1966).
22. Dawkins, A. W., Grove, J. F., and Tidd, B. K., *Chem. Commun.* p. 27 (1965).
23. Dockerill, B., Hanson, J. R., and Siverns, M., *Phytochemistry* **17,** 427 (1978).
24. Ellison, R. A., and Kotsonis, F. N., *Appl. Microbiol.* **27,** 423 (1974).
25. Ellison, R. A., and Kotsonis, F. N., *J. Org. Chem.* **41,** 576 (1976).
26. Eppley, R. M., and Bailey, W. J., *Science* **181,** 758 (1973).
27. Eppley, R. M., Mazzola, E. P., Highet, R. J., and Bailey, W. J., *J. Org. Chem.* **42,** 240 (1977).
27a. Fetz, E., Böhner, B., and Tamm, C., *Helv. Chim. Acta* **49,** 1669 (1965).
28. Fishman, J., Jones, E. R. H., Lowe, G., and Whiting, M. C., *J. Chem. Soc.* p. 3948 (1960).
29. Flury, E., Mauli, R., and Sigg, H. P., *Chem. Commun.* p. 26 (1965).
30. Freeman, G. G., *J. Gen. Microbiol.* **12,** 213 (1955).
31. Freeman, G. G., Gill, J. E., and Waring, W. S., *J. Chem. Soc.* p. 1105 (1959).
32. Freeman, G. G., and Morrison, R. I., *Nature (London)* **162,** 20 (1948).
33. Fujimoto, Y., Morita, Y., and Tatsuno, T., *Chem. Pharm. Bull.* **20,** 1194 (1972).
34. Fujimoto, Y., Yokuna, S., Nakamura, T., Morikawa, T., and Tatsuno, T., *Tetrahedron Lett.* p. 2523 (1974).
35. Gardner, D., Glen, A. T., and Turner, W. B., *J. Chem. Soc., Perkin Trans. 1* p. 2576 (1972).
36. Gilgan, M. W., Smalley, E. B., and Strong, F. M., *Arch. Biochem. Biophys.* **114,** 1 (1966).
37. Glaz, E. T., Csanyi, E., and Gyimesi, J., *Nature (London)* **212,** 617 (1966).
38. Godtfredsen, W. O., and Vangedal, S., *Proc. Chem. Soc.* p. 188 (1964).
39. Godtfredsen, W. O., and Vangedal, S., *Acta Chem. Scand.* **19,** 1088 (1965).

40. Grove, J. F., *J. Chem. Soc. C* p. 378 (1970).
41. Grove, J. F., and Mortimer, P. H., *Biochem. Pharmacol.* **18,** 1473 (1969).
42. Gutzwiller, J., Mauli, R., Sigg, H. P., and Tamm, C., *Helv. Chim. Acta* **47,** 2234 (1964).
43. Gutzwiller, J., and Tamm, C., *Helv. Chim. Acta* **48,** 157 (1965).
44. Gutzwiller, J., and Tamm, C., *Helv. Chim. Acta* **48,** 177 (1965).
45. Gutzwiller, J., and Tamm, C., *Helv. Chim. Acta* **46,** 1786 (1963).
46. Gyimesi, J., *Acta Chim. Acad. Sci. Hung.* **45,** 323 (1965).
47. Gyimesi, J., and Melera, A., *Tetrahedron Lett.* p. 1665 (1967).
48. Hanson, J. R., Marten, T., and Siverns, M., *J. Chem. Soc., Perkin Trans 1,* p. 1033 (1974).
49. Härri, E., Loeffler, W., Sigg, H. P., Stahelin, H., Stoll, C., Tamm, C., and Wiesinger, E., *Helv. Chim Acta* **45,** 839 (1062).
50. Ikediobi, C. O., Hsu, I. C., Bamburg, J. R., and Strong, F. M., *Anal. Biochem.* **43,** 327 (1971).
51. Ilus, T., Ward, P., Nummi, M., Adlercreutz, H., and Gripenberg, J., *Phytochemistry* **16,** 1839 (1977).
52. Ishii, K., *Phytochemistry* **14,** 2469 (1975).
53. Ishii, K., Pathre, S. V., and Mirocha, C. J., *J. Agric. Food Chem.* **26,** 649 (1978).
54. Ishii, K., Sakai, K., Ueno, Y., Tsunoda, H., and Enomoto, M., *Appl. Microbiol.* **22,** 718 (1971).
55. Jones, E. R. H., and Lowe, G., *J. Chem. Soc.* p. 3959 (1960).
56. Knoll, W., and Tamm, C., *Helv. Chim. Acta* **58,** 1162 (1975).
57. Kotsonis, F. N., Ellison, R. A., and Smalley, E. B., *Appl. Microbiol.* **30,** 493 (1975).
58. Kupchan, S. M., Jarvis, B. B., Dailey, R. G., Jr., Bright, W., Bryan, R. F., and Shizuri, Y., *J. Am. Chem. Soc.* **98,** 7092 (1976).
59. Lansden, J. A., Cole, R. J., Dorner, J. W., Cox, R. H., and Cutler, H. G., *J. Agric. Food Chem.* **26,** 246 (1978).
60. Loevens, Kemiske Fabrik Produktionsaktieselskab, Netherlands Patent 302, 527 (1964); *Chem. Abstr.* **62,** 1050f (1965).
61. Masouka, N., and Kawikawa, T., *Tetrahedron Lett.* p. 1691 (1976).
62. Matsumoto, M., Minato, H., Tori, K., and Ueyama, M., *Tetrahedron Lett.* p. 4093 (1977).
63. Matsumoto, M., Minato, H., Uotani, N., Matsumoto, K., and Kondo, E., *J. Antibiot.* **30,** 681 (1977).
64. McPhail, A. T., and Sim, G. A., *Chem. Commun.* p. 350 (1965).
65. Minato, H., Katayama, T., and Tori, K., *Tetrahedron Lett.* 2579 (1975).
66. Morooka, N., Nakano, N., Nakazawa, S., and Tsunoda, H., *Nippon Nogeikagaku Kaishi* **45,** 151 (1971).
67. Morooka, N., Uratsuji, N., Yoshizawa, T., and Yamamoto, H., *J. Food Hyg. Soc. Jpn.* **13,** 368 (1972).
68. Nespiak, A., Kocor, M., and Siewinski, A., *Nature (London)* **192,** 138 (1961).
69. Okuchi, M., Itoh, M., Kaneko, Y., and Doi, S., *J. Agric. Biol. Chem. Jpn.* **32,** 394 (1968).
70. Pathre, S. V., Mirocha, C. J., Christensen, C. M., and Behrens, J., *J. Agric. Food Chem.* **24,** 97 (1976).
71. Prentice, N., and Dickson, A. D., *Biotechnol. Bioeng.* **10,** 413 (1968).
71a. Rissom, T., Jakobsen, H. J., Rastrup-Andersen, N., and Lorck, H., *Acta Chem. Scand., Ser. B* **32,** 499 (1978).
72. Sato, N., Ueno, Y., and Enomoto, M., *Jpn. J. Pharmacol.* **25,** 263 (1975).
73. Schindler, D., *Nature (London)* **249,** 38 (1974).
74. Sigg, H. P., Mauli, R., Flury, E., and Hauser, D., *Helv. Chim. Acta* **48,** 962 (1965).
75. Smalley, E. B., *J. Am. Vet. Med. Assoc.* **163,** 1278 (1973).
76. Smalley, E. B., and Strong, F. M., *in* "Mycotoxins" (I. F. H. Purchase, ed.), p. 199. Am. Elsevier, New York, 1974.
77. Snatzke, G., and Tamm, C., *Helv. Chim. Acta* **50,** 1618 (1967).

78. Steyn, P. S., Vleggaar, R., Rabie, C. J., Kriek, N. P. J., and Harington, J. S., *Phytochemistry* **17,** 949 (1978).
79. Szathmary, C. I., Mirocha, C. J., Palyusik, M., and Pathre, S. V., *Appl. Environ. Microbiol.* **32,** 579 (1976).
80. Tamm, C., *Angew. Chem., Int. Ed. Engl.* **5,** 523 (1966).
81. Tamm, C., *Fortschr. Chem. Org. Naturst.* **31,** 63 (1974).
82. Tanaka, T., Ogawa, K., Toyasaki, N., Matsuda, Y., Matsuki, Y., and Ueno, Y., *Proc. Jpn. Assoc. Mycotoxicol.* p. 16 (1978).
83. Tatsuno, T., Fujimoto, Y., and Morita, Y., *Tetrahedron Lett.* p. 33, 2823 (1969).
84. Tatsuno, T., Morita, Y. , Tsunoda, H., and Umeda, M., *Chem. Pharm. Bull.* **18,** 1485 (1970).
85. Tatsuno, T., Ohtsubo, K., and Saito, M., *Pure Appl. Chem.* **35,** 309 (1973).
86. Tatsuno, T., Saito, M., Enomoto, M., and Tsunoda, H., *Chem. Pharm. Bull.* **16,** 2519 (1968).
87. Tidd, B. K., *J. Chem. Soc. C* p. 218 (1967).
88. Traxler, P., and Tamm, C., *Helv. Chim. Acta* **53,** 1846 (1970).
89. Ueno, Y., and Fukushima, K., *Experientia* **24,** 1032 (1968).
90. Ueno, Y., Ishii, K., Sakai, K., Kanaeda, S., Tsunoda, H., Tanaka, T., and Enomoto, M., *Jpn. J. Exp. Med.* **42,** 187 (1972).
91. Ueno, Y., Ishii, K., Sawano, M., Ohtsubo, K., Matsuda, Y., Tanaka, T., Kurata, H., and Ichinoe, M., *Jpn. J. Exp. Med.* **47,** 177 (1977).
92. Ueno, Y., Ishikawa, Y., Amakai, K., Nakajima, M., Saito, M., Enomoto, M., and Ohtsubo, K., *Jpn. J. Exp. Med.* **40,** 33 (1970).
93. Ueno, Y., Ishikawa, Y., Saito-Amakai, K., and Tsunoda, H., *Chem. Pharm. Bull.* **18,** 304 (1970).
94. Ueno, Y., Nakajima, M., Sakai, K., Ishii, K., Sato, N., and Shimada, N., *J. Biochem. (Tokyo)* **74,** 285 (1973).
95. Ueno, Y., Sato, N., and Ishii, K., *Jpn. J. Pharmacol.* **245,** 40 (1974).
96. Ueno, Y., Ueno, I., Amakai, K., Ishikawa, Y., Tsunoda, H., Okubo, K., Saito, M., and Enomoto, M., *Jpn. J. Exp. Med.* **41,** 507 (1971).
97. Ueno, Y., Ueno, I., Tatsuno, T., Ohokubo, K., and Tsunoda, H., *Experientia* **25,** 1062 (1969).
98. Vesonder, R. F., Ciegler, A., and Jensen, A. H., *Appl. Microbiol.* **26,** 1008 (1973).
99. Vesonder, R. F., Ciegler, A., Jensen, A. H., Rohwedder, W. K., and Weisleder, D., *Appl. Environ. Microbiol.* **31,** 280 (1976).
100. Wei, C., Hansen, B. S., Vaughan, M. H., Jr., and McLaughlin, C. S., *Proc. Natl. Acad. Sci. U.S.A.* **71,** 713 (1974).
101. Wei, C., and McLaughlin, C. S., *Biochem. Biophys. Res. Commun.* **57,** 838 (1974).
102. Yoshizawa, T., and Morooka, N., *J. Agric. Biol. Chem. Jpn.* **37,** 2933 (1973).
103. Yoshizawa, T., and Morooka, N., *J. Food Hyg. Soc. Jpn.* **15,** 261 (1974).
104. Yoshizawa, T., and Morooka, N., *Appl. Microbiol.* **29,** 54 (1975).
105. Zurcher, W., and Tamm, C., *Helv. Chim. Acta* **49,** 2594 (1966).

The Cytochalasins

6

The cytochalasins are a rapidly growing group of fungal metabolites produced by several different and unrelated fungal species. They have been assigned different trivial names given by the different laboratories which named them after the different fungal sources. These names were zygosporins, phomins, chaetoglobosins, kodocytochalasins, paspalins, and cytochalasins. The latter trivial name "cytochalasin," which was given by English scientists because of the novel effects on mammalian cells, has been adopted for the basic skeleton of the class of compounds. The cytochalasins can be divided into two general groups based on whether they contain a phenylalanine or tryptophan moiety at position 10 of the perhydroisoindole ring system. Those that contain a tryptophan moiety generally have been referred to as the chaetoglobosins. Natori[28] has classified the cytochalasins into several groups based on the substitutent at position 10 (phenyl or indol-3-yl), the type of the six-membered ring in the perhydroisoindole ring system (four different types), and the structure of the macrocyclic ring after C_{16}.[28] The cytochalasins generally can be described as a phenylalanine or tryptophan moiety linked to a perhydroisoindole moiety which is in turn linked to a C^{16}–C_{18} polyketide ring system containing a carbocyclic (cytochalasin C), a lactone (cytochalasin A), or a cyclic carbonate (cytochalasin E) moiety.

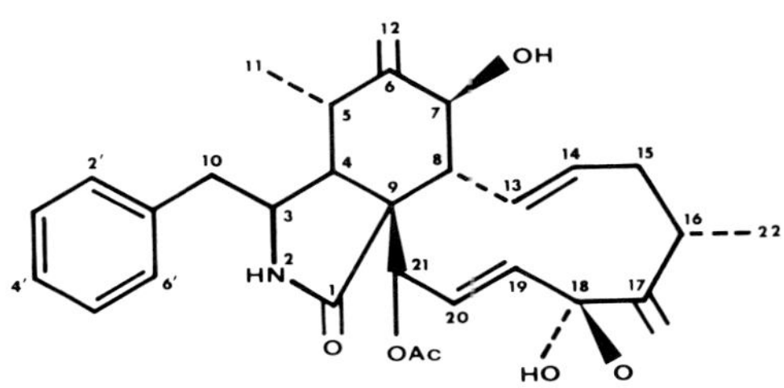

[11]Cytochalasin

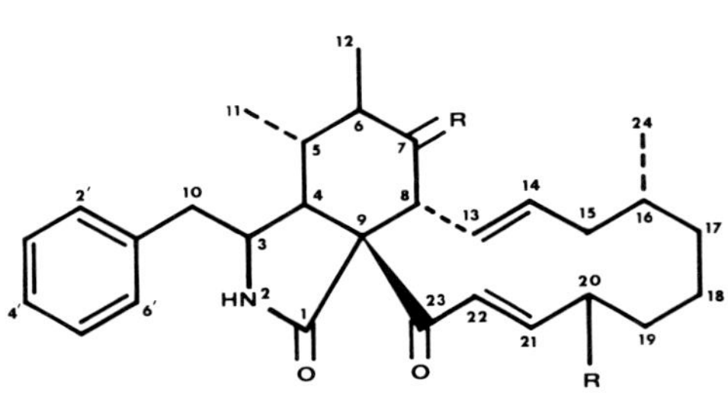

[13]Cytochalasin

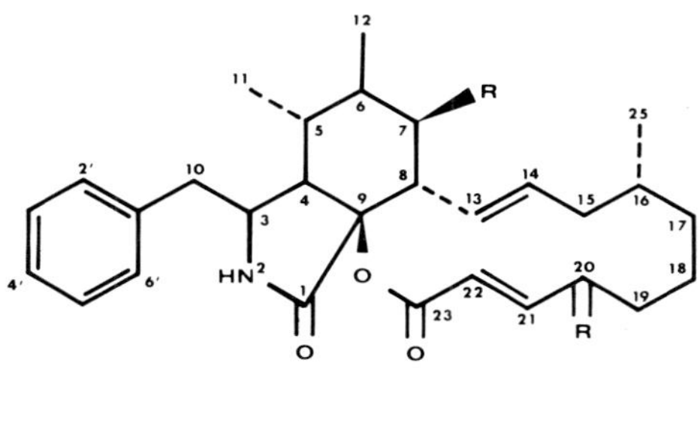

24-Oxa-[14]cytochalasin

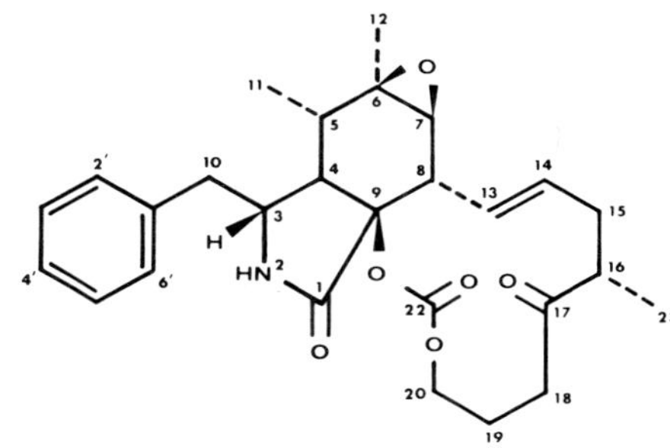

21,23-Dioxa-[13]cytochalasin

The cytochalasins generally possess the following distinctive biological effects: inhibition of cytoplasmic cleavage but not nuclear division in cell culture, resulting in polynucleate cells; inhibition of cell movement; and the novel effect of nuclear extrusion in cell culture leading to viable nuclei isolated from the cytoplasm. More recently, some of these compounds have been recognized for their potential as mycotoxins.[18]

Cytochalasins	Molecular weight	Molecular formula
Cytochalasin group		
Cytochalasin E	495.2256	$C_{28}H_{33}O_7N$
Cytochalasin G	474.2518	$C_{29}H_{34}O_4N_2$
Cytochalasin A	477.2515	$C_{29}H_{35}O_5N$
Cytochalasin B	479.2671	$C_{29}H_{37}O_5N$
Cytochalasin F	479.2671	$C_{29}H_{37}O_5N$
Cytochalasin H	493.2828	$C_{30}H_{39}O_5N$
Cytochalasin C	507.2620	$C_{30}H_{37}O_6N$
Cytochalasin D	507.2620	$C_{30}H_{37}O_6N$
Chaetoglobosin group		
Chaetoglobosin A	528.2623	$C_{32}H_{36}O_5N_2$
Chaetoglobosin B	528.2623	$C_{32}H_{36}O_5N_2$
Chaetoglobosin C	528.2623	$C_{32}H_{36}O_5N_2$
Chaetoglobosin D	528.2623	$C_{32}H_{36}O_5N_2$
Chaetoglobosin E	530.2779	$C_{32}H_{38}O_5N_2$
Chaetoglobosin F	530.2779	$C_{32}H_{38}O_5N_2$
Chaetoglobosin G	528.2620	$C_{32}H_{36}O_5N_2$
Chaetoglobosin J	512.2500	$C_{32}H_{36}O_4N_2$
Chaetoglobosin K	556.2934	$C_{34}H_{40}O_5N_2$
Deoxaphomin	463.2722	$C_{29}H_{37}O_4N$
Proxiphomin	431.2824	$C_{29}H_{37}O_2N$
Protophomin	445.2616	$C_{29}H_{35}O_3N$
Zygosporin D	465.2515	$C_{28}H_{35}O_5N$
Zygosporin E	491.2722	$C_{30}H_{37}O_5N$
Zygosporin F	549.2725	$C_{32}H_{39}O_7N$
Zygosporin G	491.2671	$C_{30}H_{37}O_5N$

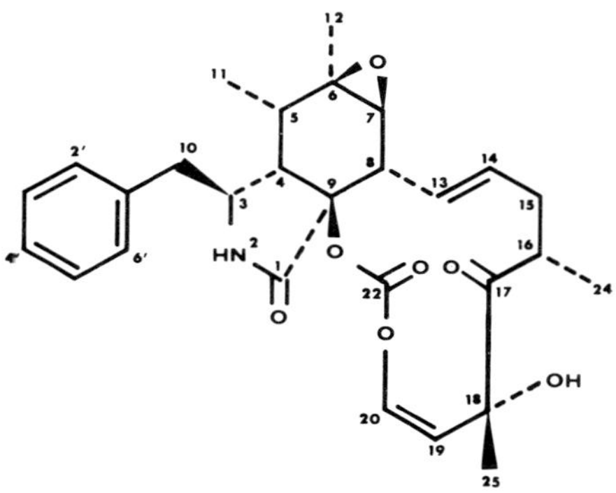

Common name	Cytochalasin E (6,7-epoxy-10-phenyl-5,6,16,18-tetramethyl-21,23-dioxa-[13]cytochalas-13,19-diene-17,22-dione)
Molecular weight	495.2256
Molecular formula	$C_{28}H_{33}O_7N$
General characteristics[11,18]	Crystals from acetone–hexane, mp 206°–208°C (dec.) $[\alpha]_D^{25}$ −25.6° (in methanol) Soluble in dichloromethane, chloroform, acetone, and methanol; sparingly soluble in water. It is unstable in chloroform solution and under acidic conditions.
UV data	Weak absorptions between 240 and 300 nm due in part to the $\pi \rightarrow \pi^*$ transition of the monosubstituted aromatic ring.
Fungal source[3,4,11]	*Rosellinia necutrix* *Aspergillus clavatus*
Toxicity data[11,13,18]	LD_{50} in rats; oral 9.1 mg/kg, IP 2.6 mg/kg. Death was due to circulatory collapse caused by massive extravascular effusion of plasma. The LD_{50} values for a single IP dose of cytochalasin E were 1-day-old rats, 0.98 mg/kg; adolescent rats, 2.60 mg/kg; mice, 4.60 mg/kg; and guinea pigs, 0.5–1.5 mg/kg. In adolescent rats dosed orally, LD_{50} was 1.3 mg/kg. Histopathologic examination showed congestive degenerative changes, necrosis of liver, kidney, spleen, pancreas, and small intestine, brain edema, pulmonary hemorrhages and injury to

vascular walls. Cytochalasin E rarely produces nuclear extrusion typical of other cytochalasins; however, it produces a unique halo around the nuclei.

TLC data[18,47]

Adsorbent: silica gel GF_{254}
Solvent: acetone–hexane, 2:3 v/v
R_f: 0.25–0.30
Detection: H_2SO_4 spraying and heat
Adsorbent: silica gel G-HR
Solvent: A: chloroform–acetone, 93:7 v/v
B: toluene–ethyl acetate–formic acid, 5:4:1 v/v/v
R_f: A: 0.10
B: 0.60
Detection: a dark purple fluorescent spot under UV light after spraying with 50% ethanolic H_2SO_4 and heating.

^{13}C NMR data

C1, 168.9 s; C3, 52.1 d; C4, 45.6 d; C5, 35.5 d; C6, 56.1 s; C7, 58.8 d; C8, 45.9 d; C9, 86.9 s; C10, 42.6 t; C11, 12.4 q; C12, 19.8 q; C13, 127.6 d; C14, 131.4 d; C15, 38.5 t; C16, 39.6 d; C17, 211.4 s; C18, 76.5 s; C19, 121.0 d; C20, 140.5 d; C22, 148.8 s; C24, 19.1 q; C25, 24.8 q; C1′, 136.1 s; C2′,6′, 130.1 d; C3′,5′, 127.9 d; C4′, 126.3 d

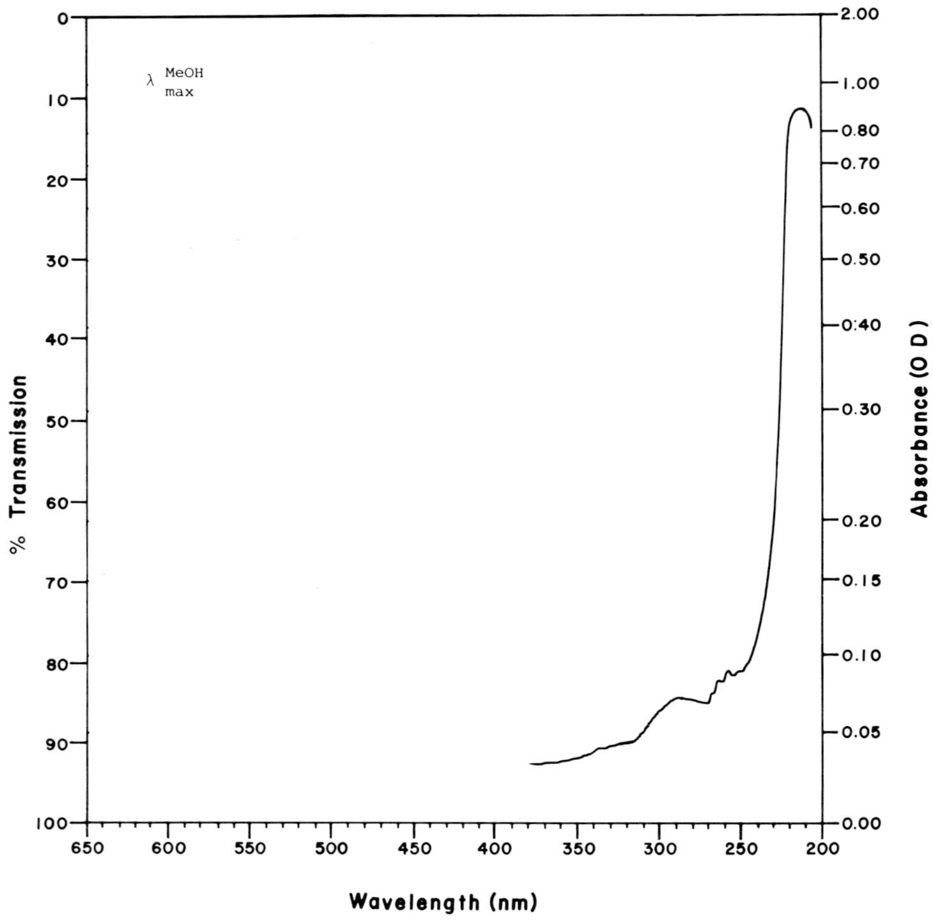
λ MeOH max
% Transmission
Absorbance (O D)
Wavelength (nm)
0
10
20
30
40
50
60
70
80
90
100
2.00
1.00
0.80
0.70
0.60
0.50
0.40
0.30
0.20
0.15
0.10
0.05
0.00
650
600
550
500
450
400
350
300
250
200

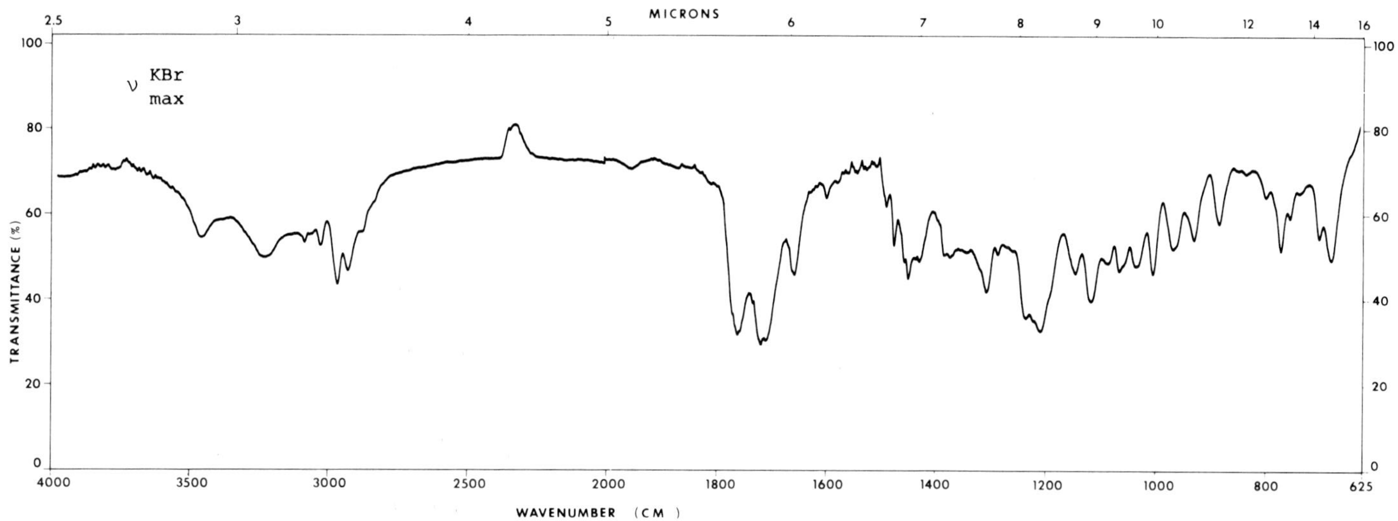

MICRONS
TRANSMITTANCE (%)
WAVENUMBER (CM)
ν KBr max

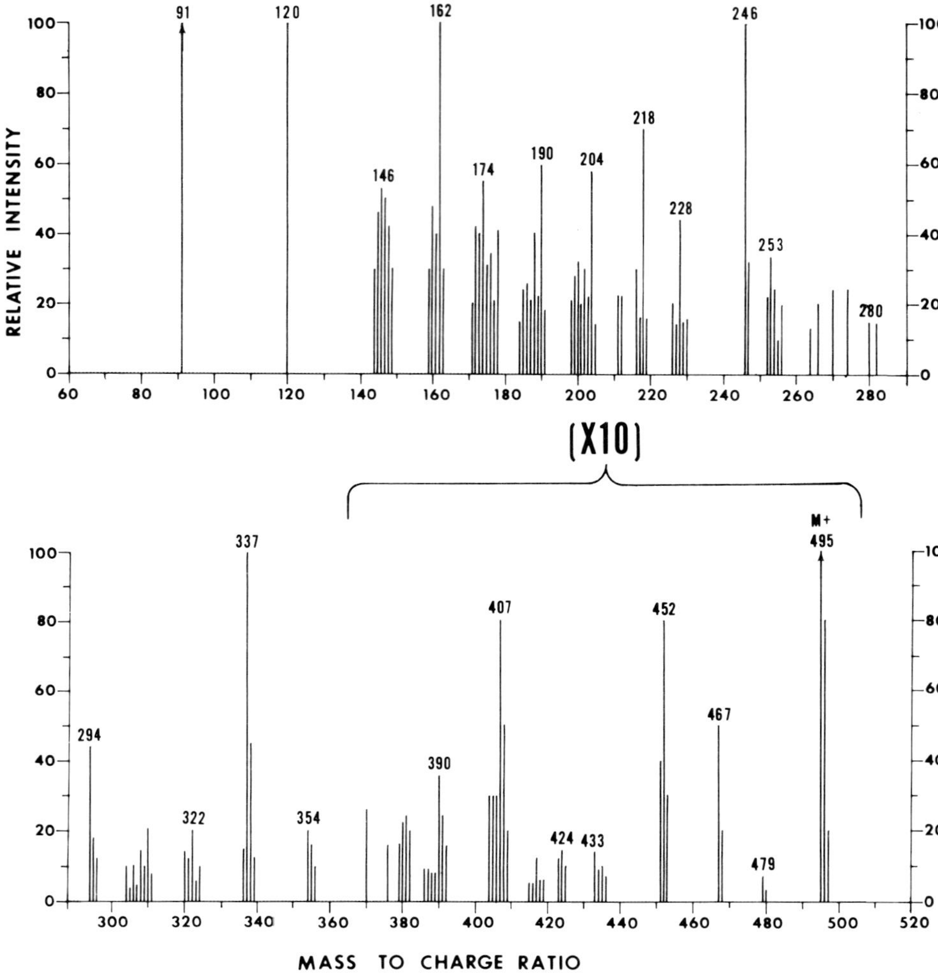
RELATIVE INTENSITY
MASS TO CHARGE RATIO
(X10)
M+
91
120
146
162
174
190
204
218
228
246
253
280
294
322
337
354
390
407
424
433
452
467
479
495

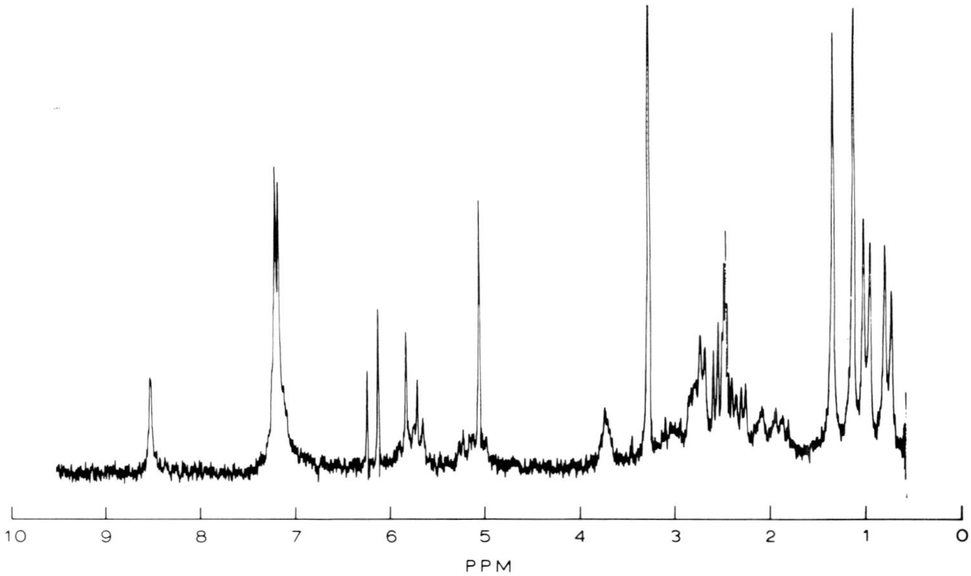

The 100 MHz ^{1}H NMR spectrum of cytochalasin E in DMSO solution.

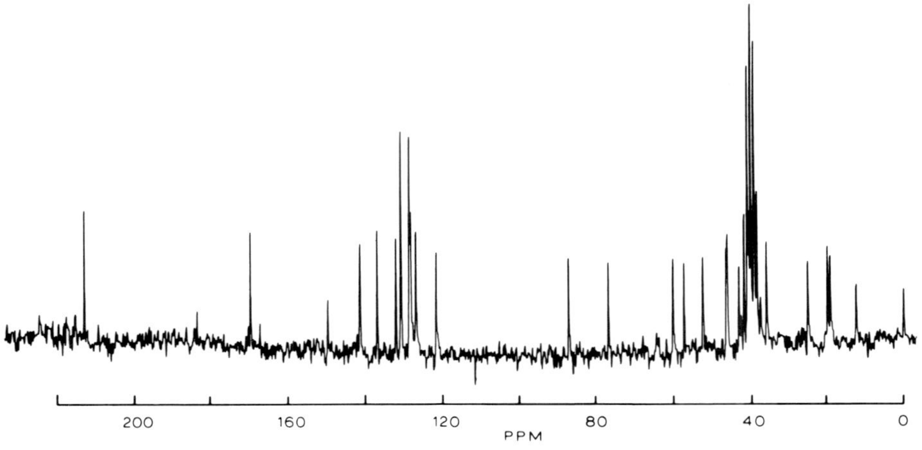

The 25 MHz ^{13}C NMR spectrum of cytochalasin E in DMSO solution.

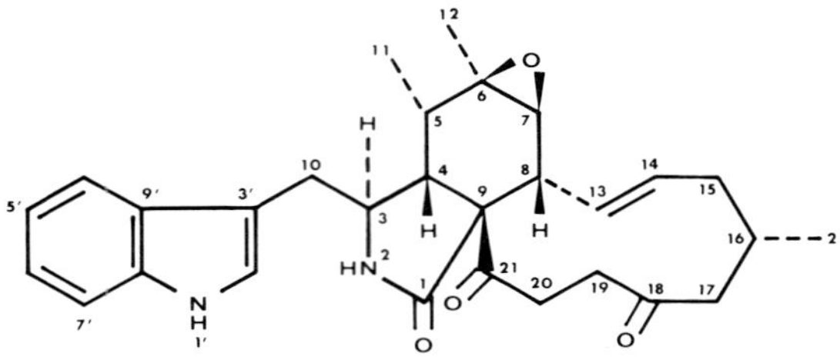

Common name	Cytochalasin G [6,7-Epoxy-10-(indol-3-yl)-5,6,16-trimethyl-[11]cytochalas-13-ene-1,18,21-trione]
Molecular weight	474.2518
Molecular formula	$C_{29}H_{34}O_4N_2$
General characteristics[12]	Needles from methanol, mp 255°–257°C $[\alpha]_D^{24}$ −99° (C = 0.35 in methanol)
UV data[12]	λ_{max}^{MeOH} nm (ϵ): 222(3,250) and 283(535)
IR data[12]	$\nu_{max}^{(Nujol)}$ cm^{-1}: 3360, 3200, 1716, 1678, 1112, 978, 878, and 740
Fungal source[12]	*Nigrosabulum* spp.
Toxicity data	Not reported
^{13}C NMR data	C1, 173.2 s; C3, 51.7 d; C4, 49.1 d; C5, 35.6 d; C6, 56.9 s; C7, 60.7 d; C8, 47.4 d; C9, 64.1 s; C10, 42.1 t; C11, 12.1 q; C12, 22.4 q; C13, 128.4 d; C14, 132.8 d; C15, 32.4 t; C16, 27.4 d; C17, 36.2 t; C18, 206.5 s; C19, 46.1 t; C20, 46.2 t; C21, 207.4 s; C22, 19.0 q; C2′, 120.7 d; C3′, 108.3 s; C4′, 118.4 d; C5′, 124.9 d; C6′, 118.4 d; C7′, 111.0 d; C8′, 136.0 s; C9′, 127.6 s

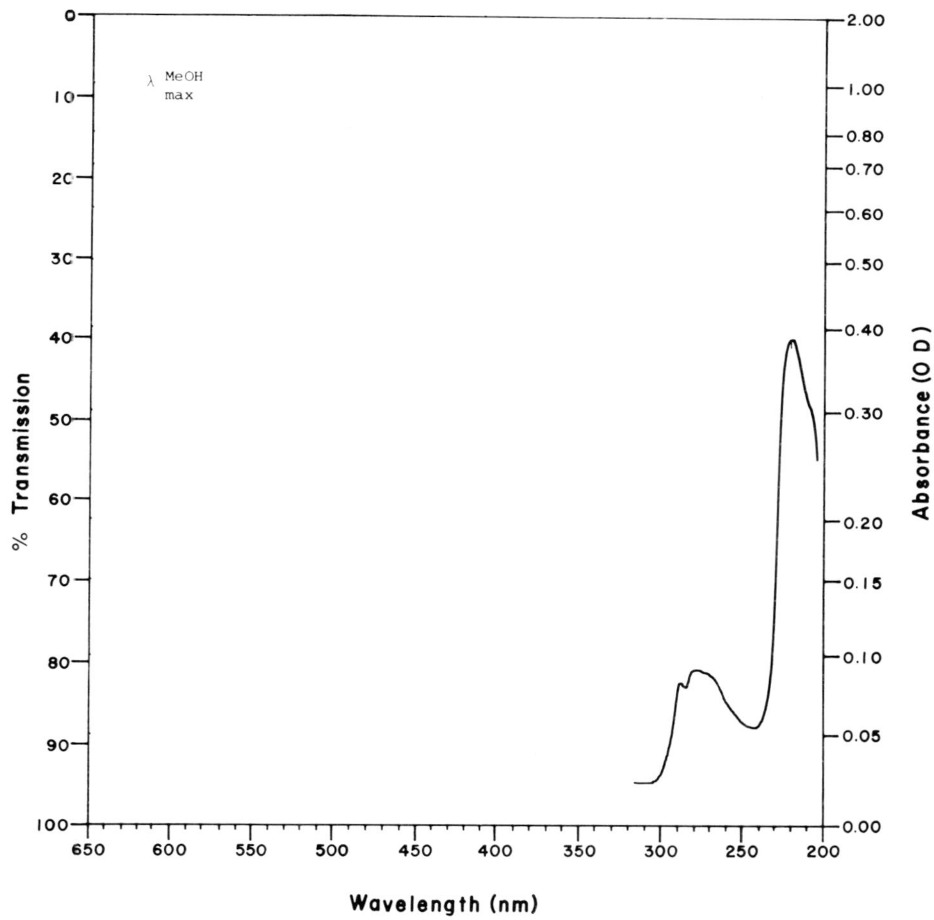
λ MeOH max
% Transmission
0
10
20
30
40
50
60
70
80
90
100
Absorbance (O D)
2.00
1.00
0.80
0.70
0.60
0.50
0.40
0.30
0.20
0.15
0.10
0.05
0.00
650
600
550
500
450
400
350
300
250
200
Wavelength (nm)

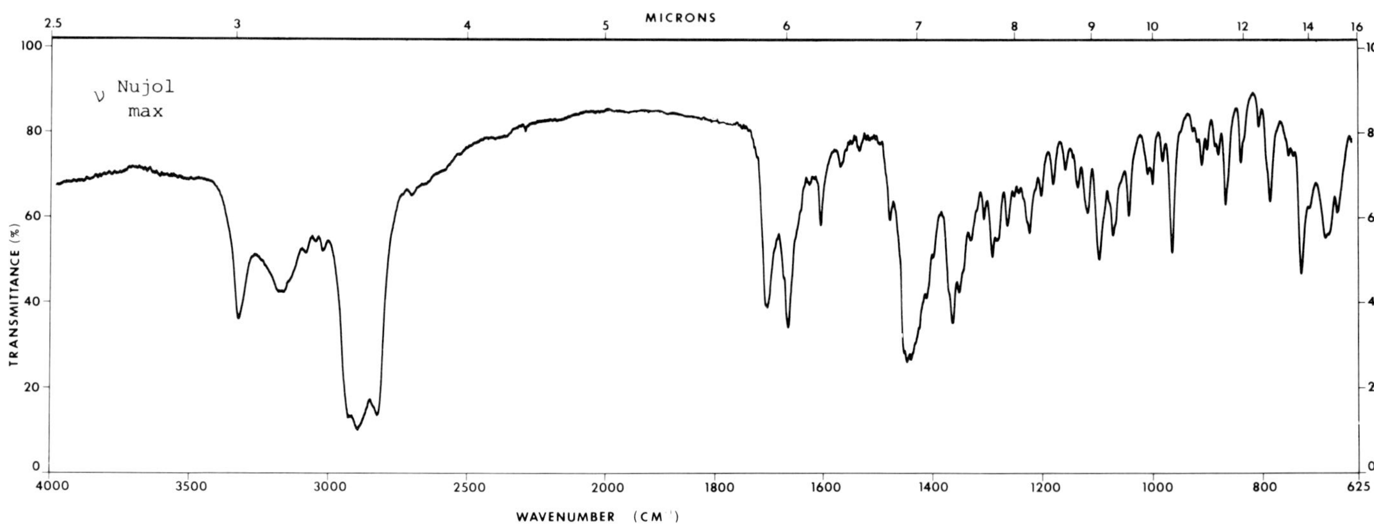
MICRONS
TRANSMITTANCE (%)
WAVENUMBER (CM⁻¹)
ν Nujol max

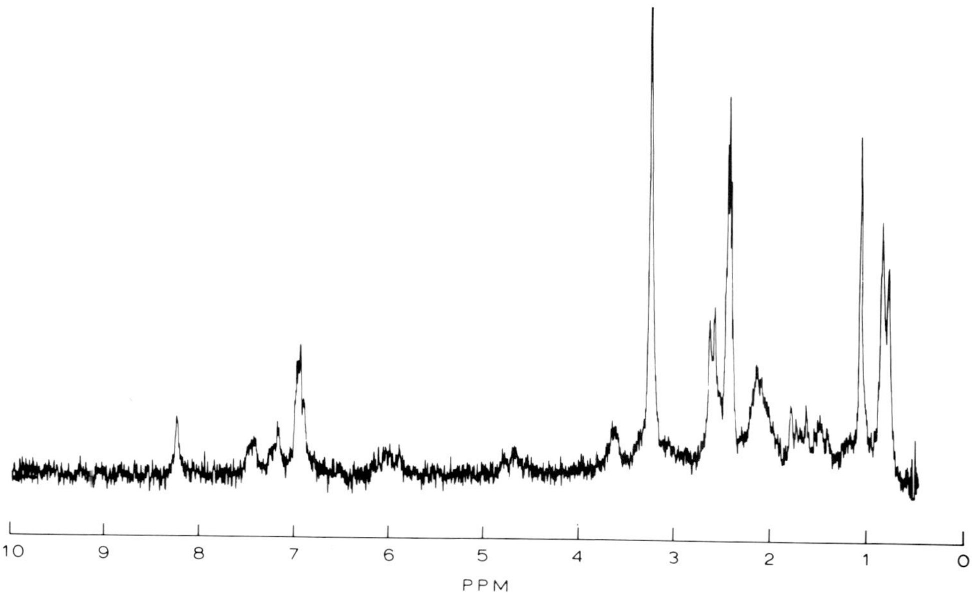

The 100 MHz [1]NMR spectrum of cytochalasin G in DMSO solution.

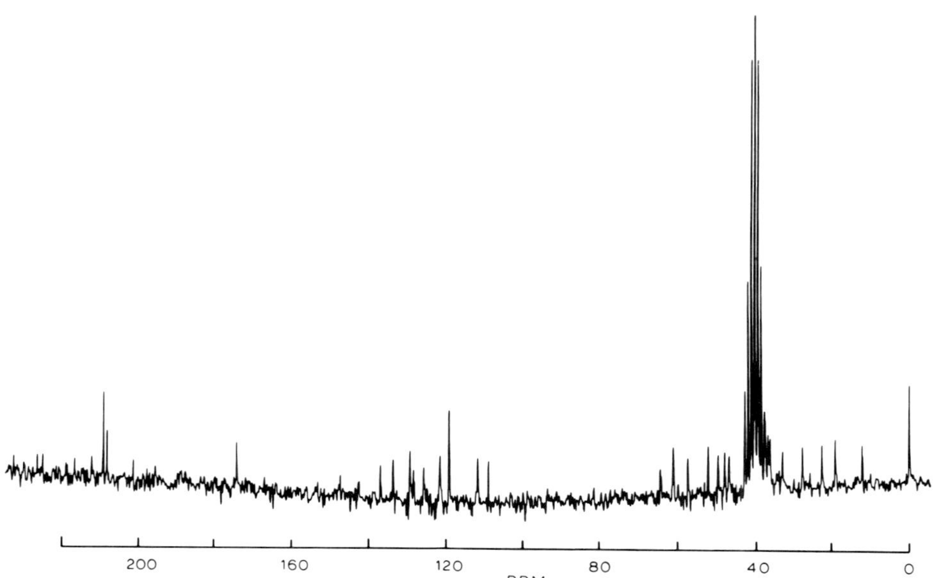

The 25 MHz ^{13}C NMR spectrum of cytochalasin G in DMSO solution.

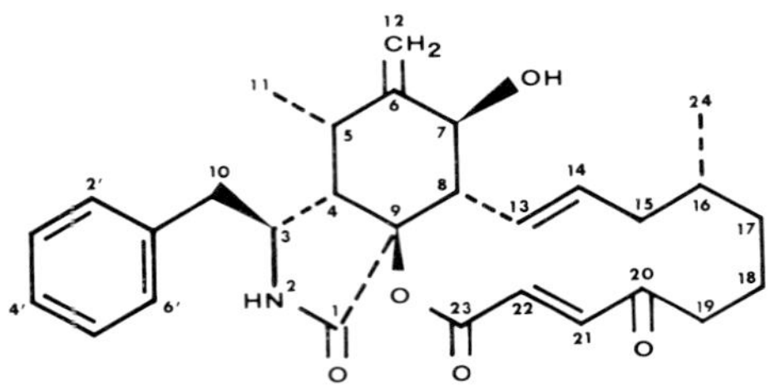

Common name	Cytochalasin A (Dehydrophomin) [7-Hydroxy-10-phenyl-5,16-dimethyl-24-oxa-[14]cytochalas-6(12),13,21-triene-20, 23-dione]
Molecular weight	477.2515
Molecular formula	$C_{29}H_{35}O_5N$
General characteristics[36]	mp 193°–195°C $[\alpha]_D^{21}$ +83.7° (*C* = 1.0 in methanol)
Fungal source[1,36]	*Helminthosporium dematioideum* *Phoma* spp.
Toxicity data[45]	Cytoxicity to HeLa cells, ED_{50} 3.2 μg/ml. Inhibits movement (reversible) and cytoplasmic cleavage in cultured cells leading to multinucleated giant cells; at higher levels nuclear extrusion occurs.
TLC data[47]	*Adsorbent:* silica gel G-HR *Solvent:* A: chloroform–acetone, 93:7 v/v B: toluene–ethyl acetate–formic acid, 5:4:1 v/v/v R_f: A: 0.29 B: 0.68 *Detection:* a blue fluorescent spot under UV light after spraying with 50% ethanolic H_2SO_4 and heating.
^{13}C NMR data	C1, 169.9 s; C3, 52.8 d; C4, 46.7 d; C5, 30.9 d; C6, 150.8 s; C7, 68.1 d; C8, 49.2 d; C9, 83.5 s; C10, 42.3 t; C11, 13.5 q; C12, 112.2 t; C13, 128.1 d; C14, 133.5 d; C15, 32.4 t; C16, 30.9 d; C17, 30.6 t; C18, 21.3 t; C19, 40.8 t; C20, 201.6 s; C21, 140.5 d; C22, 127.9 d; C23, 163.5 s; C24, 20.2 q; C1′, 137.4 s; C2′,6′, 129.5 d; C3′,5′, 128.1 d; C4′, 126.2 d

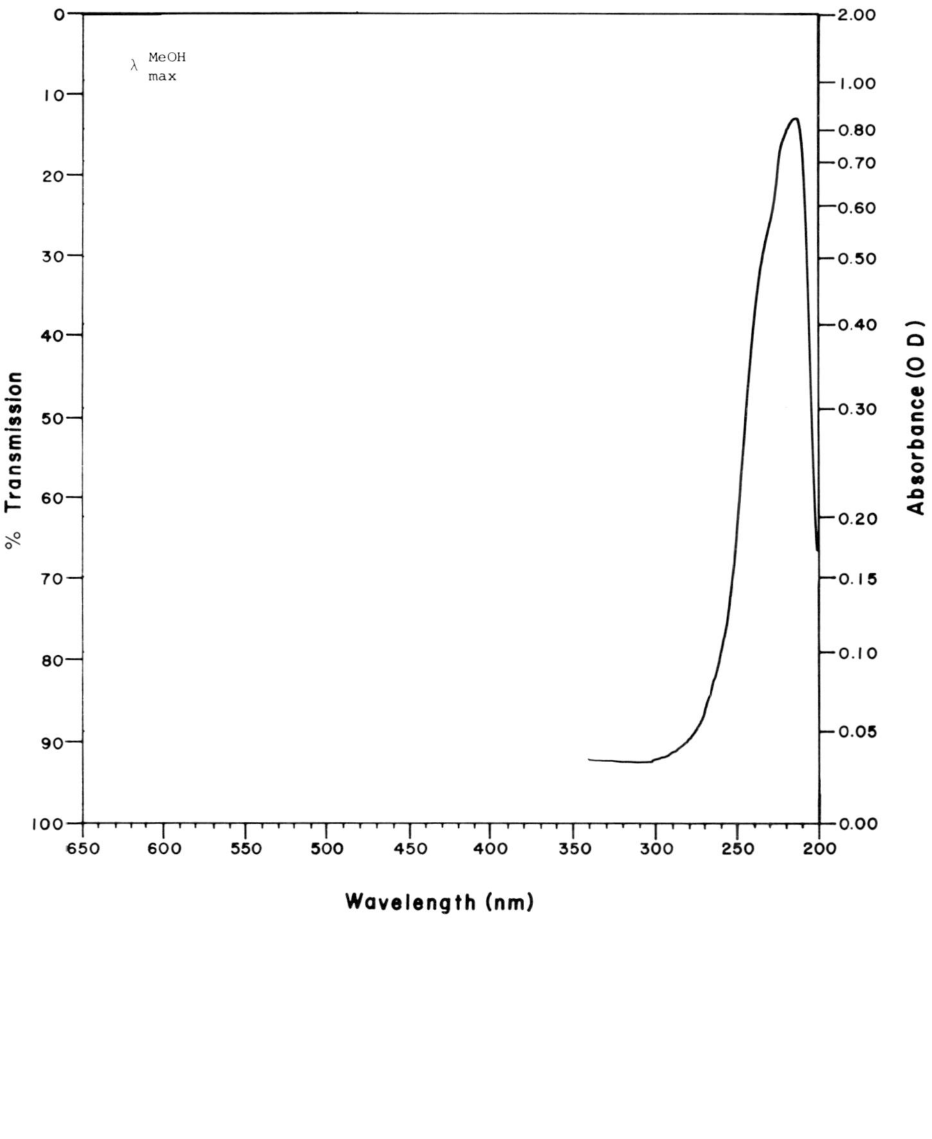
λ MeOH max
% Transmission
0
10
20
30
40
50
60
70
80
90
100
Absorbance (O D)
2.00
1.00
0.80
0.70
0.60
0.50
0.40
0.30
0.20
0.15
0.10
0.05
0.00
650
600
550
500
450
400
350
300
250
200
Wavelength (nm)

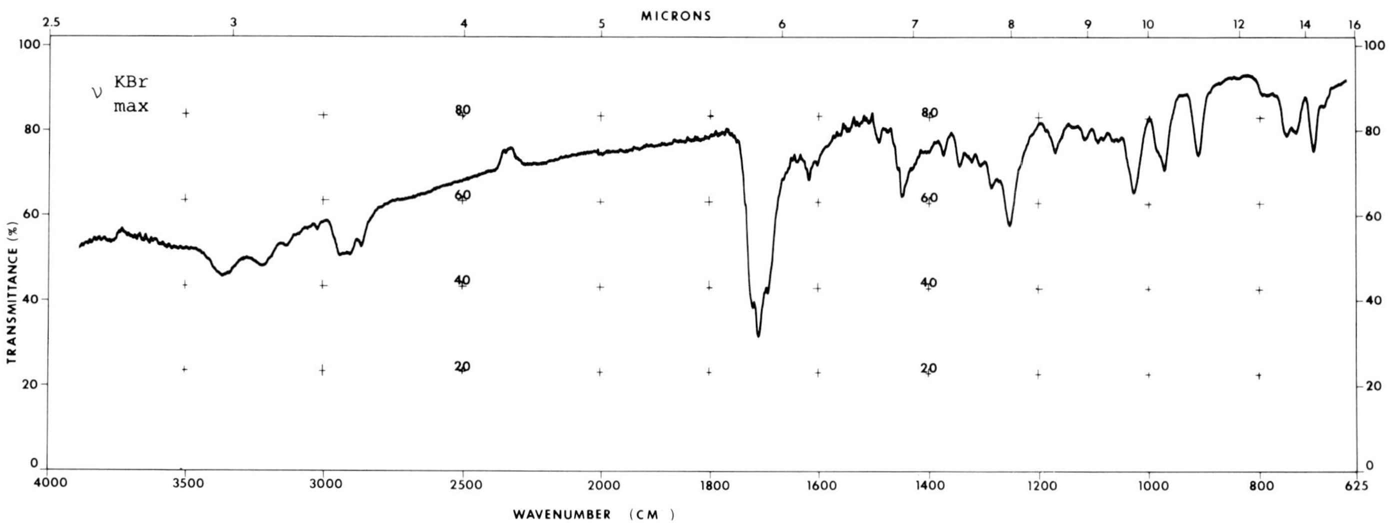

MICRONS
2.5
3
4
5
6
7
8
9
10
12
14
16
ν_{max}^{KBr}
TRANSMITTANCE (%)
100
80
60
40
20
0
WAVENUMBER (CM)
4000
3500
3000
2500
2000
1800
1600
1400
1200
1000
800
625

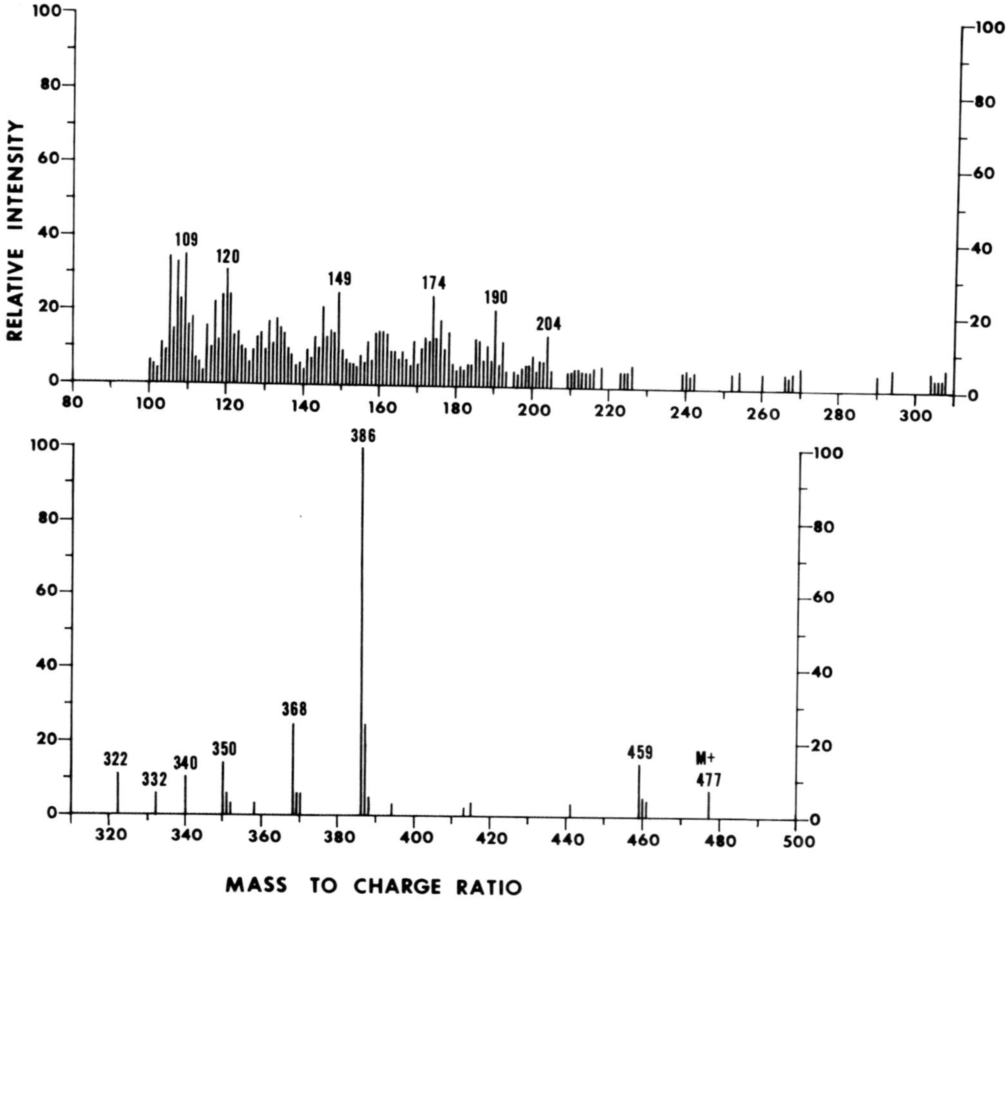
RELATIVE INTENSITY
MASS TO CHARGE RATIO
109
120
149
174
190
204
322
332
340
350
368
386
459
M+
477

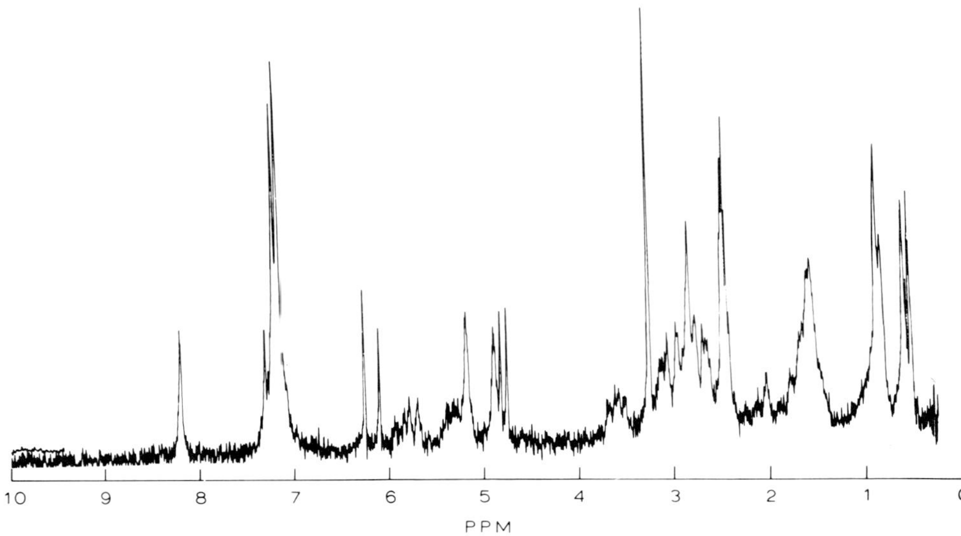

The 100 MHz ^1H NMR spectrum of cytochalasin A in DMSO solution.

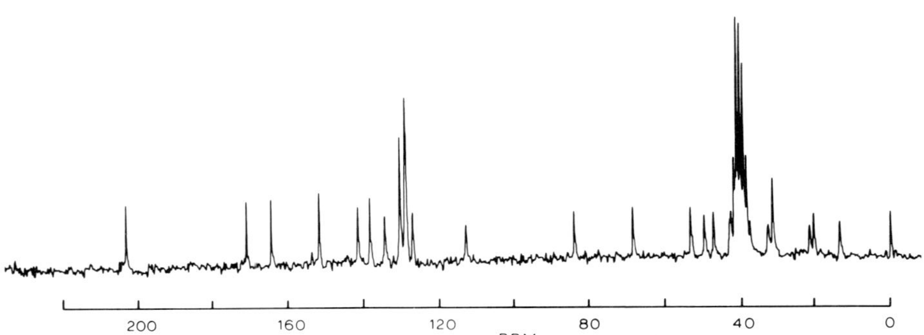

The 25 MHz ^{13}C NMR spectrum of cytochalasin A in DMSO solution.

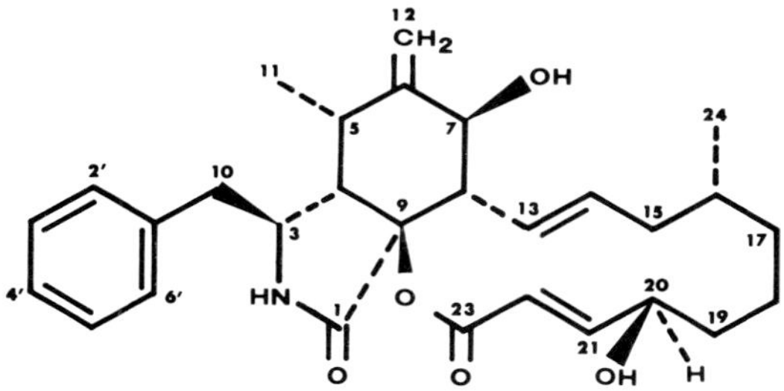

Common name	Cytochalasin B (Phomin) [7,20-Dihydroxy-10-phenyl-5,16-dimethyl-24-oxa-[14]cytochalas-6(12),13,21-trien-23-one]
Molecular weight	479.2671
Molecular formula	$C_{29}H_{37}O_5N$
General characteristics[1,34]	mp 221°–223°C $[\alpha]_D^{21}$ +86.7° (C = 0.9 in methanol) mp 218°C, from chloroform solution $[\alpha]_D^{25}$ +84° (C = 1.0 in ethanol)
UV data[34]	λ_{max} nm (ϵ): 220(44,668); 257(sh), 265(sh), and 269(sh)
Fungal source[1,34,35]	*Helminthosporium dematioideum* *Hormiscium* spp. *Phoma* spp.
Toxicity data[6,7,16,17,19,21–23,27,33,37,42,45,48]	Cytochalasin B inhibits cellular motility of moving L cells. Cytoplasmic cleavage is inhibited without interference with nuclear division, resulting in binucleate and multinucleate cells. The effect is reversible. Mammalian cells can be enucleated by treatment with cytochalasin B (10 μg/ml) followed by centrifiguration. Cytochalasin B has been reported to inhibit the following cellular functions: release of growth hormone; thyroid secretion; phagocytosis; platelet aggregation and clot retraction; and hexase transport. Cytotoxicity to HeLa cells, ED_{50} between 1 and 2.5 μg/ml.
TLC data[47]	*Adsorbent:* silica gel G-HR *Solvent:* A: chloroform–acetone, 93:7 v/v B: toluene–ethyl acetate–formic acid, 5:4:1 v/v/v

R_f: A: 0.11
B: 0.60

Detection: a blue fluorescent spot under UV light after spraying with 50% ethanolic H_2SO_4 and heating.

Other significant references
X-Ray structure[24]

^{13}C NMR data
C1, 170.6 s; C3, 52.3 d; C4, 46.5 d; C5, 30.4 d; C6, 150.9 s; C7, 69.4 d; C8, 47.5 d; C9, 83.3 s; C10, 42.9 t; C11, 13.2 q; C12, 112.2 t; C13, 128.4 d; C14, 133.0 d; C15, 34.8 t; C16, 30.9 d; C17, 34.2 t; C18, 20.1 t; C19, 41.6 t; C20, 68.6 d; C21, 153.2 d; C22, 118.1 d; C23, 164.0 s; C24, 20.2 q; C1′, 136.9 s; C2′,6′, 129.6 d; C3′.5′, 128.1 d; C4′, 126.3 d

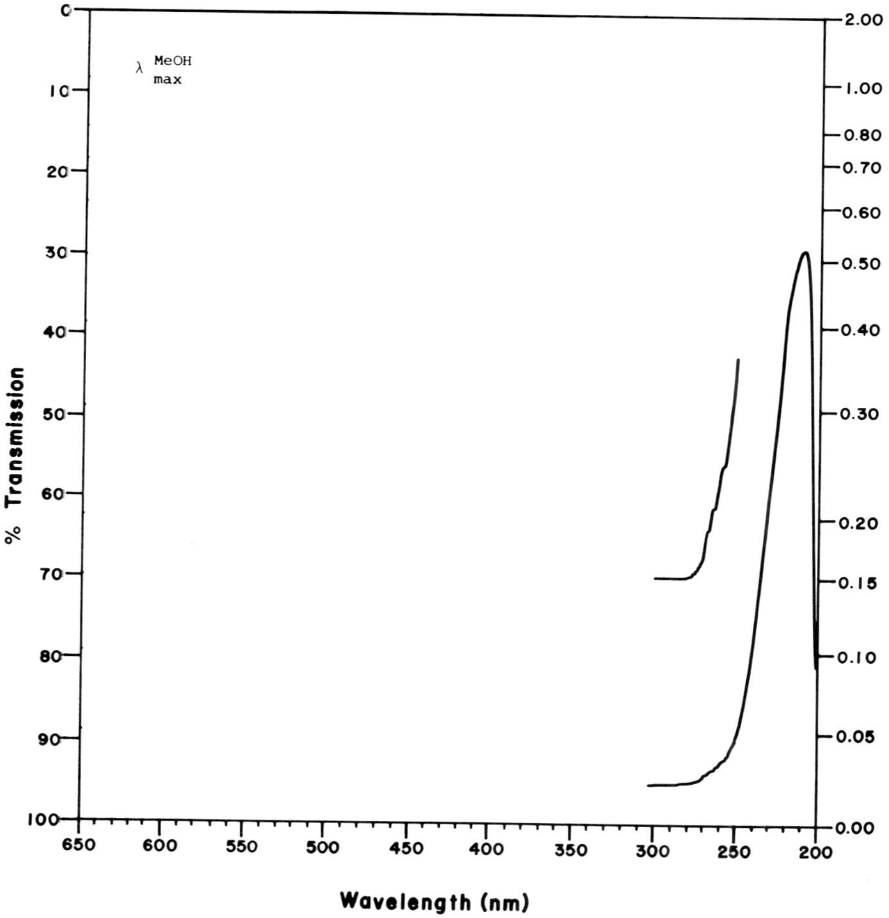
λ MeOH max
% Transmission
0
10
20
30
40
50
60
70
80
90
100
650
600
550
500
450
400
350
300
250
200
Wavelength (nm)
Absorbance (O D)
2.00
1.00
0.80
0.70
0.60
0.50
0.40
0.30
0.20
0.15
0.10
0.05
0.00

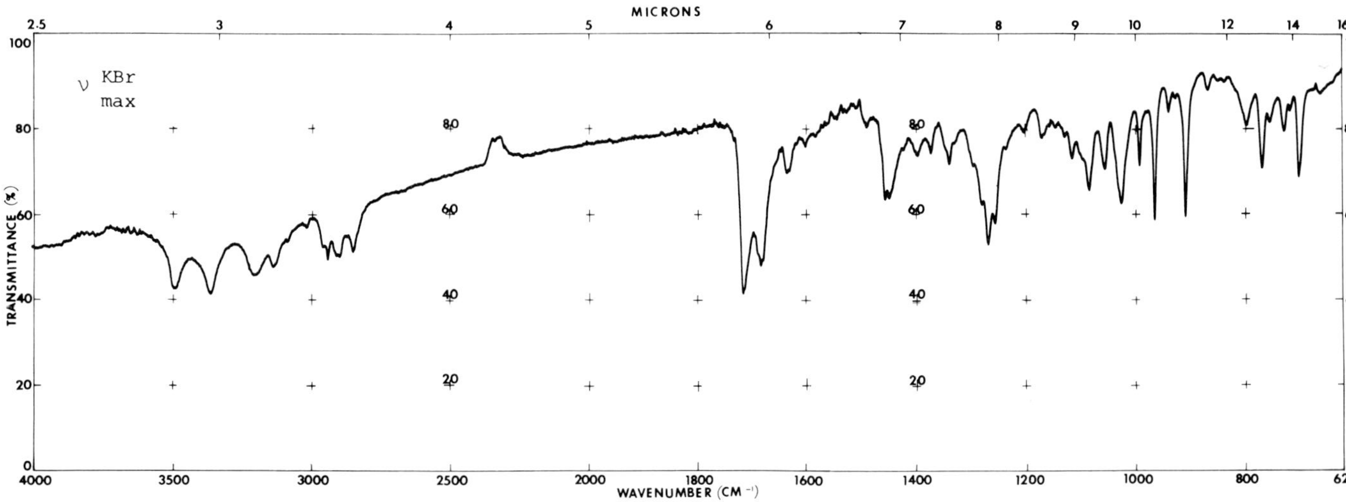
MICRONS
ν KBr max
TRANSMITTANCE (%)
WAVENUMBER (CM⁻¹)

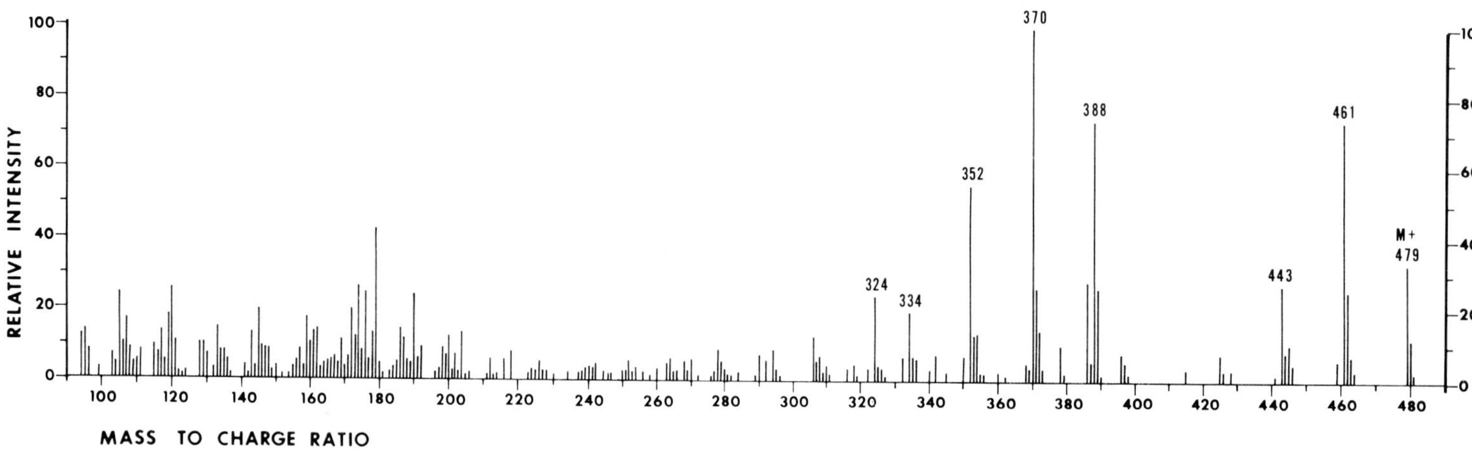
RELATIVE INTENSITY
M+ 479
461
443
388
370
352
334
324
MASS TO CHARGE RATIO

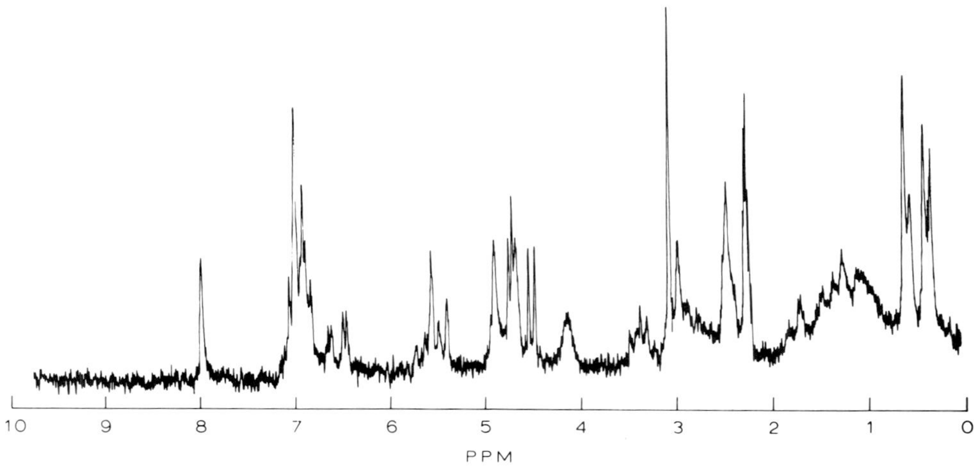

The 100 MHz ^1H NMR spectrum of cytochalasin B in DMSO solution.

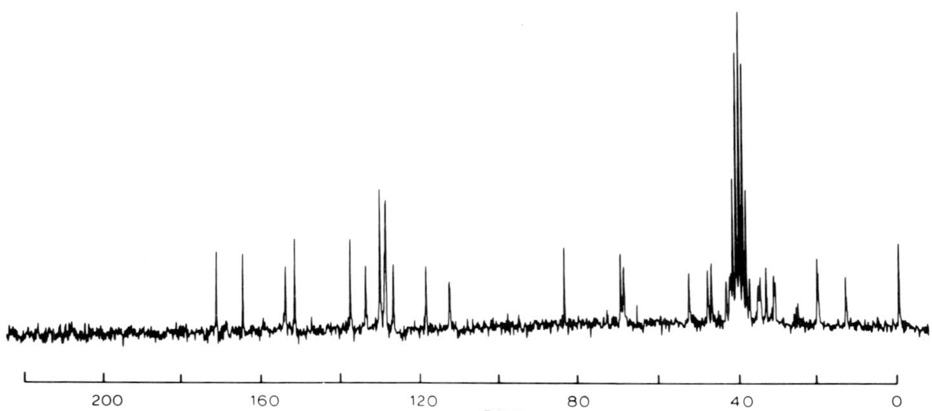

The 25 MHz ^{13}C NMR spectrum of cytochalasin B in DMSO solution.

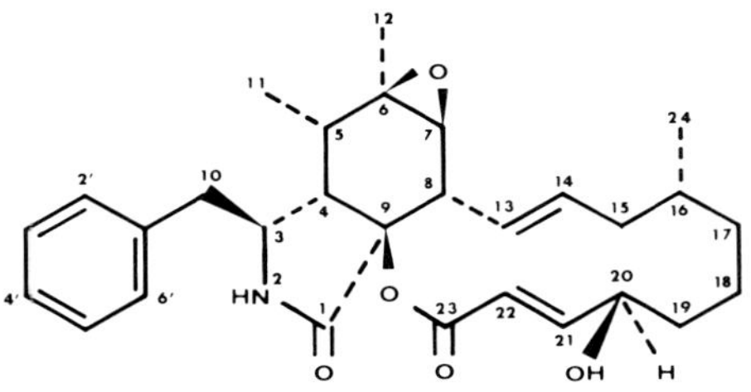

Common name	Cytochalasin F [6,7-Epoxy-20-hydroxy-10-phenyl-5,6,16-trimethyl-24-oxa-[14]cytochalas-13,21-dien-23-one]
Molecular weight	479.2671
Molecular formula	$C_{29}H_{37}O_5N$
Fungal source[3,4]	*Helminthosporium dematioideum*
Toxicity data	Toxicity unknown
Other significant references	Structure elucidation[11]

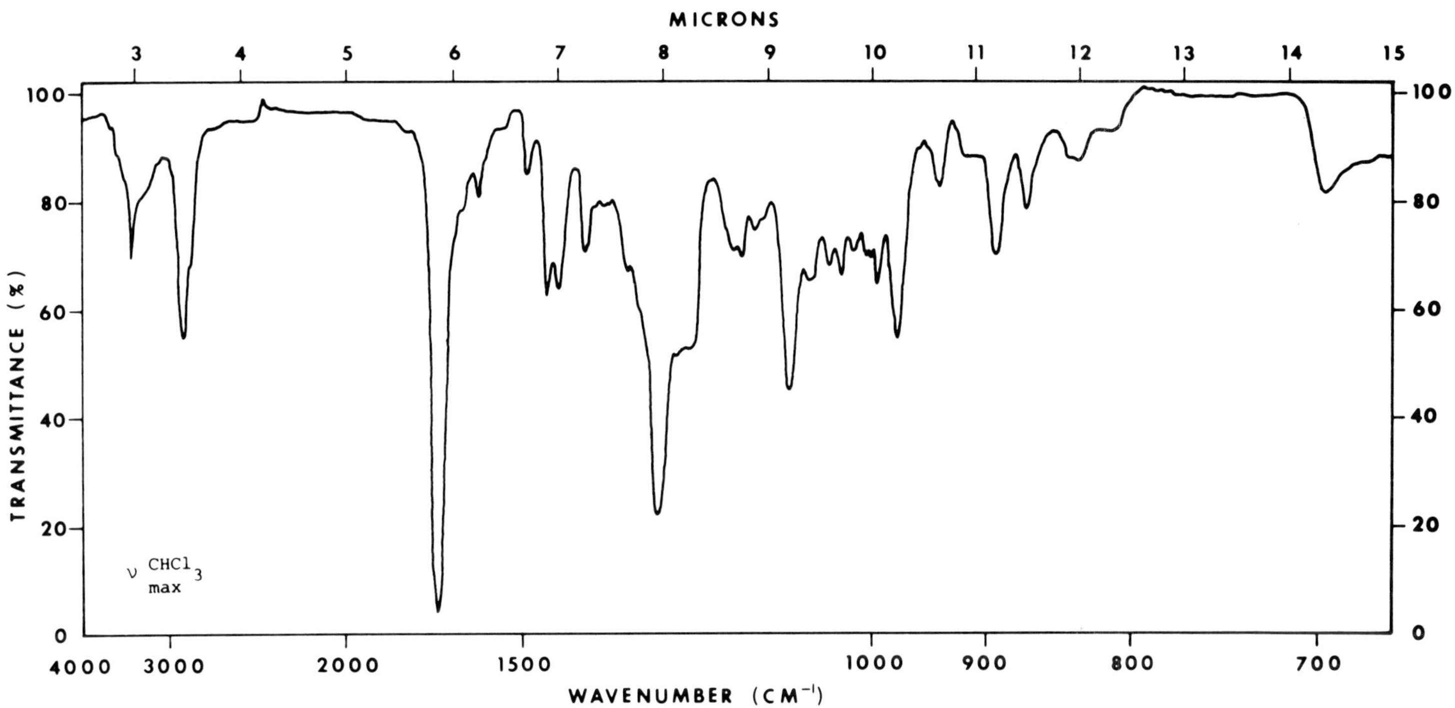
MICRONS
3
4
5
6
7
8
9
10
11
12
13
14
15
TRANSMITTANCE (%)
100
80
60
40
20
0
$\nu_{max}^{CHCl_3}$
4000
3000
2000
1500
1000
900
800
700
WAVENUMBER (CM^{-1})

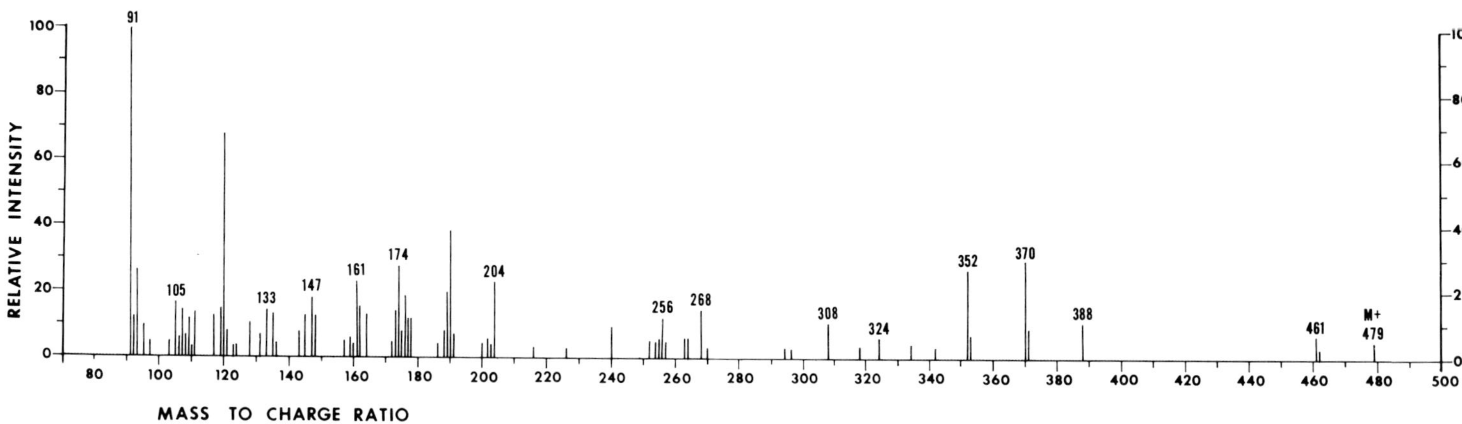

RELATIVE INTENSITY
MASS TO CHARGE RATIO
91
105
133
147
161
174
204
256
268
308
324
352
370
388
461
M+
479

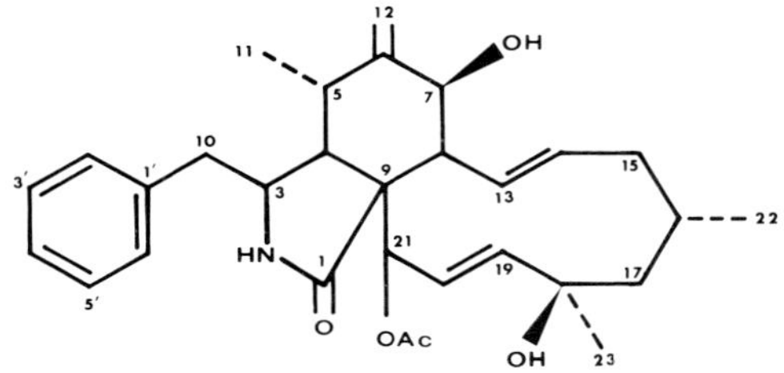

Common name	Cytochalasin H (Kodocytochalasin-1) [7,18-Dihydroxy-5,16,18-trimethyl-10-phenyl-[11]cytochalas-21-acetoxy-6(2),13,19-triene]
Molecular weight	493.2828
Molecular formula	$C_{30}H_{39}O_5N$
General characteristics[31,32,47]	Crystals from ethyl ether, mp 255°–257°C Colorless needles from methanol, mp 268°–269°C $[\alpha]_D^{25}$ +63° (C = 0.103, in methanol) White needles from chloroform–ethyl ether mp 258°–263°C $[\alpha]_D$ −9.0° (in chloroform); +91.2° (in methanol)
UV data[47]	End absorption and weak $\pi \rightarrow \pi^*$ transition due to aromatic ring
Fungal source[30,31,47]	*Phomopsis* spp. *Phomopsis paspalli*
Toxicity data[47]	LD_{50} 12.5 mg/kg dosed orally to day-old cockerels. A plant growth inhibitor; inhibits floral development in tobacco
TLC data[47]	*Adsorbent:* silica gel G-HR *Solvent:* A: toluene–ethyl acetate–formic acid, 5:4:1 v/v/v B: chloroform–acetone, 93:7 v/v R_f: A: 0.62 B: 0.13 *Detection:* 50% ethanolic H_2SO_4 sprayed alone or in combination with 3% ethanolic phosphomolybdic acid followed by heating for 5 min at 100°C

Other significant references X-Ray crystallography[8,25]

^{13}C NMR data C1, 174.4 s; C3, 57.8 d; C4, 50.3 d; C5, 32.9 d; C6, 148.2 s; C7, 69.8 d; C8, 47.3 d; C9, 51.9 s; C10, 45.6 t; C11, 14.0 q; C12, 113.8 t; C13, 127.2 d; C14, 138.3 d; C15, 42.9 t; C16, 31.2 d; C17, 57.8 t; C18, 74.2 s; C19, 127.0 d; C20, 138.5 d; C21, 77.5 d; C22, 26.5 q; C23, 28.5 q; C1′, 137.4 s; C2′,6′, 129.1 d; C3′,5′, 128.9 d; C4′, 125.9 d; $CH_3\underline{C}$ = 0, 170.0 s; $\underline{C}H_3$ C = 0, 20.8 q

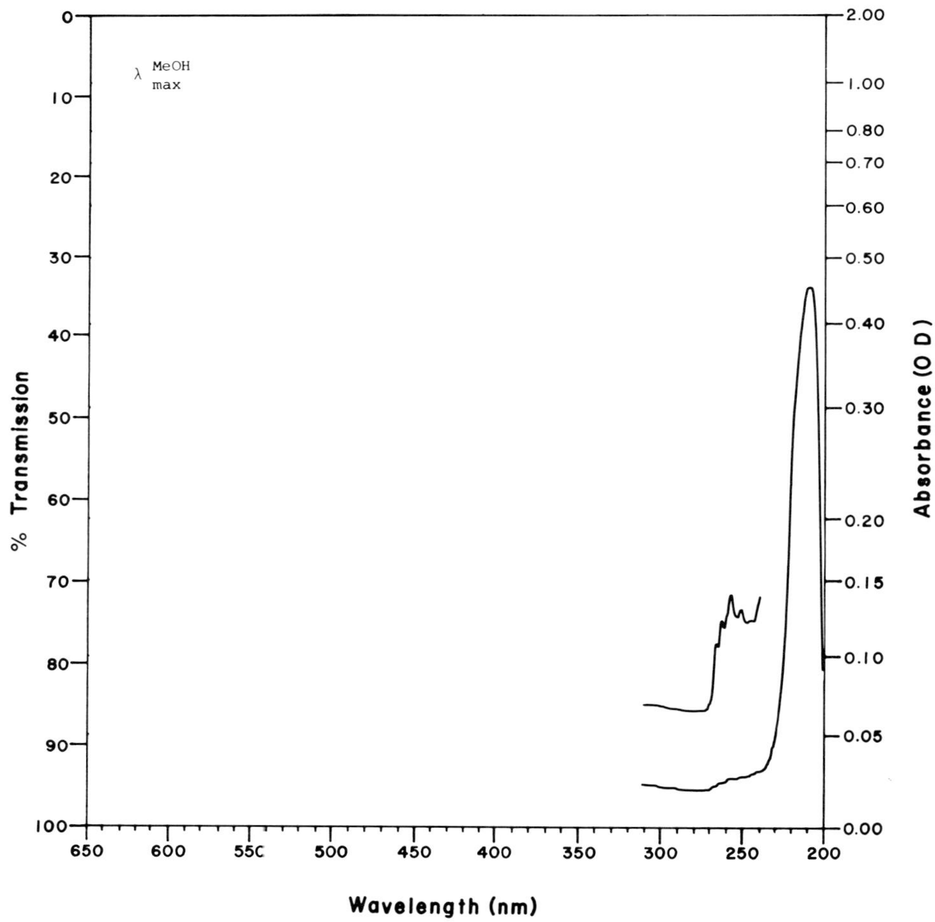

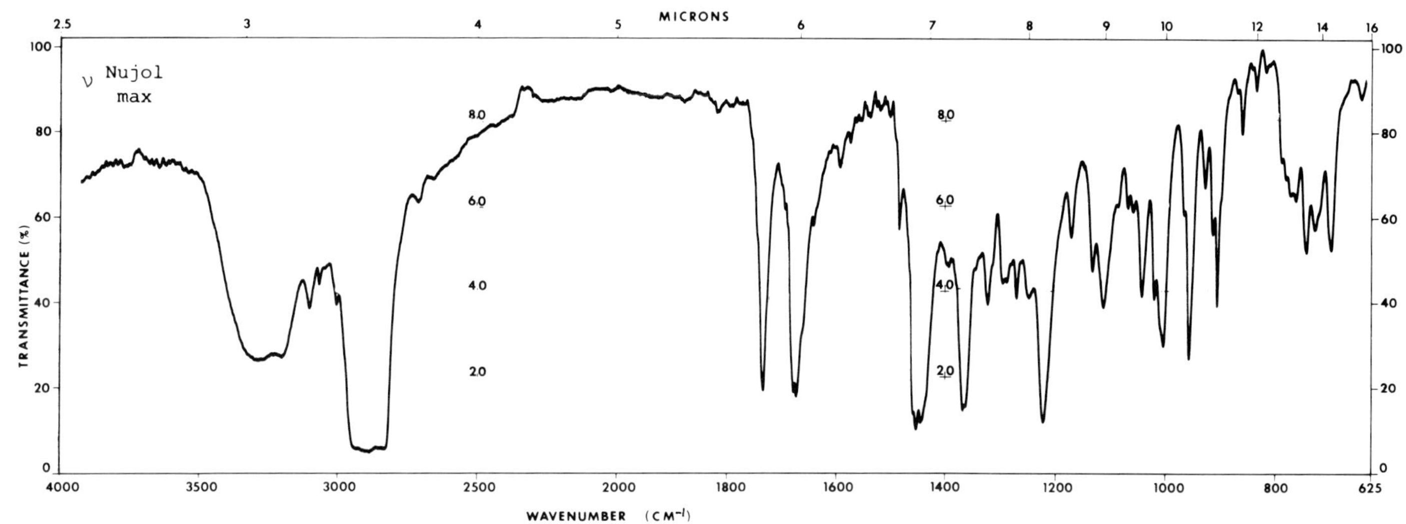
MICRONS
WAVENUMBER (CM⁻¹)
TRANSMITTANCE (%)
ν Nujol max

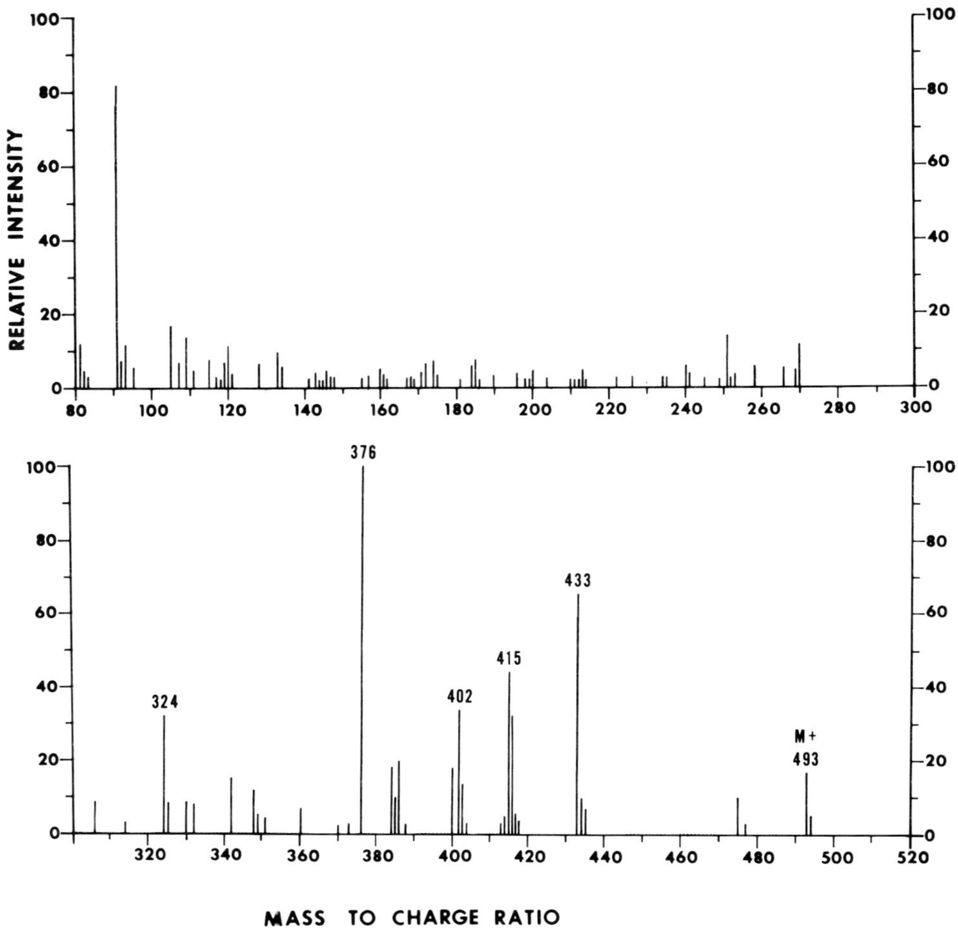
RELATIVE INTENSITY
MASS TO CHARGE RATIO
376
433
415
402
324
M+
493

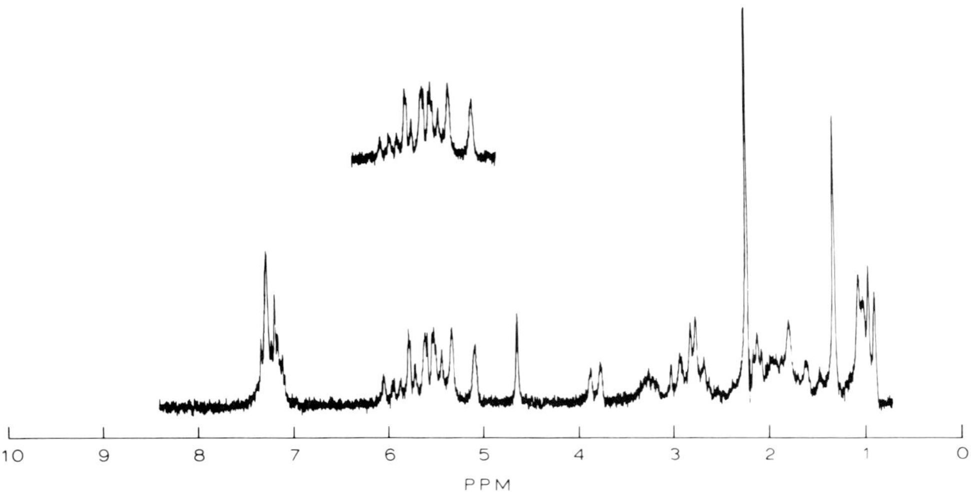

The 100 MHz ^1H NMR spectrum of cytochalasin H in $CDCl_3$ solution.

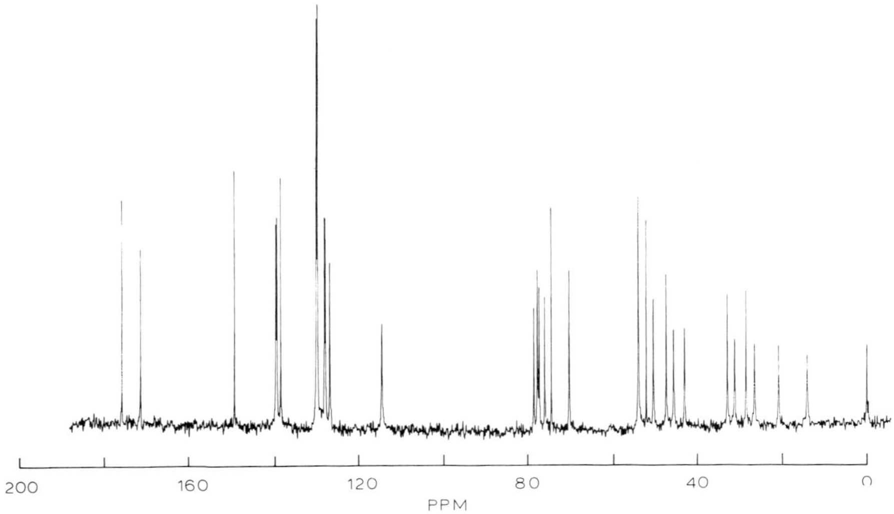

The 25 MHz ^{13}C NMR spectrum of cytochalasin H in $CDCl_3$ solution.

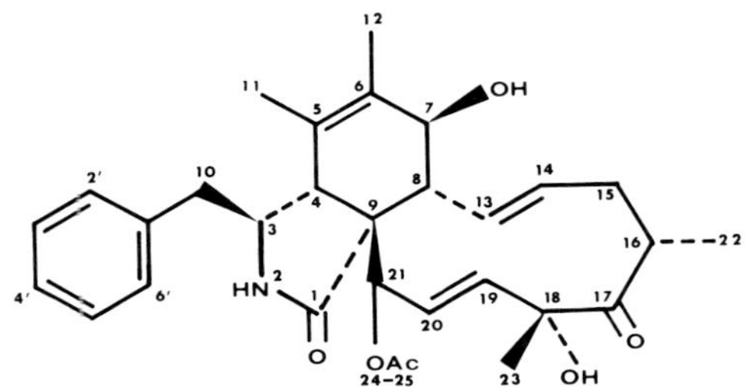

Common name	Cytochalasin C [7,18-Dihydroxy-10-phenyl-5,6,16,18-tetramethyl-[11]cytochalas-21-acetoxy-5-13,19-trien-17-one]
Molecular weight	507.2620
Molecular formula	$C_{30}H_{37}O_6N$
General characteristics[2,5]	mp 260°C (dec.) $[\alpha]_D^{28}$ $-14.7°$ (C = 0.82 in dioxane)
UV data	End absorption
Fungal source[2,5]	*Metarrhizium anisopliae*
Toxicity data[14]	Cytochalasin C shows essentially the same activity as cytochalasin B except it is about 10 times more potent in mammalian cell culture (inhibits movement and cytoplasmic cleavage); at higher doses causes nuclear extrusion which may lead to total enucleation.
TLC data[47]	*Adsorbent:* silica gel G-HR *Solvent:* A: chloroform–acetone, 93:7 v/v B: toluene–ethyl acetate–formic acid, 5:4:1 v/v/v R_f: A: 0.09 B: 0.55 *Detection:* a dull yellow fluorescent spot under UV light after spraying with 50% ethanolic H_2SO_4 and heating.
^{13}C NMR data	C1, 173.9 s; C3, 48.9 d; C4, 59.6 d; C5, 125.4 s; C6, 133.0 s; C7, 67.8 d; C8, 48.5 d; C9, 52.5 s; C10, 43.6 t; C11, 14.1 q; C12, 16.4 q; C13, 131.0 d; C14, 131.5 d; C15, 37.1 t; C16, 40.4 d; C17, 209.8 s; C18, 77.5

s; C19, 127.5 d; C20, 131.7 d; C21, 74.9 d; C22, 19.2 q; C23, 24.6 q; C1′, 137.6 s; C2′,6′, 129.2 d; C3′,5′, 128.4 d; C4′, 126.4 d; $CH_3\underline{C=O}$, 170.1 s; $\underline{CH_3}C=O$, 20.5 q

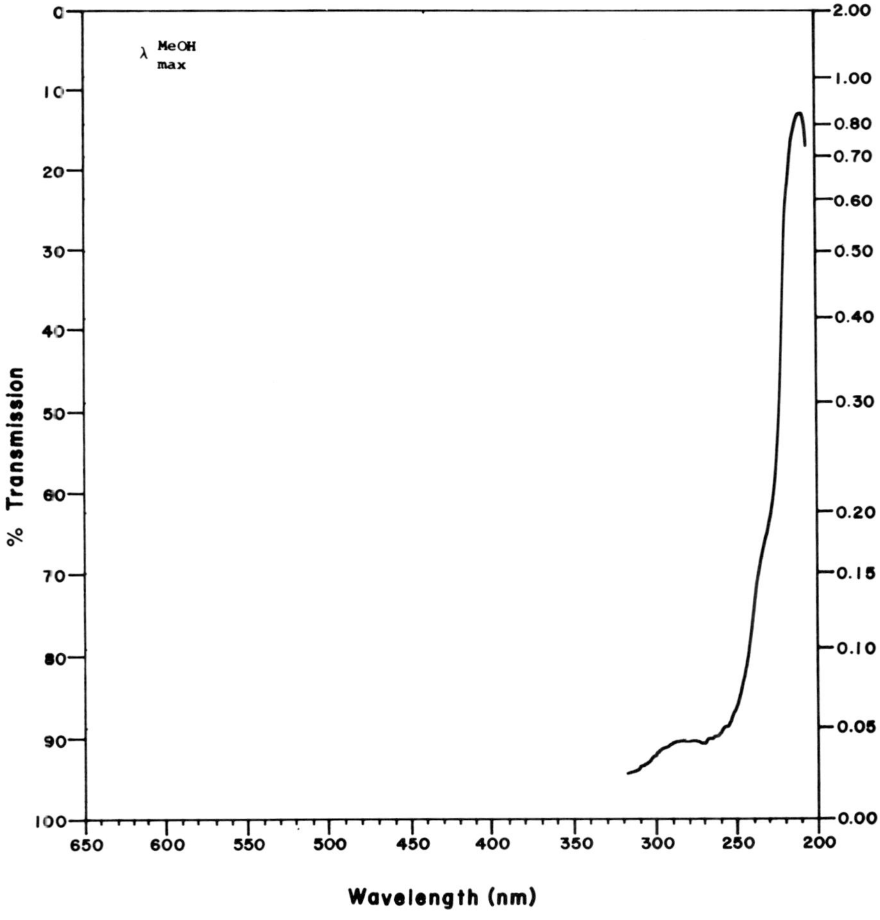

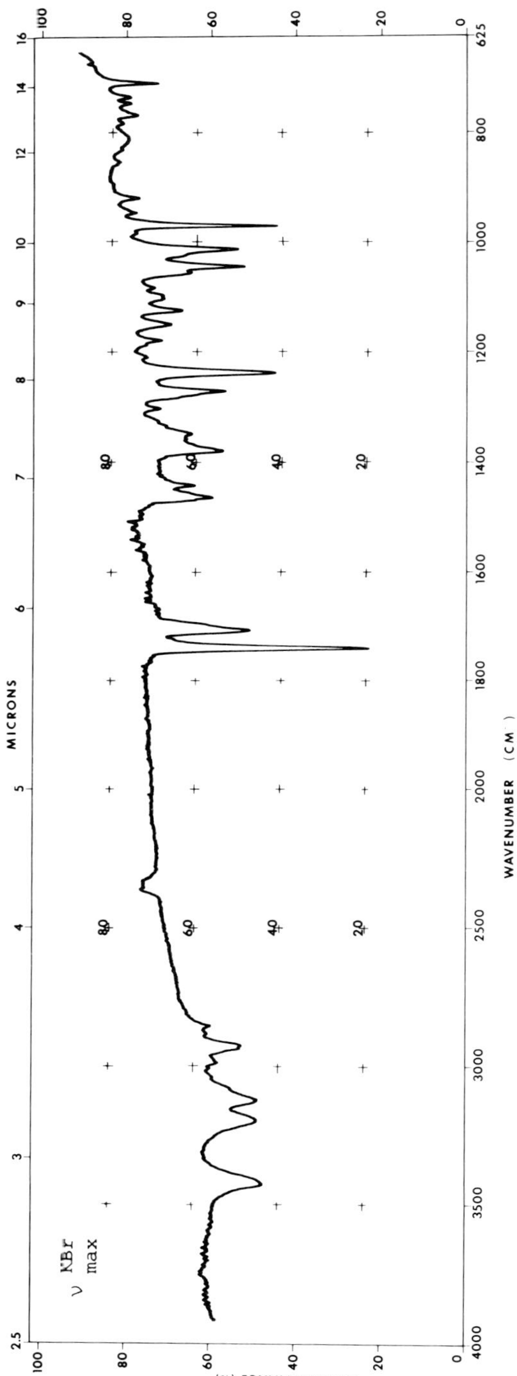
MICRONS
WAVENUMBER (CM⁻¹)
TRANSMITTANCE (%)
ν KBr max

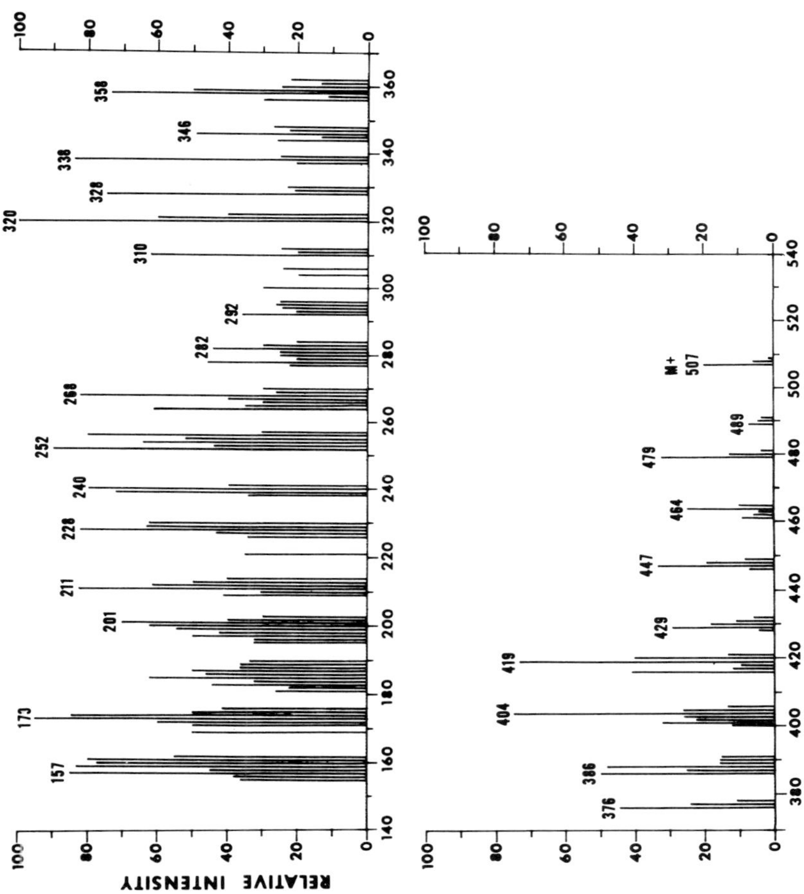
RELATIVE INTENSITY
MASS TO CHARGE RATIO
157
173
201
211
228
240
252
268
282
292
310
320
328
338
346
358
376
386
404
419
429
447
464
479
489
M+
507

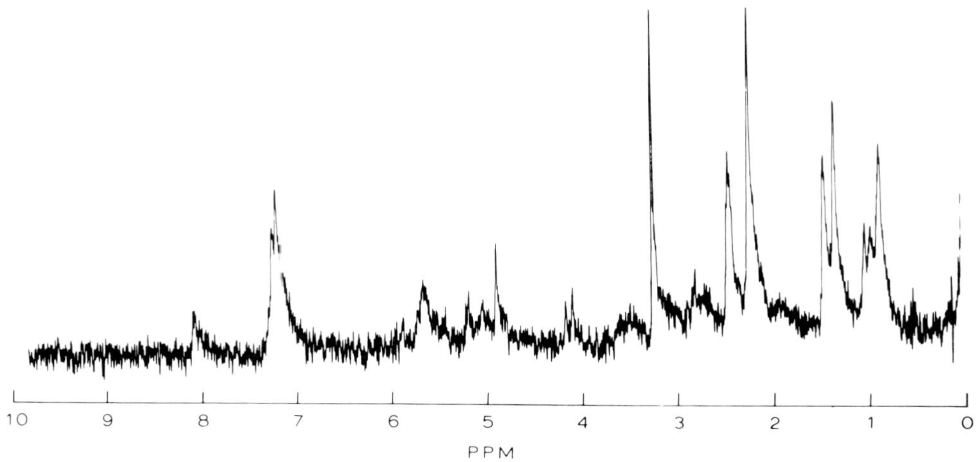

The 100 MHz ^{1}H NMR spectrum of cytochalasin C in DMSO solution.

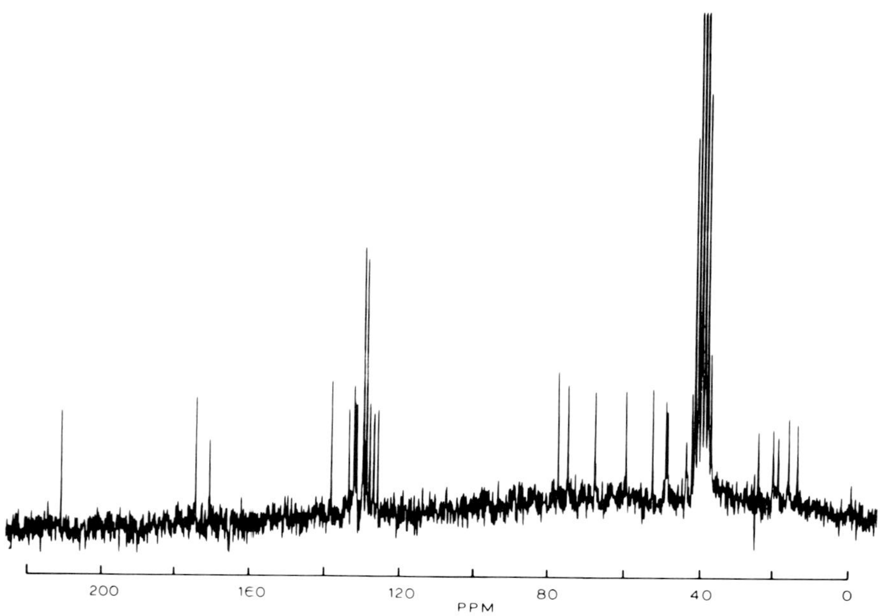

The 25 MHz ^{13}C NMR spectrum of cytochalasin C in DMSO solution.

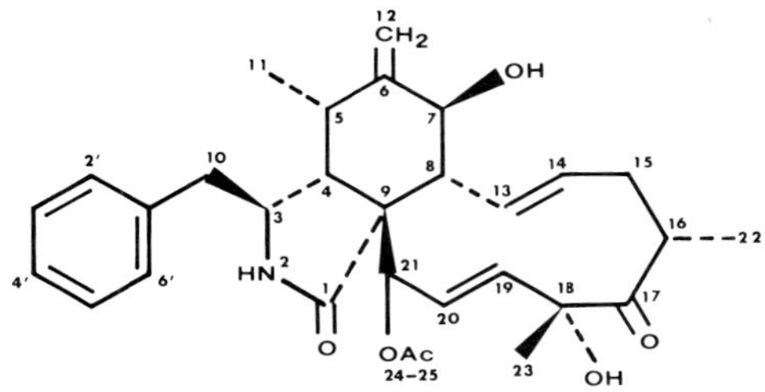

Common name	Cytochalasin D (Zygosporin A) [7,18-Dihydroxy-10-phenyl-5,16,18-trimethyl-[11]cytochalas-21-acetoxy-6(12),13,19-trien-17-one
Molecular weight	507.2620
Molecular formula	$C_{30}H_{37}O_6N$
General characteristics[2,5]	Needles from acetone–petroleum ether, mp 267°–271°C $[\alpha]_D^{25}$ −7.5° (*C* = 0.55 in dioxane)
UV data	End absorption
Fungal source[2,5,20,26]	*Metarrhizium anisopliae* *Zygosporium masonii*
Toxicity data[14]	Cytochalasin D shows essentially similar activity to cytochalasin B, except it is about 10 times more potent in mammalian cells (inhibits movement and cytoplasmic cleavage); at higher doses causes nuclear extrusion which may lead to total enucleation.
TLC data[47]	*Adsorbent:* silica gel G-HR *Solvent:* A: chloroform–acetone, 93:7 v/v B: toluene–ethyl acetate–formic acid, 5:4:1 v/v/v R_f: A: 0.06 B: 0.51 *Detection:* a bright yellow fluorescent spot under UV light after spraying with 50% ethanolic H_2SO_4 and heating.
Other significant references	Biosynthesis[46]

^{13}C NMR data C1, 173.4 s; C3, 52.4 d; C4, 47.6 d; C5, 31.5 d; C6, 150.6 s; C7, 70.0 d; C8, 46.0 d; C9, 53.1 s; C10, 43.6 t; C11, 12.7 q; C12, 111.3 t; C13, 130.7 d; C14, 131.6 d; C15, 36.8 t; C16, 40.2 d; C17, 209.8 s; C18, 77.4 s; C19, 126.9 d; C20, 132.0 d; C21, 76.3 d; C22, 19.1 q; C23, 24.5 q; C1′, 136.9 s; C2′,6′, 129.5 d; C3′,5′, 128.1 d; C4′, 126.4 d; $CH_3\underline{C=O}$, 169.8, s; $CH_3C=O$, 20.4 q

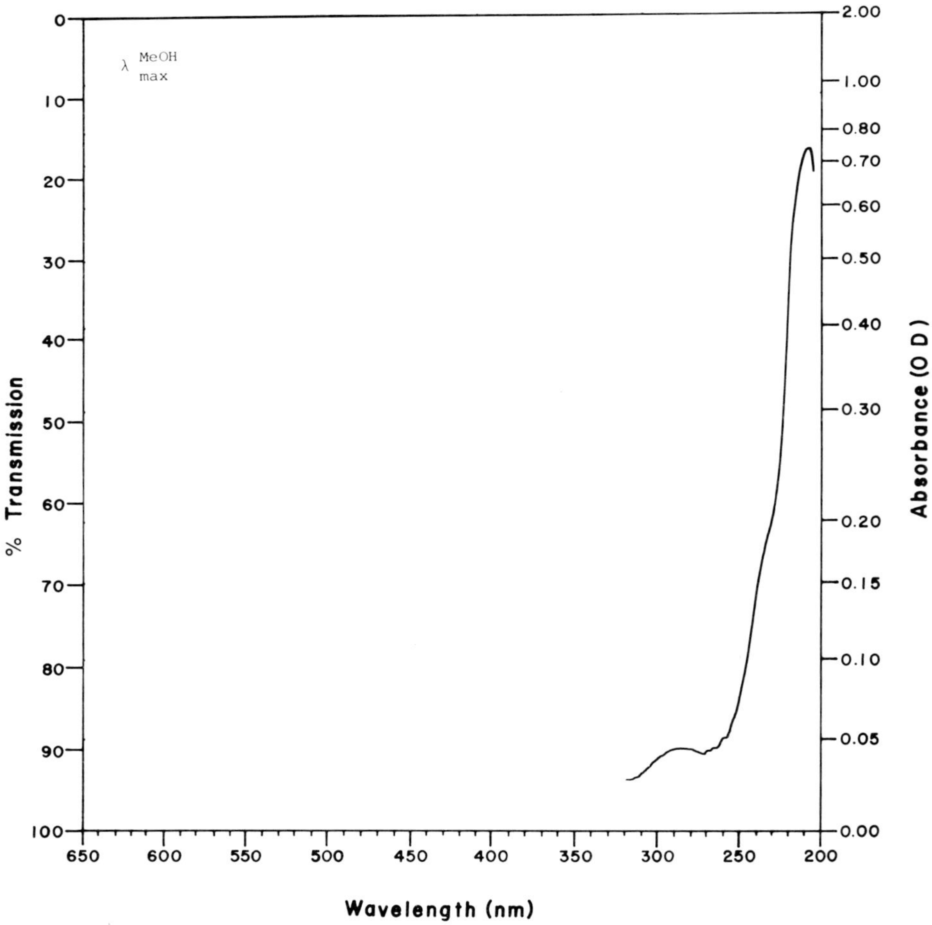

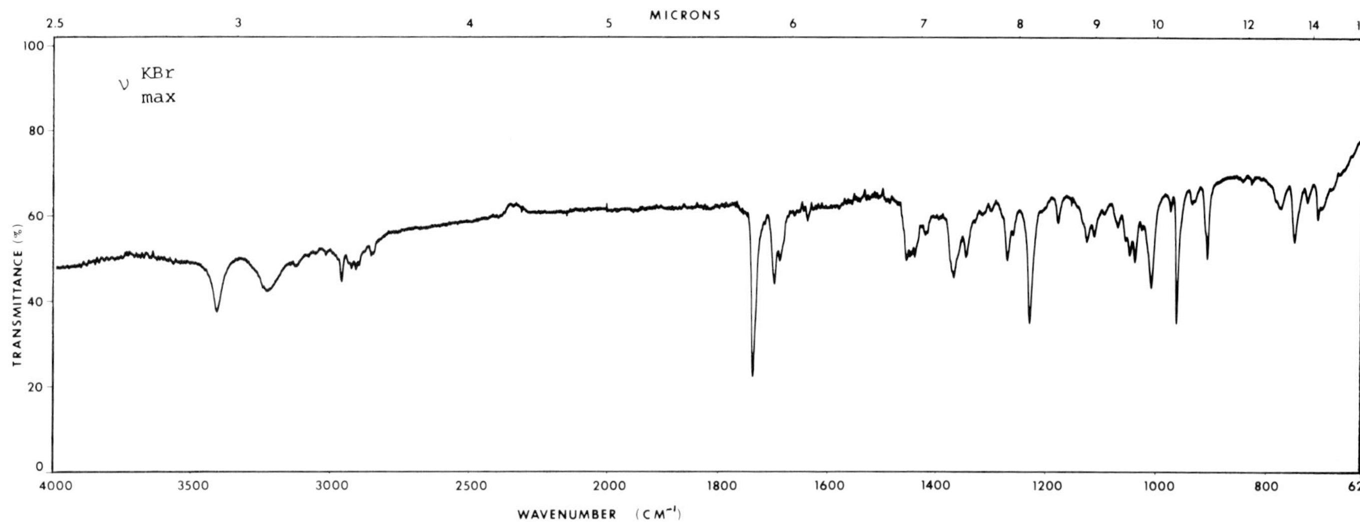

MICRONS
2.5
3
4
5
6
7
8
9
10
12
14
16
ν_{max}^{KBr}
TRANSMITTANCE (%)
100
80
60
40
20
0
4000
3500
3000
2500
2000
1800
1600
1400
1200
1000
800
625
WAVENUMBER (CM⁻¹)

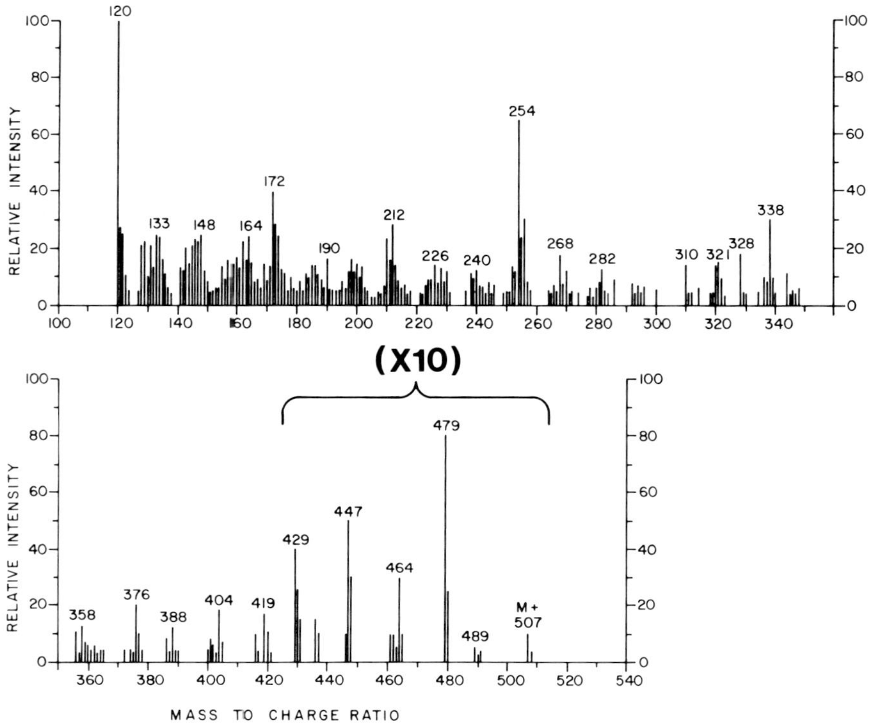

RELATIVE INTENSITY
MASS TO CHARGE RATIO
(X10)
M+
507

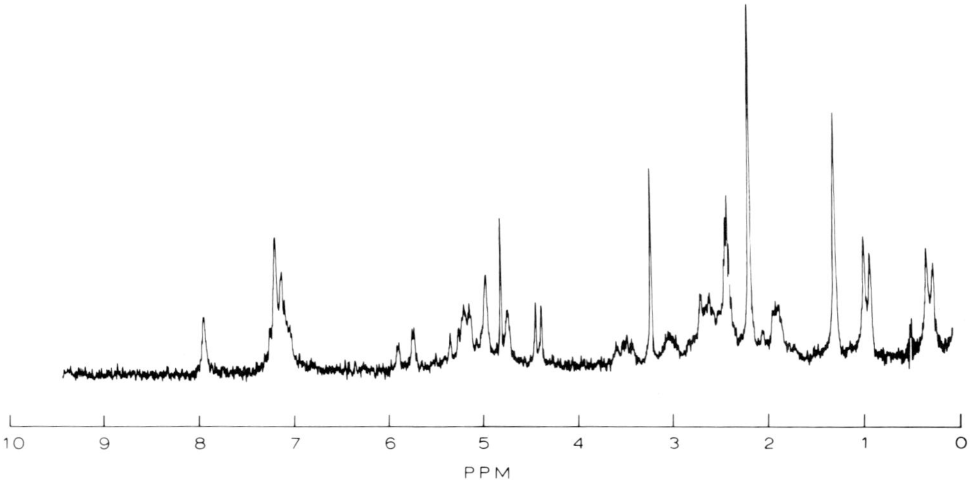

The 100 MHz ^{1}H NMR spectrum of cytochalasin D in DMSO solution.

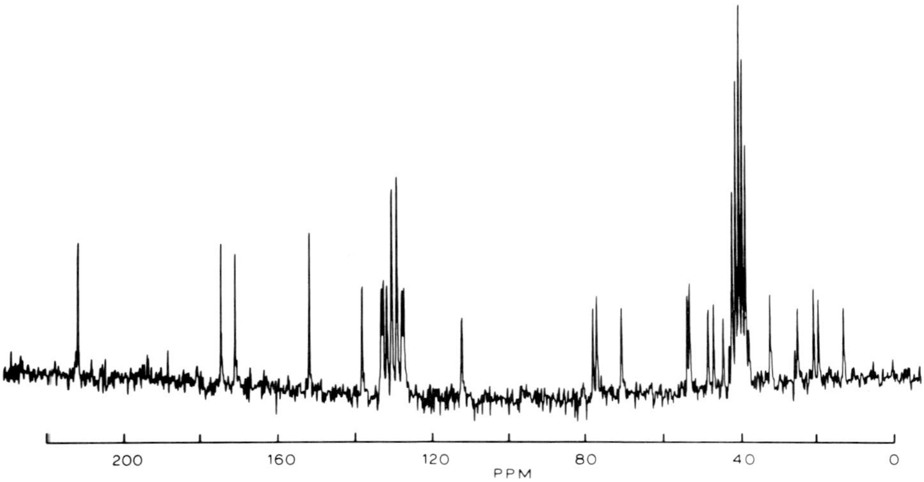

The 25 MHz ^{13}C NMR spectrum of cytochalasin D in DMSO solution.

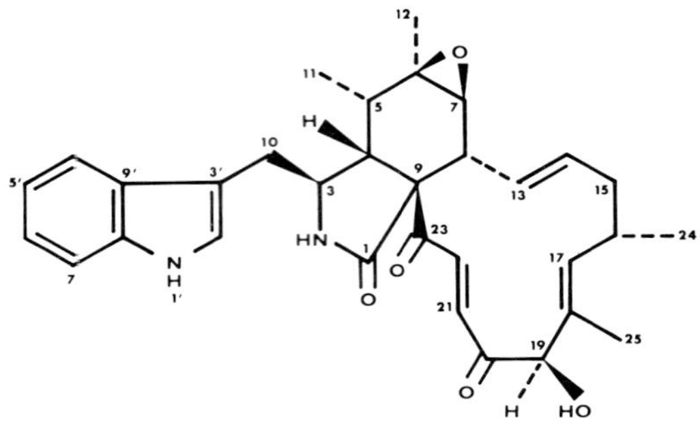

Common name	Chaetoglobosin A [6,7-Epoxy-10-(indol-3-yl)-19-hydroxy-5,6,16,18-tetramethyl-[13]cytochalas-13,17,21-triene-20,23-dione]
Molecular weight	528.2623
Molecular formula	$C_{32}H_{36}O_5N_2$
General characteristics[40]	Crystals from dichloromethane, mp 168°–170°C $[\alpha]_D$ −270° (in methanol)
UV data[40]	λ_{max}^{EtOH} nm (ϵ): 223(40,700), 274(6,600), 282(6,600), and 291(5,300)
Fungal source[43]	*Chaetomium globosum*
Toxicity data[28,29,45]	Cytotoxicity to HeLa cells, ED_{50} between 3.2 and 10 μg/ml. Causes inhibition of cytoplasmic cleavage resulting in polynucleate cells. LD_{50} in male mice dosed subcutaneously, 6.5 mg/kg; in female mice, 17.8 mg/kg.
TLC data[47]	*Adsorbent:* silica gel G-HR *Solvent:* A: chloroform–acetone, 93:7 v/v B: toluene–ethyl acetate–formic acid, 5:4:1 v/v/v R_f: A: 0.05 B: 0.65 *Detection:* a light olive gray fluorescent spot under UV light after spraying with 50% ethanolic H_2SO_4 and heating.
Other significant references	X-Ray crystallography[43]

^{1}H NMR data[40]

H2, 5.85; H3, 3.81 (*J* = 3.0, 4.3, 7.5); H4, 3.03 (*J* = 3.0, 5.0); H5, 1.85 (*J* = 5.0, 7.0); H7, 2.78 (*J* = 5.0); H8, 2.14 (*J* = 5.0, 9.5); H10, 2.63 (*J* = 1.5, 7.5, 15); H11, 1.24 (*J* = 7.0); H12, 1.29; H13, 6.05 (*J* = 9.5, 15); H14, 5.20 (*J* = 4.0, 10.0, 15); H15, 1.8–2.4; H16, 2.42 (J = 6.5, 9.0); H17, 5.57 (J = 9.0); H19, 5.01; H21, 7.72 (*J* = 16.5); H22, 6.50 (*J* = 16.5); H24, 1.00 (*J* = 6.5); H25, 1.31; 19OH, 3.84; H1′, 8.21; H2′, 6.94; H4′–7′, 7.0–7.5

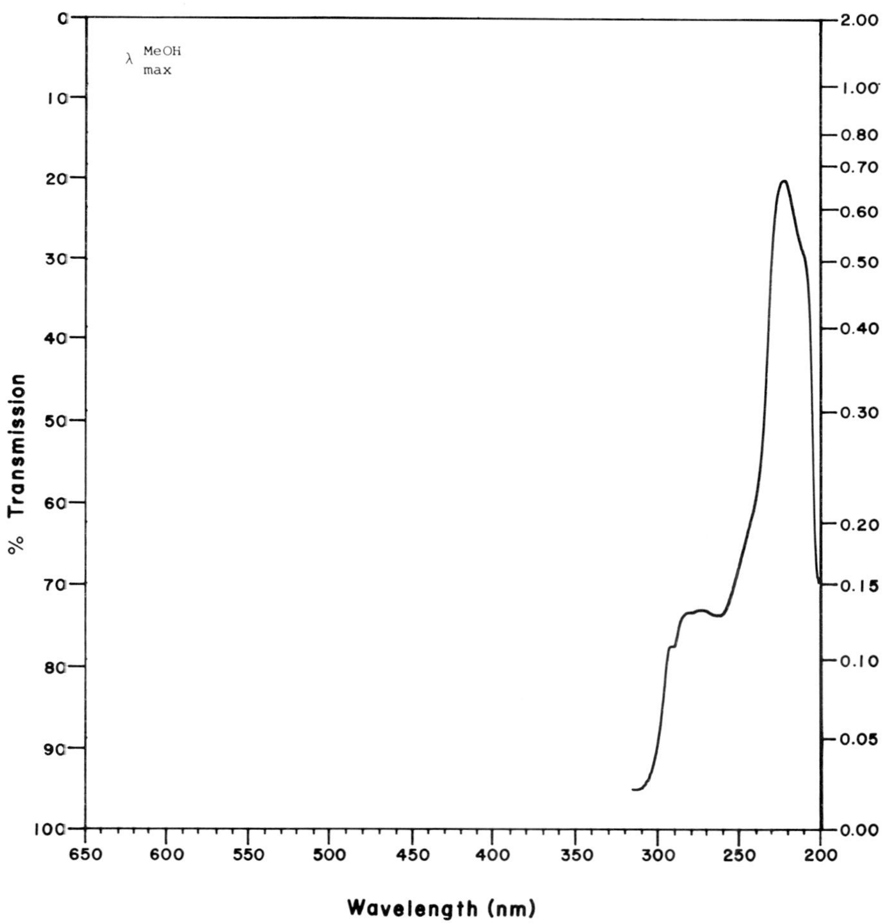

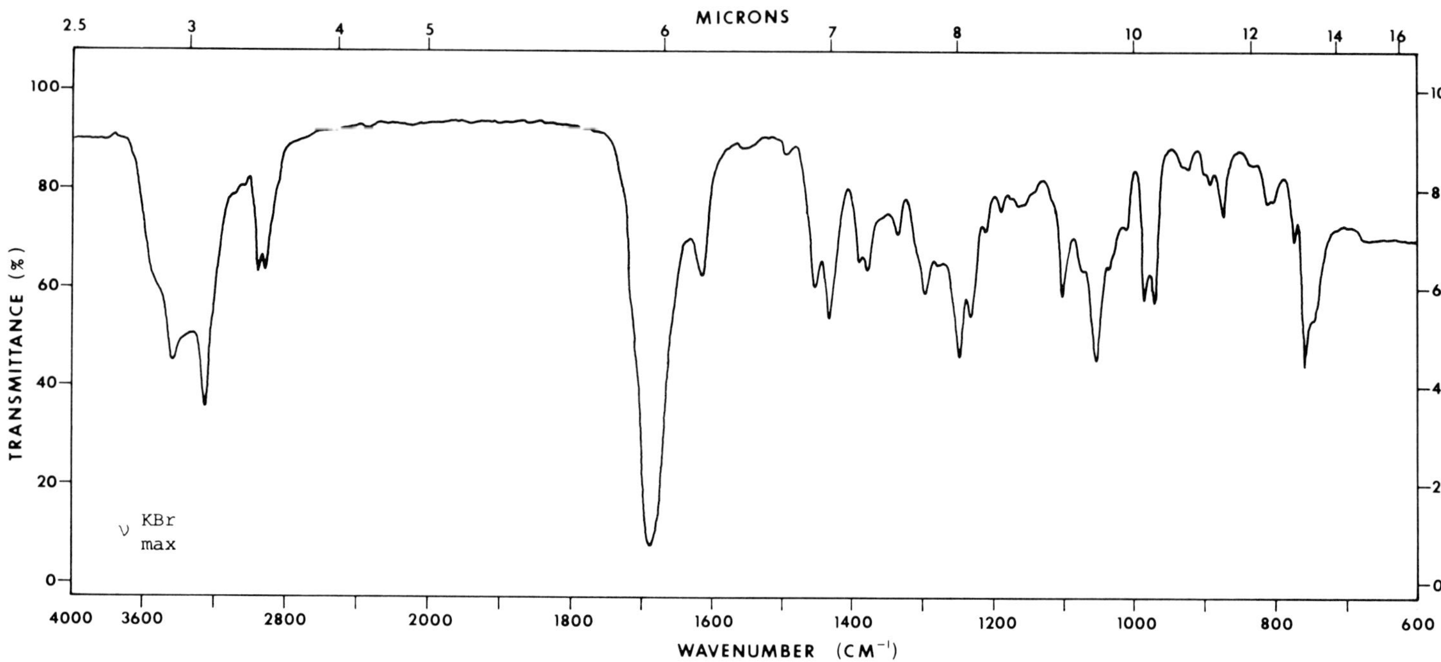
MICRONS
2.5
3
4
5
6
7
8
10
12
14
16
TRANSMITTANCE (%)
100
80
60
40
20
0
ν_{max}^{KBr}
4000
3600
2800
2000
1800
1600
1400
1200
1000
800
600
WAVENUMBER (CM^{-1})

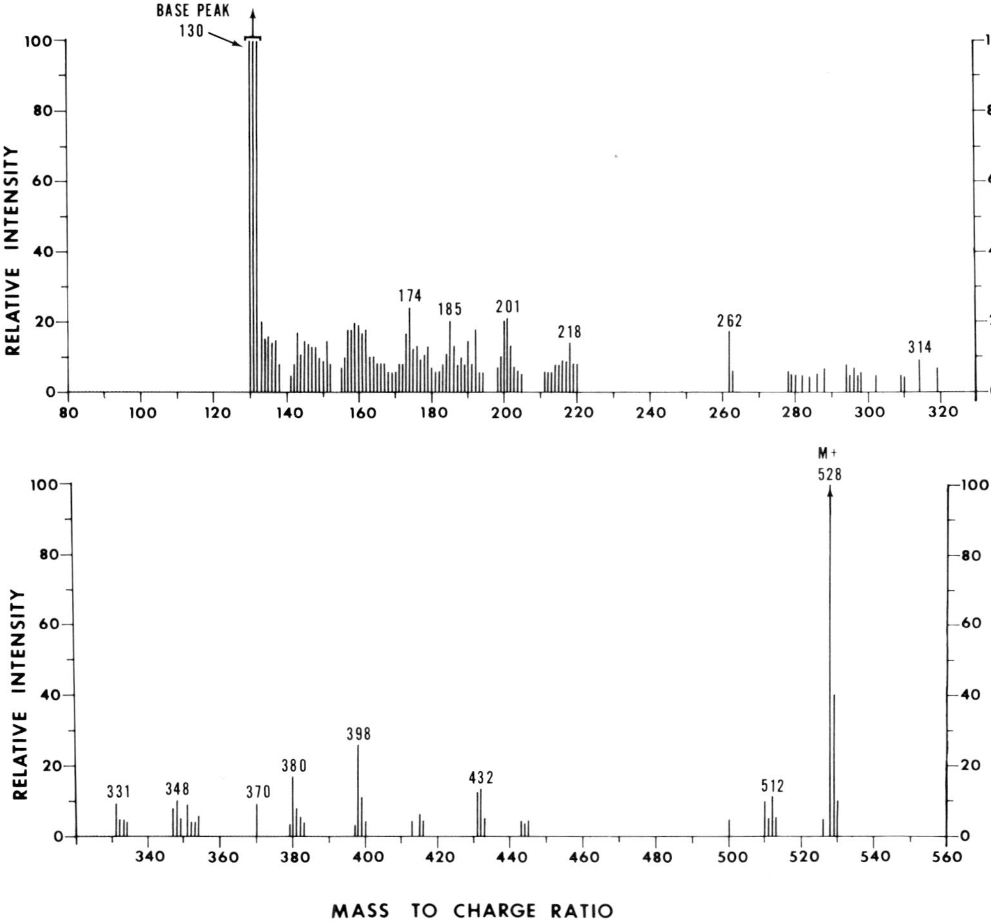

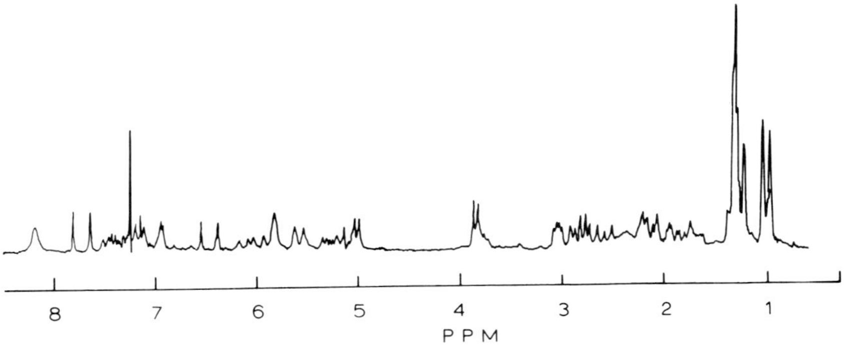

The 100 MHz ^{1}H NMR spectrum of chaetoglobosin A in $CDCl_3$ solution.

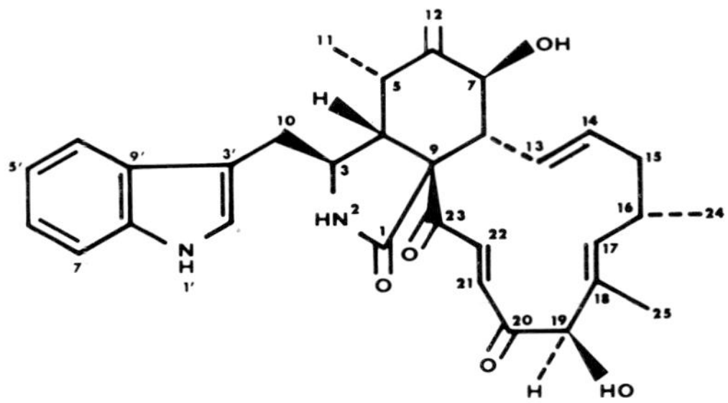

Common name	Chaetoglobosin B [7,19-Dihydroxy-10-(indol-3-yl)-5,16,18-trimethyl-[13]cytochalas-6(12),13,17,21-tetraene-20,23-dione]
Molecular weight	528.2623
Molecular formula	$C_{32}H_{36}O_5N_2$
General characteristics[40]	Crystals from benzene, mp 186°–187°C $[\alpha]_D$ −176° (in methanol)
UV data[28]	λ_{max}^{EtOH} nm (ϵ): 222(43,600), 274(7,900), 281(7,900), and 290(6,700)
Fungal source[40]	*Chaetomium globosum*
Toxicity data[28,45]	Cytotoxicity to HeLa cells, ED_{50} between 3.2 and 10 μg/ml; causes inhibition of cytoplasmic cleavage, resulting in polynucleate cells.
TLC data[47]	*Adsorbent:* silica gel G-HR *Solvent:* A: chloroform–acetone, 93:7 v/v B: toluene–ethylacetate–formic acid, 5:4:1 v/v/v R_f: A: 0.05 B: 0.61 *Detection:* a dark olive-gray fluorescent spot under UV light after spraying with 50% ethanolic H_2SO_4 and heating.
^{1}H NMR data[40]	H4, 3.40; H7, 3.53; H8, 2.06; H11, 1.64; H12, 1.73. The remainder of the protons are the same as in chaetoglobosin A.

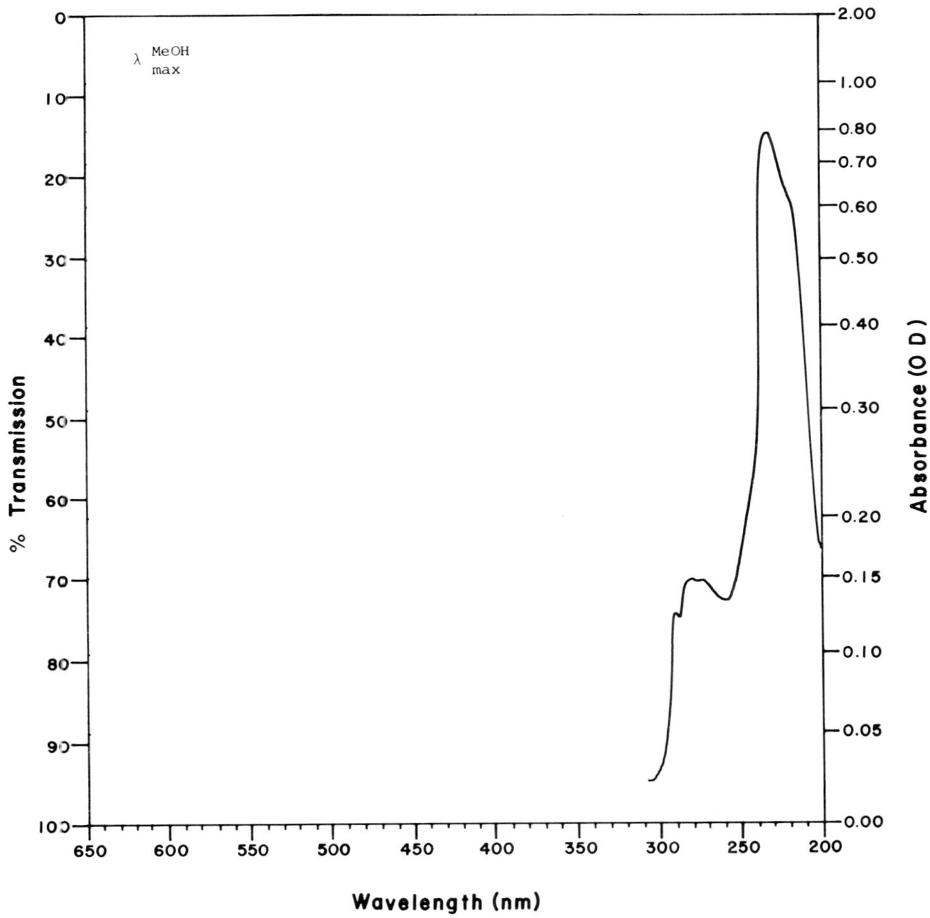
λ MeOH max
% Transmission
Absorbance (O D)
Wavelength (nm)
0
10
20
30
40
50
60
70
80
90
100
2.00
1.00
0.80
0.70
0.60
0.50
0.40
0.30
0.20
0.15
0.10
0.05
0.00
650
600
550
500
450
400
350
300
250
200

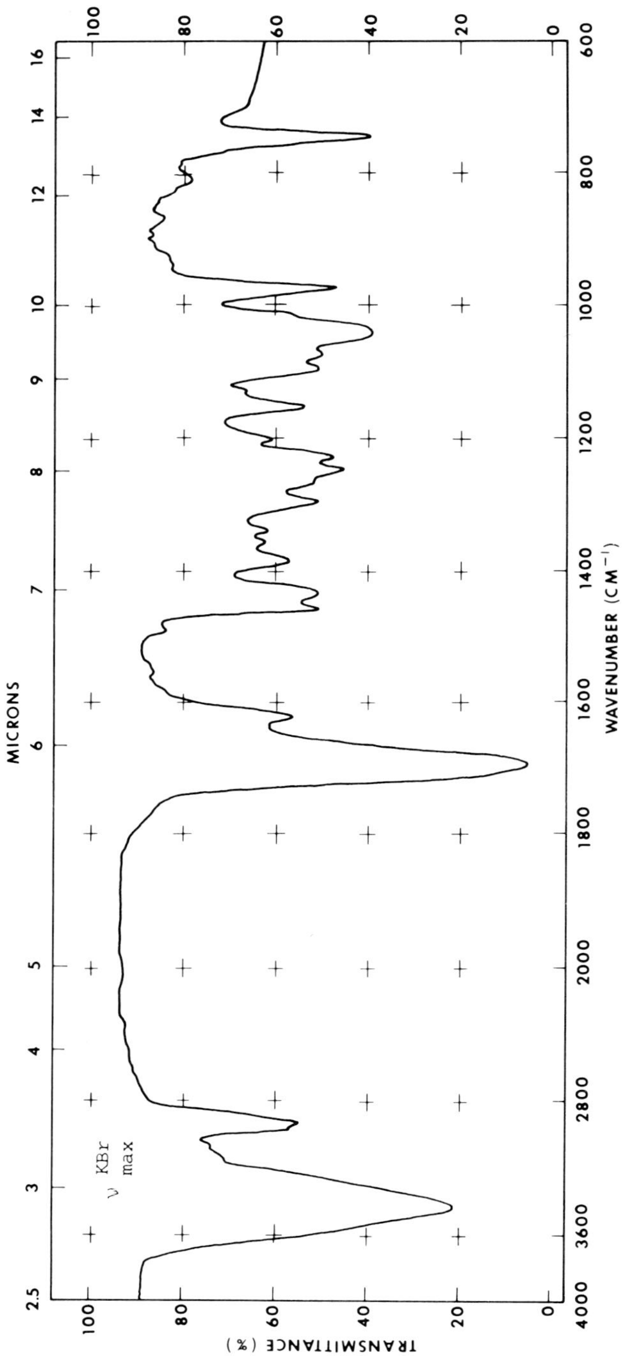
MICRONS
2.5 3 4 5 6 7 8 9 10 12 14 16
WAVENUMBER (CM^{-1})
4000 3600 2800 2000 1800 1600 1400 1200 1000 800 600
TRANSMITTANCE (%)
100 80 60 40 20 0
ν_{max}^{KBr}

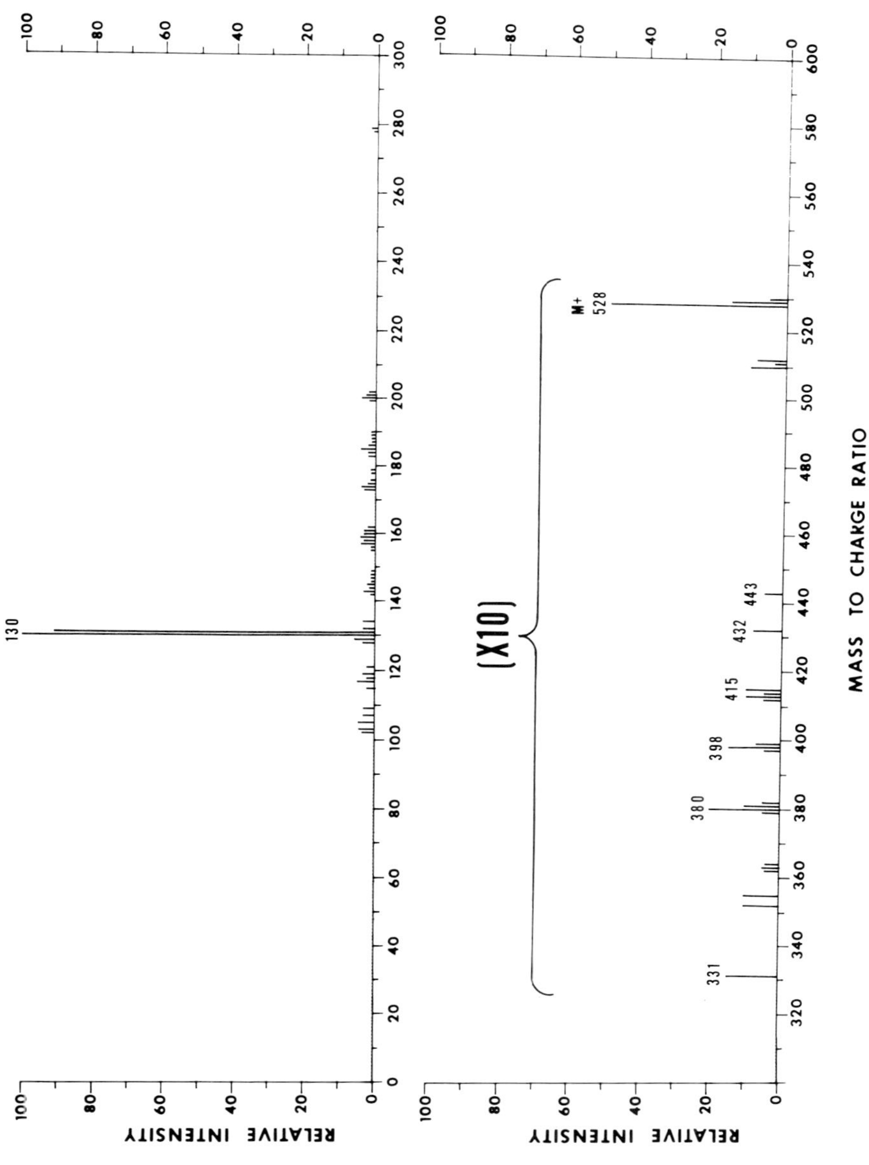
130
(X10)
M+
528
443
432
415
398
380
331
MASS TO CHARGE RATIO
RELATIVE INTENSITY
RELATIVE INTENSITY

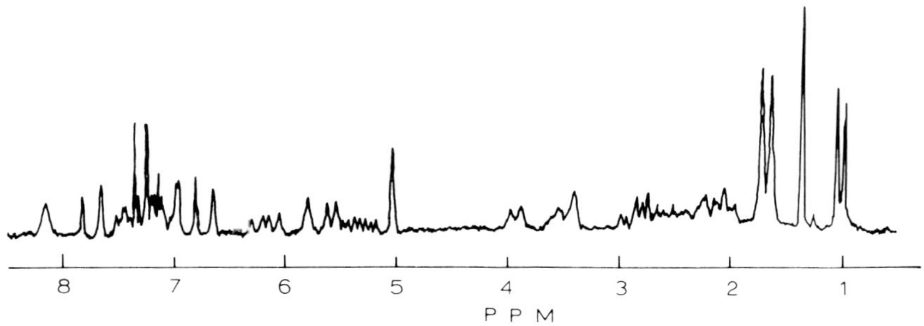

The 100 MHz ^{1}H NMR spectrum of chaetoglobosin B in $CDCl_3$ solution.

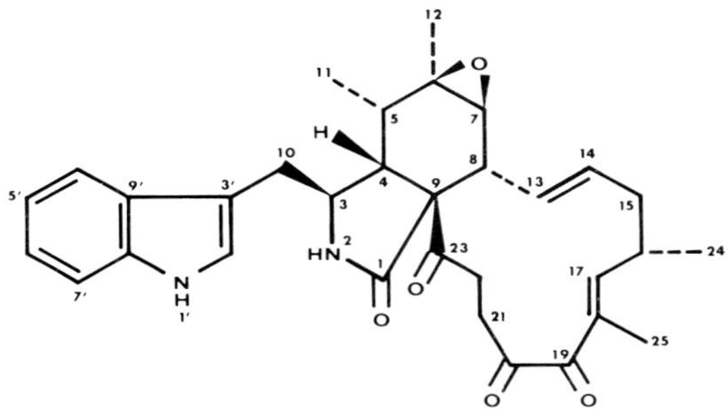

Common name	Chaetoglobosin C [6,7-Epoxy-10-(indol-3-yl)-5,6,16,18-tetramethyl-[13]cytochalas-13,17-diene-19,20,23-trione]
Molecular weight	528.2623
Molecular formula	$C_{32}H_{36}O_5N_2$
General characteristics[40,44]	Colorless leaflets from acetone, mp 260°–263°C $[\alpha]_D$ −30° (in methanol)
UV data[40]	λ_{max}^{EtOH} nm (ϵ): 222(36,300), 273(6,700), 281(6,700), and 291(5,700)
Fungal source[40,44]	*Chaetomium globosum* *C. cochliodes* *Penicillium aurantio-virens*
Toxicity data[28,45]	Cytotoxicity to HeLa cells, ED^{50} between 10 and 32 μg/ml. Oral LD_{50} determination was not successful due to insolubility of purified toxin; crude extracts were highly toxic.
TLC data[47]	*Adsorbent:* silica gel G-HR *Solvent:* A: chloroform–acetone, 93:7 v/v B: toluene–ethyl acetate–formic acid, 5:4:1 v/v/v R_f: A: 0.13 B: 0.71 *Detection:* a slate gray fluorescent spot under UV light after spraying with 50% ethanolic H_2SO_4 and heating.
Other significant references	X-Ray crystallography[44]

^{13}C NMR data C1, 173.8 s; C3, 52.3 d; C4, 48.4 d; C5, 36.1 d; C6, 56.6 s; C7, 60.3 d; C8, 48.4 d; C9, 62.2 s; C10, 39.5 t; C11, 9.9 q; C12, 19.3 q; C13, 127.0 d; C14, 133.1 d; C15, 32.6 t; C16, 37.0 d; C17, 155.8 d; C18, 130.9 s; C19, 196.1 s; C20, 205.2 s; C21, 31.9 t; C22, 31.9 t; C23, 208.1 s; C24, 12.3 q; C25, 19.1 q; C2′, 120.9 d; C3′, 108.0 s; C4′, 118.6 d; C5′, 125.1 d; C6′, 118.3 d; C7′, 111.3 d; C8′, 135.8 s; C9′, 127.6 s

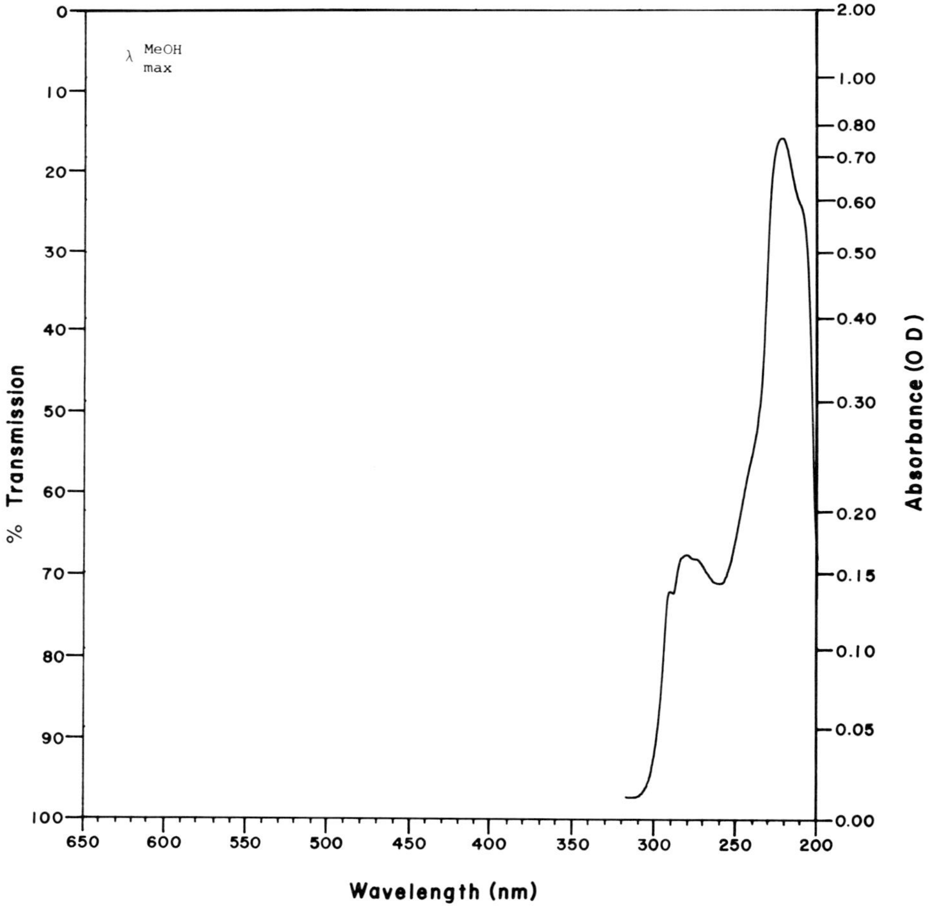

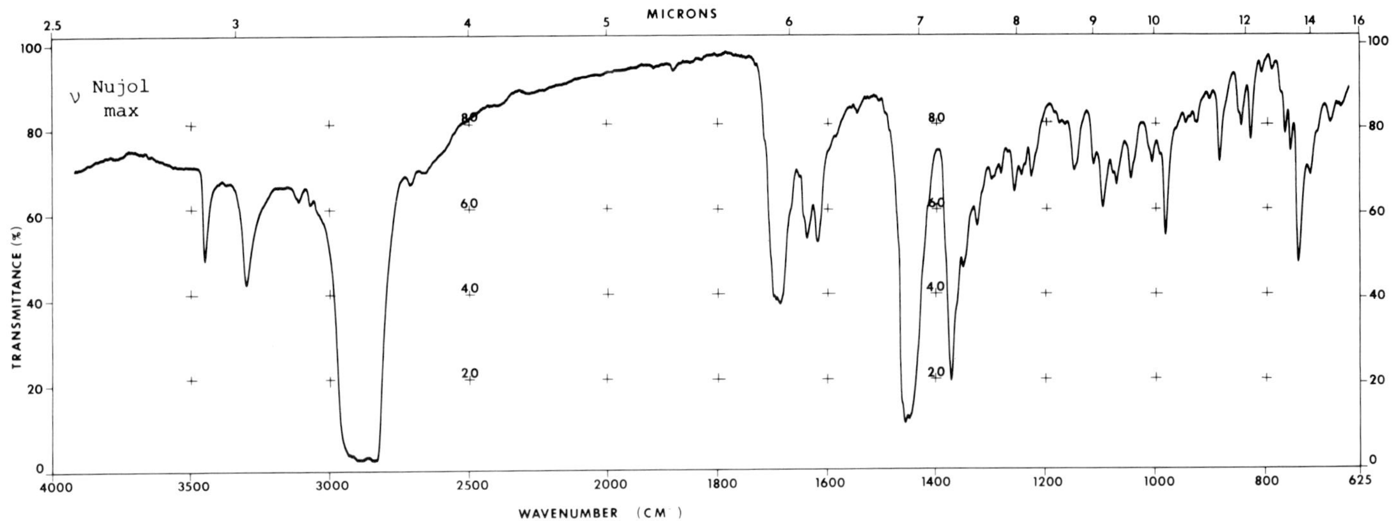
MICRONS
TRANSMITTANCE (%)
WAVENUMBER (CM⁻¹)
ν Nujol max

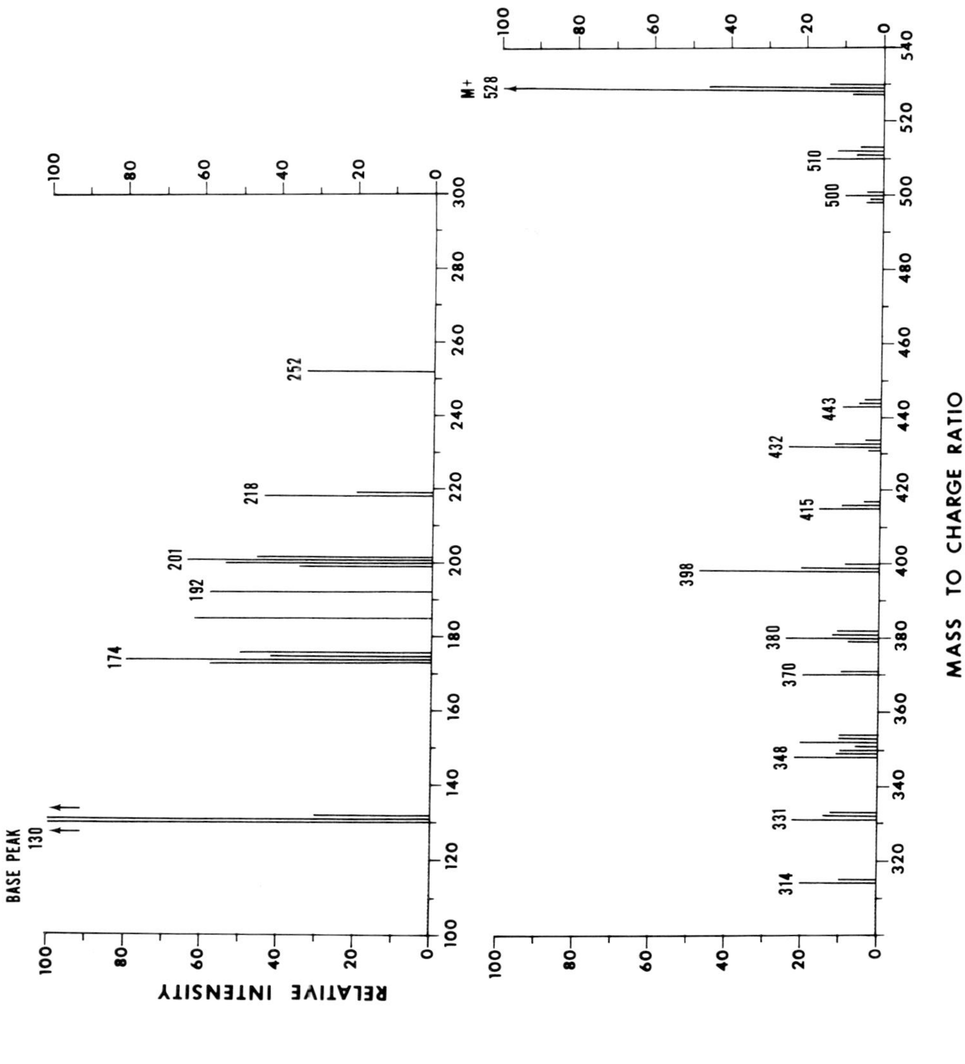
BASE PEAK
130
174
192
201
218
252
M+
528
510
500
443
432
415
398
380
370
348
331
314
RELATIVE INTENSITY
MASS TO CHARGE RATIO

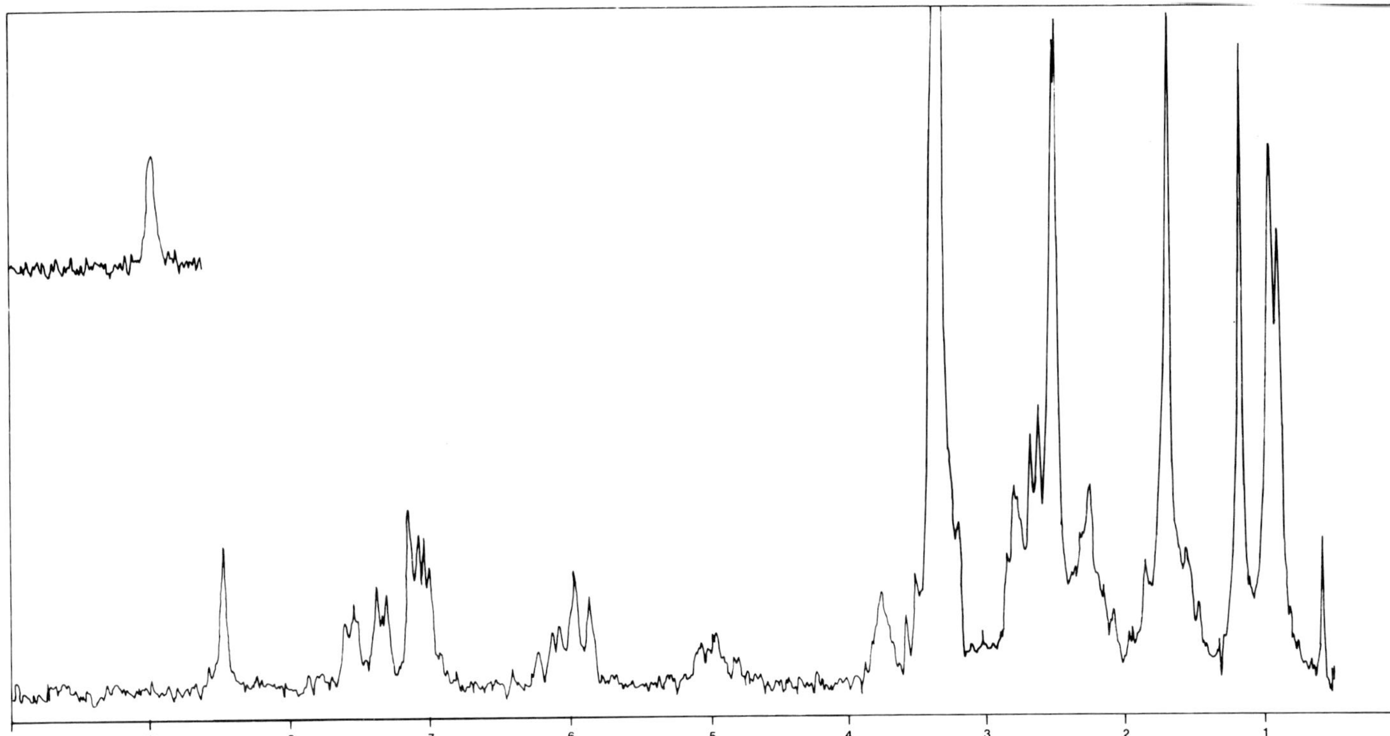

The 100 MHz ^{1}H NMR spectrum of chaetoglobosin C in DMSO solution.

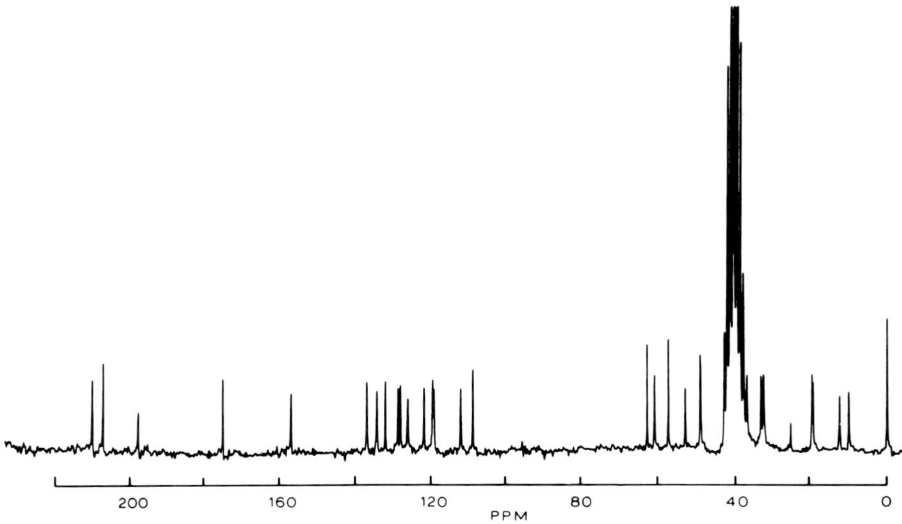

The 25 MHz ^{13}C NMR spectrum of chaetoglobosin C in DMSO solution.

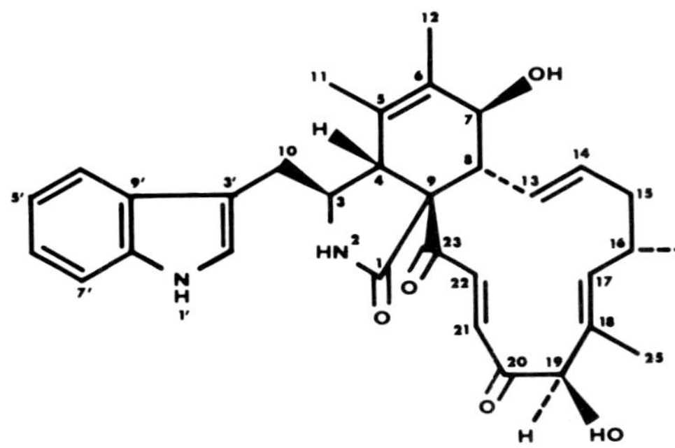

Common name	Chaetoglobosin D [5,6,16,18-tetramethyl-7,19-dihydroxy-10-(indol-3-yl)-[13]cytochalas-6(12),13,17,21-tetraene]
Molecular weight	528.2623
Molecular formula	$C_{32}H_{36}O_5N_2$
General characteristics[40,45]	Pale yellow prisms from chloroform solution, mp 216°C $[\alpha]_D$ −269° (in methanol)
UV data[40]	λ_{max}^{EtOH} nm (ϵ): 221(43,600), 273(9,100), 281(9,100), and 290(7,500)
IR data[40]	ν_{max}^{KBr} cm^{-1}: 3421, 3280, 1686, 1606, 972, 908, and 750
Mass spectrum	Molecular ion (M^+) 528.261; base peak 130.037 ($C_9H_8N_1$)
Fungal source[40,45]	*Chaetomium globosum*
Toxicity data[28]	Cytotoxicity to HeLa cells, ED_{50} between 3.2 and 10 μg/ml.
^{1}H NMR data[39]	H4, 3.30; H5, 3.14; H7, 4.59; H8, 2.92; H11, 1.23; H12, 5.21, 5.66; H13, 6.53; H14, 5.44. The remainder of the protons are the same as in chaetoglobosin A.

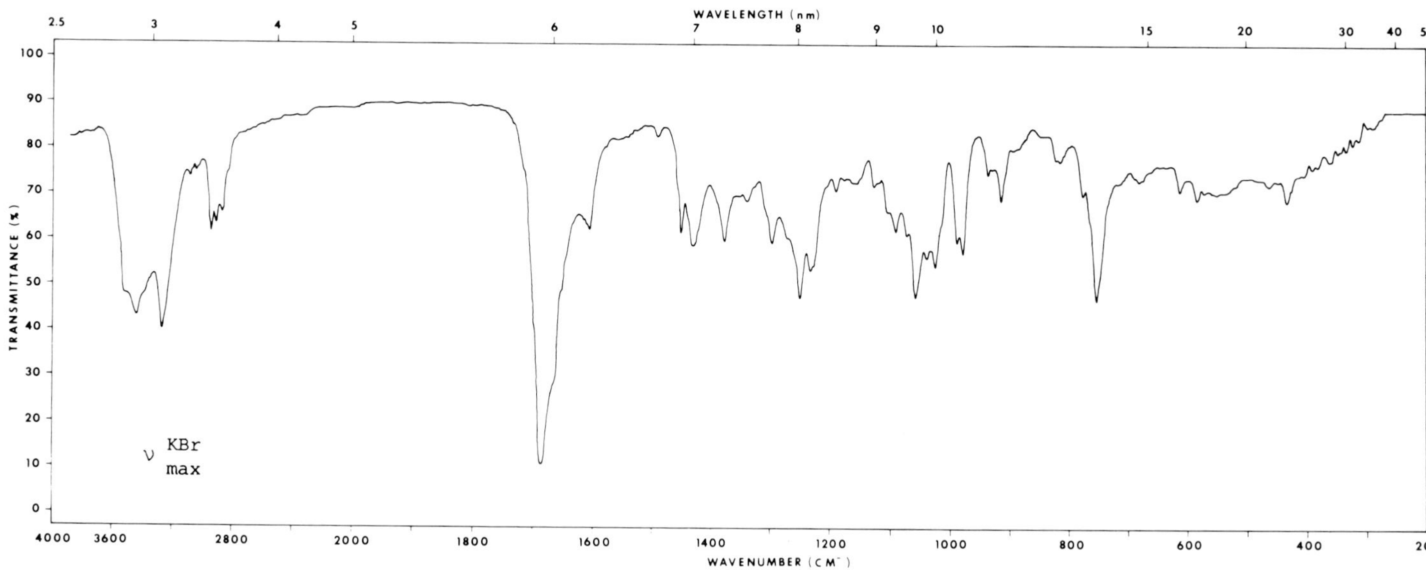
WAVELENGTH (nm)
2.5
3
4
5
6
7
8
9
10
15
20
30
40
50
TRANSMITTANCE (%)
100
90
80
70
60
50
40
30
20
10
0
4000
3600
2800
2000
1800
1600
1400
1200
1000
800
600
400
200
WAVENUMBER (CM⁻)
ν KBr max

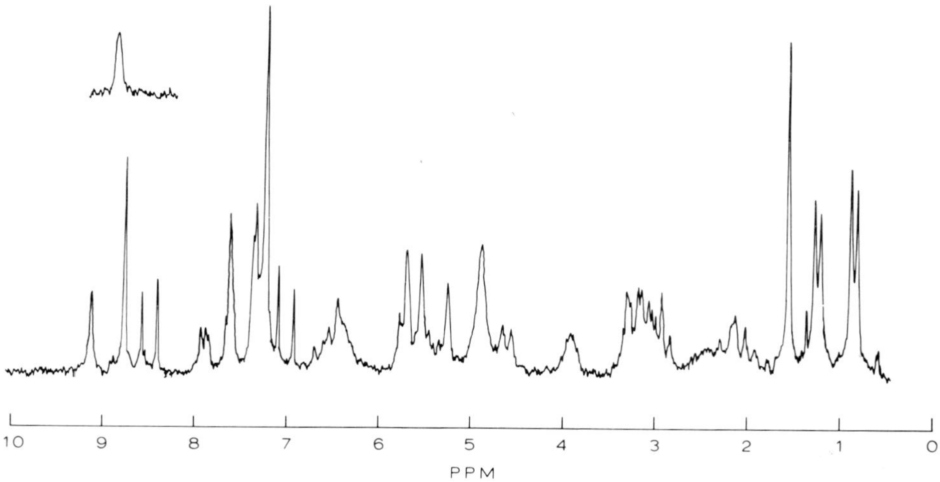

The 100 MHz ^{1}H NMR spectrum of chaetoglobosin D in pyridine-d_5 solution.

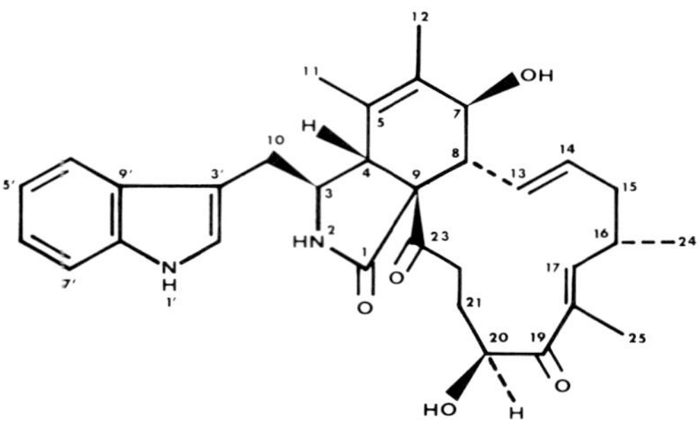

Common name	Chaetoglobosin E [7,20-Dihydroxy-10-(indol-3-yl)-5,6,16,18-tetramethyl-[13]cytochalas-13,17-diene-19,23,dione]
Molecular weight	530.2779
Molecular formula	$C_{32}H_{38}O_5N_2$
General characteristics[39,41]	Colorless needles from methanol, mp 279°–280°C $[\alpha]_D$ +158° (in methanol)
UV data[39,41]	λ_{max}^{EtOH} nm (ϵ): 221(56,200), 275(7,000), 281(7,000), and 291(6,300)
IR data[39,41]	ν_{max}^{KBr} cm^{-1}: 3410, 1704, 1676, 1048, and 746
Mass spectrum[39,41]	M+ 530.275; base peak 130.131 ($C_9H_8N_1$)
Toxicity data[28]	Cytotoxicity to HeLa cells, ED_{50} between 3.2 and 10 μg/ml.
^1H NMR data[39]	H7, 4.56; 7OH, 5.97; H11, 1.49; H12, 1.90; H17, 6.34; H20, 5.29; 20OH, 6.10; H21, 2.41; H22, 3.41. The remainder of the protons are the same as in chaetoglobosin A.

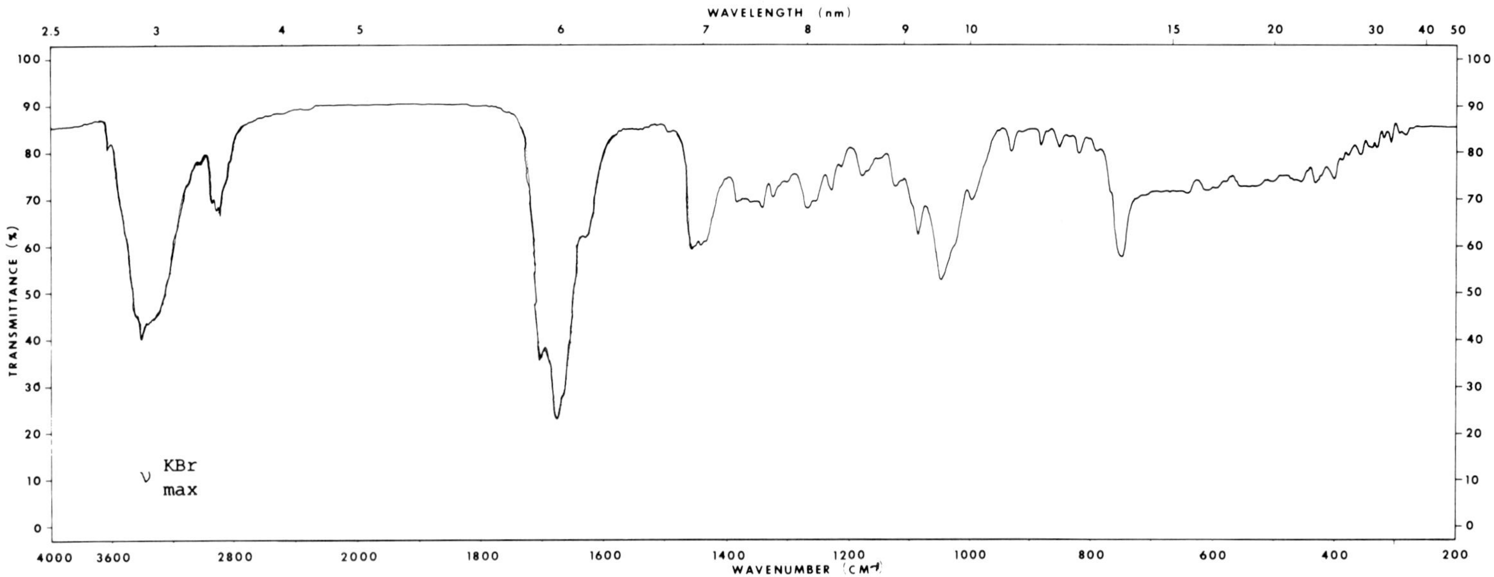
WAVELENGTH (nm)
TRANSMITTANCE (%)
WAVENUMBER (CM⁻¹)
ν KBr max

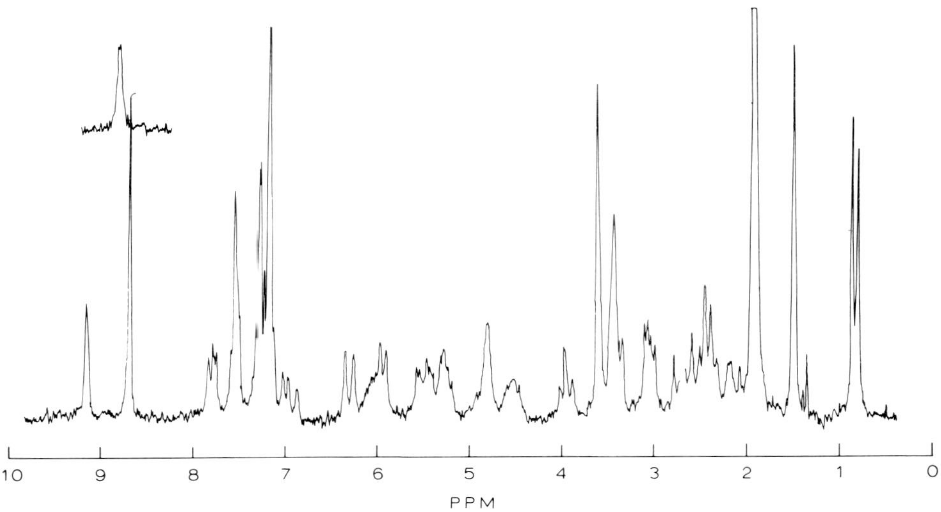

The 100 MHz ^{1}H NMR spectrum of chaetoglobosin E in pyridine-d_5 solution.

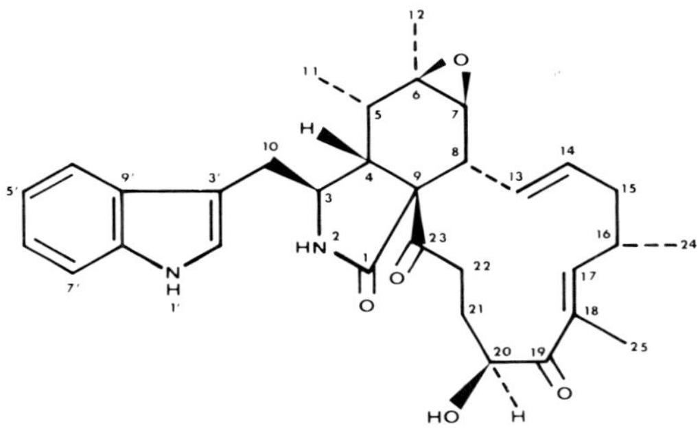

Common name	Chaetoglobosin F [6,7-Epoxy-10-(indol-3-yl)-5,6,16,18-tetramethyl-[13]cytochalas-13,17-diene-19,23-dione]
Molecular weight	530.2779
Molecular formula	$C_{32}H_{38}O_5N_2$
General characteristics[39,45]	Colorless leaflets from benzene, mp 177°–178°C $[\alpha]_D$ −69° (in chloroform)
UV data[39,45]	λ_{max}^{EtOH} nm (ϵ): 222(47,800), 276(6,900), 283(6,700), and 292(6,000)
IR data[39,45]	ν_{max}^{KBr} cm^{-1}: 3346, 1676, 1618, 979, 879, and 740
Mass spectrum	M+ 530.275; base peak 130.107 ($C_9H_8N_1$)
Fungal source[45]	*Chaetomium globosum*
Toxicity data[45]	Cytotoxicity to HeLa cells, ED_{50} 10 μg/ml.
^1H NMR data[39]	H5, 2.01; H7, 3.12; H11, 0.97; H12, 1.25; H17, 6.26; H20, 5.14; H21, 2.27; H22, 2.95; H25, 1.93. The remainder of the protons are the same as in chaetoglobosin A.

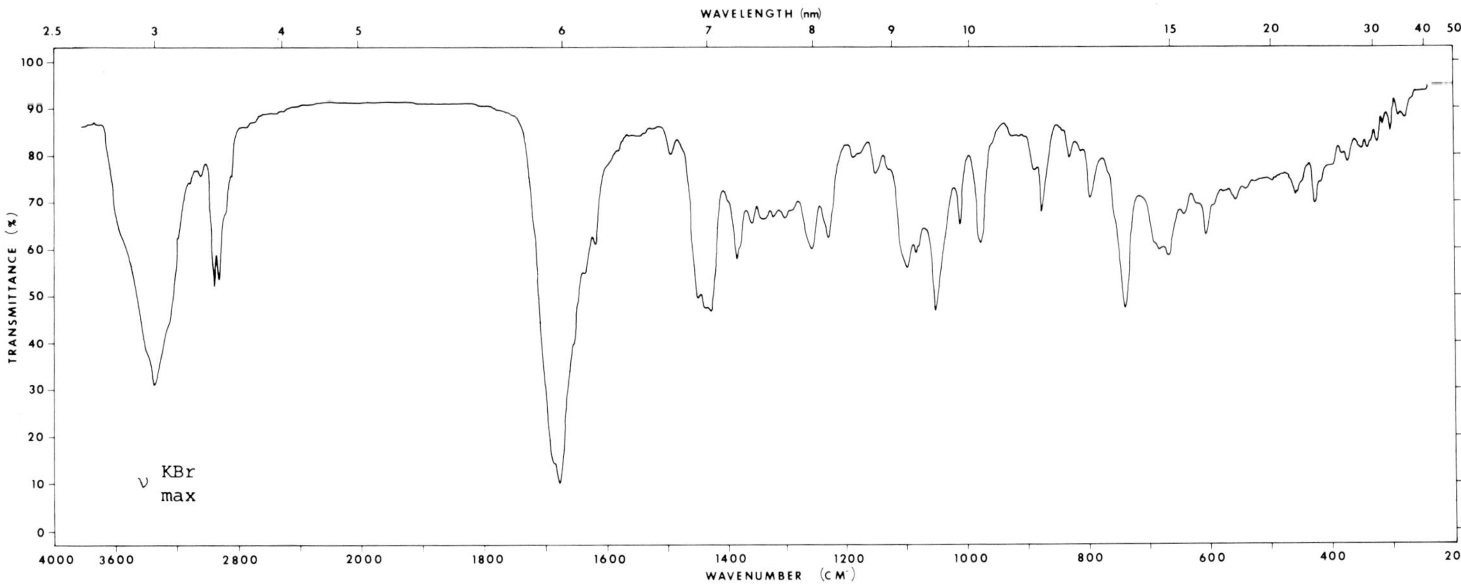

WAVELENGTH (nm)
2.5
3
4
5
6
7
8
9
10
15
20
30
40
50
TRANSMITTANCE (%)
100
90
80
70
60
50
40
30
20
10
0
ν KBr max
4000
3600
2800
2000
1800
1600
1400
1200
1000
800
600
400
200
WAVENUMBER (CM⁻)

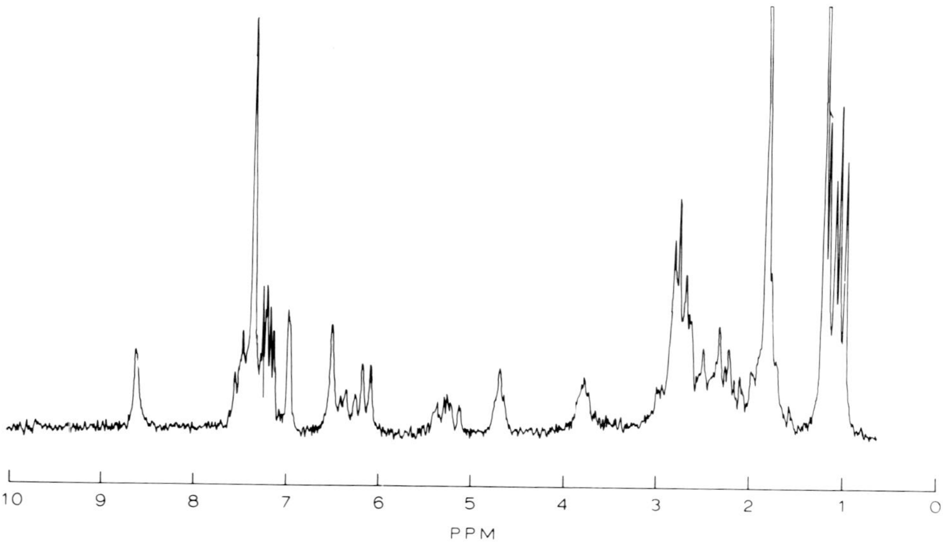

The 100 MHz ^{1}H NMR spectrum of chaetoglobosin F in pyridine-d_5 solution.

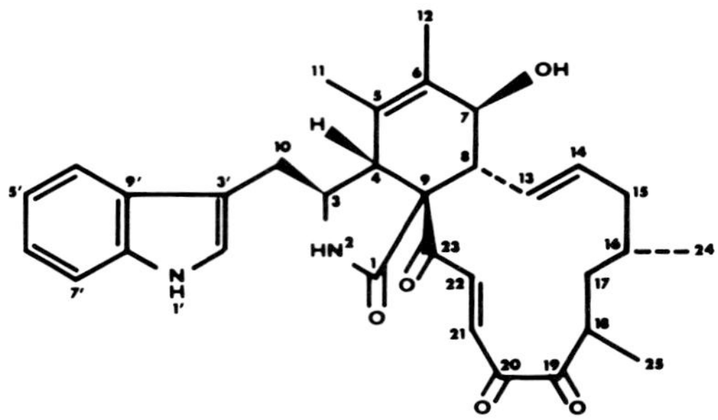

Common name	Chaetoglobosin G
Molecular weight	528.262
Molecular formula	$C_{32}H_{36}O_5N_2$
General characteristics[38]	Colorless leaflets from methanol, mp 251°–253°C $[\alpha]_D$ +89°
UV data[38]	λ_{max}^{EtOH} nm (ϵ): 222(32,300), 275(6,000), 282(6,000), and 291(5,300)
IR data[38]	ν_{max}^{KBr} cm^{-1}: 3455, 3300, 1713, 1693, 1646, 1623, 987, 948, and 741
Fungal source[38]	*Chaetomium globosum*
Toxicity data[38]	Cytotoxicity to cultured HeLa cells at *ca.* 3 μg/ml, forms multinucleated cells.
^1H NMR data[38]	Perhydroisoindolone protons same as in chaetoglobosin B and E. Thirteen-membered ring protons are the same as in chaetoglobsin C.

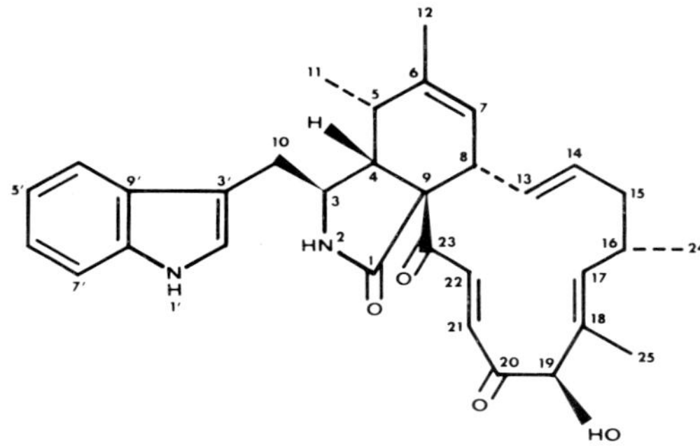

Common name	Chaetoglobosin J
Molecular weight	512.2500
Molecular formula	$C_{32}H_{36}O_4N_2$
General characteristics[38]	Pale yellow prisms from benzene, mp 149°–151°C
UV data[38]	λ_{max}^{EtOH} nm (ϵ): 224(47,800), 270(7,000), 280(7,000), and 290(6,000)
IR data[38]	ν_{max}^{KBr} cm^{-1}: 3412, 3273, 1683, 1639, 1612, 980, 975, 925, and 750
Fungal source[38]	*Chaetomium globosum*
Toxicity data[38]	Cytotoxicity to cultured HeLa cells at *ca.* 3 μg/ml. Unlike other members of the group, chaetoglobosin J does not form multinucleated cells.
^{1}*H NMR data*[38]	H4, 3.12; H5, 2.46; H7, 5.30; H8, 2.56; H11, 1.42; H12, 1.80. The remainder of the protons are the same as in chaetoglobosin A.

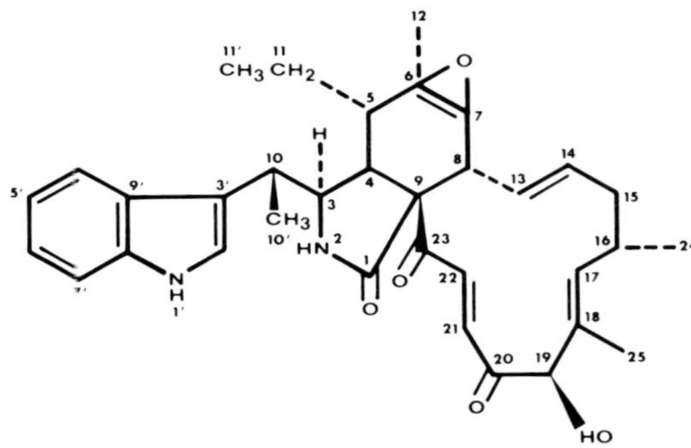

Common name	Chaetoglobosin K
Molecular weight	556.2934
Molecular formula	$C_{34}H_{40}O_5N_2$
General characteristics[15]	Rectangular trapezoids from acetone, mp 264°–266°C
UV data[15]	λ_{max}^{EtOH} nm (ϵ): 219(44,600); also absorptions at 274, 282, and 290 nm in concentrated solutions
Fungal source[15]	*Diplodia macrospora*
Toxicity data[15]	Significant growth inhibitor of wheat coleoptiles, activity down to 10^{-7} *M*.
^{13}C NMR data	C1, 173.2 s; C3, 56.7 d; C4, 49.2 d; C5, 36.3 d; C6, 57.2 s; C7, 61.6 d; C8, 44.0 d; C9, 63.9 s; C10, 44.0 d; C10′, 10.6 q; C11, 21.7 t; C11′, 13.8 q; C12, 19.6 q; C13, 133.2 d; C14, 128.2 d; C15, 41.9 t; C16, 31.9 d; C16′, 12.6 q; C17, 140.1 d; C18, 125.8 s; C18′, 20.9 q; C19, 81.6 d; C20, 190.0 s; C21, 130.9 d; C22, 136.4 d; C23, 193.9 s; C2′, 122.3 d; C3′, 116.5 s; C19′, 132.2 s; C4′, 118.4 d; C5′, 119.7 d; C6′, 121.7 d; C7′, 111.5 d; C8′, 136.4 s
^{1}H NMR data	H2, 6.23; H3, 3.89 (*J* = 4.4); H4, 3.14; H5, 1.90; H7, 2.74 (*J* = 5.3); H8, 2.16; H10, 3.14; H10′, 1.16 (*J* = 7.3); H11, 1.51; H11′, 1.01 (*J* = 7.0); H12, 1.29; H13, 6.09 (*H* = 10.2, 15.6); H14, 5.01; H15, 2.16; H16, 2.35; H16′, 1.12 (*J* = 7.1); H17, 5.56 (*J* = 8.4); H18′, 1.29; H19, 4.99 (*J* = 4.5); H21, 7.65 (*J* = 16.5); H22, 6.49; H1′, 8.53; H2′, 6.90; H4–7, 7.0–7.5

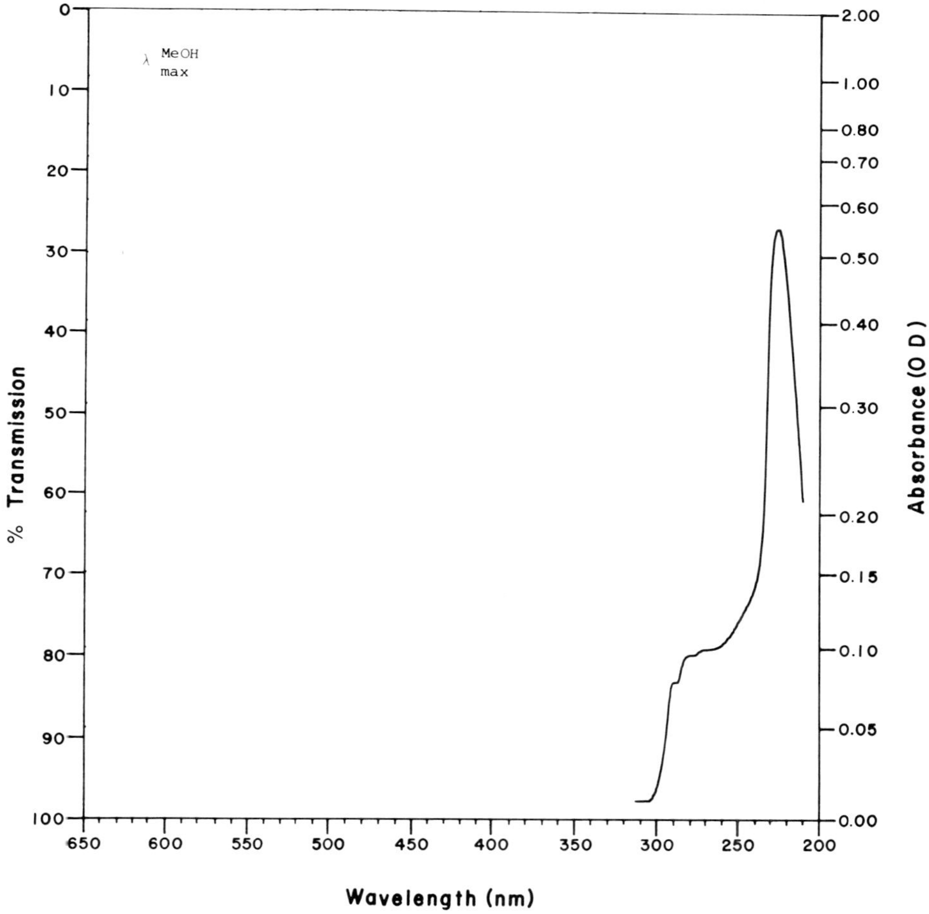

λ MeOH max
% Transmission
Absorbance (O D)
Wavelength (nm)
0
10
20
30
40
50
60
70
80
90
100
2.00
1.00
0.80
0.70
0.60
0.50
0.40
0.30
0.20
0.15
0.10
0.05
0.00
650
600
550
500
450
400
350
300
250
200

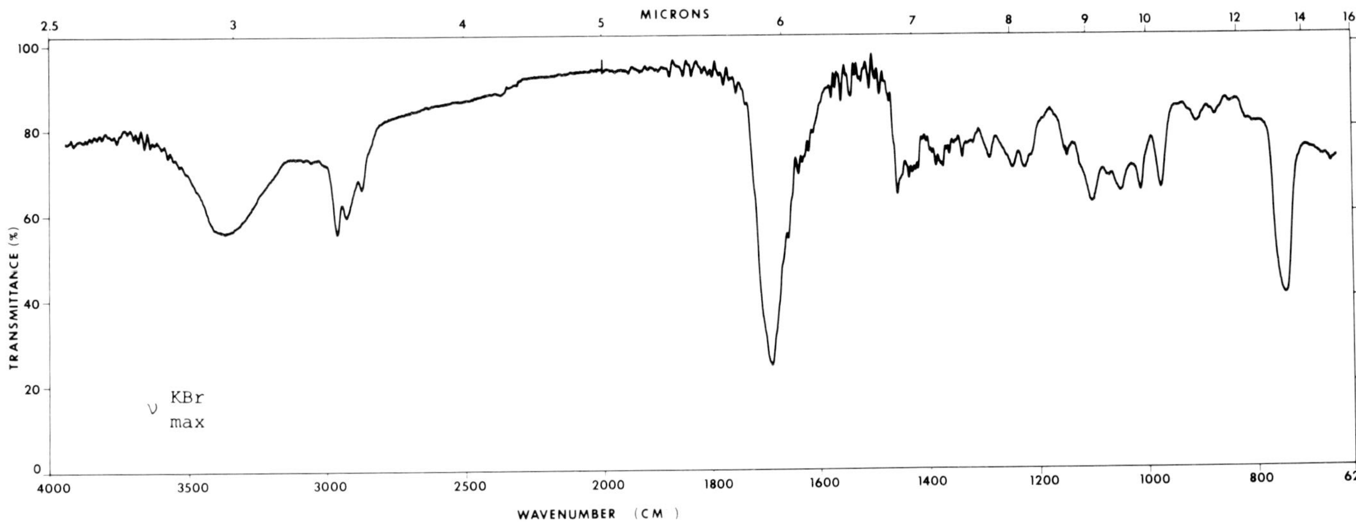
MICRONS
TRANSMITTANCE (%)
WAVENUMBER (CM)
ν KBr max

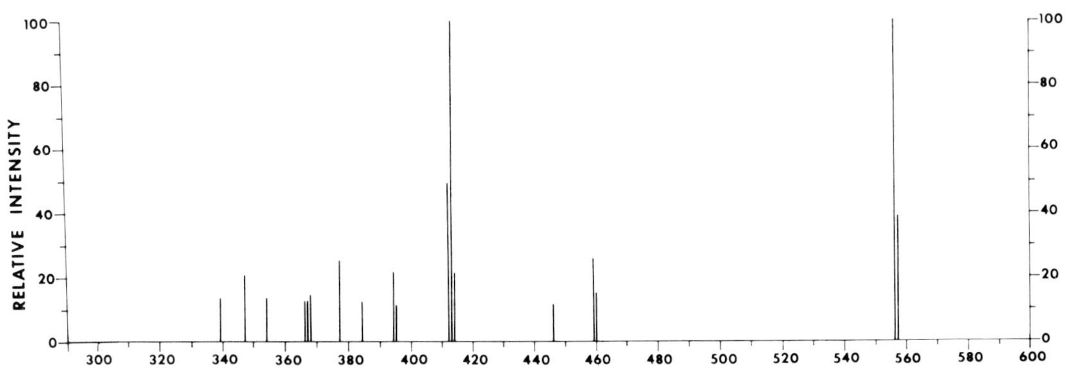
RELATIVE INTENSITY

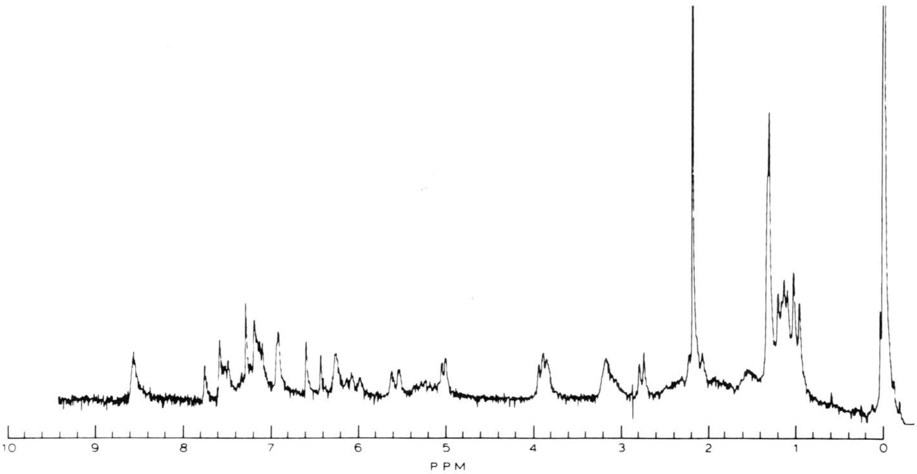

The 100 MHz ^{1}H NMR spectrum of chaetoglobosin K in $CDCl_3$ solution.

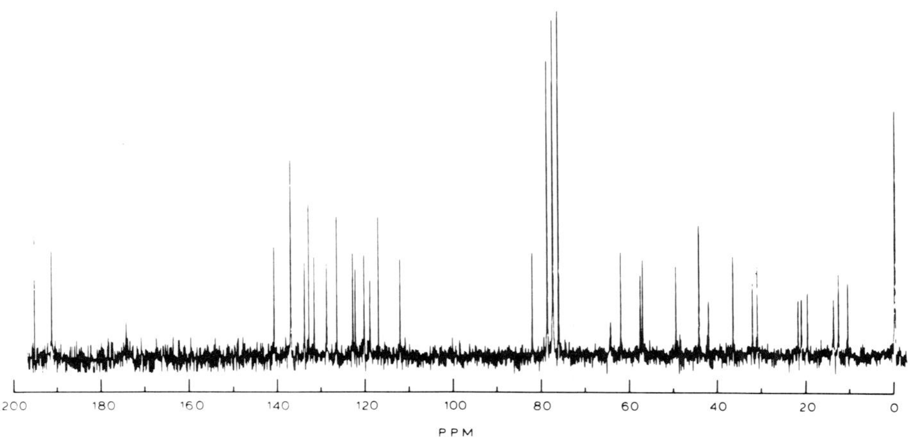

The 25 MHz ^{13}C NMR spectrum of chaetoglobosin K in $CDCl_3$ solution.

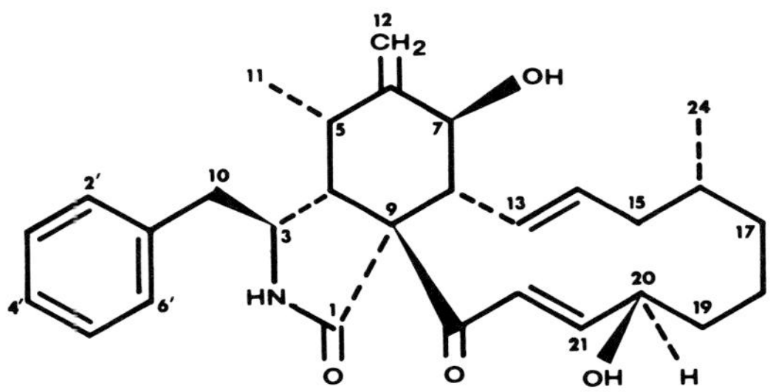

Common name	Deoxaphomin [(7*S*,16*R*,20*R*)-7,20-Dihydroxy-5,16-dimethyl-10-phenyl-[13]cytochalas-6(12),13^t,21^t-triene-1,23-dione]
Molecular weight	463.2722
Molecular formula	$C_{29}H_{37}O_4N$
General characteristics[9]	Amorphous $[\alpha]_D^{24}$ −50.8° (*C* = 0.178 in chloroform)
UV data[9]	λ_{max}^{MeOH} nm (ϵ): 217(14,100) and 234(8,700)
IR data[9]	$\nu_{max}^{CH_2Cl_2}$ cm^{-1}: 3590, 3540, 3410, 3060, 3020, 1710–1670, 1622, 1600, 1490, and 1020
Fungal source[9]	*Phoma* spp.
Toxicity data[9]	Biological activity comparable to phomin, cytostatic activity (*in vitro*) of P-185 mastocyte cells, ED_{50} 0.29 mg/liter; in target cell destruction test treatments of 10, 1, and 0.1 μg/liter gave inhibition of 100, 76, and 9%, respectively.
^1H NMR data[9]	H2, 6.36; H3, 2.81; H4, 3.15; H5, 3.44; H7, 3.97; H8, 2.50; H10, 2.60; H11, 0.94; H12, 5.09, 5.29; H13, 6.18; H14, 5.40; H15–19, 1.0–2.0; H20, 4.20; H21, 6.77; H22, 7.20; H24, 0.98; H2′–6′, 7.20; OH, 3.10,2.35

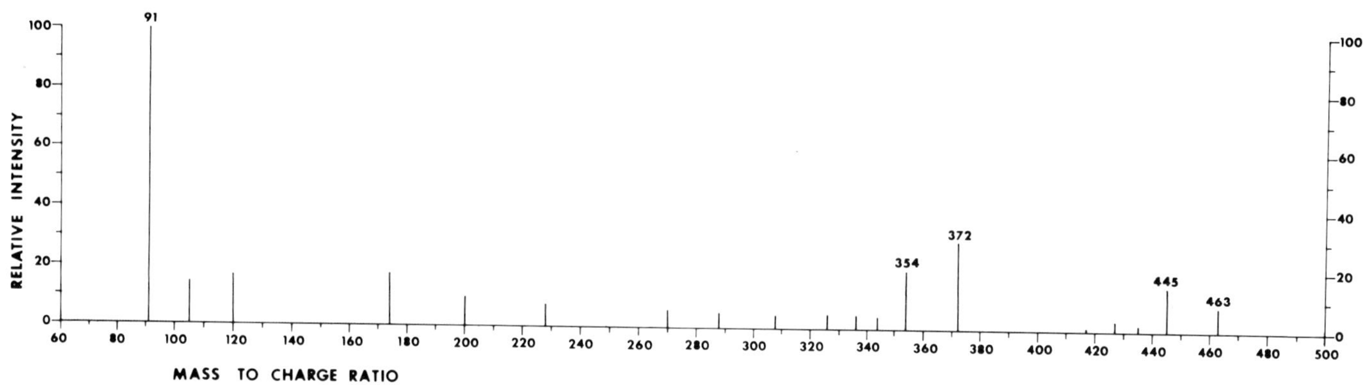
91
354
372
445
463
RELATIVE INTENSITY
MASS TO CHARGE RATIO

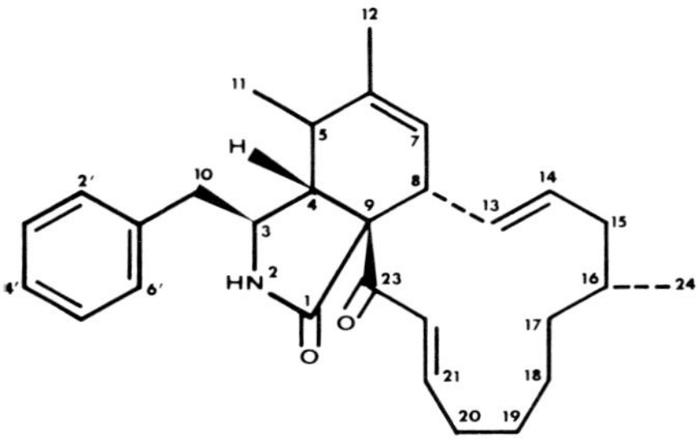

Common name	Proxiphomin [5,6,16-Trimethyl-10-phenyl-[13]-cytochalasa-6(7),13^t,21^t-triene-1,23-dione]
Molecular weight	431.2824
Molecular formula	$C_{29}H_{37}O_2N$
General characteristics[10]	Amorphous $[\alpha]_D^{24}$ −140° (C = 0.156 in chlororform)
UV data[10]	λ_{max}^{EtOH} nm (ϵ): 206(22,900), 217(12,300), and 243(10,700)
IR data[10]	$\nu_{max}^{CCl_4}$ cm^{-1}: 3420, 3200, 3080, 3060, 3020, 1700–1670, 1610, and 1490
Fungal source[10]	*Phoma* spp.
Toxicity data	Biological activity not reported
^1H NMR data[10]	H2, 5.84; H3, 3.24; H4, 3.24; H7, 5.44; H10, 2.52, 2.85; H11, 1.22; H12, 1.80; H13, 6.30; H14, 5.24; H15–18, 1.0–3.0; H21, 6.80; H22, 7.26; H24, 0.95; H2′–6′, 7.26

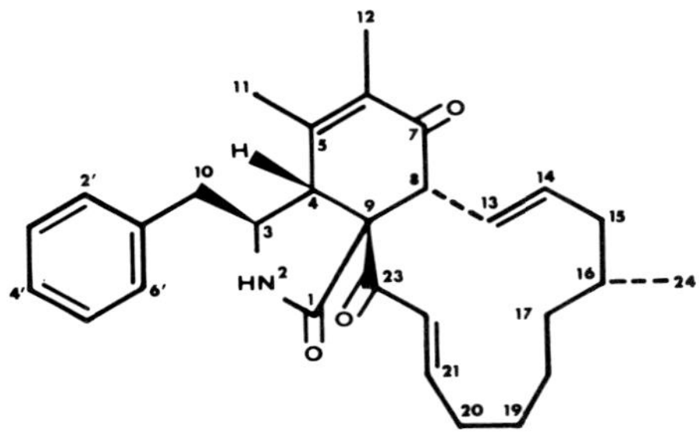

Common name	Protophomin [5,6,16-Trimethyl-10-phenyl-[13]-cytochalasa-5(6),13^t,21^t-triene-1,7,23-trione]
Molecular weight	445.2616
Molecular formula	$C_{29}H_{35}O_3N$
General characteristics[10]	Crystals from acetone, mp 252°–256°C $[\alpha]_D^{24}$ −112° (C = 0.42 in chloroform)
UV data[10]	λ_{max}^{EtOH} nm (ϵ): 207(14,100) and 245(13,200)
IR data[10]	ν_{max}^{KBr} cm^{-1}: 3600–3200, 1700, 1680, 1660, 1610, and 1490
Fungal source[10]	*Phoma* spp.
Toxicity data	Biological activity not known
^1H NMR data[10]	H2, 6.53; H3, 3.50; H4, 3.71; H8, 3.13; H10, 2.53, 2.70; H11, 1.53; H12, 1.70; H13, 6.24; H14, 5.09; H15–19, 1.0–3.0; H21, 7.07; H22, 6.73; H24, 0.84; H2′–6′, 7.24

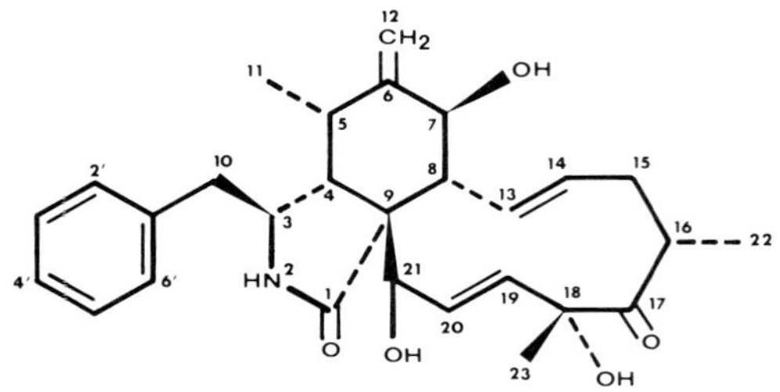

Common name	Zygosporin D
Molecular weight	465.2515
Molecular formula	$C_{28}H_{35}O_5N$
General characteristics[26]	Colorless plates from methanol, mp 180°–190°C $[\alpha]_D^{23}$ −14.9° (C = 0.759 in dioxane)
IR data[26]	ν_{max} cm^{-1}: 3400, 1700, 1127, 1107, 1075, 1005, 968, and 910
Fungal source[26]	*Zygosporium masonii*
Toxicity data[26]	Cytotoxicity to HeLa cells.
TLC data[26]	*Adsorbent:* silica gel *Solvent:* chloroform–methanol, 10:1 v/v R_f: 0.50 *Detection:* not reported

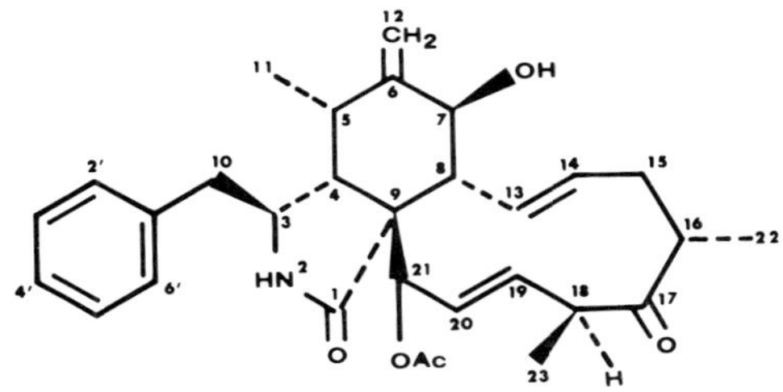

Common name	Zygosporin E
Molecular weight	491.2722
Molecular formula	$C_{30}H_{37}O_5N$
General characteristics[26]	Colorless needles from acetone–*n*-hexane, mp 218°–223.5°C $[\alpha]_D^{24}$ +6.2° (*C* = 0.971 in dioxane)
IR data[26]	ν_{max} cm^{-1}: 3524, 3420, 1743, 1703, 1113, 1010, 973, and 912
Fungal source[26]	*Zygosporium masonii*
Toxicity data[26]	Cytotoxic to HeLa cells
TLC data[26]	*Adsorbent:* silica gel *Solvent:* ethyl acetate R_f: 0.48 *Detection:* not reported

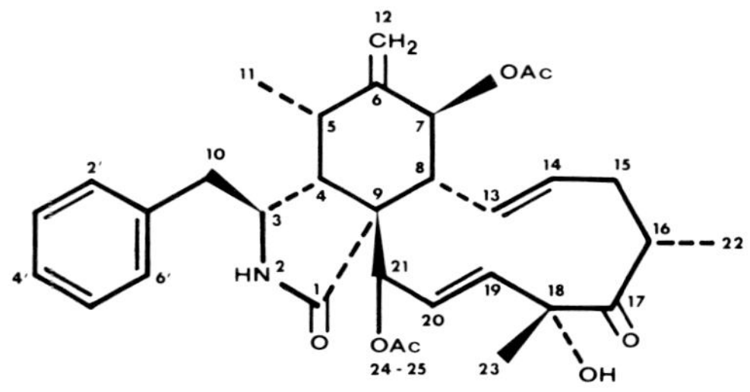

Common name	Zygosporin F (Cytochalasin D monoacetate)
Molecular weight	549.2725
Molecular formula	$C_{32}H_{39}O_7N$
General characteristics[26]	Colorless prisms from diisopropyl ether, mp 126°–129°C $[\alpha]_D$ −12.0° (*C* = 0.775 in dioxane)
IR data[26]	ν_{max} cm^{-1}: 3420, 1740, 1704, 1113, 1043, 1009, 965, and 910
Fungal source[26]	*Zygosporium masonii*
Toxicity data[26]	Cytotoxic to HeLa cells
TLC data[26]	*Adsorbent:* silica gel *Solvent:* toluene–methanol, 10:1 v/v R_f: 0.28 *Detection:* not reported

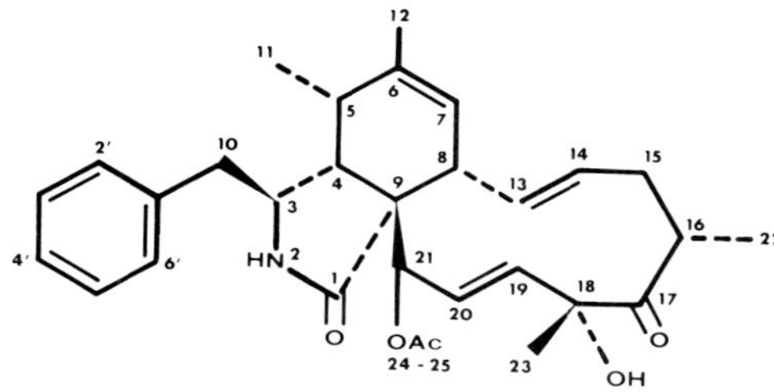

Common name	Zygosporin G
Molecular weight	491.2671
Molecular formula	$C_{30}H_{37}O_5N$
General characteristics[26]	Colorless prisms from diisopropyl ether, mp 115°–125°C $[\alpha]_D^{24}$ −82.0° (*C* = 0.870 in dioxane)
IR data[26]	ν_{max} cm^{-1}: 3410, 1743, 1701, 1122, 1045, 1008, and 965 $\nu_{max}^{(Nujol)}$ cm^{-1}: 3430, 3285, 1742, 1693, 1228, 1121, 1044, 1007, 965, 745, and 703
Fungal source[26]	*Zygosporium masonii*
Toxicity data[26]	Cytotoxic to HeLa cells.
TLC data[26]	*Adsorbent:* silica gel *Solvent:* toluene–methanol, 10:1 v/v R_f: 0.35 *Detection:* not reported

REFERENCES

1. Aldridge, D. C., Armstrong, J. J., Speake, R. N., and Turner, W. B., *J. Chem. Soc.* p. 1667 (1967).
2. Aldridge, D. C., Armstrong, J. J., Speake, R. N., and Turner, W. B., *Chem. Commun.* p. 26 (1967).
3. Aldridge, D. C., Burrows, B. F., and Turner, W. B., *Chem. Commun.* p. 148 (1972).
4. Aldridge, D. C., Greatbanks, D., and Turner, W. B., *Chem. Commun.* p. 551 (1973).
5. Aldridge, D. C., and Turner, W. B., *J. Chem. Soc. C* p. 923 (1969).
6. Allison, A. C., Davies, P., and DePetris, S., *Nature (London), New Biol.* **232,** 1536 (1971).
7. Axline, S. G., and Reaven, E., *J. Clin. Invest.* **51,** 6a (1972).
8. Beno, M. A., Cox, R. H., Wells, J. M., Cole, R. J., Kirksey, J. W., and Christoph, G. G., *J. Am. Chem. Soc.* **99,** 4123 (1977).
9. Binder, M., and Tamm, C.,*Helv. Chim. Acta* **56,** 966 (1973a).
10. Binder, M., and Tamm, C.,*Helv. Chim. Acta* **56,** 2387 (1973b).
11. Büchi, G., Kitaura, Y., Yuan, S., Wright, H. E., Clardy, J., Demain, A. L., Glinsukon, T., Hunt, N., and Wogan, G. N., *J. Am. Chem. Soc* **95,** 5423 (1973).
12. Cameron, A. F., Freer, A., Hesp, B., and Strawson, C. J., *J. Chem. Soc. Perkin Trans. 2* p. 1741 (1974).
13. Carter, S. B., *Endeavour* **31,** 77 (1972).
14. Carter, S. B., *Nature (London)* **213,** 261 (1967).
15. Cutler, H. G., personal communication.
16. Davis, A. T., Estensen, R. D., and Ouie, P. G., *Proc. Soc. Exp. Biol. Med.* **137,** 161 (1971).
17. Estensen, R. D., and Plagemann, P. G., *Proc. Natl. Acad. Sci. U.S.A.* **69,** 1430 (1972).
18. Glinsukon, T., Shank, R. C., Wogan, G. N., and Newberne, P. M., *Toxicol. Appl. Pharmacol.* **32,** 135 (1975).
19. Haslam, R. J., *Biochem. J.* **127,** 34P (1972).
20. Hayakawa, S., Matsushima, T., Kimura, T., Minato, H., and Katagiri, K., *J. Antibiot.* **21,** 523 (1968).
21. Kletzien, R. F., Perdue, J. F., and Springer, A., *J. Biol. Chem.* **247,** 2964 (1972).
22. Krishan, A., *J. Cell Biol.* **54,** 657 (1972).
23. Majno, G., Bouvier, C. A., Gabbiani, G., Ryan, C. B., and Statkov, P., *Thromb. Diath. Haemorrh.* **28,** 49 (1972).
24. McLaughlin, G. M., Sim, G. A., Kiechel, J. R., and Tamm, C., *Chem. Commun.* p. 1398 (1970).
25. McMillan, J. A., Chiang, C. C., Greensley, M. K., Paul, I. C., Patwardhan, S. A., Dev, S., Beno, M. A., and Christoph, G. G., *J. Chem. Soc., Chem. Commun.* **4,** 105 (1977).
26. Minato, H., and Matsumoto, M., *J. Chem. Soc. C* p. 38 (1970).
27. Mizel, S. B., and Wilson, L., *J. Biol. Chem.* **247,** 4102 (1972).
28. Natori, S., *in* "Mycotoxins in Human and Animal Health" (J. V. Rodricks, C. W. Hesseltine, and M. A. Mehlman, eds.), Pathotox Publ., Park Forest South, Illinois, 1977.
29. Ohtsubo, K., Saito, M., Sekita, S., Yoshihira, K., and Natori, S., *Jpn. J. Exp. Med.* **48,** 105 (1978).
30. Padhye, S. B., Joshi, C. R., and Pendse, G. S., *J. Chromatogr.* **108,** 405 (1975).
31. Patwardhan, S. A., Pandey, R. C., and Dev, S., *Phytochemistry* **13,** 1985 (1974).
32. Pendse, G. S., *Experientia* **30,** 107 (1974).
33. Prescott, D. M., Myerson, D., and Wallace, *J., Exp. Cell Res.* **71,** 480 (1972).
34. Pribela, A., Tomko, J., and Dolejs, L., *Chem. Zvesti* **28,** 710 (1974).
35. Rothweiler, W., and Tamm, C., *Experientia* **22,** 750 (1966).
36. Rothweiler, W., and Tamm, C., *Helv. Chim. Acta* **53,** 696 (1970).
37. Schofield, J. G., *Nature (London) New Biol.* **234,** 215 (1971).

38. Sekita, S., Yoshihira, K., Natori, S., and Kuwano, H., *Tetrahedron Lett.* **32,** 2771 (1977).
39. Sekita, S., Yoshihira, K., Natori, S., and Kuwano, H., *Tetrahedron Lett.* p. 1351 (1976).
40. Sekita, S., Yoshihira, K., Natori, S., and Kuwano, H., *Tetrahedron Lett.* p. 2109 (1973).
41. Sekita, S., Yoshihira, K., Natori, S., Kuwano, H., and Umeda, M., *Symp. Chem. Nat. Prod., 20th, 1976* Symposium papers, p. 396 (1976).
42. Shepro, D., Belamarich, F. A., Robblee, L., and Chao, F. C., *J. Cell Biol.* **47,** 544 (1970).
43. Silverton, J. V., Akiyama T., Kabuto, C., Sekita, S., Yoshihira, K., and Natori, S., *Tetrahedron Lett.* p. 1349 (1976).
44. Springer, J. P., Clardy, J., Wells, J. M., Cole, R. J., Kirksey, J. W., MacFarlane, R. D., and Torgerson, D. F., *Tetrahedron Lett.* p. 1355 (1976).
45. Umeda, M., Ohtsubo, K., Saito, M., Sekita, S., Yoshihira, K., Natori, S., Udagawa, S., Sakabe, F., and Kurata, H., *Experientia* **31,** 435 (1975).
46. Vederas, J. C., and Tanm, C., *Helv. Chim. Acta* **59,** 558 (1976).
47. Wells, J. M., Cutler, H. G., and Cole, R. J., *Can. J. Microbiol.* **22,** 1137 (1976).
48. Williams, J. A., and Wolff, J., *Biochem. Biophys. Res. Commun.* **44,** 422 (1971).

7

Rubratoxins

Evidence from field observations in conjunction with laboratory studies has strongly implicated *Penicillium rubrum* in the etiology of certain naturally occurring diseases of pigs and cattle ingesting molded corn.[1,2,18,20–22] These observations prompted scientists to isolate and identify the toxic agents produced by *P. rubrum*. Recently, two major toxins produced by *P. rubrum*, rubratoxin A and rubratoxin B, have been isolated and identified as complex nonadrides with anhydride and lactone rings. Rubratoxin B is the more toxic and the more prevalent member found in crude extracts of *P. rubrum*. Rubratoxin B has various effects on animals; however, it is mainly hepatotoxic and nephrotoxic. Rubratoxin A is acutely toxic; no chronic studies have been done. Evidence has been presented that the rubratoxins may be more economically important by acting in concert with or by synergizing the effects of other toxic metabolites such as the aflatoxins.[9]

Three reviews covering the subject have been published.[3,10,16]

Rubratoxin group	Molecular weight	Molecular formula
Rubratoxin B	518.1786	$C_{26}H_{30}O_{11}$
Rubratoxin A	520.1942	$C_{26}H_{32}O_{11}$

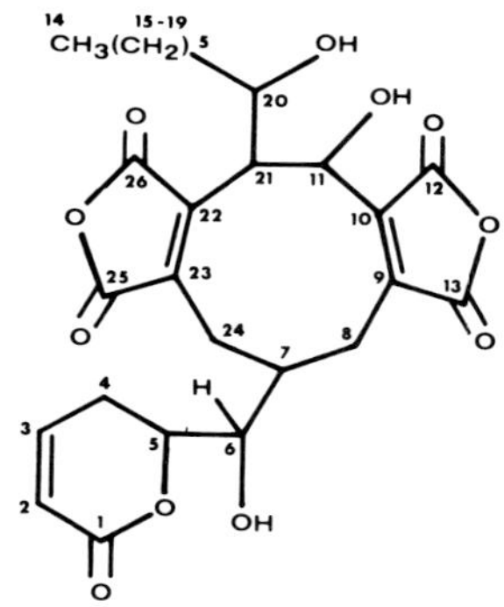

Common name	Rubratoxin B
Molecular weight	518.1786 (mass spectral analysis shows 474 *m/e* due to $M^+ - CO_2$)
Molecular formula	$C_{26}H_{30}O_{11}$
General characteristics[11,14]	Shape of crystals variable, depending on solvent and conditions; fine needlelike laths from benzene–ethyl acetate, mp 168°–170°C (dec.) $[\alpha]_D^{20}$ +67° (C = 2.0 in acetone) pK_a = 4.69
UV data[14]	$\lambda_{max}^{CH_3CN}$ nm (ϵ): 251(9,700)
Fungal source[9,15,18,22]	*Penicillium rubrum* Stoll (NRRL 3290) *P. purpurogenum* Stoll
Toxicity data[4–7,10,22,23]	The LD_{50} to mice dosed IP is 3.0 mg/kg (propylene glycol carrier). The acute oral LD_{50} in day-old chicks was 83.2 mg/kg. The most characteristic gross sign in birds receiving a lethal dose was red mottled livers. A dose of 500 μg/g of diet was required to cause a significant decrease in body weight; 1000 μg/g was required for a significant hypertrophy of the liver and atrophy of the bursa of Fabricius. Spontaneous hemorrhaging was not observed. Rubratoxin B has also shown antiprotozoal activity, inhibition of *in vitro* oxygen uptake of Krebs cycle intermediates, and acute toxicity in dogs. In the latter, plasma enzyme tests indicated hepatic cell necrosis; other clinical signs were anorexia followed by dehydration, somnolence, diarrhea, and jaundice. Rubratoxin B is nephrotoxic

in dogs. Studies with dogs by Hayes and Williams[7] showed a striking similarity between experimentally induced lesions when rubratoxin B and aflatoxin B_1 were administered together and those described in natural outbreaks of hepatitis X. Rubratoxin B markedly inhibited the hepatic cytochrome *P*-450-dependent monooxygenase system when measured 1, 6, 48, 72, and 96 hr after a single exposure. At 48 hr, dose levels as low as 0.25 mg/kg rubratoxin B inhibited pentobarbital hydroxylase and ethyl morphine demethylase. Single exposure of male mice to rubratoxin B resulted in depression in activity of the cytochrome *P*-450-dependent monooxygenase enzyme system and that after maximum inhibition by a single dose at 48 hr, the trend was for the enzymes to return to control values.

TLC data[8]

Adsorbent: silica gel HF_{254}
Solvent: glacial acetic acid–methanol–chloroform, 2:20:80 v/v/v
R_f: 0.54–0.59*; 0.56
Detection: a dark spot against a green fluorescent background when examined under shortwave UV light.

Other significant references

Isolation[8,18,22]; structure[12–14]; chemistry[11]; review[3]; HPLC[19]

^{13}C NMR Shifts

C14.2 q; C23.2 t; C25.3 t; C26.3 t; C26.9 t; C32.4 t; C36.9 t (2); C48.6 d; C65.0 d; C70.3 d; C77.3 d; C79.6 d; C121.0 d; C139.9 s; C143.7 s; C144.0 s; C145.8 d; C147.8 s; C163.1 s; C166.9 s

* An artifact develops if the plates are exposed to the atmosphere after spotting for 2–3 min before development. It occurs at R_f 0.08–0.14.

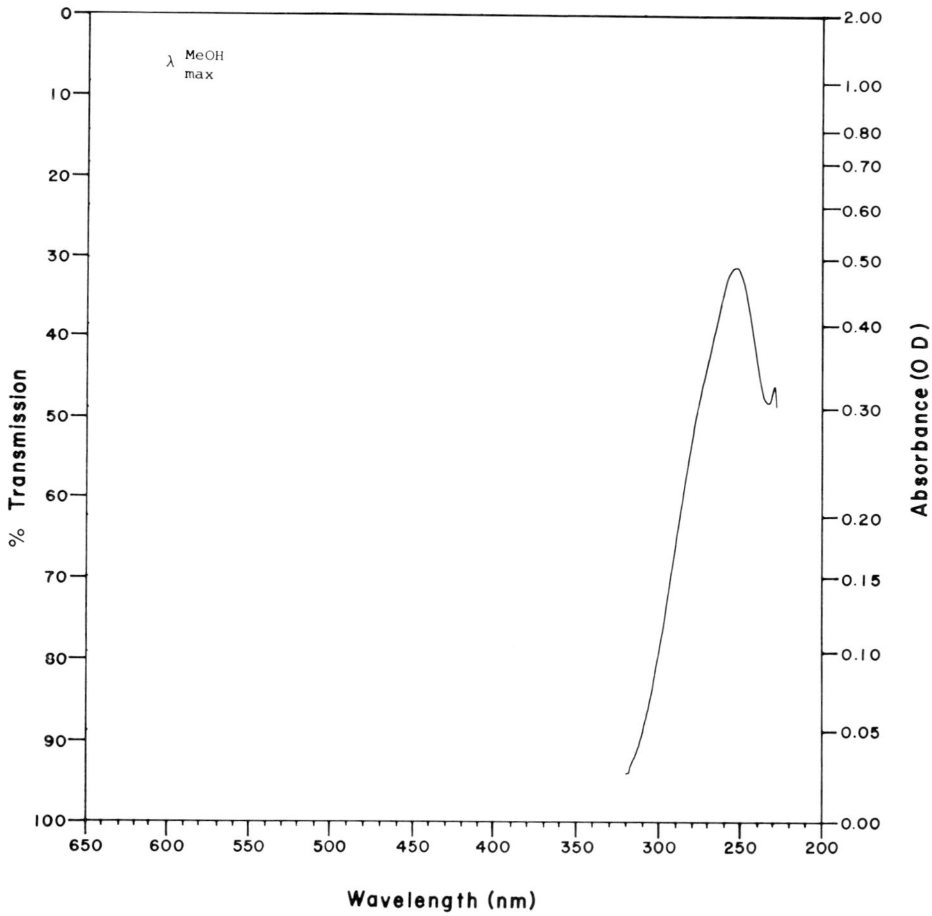
λ MeOH max
% Transmission
Absorbance (O D)
Wavelength (nm)
0
10
20
30
40
50
60
70
80
90
100
2.00
1.00
0.80
0.70
0.60
0.50
0.40
0.30
0.20
0.15
0.10
0.05
0.00
650
600
550
500
450
400
350
300
250
200

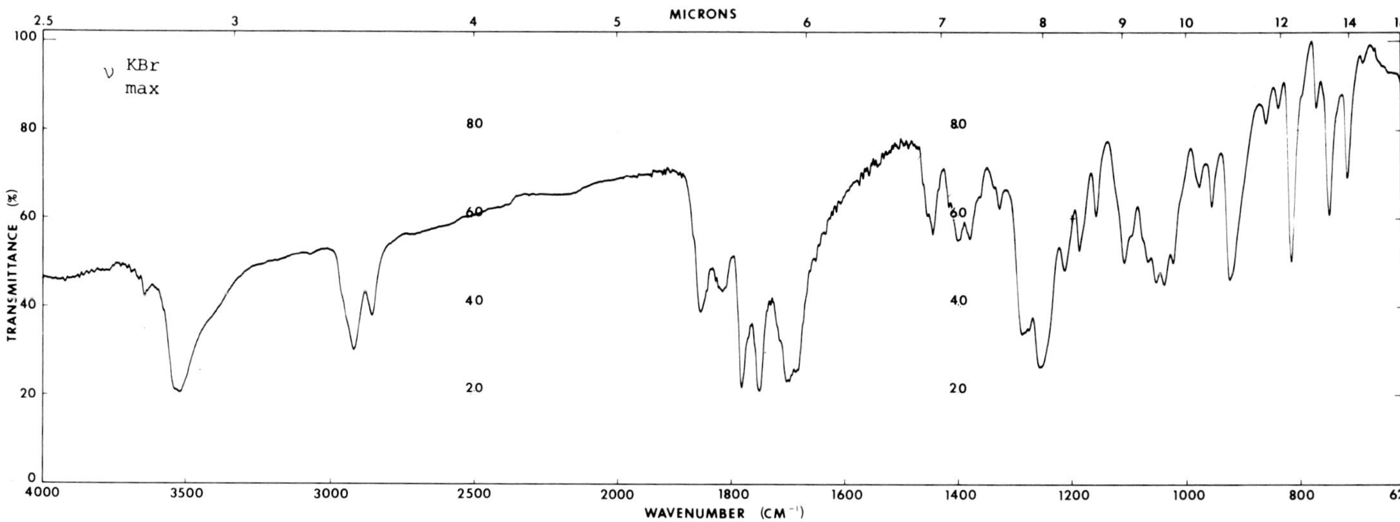
MICRONS
ν KBr max
TRANSMITTANCE (%)
WAVENUMBER (CM⁻¹)

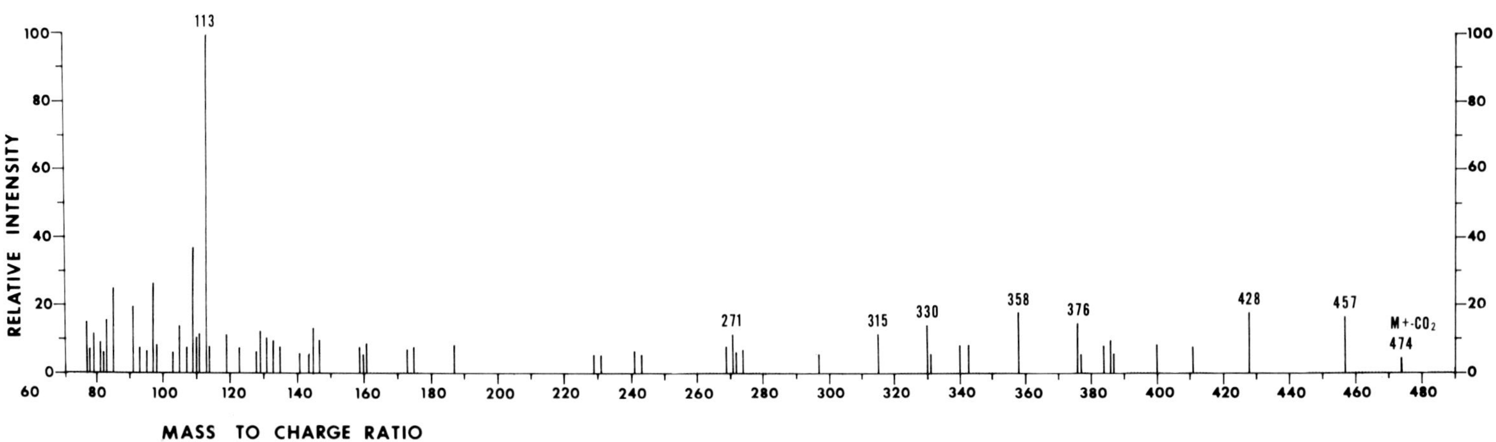
113
271
315
330
358
376
428
457
M+-CO2
474
RELATIVE INTENSITY
MASS TO CHARGE RATIO

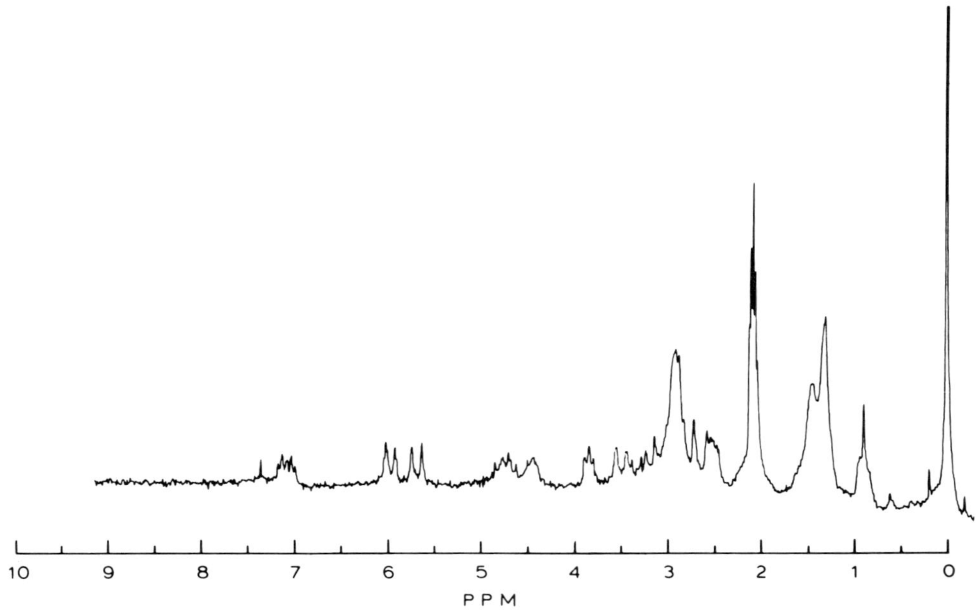

The 100 MHz ^{1}H NMR spectrum of rubratoxin B in acetone solution.

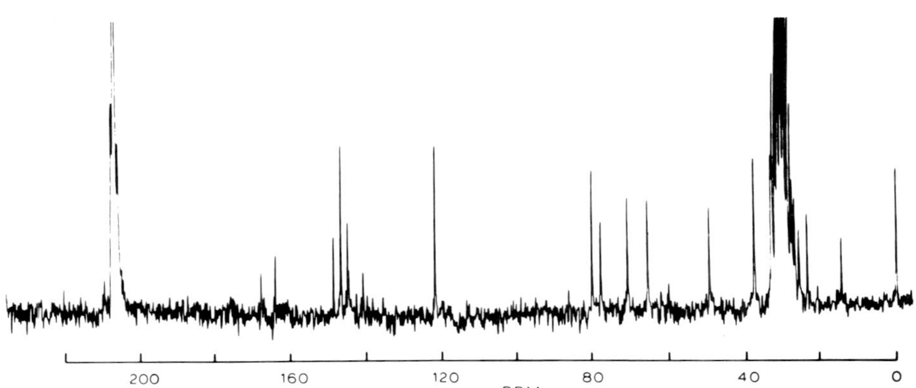

The 25 MHz[13]C NMR spectrum of rubratoxin B in acetone solution.

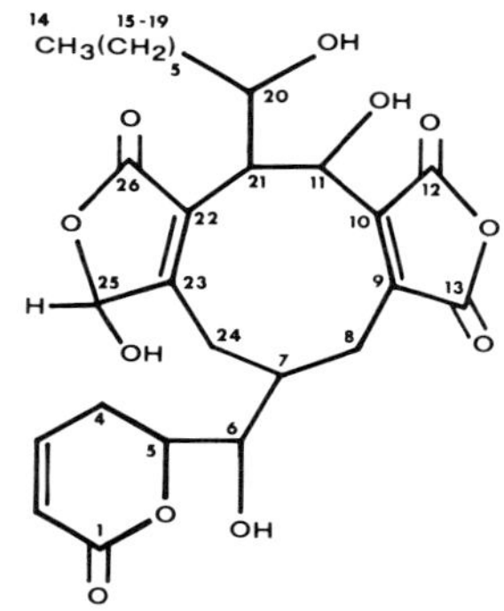

Common name	Rubratoxin A
Molecular weight	520.1942 (mass spectral analysis shows 458.1932 due to $M^+ - CO_2$ and H_2O)
Molecular formula	$C_{26}H_{32}O_{11}$
General characteristics[11,14]	Needles from ethyl acetate, mp 204°–206°C (dec.) $[\alpha]_D$ +87° (C = 2.0 in acetone)
UV data[14]	$\lambda_{max}^{CH_3CN}$ nm (ϵ): 204(31,900), inflection at 225 and 252(sh)(4,430)
Fungal source[15,22]	*Penicillium rubrum* (NRRL A-11785) *P. purpurogenum* Stoll Rubratoxin A occurs as the minor metabolite along with rubratoxin B, the major toxic metabolite from the above fungi.
Toxicity data[10]	The LD_{50} to mice dosed IP was 6.6 mg/kg (propylene glycol carrier)
TLC data[8]	*Adsorbent:* silica gel HF_{254} *Solvent:* glacial acetic acid–methanol–chloroform, 2:20:80 v/v/v R_f: 0.70 *Detection:* a dark spot against a green fluorescent background when examined under shortwave UV light.
Other significant references	Isolation[8,18]; structure[12–14]; chemistry[11]
^1H NMR data	No assignment of ^1H spectrum

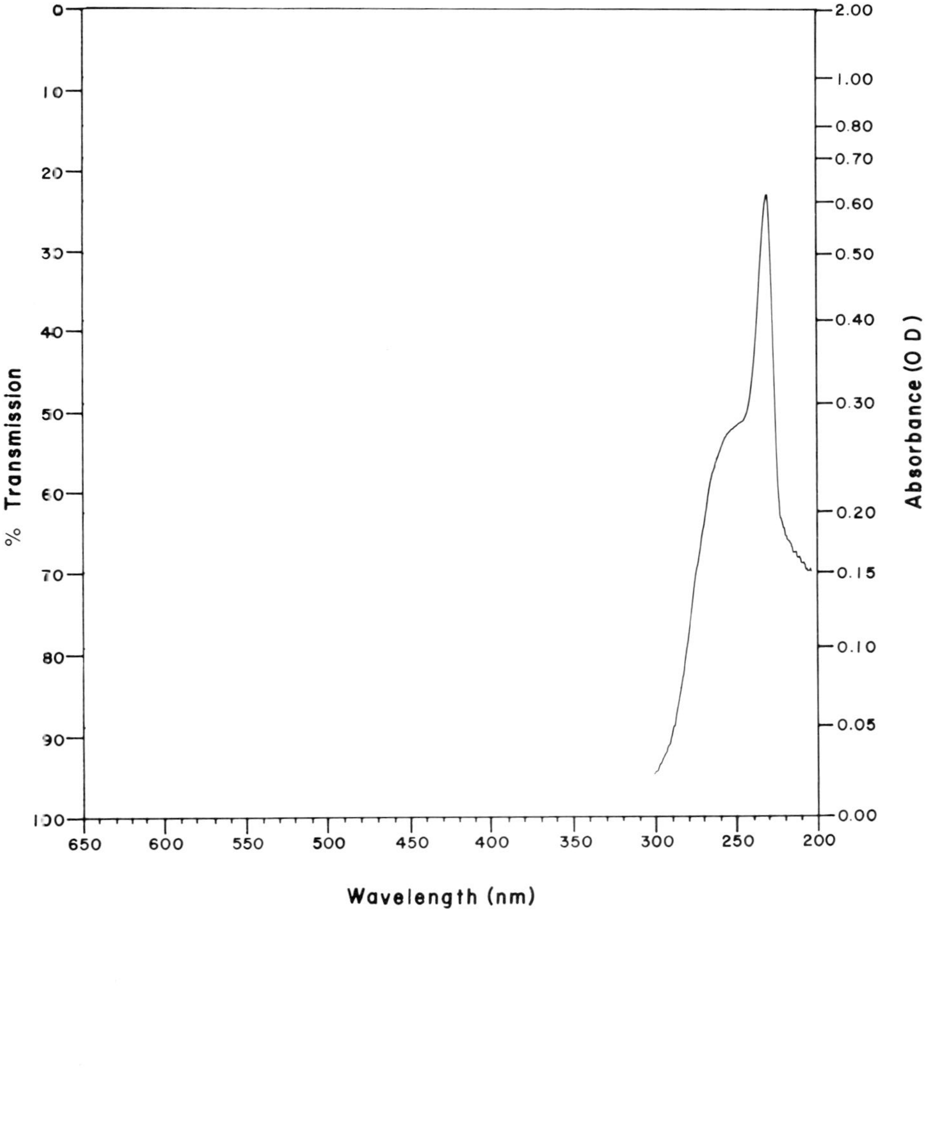
% Transmission
0
10
20
30
40
50
60
70
80
90
100
Absorbance (O D)
2.00
1.00
0.80
0.70
0.60
0.50
0.40
0.30
0.20
0.15
0.10
0.05
0.00
650
600
550
500
450
400
350
300
250
200
Wavelength (nm)

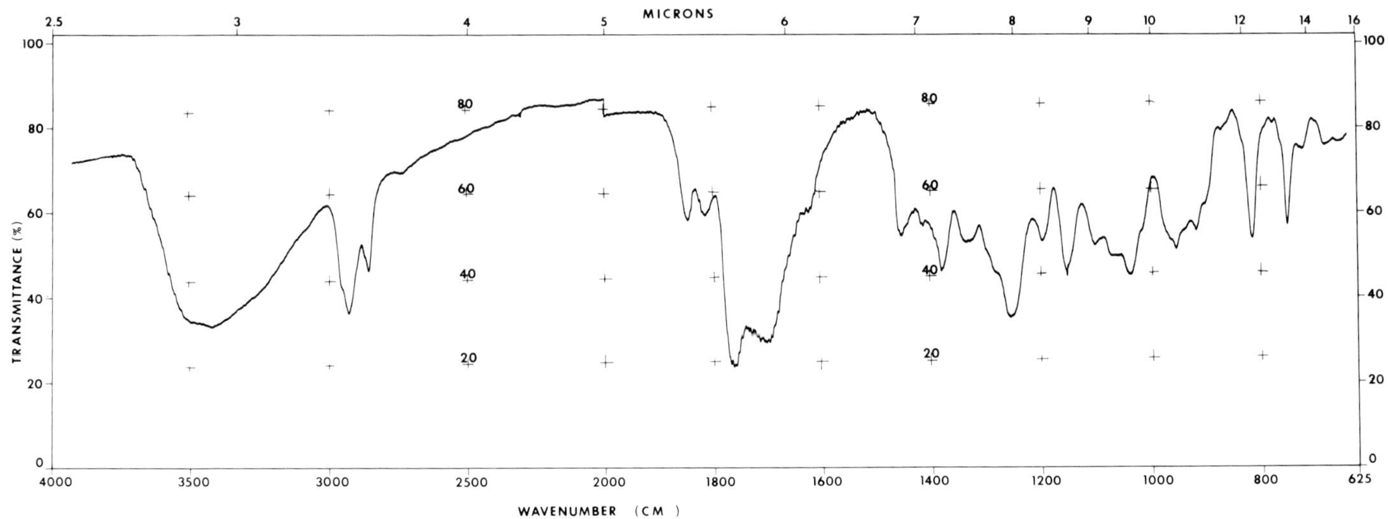
MICRONS
TRANSMITTANCE (%)
WAVENUMBER (CM)

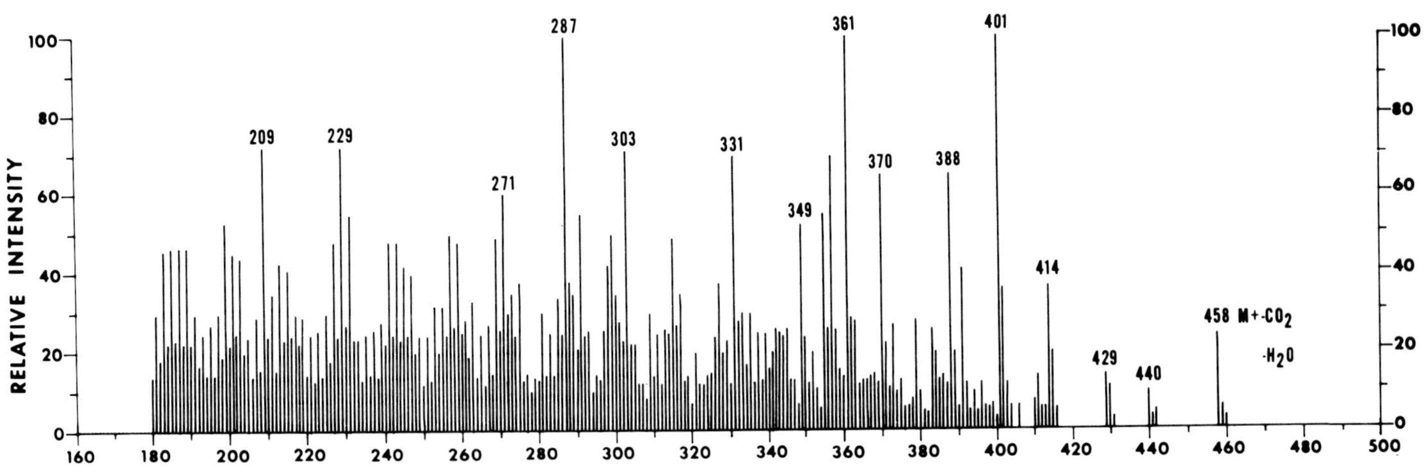
RELATIVE INTENSITY
209
229
271
287
303
331
349
361
370
388
401
414
429
440
458 M+-CO2
-H2O

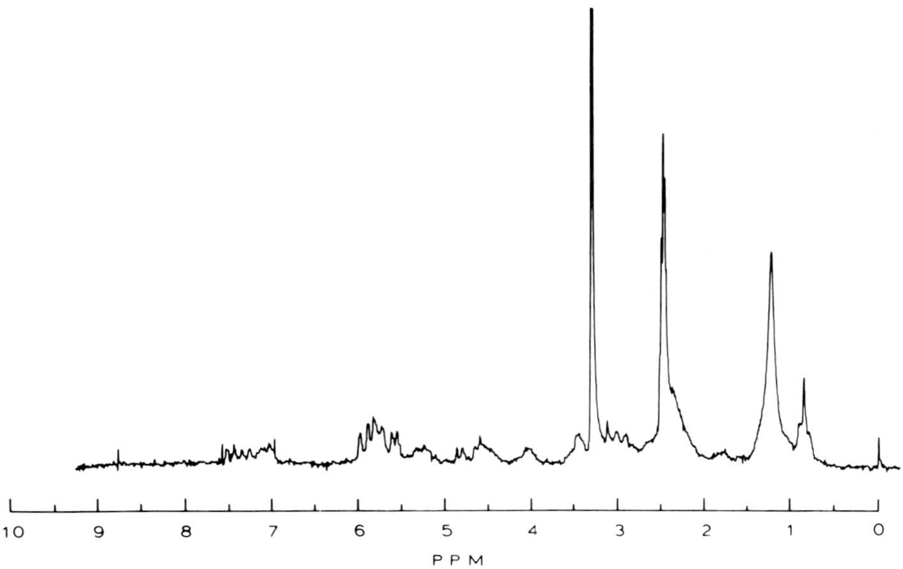

The 100 MHz ^1H NMR spectrum of rubratoxin A in DMSO solution.

REFERENCES

1. Burnside, J. E., Sippel, W. L., Forgacs, J., Carll, W. T., Atwood, M. B., and Doll, E. R., *Am. J. Vet. Res.* **18,** 817 (1957).
2. Forgacs, J., Koch, H., Carll, W. T., and White-Stevens, R. M., *Am. J. Vet. Res.* **19,** 744 (1958).
3. Hayes, A. W., *in* "Mycotoxins in Human and Animal Health," (J. V. Rodricks, C. W. Hesseltine, and M. A. Mehlman, eds.), p. 507. Pathotox Publ. Inc. Park Forest South, Illinois, 1976.
4. Hayes, A. W., *Antimicrob. Agents & Chemother.* **4,** 80 (1973).
5. Hayes, A. W., and Hannan, C. J., *Toxicol. Appl. Pharmacol.* **25,** 30 (1973).
6. Hayes, A. W., Neville, J. A., and Hollingsworth, E. B., *Toxicol. Appl. Pharmacol.* **25,** 606 (1973).
7. Hayes, A. W., and Williams, W. L., *J. Environ. Pathol. Toxicol.* **1,** 59 (1977).
8. Hayes, A. W., and Wilson, B. J., *Appl. Microbiol.* **16,** 1163 (1968).
9. Moss, M. O., and Hill, I. W., *Mycopathol. Mycol. Appl.* **40,** 80 (1970).
10. Moss, M. O., *in* "Microbial Toxins" (A. Ciegler, S. Kadis, and S. J. Ajl, eds.), Vol. 6, p. 381. Academic Press, New York, 1971.
11. Moss, M. O., Robinson, F. V., and Wood, A. B., *J. Chem. Soc.* p. 619 (1971).
12. Moss, M. O., Robinson, F. V., Wood, A. B., and Morrison, A., *Chem. Ind.* (*London*) p. 755 (1967).
13. Moss, M. O., Robinson, F. V., Wood, A. B., Paisley, H. M., and Fenney, J., *Nature* (*London*) **220,** 767 (1968).
14. Moss, M. O., Wood, A. B., and Robinson, F. V., *Tetrahedron Lett.* p. 367 (1969).
15. Natori, S., Sakaki, S., Kurata, H., Udagawa, S., Ichinoe, M., Saito, M., Umeda, M., and Ohtsubo, K., *Appl. Microbiol.* **19,** 613 (1970).
16. Newberne, P. M., *in* "Mycotoxins" (I. F. H. Purchase, ed.), p. 163. Elsevier, Am. New York, 1974.
17. Siroj, M. Y., and Hayes, A. W., *Toxicol. Appl. Pharmacol.* **48,** 351–359 (1979).
18. Townsend, R. J., Moss, M. O., and Peck, H. M., *J. Pharm. Pharmacol.* **18,** 473 (1966).
19. Unger, P. D., and Hayes, A. W., *J. Chromatogr.* **153,** 115 (1978).
20. Wilson, B. J., and Harbison, R. D., *J. Am. Vet. Med. Assoc.* **163,** 1274 (1973).
21. Wilson, B. J., and Wilson, C. H., *J. Bacteriol.* **83,** 693 (1962).
22. Wilson, B. J., and Wilson, C. H., *J. Bacteriol.* **84,** 283 (1962).
23. Wyatt, R. D., and Hamilton, P. B., *Poult. Sci.* **51,** 1383 (1972).

8

Tremorgen Group

The increased research on mycotoxins in general over the past decade has led to the discovery of fungal metabolites that are capable of eliciting sustained or intermittent tremors in vertebrate animals. Before the discovery of tremorgenic mycotoxins, chemicals known to induce tremorgenic effects in animals were rare. The tremorgenic mycotoxins reported to date can be conveniently placed into 4 groups based on chemical similarity. These are the penitrem group, the fumitremorgin-verruculogen group, the paspalitrem group, and the tryptoquivaline group. All of the tremorgens in the above groups have in common an indole moiety derived from tryptophan. Fungi capable of producing tremorgenic metabolites belong to the genera *Penicillium, Aspergillus,* and *Claviceps*.

The basic chemical structures with stereochemistry have been elucidated in all the groups except the penitrem group. The chemical structures of penitrem A and related metabolites have not yet been elucidated.

Some members of the paspalitrem group have been strongly implicated in the etiology of the "Paspalum staggers" syndrome. It has been shown that paspalinine and two closely related tremorgenic metabolites, paspalitrem A and B, are present in *C. paspali* sclerotia from toxic pastures.[22]

Verruculogen and penitrem A appear to cause tremors by inhibiting natural neurons which inhibit the α-motor cells of the anterior horn. In efforts to determine the mode of action of fungal tremorgens, it has become apparent that they provide useful tools in the study of central nervous system function.

Also included in the tremorgenic mycotoxin group are several metabolites that are chemically related but not tremorgenic. These include cyclopiazonic acid and congers, austamide, echinulin, preechinulin, deoxybrevianamide E, tenuazonic acid, etc.

Three reviews on the subject of tremorgenic mycotoxins have appeared.[11,17,18]

Tremorgens and related metabolites	Molecular weight	Molecular formula
Fumitremorgin Group		
Fumitremorgin A	579.2943	$C_{32}H_{41}O_7N_3$
Fumitremorgin B	479.2420	$C_{27}H_{33}O_5N_3$
Fumitremorgin C (SM-Q)	379.1895	$C_{22}H_{25}O_3N_3$
Verruculogen	511.2318	$C_{27}H_{33}O_7N_3$
15-Acetoxyverruculogen	569.2372	$C_{29}H_{35}O_9N_3$
TR-2	429.1899	$C_{22}H_{27}O_6N_3$
Penitrem Group		
Penitrem A	633.2807	$C_{37}H_{44}O_6NCl$
Penitrem B	583.3296	$C_{37}H_{45}O_5N$
Paspalitrem Group		
Paxilline	435.2409	$C_{27}H_{33}O_4N$
Paspaline	421.2980	$C_{28}H_{39}O_2N$
Paspalicine	417.2303	$C_{27}H_{31}O_3N$
Paspalinine	433.2252	$C_{27}H_{31}O_4N$
Paspalitrem A	501.2879	$C_{32}H_{39}O_4N$
Paspalitrem B	517.2827	$C_{32}H_{39}O_5N$
Aflatrem	501.2879	$C_{32}H_{39}O_4N$
Tryptoquivaline group		
Tryptoquivaline	546.2114	$C_{29}H_{30}O_7N_4$
Nortryptoquivalone (tryptoquivalone)	488.1695	$C_{26}H_{24}O_6N_4$
Nortryptoquivaline	532.1958	$C_{28}H_{28}O_7N_4$
Deoxytryptoquivaline	530.2165	$C_{29}H_{30}O_6N_4$
Deoxynortryptoquivalone	472.1746	$C_{26}H_{24}O_5N_4$
Deoxynortryptoquivaline	516.2008	$C_{28}H_{28}O_6N_4$
Tryptoquivaline E	418.1277	$C_{22}H_{18}O_5N_4$
Tryptoquivaline F	402.1328	$C_{22}H_{18}O_4N_4$
Tryptoquivaline G	432.1433	$C_{23}H_{20}O_5N_4$
Tryptoquivaline H	418.1277	$C_{22}H_{18}O_5N_4$
Tryptoquivaline I	502.1852	$C_{27}H_{26}O_6N_4$
Tryptoquivaline J	402.1328	$C_{22}H_{18}O_4N_4$
Tryptoquivaline L	432.1433	$C_{23}H_{20}O_5N_4$
Tryptoquivaline M	532.1958	$C_{28}H_{28}O_7N_4$
Tryptoquivaline N	472.1746	$C_{26}H_{24}O_5N_4$
Related nontremorgenic metabolites		
Deoxybrevianamide E	351.1946	$C_{21}H_{25}O_2N_3$
Preechinulin	325.1790	$C_{19}H_{23}O_2N_3$
Neoechinulin	391.1895	$C_{23}H_{25}O_3N_3$
Neoechinulin E	323.1267	$C_{18}H_{17}O_3N_3$
Neoechinulin D	391.2259	$C_{24}H_{29}O_2N_3$
Cryptoechinulin G	457.2718	$C_{29}H_{35}O_2N_3$
Neoechinulin A	323.1633	$C_{19}H_{21}O_2N_3$
Neoechinulin B	321.1477	$C_{19}H_{19}O_2N_3$
Neoechinulin C	389.2103	$C_{24}H_{27}O_2N_3$
Isoechinulin A	391.2259	$C_{24}H_{29}O_2N_3$
Isoechinulin B	389.2103	$C_{24}H_{27}O_2N_3$
Isoechinulin C	405.2051	$C_{24}H_{27}O_3N_3$
Cryptoechinulin A	389.2103	$C_{24}H_{27}O_2N_3$
Echinulin	461.3042	$C_{29}H_{39}O_2N_3$
Austamide	363.1582	$C_{21}H_{21}O_3N_3$
Tetramic acid group		
Tenuazonic acid	197.1051	$C_{10}H_{15}O_3N$
Cyclopiazonic acid	336.1473	$C_{20}H_{20}O_3N_2$
Cyclopiazonic acid imine	335.1633	$C_{20}H_{21}O_2N_3$
Bissecodehydrocyclopiazonic acid	338.1630	$C_{20}H_{22}O_3N_2$

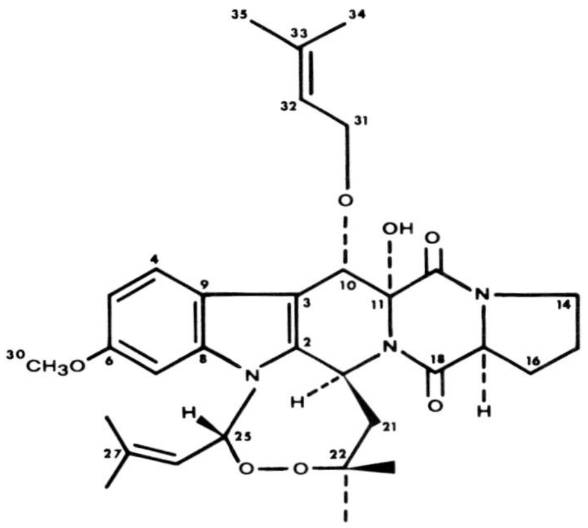

Common name	Fumitremorgin A
Molecular weight[100,101]	579.2943
Molecular formula	$C_{32}H_{41}O_7N_3$
General characteristics[99–101]	Colorless prisms from methanol, mp 206°–209°C $[\alpha]_D^{10}$ +61° (acetone) Soluble in chloroform and ethyl acetate; slightly soluble in methanol and ethanol; unstable to prolonged exposure to light.
UV data[97]	λ_{max}^{EtOH} nm (ϵ): 226(31,700),277(5,300), and 296(4,900)
Fungal source[77,101]	*Aspergillus fumigatus* Fres. *A. caespitosus*
Toxicity data[101]	1 mg dosed IP to mice caused perceptible tremors; 5 mg dosed IP to mice caused sustained tremors and 70% mortality.
TLC data	*Adsorbent:* silica gel G-HR *Solvent:* A: chloroform–acetone, 93:7 v/v B: toluene–ethyl acetate–formic acid, 5:4:1 v/v/v R_f: A: 0.30 B: 0.65 *Detection:* slate gray–blue spot in visible light or mustard-colored spot under UV light; develops immediately after spraying with 50% ethanolic H_2SO_4.

NOTE: The color response and R_f's of verruculogen are identical to those of fumitremorgin A in the above conditions.

Other significant references X-Ray structure[33]

^{1}H NMR data[99] H4, 6.59; H5, 6.82; H7, 7.68; H10, 5.50; 11OH, 4.48; H14, 3.63; H15, 1.60–2.20; H16, 1.60–2.20; H17, 6.13; H20, 5.06; H21, 2.40; H23, 0.99; H24, 2.00; H25, 6.62; H26, 5.02; H28, 1.71; H29, 1.71; H30, 3.84; H31, 4.71; H32, 5.60; H34, 1.81; H35, 1.81

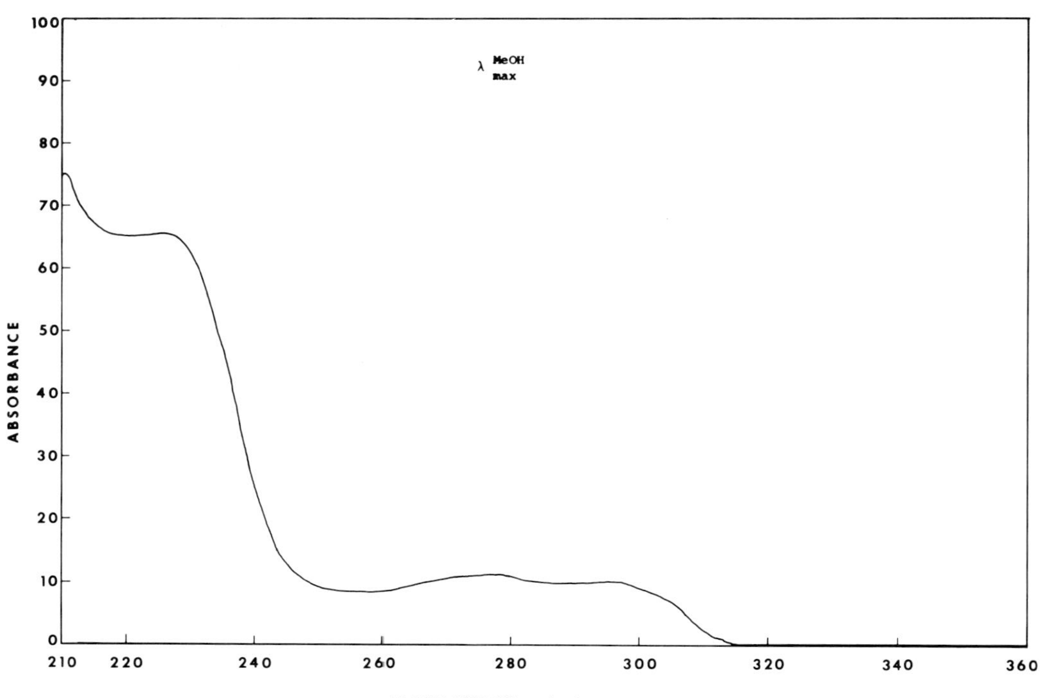

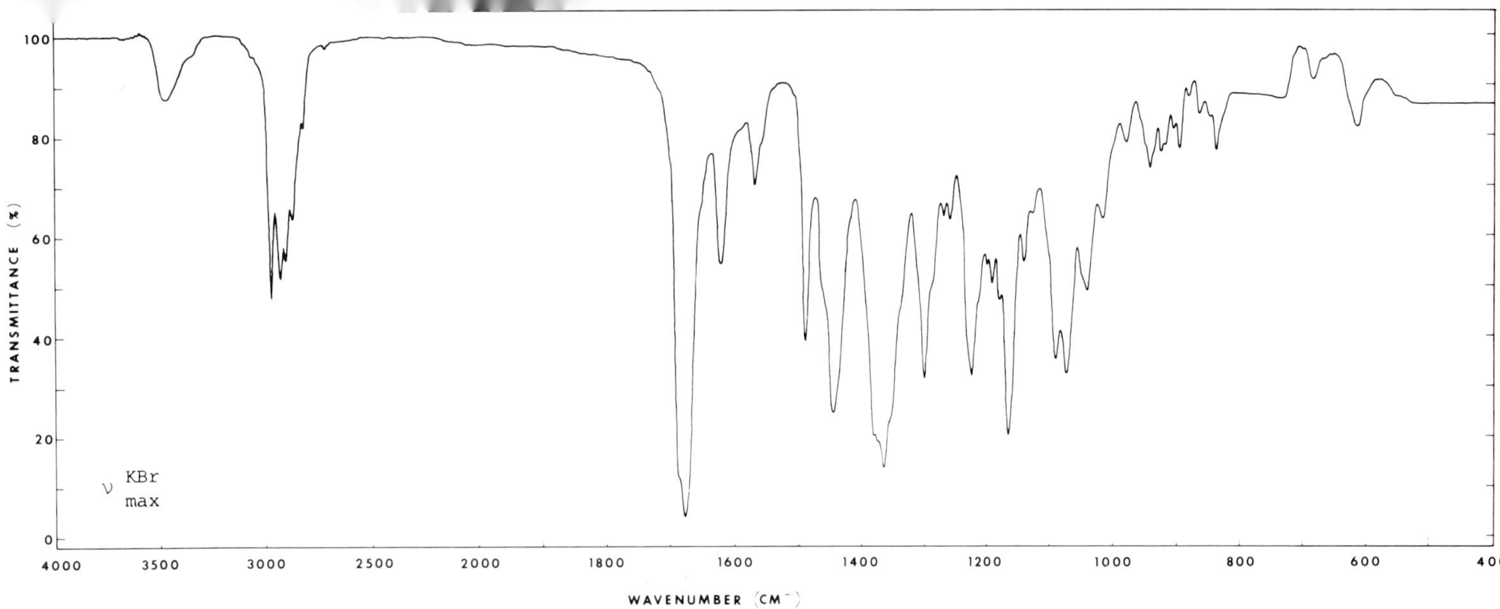

TRANSMITTANCE (%)
100
80
60
40
20
0
ν KBr max
4000
3500
3000
2500
2000
1800
1600
1400
1200
1000
800
600
400
WAVENUMBER (CM⁻)

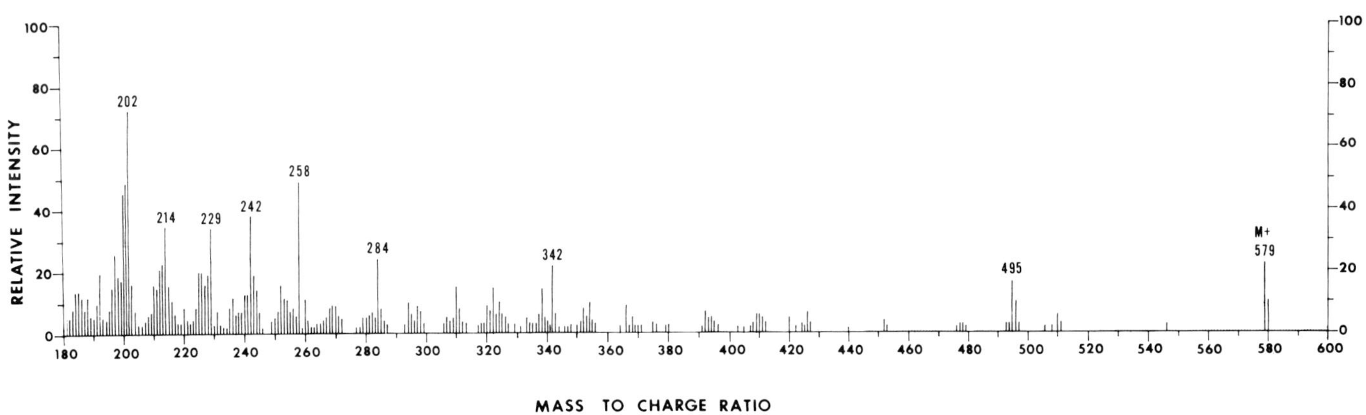

RELATIVE INTENSITY
100
80
60
40
20
0
202
214
229
242
258
284
342
495
M+
579
180
200
220
240
260
280
300
320
340
360
380
400
420
440
460
480
500
520
540
560
580
600
MASS TO CHARGE RATIO

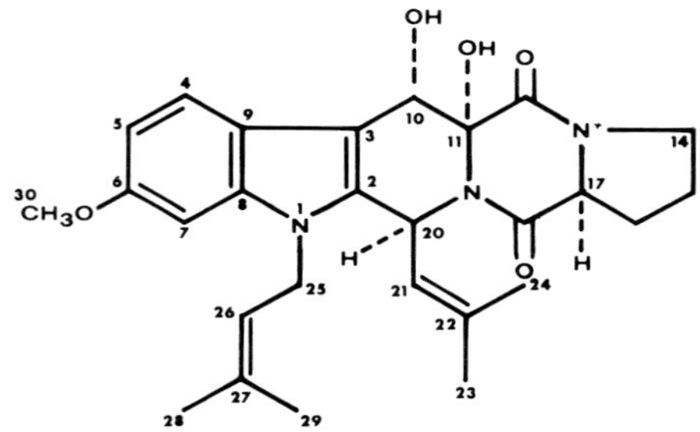

Common name[32]	Fumitremorgin B (Lanosulin)
Molecular weight	479.2420
Molecular formula	$C_{27}H_{33}O_5N_3$
General characteristics[100]	Colorless needles from methanol, mp 211°–212°C
UV data[97]	λ_{max}^{EtOH} nm (ϵ): 228(36,400), 278(7,500), and 297(8,400)
Fungal source[32,77,101]	*Aspergillus fumigatus* Fres. *A. caespitosus* (NRRL 1929)
Toxicity data[101]	Fumitremorgin B dosed at 1 mg/mouse (IP) caused perceptible tremors; at 5 mg/mouse (IP), it caused sustained tremors and 70% mortality.
TLC data[77]	*Adsorbent:* silica gel 60 HR *Solvent:* A: diethyl ether B: acetone-methylene chloride, 5:95 v/v R_f: A: 0.67 B: 0.38 *Detection:* Bright blue fluorescent spot under UV light after spraying with 20% ethanolic $AlCl_3$ and heating for 10 min at 100°C.
Other significant references	X-Ray structure.[98] Possible role in staggers syndrome.[39,73]
^{13}C *NMR data*	C2, 130.8 s; C3, 104.2 s; C4, 121.1 d; C5, 104.0 d; C6, 155.8 s; C7, 93.7 d; C8, 137.6 s; C9, 120.3 s; C10, 68.8 d; C11, 82.8 d; C12, 170.0 s; C14, 48.9 t; C15, 22.5 t; C16, 28.9 t; C17, 58.6 d; C18, 165.8 s; C20, 58.2 d; C21, 122.7 d; C22, 134.8 s; C23, 25.5 q; C24, 18.3 q; C25, 45.1 t; C26, 120.0 d; C27, 134.4 s; C28, 25.5 q; C29, 18.3 q; C30, 55.6 q

^{1}H NMR data H4, 7.80 (J = 9.0); H5, 6.73 (J = 9.0, 2.0); H7, 6.64 (J = 2.0); H10, 5.72; H14, 3.80; H15, 2.00; H16, 2.00; H17, 4.50; H20, 5.95; H21, 4.70; H23, 1.67; H24, 1.97; H25, 4.50; H26, 4.50; H28, 1.60; H29, 1.81; H30, 3.82

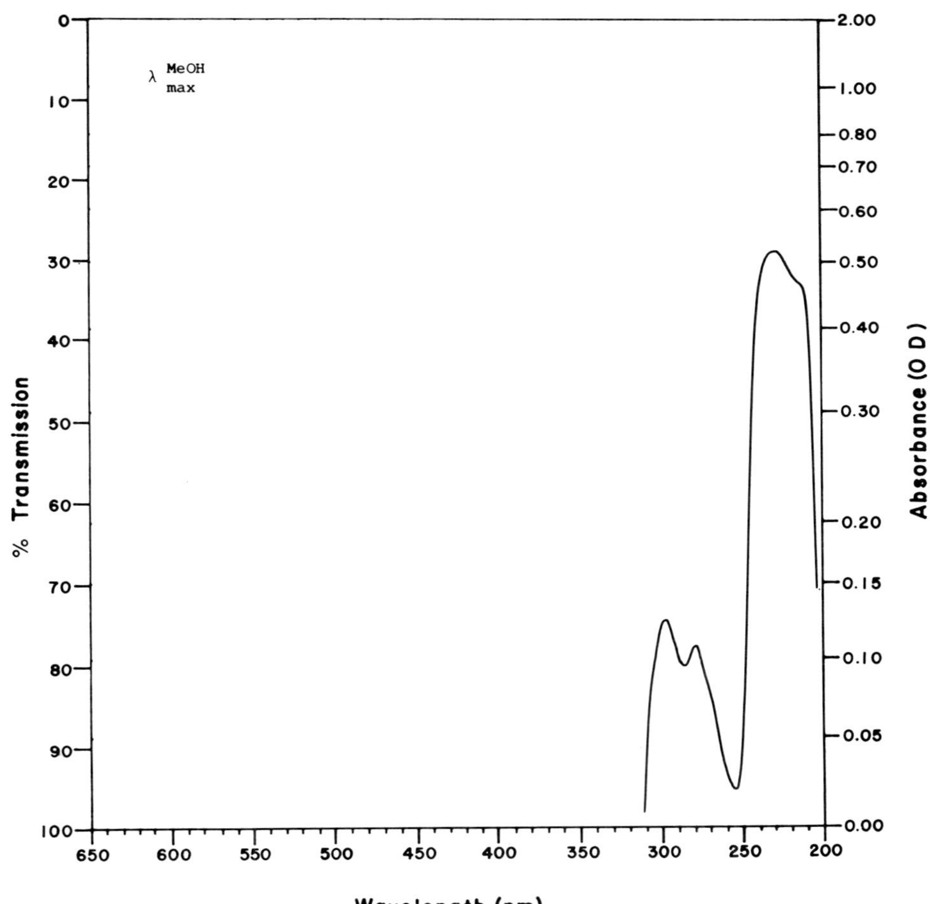

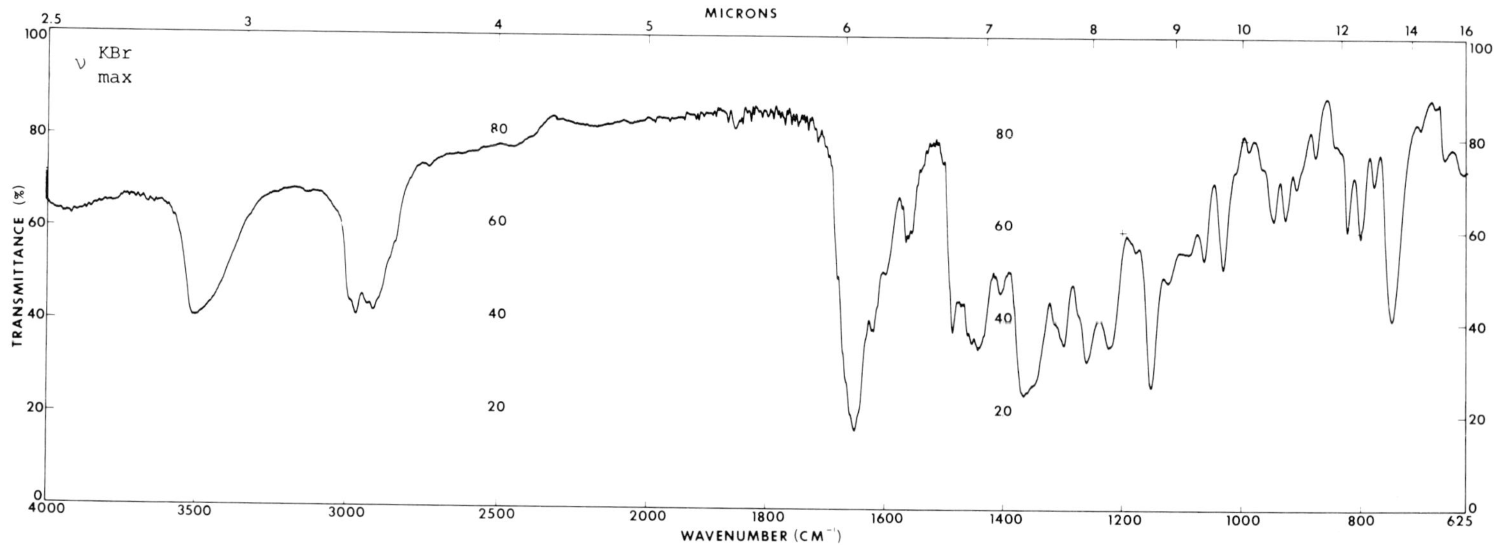
MICRONS
ν KBr max
TRANSMITTANCE (%)
WAVENUMBER (CM⁻¹)

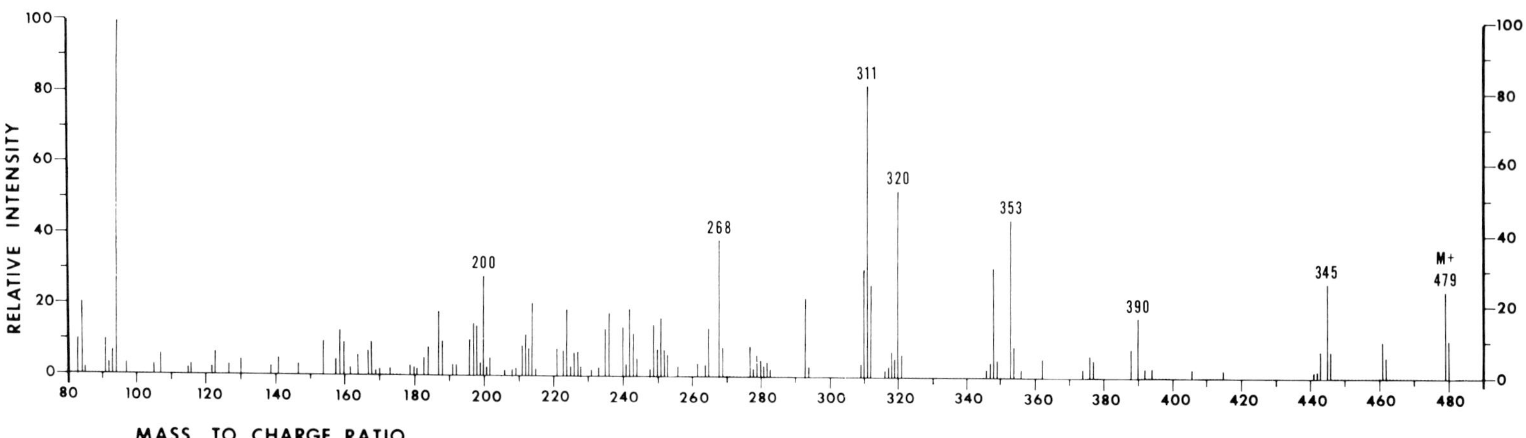
311
320
353
268
200
390
345
M+ 479
RELATIVE INTENSITY
MASS TO CHARGE RATIO

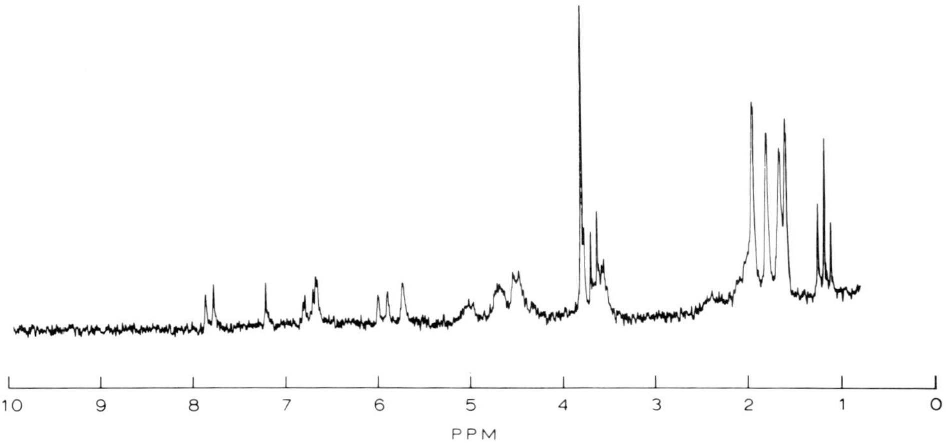

The 100 MHz ^{1}H NMR spectrum of fumitremorgin B in $CDCl_3$ solution.

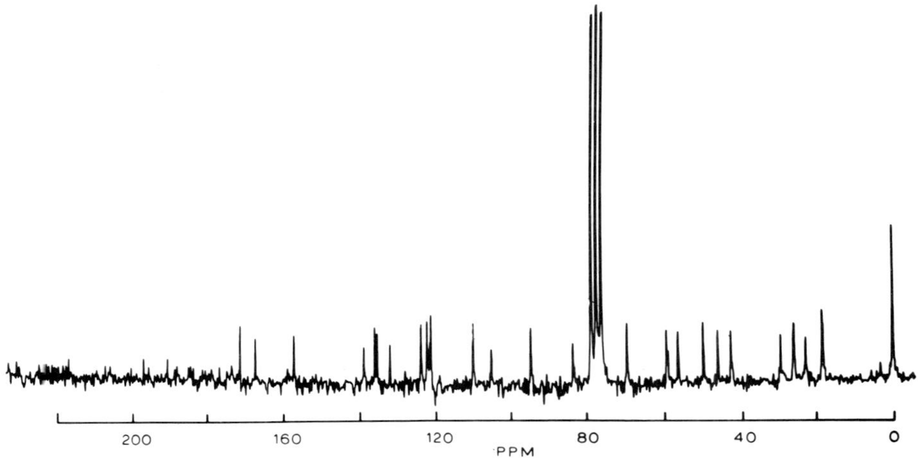

The 25 MHz ^{13}C NMR spectrum of fumitremorgin B in $CDCl_3$ solution.

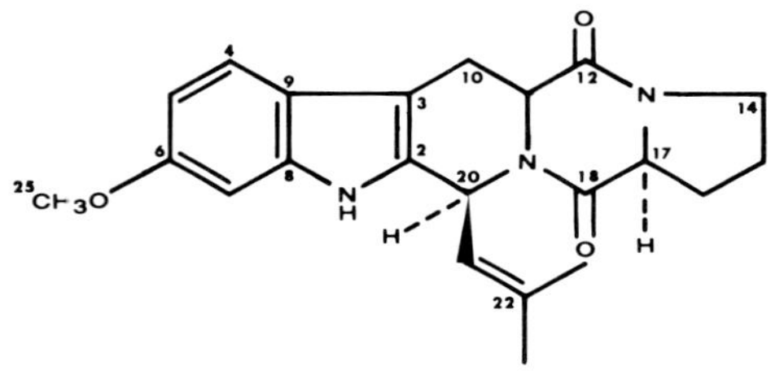

Common name	Fumitremorgin C (SM-Q)
Molecular weight	379.1895
Molecular formula[13]	$C_{22}H_{25}O_3N_3$
General characteristics	Crystals from ethyl acetate, mp 125°–130°C
UV data	λ_{max}^{MeOH} nm: 224,272, and 294 (ϵ_{max} not reported)
Fungal source[22]	*Aspergillus fumigatus* Fres.
Toxicity data[22]	Tremorgenic to day-old cockerels at levels down to 25 mg/kg (oral).
TLC data[22]	*Adsorbent:* silica gel G-HR *Solvent:* toluene–ethyl acetate–formic acid, 5:4:1 v/v/v R_f: 0.55 *Detection:* bright orange spot develops immediately after spraying with 50% ethanolic H_2SO_4 and minimal heating
^{13}*C NMR data*	C2, 132.1 s; C3, 106.2 s; C4, 118.8 d; C5, 109.4 d; C6, 156.4 s; C7, 95.2; C8, 136.9 s; C9, 120.7 s; C10, 45.4 t; C11, 51.0 d; C12, 169.4 s; C14, 45.4 t; C15, 23.0 t; C16, 28.6 t; C17, 59.2 d; C18, 165.6 s; C20, 56.8 d; C21, 124.1 d; C22, 133.9 s; C23, 18.1 q; C24, 25.7 q; C25, 55.7 q
^{1}*H NMR data*	1-NH, 8.32; H4, 7.48 (J = 9.0); H5, 6.86 (J = 9.0, 2.0); H7, 6.92 (J = 2.0); H17, 4.16; H20, 4.98 (J = 9.0); H21, (6.06) (J = 9.0); H23, 2.07; H24, 1.72; H25, 3.91

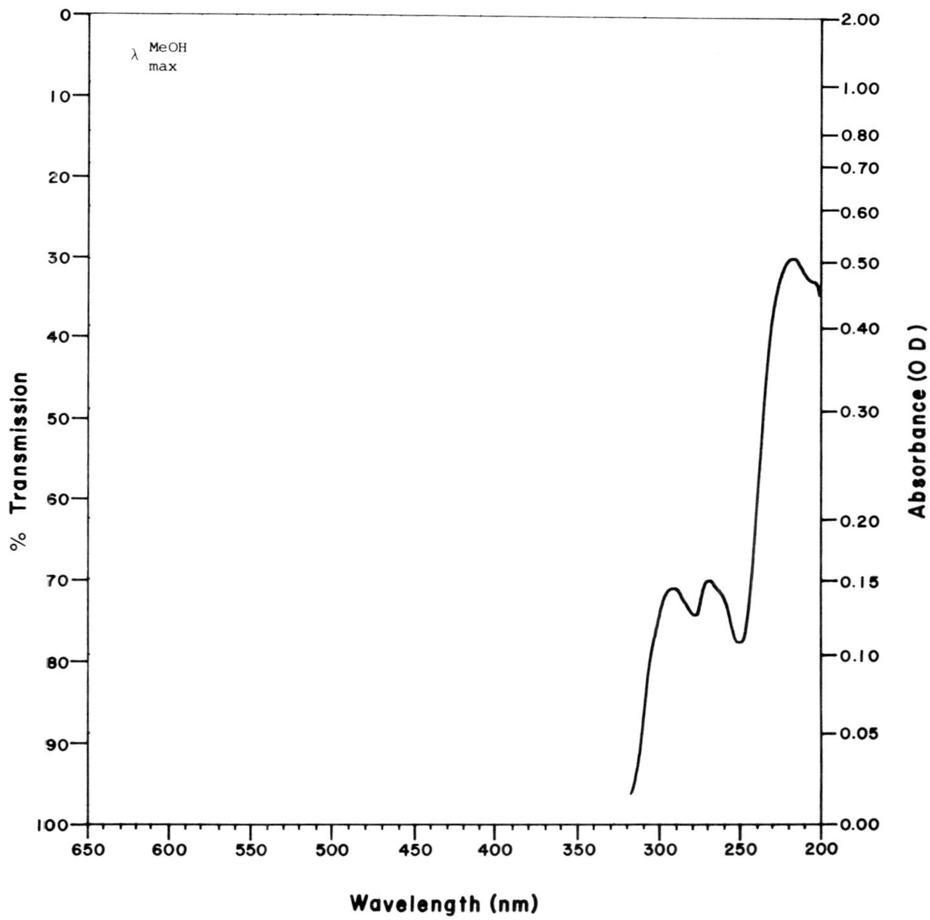
λ MeOH max
% Transmission
Absorbance (O D)
Wavelength (nm)
0
10
20
30
40
50
60
70
80
90
100
2.00
1.00
0.80
0.70
0.60
0.50
0.40
0.30
0.20
0.15
0.10
0.05
0.00
650
600
550
500
450
400
350
300
250
200

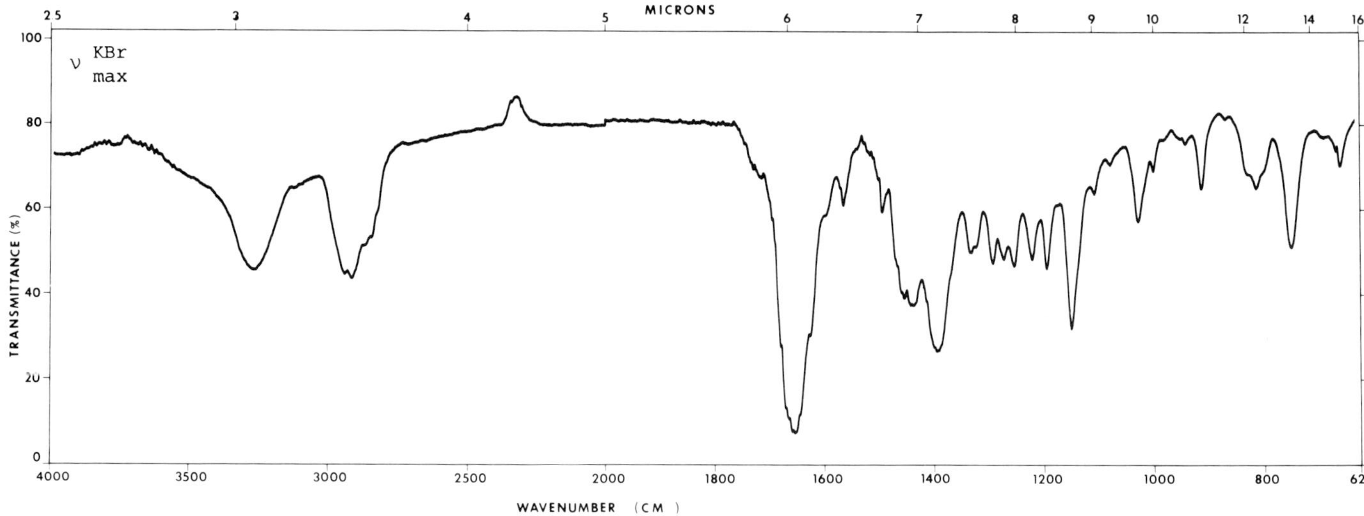
MICRONS
2.5
3
4
5
6
7
8
9
10
12
14
16
ν KBr max
TRANSMITTANCE (%)
100
80
60
40
20
0
4000
3500
3000
2500
2000
1800
1600
1400
1200
1000
800
625
WAVENUMBER (CM)

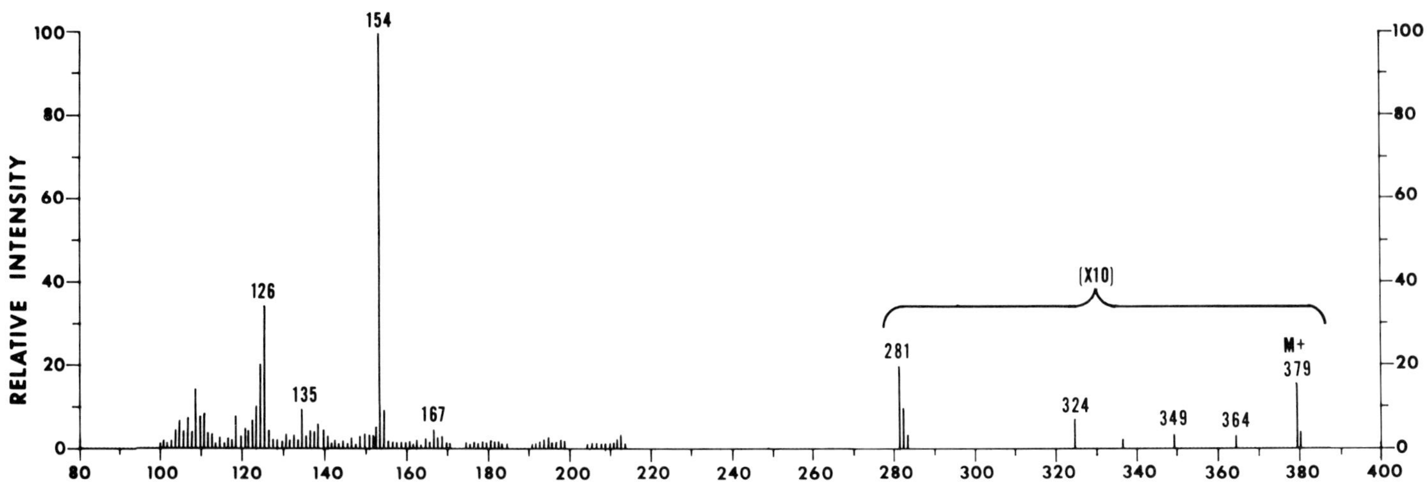
RELATIVE INTENSITY
100
80
60
40
20
0
154
126
135
167
(X10)
281
324
349
364
M+
379
80
100
120
140
160
180
200
220
240
260
280
300
320
340
360
380
400

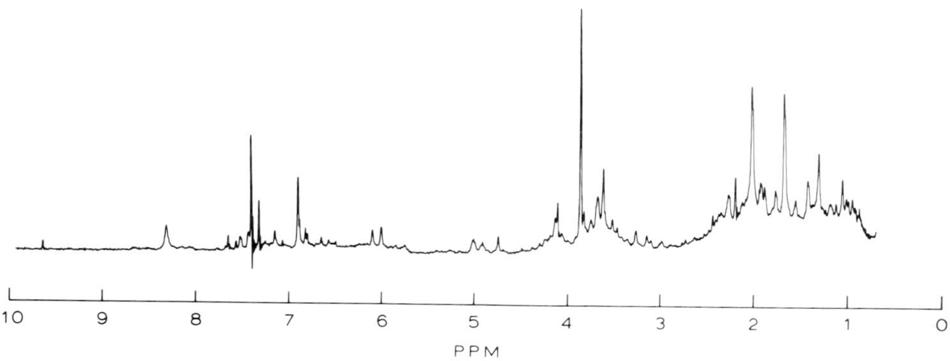

The 100 MHz ^{1}H NMR spectrum of fumitremorgin C in $CDCl_3$ solution.

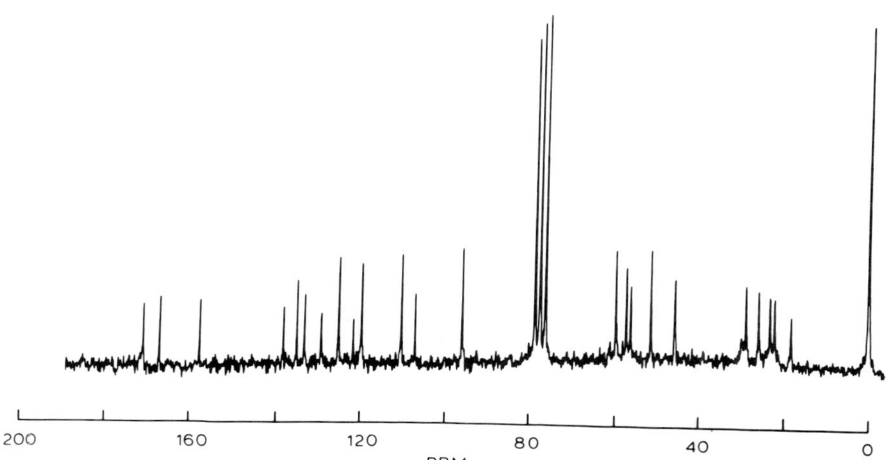

The 25 MHz ^{13}C NMR spectrum of fumitremorgin C in $CDCl_3$ solution.

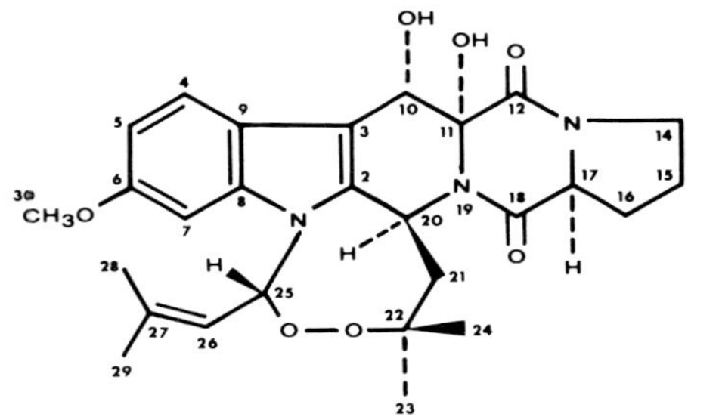

Common name — Verruculogen

Molecular weight — 511.2318

Molecular formula[23] — $C_{27}H_{33}O_7N_3$

General characteristics[23] — Colorless plates from benzene–ethyl alcohol 1:1 v/v, mp 233°–235°C (dec.)
$[\alpha]_D$ −27.7° (in chloroform)
Soluble in benzene, ethyl acetate, and acetone; slightly soluble in ethanol; very soluble in $CHCl_3$
Verruculogen is unstable to prolonged exposure to light.

UV data[23] — λ_{max}^{EtOH} nm (ϵ): 226(47,500), 277(11,000), and 295(9,750)

CD data[23] — $\Delta\epsilon$ (in ethanol solution) nm: 290(+0.16) and 265(+0.56)

Fungal source[22,23,25,26,38,77,97,104,105] — *Penicillium verruculosum* Peyronel (ATCC 24640; NRRL 5881)
P. paraherquei
P. piscarium Westling
P. janthinellum
Aspergillus caespitosus (NRRL 1929)
A. fumigatus

Toxicity data[23,48,49] — The LD_{50} of verruculogen to mice was 2.4 mg/kg (IP) and 126.7 mg/kg (oral); to day-old cockerels, 15.2 mg/kg (IP) and 265.5 mg/kg (oral).

The ED_{50} for tremor response in mice was 0.39 mg/kg (IP); in day-old cockerels it was 0.33 mg/kg (IP). Tremors are sustained at higher levels, intermittent at lower levels, and enhanced by enforced movement. Verruculogen administration produced a decrease in γ-aminobutyric acid levels in mouse central nervous system; verruculogen induced tremor is mediated by a loss of inhibitory GABA function.

TLC data[23]

Adsorbent: silica gel G-HR
Solvent: A: chloroform–acetone, 93:7 v/v
B: toluene–ethyl acetate–formic acid, 5:4:1 v/v/v
R_f: A: 0.30
B: 0.65
Detection: slate gray-blue spot in visible light or mustard-colored spot under UV light; develops immediately after spraying with 50% ethanolic H_2SO_4.
NOTE: The R_f's and color response of fumitremorgin A are identical to verruculogen in the above conditions.

Other significant references

X-Ray structure[34]

^{13}C NMR data

C2, 131.6 s; C3, 105.7 s; C4, 121.7 d; C5, 109.3 d; C6, 156. 4 s; C7, 93.9 d; C8, 136.3 s; C9, 121.1 s; C10, 68.7 d; C11, 82.6 s; C12, 170.7 s; C14, 51.3 t; C15, 22.7 t; C16, 29.1 t; C17, 58.7 d; C18, 166.2 s; C20, 48.9 d; C21, 45.3 t; C22, 82.1 s; C23, 27.1 q; C24, 25.6 q; C25, 85.8 q; C26, 118.7 d; C27, 142.9 s; C28, 18.8 q; C29, 24.2 q; C30, 55.7 q

^{1}H NMR data

H4, 7.90 (J = 9.0); H5, 6.80 (J = 9.0, 2.0); H7, 6.60 (J = 2.0); H10, 5.65; 10-OH, 4.80; H14, 3.65; H15, 2.00; H16, 2.00; H17, 4.50; H20, 6.05; H21, 2.00; H23, 1.02; H24, 1.80; H25, 6.70 (J = 8.0); H26, 5.05; H28, 1.80; H29, 2.00; H30, 3.86

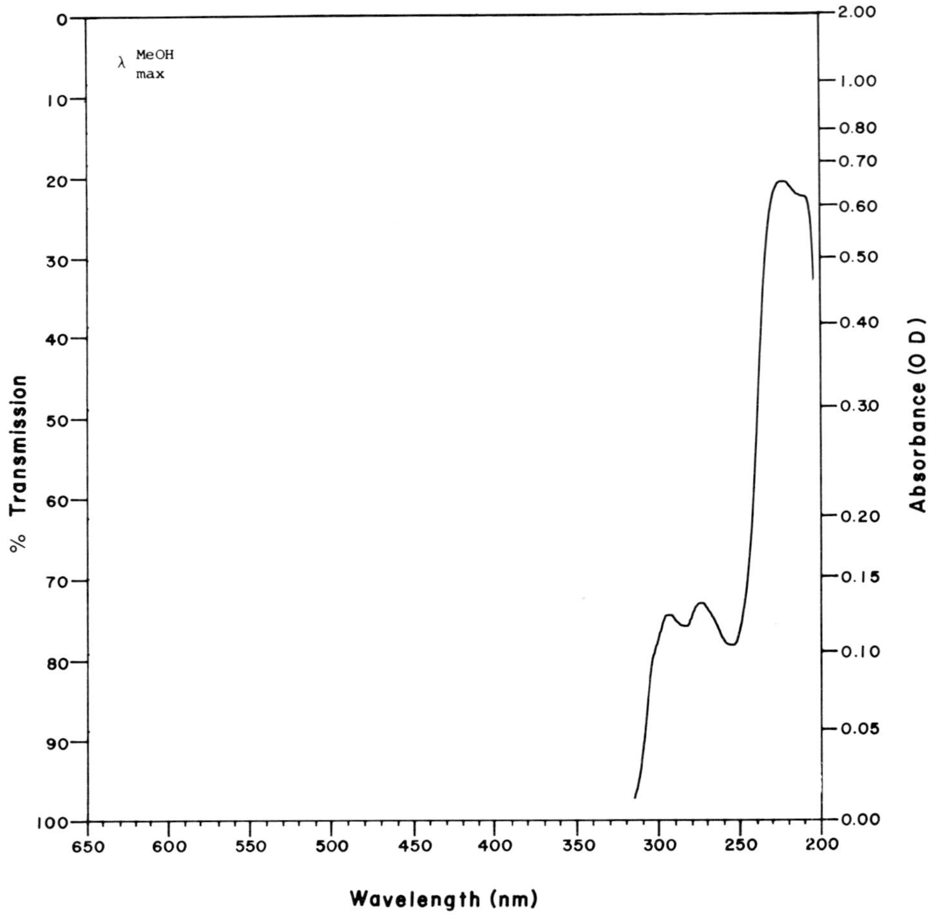

λ MeOH max
% Transmission
0
10
20
30
40
50
60
70
80
90
100
Absorbance (O D)
2.00
1.00
0.80
0.70
0.60
0.50
0.40
0.30
0.20
0.15
0.10
0.05
0.00
650
600
550
500
450
400
350
300
250
200
Wavelength (nm)

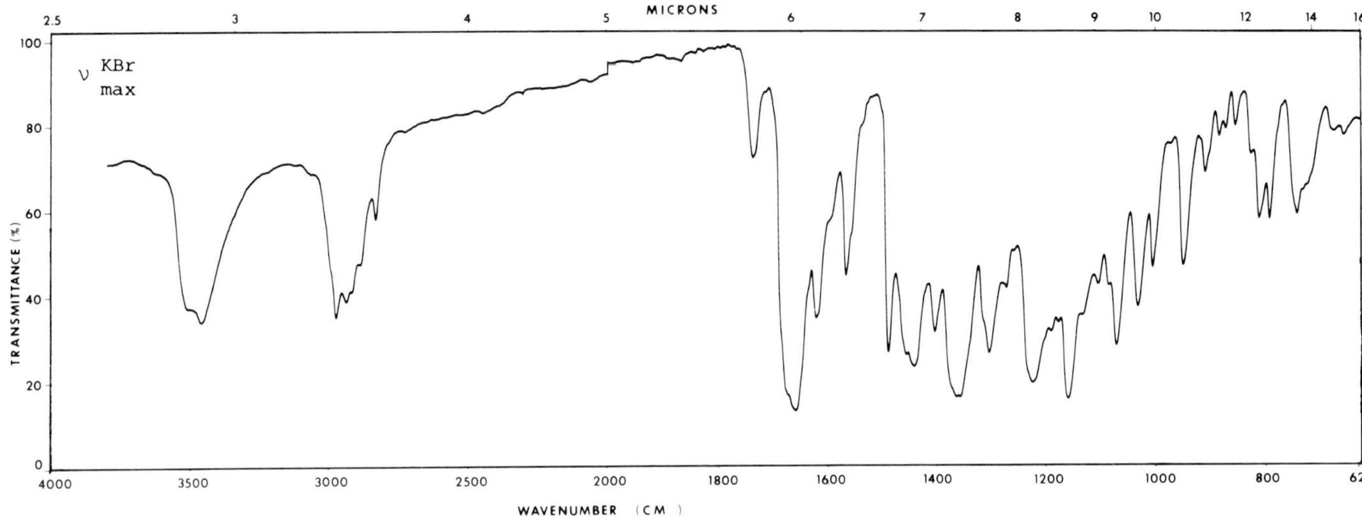
MICRONS
ν KBr max
TRANSMITTANCE (%)
WAVENUMBER (CM)

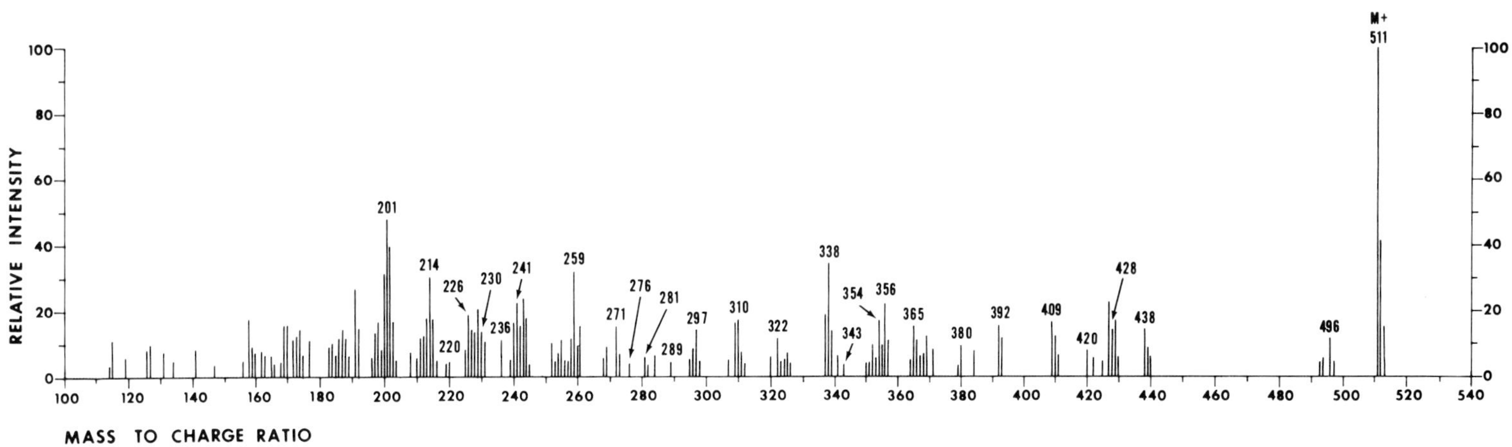
M+ 511
201
214
226
230
236
241
259
220
271
276
281
289
297
310
322
338
343
354
356
365
380
392
409
420
428
438
496
RELATIVE INTENSITY
MASS TO CHARGE RATIO

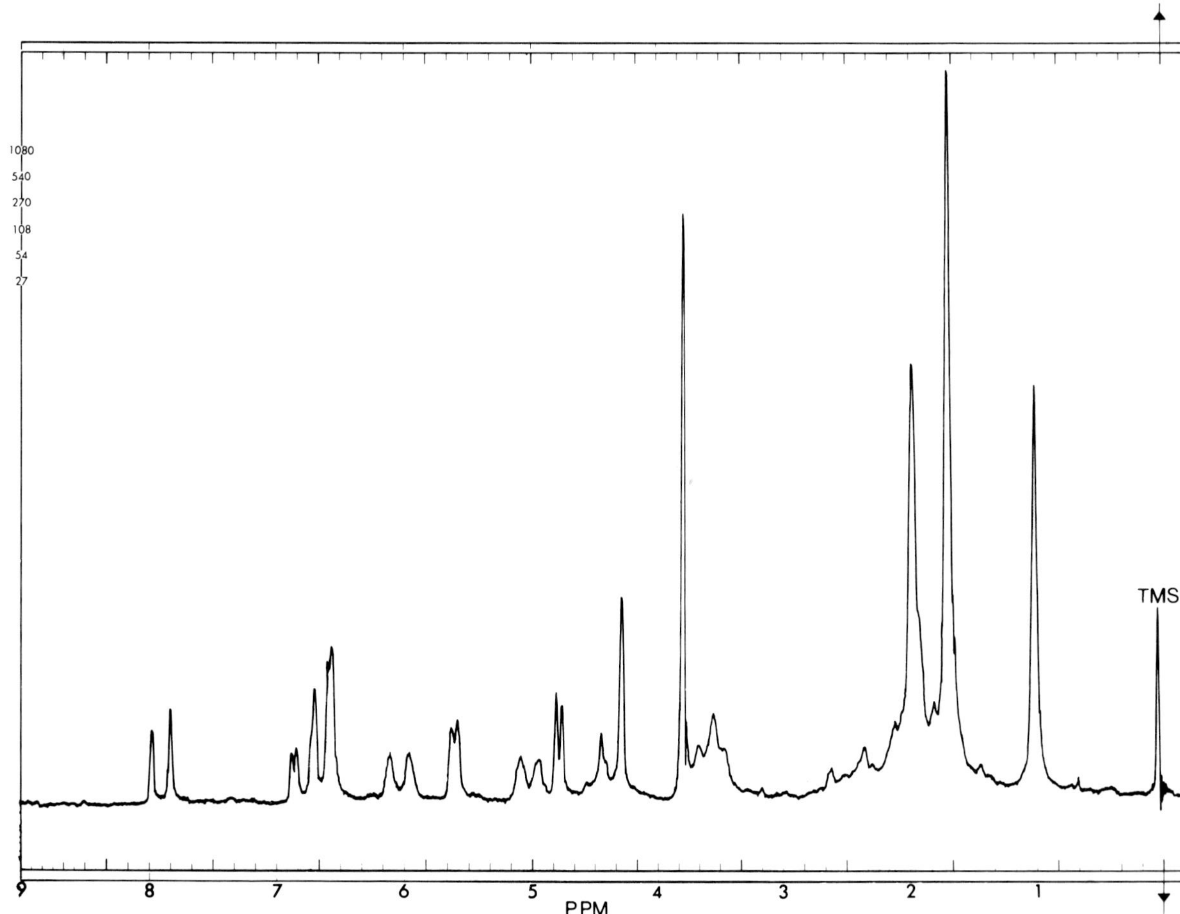

The 60 MHz ^1H NMR spectrum of verruculogen in $CDCl_3$ solution.

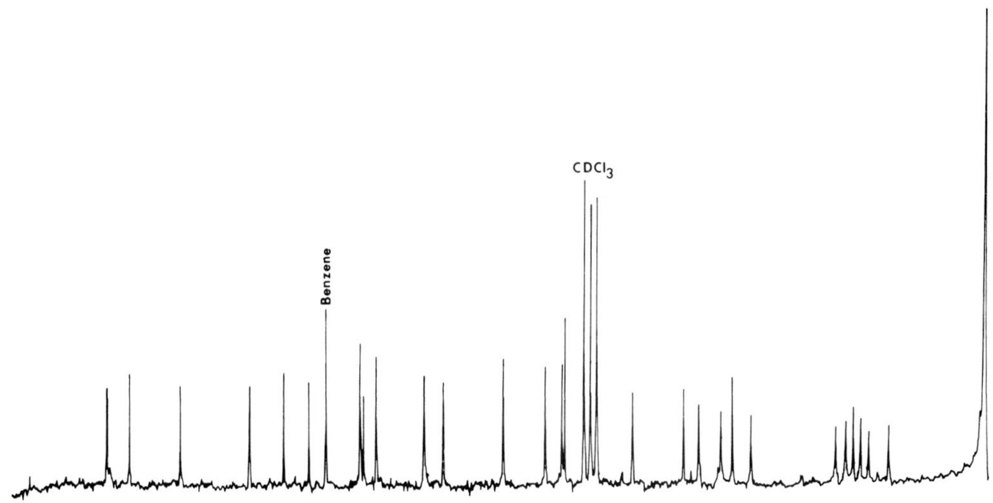

The 60 MHz ^{13}C NMR spectrum of verruculogen in $CDCl_3$ solution.

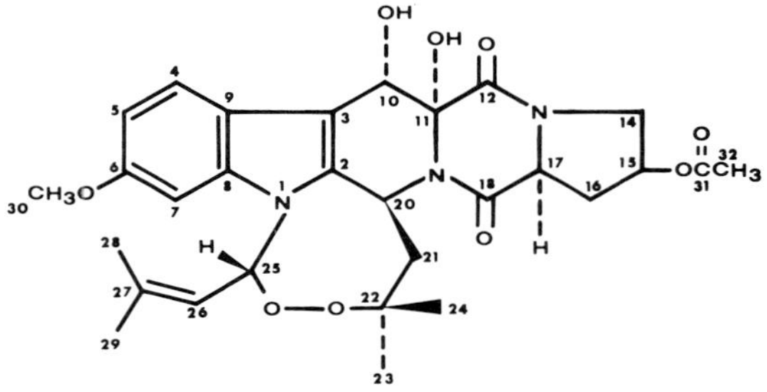

Common name	15-Acetoxyverruculogen
Molecular weight	569.2372
Molecular formula	$C_{29}H_{35}O_9N_3$
General characteristics[16]	mp 217°–218°C (dec.)
UV data[91]	λ_{max}^{MeOH} nm: 227, and 295 (ϵ_{max} not reported)
Fungal source[91]	*Penicillium verruculosum*
Toxicity data[16]	Tremorgenic but activity less than that observed for comparable amount of verruculogen.
TLC data[16]	*Adsorbent:* silica gel G-HR *Solvent:* toluene–ethyl acetate–formic acid, 5:4:1 v/v/v R_f: 0.66 *Detection:* slate-gray spot develops immediately after spraying with 50% ethanolic H_2SO_4

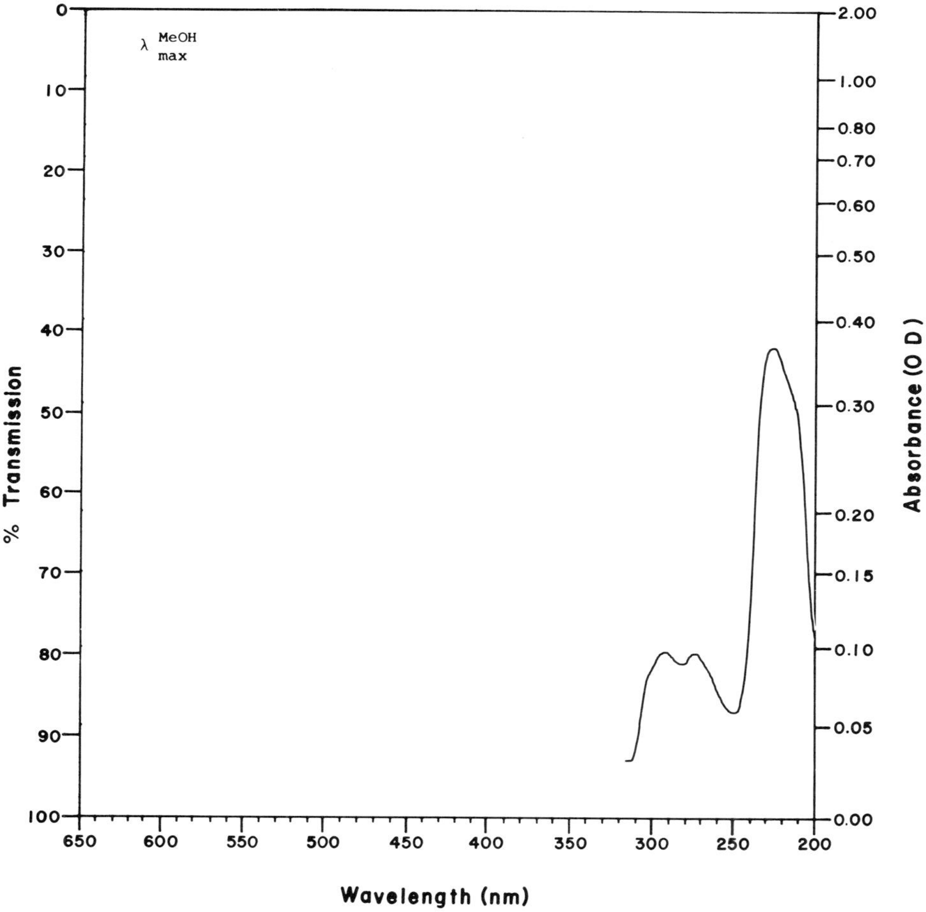

λ MeOH max
% Transmission
0
10
20
30
40
50
60
70
80
90
100
Absorbance (O D)
2.00
1.00
0.80
0.70
0.60
0.50
0.40
0.30
0.20
0.15
0.10
0.05
0.00
650
600
550
500
450
400
350
300
250
200
Wavelength (nm)

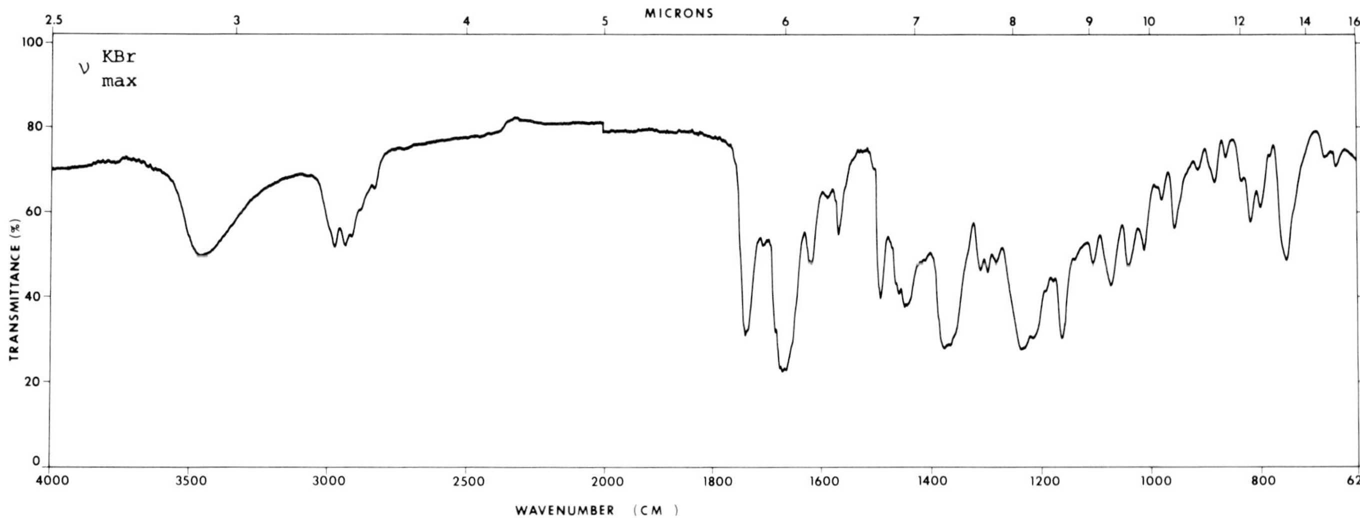
MICRONS
2.5
3
4
5
6
7
8
9
10
12
14
16
ν KBr max
TRANSMITTANCE (%)
100
80
60
40
20
0
4000
3500
3000
2500
2000
1800
1600
1400
1200
1000
800
625
WAVENUMBER (CM)

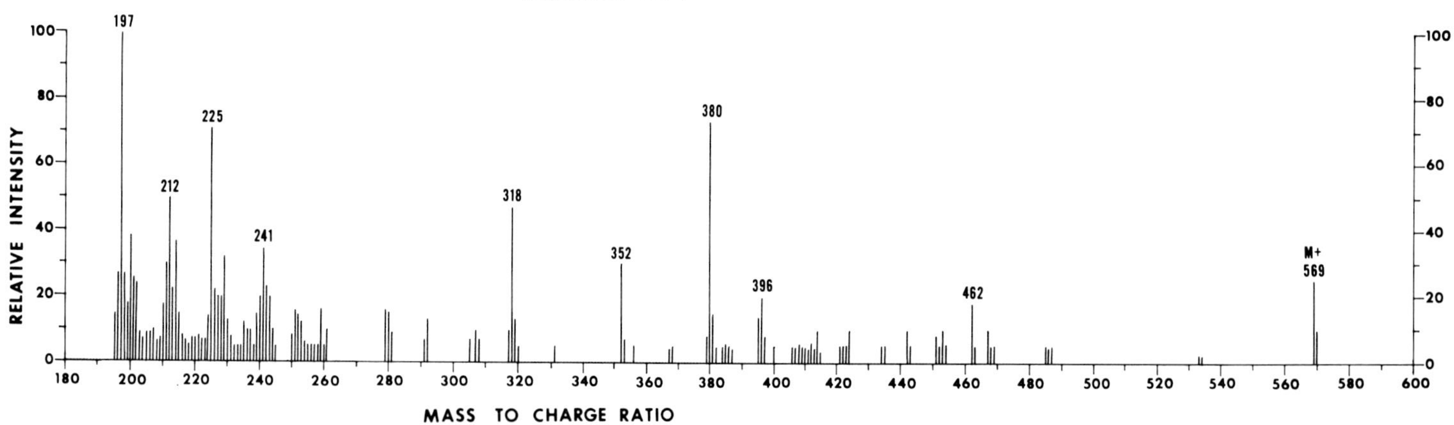
RELATIVE INTENSITY
100
80
60
40
20
0
197
212
225
241
318
352
380
396
462
M+
569
180
200
220
240
260
280
300
320
340
360
380
400
420
440
460
480
500
520
540
560
580
600
MASS TO CHARGE RATIO

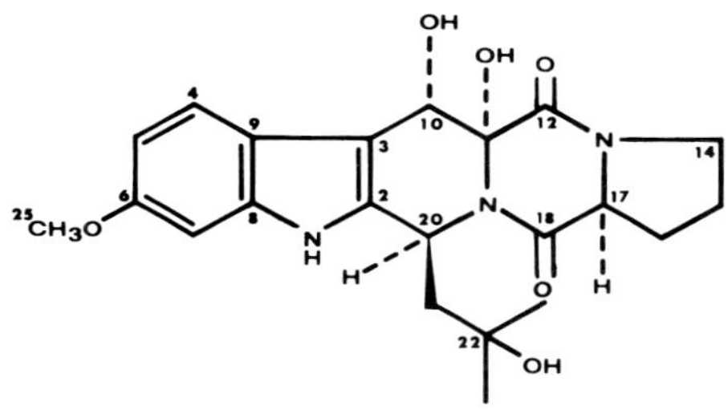

Common name	TR-2
Molecular weight	429.1899
Molecular formula	$C_{22}H_{27}O_6N_3$
General characteristics[21]	Crystals from benzene–ethyl acetate, 95:5 v/v mp 150°–152°C
UV data[21]	λ_{max}^{EtOH} nm (ϵ): 224(37,400), 268(6,830), and 294(7,540)
Fungal source[20–22]	*Penicillium verruculosum* Peyronel *Aspergillus fumigatus* Fres.
Toxicity data[21]	Perceptible tremors in day-old cockerels dosed orally at levels down to 12.5 mg/kg.
TLC data[22]	*Adsorbent:* silica gel G-HR *Solvent:* toluene–ethyl acetate–formic acid, 5:4:1 v/v/v R_f: 0.46 *Detection:* light brown fluorescent spot after spraying with 50% ethanolic H_2SO_4 and heating at 100°C for 5 min.
^{13}*C NMR data*	C2, 131.8 s; C3, 105.1 s; C4, 121.2 d; C5, 109.8 d; C6, 156.6 s; C7, 95.2; C8, 137.2 s; C9, 120.8 s; C10, 68.8 d; C11, 83.4 d; C12, 171.4 s; C14, 49.9 t; C15, 22.7 t; C16, 29.6 t; C17, 58.9 d; C18, 166.2 s; C20, 57.4 d; C21, 45.3 t; C22, 71.2 s; C23, 31.9 q; C24, 29.7 q; C25, 55.7 q
^{1}*H NMR data*	H4, 7.60 (J = 9.0); H5, 6.60 (J = 9.0, 3.0); H7, 6.87; H10, 5.52; 10-OH, 5.00; 11-OH, 4.20; H14, 3.54; H15, 1.85; H16, 1.85; H17, 4.33; H20, 5.37; H21, 1.85; H23, 1.05; H24, 1.30; 22-OH, 5.94; H25, 3.71

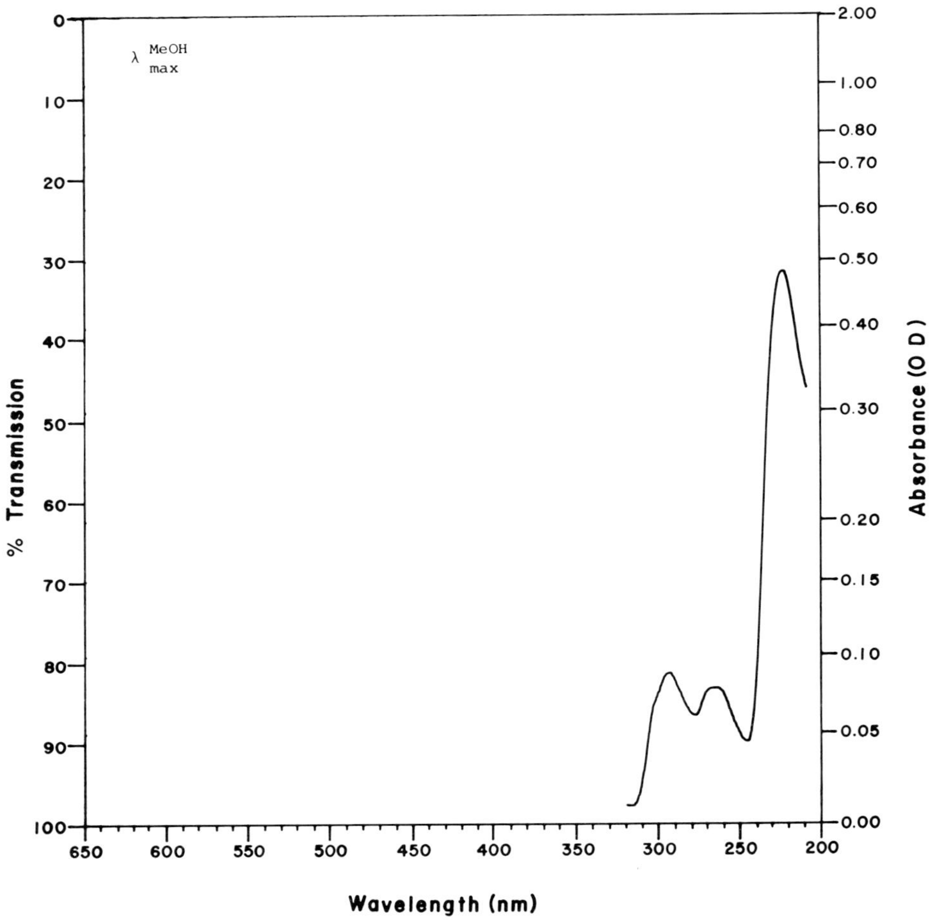
λ MeOH max
% Transmission
0
10
20
30
40
50
60
70
80
90
100
Absorbance (O D)
2.00
1.00
0.80
0.70
0.60
0.50
0.40
0.30
0.20
0.15
0.10
0.05
0.00
650 600 550 500 450 400 350 300 250 200
Wavelength (nm)

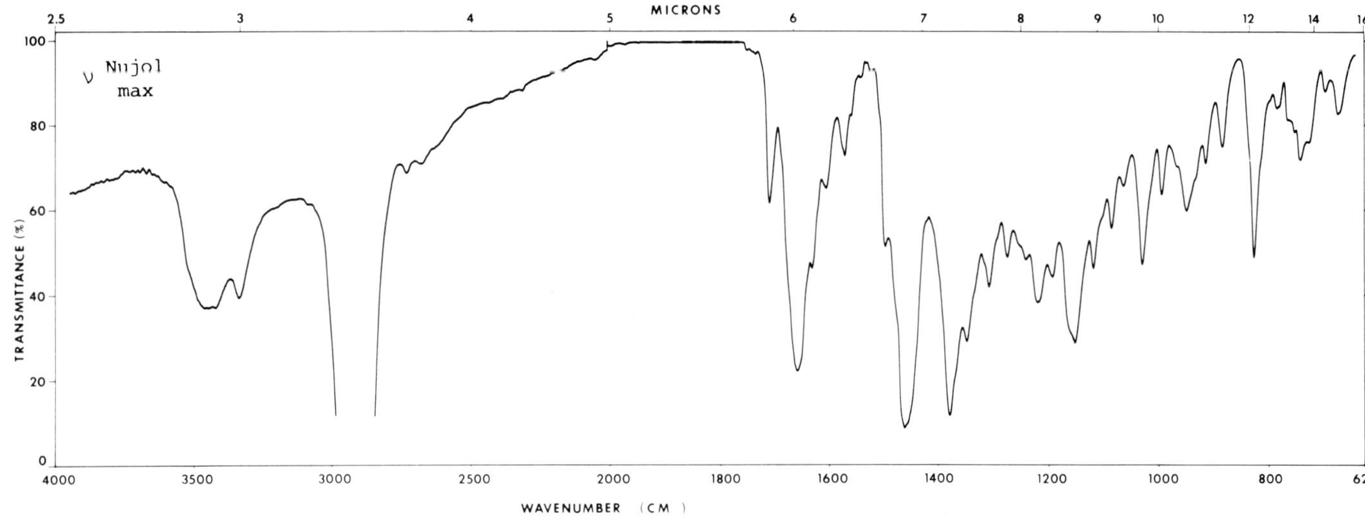
MICRONS
2.5
3
4
5
6
7
8
9
10
12
14
16
ν Nujol max
TRANSMITTANCE (%)
100
80
60
40
20
0
4000
3500
3000
2500
2000
1800
1600
1400
1200
1000
800
625
WAVENUMBER (CM)

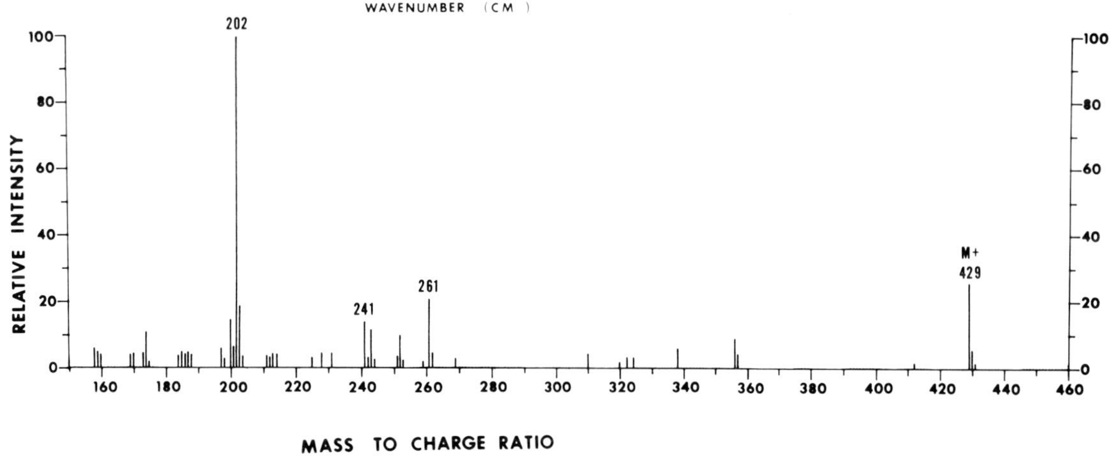
202
241
261
M+
429
RELATIVE INTENSITY
100
80
60
40
20
0
160
180
200
220
240
260
280
300
320
340
360
380
400
420
440
460
MASS TO CHARGE RATIO

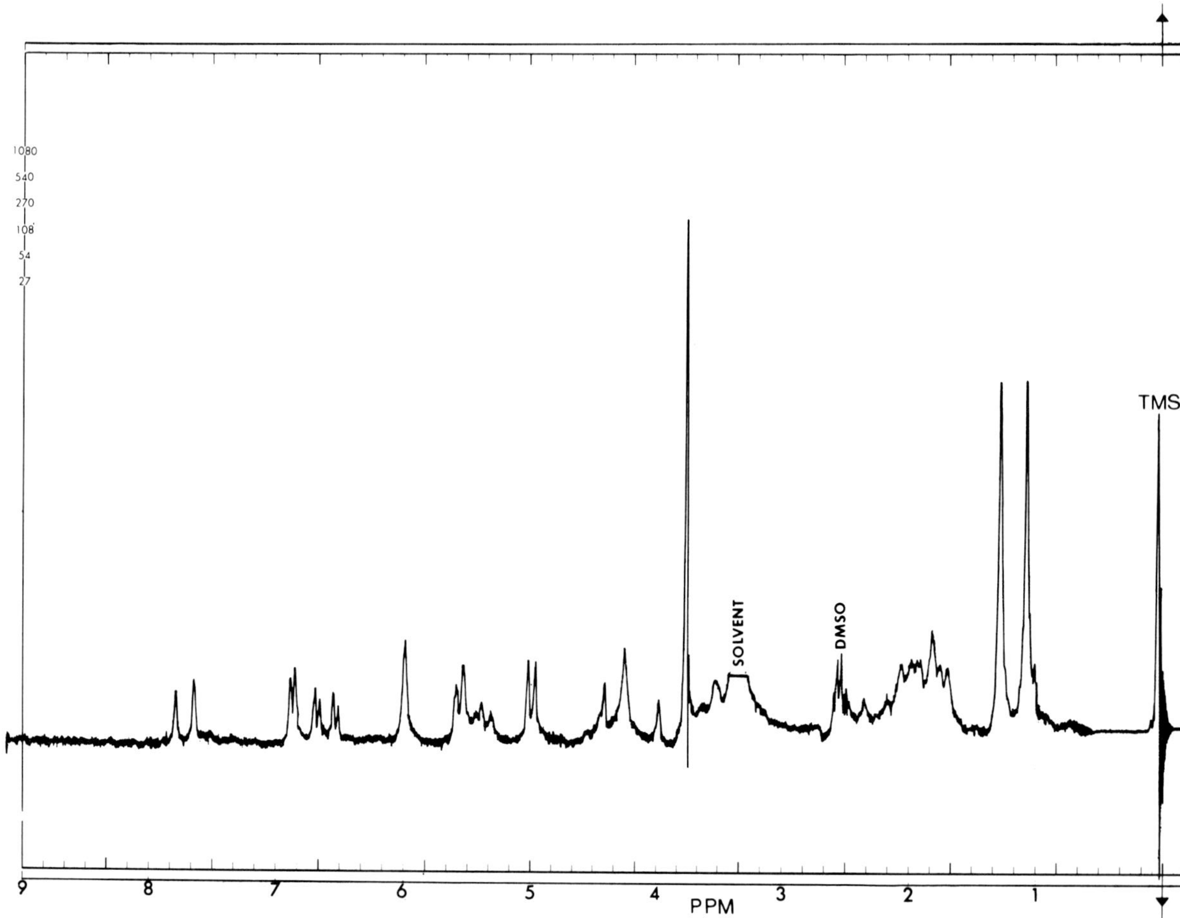

The 60 MHz ^{1}H NMR spectrum of TR-2 in DMSO solution.

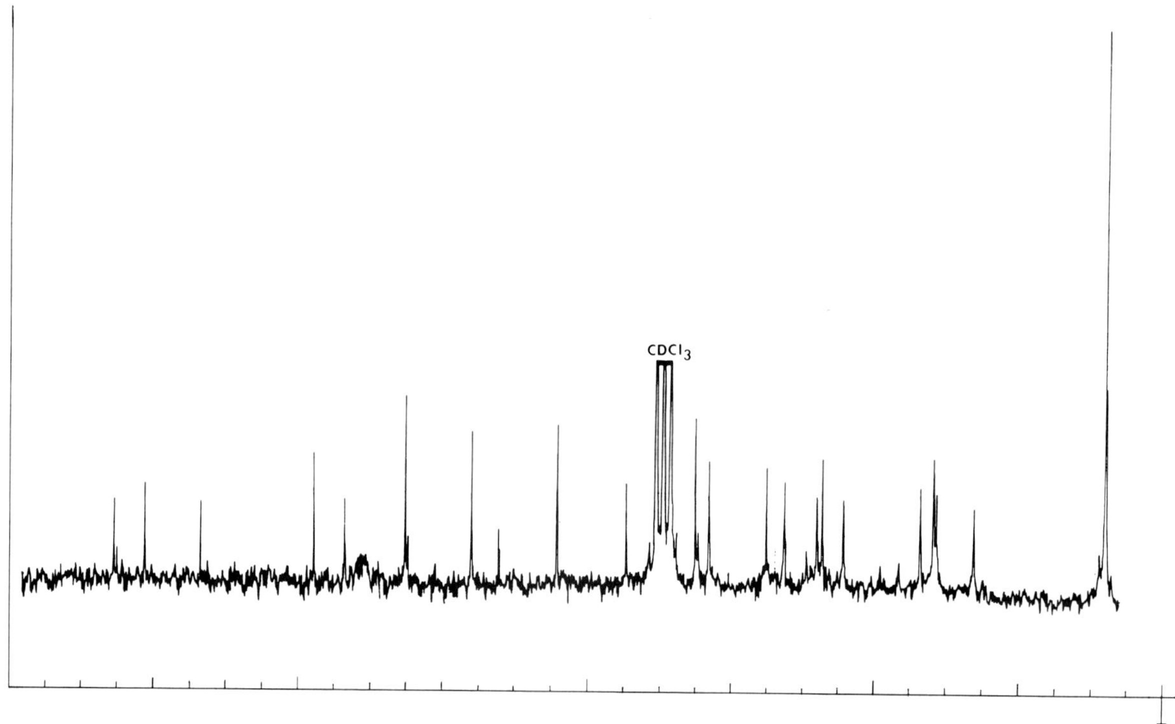

The 25 MHz ^{13}C NMR spectrum of TR-2 in $CDCl_3$ solution.

Common name	Penitrem A*
Molecular weight[96]	633.2807
Molecular formula[96]	$C_{37}H_{44}O_6NCl$
General characteristics[11]	Microcrystals from ethyl ether, petroleum ether, mp 180°–200°C (dec.) Fine needles from ethanol–water, mp 237°–239°C (dec.) Penitrem A is soluble in ethyl ether, chloroform, acetone, methanol; sparingly soluble in water, aqueous HCl and NaOH. It is unstable in acid, rapidly forming a blue color. It forms a monoacetate; mp 203°–205°C (dec.)
UV data[11]	λ_{max}^{MeOH} nm (ϵ): 233(37,000) and 295(11,600)
Fungal source[50]	*Penicillium cyclopium* *P. crustosum* *P. palitans* (NRRL 3468) *P. puberulum* *P. spinulosum*
Toxicity data[10,28,44,5C,81,85]	Penitrem A produced perceptible tremors in mice dosed IP at 250 μg/kg. Its LD_{50} in mice dosed IP was 1.05 mg/kg. It has been shown that penitrem A induces tremors by acting as an antagonist to production of glycine in the central nervous system. Glycine is a known inhibitor of interneurons in the spinal cord.
TLC data[10,50]	*Adsorbent:* silica gel G *Solvent:* chloroform–methanol, 95:5 v/v R_f: 0.4 *Detection:* visualized as green spot when sprayed with 2% $FeCl_3$ in BuOH followed by gentle heating. Visualized as blue spot when sprayed with 50% ethanolic H_2SO_4 and heated.
Other significant references	Possible role in staggers syndrome[31,73]

* The structure of penitrem A is not known.

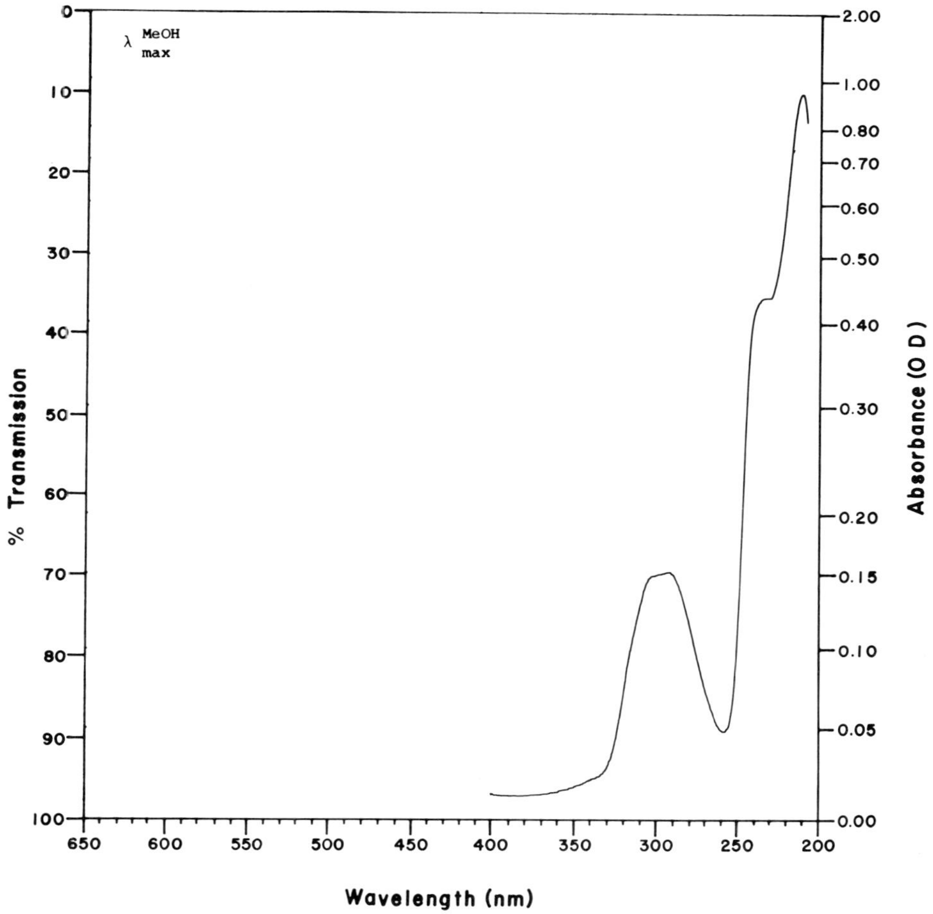

λ MeOH max
% Transmission
0
10
20
30
40
50
60
70
80
90
100
Absorbance (O D)
2.00
1.00
0.80
0.70
0.60
0.50
0.40
0.30
0.20
0.15
0.10
0.05
0.00
650
600
550
500
450
400
350
300
250
200
Wavelength (nm)

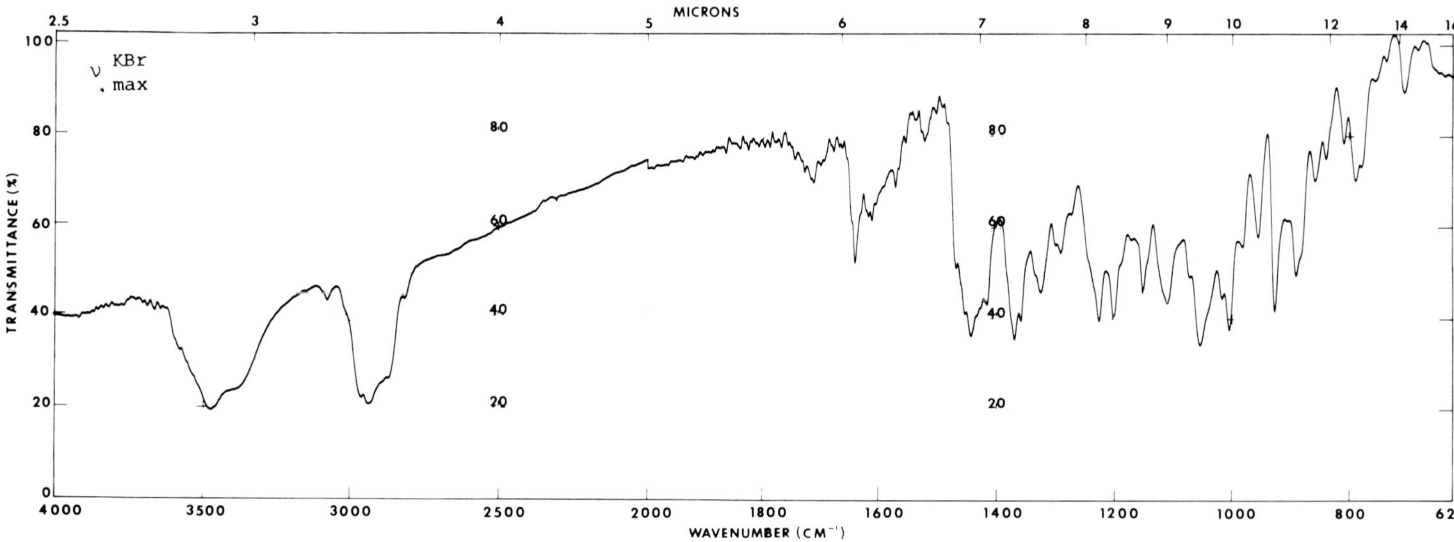
MICRONS
ν^{KBr}_{max}
TRANSMITTANCE (%)
WAVENUMBER (CM⁻¹)
2.5
3
4
5
6
7
8
9
10
12
14
16
4000
3500
3000
2500
2000
1800
1600
1400
1200
1000
800
625
0
20
40
60
80
100

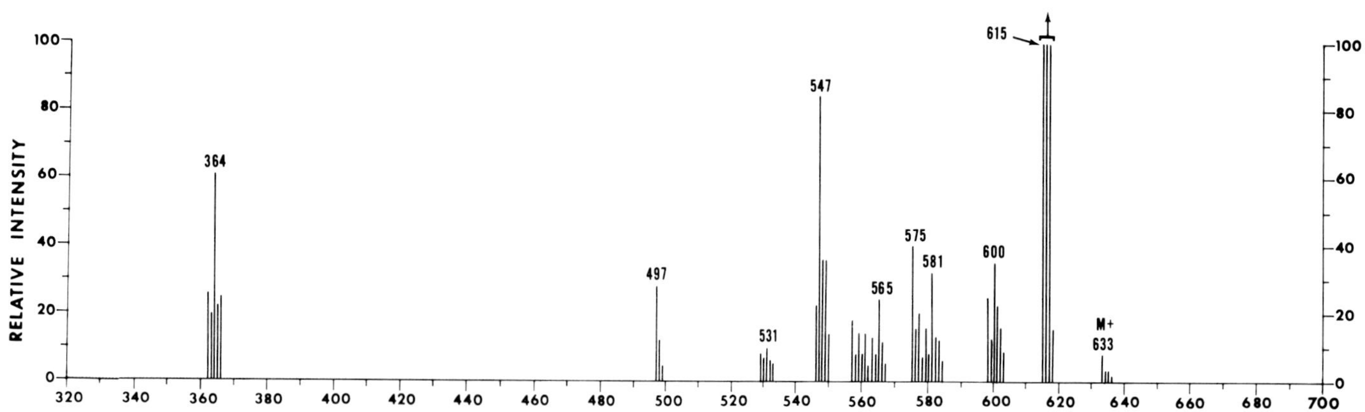
RELATIVE INTENSITY
364
497
531
547
565
575
581
600
615
M+
633
320
340
360
380
400
420
440
460
480
500
520
540
560
580
600
620
640
660
680
700
0
20
40
60
80
100

Common name	Penitrem B*
Molecular weight	583.3296
Molecular formula	$C_{37}H_{45}O_5N$
General characteristics[50]	Penitrem B crystallizes from ethanol–water to form fine needles, mp 185°–195°C. It is soluble in ethyl ether, chloroform, acetone, methanol; sparingly soluble in water, aqueous HCl and NaOH. It is unstable in acidic solution, rapidly forming a blue color.
UV data[50]	λ_{max}^{MeOH} nm (ϵ): 227(38,450), 286(13,100), and 297(11,000)
Fungal source[50]	*Penicillium palitans* (NRRL 3468)
Toxicity data[50]	Penitrem B produced perceptible tremors in mice dosed IP at 1.3 mg/kg. The LD_{50} of penitrem B in mice dosed IP was 5.84 mg/kg.
TLC data[50]	*Adsorbent:* silica gel G-HR *Solvent:* chloroform–acetone, 93:7 v/v R_f: 0.60 *Detection:* green color after spraying with 1% $FeCl_3$ in BuOH and gentle heating with a warm air stream.

* The structure of penitrem B is not known.

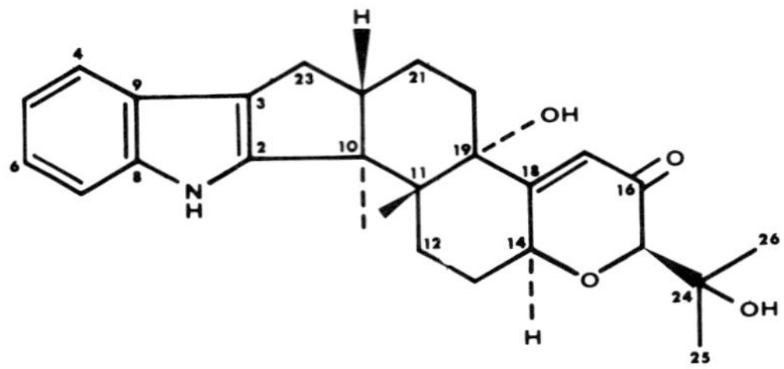

Common name	Paxilline
Molecular weight	435.2409
Molecular formula	$C_{27}H_{33}O_4N$
General characteristics[24]	Clear cubes from acetone, mp 252°C (paxilline crystallizes with 1 molecule of acetone H-bonded to the hydrogen of the indole nitrogen)
UV data[24]	λ_{max}^{MeOH} nm (ϵ): 230(41,500) and 281(8,000)
CD data[24]	$\Delta\epsilon$ nm: 335(+1.2) and 300(+3.3)
Fungal source[24]	*Penicillium paxilli* Bainier
Toxicity data[24]	In day-old cockerels dosed orally: at 25 mg/kg (1 mg/cockerel) tremors were intermittent; at 100 mg/kg (4 mg/cockerel) tremors were more sustained. Mortality was observed at higher dosage levels. In mice dosed intraperitoneally, intermittent tremors were observed at levels down to 35 mg/kg. Tremors were enhanced when the animal tried to move or was forced to move. Mice were noticeably hypersensitive to sound stimuli. No mortality was observed in mice dosed up to 227 mg/kg. Mice continued to feed or attempt to feed.
TLC data[24]	*Adsorbent:* silica gel G-HR *Solvent:* toluene–ethyl acetate–formic acid, 5:4:1 v/v/v R_f: 0.75 *Detection:* A dark blue spot with 50% ethanolic H_2SO_4 spray; a greenish-gray spot with 3% ethanolic phosphomolybdic acid.
Other significant references	X-Ray structure[82]

^{13}*C NMR data*

C2, 152.1 s; C3, 114.5 s; C4, 117.2 d; C5, 118.4 d; C6, 118.8 d; C7, 111.4 d; C8, 139.3 s; C9, 124.1 s; C10, 50.0 s; C11, 39.9 s; C12, 20.6 t*; C13, 25.8 t*; C14, 72.2 d; C15, 83.1 d; C16, 196.9 s; C17, 117.9 d; C18, 169.9 s; C19, 75.4 s; C20, 26.7 t*; C21, 28.1 t*; C22, 49.1 d; C23, 32.4 t; C24, 70.8 s; C25, 25.7 q†; C26, 25.8 q†; C27, 18.6 q; C28, 16.2 q

* Assignments uncertain at this time.
† Assignments may be reversed.

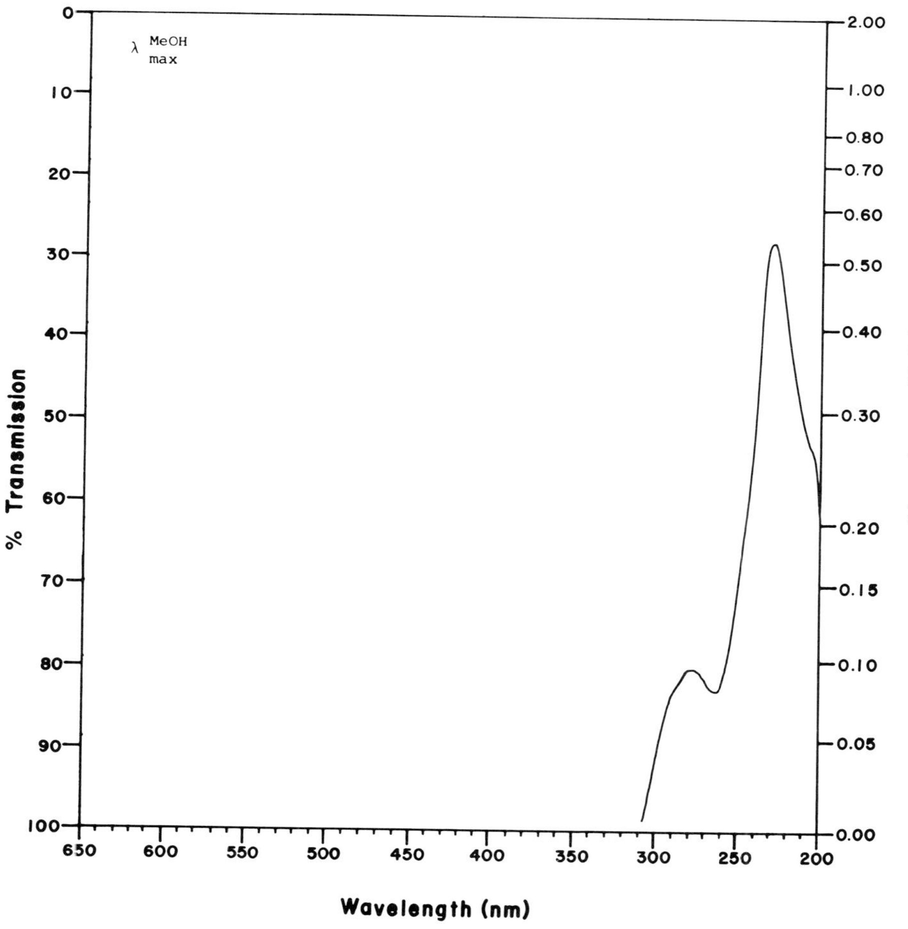

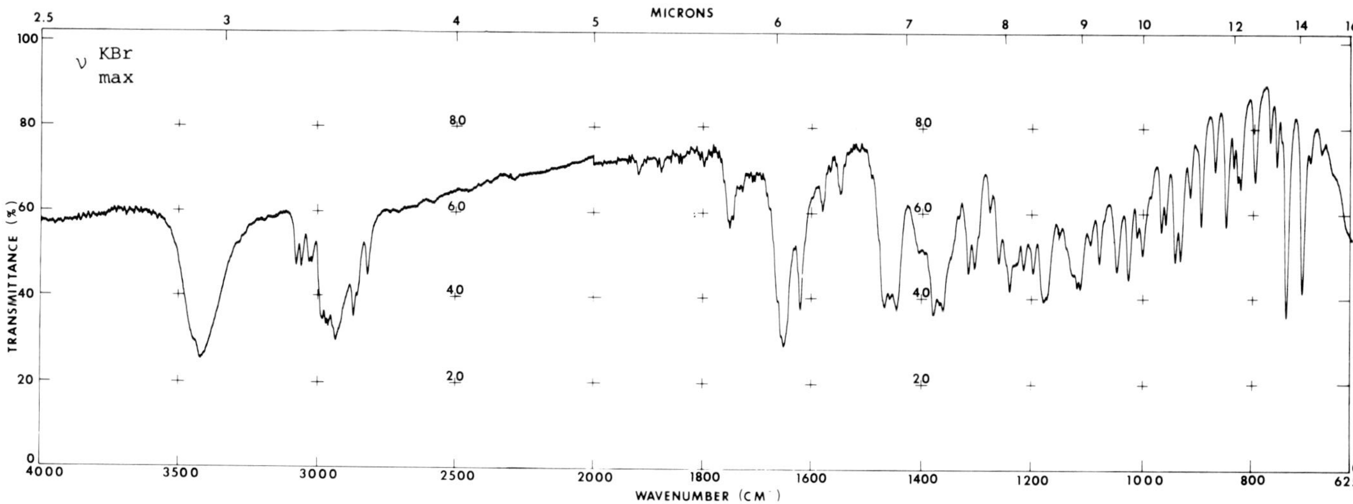
MICRONS
ν KBr max
TRANSMITTANCE (%)
WAVENUMBER (CM⁻¹)

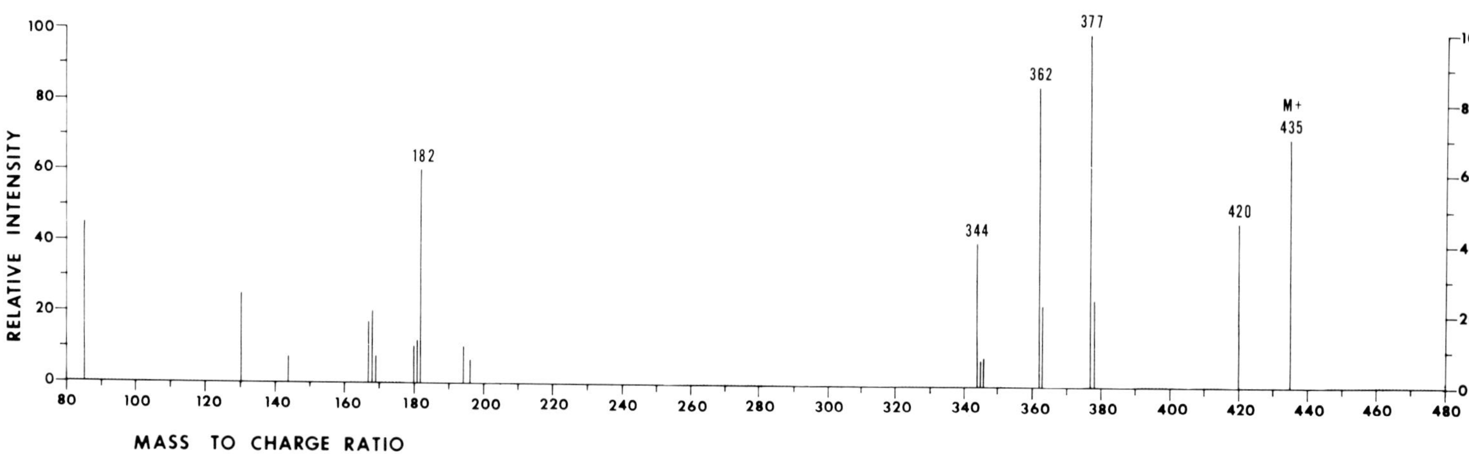
182
344
362
377
420
M+
435
RELATIVE INTENSITY
MASS TO CHARGE RATIO

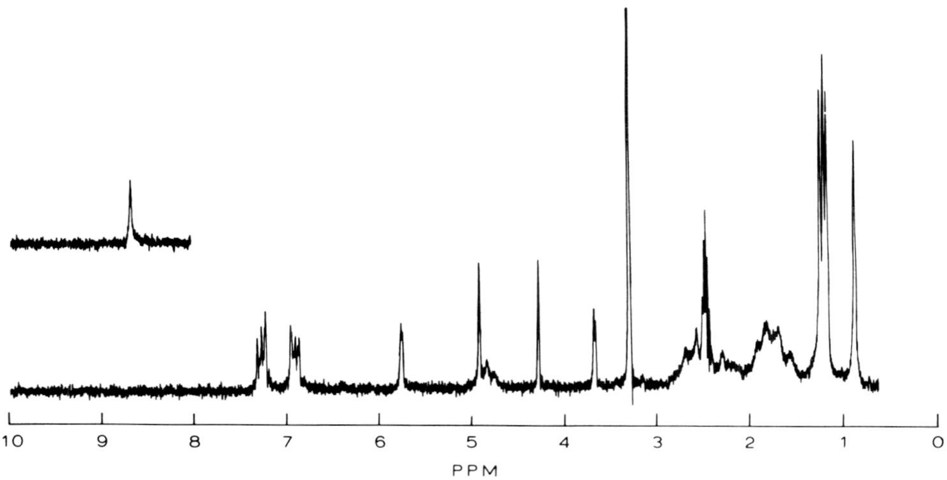

The 100 MHz ^{1}H NMR spectrum of paxilline in DMSO solution.

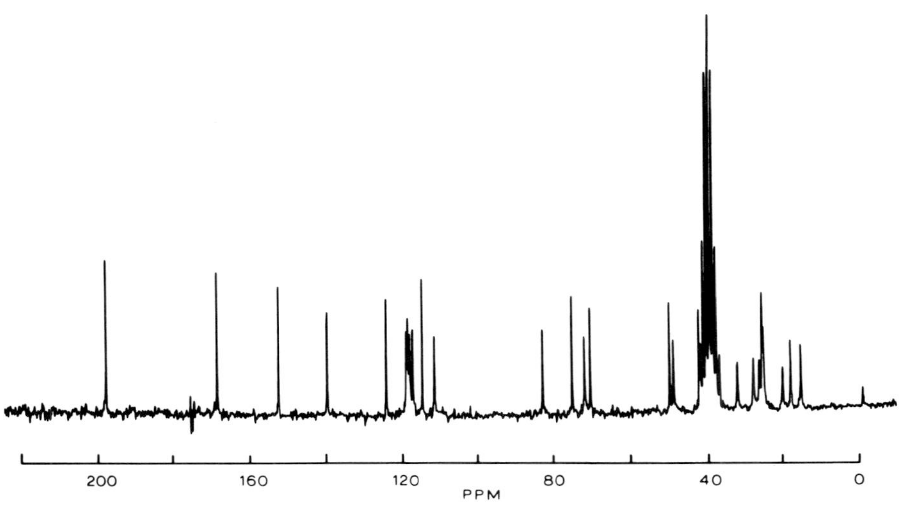

The 25 MHz ^{13}C NMR spectrum of paxilline in DMSO solution.

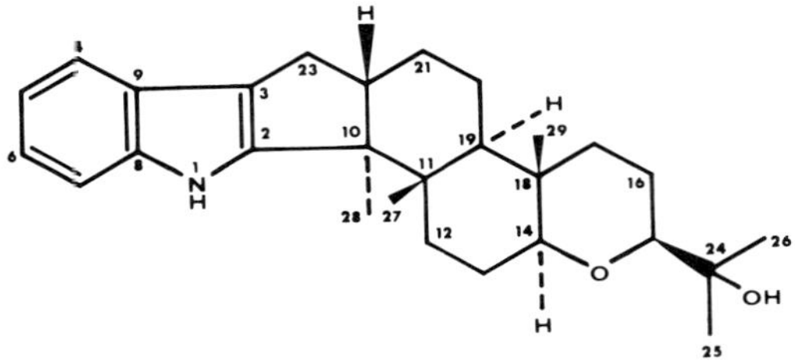

Common name	Paspaline
Molecular weight	421.2980
Molecular formula	$C_{28}H_{39}O_2N$
General characteristics[35]	Crystals from methanol, mp 254°C sublimed sample, mp 264°C $[\alpha]_D$ −23° (C = 0.36)
UV data[35]	λ_{max} nm (ϵ): 228–231(25,118), 282(7,943), and 291(6,309)
Fungal source[35,43,57]	*Claviceps paspali* Stammes
Toxicity data[17]	Nontremorgenic in mice dosed IP up to 500 mg/kg; chemically related to tremorgens paxilline and paspalinine
TLC data[16]	*Adsorbent:* silica gel G-HR *Solvent:* toluene–ethyl acetate–formic acid, 5:4:1 v/v/v R_f: 0.76 *Detection:* dull fluorescent spot after spraying with 50% ethanolic H_2SO_4 and heating at 100°C for 5 min.
Other significant references	X-Ray structure[83]; biosynthesis[1]
^{13}C NMR data	C2, 150.7 s; C3, 118.3 s; C4, 118.3 d; C5, 119.4 d; C6, 120.4 d; C7, 111.3 d; C8, 139.9 s; C9, 125.1 s; C10, 52.9 s; C11, 42.2 s; C12, 21.9 t*; C13, 24.6 t*; C14, 84.7 d; C15, 85.7 d; C16, 37.7 t; C17, 25.2 t*; C18, 36.4 s; C19, 46.4 d; C20, 21.9 t*; C21, 27.5 t*; C22, 48.7 d; C23, 33.8 t; C24, 71.9 s; C25, 33.7 q†; C26, 26.1 q†; C27, 19.9 q; C28, 14.5 q; C29, 12.7 q

* Assignments uncertain at this time.

† Assignments may be reversed.

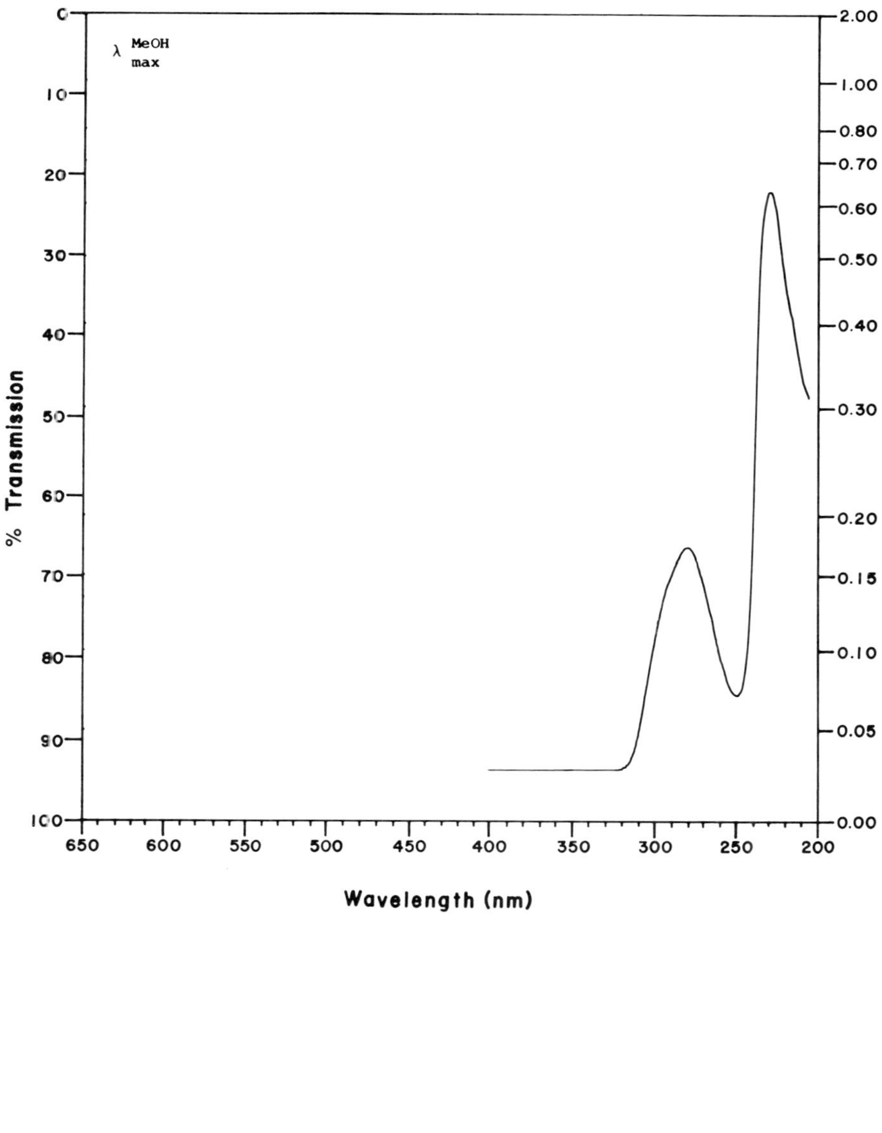

λ MeOH max
% Transmission
Absorbance (O D)
Wavelength (nm)
0
10
20
30
40
50
60
70
80
90
100
2.00
1.00
0.80
0.70
0.60
0.50
0.40
0.30
0.20
0.15
0.10
0.05
0.00
650
600
550
500
450
400
350
300
250
200

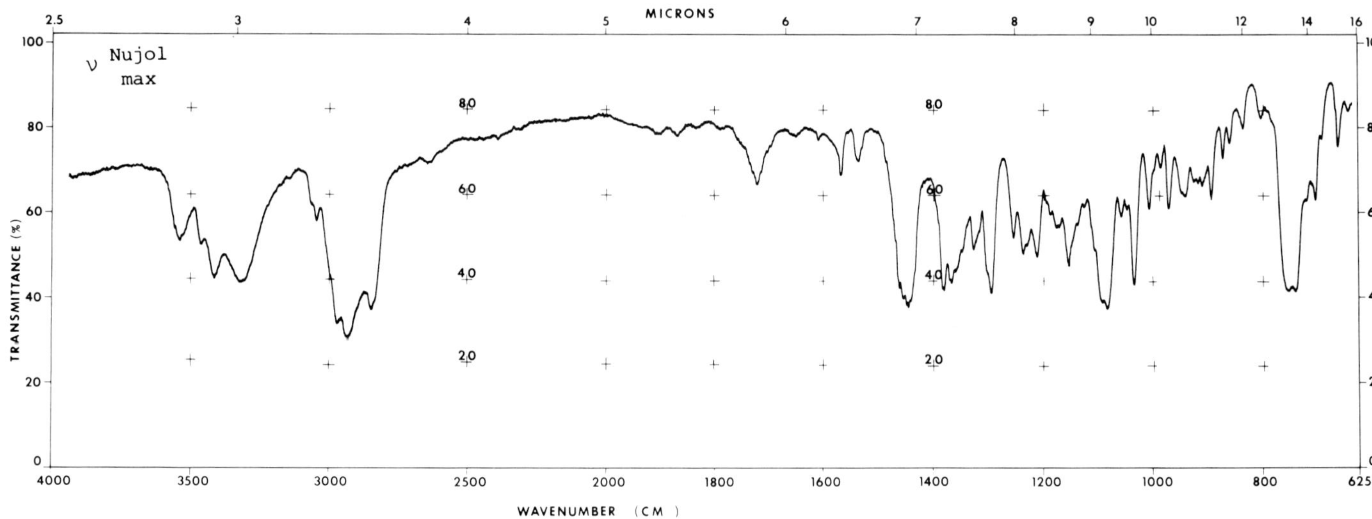
MICRONS
ν Nujol max
TRANSMITTANCE (%)
WAVENUMBER (CM)

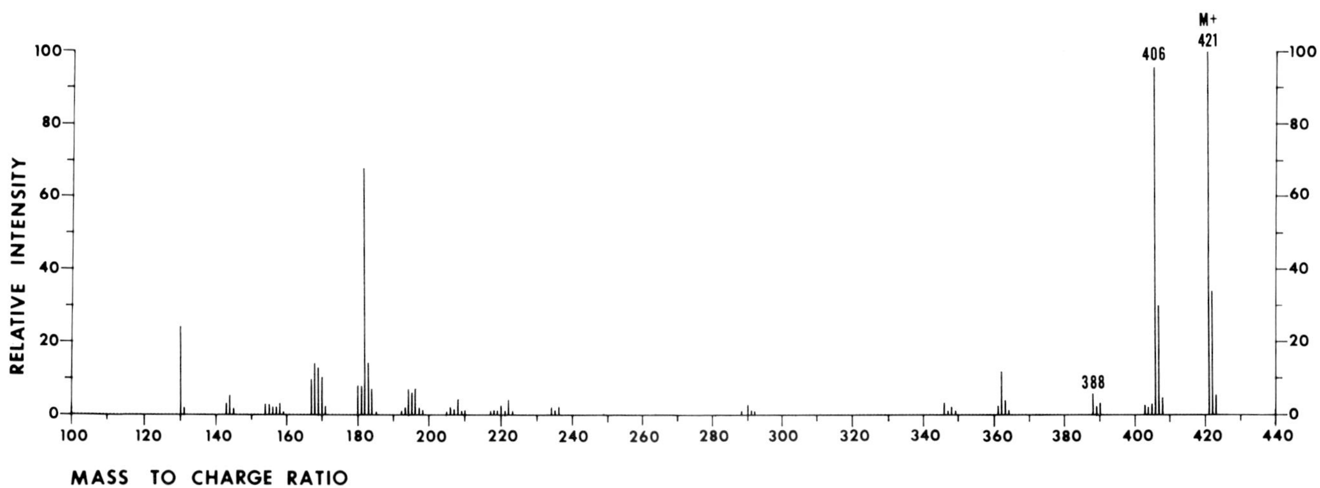
M+
421
406
388
RELATIVE INTENSITY
MASS TO CHARGE RATIO

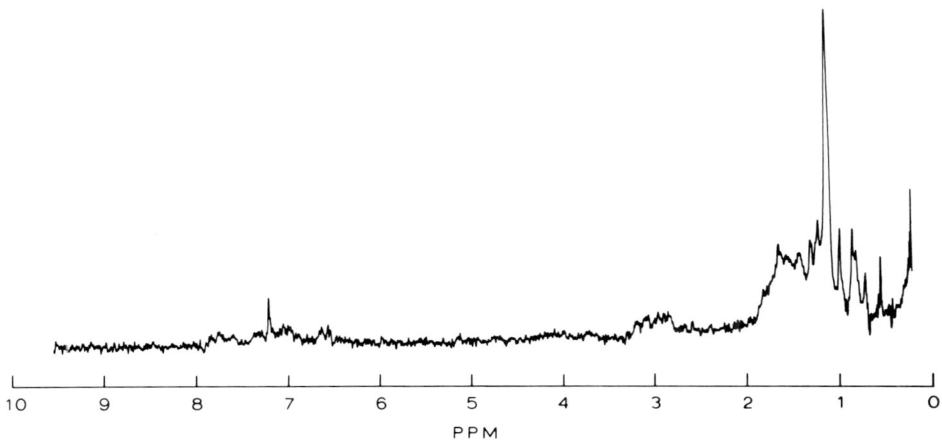

The 100 MHz ^1H NMR spectrum of paspaline in $CDCl_3$/DMSO solution.

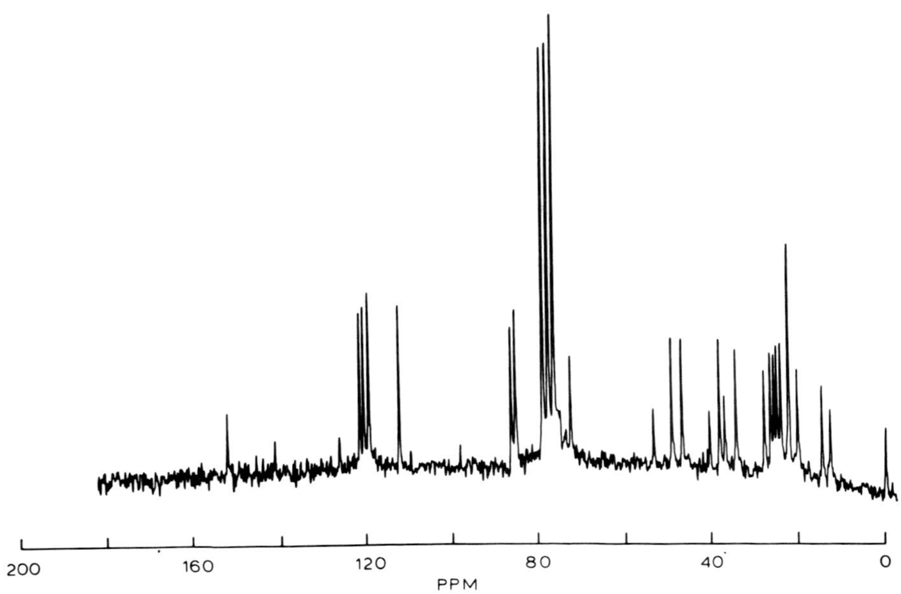

The 25 MHz ^{13}C NMR spectrum of paspaline in $CDCl_3$/DMSO solution.

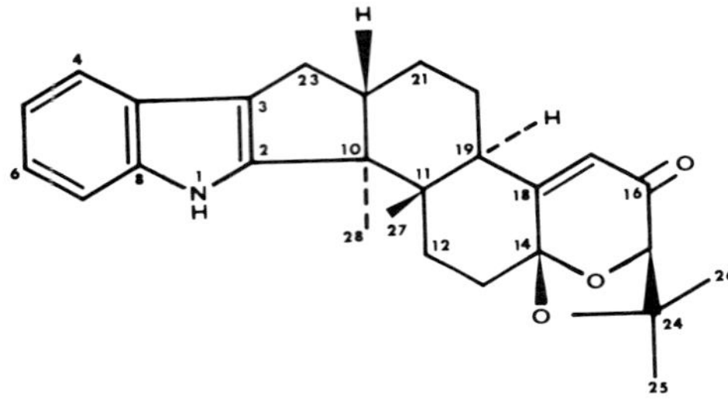

Common name	Paspalicine
Molecular weight	417.2303
Molecular formula	$C_{27}H_{31}O_3N$
General characteristics	Large needles from acetone, mp ~240°C (dec.) $[\alpha]_D$ +173° (C + 0.5) Crystals from methanol, mp ~230°C
UV data[35]	λ_{max} nm (ϵ): 231(47,863), 250(17,782), and 275(12,589)
CD data[35]	$\Delta\epsilon$ +11.8
Fungal source[35]	*Claviceps paspali* Stammes
Toxicity data[17]	Nontremorgenic relative of paspalinine. No clinical signs in mice dosed (IP) up to 500 mg/kg
TLC data[16]	*Adsorbent:* silica gel G-HR *Solvent:* toluene–ethyl acetate–formic acid, 5:4:1 v/v/v R_f: 0.85 *Detection:* dark spot with fluorescent halo under UV light after spraying with 50% ethanolic H_2SO_4 and heating at 100°C for 5 min.
Other significant references	X-Ray structure[83]
^{13}*C NMR data*	C2, 149.4 s; C3, 118.4 s; C4, 118.6 d; C5, 119.9 d; C6, 120.8 d; C7, 111.6 d; C8, 140.2 s; C9, 125.3 s; C10, 51.5 s; C11, 39.9 s; C12, 21.6 t*; C13, 27.7 t*; C14, 104.4 s; C15, 88.4 s; C16, 197.6 s; C17, 118.4 d; C18, 171.8 s; C19, 37.5 d; C20, 28.6 t*; C21, 29.4 t*; C22, 48.9 d; C23, 32.1 t; C24, 78.2 s; C25, 23.8 q; C26, 29.0 q; C27, 23.2 q; C28, 14.9 q

* Assignments uncertain at this time.

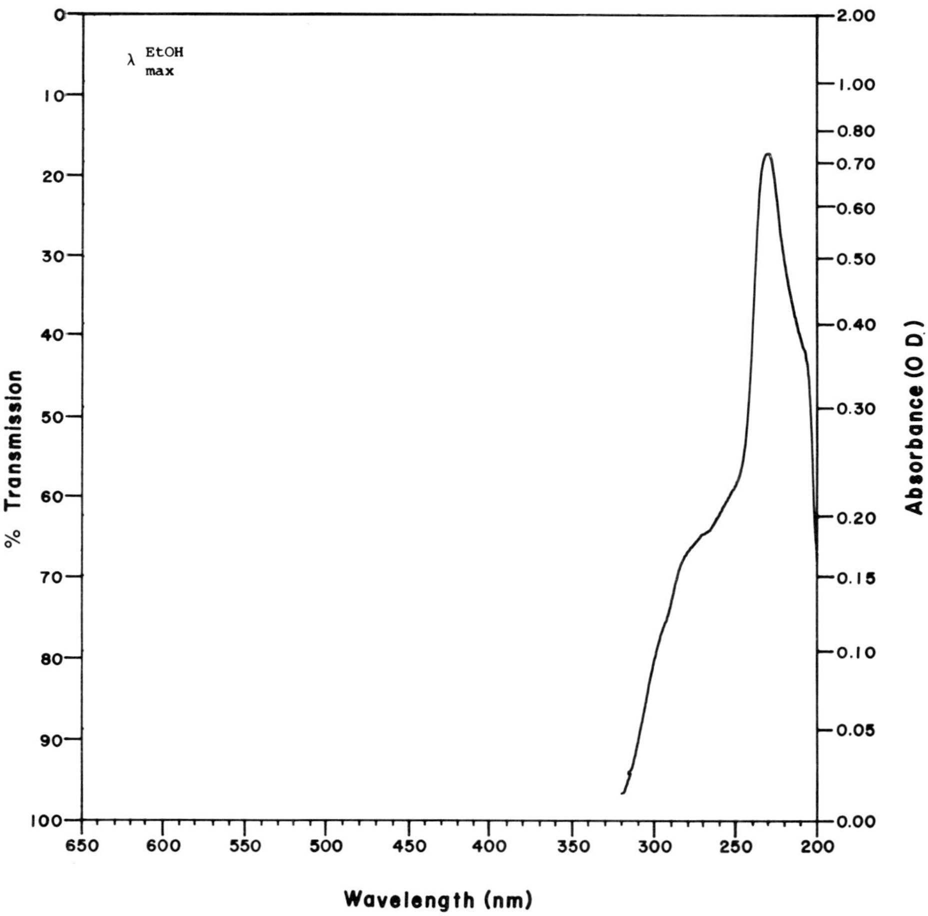
λ EtOH max
% Transmission
Absorbance (O D)
Wavelength (nm)
0
10
20
30
40
50
60
70
80
90
100
2.00
1.00
0.80
0.70
0.60
0.50
0.40
0.30
0.20
0.15
0.10
0.05
0.00
650
600
550
500
450
400
350
300
250
200

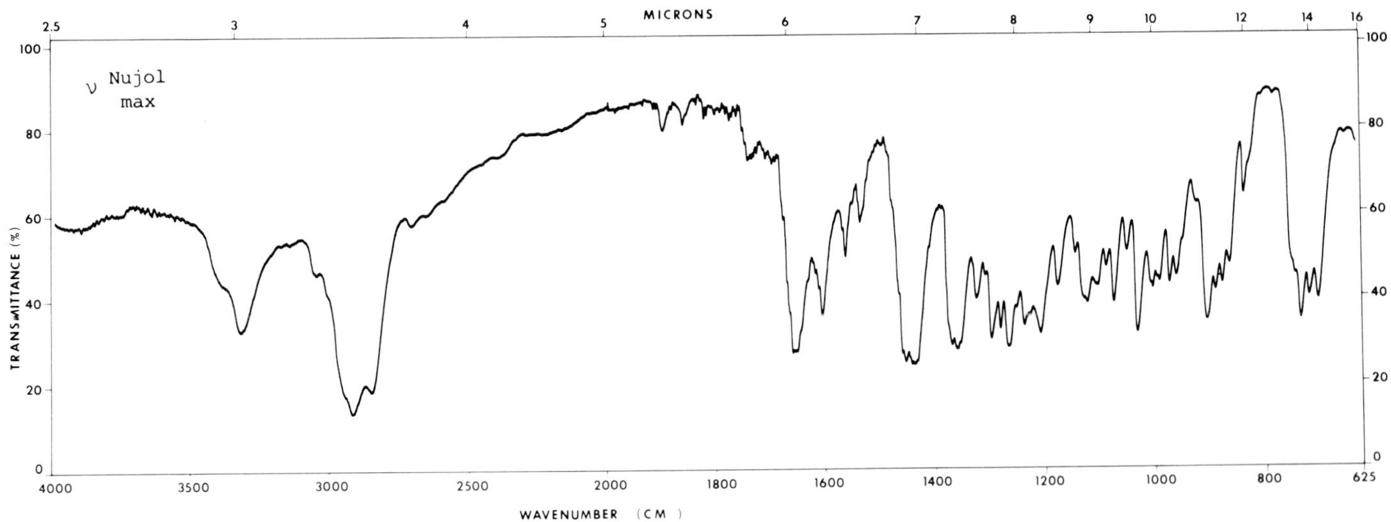
MICRONS
ν Nujol max
TRANSMITTANCE (%)
WAVENUMBER (CM)

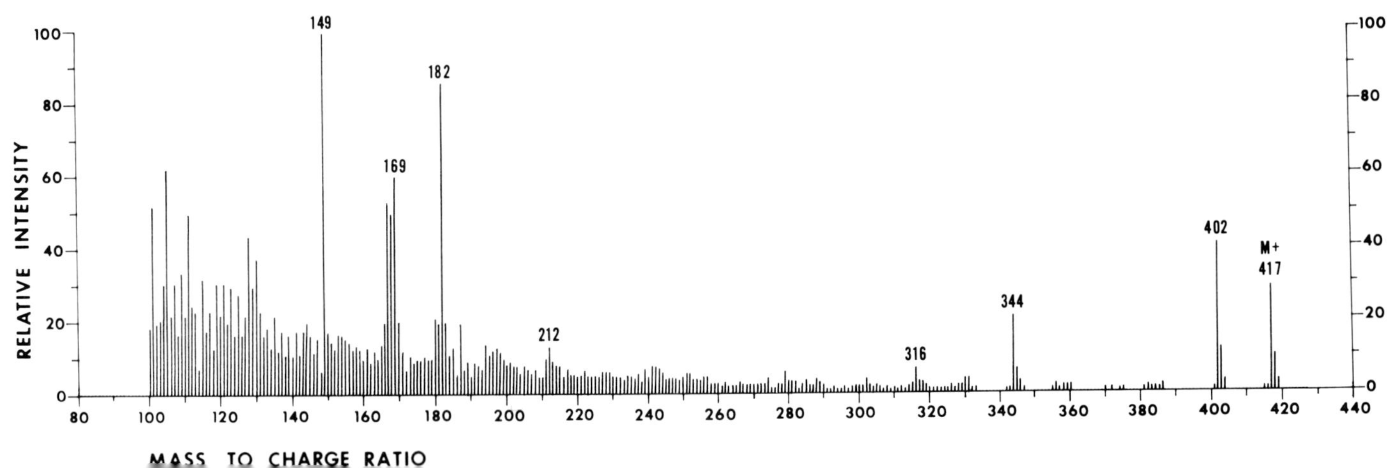
149
182
169
212
316
344
402
M+
417
RELATIVE INTENSITY
MASS TO CHARGE RATIO

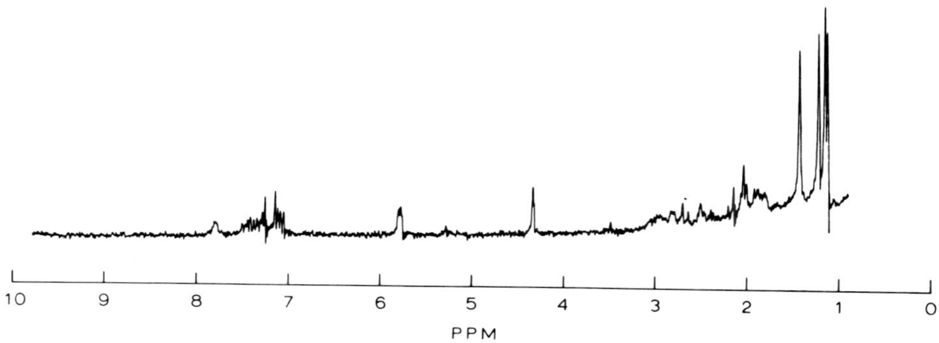

The 100 MHz ^1H NMR spectrum of paspalicine in $CDCl_3$ solution.

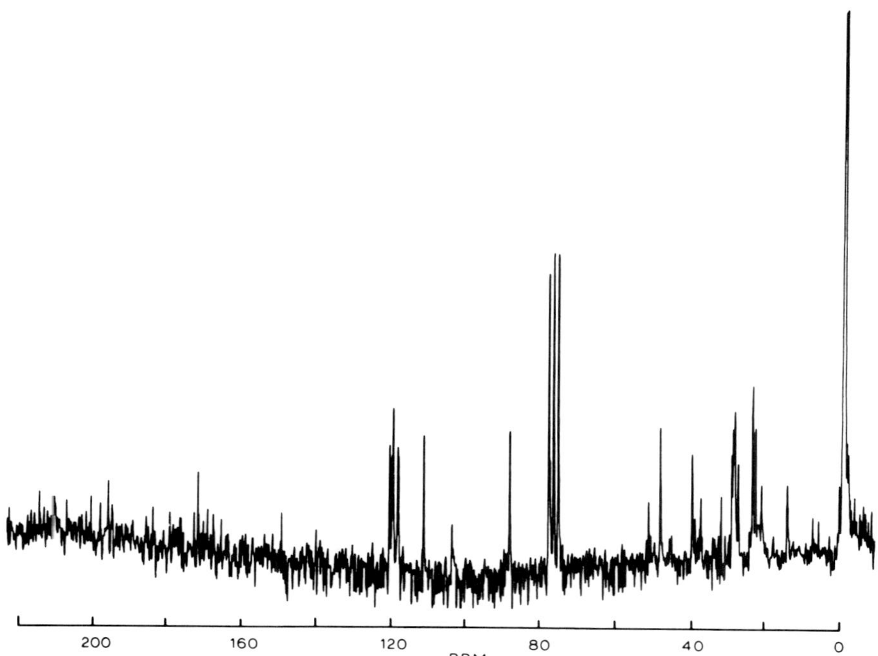

The 25 MHz ^{13}C NMR spectrum of paspalicine in $CDCl_3$ solution.

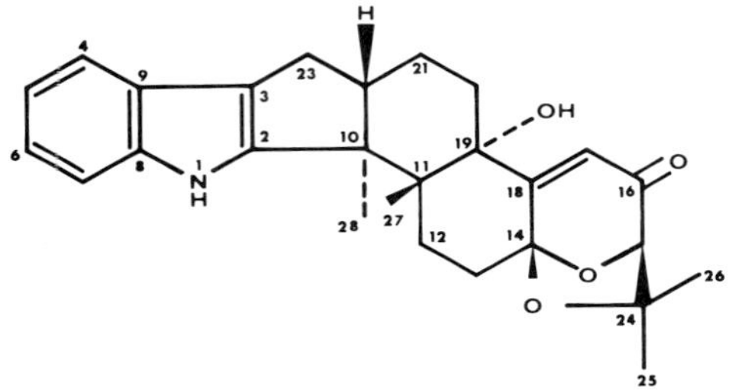

Common name[4,12,37]	Paspalinine
Molecular weight	433.2252
Molecular formula	$C_{27}H_{31}O_4N$
UV data[19]	λ_{max}^{MeOH} nm (ϵ): 232(39,800) and 275(10,500)
Fungal source[4]	*Claviceps paspali* Stammes sclerotia
Toxicity data[19]	ED_{50} for tremor > 14 mg/kg IP in mice. Other clinical signs: uncoordinated movement, diarrhea, rough hair coat and hypersensitivity to sound stimuli. Gross clinical signs very similar to those produced by paxilline.
TLC data[19]	*Adsorbent:* silica gel G-HR *Solvent:* chloroform–acetone, 93:7 v/v R_f: 0.52 *Detection:* blue spot after spraying with 50% ethanolic H_2SO_4 and heating at 100°C for 5 min; fluorescent under UV light.
Other significant references	Review of paspalum staggers syndrome[29,41,60]
^{13}*C NMR data*	C2, 152.1 s; C3, 117.1 s; C4, 118.4 d; C5, 119.5 d; C6, 120.3 d; C7, 111.5 d; C8, 139.9 s; C9, 125.2; C10, 51.5 s; C11, 39.9 s; C12, 21.1 t*; C13, 26.3 t; C14, 104.5 s; C15, 87.9 s; C16, 197.2 s; C17, 117.5 d; C18, 169.9 s; C19, 77.4 s; C20, 27.4 t*; C21, 28.2 t*; C22, 48.6 d; C23, 32.8 t; C24, 78.6 s; C25, 23.5 q; C26, 28.8 q; C27, 23.0 q; C28, 16.3 q

* Assignments uncertain at this time.

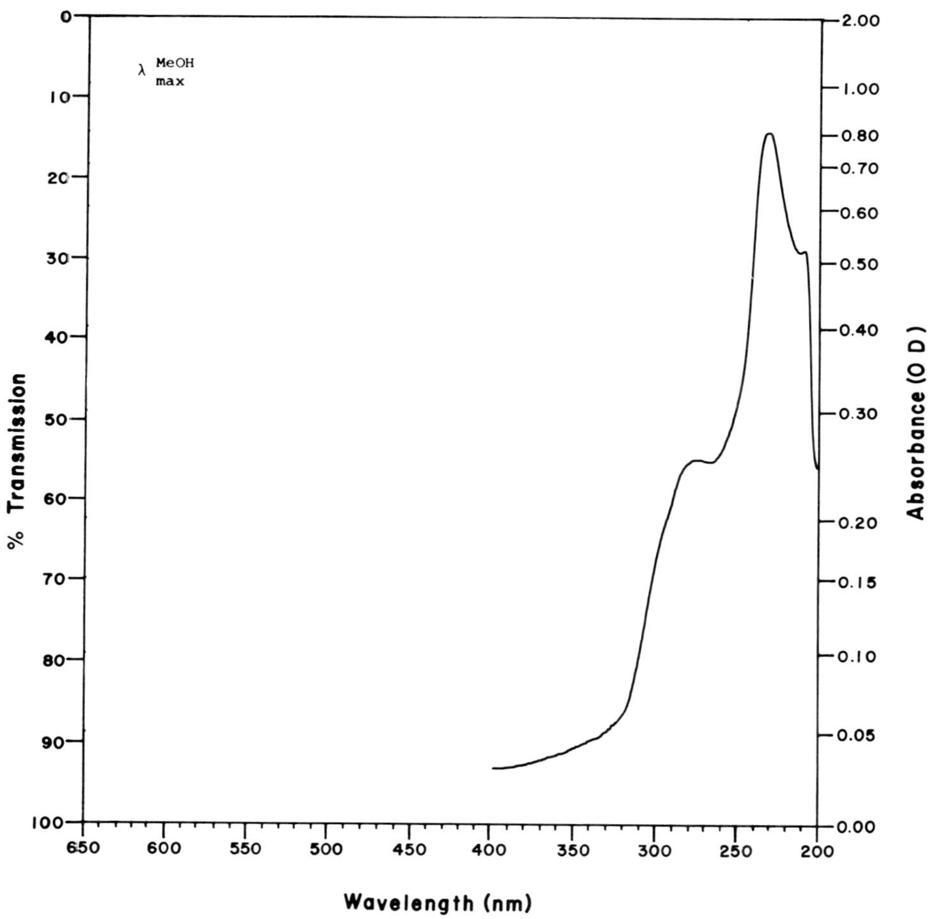

λ MeOH max
% Transmission
Absorbance (O D)
Wavelength (nm)
0
10
20
30
40
50
60
70
80
90
100
2.00
1.00
0.80
0.70
0.60
0.50
0.40
0.30
0.20
0.15
0.10
0.05
0.00
650
600
550
500
450
400
350
300
250
200

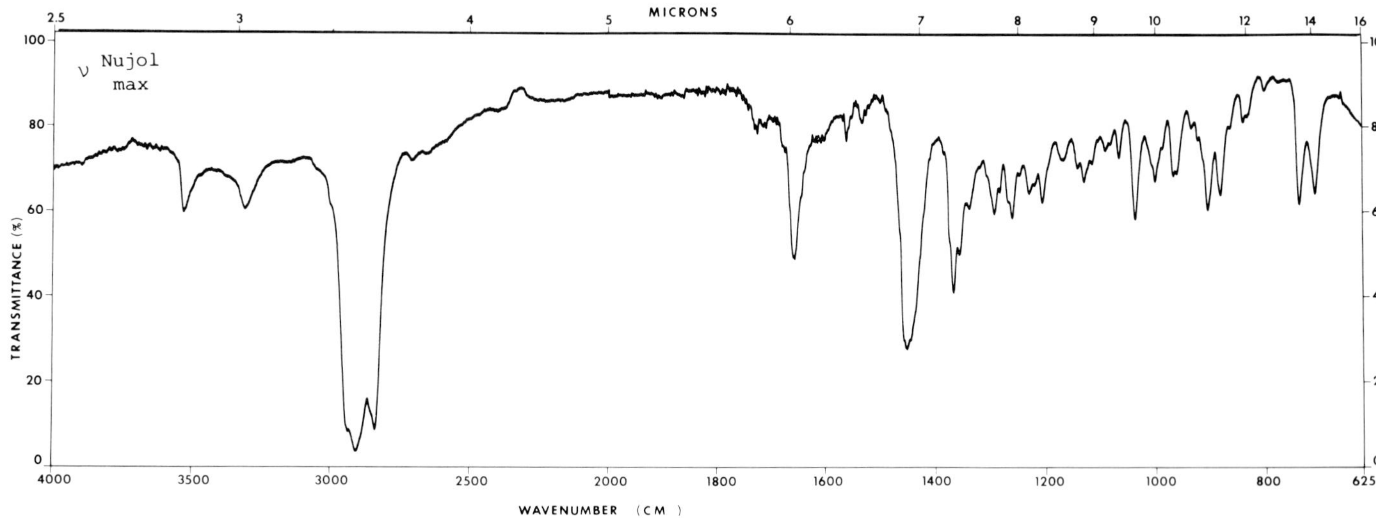

MICRONS
ν_{max}^{Nujol}
TRANSMITTANCE (%)
WAVENUMBER (CM)

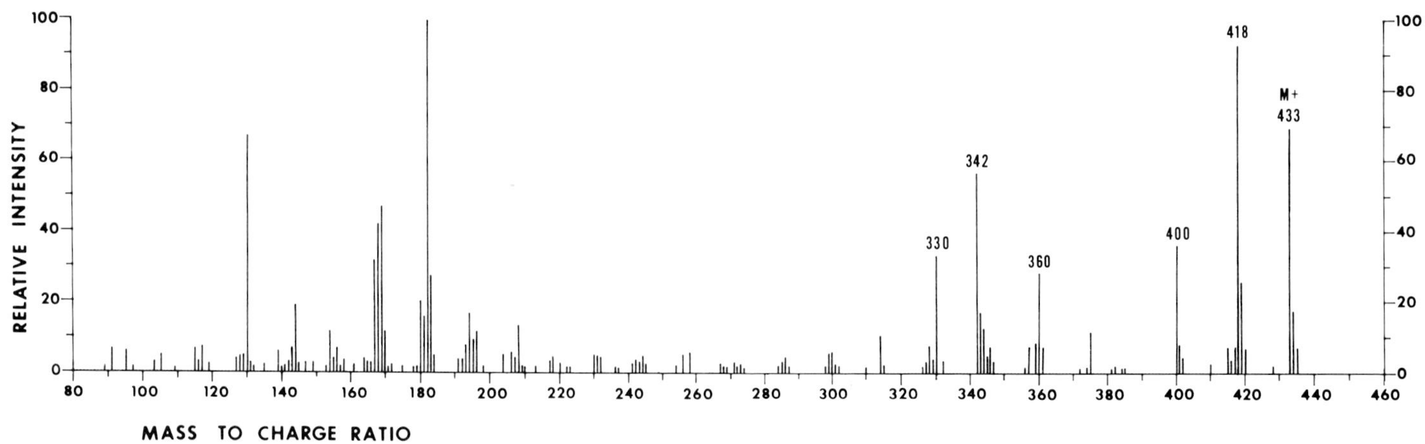

RELATIVE INTENSITY
MASS TO CHARGE RATIO
330
342
360
400
418
M+
433

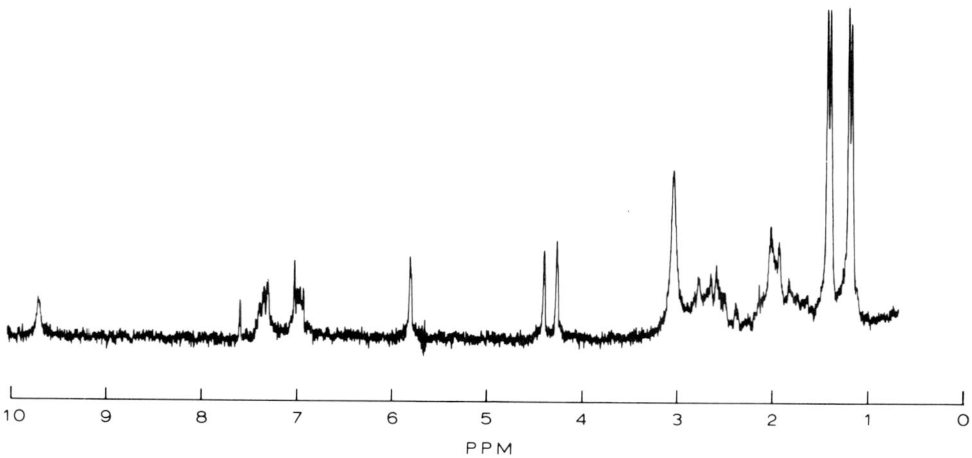

The 100 MHz ^{1}H NMR spectrum of paspalinine in $CDCl_3$ solution.

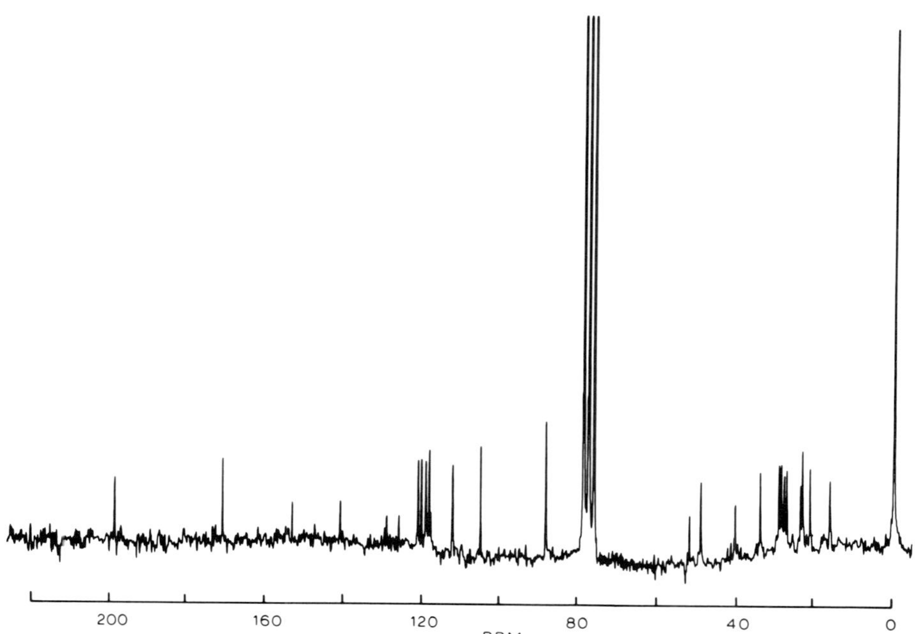

The 25 MHz ^{13}C NMR spectrum of paspalinine in $CDCl_3$ solution.

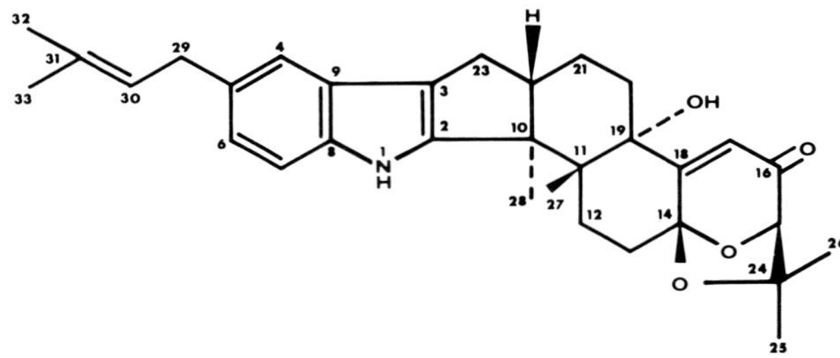

Common name[36]	Paspalitrem A (3-Methyl-2-butenylpaspalinine)
Molecular weight	501.2879
Molecular formula	$C_{32}H_{39}O_4N$
UV data[22]	λ_{max}^{MeOH} nm: 234 and 278 (ϵ_{max} not reported)
Fungal source[22]	*Claviceps paspali* Stammes sclerotia
Toxicity data[22]	ED_{50} in mice dosed IP was < 14.0 mg/kg. Clinical signs were intermittent tremors, diarrhea, rough hair coat, and hypersensitivity to sound stimuli. Gross clinical signs very similar to those produced by paxilline.
TLC data[22]	*Adsorbent:* silica gel G-HR *Solvent:* chloroform–acetone, 93:7 v/v R_f: 0.60 *Detection:* blue spot after spraying with 50% ethanolic H_2SO_4 and heating at 100°C for 5 min; fluorescent under UV light.
^{13}*C NMR data*	C2, 151.2 s; C3, 116.7 s; C4, 117.6 d; C5, 127.9 s; C6, 120.9 d; C7, 109.4 d; C8, 139.8 s; C9, 124.6 s; C10, 51.2 s; C11, 39.9 s; C12, 21.1 t*; C13, 27.0 t*; C14, 104.4 s; C15, 88.0 s; C16, 197.3 s; C17, 117.6 d; C18, 169.8 s; C19, 77.6 s; C20, 28.3 t*; C21, 29.4 t*; C22, 48.7 d; C23, 33.8 t; C24, 78.7 s; C25, 23.6 q; C26, 28.9 q; C27, 23.1 q; C28, 16.3 q; C29, 32.0 t; C30, 123.7 d; C31, 133.0 s; C32, 18.0 q; C33, 25.8 q

* Assignments uncertain at this time.

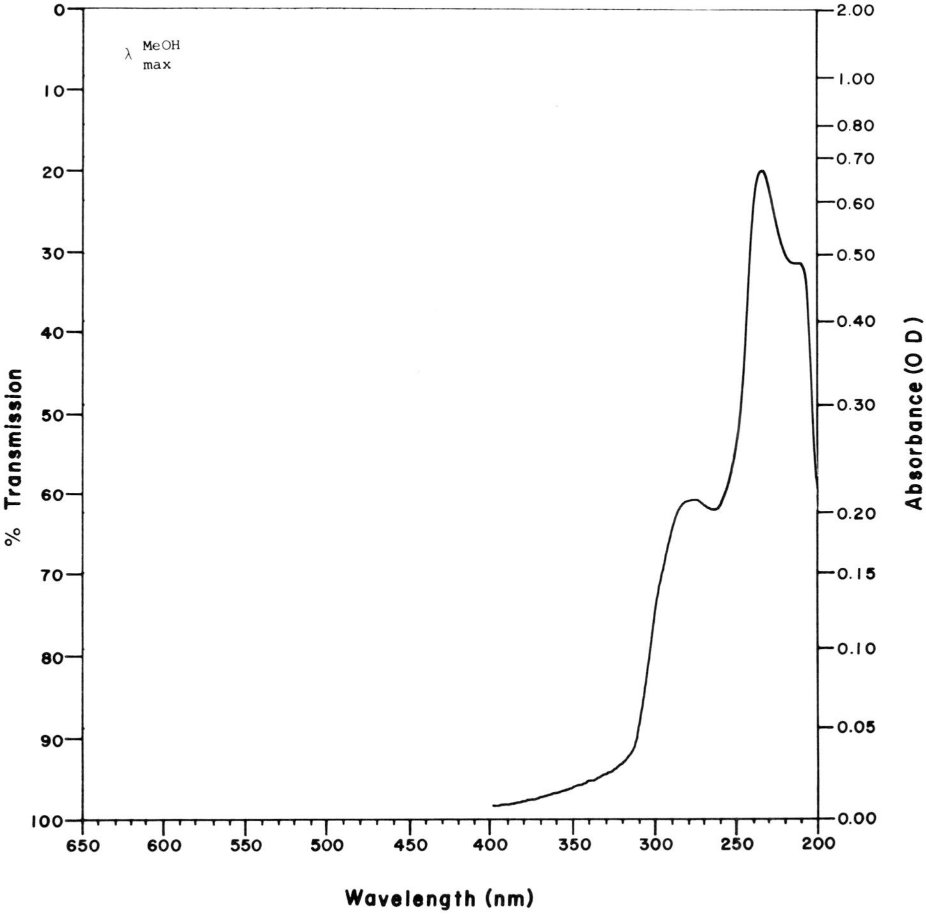

λ MeOH max
% Transmission
0
10
20
30
40
50
60
70
80
90
100
Absorbance (O D)
2.00
1.00
0.80
0.70
0.60
0.50
0.40
0.30
0.20
0.15
0.10
0.05
0.00
650
600
550
500
450
400
350
300
250
200
Wavelength (nm)

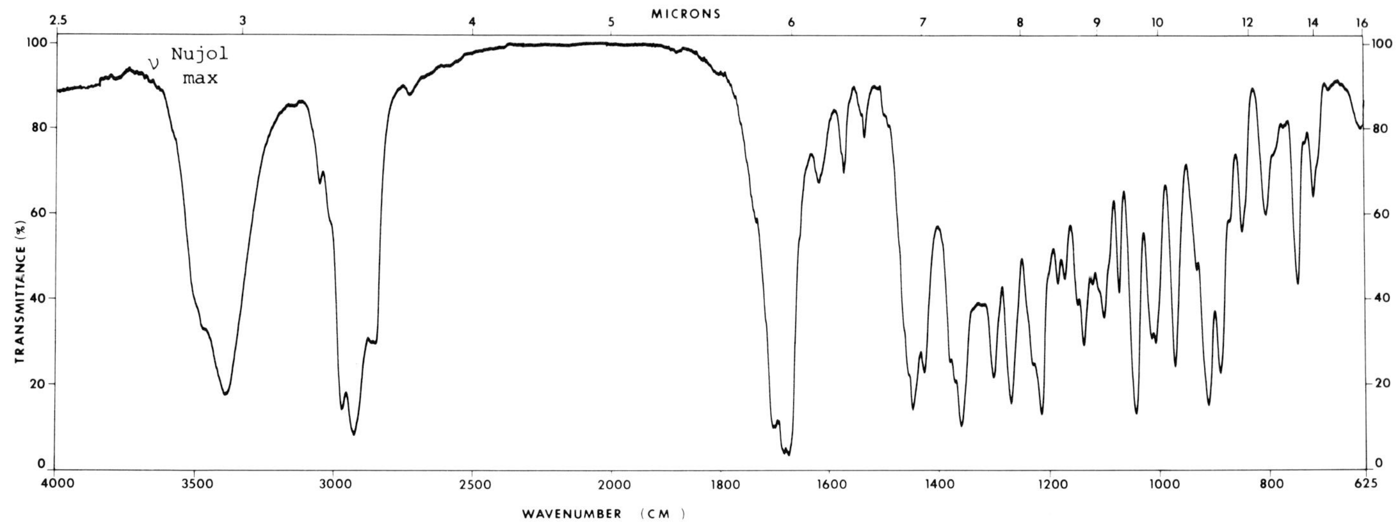
MICRONS
ν Nujol max
TRANSMITTANCE (%)
WAVENUMBER (CM)

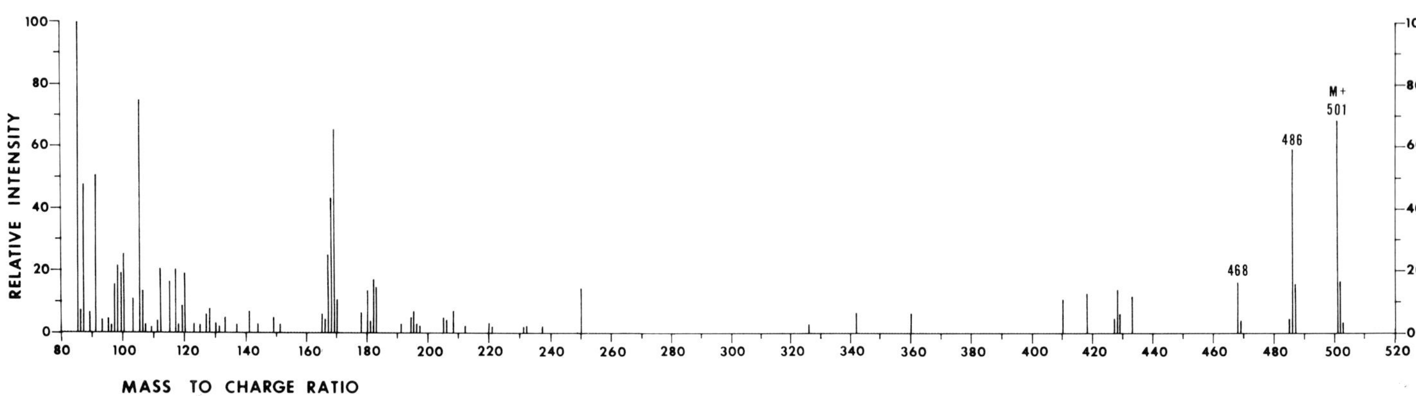
RELATIVE INTENSITY
M+ 501
486
468
MASS TO CHARGE RATIO

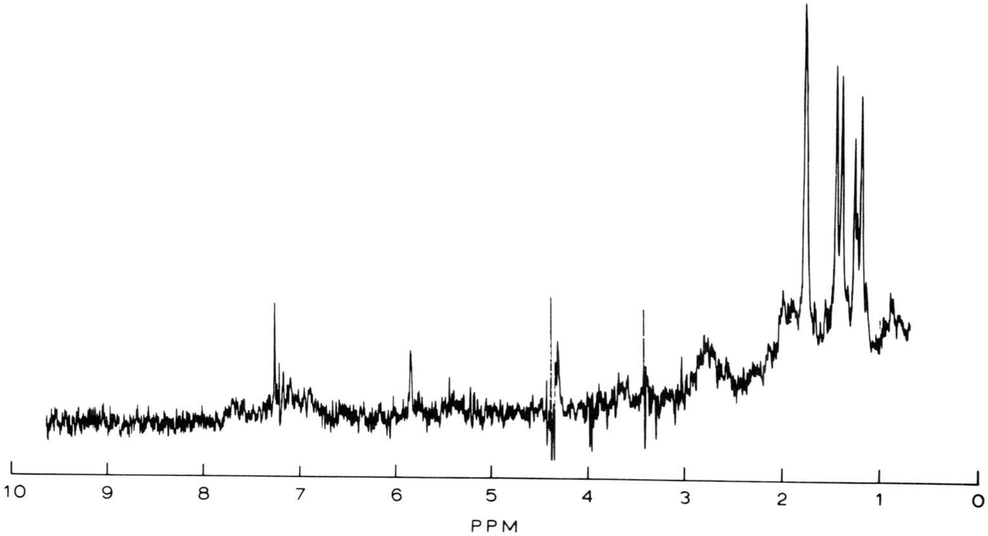

The 100 MHz ^{1}H NMR spectrum of paspalitrem A in $CDCl_3$ solution.

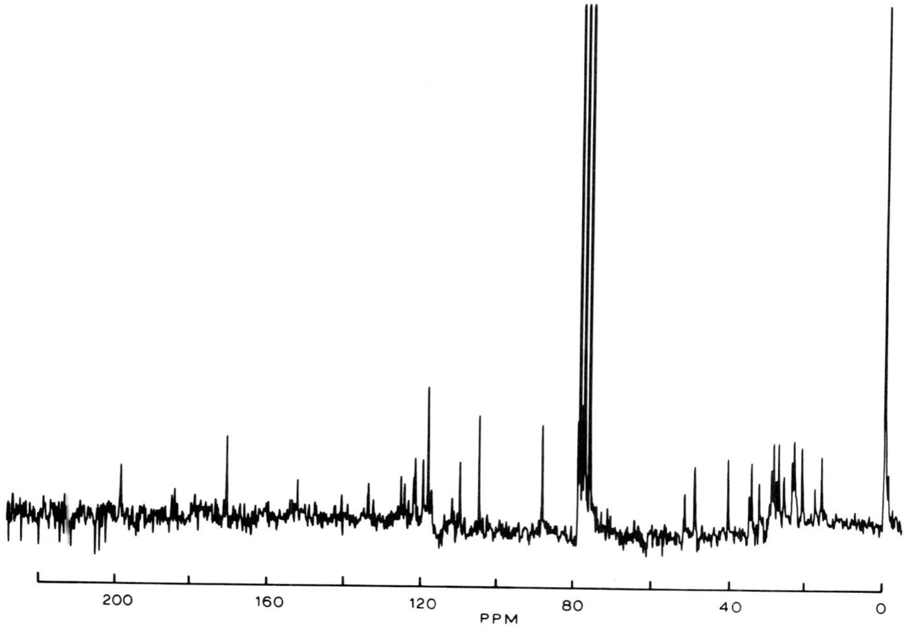

The 25 MHz ^{13}C NMR spectrum of paspalitrem A in $CDCl_3$ solution.

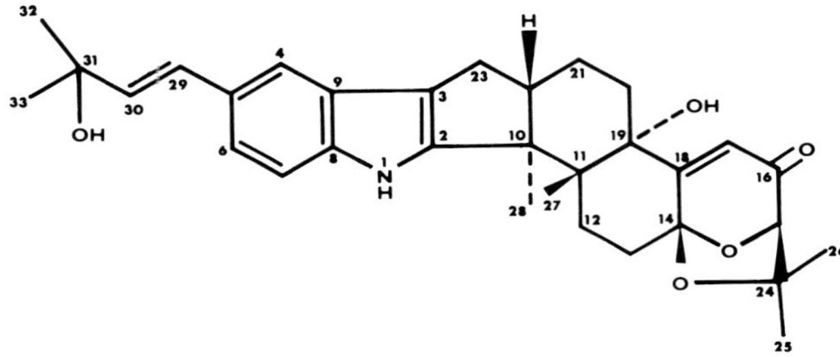

Common name[36]	Paspalitrem B (3-Methyl-3-hydroxy-1-butenyl-paspalinine)
Molecular weight	517.2827
Molecular formula	$C_{32}H_{39}O_5N$
General characteristics[22]	Paspalitrem B readily forms a methylated derivative in methanol solution and acidic conditions. The methylation occurs with the tertiary hydroxyl group located on the 3-methyl-3-hydroxy-1-butenyl moiety.
UV data[22]	λ_{max}^{MeOH} nm: 227, 248, 305, and 336 (ϵ_{max} not reported)
Fungal source[22]	*Claviceps paspali* Stammes sclerotia
Toxicity data[22]	ED_{50} for tremorgenic activity was < 14 mg/kg in mice dosed IP; other clinical signs were diarrhea, rough hair coat, and hypersensitivity to sound stimuli. Gross clinical signs very similar to those produced by paxilline.
TLC data[22]	*Adsorbent:* silica gel G-HR *Solvent:* chloroform–acetone, 93:7 v/v R_f: 0.20 *Detection:* spray with 50% ethanolic H_2SO_4 and heat at 100°C for 5 min. Blue-green spot after spraying; turns blue after heating; fluorescent under UV light.
^{13}*C NMR data*	C2, 152.3 s; C3, 116.4 s; C4, 116.0 d; C5, 128.6 s; C6, 120.9 d; C7, 110.5 d; C8, 140.3 s; C9, 124.2 s; C10, 51.2 s; C11, 39.9 d; C12, 21.1 t*; C13, 27.0 t*;

* Assignments uncertain at this time.

C14, 104.4 s; C15, 87.9 s; C16, 197.3 s; C17, 117.6 d; C18, 169.7 s; C19, 77.5 s; C20, 28.3 t*; C21, 29.9 t*; C22, 48.6 d; C23, 33.8 t; C24, 78.7 s; C25, 23.6 q; C26, 28.8 q; C27, 23.0 q; C28, 16.3 q; C29, 124.7 d; C30, 137.5 d; C31, 71.2 s; C32, 29.9 q; C33, 29.9 q

* Assignments uncertain at this time.

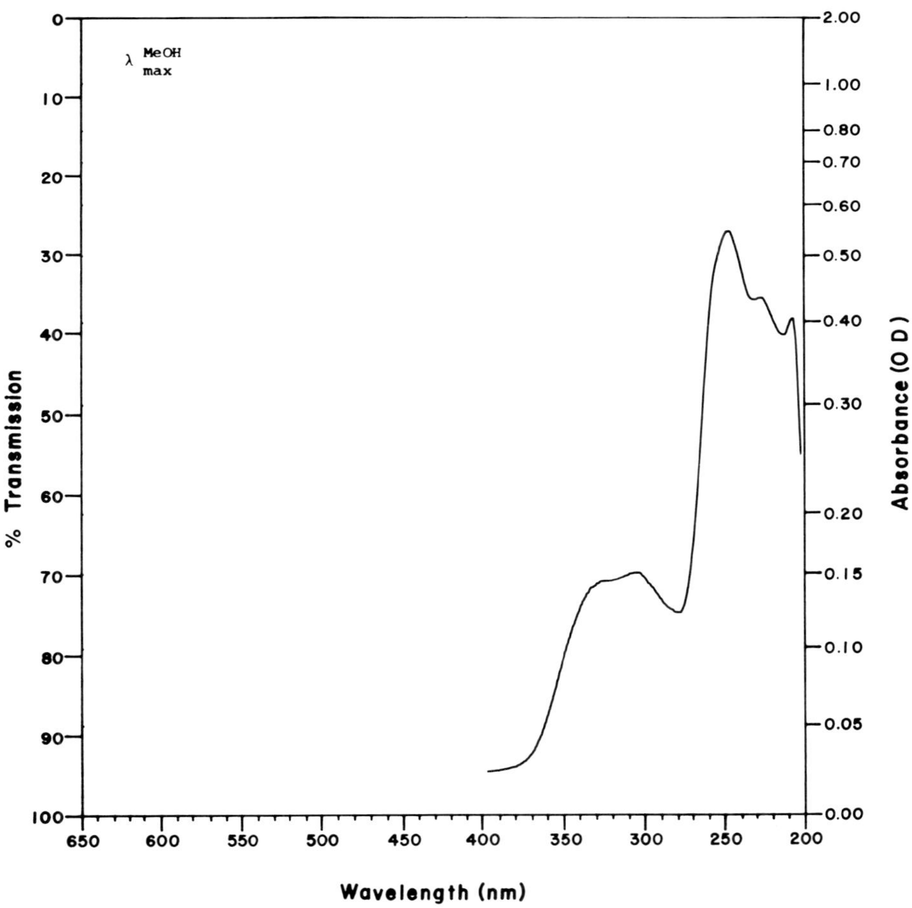

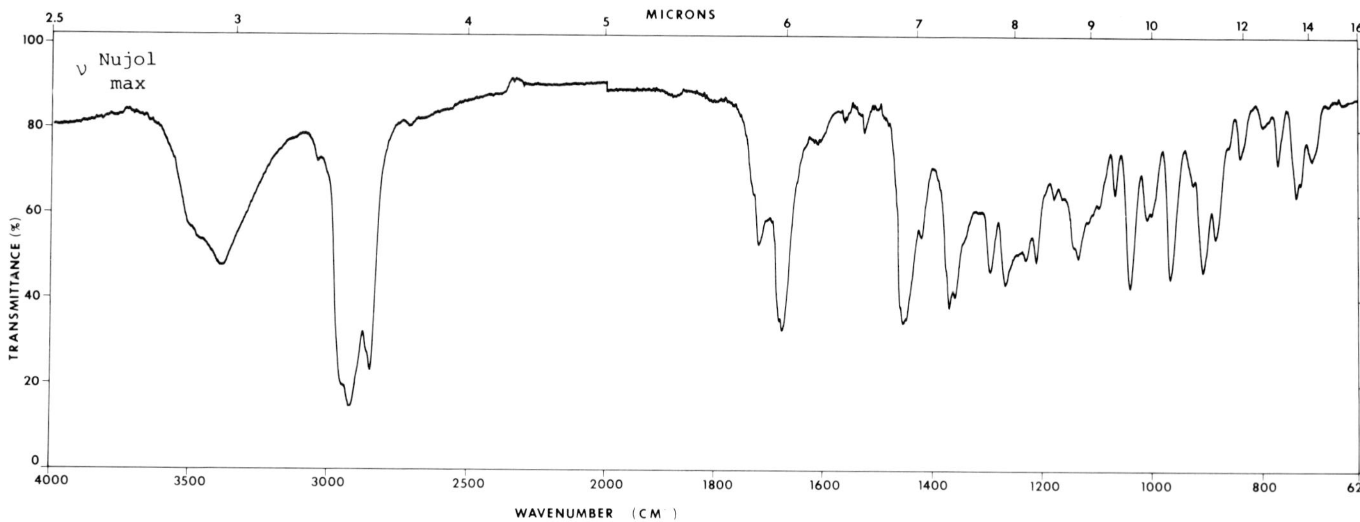

MICRONS
ν Nujol max
TRANSMITTANCE (%)
WAVENUMBER (CM⁻¹)

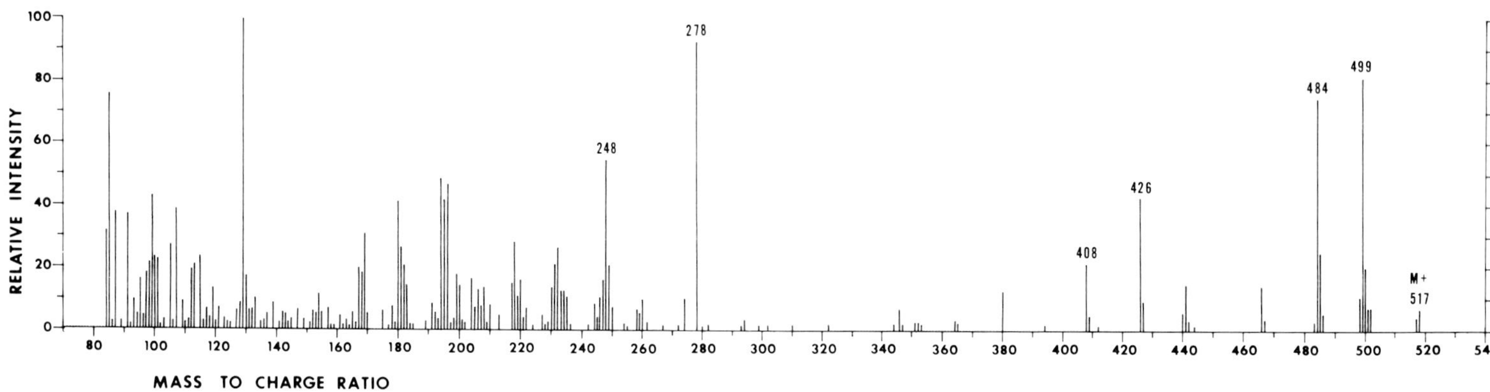

278
248
408
426
484
499
M+ 517
RELATIVE INTENSITY
MASS TO CHARGE RATIO

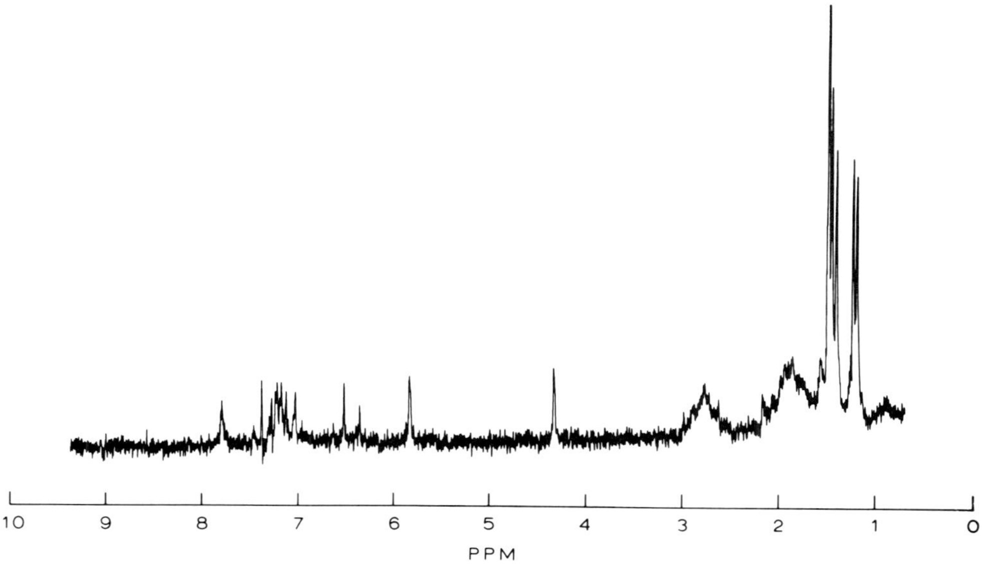

The 100 MHz ^{1}H NMR spectrum of paspalitrem B in $CDCl_3$ solution.

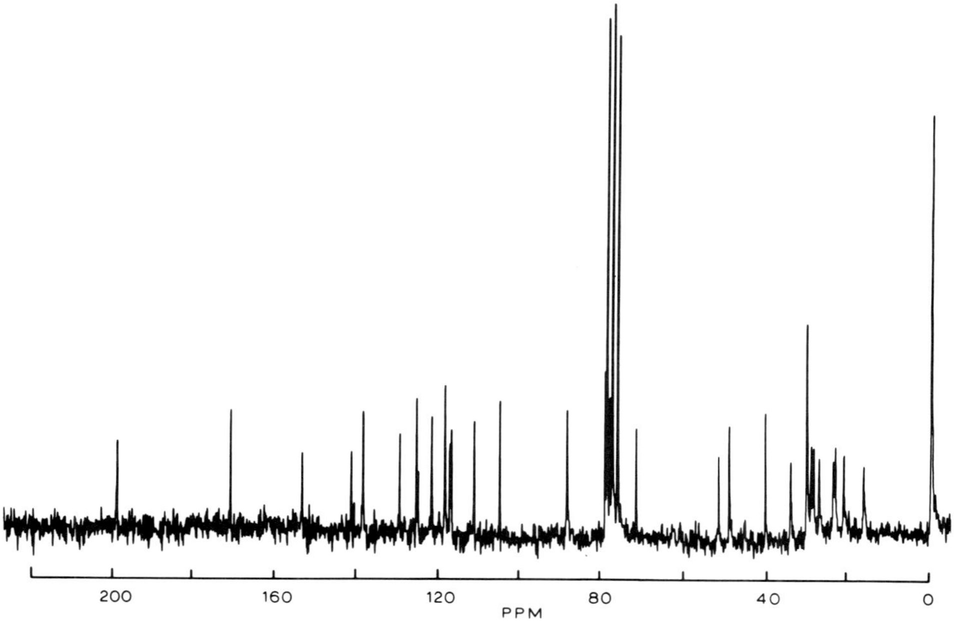

The 25 MHz ^{13}C NMR spectrum of paspalitrem B in $CDCl_3$ solution.

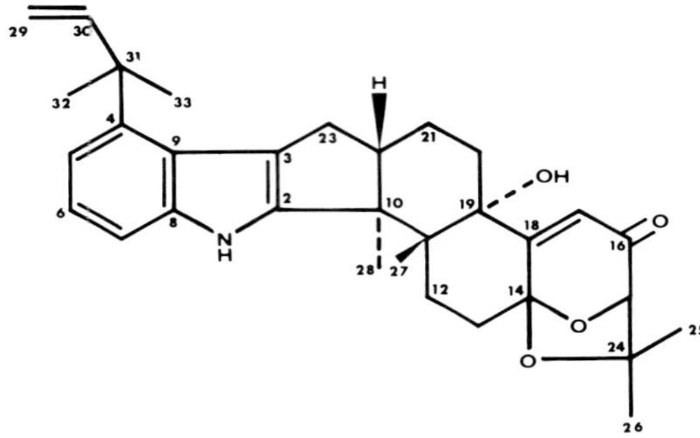

Common name[36,95]	Aflatrem (α,α-Dimethylallyl paspalinine)
Molecular weight[36]	501.2879
Molecular formula	$C_{32}H_{39}O_4N$
General characteristics[16]	Colorless needles from ethyl ether–acetone, mp 222°–224°C
UV data[36]	λ_{max}^{EtOH} nm (ε): 231(27,750), 282(9,050), and 292(sh)(7,850)
Fungal source[95]	*Aspergillus flavus*
Toxicity data[95]	Tremorgenic when dosed IP or orally to mice, guinea pigs, and rats. A dose of 1 mg of partially purified toxin to mice via stomach tube initially resulted in inactivity but animals responded to auditory and tactile stimuli. Tremors became marked when animals attempted to move. Trembling became more pronounced in 1–2 hr after dosing and continued for several hours. Marked improvement was usually evident the day after treatment with the toxin.
TLC data[16]	*Adsorbent:* silica gel G-HR *Solvent:* chloroform–acetone, 93:7 v/v R_f: 0.65 *Detection:* blue spot after spraying with H_2SO_4 followed by 1% dimethyl aminobenzaldehyde (ethanolic) and heating at 100°C for 1–2 min.
^{13}C NMR data	C2, 151.8 s; C3, 115.1 s; C4, 140.6 s; C5, 115.8 d; C6, 118.9 d; C7, 110.6 d; C8, 139.3 s; C9, 123.3 s; C10, 50.5 s; C11, 39.4 s; C12, 21.0 t; C13, 26.4 t;

C14, 104.5 s; C15, 87.6 d; C16, 197.0 s; C17, 116.9 d; C18, 170.0 s; C19, 77.6 s; C20, 28.2 t; C21, 32.6 t; C22, 48.0 d; C23, 33.8 t; C24, 78.0 s; C25, 28.8 q; C26, 22.9 q; C27, 22.9 q; C28, 16.0 q; C29, 111.1 t; C30, 140.6 d; C31, 41.3 s; C32, 29.3 q; C33, 29.5 q

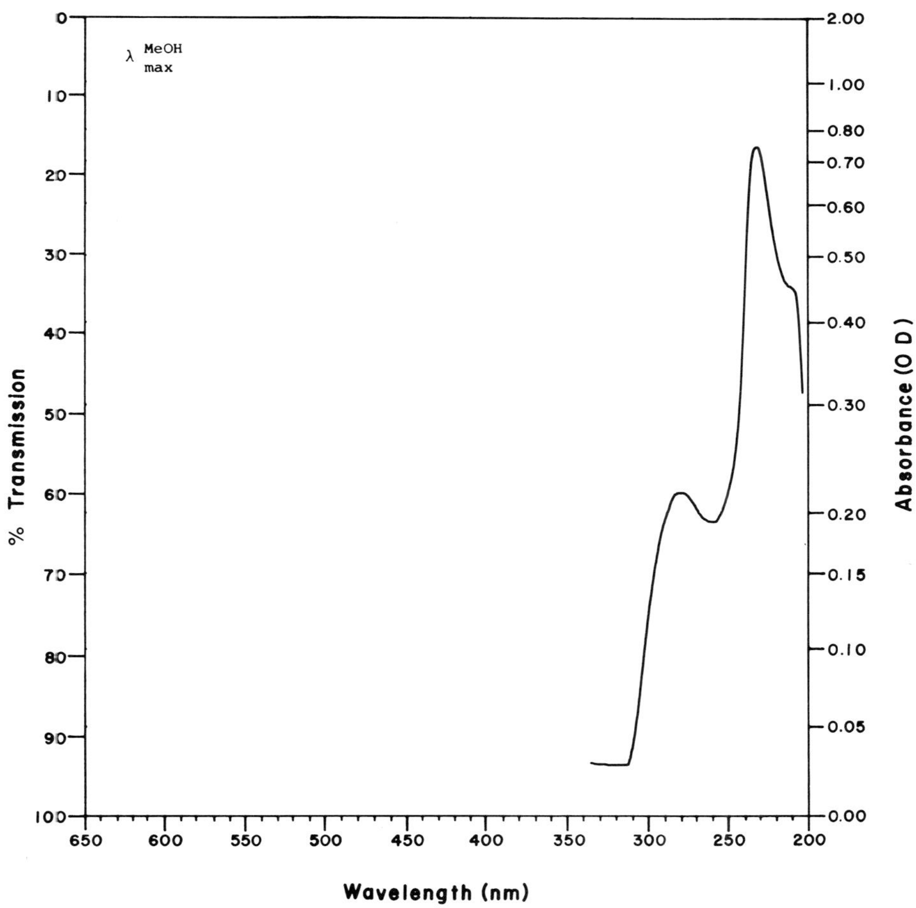

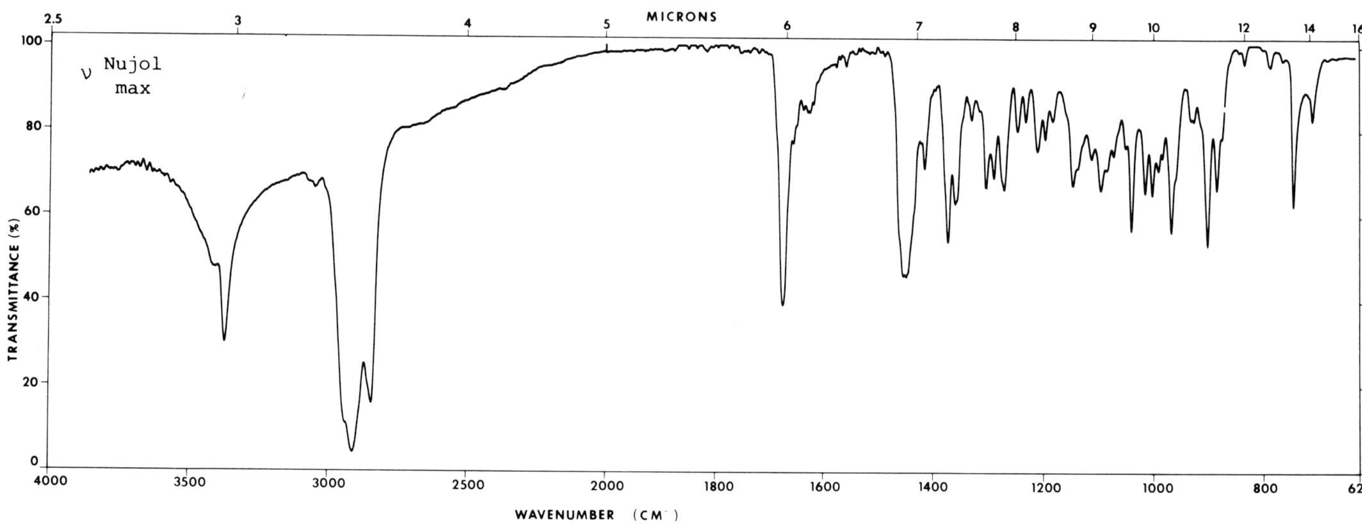

MICRONS
2.5
3
4
5
6
7
8
9
10
12
14
16
TRANSMITTANCE (%)
100
80
60
40
20
0
WAVENUMBER (CM⁻¹)
4000
3500
3000
2500
2000
1800
1600
1400
1200
1000
800
625
ν_{max}^{Nujol}

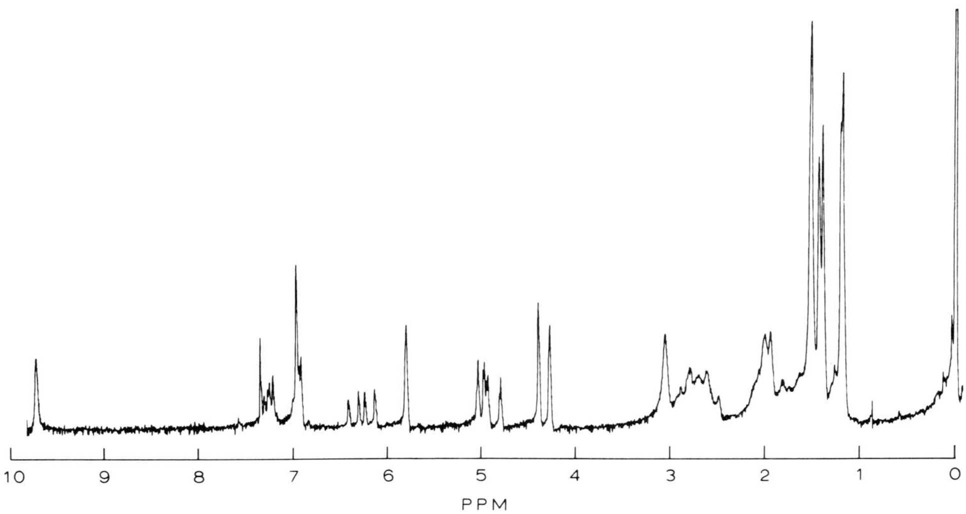

The 100 MHz ^1H NMR spectrum of aflatrem in $CDCl_3$/DMSO solution.

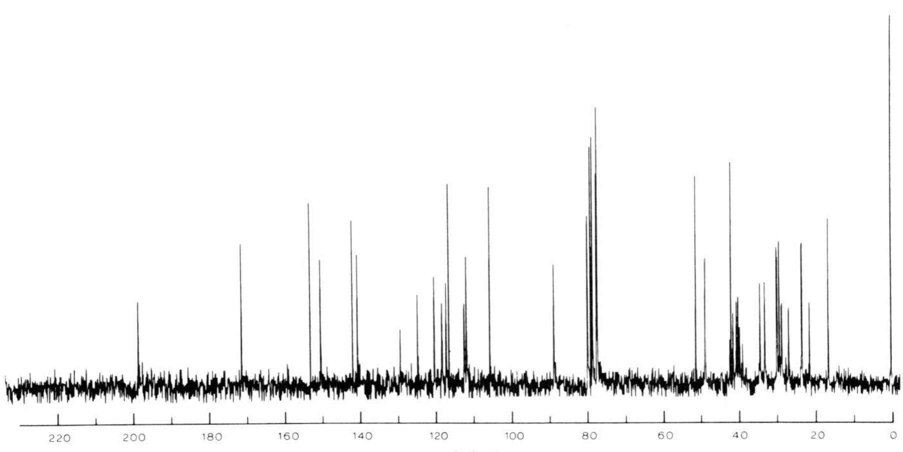

The 25 MHz ^{13}C NMR spectrum of aflatrem in $CDCl_3$/DMSO solution.

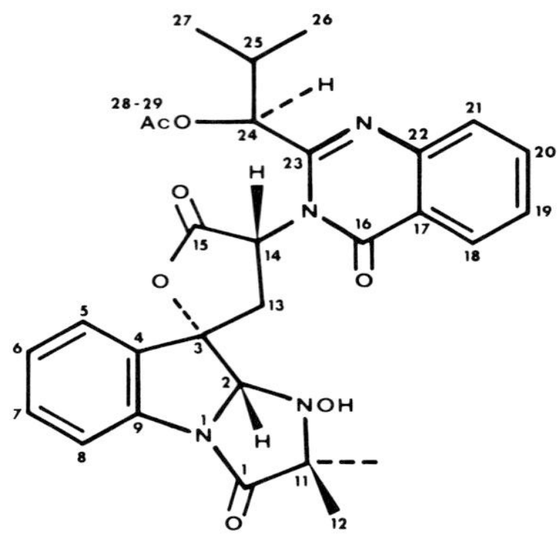

Common name[68b]	Tryptoquivaline (Tryptoquivaline A and C)
Molecular weight[6,14]	546.2114
Molecular formula	$C_{29}H_{30}O_7N_4$
General characteristics[6,14,84]	Colorless prisms from dichloromethane–hexane, mp 155°–157°C $[\alpha]_D^{25}$ 130° (C = 0.22 in chloroform)
UV data[6]	λ_{max}^{EtOH} nm (ϵ): 228(42,200), 232(sh)(40,200), 252(sh)(19,500), 268(10,900), 279(9,500), 307(3,700), and 319(3,000)
CD data[6]	$\Delta\epsilon$ (in 95% ethanol) nm: 258(−5.3), 282(0), 305(2.2), and 327(0)
Fungal source[6,14]	*Aspergillus clavatus* (NRRL 5890)
Toxicity data[42]	Tryptoquivaline (dosed IP, up to 50 μl DMSO solution) induced persistent tremors in rats.
TLC data[14]	*Adsorbent:* silica gel GF_{254} *Solvent:* benzene–acetone, 9:1 v/v R_f: not reported *Detection:* not reported
^1H NMR data[6]	($CDCl_3$) H2, 5.00; H5–8, 7.12–7.90; H11, 1.50; H12, 1.52; H13a, 3.10 (J = 10); H13b, 3.15 (J = 10); H14, 5.70 (J = 10); H18, 8.22 (J = 8); H19–21, 7.12–7.90; H24, 5.61 (J = 9); H25, 2.63; H26, 1.03; H27, 1.17; CH_3CO, 2.19; NOH, 3.63,4.04

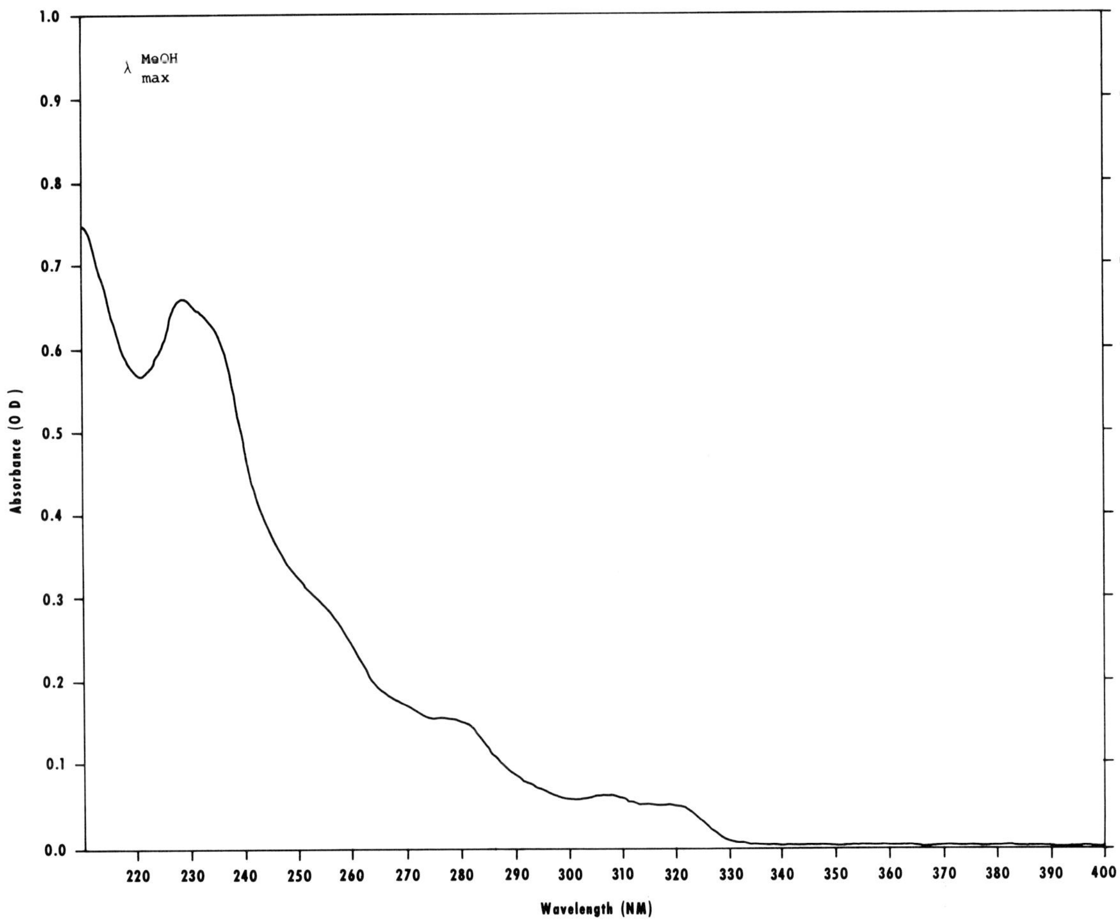
λ MeOH max
Absorbance (O D)
Wavelength (NM)
1.0
0.9
0.8
0.7
0.6
0.5
0.4
0.3
0.2
0.1
0.0
220
230
240
250
260
270
280
290
300
310
320
330
340
350
360
370
380
390
400

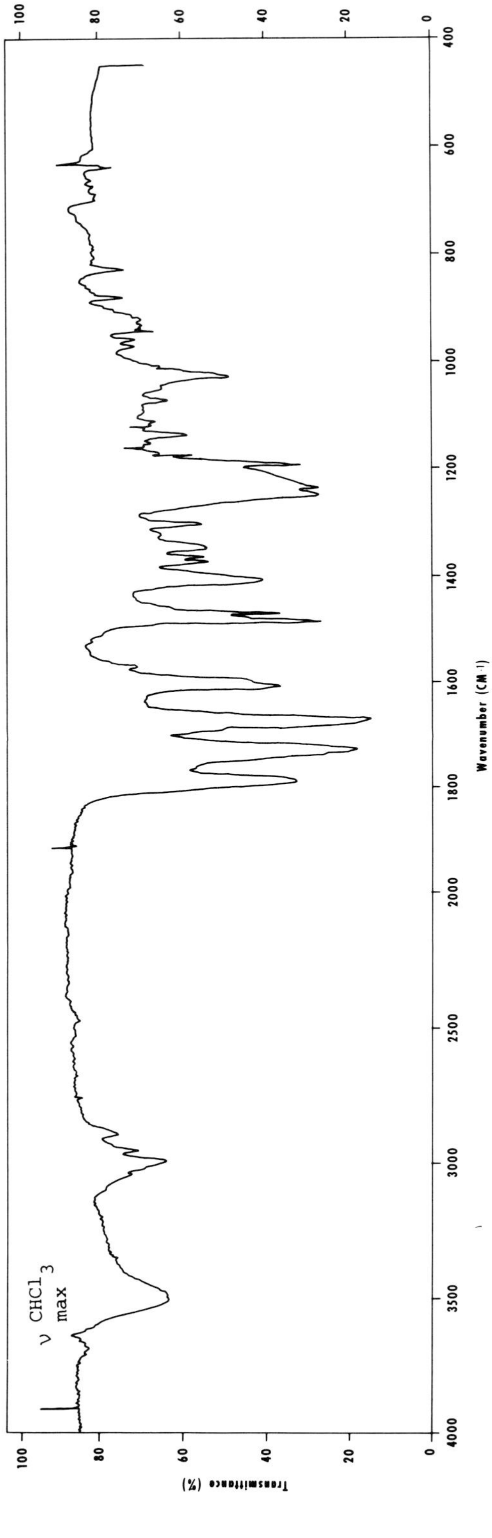

ν CHCl3 max
Transmittance (%)
Wavenumber (CM-1)
100
80
60
40
20
0
4000
3500
3000
2500
2000
1800
1600
1400
1200
1000
800
600
400

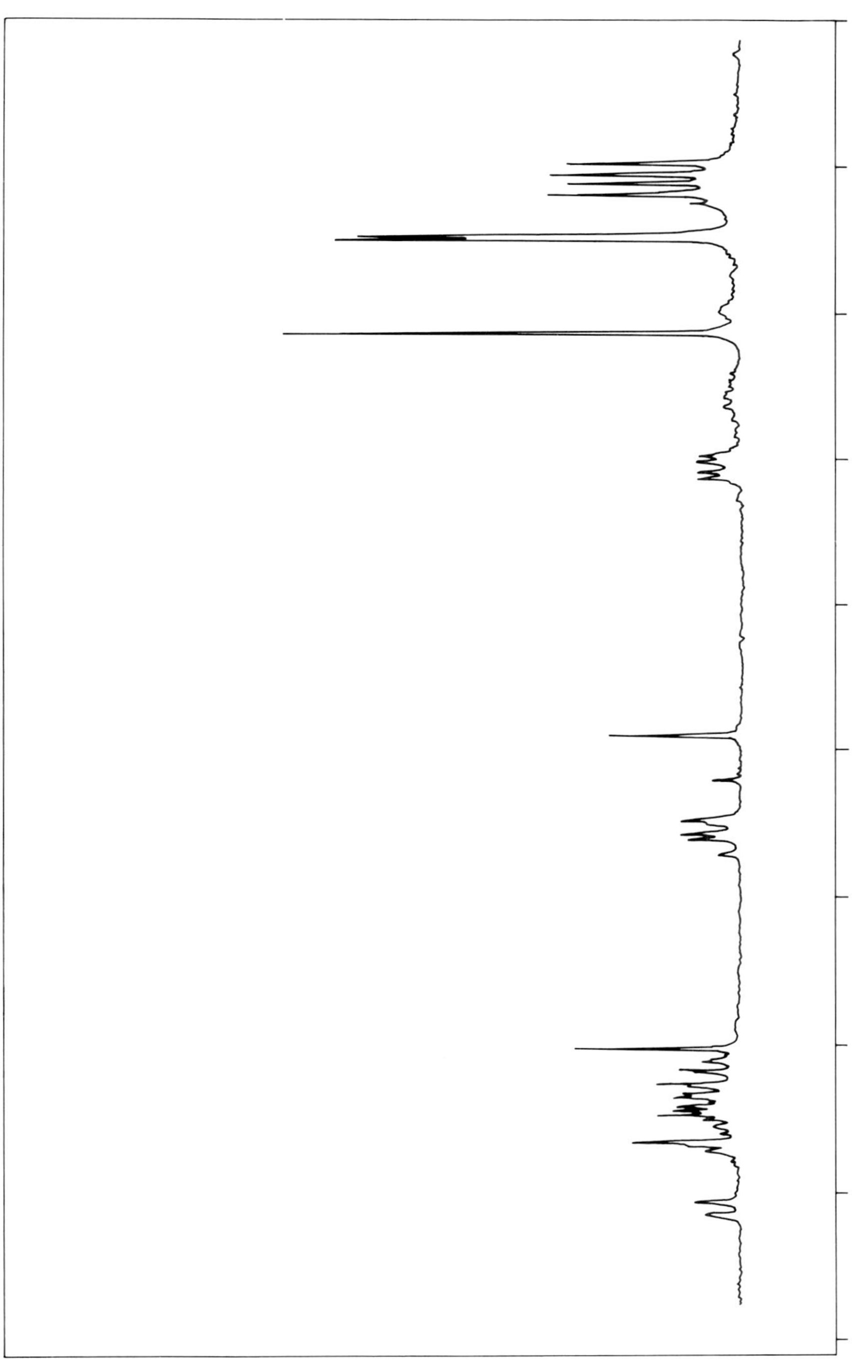
0PPM
1.0
2.0
3.0
4.0
5.0
6.0
7.0
8.0
9.0

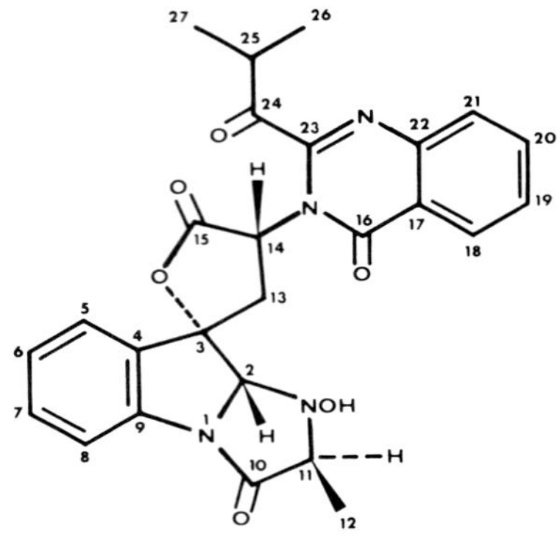

Common name	Nortryptoquivalone (Tryptoquivalone; Tryptoquivaline B)
Molecular weight[6,14]	488.1695
Molecular formula	$C_{26}H_{24}O_6N_4$
General characteristics[6,14]	Colorless prisms from dichloromethane hexane, mp 208°–209°C $[\alpha]_D^{25}$ 255° (C = 0.30 in chloroform)
UV data[6,14]	λ_{max}^{EtOH} nm (ϵ): 228(sh)(31,300), 232(31,800), 253(sh)(16,900), 278(7,600), 289(6,500), 304(5,400), and 316(sh)(4,500)
CD data[6]	$\Delta\epsilon$ (in 95% ethanol) nm: 256(−3.8), 273(0), 298(3.6), and 342(0)
Fungal source[6,14]	*Aspergillus clavatus* (NRRL 5890)
Toxicity data[42]	Nortryptoquivalone dosed IP (up to 50 μl DMSO) induced persistent tremors in rats.
¹H NMR data[6]	($CDCl_3$) H2, 5.24; H5–8, 7.11–7.74; H11, 1.53; H12, 1.53; H13a, 3.06 (J = 10); H13b, 3.06 (J = 10); H14, 5.65 (J = 10); H18, 8.20; H19–21, 7.11–7.74; H24, 5.52 (J = 9); H25, 2.54; H26, 1.04 (J = 7); H27, 1.20 (J = 7); CH_3CO, 2.16

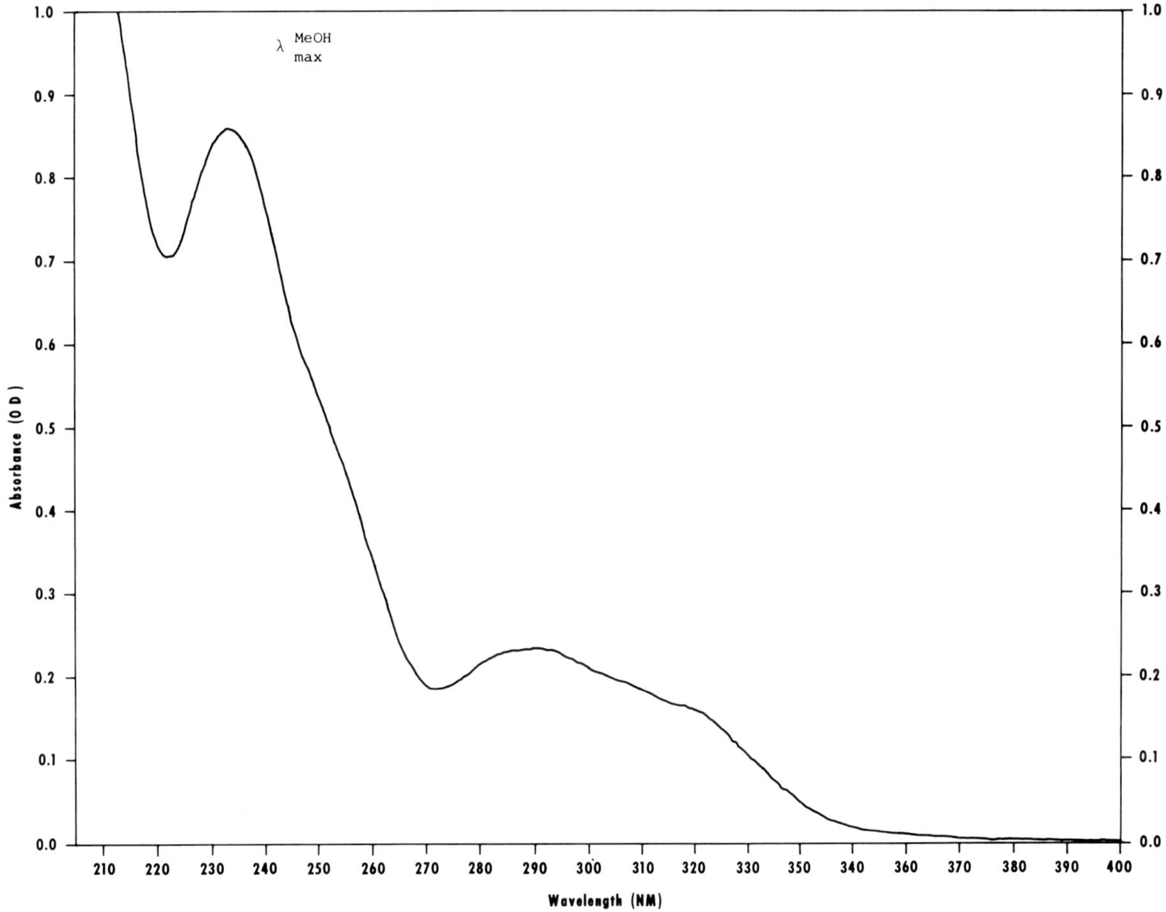

λ_{max}^{MeOH}
Absorbance (O D)
Wavelength (NM)

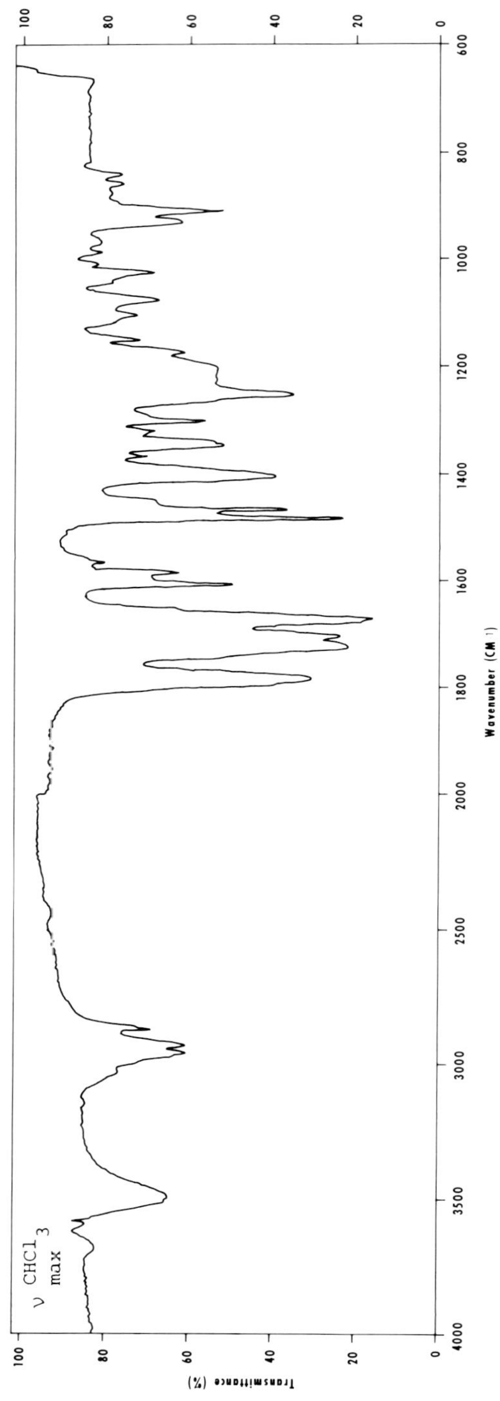
ν CHCl3 max
Wavenumber (CM-1)
Transmittance (%)
4000
3500
3000
2500
2000
1800
1600
1400
1200
1000
800
600
0
20
40
60
80
100

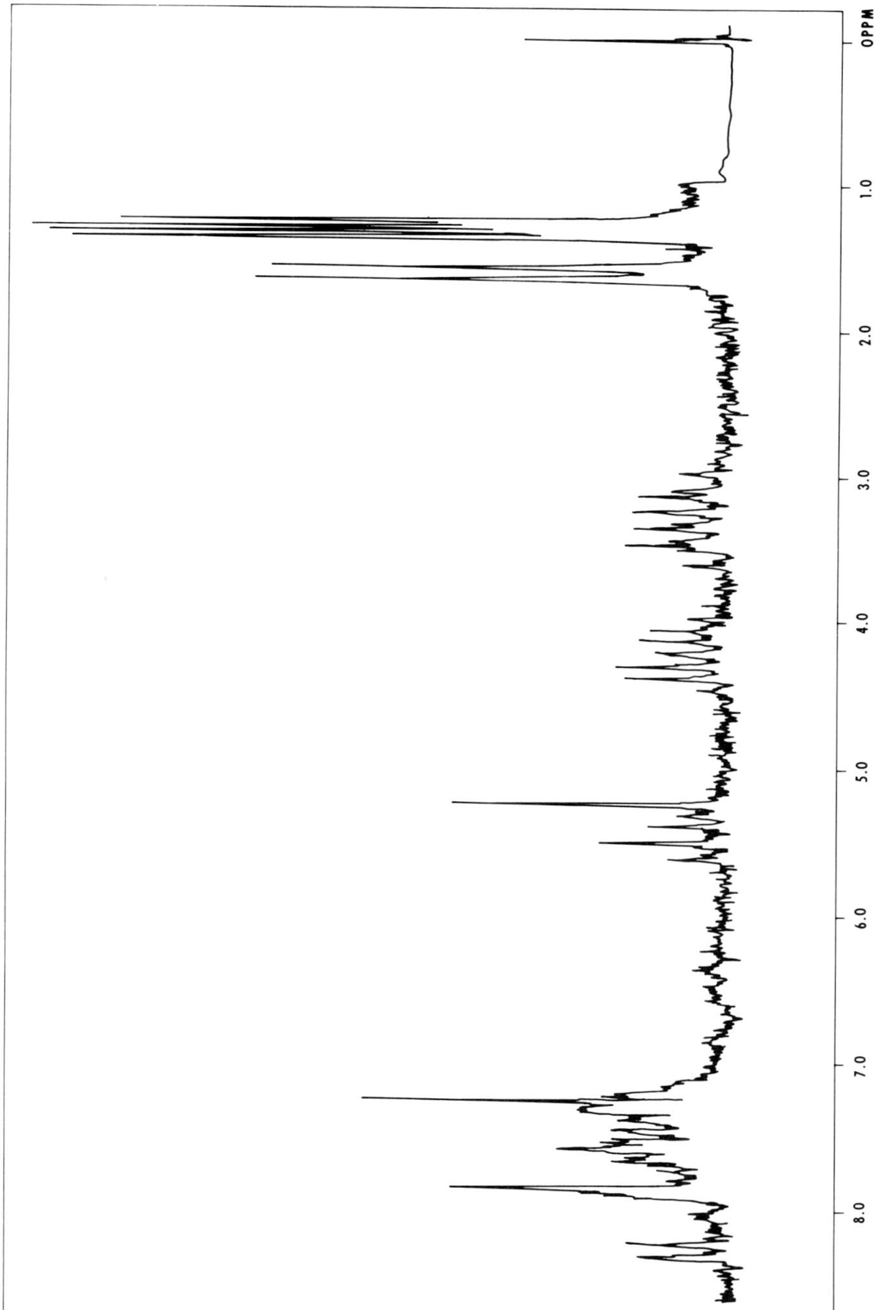
0PPM
1.0
2.0
3.0
4.0
5.0
6.0
7.0
8.0
9.0

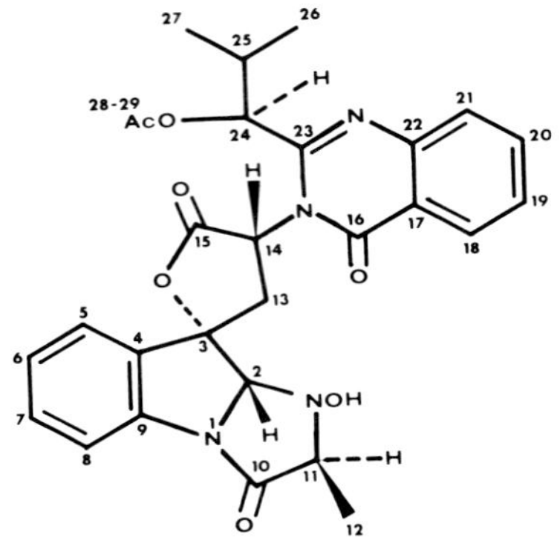

Common name[68b]	Nortryptoquivaline (Tryptoquivaline D)
Molecular weight	532.1958
Molecular formula	$C_{28}H_{28}O_7N_4$
General characteristics[6]	Colorless prisms from dichloromethane hexane, mp 256°–258°C $[\alpha]_D^{25}$ 170° (C = 0.64 in chloroform)
UV data[6]	λ_{max}^{EtOH} nm (ϵ): 228(43,600), 233(sh)(42,000), 254(sh)(18,700), 267(sh)(11,900), 279(10,200), 306(4,500), and 319(3,500)
CD data[6]	$\Delta\epsilon$ (taken in 95% ethanol) nm: 254(−5.4), 287(0), 307(2.1), and 325(0)
IR data[6]	$\nu_{max}^{CHCl_3}$ cm^{-1}: 3490, 2980, 2940, 2880, 1790, 1728, 1670, 1610, 1485, 1471, and 1410
Fungal source[6,68b]	*Aspergillus clavatus* (NRRL 5890) *A. fumigatus*
Toxicity data	Tremorgenic activity not reported in detail; however, metabolite is related to the known tremorgens tryptoquivaline and nortryptoquivalone.
^{1}H NMR data[6]	($CDCl_3$) H2, 5.10; H5–8, 7.01–7.79; H11, 4.28 (J = 7); H12, 1.58 (J = 7); H13a, 2.94 (J = 10,13); H13b, 3.18 (J = 10,13); H14, 5.65 (J = 10); H18, 8.12; H19–21, 7.01–7.79; H24, 5.54 (J = 9); H25, 2.57; H26, 1.02 (J = 7); H27, 1.16 (J = 7); CH_3CO, 2.16

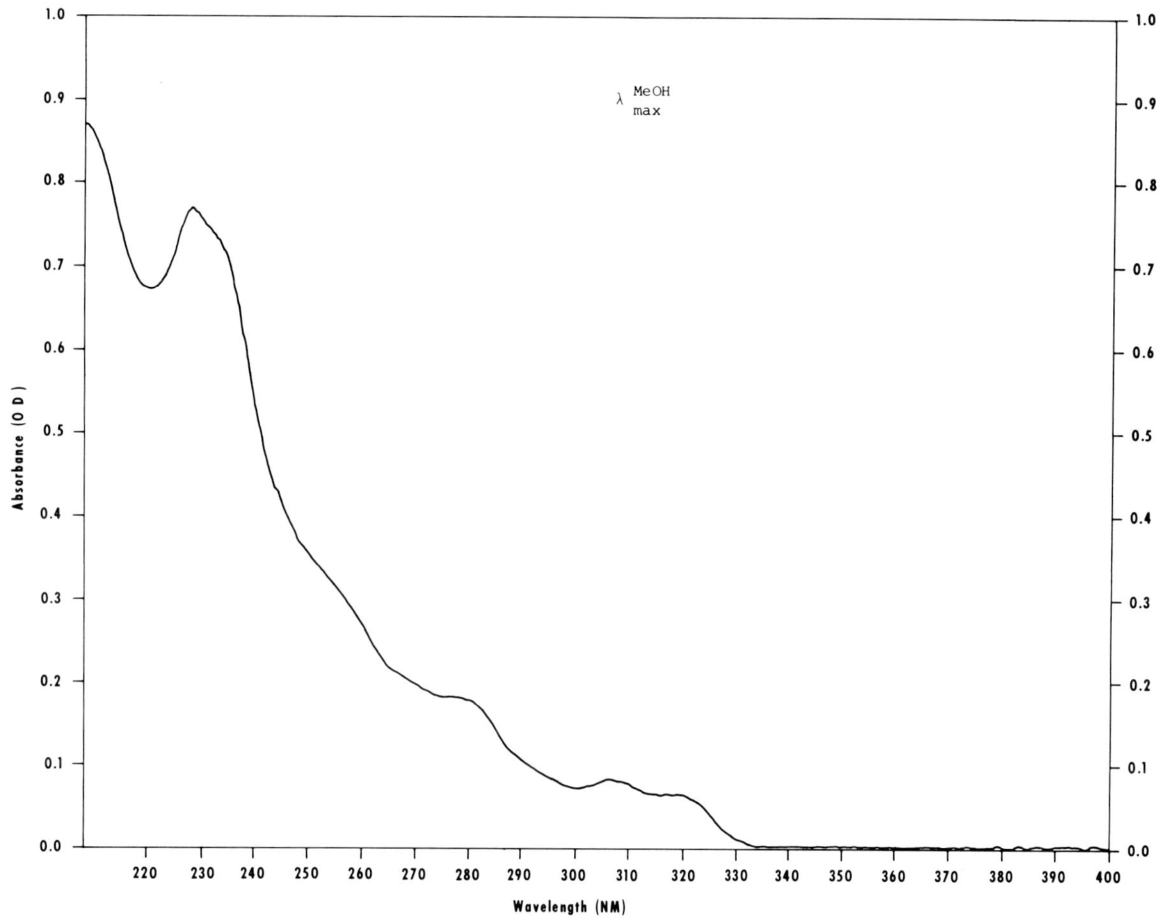
λ MeOH max
Absorbance (O D)
Wavelength (NM)

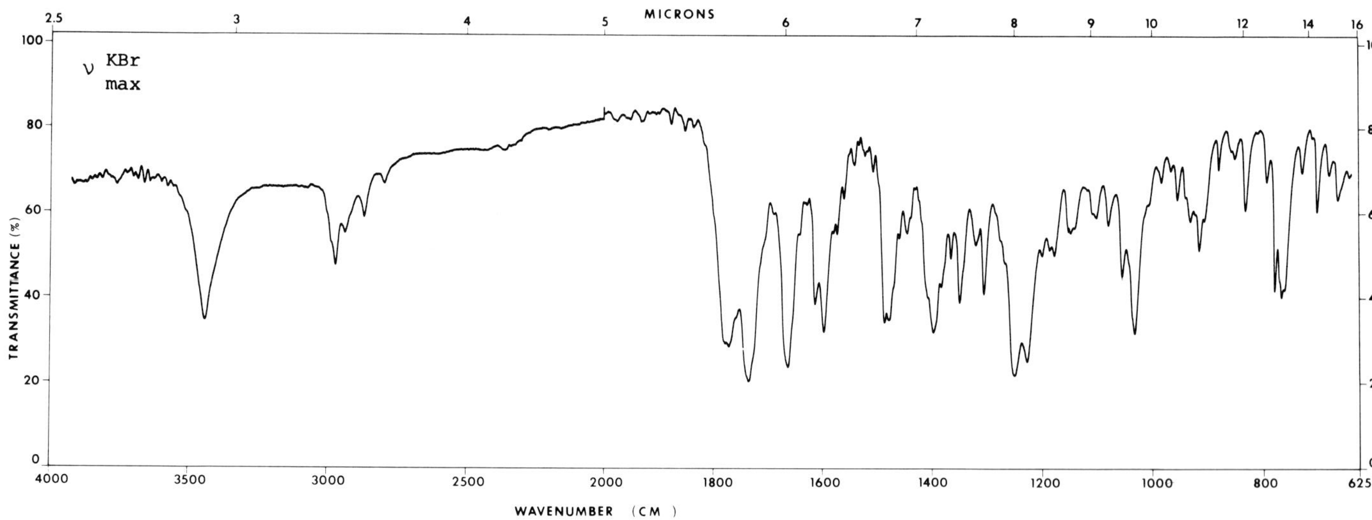

MICRONS
2.5
3
4
5
6
7
8
9
10
12
14
16
ν KBr
max
TRANSMITTANCE (%)
100
80
60
40
20
0
4000
3500
3000
2500
2000
1800
1600
1400
1200
1000
800
625
WAVENUMBER (CM)

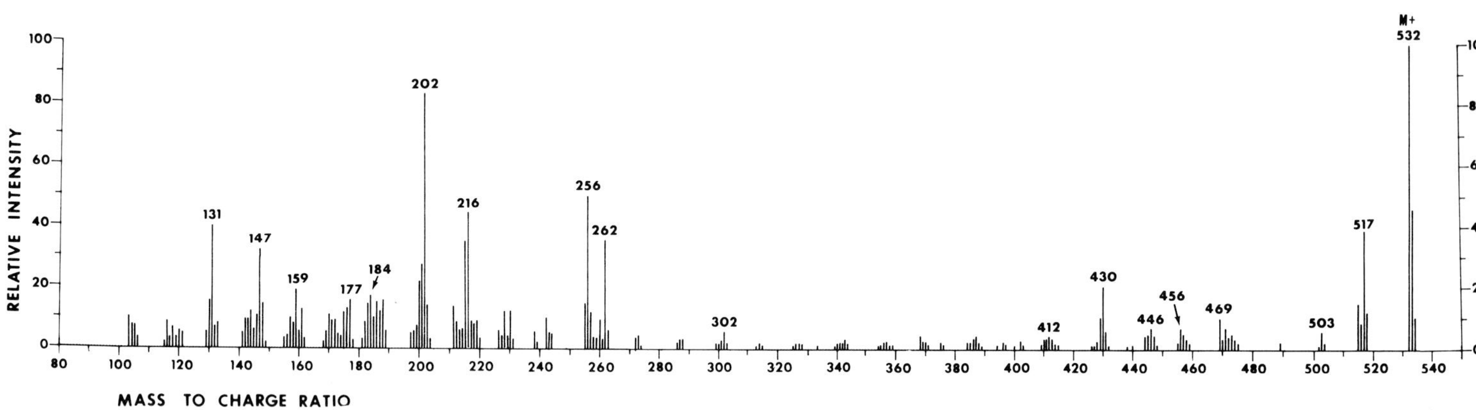

RELATIVE INTENSITY
100
80
60
40
20
0
M+
532
517
503
469
456
446
430
412
302
262
256
216
202
184
177
159
147
131
80
100
120
140
160
180
200
220
240
260
280
300
320
340
360
380
400
420
440
460
480
500
520
540
MASS TO CHARGE RATIO

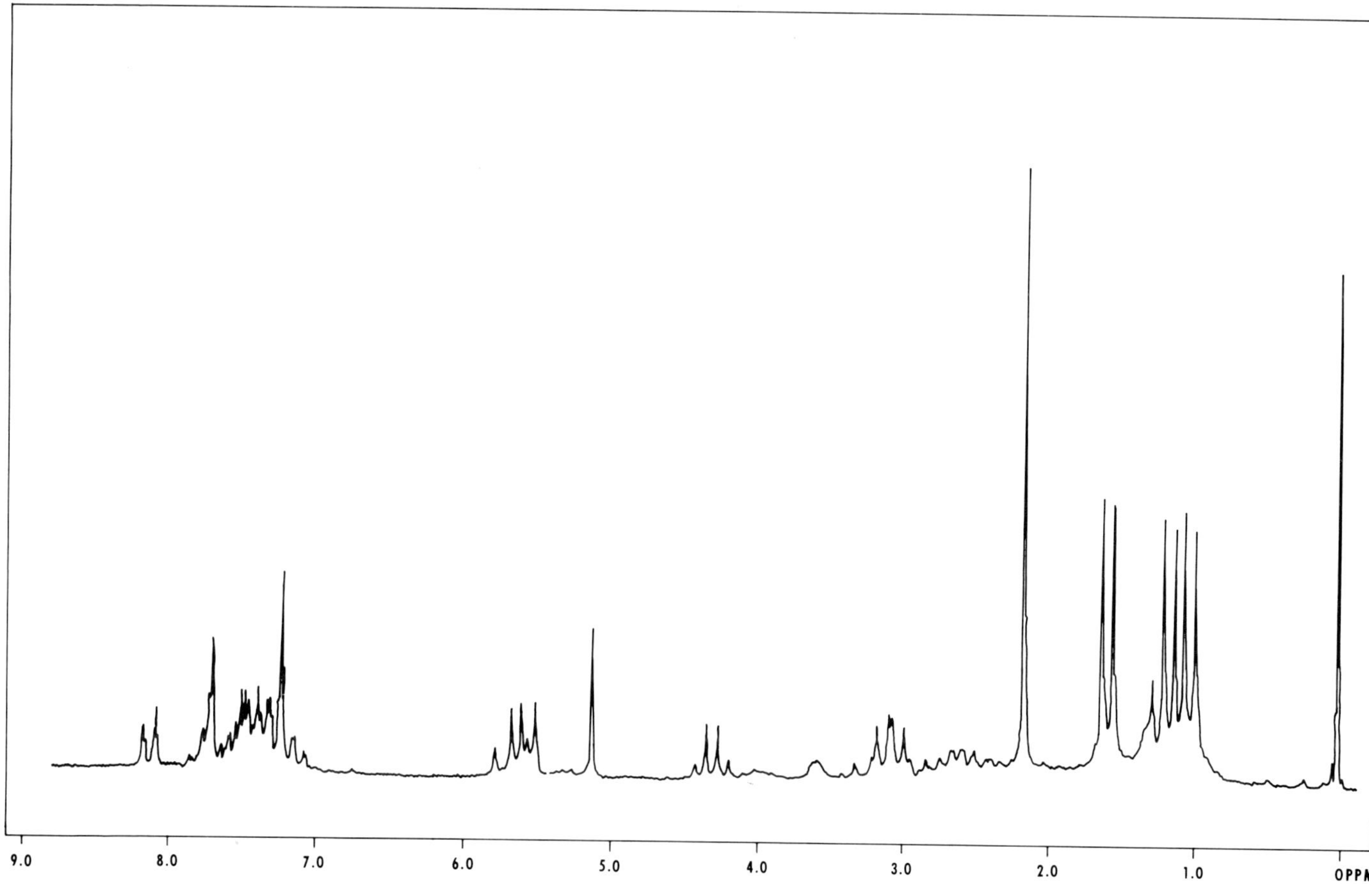

The 100 MHz ^1H NMR spectrum of tryptoquivaline in $CDCl_3$.

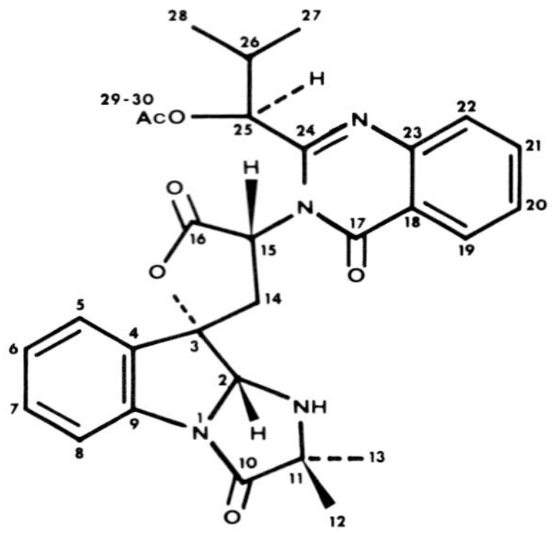

Common name	Deoxytryptoquivaline
Molecular weight	530.2165
Molecular formula	$C_{29}H_{30}O_6N_4$
General characteristics[6]	White needles from dichloromethane–hexane, mp 150°–152°C $[\alpha]_D^{25}$ 56.8° (C = 0.78 in chloroform)
UV data[6]	λ_{max}^{EtOH} nm (ϵ): 227(44,500), 232(sh)(41,900), 252(sh)(18,500), 267(sh)(12,000), 278(sh)(10,300), 304(3,300), and 318(sh)(2,700)
IR data[6]	$\nu_{max}^{CHCl_3}$ cm^{-1}: 3360, 3310, 2980, 2935, 2875, 1790, 1720, 1676, 1604, 1483, and 1469
Fungal source[6,14]	*Aspergillus clavatus*
Toxicity data	Details of tremorgenic activity not reported; however, metabolite is chemically related to known tremorgens tryptoquivaline and nortryptoquivalone.
^1H NMR data[6]	($CDCl_3$) H2, 5.24; H5–8, 7.11–7.74; H11, 1.53; H12, 1.53; H13a, 3.06 (J = 10); H13b, 3.06 (J = 10); H14, 5.65 (J = 10); H18, 8.20; H19–21, 7.11–7.74; H24, 5.52 (J = 9); H25, 2.54; H26, 1.04 (J = 7); H27, 1.20 (J = 7); CH_3CO, 2.16

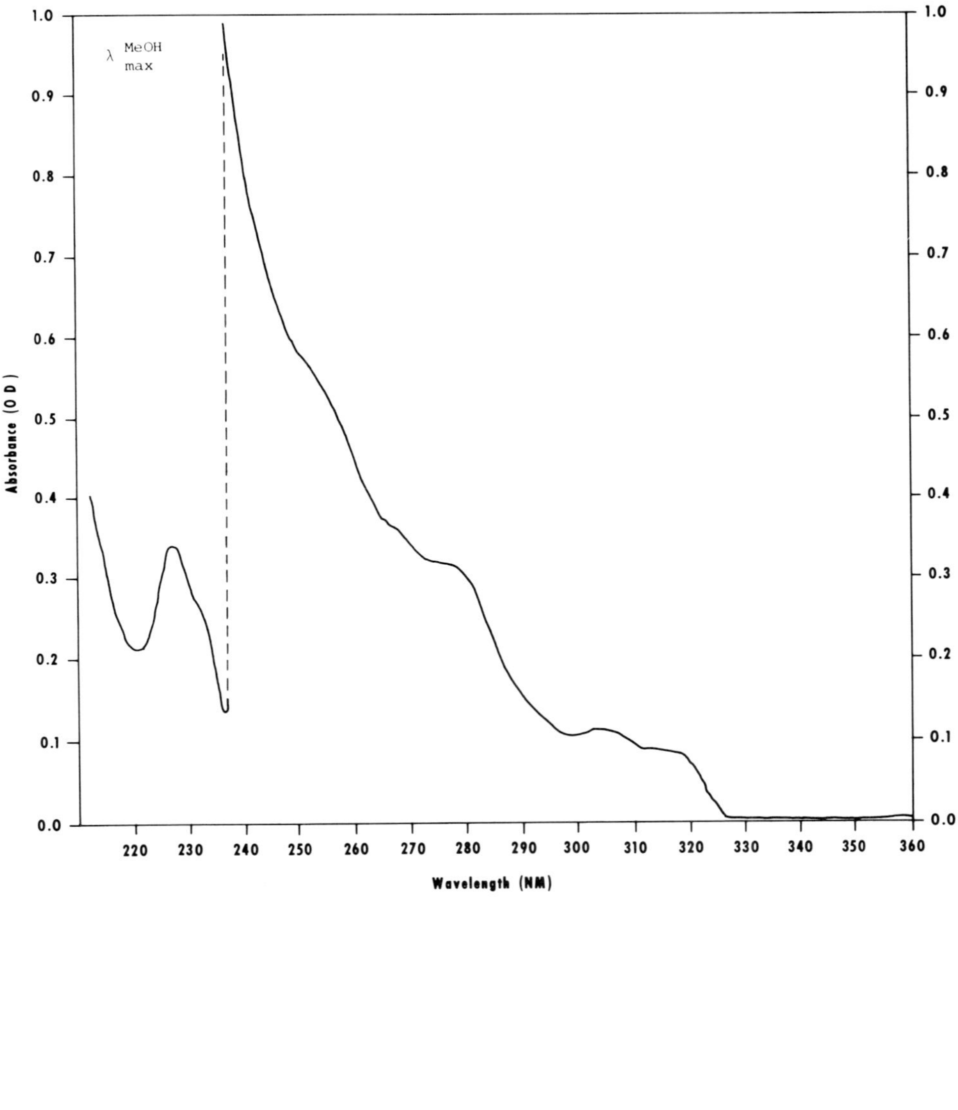
λ MeOH max
Absorbance (O D)
1.0
0.9
0.8
0.7
0.6
0.5
0.4
0.3
0.2
0.1
0.0
220 230 240 250 260 270 280 290 300 310 320 330 340 350 360
Wavelength (NM)

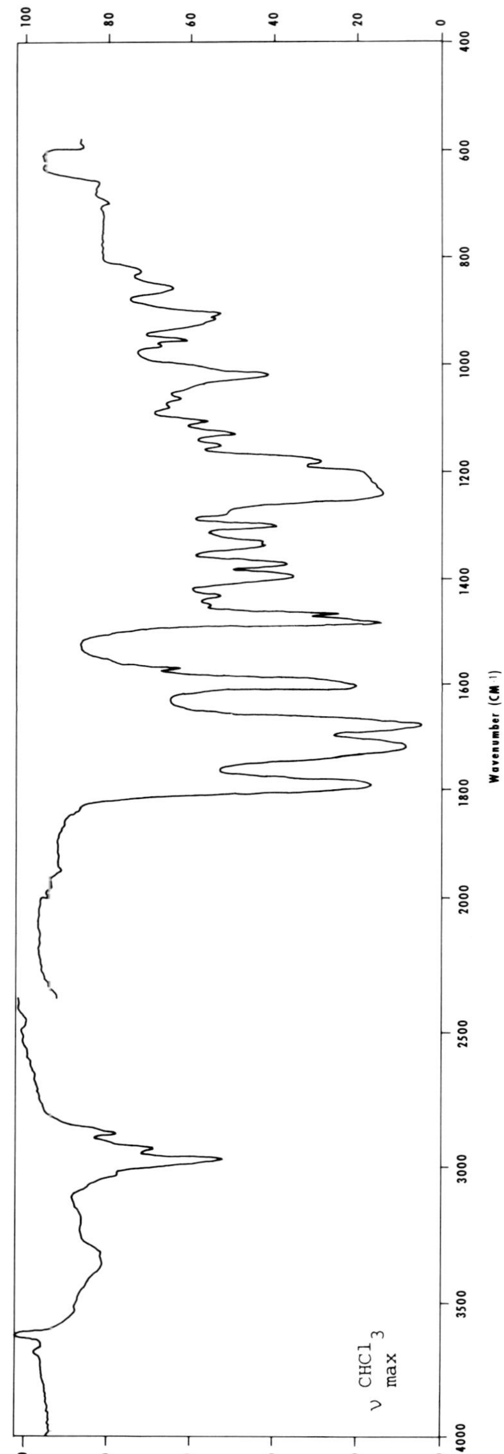

Transmittance (%)
Wavenumber (CM⁻¹)
ν CHCl3 max

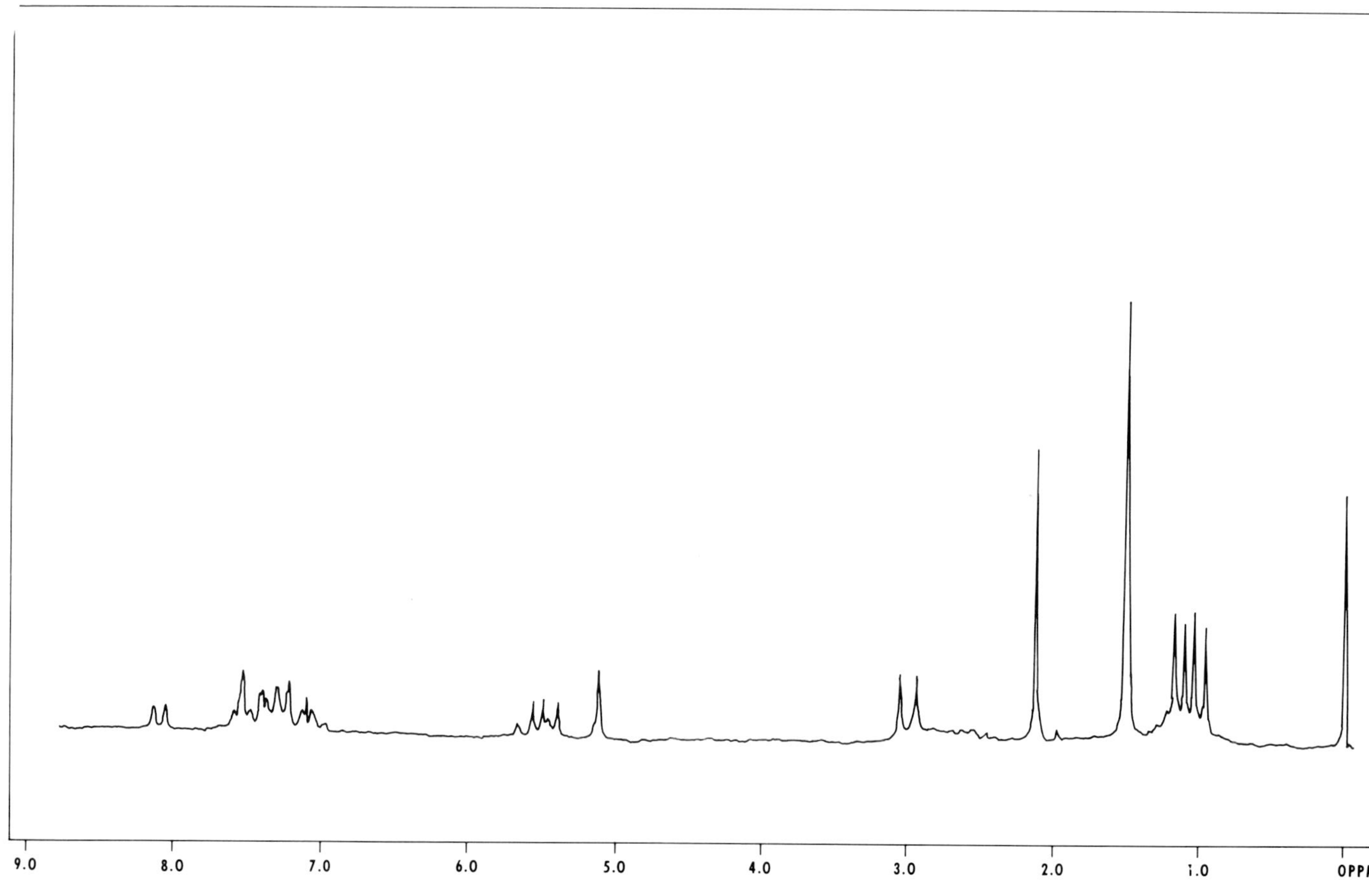

The 100 MHz ^{1}H NMR spectrum of deoxytryptoquivaline in $CDCl_3$/TMS.

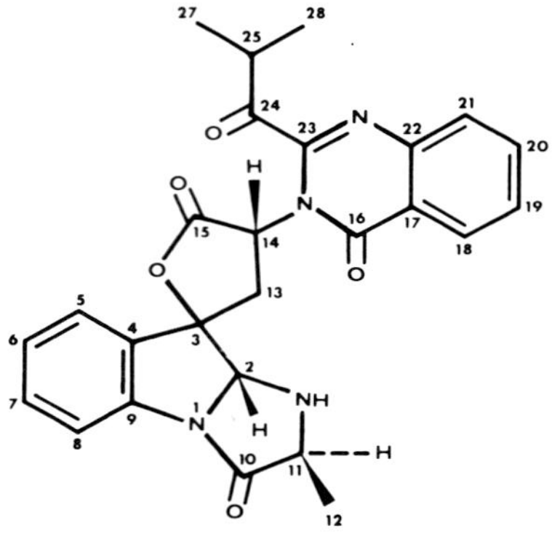

Common name	Deoxynortryptoquivalone
Molecular weight	472.1746
Molecular formula	$C_{26}H_{24}O_5N_4$
General characteristics[6]	Fine white needles from diethyl ether, mp 192°–193°C $[\alpha]_D^{25}$ 171° (C = 0.79 in chloroform)
UV data[6]	λ_{max}^{EtOH} nm (ϵ): 232(32,400), 288(9,250), and 320(sh)(6,250)
IR data[6]	$\nu_{max}^{CHCl_3}$ cm^{-1}: 3360, 2980, 2935, 2880, 1790, 1705, 1680, 1610, 1588, 1484, and 1469
Fungal source[6]	*Aspergillus clavatus*
Toxicity data	Details of tremorgenic activity not reported; however, the metabolite is related to the known tremorgens tryptoquivaline and nortryptoquivalone
^{1}H NMR data[6]	($CDCl_3$) H2, 5.36; H5–8, 7.07–7.84; H11, 4.12 (J = 7); H12, 1.56 (J = 7); H13a, 3.02 (J = 10,13); H13b, 3.32 (J = 10,13); H14, 5.48 (J = 10); H18, 8.24, H19–21, 7.07–7.84; H25, 4.08; H26, 1.26 (J = 7); H27, 1.30 (J = 7)

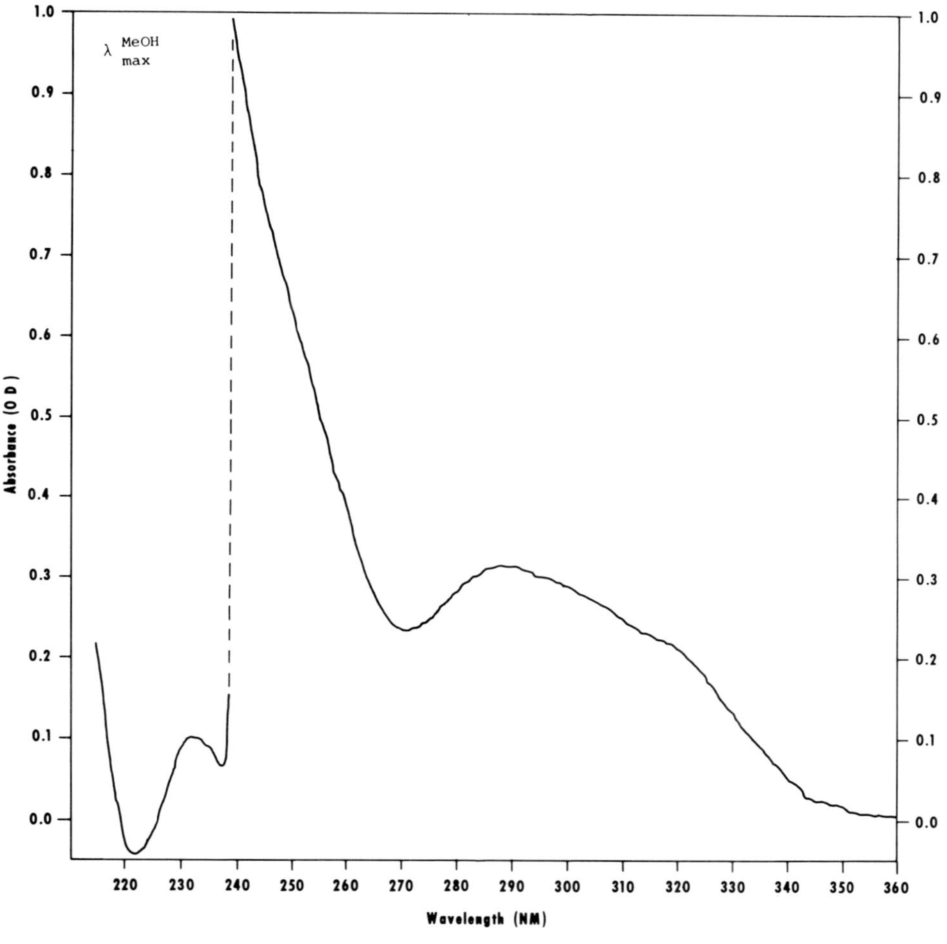

λ MeOH max
Absorbance (O D)
1.0
0.9
0.8
0.7
0.6
0.5
0.4
0.3
0.2
0.1
0.0
220 230 240 250 260 270 280 290 300 310 320 330 340 350 360
Wavelength (NM)

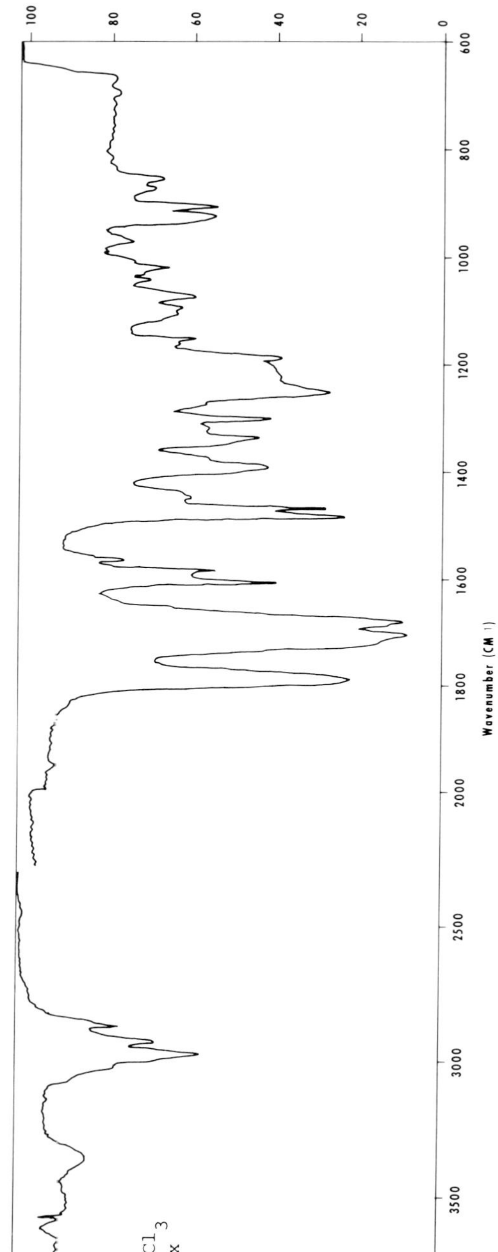
$\nu_{max}^{CHCl_3}$
Wavenumber (CM⁻¹)
Transmittance (%)

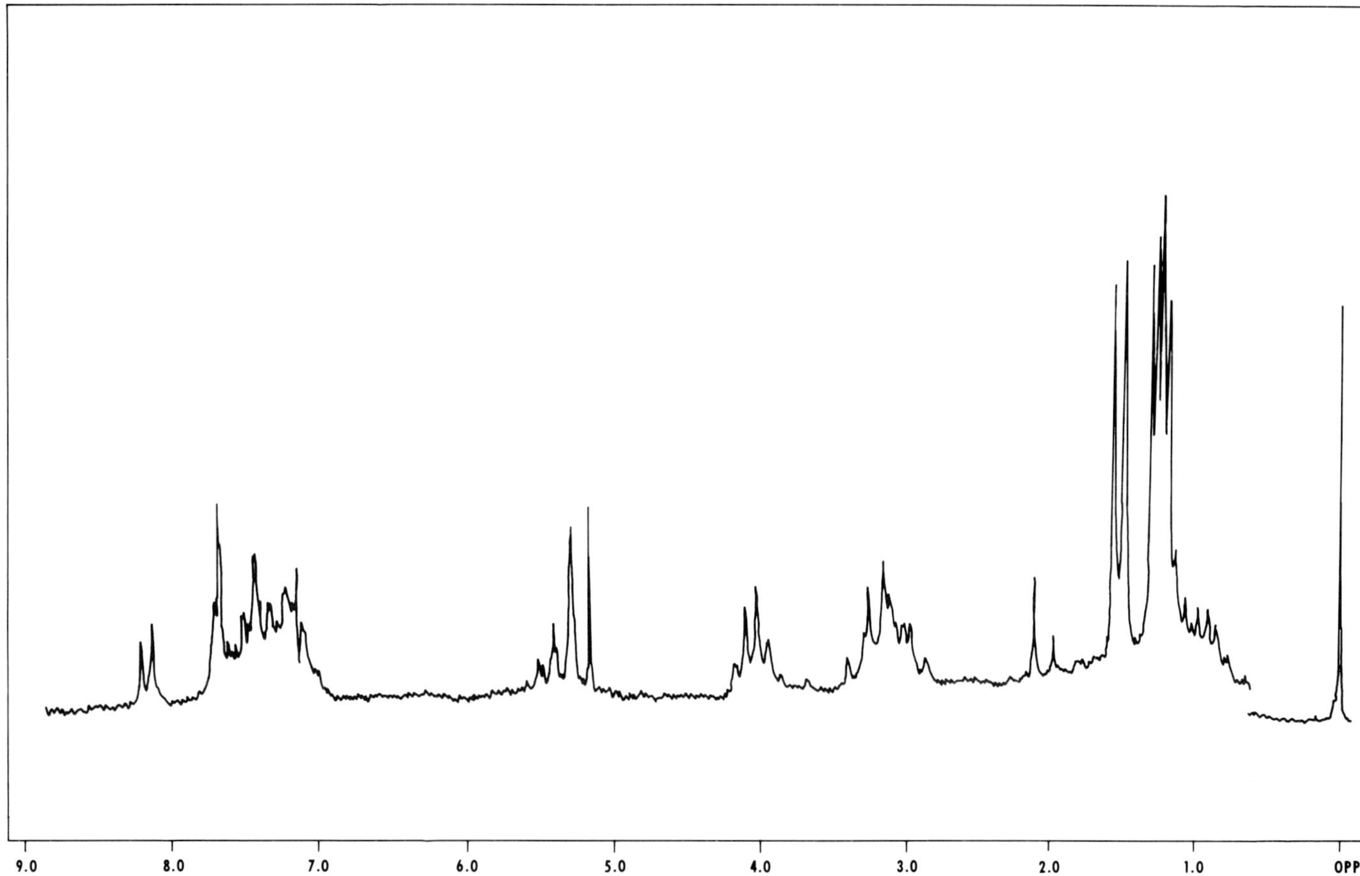

The 100 MHz ^{1}H NMR spectrum of deoxynortryptoquivalone in $CDCl_3$/TMS.

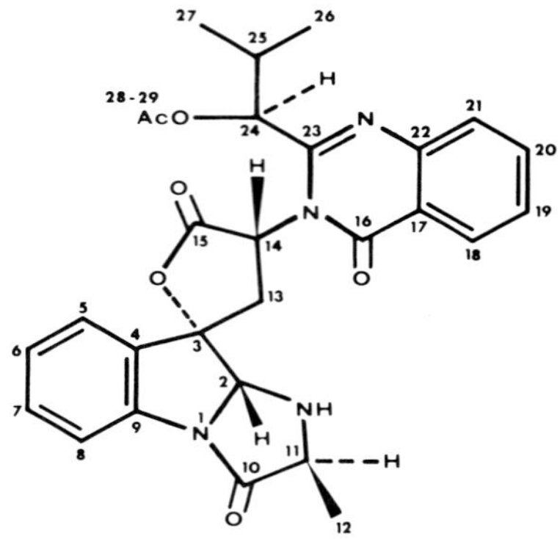

Common name	Deoxynortryptoquivaline
Molecular weight	516.2008
Molecular formula	$C_{28}H_{28}O_6N_4$
General characteristics[6]	Colorless prisms from diethyl ether, mp 158°–160°C $[\alpha]_D^{25}$ 69.5° (C = 0.82 in chloroform)
UV data[6]	λ_{max}^{EtOH} nm (ϵ): 228(43,900), 233(sh)(40,100), 254(sh)(15,600), 268(11,700), 278(sh)(10,500), 305(4,100), and 317(3,300)
IR data[6]	$\nu_{max}^{CHCl_3}$ cm^{-1}: 3360, 2975, 2935, 2880, 1790, 1724, 1676, 1607, and 1483
Fungal source[6]	*Aspergillus clavatus*
Toxicity data	Details of tremorgenic activity not reported; however, the metabolite is related to the known tremorgens tryptoquivaline and nortryptoquivalone
^1H NMR data[6]	($CDCl_3$) H2, 5.22; H5–8, 7.02–7.74; H11, 4.12 (J = 7); H12, 1.55 (J = 7); H13a, 2.86 (J = 10,13); H13b, 3.07 (J = 10,13); H14, 5.65 (J = 10); H18, 8.14; H19–21, 7.02–7.74; H24, 5.55 (J = 9); H25, 2.58; H26, 1.00 (J = 7); H27, 1.15 (J = 7); CH_3CO, 2.16

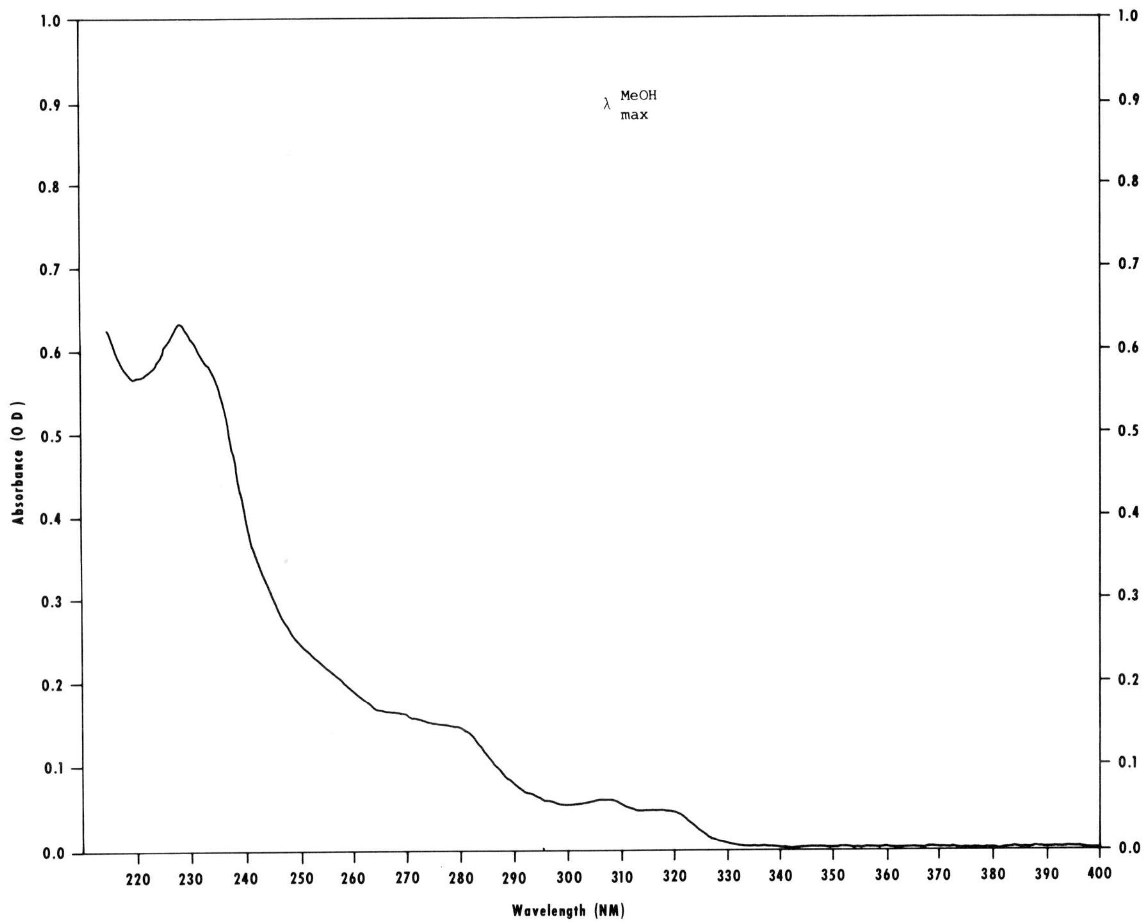
λ MeOH max
Absorbance (O D)
Wavelength (NM)
1.0
0.9
0.8
0.7
0.6
0.5
0.4
0.3
0.2
0.1
0.0
220
230
240
250
260
270
280
290
300
310
320
330
340
350
360
370
380
390
400

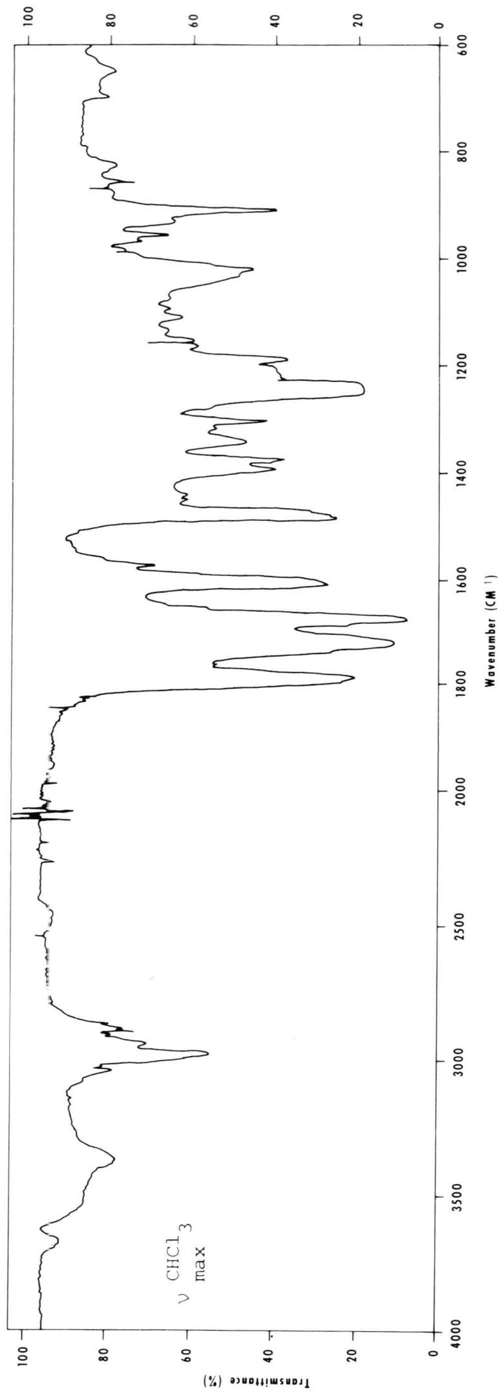

Transmittance (%)
Wavenumber (CM⁻¹)
ν CHCl3 max

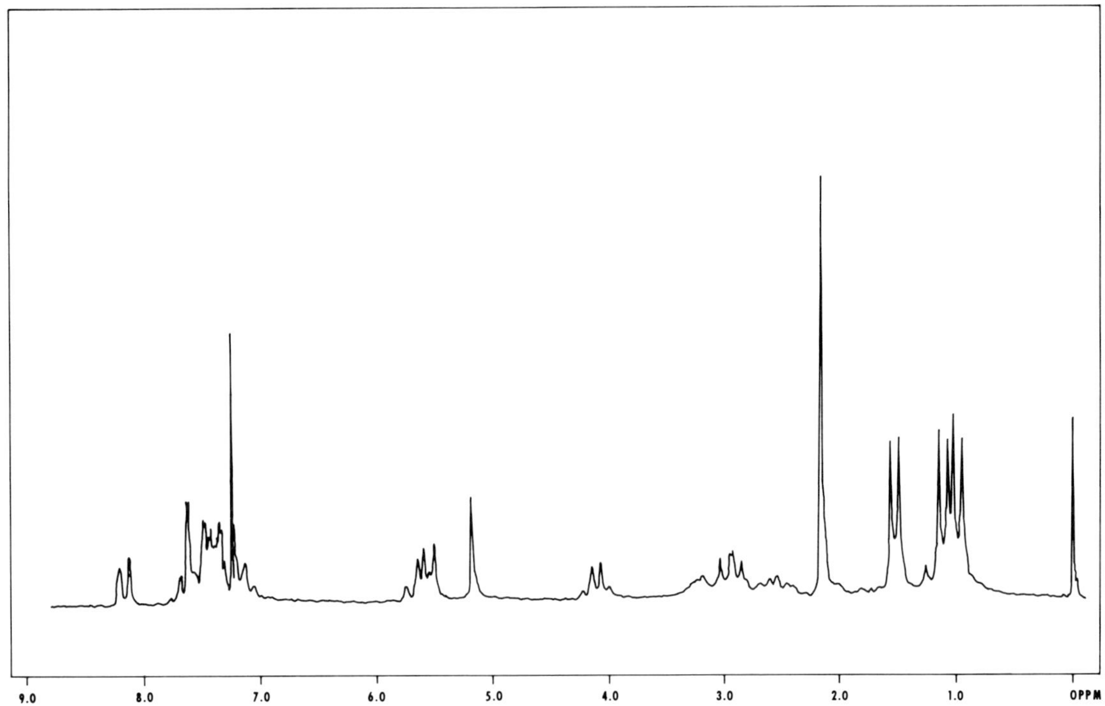

The 100 MHz ^{1}H NMR spectrum of deoxynortryptoquivaline in $CDCl_3$.

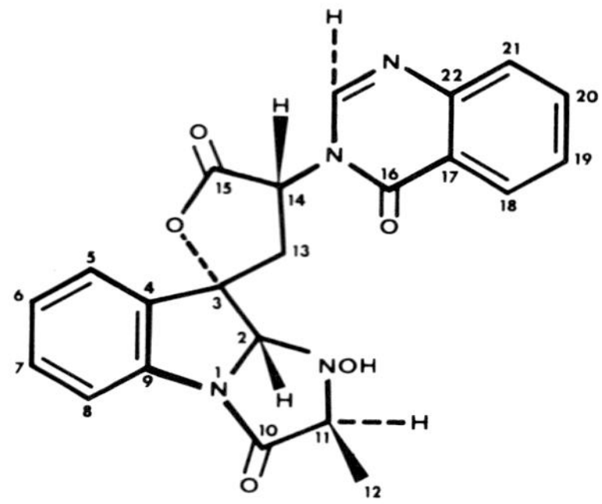

Common name	Tryptoquivaline E
Molecular weight	418.1277
Molecular formula	$C_{22}H_{18}O_5N_4$
General characteristics[103]	Colorless feathers from acetone, mp ~257°C (dec.) $[\alpha]_D^{15.5}$ +257° (C = 0.009 in chloroform)
UV data[103]	λ_{max}^{MeOH} nm (ϵ): 225.5(32,300), 232(sh)(29,800), 254(sh)(16,000), 265.5(sh)(11,700), 275.5(sh)(8,400), 291(sh)(3,500), 303(3,000), and 315(2,600)
IR data[103]	ν_{max}^{KBr} cm^{-1}: 3430, 1780, 1740, 1732, 1677, 1658, and 1614
Fungal source[103]	*Aspergillus fumigatus*
Toxicity data[103]	Related to tremorgen tryptoquivaline; however, not demonstrated to be tremorgenic.
^{1}H NMR data[102]	(pyridine) H2, 5.42; H5–8, 6.83–7.80; H11, 4.26 (J = 7); H12, 1.63 (J = 7); H13a, 3.38 (J = 10,13); H13b, 3.56 (J = 10,13); H14, 6.48 (J = 10); H18, 8.17; H19–21, 6.83–7.80; H23, 8.59; NOH, 10.42

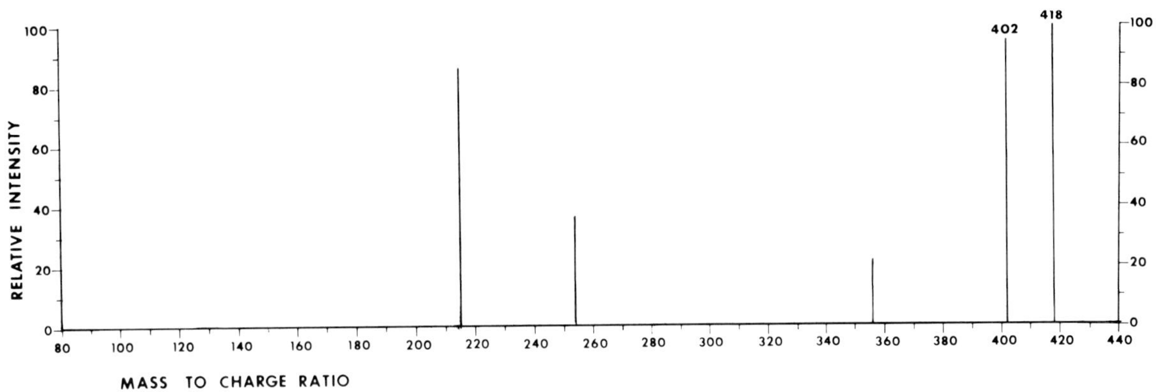

418
402
100
80
60
40
20
0
80
100
120
140
160
180
200
220
240
260
280
300
320
340
360
380
400
420
440
RELATIVE INTENSITY
MASS TO CHARGE RATIO

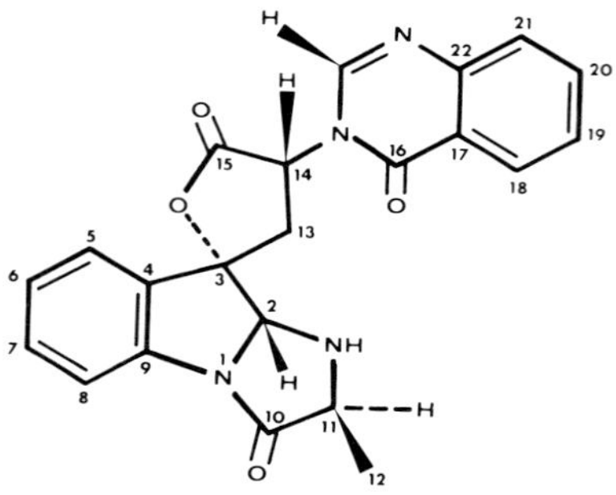

Common name	Tryptoquivaline F
Molecular weight	402.1328
Molecular formula	$C_{22}H_{18}O_4N_4$
General characteristics[103]	Colorless fine needles from methanol, mp ~277°C (dec.) $[\alpha]_D^{15.5}$ −109° (C = 0.006 in chloroform)
UV data[103]	λ_{max}^{MeOH} nm (ϵ): 226(33,400), 232(sh)(31,200), 255(sh)(12,400), 265(sh)(11,100), 276(sh)(8,100), 290(sh)(3,100), 303(2,800), and 315(2,200)
IR data[103]	ν_{max}^{KBr} cm^{-1}: 3365, 1775, 1725, 1664, and 1606
Fungal source[103]	*Aspergillus fumigatus*
Toxicity data[103]	Related to tremorgen tryptoquivaline; however, not demonstrated to be tremorgenic.
^{1}H NMR data[102] *for Tryptoquivaline F acetate ($CDCl_3$)*	($CDCl_3$) H2, 5.65; H5–8, 7.12–8.07; H11, 4.44 (J = 7); H12, 1.63 (J = 7); H13a, 3.19 (J = 9,13); H13b, 3.33 (J = 9,13); H14, 5.40 (J = 9); H18, 8.21 (J = 7); H19–21, 7.12–8.07; H23, 8.04; CH_3CO, 2.15

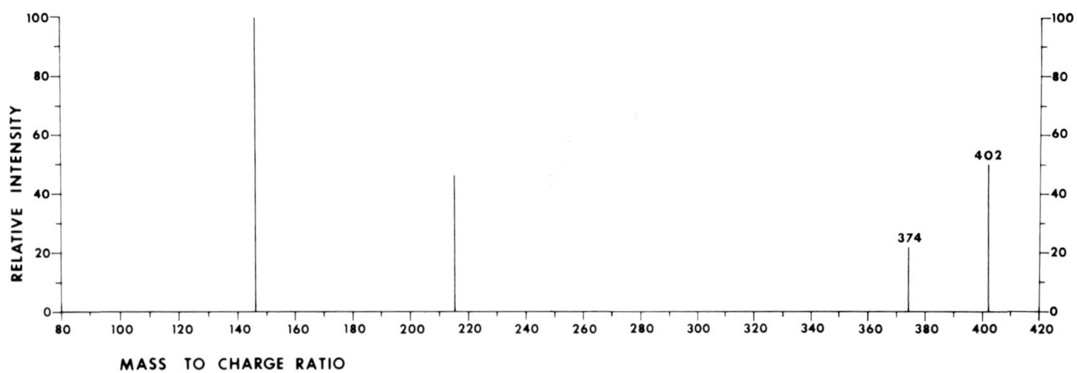
RELATIVE INTENSITY
MASS TO CHARGE RATIO
374
402
0
20
40
60
80
100
80
100
120
140
160
180
200
220
240
260
280
300
320
340
360
380
400
420

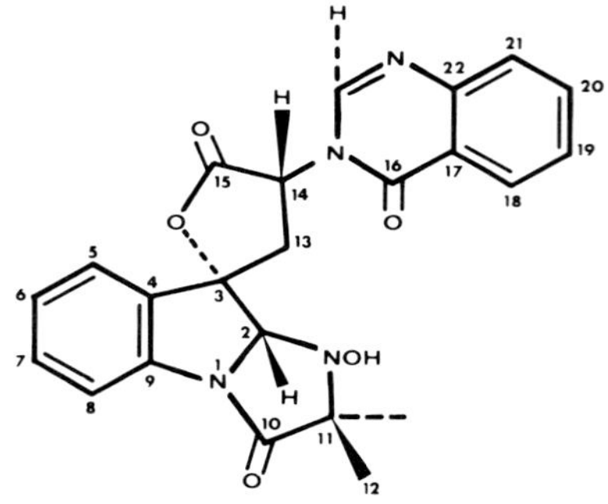

Common name	Tryptoquivaline G
Molecular weight	432.1433
Molecular formula	$C_{23}H_{20}O_5N_4$
General characteristics[103]	Colorless prisms from acetone, mp 240°–241.5°C (dec.) $[\alpha]_D^{11}$ +215° (C = 0.011 in acetone)
UV data[103]	λ_{max}^{MeOH} nm (ϵ): 226(34,300), 232(sh)(31,700), 253(sh)(17,400), 265(sh)(11,800), 275(sh)(8,200), 291(sh)(3,700), 302(3,000), and 315(2,500)
IR data[103]	ν_{max}^{KBr} cm^{-1}: 3470, 1778, 1738, 1662, and 1610
Fungal source[103]	*Aspergillus fumigatus*
Toxicity data[103]	Related to tremorgen tryptoquivaline; however, not demonstrated to be tremorgenic.
^1H NMR data[102]	(pyridine), H2, 5.30; H5–8, 6.96–7.96; H11, 4.26 (J = 7); H12, 1.50, 1.62; H13a, 3.43 (J = 10,14); H13b, 3.64; H14, 6.56 (J = 10); H18, 8.23; H19–21, 6.96–7.96; H23, 8.66; NOH, 10.61

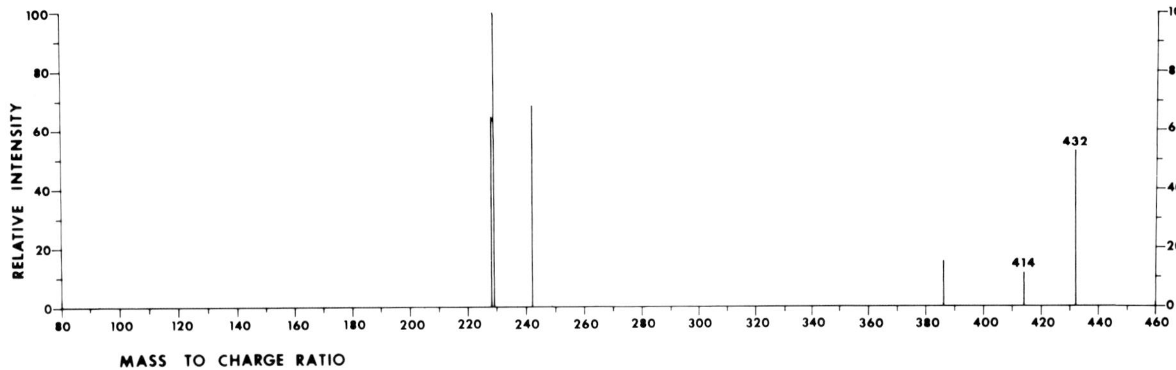

RELATIVE INTENSITY
100
80
60
40
20
0
80
100
120
140
160
180
200
220
240
260
280
300
320
340
360
380
400
420
440
460
414
432
MASS TO CHARGE RATIO

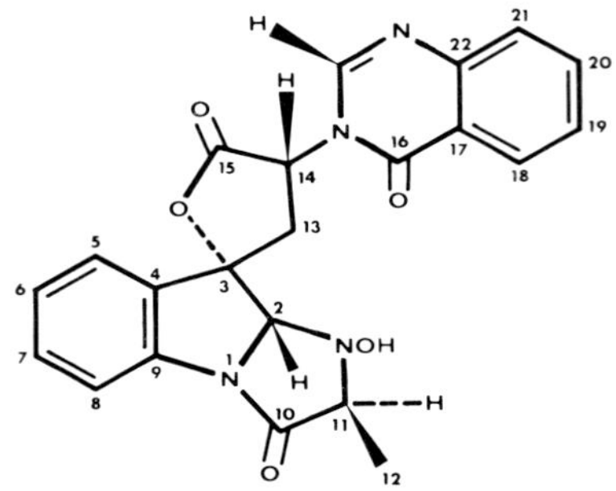

Common name	Tryptoquivaline H
Molecular weight	418.1277
Molecular formula	$C_{22}H_{18}O_5N_4$
General characteristics[103]	Colorless fine needles from methanol, mp ~274°C (dec.) $[\alpha]_D^{11}$ −155° (C = 0.021 in acetone)
UV data[103]	λ_{max}^{MeOH} nm (ϵ): 226(33,100), 232(sh)(30,900), 255(sh)(16,600), 266(sh)(11,300), 276(sh)(8,500), 291(sh)(3,600), 303(3,100), and 315(2,500)
IR data[103]	ν_{max}^{KBr} cm^{-1}: 3430, 1780, 1742, 1734, 1667, and 1609
Fungal source[103]	*Aspergillus fumigatus*
Toxicity data[103]	Related to tremorgen tryptoquivaline; however, not demonstrated to be tremorgenic.
^{1}H NMR data[102]	(pyridine), H2, 5.25; H5–8, 6.68–7.88; H11, 3.98 (J = 8); H12, 1.45 (J = 8); H13a, 2.96 (J = 10,13); H13b, 3.60 (J = 10,13); H14, 5.76 (J = 10); H18, 8.11; H19–21, 6.68–7.88; H23, 8.38; NOH, 10.60

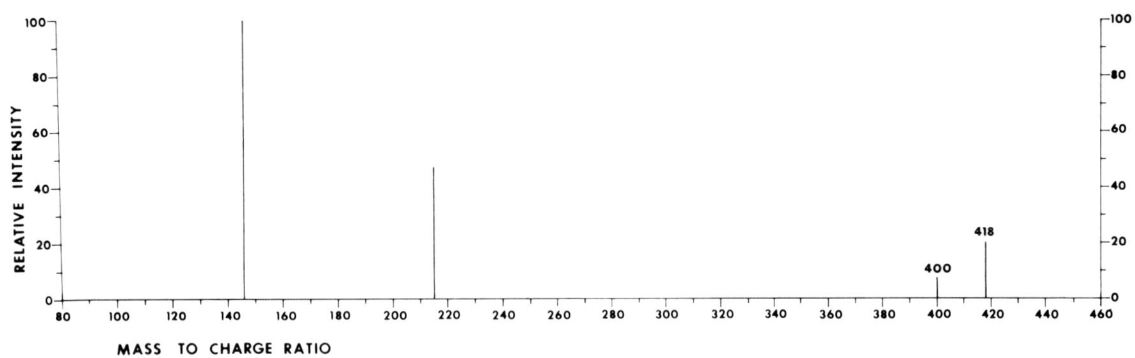
RELATIVE INTENSITY
MASS TO CHARGE RATIO
400
418
0
20
40
60
80
100
80
100
120
140
160
180
200
220
240
260
280
300
320
340
360
380
400
420
440
460

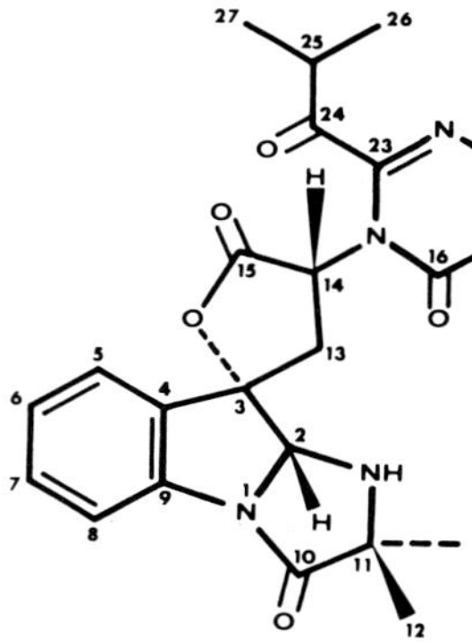

Common name	Tryptoquivaline I
Molecular weight	502.1852
Molecular formula	$C_{27}H_{26}O_6N_4$
General characteristics[103]	Colorless leaflets from methylene–methanol, mp 232°–235.5°C (dec.) $[\alpha]_D^{14}$ +239° (C = 0.16 in chloroform)
UV data[103]	λ_{max}^{MeOH} nm (ϵ): 235(31,700), 250(sh)(21,400), 292(9,600), and 321(sh)(6,100)
IR data[103]	ν_{max}^{KBr} cm^{-1}: 3480, 1780, 1732, 1710, 1675, and 1609
Fungal source[103]	*Aspergillus fumigatus*
Toxicity data[103]	Related to tremorgen tryptoquivaline; however, not demonstrated to be tremorgenic.
^{1}H NMR data[102]	($CDCl_3$), H2, 4.99; H5–8, 7.00–7.94; H12, 1.49; H13a, 3.06 (J = 10,14); H13b, 3.39 (J = 10,14); H14, 5.47 (J = 10); H18, 8.24 (J = 7); H19–21, 7.00–7.94; H25, 4.07 (J = 7); H26, 1.22; H27, 1.28; NOH, 7.01

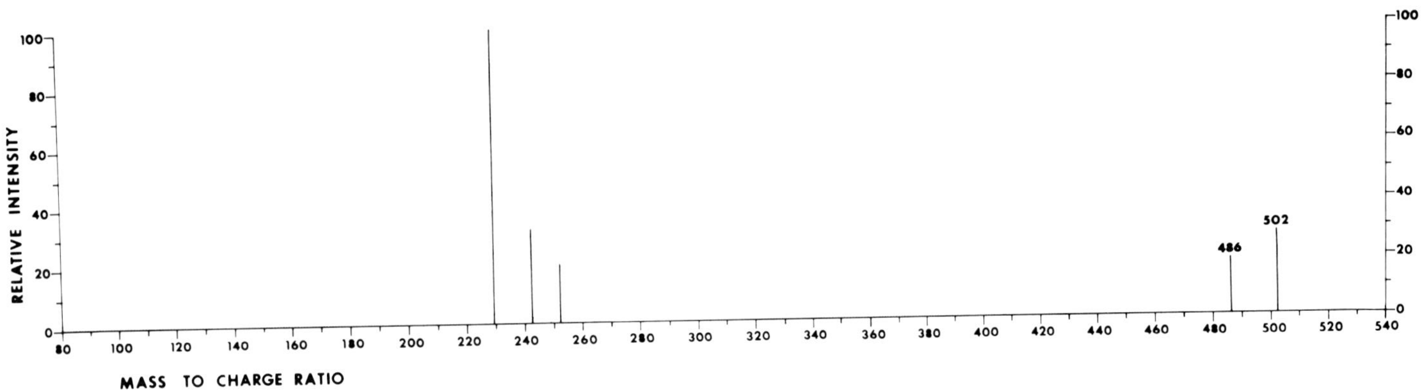
RELATIVE INTENSITY
100
80
60
40
20
0
80
100
120
140
160
180
200
220
240
260
280
300
320
340
360
380
400
420
440
460
480
500
520
540
486
502
MASS TO CHARGE RATIO

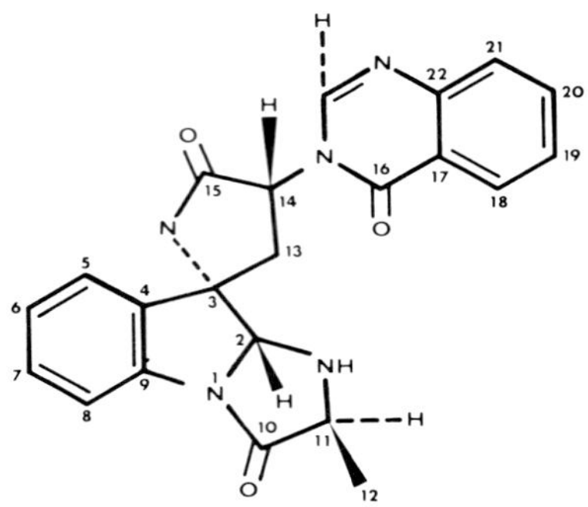

Common name	Tryptoquivaline J
Molecular weight	402.1328
Molecular formula	$C_{22}H_{18}O_4N_4$
General characteristics[103]	Colorless fine needles from acetone–methanol, mp 254°–258°C (dec.) $[\alpha]_D^{14}$ + 135° (C = 0.024 in acetone)
UV data[103]	λ_{max}^{MeOH} nm (ϵ): 225.5(41,100), 231(sh)(38,000), 253(sh)(16,200), 264(sh)(12,800), 275(sh)(9,700), 290(sh)(4,200), 302(3,900), and 310(3,100)
IR data[103]	ν_{max}^{KBr} cm^{-1}: 3375, 1780, 1713, 1670, and 1610
Fungal source[103]	*Aspergillus fumigatus*
Toxicity data[103]	Related to tremorgen tryptoquivaline; however, not demonstrated to be tremorgenic.
^1H NMR data[102]	(DMSO) H2, 5.41 (J = 6); H5–8, 7.20–8.02; H11, 3.87 (J = 7); H12, 1.41 (J = 7); H13a, 3.08; H13b, 3.10; H14, 6.01 (J = 10); H18, 8.17; H19–21, 7.20–8.02; H23, 8.49; NH, 3.76

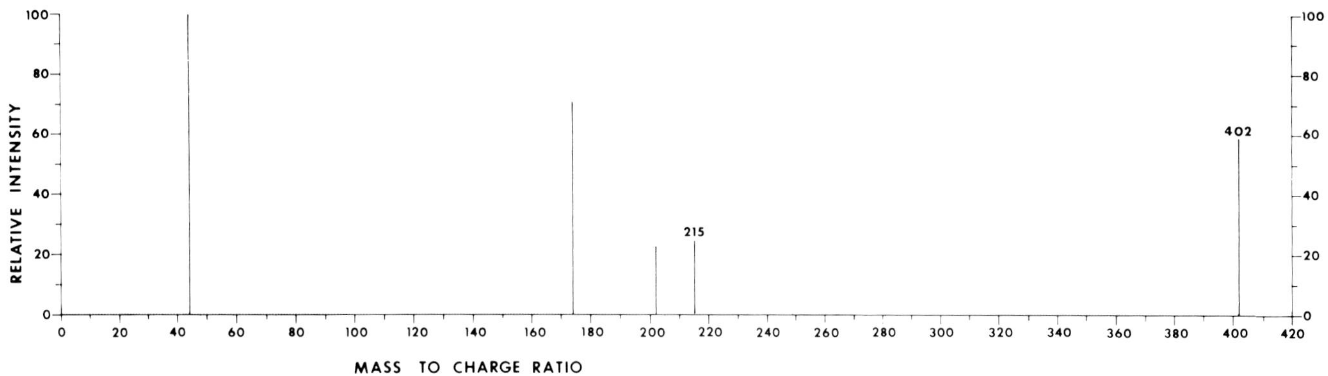
RELATIVE INTENSITY
MASS TO CHARGE RATIO
215
402
0
20
40
60
80
100
120
140
160
180
200
220
240
260
280
300
320
340
360
380
400
420

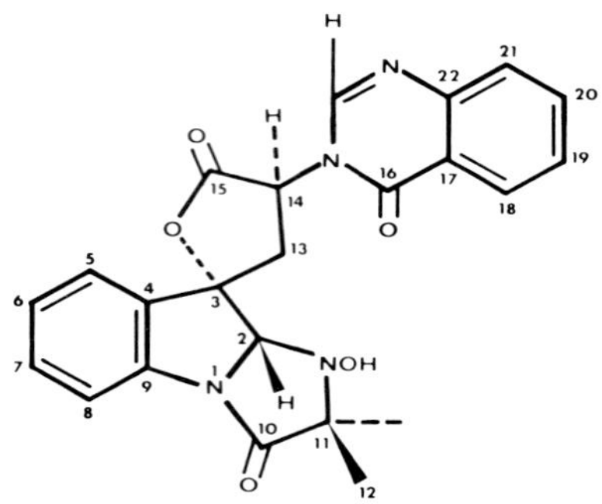

Common name	Tryptoquivaline L
Molecular weight	432.1433
Molecular formula	$C_{23}H_{20}O_5N_4$
General characteristics[104]	Colorless leaflets from acetone, mp 265°–268°C $[\alpha]_D^{23}$ −229° (C = 0.03 in dimethyl sulfoxide)
UV data[104]	λ_{max}^{MeOH} nm (ϵ): 216(sh)(33,900), 226(33,000), 231(sh)(30,800), 252(sh)(17,400), 264(sh)(12,300), 274(sh)(8,400), 290(sh)(3,700), 302(3,000), and 315(2,400)
IR data[104]	ν_{max}^{KBr} cm^{-1}: 3230, 1784, 1749, 1670, 1616, 1485, 1260, and 1200
Fungal source[104]	*Aspergillus fumigatus*
Toxicity data	Toxicity unknown
^1H NMR data[104]	(DMSO) H2, 5.21; H5–8, 7.24–8.00; H12, 1.26, 1.36; H13, 3.03, 3.43; H14, 5.57; H18, 8.23; H19–21, 7.24–8.00; H23, 8.53; NOH, 8.74

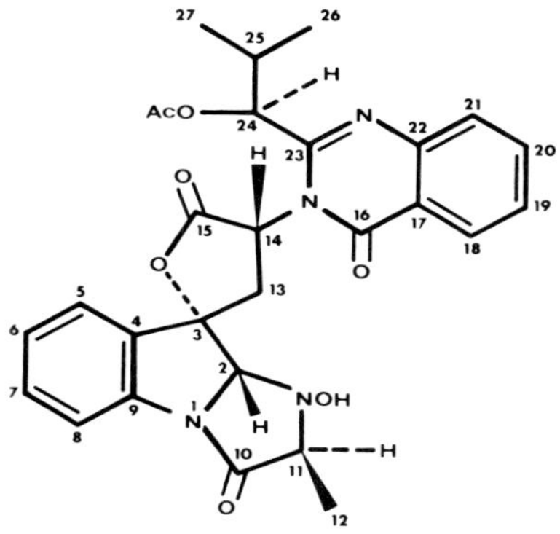

Common name	Tryptoquivaline M
Molecular weight	532.1958
Molecular formula	$C_{28}H_{28}O_7N_4$
General characteristics[104]	Colorless plates from methanol–water, mp 157°–164°C $[\alpha]_D^{24}$ −154° (C = 0.50 in chloroform)
UV data[104]	λ_{max}^{MeOH} nm (ϵ): 228(32,700), 232(30,700), 255(15,400), 278(9,300), 305(2,900), and 317(2,700)
IR data[104]	ν_{max}^{KBr} cm^{-1}: 3400, 1788, 1725, 1678, 1600, 1480, 1464, and 1210
Fungal source[104]	*Aspergillus fumigatus*
Toxicity data	Toxicity unknown
^1H NMR data[104]	H2, 5.22; H5–8, 7.12–7.88. H11, 4.15 (J = 7); H12, 1.55 (J = 7); H13, 2.68, 3.65 (J = 11,10); H14, 5.87 (J = 10); H18, 8.24 (J = 8); H19–21, 7.12–7.88; H24, 5.54 (J = 10); H25, 2.74; H26, 0.92 (J = 6); H27, 1.04 (J = 6); CH_3CO, 2.16; OH, 6.86

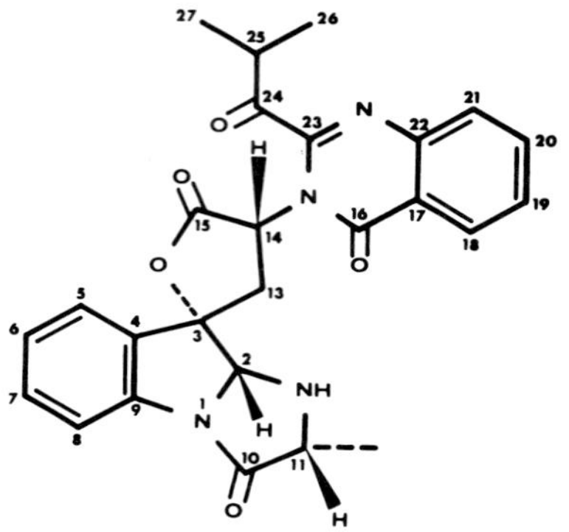

Common name	Tryptoquivaline N
Molecular weight	472.1746
Molecular formula	$C_{26}H_{24}O_5N_4$
General characteristics[104]	Colorless needles from methanol, mp 193°–197°C $[\alpha]_D^{24}$ +127° (C = 0.66 in dimethyl sulfoxide)
UV data[104]	λ_{max}^{MeOH} nm (ϵ): 232(32,300), 251(sh)(17,900), 291(9,000), and 320(sh)(6,100)
IR data[104]	ν_{max}^{KBr} cm^{-1}: 3360, 1780, 1722, 1705, 1680, 1607, 1481, and 1250
Fungal source[104]	*Aspergillus fumigatus*
Toxicity data	Toxicity unknown

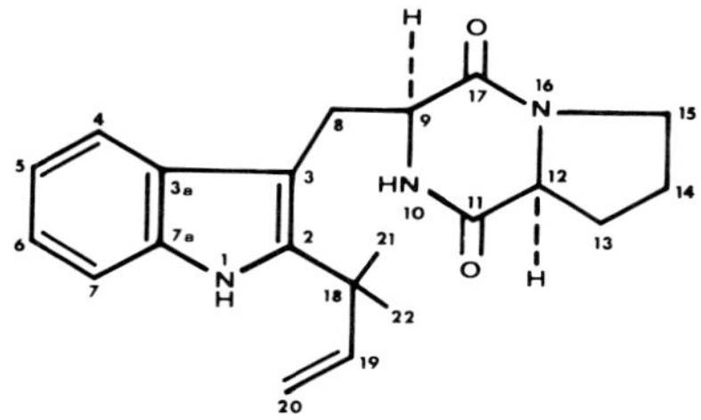

Common name	Deoxybrevianamide E [L-Prolyl-2-(1′,1′-dimethylallyl)-L-tryptophyldiketopiperazine]
Molecular weight	351.1946
Molecular formula	$C_{21}H_{25}O_2N_3$
General characteristics[87]	Homogeneous powder from benzene $[\alpha]_D^{22}$ −59° (*C* = 1.2 in chloroform)
UV data[87]	λ_{max}^{EtOH} nm (ϵ): 225(32,300), 275(sh)(7,000), 283(8,100), and 291(7,000)
IR data[87]	$\nu_{max}^{CHCl_3}$ cm^{-1}: 3481, 3458, 3366, 1670, and 1665
Fungal source[5,87]	*Aspergillus ustus* (Bainier) Thom and Church (C.S.I.R. 1128; NRRL 5856) *Penicillium italicum* Wehmer
Toxicity data	Toxicity unknown
TLC data[78]	*Adsorbent:* silica gel *Solvent:* chloroform–acetone, 9:1 v/v R_f: 0.75 *Detection:* blue spot after spraying with 65% H_2SO_4 and heating at ca. 120°C for 10 min.
Other significant references	Synthesis[74]

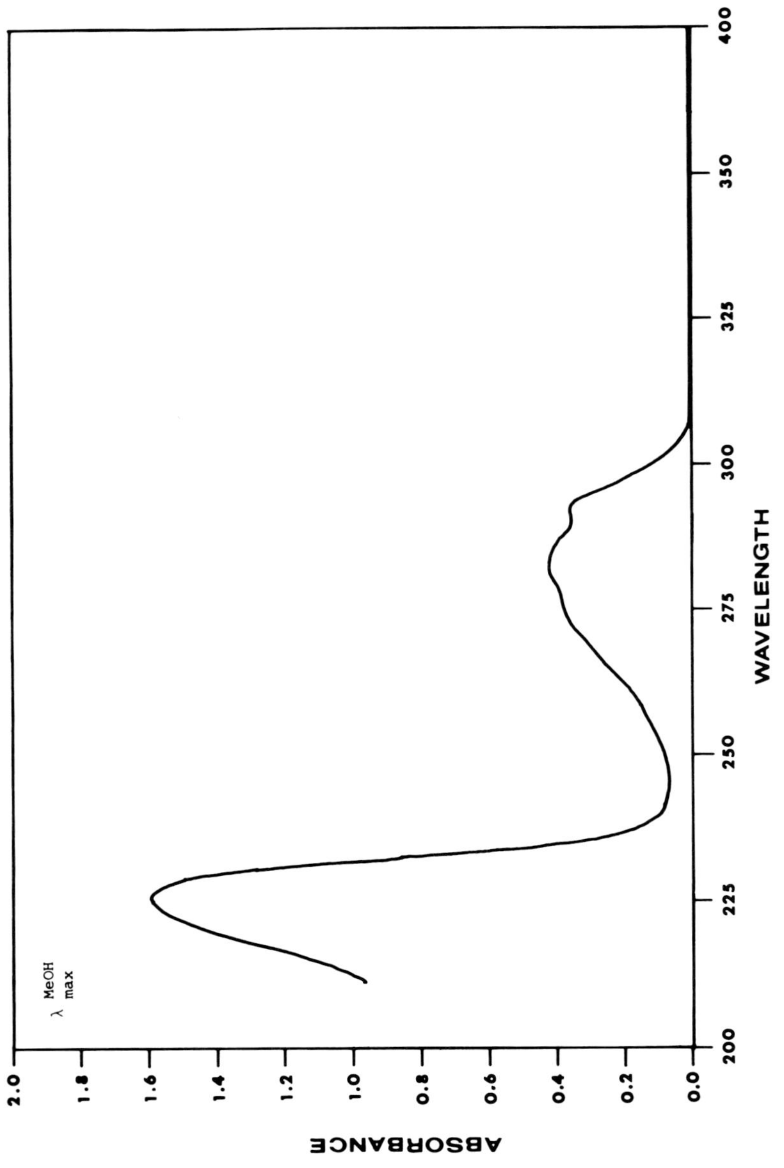
λ MeOH max
ABSORBANCE
2.0
1.8
1.6
1.4
1.2
1.0
0.8
0.6
0.4
0.2
0.0
200
225
250
275
300
325
350
400
WAVELENGTH

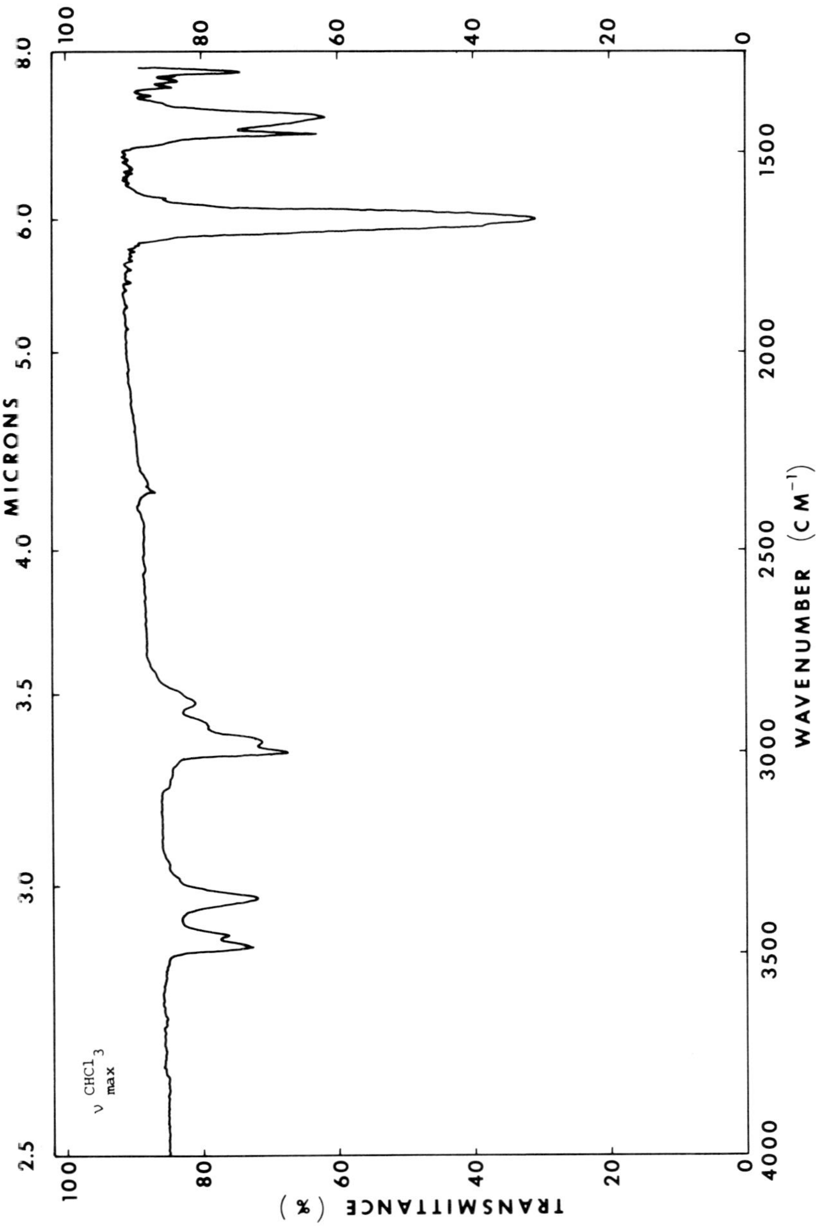
MICRONS
WAVENUMBER (CM^{-1})
TRANSMITTANCE (%)
$\nu_{max}^{CHCl_3}$

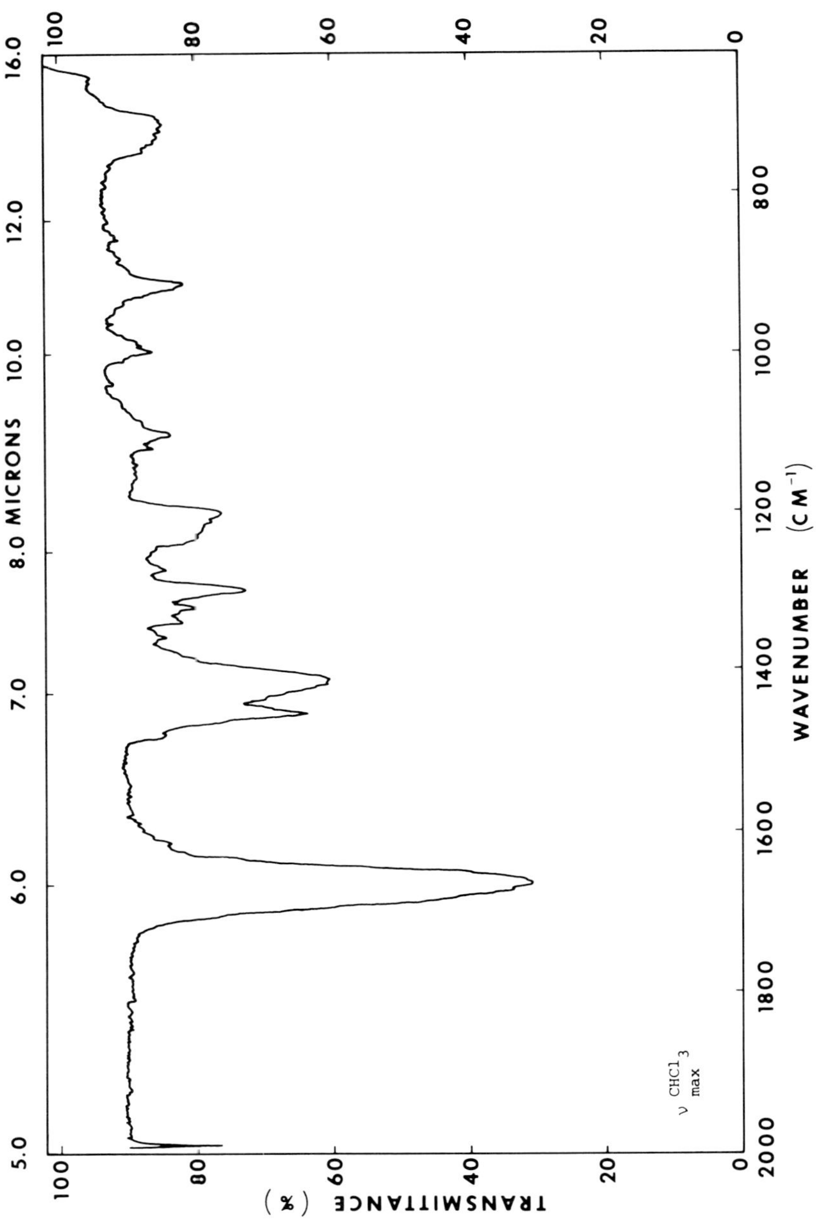
MICRONS
5.0
6.0
7.0
8.0
10.0
12.0
16.0
TRANSMITTANCE (%)
0
20
40
60
80
100
WAVENUMBER (CM⁻¹)
2000
1800
1600
1400
1200
1000
800
ν CHCl3 max

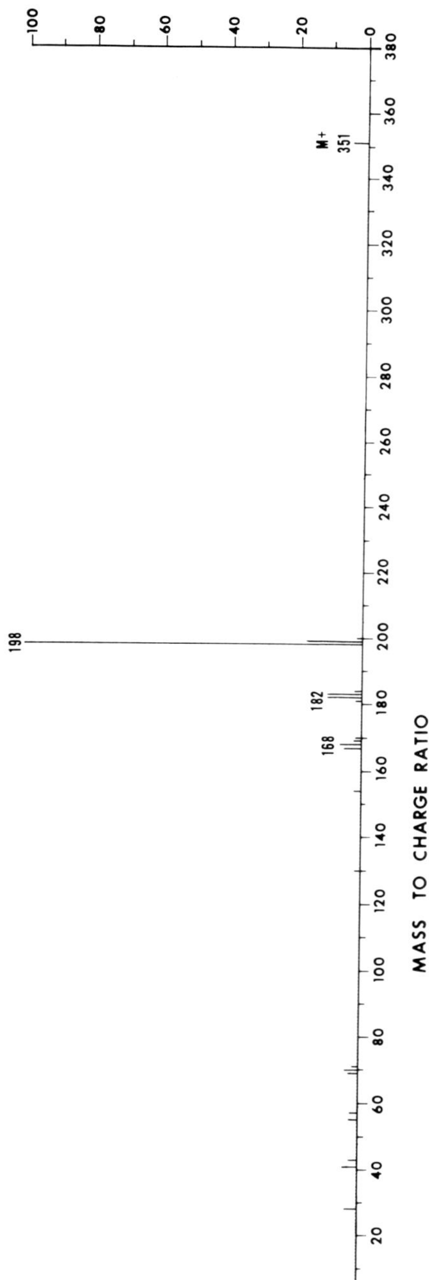
RELATIVE INTENSITY
MASS TO CHARGE RATIO
198
182
168
M+
351

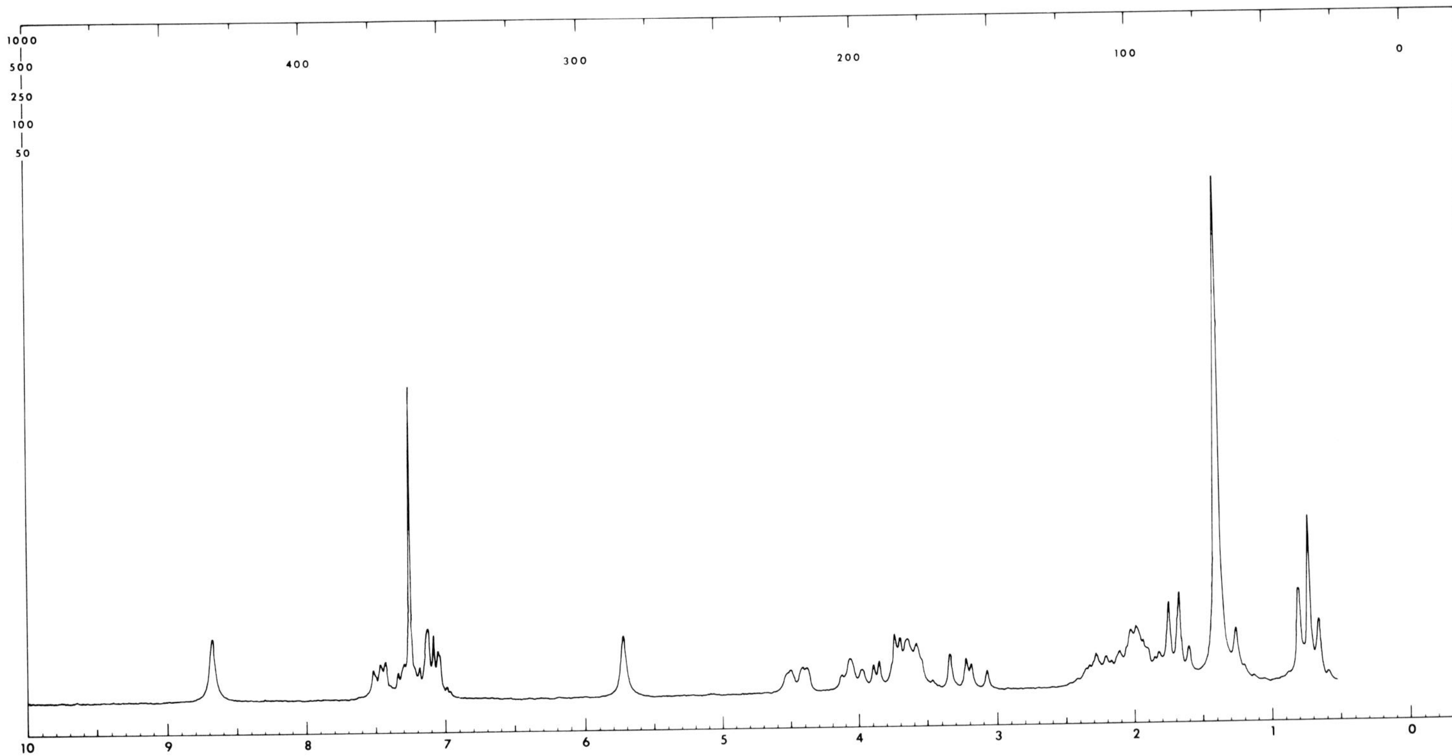

The 100 MHz ^{1}H NMR spectrum of deoxybrevianamide E in $CDCl_3$ solution.

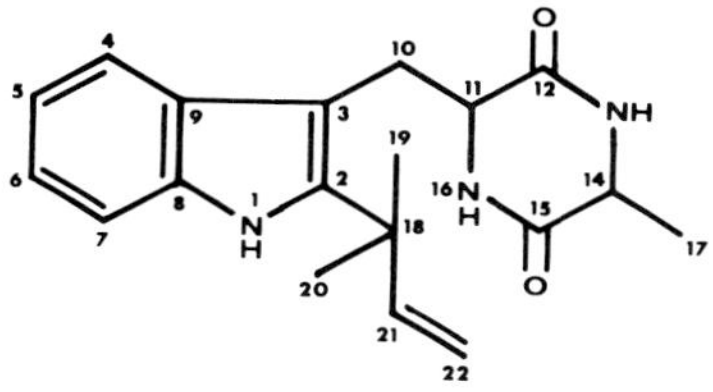

Common name	Preechinulin [(+)-L-Alanyl(1,1-dimethylallyl)tryptophan anyhydride]
Molecular weight	325.1790
Molecular formula	$C_{19}H_{23}O_2N_3$
General characteristics[51,90]	Fine white needles from methylene chloride–benzene, mp 294°–297°C $[\alpha]_D^{24}$ +50° (C = 0.38 in acetic acid) $[\alpha]_D^{20}$ +22.3° (in acetic acid)
UV data[51]	λ_{max}^{EtOH} nm (ϵ): 225(32,500), 283(7,750), and 291(7,000)
IR data[51]	ν_{max}^{KCl} cm^{-1}: 3380, 3210, 1670, 1460, 993, and 910
Fungal source[51,90]	*Aspergillus amstelodami* *A. chevalieri*
Toxicity data	Toxicity unknown
Other significant references	Biosynthesis[3]; synthesis[51]

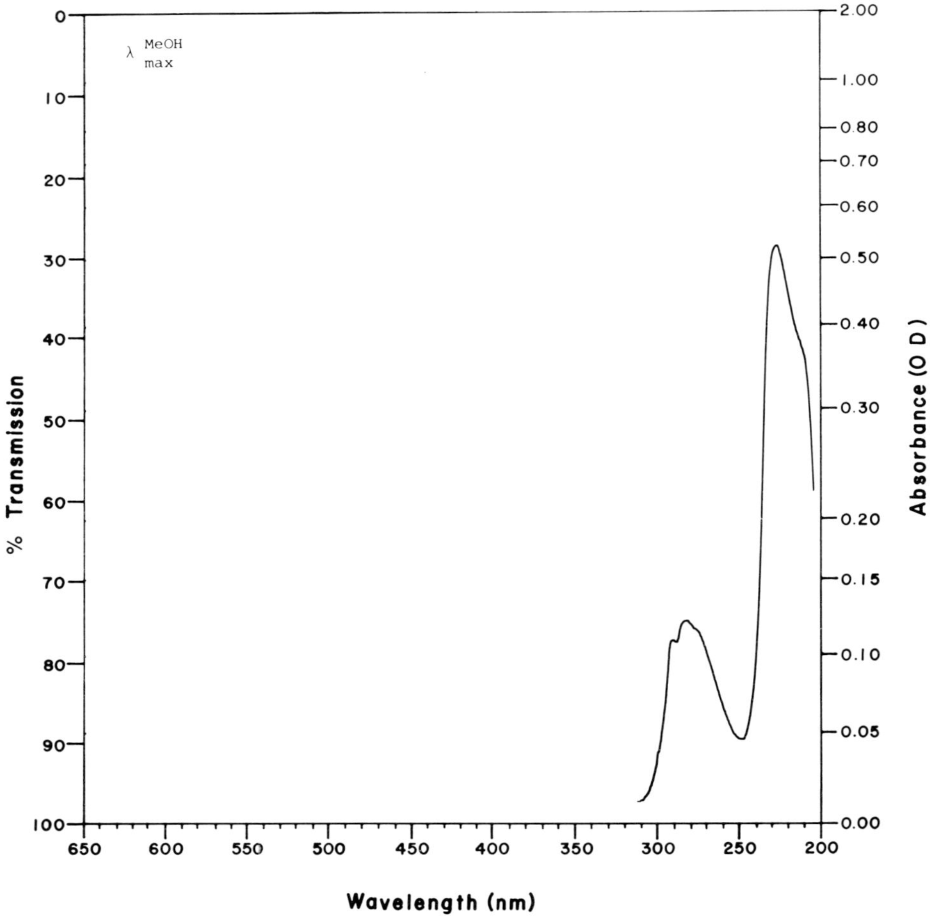
λ MeOH max
% Transmission
Absorbance (O D)
Wavelength (nm)
0
10
20
30
40
50
60
70
80
90
100
650
600
550
500
450
400
350
300
250
200
2.00
1.00
0.80
0.70
0.60
0.50
0.40
0.30
0.20
0.15
0.10
0.05
0.00

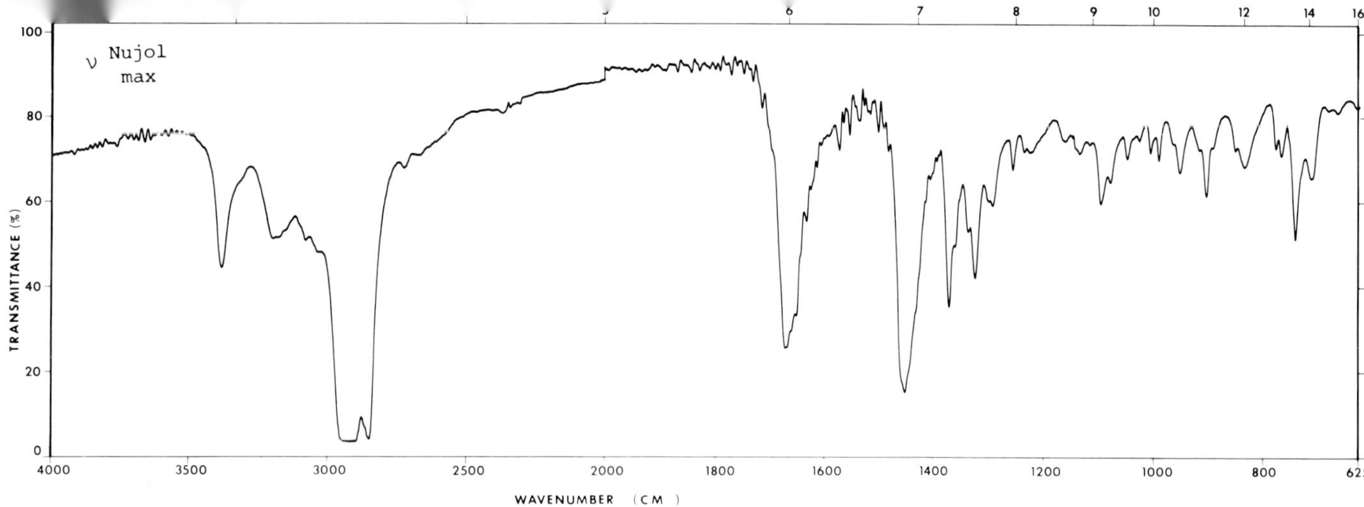

ν Nujol max
TRANSMITTANCE (%)
WAVENUMBER (CM)

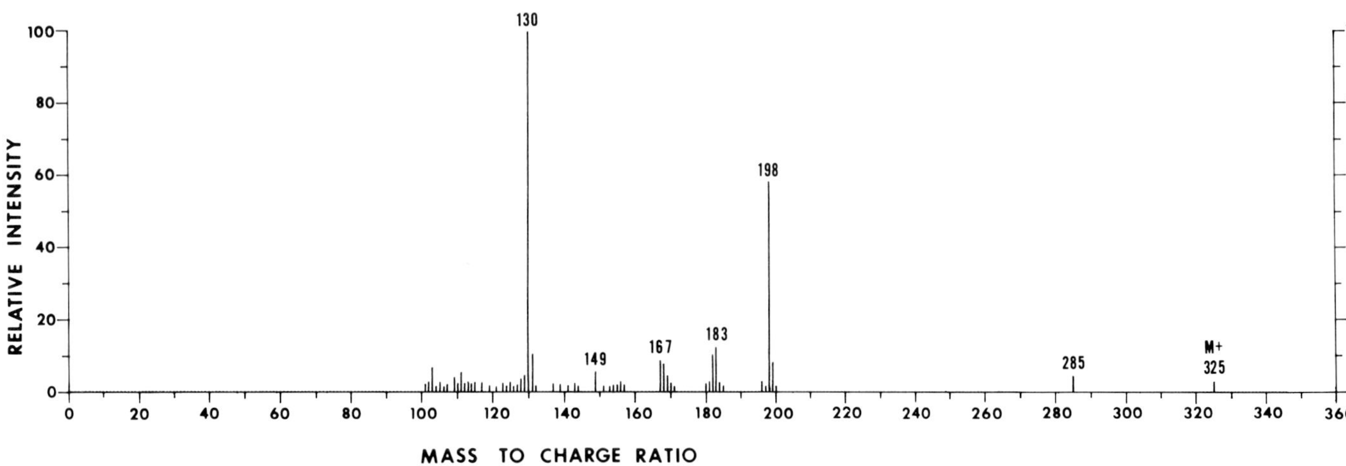

130
149
167
183
198
285
M+ 325
RELATIVE INTENSITY
MASS TO CHARGE RATIO

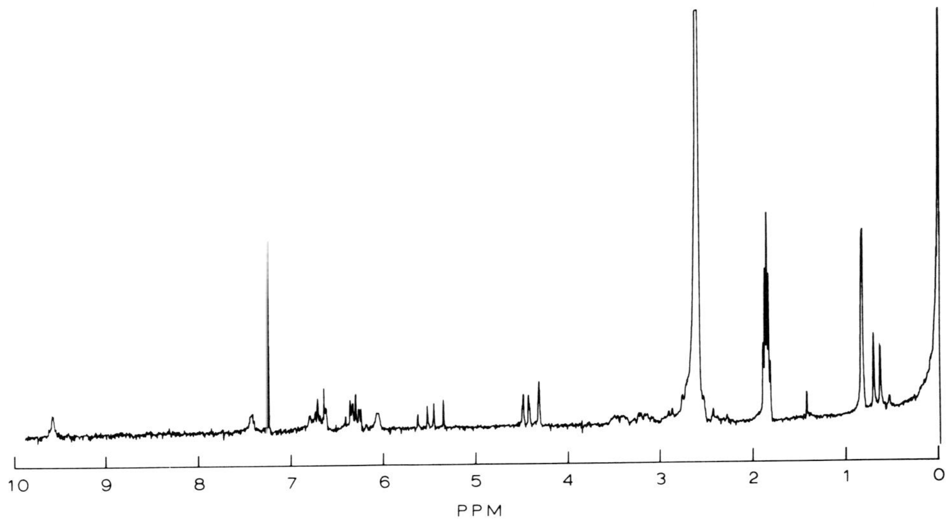

The 100 MHz ^{1}H NMR spectrum of preechinulin in $CDCl_3$/DMSO solution.

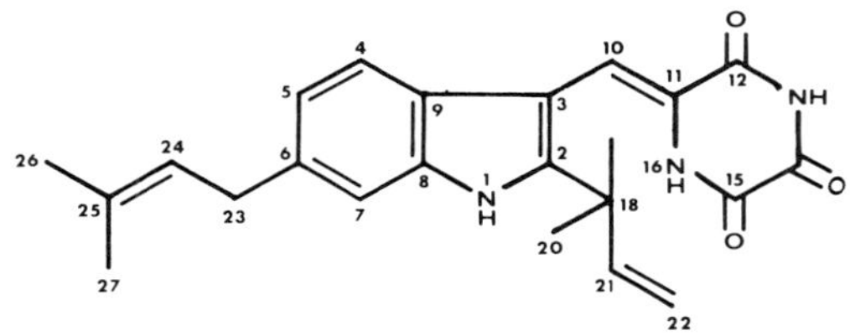

Common name	Neoechinulin
Molecular weight	391.1895
Molecular formula	$C_{23}H_{25}O_3N_3$
UV data[62]	λ_{max}^{EtOH} nm (ϵ): 231(32,300), 287(13,100), and 420(9,700)
Fungal source[62]	*Aspergillus amstelodami* (Mangin) Thom and Church
Toxicity data	Toxicity unknown
TLC data[62]	*Adsorbent:* silica gel HF_{254} *Solvent:* A: ethyl acetate–hexane, 2:1 v/v B: benzene-ethyl acetate, 1:1 v/v C: benzene-ethyl acetate, 2:1 v/v R_f: A: 0.75 B: 0.50 C: 0.30 *Detection:* not reported
^{13}*C NMR data*[4a,7]	C2, 131.1; C3, 103.4; C4, 116.3; C5, 119.2; C6, 124.0; C7, 110.8; C8, 134.6; C9, 123.1; C10, 120.9; C11, 145.7; C12, 152.0; C14, 157.0; C15, 160.2; C18, 38.9; C19, 27.7; C20, 27.7; C21, 145.1; C22, 111.9; C23, 33.9; C24, 124.0; C25, 135.7; C26, 25.4; C27, 17.6
^{1}*H NMR data*[4a,7]	H4, 7.11; H5, 6.83; H7, 7.51; H19, 1.57; H20, 1.57; H21, 6.10; H22, 5.12; 5.11; H23, 3.42; H24, 5.38; H26, 1.75; H27, 1.75; NH, 9.59; 11.06; 12.07

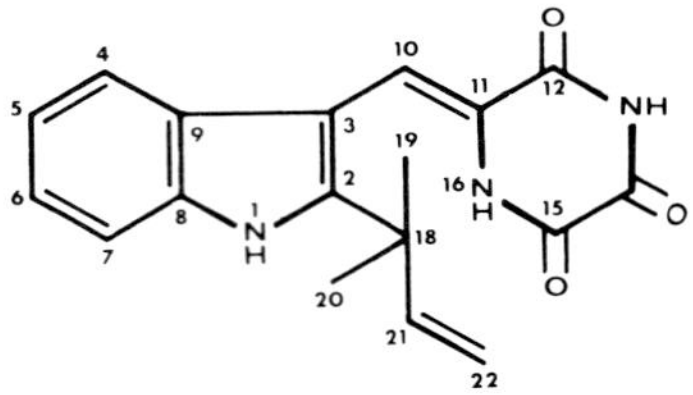

Common name	Neoechinulin E
Molecular weight	323.1267
Molecular formula	$C_{18}H_{17}O_3N_3$
General characteristics[62]	Orange-red crystals from methanol, mp 275°C
UV data[62]	λ_{max}^{EtOH} nm (ϵ): 228(37,100), 281(9,100), and 410(8,100)
IR data[62]	ν_{max}^{KBr} cm^{-1}: 3330, 3150, 3020, 2840, 1740, 1690, and 1600
Fungal source[62]	*Aspergillus amstelodami*
Toxicity data	Toxicity unknown
TLC data[62]	*Adsorbent:* silica gel HF_{254} *Solvent:* A: ethyl acetate–hexane, 2:1 v/v B: benzene–ethyl acetate, 1:1 v/v C: benzene–ethyl acetate, 2:1 v/v R_f: A: 0.51 B: 0.36 C: 0.17 *Detection:* visible light
^{1}*H NMR data*[62]	H1, 8.64; H4–7, 6.90–7.55; H10, 7.40; H13, 10.58; H16, 10.58; H19, 1.54; H20, 1.54; H21, 6.06; H23, 5.06; H24, 5.20

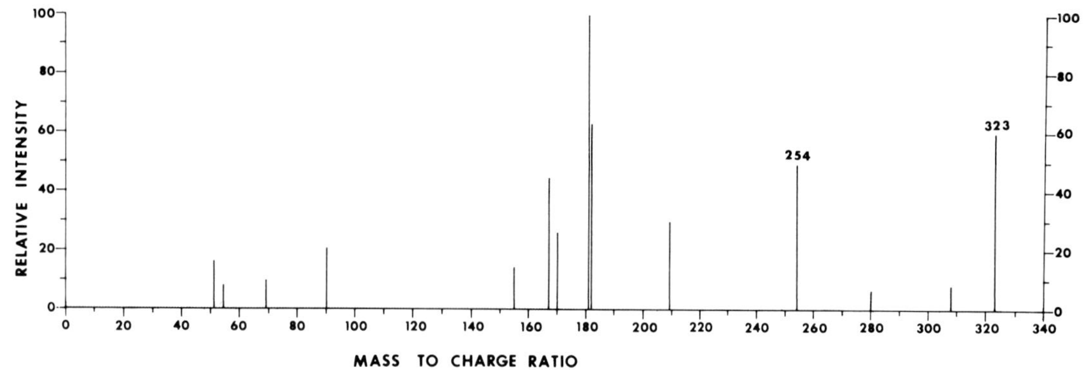
RELATIVE INTENSITY
MASS TO CHARGE RATIO
254
323
0
20
40
60
80
100
120
140
160
180
200
220
240
260
280
300
320
340

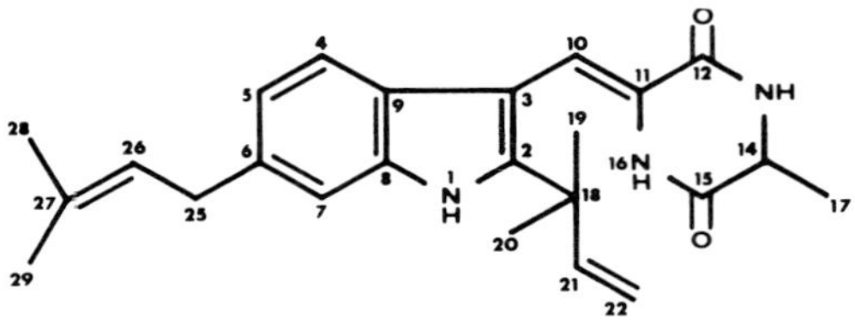

Common name	Neoechinulin D
Molecular weight	391.2259
Molecular formula	$C_{24}H_{29}O_2N_3$
General characteristics[62]	Ivory crystals from methanol, mp 223°–225°C
UV data[62]	λ_{max}^{EtOH} nm (ϵ): 231(34,600), 265(11,400), 296(10,400), and 345(10,900)
IR data[62]	ν_{max}^{KBr} cm^{-1}: 3340, 2940, 1670, and 1630
Fungal source[68]	*Aspergillus amstelodami* *A. ruber*
Toxicity data	Toxicity unknown
TLC data[62]	*Adsorbent:* silica gel HF_{254} *Solvent:* A: ethyl acetate–hexane, 2:1 v/v B: benzene–ethyl acetate, 1:1 v/v C: benzene–ethyl acetate, 2:1 v/v R_f: A: 0.37 B: 0.23 C: 0.12 *Detection:* not reported
^{1}*H NMR data*[62]	H1, 8.32; H4,5,7, 6.95–7.40; H10, 7.01; H13, 10.38; H14, 4.25; H16, 10.62; H17, 1.53; H19, 1.48; H20, 1.48; H21, 6.11; H23, 5.06; H24, 5.08; H25, 3.40; H26, 5.33; H28, 1.72; H29, 1.72

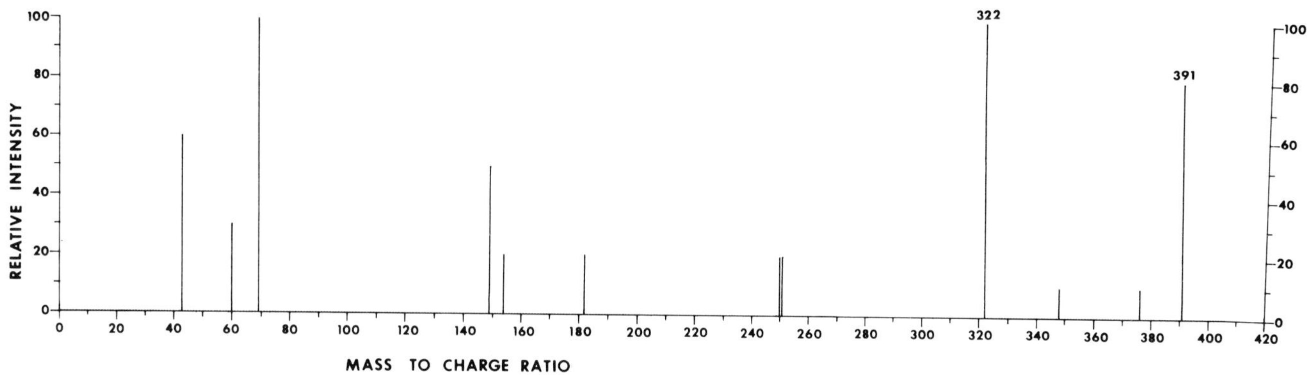

322
391
RELATIVE INTENSITY
0
20
40
60
80
100
0
20
40
60
80
100
120
140
160
180
200
220
240
260
280
300
320
340
360
380
400
420
MASS TO CHARGE RATIO

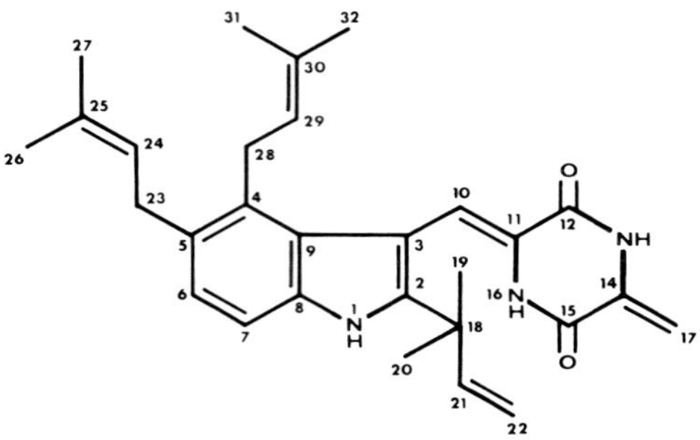

Common name	Cryptoechinulin G
Molecular weight	457.2718
Molecular formula	$C_{29}H_{35}O_2N_3$
General characteristics[40]	Amorphous solid, optically inactive
Fungal source[40]	*Aspergillus ruber*
Toxicity data	Toxicity unknown
^{13}*C NMR data*[40]*	C2, 126.6; C3, 101.3; C4, 128.9; C5, 130.3; C6, 123.0; C7, 109.5; C8, 134.5; C9, 130.2; C10, 114.6; C11, 142.2; C12, 156.6; C14, 134.6; C15, 156.1; C17, 100.4; C21, 145.4; C22, 111.1; C24, 124.9; C25, 130.2; C29, 124.7; C30, 130.2

* Data for saturated carbons not reported.

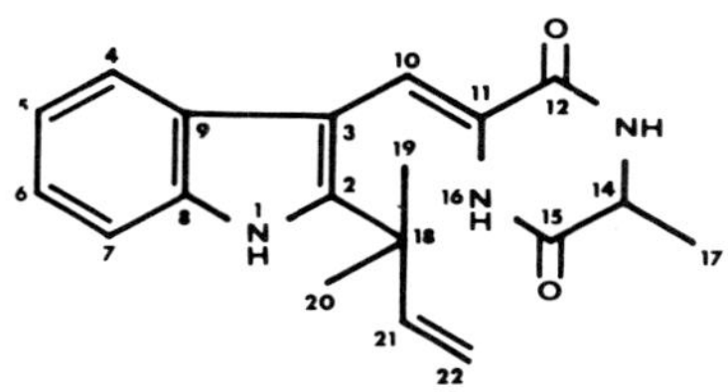

Common name	Neoechinulin A
Molecular weight	323.1633
Molecular formula	$C_{19}H_{21}O_2N_3$
General characteristics[62]	Ivory crystals from methanol, mp 264°–265°C
UV data[62]	λ_{max}^{EtOH} nm (ϵ): 229(23,300), 286(11,700), 292(10,400), and 338(9,500)
IR data[62]	ν_{max}^{KBr} cm^{-1}: 3360, 2980, 1670, and 1630
Fungal source[62]	*Aspergillus ruber* *A. amstelodami*
Toxicity data	Toxicity unknown
TLC data[62]	*Adsorbent:* silica gel HF_{254} *Solvent:* A: ethyl acetate–hexane, 2:1 v/v B: benzene–ethyl acetate, 1:1 v/v C: benzene–ethyl acetate, 2:1 v/v R_f: A: 0.24 B: 0.17 C: 0.09 *Detection:* not reported
^{1}H NMR data[62]	H1, 8.30; H4, 7.00–7.50; H5, 7.00–7.50; H6, 7.00–7.50; H7, 7.00–7.50; H10, 6.93; H13, 8.30; H14, 4.15; H16, 10.96; H17, 1.42; H19, 1.50; H20, 1.50; H21, 6.12; H23, 5.04; H24, 5.06
^{13}C NMR data[68b]	C2, 144.1; C3, 103.3; C4, 118.7; C5, 120.8; C6, 119.5; C7, 110.2; C8, 135.2; C9, 126.0; C10, 111.6; C11, 124.8; C12, 159.7; C14, 50.5; C15, 166.3; C17, 17.9; C18, 40.2; C19, 27.5; C20, 27.5; C21, 145.2; C22, 111.6

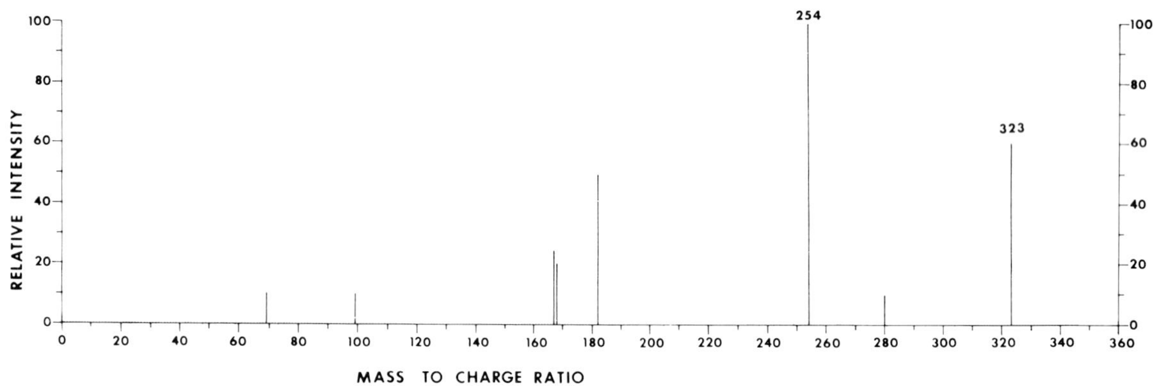
254
323
RELATIVE INTENSITY
MASS TO CHARGE RATIO

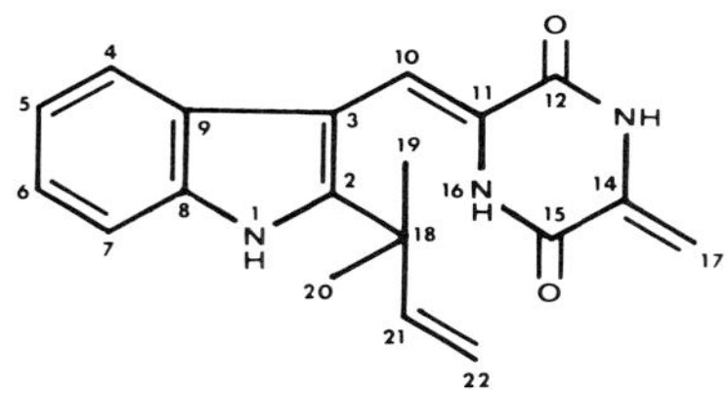

Common name	Neoechinulin B
Molecular weight	321.1477
Molecular formula	$C_{19}H_{19}O_2N_3$
General characteristics[62]	Yellow crystals from methanol, mp 234°–236°C
UV data[62]	λ_{max}^{EtOH} nm (ϵ): 228(28,100), 273(19,000), 284(18,100), and 374(10,400)
IR data[62]	ν_{max}^{KBr} cm^{-1}: 3360, 2980, 2925, 1680, and 1645
Fungal source[62]	*Aspergillus amstelodami*
Toxicity data	Toxicity unknown
TLC data[62]	*Adsorbent:* silica gel HF_{254} *Solvent:* A: ethyl acetate–hexane, 2:1 v/v B: benzene–ethyl acetate, 1:1 v/v C: benzene–ethyl acetate, 2:1 v/v R_f: A: 0.77 B: 0.46 C: 0.26 *Detection:* not reported
^{1}*H NMR data*[62]	H1, 8.36; H4–7, 6.70–7.50; H10, 6.85; H13, 10.61; H16, 10.86; H17, 5.08, 5.34; H19, 1.50; H20, 1.50; H21, 6.07; H23, 5.07; H24, 5.09

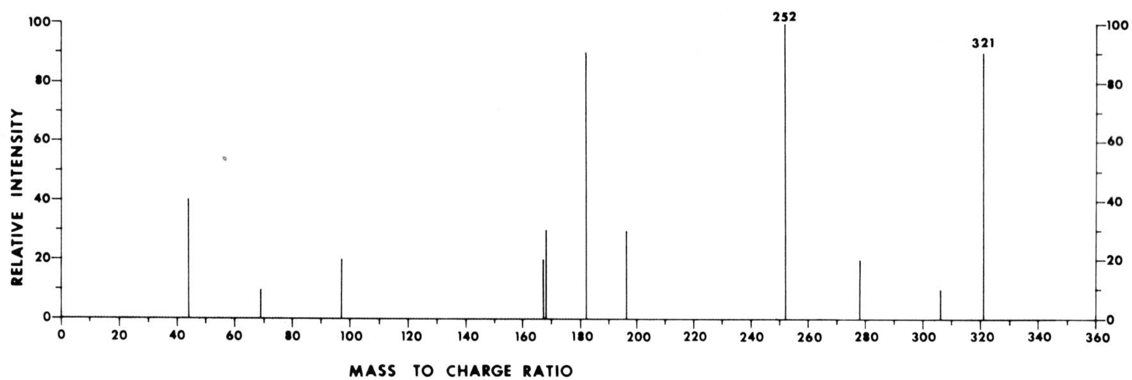

252
321
RELATIVE INTENSITY
MASS TO CHARGE RATIO

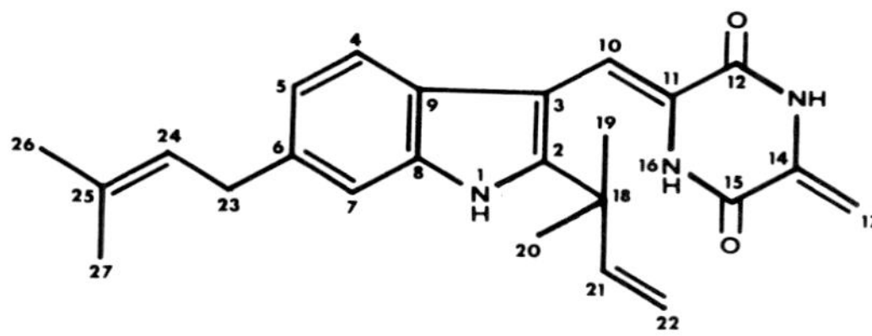

Common name	Neoechinulin C
Molecular weight	389.2103
Molecular formula	$C_{24}H_{27}O_2N_3$
General characteristics[62]	Yellow crystals from light petroleum–benzene 1:1 v/v, mp 205°–207°C
UV data[62]	λ_{max}^{EtOH} nm (ϵ): 231(34,600), 275(22,300), 290(18,600), and 380(12,000)
IR data[62]	ν_{max}^{KBr} cm^{-1}: 3350, 2970, 2925, 1680, and 1640
Fungal source[62]	*Aspergillus amstelodami*
Toxicity data	Toxicity unknown
TLC data[62]	*Adsorbent:* silica gel HF_{254} *Solvent:* A: ethyl acetate–hexane, 2:1 v/v B: benzene–ethyl acetate, 1:1 v/v C: benzene–ethyl acetate, 2:1 v/v R_f: A: 0.80 B: 0.53 C: 0.39 *Detection:* not reported
^{1}*H NMR data*[62]	H1, 8.66; H4,5,7, 6.80–7.40; H10, 7.01; H13, 10.79; H16, 10.96; H17, 4.96, 5.27; H19, 1.48; H20, 1.48; H21, 6.07; H23, 5.06; H24, 5.08; H25, 3.40; H26, 5.35; H28, 1.76; H29, 1.76

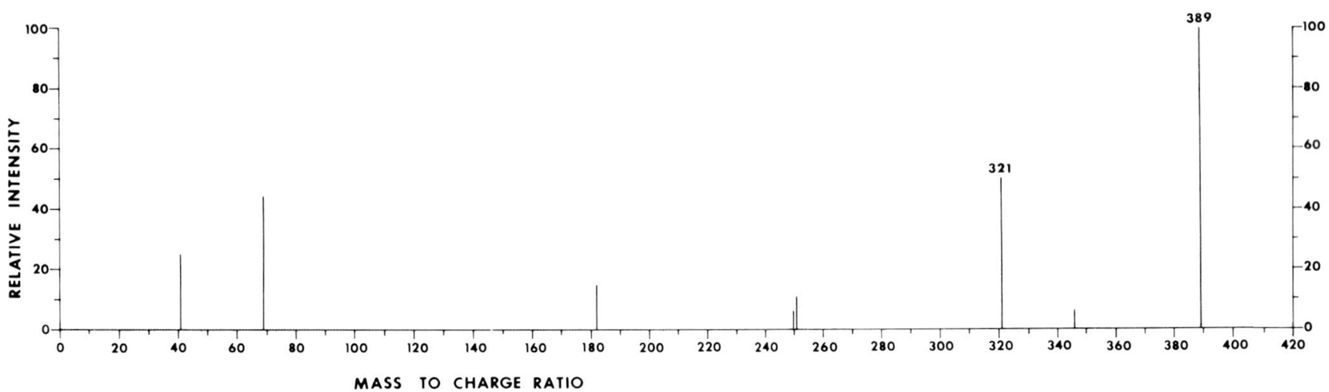
389
321
RELATIVE INTENSITY
MASS TO CHARGE RATIO
0
20
40
60
80
100
120
140
160
180
200
220
240
260
280
300
320
340
360
380
400
420

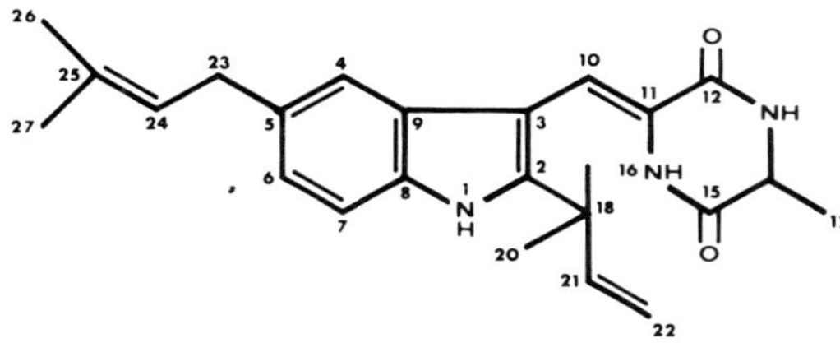

Common name	Isoechinulin A
Molecular weight	391.2259
Molecular formula	$C_{24}H_{29}O_2N_3$
UV data[68a]	λ_{max}^{EtOH} nm (ϵ): 227(31,500), 289(8,900), and 341(9,400)
IR data[68a]	ν_{max}^{KBr} cm^{-1}: 3260, 1675, and 1633
Fungal source[68a]	*Aspergillus ruber*
Toxicity data[68a]	Isoechinulin A inhibited growth of silkworm larvae
Other significant references	^{13}C NMR[68b]
^{1}*H NMR data*[68a]	H1, 10.21; H4, 7.10; H6, 6.99; H7, 7.34; H10, 7.15; H13, 7.92; H14, 4.28; H16, 7.53; H17, 1.55; H19, 1.55; H20, 1.55; H21, 6.16; H23, 5.13; H24, 5.13; H25, 3.40; H26, 5.38; H28, 1.71; H29, 1.71
^{13}*C NMR data*[68b]	C2, 143.9; C3, 102.8; C4, 123.2*; C5, 132.9†; C6, 117.9; C7, 111.1‡; C8, 134.6; C9, 126.4; C10, 112.1‡; C11, 124.3; C12, 160.3; C14, 51.6; C15, 165.8; C17, 20.9; C18, 39.3; C19, 27.4; C20, 27.4; C21, 144.1; C22, 113.1; C23, 34.6; C24, 124.0*; C25, 132.0†; C26, 25.8; C27, 17.9

* Assignments may be interchanged.
† Assignments may be interchanged.
‡ Assignments may be interchanged.

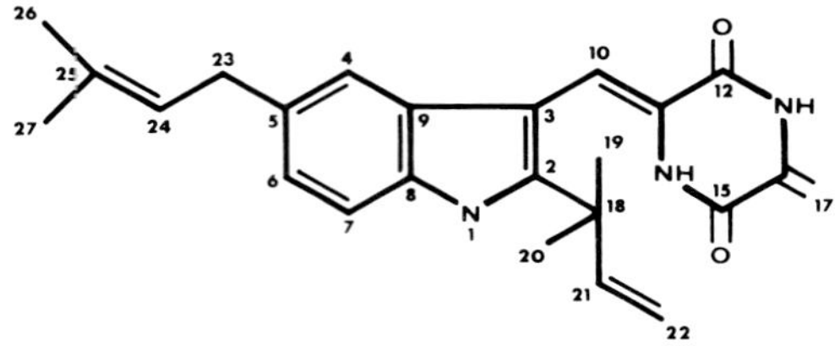

Common name	Isoechinulin B
Molecular weight	389.2103
Molecular formula	$C_{24}H_{27}O_2N_3$
UV data[68a]	λ_{max}^{MeOH} nm (ϵ): 228(27,900), 272(18,700), 285(sh)(16,600), and 370(10,400)
IR data[68a]	ν_{max}^{KBr} cm^{-1}: 3310, 1678, and 1643
Fungal source[68a]	*Aspergillus ruber*
Toxicity data	Toxicity unknown
Other significant references	^{13}C NMR[68b]
^{1}*H NMR data*[68a]	H1, 10.23; H4, 7.00–7.20; H6, 6.96; H7, 7.30; H10, 7.00–7.20; H13, 9.80; H16, 8.08; H17, 5.00–5.40; H19, 1.54; H20, 1.54; H21, 6.13; H23, 5.00–5.40; H24, 5.00–5.40; H25, 3.38; H26, 5.00–5.40; H28, 1.68; H29, 1.68
^{13}*C NMR data*[68b]	C2, 144.6; C3, 102.7; C4, 123.2*; C5, 133.0†; C6, 118.0; C7, 111.3‡; C8, 134.7; C9, 126.0; C10, 113.5‡; C11, 124.0; C12, 158.5; C14, 133.6; C15, 156.1; C17, 103.0; C18, 39.2; C19, 27.4; C20, 27.4; C21, 144.2; C23, 34.5; C24, 124.0*; C25, 132.0†; C26, 25.7; C27, 17.8

* Assignments may be interchanged.
† Assignments may be interchanged.
‡ Assignments may be interchanged.

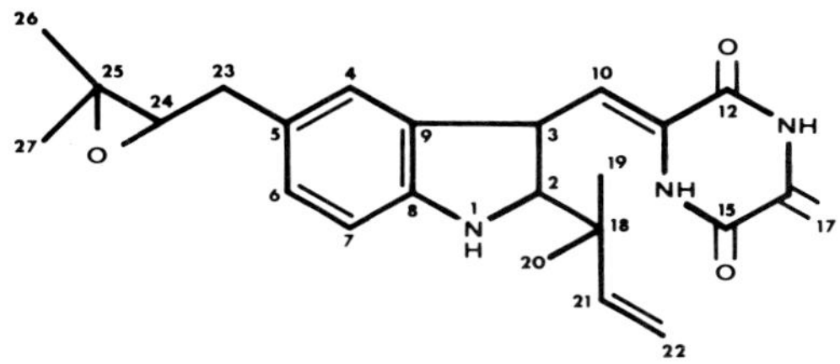

Common name	Isoechinulin C
Molecular weight	405.2051
Molecular formula	$C_{24}H_{27}O_3N_3$
UV data[68a]	λ_{max}^{MeOH} nm (ϵ): 229(26,700), 272(18,000), 286(sh)(15,800), and 371(9,700)
IR data[68a]	ν_{max}^{KBr} cm^{-1}: 3250, 1676, and 1644
Fungal source[68a]	*Aspergillus ruber*
Toxicity data	Toxicity unknown
Other significant references	^{13}C NMR[68b]
^{1}*H NMR data*[68a]	H1, 10.21; H4, 7.02; H6, 7.02; H7, 7.31; H10, 7.07; H13, 9.73; H16, 8.12; H17, 5.00; 5.34; H19, 1.56; H20, 1.56; H21, 6.11; H23, 5.08; H24, 5.05; H25, 2.70–3.10; H26, 2.70–3.10; H28, 1.22; H29, 1.33
^{13}*C NMR data*[68b]	C2, 140.2; C3, 104.5; C4, 124.5; C5, 131.2; C6, 119.9; C7, 112.8*; C8, 136.4; C9, 127.8; C10, 112.9*; C11, 126.4; C12, 158.3; C14, 135.5; C15, 157.0; C17, 100.6; C18, 40.6; C19, 28.4; C20, 28.4; C21, 146.2; C22, 112.9*; C23, 36.7; C24, 65.9; C25, 59.4; C26, 25.5; C27, 19.6

* Assignments may be interchanged.

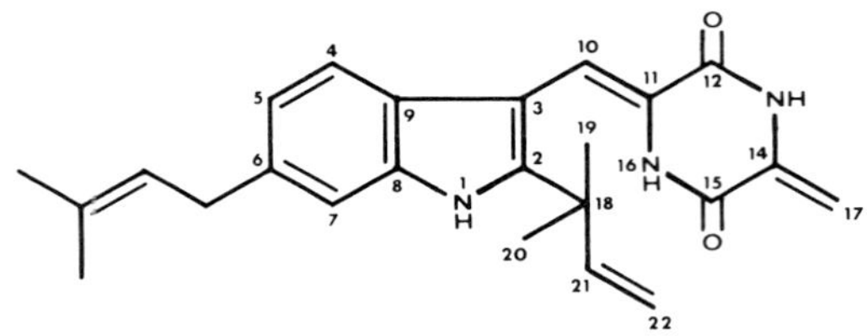

Common name	Cryptoechinulin A
Molecular weight	389.2103
Molecular formula	$C_{24}H_{27}O_2N_3$
General characteristics[7,8]	Optically inactive, forms yellowish crystals from benzene, mp 190°–192°C
UV data[7,8]	λ_{max}^{EtOH} nm (ϵ): 231(37,000), 273(24,100), 283(sh)(21,400), and 380(14,000)
Fungal source[7]	*Aspergillus amstelodami*
Toxicity data	Toxicity unknown
^{13}C NMR data[7]	C2, 131.1; C3, 102.8; C4, 112.0; C5, 118.5; C6, 124.0; C7, 110.9; C8, 134.4; C9, 124.0; C10, 120.9; C11, 142.2; C12, 155.7; C14, 134.8; C15, 157.1; C17, 100.0; C18, 38.6; C19, 27.5; C20, 27.5; C21, 144.1; C22, 111.7; C23, 33.9; C24, 124.0; C25, 135.6; C26, 25.4; C27, 17.6

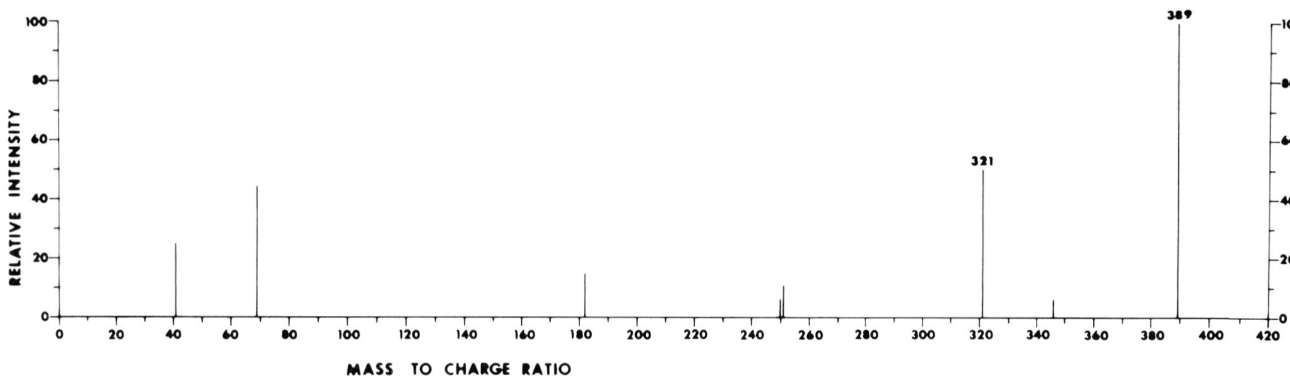

389
321
RELATIVE INTENSITY
100
80
60
40
20
0
0 20 40 60 80 100 120 140 160 180 200 220 240 260 280 300 320 340 360 380 400 420
MASS TO CHARGE RATIO

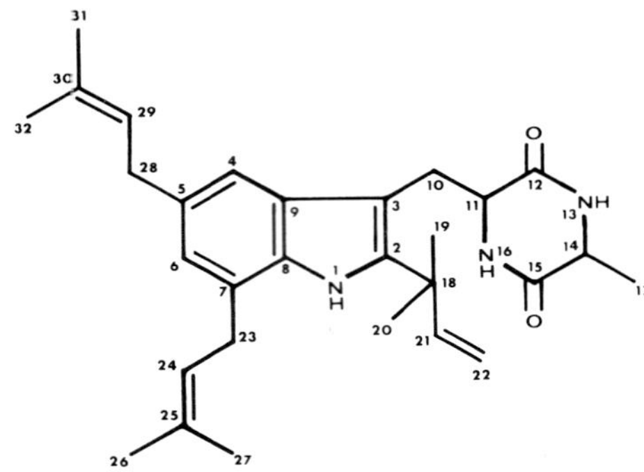

Common name	Echinulin (3{[2-(1,1-dimethylallyl)-5,7-bis(3-methyl-2-butenyl)indol-3-yl]methyl}-6-methyl-2,5-piperazinedione)
Molecular weight	461.3042
Molecular formula	$C_{29}H_{39}O_2N_3$
General characteristics[9,59,90]	Needles from butanol, mp 242°–243°C. Soluble in glacial acetic acid, chloroform, pyridine, dioxane; slightly soluble in warm alcohol, butanol, benzene, ethyl ether, acetone, and carbon tetrachloride.
UV data[9]	λ_{max}^{EtOH} nm (ϵ): 230(39,810), 279(9,549), and 286(9,120)
Fungal source[59,90]	*Aspergillus amstelodami* *A. chevalieri* (Mangin) Thom and Church *A. echinulatus*
Toxicity data	Toxicity unknown
TLC data[62]	*Adsorbent:* silica gel HF_{254} *Solvent:* A: ethyl acetate–hexane, 2:1 v/v B: benzene–ethyl acetate, 1:1 v/v C: benzene–ethyl acetate, 2:1 v/v R_f: A: 0.14 B: 0.12 C: 0.05 *Detection:* not reported
Other significant references	Chemistry[9]; biochemistry[2]; biosynthesis.[3]
^{13}*C NMR data*[68b]	C2, 141.1 s; C3, 104.5 s; C4, 121.5 d; C5, 130.4 s; C6, 114.5 d; C7, 124.4 s; C8, 131.8 s; C9, 130.4 s;

C10, 29.9 t; C11, 55.2 d; C12, 167.4 s; C14, 50.4 d; C15, 167.8 s; C17, 20.2 q; C18, 38.9 s; C19, 27.9 q; C20, 27.9 q; C21, 145.9 d; C22, 111.2 t; C23, 30.4 t; C24, 122.3 d; C25, 130.4 s; C26, 25.6 q; C27, 17.7 q; C28, 34.3 t; C29, 123.2 d; C30, 131.9 s; C31, 25.6 q; C32, 17.7 q

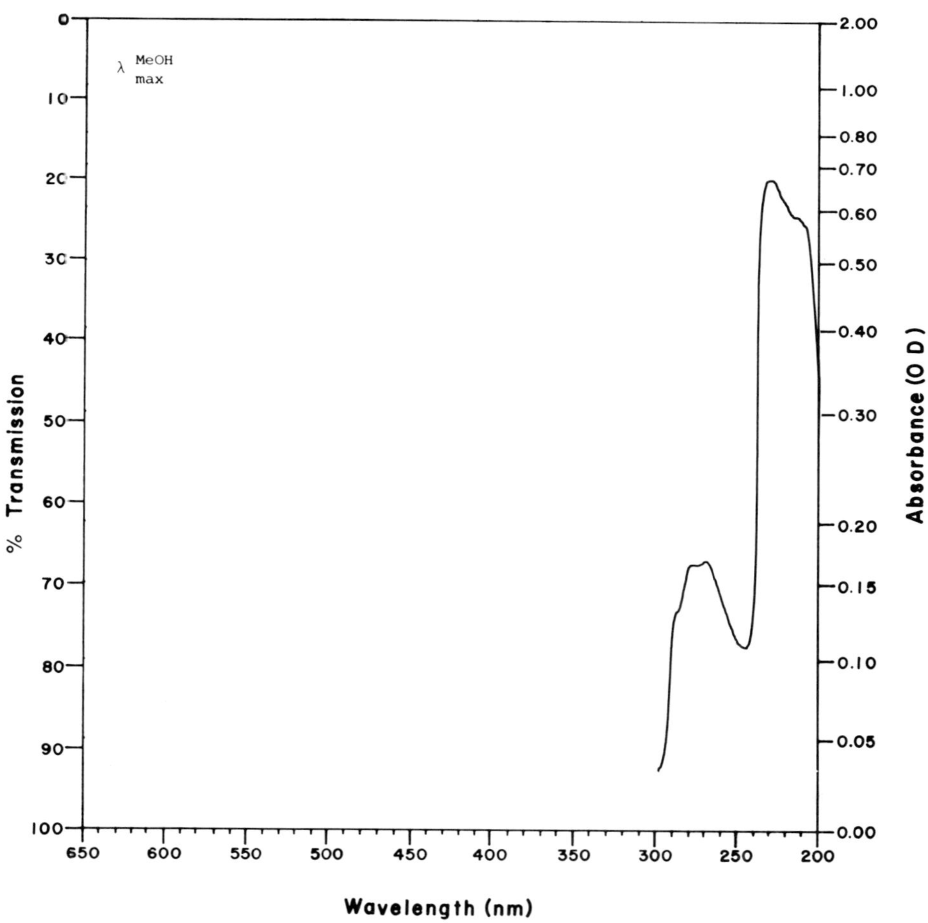

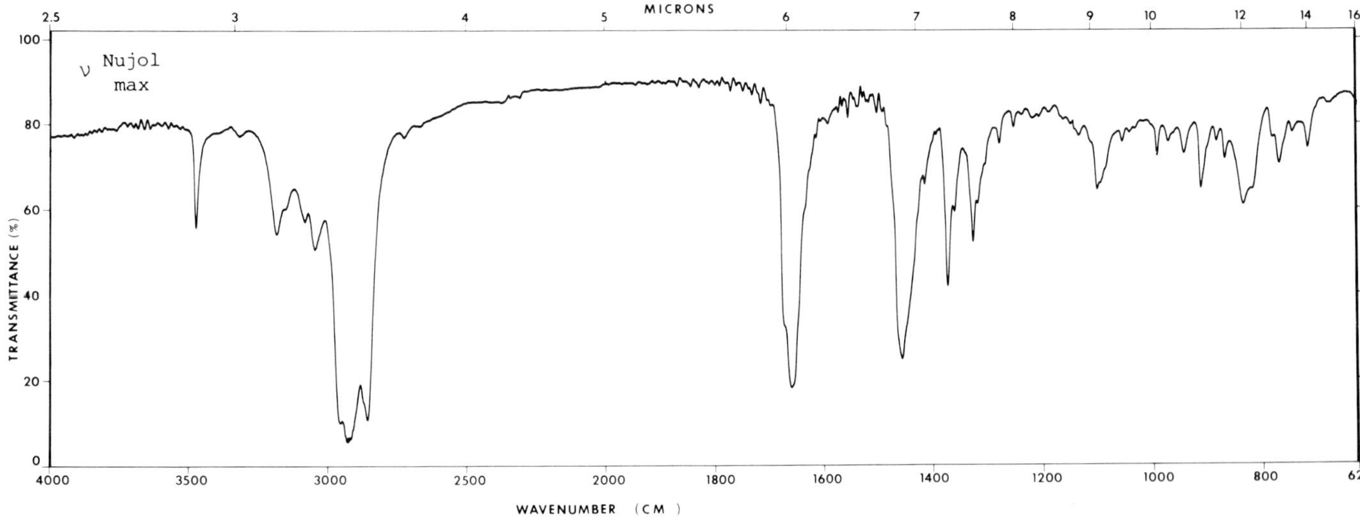
MICRONS
2.5
3
4
5
6
7
8
9
10
12
14
16
ν Nujol max
TRANSMITTANCE (%)
100
80
60
40
20
0
4000
3500
3000
2500
2000
1800
1600
1400
1200
1000
800
625
WAVENUMBER (CM)

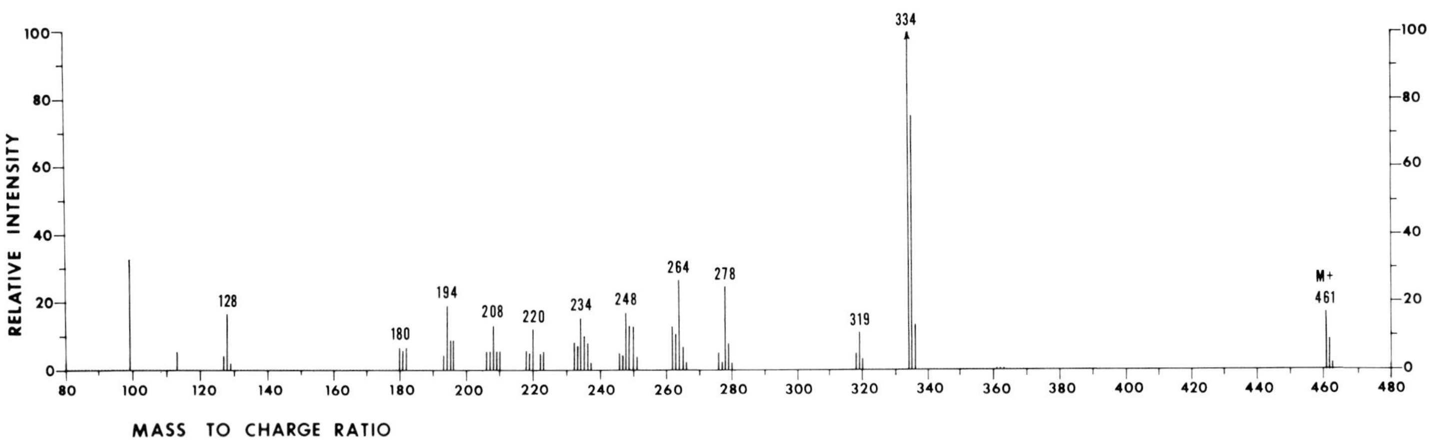
RELATIVE INTENSITY
100
80
60
40
20
0
334
M+
461
128
180
194
208
220
234
248
264
278
319
80
100
120
140
160
180
200
220
240
260
280
300
320
340
360
380
400
420
440
460
480
MASS TO CHARGE RATIO

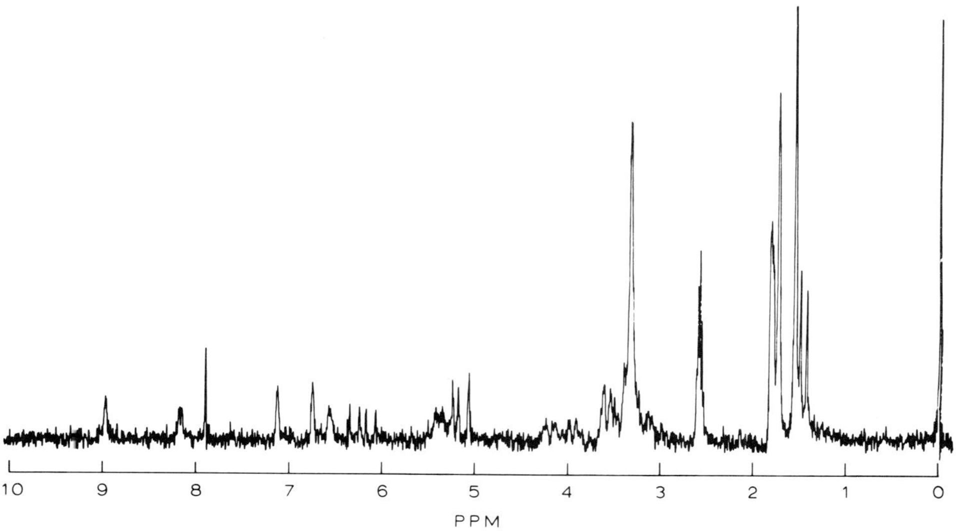

The 100 MHz ^1H NMR spectrum of echinulin in $CDCl_3$/DMSO solution.

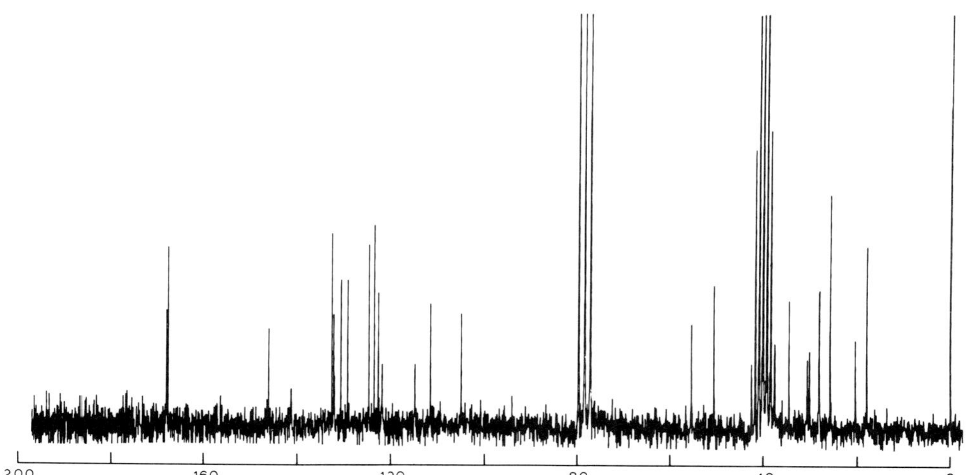

The 25 MHz ^{13}C NMR spectrum of echinulin in $CDCl_3$/DMSO solution.

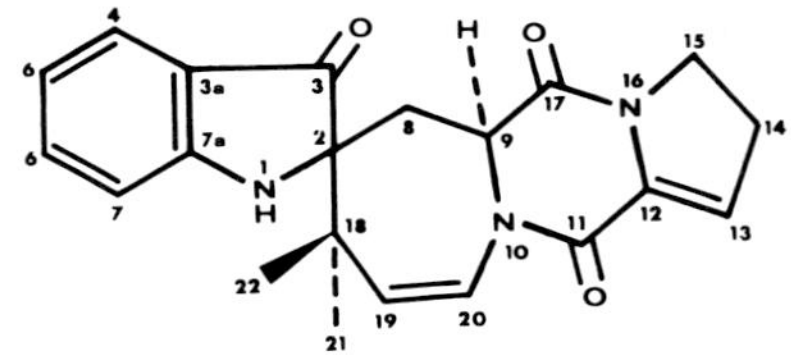

Common name	Austamide
Molecular weight[86]	363.1582
Molecular formula	$C_{21}H_{21}O_3N_3$
General characteristics[86]	A yellow amorphous compound or homogeneous powder $[\alpha]_D^{20}$ +152° (C = 1 in ethanol)
UV data[86]	λ_{max}^{EtOH} nm (ϵ): 234(26,302), 256(117), 268(sh)(1,096), 282(8,709), and 392(2,691) (unchanged upon addition of acid or base)
CD data[86]	$\Delta\epsilon$ (in methanol) nm: 390(+2.0), 285(+1.8), 234(20–25)
IR data[86]	$\nu_{max}^{CHCl_3}$ cm^{-1}: 3420, 1700, 1680, and 1650
Fungal source[86]	*Aspergillus ustus* (Bainier) Thom and Church (C.S.I.R. 1128; NRRL 5856)
Toxicity data[87]	Austamide was one of the active components in extracts of *A. ustus* (cultured on maize) causing acute toxicosis in day old ducklings.
TLC data[86,87]	*Adsorbent:* silica gel *Solvent:* chloroform–methanol, 97:3 v/v R_f: not reported *Detection:* not reported
Other significant references	X-Ray structure[15]
^{13}C Chemical shifts	23.6, 26.1, 28.0, 41.9, 42.0, 45.5, 62.5, 70.7, 111.9, 118.7, 120.3, 120.4, 124.4, 125.4, 128.3, 132.2, 137.3, 154.7, 160.1, 161.1, 200.8

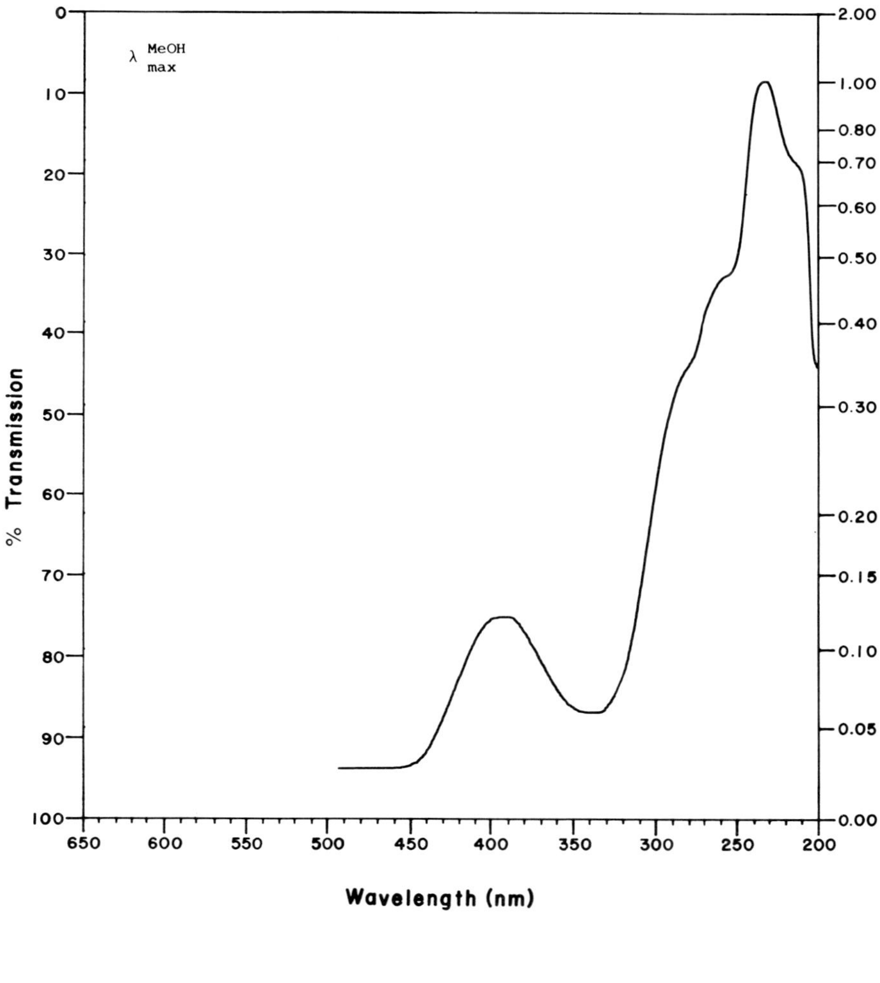
λ MeOH max
% Transmission
Absorbance (O D)
Wavelength (nm)
0
10
20
30
40
50
60
70
80
90
100
2.00
1.00
0.80
0.70
0.60
0.50
0.40
0.30
0.20
0.15
0.10
0.05
0.00
650
600
550
500
450
400
350
300
250
200

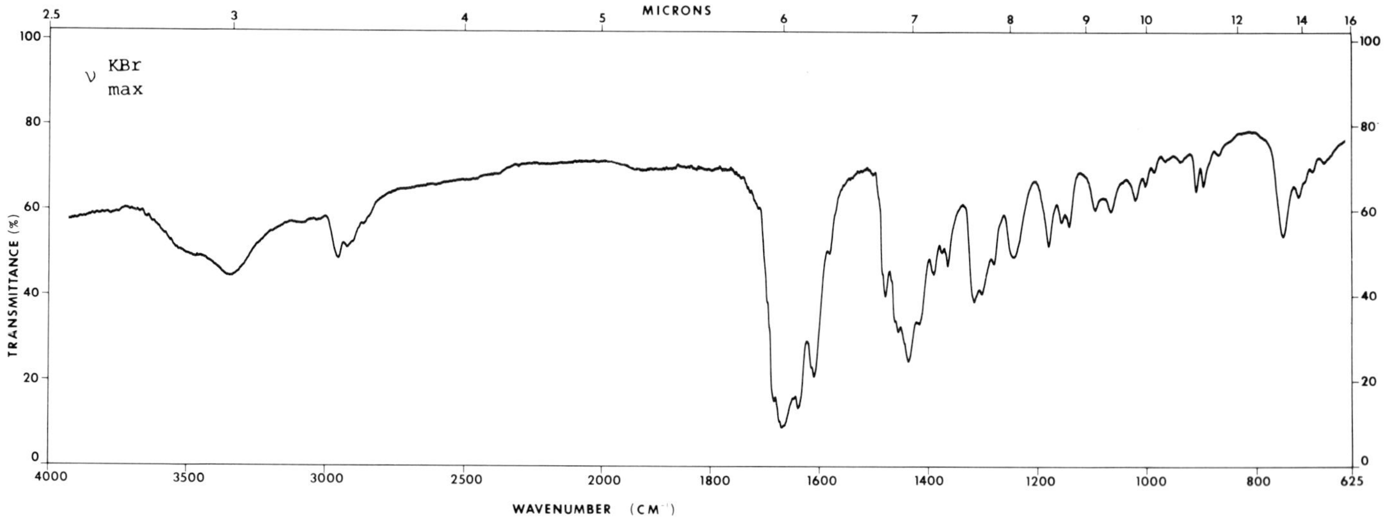
MICRONS
ν KBr max
TRANSMITTANCE (%)
WAVENUMBER (CM⁻¹)

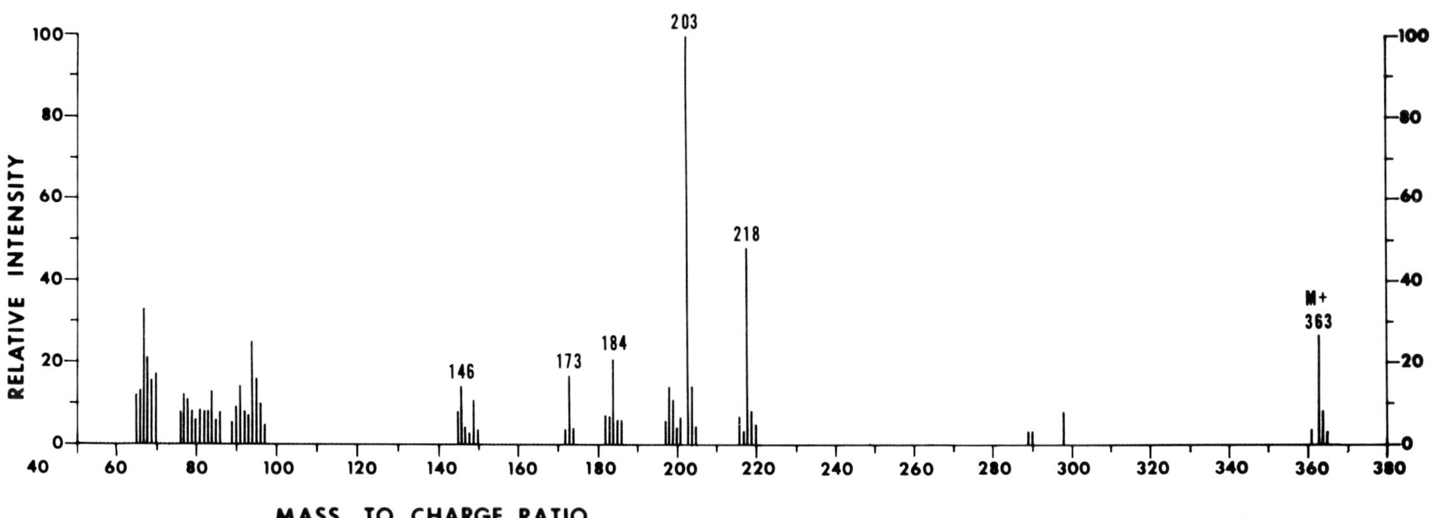
RELATIVE INTENSITY
146
173
184
203
218
M+
363
MASS TO CHARGE RATIO

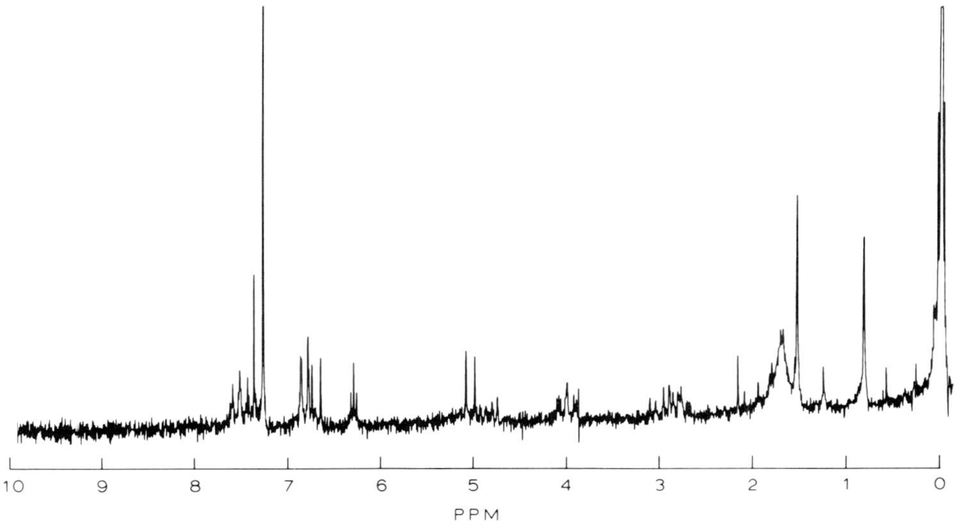

The 100 MHz ^1H NMR spectrum of austamide in $CDCl_3$ solution.

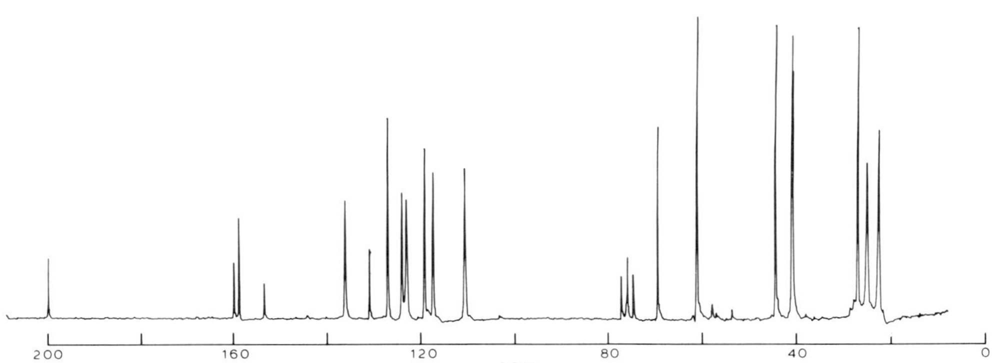

The 25 MHz ^{13}C NMR spectrum of austamide in $CDCl_3$ solution.

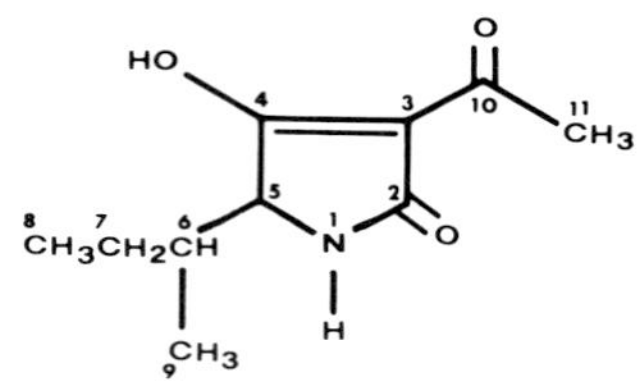

Common name	Tenuazonic acid (3-Acetyl-5-*sec*-butyltetramic acid)
Molecular weight	197.1051
Molecular formula	$C_{10}H_{15}O_3N$
General characteristics[88]	Tenuazonic acid is a homogenous oil; Ca and Mg tenuazonate occurs as an amorphous powder. The powder is insoluble in water; slightly soluble in chloroform, and very soluble in methanol. It gives a red-brown color with $FeCl_3$; yellow with 2,4-DNP; gray brown with 1% $Ce(SO_4)_2$ in 6*N* H_2SO_4
UV data[75,88,89]	λ_{max}^{EtOH} nm (ϵ): 217(5,248) and 277(13,489) $\lambda_{max}^{H_2O}$ nm (ϵ): 239(11,220) and 279(14,454) $\lambda_{max}^{0.09\,N\ NaOH}$ nm (ϵ): 240(11,748) and 279(14,791) λ_{max}^{HCl} nm (ϵ): 220(6,309) and 277(12,589)
CD data[88]	$\Delta\epsilon$ (in methanol) nm: 235 (-3.7) and 275 (-6.9)
Fungal source[30,52,53,55,64,75,88,89,92]	*Pyricularia oryzae* Cav. *Alternaria tenuis* Auct. *A. longipes* *A. kikuchiana* *A. mali* *A. alternata* *A. tenuissima* *Phoma sorghina* Sacc. *Aspergillus* spp. *Sphacropsidales* spp.
Toxicity data[54,58,65,67,72,79,80,93,94]	*Phytotoxicity:* Tenuazonic acid has been found as the major vivotoxin in naturally infected rice plants at 2.6 mg/kg of plant tissue. It exhibited conspicuous stunting on seedling growth of several species and inhibited the growth of suspended cells of soybean and rice. Biochemically, it inhibited pro-

tein and nucleic acid synthesis. It has also been isolated as vivotoxin in tobacco brown-spot disease.

Animal toxicity: Biological activities reported for tenuazonic acid are: antitumor, antiviral, inhibitor of protein synthesis *in vivo* and *in vitro* and exhibited a protective effect on *in vivo* cell death of cells exposed to 1-β-D-arabinofuranosylcytosine. The protective effect was attributed to inhibition of protein synthesis.

Rabie *et al.*[72] have implicated a mycotoxin from *P. sorghina* in the etiology of onyalai, a hematologic disorder which occurs widely among the black African population south of the Sahara. The role of tenuazonic acid in this disease is unknown.

TLC data[88]

Adsorbent: silica gel
Solvent: chloroform–methanol, 8.5:1.5 v/v
R_f: 0.30 (tenuazonates)
Detection: red-brown color after spraying with ethanolic ferric chloride

Other significant references

Chemistry[9,88]

^{13}C NMR data

C1, 15.9; C2, 198.1; C3, 103.9; C4, 175.2; C5, 187.0; C6, 67.2; C7, 38.2; C8, 20.2; C9, 24.8; C10, 12.1

^{1}H NMR data

H1, 2.30; H6, 3.94; H7, 2.48; H8, 1.18; H9, 2.48; H10, 1.00

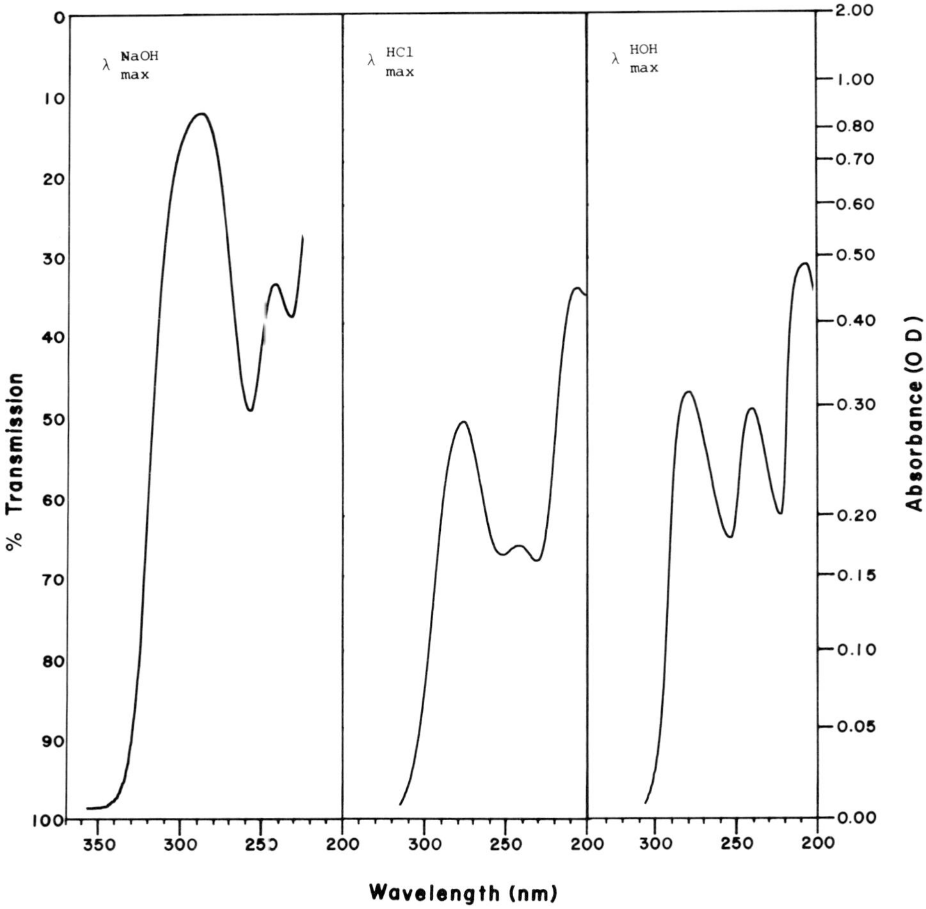
λ NaOH max
λ HCl max
λ HOH max
% Transmission
Absorbance (O D)
Wavelength (nm)
0
10
20
30
40
50
60
70
80
90
100
2.00
1.00
0.80
0.70
0.60
0.50
0.40
0.30
0.20
0.15
0.10
0.05
0.00
350
300
250
200
300
250
200
300
250
200

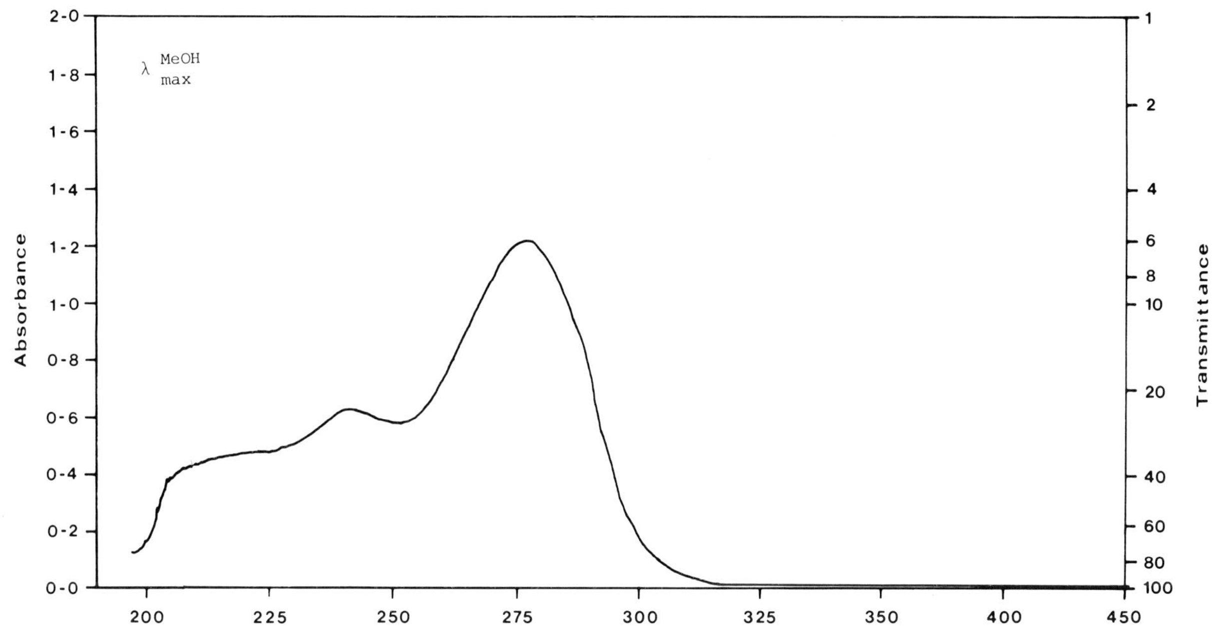

λ MeOH max
Absorbance
2-0
1-8
1-6
1-4
1-2
1-0
0-8
0-6
0-4
0-2
0-0
200
225
250
275
300
325
350
400
450
Transmittance
1
2
4
6
8
10
20
40
60
80
100

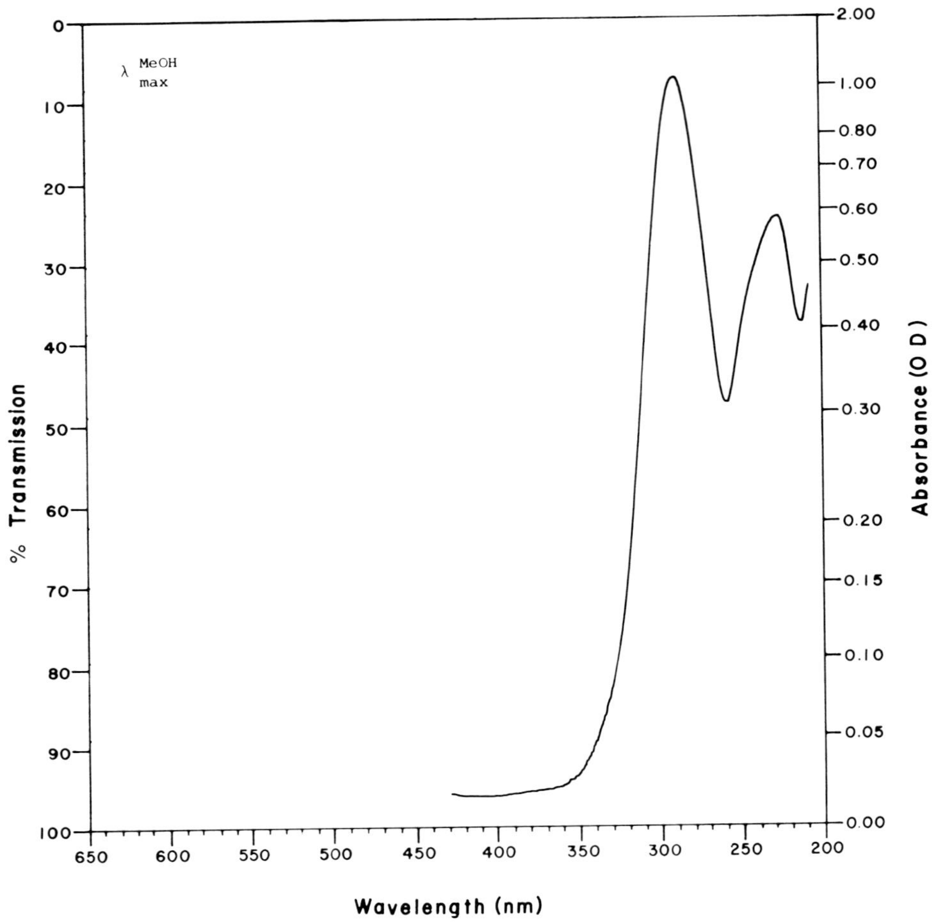

Tenuazonic acid (copper complex).

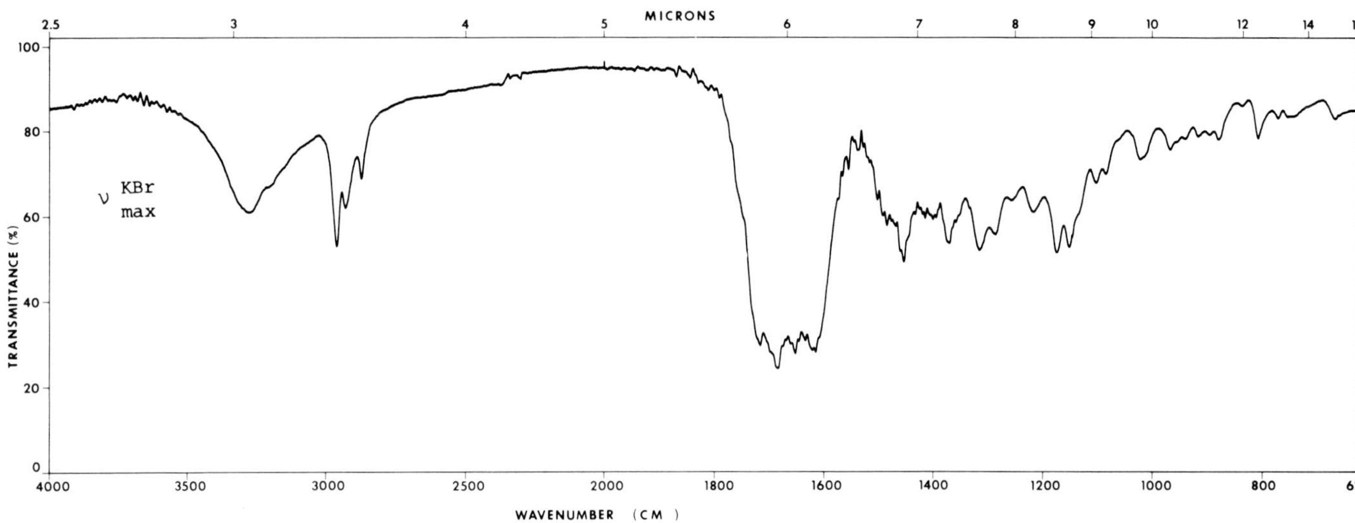
MICRONS
2.5
3
4
5
6
7
8
9
10
12
14
16
100
80
60
40
20
0
TRANSMITTANCE (%)
ν KBr max
4000
3500
3000
2500
2000
1800
1600
1400
1200
1000
800
625
WAVENUMBER (CM)

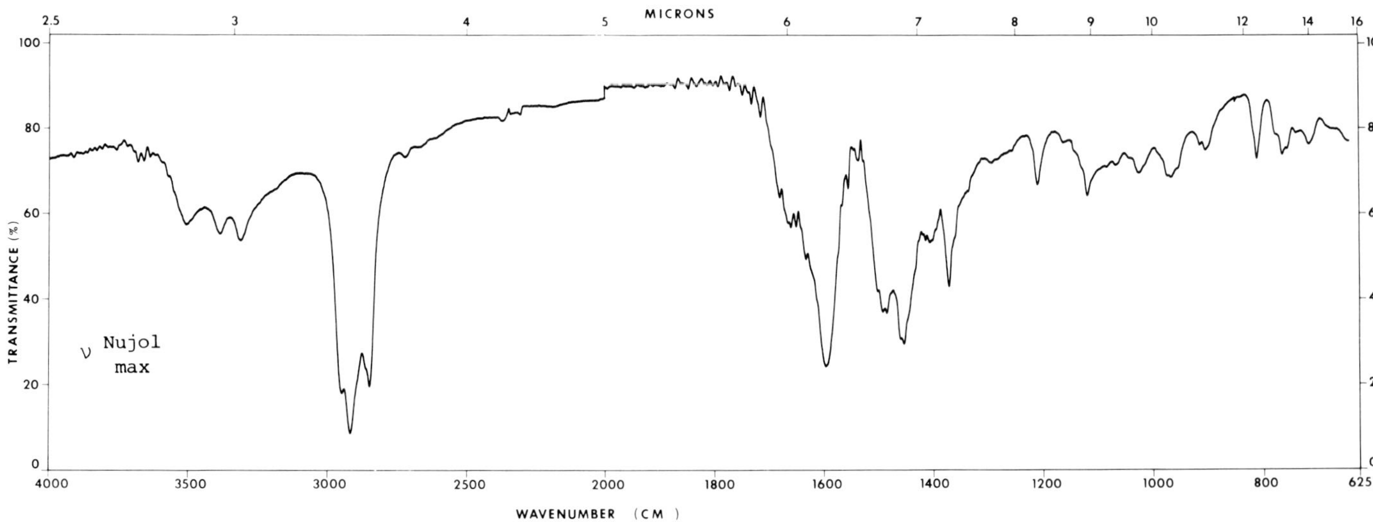

Tenuazonic acid (copper complex).

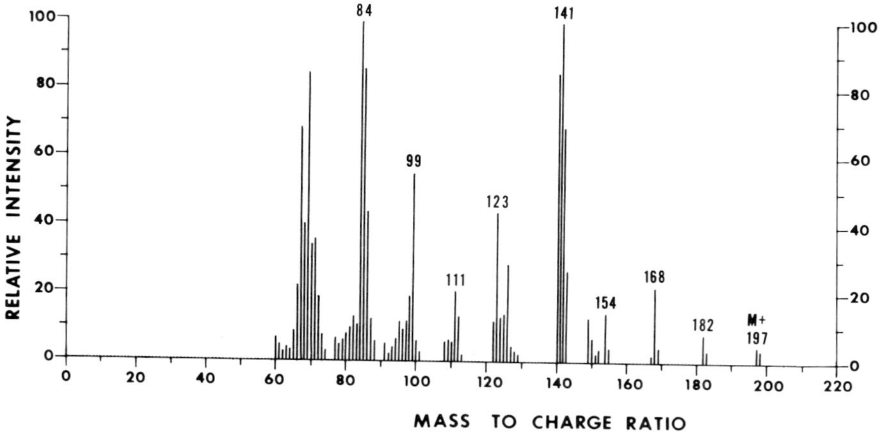
RELATIVE INTENSITY
MASS TO CHARGE RATIO
84
141
99
123
111
154
168
182
M+
197
0
20
40
60
80
100
120
140
160
180
200
220

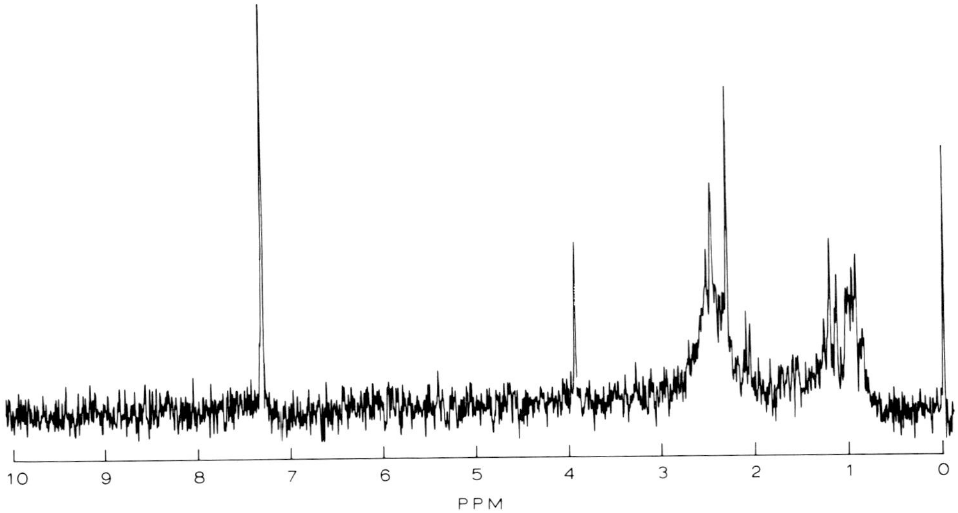

The 100 MHz ^{1}H NMR spectrum of tenuazonic acid in $CDCl_3$ solution.

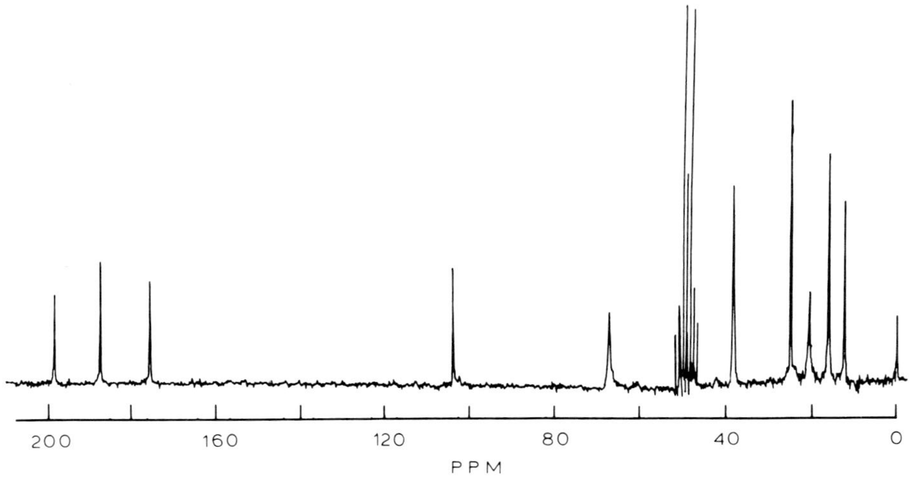

The 25 MHz ^{13}C NMR spectrum of tenuazonic acid in methanol-d_4 solution.

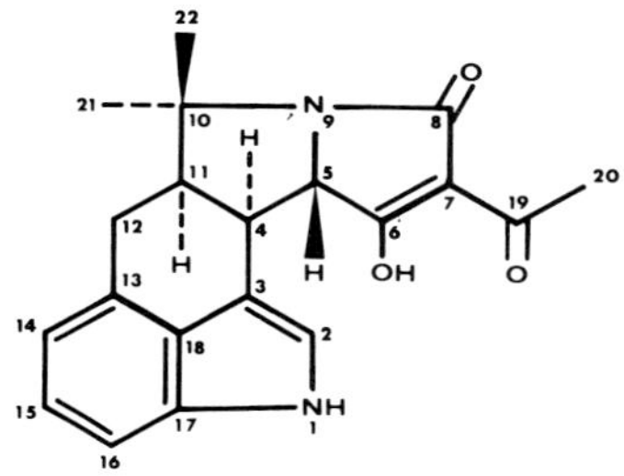

Common name	Cyclopiazonic acid
Molecular weight	336.1473
Molecular formula	$C_{20}H_{20}O_3N_2$
General characteristics[46,69]	Crystals from methanol, mp 245°–246°C $[\alpha]_D^{20}$ −36.1° (C = 0.36 in pyridine)
UV data[46]	λ_{max} nm (ϵ): 225(39,810), 253(16,595), 275(sh)(19,054), 284(20,417), and 292(sh)(17,378)
CD data[46]	$\Delta\epsilon$ (C = 1.5) nm: 320(+1.06), 298(+4.24), 286(0), 273(−9.53), 254(+1.06), 228(−40.27), and 220(−22.28)
Fungal source[46,69]	*Penicillium cyclopium* Westling (NRRL 3523) *P. puberulum* *Aspergillus versicolor* (Vuill.) Tiraboschi *A. flavus*
Toxicity data[69]	Acutely toxic to rats: Intraperitoneal injection produces hyperesthesia and convulsions followed by death in about 30 min. The LD_{50} in males is 2.3 mg/kg. Oral administration results in death in 1–5 days with an LD_{50} of 36 mg/kg in males and 63 mg/kg in females. The lesions produced include degenerative changes and necrosis in the liver, spleen, exocrine and endocrine pancreas, kidney, salivary glands, myocardium, and skeletal muscle. An unusual lesion, seen in the bile ducts and other ducts, was enlargement of the nucleus with margination of the chromatin and subsequent enlargement of the cells lining the ducts. It appears that cyclopiazonic acid produces focal necrosis in most

organs at high dosage and, at lower dosage, affects ducts or organs (such as the islets of Langerhans) originating from ducts.

Cyclopiazonic acid administered orally to rats has no immediate effect. Since it behaves as a monobasic acid, it is insoluble in aqueous solvents below pH 7.0 and is, therefore, probably not readily absorbed.

TLC data[63]

Adsorbent: silica gel F_{254} coated with 8% oxalic acid
Solvent: chloroform–methyl isobutyl ketone, 4:1 v/v
R_f: 0.5
Detection: spraying with dimethylaminobenzaldehyde and 50% ethanolic H_2SO_4 followed by heating for 5 min at 100°C.

Other significant references

Biochemistry[76]; biosynthesis[47,63]

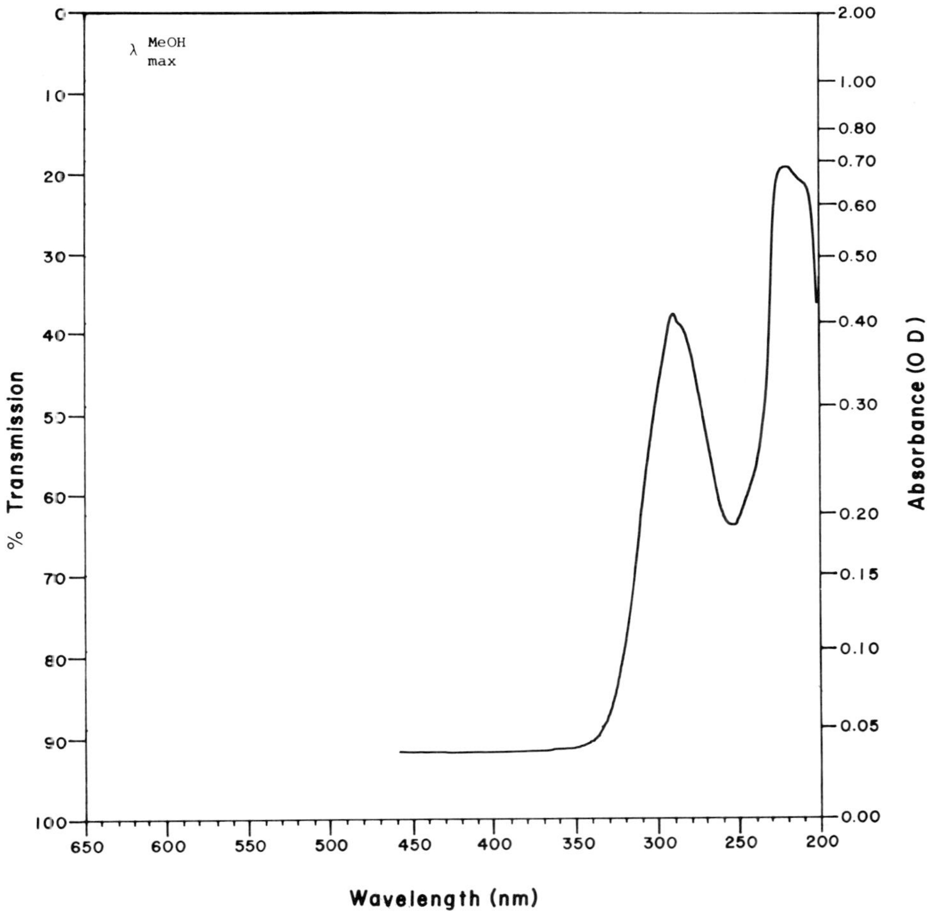

λ MeOH max
% Transmission
0
10
20
30
40
50
60
70
80
90
100
Absorbance (O D)
2.00
1.00
0.80
0.70
0.60
0.50
0.40
0.30
0.20
0.15
0.10
0.05
0.00
650
600
550
500
450
400
350
300
250
200
Wavelength (nm)

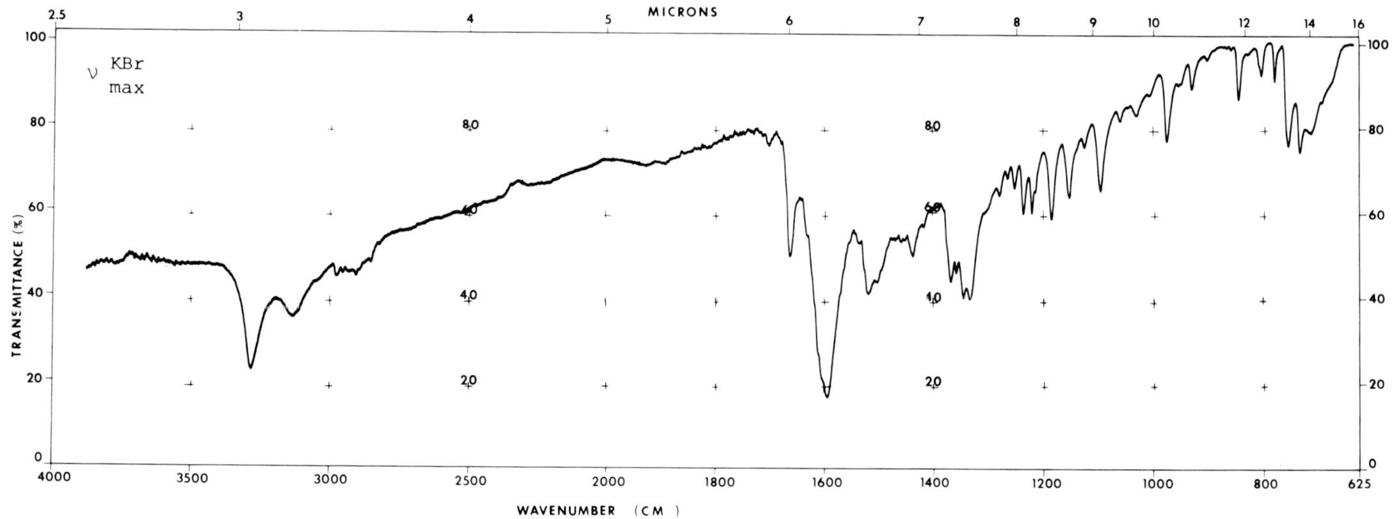
MICRONS
ν KBr max
TRANSMITTANCE (%)
WAVENUMBER (CM)
M+ 336

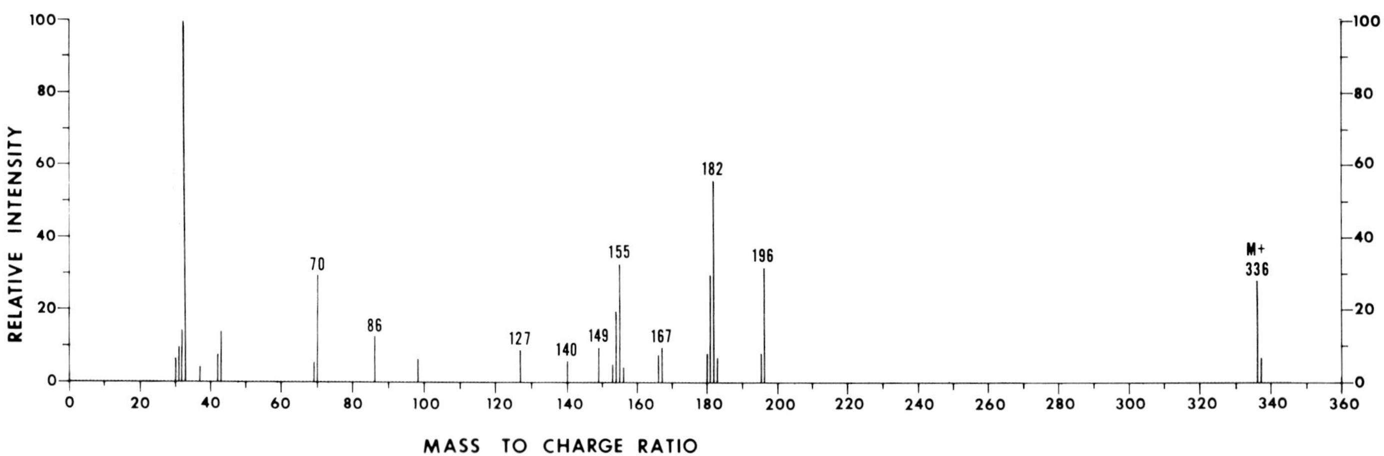
RELATIVE INTENSITY
70
86
127
140
149
155
167
182
196
M+
336
MASS TO CHARGE RATIO

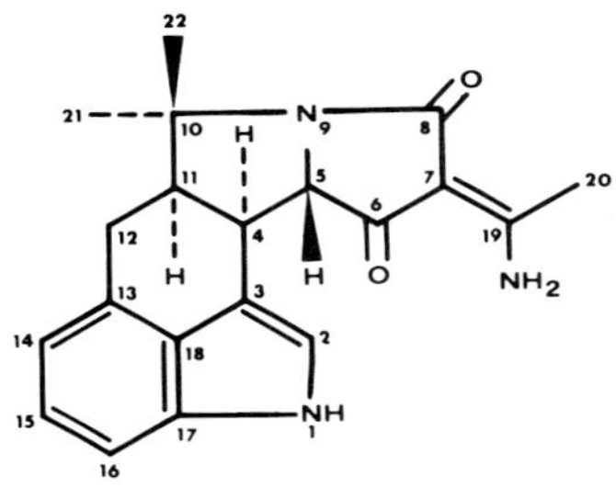

Common name	Cyclopiazonic acid imine
Molecular weight	335.1633
Molecular formula	$C_{20}H_{21}O_2N_3$
General characteristics[46,69]	Crystals from methanol, mp 277°–278°C $[\alpha]_D^{20}$ −11.1°
UV data[46,69]	λ_{max} nm (ϵ): 224(33,018), 244(sh)(10,715), 275(sh)(14,454), 286(sh)(20,892), and 293(22,387)
Fungal source[46,69]	*Penicillium cyclopium* Westling *Aspergillus versicolor* (Vuill.) Tiraboschi
Toxicity data	Relatively nontoxic compared to cyclopiazonic acid.

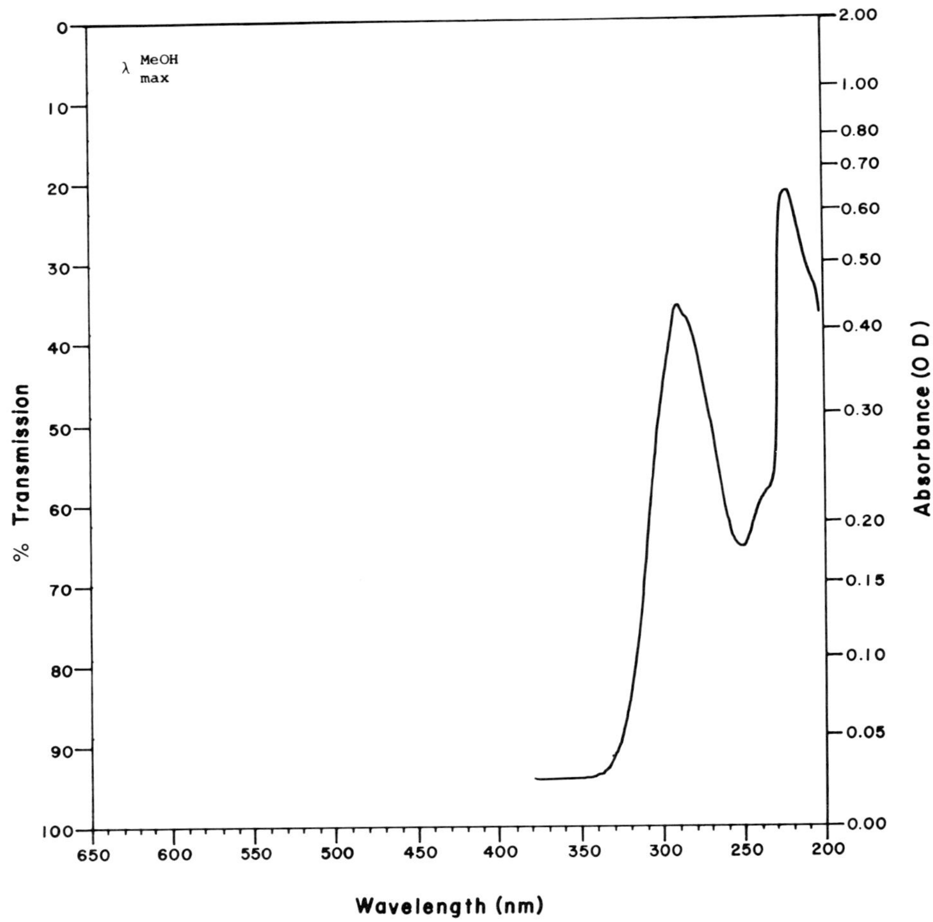
λ MeOH max
% Transmission
Absorbance (O D)
Wavelength (nm)
0
10
20
30
40
50
60
70
80
90
100
2.00
1.00
0.80
0.70
0.60
0.50
0.40
0.30
0.20
0.15
0.10
0.05
0.00
650
600
550
500
450
400
350
300
250
200

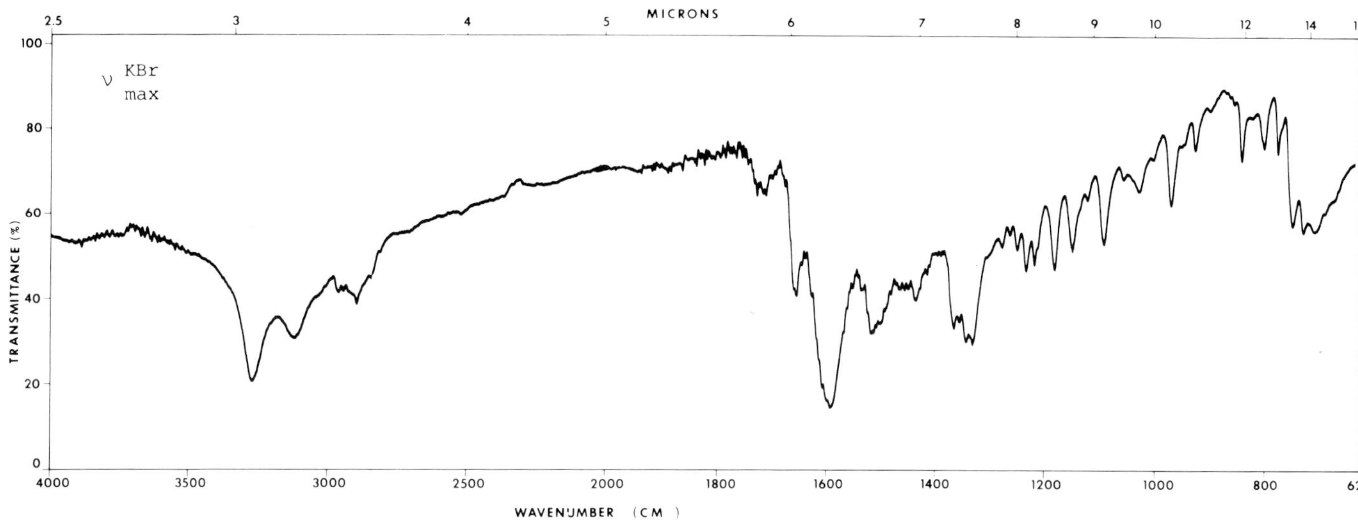
MICRONS
2.5
3
4
5
6
7
8
9
10
12
14
16
100
80
60
40
20
0
TRANSMITTANCE (%)
ν KBr max
4000
3500
3000
2500
2000
1800
1600
1400
1200
1000
800
625
WAVENUMBER (CM)

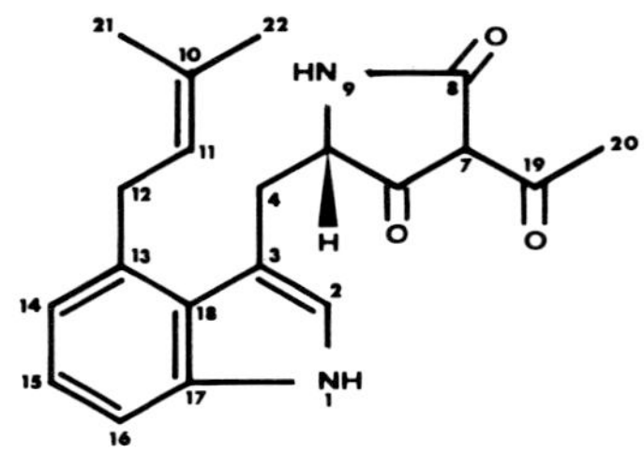

Common name	Bissecodehydrocyclopiazonic acid
Molecular weight	338.1630
Molecular formula	$C_{20}H_{22}O_3N_2$
General characteristics[46,69]	Crystals from methanol, mp 168°–169°C $[\alpha]_D^{20}$ −186° (*C* = 0.42 in pyridine)
UV data[46,69]	λ_{max} nm (ϵ): 225(35,481), 276(19,054), and 296(sh)(11,748)
CD data[46,69]	$\Delta\epsilon$ nm: 280(ca −5.0) and 325(ca −1.0)
Fungal source[46,69]	*Penicillium cyclopium* Westling *Aspergillus versicolor* (Vuill.) Tiraboschi
Toxicity data[46]	Relatively nontoxic compared to cyclopiazonic acid

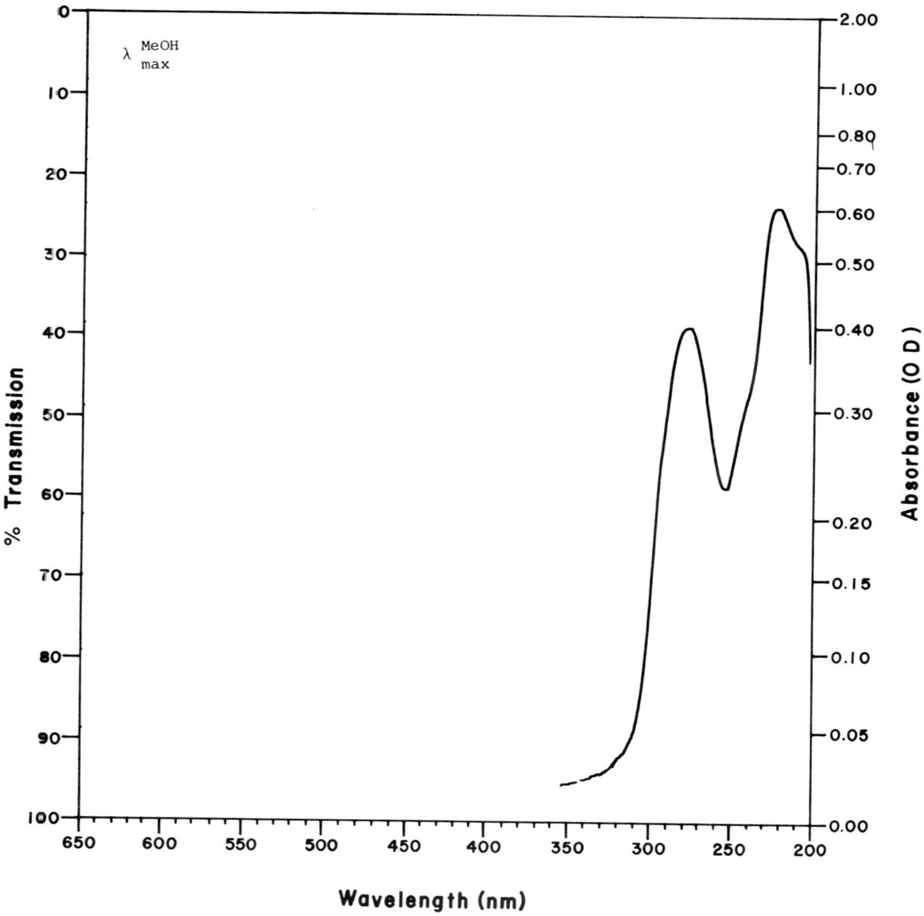
λ MeOH max
% Transmission
0
10
20
30
40
50
60
70
80
90
100
Absorbance (O D)
2.00
1.00
0.80
0.70
0.60
0.50
0.40
0.30
0.20
0.15
0.10
0.05
0.00
650
600
550
500
450
400
350
300
250
200
Wavelength (nm)

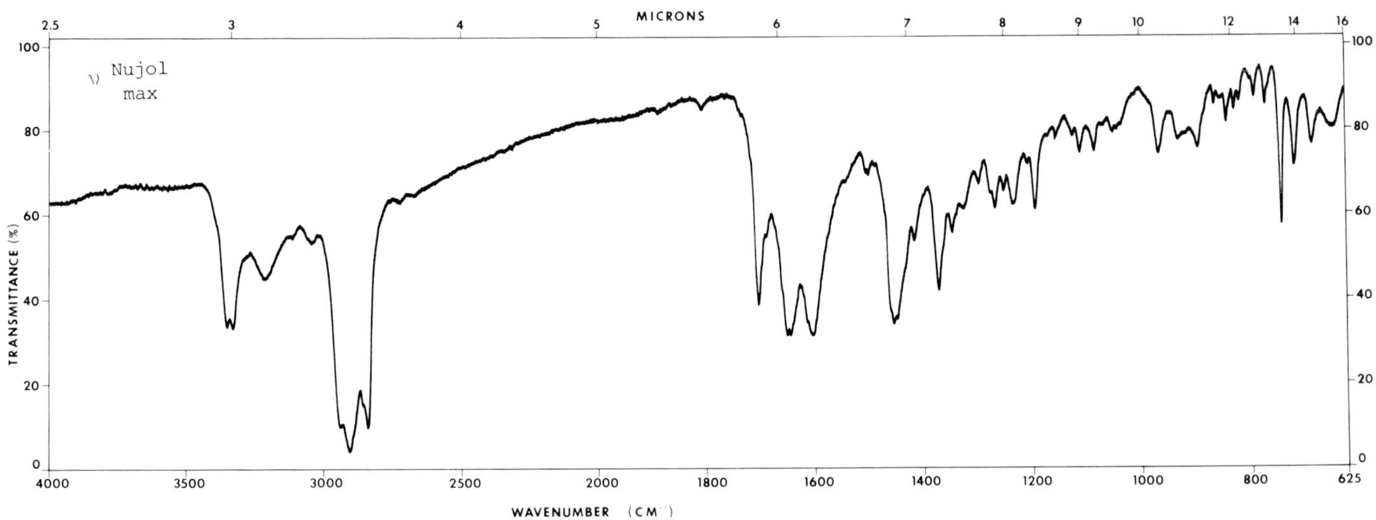

MICRONS
2.5
3
4
5
6
7
8
9
10
12
14
16
100
80
60
40
20
0
TRANSMITTANCE (%)
ν Nujol
max
4000
3500
3000
2500
2000
1800
1600
1400
1200
1000
800
625
WAVENUMBER (CM⁻¹)

REFERENCES

1. Acklin, W., Weibel, F., and Arigoni, D., *Chimia* **31,** 2, 63 (1977).
2. Allen, C. M., Jr., *Biochemistry* **11,** 2154 (1972).
3. Allen, C. M., Jr., *J. Am. Chem. Soc.* **95,** 2386 (1973).
4. Aragonne, D., Acklin, W., and Fehr, T., *Helv. Chim. Acta* (in press).

4a. Barbetta, M., Casnati, G., Pochini, A., and Selva, A., *Tetrahedron Lett.* p. 4457 (1969).

5. Birch, A. J., and Wright, J. J., *Tetrahedron* **26,** 2329 (1970).
6. Büchi, G., Luk, K. C., Kobbe, B., and Townsend, J. M., *J. Org. Chem.* **42,** 224 (1977).
7. Cardillo, R., Fuganti, C., Gatti, G., Ghiringhelli, D., and Grasselli, P., *Tetrahedron Lett.* p. 3163 (1974).
8. Cardillo, R., Fuganti, C., Ghiringhelli, D., Grasselli, P., and Gatti, G., *J. Chem. Soc., Chem. Commun.* p. 778 (1975).
9. Casnati, G., Piozzi, F., Quilico, A., and Cardani, C., *Gazz. Chim. Ital.* **90,** 476 (1960).
10. Ciegler, A., *Appl. Microbiol.* **18,** 128 (1969).
11. Ciegler, A., Vesonder, R. F., and Cole, R. J., *Adv. Chem. Ser.* **149,** 163 (1976).
12. Clardy, J., personal communication.
13. Clardy, J., and Cole, R. J., personal communication.
14. Clardy, J., Springer, J. P., Büchi, G., Matsuo, K., and Wightman, R., *J. Am. Chem. Soc.* **97,** 663 (1975).
15. Coetzer, J., and Steyn, P. S., *Acta Crystallogr., Sect. B* **29,** 685 (1973).
16. Cole, R. J., personal communication.
17. Cole, R. J., Tremorgenic mycotoxins. *In* "Mycotoxins in Human and Animal Health" (J. V. Rodricks, C. W. Hesseltine, and M. A. Mehlman, eds.), p. 583. Pathotox Publ. Inc., Park Forest South, Illinois, 1976.
18. Cole, R. J., Tremorgenic mycotoxins: Current status. *In* "Antinutrients and Natural Toxicants in Foods." Food and Nutrition Press, Inc. (in press).
19. Cole, R. J., Dorner, J. W., Lansden, J. A., Cox, R. H., Cunfer, B., Nicholson, S. S., and Bedell, D. M., *J. Agric. Food. Chem.* **25,** 826 (1977).
20. Cole, R. J., and Kirksey, J. W., *J. Agric. Food Chem.* **21,** 927 (1973).
21. Cole, R. J., Kirksey, J. W., Cox, R. H., and Clardy, J., *J. Agric. Food Chem.* **23,** 1015 (1975).
22. Cole, R. J., Kirksey, J. W., Dorner, J. W., Wilson, D. M., Johnson, J., Jr., Johnson, N., Bedell, D. M., Springer, J. P., Chexal, K. K., Clardy, J., and Cox, R. H., *J. Agric. Food Chem.* **25,** 826 (1977).
23. Cole, R. J., Kirksey, J. W., Moore, J. H., Blankenship, B. R., Diener, U. L., and Davis, N. D., *Appl. Microbiol.* **24,** 248 (1972).
24. Cole, R. J., Kirksey, J. W., and Wells, J. M., *Can. J. Microbiol.* **20,** 1159 (1974).
25. Cole, R. J., Kirksey, J. W., Dorner, J. W., Wilson, D. M., Johnson, J., Jr., Bedell, D., Springer, J. P., Chexal, K. K., Clardy, J., and Cox, R. H., *Ann. Nutr. Aliment.* **31,** 685 (1977).
26. Culvenor, C. C. J., personal communication.
27. Cysewski, S. J., *Proc. Intersect. Congr. IAMS, 1st, 1900* Vol. 4, p. 373 (1975).
28. Cysewski, S. J., Baetz, A. L., and Pier, A. C., *Am. J. Vet. Sci.* **36,** 53 (1975).
29. Cysewski, S. J., *J. Am. Vet. Med. Assoc.* **163,** 1291 (1973).
30. Davis, N. D., Diener, U. L., and Morgan-Jones, G., *Appl. Environ. Microbiol.* **34,** 155 (1977).
31. Di Menna, M. E., Mantle, P. G., and Mortimer, P. H., *N. Z. Vet. J.* **24,** 45 (1976).
32. Dix, D. T., Martin, J., and Moppett, C. E., *J. Chem. Soc., Chem. Commun.* p. 1168 (1972).
33. Eickman, N., Clardy, J., Cole, R. J., and Kirksey, J. W., *Tetrahedron Lett.* p. 1051 (1975).

34. Fayos, J., Lokensgard, D., Clardy, J., Cole, R. J., and Kirksey, J. W., *J. Am. Chem. Soc.* **96,** 6785 (1974).
35. Fehr, T., and Acklin, W., *Helv. Chim. Acta* **49,** 1907 (1966).
36. Gallagher, R. T., Clardy, J., and Wilson, B. J., *Tetrahedron Lett.* **21,** 239 (1980).
37. Gallagher, R. T., Finer, J., Clardy, J., Leutwiler, A., Weibel, F., Acklin, W., and Arigoni, D., *Tetrahedron Lett.* (in press).
38. Gallagher, R. T., and Latch, G. C. M., *Appl. Environ. Microbiol.* **33,** 730 (1977).
39. Gallagher, R. T., Keogh, R. G., Latch, G. C. M., and Reid, C. S. W., *N. Z. J. Agric. Res.* **20,** 431 (1977).
40. Gatti, G., Cardillo, R., and Fuganti, C., *Tetrahedron Lett.* p. 2605 (1978).
41. Gieger, M., and Barrentine, B. F., *J. Am. Chem. Soc.* **61,** 966 (1939).
42. Glinsukon, T., Yuan, S. S., Wightman, R., Kitaura, Y., Büchi, G., and Shank, R. C., *Plant Foods Man* **1,** 113 (1974).
43. Gysi, R. P., Dissertation No. 4990. Eidgenössischen Technischen Hochschule, Zurich, Switzerland (1973).
44. Hayey, A. W., Presley, D. B., and Neville, J. A., *Toxicol. Appl. Pharmacol.* **35,** 311 (1976).
45. Holzapfel, C. W., *Tetrahedron* **24,** 2101 (1968).
46. Holzapfel, C. W., Hutchison, R. D., and Wilkins, D. C., *Tetrahedron* **26,** 5239 (1970).
47. Holzapfel, C. W., and Wilkins, D. C., *Phytochemistry* **10,** 351 (1971).
48. Hotujac, Lj., and Stern, P., *Acta Med. Jugosl.* **28,** 223 (1974).
49. Hotujac, Lj., Muftić, R. H., and Filipović, N., *Pharmacology* **14,** 297 (1976).
50. Hou, C. T., Ciegler, A., and Hesseltine, C. W., *Can. J. Microbiol.* **17,** 599 (1971).
51. Houghton, E., and Saxton, J. E., *J. Chem. Soc. C* p. 1003 (1969).
52. Iwasaki, S., Muro, H., Nozoe, S., Okuda, S., and Sato, Z., *Tetrahedron Lett.* p. 13 (1972).
53. Janardhanan, K. K., and Husain, A., *Indian J. Exp. Biol.* **13,** 321 (1975).
54. Kaczka, T., Gitterman, C. D., Dulaney, E. L., Smith, M. C., Hendlin, D., Woodruff, H. B., and Folkers, K., *Biochem. Biophys. Res. Commun.* **14,** 54 (1964).
55. Kinoshita, T., Renbutsu, Y., Khan, I. D., Kohmoto, K., and Nishimura, S., *Ann. Phytopathol. Soc. Jpn.* **38,** 397 (1972).
56. Lanigan, G. W., Payne, A. L., and Cockrum, P. A., personal communication.
57. Leutwiler, A., Dissertation No. 5163. Eidgenössischen Technischen Hochschule, Zurich, Switzerland (1973).
58. Lieberman, N. W., Verbin, R. S., Landay, M., Liang, H., Farber, E., Lee, T., and Sear, R., *Cancer Res.* **30,** 942 (1970).
59. MacDonald, J. C., and Slater, G. P., *Can. J. Microbiol.* **12,** 455 (1966).
60. Mantle, P. G., Mortimer, P. H., and White, E. P., *Res. Vet. Sci.* **24,** 49 (1978).
61. Marchelli, R., Dossena, A., and Casnati, G., *J. Chem. Soc., Chem. Commun.* p. 779 (1975).
62. Marchelli, R., Dossena, A., Pochini, A., and Dradi, E., *J. Chem. Soc., Perkin Trans. I* p. 713 (1977).
63. McGrath, R. M., Steyn, P. S., Ferreira, N. P., and Neethling, D. C., *Bioorg. Chem.* **4,** 11 (1976).
64. Meronuck, R. A., Steele, J. A., Mirocha, C. J., and Christensen, C. M., *Appl. Microbiol.* **23,** 613 (1972).
65. Mikami, Y., Nishijima, N., Iimura, H., Suzuki, H., and Tamura, S., *Agric. Biol. Chem.* (*Tokyo*) **35,** 611 (1971).
66. Mikami, Y., Nishijima, Y., and Suzuki, A., *J. Agric. Chem. Soc. Jpn.* **46,** 473 (1972).
67. Miller, F. A., Rightsel, W. A., Sloan, B. J., Ehrlich, J., French, J. C., and Bartz, Q. R., *Nature* (*London*) **200,** 1338 (1963).
68. Nagasawa, H., Isogai, A., Ikada, K., Sato, S., Murakoshi, A., Suzuki, A., and Tamuro, S., *Agric. Biol. Chem.* **39,** 1901 (1975).
68a. Nagasawa, H., Isogai, A., Suzuki, A., and Tamura, S., *Tetrahedron Lett.* p. 1601 (1976).
68b. Nagasawa, H., Isogai, A., Suzuki, A., and Tamura, S., *Agric. Biol. Chem.* **43,** 1759 (1979).

69. Ohmomo, S., Sugita, M., and Abe, M., *J. Agric. Chem. Soc. Jpn.* **47,** 57 (1973).
70. Patterson, D. S. P., Roberts, B. A., Shreeve, B. J., MacDonald, S. W., and Hayes, A. N., *Appl. Environ. Microbiol.* **37,** 172 (1979).
71. Purchase, I. F. H., *Toxicol. Appl. Pharmacol.* **18,** 114 (1971).
72. Rabie, C. J., Van Rensburg, S. J., Van der Watt, J. J., and Lubben, A., *S. Afr. Med. J.* **49,** 1647 (1975).
73. Richard, J. L., personal communication (1978).
74. Ritchie, R., and Saxton, J. E., *J. Chem. Soc., Chem. Commun.* p. 611 (1975).
75. Rosett, T., Sankhala, R. H., Stickings, C. E., Taylor, M. E. U., and Thomas, R., *Biochem. J.* **67,** 390 (1957).
76. Schabort, J. C., Wilkens, D. C., Holzapfel, C. W., Potgieter, D. J. J., and Neitz, A. W., *Biochim. Biophys. Acta* **250,** 311 (1971).
77. Schroeder, H. W., Cole, R. J., Hein, J., Jr., and Kirksey, J. W., *Appl. Microbiol.* **29,** 857 (1975).
78. Scott, P. M., Kennedy, P. C., Harwig, J., and Chen, Y.-K., *Appl. Microbiol.* **28,** 892 (1974).
79. Shigeura, H. T., and Gordon, C. N., *Biochemistry* **2,** 1132 (1962).
80. Smith, E. R., Fredrickson, T. N., and Hadidian, Z., *Cancer Chemother. Rep.* **52,** 579 (1968).
81. Sobotka, T. J., Brodie, R. E., and Spaid, S. L., *Pharmacology* **16,** 287 (1978).
82. Springer, J. P., Clardy, J., Wells, J. M., Cole, R. J., and Kirksey, J. W., *Tetrahedron Lett.* p. 2531 (1975).
83. Springer, J. P., and Clardy, J., *Helv. Chim. Acta* **21,** 231 (1980).
84. Springer, J. P., *Tetrahedron Lett.* p. 339 (1979).
85. Stern, P., *Jugosl. Physiol. Pharmacol. Acta* **7,** 187 (1971).
86. Steyn, P. S., *Tetrahedron Lett.* p. 3331 (1971).
87. Steyn, P. S., *Tetrahedron* **29,** 107 (1973).
88. Steyn, P. S., and Rabie, C. J., *Phytochemistry* **15,** 1977 (1979).
89. Stickings, C. E., *Biochem. J.* **72,** 332 (1959).
90. Stipanovic, R. D., and Schroeder, H. W., *Trans. Br. Mycol. Soc.* **66,** 178 (1976).
91. Tenabe, M., and Clardy, J., personal communication.
92. Umetsu, N., Kaji, J., and Tamari, K., *Agric. Biol. Chem.* **36,** 859 (1972).
93. Umetsu, N., Kaji, J., and Tamari, K., *Agric. Biol. Chem.* **37,** 451 (1973).
94. Umetsu, N., Muramatsu, T., Honda, H., and Tamari, K., *Agric. Biol. Chen.* **38,** 791 (1974).
95. Wilson, B. J., and Wilson, C. H., *Science* **144,** 177 (1964).
96. Wilson, B. J., Wilson, C. H., and Hayes, A. W., *Nature (London)* **220,** 77 (1968).
97. Yamazaki, M., personal communication.
98. Yamazaki, M., Fujimoto, H., Akiyama, T., Sankawa, U., and Iitaka, Y., *Tetrahedron Lett.* p. 27 (1975).
99. Yamazaki, M., Fujimoto, H., and Kawasaki, T., *Tetrahedron Lett.* p. 1241 (1975).
100. Yamazaki, M., Sasago, K., and Miyaki, K., *J. Chem. Soc., Chem. Commun.* p. 408 (1974).
101. Yamazaki, M., Suzuki, S., and Miyaki, K., *Chem. Pharm. Bull.* **19,** 1739 (1971).
102. Yamazaki, M., Fujimoto, H., and Okuyama, E., *Chem. Pharm. Bull.* **25,** 2554 (1977).
103. Yamazaki, M., Fujimoto, H., and Okuyama, E., *Chem. Pharm. Bull.* **26,** 111 (1978).
104. Yamazaki, M., Okuyama, E., and Maebayashi, Y., *Chem. Pharm. Bull.* **27,** 1611 (1979).
105. Yashizaga, T., Morooka, N., Sawada, Y., and Udagawa, S., *Appl. Environ. Microbiol.* **32,** 441 (1976).

9

Toxic Lactones

Chemically, the mycotoxins patulin, penicillic acid, and ascladiol are relatively noncomplex in nature. They all contain a five-membered cyclic lactone ring. Patulin and penicillic acid have been studied extensively. They are produced by several fungi, mostly belonging to the genera *Aspergillus* and *Penicillium*. This, in conjunction with the reports that they are carcinogenic, makes them potential health hazards to animals and man, and they merit further study and evaluation. Both have been found to occur naturally.

Wilson has recently reviewed the subject.[40]

Toxic lactones	Molecular weight	Molecular formula
Patulin	154.0266	$C_7H_6O_4$
Isopatulin	154.0266	$C_7H_6O_4$
Ascladiol	156.0420	$C_7H_8O_4$
Penicillic acid	170.0579	$C_8H_{10}O_4$

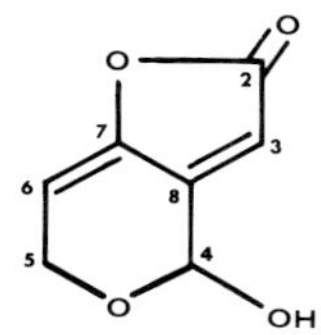

Common name

Patulin (Claviformin; Clavitin; Clavicin; Expansion; Penicidin; Mycoin; Leucopin; Tercinin) (4-Hydroxy-4*H*-furo[3,2-*c*]pyran-2(6*H*)-one)

Molecular weight

154.0266

Molecular formula

$C_7H_6O_4$

General characteristics[41,42]

Crystals from benzene, mp 110°–111°C
$[\alpha]_D$ 0°
Patulin is soluble in water, alcohols, acetone, ethyl acetate, chloroform; sparingly soluble in ethyl ether, benzene; insoluble in petroleum ether. It is somewhat unstable in polar solvents such as water and methanol and loses its biological potency in alkaline conditions.
Acetyl derivative, prisms from methanol, mp 118°–120°C;
Phenylhydrazone, orange plates, mp 152°–153°C;
Oxime, crystals from water, mp 152°–153°C (dec.);
Isopatulin, prisms from ethyl acetate, mp 88°–90°C

UV data[42]

λ_{max} nm (ϵ): 276(16,600)
Isopatulin:
λ_{max}^{MeOH} nm (ϵ): 269(12,800)

Fungal source[3–5,17,40]

Penicillium expansum (*P. leucopus*)
P. patulum Bainier (*P. urticae; P. griseofulvum*)
P. claviforme
P. equinum (*P. terrestre*)
P. novae-zeelandiae
P. lapidosum
P. granulatum (*P. divergens*)
P. lanosum
P. melinii
P. cyclopium
P. cyaneo-fulvum
P. roqueforti
Aspergillus clavatus

A. giganteus
A. terreus
Byssochlamys nivea (*Gymnoascus* spp.)

Toxicity data[6,8,11–13,20,21,30,31,37,40]

Patulin was originally isolated because of its antibiotic properties against gram-positive and gram-negative bacteria. However, it was soon found to be too toxic for use as an antimicrobial agent in man. The LD_{50} of patulin in White Leghorn cockerels was 170 mg/kg. Sublethal doses produced hemorrhaging in the digestive tract. LD_{50} (IP) in mice was 5 mg/kg. Patulin has been reported to be carcinogenic and a potent mitotic toxin in addition to being toxic to many protozoa, fungi, mammals, plants, HeLa cells, and viruses. Patulin has been found contaminating apple juice and has been implicated in the death of cattle.
NOTE: See current reviews by Wilson,[40] Scott,[30] and Ciegler.[9]

TLC data[30]

Adsorbent: silica gel G-HR
Solvent: toluene–ethyl acetate–formic acid, 5:4:1 v/v/v
R_f: 0.58
Detection: brown spot after spraying with 50% ethanolic H_2SO_4 and heating at 100°C for 5 min.
NOTE: For extensive review of the TLC of patulin, see Scott.[30]

GLC data[15,27]

Solid support: Gas Chrom-9
Liquid phase: 10% Dc-200
Column temp: 175°C
Detector temp: 190°C
Injector temp: 190°C
Relative retention time: (relative to β-BHC and analyzed as TMS derivative): 0.56
Solid support: Gas Chrom-Q, 100-120 mesh
Liquid phases: 3% Dexsil 300
3% OV-17
3% OV-25
NOTE: For extensive discussion of relative retention times, conditions, etc., refer to Pero.[27]

Other significant references

Structure[41,42]; synthesis[43]; biosynthesis[14,29]; high pressure liquid chromatography[39]; review[9]

^{13}C NMR data C2, 168.5 s; C3, 109.8 d; C4, 88.4 d; C5, 59.1 t; C6, 107.9 d; C7, 151.5 s; C8, 146.1 s

^{1}H NMR data H3, 6.96; H4, 5.90; H5a, 4.32; H5b, 4.66; H6, 5.90; 4-OH, 5.94

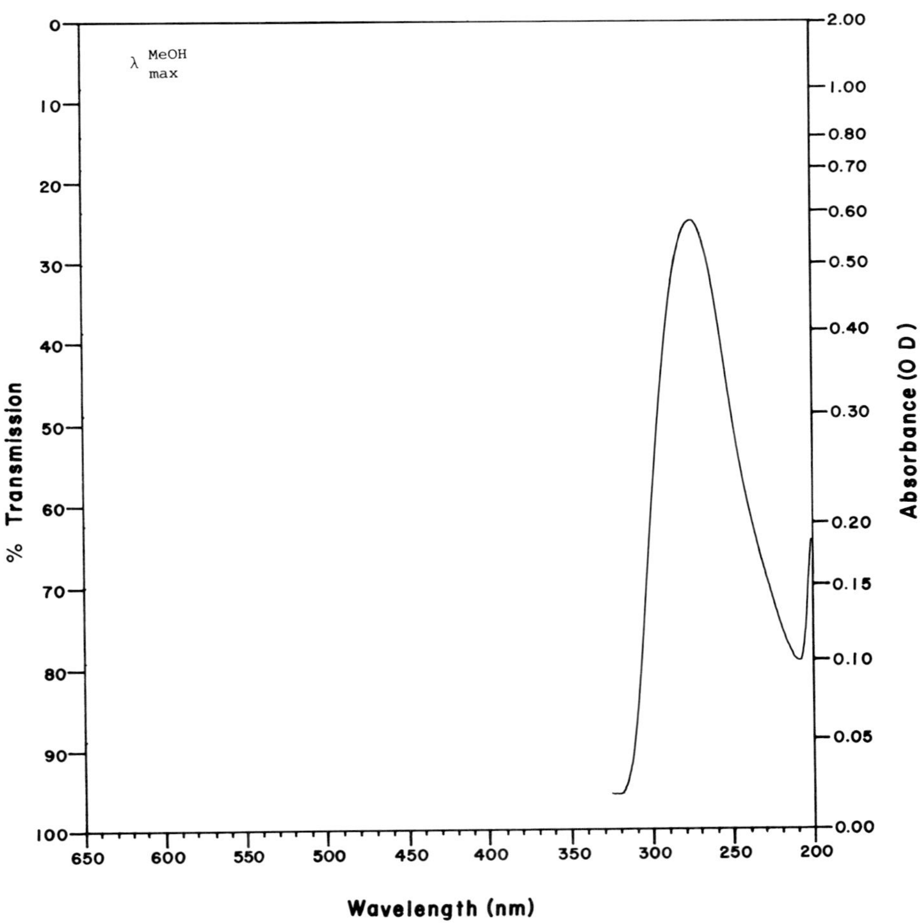

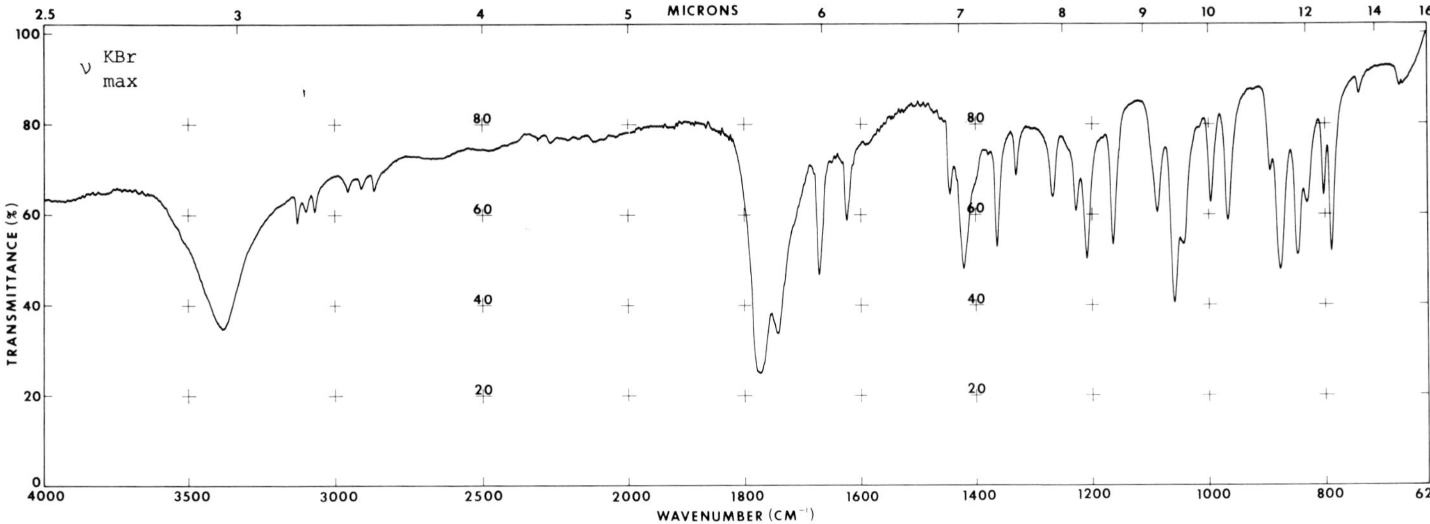

MICRONS
2.5
3
4
5
6
7
8
9
10
12
14
16
ν KBr max
TRANSMITTANCE (%)
WAVENUMBER (CM⁻¹)
4000
3500
3000
2500
2000
1800
1600
1400
1200
1000
800
625

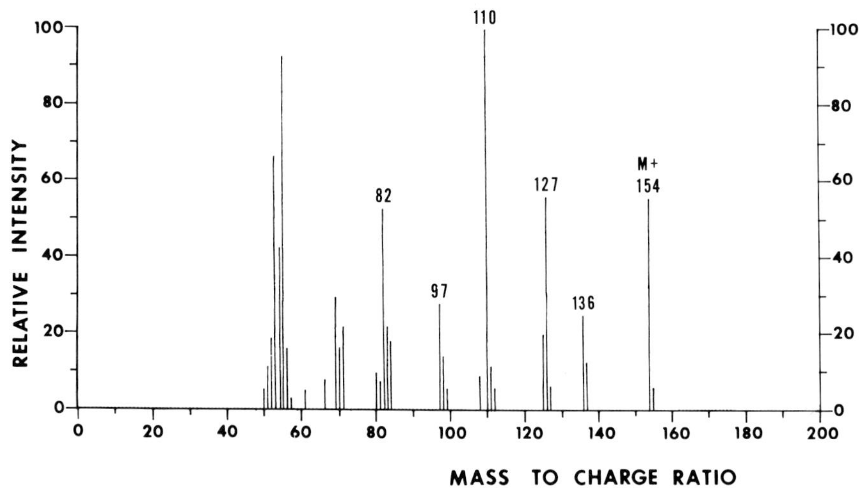

RELATIVE INTENSITY
MASS TO CHARGE RATIO
82
97
110
127
136
M+
154

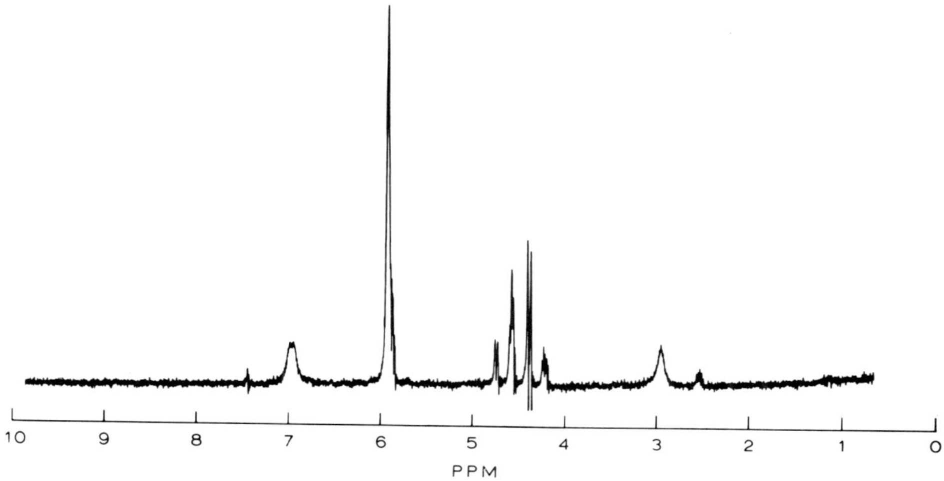

The 100 MHz ^1H NMR spectrum of patulin in $CDCl_3$/DMSO solution.

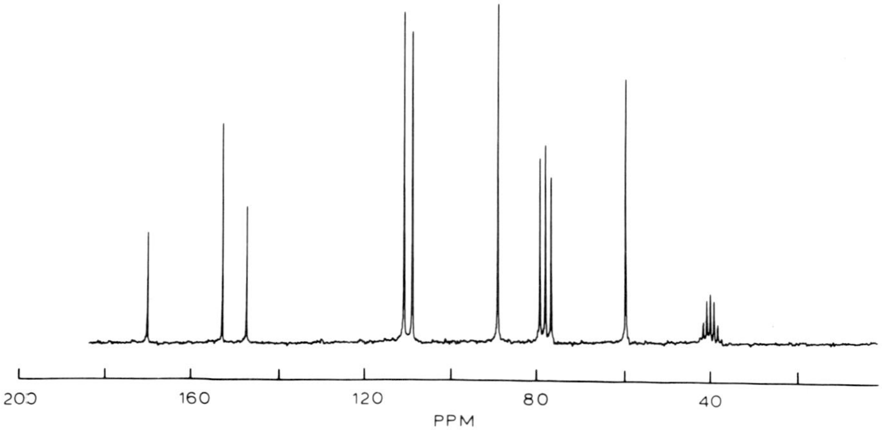

The 25 MHz ^{13}C NMR spectrum of patulin in $CDCl^3$/DMSO solution.

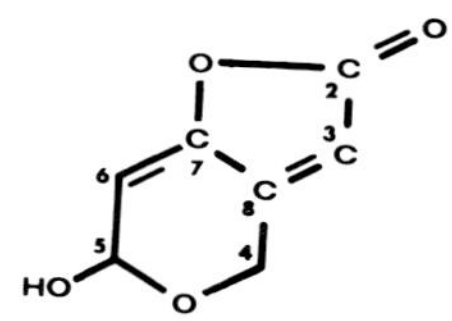

Common name	Isopatulin (5-Hydroxy-4*H*-furo[3,2-*c*]pyran-2(6*H*)-one)
Molecular weight	154.0266
Molecular formula	$C_7H_6O_4$
General characteristics[32]	Crystals from ethyl acetate, mp 88°–90°C $[\alpha]_D$ 0°
UV data[32]	λ_{max}^{MeOH} nm: 269 (ϵ_{max} not reported)
IR data[32]	ν_{max}^{Nujol} cm^{-1}: 3400, 1780, 1755, 1675, 1610, 1440, 1400, 1280, 1225, 1175, 1110, 1070, 1040, 1000, 980, 900, 845, 820, and 795
Fungal source[32]	*Penicillium urticae* (NRRL 2159A)
Toxicity data	Toxicity unknown
TLC data[32]	*Adsorbent:* silica gel *Solvent:* chloroform–ethyl acetate–ethyl ether, 55:2:1 v/v/v R_f: 0.30 *Detection:* visible under UV light
Mass spectroscopy data	*m/e:* 154, 136, 126, 112, 97, 67, 39
^{1}*H NMR data*[32]	(acetone-d_6) H3, 6.01; H4a, 4.74 (J = 16.9, 1.2); H4b, 5.04 (J = 16.9, 2.2); H5, 5.91; H6, 5.72; 5OH, 3.08
^{13}*C NMR data*[32]	C2, 169.6 s; C3, 109.0 d; C4, 58.2 t; C5, 90.4 d; C6, 110.8 d; C7, 153.4 s; C8, 150.2 s

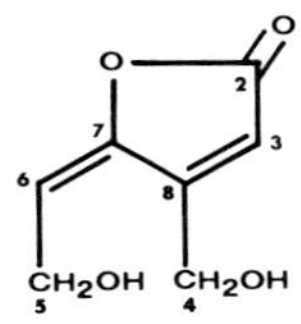

Common name	Ascladiol
Molecular weight	156.0420
Molecular formula	$C_7H_8O_4$
General characteristics[35]	Crystals from chloroform–acetone, mp 65°–66°C
UV data[35]	λ_{max}^{EtOH} nm: 271 (ϵ_{max} not reported)
Fungal source[35]	*Aspergillus clavatus*
Toxicity data[35]	The acute toxicity (IP) was reported to be only ¼ that of patulin in mice.
TLC data[10]	*Adsorbent:* silica gel G-HR *Solvent:* toluene–ethyl acetate–formic acid, 5:4:1 v/v/v R_f: 0.30 (Patulin has R_f 0.58 in this same system.) *Detection:* light brown fluorescent spot after spraying with 50% ethanolic H_2SO_4 and heating at 110° for 5 min.
^{1}H NMR data[35]	(acetone-d_6) H3, 6.29 q; H4, 4.74 s; H5, 4.30 m; H6, 5.87 m; OH, 4.74

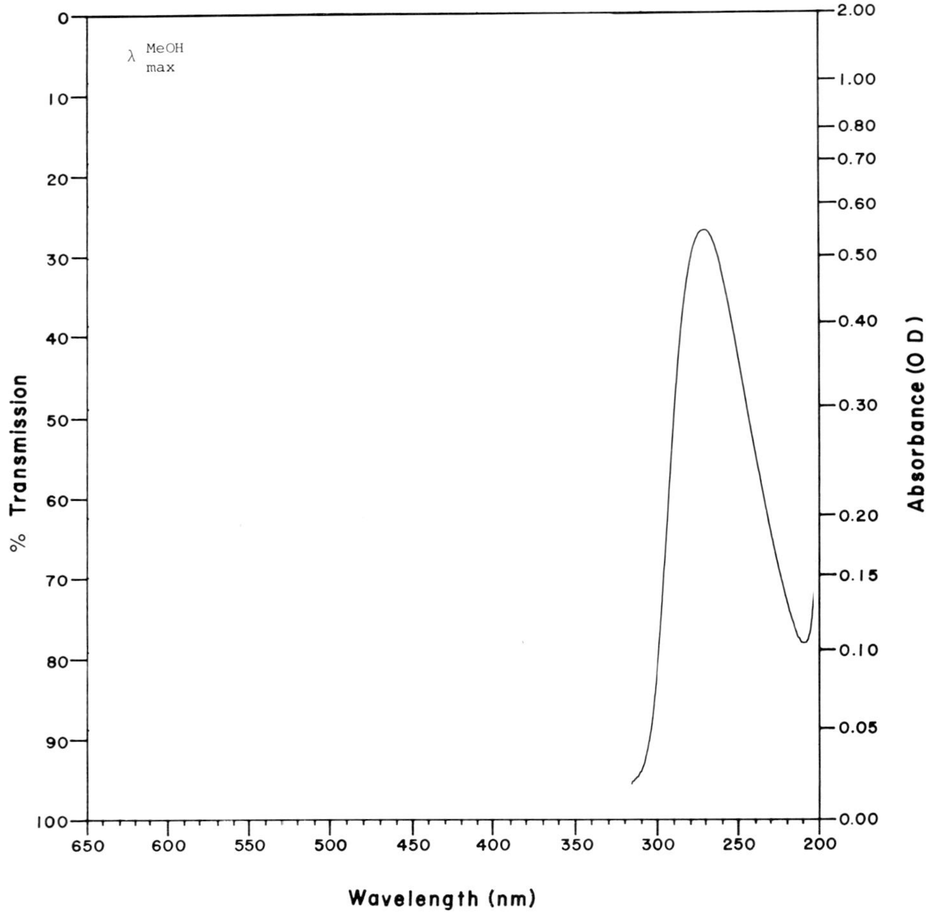

λ_{max}^{MeOH}
% Transmission
Absorbance (O D)
Wavelength (nm)
0
10
20
30
40
50
60
70
80
90
100
2.00
1.00
0.80
0.70
0.60
0.50
0.40
0.30
0.20
0.15
0.10
0.05
0.00
650
600
550
500
450
400
350
300
250
200

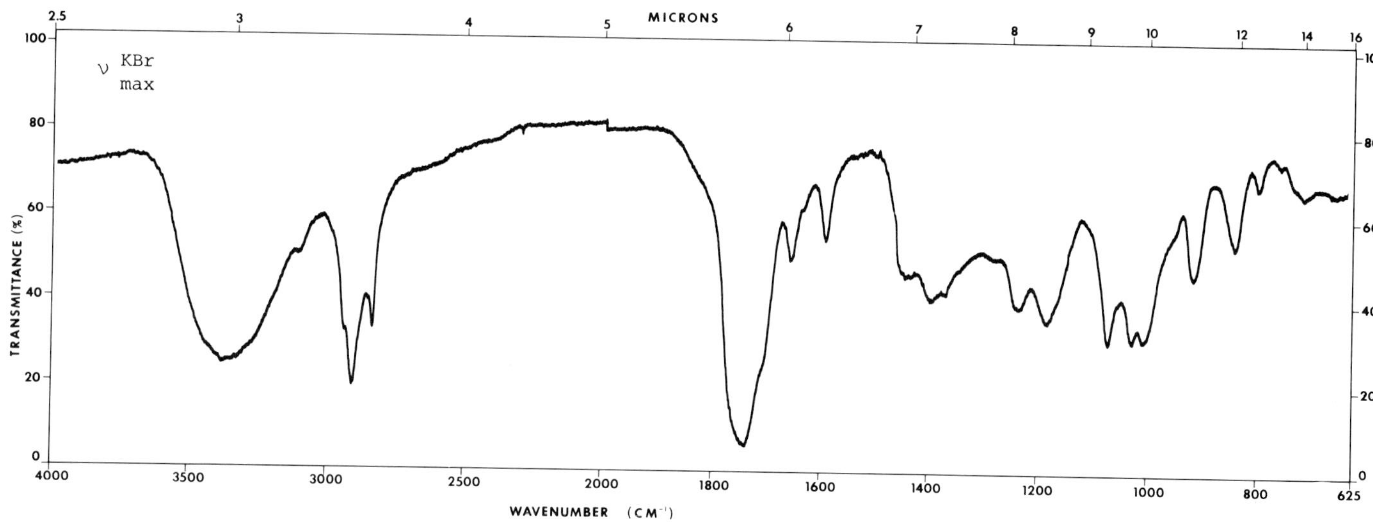

MICRONS
2.5
3
4
5
6
7
8
9
10
12
14
16
ν KBr max
TRANSMITTANCE (%)
100
80
60
40
20
0
4000
3500
3000
2500
2000
1800
1600
1400
1200
1000
800
625
WAVENUMBER (CM⁻¹)

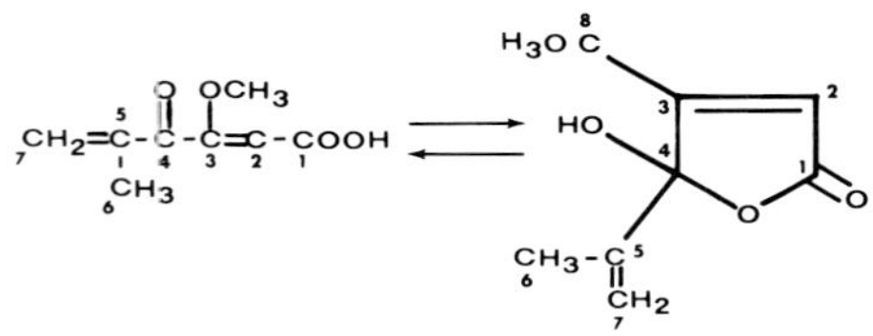

Common name	Penicillic acid (3-Methoxy-5-methyl-4-oxo-2,5-hexadienoic acid; 2-keto-β-methoxy-δ-methylene-Δα-hexenoic acid)
Molecular weight	170.0579
Molecular formula	$C_8H_{10}O_4$
General characteristics[7]	Crystals from petroleum ether, mp 86°–87°C; from H_2O, 58°–64°C $[\alpha]_D$ 0° Methyl ester, colorless prisms, mp 35°C Penicillic acid can occur as keto form or lactone form.
UV data[28]	λ_{max}^{MeOH} nm (ϵ): 221(10,500)
Fungal source[40]	*Penicillium lividum* *P. puberulum* *P. griseum* *P. simplicissimum* *P. cyclopium* *P. thomii* *P. roqueforti* (*P. suavolens*) *P. martensii* *P. fenelliae* *P. aurantio-virens* *P. janthinellum* *P. viridicatum* *P. palitans* *P. baarnense* *P. madriti* *P. lilacinum* *P. canescens* *P. chrysogenum*

P. olivino-viride
Aspergillus ochraceus
A. sulphureus
A. melleus (*A. quercinus*)
A. sclerotiorum
A. alliaceus
A. ostianus
Paecilomyces ehrlichii

Toxicity data[2,16,18,22–26,34,38,44]

Subcutaneous injection of penicillic acid in mice at ca. 200–300 mg/kg was lethal. The LD_{50} to mice dosed SC was 110 mg/kg; IV was 250 mg/kg. The LD_{50} for rabbits dosed SC was 100–200 mg/kg. Penicillic acid has been shown to have antibacterial (mostly gram-negative), antiviral, antitumor, and cytotoxic effects.

TLC data[15]

Adsorbent: Kieselgel G
Solvent: benzene–methanol–acetic acid, 18:1:1 v/v/v
R_f: 0.25 (Patulin has same R_f in this system)
Detection: fluoresces under UV light after exposure to NH_3; yellow fluorescence with NH_3 and phenylhydrazine

GLC data[15,27,36]

Solid support: Gas Chrom-Q
Liquid phase: 10% Dc-200
Column temp: 175°C
Detector temp: 190°C
Injector temp: 190°C
Relative retention time
(relative to β-BHC and were analyzed as TMS derivatives): 0.34
(relative retention time for patulin in same system): 0.56

Solid support: Gas Chrom-Q, 100–102 mesh
Liquid phase: 3% Dexsil 300
3% OV-17
3% OV-25
NOTE: For extensive discussion of relative retention times, conditions, etc., refer to Pero.[27] Additional GLC analyses see Thorp.[36]

Other significant references

^1H NMR[33]; ^{13}C NMR[33]; synthesis[28]; infrared studies[19]

^{13}*C NMR data*[33]	C1, 169.5 s; C2, 89.6 d; C3, 178.9 s; C4, 102.1 s; C5, 139.8 s; C6, 16.9 q; C7, 114.0 t; C8, 59.6 q
^{1}*H NMR data*[33]	H2, 5.09; H6, 1.76; H7a, 5.18; H7b, 5.44; H8, 3.90

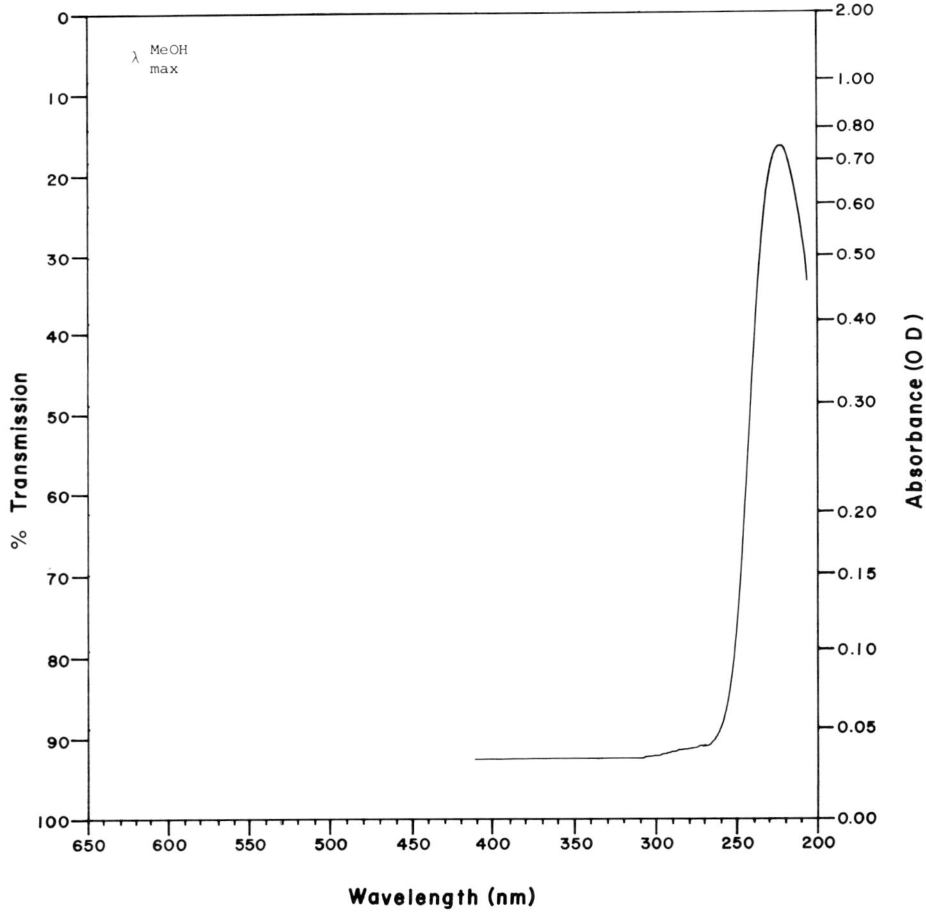

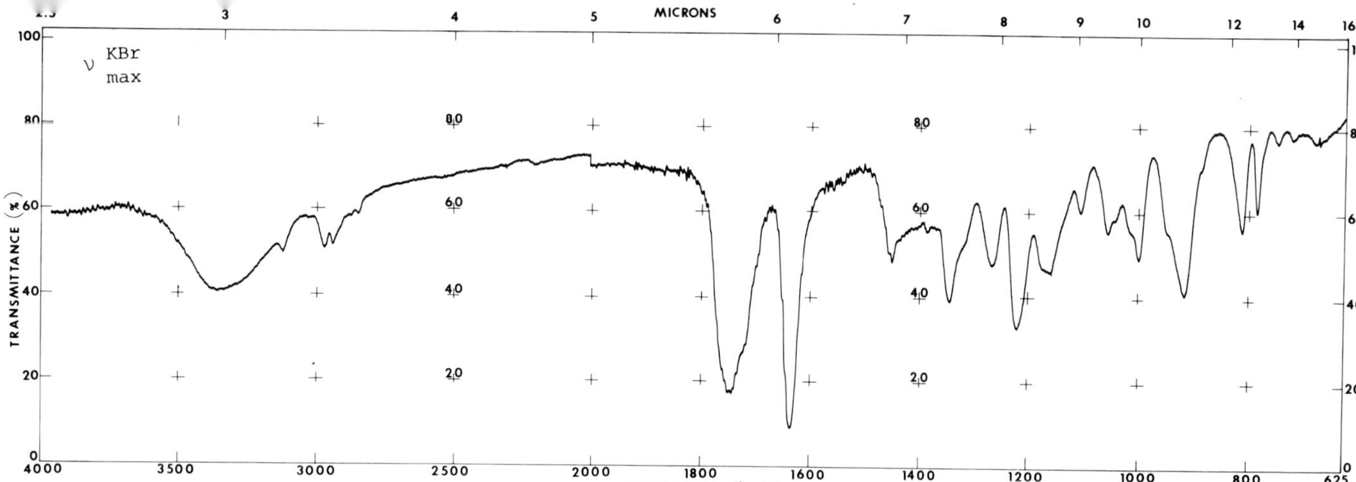

PENICILLIC ACID

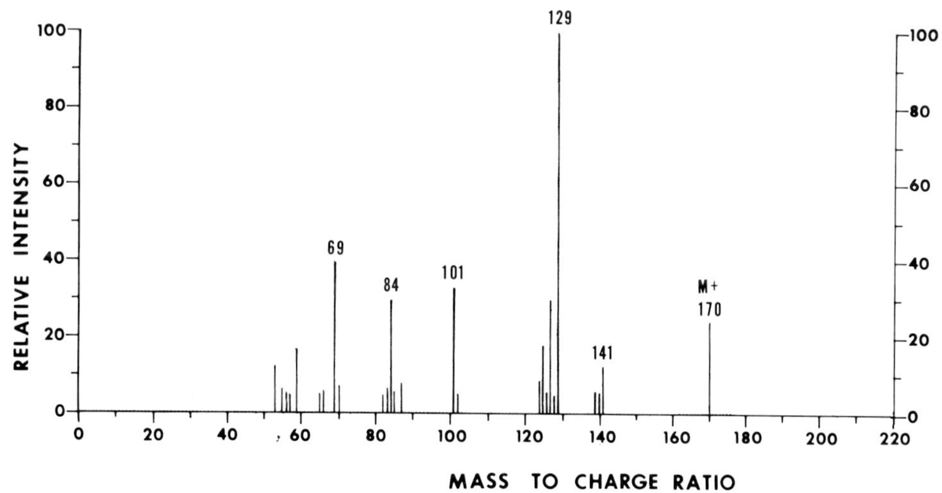

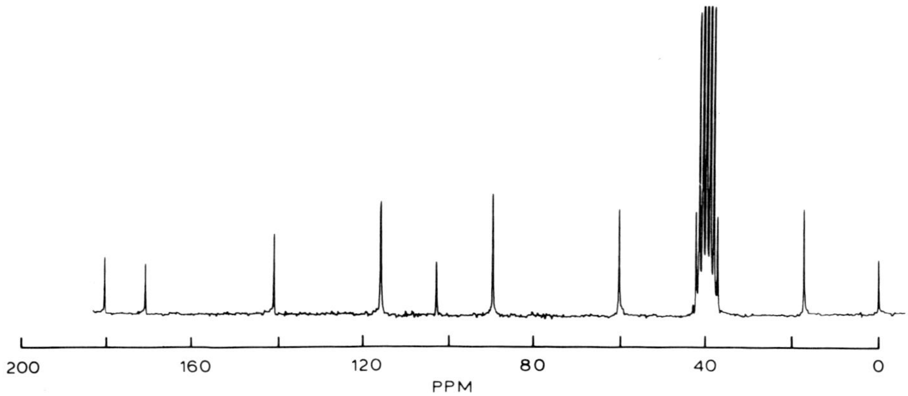

The 25 MHz ^{13} C NMR spectrum of penicillic acid in DMSO solution.

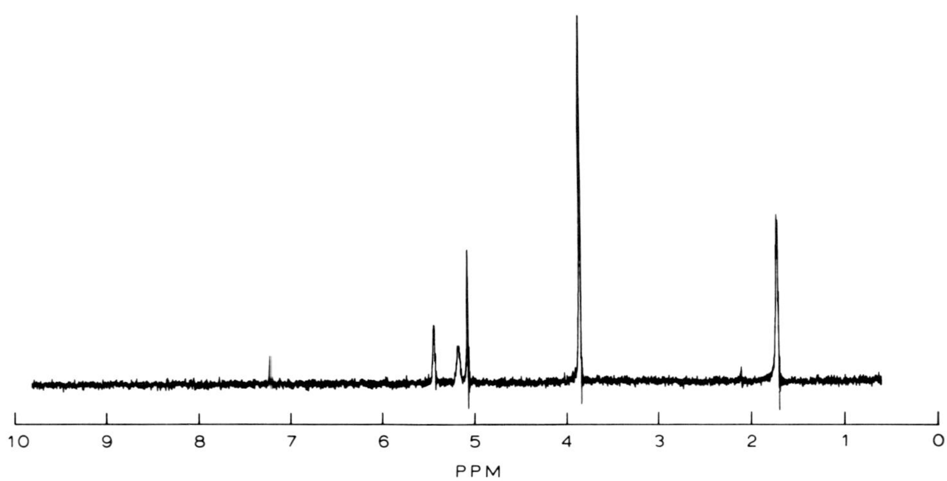

The 25 MHz ^{13}C NMR spectrum of penicillic acid in $CDCl_3$ solution.

REFERENCES

1. Al-Rawi, J. M. A., Elvidge, J. A., Jaiswal, D. K., Jones, J. R., and Thomas, R., *J. Chem. Soc., Chem. Commun.* p. 220 (1974).
2. Alsberg, C. L., and Black, O. F., *U.S. Dept. Agric., Bur. Plant Ind. Bull.* No. 270 (1913).
3. Anslow, W. K., Raistrick, H., and Smith, G., *J. Soc. Chem. Ind., London* **62,** 236 (1943).
4. Berestetskii, O. A., Hadkernichnii, S. P., and Patika, V. F., *Khim. Prir. Soedin.* 420 (1974).
5. Birkinshaw, J. H., Bracken, A., Michael, S. E., and Raistrick, H., *Biochem. J.* **37,** 726 (1943).
6. Birkinshaw, J. H., Bracken, A., Michael, S. E., and Raistrick, H., *Lancet* **245,** 625 (1943).
7. Birkinshaw, J. H., Oxford, A. E., and Raistrick, H., *Biochem. J.* **30,** 394 (1936).
8. Broom, W. A., Bulbring, E., Chapman, C. J., Hampton, J. W. F., Thomson, A. M., Unger, J., Wein, R., and Woolfe, G., *Br. J. Exp. Pathol.* **25,** 195 (1944).
9. Ciegler, A., *in* "Mycotoxins in Human and Animal Health" (J. V. Rodricks, C. N. Hesseltine, and M. A. Mehlman, eds.), p. 609. Pathotox Publ., Inc., Park Forest South, Illinois, 1977.
10. Cole, R. J., personal communication.
11. Dickens, F., and Jones, H. E. H., *Br. J. Cancer* **15,** 85 (1961).
12. Dustin, P., Jr., Pharmacol. Rev. **15,** 449 (1963).
13. Enomoto, M., and Saito, M., *Annu. Rev. Microbiol.* **26,** 279 (1972).
14. Forrester, P. I., and Gaucher, G. M., *Biochemistry* **11,** 1102 (1972).
15. Fujimoto, Y., Suzuki, T., and Hoshino, Y., *J. Chromatogr.* **105,** 99 (1975).
16. Hall, E. A., Kavanagh, F., and Asheshov, I. N., *Antibiot. Chemother.* (*Washington, D.C.*) **1,** 369 (1951).
17. Harwig, J., Blanchfield, B. J., and Scott, P. M., *Can. Inst. Food Sci. Technol. J.* **11,** 149 (1978).
18. Kawasaki, I., Oki, T., Umeda, M., and Saito, M., *Jpn. J. Exp. Med.* **42,** 327 (1972).
19. Kovac, S., and Solcaniova, E., *Tetrahedron* **25,** 3617 (1969).
20. Koyyboki, T., Kowszyk-Gindifer, Z., and Kurylowicz, W., "Antibiotic Origin, Nature, and Properties," Vol. II, p. 1223. Permagon, Oxford, 1967.
21. Lovett, J., *Poult. Sci.* **51,** 2097 (1972).
22. Murnaghan, M. F., *J. Pharmacol.* **88,** 119 (1946).
23. Natori, S., Sakaki, S., Kurata, H., Udagawa, S., Ichinoe, M., Saito, M., and Umeda, M., *Chem. Pharm. Bull.* **18,** 2259 (1970).
24. Oxford, A. E., *Biochem. J.* **36,** 438 (1942).
25. Oxford, A. E., *Chem. Ind.* (*London*) p. 48 (1942).
26. Oxford, A. E., Raistrick, H., and Smith, G., *Chem. Ind.* (*London*) p. 22 (1942).
27. Pero, R. W., Harvan, D., and Owens, R. G., *J. Chromatogr.* **65,** 501 (1972).
28. Raphael, R. A., *J. Chem. Soc.* p. 1508 (1948).
29. Scott, A. I., Zamir, L., Phillips, G. T., and Yalpani, M., *Bioorg. Chem.* **2,** 124 (1973).
30. Scott, P. M., *in* "Mycotoxins" (I. F. H. Purchase, ed.), p. 383. Elsevier, Amsterdam, 1974.
31. Scott, P. M., Miles, W. F., Toft, P., and Dube, J. G., *J. Agric. Food Chem.* **20,** 450 (1972).
32. Sekiguchi, J., Gaucher, G. M., and Yamada, Y., *Tetrahedron Lett.* p. 41 (1979).
33. Seto, H., Cary, L. W., and Tenabe, M., *J. Antibiot.* **27,** 558 (1974).
34. Suzuki, S., Kimura, T., Saito, F., and Ando, K., *Agric. Biol. Chem.* **35,** 287 (1971).
35. Suzuki, T., Takeda, M., and Tanabe, H., *Chem. Pharm. Bull.* **19,** 1786 (1971).
36. Thorpe, C. W., and Johnson, R. L., *J. Assoc. Off. Anal. Chem.* **57,** 861 (1974).
37. Ukai, T., Yamamoto, Y., and Yamamoto, T., *J. Pharm. Soc. Jpn.* **74,** 450 (1954).
38. Umeda, M., *Jpn. J. Exp. Med.* **41,** 195 (1971).
39. Ware, G. M., Thorpe, C. W., and Pohland, A. E., *J. Assoc. Off. Anal. Chem.* **57,** 1111 (1975).
40. Wilson, D. M., *Adv. Chem. Ser.* **149,** 90 (1976).

41. Woodward, R. B., and Singh, G., *Experientia* **6,** 238 (1950).
42. Woodward, R. B., and Singh, G., *J. Am. Chem. Soc.* **71,** 758 (1949).
43. Woodward, R. B., and Singh, G., *J. Am. Chem. Soc.* **72,** 1428 (1950).
44. Zeller, A., and Matthes, T., *Dtsch. Gesundheitswes.* **1,** 499 (1946).

Roquefortines

10

The following alkaloids are of interest as potentially significant mycotoxins since the fungi that produce them are commonly found in fermenting silage and used in the production of Roquefort cheese and other types of blue cheese. Roquefortine A and B (isofumigaclavine A and B) are stereoisomers of fumigaclavine A and B, metabolites commonly produced by *Aspergillus fumigatus*. Roquefortine (roquefortine C), an unrelated alkaloid, was found naturally occurring in blue cheese from seven countries (up to 6.8 ppm).[11] It was usually accompanied by isofumigaclavine A (up to 4.7 ppm) and traces of isofumigaclavine B. Fumigaclavines A and C were produced by isolates of *A. fumigatus* involved in a toxic syndrome of cattle ingesting molded silage.[6]

The rugulovasine group of alkaloids was reported from cultures of *P. islandicum* and *P. concavo-rugulosum*. Chlororugulovasines A and B are reportedly the first naturally occurring halogenated ergoline alkaloids. The diasteriomeric rugulovasines and chlororugulovasines can be interconverted upon warming in polar solvents.[4,16]

Roquefortine group	Molecular weight	Molecular formula
Chlororugulovasine A	302.0821	$C_{16}H_{15}O_2N_2Cl$
Chlororugulovasine B	302.0821	$C_{16}H_{15}O_2N_2Cl$
Rugulovasine A	268.1211	$C_{16}H_{16}O_2N_2$
Rugulovasine B	268.1211	$C_{16}H_{16}O_2N_2$
Fumigaclavine A (SM-2)	298.1681	$C_{18}H_{22}O_2N_2$
Roquefortine A (Isofumigaclavine A)	298.1681	$C_{18}H_{22}O_2N_2$
Fumigaclavine B	256.1575	$C_{16}H_{20}ON_2$
Roquefortine B (Isofumigaclavine B)	256.1575	$C_{16}H_{20}ON_2$
Fumigaclavine C (SM-1)	366.2307	$C_{23}H_{30}O_2N_2$
Roquefortine or Roquefortine C	389.1851	$C_{22}H_{23}O_2N_5$

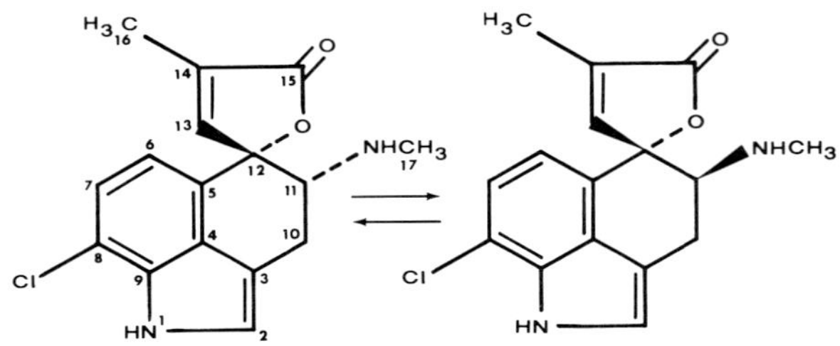

8-Chlororugulovasine A ⇄ 8-Chlororugulovasine B

Common name	8-Chlororugulovasine A
Molecular weight	302.0821 (mass spectrum identical to that of 8-chlororugulovasine B)
Molecular formula	$C_{16}H_{15}O_2N_2Cl$
UV data[4]	λ_{max}^{MeOH} nm (ϵ): 225(21,400), 280(sh)(4,600), 292(5,200), and 298(5,400)
Fungal source[5]	*Penicillium islandicum* Sopp
Toxicity data[5]	LD_{50} in day old cockerels dosed orally was 75–125 mg/kg.
TLC data[4]	*Adsorbent:* silica gel PF_{254} *Solvent:* chloroform–acetone–methanol, 93:7:5 v/v/v R_f: 0.27 *Detection:* rose-colored spot after spraying with 50% ethanolic H_2SO_4 followed immediately by spraying with ninhydrin and heating at 100°C for 5–10 min.
Other significant references	X-Ray crystallography and stereochemistry[4]
^{13}C NMR data	C2, 120.6 d; C3, 109.9 s; C4, 124.8 s; C5, 127.4 s; C6, 116.0 d; C7, 122.3 d; C8, 116.5 s; C9, 131.1 s*; C10, 24.3 t; C11, 63.6 d; C12, 87.4 s; C13, 150.2 d; C14, 129.8 s*; C15, 173.5 s; C16, 10.8 q; C17, 34.8 q
^{1}H NMR data	H1, 8.60; H2, 7.00; H6, 6.77 (J = 7.6); H7, 7.11 (J = 7.6); H10, 3.30; H11, 3.30; 11NH, 1.43; H13, 7.35 (J = 1.5); H16, 2.02 (J = 1.5); H17, 2.43

* Assignments may be interchanged.

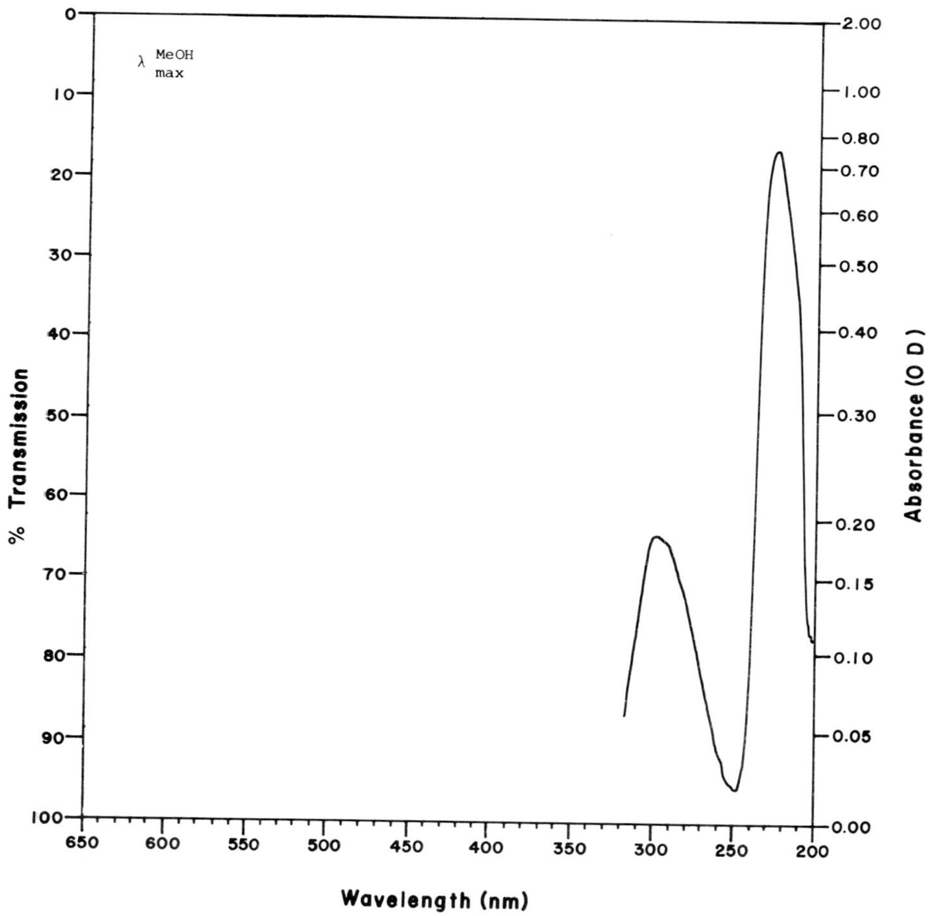

λ MeOH max
% Transmission
0
10
20
30
40
50
60
70
80
90
100
Absorbance (O D)
2.00
1.00
0.80
0.70
0.60
0.50
0.40
0.30
0.20
0.15
0.10
0.05
0.00
650
600
550
500
450
400
350
300
250
200
Wavelength (nm)

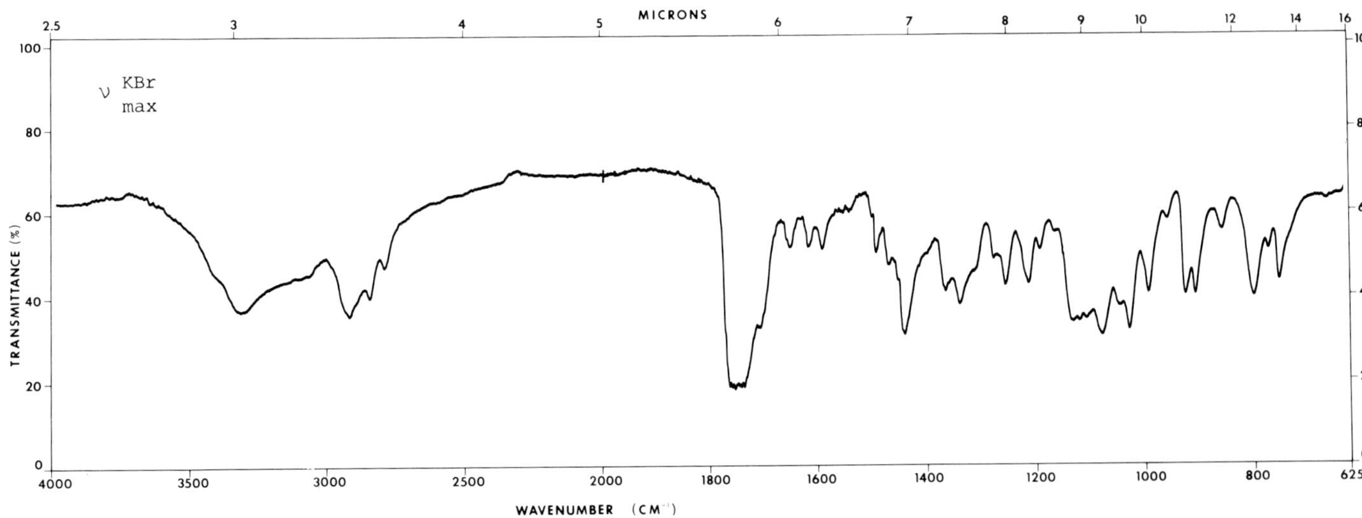
MICRONS
2.5
3
4
5
6
7
8
9
10
12
14
16
ν KBr max
TRANSMITTANCE (%)
100
80
60
40
20
0
4000
3500
3000
2500
2000
1800
1600
1400
1200
1000
800
625
WAVENUMBER (CM⁻¹)

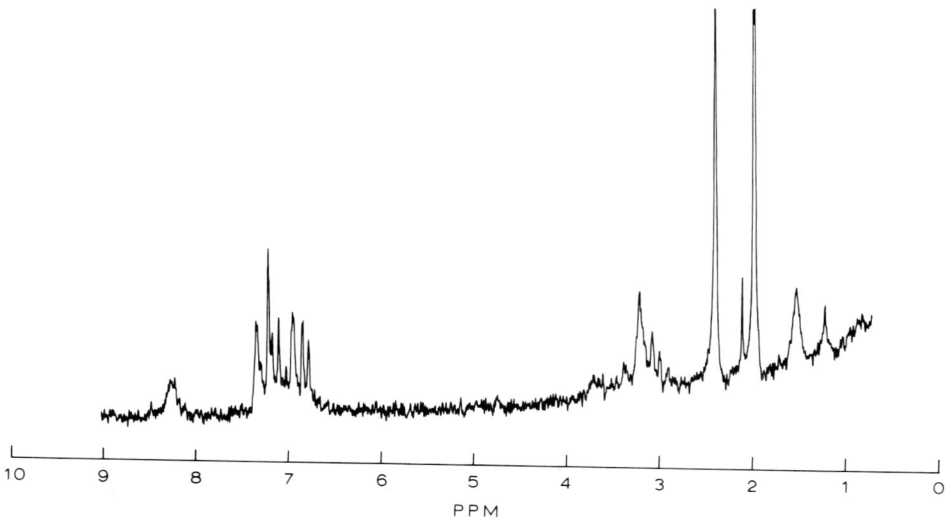

The 100 MHz ^1H NMR spectrum of chlororugulovasine A in $CDCl_3$ solution.

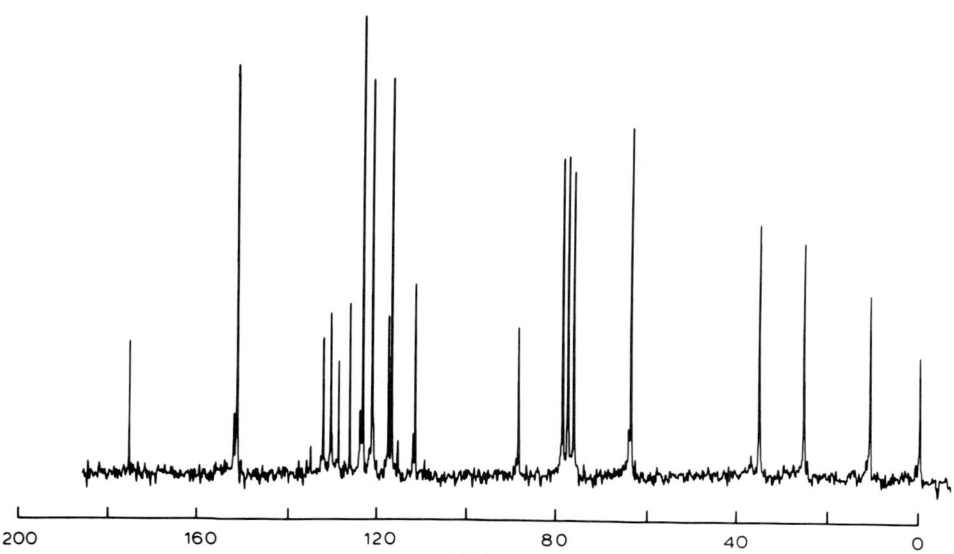

The 25 MHz ^{13}C NMR spectrum of chlororugulovasine A in $CDCl_3$ solution.

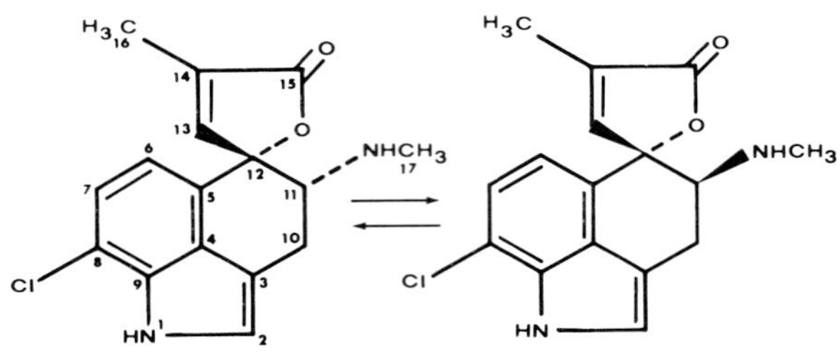

Common name	8-Chlororugulovasine B
Molecular weight	302.0821
Molecular formula	$C_{16}H_{15}O_2N_2Cl$
UV data[4]	λ_{max}^{MeOH} nm (ϵ): 224(22,400), 278(4,700), 288(5,100), and 295(5,400)
Fungal source[5]	*Penicillium islandicum* Sopp
Toxicity data[5]	LD_{50} of chlororugulovasine B in preliminary studies showed it to be as toxic as chlororugulovasine A in day-old cockerels dosed orally (75–125 mg/kg).
TLC data[4]	*Adsorbent:* silica gel PF_{254} *Solvent:* chloroform–acetone–methanol, 73:7:5 v/v/v R_f: 0.44 *Detection:* rose-colored spot after spraying with 50% ethanolic H_2SO_4 followed immediately with ninhydrin spraying and heating at 100°C for 5–10 min.
Other significant references	X-Ray crystallography and stereochemistry[4]
^{13}C NMR data	C2, 120.1 d; C3, 110.6 s; C4, 124.9 s; C5, 127.5 s; C6, 115.8 d; C7, 122.0 d; C8, 116.5 s; C9, 130.9 s*; C10, 25.1 t; C11, 63.0 d; C12, 87.8 s; C13, 149.9 d; C14, 129.3 s*; C15, 173.8 s; C16, 10.7 q; C17, 34.9 q
^1H NMR Assignments	H1, 8.89; H2, 7.09; H6, 6.73 (J = 7.6); H7, 7.05 (J = 7.6); H10, 3.07; H11, 3.07; 11NH, 1.61; H13, 6.78 (J = 1.5); H16, 2.01 (J = 1.5); H17, 2.44

* Assignments may be interchanged.

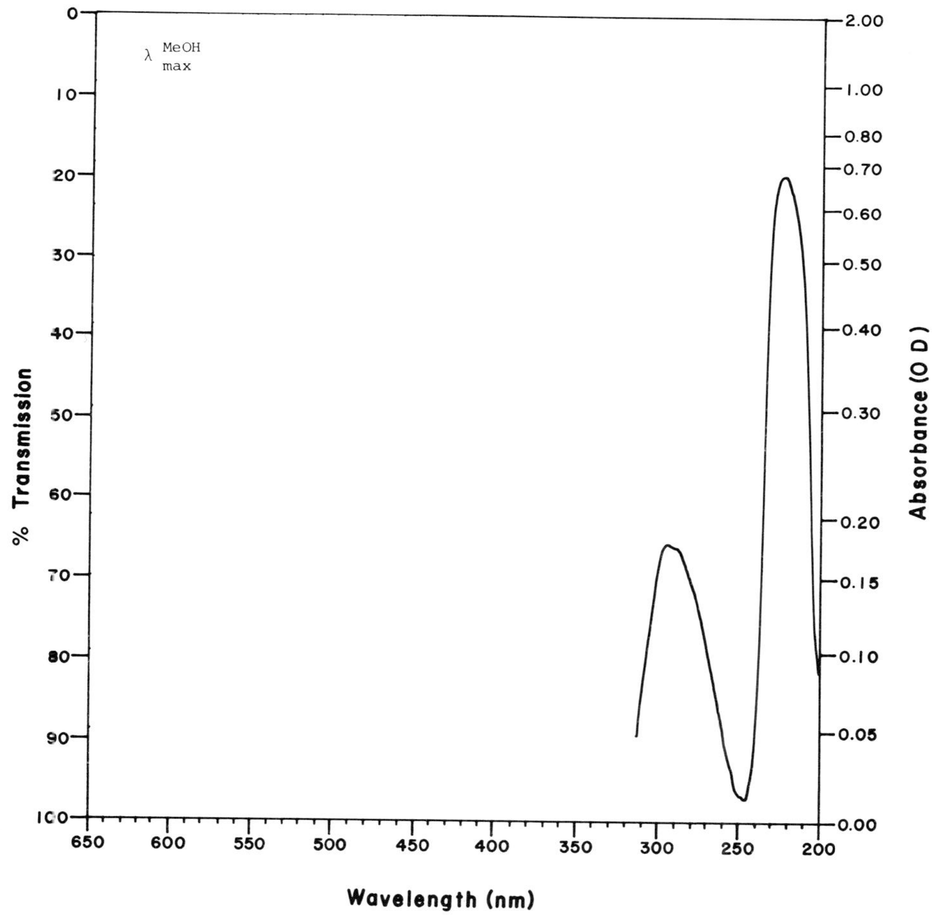
λ MeOH max
% Transmission
0
10
20
30
40
50
60
70
80
90
100
Absorbance (O D)
2.00
1.00
0.80
0.70
0.60
0.50
0.40
0.30
0.20
0.15
0.10
0.05
0.00
650
600
550
500
450
400
350
300
250
200
Wavelength (nm)

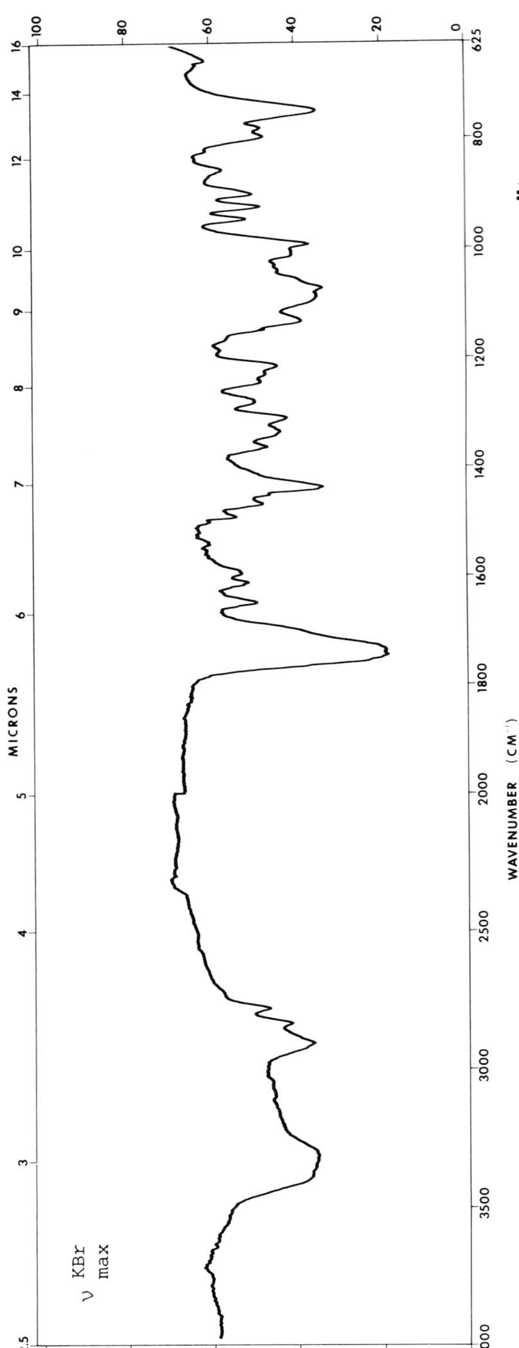

MICRONS
WAVENUMBER (CM⁻¹)
TRANSMITTANCE (%)
ν KBr max

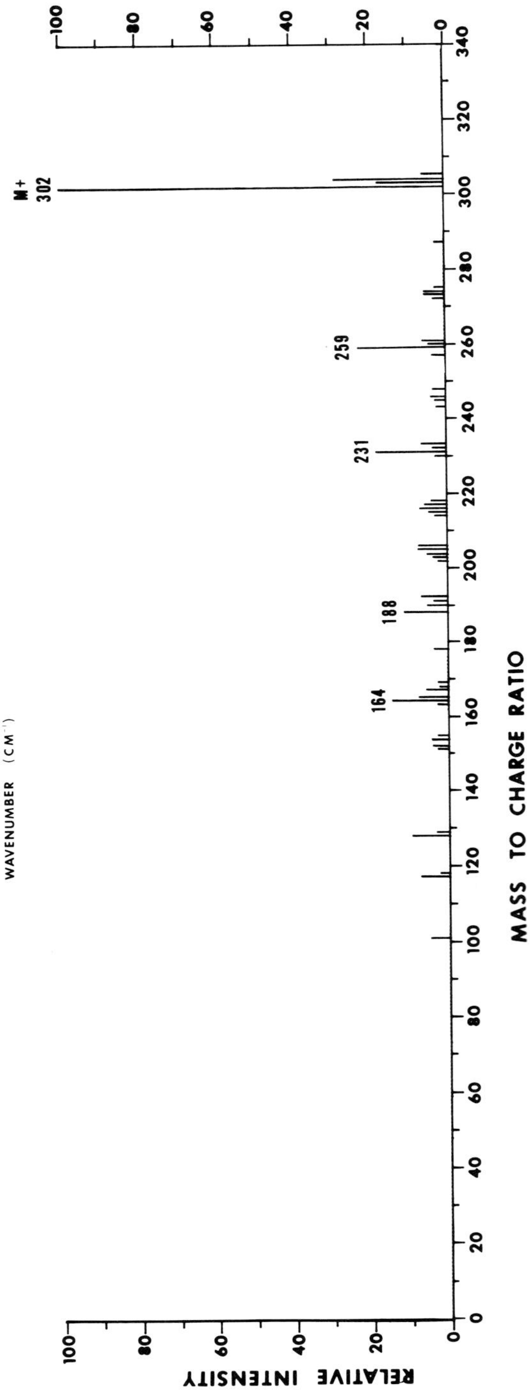

M+
302
259
231
188
164
MASS TO CHARGE RATIO
RELATIVE INTENSITY

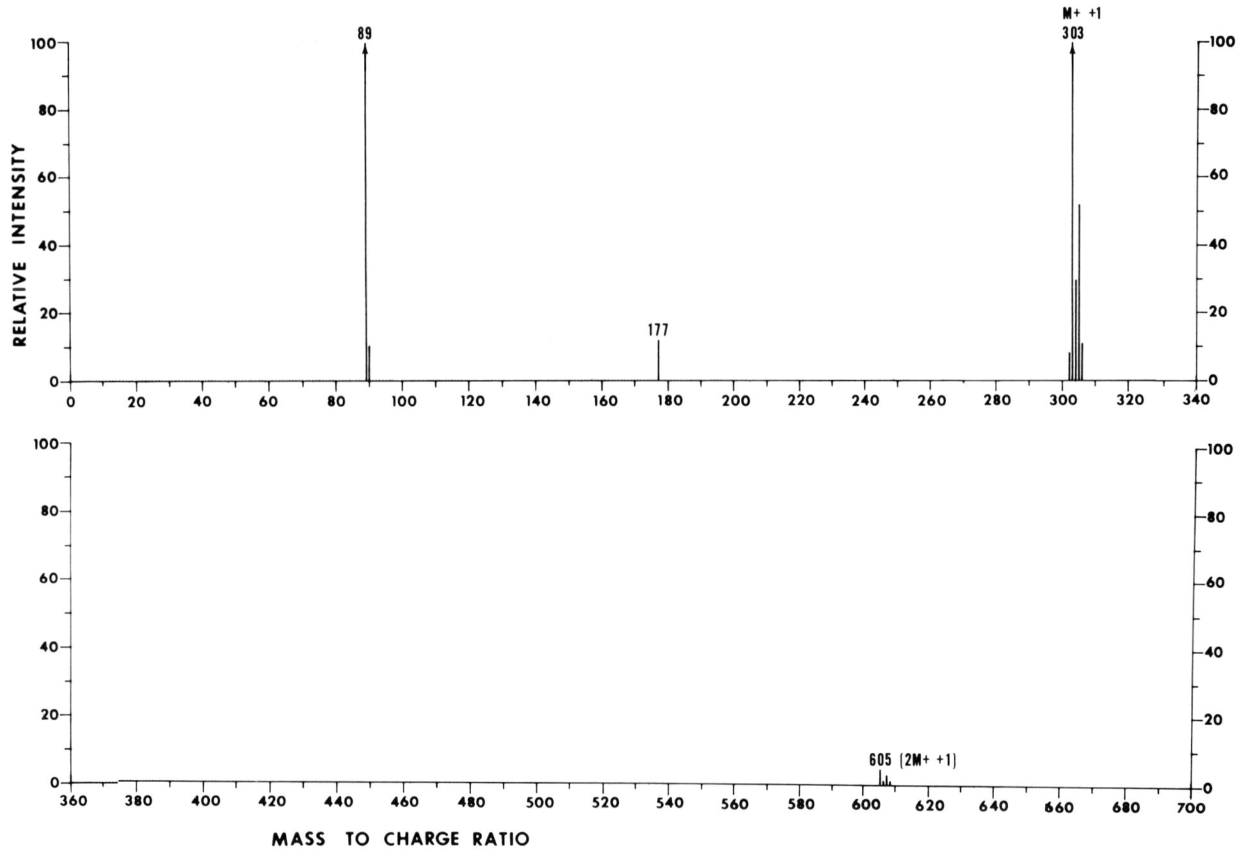

Chemical ionization mass spectrum. Isobutane used as ionizing gas.

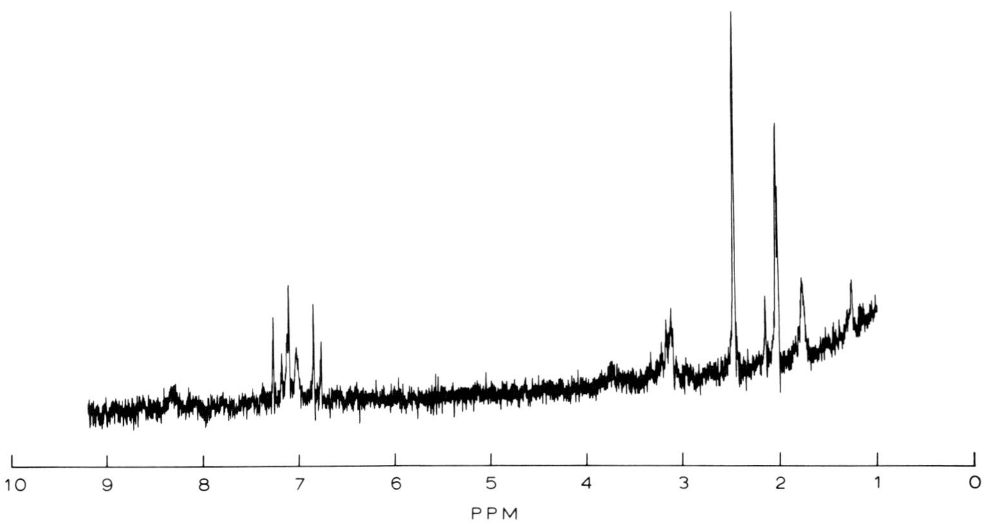

The 100 MHz ^{1}H NMR spectrum of chlororugulovasine B in $CDCl_3$ solution.

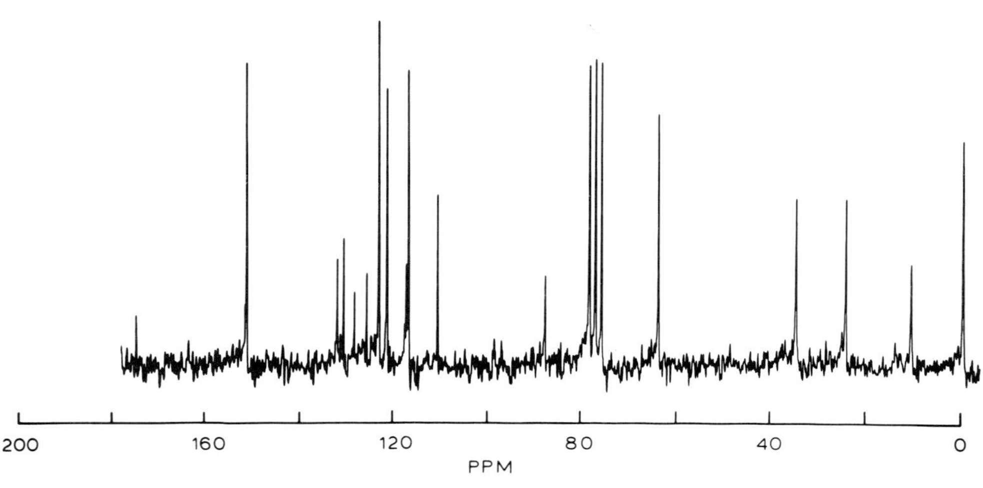

The 25 MHz ^{13}C NMR spectrum of chlororugulovasine B in $CDCl_3$ solution.

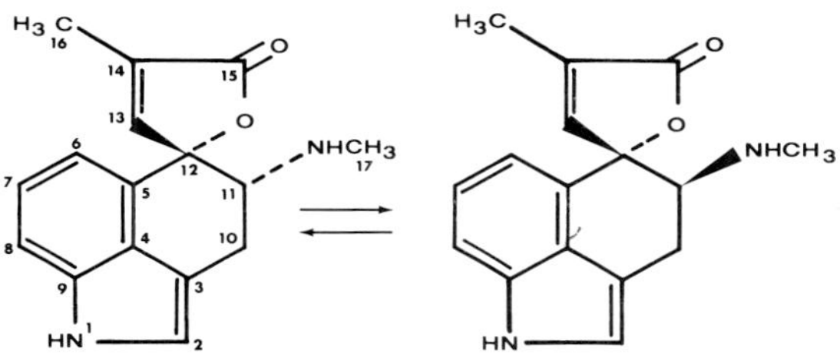

Rugulovasine A ⇄ Rugulovasine B

Common name	Rugulovasine A
Molecular weight	268.1211
Molecular formula	$C_{16}H_{16}O_2N_2$
General characteristics[1,4,16]	mp 138°C (dec.) $[\alpha]^{22}_{436}$ −3.0° (C = 1.0 in pyridine) $[\alpha]^{22}_{D}$ 0.0° Rugulovasine A can be interconverted with its diastereomer, rugulovasine B upon warming in polar solvent.
UV data[1]	λ_{max}^{EtOH} nm (ϵ): 224(23,400), 277(5,000), 288(6,000), and 295(6,000)
Fungal source[1,4]	*Penicillium islandicum* Sopp *P. concavo-rugulosum* *P. verrucosum* Dierckx
Toxicity data[1]	Rugulovasine A exhibited hypotensive action on chloralose-urethanized cats; the minimum effective dose (IV) was 0.2–0.5 mg/kg.
TLC data[4]	*Adsorbent:* silica gel PF_{254} *Solvent:* chloroform–acetone–methanol, 93:7:5 v/v/v R_f: 0.22 *Detection:* rose-colored spot after spraying with 50% ethanolic H_2SO_4 followed immediately by spraying with ninhydrin and heating at 100°C for 5–10 min.
^{13}C NMR data	C2, 119.6 d; C3, 109.8 s; C4, 126.4 s*; C5, 128.2 s*; C6, 114.8 d; C7, 122.8 d; C8, 111.4 d; C9, 133.9 s; C10, 25.4 t; C11, 63.7 d; C12, 88.5 s; C13, 150.6 d; C14, 129.2 s; C15, 174.3 s; C16, 10.8 q; C17, 35.1 q

* Assignments may be interchanged.

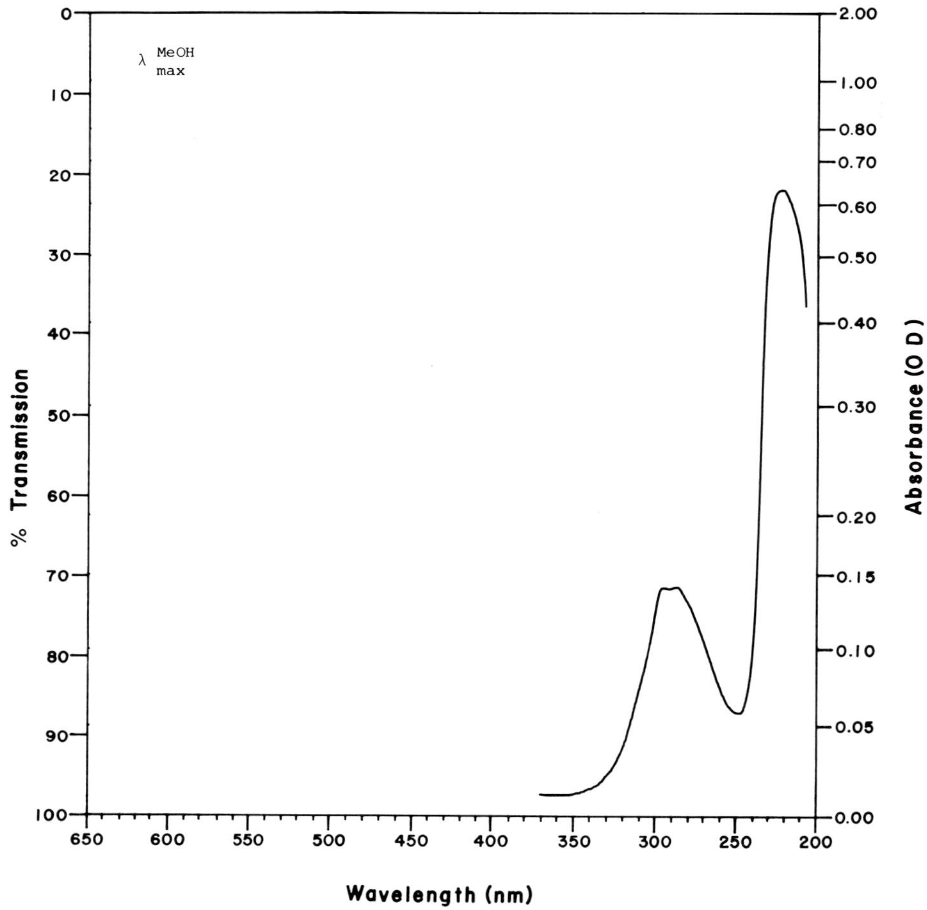
λ MeOH max
% Transmission
Absorbance (O D)
Wavelength (nm)
0
10
20
30
40
50
60
70
80
90
100
2.00
1.00
0.80
0.70
0.60
0.50
0.40
0.30
0.20
0.15
0.10
0.05
0.00
650
600
550
500
450
400
350
300
250
200

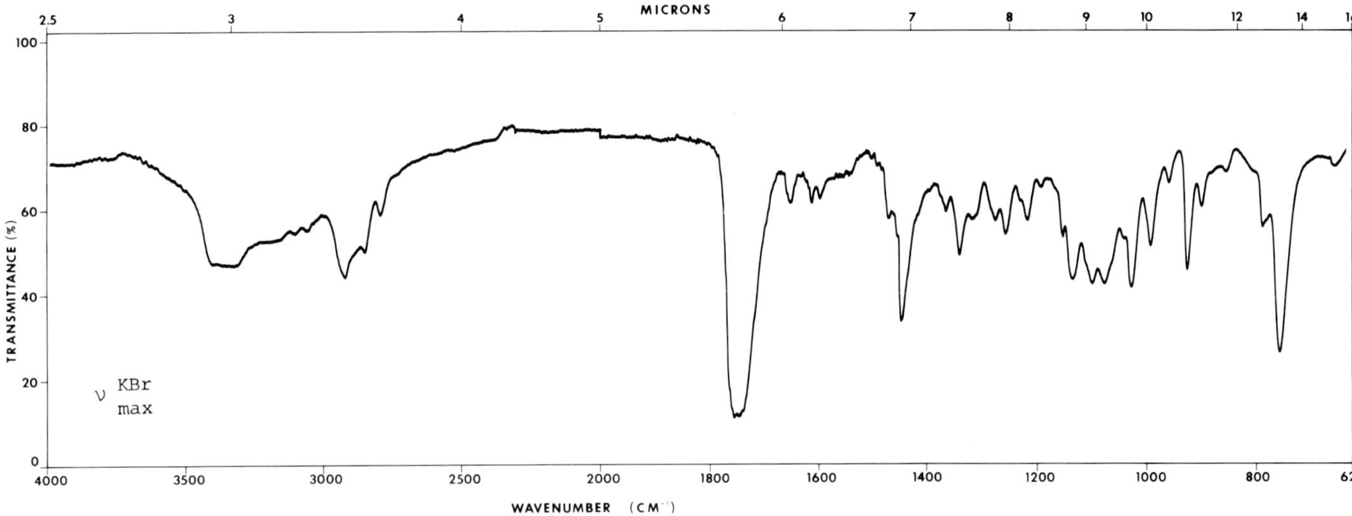
MICRONS
TRANSMITTANCE (%)
ν KBr max
WAVENUMBER (CM⁻¹)

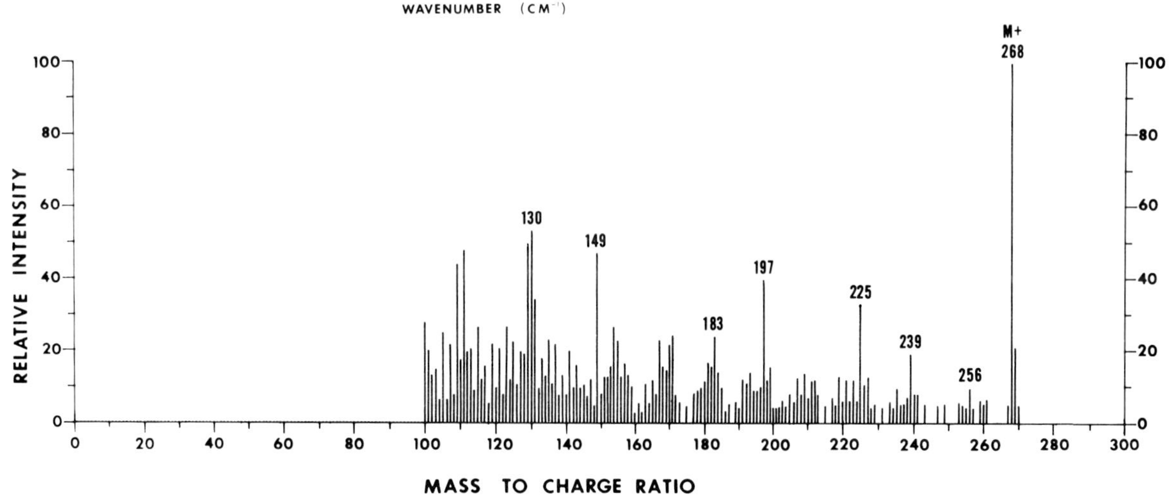
M+
268
130
149
197
183
225
239
256
RELATIVE INTENSITY
MASS TO CHARGE RATIO

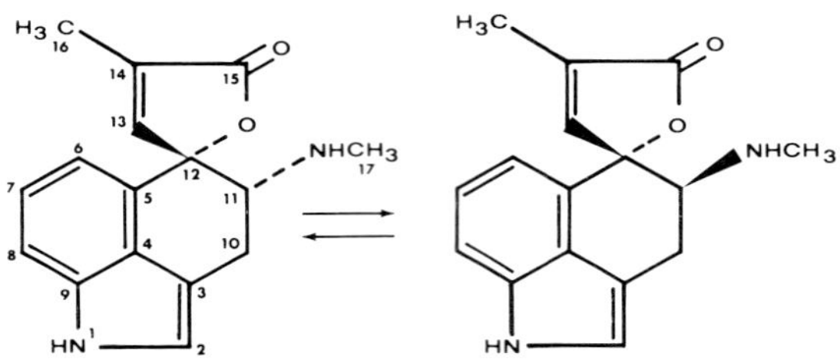

Rugulovasine A ⇄ Rugulovasine B

Common name	Rugulovasine B
Molecular weight	268.1211
Molecular formula	$C_{16}H_{16}O_2N_2$
General characteristics[1,16]	Colorless resinous oil $[\alpha]^{22}_{436}$ +1.4° (C = 1.0 in pyridine) $[\alpha]^{22}_{D}$ 0.0°
UV data[1,16]	λ^{EtOH}_{max} nm (ϵ): 224(14,500), 277(4,800), 288(5,400), and 295(5,200)
Fungal source[1,5]	*Penicillium islandicum* Sopp *P. concavo-rugulosum*
Toxicity data[1]	Rugulovasine B exhibited hypotensive action in experimental animals; in chloralose-urethanized cats, the minimum effective dose (IV) was 0.025–0.05 mg/kg.
TLC data[4]	*Adsorbent:* silica gel PF_{254} *Solvent:* chloroform–acetone–methanol, 93:7:5 v/v/v R_f: 0.34 *Detection:* rose-colored spot after spraying with 50% ethanolic H_2SO_4 followed immediately by spraying with ninhydrin and heating at 100°C for 5–10 min.
^{13}C NMR data	C2, 120.0 d; C3, 108.4 s; C4, 125.8 s*; C5, 126.1 s*; C6, 114.4 d; C7, 122.7 d; C8, 111.2 d; C9, 133.9 s; C10, 24.3 t; C11, 63.6 d; C12, 88.2 s; C13, 150.9 d; C14, 129.2 s; C15, 174.0 s; C16, 10.7 q; C17, 34.7 q
^{1}H NMR data	H1, 8.68; H2, 7.00; H6, 6.83; H7, 7.2: H8, 7.2; H10, 3.25; H11, 3.25; 11NH, 1.70; H13, 6.98; H16, 2.02; H17, 2.43

* Assignments may be interchanged.

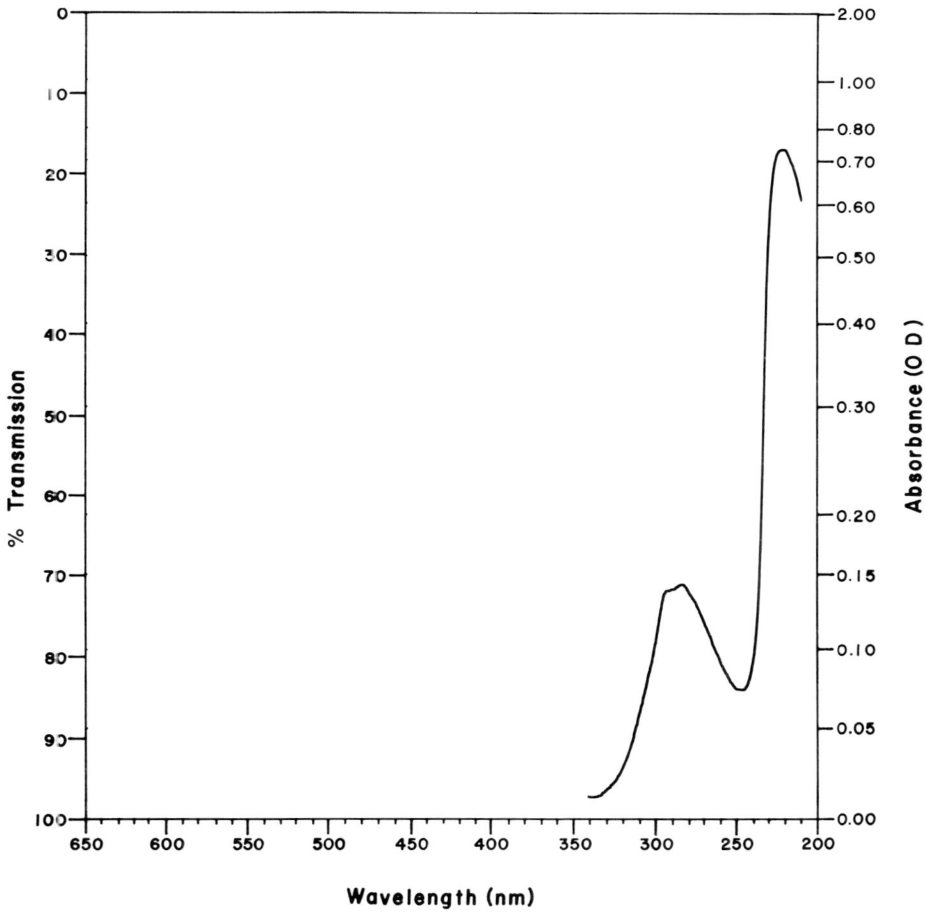
% Transmission
Absorbance (O D)
Wavelength (nm)
0
10
20
30
40
50
60
70
80
90
100
2.00
1.00
0.80
0.70
0.60
0.50
0.40
0.30
0.20
0.15
0.10
0.05
0.00
650
600
550
500
450
400
350
300
250
200

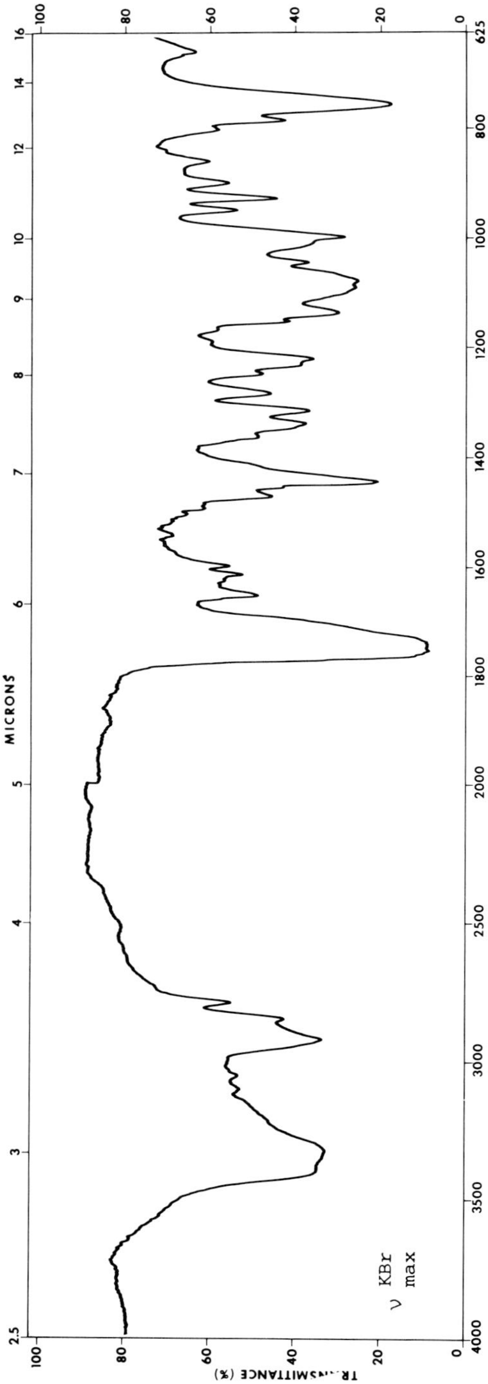
MICRONS
WAVENUMBER (CM⁻¹)
TRANSMITTANCE (%)
ν KBr max

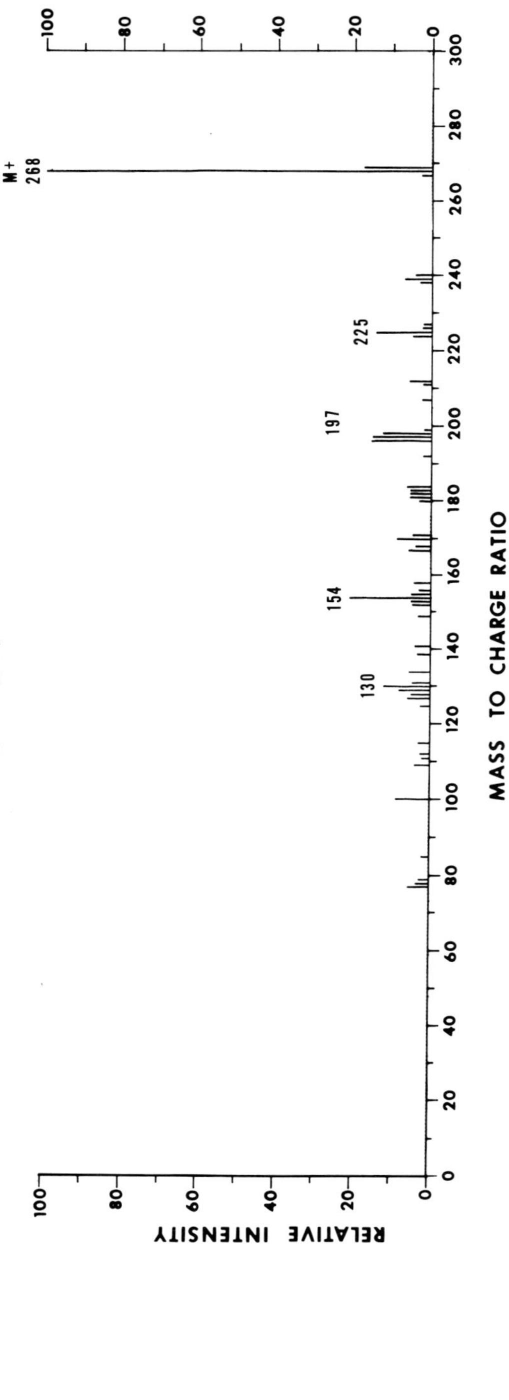
M+
268
225
197
154
130
MASS TO CHARGE RATIO
RELATIVE INTENSITY

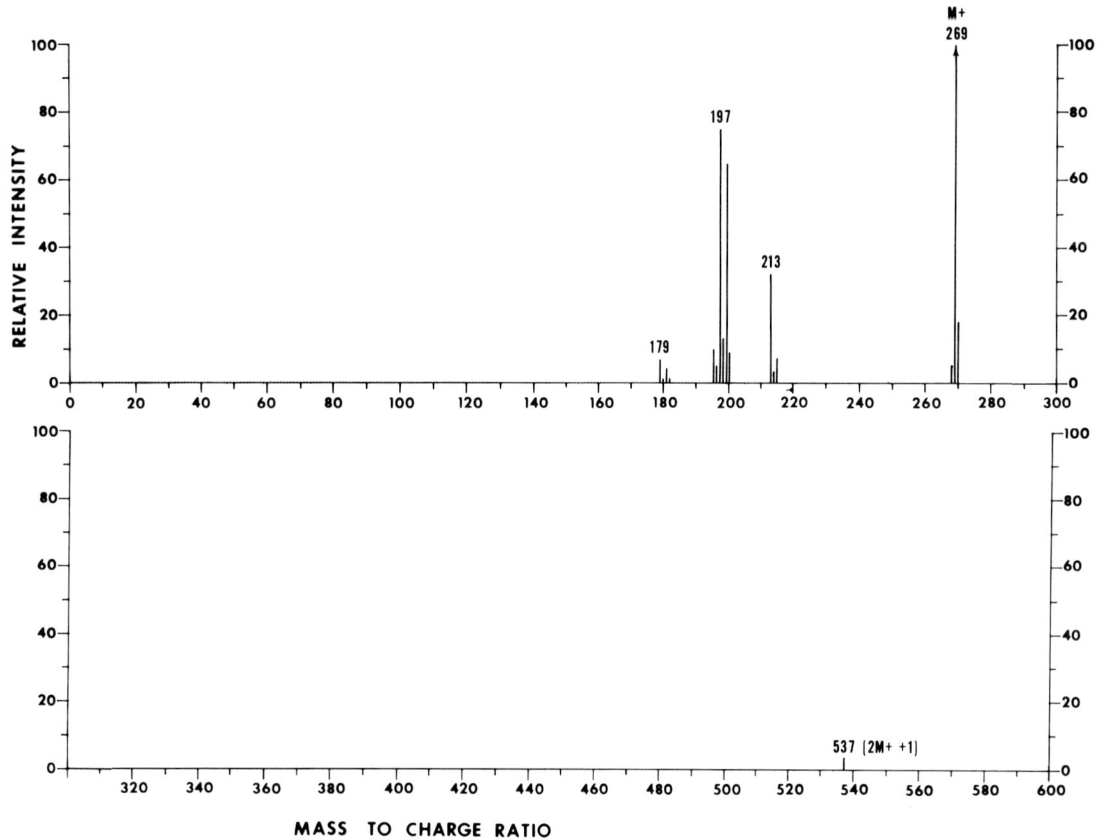

Chemical ionization spectrum using isobutane as ionizing gas.

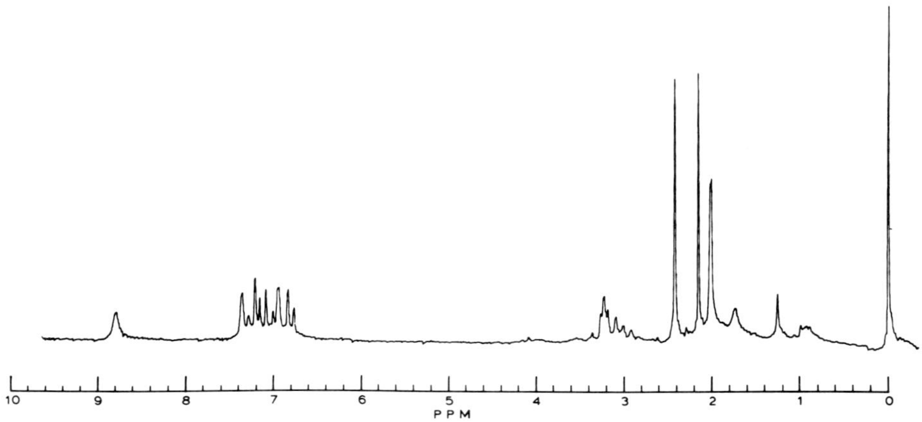

The 100 MHz ^{1}H NMR spectrum of rugulovasine B in $CDCl_3$ solution.

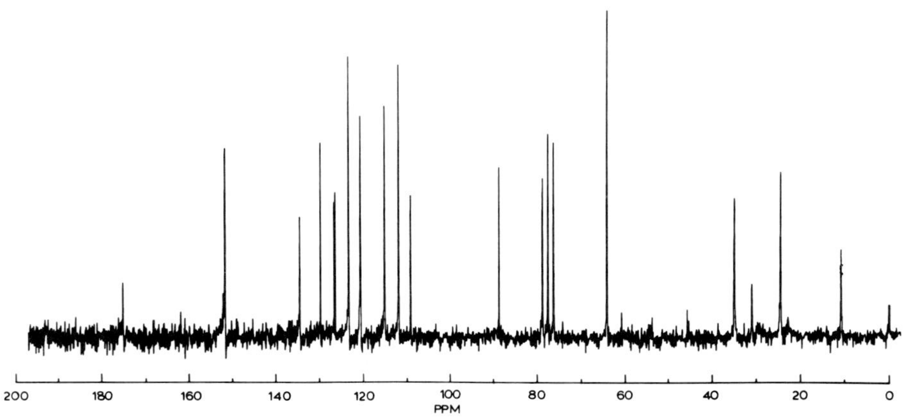

The 25 MHz ^{13}C NMR spectrum of rugulovasine B in $CDCl_3$ solution.

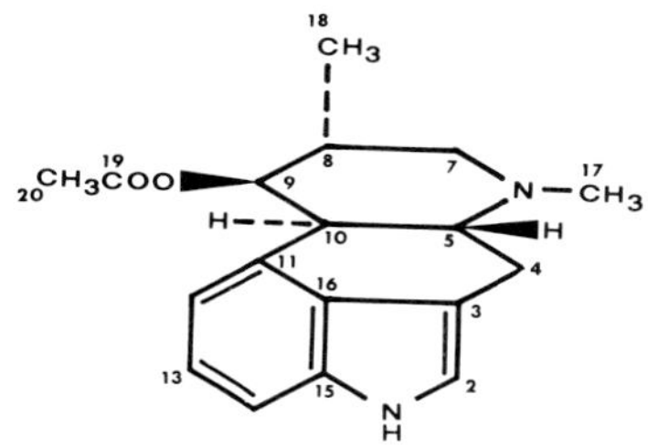

Common name	Fumigaclavine A (9-Acetoxy-6,8-dimethylergoline)
Molecular weight	298.1681
Molecular formula	$C_{18}H_{22}O_2N_2$
General characteristics[14]	Colorless needles, mp 84°85°C (crystals from aqueous methanol) $[\alpha]_{5461}$ −56.7° (pyridine) Hydrochloride: prisms from ethanol, mp 304°–305° (dec.) $[\alpha]^{22}_{5461}$ −56.7° (C = 1.5 in methanol)
UV data[14]	λ_{max} nm: 225, 275, 282, and 293 (ϵ_{max} not reported)
Fungal source[2,6,14,15]	*Aspergillus fumigatus* Fres.
Toxicity data	Toxicity unknown
TLC data[6]	*Adsorbent:* silica gel G-HR *Solvent:* toluene–ethyl acetate–formic acid, 4:4:1 v/v/v R_f: 0.05 *Detection:* purple spot after spraying with 50% ethanolic H_2SO_4 and heating at 100°C for 5–10 min.
Other significant references	X-Ray crystallography[6]
^{13}C NMR data	C2, 117.7 d; C3, 111.6 s; C4, 26.8 t; C5, 61.9 d; C7, 57.9 t; C8 33.2 d; C9, 71.6 d; C10, 39.8 d; C11, 129.9 s; C12, 112.9 d; C13, 123.2 d; C14, 108.3 d; C15, 133.6 s; C16, 126.8 s; C17, 43.6 q; C18, 16.8 q; C19, 170.8 s; C20, 21.2 q
^{1}H NMR data	1-NH, 7.94; H2, 6.86; H9, 5.70; H12, 7.12; H13, 6.76; H14, 7.16; H17, 2.47; H18, 1.35 (J = 7.0); H20, 1.90

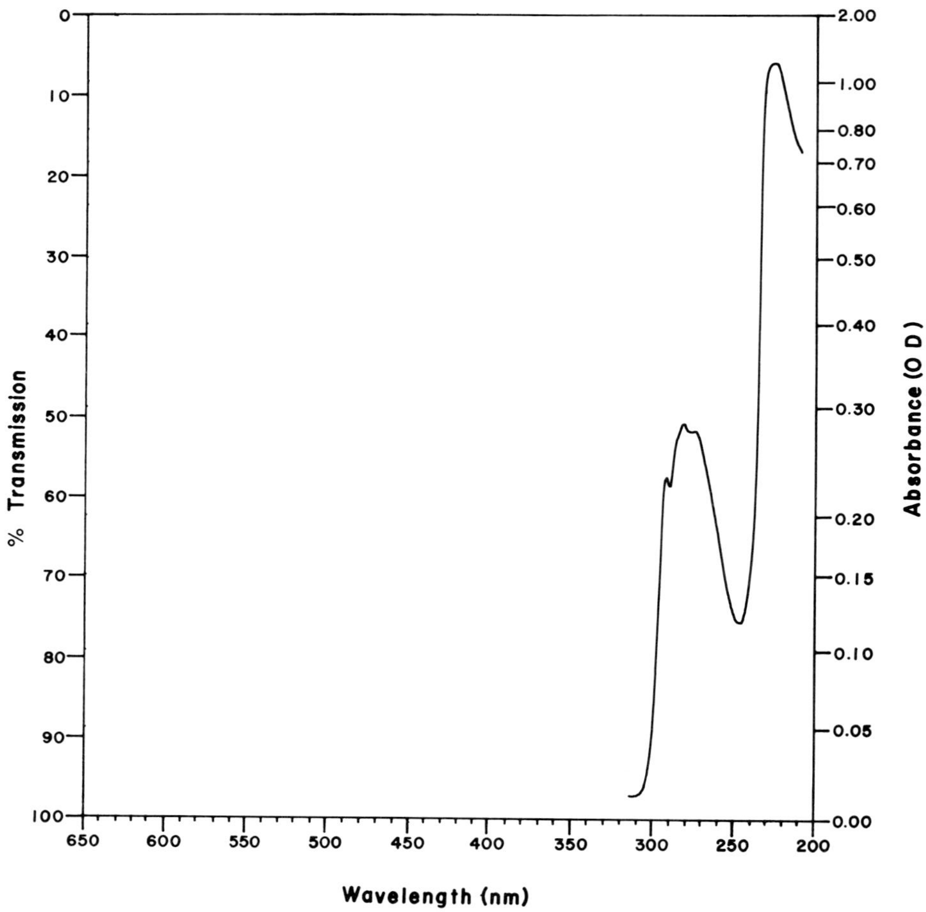
% Transmission
0
10
20
30
40
50
60
70
80
90
100
Absorbance (O D)
2.00
1.00
0.80
0.70
0.60
0.50
0.40
0.30
0.20
0.15
0.10
0.05
0.00
650
600
550
500
450
400
350
300
250
200
Wavelength (nm)

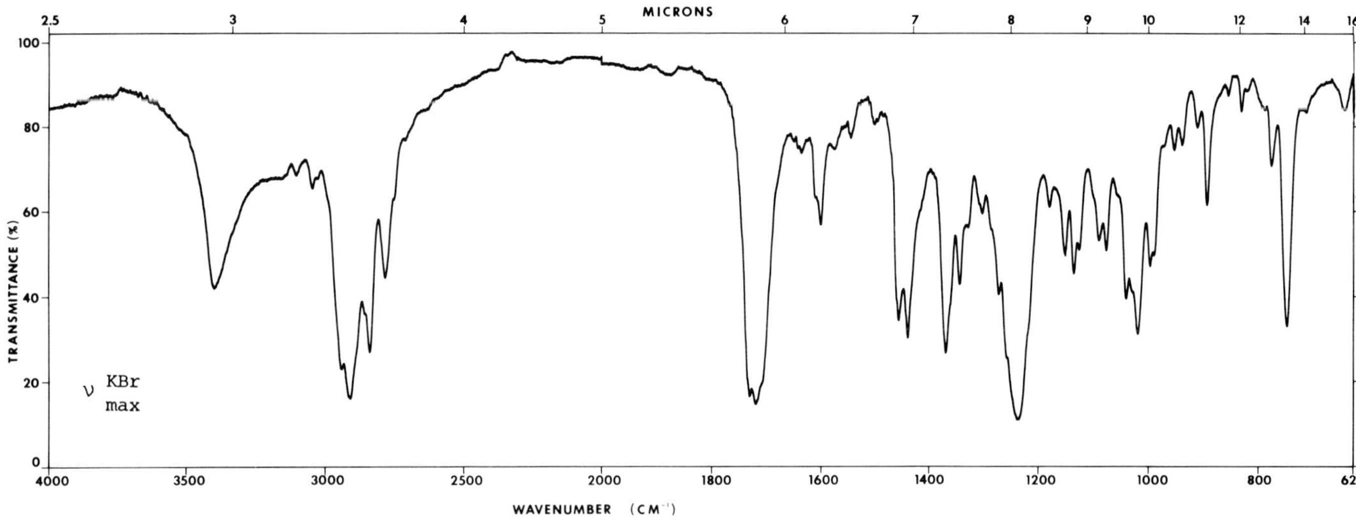
MICRONS
TRANSMITTANCE (%)
WAVENUMBER (CM⁻¹)
ν KBr max

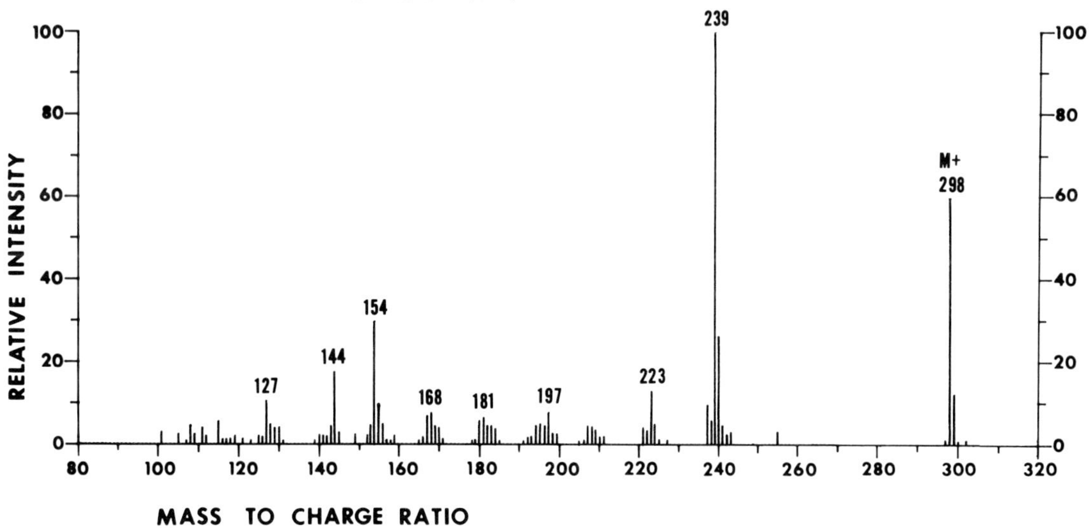
RELATIVE INTENSITY
MASS TO CHARGE RATIO
127
144
154
168
181
197
223
239
M+
298

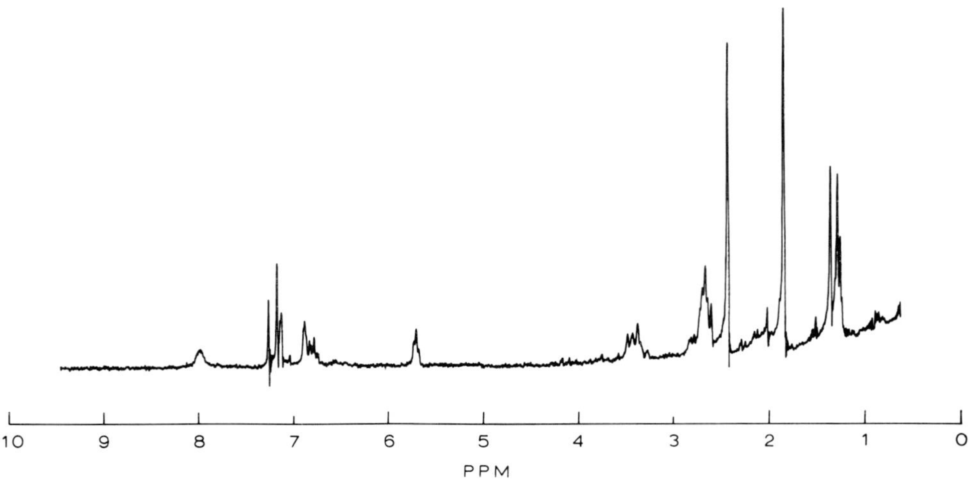

The 100 MHz ^{1}H NMR spectrum of fumigaclavine A in $CDCl_3$ solution.

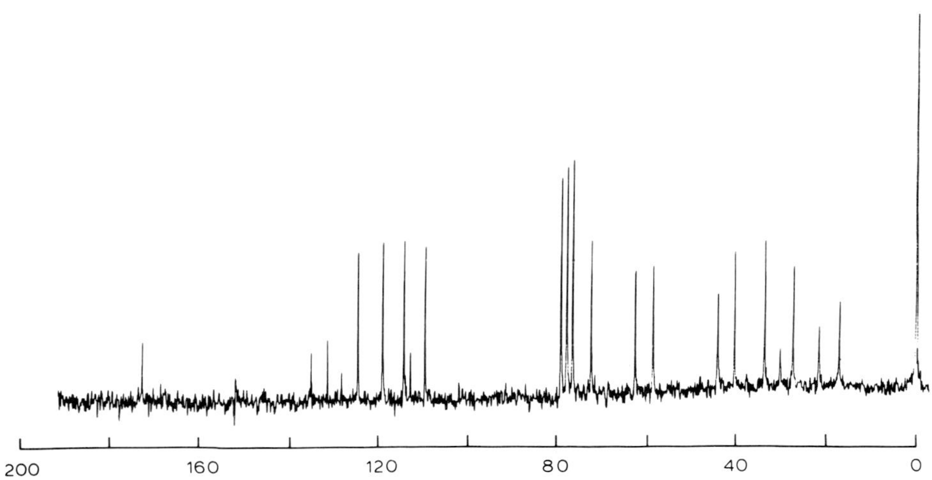

The 25 MHz ^{13}C NMR spectrum of fumigaclavine A in $CDCl_3$ solution.

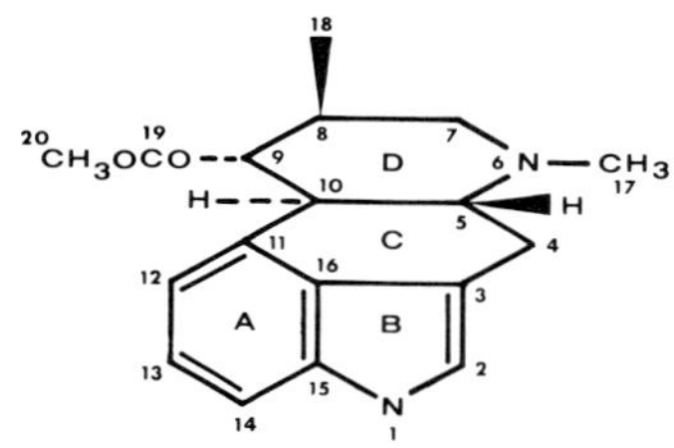

Common name[8,9,12]	Roquefortine A (Isofumigaclavine A) (9α-Acetoxy-6,8β-dimethylergoline)
Molecular weight	298.1681
Molecular formula	$C_{18}H_{22}O_2N_2$
General characteristics[12]	Crystals from benzene, mp 190°–193°C $[\alpha]_D^{22}$ −54.1° (C = 0.67 in chloroform)
UV data[12]	λ_{max}^{MeOH} nm (ϵ): 226(46,300), 277(sh)(6,000), 283(6,500), and 293(5,600)
Fungal source[8]	*Penicillium roqueforti* Thom
Toxicity data[12]	Roquefortine A exhibited a variety of weak pharmacological actions, such as muscle relaxant, antidepressant, local anesthetic, etc., in experimental animals. Its LD_{50} in mice dosed IP was 340 mg/kg.
TLC data[12]	*Adsorbent:* silica gel F 1500/LS 254 (Schleicher and Schull) *Solvent:* chloroform–methanol–28% ammonium hydroxide, 85:15:1 v/v/v R_f: 0.62 *Detection:* a mauve spot after spraying with 50% H_2SO_4 and heating at about 100°C for 10 min.
^{13}C NMR data	C2, 119.1 d; C3, 110.7 s; C4, 28.4 t; C5, 62.3 d; C7, 55.4 t; C8, 37.5 d; C9, 79.6 d; C10, 38.5 d; C11, 131.5 s; C12, 113.9 d; C13, 122.8 d; C14, 110.0 d; C15, 134.2 s; C16, 113.9 s; C17, 45.9 q; C18, 16.4 q; C19, 171.5 s; C20, 22.2 q
^1H NMR data	1-NH, 10.76; H2, 7.07; H9, 5.06; H12, 6.70 (J = 6.7); H13, 7.07; H14, 7.25 (J = 8.5); H17, 2.40; H18, 0.95 (J = 4.0); H20, 2.28

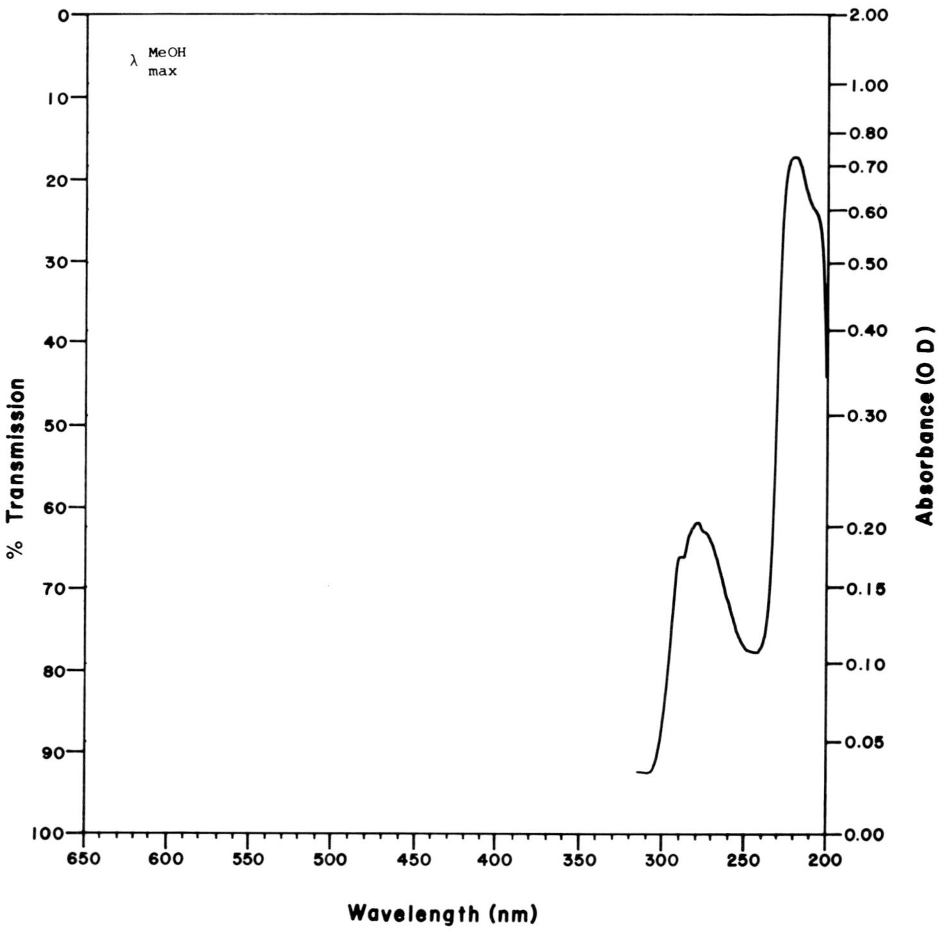
λ MeOH max
% Transmission
Absorbance (O D)
Wavelength (nm)
0
10
20
30
40
50
60
70
80
90
100
2.00
1.00
0.80
0.70
0.60
0.50
0.40
0.30
0.20
0.15
0.10
0.05
0.00
650
600
550
500
450
400
350
300
250
200

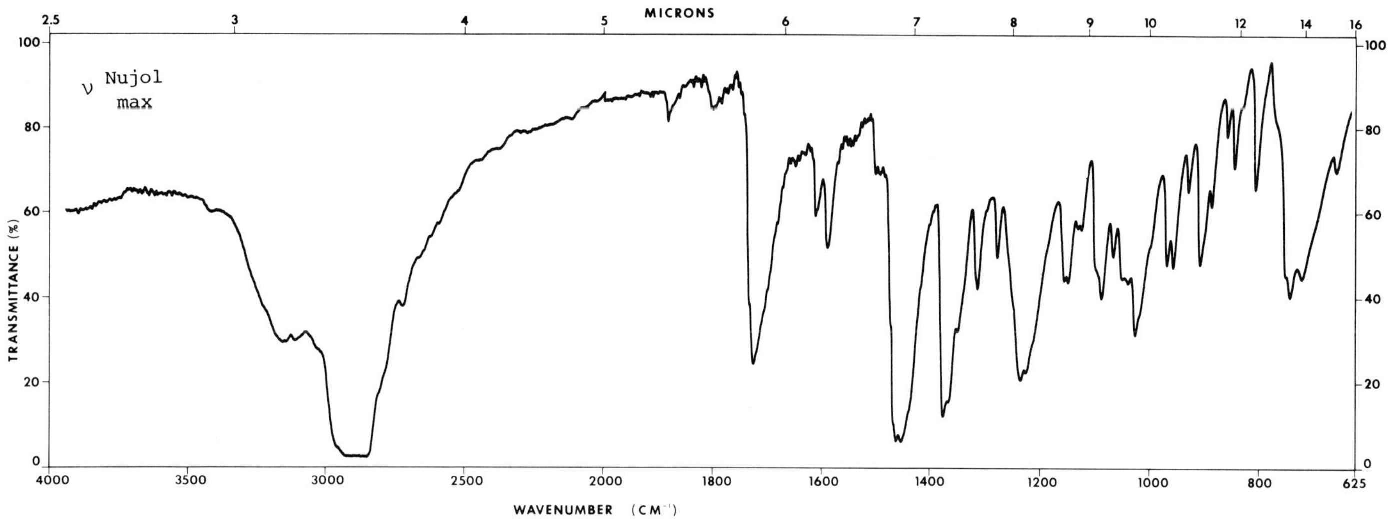
MICRONS
TRANSMITTANCE (%)
ν Nujol max
WAVENUMBER (CM⁻¹)

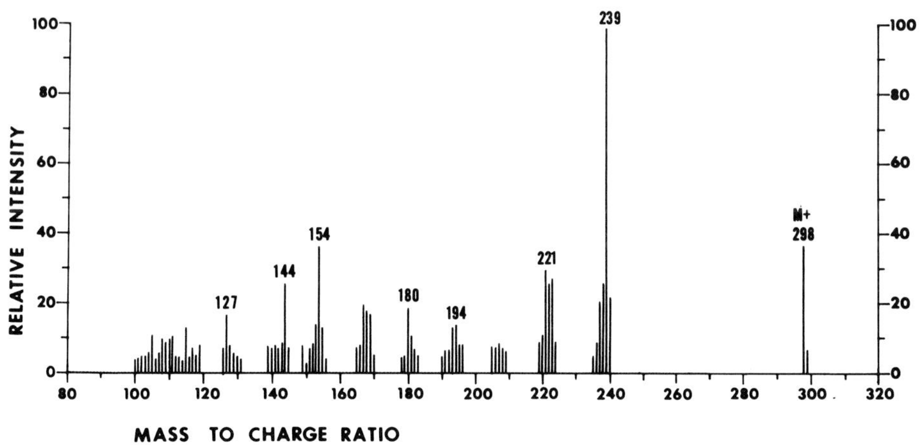
RELATIVE INTENSITY
127
144
154
180
194
221
239
M+
298
MASS TO CHARGE RATIO

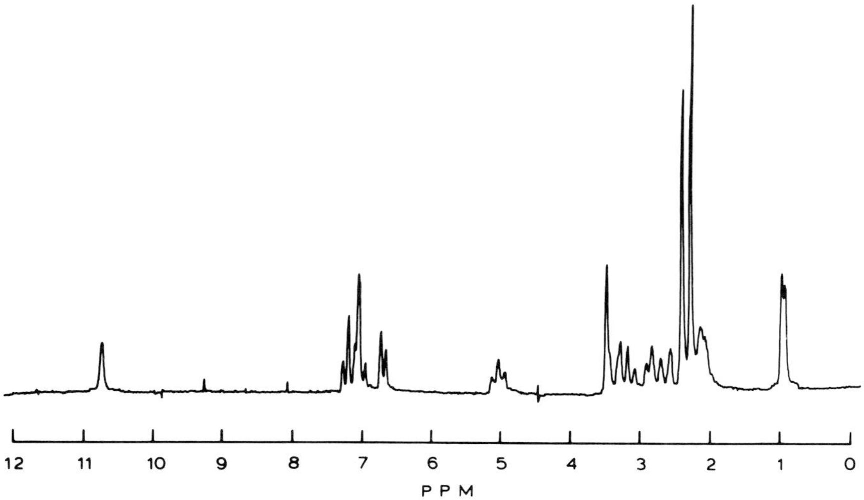

The 100 MHz ^{1}H NMR spectrum of roquefortine A in DMSO solution.

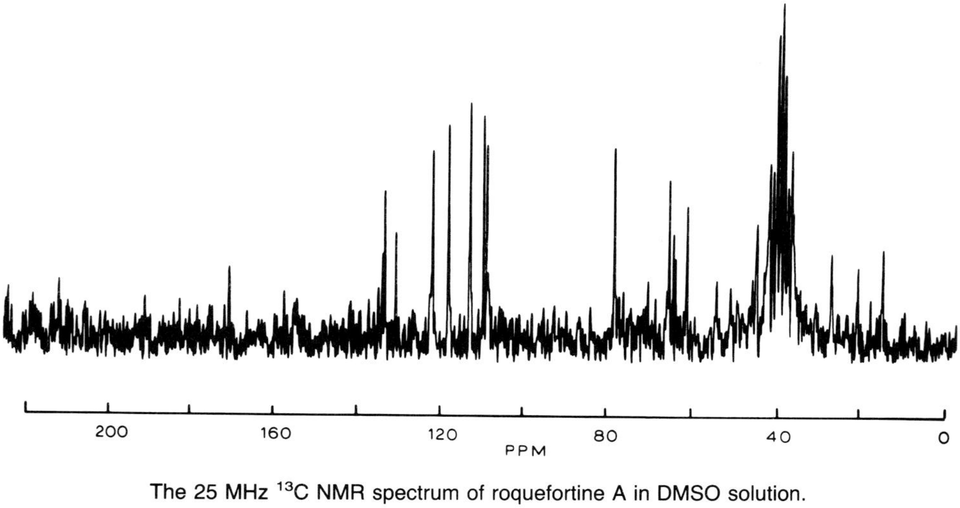

The 25 MHz ^{13}C NMR spectrum of roquefortine A in DMSO solution.

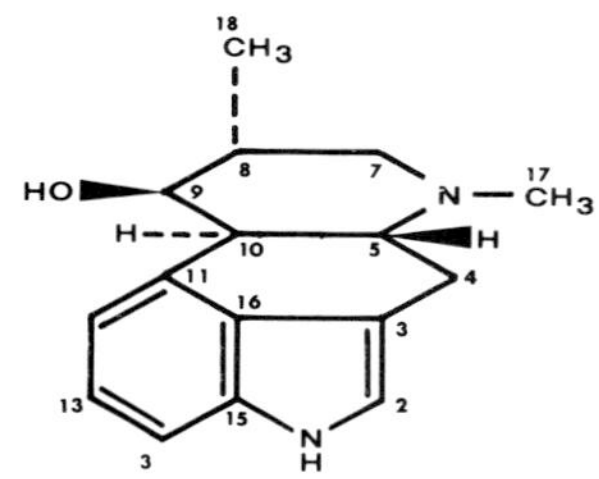

Common name	Fumigaclavine B (6,8-Dimethylergolin-9-ol)
Molecular weight	256.1575
Molecular formula	$C_{16}H_{20}ON_2$
General characteristics[14]	Needles from aqueous methanol, 244°–245°C (at 260°C, the melt solidified and remelted at 265°–267°C) $[\alpha]^{22}_{5461}$ −6.3° (C = 1.2 in methanol); −113° (C = 0.6 in pyridine)
UV data[14]	λ_{max} nm (ϵ): 225(30,900), 275(6,200), 282(6,600), and 293(5,200)
Fungal source[2,6,14,15]	*Aspergillus fumigatus* Fres. *Rhizopus arrhizus* Fischer
Toxicity data	Toxicity unknown
TLC data[6]	*Adsorbent:* silica gel G-HR *Solvent:* toluene–ethyl acetate–formic acid, 5:4:1 v/v/v R_f: origin *Detection:* purple spot after spraying with 50% ethanolic H_2SO_4 and heating at 100°C for 5–10 min.
Other significant references	X-Ray crystallography[6]; stereochemistry[3]
^{13}C NMR data	C2, 117.6 d; C3, 111.6 s; C4, 26.2 t; C5, 60.9 d; C7, 57.4 t; C8, 34.0 d; C9, 69.0 d; C10, 41.6 d; C11, 129.5 s; C12, 112.2 d; C13, 123.0 d; C14, 108.7 d; C15, 133.5 s; C16, 126.9 s; C17, 43.4 q; C18, 16.9 q
^{1}H NMR data	1-NH, 8.04; H2, 6.96; H9, 4.54; H11, 6.87; H12, 7.18; H13, 7.21; H17, 2.43; H18, 1.30

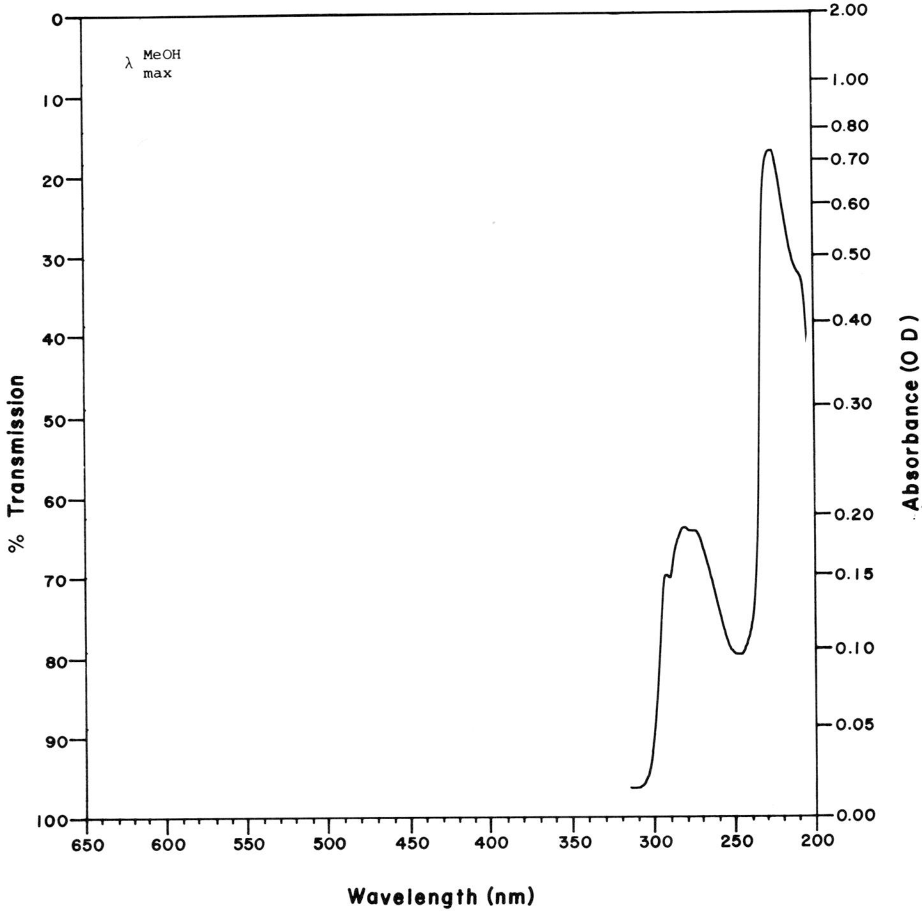
λ MeOH max
% Transmission
0
10
20
30
40
50
60
70
80
90
100
Absorbance (O D)
2.00
1.00
0.80
0.70
0.60
0.50
0.40
0.30
0.20
0.15
0.10
0.05
0.00
650 600 550 500 450 400 350 300 250 200
Wavelength (nm)

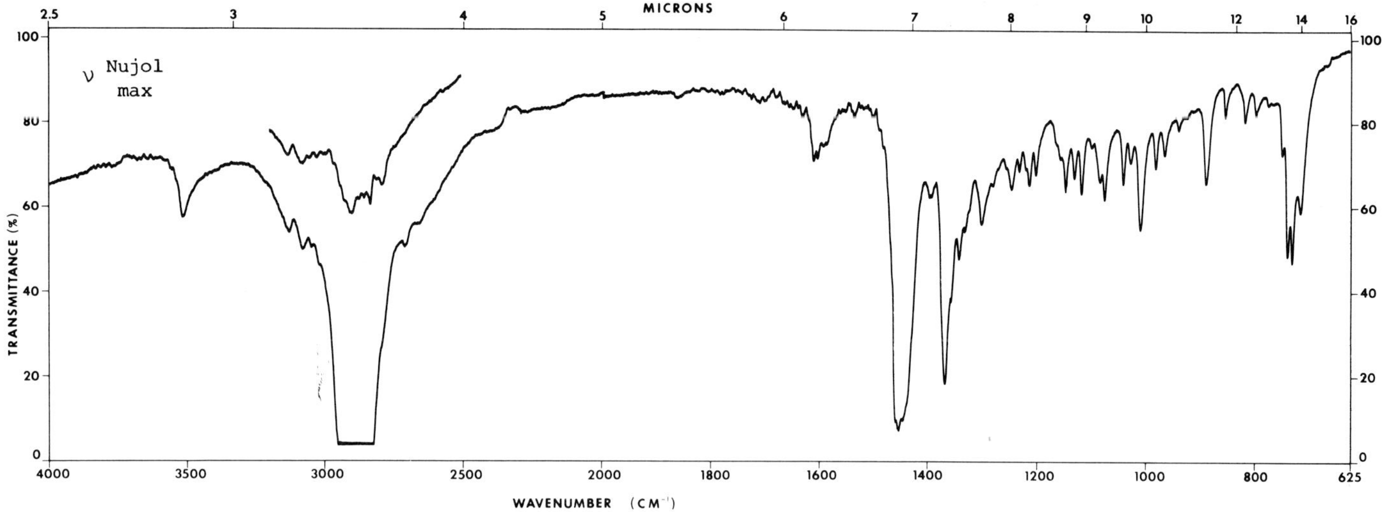
MICRONS
ν Nujol max
TRANSMITTANCE (%)
WAVENUMBER (CM⁻¹)

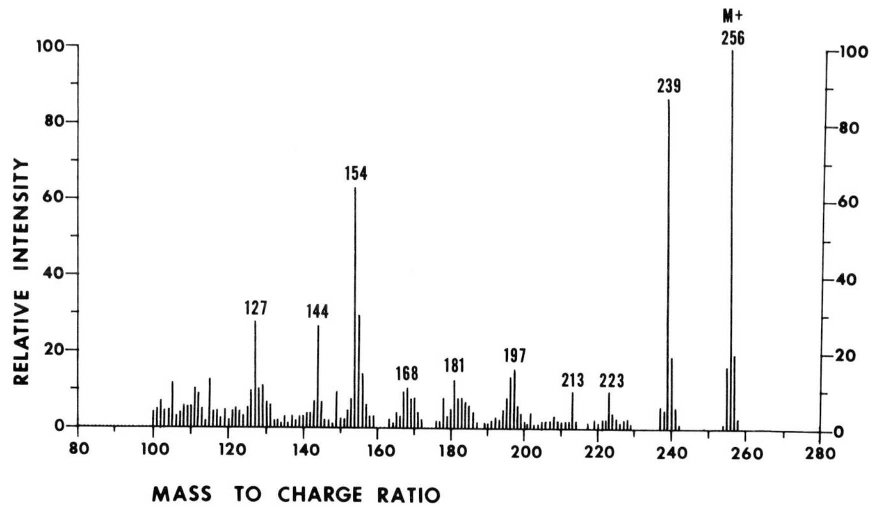
RELATIVE INTENSITY
127
144
154
168
181
197
213
223
239
M+
256
MASS TO CHARGE RATIO

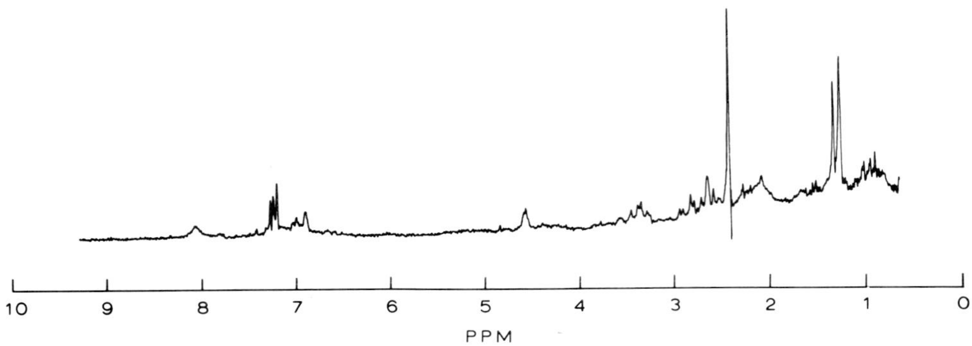

The 100 MHz ^{1}H NMR spectrum of fumigaclavine B in $CDCl_3$ solution.

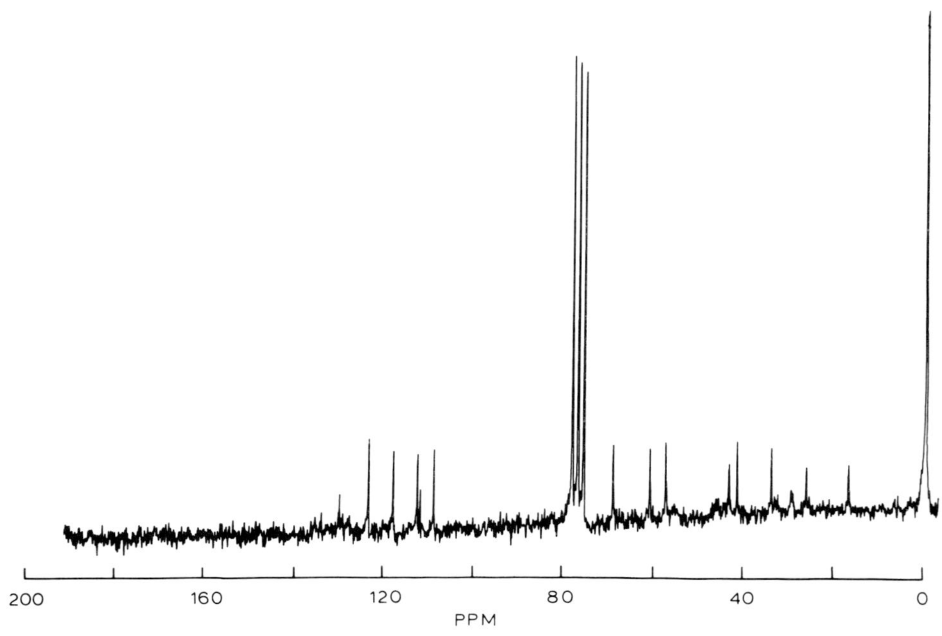

The 25 MHz ^{13}C NMR spectrum of fumigaclavine B in $CDCl_3$ solution.

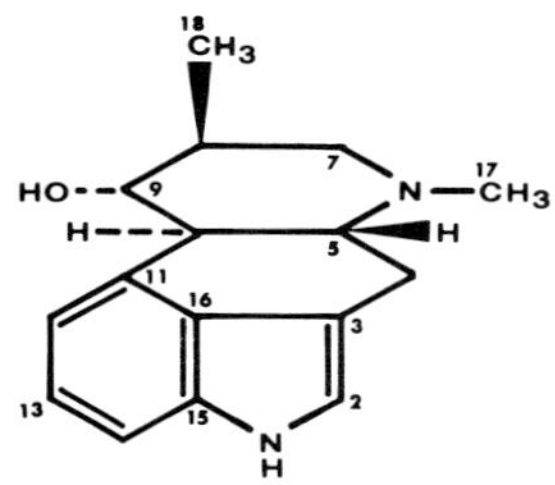

Common name	Roquefortine B (Isofumigaclavine B) (6,8β-Dimethylergolin-9α-ol)
Molecular weight	256.1575
Molecular formula	$C_{16}H_{20}ON_2$
General characteristics[8,12]	mp 278°–282°C (dec.) Colorless needles from methanol, mp 222°–225°C (760 mm Hg) $[\alpha]_D^{15}$ −147° (C = 0.5 in pyridine)
UV data[8,10,12]	λ_{max}^{EtOH} nm (ϵ): 224(44,700), 275(6,200), 282(7,000), and 294(6,000) λ_{max}^{EtOH} nm (ϵ): 225(25,100), 277(sh)(4,300) 283(4,600), and 293(3,800)
Fungal source[8]	*Penicillium roqueforti*
Toxicity data[8]	The LD_{50} of roquefortine B dosed IP to mice was 1000 mg/kg.
TLC data[12]	*Adsorbent:* silica gel F 1500/LS254 thin layer sheets *Solvent:* chloroform–methanol–28% ammonium hydroxide, 85:15:1 v/v/v R_f: 0.31 *Detection:* a mauve spot after spraying with 50% H_2SO_4 and heating at about 100°C for 10 min.

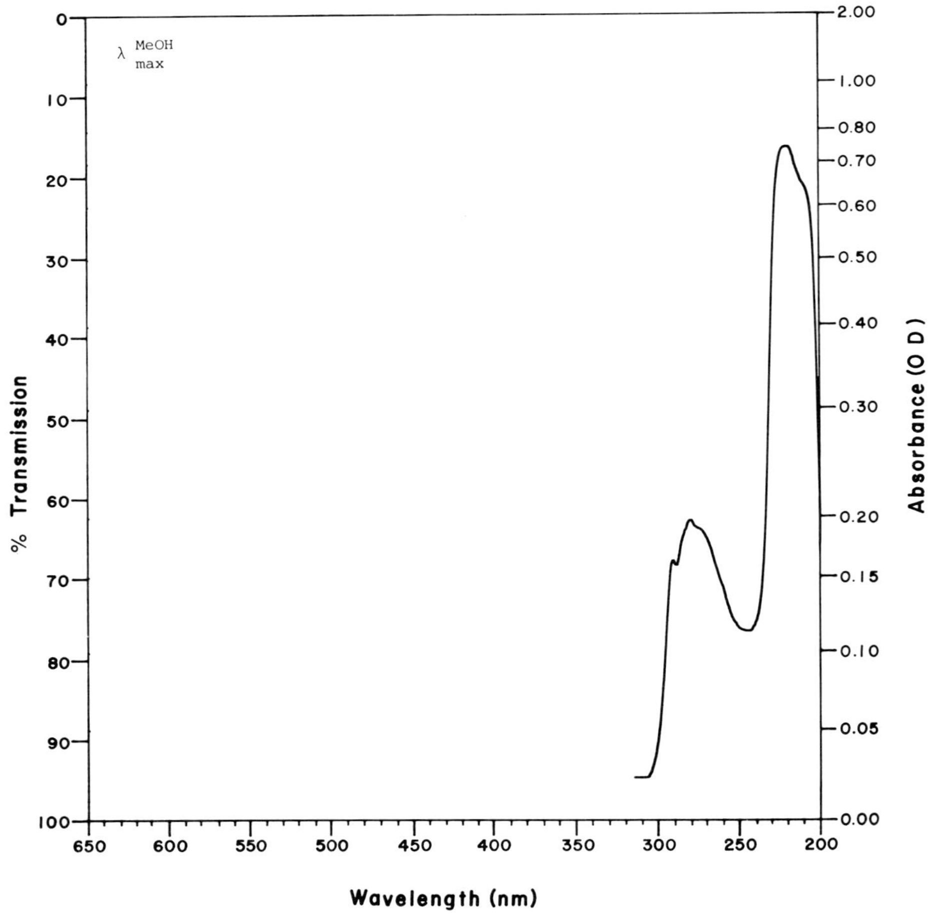
λ MeOH max
% Transmission
0
10
20
30
40
50
60
70
80
90
100
Absorbance (O D)
2.00
1.00
0.80
0.70
0.60
0.50
0.40
0.30
0.20
0.15
0.10
0.05
0.00
650
600
550
500
450
400
350
300
250
200
Wavelength (nm)

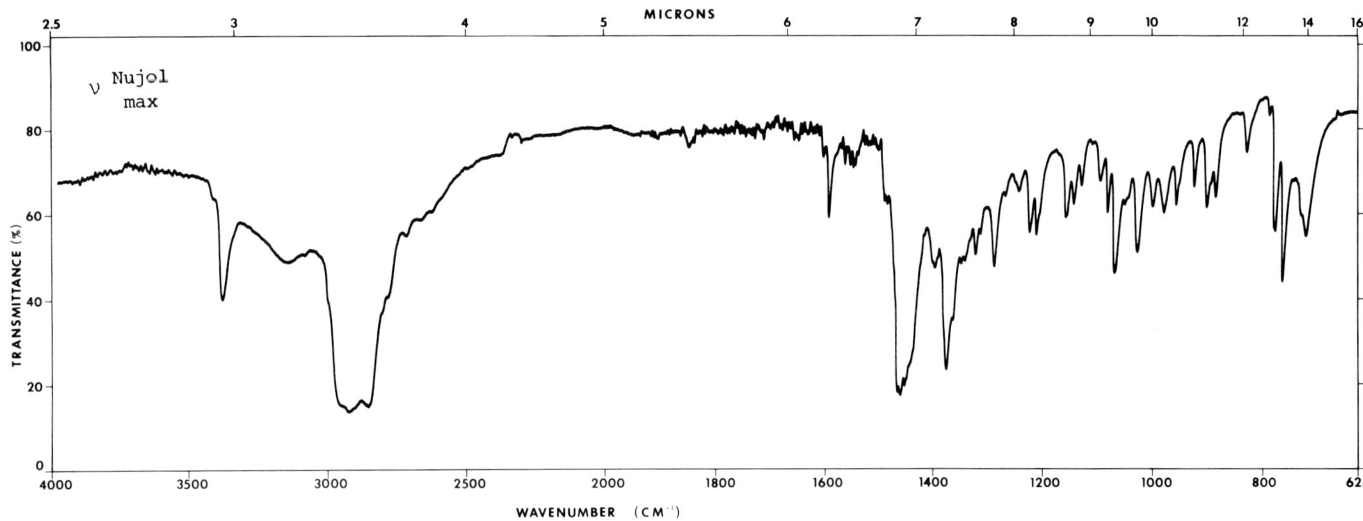
MICRONS
TRANSMITTANCE (%)
WAVENUMBER (CM⁻¹)
ν Nujol max

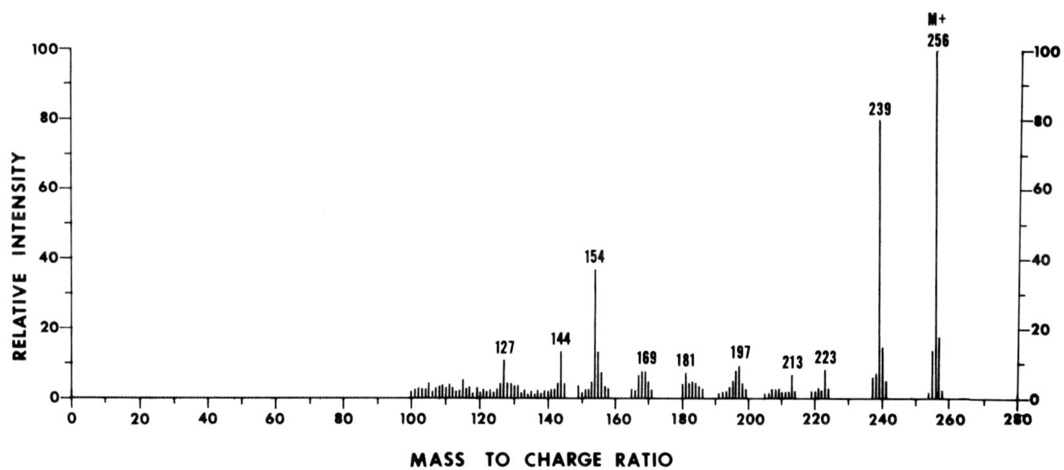
RELATIVE INTENSITY
MASS TO CHARGE RATIO
M+ 256
239
154
144
127
169
181
197
213
223

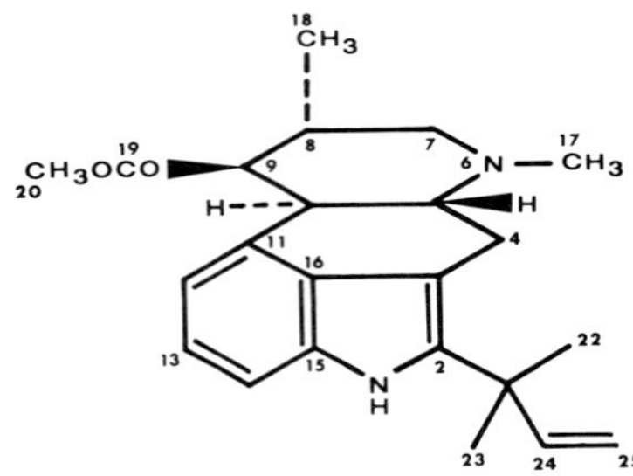

Common name[6,15]	Fumigaclavine C (2-Dimethylallyl-9-acetoxy-6,8-dimethylergoline)
Molecular weight[2]	366.2307
Molecular formula	$C_{23}H_{30}O_2N_2$
General characteristics[6,15]	Crystals from ethanol, acetone, or ethyl acetate (prisms), mp 191°–193°C Needles from methanol, mp 198°C $[\alpha]_D^{22}$ −90° $[\alpha]_{5461}^{22}$ −119° (C = 1.0 in chloroform) $[\alpha]_D^{22}$ −132° $[\alpha]_{5461}^{22}$ −160° (C = 1.0 in pyridine)
UV data[6,15]	λ_{max} nm: 229, 284, and 292 (ϵ_{max} not reported) λ_{max}^{MeOH} nm (ϵ): 225(34,700), 227(10,200), 283(11,000), and 292(9,500)
Fungal source[2,6,15]	*Aspergillus fumigatus* Fres.
Toxicity data[6]	The LD_{50} of fumigaclavine C dosed orally to day-old cockerels was 150 mg/kg. Histopathology of surviving cockerels showed vaculation of the hepatic parenchymal cells that may have been directly related to the toxic action of the toxin or due to anorexia.
TLC data[6]	*Adsorbent:* silica gel G-HR *Solvent:* toluene–ethyl acetate–formic acid, 5:4:1 v/v/v R_f: 0.14 *Detection:* blue spot after spraying with 50% ethanolic H_2SO_4 and heating at 100°C for 5–10 min.

Other significant references	X-Ray crystallography[6]
^{13}C NMR data	C2, 131.1 s; C3, 106.2 s; C4, 28.1 t; C5, 61.6 d; C7, 57.8 t; C8, 33.1 d; C9, 71.4 d; C10, 39.4 d; C11, 129.1 s; C12, 112.8 d; C13, 122.2 d; C14, 107.6 d; C15, 136.6 s; C16, 128.0 s; C17, 43.5 q; C18, 16.7 q; C19, 170.8 s; C20, 21.2 q; C21, 39.1 s; C22, 27.3 q; C23, 27.4 q; C24, 145.7 d; C25, 111.8 t
^{1}H NMR data	1-NH, 7.76; H9, 5.67; H12, 7.04; H13, 6.72; H14, 7.08; H17, 2.46; H18, 1.35; H20, 1.90; H22, 1.56; H23, 1.56; H24, 6.10; H25a, 5.05; H25b, 5.20

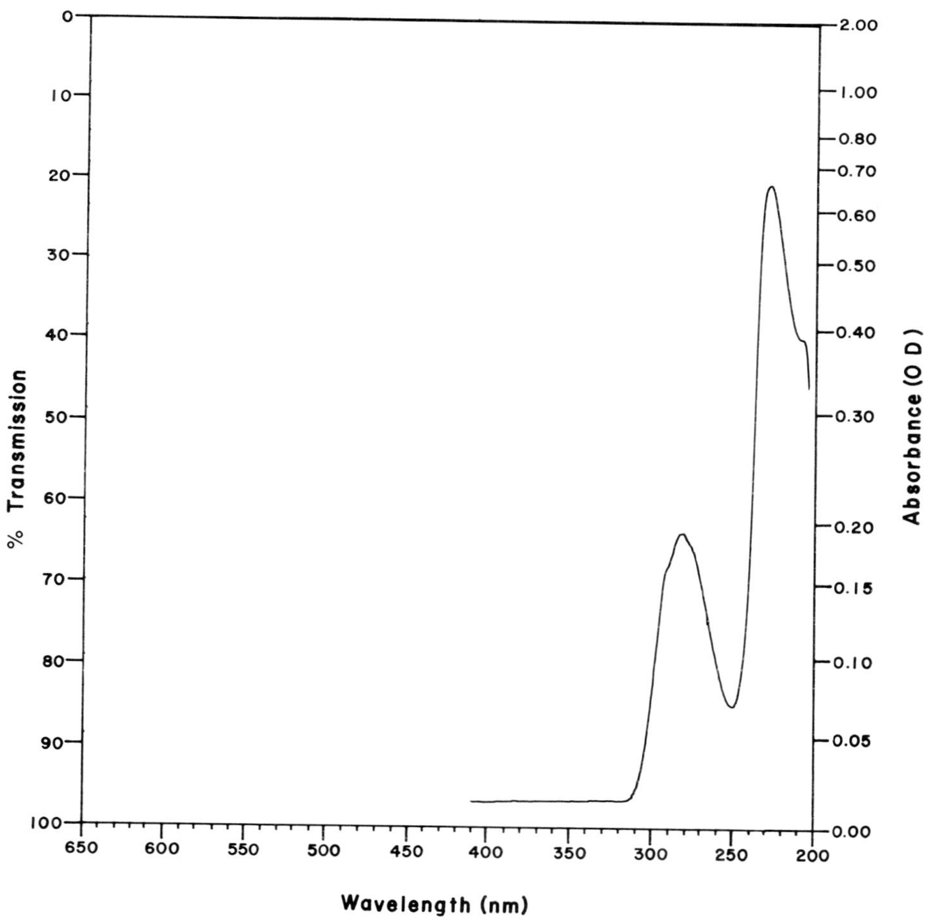

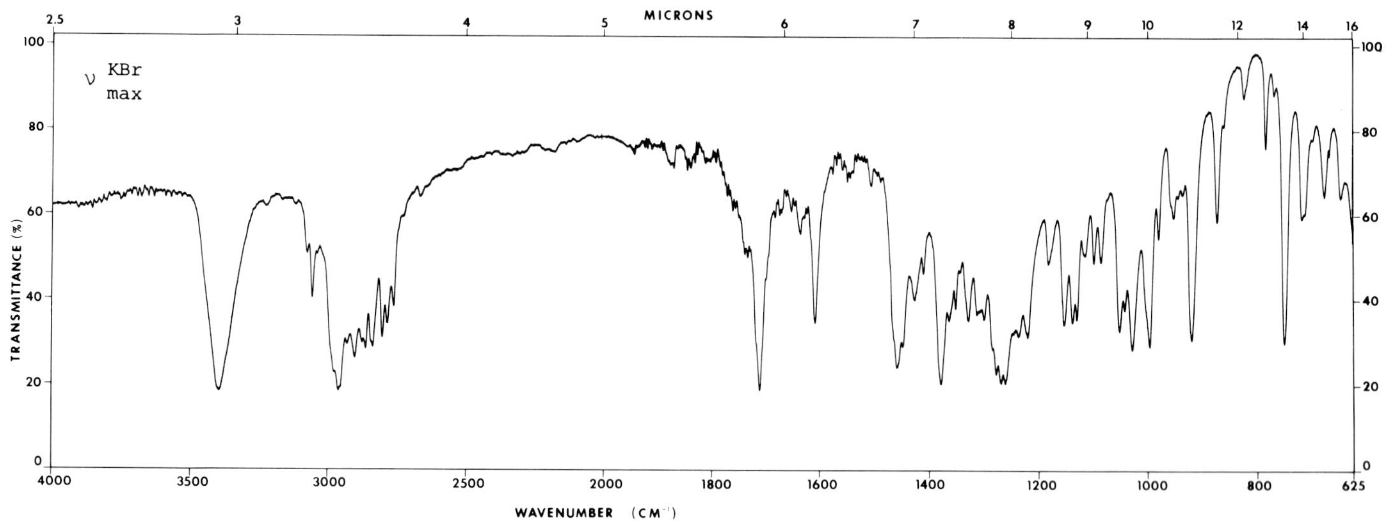
MICRONS
TRANSMITTANCE (%)
ν KBr max
WAVENUMBER (CM⁻¹)

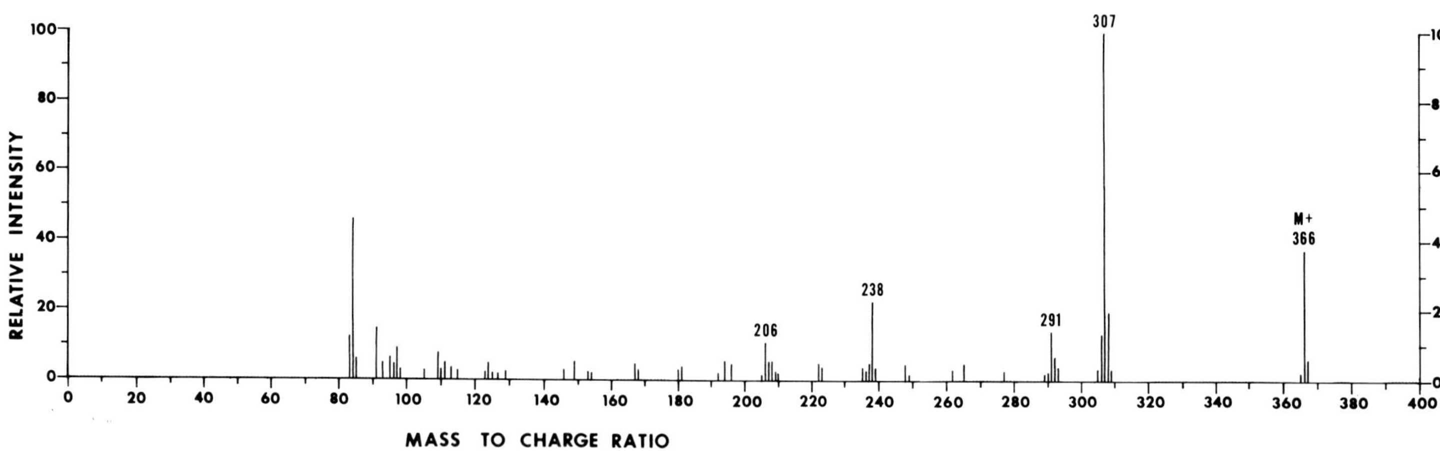
RELATIVE INTENSITY
MASS TO CHARGE RATIO
307
M+ 366
291
238
206

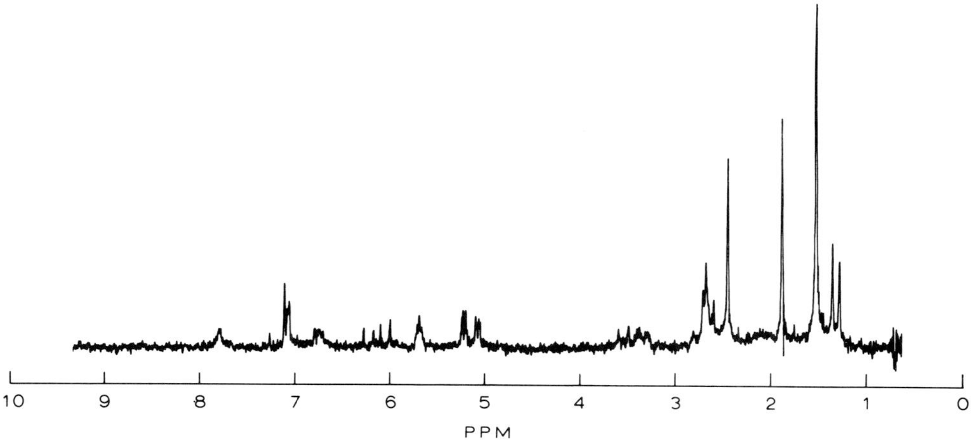

The 100 MHz ^{1}H NMR spectrum of fumigaclavine C in $CDCl_3$ solution.

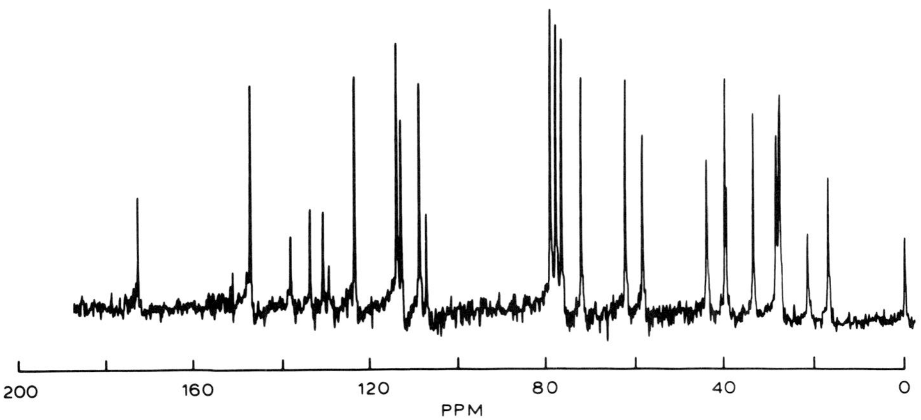

The 25 MHz ^{13}C NMR spectrum of fumigaclavine C in $CDCl_3$ solution.

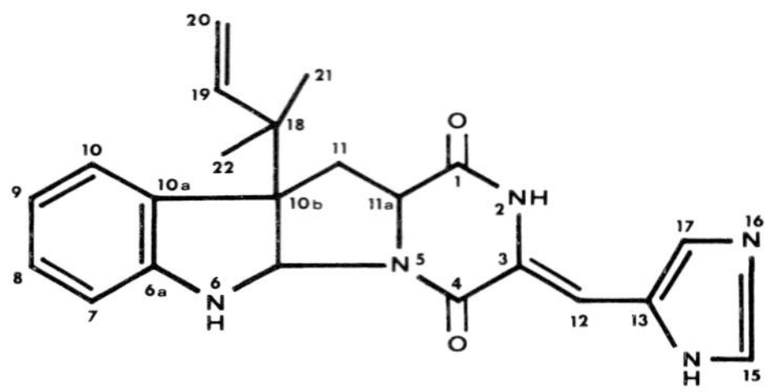

Common name[8,9,12]
Roquefortine (Roquefortine C)
(10β-(1,1-Dimethyl-2-propenyl)-3-(imidazol-4-ylmethylene)-5α,10β,11,11α-tetrahydro-2*H*-pyrazino[1′,2′:1,5]pyrrolo[2,3-*b*]indole-1,4-(3*H*,6*H*)-dione)

Molecular weight
389.1851

Molecular formula
$C_{22}H_{23}O_2N$

General characteristics[9,12]
Colorless needles from methanol–water, mp 195°–200°C (dec.)
$[\alpha]_D^{22}$ −703° (*C* = 1.0 in chloroform)

UV data[12]
λ_{max}^{EtOH} nm (ε): 209(29,500), 240(16,200), and 328(27,000)

Fungal source[9,12]
Penicillium roqueforti

Toxicity data[7,12]
The LD_{50} (IP, male mice) was 15–20 mg/kg. Doses of 50–100 mg/kg caused prostration and an atonic posture. The prostration was interrupted for a few seconds by opisthotonoid seizures which could be brought about by noise or a tap on the cage. Death followed within a few hours. At doses of 10 mg/kg, convulsions were replaced by simple contractions and the state of prostration by equilibrium disorders during movement.

TLC data[12]
Adsorbent: silica gel F 1500/LS 254
Solvent: chloroform–methanol–28% ammonium hydroxide, 90:10:1 v/v/v
R_f: 0.46
Detection: a blue spot after spraying with 50% H_2SO_4 and heating at 110°C for 10 min.

Other significant references
Configuration of 3,12 double bond[13]

¹³C NMR data[12] C1, 166.2 s; C3, 123.7 s; C4, 159.8 s; C5a, 78.9 d; C6a, 151.9 s; C7, 110.3 d; C8, 129.8 d; C9, 118.9 d; C10, 125.6 d; C10a, 129.4 s; C10b, 62.1 d; C11, 38.0 t; C11a, 59.4 d; C12, 109.6 d; C13, 125.6 d; C15, 137.3 d; C17, 134.5 d; C18, 41.7 s; C19, 144.9 d; C20, 114.2 t; C21, 23.0 q; C22, 23.3 q;

¹H NMR data H5a, 5.71; H7–10, 6.58–7.30; H11, 2.56; H11a, 4.00; H12, 6.48; H15, 7.73; H17, 7.20; H19, 6.12; H20, 5.10; H21, 1.06; H22, 1.17

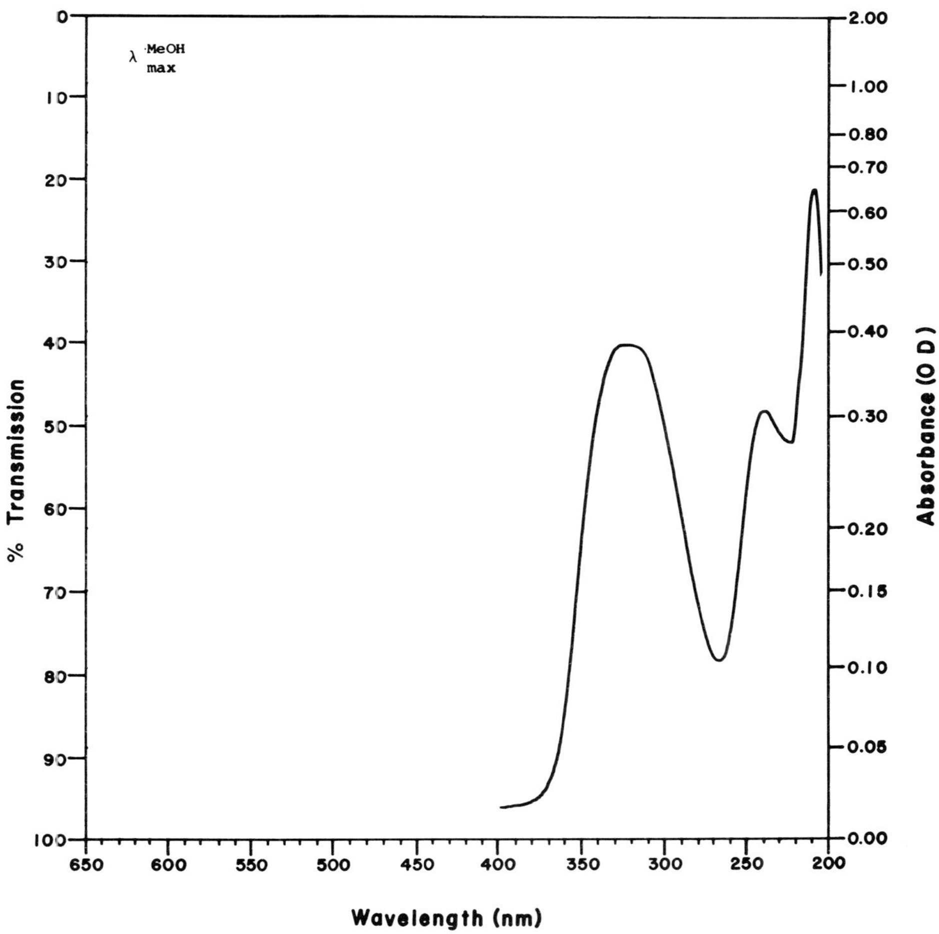

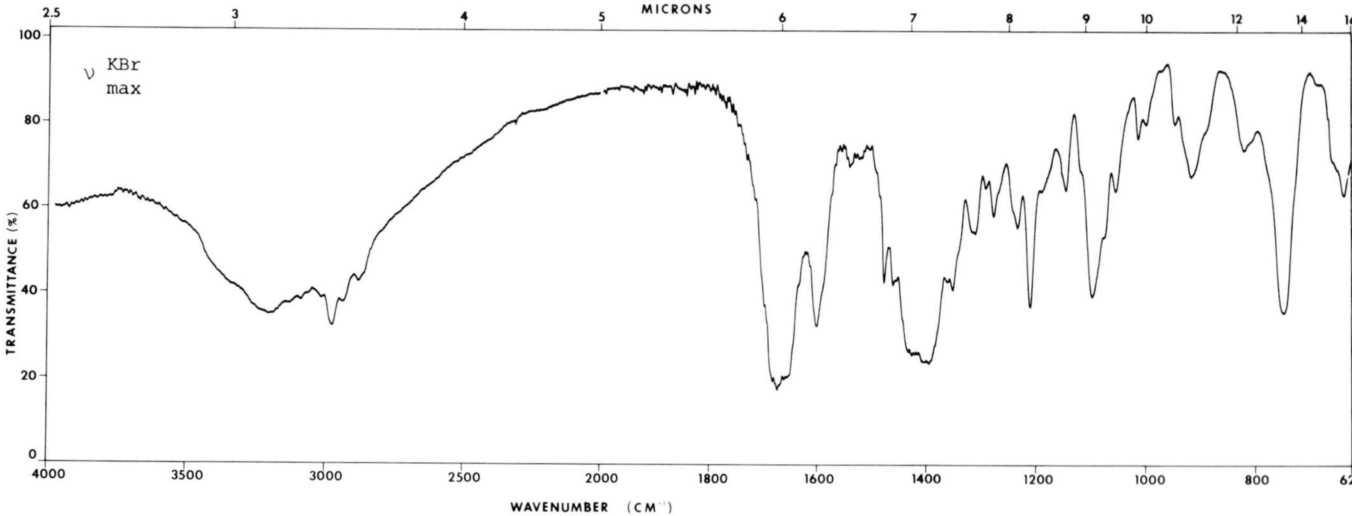
MICRONS
2.5
3
4
5
6
7
8
9
10
12
14
16
ν KBr max
TRANSMITTANCE (%)
100
80
60
40
20
0
4000
3500
3000
2500
2000
1800
1600
1400
1200
1000
800
625
WAVENUMBER (CM⁻¹)

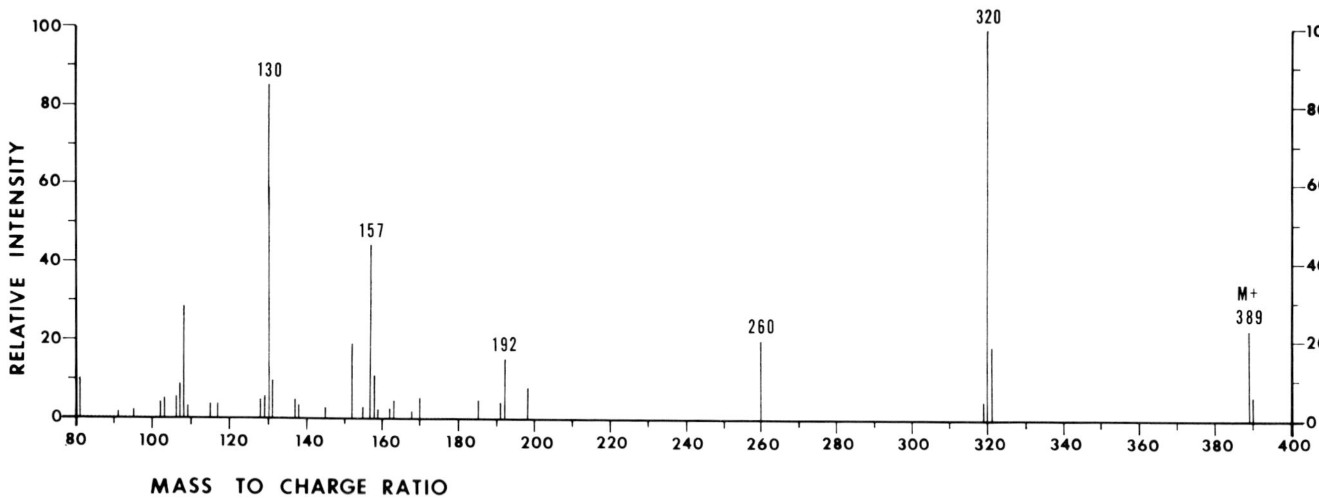
RELATIVE INTENSITY
100
80
60
40
20
0
130
157
192
260
320
M+
389
80
100
120
140
160
180
200
220
240
260
280
300
320
340
360
380
400
MASS TO CHARGE RATIO

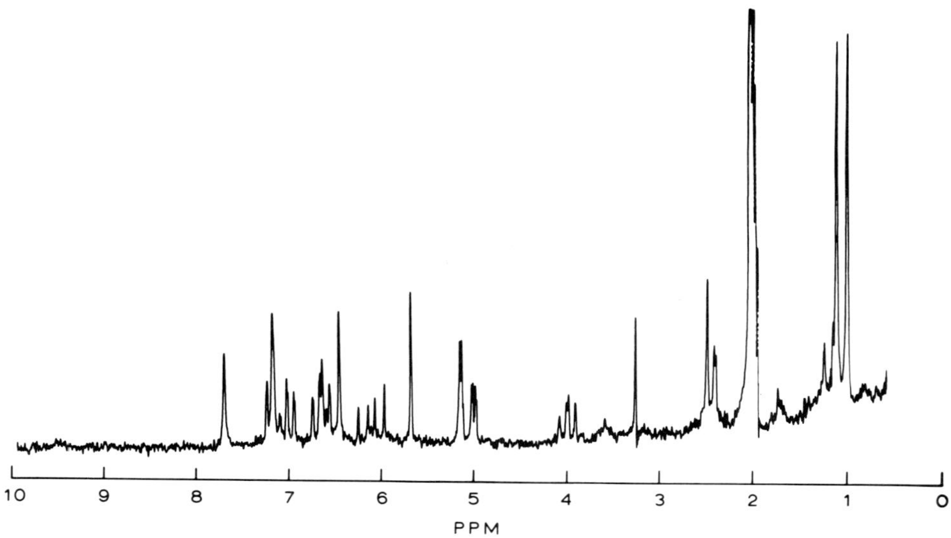

The 100 MHz ^{1}H NMR spectrum of roquefortine C in acetone solution.

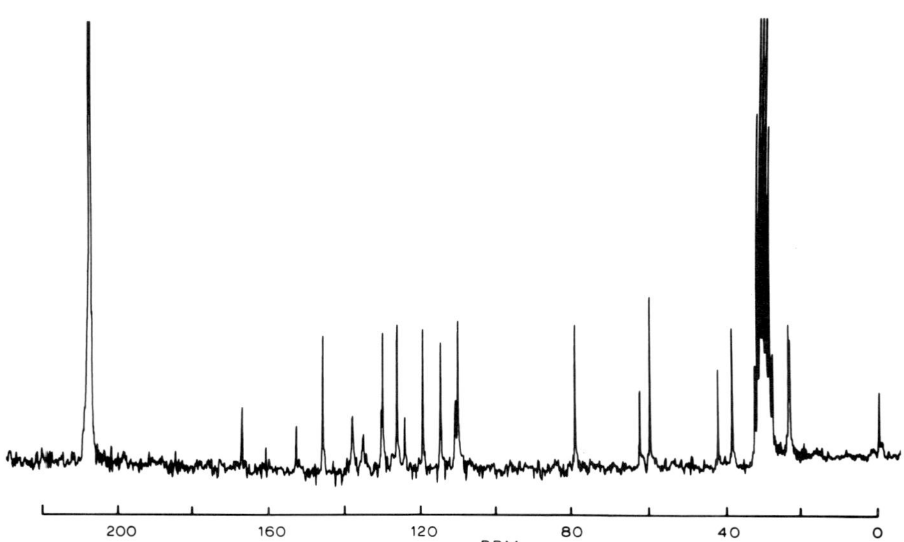

The 25 MHz ^{13}C NMR of roquefortine C in acetone solution.

REFERENCES

1. Abe, M., Ohmomo, S., Ohashi, T., and Tabuchi, T., *Agric. Biol. Chem.* **33,** 469 (1969).
2. Abe, M., Yamatodani, S., Yamano, T., Kozy, Y., and Yamada, S., *Nippon Nogei Kagaku Kaishi* **41,** 72 (1967).
3. Bach, N. J., Boaz, H. E., Kornfeld, E. C., Chang, C.-J., Floss, H. G., Hagaman, E. W., and Wenkert, E., *J. Org. Chem.* **39,** 1272 (1974).
4. Cole, R. J., Kirksey, J. W., Clardy, J., Eickman, N., Weinreb, S. M., Singh, P., and Kim, D., *Tetrahedron Lett.* p. 3849 (1976).
5. Cole, R. J., Kirksey, J. W., Cutler, H. G., Wilson, D. M., and Morgan-Jones, G., *Can. J. Microbiol.* **22,** 741 (1976).
6. Cole, R. J., Kirksey, J. W., Dorner, J. W., Wilson, D. M., Johnson, J., Jr., Johnson, N., Bedell, D., Springer, J. P., Chexal, K. K., Clardy, J., and Cox, R. H., *J. Agric. Food Chem.* **25,** 826 (1977).
7. Experiments conducted at the Institut de Recherches Scientifiques sur le cancer, Villejuif, France by C. Frayssinet and Christiane Frayssinet.
8. Ohmomo, S., Sato, T., Utagawa, T., and Abe, M., *Agric. Biol. Chem.* **39,** 1333 (1975).
9. Ohmomo, S., Sato, T., Utagawa, T., and Abe, M., *J. Agric. Chem. Soc. Jpn.* **49,** 53 (1975).
10. Scott, P. M., personal communication.
11. Scott, P. M., and Kennedy, P. C., *J. Agric. Food Chem.* **24,** 865 (1976).
12. Scott, P. M., Merrien, M. A., and Polonsky, J., *Experientia* **32,** 140 (1976).
13. Scott, P. M., Polonsky, J., and Merrien, M. A., *J. Agric. Food Chem.* **27,** 201 (1979).
14. Spilsbury, J. F., and Wilkinson, S., *J. Chem. Soc.* p. 2085 (1961).
15. Yamano, T., Kishino, K., Yamatodani, S., and Abe, M., *Annu. Rep. Takeda Res. Lab.* **21,** 95 (1962).
16. Yamatodani, S., Asahi, Y., Matsukura, A., Ohmomo, S., and Abe, M., *Agric. Biol. Chem.* **34,** 485 (1970).

11

Epipolythiopiperazine-3,6-diones

The epipolythiopiperazine-3,6-diones represent an economically important group of fungal metabolites produced by a variety of unrelated fungi. At present, the most notable are the sporidesmins produced by *Pithomyces chartarum*. The latter have been strongly implicated in the etiology of facial eczema, an important disease of ruminants. The disease has been studied most intensively in New Zealand because of its prevalence there. The suspect fungus is a saprophyte which grows on dead and decaying plant material. It is capable of producing a complex mixture of closely related toxic metabolites called the sporidesmins (sporidesmins A through G). The sporidesmins are characterized by a 3,6-epidithia-2.5-dioxopiperazine moiety and an indolopyrrolopyrazine moiety. They differ from one another by the substituents on the indolopyrrolopyrazine moiety and the number of sulfur atoms across the diketopiperazine ring.

Gliotoxin, a related epipolythiopiperazine-3,6-dione, was originally isolated because of its antibiotic properties. It has subsequently been shown to be acutely toxic to vertebrate animals.

Chetomin is a complex member of this group that is biologically active against gram-positive bacteria and viruses.[52,62]

Hyalodendrin and related metabolites, the newest members of the class of epipolythiopiperazine-3,6-diones, are characterized by a benzyldiketopiperazine moiety. Hyalodendrin also possesses broad spectrum antifungal activity; however, vertebrate toxicity is unknown.

Several comprehensive reviews covering the various aspects of these novel metabolites have been published.[3,11,28,52,60]

Epipolythiopiperazine-3,6-diones	Molecular weight	Molecular formula
Gliotoxin	326.0393	$C_{13}H_{14}O_4N_2S_2$
Gliotoxin acetate	368.0365	$C_{15}H_{16}O_5N_2S_2$
Dehydrogliotoxin	324.0232	$C_{13}H_{12}O_4N_2S_2$
Hyalodendrin	324.0595	$C_{14}H_{16}O_3N_2S_2$
Hyalodendrin tetrasulfide	388.0035	$C_{14}H_{16}O_3N_2S_4$
Bisdethiodi(methylthio)hyalodendrin	354.1063	$C_{16}H_{22}O_3N_2S_2$
Sporidesmin	473.0479	$C_{18}H_{20}O_6N_3S_2Cl$
Sporidesmin B	457.0530	$C_{18}H_{20}O_5N_3S_2Cl$
Sporidesmin C acetate	589.0402	$C_{22}H_{24}O_8N_3S_3Cl$
Sporidesmin D	503.0940	$C_{20}H_{26}O_6N_3S_2Cl$
Sporidesmin E	505.0199	$C_{18}H_{20}O_6N_3S_3Cl$
Sporidesmin F	455.0917	$C_{19}H_{22}O_6N_3SCl$
Sporidesmin G	536.9912	$C_{18}H_{20}O_6N_3S_4Cl$
Sporidesmin H	339.1219	$C_{18}H_{17}O_4N_3$
Sporidesmin J	459.0326	$C_{17}H_{18}O_6N_3S_2Cl$
Chaetocin	696.0938	$C_{30}H_{28}O_6N_6S_4$
Chetomin	710.1094	$C_{31}H_{30}O_6N_6S_4$
Verticillin A	696.0938	$C_{30}H_{28}O_6N_6S_4$
Aranotin	462.0496	$C_{20}H_{18}O_7N_2S_2$
Acetylaranotin (LL-S88 α)	504.0658	$C_{22}H_{20}O_8N_2S_2$
Bisdethiodi(methylthio)acetylaranotin (LL-S88 β)	534.1120	$C_{24}H_{26}O_8N_2S_2$
Apoaranotin	446.0547	$C_{20}H_{18}O_6N_2S_2$
Bisdethiodi(methylthio)acetylapoaranotin	518.1171	$C_{24}H_{26}O_7N_2S_2$

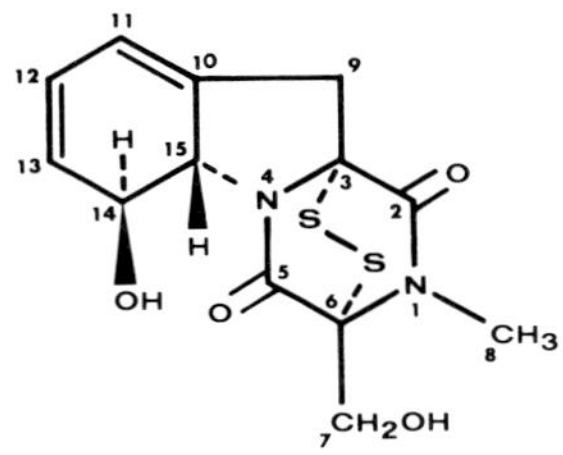

Common name	Gliotoxin
Molecular weight	326.0393
Molecular formula	$C_{13}H_{14}O_4N_2S_2$
General characteristics[23,24]	mp 221°C $[\alpha]_{589}$ nm −840° (C = 0.103 in chloroform); −936° (C = 0.0675 in dioxane)
UV data[5,17]	λ_{max}^{EtOH} nm (ϵ): 216(2,900) and 272(6,300)
CD data[17]	$\Delta\epsilon$ (methanol solution) nm: 234(−30) and 270(+7.7)
Fungal source[23,63]	*Penicillium terlikowskii* *P. obscurum* *Aspergillus terreus* *A. chevalieri* *A. fumigatus* *Gliocladium fimbriatum* (*Trichoderma viride*) (NRRL 1828 and NRRL 75)
Toxicity data[23,32]	Gliotoxin has potent antibiotic properties against bacteria, fungi, and RNA viruses. It was acutely toxic to rabbits (LD_{50} = 45 mg/kg), mice (LD_{50} = 50 mg/kg), and rats (LD_{50} = 50–65 mg/kg).
Other significant references	Chemistry[8]; stereochemistry[18]; structure elucidation[5]; X-ray crystallography[4,14]; synthesis[15]; biosynthesis[9]
^{13}C NMR data	C2, 165.1*s; C3, 77.3 s; C5, 165.9*s; C6, 77.3 s; C7, 60.5 t; C8, 27.5 q; C9, 36.6 t; C10, 130.7 s; C11, 123.3 d; C12, 120.1 d; C13, 129.8 d; C14, 73.1 d; C15, 69.7 d
^{1}H NMR data	H7, 4.80; H8, 3.22; H9a, 4.18 (J = 13.0); H9b, 4.44 (J = 13.0); H11,12,13, 5.90; H14, 3.72 (J = 18.0); H15, 2.90 (J = 18.0)

* Assignments may be reversed.

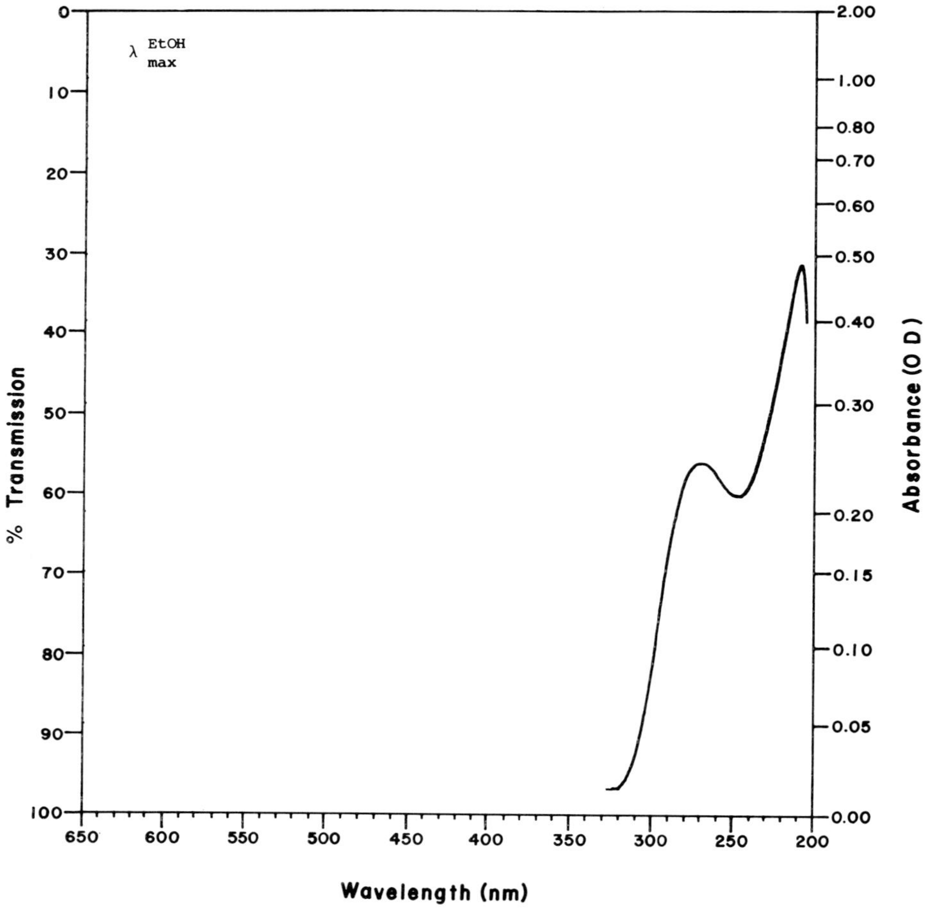
λ EtOH max
% Transmission
0
10
20
30
40
50
60
70
80
90
100
Absorbance (O D)
2.00
1.00
0.80
0.70
0.60
0.50
0.40
0.30
0.20
0.15
0.10
0.05
0.00
650
600
550
500
450
400
350
300
250
200
Wavelength (nm)

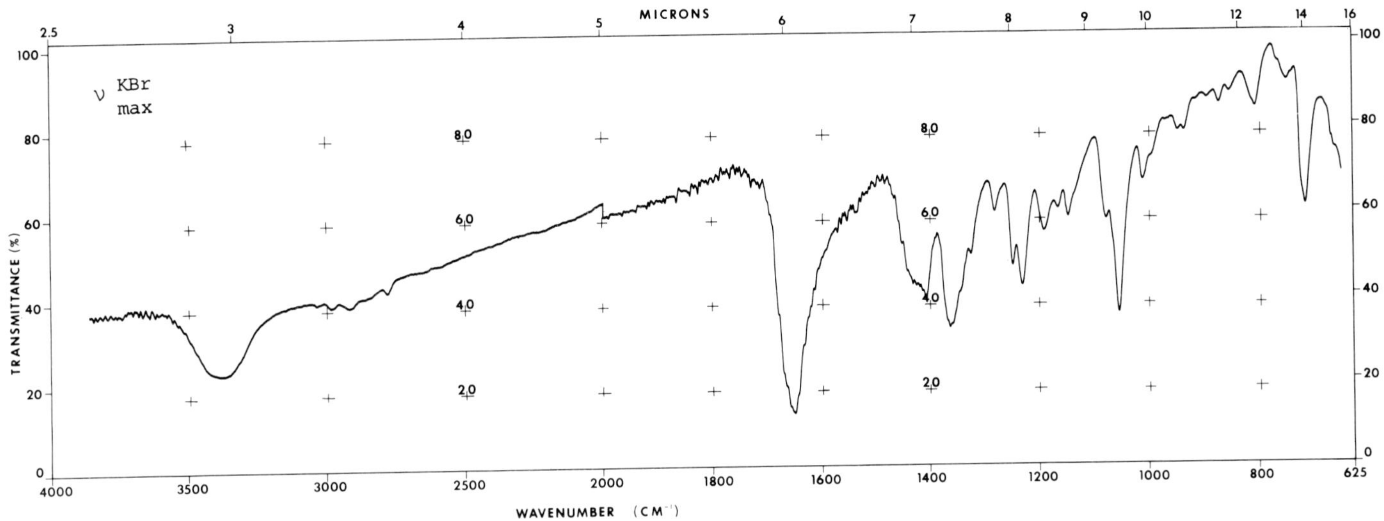
MICRONS
ν KBr max
TRANSMITTANCE (%)
WAVENUMBER (CM⁻¹)

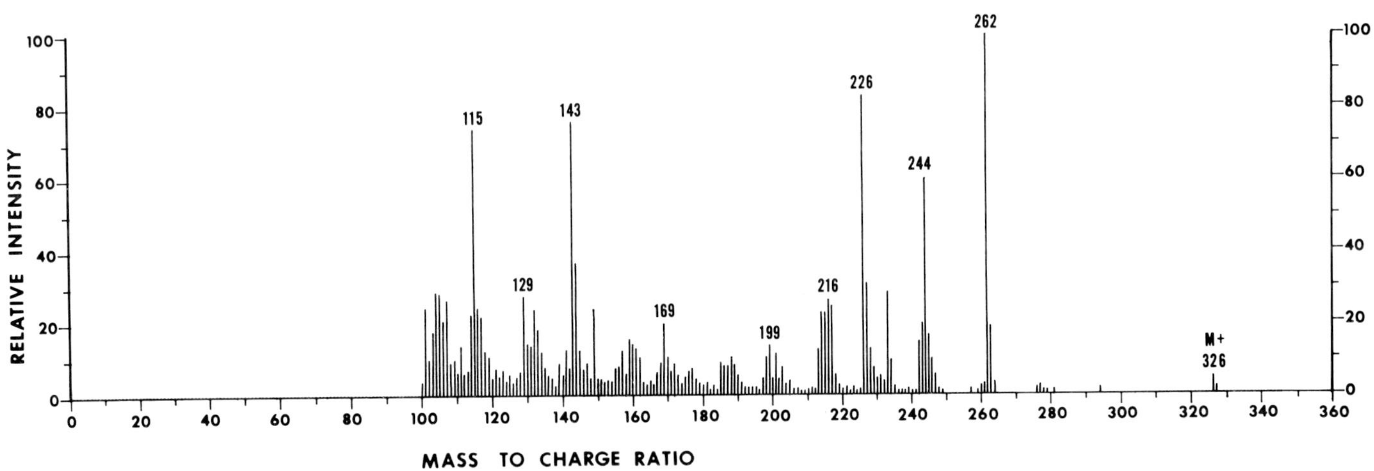
RELATIVE INTENSITY
115
129
143
169
199
216
226
244
262
M+
326
MASS TO CHARGE RATIO

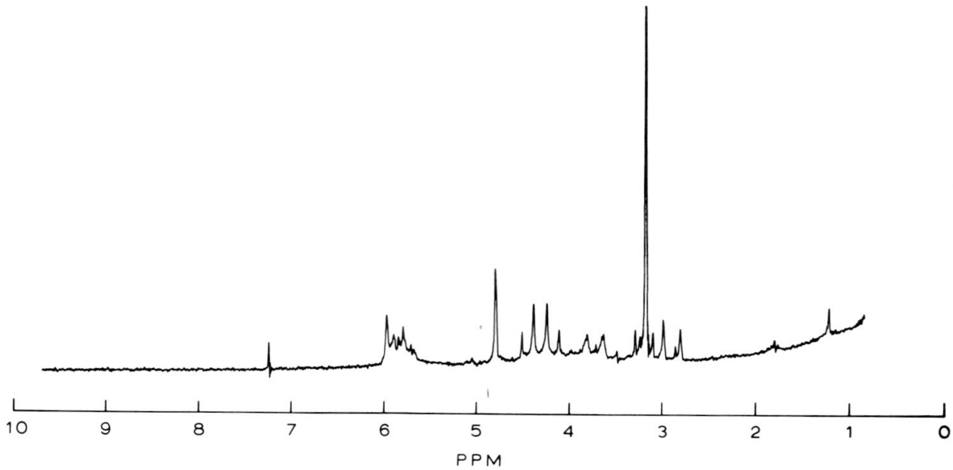

The 100 MHz ^1H NMR spectrum of gliotoxin in $CDCl_3$ solution.

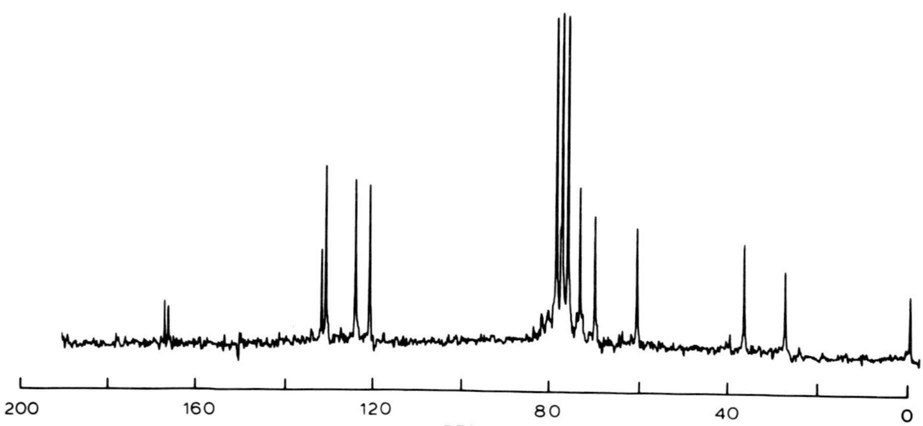

The 25 MHz ^{13}C NMR spectrum of gliotoxin in $CDCl_3$ solution.

Common name	Gliotoxin acetate
Molecular weight	368.0365
Molecular formula	$C_{15}H_{16}O_5N_2S_2$
General characteristics[22,60]	mp 162°–163°C; 170°–172°C $[\alpha]_D^{19}$ −720° (C = 0.6 in chloroform)
UV data[22,60]	λ_{max} nm (ϵ): 268(6,300)
IR data[22,60]	ν_{max}^{KBr} cm^{-1}: 1735, 1690, 1675, 1425, 1370, 1230, 1085, 1050, 825, 720, and 650
Fungal source[22,60]	*Penicilliun terlikowskii*

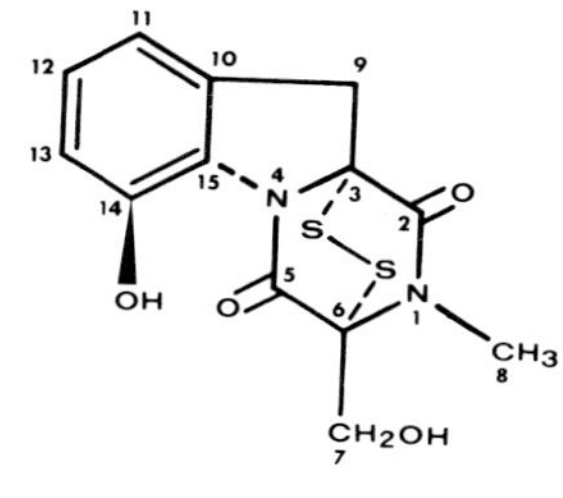

Common name	Dehydrogliotoxin
Molecular weight	324.0232
Molecular formula	$C_{13}H_{12}O_4N_2S_2$
General characteristics[29,51]	Crystals from toluene, mp 188°–189°C $[\alpha]_{589}$ nm −1310° (C = 0.23 in chloroform)
UV data[29,51]	λ_{max}^{EtOH} nm (ϵ): 214(21,800), 272(5,300), and 300(4,600)
IR data[29,51]	ν_{max}^{KBr} cm^{-1}: 3580 and 1685
Fungal source[29,51]	*Penicillium terlikowskii*
Toxicity data[29,51]	Antibiotic activity against *Bacillus subtilis* at concentrations similar to those of gliotoxin.

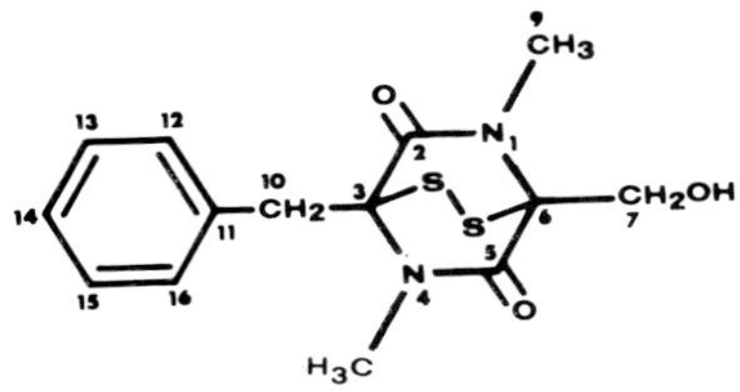

Common name	Hyalodendrin (3-Benzyl-6-hydroxymethyl-1,4-dimethyl-3,6-epidithiopiperazine-2,5-dione)
Molecular weight	324.0595 Mass spectral analysis shows base peak *m/e* 260 due to $M^+ - S_2$
Molecular formula	$C_{14}H_{16}O_3N_2S_2$
General characteristics[54,58]	Cyrstals (prisms) from ethyl ether, mp 101°C $[\alpha]_D^{22}$ +26° (*C* = 1.06 in chloroform)
UV data[54,58]	λ_{max}^{EtOH} nm (ϵ): <220(ca. 1000) and 260(slight inflection)
CD data[54]	$\Delta\epsilon$ (*C* = 0.376 in methanol, 1 mm cell) nm: 234(22.5, 252(0), 264(8.36), and 310(0)
Fungal source[58]	*Hyalodendron* spp.
Toxicity data[54]	Hyalodendrin shows *in vitro* antimicrobial activity against a broad spectrum of fungi associated with disease in plants and trees, and decay in wood products. Also effective against several microorganisms pathogenic to man. LD_{50} in adult Swiss albino mice (IP) 75 mg/kg.
TLC data[54]	*Adsorbent:* Merck GF_{254} *Solvent:* benzene–acetone, 94:6 R_f: 0.60 *Detection:* UV irradiation
Other significant references	Chemistry[55,57]; synthesis[56]; IR data[58]
^{13}C NMR data	C2, 166.9*s; C3, 75.3 s; C5, 168.0*s; C6, 75.3 s; C7, 61.3 t; C8, 27.5 q; C9, 28.6 q; C10, 36.9 t; C11, 134.1 s; C12,16, 129.1 d; C13,15, 128.7 d; C14, 127.4 d

* Assignments may be reversed.

^{1}H NMR data H7a, 4.38 (J = 12.0); H7b, 4.42 (J = 12.2); H8, 3.22; H9, 2.98; H10a, 3.60 (J = 16.0); H10b, 4.08 (J = 16.0); H12–16, 7.26

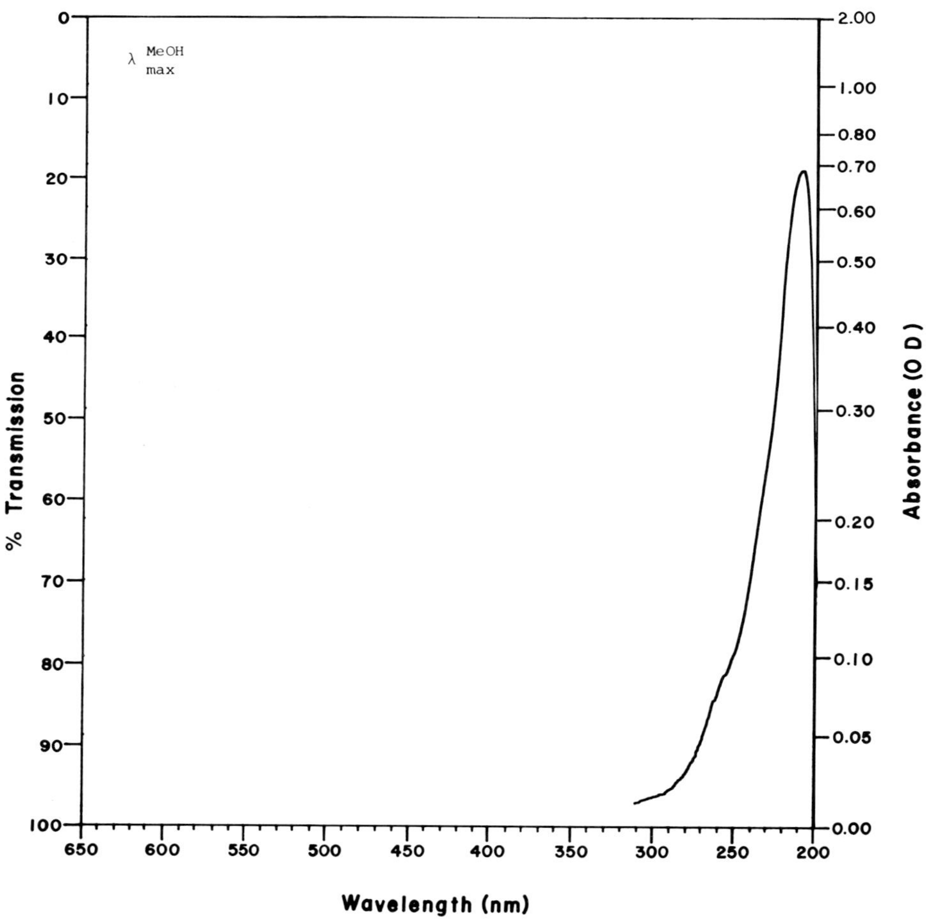

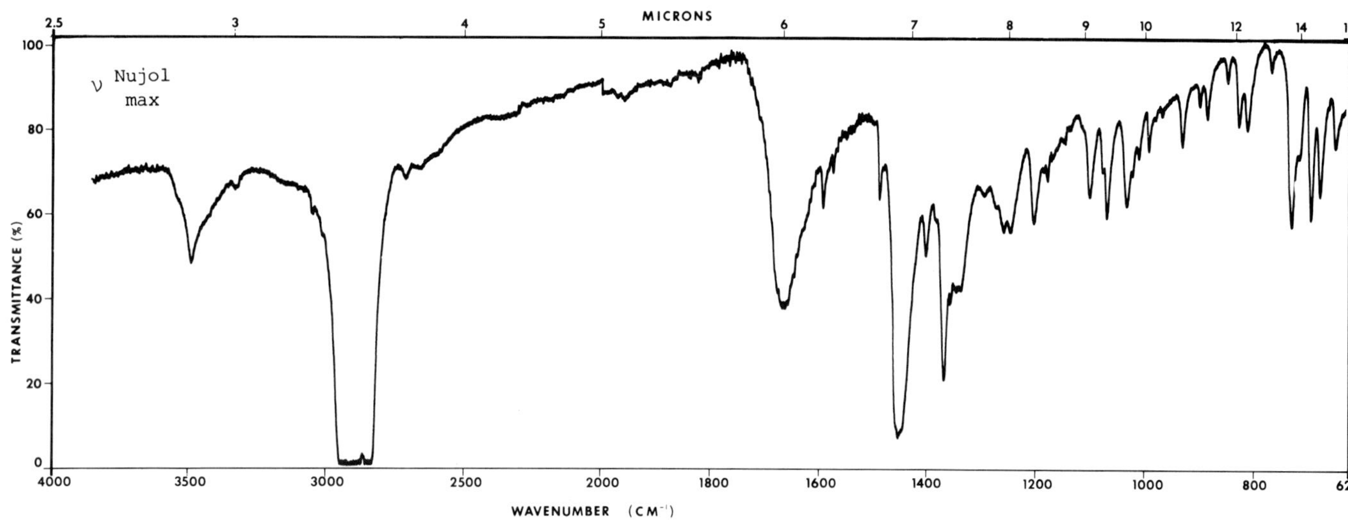
MICRONS
2.5
3
4
5
6
7
8
9
10
12
14
16
ν Nujol max
TRANSMITTANCE (%)
100
80
60
40
20
0
4000
3500
3000
2500
2000
1800
1600
1400
1200
1000
800
625
WAVENUMBER (CM⁻¹)

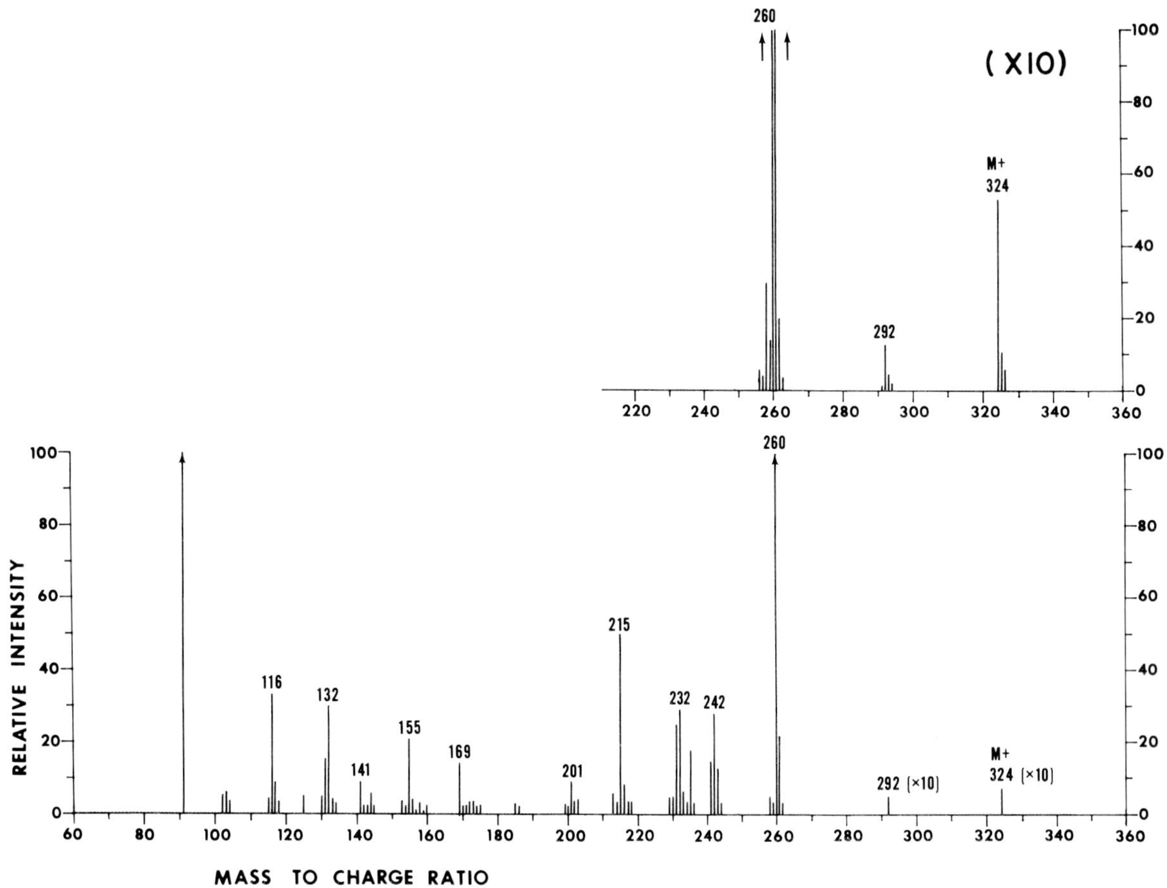

260
(X10)
M+
324
292
220
240
260
280
300
320
340
360
100
80
60
40
20
0
RELATIVE INTENSITY
116
132
141
155
169
201
215
232
242
260
292 (×10)
M+
324 (×10)
60
80
100
120
140
160
180
200
MASS TO CHARGE RATIO

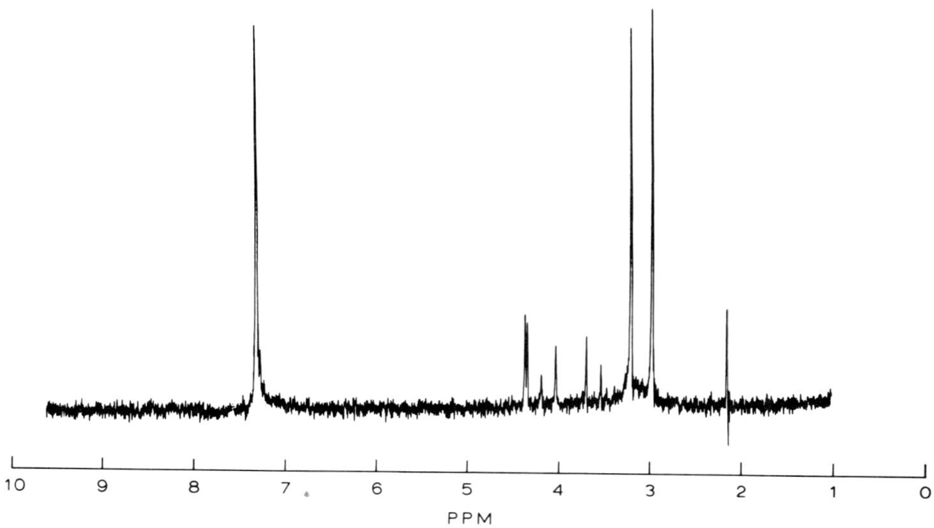

The 100 MHz ^1H NMR spectrum of hyalodendrin in $CDCl_3$ solution.

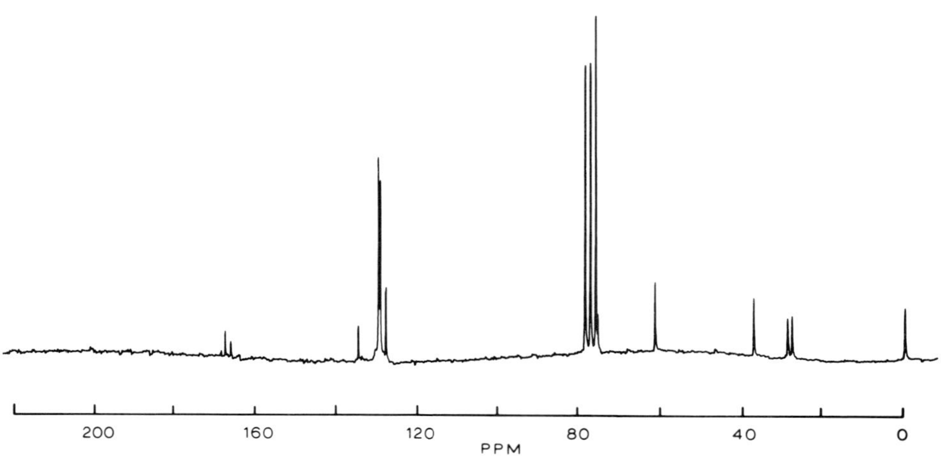

The 25 MHz ^{13}C NMR spectrum of hyalodendrin in $CDCl_3$ solution.

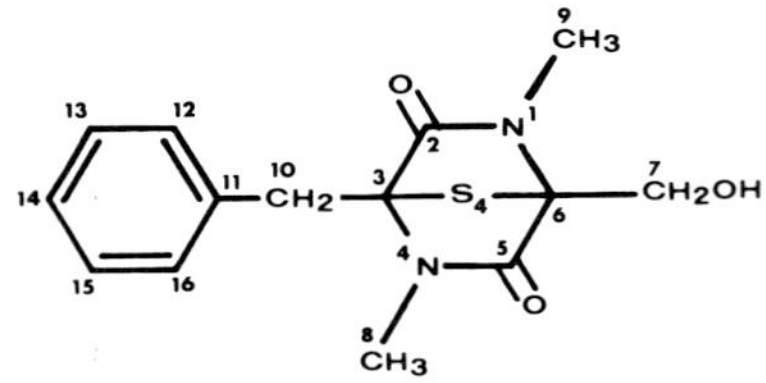

Common name	Hyalodendrin tetrasulfide [(3*S*,6*S*)3-Benzyl-6-hydroxymethyl-1,4-dimethyl-3,6-epitetra-thiopiperazine-2,5-dione]
Molecular weight	388.0035
Molecular formula	$C_{14}H_{16}O_3N_2S_4$
General characteristics[57]	Crystals from benzene–cyclohexane, mp 133°–139°C
CD data[57]	$\Delta\epsilon$ (2.8×10^{-3} *M* in methanol, 1 mm cell) nm: 370(0), 328(+0.9), 316(0), 297(−4.4), 287(0), 263(+13.6), 225(+12.1), 230(0), and <220 ($-ve_{max}$)
IR data[57]	ν_{max}^{KBr} cm^{-1}: 3400, 1666, 1648, 1375, 1060, 742, 735, 701, and 657
Fungal urce[57]	*Hyalodendron* spp. Unidentified NRRL 3888 The enantiomeric metabolite with *R* configuration at positions 3 and 6 are produced by *Penicillium turbatum*. Gliovictin from *Heleminthosporium victoriae* is likewise enantiomeric (3*R*,6*R*).
TLC data[57]	*Adsorbent:* Kieselgel *Solvent:* benzene–acetone, 9:1 v/v R_f: 0.5 *Detection:* not reported

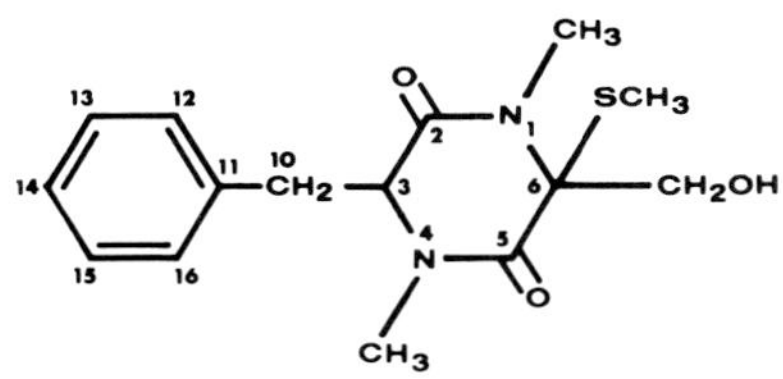

Common name	Bisdethiodi(methylthio)hyalodendrin [3-Benzyl-6-hydroxymethyl-1,4-dimethyl-2,5-dioxo-3,6-di(methylthio)piperazine]
Molecular weight	354.1063
Molecular formula	$C_{16}H_{22}O_3N_2S_2$
General characteristics[55]	Colorless needles from dichloromethane–cyclohexane, mp 140°–140.5° C $[\alpha]_D^{23}$ +64° (C = 1.071 in chloroform)
UV data[55]	λ_{max}^{EtOH} nm (ϵ): <220 and 260(900)
IR data[55]	ν_{max}^{KBr} cm^{-1}: 3375, 1660, 1644, 1636, 1499, 1375, 735, and 700
Fungal source[55]	*Hyalodendron* spp.

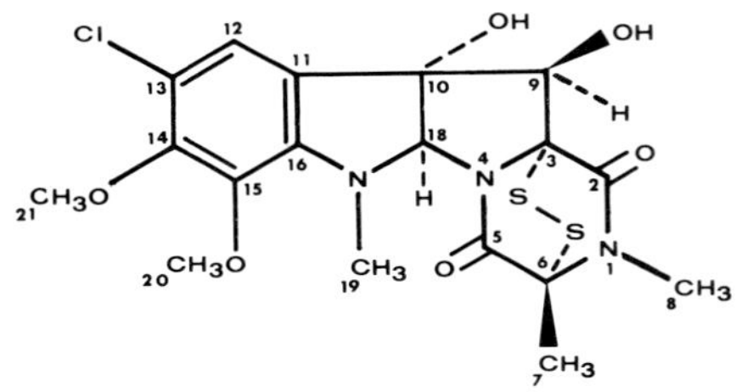

Common name	Sporidesmin (Sporidesmin A)
Molecular weight	473.0479
Molecular formula	$C_{18}H_{20}O_6N_3S_2Cl$
General characteristics[19]	mp 179°C; benzene solvate, mp 110°–120°C $[\alpha]_D^{23}$ −185° (*C* = 1.1 in methanol); +38° (*C* = 1.4 in chloroform) Diacetate, mp 170°–171°C $[\alpha]_D^{22}$ +43.4° (*C* = 0.51 in chloroform)
UV data[19]	$\lambda_{max}^{Et_2O}$ nm (ϵ): 218(39,800), 254(13,100), and 302(2,800)
CD data[19]	$\Delta\epsilon$ (in methanol) nm: 365(0.3), 300(+2.59), 268(5.88), and 232(−34.0) $\Delta\epsilon$ (in ethanol) nm: 360(+0.31), 300(+1.80), and 235(−20.9)
Fungal source[3]	*Pithomyces chartarum* (Berk. and Curt.) M.B. Ellis Syn. Syn. *Sporidesmium bakeri* Sydon: *Scheleobrachea echinulata* (Speg.) Hughes *Piricauda chartarum* (Berk. and Curt.) R. T. Moore *Sporidesmium chartarum* Berk. and Curt., and *Sporidesmium echinulatum* Speg.
Toxicity data[11,36–40,45,46]	*Mice:* a lethal dose of sporidesmin was in the range of 200–300 mg/kg. Typical clinical signs were pleural effusions, pulmonary edema, icterus, and hepatic lesions. Some mice exhibited signs of progressive anemia. *Rats:* The LD_{50} in weanling female rats dosed IP was approximately 8 mg/kg, depending on their diet. Typical clinical signs in rats dosed with spor-

idesmin were a general inflammatory response marked by ascites and/or pleural effusions, increased capilla permeability and diarrhea.
Sheep: Sporidesmin and other closely related metabolites of *P. chartarum* are strongly implicated in the etiology of a disease syndrome in sheep characterized by hepatic damage, ultimately expressed as photosensitivity and facial eczema. A dose of 0.5–1 mg/kg orally, SC, or IV induced severe disease. At 1 mg/kg given orally to female lambs, 80% of the animals died in 24 days; at 0.5 mg/kg, 12 died in the same period. The clinical response to a single oral dose resulted in anorexia and diarrhea initially, followed shortly by a rise in serum bile acids conjugated with taurine. Other effects were lesions in the walls of the blood vessels and accumulation of lipid in the liver. After about 10 days, many animals became photosensitive and icteric.

TLC data[19]

Adsorbent: silica gel F_{254} (Merck)
Solvent: A: benzene–ethyl acetate, 4:1 v/v
B: chloroform–methanol, 19:1 v/v
R_f: A: 0.38
B: 0.57
Detection: spray with 5% aqueous silver nitrate or observe under reflected short wave UV light.

Other significant references

X-Ray crystallography[13,30]; synthesis[26,27,41]; structure[19]; isolation[59]; mass spectroscopy[53]; review[3,11,28,52,60]; NMR[49]

^{13}C NMR data[49]

C2, 166.7 (169.1)*; C3, 77.7 (77.0); C5, 164.9 (168.0); C6, 74.5 (75.5); C7, 18.2 (18.4); C8, 27.3 (27.2); C9, 82.5 (74.9); C10, 90.4 (91.0); C11, 126.9 (121.4); C12, 121.0 (118.7); C13, 118.8 (119.7); C14, 151.6 (152.0); C15, 145.8 (145.6); C16, 140.8 (140.8); C18, 95.3 (93.4); C19, 39.2 (38.9); C20, 60.9 (60.6); C21, 60.9 (60.9); *C*H_3CO, (19.7, 21.6); CH_3*C*O, (168.0, 169.1)

^{1}H NMR data[49]

H7, 2.06; H8, 3.30; H9, 4.60; H10, 4.85; H12, 7.10; H18, 5.32; H19, 3.07; H20, 3.83; H21, 3.88

* Data in parentheses for sporidesmin diacetate.

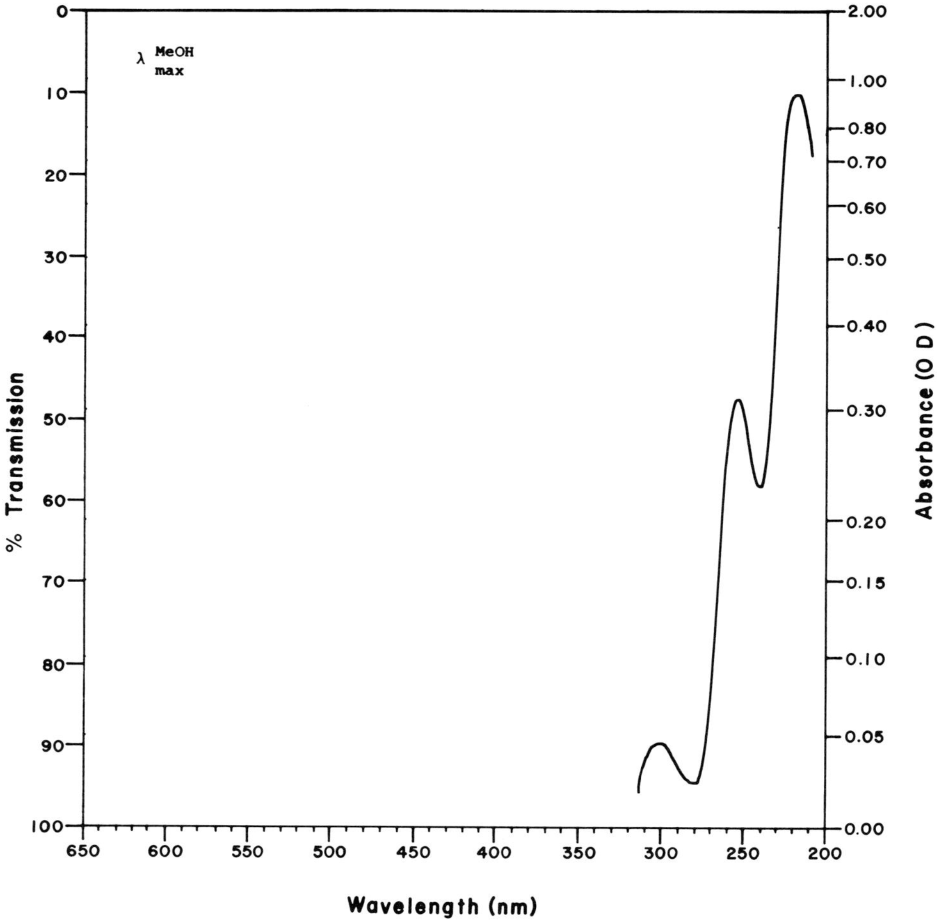
λ MeOH max
% Transmission
0
10
20
30
40
50
60
70
80
90
100
Absorbance (O D)
2.00
1.00
0.80
0.70
0.60
0.50
0.40
0.30
0.20
0.15
0.10
0.05
0.00
650
600
550
500
450
400
350
300
250
200
Wavelength (nm)

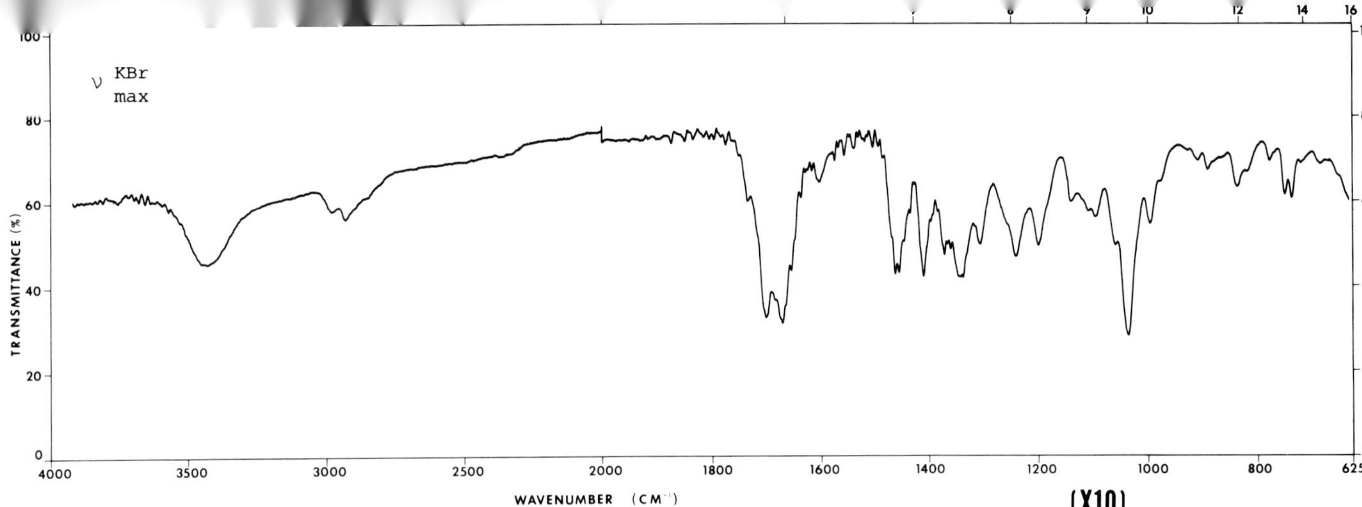
ν KBr max
TRANSMITTANCE (%)
WAVENUMBER (CM⁻¹)

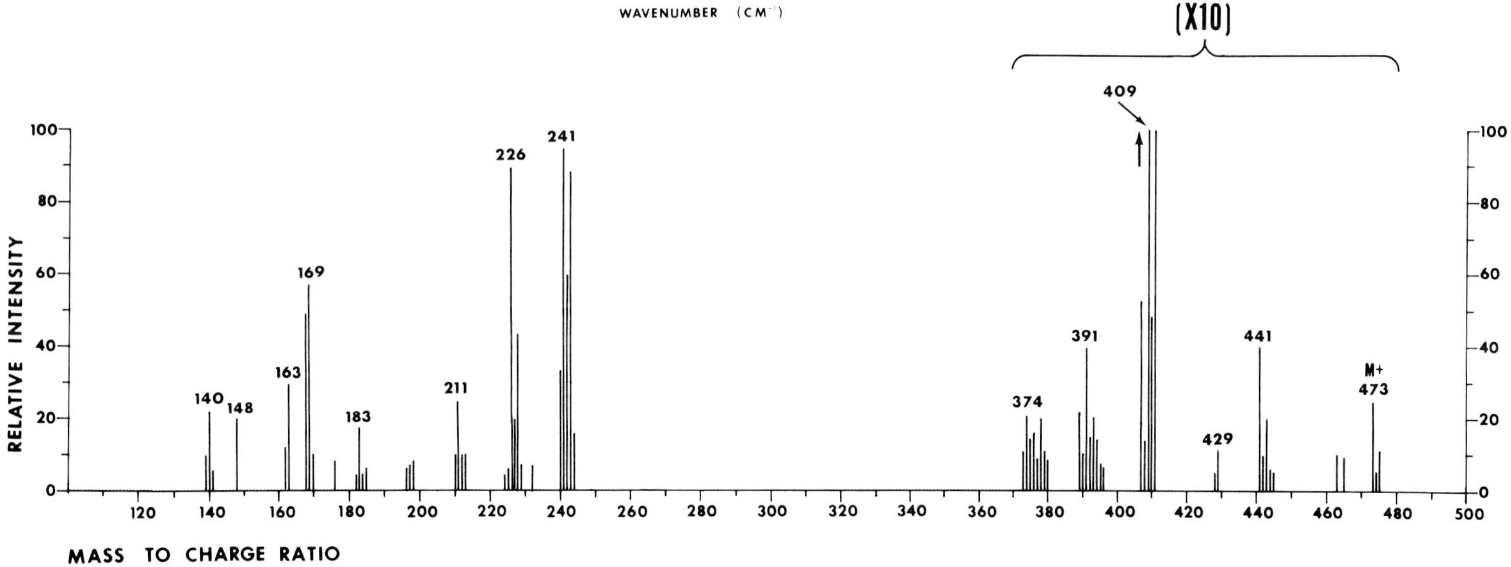
(X10)
409
226
241
169
163
140
148
183
211
374
391
441
429
M+
473
RELATIVE INTENSITY
MASS TO CHARGE RATIO

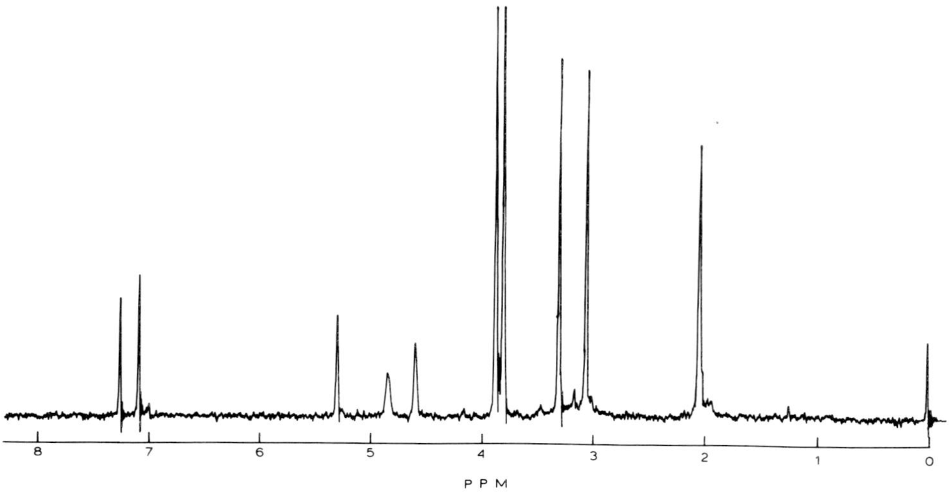

The 100 MHz ^1H NMR spectrum of sporidesmin in $CDCl_3$ solution.

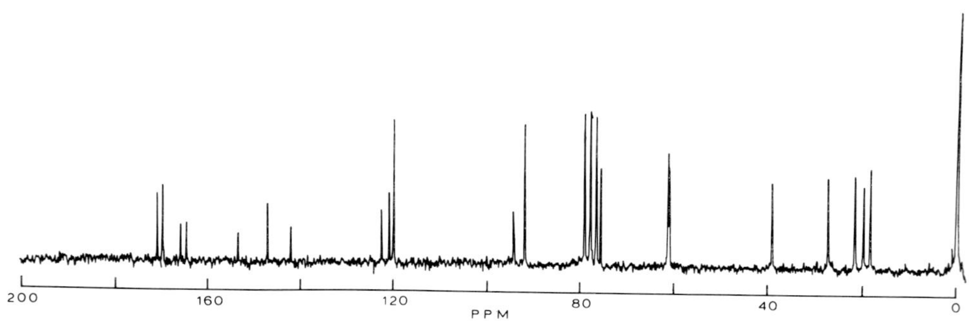

The 25 MHz ^{13}C NMR spectrum of sporidesmin diacetate in $CDCl_3$ solution.

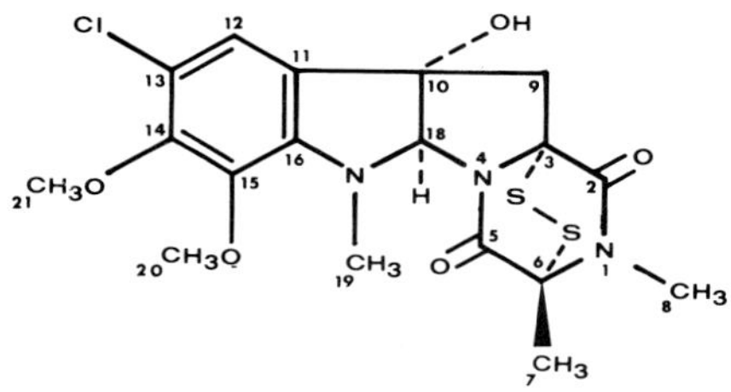

Common name	Sporidesmin B
Molecular weight	457.0530
Molecular formula	$C_{18}H_{20}O_5N_3S_2Cl$
General characteristics[50]	Colorless needles from aqueous acetone, mp 183°C $[\alpha]_D^{21}$ −27° (C = 1.0 in methanol) +12° (C = 0.75 in chloroform)
UV data[50]	$\lambda_{max}^{Et_2O}$ nm (ϵ): 218(31,600), 256(12,000), and 307(2,500)
IR data[50]	ν_{max}^{Nujol} cm^{-1}: 1697, 1666, and in chloroform: 1705
Fungal source[50]	*Pithomyces chartarum*
Toxicity data	Not reported
Other significant references	Synthesis[41]; mass spectroscopy[53]
^1H NMR data[50]	H7, 2.03; H8, 3.03; H9, 2.80 (J = 16.1), 3.20 (J = 16.1); H12, 7.03; H18, 6.40; H19, 3.24; H20, 3.81*; H21, 3.86*; 10-OH, 3.53

* Assignments may be reversed.

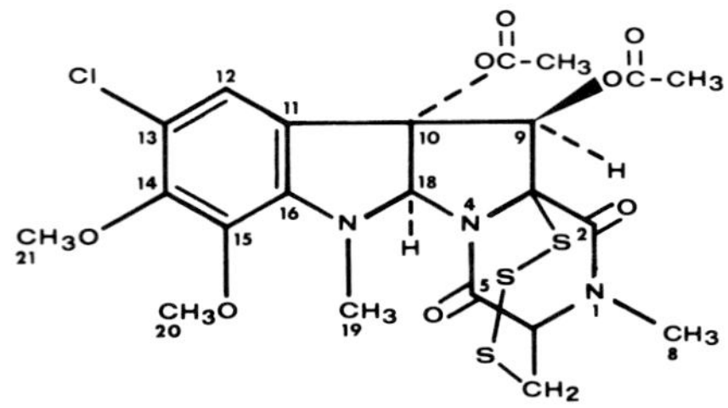

Common name	Sporidesmin C diacetate
Molecular weight	589.0402
Molecular formula	$C_{22}H_{24}O_8N_3S_3Cl$
General characteristics[20]	Needles from isopropanol, mp 230°–240° (dec.) after prior melting and resolidification at 195°–200°C $[\alpha]_D$ −215° (C = 0.46 in chloroform)
UV data[20]	λ_{max}^{MeOH} nm (ϵ): 222(22,300), 255(9,900), and 309(2,900)
Fungal source[20]	*Pithomyces chartarum*

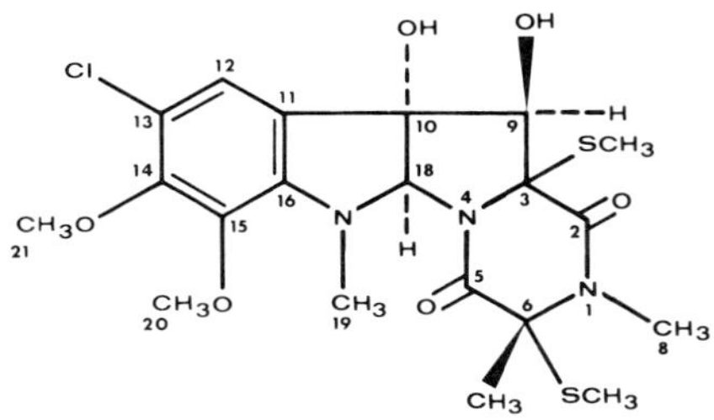

Common name	Sporidesmin D
Molecular weight	503.0940
Molecular formula	$C_{20}H_{26}O_6N_3S_2Cl$
General characteristics[21]	Colorless needles from ether as etherate, mp 110°–120°C Ethanolate from ethanol, mp 105°–107°C $[\alpha]_D^{23}$ 58° (C = 0.11 in chloroform)
UV data[21]	λ_{max}^{MeOH} nm (ϵ): 216(28,100), 252(10,000), and 300(2,000)
IR data[21]	ν_{max}^{KBr} cm^{-1}: 3450, 3330, 1680, 1665, and 1605
Fungal source	*Pithomyces chartarum*
^{13}C NMR data[49]	C2, 167.1; C3, 73.1; 3-SCH_3, 14.5*; C5, 165.0; C6, 69.7; 7-CH_3, 25.5; 7-SCH_3, 15.6*; C8, 28.8; C9, 80.4; C10, 90.3; C11, 128.3; C12, 121.8; C13, 119.1; C14, 151.3; C15, 146.1; C16, 141.0; C18, 96.0; C19, 41.3; C20, 60.3†; C21, 61.0†
^1H NMR data[21]	7-CH_3, 1.87; 7-SCH_3, 2.32*; 3-SCH_3, 2.40*; H9, 4.67; H12, 7.10, H18, 5.30; H19, 3.38; H8, 3.08; H20, 3.81‡; H21, 3.88‡; OH, 3.05; 3.95

* Assignments may be reversed.
† Assignments may be reversed.
‡ Assignments may be reversed.

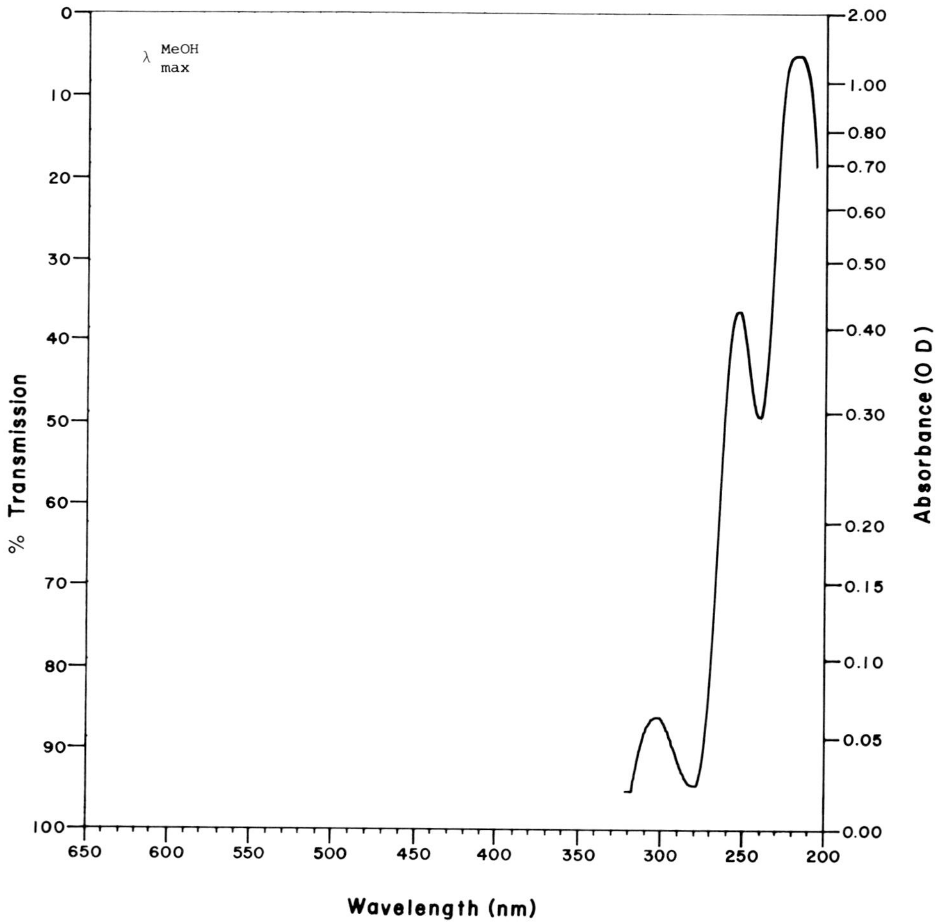

λ MeOH max
% Transmission
0
10
20
30
40
50
60
70
80
90
100
Absorbance (O D)
2.00
1.00
0.80
0.70
0.60
0.50
0.40
0.30
0.20
0.15
0.10
0.05
0.00
650
600
550
500
450
400
350
300
250
200
Wavelength (nm)

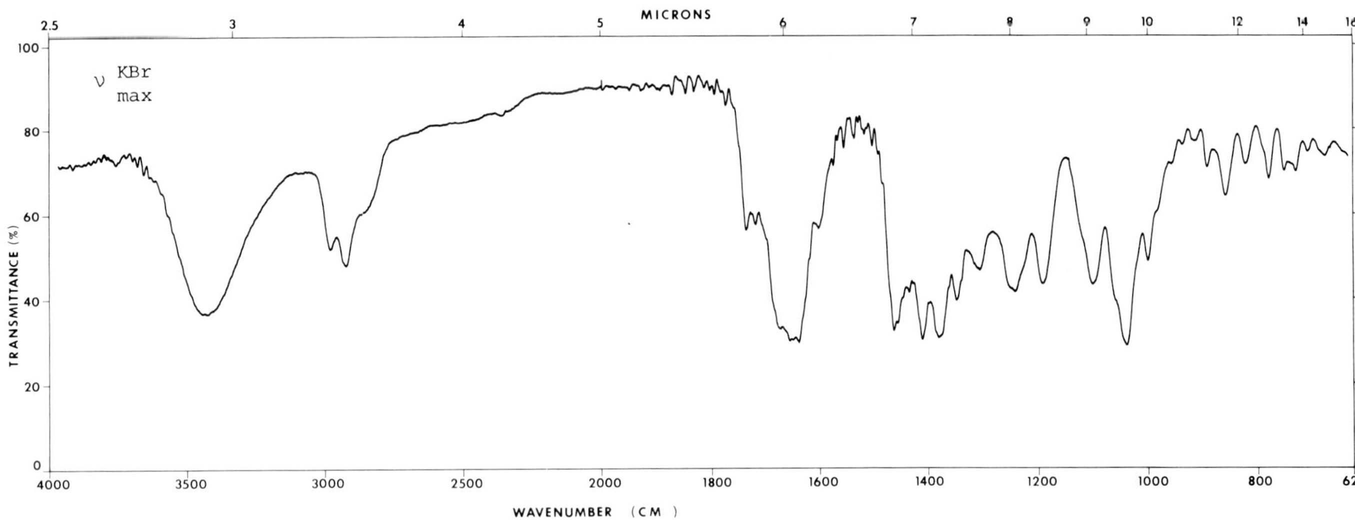
MICRONS
ν KBr max
TRANSMITTANCE (%)
WAVENUMBER (CM)

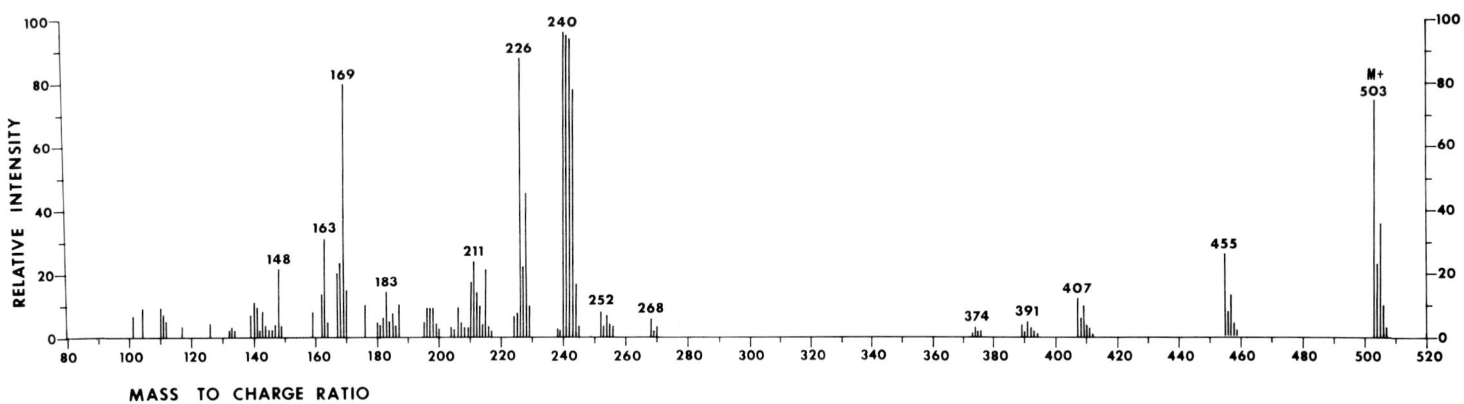
RELATIVE INTENSITY
148
163
169
183
211
226
240
252
268
374
391
407
455
M+
503
MASS TO CHARGE RATIO

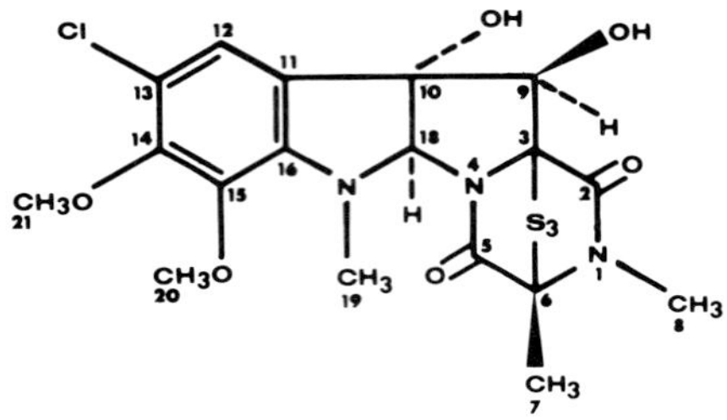

Common name	Sporidesmin E
Molecular weight	505.0199
Molecular formula	$C_{18}H_{20}O_6N_3S_3Cl$
General characteristics[48]	Sporidesmin E ethanolate from ethanol, mp 180°–187°C $[\alpha]_D^{20}$ −131° (C = 0.065 in chloroform) Etherate from ethyl ether, mp 180°–185°C
UV data[48]	λ_{max}^{MeOH} nm (ϵ): 217(33,100), 252(16,500), and 295(3,100)
IR data[48]	ν_{max}^{KBr} cm^{-1}: 3325, 1690, and 1655
Fungal source[48]	*Pithomyces chartarum*
^{1}H NMR data[48]*	H7, 2.00, 1.95; H8, 3.14, 3.02; H9, 4.62, 4.52; H12, 7.08, 7.07; H18, 5.41, 5.29; H19, 3.50, 3.33; H20, 21, 3.90, 3.87, 3.83

* Ether of solvation at 1.33, 1.08, and 3.86–3.50 ppm.

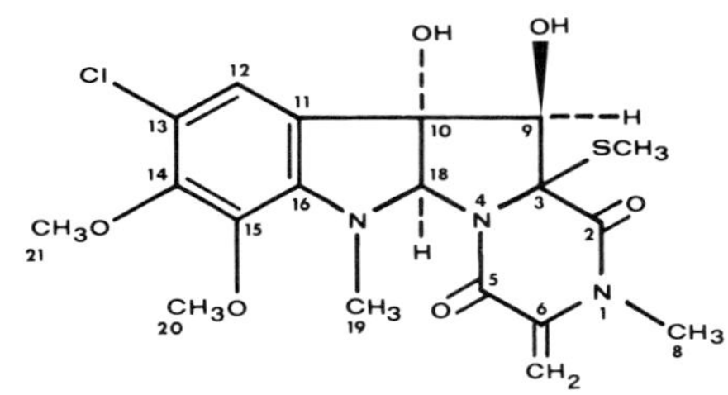

Common name	Sporidesmin F
Molecular weight	455.0917
Molecular formula	$C_{19}H_{22}O_6N_3SCl$
General characteristics[21]	Amorphous solid from ethyl acetate–light petroleum, mp 65°–75°C
UV data[21]	λ_{max}^{MeOH} nm (ϵ): 216(28,800), 250(13,800), and 298(2,000)
IR data[21]	ν_{max}^{KBr} cm^{-1}: 3430, 1690, and 1615, and 1605
Fungal source[21]	*Pithomyces chartarum*
Toxicity data	Toxicity unknown
^{1}H NMR data[21]	3-SCH_3, 2.31; H7, 5.88, 5.98; H8, 3.24; H9, 4.63; H12, 7.09; H18, 5.46; H19, 3.40; H20, 3.82*; H21, 3.88*

* Assignments may be reversed.

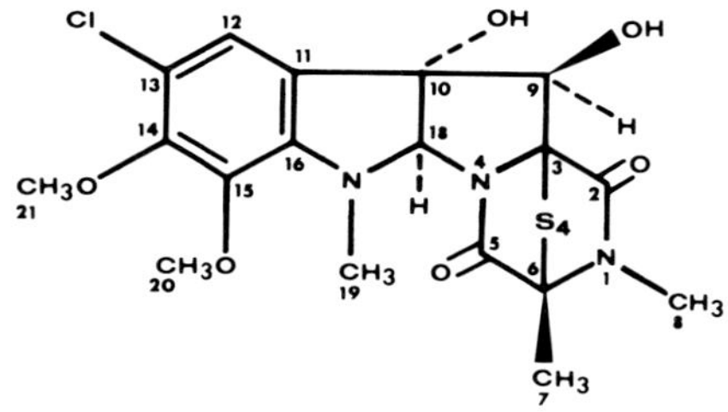

Common name	Sporidesmin G (9-Chloro-3,11*a*-epitetrathio-2,3,5a,6,10*b*,11-hexahydro-10*b*,11-dihydroxy-7,8-dimethyoxy-2,3,6-trimethyl-11a*H*-pyrazeno-[1′,2′:1,5]pyrrolo[2,3-6]indole-1,4-dione)
Molecular weight	536.9912
Molecular formula	$C_{18}H_{20}O_6N_3S_4Cl$
General characteristics[12]	Crystals from ether, mp 148°–153°C $[\alpha]_D^{20}$ −217° (*C* = 0.023 in chloroform)
UV data[12]	λ_{max}^{MeOH} nm (ϵ): 216(44,700), 250(11,500), and 298(3,000)
IR data$_{12}$	$\nu_{max}^{CHCl_3}$ cm^{-1}: 3570, 3520, 1690, and 1660 ν_{max}^{KBr} cm^{-1}: 1210, 825, 725, 585, 510, 505, 480, 410, and 395
Fungal source[12]	*Pithomyces chartarum*
^{1}*H NMR data*[12]	H7, 1.97; H8, 3.07; H9, 4.65; H12, 7.18; H18, 5.10; H19, 3.40; H20, 3.84*; ·H21, 3.90*

* Assignments may be reversed.

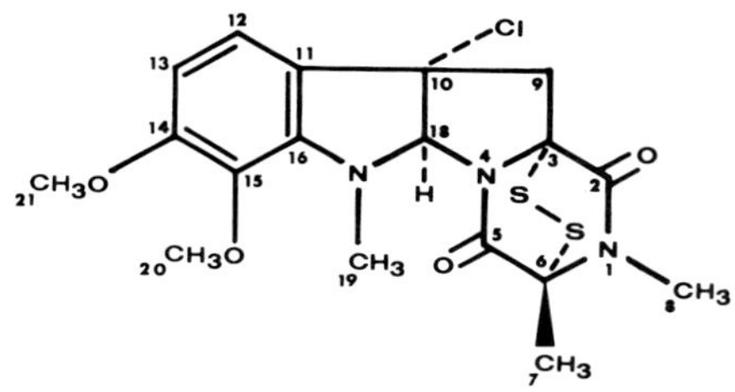

Common name	Sporidesmin H
Molecular weight	339.1219
Molecular formula	$C_{18}H_{17}O_4N_3$
General characteristics[48a]	Colorless amorphous solid from diethyl ether–light petroleum, mp 150°–152°C
UV data[48a]	λ_{max}^{MeOH} nm (ϵ): 216(23,400), 252(11,200), and 290(6,400)
IR data[48a]	ν_{max}^{KBr} cm^{-1}: 1700
Fungal source[48a]	*Pithomyces chartarum*
Toxicity data	Toxicity unknown
^1H NMR data[48a]	H7, 2.02; H8, 3.03; H9a, 2.70 (J = 16); H9b, 3.27 (J = 16); H12, 6.98 (J = 8); H13, 6.43 (J = 8); H18, 5.38; H19, 3.30; H20, 3.75*; H21, 3.83*

* Assignments may be reversed.

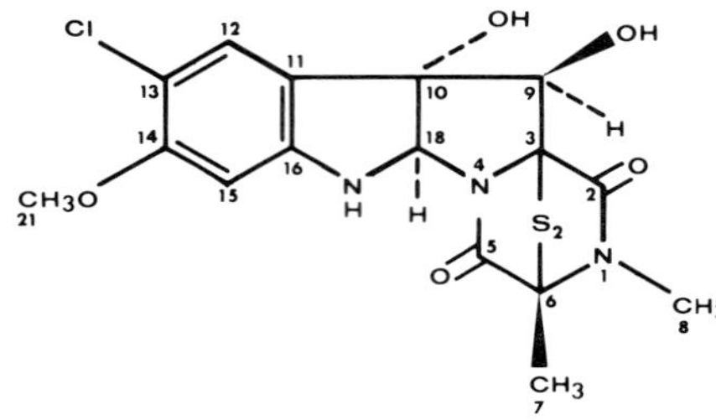

Common name	Sporidesmin J
Molecular weight	459.0326
Molecular formula	$C_{17}H_{18}O_6N_3S_2Cl$
General characteristics[48a]	Crystals from ethyl ether, mp 168°–169°C $[\alpha]_D^{19}$ +43° (*C* = 0.25 in chloroform)
Fungal source[48a]	*Pithomyces chartarum*
Toxicity data	Toxicity unknown
^{1}*H NMR data*[48a]	H7, 2.03; H8, 3.06; H9, 4.74; H12, 7.14; H17, 5.17; H18, 5.67; H19, 3.85*; H20, 3.88*

* Assignments may be reversed.

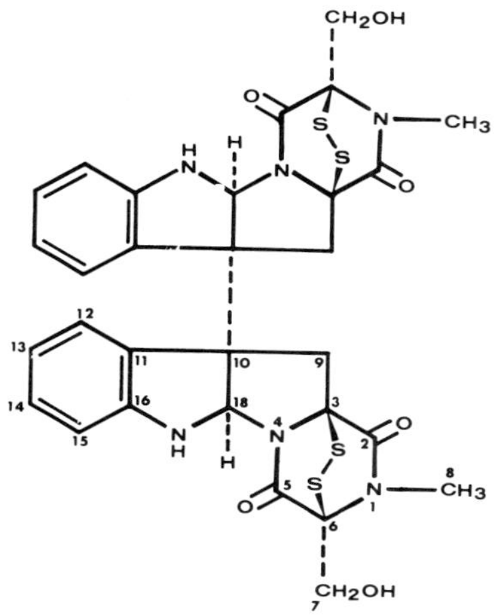

Common name	Chaetocin
Molecular weight	696.0938
Molecular formula	$C_{30}H_{28}O_6N_6S_4$
General characteristics[17]	Crystals from pyridine–methanol, mp 240°C $[\alpha]_D^{20}$ +379° (C = 1.00 in pyridine); +789° (C = 1.00 in dimethyl sulfoxide) Diacetate: crystals from acetone, mp 220°–225°C (dec.) $[\alpha]_D^{20}$ +679° (C = 0.99 in chloroform)
UV data[17]	λ_{max}^{DMSO} nm (ϵ): 306(6,000)
CD data[17]	$\Delta\epsilon$ (diacetate) nm: 239(+73.8), 270(−6.5), and 304(+14.3)(in methanol)
IR data[17]	ν_{max}^{KBr} cm^{-1}: 3520, 3420, 3330, 1680, 1655, and 1605
Fungal source[17]	*Chaetomium minutum*
Toxicity data[17]	Antibacterial and cytostatic activity
Other significant references	X-Ray structure[17]
^{13}C NMR data[16,17]	C2, 160.6; C3, 77.1; C5, 165.4; C6, 73.4; C7, 58.6; C8, 27.5; C9, 40.0; C10, 59.5; C11, 127.8; C12, 129.4; C13, 117.9; C14, 124.6; C15, 109.1; C16, 149.9; C18, 79.5
^{1}H NMR data[16,17]	H7, 4.18; H8, 3.02; H9, 2.92, 3.64; H12–15, 6.5–7.5; H17, 6.5–7.5; H15, 5.30; 7-OH, 5.80

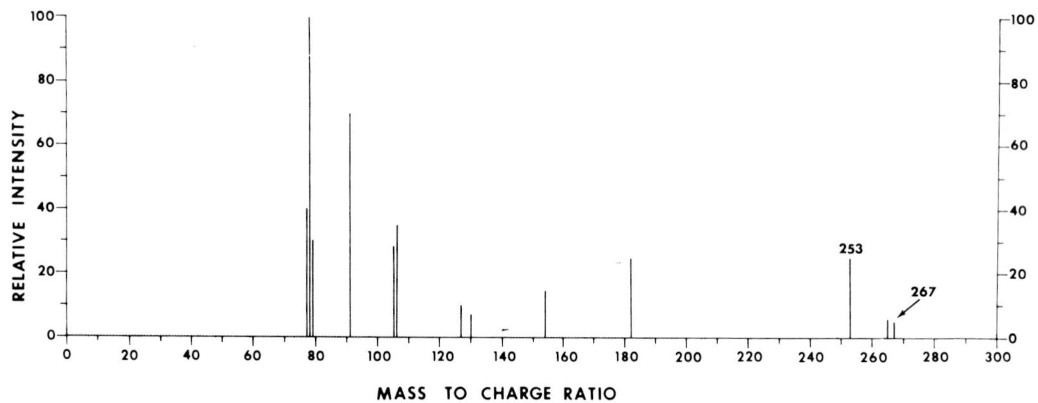
RELATIVE INTENSITY
MASS TO CHARGE RATIO
253
267

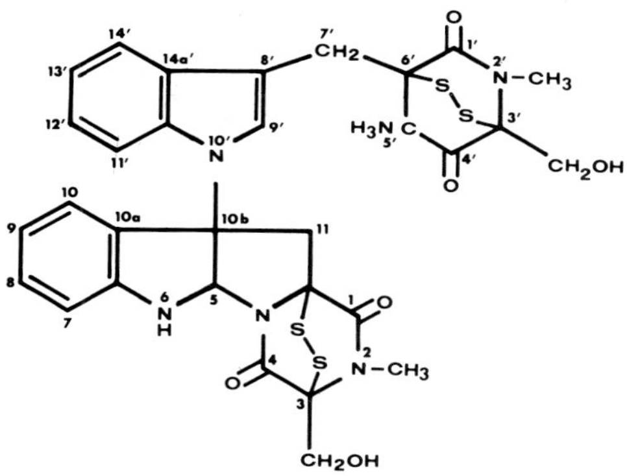

Common name	Chetomin
Molecular weight	710.1094
Molecular formula	$C_{31}H_{30}O_6N_6S_4$
General characteristics[7,31]	Amorphous glasslike solid $[\alpha]_D^{24}$ +258° (C = 0.095 in chloroform)
UV data[7,31,52]	λ_{max} nm (ϵ): 275(9,100), 284(9,600), and 295(8,250)
Fungal source[7,31,60]	*Chaetomium cochliodes* *C. globosum*
Toxicity data[1,2,6,7,31,52,60,62]	The toxicity of chetomin in sheep is similar to that of sporidesmin except for the absence of the relapse phase (icterus and facial eczema), and for being 50 times less active than sporidesmin when given orally. Chetomin also shows antibiotic activity against gram-positive bacteria and viruses (minimum inhibitory concentration vs. *Bacillus subtilis,* 0.1 μg/ml). Chetomin is thought to be partly involved in the etiology of ill-thrift syndrome of sheep in Nova Scotia. At 0.02 μg/ml, it inhibited protein synthesis in HeLa cells. Its oral LD_{50} in rats was 75 mg/kg; in turkeys, it was 30 mg/kg and in lambs dosed IP, it was 5 mg/kg.
^{13}C NMR data[7]	C1, 165.6 s; C3, 76.3 s; C4, 163.0 s; C5, 80.3 d; C6a, 148.4 s; C7, 111.0 d; C8, 124.9 d; C9, 120.2 d; C10, 131.2 d; C10a, 126.7 s; C10b, 73.6 s; C11, 42.7 t; C11a, 73.6 s; C12, 60.4 t*; C13, 27.5 q; C1′, 165.6

* Assignments may be reversed.

	s; C3′, 76.6 s; C4′, 166.7 s; C6′, 74.9 s; C7′ 27.1 t; C8′, 107.8 s; C9′, 127.2 d; C10a′, 134.1 s; C11′, 111.9 d; C12′, 122.8 d; C13′, 120.6 d; C14′, 119.1 d; C14a′, 130.3 s; C15′, 61.6 t[b]; C16′, 27.5 q; C17′, 28.3 q
^{1}H NMR data[7]	H5, 6.17; H6, 5.44; H7, 6.75 (*J* = 7.5); H8, 7.23 (*J* = 7.3); H9, 6.90 (*J* = 8.0); H10, 7.22; H11, 4.49, 3.09 (*J* = 15.4); H12, 4.72 (*J* = 12.7); H13, 2.93; H7′, 2.62 (*J* = 15.5); H9′, 7.34; H11′, 7.29 (*J* = 1.2, 0.5, 8.0); H12′, 7.19 (*J* = 7.0, 1.2); H13′, 7.19 (*J* = 7.8); H14′, 7.64; H15′, 4.72 (*J* = 12.7); H16′, 3.14*; H17′, 3.15*

* Assignments may be reversed.

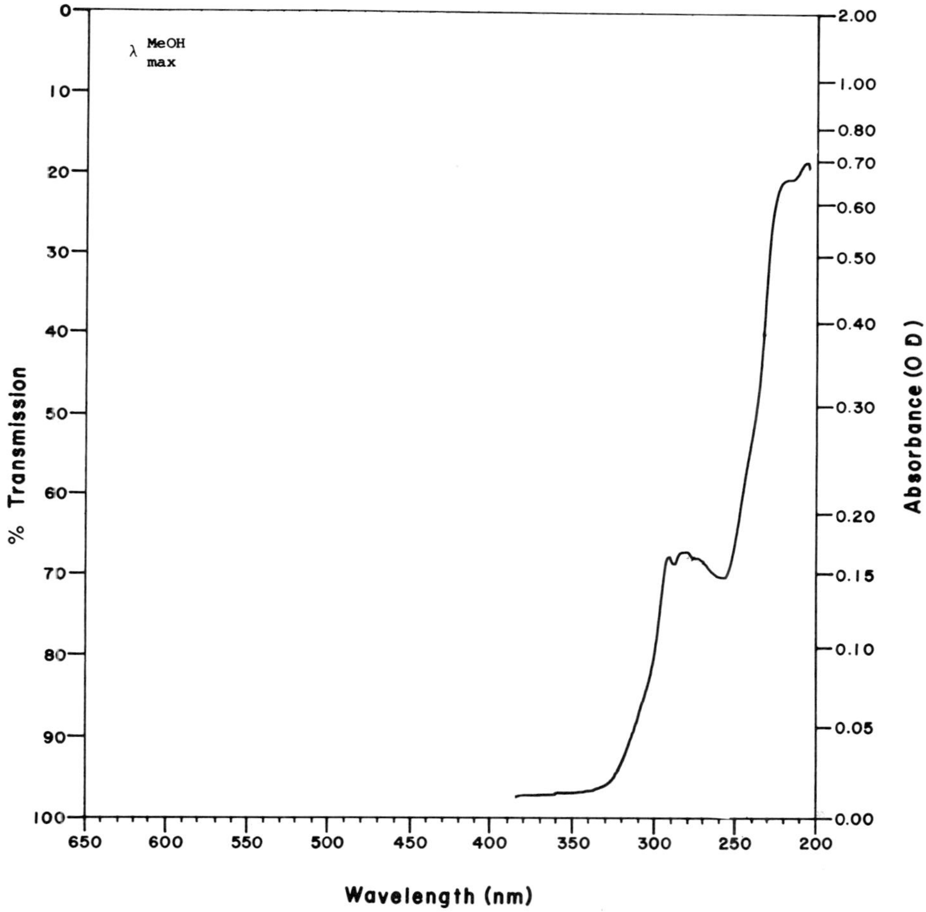
λ MeOH max
% Transmission
Absorbance (O D)
Wavelength (nm)
0
10
20
30
40
50
60
70
80
90
100
2.00
1.00
0.80
0.70
0.60
0.50
0.40
0.30
0.20
0.15
0.10
0.05
0.00
650
600
550
500
450
400
350
300
250
200

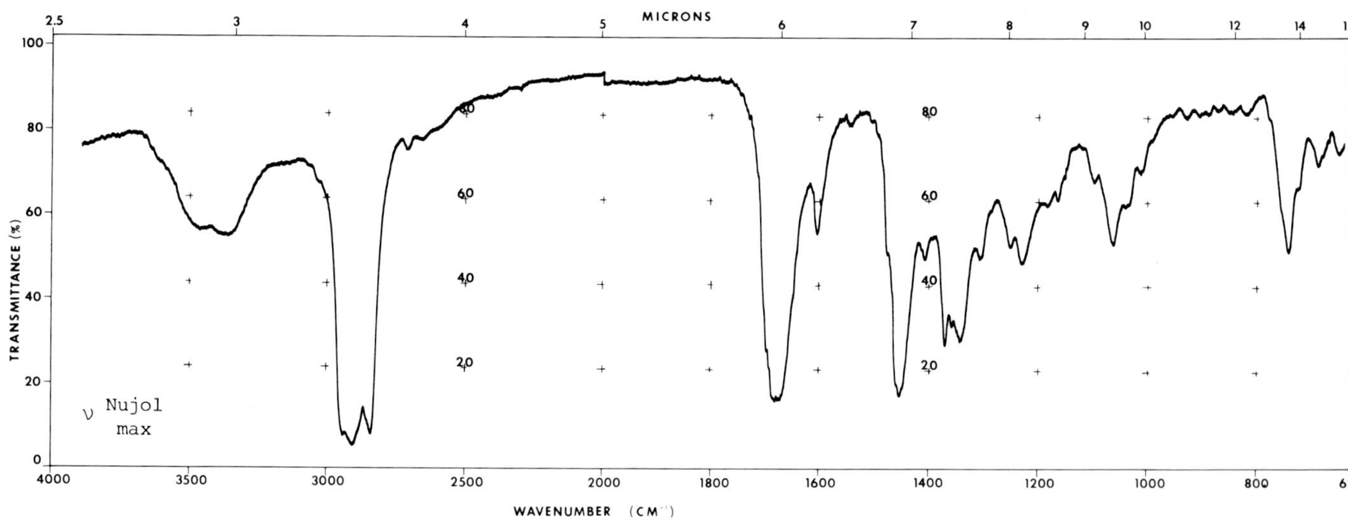
MICRONS
TRANSMITTANCE (%)
WAVENUMBER (CM⁻¹)
ν Nujol max

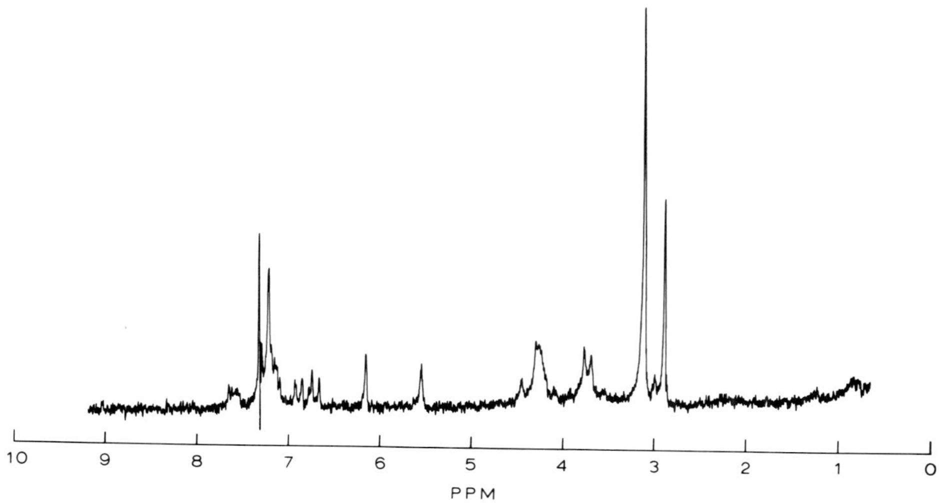

The 100 MHz ^1H NMR spectrum of chetomin in $CDCl_3$ solution.

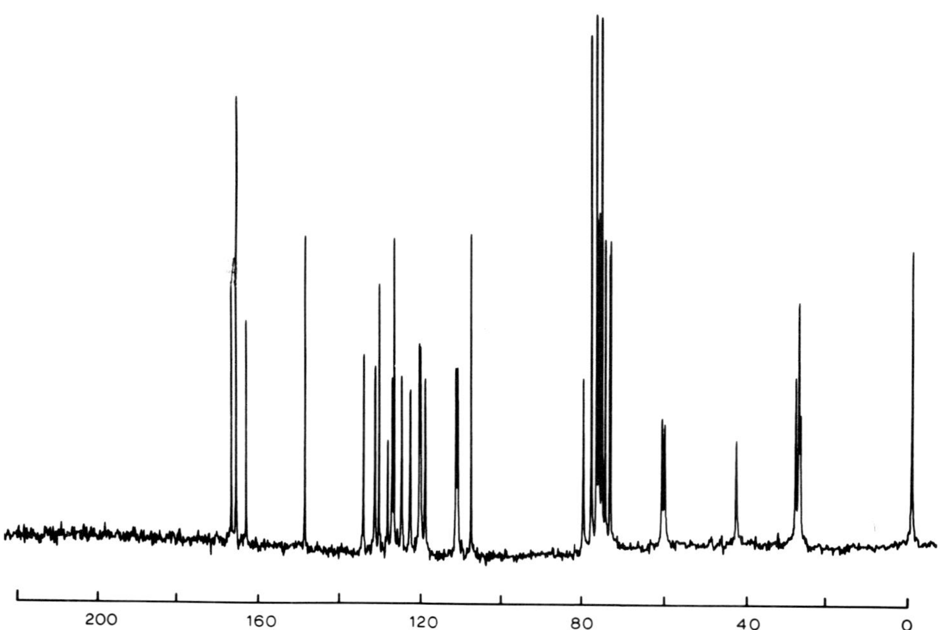

The 25 MHz ^{13}C NMR spectrum of chetomin in $CDCl_3$ solution.

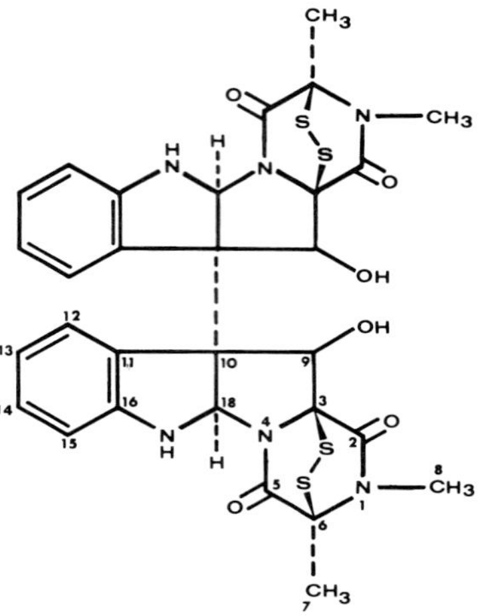

Common name	Verticillin A
Molecular weight	696.0938
Molecular formula	$C_{30}H_{28}O_6N_6S_4$
General characteristics[25,34]	Pale yellow plates from chloroform (chloroform solvate), mp 199°–213°C (dec.) Pale yellow needles from pyridine (pyridine solvate), mp 202°–217°C (dec.) Pale yellow amorphous powder from tetrahydrofuran, mp 203°–214°C Acetate, yellow powder from chloroform–methanol mp 220°–243°C (dec.) $[\alpha]_D$ +727.5°
UV data[25,34]	$\lambda_{max}^{dioxane}$ nm (ϵ): 306(6,300)
IR data[25,34]	ν_{max}^{Nujol} cm^{-1}: 3420, 3335, 1703, 1694, 1675, 1608, 1594, 1350, 1300, 1246, 1202, 1141, 1092, 1064, 982, 753, and 745
Fungal source[25,34]	A species of *Verticillium,* an imperfect fungus isolated from the basocarp of *Coltricia cinnamomea* (*Polystictus cinnamomeus*)
Toxicity data[25,34]	It is cytotoxic to HeLa cells, ED_{50} 0.2 μg/ml; showed antitumor effects against Ehrlich asites tumor; showed antiviral activity against polio-virus (Type 1). The acute toxicity (LD_{50}) in mice was 7.6 mg/kg by intraperitoneal administration (delayed toxicity).

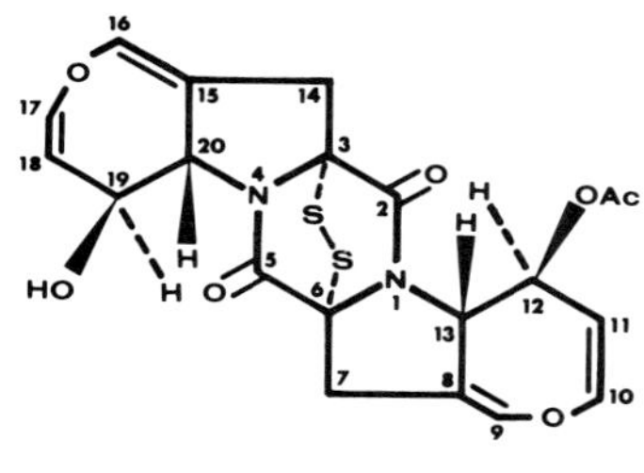

Common name	Aranotin
Molecular weight	462.0496
Molecular formula	$C_{20}H_{18}O_7N_2S_2$
General characteristics[42–44,60]	Crystals, mp 198°–200°C (dec.)
Fungal source[42–44,60]	*Arachniotus aureus* (Eidam) Schroeter (NRRL 3205)

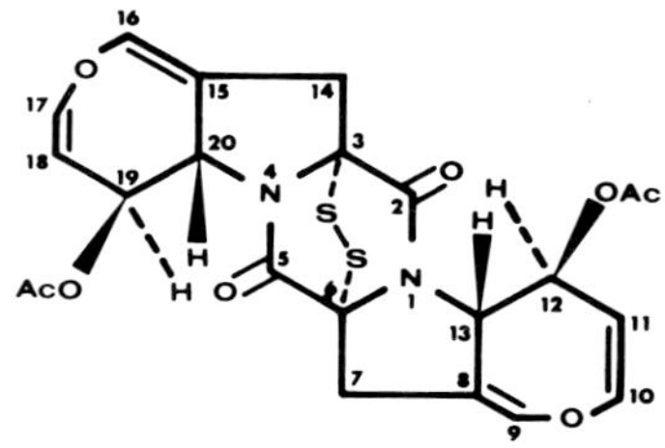

Common name — Acetylaranotin (LL-S88 α)

Molecular weight — 504.0658

Molecular formula — $C_{22}H_{20}O_8N_2S_2$

General characteristics[32,33,42–44] — Yellow tetragonal bipyrimidal crystals from ethanol, mp 215°–230°C (dec.)
$[\alpha]_D^{26}$ −549.7° (in chloroform)

UV data[32,33,42–44] — λ_{max}^{MeOH} end absorption with shoulders at nm: 270(1,800) and 225(10,200)

IR data[32,33,42–44] — ν_{max}^{KBr} cm^{-1}: 1735, 1715, 1365, 1340, 1225, 1140, 1040, and 1020

Fungal source[32,33,41,42,44] — *Aspergillus terreus* (NRRL 3319)
Arachniotus aureus (Eidam) Schroeter (NRRL 3205)

Toxicity data[32,33,42–44] — Biologically active in tissue culture against strains of rhino-Coxsackie, polio- and parainfluenza viruses. Completely blocks viral RNA synthesis without effect on cellular RNA synthesis.

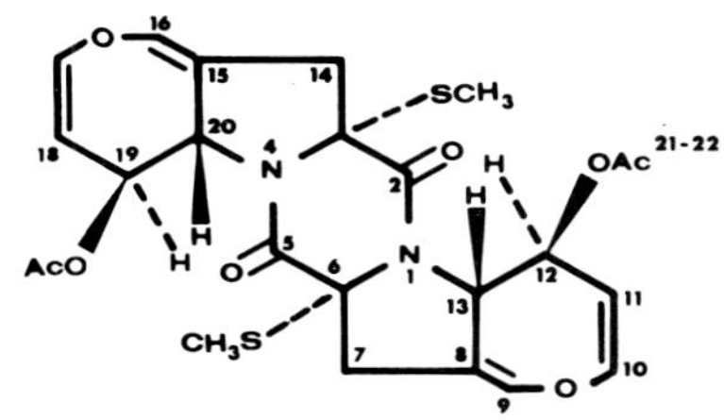

Common name	Bisdethiodi(methylthio)acetylaranotin (LL-S88 β)
Molecular weight	534.1120
Molecular formula	$C_{24}H_{26}O_8N_2S_2$
General characteristics[35,42–44]	White platelets from ethanol, mp 213°–236°C (dec.)
UV data[35,42–44]	λ_{max}^{MeoH} End absorption with shoulder at nm: 225(1,900) and 220(14,100)
Fungal source[35,42–44]	*Arachniotus aureus* (Eidam) Schroeter
Toxicity data[35,42–44]	No antiviral activity has been observed for this metabolite.
Other significant references	X-Ray structure[10,35]

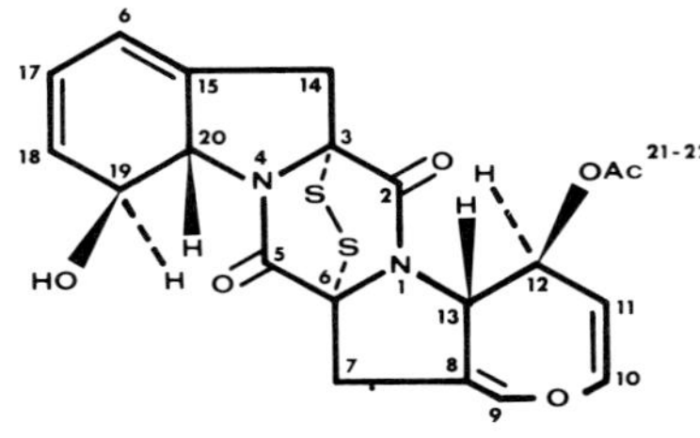

Common name	Apoaranotin
Molecular weight	446.0547
Molecular formula	$C_{20}H_{18}O_6N_2S_2$
General characteristics[43,60]	mp 200°–205°C; acetate mp 177°–180°C (dec.) $[\alpha]_D^{26}$ −492° (in chloroform)
UV data[43,60]	λ_{max}^{EtOH} nm (ϵ): 265(3,190)
CD data[43,60]	$\Delta\epsilon$ (in methanol) nm: 229(−49.0)
IR data[43,60]	ν_{max}^{Nujol} cm^{-1}: 3350, 1725, 1660, 1245, and 1145
Fungal source[43,60]	*Arachniotus aureus* (Eidam) Schroeter
Toxicity data[43,60]	Apoaranotin shows antiviral activity against a Coxsackie A21 virus infection in mice.

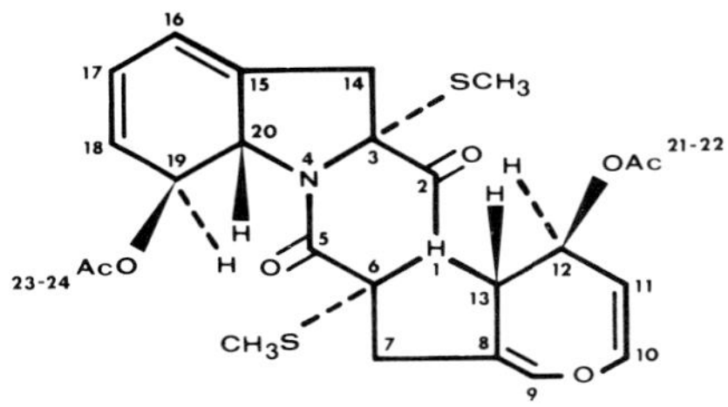

Common name	Bisdethiodi(methylthio)acetylapoaranotin
Molecular weight	518.1171
Molecular formula	$C_{24}H_{26}O_7N_2S_2$
General characteristics[43,44]	$[\alpha]_D^{26}$ $-175°$ (in chloroform)
UV data[43,44]	λ_{max}^{EtOH} nm (ϵ): 265(3,880)
CD data[43,44]	$\Delta\epsilon$ (in methanol) nm: 347(-0.72), 320(-0.42), 270($+5.3$), and 229(-49.0)
Toxicity data	Not reported
Fungal source[43,44]	*Arachniotus aureus* (Eidam) Schroeter

REFERENCES

1. Argoudelis, A. D., and Reusser, F., *J. Antibiot.* **24,** 383 (1971).
2. Argoudelis, A. D., and Reusser, F., *J. Antibiot.* **25,** 171 (1972).
3. Atherton, L. G., Brewer, D., and Taylor, A., *in* "Mycotoxins" (I. F. H. Purchase, ed.), p. 29. Elsevier, Amsterdam, 1974.
4. Beecham, A. F., Fridrichsons, J., and Mathieson, A. McL., *Tetrahedron Lett.* p. 3131 (1966).
5. Bell, M. R., Johnson, J. R., Wildi, B. S., and Woodward, R. B., *J. Am. Chem. Soc.* **80,** 1001 (1958).
6. Brewer, D., Duncan, J. M., Jerram, W. A., Leach, C. K., Safe, S., Taylor, A., Vining, L. C., Archibald, R. McG., Stevenson, R. G., Mirocha, C. J., and Christensen, C. M., *Can. J. Microbiol.* **18,** 1129 (1972).
7. Brewer, D., McInnes, A. G., Smith, D. G., Taylor, A., Walter, J. A., Loosli, H. R., and Kis, Z. L., *J. Chem. Soc., Perkin Trans. 1* p. 1248 (1978).
8. Bruce, W. F., Dutcher, J. D., Johnson, J. R., and Miller, L. L., *J. Am. Chem. Soc.* **66,** 614 (1944).
9. Bu'Lock, J. D., and Leich, C., *J. Chem. Soc., Chem. Commun.* p. 628 (1975).
10. Cosulich, D. B., Nelson, N. R., and van der Hende, J. H., *J. Am. Chem. Soc.* **90,** 6519 (1968).
11. Done, J., Mortimer, P. H., and Taylor, A., *Res. Vet. Sci.* **3,** 161 (1962).
12. Francis, E., Rahman, R., Safe, S., and Taylor, A., *J. Chem. Soc., Perkin Trans. 1* p. 470 (1972).
13. Fridrichsons, J., and Mathieson, A. McL., *Tetrahedron Lett.* p. 1265 (1962).
14. Fridrichsons, J., and Mathieson, A. McL., *Acta Crystallogr.* **23,** 439 (1967).
15. Fukuyama, T., and Kishi, Y., *J. Am. Chem. Soc.* **98,** 6723 (1976).
16. Hauser, D., Loosli, H. R., and Niklaus, P., *Helv. Chim. Acta* **55,** 2182 (1972).
17. Hauser, D., Weber, H. P., and Sigg, H. P., *Helv. Chim. Acta* **53,** 1061 (1970).
18. Herrmann, H., Hodges, R., and Taylor, A., *J. Chem. Soc.* p. 4315 (1964).
19. Hodges, R., Ronaldson, J. W., Shannon, J. S., Taylor, A., and White, E. P., *J. Chem. Soc.* p. 26 (1964).
20. Hodges, R., and Shannon, J. S., *Aust. J. Chem.* **19,** 1059 (1966).
21. Jamieson, W. D., Rahman, R., and Taylor, A., *J. Chem. Soc. C* p. 1564 (1969).
22. Johnson, J. R., and Buchanan, J. B., *J. Am. Chem. Soc.* **75,** 2103 (1953).
23. Johnson, J. R., Bruce, W. F., and Dutcher, J. D., *J. Am. Chem. Soc.* **65,** 2005 (1943).
24. Johnson, J. R., McCrone, W. C., and Bruce, W. F., *J. Am. Chem. Soc.* **66,** 501 (1944).
25. Katagiri, K., Sato, K., Hayakawa, S., Matsushima, T., and Minato, H., *J. Antibiot., Ser. A* **23,** 420 (1970).
26. Kishi, Y., Fukuyama, T., and Nakatsuka, S., *J. Am. Chem. Soc.* **95,** 6492 (1973).
27. Kishi, Y., Nakatsuka, S., Fukuyama, T., and Havel, M., *J. Am. Chem. Soc.* **95,** 6493 (1973).
28. Leigh, C., and Taylor, A., *Adv. Chem. Ser.* **149,** 228 (1976).
29. Lowe, G., Taylor, A., and Vining, L. C., *J. Chem. Soc.* p. 1799 (1966).
30. Mathieson, A. McL., *Tetrahedron Lett.* **2,** 81 (1963).
31. McInnes, A. G., Taylor, A., and Walter, J. A., *J. Am. Chem. Soc.* **98,** 6741 (1976).
32. Miller, P. A., Milstrey, K. P., and Trown, P. W., *Science* **159,** 431 (1968).
33. Miller, P. A., Trown, P. W., Fulmor, W., Morton, G. O., and Karliner, J., *Biochem. Biophys. Res. Commun.* **33,** 219 (1968).
34. Minato, H., Matsumoto, M., and Katayama, T., *Chem. Commun.* p. 44 (1971).
35. Moncrief, J. W., *J. Am. Chem. Soc.* **90,** 6517 (1968).
36. Mortimer, P. H., *N. Z. J. Agric. Res.* **13,** 437 (1970).
37. Mortimer, P. H., and Taylor, A., *Res. Vet. Sci.* **3,** 147 (1962).
38. Mortimer, P. H., *Res. Vet. Sci.* **3,** 269 (1962).
39. Mortimer, P. H., *Res. Vet. Sci.* **4,** 166 (1963).

40. Mortimer, P. H., Taylor, A., and Shorland, F. B., *Nature* (*London*) **194,** 559 (1962).
41. Nakasuka, S., Fukuyama, T., and Kishi, Y., *Tetrahedron Lett.* p. 1549 (1974).
42. Narajan, R., Huckstep, L. L., Lively, D. H., DeLong, D. C., Marsh, M. M., and Neuss, N., *J. Am. Chem. Soc.* **90,** 2980 (1968).
43. Neuss, N., Boeck, L. D., Brannon, D. R., Cline, J. C., DeLong, D. C., Gorman, M., Huckstep, L. L., Lively, D. H., Mabe, J., Marsh, M. M., Molloy, B. B., Nagarajan, R., Nelson, J. D., and Stark, M. M., *Antimicrob. Agents Chemother.* p. 213 (1968).
44. Neuss, N., Nagarajan, R., Molloy, B. B., and Huckstep, L. L., *Tetrahedron Lett.* p. 4467 (1968).
45. Peters, J. A., *Nature* (*London*) **200,** 286 (1963).
46. Peters, J. A., and Smith, L. M., *Biochem. J.* **92,** 379 (1964).
47. Przybulska, M., and Gopalakrishna, E. M., *Acta Crystallogr., Sect. B* **30,** 597 (1974).
48. Rahman, R., Safe, S., and Taylor, A., *J. Chem. Soc. C* p. 1665 (1969).
48a. Rahman, R., Safe, S., and Taylor, A., *J. Chem. Soc., Perkin Trans. 1* p. 1476 (1978).
49. Ronaldson, J. W., *Aust. J. Chem.* **29,** 2307 (1976).
50. Ronaldson, J. W., Taylor, A., White, E. P., and Abraham, R. J., *J. Chem. Soc.* p. 3172 (1963).
51. Safe, S., and Taylor, A., *J. Chem. Soc. C* p. 432 (1970).
52. Safe, S., and Taylor, A., *J. Chem. Soc., Perkin Trans. 1* p. 472 (1972).
53. Shannon, J. S., *Tetrahedron Lett.* p. 801 (1963).
54. Stillwell, M. A., Magasi, L. P., and Strunz, G. M., *Can. J. Microbiol.* **20,** 759 (1974).
55. Strunz, G. M., Heissner, C. J., Kakushima, M., and Stillwell, M. A., *Can. J. Chem.* **52,** 366 (1974).
56. Strunz, G. M., and Kakushima, M., *Experientia* **30,** 719 (1974).
57. Strunz, G. M., Kakushima, M., and Stillwell, M. A., *Can. J. Chem.* **53,** 295 (1975).
58. Strunz, G. M., Kakushima, M., Stillwell, M. A., and Heissner, C. J., *J. Chem. Soc., Perkin Trans. 1* p. 2600 (1973).
59. Synge, R. L. M., and White, E. P., *Chem. Ind.* (*London*) p. 1456 (1959).
60. Taylor, A., *in* "Microbial Toxins" (S. Kadis, A. Ciegler, and S. J. Ajl, ed.), Vol. 7, p. 337. Academic Press, New York, 1971.
61. Thornton, R. H., and Percival, J. C., *Nature* (*London*) **183,** 63 (1959).
62. Trown, P. W., *Biochem. Biophys. Res. Commun.* **33,** 402 (1968).
63. Wilkinson, S., and Spilsbury, J. F., *Nature* (*London*) **66,** 619 (1965).

12

Alternaria Toxins

Although *Alternaria* spp. can be found on various agricultural commodities such as wheat, tobacco, corn, peanuts, sorghum, barley, oats, alfalfa, and grasses, their mycotoxin-producing potential has not been exhaustively studied.[5] Studies to date have produced the identity of several toxic metabolites belonging to at least three classes of chemicals. *Alternaria tenuis* is capable of producing the toxic tetramic acid, tenuazonic acid (included in a different section). The second class of chemicals is characterized by the dibenzo[*a*]pyrone moiety. Four members of this class have been isolated and identified. These are alternariol, alternariol monomethyl ether, altenuene, and altenuisol. The third structural class is represented by altertoxin I and altertoxin II, structurally unknown toxic metabolites of *Alternaria* spp.

The *Alternaria* metabolites are acutely toxic, cytotoxic, and teratogenic. Both alternariol and alternariol monomethyl ether have been shown to be synergistic against bacteria.

Two reviews currently appear in the literature on the subject of *Alternaria* toxins.[5,12]

Alternaria toxins	Molecular weight	Molecular formula
Alternariol	258.0528	$C_{14}H_{10}O_5$
Altenuisol	274.0476	$C_{14}H_{10}O_6$
Alternariol monomethyl ether	272.0684	$C_{15}H_{12}O_5$
Altenuene	292.0946	$C_{15}H_{16}O_6$
Altenusin	290.0790	$C_{15}H_{14}O_6$
Dehydroaltenusin	288.0633	$C_{15}H_{12}O_6$
Altertoxin I	352.0946	$C_{20}H_{16}O_6$
Altertoxin II	350.0790	$C_{20}H_{14}O_6$

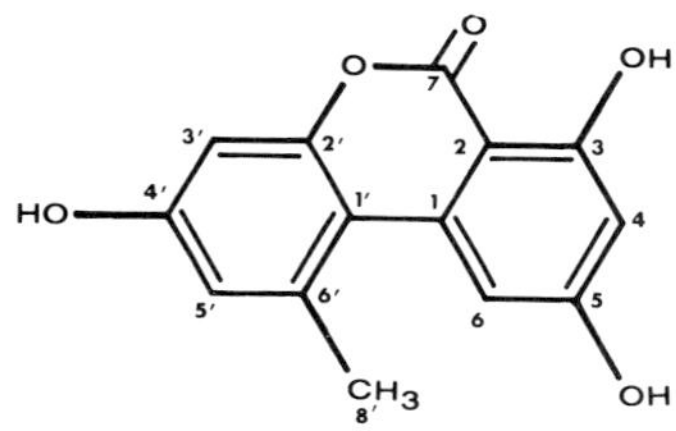

Common name	Alternariol (3,4′,5-Trihydroxy-6′-methyl-dibenzo[*a*]pyrone)
Molecular weight	258.0528
Molecular formula	$C_{14}H_{10}O_5$
General characteristics[2,12,18]	Crystals from ethanol, mp 350°C (dec.); sublimes without decomposition at 250°C in a high vacuum. $[\alpha]_D^{20}$ 0° (C = 0.8 in ethanol) Triacetate from ethanol, mp 167°–169°C
UV data[2]	λ_{max}^{EtOH} nm (ϵ): 218, 258(38,000), 302, and 330
Fungal source[2,12,18]	*Alternaria tenuis* (NRRL 5255; 5872; 5873) *A. dauci* (Kuhn) Groves and Skolko *A. cucumerina* Alternariol and alternariol monomethyl ether have been found in tobacco, grain sorghum, and pecans.[5,15,16]
Toxicity data[4,11]	Alternariol completely suppresses growth of *Staphylococcus aureus* at concentration of 1:40,000; *E. coli* at concentration of 1:20,000. The ID_{50} to HeLa cells was 6 μg/ml. In mice dosed IP (DMSO carrier) at 200 mg/kg, 3 of 10 mice died; mice were often sedated within a few minutes of dosing; the eyes of mice recovering were dull; there were occasional stomach spasms and periodic panting.
TLC data[10]	*Adsorbent:* silica gel G *Solvent:* A: benzene–ethanol, 95:5 v/v B: chloroform–methanol, 95:5 v/v R_f: A: 0.21 B: 0.19 *Detection:* blue fluorescent spot under UV light

GLC data[8]	*Solid support:* Gas Chrom Q *Liquid phase:* OV-17 (3%) *Column temp:* 100°–250°C at 8°/min rise *Injector temp:* 300°C *Detector temp:* 300°C *Retention time:* about 30 min
Other significant references	Synthesis[3,17]
^{13}C NMR data	C2, 97.6 s; C4, 100.9 d; C6, 104.2 d; C7, 165.0 s; C3′, 101.7 d; C5′, 117.3 d; C6′, 137.8 s; C8′, 25.4 q; 128.6 s; 138.2 s; 152.6 s; 158.0 s; 164.3 s; 165.0 s
^{1}H NMR data	H4, 6.30 (J = 2.0); H6, 7.14 (J = 2.8); H3′, 7.65; H5′, 6.58; H8′, 2.66

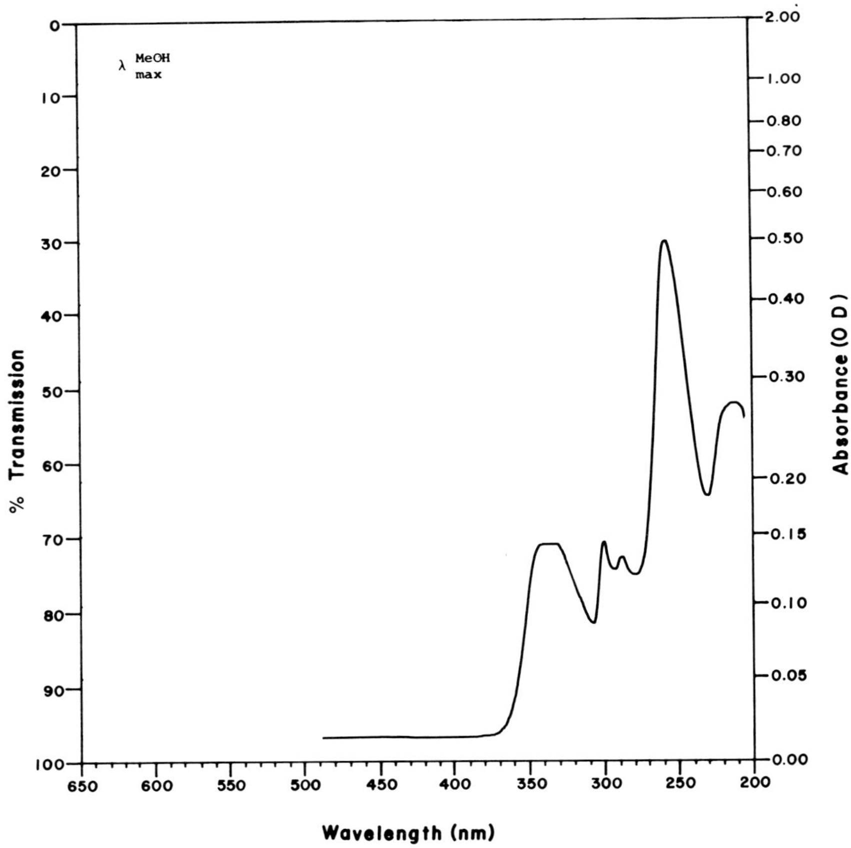

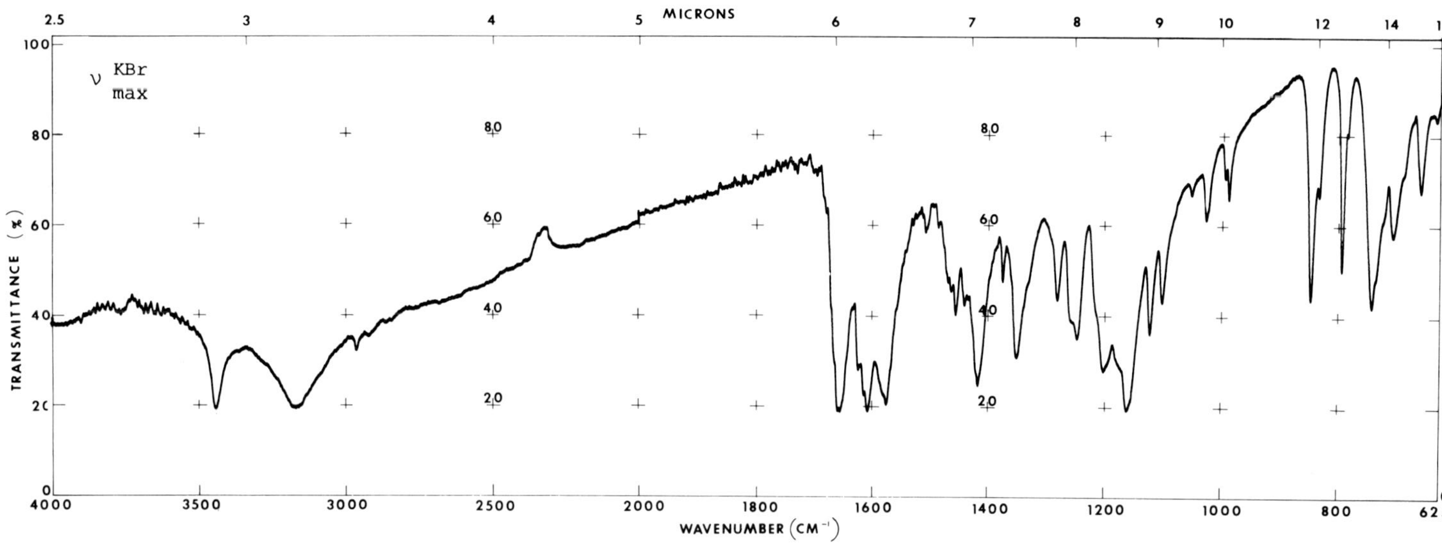
MICRONS
2.5
3
4
5
6
7
8
9
10
12
14
16
ν_{max}^{KBr}
TRANSMITTANCE (%)
100
80
60
40
20
0
4000
3500
3000
2500
2000
1800
1600
1400
1200
1000
800
625
WAVENUMBER (CM⁻¹)

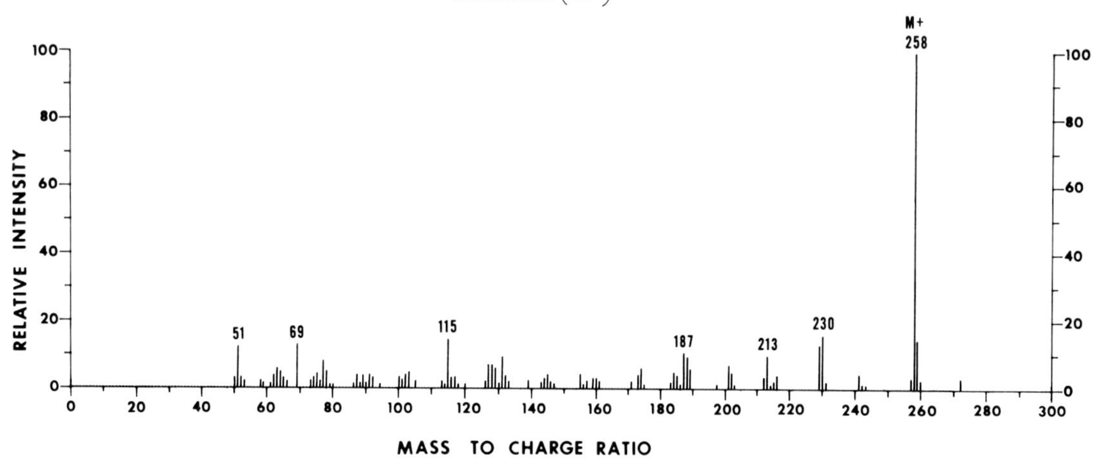
RELATIVE INTENSITY
100
80
60
40
20
0
51
69
115
187
213
230
M+
258
0
20
40
60
80
100
120
140
160
180
200
220
240
260
280
300
MASS TO CHARGE RATIO

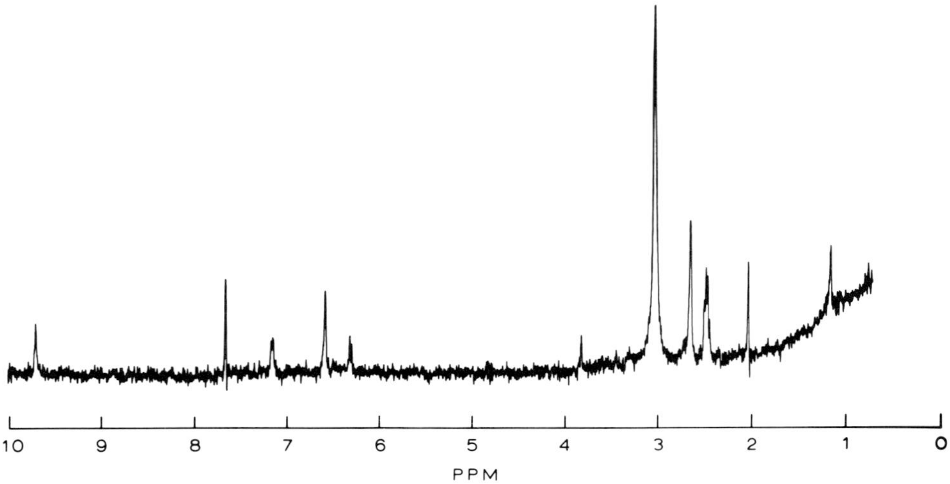

The 100 MHz ^{1}H NMR spectrum of alternariol in $CDCl_3$/DMSO solution.

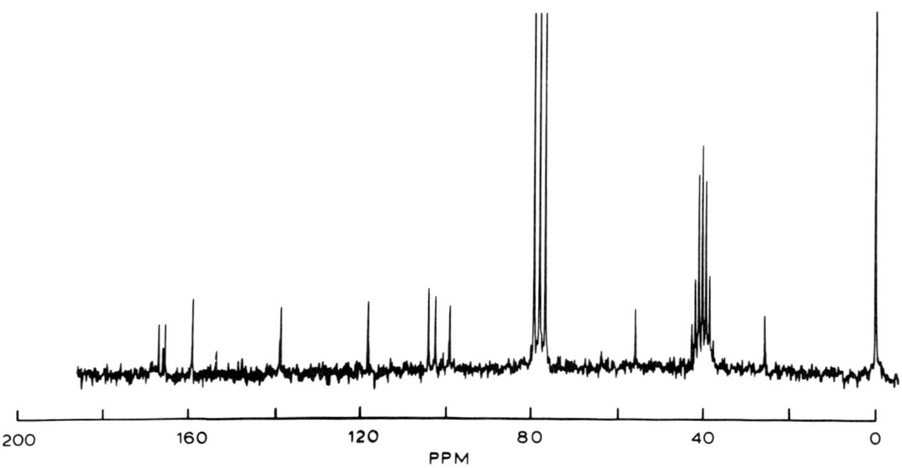

The 25 MHz ^{13}C NMR spectrum of alternariol in $CDCl_3$/DMSO solution.

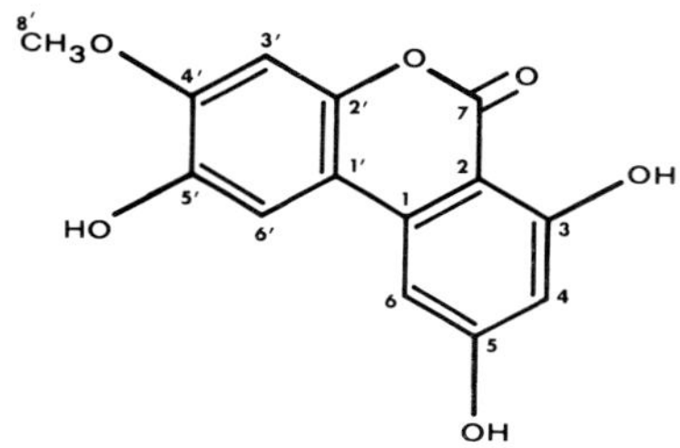

Common name	Altenuisol (Altertenuol) (3,5,5′-Trihydroxy-4′-methoxydibenzo[*a*]pyrone)
Molecular weight	274.0476
Molecular formula	$C_{14}H_{10}O_6$
General characteristics[7]	Crystallizes from hot acetic acid as slender needles, mp 277°–282°C Soluble in ethanol, acetone, ethyl acetate; sparingly soluble in chloroform, diethyl ether, insoluble in benzene and hexane. Triacetate, mp 210°–213°C
UV data[7]	λ_{max}^{EtOH} nm (ϵ): 216(9,000), 256(11,000), and 278(sh)(3,500) $\lambda_{max}^{EtOH+NaOH}$ nm (ϵ): 208(27,400) and 264(9,400)
Fungal source[7]	*Alternaria tenuis* (NRRL 5872)
Toxicity data[7]	The ID_{50} of altenuisol to HeLa cells was 8 μg/ml; it inhibited the bacterium *Bacillus mycoides* at 200-5 μg per assay disc on agar plates.
TLC data[7]	*Adsorbent:* silica gel G *Solvent:* 20% tetrahydrofuran–benzene R_f: 0.33 (alternariol = 0.54; alternariol monomethyl ether = 0.72; altenuene = 0.11) *Detection:* streaking blue fluorescent spot under UV light
^1H NMR data	H4, 6.47; H6, 6.92; H3′, 6.70; H6′, 7.38; H8′, 3.87; 3-OH, 11.84; 5,5′-OH, 11.62

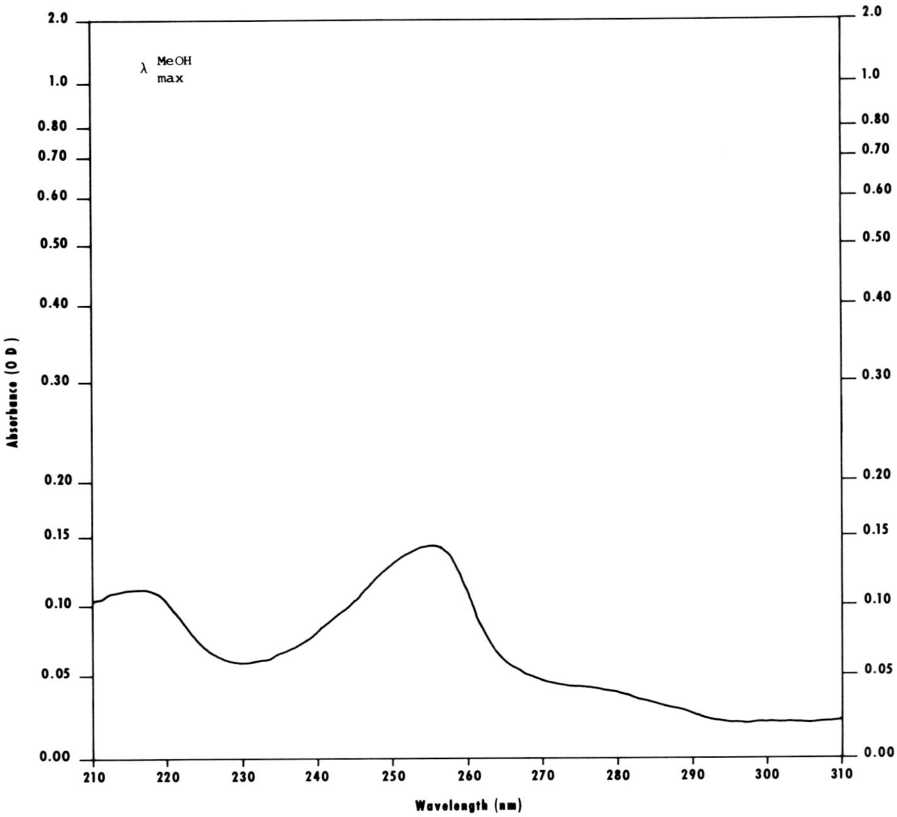
λ MeOH max
2.0
1.0
0.80
0.70
0.60
0.50
0.40
0.30
0.20
0.15
0.10
0.05
0.00
Absorbance (O D)
210
220
230
240
250
260
270
280
290
300
310
Wavelength (nm)

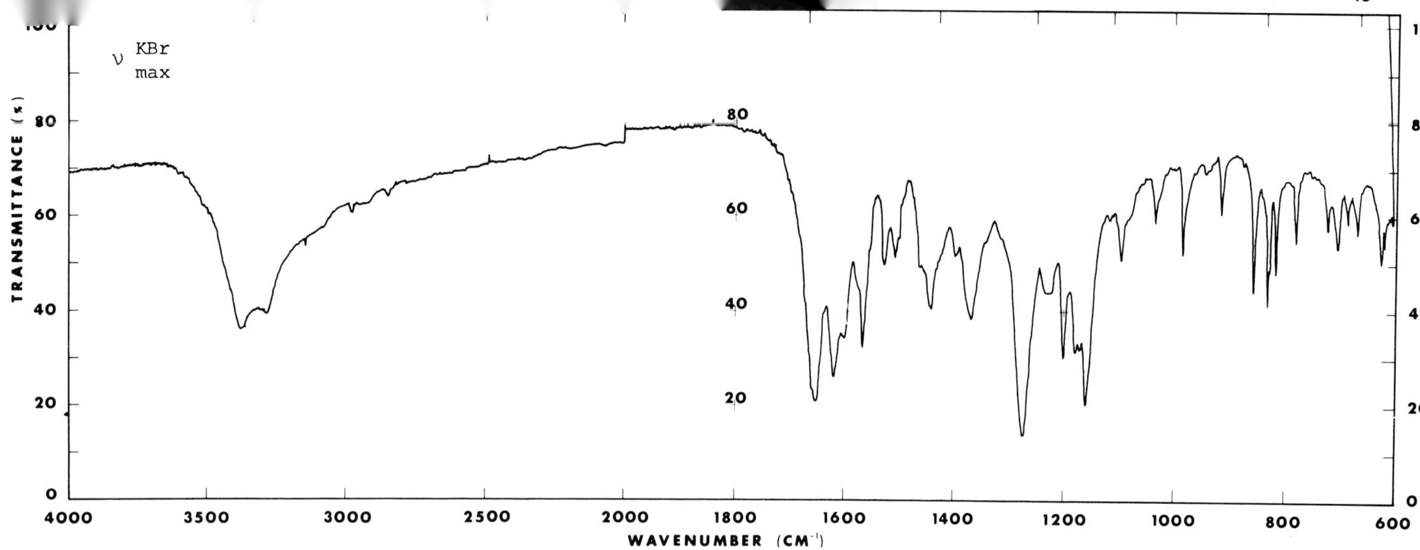
ν KBr max
TRANSMITTANCE (%)
WAVENUMBER (CM⁻¹)
4000
3500
3000
2500
2000
1800
1600
1400
1200
1000
800
600
0
20
40
60
80
100

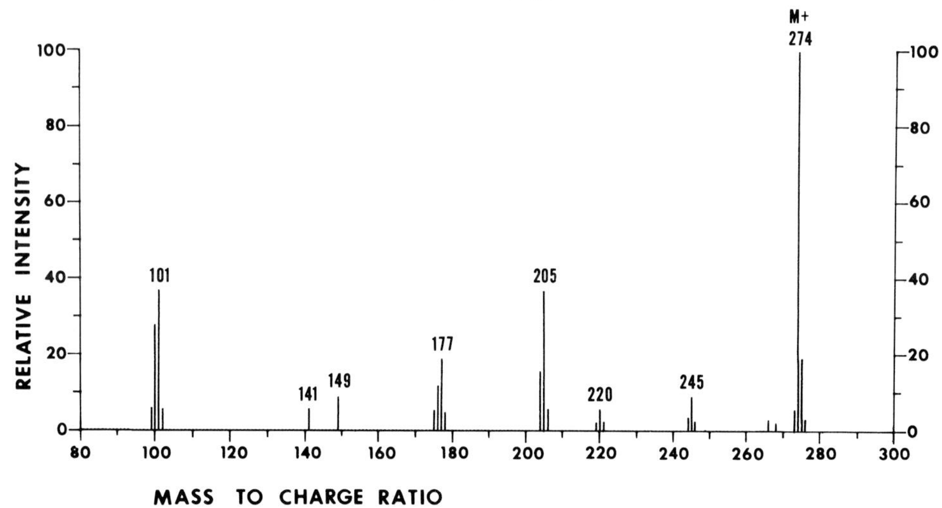
M+
274
101
141
149
177
205
220
245
RELATIVE INTENSITY
MASS TO CHARGE RATIO
80
100
120
140
160
180
200
220
240
260
280
300

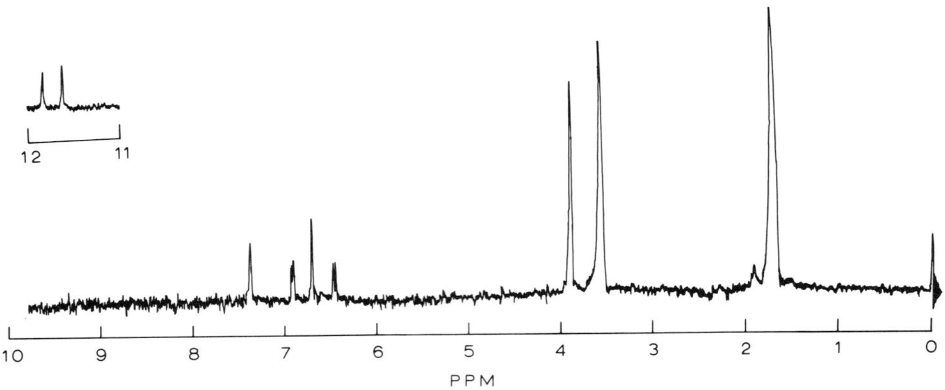

The 100 MHz ^{1}H NMR spectrum of altenuisol in THF-d_8 solution.

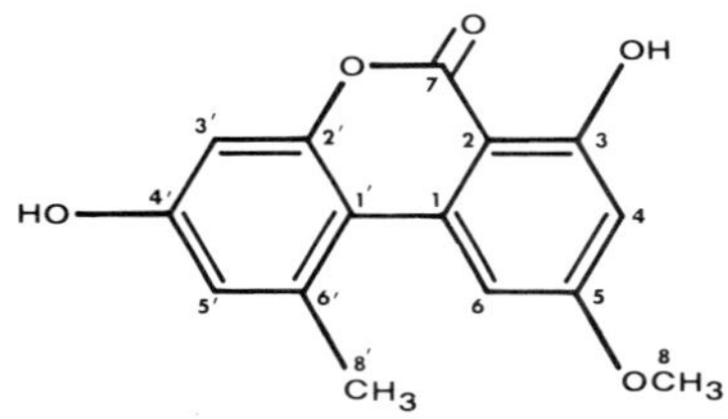

Common name	Alternariol monomethyl ether (3,4′-Dihydroxy-5-methyoxy-6′-methyldibenzo[*a*]pyrone)
Molecular weight	272.0684
Molecular formula	$C_{15}H_{12}O_5$
General characteristics[8,12]	Crystals from ethanol, mp 267°C (dec.) Sublimes without decomposition at 180°–200°C in high vacuum. $[\alpha]_D^{20}$ 0° (C = 0.8 in ethanol)
UV data[16]	λ_{max} nm: 335-342 (broad), 301, 290, 257, and 230 (ϵ_{max} not reported)
Fungal source[2,12,18]	*Alternaria tenuis* (NRRL 5831) *A. dauci* (Kuhn) Groves and Skolko *A. cucumerina* Alternariol monomethyl ether and alternariol have been found in tobacco, grain sorghum, and pecans.[5,15,16]
Toxicity data[4,11]	Alternariol monomethyl ether showed cytotoxic activity to bacteria and mammalian cells. Inhibition to *Bacillus mycoides* was observed at 500 μg/disc; the ID_{50} to HeLa cells was 8–14 μg/ml. In mice dosed IP (DMSO carrier) at 400 mg/kg, 1 of 10 mice died; mice were often sedated within a few minutes of dosing; the eyes of the mice were dull, there were occasional stomach spasms and periodic panting.
TLC data[10]	*Adsorbent:* silica gel G *Solvent:* A: benzene–ethanol, 95:5 v/v B: chloroform–methanol, 95:5 v/v R_f: A: 0.39 B: 0.54 *Detection:* blue fluorescent spot under UV light

GLC data[8]	*Solid support:* Gas Chrom Q *Liquid phase:* OV-17 (3%) *Column temp:* 100°–250°C at 8° min rise *Detector temp:* 300°C *Injector temp:* 300°C *Retention time:* approximately 31 min
Other significant references	Synthesis[17]
^{13}C NMR data	C2, 98.7 s; C4, 98.5 d; C6, 103.4 d; C7, 165.0 s; C8, 55.4 q; C3′, 101.9 d; C5′, 117.5 d; C6′, 137.8 s; C8′, 25.4 q; 137.9 s; 146.7 s; 152.6 s; 158.2 s; 164.4 s; 166.0 s
^{1}H NMR data	H4, 6.29; H6, 7.12; H8, 3.84; H3′, 7.60; H5′, 6.60; H8′, 2.68

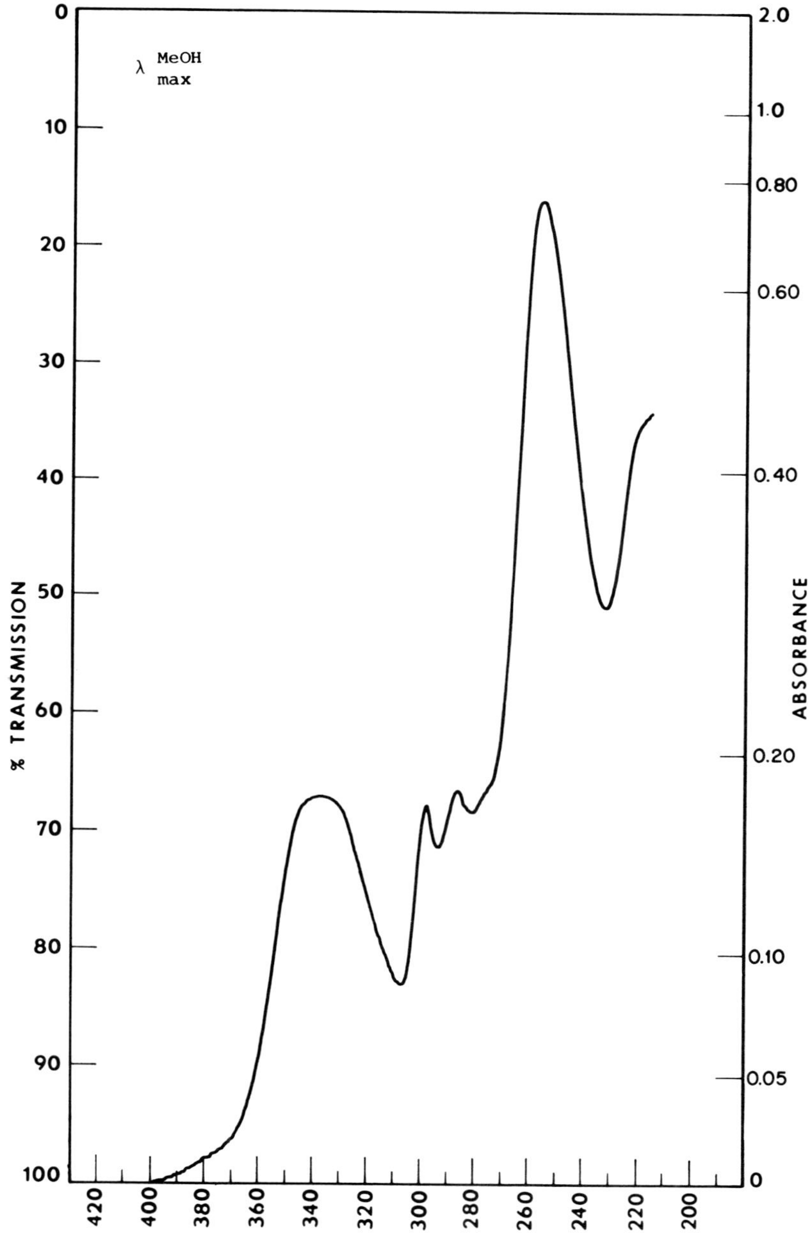
λ MeOH max
% TRANSMISSION
ABSORBANCE
0
10
20
30
40
50
60
70
80
90
100
2.0
1.0
0.80
0.60
0.40
0.20
0.10
0.05
0
420
400
380
360
340
320
300
280
260
240
220
200

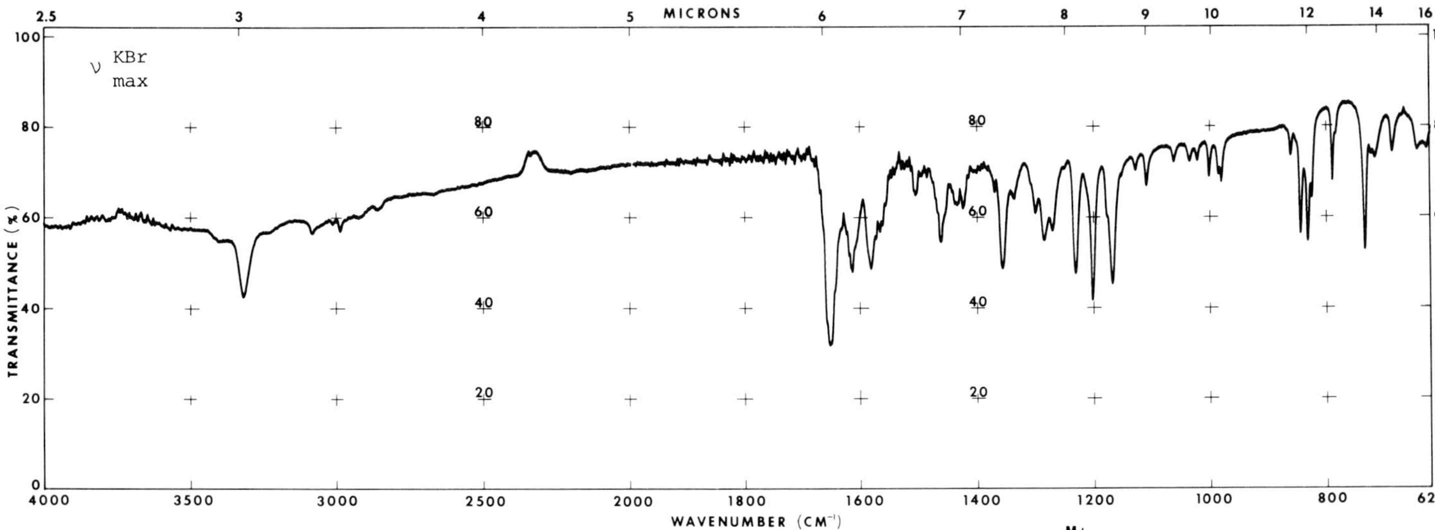

MICRONS
2.5 3 4 5 6 7 8 9 10 12 14 16
ν KBr max
TRANSMITTANCE (%)
100 80 60 40 20 0
4000 3500 3000 2500 2000 1800 1600 1400 1200 1000 800 625
WAVENUMBER (CM⁻¹)

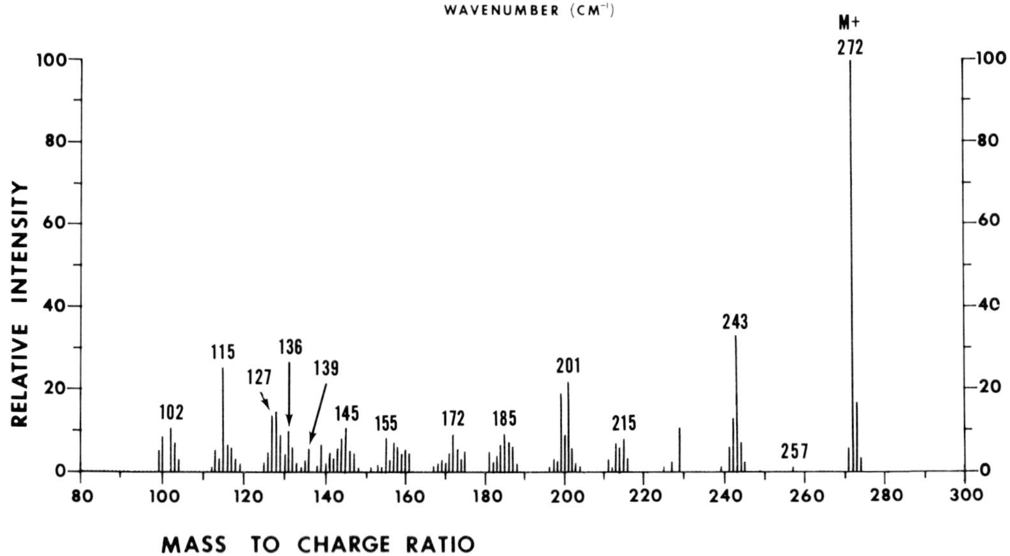

RELATIVE INTENSITY
100 80 60 40 20 0
M+
272
102
115
127
136
139
145
155
172
185
201
215
243
257
80 100 120 140 160 180 200 220 240 260 280 300
MASS TO CHARGE RATIO

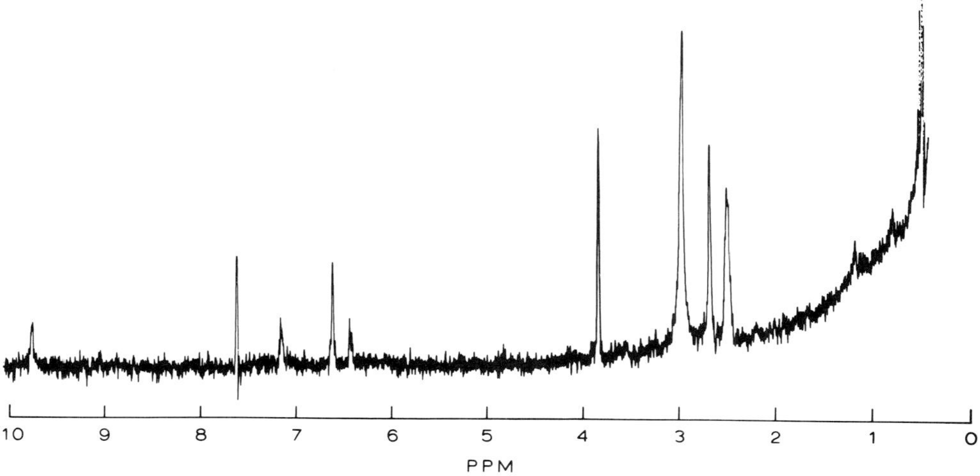

The 100 MHz ^1H NMR spectrum of alternariol monomethyl ether in $CDCl_3$/DMSO solution.

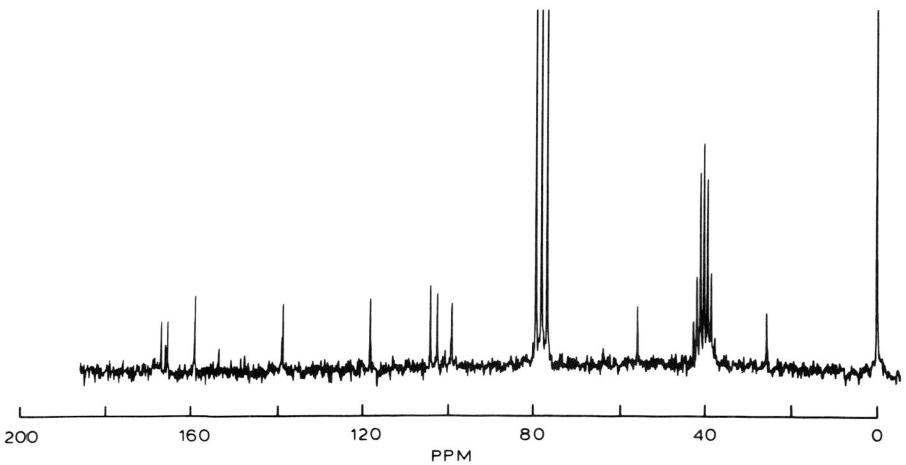

The 25 MHz ^{13}C NMR spectrum of alternariol monomethyl ether in $CDCl_3$/DMSO solution.

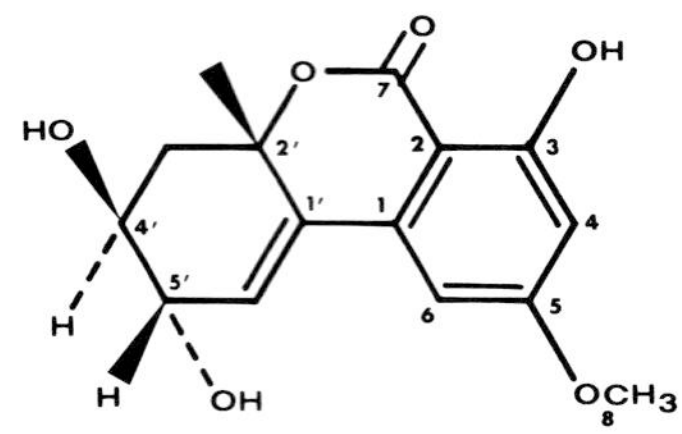

Common name	Altenuene (2′,3′,4′,5′-Tetrahydro-3,4′β, 5′α-trihydroxy-5-methoxy-2′β-methyldibenzo[*a*]-pyrone)
Molecular weight	292.0946
Molecular formula	$C_{15}H_{16}O_6$
General characteristics[9]	Crystallizes from acetone–hexane as colorless prisms, mp 190°–191°C Readily soluble in acetone, 1 *M* NaOH, methanol, pyridine, tetrahydrofuran; slightly soluble in diethyl ether, chloroform; insoluble in benzene, hexane, 1 *M* HCl, aqueous $NaHCO_3$ and water.
UV data[9]	λ_{max}^{EtOH} nm (ϵ): 319(6,600), 278(10,000), and 240(30,000) λ_{max}^{EtOH} nm (ϵ): 253(6,600), 257(14,000), 233(sh) (25,000), and 215(35,000)
Fungal source[10]	*Alternaria tenuis* (NRRL 5872; 5873)
Toxicity data[4,11]	Altenuene showed cytotoxic activity to five bacteria by the paper-disc agar-plate method and cytotoxicity to mammalian cells (toxicity to *Bacillus mycoides,* 125 μg/disc; toxicity to HeLa cells, ID_{25} 28 μg/ml). In mice dosed IP (DMSO carrier) at 50 mg/kg, 1 of 3 mice died.
TLC data[10]	*Adsorbent:* silica gel G *Solvent:* A: benzene–ethanol, 95:5 v/v B: chloroform–methanol, 95:5 v/v R_f: A: 0.14 B: 0.13 *Detection:* blue fluorescent spot under UV light

GLC data[8]	*Solid support:* Gas Chrom Q *Liquid phase:* OV-17 (3%) *Column temp:* 100°–250°C at 8°/min rise *Injector temp:* 300°C *Detector temp:* 300°C *Retention time:* about 24 min
Other significant references	X-Ray crystallography[6]
^{13}C NMR data	C4, 100.4 d; C6, 102.4 d; C7, 165.7 s; C8, 55.5 q; C2′, 81.0 s; C3′, 40.2 t; C4′, 69.6 d; C5′, 72.4 d; C6′, 132.5 d; C7′, 27.8 d; 130.9 s; 139.1 s; 143.6 s; 165.7 s; 168.7 s
^{1}H NMR data	H4, 6.15 (J = 2.0); H6, 6.53 (J = 2.0); H8, 3.87; H3′a, 1.78 (J = 10.0, 15.0); H3′b, 2.54 (J = 4.0, 15.0); H4′, 3.78; H5′, 4.12 (J = 2.0); H6′, 6.41 (J = 2.0, 6.5); H7′, 1.51

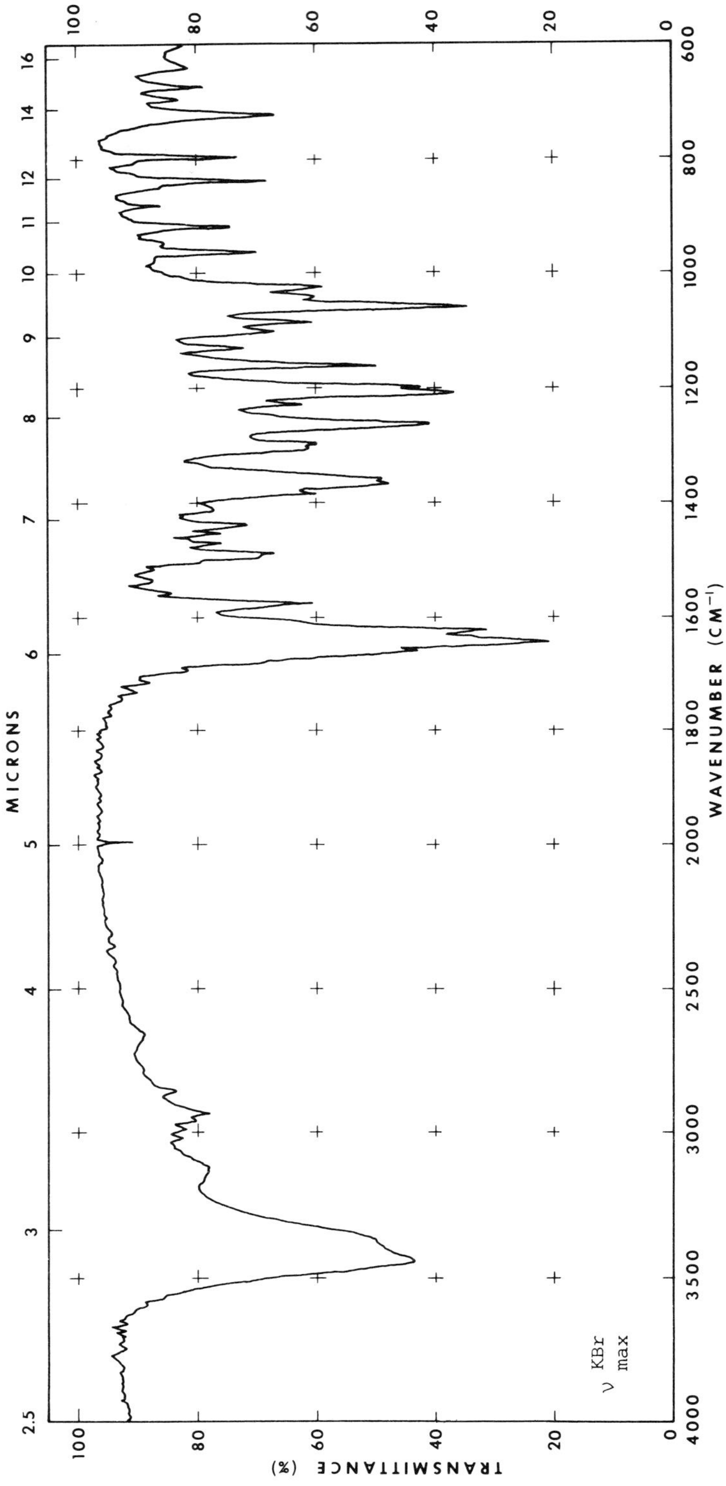
MICRONS
WAVENUMBER (CM⁻¹)
TRANSMITTANCE (%)
ν KBr max
2.5
3
4
5
6
7
8
9
10
11
12
14
16
4000
3500
3000
2500
2000
1800
1600
1400
1200
1000
800
600
100
80
60
40
20
0

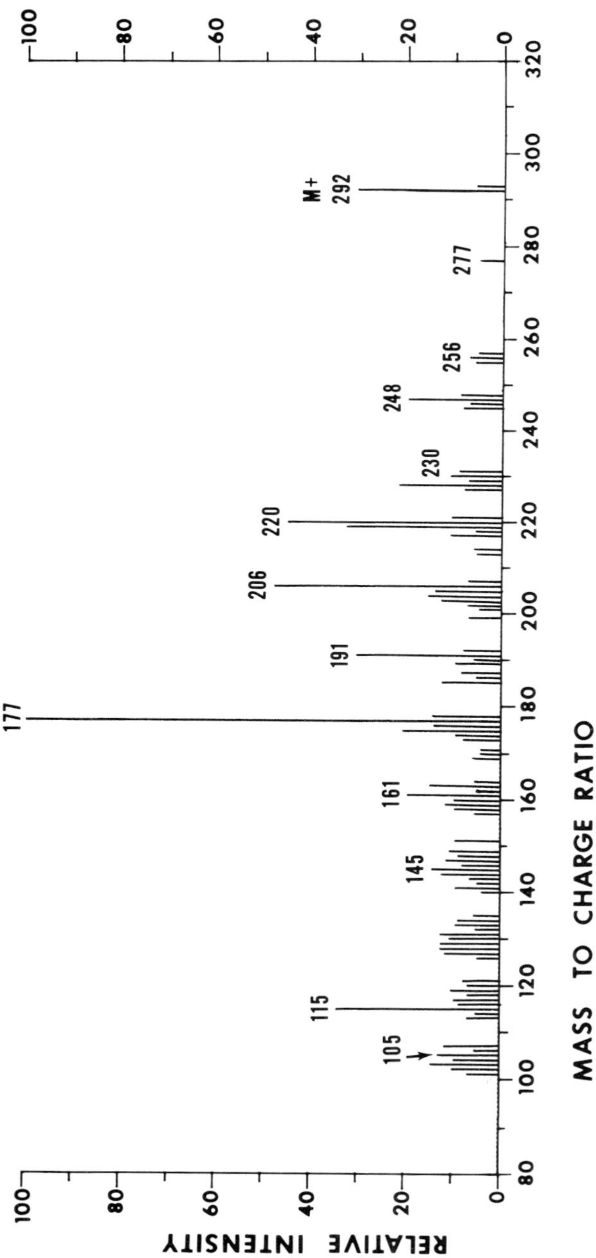
RELATIVE INTENSITY
MASS TO CHARGE RATIO
0
20
40
60
80
100
80
100
120
140
160
180
200
220
240
260
280
300
320
105
115
145
161
177
191
206
220
230
248
256
277
M+
292

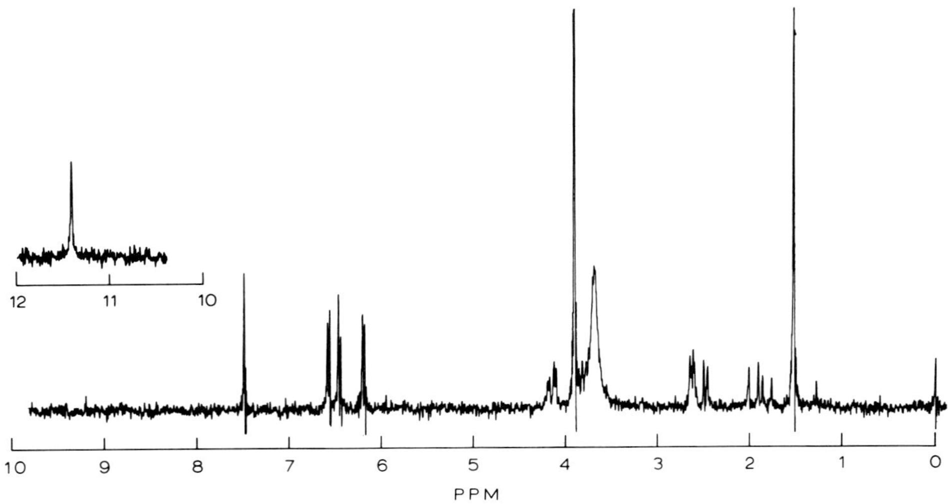

The 100 MHz ^{1}H NMR spectrum of altenuene in $CDCl_3$/DMSO solution.

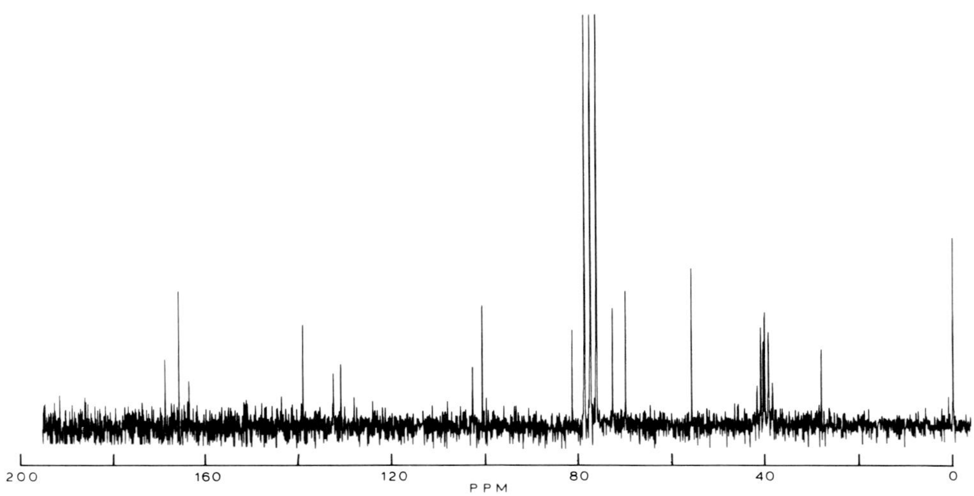

The 25 MHz ^{13}C NMR spectrum of altenuene in $CDCl_3$/DMSO solution.

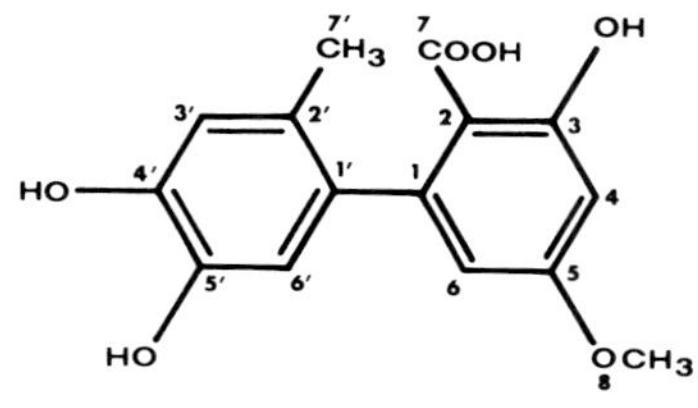

Common name	Altenusin
Molecular weight	290.0790
Molecular formula	$C_{15}H_{14}O_6$
General characteristics[1,14]	Colorless prisms from chloroform, mp 202°–203°C Undried crystals melt at 95°C, resolidify and melt at 202°–203°C Soluble in ethyl ether, ethanol, methanol, and aqueous $NaHCO_3$
UV data[1,14]	λ_{max}^{EtOH} nm: 217, 247(sh), and 290 (ϵ_{max} not reported)
IR data[1,14]	ν_{max}^{KCl} cm^{-1}: 3350, 1635, 1607, 1520, 1460, 1440, 1365, 1255, 1230, 1205, 1158, 1075, 1040, 980, 875, 860, and 800
Fungal source[1,14]	*Alternaria tenuis* Auct.
Toxicity data	Toxicity unknown
^1H NMR data[1]	(DMSO-d_6) H4, 6.15 (J = 2.5); H6, 6.46; H8, 3.77; H2′, 6.54; H5′, 6.46; H7′, 1.88; OH, 5.0–6.0; 8.60

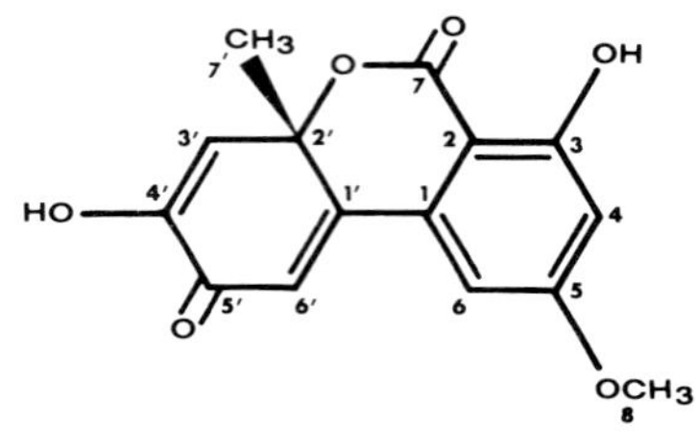

Common name	Dehydroaltenusin
Molecular weight	288.0633
Molecular formula	$C_{15}H_{12}O_6$
General characteristics[1,13]	Yellow plate crystals from acetone, mp 190°–193°C (dec.)
UV data[1,13]	λ_{max}^{EtOH} nm (ϵ): 217(33,800), 249(12,600), and 300(7,500)
IR data[1,13]	ν_{max}^{KCl} cm^{-1}: 3380, 1655, 1645(sh), 1625, 1580, 1275, 1230, 1210, 1165, 1080, 1032, 975, 856, 800, and 750
Fungal source[1,13]	*Alternaria tenuis*
Toxicity data	Toxicity unknown
^{1}*H NMR data*[1]	(acetone-d_6) H4, 6.68 (J = 2.5); H6, 7.05 (J = 2.5); H8, 4.00; H3′, 6.23; H6′, 6.85; H7′, 1.76; OH, 8.00; 11.32

Common name	Altertoxin I*
Molecular weight	352.0946
Molecular formula	$C_{20}H_{16}O_6$
General characteristics[4,11]	A yellow amorphous solid with an undefined melting point, ca 210°C (dec.) Bright yellow fluorescence under UV light.
Fungal source[4,11]	*Alternaria tenuis* *A. mali*
Toxicity data[4,11]	The ID_{50} of altertoxin I to HeLa cells was 20 μg/ml; toxicity to *Bacillus mycoides* was 250 μg/disc on agar plates. In mice dosed IP (DMSO carrier), an LD_{50} was between 100–200 mg/kg; at 200 mg/kg, an LD_{100} was observed.
^{13}C *NMR data*	C33.9 t; C35.0 t; C47.7 t; C51.9 d; C65.2 d; C68.5 s; C113.9 s; C116.1 d; C116.8 s; C118.2 d; C123.3 s; C124.5 s; C131.9 d; C132.0 d; C137.3 s; C139.9 s; C161.1 s; C161.5 s; C203.4 s; C205.5 s

* Structure not known.

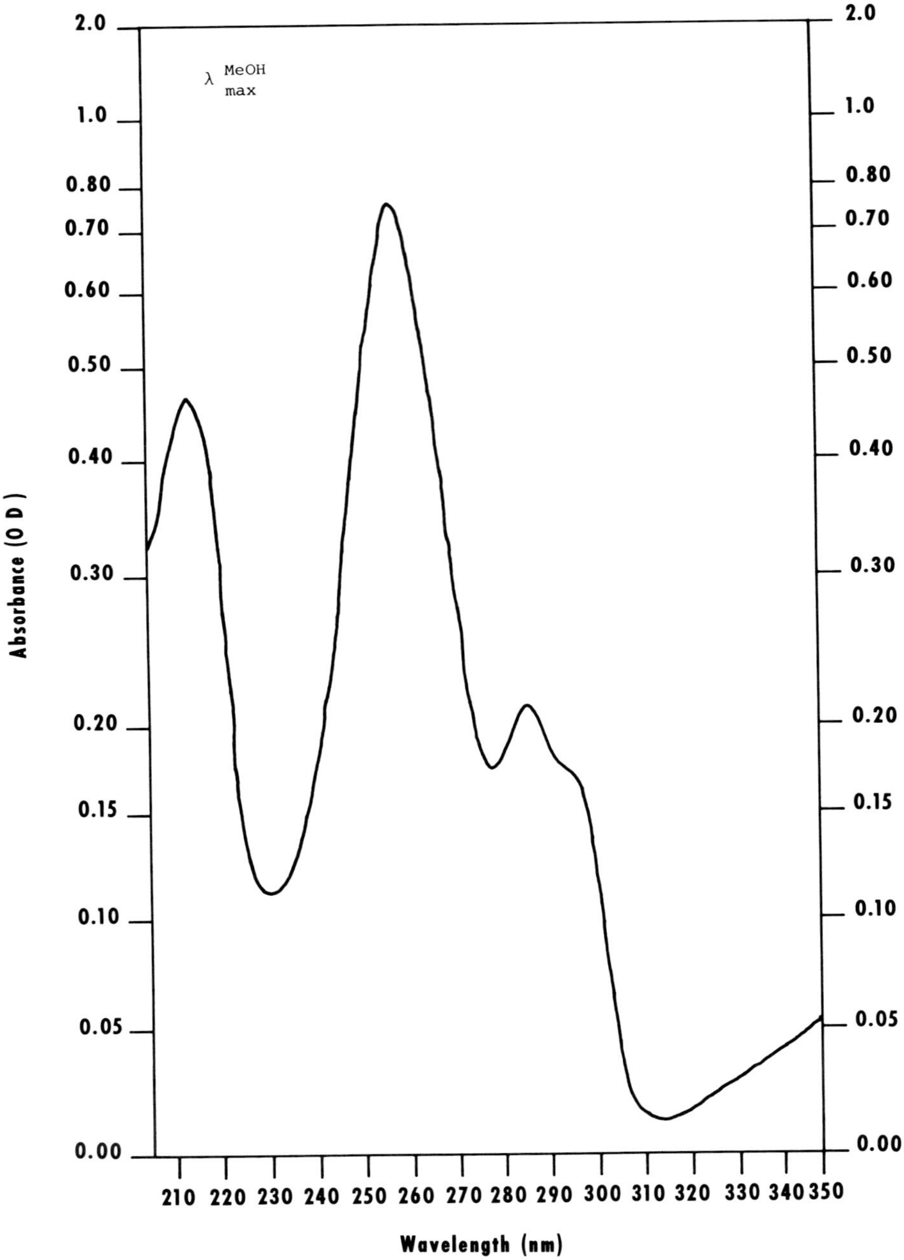
λ MeOH max
Absorbance (O D)
2.0
1.0
0.80
0.70
0.60
0.50
0.40
0.30
0.20
0.15
0.10
0.05
0.00
210 220 230 240 250 260 270 280 290 300 310 320 330 340 350
Wavelength (nm)

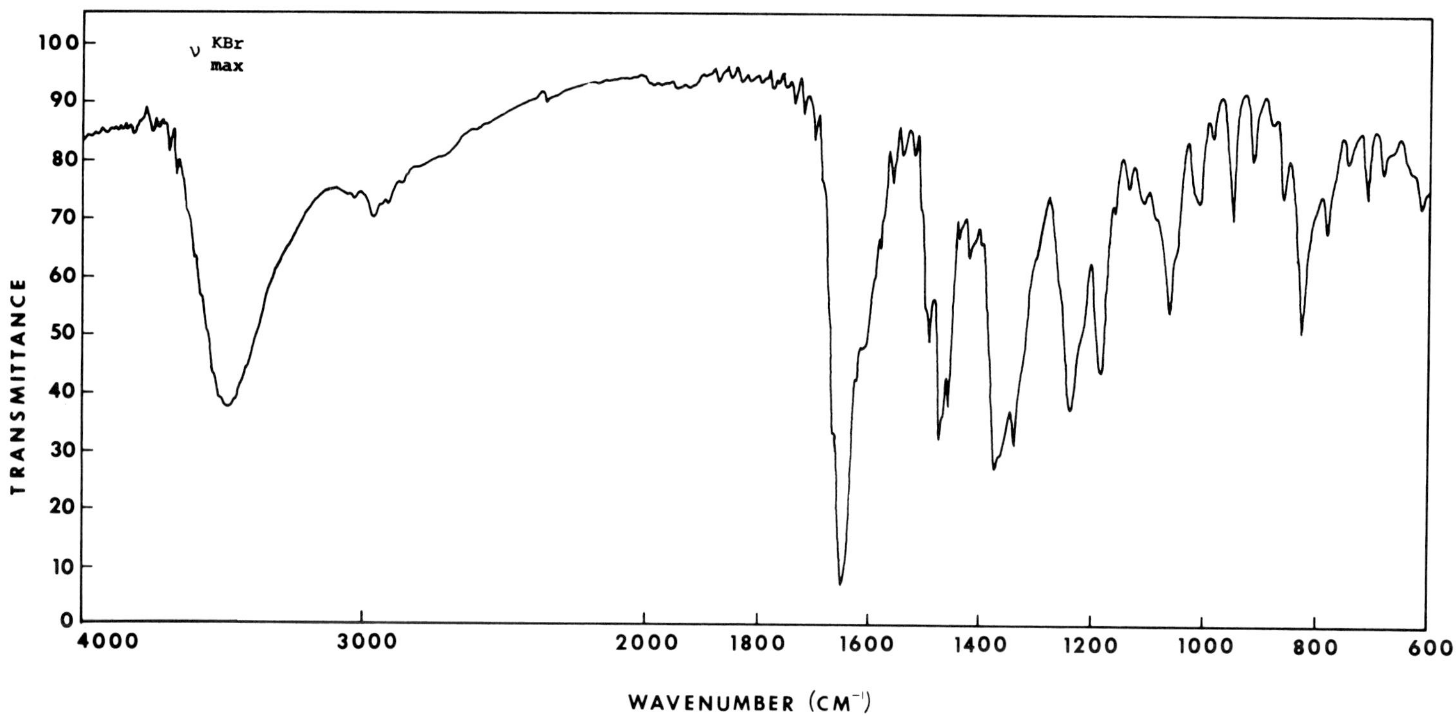

ν KBr max
100
90
80
70
60
50
40
30
20
10
0
TRANSMITTANCE
4000
3000
2000
1800
1600
1400
1200
1000
800
600
WAVENUMBER (CM⁻¹)

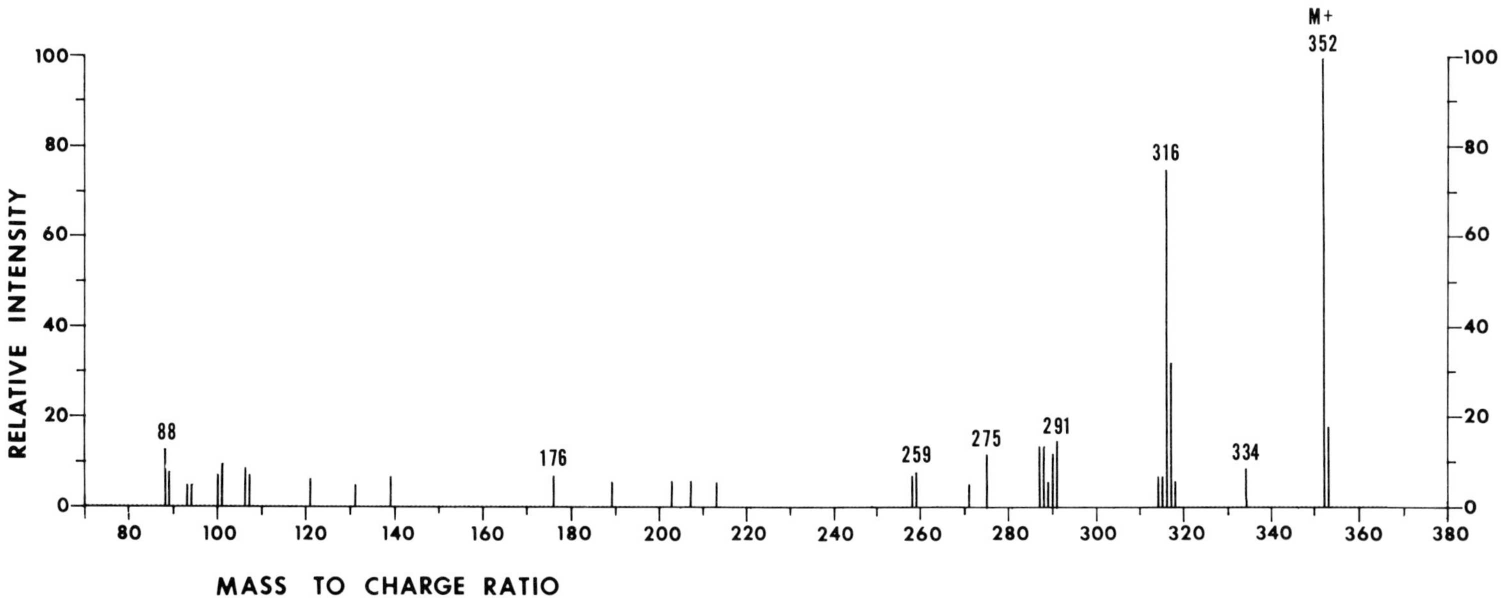
M+
352
316
334
291
275
259
176
88
RELATIVE INTENSITY
0
20
40
60
80
100
80
100
120
140
160
180
200
220
240
260
280
300
320
340
360
380
MASS TO CHARGE RATIO

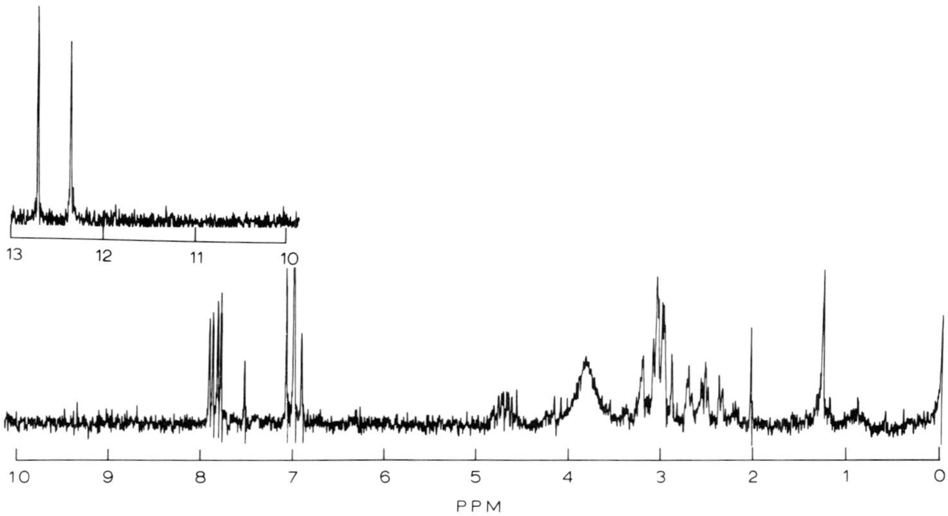

The 100 MHz ^{1}H NMR spectrum of altertoxin I in $CDCl_3$/DMSO solution.

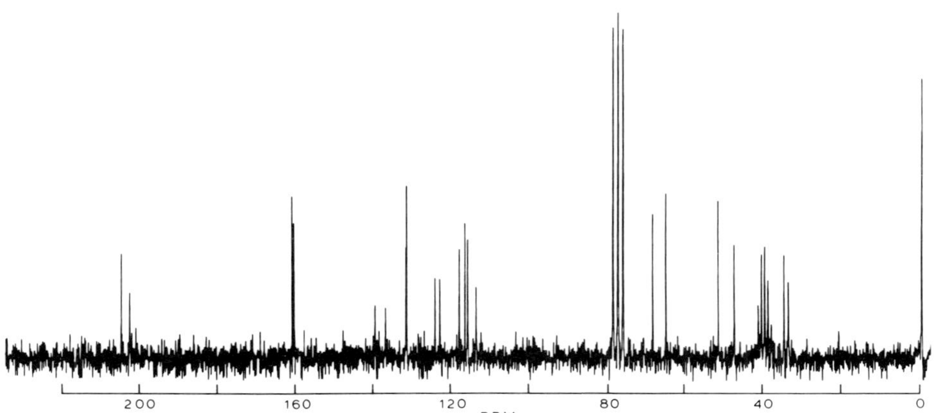

The 25 MHz ^{13}C NMR spectrum of altertoxin I in $CDCl_3$/DMSO solution.

Common name	Altertoxin II*
Molecular weight	350.0790
Molecular formula	$C_{20}H_{14}O_6$
General characteristics[4,11]	Orange crystalline solid, mp 185°–195°C (dec.) A dark quenching spot under UV light
Fungal source[4,11]	*Alternaria tenuis* *A. mali*
Toxicity data[4,11]	The ID_{50} of altertoxin II to HeLa cells was 0.5 μg/ml; inhibition of *Bacillus mycoides* was observed at 250 μg/disc on agar plates. In mice dosed IP (DMSO carrier), an LD_{50} was between 100–200 mg/kg; at 200 mg/kg, an LD_{100} was observed.

* Structure not known.

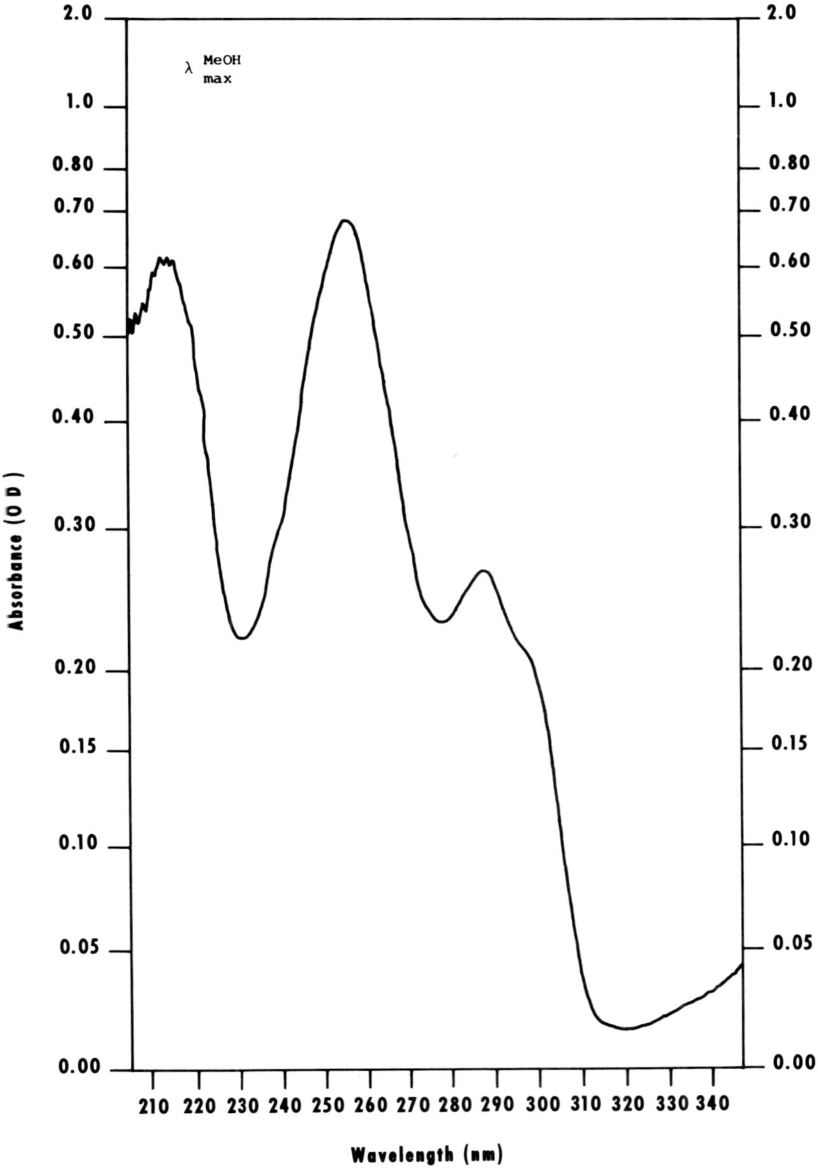
λ MeOH max
2.0
1.0
0.80
0.70
0.60
0.50
0.40
0.30
0.20
0.15
0.10
0.05
0.00
Absorbance (O D)
210 220 230 240 250 260 270 280 290 300 310 320 330 340
Wavelength (nm)

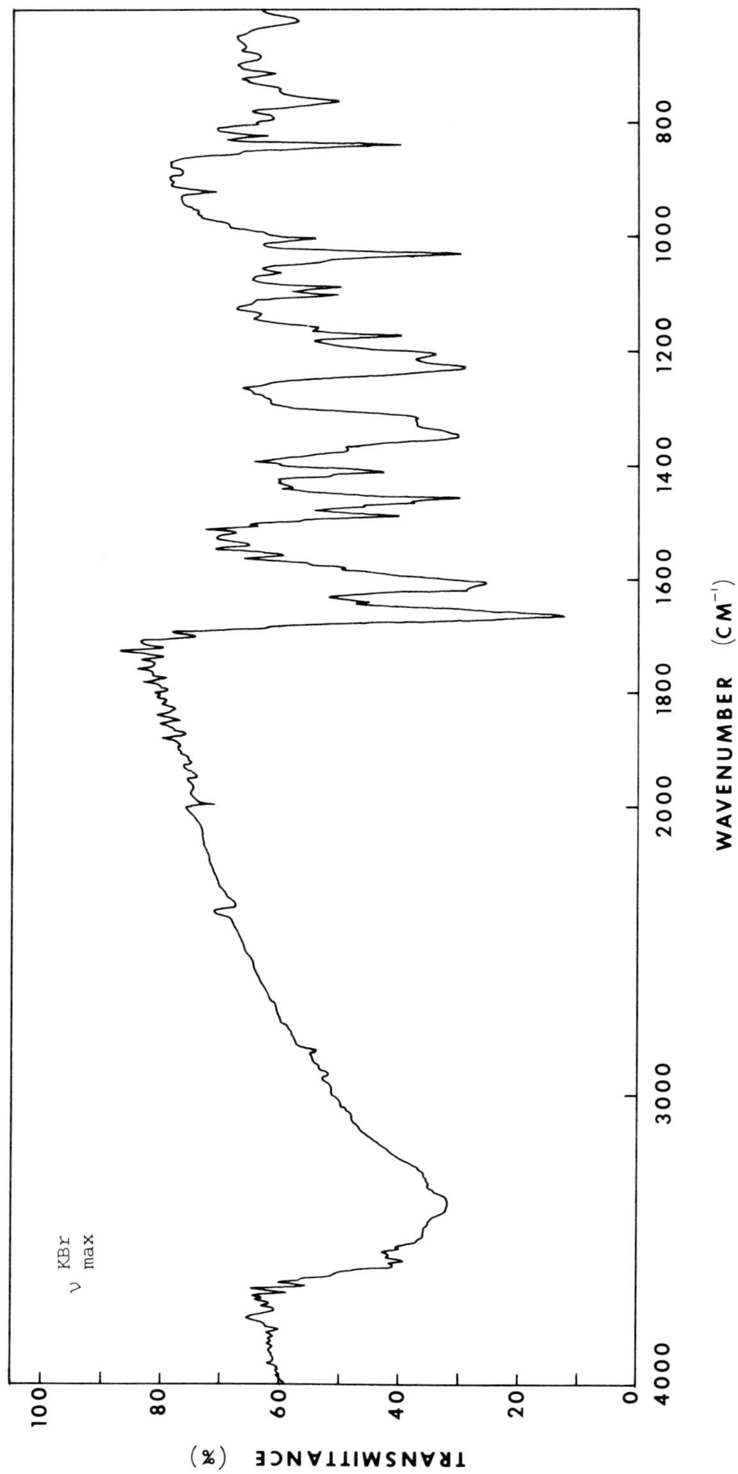

ν KBr max
TRANSMITTANCE (%)
100
80
60
40
20
0
4000
3000
2000
1800
1600
1400
1200
1000
800
WAVENUMBER (CM^{-1})

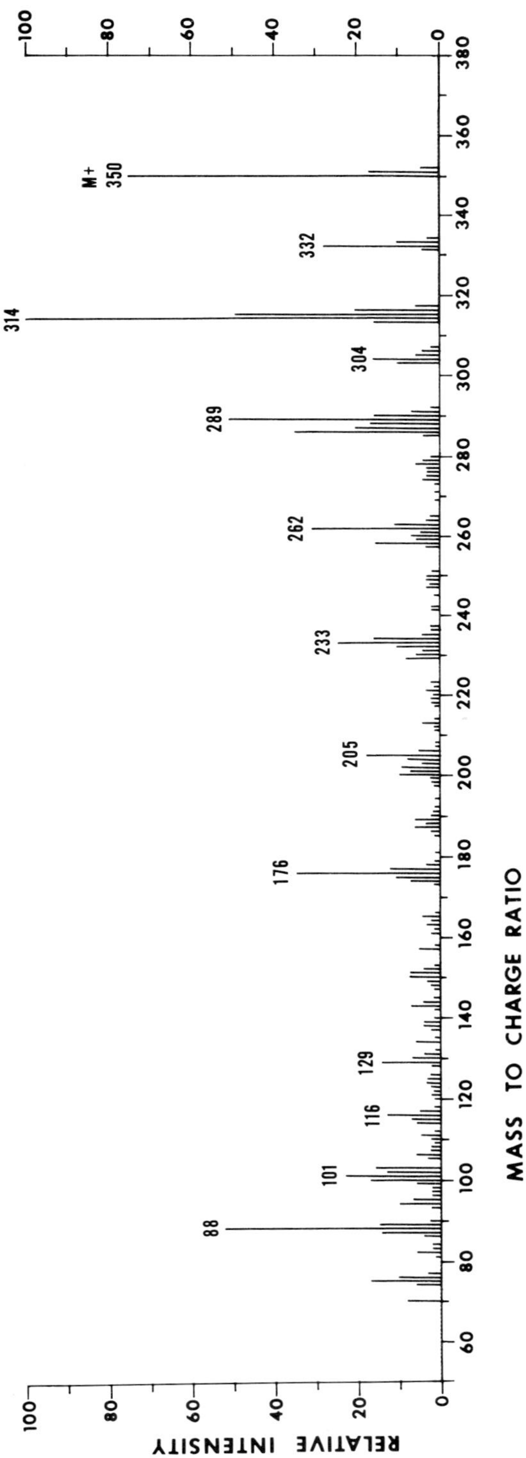
M+
350
332
314
304
289
262
233
205
176
129
116
101
88
MASS TO CHARGE RATIO
RELATIVE INTENSITY
60
80
100
120
140
160
180
200
220
240
260
280
300
320
340
360
380
0
20
40
60
80
100

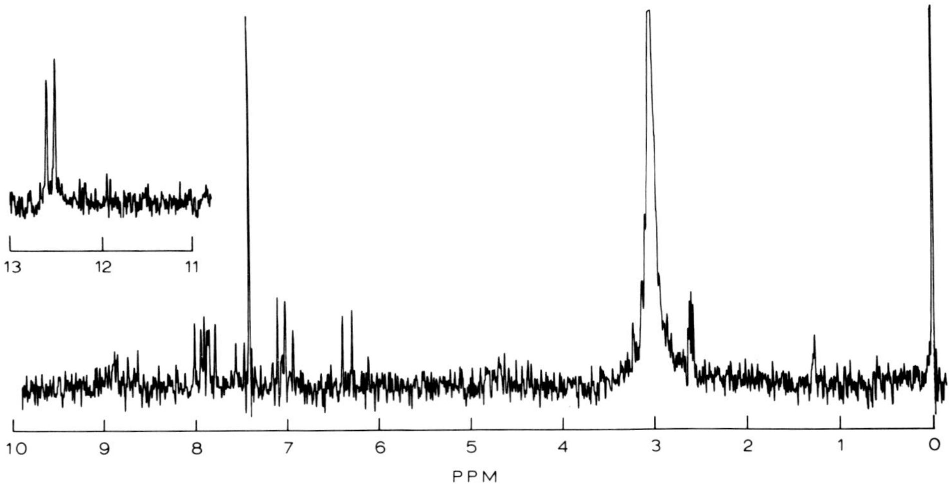

The 100 MHz ^1H NMR spectrum of altertoxin II in $CDCl_3$/DMSO solution.

REFERENCES

1. Coombe, R. G., *Aust. J. Chem.* **23,** 2343 (1970).
2. Freeman, G. G., *Phytochemistry* **5,** 719 (1966).
3. Harris, T. M., and Hay, J. V., *J. Am. Chem. Soc.* **99,** 1631 (1977).
4. Harvan, D., and Pero, R. W., *Adv. Chem. Ser.* **149,** 344 (1976).
5. Lucas, G. B., Pero, R. W., Snow, J. P., and Harvan, D., *J. Agric. Food Chem.* **19,** 1274 (1971).
6. McPhail, A. T., Miller, R. W., Harvan, D., and Pero, R. W., *J. Chem. Soc., Chem. Commun.* p. 682 (1973).
7. Pero, R. W., Harvan, D., and Blois, M. C., *Tetrahedron Lett.* p. 945 (1973).
8. Pero, R. W., and Harvan, D., *J. Chromatogr.* **80,** 255 (1973).
9. Pero, R. W., Owens, R. G., Dale, S. W., and Harvan, D., *Biochim. Biophys. Acta* **230,** 170 (1971).
10. Pero, R. W., Owens, R. G., and Harvan, D., *Anal. Biochem.* **43,** 80 (1971).
11. Pero, R. W., Posner, H., Blois, M., Harvan, D., and Spalding, J. W., *Environ. Health Perspect.* **4,** 87 (1973).
12. Raistrick, H., Stickings, C. E., and Thomas, R., *Biochem. J.* **55,** 421 (1953).
13. Rogers, D., and Williams, D. J., *Chem. Commun.* p. 393 (1971).
14. Rosett, T., Sankhala, R. H., Stickings, C. E., Taylor, M. E. U., and Thomas, R., *Biochem. J.* **67,** 390 (1957).
15. Schroeder, H. W., and Cole, R. J., *J. Agric. Food Chem.* **25,** 204 (1977).
16. Seitz, L. M., Mohr, H. E., Sauer, D. B., Burrough, R., and Paukstelis, J. V., *J. Agric. Food Chem.* **23,** 1 (1975).
17. Sóti, F., Incze, M., Kajtár-Peredy, M., Baitz-Gács, E., Imre, L., and Farkas, L., *Chem. Ber.* **110,** 979 (1977).
18. Starratt, A. N., and White, S. A., *Phytochemistry* **7,** 1883 (1968).

13

Secalonic Acids

The secalonic acids are a group of closely related fungal metabolites isolated originally from the sclerotia of *Claviceps purpurea* (ergochromes) and more recently isolated from *Aspergillus ochraceus* (secalonic acid A), *A. aculeatus* (secalonic acids D and F), *Pyrenochaeta terrestris* (secalonic acids A, B, and G), and *Penicillium oxalicum* (secalonic acid D).[2,8,10] Chemically, they are xanthone dimers having identical molecular weights (638.1635) and molecular formulas ($C_{32}H_{30}O_{14}$). They differ in their stereochemistry at positions 5, 6, 10a, 5′, 6′, and 10a′. The xanthone monomers are linked at the 2,2′ positions.

Secalonic acid D has been shown to be toxic to mice, and secalonic acid A reportedly showed antibacterial and phlogistic activity.[8–10]

Secalonic acid group	Molecular weight	Molecular formula
Secalonic acid A	638.1635	$C_{32}H_{30}O_{14}$
Secalonic acid B (E)	638.1635	$C_{32}H_{30}O_{14}$
Secalonic acid C	638.1635	$C_{32}H_{30}O_{14}$
Secalonic acid D	638.1635	$C_{32}H_{30}O_{14}$
Secalonic acid F	638.1635	$C_{32}H_{30}O_{14}$
Secalonic acid G	638.1635	$C_{32}H_{30}O_{14}$

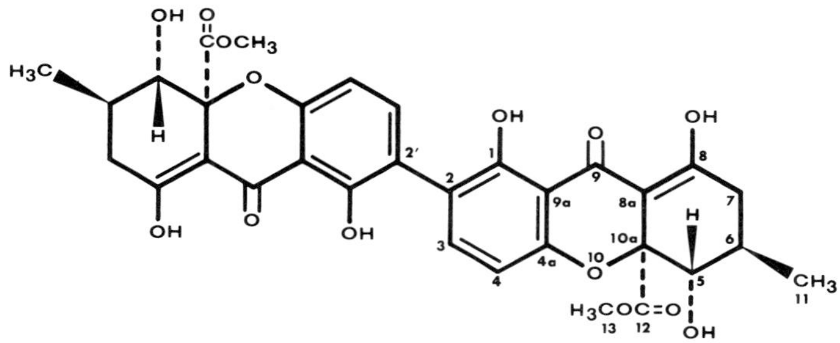

Common name	Secalonic acid A (Ergochrome EE)
Molecular weight	638.1635
Molecular formula	$C_{32}H_{30}O_{14}$
General characteristics[10]	Yellow crystals from chloroform, mp 246°–248°C; from acetone, mp 243°C $[\alpha]_D^{20}$ −75° (in chloroform), −202° (in pyridine)
UV data[10]	λ_{max} nm (ϵ): 252(17,200) and 346(31,000)
Fungal source[2,6,10]	*Claviceps purpurea* (sclerotia) *Aspergillus ochraceus* *Phoma terrestris* *Pyrenochaeta terrestris* Lichen source—*Parmelia entotheiochroa*
Toxicity data[10]	The LD_{100} in mice dosed IP was 100 mg/kg; at 50 mg/kg, 6 of 8 mice died. No mortalities were observed with oral doses up to 250 mg/kg in mice. It was nontoxic to chick embryos.
TLC data[8,9]	*Adsorbent:* silica gel impregnated with 6% oxalic acid *Solvent:* chloroform–4-methyl-2-pentanone, 9:1 v/v R_f: 0.23 *Detection:* 1% ceric sulfate in 6 *N* H_2SO_4 spray
Other significant references	Mass spectroscopy[4]; biosynthesis[7]; ^{1}H NMR[3]; X-ray diffraction[5]
^{13}*C NMR data*[7]	C1,1′, 159.9; C2,2′, 118.1; C3,3′, 140.8; C4,4′, 107.8; C4a,4a′, 159.9; C5,5′, 76.6; C6,6′, 30.5; C7,7′, 36.7; C8,8′, 178.6; C8a,8a′, 102.6; C9,9′,

	187.8; C9a,9a′, 107.5; C10a,10a′, 86.2; C11,11′, 18.4; C12,12′, 171.1; C13,13′, 52.8
^{1}H NMR data[4]	H3, 7.74 (*J* = 8); H4, 6.60 (*J* = 8); H5, 3.88; H6, 2.74; H7, 2.74; H11, 1.16 (*J* = 6); H13, 3.68; 1-OH, 13.78; 5-OH, 1.54; 8-OH, 11.70

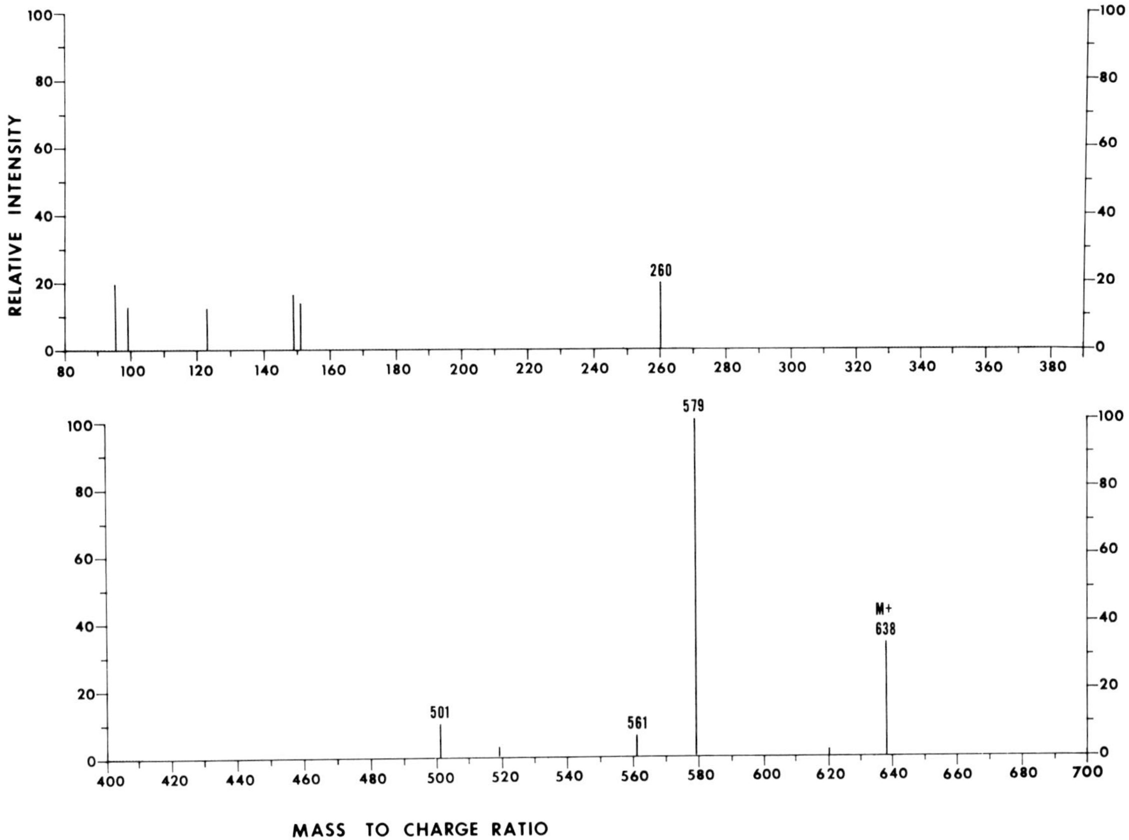
RELATIVE INTENSITY
260
579
501
561
M+
638
MASS TO CHARGE RATIO

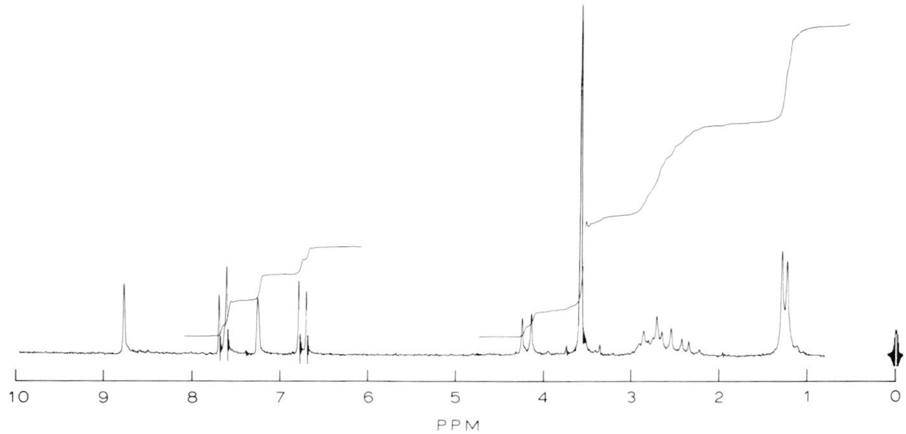

The 100 MHz ¹H NMR spectrum of secalonic acid A in pyridine-d_5 solution.

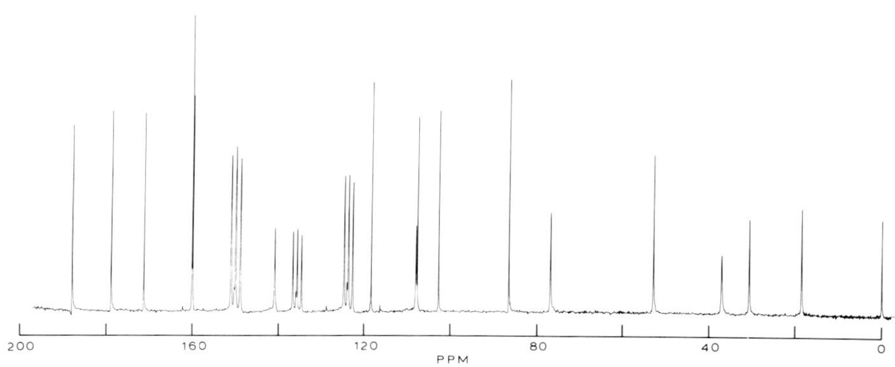

The 25 MHz ¹³C NMR spectrum of secalonic acid A in pyridine-d_5 solution.

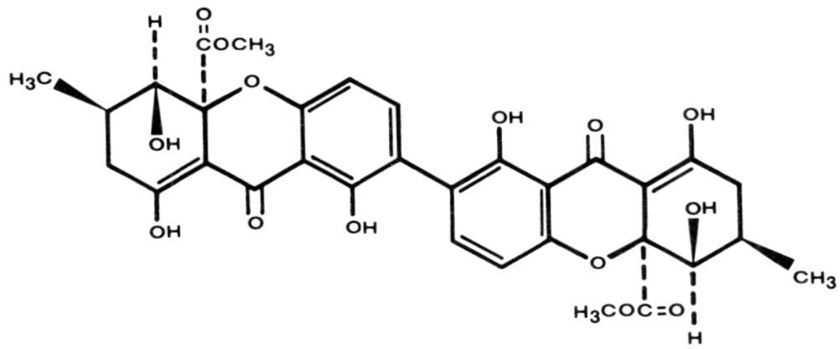

Common name	Secalonic acid B (E)
Molecular weight	638.1635
Molecular formula	$C_{32}H_{30}O_{14}$
General characteristics[8,9]	mp 259°–262°C
Fungal source[2,6]	*Claviceps purpurea* (sclerotia) *Pyrenochaeta terrestris*
Toxicity data	Toxicity unknown
TLC data[9]	*Adsorbent:* silica gel impregnated with 6% oxalic acid *Solvent:* chloroform–4-methyl-2-pentanone, 9:1 v/v R_f: 0.46 *Detection:* 1% ceric sulfate in 6 *N* H_2SO_4 spray
Other significant references	Mass spectroscopy[4]; ^{1}H NMR[3]
^{13}*C NMR data**	(pyridine-d_5) C1, 160.1; C2, 118.6; C3, 140.4; C4, 107.6; C4a, 158.5; C5, 71.4; C6, 29.5; C7, 33.6; C8, 180.7; C8a, 101.5; C9, 188.4; C9a, 107.4; C10a, 86.1; C11, 17.8; C12, 172.2; C13, 53.2
^{1}*H NMR data**	H3, 7.53 (J = 8.5); H4, 6.71; H5, 4.58 (J = 0.8); H6, 2.25 (J = 0.8, 5.8, 11.0); H7eq, 2.46 (J = 5.8, 18.6); H7ax, 2.77 (J = 11.0, 18.6); H11, 1.19 (J = 6.4); H13, 3.57; OH's, 11–13

* Spectra and assignments for secalonic acid E. Courtesy of Dr. A. G. McInnes.

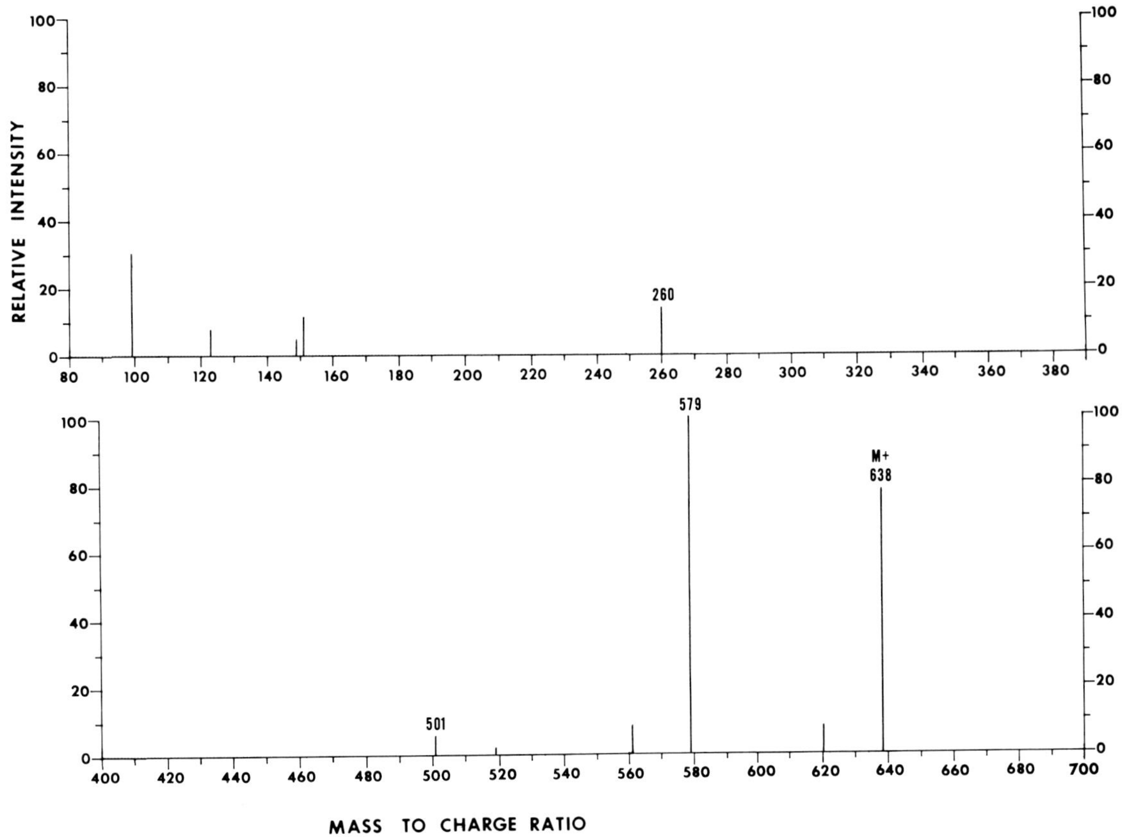
RELATIVE INTENSITY
260
579
M+
638
501
MASS TO CHARGE RATIO

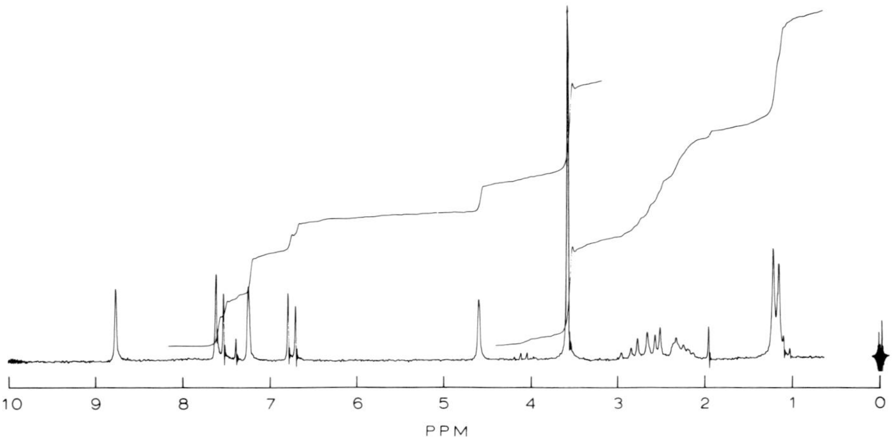

The 100 MHz ^{1}H NMR spectrum of secalonic acid E in pyridine-d_5 solution.

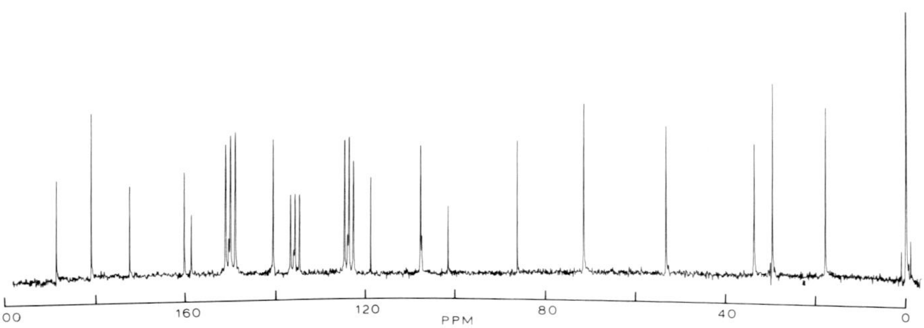

The 25 MHz ^{13}C NMR spectrum of secalonic acid E in pyridine-d_5 solution.

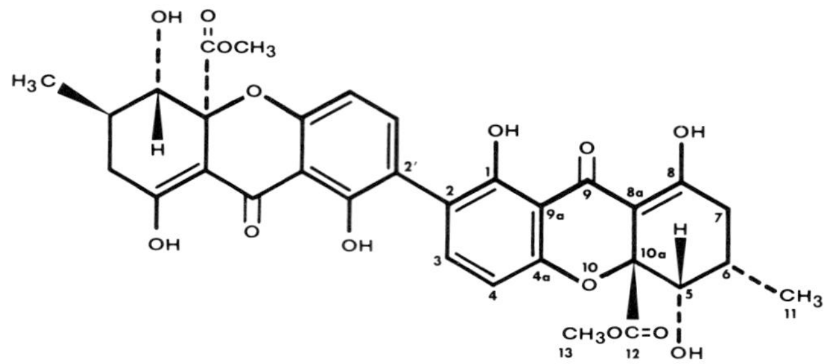

Common name	Secalonic acid C
Molecular weight	638.1635
Molecular formula	$C_{32}H_{30}O_{14}$
General characteristics[2]	mp 187°–189°C
Fungal source[2]	*Claviceps purpurea* (sclerotia)
Toxicity data	Toxicity unknown
Other significant references	Mass spectroscopy[4]

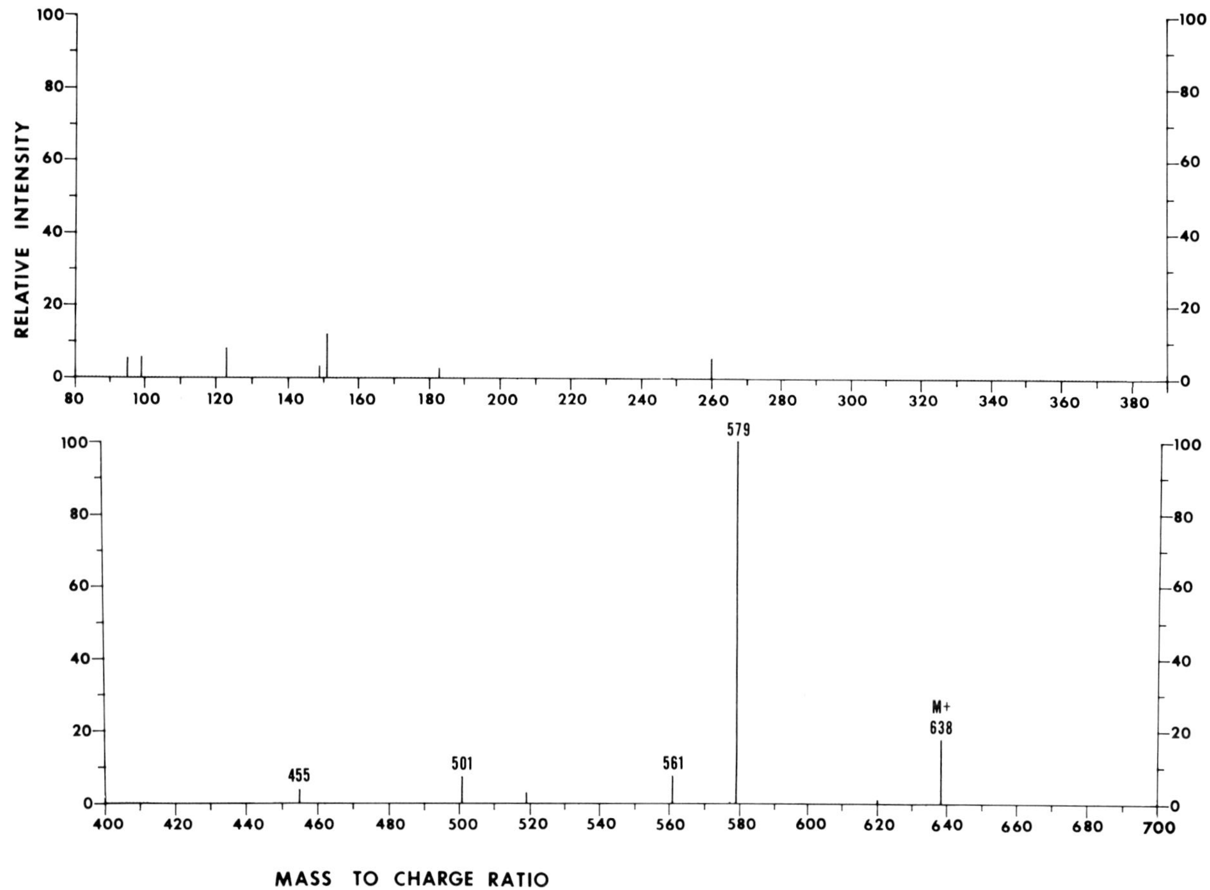
RELATIVE INTENSITY
MASS TO CHARGE RATIO
455
501
561
579
M+
638

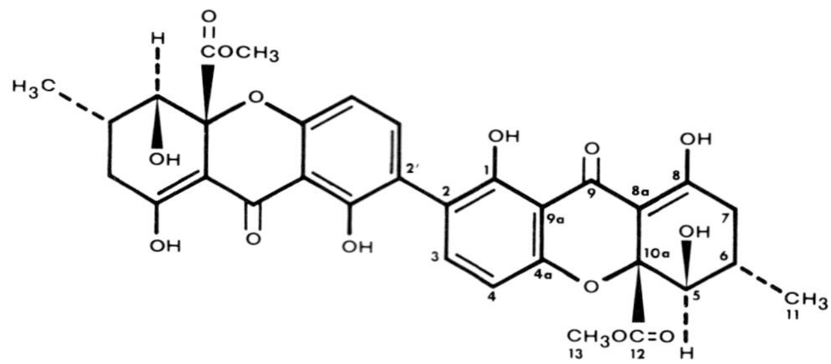

Common name Secalonic acid D

Molecular weight 638.1635

Molecular formula $C_{32}H_{30}O_{14}$

General characteristics[8,9] Pale yellow crystals from chloroform, mp 253°–255°C
Secalonic acid D is the optical antipode of secalonic acid A.
$[\alpha]_D^{20}$ +80° (in chloroform)
+192° (in pyridine)

UV data[9] λ_{max}^{EtOH} nm (ϵ): 337(31,600) and 248(18,200)

CD data[9] $\Delta\epsilon$ (C = 2.0 in dioxane) nm: 400(0), 334(+16.8), 300-290(0), 270(−5.9), 260(−2.4), 255(−60), 220(−45.20)

ORD data[9] [ψ] (C = 3.0 in chloroform) nm: 450(+2,050), 350(+20,500), 337(0), 312(−43,000), 280(−55,200), 265(−51,000), and 240(−90,000)

Fungal source[2,8,10] *Penicillium oxalicum*
Claviceps purpurea (sclerotia)
Aspergillus aculeatus Iizuka

Toxicity data[8,9] The LD_{50} of secalonic acid D dosed IP to mice in an aqueous 5-methyl cellulose suspension was 42 mg/kg (both male and female mice).

TLC data[9] *Adsorbent:* silica gel impregnated with 6% oxalic acid
Solvent: chloroform–4-methyl-2-pentanone, 9:1 v/v
R_f: 0.23
Detection: 1% ceric sulfate in 6 N H_2SO_4 spray

Other significant references Mass spectroscopy[4]

^{13}C NMR data C1,1′, 158.3 s*; C2,2′, 116.9 s; C3,3′, 139.7 d; C4,4′, 106.9 d; C4a,4a′, 157.9 s*; C5,5′, 74.9 d; C6,6′, 29.7 d; C7,7′, 35.6 t; C8,8′, 177.4 s; C8a,8a′, 101.3 s; C9,9′, 185.9 s; C9a,9a′, 105.8 s; C10a, 10a′, 84.8 s; C11,11′, 17.6 q; C12,12′, 169.4 s; C13,13′, 52.5 q

^{1}H NMR data H3, 7.46 (J = 8); H4, 6.62 (J = 8); H5, 3.80; H6, 2.0–3.0; H7, 2.0–3.0; H11, 1.05; H13, 3.60; 1-OH, 13.72; 5-OH, 6.00; 8-OH, 11.65

* Assignments may be reversed.

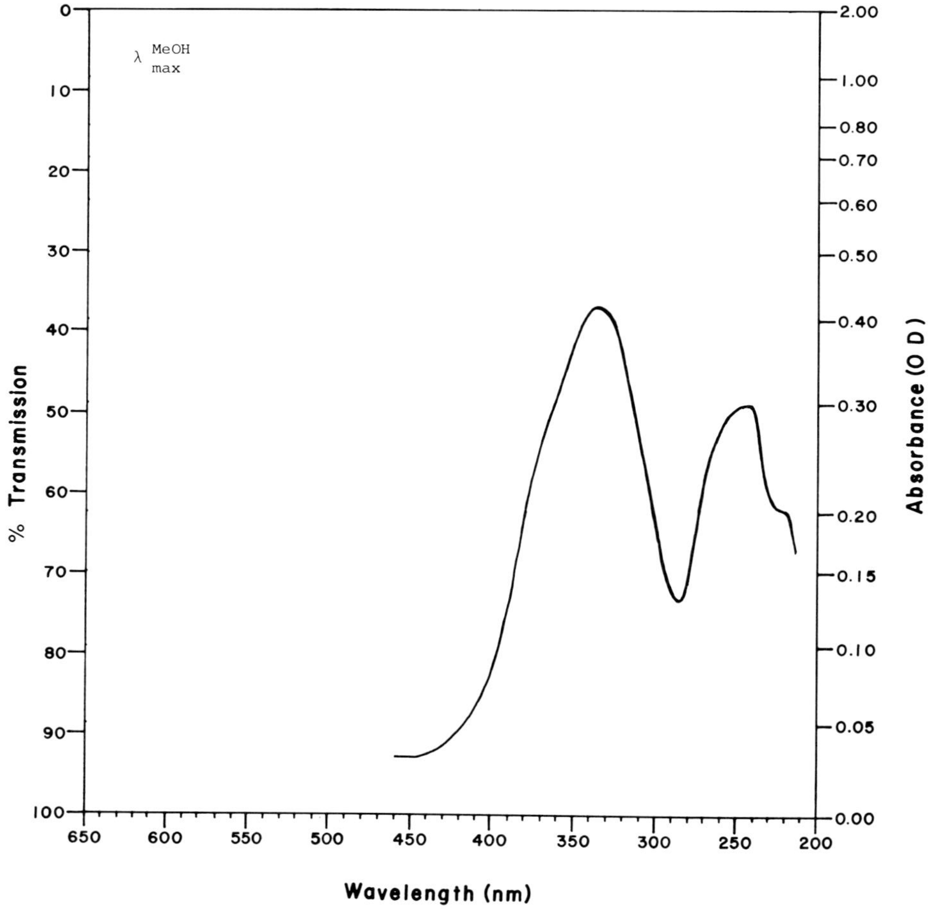

λ_{max}^{MeOH}
% Transmission
Absorbance (O D)
Wavelength (nm)
0
10
20
30
40
50
60
70
80
90
100
2.00
1.00
0.80
0.70
0.60
0.50
0.40
0.30
0.20
0.15
0.10
0.05
0.00
650
600
550
500
450
400
350
300
250
200

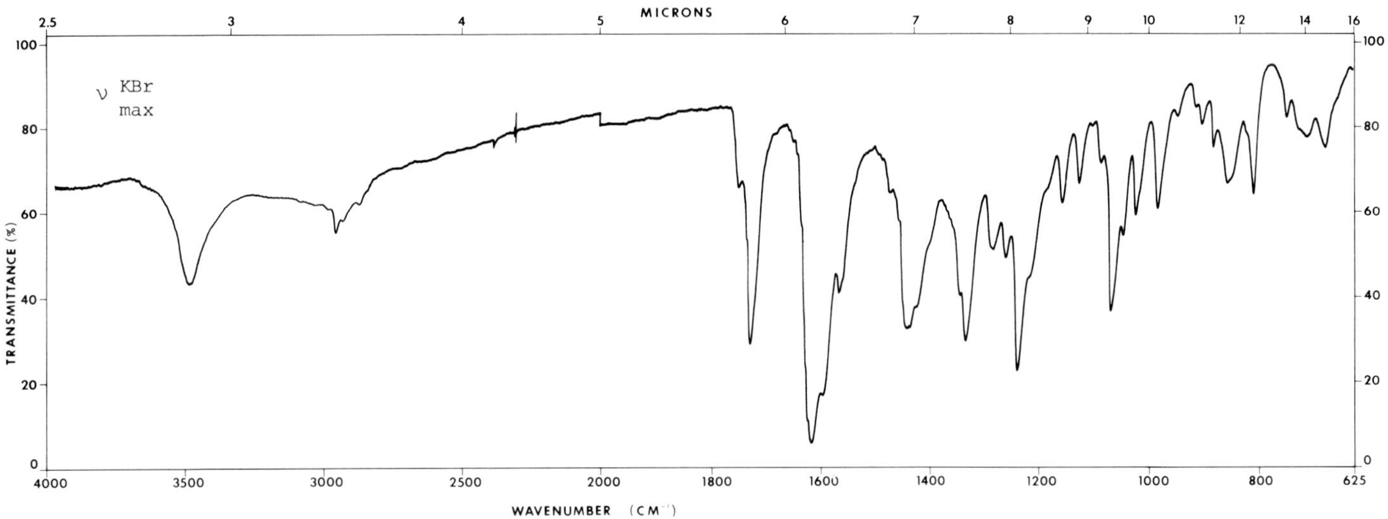
MICRONS
WAVENUMBER (CM⁻¹)
TRANSMITTANCE (%)
ν KBr max

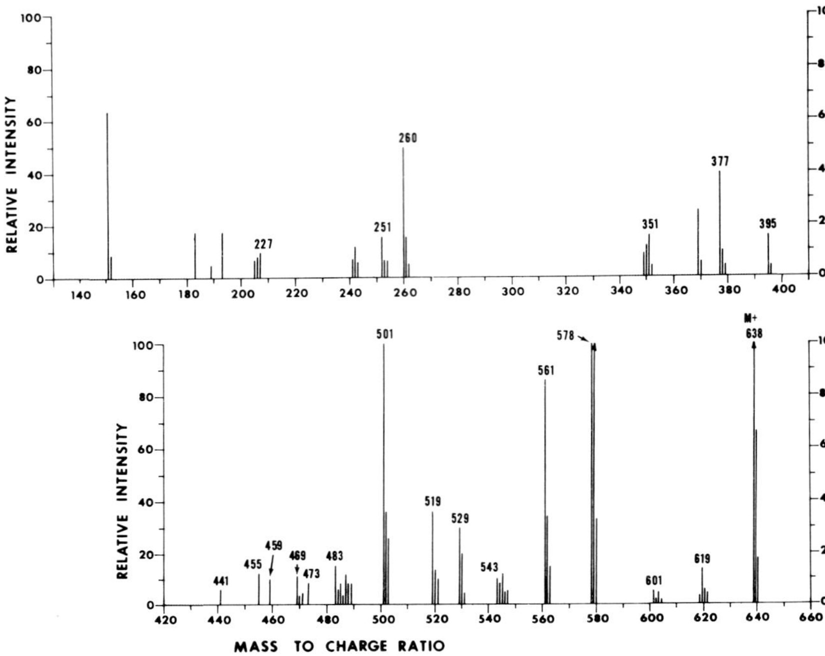
RELATIVE INTENSITY
227
251
260
351
377
395
RELATIVE INTENSITY
441
455
459
469
473
483
501
519
529
543
561
578
M+
638
601
619
MASS TO CHARGE RATIO

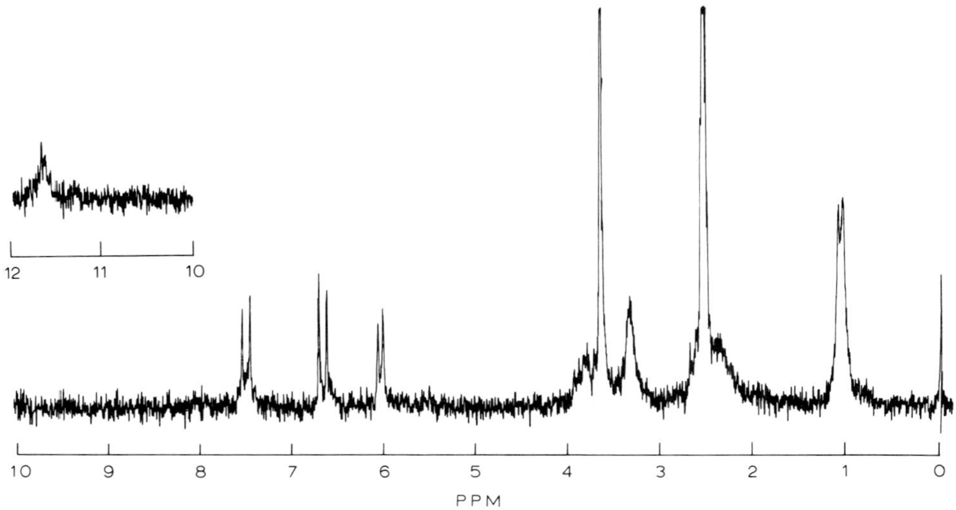

The 100 MHz ^{1}H NMR spectrum of secalonic acid D in DMSO solution.

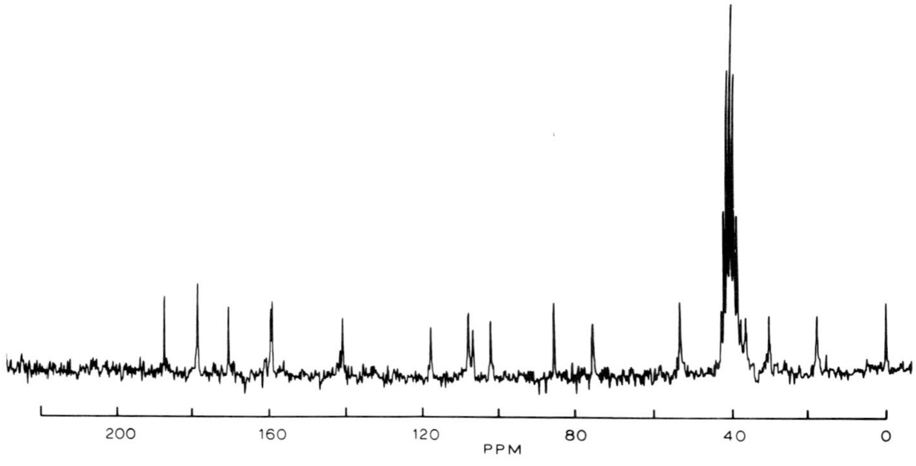

The 25 MHz ^{13}C NMR spectrum of secalonic acid D in DMSO solution.

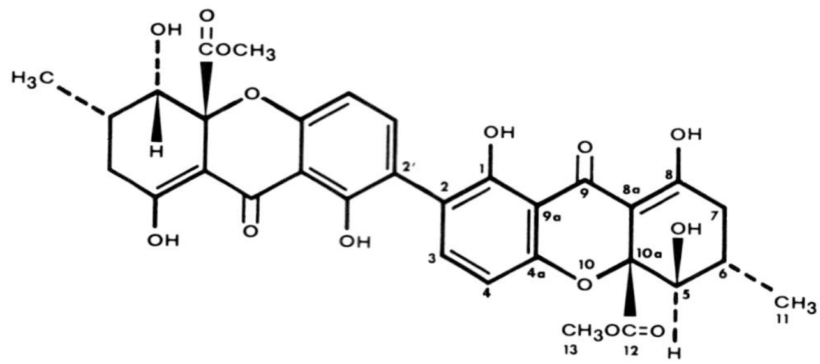

Common name	Secalonic acid F
Molecular weight	638.1635
Molecular formula	$C_{32}H_{30}O_{14}$
General characteristics[1]	Yellow needles from benzene–cyclohexane, mp 218°–221°C $[\alpha]_D^{20}$ +202° $[\alpha]_{578}^{20}$ +214° (C = 0.13 in pyridine)
UV data[1]	λ_{max}^{EtOH} nm (ϵ): 236(19,250), 263(17,300), and 388(37,000)
CD data	$\Delta\epsilon$ (C = 4.8 × 10^{-2} mg/ml in dioxane) nm: 400(0), 332(+17), 275–260(0), 223(−43), 215(−25)
Fungal source[1]	*Aspergillus aculeatus* Iizuka
Toxicity data[1]	Secalonic acid F showed antimicrobial activity against *Bacillus megaterium;* details of vertebrate toxicity not reported.
TLC data[1]	*Adsorbent:* silica gel GF_{254} containing 6% tartaric acid *Solvent:* 2-pentanone–chloroform, 2:8 v/v R_f: 0.29 (secalonic acid D had R_f 0.17 in this system) *Detection:* red-brown spot with ferric chloride spray
^{1}H NMR data[1]	H3,3′, 7.35, 7.39 (J = 9); H4,4′, 6.52, 6.58 (J = 9); H5,5′, 3.87, 4.09; H6,6′, 2.0–3.0; H7,7′, 2.0–3.0; H11,11′, 1.14 (J = 7); H13,13′, 3.67; 1,1′-OH, 13.70, 13.88; 5,5′-OH, 2.67, 3.86; 8,8′-OH, 11.65, 11.80

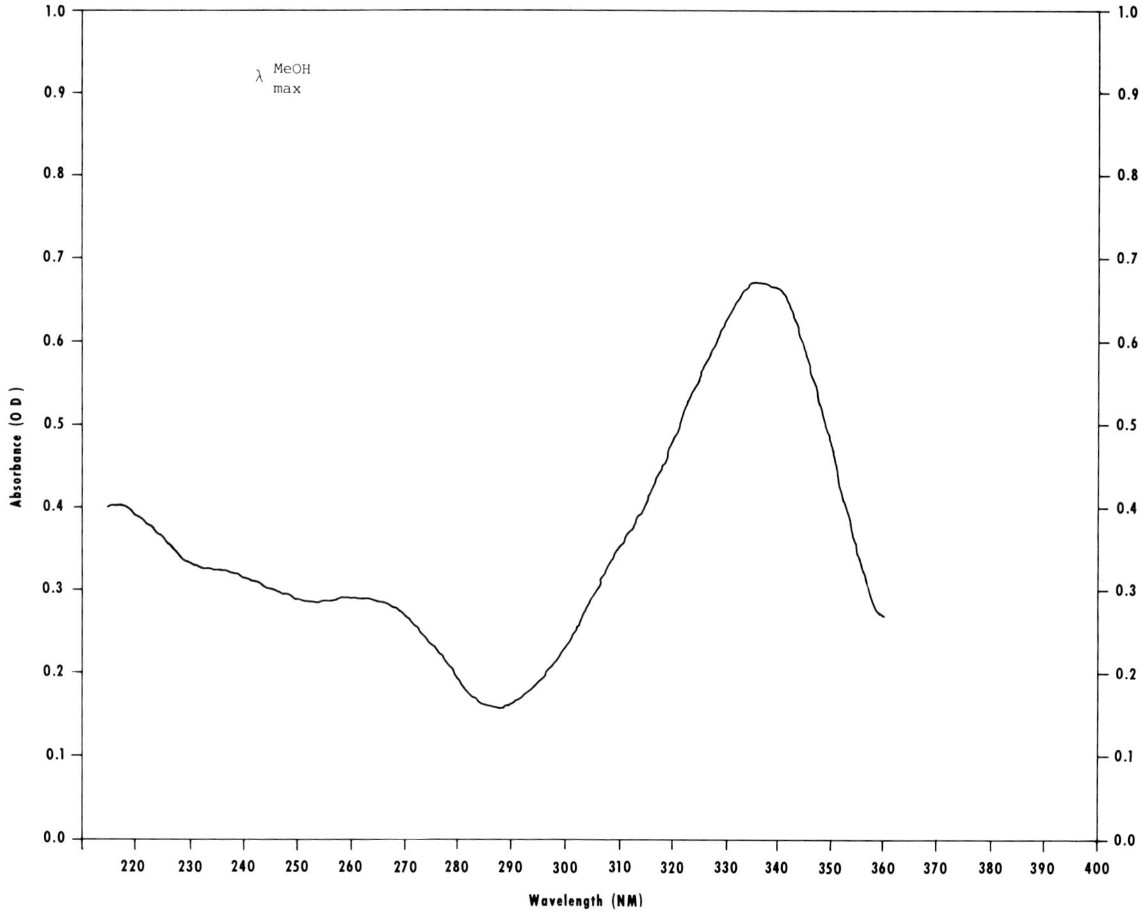

λ MeOH max
Absorbance (O D)
Wavelength (NM)
1.0
0.9
0.8
0.7
0.6
0.5
0.4
0.3
0.2
0.1
0.0
220
230
240
250
260
270
280
290
300
310
320
330
340
350
360
370
380
390
400

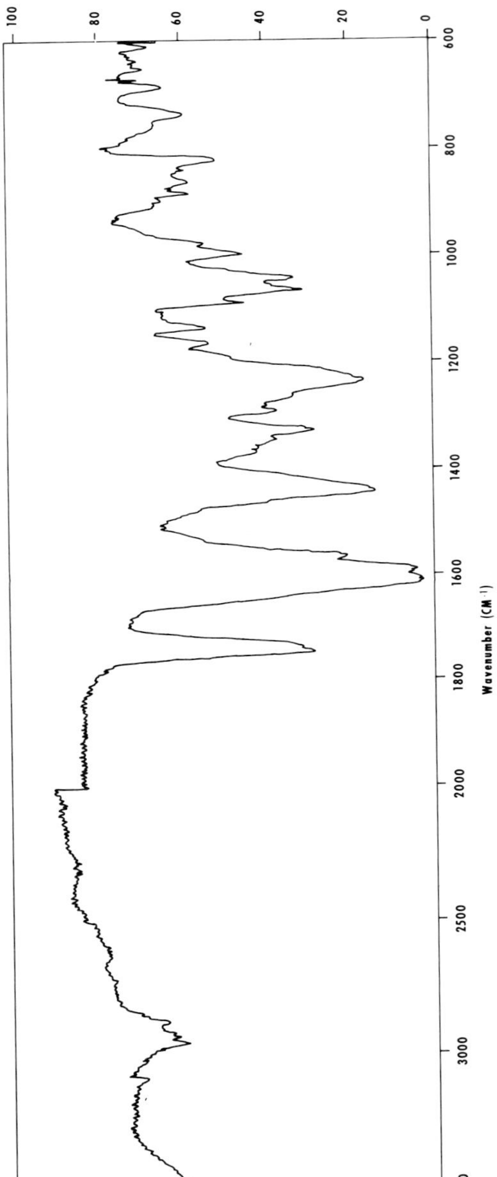

Transmittance (%)
Wavenumber (CM⁻¹)
ν KBr max

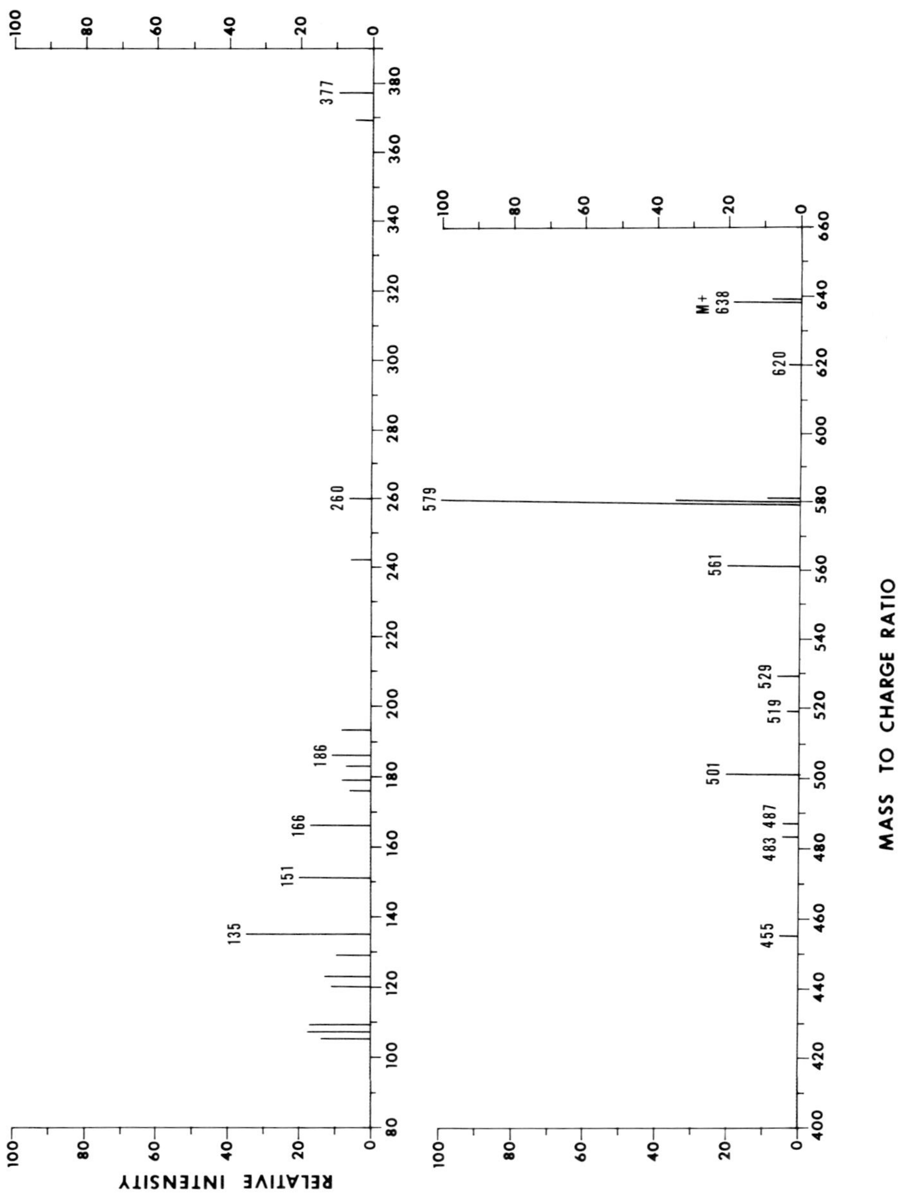
RELATIVE INTENSITY
MASS TO CHARGE RATIO
135
151
166
186
260
377
455
483
487
501
519
529
561
579
620
M+
638

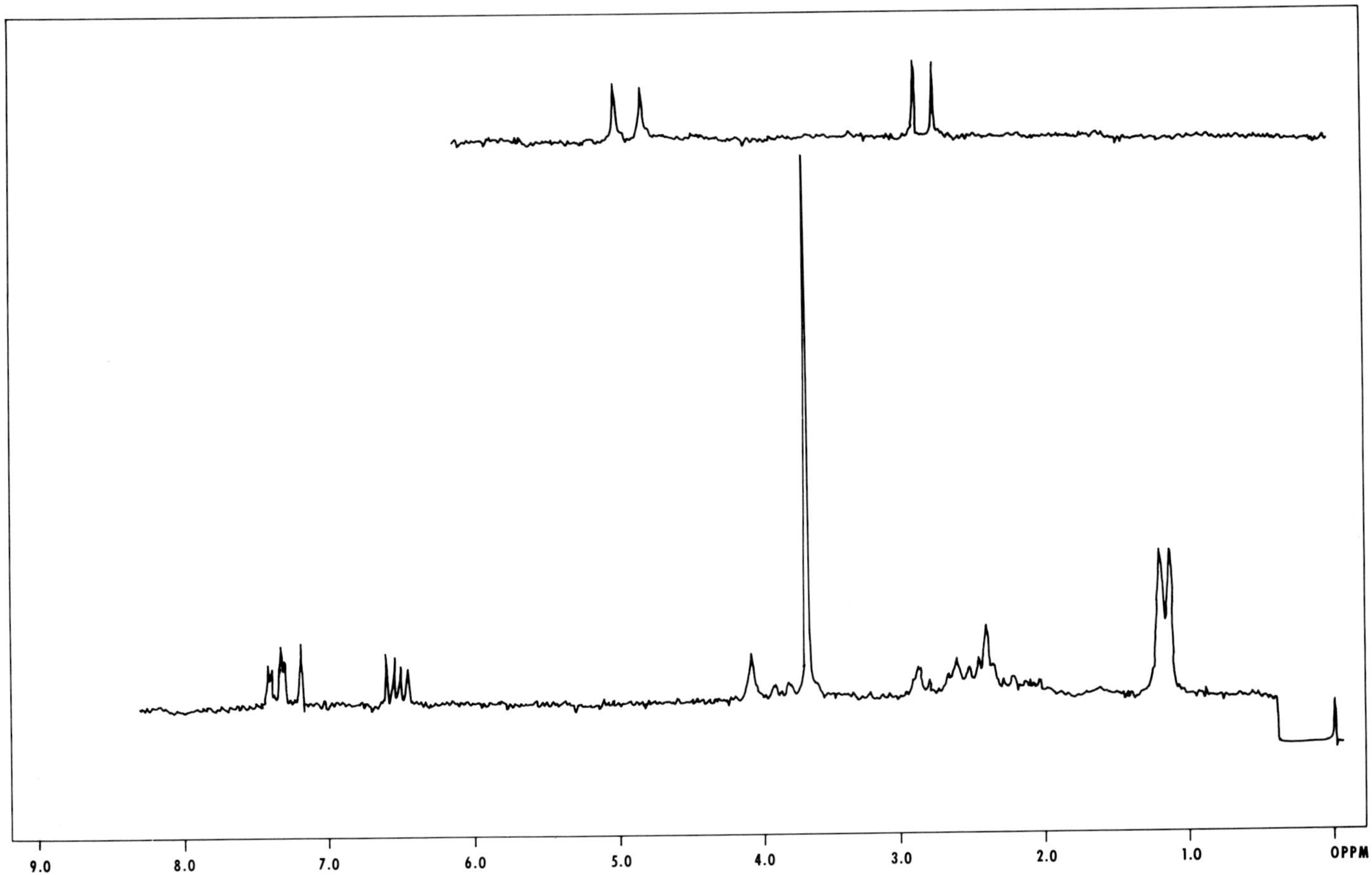

The 100 MHz ^{1}H NMR spectrum of secalonic acid F in $CDCl_3$.

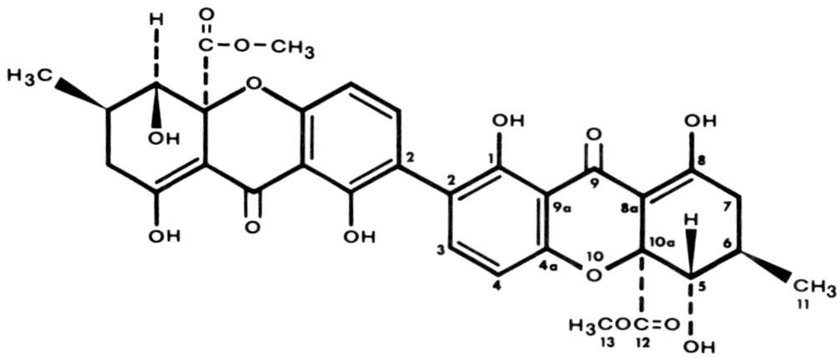

Common name	Secalonic acid G
Molecular weight	638.1635
Molecular formula	$C_{32}H_{30}O_{14}$
General characteristics[6]	Yellow needles from ethanol, mp 206°–207°C $[\alpha]_D^{20}$ −106.7° (C = 1.055 in chloroform) −201.6° (C = 1.105 in pyridine)
UV data[6]	$\lambda_{max}^{dioxane}$ nm (ϵ): 240(21,900), 265(19,000), and 338(37,200)
IR data[6]	ν_{max}^{KBr} cm^{-1}: 3600, 3550, 3450, 1735, and 1600
Fungal source[6]	*Pyrenochaeta terrestris*
Toxicity data	Toxicity unknown
*^{13}C NMR data**	(pyridine-d_5) C1, 160.0; C2, 118.1; C3, 140.7; C4, 107.8; C4a, 159.9; C5, 76.6; C6, 30.6; C7, 36.9; C8, 178.7; C8a, 102.6; C9, 187.7; C9a, 107.5; C10a, 86.3; C11, 18.5; C12, 171.2; C13, 52.7; C1′, 160.0; C2′, 118.8; C3′, 140.4; C4′, 107.7; C4a′, 158.5; C5′, 71.4; C6′, 29.5; C7′, 33.6; C8′, 180.8; C8a′, 101.5; C9′, 188.3; C9a′, 107.4; C10a′, 86.1; C11′, 17.8; C12′, 172.3; C13′, 53.2
*^{1}H NMR data**	H3, 7.58 (J = 8.5); H4, 6.74; H5, 4.19 (J = 10.3); H6, 2.80 (J = 6.0, 10.3, 11.0); H7ax, 2.76 (J = 11.0, 19.0); H7eq, 2.44 (J = 6.0, 19.0); H11, 1.26 (J = 6.0); H13, 3.57; H3′, 7.56 (J = 8.5); H4′, 6.70; H5′, 4.58; H6′, 2.25 (J = 5.8, 11.0); H7′eq, 2.46 (J = 5.8, 18.6); H7′ax, 2.76 (J = 11.0, 18.6); H11′, 1.19; H13′, 3.60

* Spectra and assignments courtesy of Dr. A. G. McInnes.

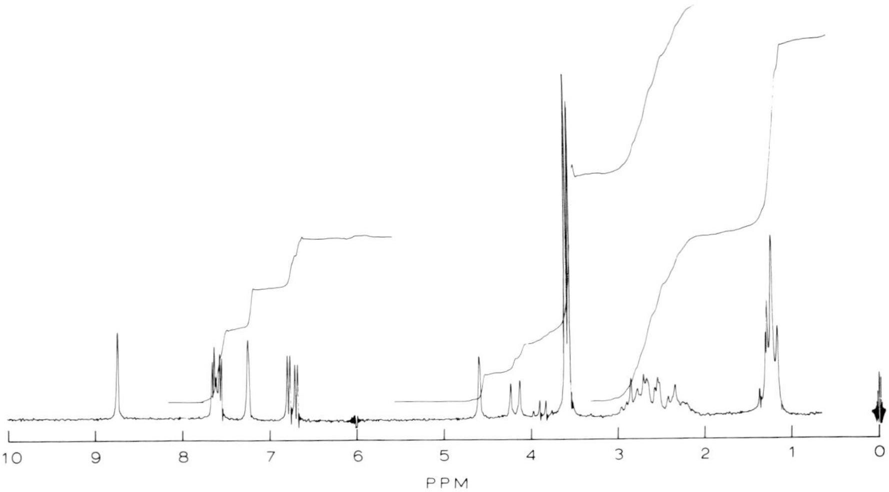

The 100 MHz ^{1}H NMR spectrum of secalonic acid G in pyridine-d_5 solution.

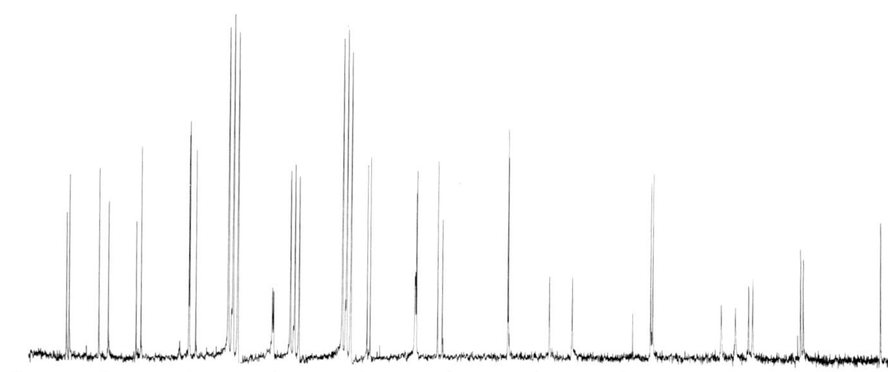

The 25 MHz ^{13}C NMR spectrum of secalonic acid G in pyridine-d_5 solution.

REFERENCES

1. Andersen, R., Büchi, G., Kobbe, B., and Demain, A. L., *J. Org. Chem.* **42,** 352 (1977).
2. Franck, B., Gottschalk, E. M., Ohnsorge, U., and Hüper, F., *Chem. Ber.* **99,** 3842 (1966).
3. Hooper, J. W., Marlow, W., Whalley, W. B., Borthwick, A. D., and Bowden, R., *J. Chem. Soc. C* p. 3580 (1971).
4. Howard, C. C., and Johnstone, R. A. W., *J. Chem. Soc., Perkin Trans. 1* p. 2033 (1973).
5. Howard, C. C., Johnstone, R. A. W., King, T. J., and Lessinger, L., *J. Chem. Soc., Perkin Trans. 1* p. 1820 (1976).
6. Kurobane, I., Vining, L. C., and McInnes, A. G., *Tetrahedron Lett.* p. 4633 (1978).
7. Kurobane, I., Vining, L. C., McInnes, A. G., Walter, J. A., and Wright, J. L. C., *Tetrahedron Lett.* p. 1379 (1978).
8. Steyn, P. S., *J. S. Afr. Chem. Inst.* **22,** 520 (1969).
9. Steyn, P. S., *Tetrahedron* **26,** 51 (1970).
10. Yamazaki, M., Maebayashi, Y., and Miyaki, K., *Chem. Pharm. Bull.* **19,** 199 (1971).

14

The Malformins

The malformins are a family of closely related cyclic pentapeptides characterized also by a disulfide bridge between two cysteine residues. They are produced by fungi belonging to the *Aspergillus niger* group. They are of biological interest because of their antibiotic activity,[14] cytotoxicity,[7] inhibition of adventitious root formation, and unique ability to cause malformations of the stems and petioles of bean plants (*Phaseolus vulgaris*) and curvature of roots of corn (*Zea mays*). More recently, they are receiving attention because of their potential as mycotoxins.[1,11]

Malformin group	Molecular weight	Molecular formula
Malformin A_1	529.2392	$C_{23}H_{39}O_5N_5S_2$
Malformin A_2	515.2236	$C_{22}H_{37}O_5N_5S_2$
Malformin B_1	529.2392	$C_{23}H_{39}O_5N_5S_2$
Malformin B_2	515.2236	$C_{22}H_{37}O_5N_5S_2$
Malformin C	529.2392	$C_{23}H_{39}O_5N_5S_2$

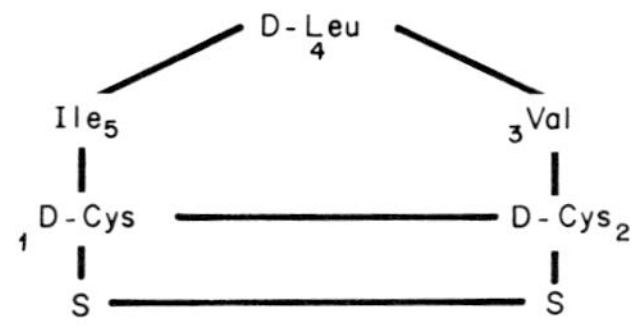

Common name Malformin A_1 (Malformin A) (The disulfide form of cyclo-D-cysteinyl-D-cysteinyl-L-valyl-D-leucyl-L-isoleucyl)

Molecular weight 529.4231

Molecular formula $C_{23}H_{39}O_5N_5S_2$

General characteristics[4] Decomposed without melting above 300°C
$[\alpha]_D^{18}$ −26.6° (C = 0.68 in methyl cellosolve)

UV data[4] End absorption

Fungal source[8–10] *Aspergillus niger* van Tieghem
A. ficuum
A. awamori
A. phoenicis

Toxicity data[4,6,7,14,18] Malformin A causes distortions on plants, possesses antibiotic activity, and inhibits mitosis and adventitious root formation. The LD_{50} for male mice (DDY strain) was 3.1 mg/kg (IP); oral > 50 mg/kg. Clinical signs were dilation with hemorrhage of the gastrointestinal tract and fading of liver and kidney.

TLC data[4,17] *Adsorbent:* silica gel
Solvent: A: water saturated ethyl acetate
B: chloroform–methanol, 9:1 v/v
R_f: A: 0.82
B: 0.57
Detection: spray with *tert*-butyl hypochlorite–KI–starch reagent or spray with 30% H_2SO_4 and heat.

Other significant references Discovery[8]; isolation[15,16]; structure[2,5,12]; conformation studies[13]; mass spectroscopy[3]

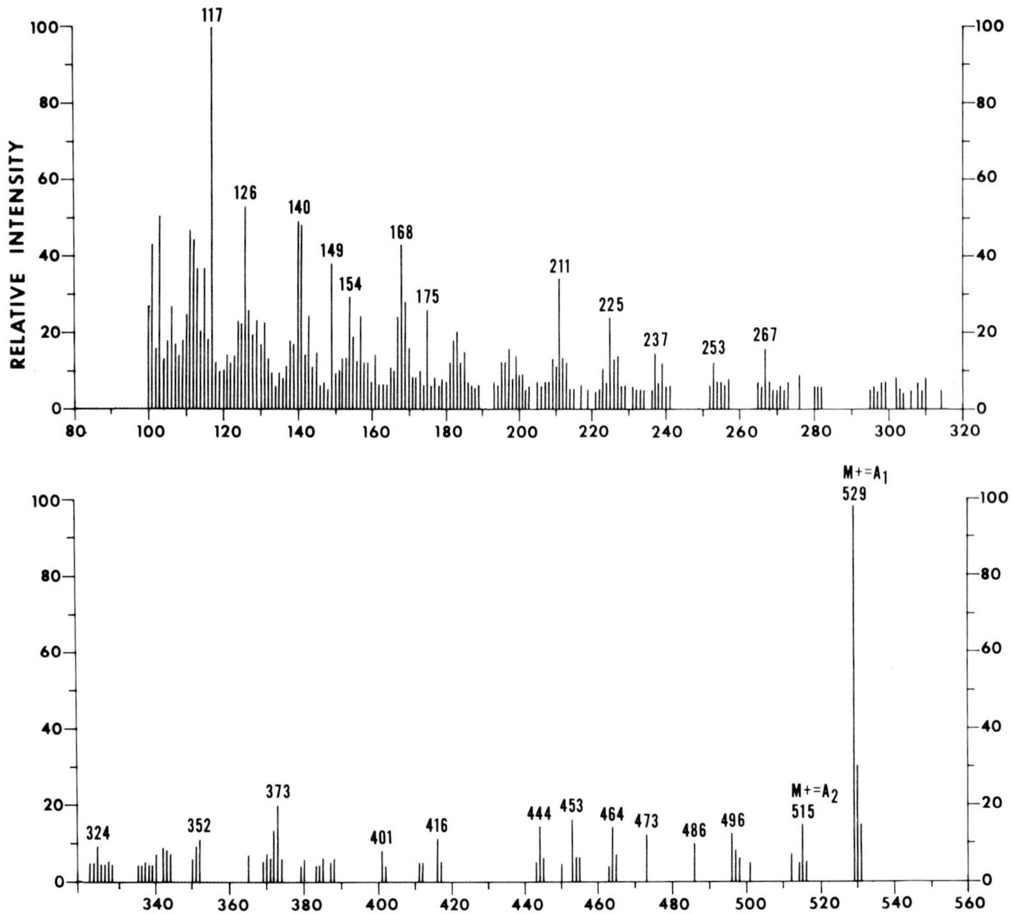

Mass spectra of malformin A_1 and A_2.

Common name	Malformin A_2* (occurs as a companion of A_1 approximately 18%)
Molecular weight	515.2236
Molecular formula	$C_{22}H_{37}O_5N_5S_2$
UV data	End absorption
Fungal source[8,17]	*Aspergillus niger*
Toxicity data	Toxicity unknown
TLC data[17]	*Adsorbent:* silica gel H *Solvent:* water-saturated ethyl acetate R_f: 0.82 *Detection:* sprayed with 30% H_2SO_4 and heated.

* Structure unknown.

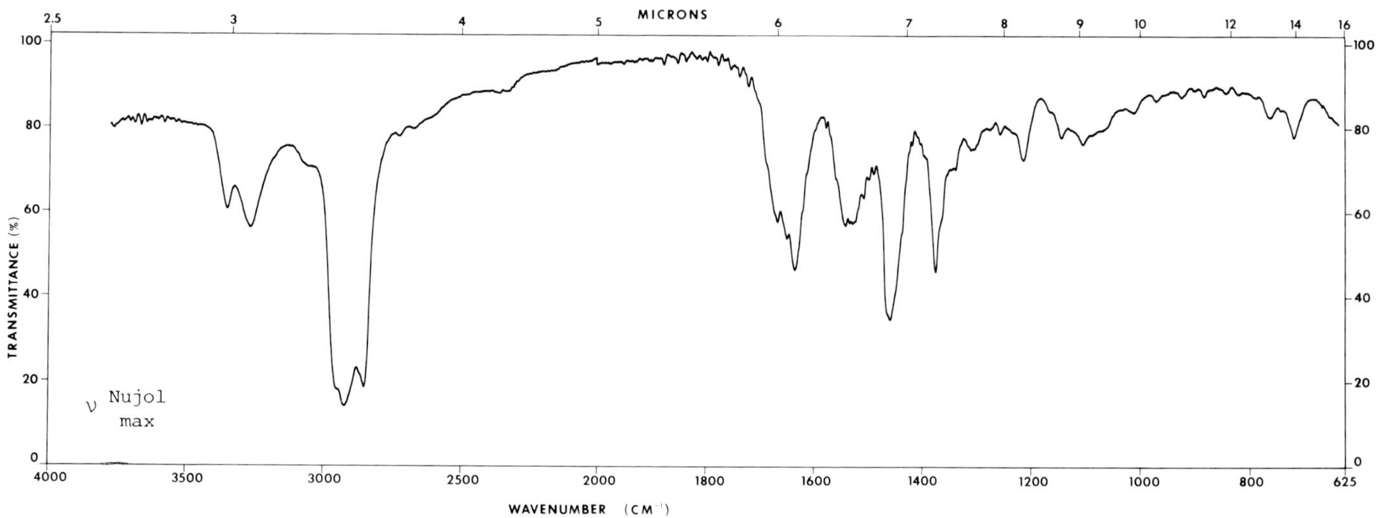
MICRONS
TRANSMITTANCE (%)
WAVENUMBER (CM⁻¹)
ν Nujol max

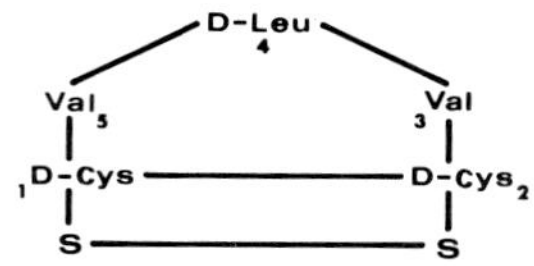

Malformin B_2

Common name[17]	Malformin B (composed of two cyclic peptides, B_1 and B_2; generally occur together in ratio of 1:1). B_1* is composed of 2 cysteine, one valine and two leucine residues B_2 is composed of 2 cysteine, two valine and one leucine residues.
Molecular weight	B_1: 529.2392 B_2: 515.2236
Molecular formula[17]	B_1: $C_{23}H_{39}O_5N_5S_2$ B_2: $C_{22}H_{37}O_5N_5S_2$
General characteristics[17]	Both decompose rapidly at > 300°C B_1: $[\alpha]_D^{25}$ $-40.3°$ (C = 0.72 in methyl cellosolve) B_2: $[\alpha]_D^{25}$ $-49.5°$ (C = 0.73 in methyl cellosolve)
UV data	End absorption
Fungal source[8–10,17]	*Aspergillus niger* *A. ficuum* *A. awamori* *A. phoenicis*
Toxicity data[17]	Similar activity as malformin A in corn roots. The optimum concentration for induction of root curvature in corn roots was 0.1 ppm. Vertebrate toxicity unknown.
TLC data[17]	*Adsorbent:* silica gel H *Solvent:* water-saturated ethyl acetate R_f: B_1: 0.84 B_2: 0.80 *Detection:* sprayed with 30% H_2SO_4 and heated.

* Structure unknown.

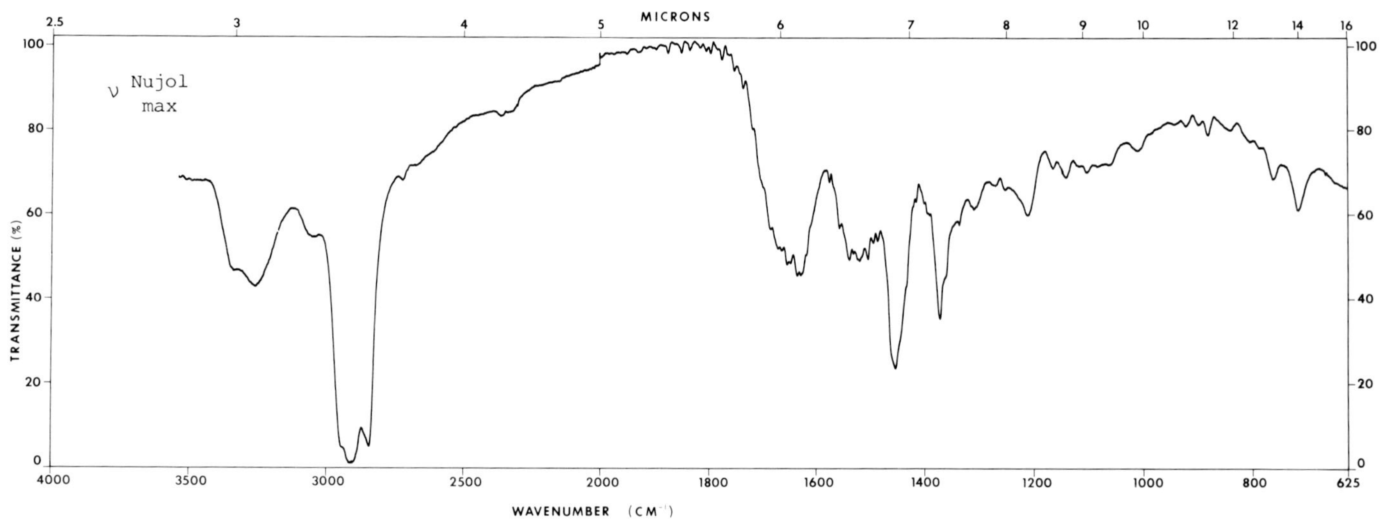

IR spectrum of malformin B_2.

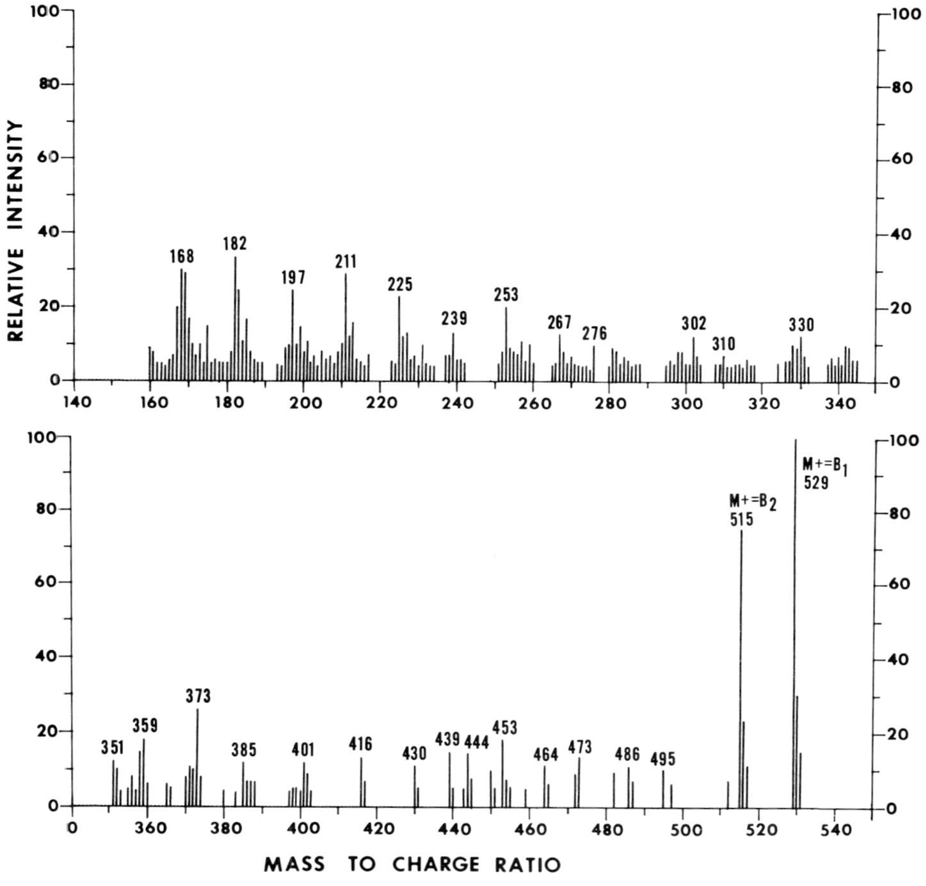

Mass spectrum of malformin B_1 and B_2.

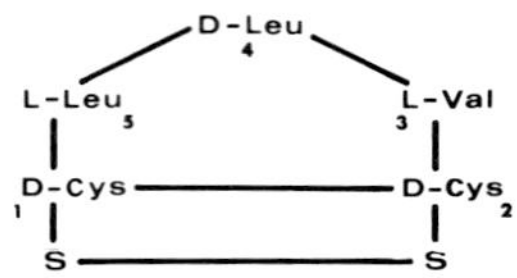

Common name	Malformin C (Cyclo-D-cysteinyl-D-cysteinyl-L-valyl-D-leucyl-L-leucyl)
Molecular weight	529.2393
Molecular formula	$C_{23}H_{39}O_5N_5S_2$
General characteristics[1]	Amorphous powder, mp >300°C (dec.) $[\alpha]_D^{25}$ −37.4° (*C* = 1.01 in dimethyl sulfoxide) $[\alpha]_D^{25}$ −37.4° (*C* = 0.52 in methyl cellosolve)
UV data[1]	End absorption only
Fungal source[1]	*Aspergillus niger*
Toxicity data[1,11]	Malformin C shows antibacterial activity against *Bacillus subtilis*. In the corn root curvature assay, the optimum concentration was 0.1 μg/ml (identical to that of malformin A). Application to the apical bud of *Phaseolus vulgaris* seedlings caused growth abnormalities identical to those produced by malformin A. The LD_{50} value in newborn rats dosed IP with DMSO carrier was 0.90 mg/kg. In 28-day-old rats, the LD_{50} was 0.87 mg/kg.
^1H NMR data	(DMSO-d_6): H18, 0.86 m; H6, 1.00–1.78 m; H1, 1.98 m; H, 3.17 m; H1, 3.46 d (*J* = 15); H5, 3.61–4.80 m; H1, 7.06 d (*J* = 11)*, 7.28 d (*J* = 9)*, 7.73 d (*J* = 8)*, 8.42 d (*J* = 6)*, 8.72 d (*J* = 5)*

* Exchangeable with D_2O.

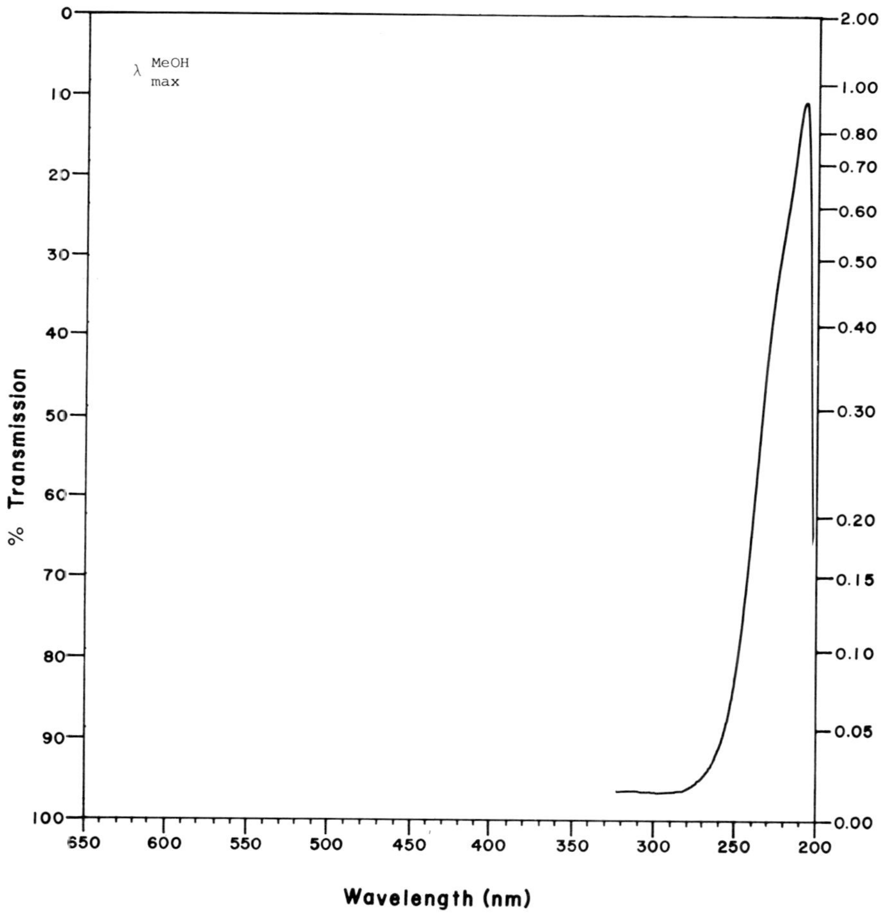
λ MeOH max
% Transmission
Absorbance (O D)
Wavelength (nm)
0
10
20
30
40
50
60
70
80
90
100
2.00
1.00
0.80
0.70
0.60
0.50
0.40
0.30
0.20
0.15
0.10
0.05
0.00
650
600
550
500
450
400
350
300
250
200

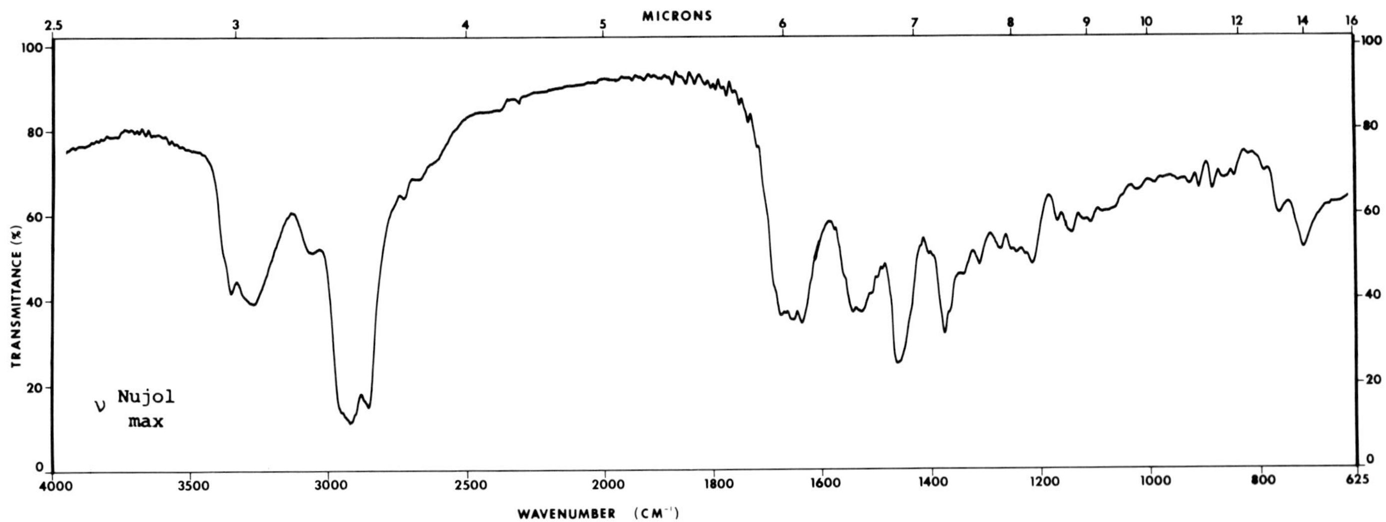
MICRONS
2.5
3
4
5
6
7
8
9
10
12
14
16
TRANSMITTANCE (%)
100
80
60
40
20
0
ν Nujol max
4000
3500
3000
2500
2000
1800
1600
1400
1200
1000
800
625
WAVENUMBER (CM⁻¹)

Malformin C

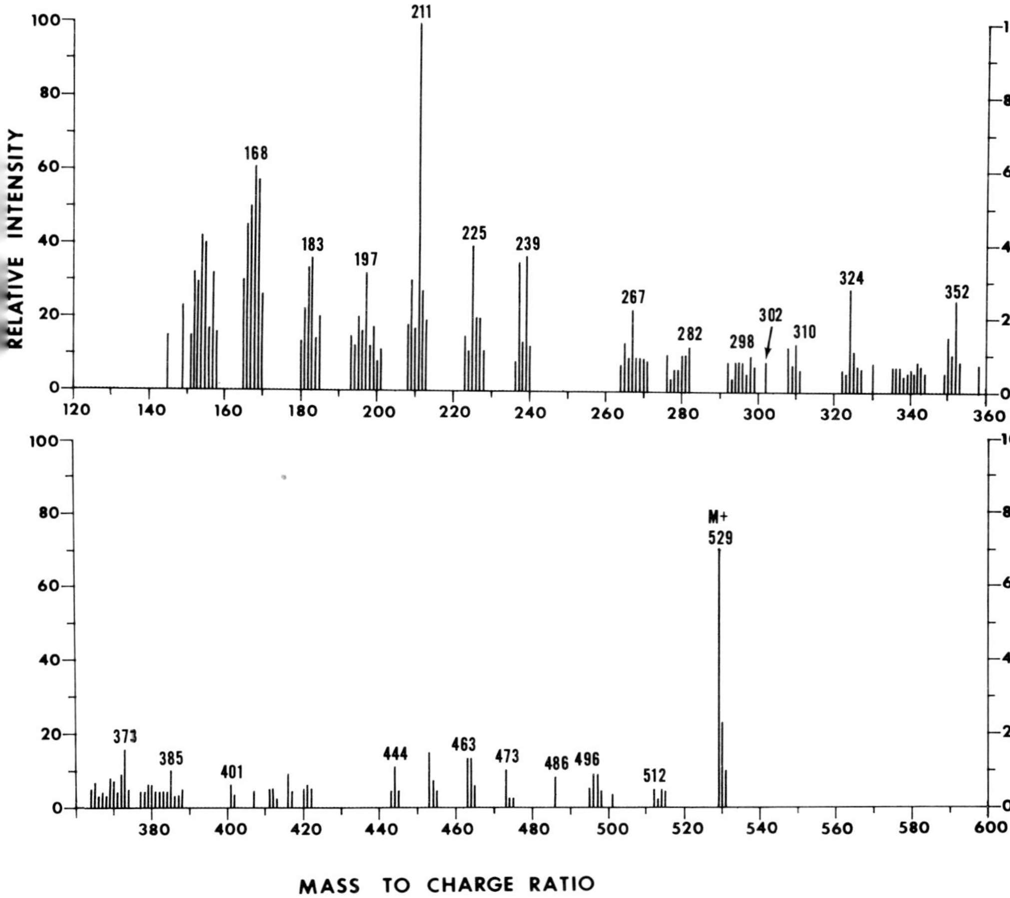
RELATIVE INTENSITY
211
168
183
197
225
239
267
282
298
302
310
324
352
M+
529
373
385
401
444
463
473
486
496
512
MASS TO CHARGE RATIO

REFERENCES

1. Anderegg, R. J., Biemann, K., Büchi, G., and Cushman, M., *J. Am. Chem. Soc.* **98,** 3365 (1976).
2. Anzai, K., and Curtis, R. W., *Phytochemistry* **4,** 263 (1965).
3. Bodanszky, M., Henes, J. B., Natarajan, S., and Stahl, G. L., *J. Antibiot.,* **29,** 549 (1976).
4. Bodanszky, M., and Stahl, G. L., *Bioorg. Chem. Ser. One* **4,** 93 (1975).
5. Bodanszky, M., and Stahl, G. L., *Proc. Natl. Acad. Sci. U.S.A.* **71,** 2791 (1974).
6. Curtis, R. W., and Fellenberg, G., *Plant Cell Physiol.* **13,** 715 (1972).
7. Curtis, R. W., *Plant Cell Physiol.* **10,** 203 (1969).
8. Curtis, R. W., *Science* **128,** 661 (1958).
9. Curtis, R. W., and Tanaka, H., *Appl. Microbiol.* **15,** 1519 (1967).
10. Iriuchijima, S., and Curtis, R. W., *Phytochemistry* **8,** 1397 (1969).
11. Kobbe, B., Cushman, M., Wogan, G. N., and Demain, A. L., *Appl. Environ. Microbiol.* **33,** 996 (1977).
12. Marumo, S., and Curtis, R. W., *Phytochemistry* **2,** 245 (1961).
13. Ptak, M., *Biopolymers* **12,** 1575 (1973).
14. Suda, S., and Curtis, R. W., *Appl. Microbiol.* **14,** 475 (1966).
15. Takahashi, N., and Curtis, R. W., *Plant Physiol.* **33,** 17 (1961).
16. Takahashi, N., and Curtis, R. W., *Plant Physiol.* **36,** 30 (1961).
17. Takeuchi, S., Senn, M., Curtis, R. W., and McLafferty, F. W., *Phytochemistry* **6,** 287 (1967).
18. Yoshizawa, T., Tsuchiya, Y., Morooka, N., and Sawada, Y., *Agric. Biol. Chem.* **39,** 1325 (1975).

Penicillium islandicum Toxins

15

It was observed that rice molded by *Penicillium islandicum* caused acute and chronic liver toxicoses in experimental animals.[6,26] These observations resulted in extensive studies concerning the nature of the toxins produced by this fungus. Subsequently, several highly toxic metabolites were isolated and identified as the toxic agents. These include several anthraquinoid pigments, a non-anthraquinoid, a nitrogen-containing red pigment (erythroskyrin), and two cyclic chlorine-containing peptides (cyclochlorotine and islanditoxin). The toxic anthraquinoid pigments can be grouped into monomeric anthraquinones such as emodin, or bianthraquinones such as luteoskyrin and rugulosin.

The two cyclic chlorine-containing pentapeptides, islanditoxin and cyclochlorotine, exhibit very similar toxic effects and chemical properties. However, different chemical structures have been proposed for these toxins. The toxic pigment erythroskyrin can be described as a β-triketonic lactam ring involving *N*-methylvaline connected to a bicyclic oxyrane ring system via a polyene moiety.

At least two reviews on the topic have been published.[6,26]

Penicillium islandicum toxins	Molecular weight	Molecular formula
Emodin	270.0528	$C_{15}H_{10}O_5$
Erythroskyrin	455.2306	$C_{26}H_{33}O_6N$
Skyrin	538.0899	$C_{30}H_{18}O_{10}$
Rugulosin	542.1212	$C_{30}H_{22}O_{10}$
Luteoskyrin	574.1110	$C_{30}H_{22}O_{12}$
Cyclochlorotine	571.1587	$C_{24}H_{31}O_7N_5Cl_2$
Islanditoxin	571.1587	$C_{24}H_{31}O_7N_5Cl_2$

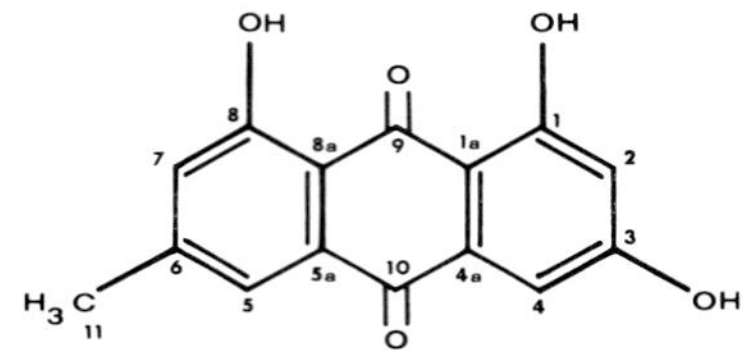

Common name	Emodin (Frangula emodin; Rheum; Archin; Frangulic acid) (1,3,8-Trihydroxy-6-methyl-anthraquinone)
Molecular weight	270.0528
Molecular formula	$C_{15}H_{10}O_5$
General characteristics[54]	Crystals from chloroform, mp 256°–257°C
UV data[3,54]	λ_{max}^{EtOH} nm (ϵ): 223(35,400), 250(18,200), 267(18,600), 290(21,800), and 442(12,500)
Fungal source[1,2,5,21,23,35,38,48,57]	*Penicillium islandicum* Sopp *P. clavariae-formis* *P. brunneum* Udagawa *P. avellaneum* (Thom et Turesson) *Aspergillus ochraceus* *A. wentii* Wehmer *Phoma foveata* Foister *Cladosporium fulvum* Cooke Also found in rhubarb root, older buckthorn (*Rhamnus frangula* L., *R. purshinana, Rumex* spp.), and other polygonaceae
Toxicity data[14,54]	The LD_{50} of emodin in day-old cockerels dosed orally was 3.7 mg/kg. Gross clinical signs were anorexia, diarrhea, accumulation of fecal material with acute epidermal irritation around the cloaca, and general debilitation. Emodin has also been shown to inhibit synchronously dividing cells of *Tetrahymena pyriformis* GL at 5 μg/ml.
TLC data[54]	*Adsorbent:* silica-7GF (Mallinckrodt) *Solvent:* A: toluene–ethyl acetate–formic acid, 5:4:1 v/v/v B: chloroform–acetone, 93:7 v/v R_f: A: 0.80 B: 0.45

	Detection: orange-red spot in visible light
Other significant references	Biosynthesis[7]
^{13}C NMR data[43]	C1, 164.3 s; C1a, 108.7 s; C2, 107.7 d; C3, 165.4 s; C4, 108.7 d; C4a, 134.8 s; C5, 120.2 d; C5a, 132.5 s; C6, 148.0 s; C7, 123.9 d; C8, 161.2 s; C8a, 112.9 s; C9, 189.4 s; C10, 180.9 s; C11, 21.4 q
^{1}H NMR data	H2, 6.54; H4, 7.07; H5, 7.39; H7, 7.07; H11, 2.48; OH, 11.95, 12.02

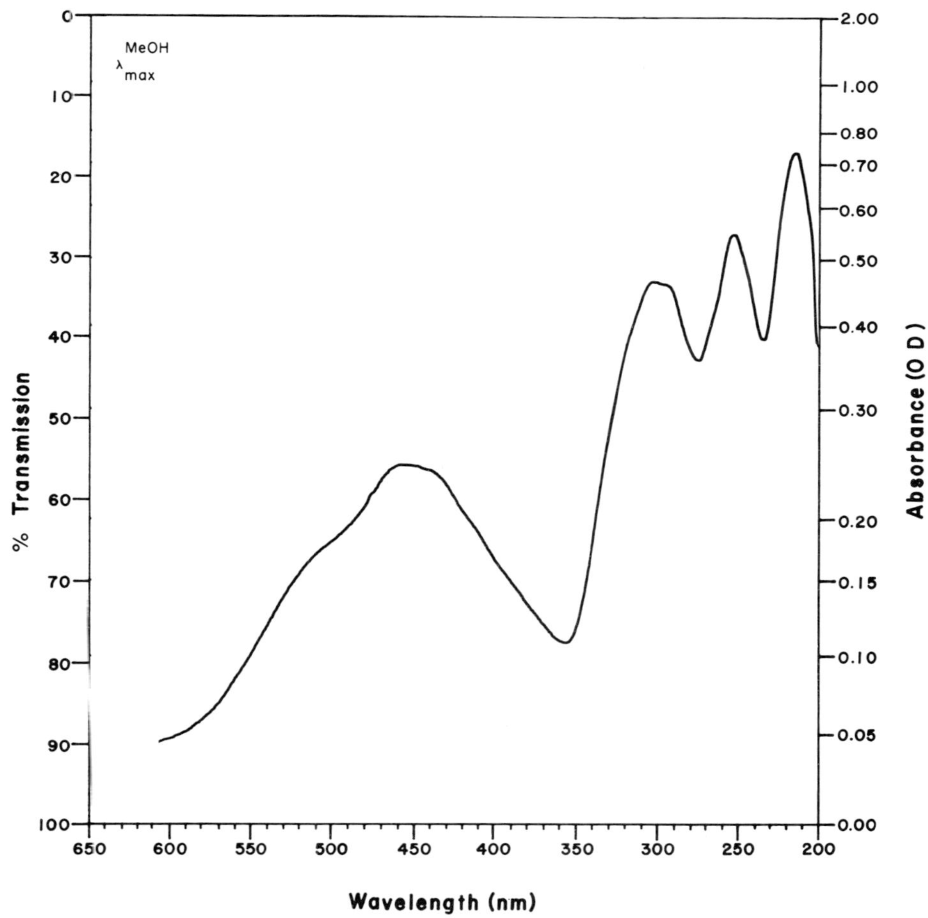

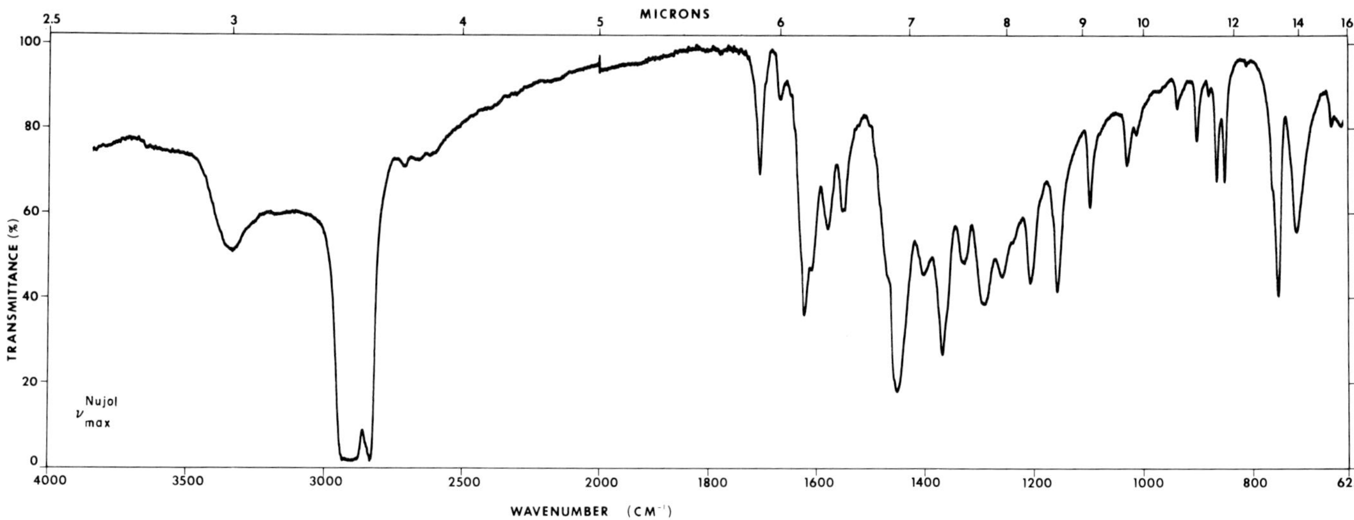
MICRONS
2.5
3
4
5
6
7
8
9
10
12
14
16
TRANSMITTANCE (%)
Nujol
ν_{max}
WAVENUMBER (CM⁻¹)
4000
3500
3000
2500
2000
1800
1600
1400
1200
1000
800
625

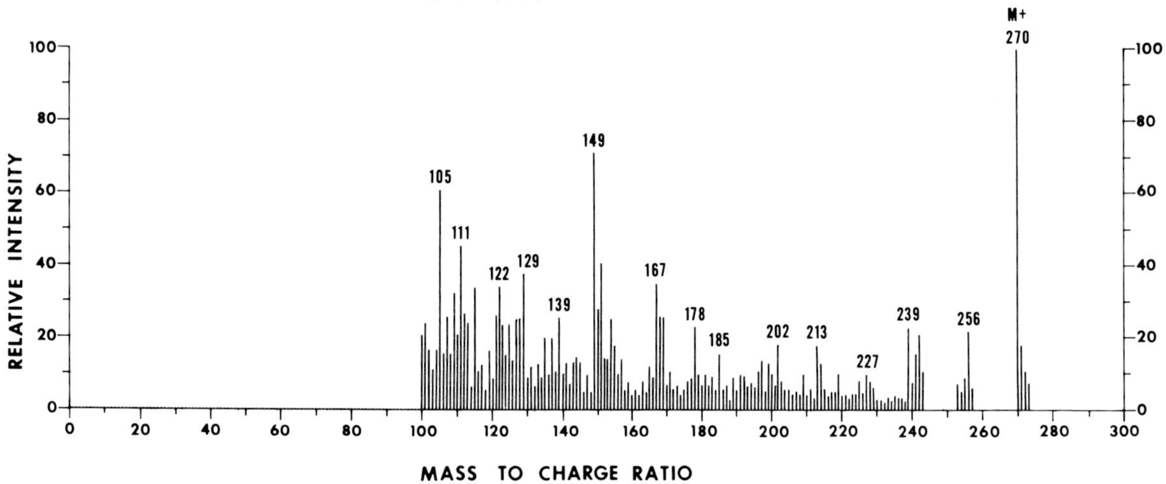
RELATIVE INTENSITY
MASS TO CHARGE RATIO
M+
270
105
111
122
129
139
149
167
178
185
202
213
227
239
256

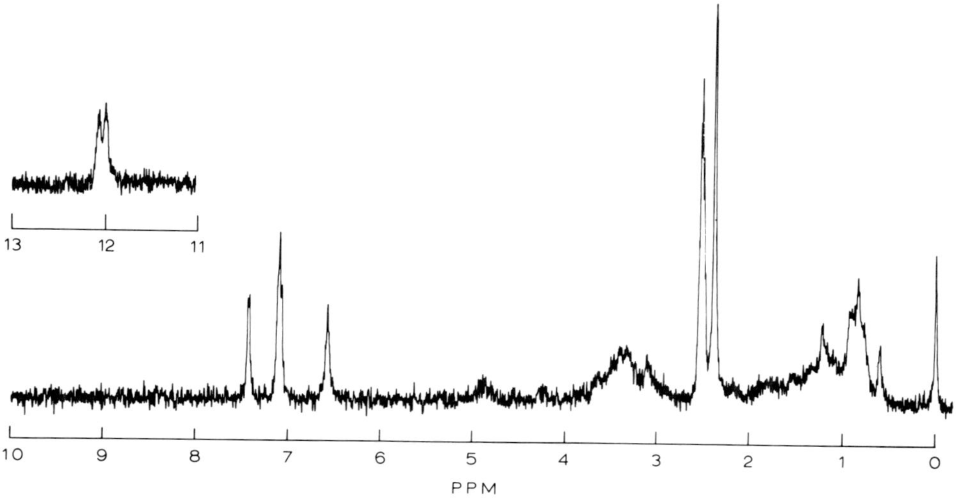

The 100 MHz ^{1}H NMR spectrum of emodin in DMSO solution.

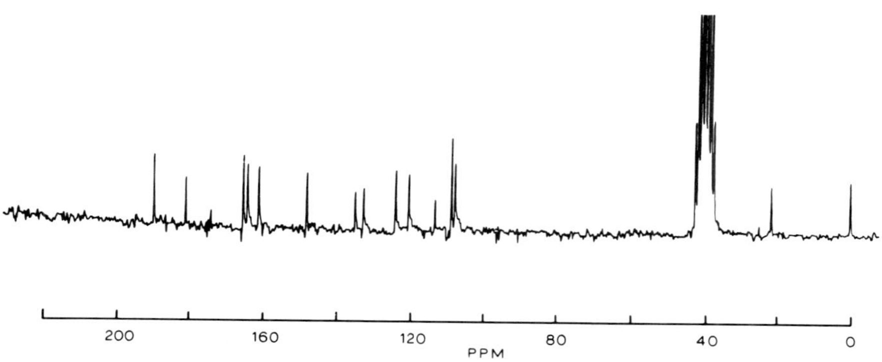

The 25 MHz ^{13}C NMR spectrum of emodin in DMSO solution.

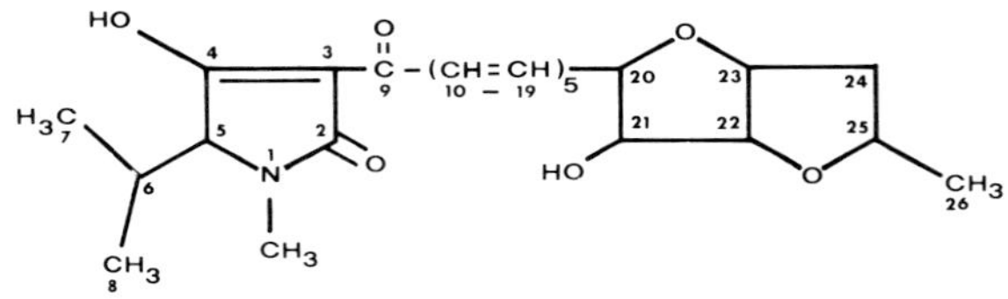

Common name	Erythroskyrin
Molecular weight	455.2306
Molecular formula	$C_{26}H_{33}O_6N$
General characteristics[9,39]	Orange-red crystals from ethanol, mp 130°–133°C $[\alpha]_D$ +46.9° Dark-red microcrystals from methanol or ethanol (effloresce on standing in air), mp 108°–110°C. Erythroskyrin is soluble in acetone, chloroform, methanol, ethanol, benzene, acetic acid, and pyridine; less soluble in ethyl ether, hexane, and petroleum ether.
UV data[39]	λ_{max}^{EtOH} nm (ϵ): 409(28,100) and 260(8,900) $\lambda_{max}^{0.1N\,NaOH}$ nm (ϵ): 392(60,000) and 260(13,800)
IR data[39]	$\nu_{max}^{CHCl_3}$ cm^{-1}: (carbonyl region) 1550, 1564, 1584, 1612, 1635, 1693, and 1010
Fungal source[9,39]	*Penicillium islandicum* Sopp (NRRL 1036)
Toxicity data[46]	The LD_{50} of erythroskyrin in 20 g mice dosed IP was 60 mg/kg while an LD_{100} was observed at 200 mg/kg. Clinical signs were paralysis, comatosis, and death. Hepatic damage, including centrolobular degeneration of the liver cells, nephrotic changes, and cellular injury of lymphatic system, spleen, and thymus were typical in mice dying 2–5 days after administration of toxin. Mice sacrificed 7 days after dosing showed less pronounced liver damage; however, an increase in mitotic cells and mobilization of Kupffers cells of the liver and slight hyperplasia of the lymph apparatus were observed.
TLC data[46]	*Adsorbent:* Kieselgel G treated with 0.1 *N* oxalic acid *Solvent:* *n*-hexane–acetone–water, 6:4:1 v/v/v R_f: 0.30 *Detection:* red spot in visible light

Other significant references	Review[26]
^{1}H NMR data[38]	H5, 3.53; H6, 1.63; H7, 0.95; H8, 1.10; H10–19, 6.30–7.1; H20, 4.75; H21, 4.50; H22, 3.79; H23, 4.06; H24, 2.25; H25, 4.25; H26, 1.27; H27, 2.98; 21OH, 2.95

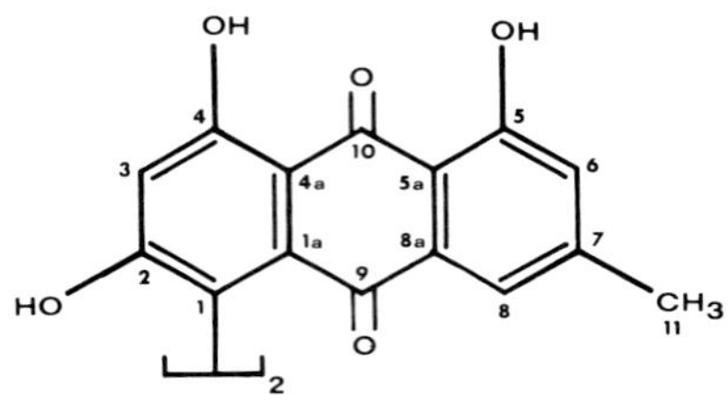

Common name	Skyrin (Endothianin) (1,1′-Bisemodin)
Molecular weight	538.0899
Molecular formula	$C_{30}H_{18}O_{10}$
General characteristics[8,34,37]	Crystallizes as either clusters of dark orange rods from ether or as hexagonal, yellow plates from acetone, mp >380°C; ethoxycarbonyl ether derivative, mp 171°–173°C; hexaacetylskyrin, thin yellow needles, mp 295°–296°C $[\alpha]_D^{22}$ 0° (in acetone) Treatment with $Na_2S_2O_4$ in alkaline solution yields reductive cleavage into 2 molecules of emodin.
Fungal source[40]	*Penicillium islandicum* Sopp (NRRL 1036 and 1175) *P. rugulosum* *P. wortmanni* *P. tardum* *P. brunneum* *Sepedonium ampullosporum* *Penicilliopsis clavariaeformis* *Endothia parasitica* *E. fluens* *E. gyrosa* *E. longirostris* *E. tropicalis* Lichen source *Acroscyphus sphaerophoroides* *Trypethliopsis bonninensis* *Physcia obscura* var. *endococcina* *Pyxine endochrysina*
Toxicity data	No apparent vertebrate toxicity
TLC data[40]	*Adsorbent:* silica gel G impregnated with 0.5 *N* oxalic acid

	Solvent: benzene–acetone, 4:1 v/v R_f: 0.48 *Detection:* yellow spot in visible light
Other significant references	^{13}C NMR[43]
^{13}C NMR data	C1, 108.9 s; C2, 164.2 s; C3, 107.0 d; C4, 164.2 s; C5, 160.9 s; C6, 123.3 d; C7, 148.0 s; C8, 120.2 d; C9, 181.8 s; C10, 189.5 s; C11, 21.4 q; C1a, 133.1 s; C4a, 123.3 s; C5a, 113.0 s; C8a, 131.2 s
^{1}H NMR data	H3, 6.69; H6, 7.07; H8, 7.22; H11, 2.35; OH's, 11.92, 12.64

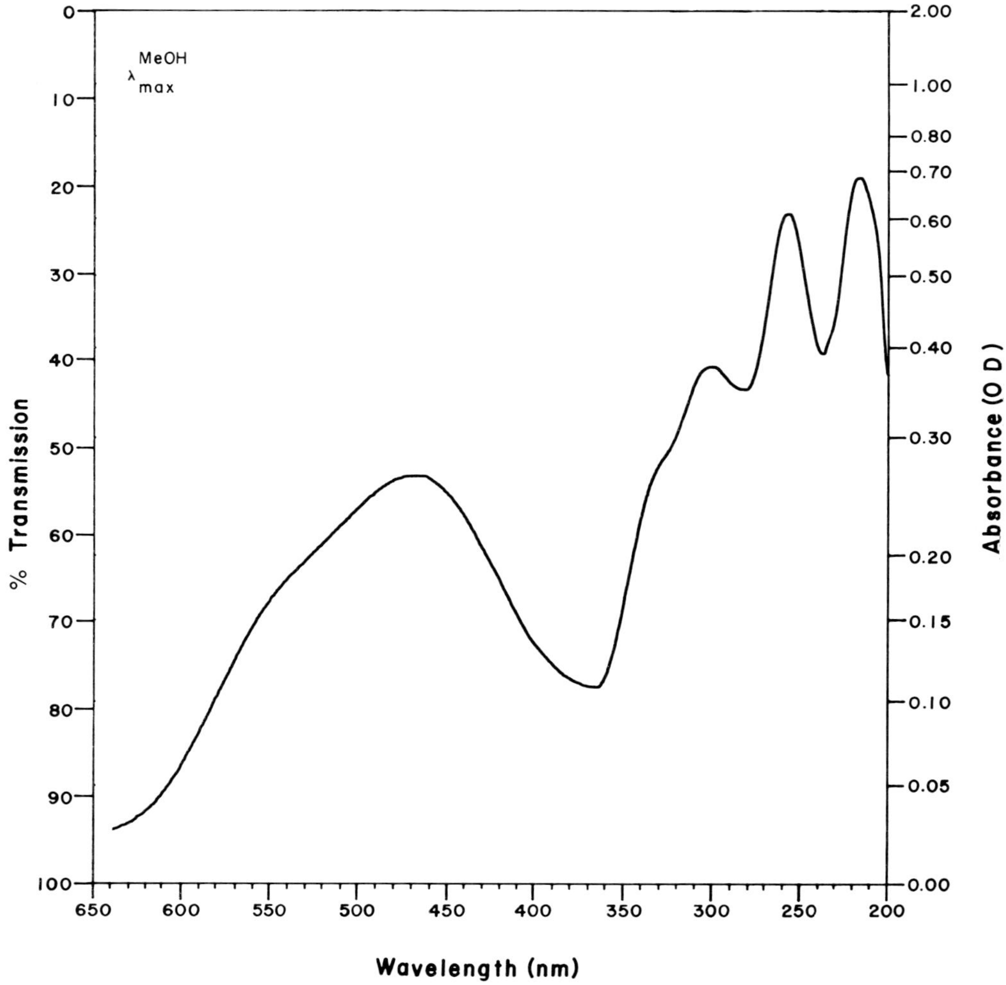

λ_{max}^{MeOH}
% Transmission
Absorbance (O D)
Wavelength (nm)
0
10
20
30
40
50
60
70
80
90
100
2.00
1.00
0.80
0.70
0.60
0.50
0.40
0.30
0.20
0.15
0.10
0.05
0.00
650
600
550
500
450
400
350
300
250
200

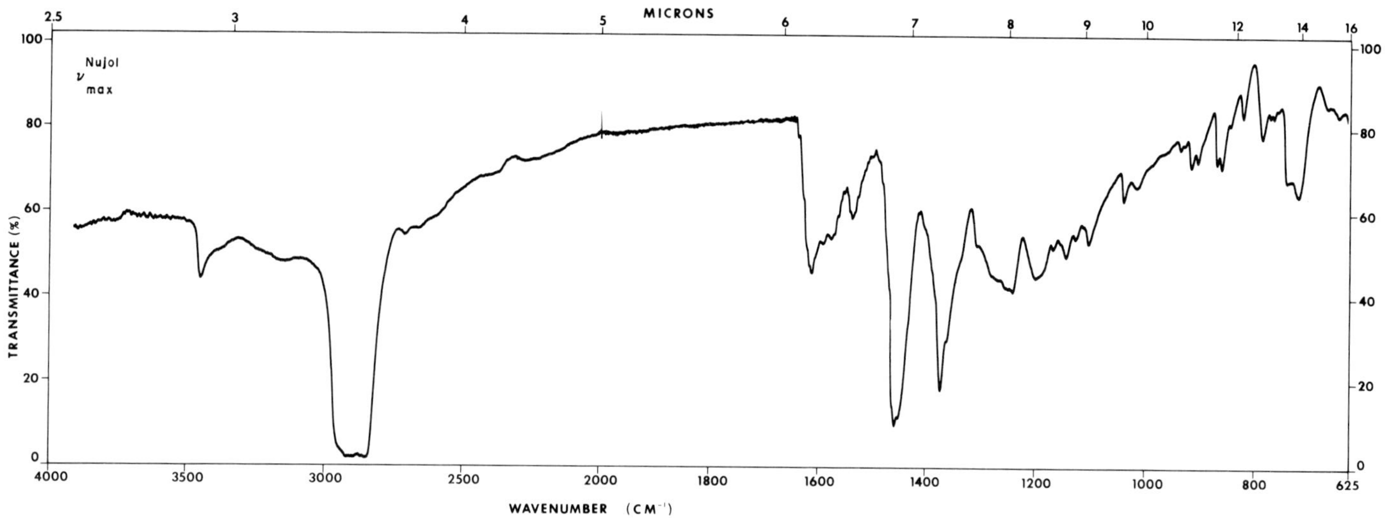

MICRONS
2.5
3
4
5
6
7
8
9
10
12
14
16
Nujol
ν max
TRANSMITTANCE (%)
100
80
60
40
20
0
WAVENUMBER (CM⁻¹)
4000
3500
3000
2500
2000
1800
1600
1400
1200
1000
800
625

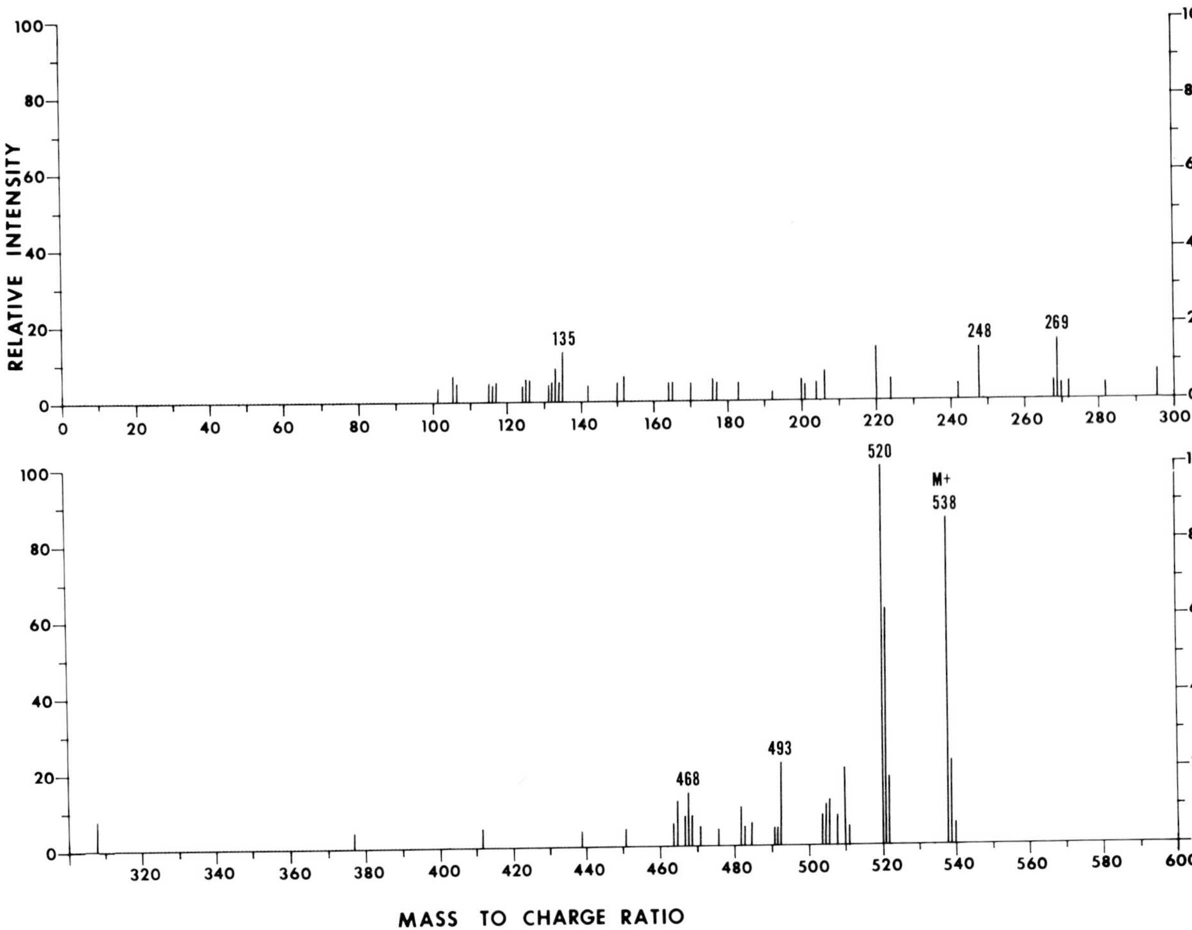
RELATIVE INTENSITY
135
248
269
520
M+
538
468
493
MASS TO CHARGE RATIO

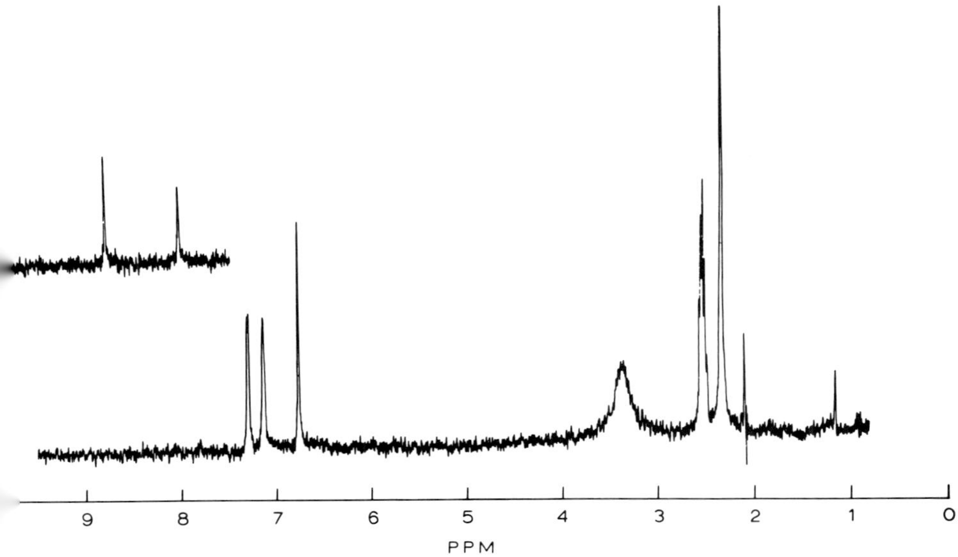

The 100 MHz ^{1}H NMR spectrum of skyrin in DMSO solution.

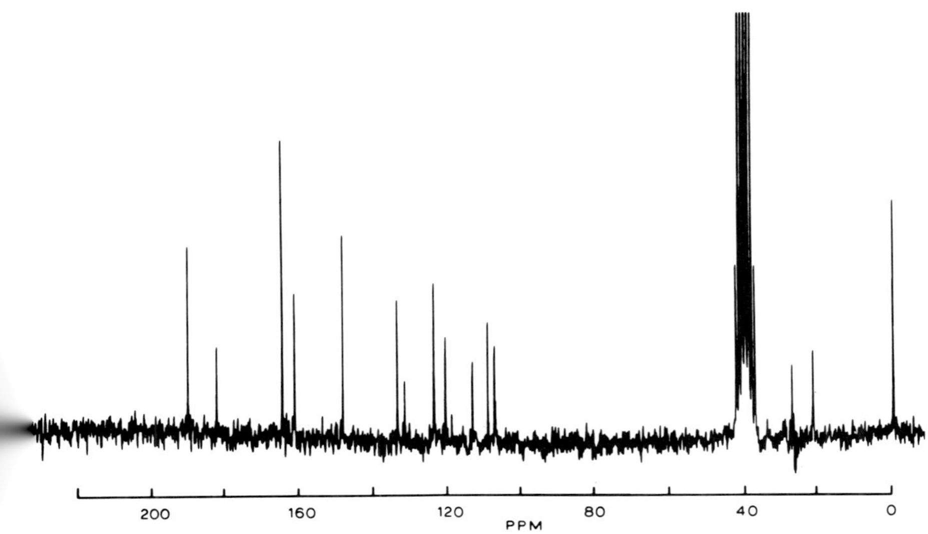

The 25 MHz ^{13}C NMR spectrum of skyrin in DMSO solution.

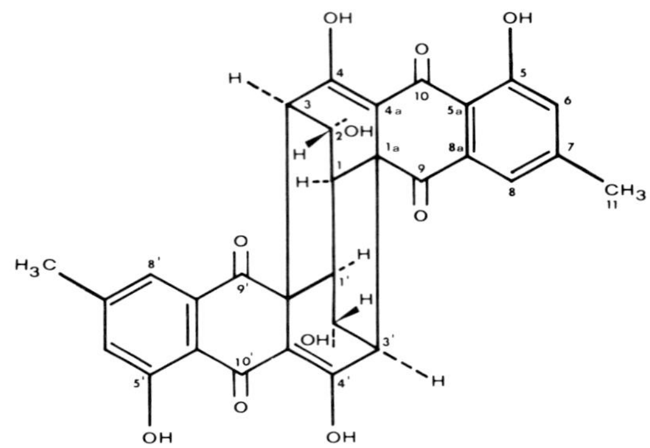

Common name[34]	(+)Rugulosin (Redicalisin) (2,2′,4,4′, 5,5′-Hexahydroxy-2,2′,3,3′-tetrahydro-7,7′-dimethyl-1,1′-bianthraquinone)
Molecular weight	542.1212
Molecular formula	$C_{30}H_{22}O_{10}$
General characteristics[41,50]	Fine yellow crystals from methanol, mp 290°C (dec.) $[\alpha]_D^{20}$ +466° (C = 0.5 in dioxane)
UV data[24,56]	λ_{max}^{MeOH} nm (ϵ): 248(27,500), 273(19,500), 302(12,000), 324(13,000), 345(14,000), 391(22,000), and 402(21,000)
CD data[24]	$\Delta\epsilon$ (in methanol): 406(+8.5), 350(−7.3), 335(−7.0), 330(−7.2), 310(−6.0), 304(−6.6), 292(−1.6), 280(−14.8), and 250(+14.8)
Fungal source[4,34,35,38,41,56]	*Penicillium rugulosum* Thom *P. variabile* *P. brunneum* *P. tardum* Thom *P. wortmanni* Klöcker *Endothia parasitica* (Murr) Anderson and Anderson *E. fluens* Shear and Stevens *E. gyrosa* *Sepedonium ampullosporum* *NOTE:* The antipode (−)rugulosin has been found in *P. islandicum* (NRRL 1175) and

Myrothecium verrucaria (Alb. and Schw.) Ditmar and Fr.

Toxicity data[4,12,28,48,50,53]

Rugulosin had an LD_{50} in male mice of 83.0 mg/kg (IP) dissolved in 0.9% NaCl; LD_{50} dissolved in olive oil was 44.0 mg/kg in female rats (IP). Characteristic signs were enlarged liver with centrolobular necrosis and fatty metamorphosis. Rats that survived over long periods following a single dose of rugulosin showed abnormal livers with perihepatitis due to postnecrotic collapse. Cellular growth in Ehrlich ascites tumor in mice was inhibited by the IP administration of 4 mg/kg/day of rugulosin, causing a prolongation of lifetime of tumor-bearing mice. Rugulosin possesses some antibiotic, antiprotozoal, and antitumor activity in addition to phototoxicity.

TLC data[40]

Adsorbent: silica gel G impregnated with 0.5 *N* oxalic acid
Solvent: benzene–hexane, 1:1 v/v
R_f: 0.25
Detection: visible light

Other significant references

X-Ray crystallography[13]; chemical conformation[25]; chemistry[29,30,36]

^{13}C NMR data

C1, 47.7 d; C2, 68.5 d; C3, 58.2 d; C4, 180.8 s; C5, 160.1 s; C6, 123.9 d; C7, 147.5 s; C8, 120.4 d; C9, 185.6 s; C10, 193.8 s; C11, 21.4 q; C1a, 55.6 s; C4a, 106.1 s; C5a, 114.1 s; C8a, 131.9 s

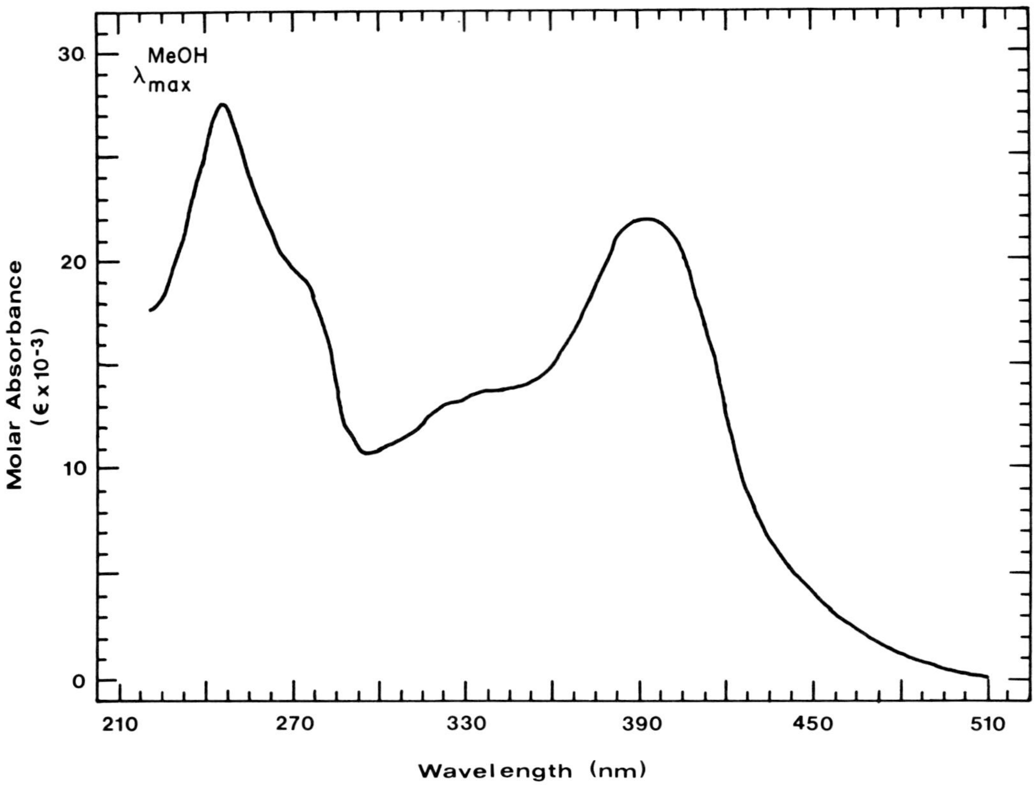

30
20
10
0
MeOH
λmax
Molar Absorbance
(ε x 10-3)
210
270
330
390
450
510
Wavelength (nm)

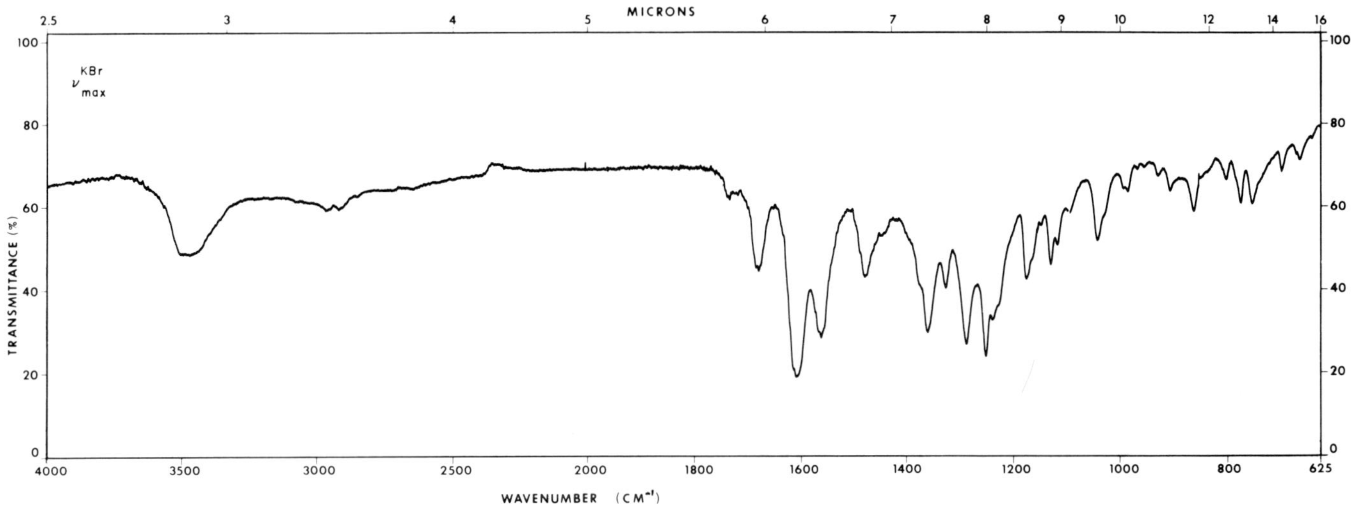
MICRONS
2.5
3
4
5
6
7
8
9
10
12
14
16
100
80
60
40
20
0
TRANSMITTANCE (%)
KBr
ν max
4000
3500
3000
2500
2000
1800
1600
1400
1200
1000
800
625
WAVENUMBER (CM⁻¹)

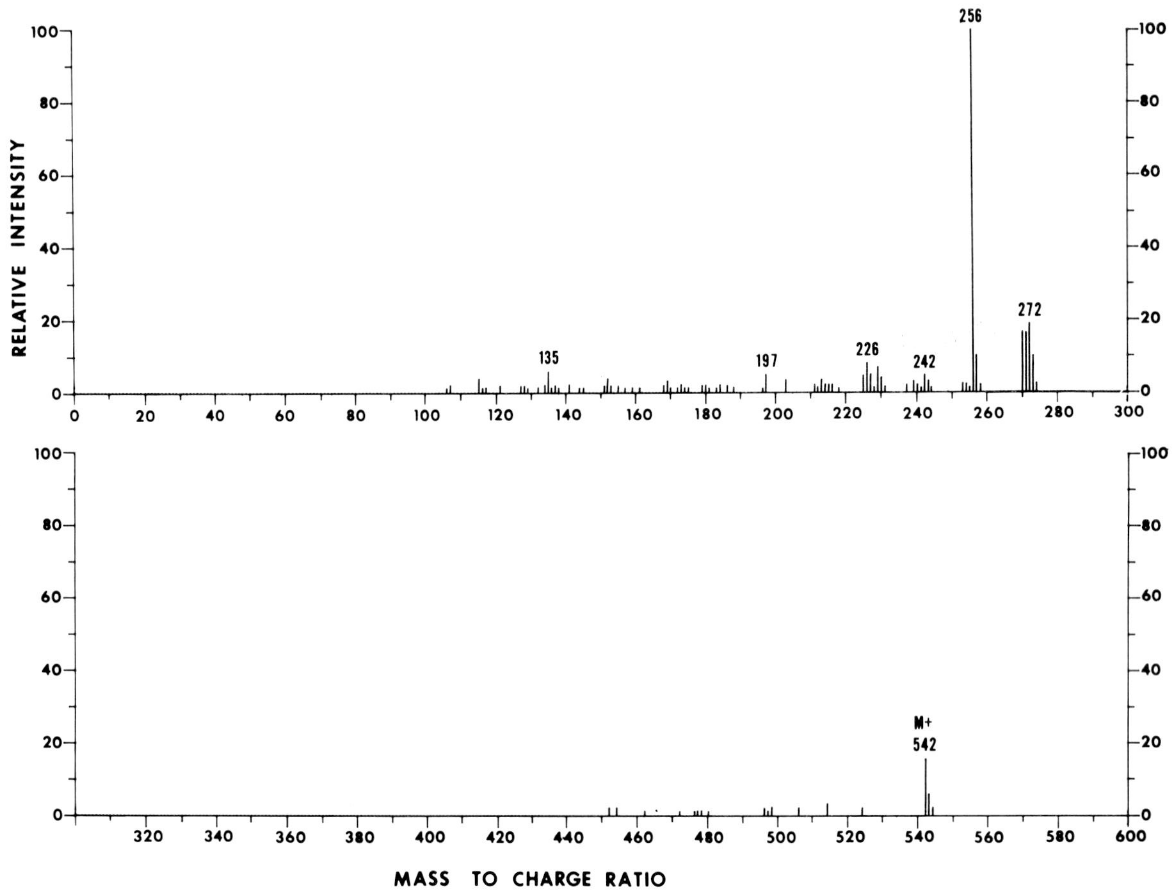
RELATIVE INTENSITY
MASS TO CHARGE RATIO
135
197
226
242
256
272
M+
542

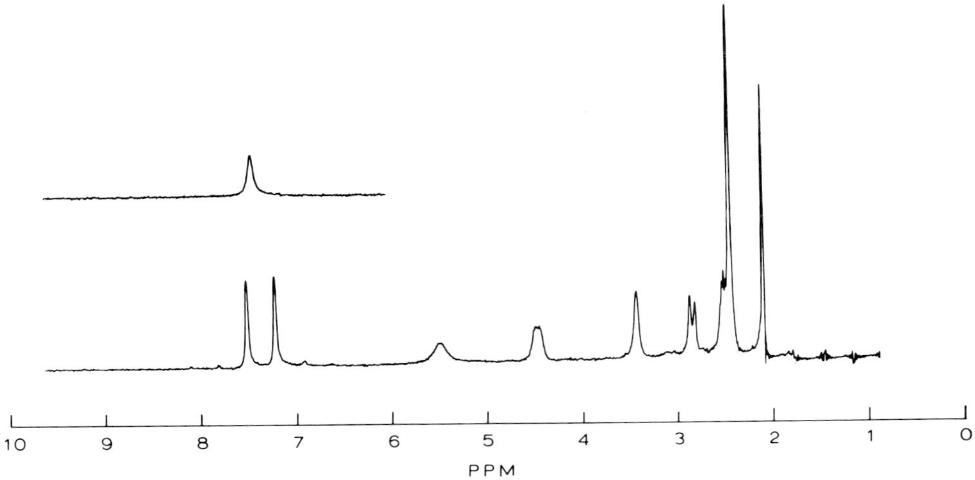

The 100 MHz ^{1}H NMR spectrum of rugulosin in DMSO solution.

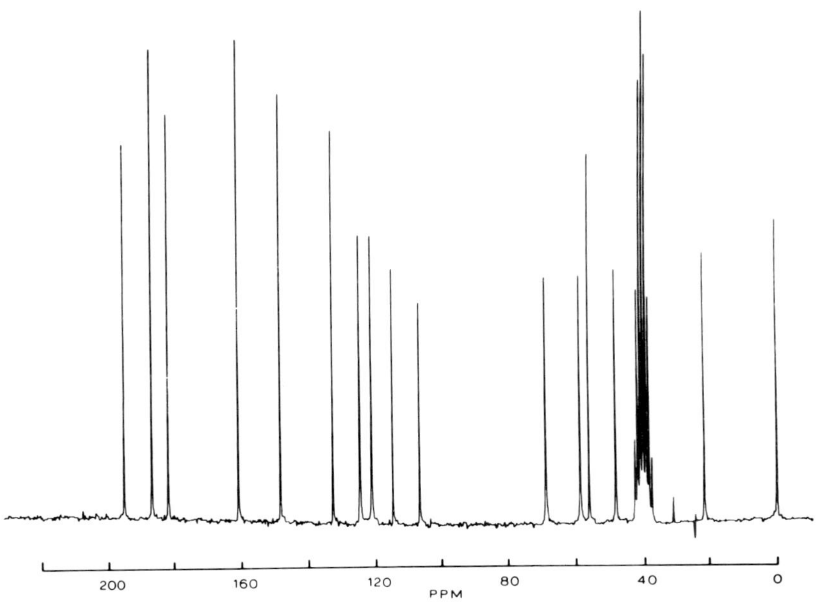

The 25 MHz ^{13}C NMR spectrum of rugulosin in DMSO solution.

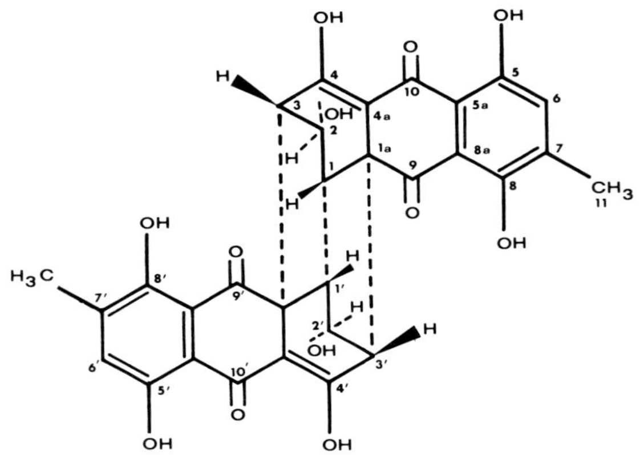

Common name[32]	(−)Luteoskyrin (Flavomycelin) (2,2′,4,4′,5,5′,8,8′-Octahydroxy-2,2′,3,3′-tetrahydro-7,7′-dimethyl-1,1′-bianthraquinone)
Molecular weight	574.1110
Molecular formula	$C_{30}H_{22}O_{12}$
General characteristics[31,33]	Yellow crystals from acetone–hexane, mp 287°C (dec.) $[\alpha]_D^{25}$ −880° (C = 0.1 in acetone) Luteoskyrin is photosensitive in several organic solvents, being converted to lumiluteoskyrin ($C_{30}H_{20}O_{12}$)
UV data[24,25,55]	$\lambda_{max}^{acetone}$ nm (ϵ): 446(28,700), 429(28,400), and 350(8,000) λ_{max}^{EtOH} nm (ϵ): 245(20,800), 275(21,300), and 430(28,800)
CD data[24]	$\Delta\epsilon$ (in methanol): 445(−17.5), 432(−16.1), 376(+3.3), 372(+3.1), 356(+7.5), 350(+5.1), 340(+7.7), 330(+5.6), 310(+9.5), 288(+22.0), 253(−18.7), and 224(+13.6)
Fungal source[22,32,42,51]	*Penicillium islandicum* Sopp (NRRL 1036) *Mycelia sterilia*
Toxicity data[26,47,49,53]	*Acute toxicity:* The acute toxicity of luteoskyrin in mice, rats, rabbits, and monkeys was characterized by hepatic lesions. Histological changes of the

liver were centrolobular necrosis and fatty degeneration of liver cells. Nuclear degeneration such as hyperchromatosis of nuclear membrane, karyorrhexis, and karyolysis were frequently found. In less-damaged liver, only single scattered liver cells were necrotic, reminiscent of the acidophilis body in viral hepatitis. Acute liver damage revealed swelling of mitochondria with irregular destruction or lytic lesions of the mitochondrial cristal, cystic dilation of endoplasmic reticulum, and positive fatty droplets via electron microscopy.

Chronic toxicity: Mice fed daily with 500 μg/10 gm body weight continuously developed liver tumors and other tumors after 6 months of feeding. The LD_{50} of luteoskyrin to mice was 40.8 mg/kg (IP), 147.0 mg/kg (SC), 221 mg/kg (PO), and 6.65 mg/kg (IV) dosed in 0.9% NaCl. Luteoskyrin acts selectively on the liver, causing atrophy, cirrhosis, chronic injury, and neoplastic changes. Only a small fraction of the administered dose is required to cause functional liver damage and morphologic damage, either acutely or chronically. Biochemically, luteoskyrin binds with DNA in the presence of magnesium ions *in vitro,* irreversibly inactivating DNA-dependent RNA polymerase.

TLC data[44]

Adsorbent: silica gel G impregnated with 0.5 *N* oxalic acid
Solvent: acetone–*n*-hexane–water, 6:3:1.5 v/v/v
R_f: 0.40
Detection: a yellow spot in visible light

Other significant references

X-Ray crystallography[13]; ^{13}C NMR[43]; absolute structure[41,43]; production[44]

^{13}C NMR data

C1, 48.8 d; C2, 68.3 d; C3, 59.8 d; C4, 180.2 s; C5, 153.6 s; C6, 128.8 d; C7, 137.8 s; C8, 155.0 s; C9, 185.1 s; C10, 200.5 s; C11, 15.9 q; C1a, 55.4 s; C4a, 106.0 s; C5a, 111.1 s; C8a, 110.6 s

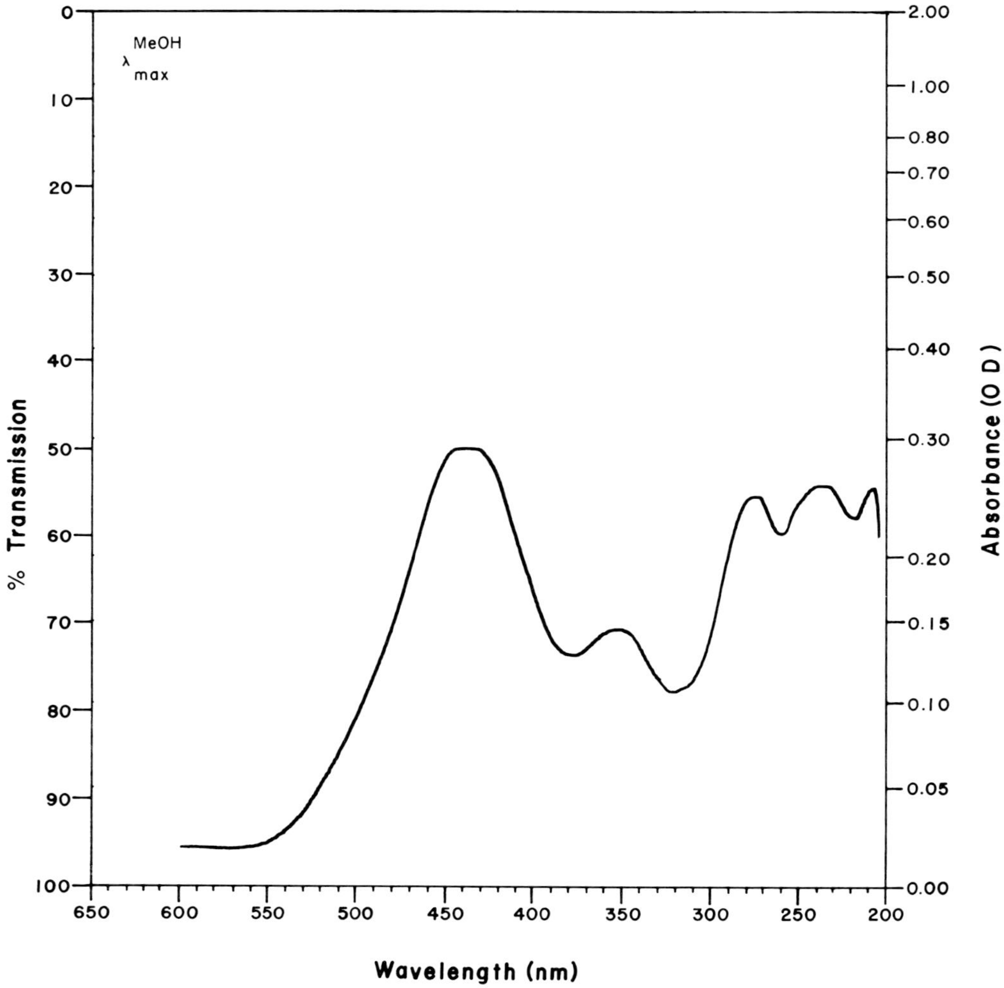

λ_{max}^{MeOH}
% Transmission
0
10
20
30
40
50
60
70
80
90
100
Absorbance (O D)
2.00
1.00
0.80
0.70
0.60
0.50
0.40
0.30
0.20
0.15
0.10
0.05
0.00
650 600 550 500 450 400 350 300 250 200
Wavelength (nm)

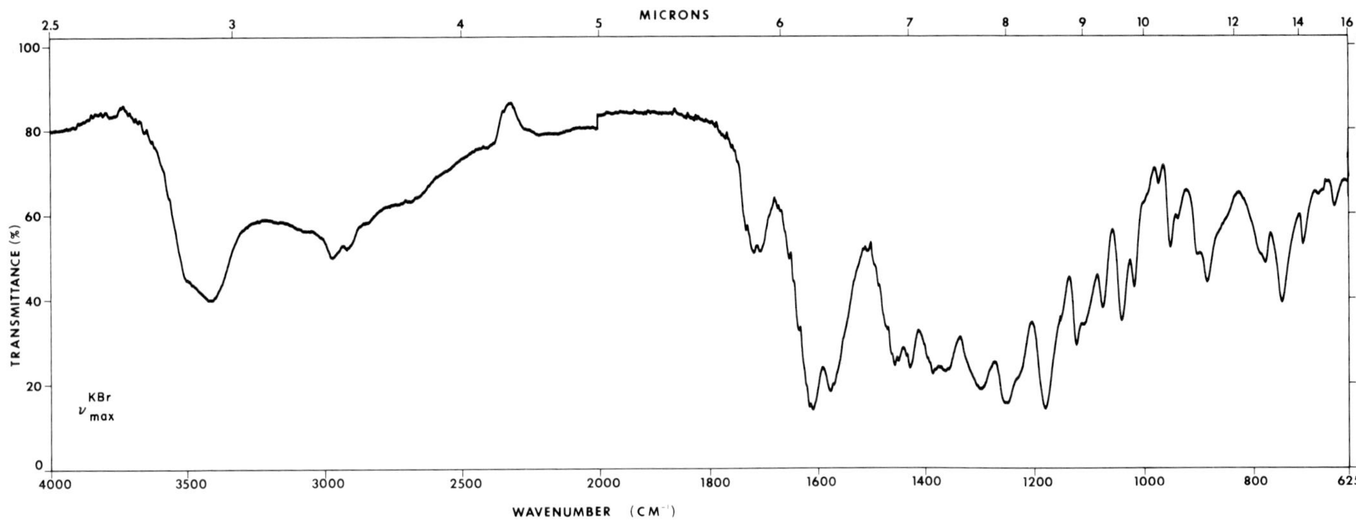
MICRONS
2.5
3
4
5
6
7
8
9
10
12
14
16
100
80
60
40
20
0
TRANSMITTANCE (%)
KBr
ν max
4000
3500
3000
2500
2000
1800
1600
1400
1200
1000
800
625
WAVENUMBER (CM⁻¹)

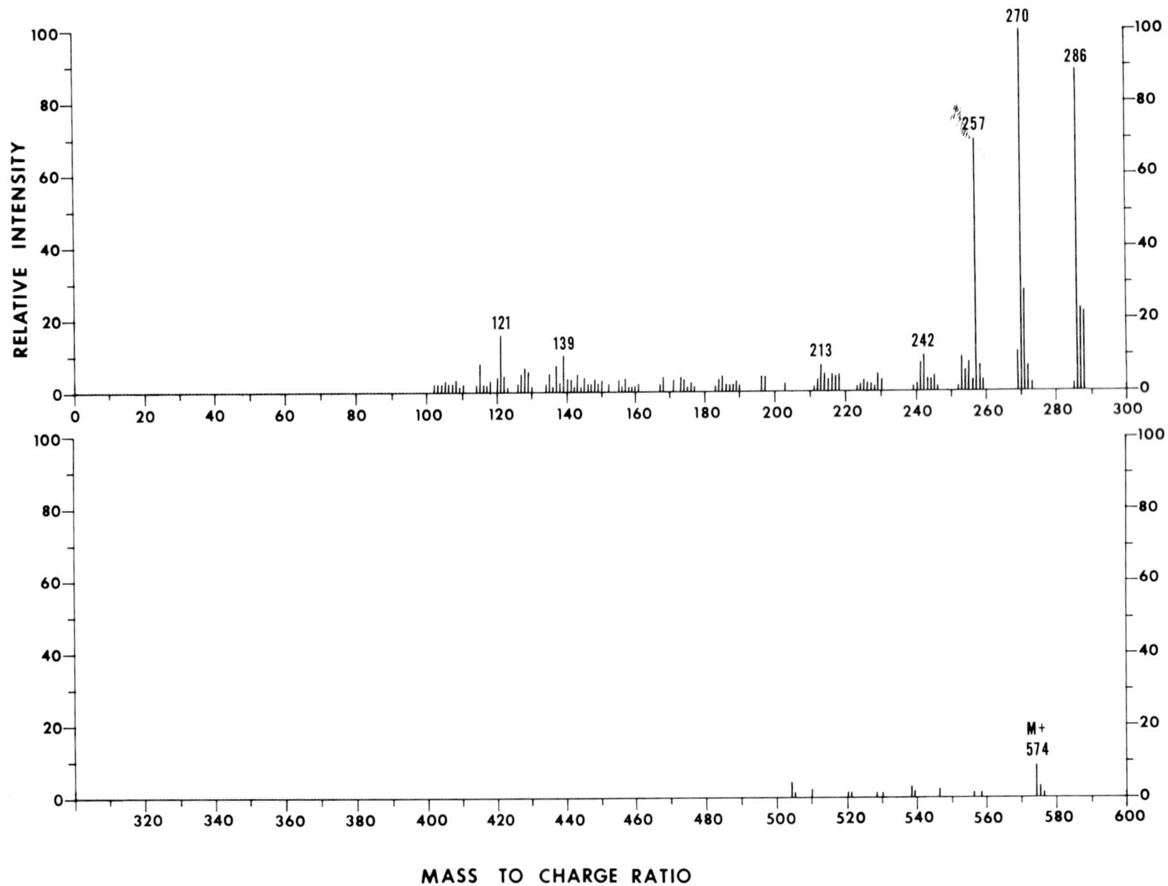

RELATIVE INTENSITY
MASS TO CHARGE RATIO
121
139
213
242
257
270
286
M+
574

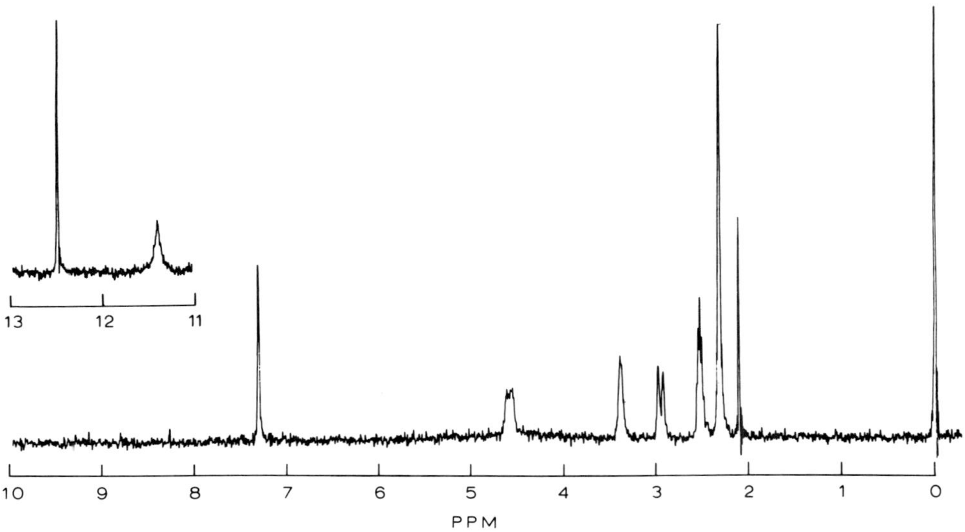

The 100 MHz ^{1}H NMR spectrum of luteoskyrin in DMSO solution.

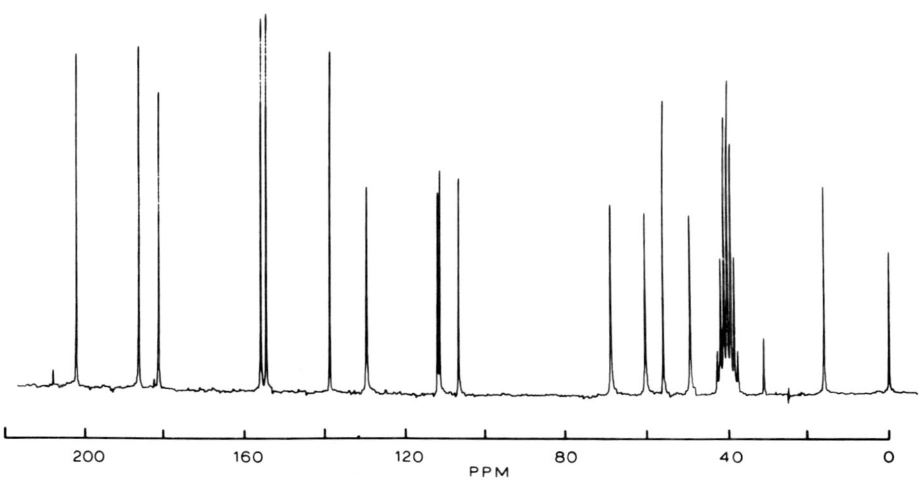

The 25 MHz ^{13}C NMR spectrum of luteoskyrin in DMSO solution.

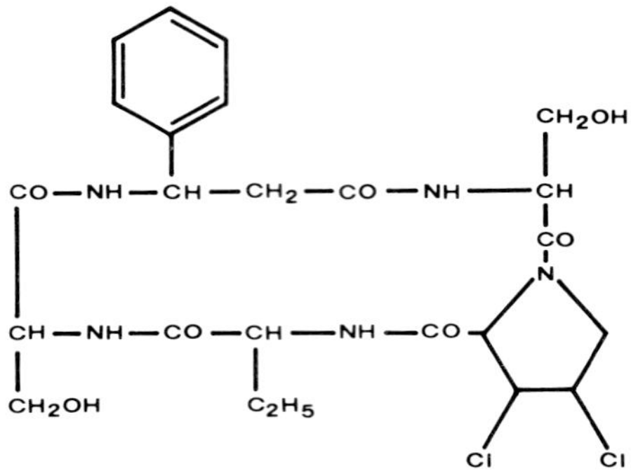

Common name	Cyclochlorotine (Chlorine-containing peptide; Cl-peptide)
Molecular weight	571.1587
Molecular formula	$C_{24}H_{31}O_7N_5Cl_2$
General characteristics[10,11]	White needles from methanol, mp 255°C (dec.) $[\alpha]_D^{16}$ −92.9° (in methanol)
UV data[10]	λ_{max} nm (ϵ): 257(230)
Fungal source[42]	*Penicillium islandicum* Sopp
Toxicity data[6,45,52]	Cyclochlorotine is a "periportal toxic agent" causing peripheral damage to the liver lobule. It accelerates liver glycogen catabolism and inhibits neogenesis *in vivo,* causing the disappearance of glycogen granules in liver. Gross effects are very rapid (liver appears pale and anemic 5 minutes after dosing), affects blood capillaries, causes necrosis of epidermis, and a low incidence of hepatoma. Liver was major target organ with fibrosis and cirrhosis typical; however, necrosis was not evident. The LD_{50} in mice was 0.47 mg/kg (SC), 0.33 mg/kg (IP), and 6.55 mg/kg (PO).
TLC data[11]	*Adsorbent:* Kieselgel G *Solvent:* *n*-butanol–acetone–water, 4:1:4 v/v/v R_f: 0.70 *Detection:* brown spot after spraying with 2 *N* H_2SO_4 and heating at 100°C for 5–10 min.
Other significant references	Synthesis[27]; X-ray crystallography[58]; chemistry[11]; production[10]; review[6]

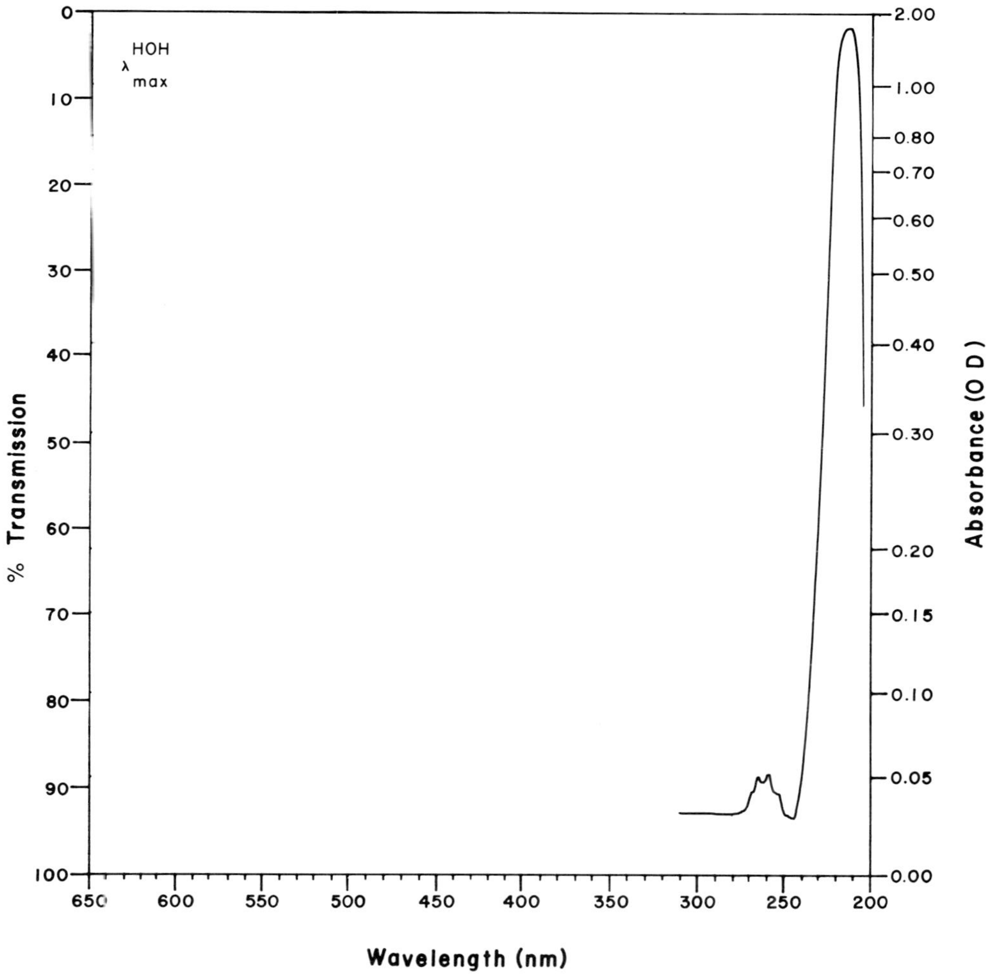
HOH
λ max
% Transmission
Absorbance (O D)
Wavelength (nm)
0
10
20
30
40
50
60
70
80
90
100
2.00
1.00
0.80
0.70
0.60
0.50
0.40
0.30
0.20
0.15
0.10
0.05
0.00
650
600
550
500
450
400
350
300
250
200

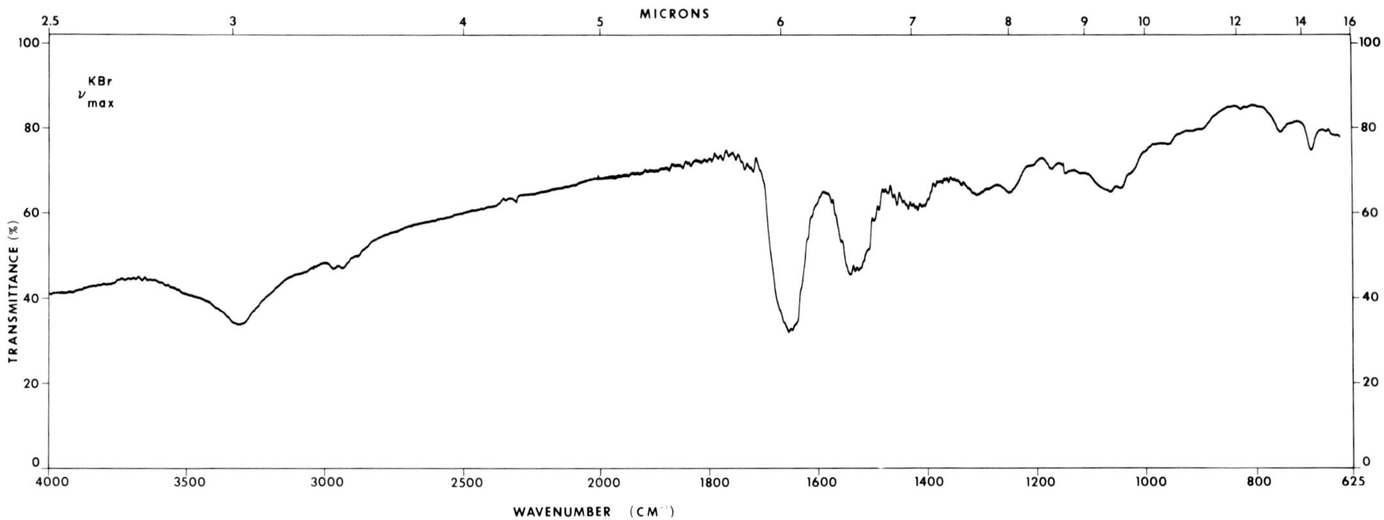
MICRONS
2.5
3
4
5
6
7
8
9
10
12
14
16
100
80
60
40
20
0
TRANSMITTANCE (%)
KBr
ν_{max}
4000
3500
3000
2500
2000
1800
1600
1400
1200
1000
800
625
WAVENUMBER (CM⁻¹)

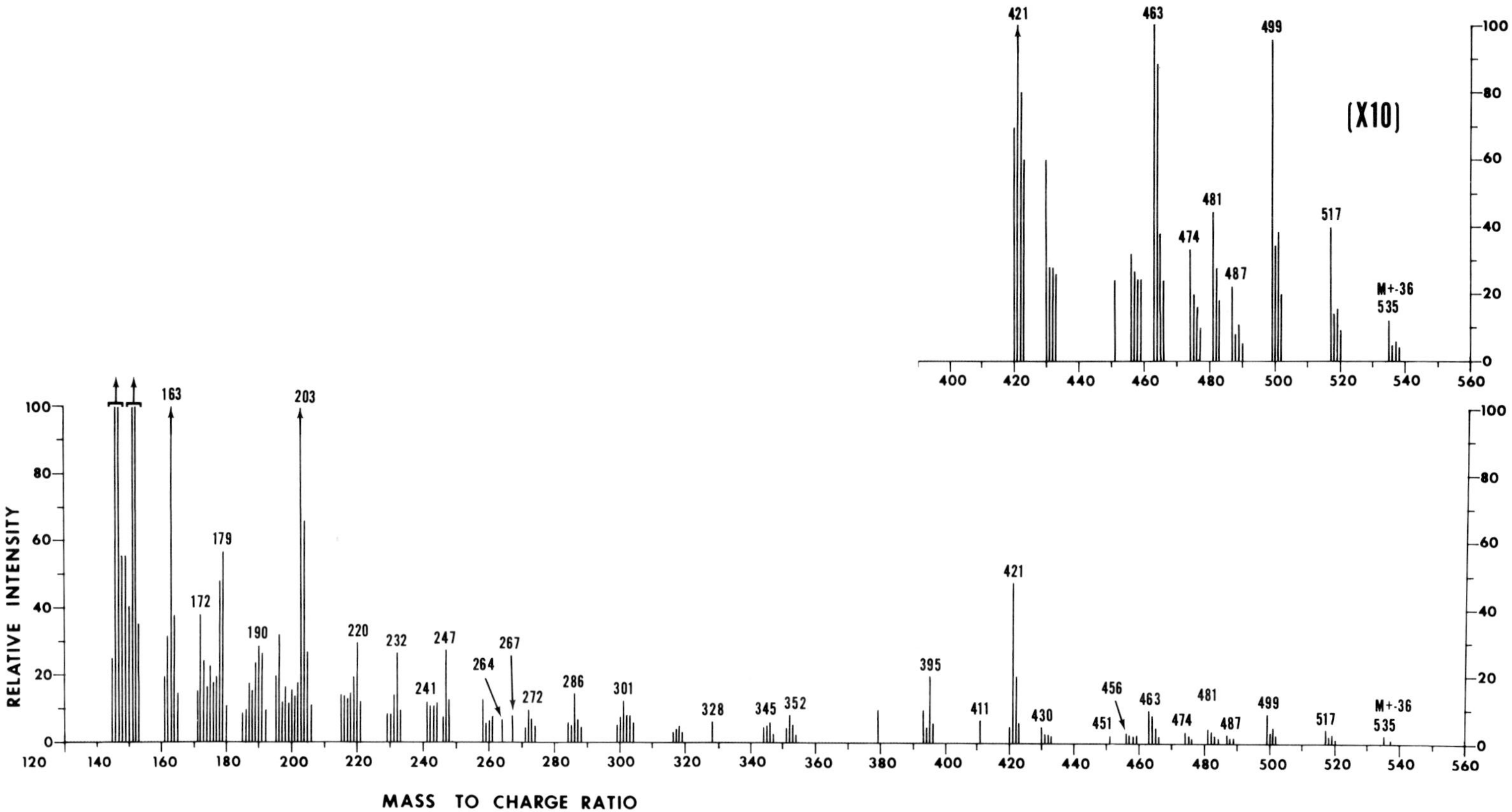

(X10)
421
463
474
481
487
499
517
M+-36
535
163
172
179
190
203
220
232
241
247
264
267
272
286
301
328
345
352
395
411
421
430
451
456
463
474
481
487
499
517
M+-36
535
RELATIVE INTENSITY
MASS TO CHARGE RATIO

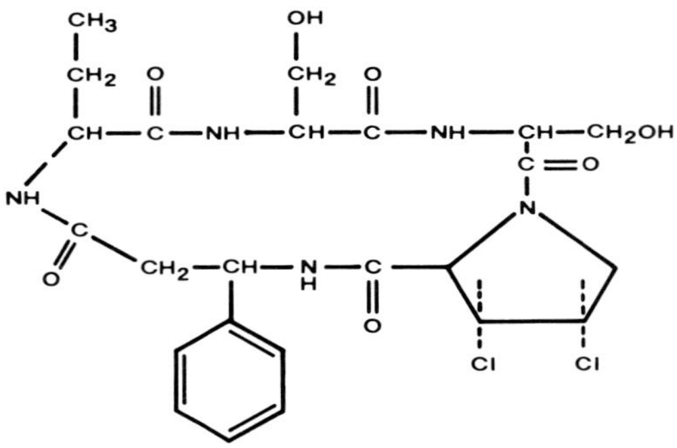

Common name	Islanditoxin
Molecular weight	571.1587
Molecular formula	$C_{24}H_{31}O_7N_5Cl_2$
General characteristics[17–20]	Colorless crystalline needles from methanol, mp 250°–251°C (dec.); from water, mp 254°C; from acetone, mp 258°C $[\alpha]_D^{21}$ −47.7° (C = 1.99 in acetic acid) Islanditoxin is very soluble in acetone, butanol, and phenol; soluble in methanol, ethanol, dioxane, and water; slightly soluble in chloroform, ethyl acetate, ether, benzene, petroleum ether, and ligroin. *NOTE:* The physical, chemical and biological characteristics of islanditoxin are very similar to cyclochlorotine; however, the proposed structure differs from cyclochlorotine in the sequence of the amino acids.
UV data[17]	$\lambda_{max}^{H_2O}$ nm (ϵ): 257(290)
IR data[15]	ν_{max}^{Nujol} cm^{-1}: 3450, 3270, 1650, 1530, and 700
Fungal source[16,19]	*Penicillium islandicum* Sopp
Toxicity data[19]	The LD_{50} of islanditoxin in mice dosed SC was 3.6 mg/kg. Histopathological examination showed degeneration of liver cells and stasis of circulating blood in the liver.
TLC data[20]	*Adsorbent:* Kieselgel G *Solvent:* *n*-butanol–acetic acid–water, 4:1:4 v/v/v R_f: 0.70 *Detection:* brown spot after spraying with 2N H_2SO_4 and heating at 100°C for 5–10 min.

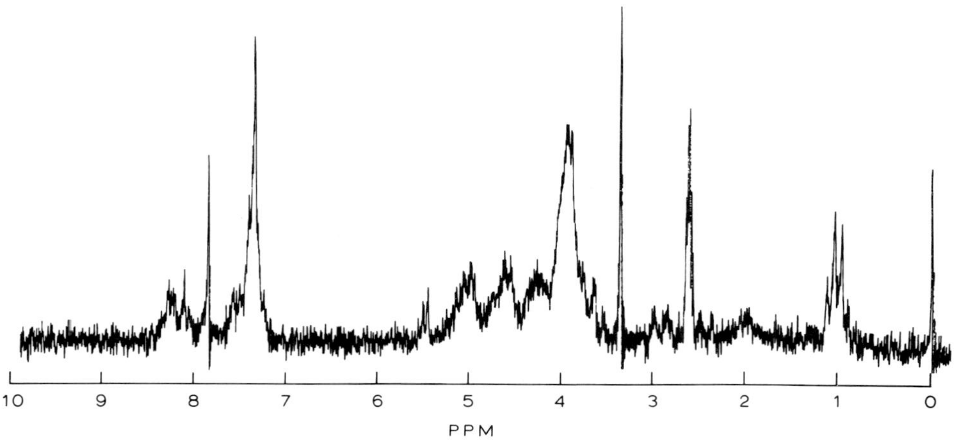

The 100 MHz ^1H NMR spectrum of islanditoxin in $CDCl_3$/DMSO solution.

REFERENCES

1. Agosti, G., Birkinshaw, J. H., and Chaplen, P., *Biochem. J.* **85,** 528 (1962).
2. Bick, I. R. C., and Rhee, C., *Biochem. J.* **98,** 112 (1966).
3. Birkinshaw, J. H., *Biochem. J.* **59,** 485 (1955).
4. Breen, J., Dacre, J. C., Raistrick, H., and Smith, G., *Biochem. J.* **60,** 618 (1955).
5. Eder, R., and Widmer, C., *Helv. Chim. Acta* **6,** 966 (1923).
6. Enomoto, M., and Ueno, I., *in* "Mycotoxins" (I. F. H. Purchase, ed.), p. 303. Elsevier, Amsterdam (1974).
7. Gatenbeck, S., *Acta Chem. Scand.* **12,** 1211 (1958).
8. Howard, B. H., and Raistrick, H., *Biochem. J.* **56,** 56 (1954).
9. Howard, B. H., and Raistrick, H., *Biochem. J.* **57,** 212 (1954).
10. Ishii, K., and Ueno, Y., *Appl. Microbiol.* **26,** 359 (1973).
11. Ishikawa, I., Ueno, Y., and Tsunoda, H., *J. Biochem. (Tokyo)* **67,** 753 (1970).
12. Itano, T., *J. Nara Med. Assoc.* **10,** 1 (1959).
13. Kobayashi, N., Itaka, Y., and Shibata, S., *Acta Crystallogr., Sect. B* **26,** 188 (1970).
14. Kogl, F., and Postowsky, J. J., *Justus Liebigs Ann. Chem.* **444,** 1 (1925).
15. Marumo, S., *Bull. Agric. Chem. Soc. Jpn.* **19,** 258 (1955).
16. Marumo, S., Miyao, K., and Matsuyama, A., *Bull. Agric. Chem. Soc. Jpn.* **19,** 262 (1955).
17. Marumo, S., *Bull. Agric. Chem. Soc. Jpn.* **23,** 428 (1959).
18. Marumo, S., Miyao, K., and Matsuyama, A., *Bull. Agric. Chem. Soc. Jpn.* **29,** 913 (1955).
19. Marumo, S., and Sumiki, Y., *J. Agric. Chem. Soc. Jpn.* **29,** 305 (1955).
20. Marumo, S., *Bull. Agric. Chem. Soc. Jpn.* **23,** 428 (1959).
21. Natori, S., Sato, F., and Udagawa, S., *Chem. Pharm. Bull.* **13,** 385 (1965).
22. Nishikawa, H., *Tokohu J. Agric. Res.* **5,** 285 (1955).
23. Ogihara, Y., Kobayashi, N., and Shibata, S., *Tetrahedron Lett.* p. 1881 (1968).
24. Pham Van Chuong, P., Bouhet, J. C., Schneider, C. R. L., and Fromageot, P., *J. Chim. Phys.* **71,** 401 (1974).
25. Pham Van Chuong, P., Bouhet, J. C., Thiery, J., and Fromageot, P., *Tetrahedron* **29,** 3533 (1973).
26. Saito, M., Enomoto, M., Tatsuno, T., and Uraguchi, K., *in* "Microbial Toxins" (A. Ciegler, S. Kadis, and S. J. Ajl, eds.), Vol. 6, p. 299. Academic Press, New York, 1971.
27. Sato, M., and Tatsuno, T., *Chem. Pharm. Bull.* **16,** 2182 (1968).
28. Schoental, R., and White, A. F., *Nature (London)* **205,** 57 (1965).
29. Shibata, S., and Ikekawa, T., *Chem. Ind. (London)* p. 360 (1962).
30. Shibata, S., and Ikekawa, T., *Chem. Pharm. Bull.* **11,** 368 (1963).
31. Shibata, S., and Kitagawa, I., *Chem. Pharm. Bull.* **9,** 352 (1961).
32. Shibata, S., Kitagawa, I., and Nishikawa, H., *Pharm. Bull.* **5,** 383 (1957).
33. Shibata, S., and Kitagawa, I., *Pharm. Bull.* **4,** 309 (1956).
34. Shibata, S., Murakami, T., Tanaka, O., Chihara, G., and Sumimoto, M., *Pharm. Bull.* **3,** 274 (1955).
35. Shibata, S., Shoji, J., Ohta, A., and Watanabe, M., *Pharm. Bull.* **5,** 380 (1957).
36. Shibata, S., Tanaka, O., Chihara, G., and Mitsuhashi, H., *Pharm. Bull.* **1,** 302 (1953).
37. Shibata, S., Tanaka, O., and Kitagawa, I., *Pharm. Bull.* **3,** 278 (1955).
38. Shibata, S., and Udagawa, S., *Chem. Pharm. Bull.* **11,** 402 (1963).
39. Shoji, J., Shibata, S., Sankawa, U., Taguchi, H., and Shibanuma, Y., *Chem. Pharm. Bull.* **13,** 1240 (1965).
40. Takeda, N., Seo, S., Ogihara, Y., Sankawa, U., Iitaka, I., Kitagawa, I., and Shibata, S., *Tetrahedron* **29,** 3703 (1973).
41. Takeda, N., Seo, S., Ogihara, Y., Sankawa, U., Iitaka, I., Kitagawa, I., and Shibata, S., *Tetrahedron* **29,** 3707 (1973).

42. Tatsuno, T., Tsukioka, M., Sakai, Y., Suzuki, Y., and Asami, Y., *Pharm. Bull.* **3,** 476 (1955).
43. Toma, F., Bouhet, J. C., Pham Van Chuong, P., Fromageot, P., Haar, W., Rüterjans, H., and Maurer, W., *Org. Magn. Reson.* **7,** 496 (1975).
44. Ueno, Y., and Ishikawa, I., *Appl. Microbiol.* **18,** 406 (1969).
45. Ueno, Y., Kaneko, M., Tatsuno, T., Ueno, I., and Uraguchi, K., *Seikagaku* **35,** 224 (1963).
46. Ueno, Y., Kato, Y., and Enomoto, M., *Jpn. J. Exp. Med.* **45,** 525 (1975).
47. Ueno, Y., Platel, A., and Fromageot, P., *Biochim. Biophys. Acta* **134,** 27 (1967).
48. Ueno, Y., and Saheki, M., *Jpn. J. Exp. Med.* **38,** 157 (1968).
49. Ueno, Y., Ueno, I., and Tatsuno, T., *Proc. Int. Congr. Biochem., 7th, 1967* Abstract, p. 625 (1967).
50. Ueno, Y., Ueno, I., Sato, N., Iitoi, Y., Saito, M., Enomoto, M., and Tsunoda, H., *Jpn. J. Exp. Med.* **41,** 177 (1971).
51. Uraguchi, K., Tatsuno, T., Sakai, Y., Sakai, F., Kobayashi, Y., Saito, M., Enomoto, M., and Miyake, M., *Jpn. J. Exp. Med.* **31,** 1 (1956).
52. Uraguchi, K., Tatsuno, T., Sakai, F., Tsukioka, M., Sakai, Y., Yonemitsu, O., Ito, H., Miyake, M., Saito, M., Enomoto, M., Shikata, T., and Ishiko, T., *Jpn. J. Exp. Med.* **31,** 17 (1961).
53. Uraguchi, K., Ueno, I., Ueno, Y., and Komai, Y., *Toxicol. Appl. Pharmacol.* **21,** 335 (1972).
54. Wells, J. M., Cole, R. J., and Kirksey, J. W., *Appl. Microbiol.* **30,** 26 (1975).
55. Yamamoto, Y., Yamamoto, T., Kanatomo, S., Tanimichi, K., and Kikui, H., *Yakugaku Zasshi* **76,** 670 (1956).
56. Yamazaki, M., Fujimoto, H., and Miyaki, K., *Yakugaku Zasshi* **92,** 101 (1972).
57. Yamazaki, M., *in* "Mycotoxins and Human Health" (I. F. H. Purchase, ed.), p. 113. Macmillan, New York, 1971.
58. Yoshioka, H., *Nippon Kagakukai Yokoshu* p. 28 (1973).

16

Sweet Potato Toxins

Although the toxic furanosesquiterpenoids produced in sweet potatoes under certain stress conditions are not mycotoxins in the strictest sense, they are included here because infection of sweet potatoes by *Fusarium solani* or *Ceratocystis fimbriata* is probably a significant mode of formation. Chemically, these compounds are composed of a furan ring substituted in the 3 position with an oxygenated pentane moiety. Although they are relatively noncomplex chemically, they are potent toxins with rather specific effects on the liver or the lungs, causing an atypical interstitial pneumonia. Chronic effects often are expressed as renal necrosis.

These stress metabolites are suspected of being the causative agents in the enzootics of livestock poisoning resulting from ingestion of mold-damaged sweet potatoes.

A review on the subject has been published.[14]

Sweet potato toxins	Molecular weight	Molecular formula
Ipomeanine	166.0629	$C_9H_{10}O_3$
1-Ipomeanol	168.0786	$C_9H_{12}O_3$
4-Ipomeanol	168.0786	$C_9H_{12}O_3$
1,4-Ipomeadiol	170.0942	$C_9H_{14}O_3$
Ipomeamaronol	266.1517	$C_{15}H_{22}O_4$
Ipomeamarone	250.1568	$C_{15}H_{22}O_3$
Dehydroipomeamarone	248.1412	$C_{15}H_{20}O_3$
4-Hydroxymyoporone	266.1517	$C_{15}H_{22}O_4$

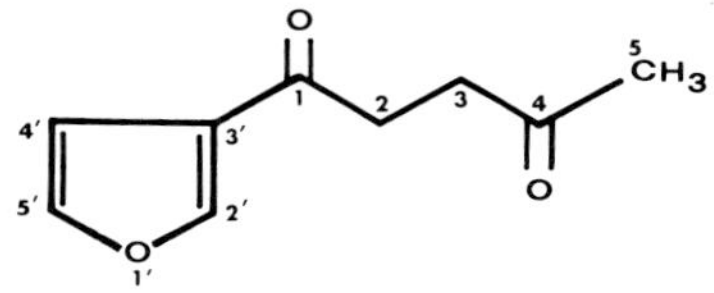

Common name	Ipomeanine [1-(3-furyl)-1,4-Pentanedione]
Molecular weight	166.0629
Molecular formula	$C_9H_{10}O_3$
General characteristics[2]	mp 41°–42°C; 117–120 bp at 2 mm Hg
UV data[2]	λ_{max}^{MeOH} nm: 211 and 251 (ϵ_{max} not reported)
Fungal source[13]	Produced on *Fusarium solani*-infected sweet potatoes and as a stress metabolite in sweet potatoes in response to general damage of tissues.
Toxicity data[2,4]	The LD_{50} of ipomeanine in mice was 26 mg/kg (oral), 25 mg/kg (IP), and 14 mg/kg (IV). Causes death due to pulmonary edemogenesis.
TLC data[5,6,14]	*Adsorbent:* silica gel, IB-F (Baker Chem. Co.) *Solvent:* benzene–methanol, 9:1 v/v R_f: not reported *Detection:* spray with Ehrlich's reagent followed by gentle heating.
GLC data[5,6,14]	*Solid support:* Gas Chrom Q, 80/100 mesh *Liquid phase:* 18% UC-W98 (Applied Science) *Retention time:* not reported
^{13}C *NMR data*	C1, 192.8 s; C2, 33.8 t; C3, 36.7 t; C4, 206.4 s; C5, 29.9 q; C2′, 146.8 d; C3′, 127.0 s; C4′, 108.9 d; C5′, 143.8 d
^{1}H *NMR data*	H2, 2.86; H3, 3.08; H5, 2.24; H2′, 8.08; H4′, 6.76; H5′, 7.44

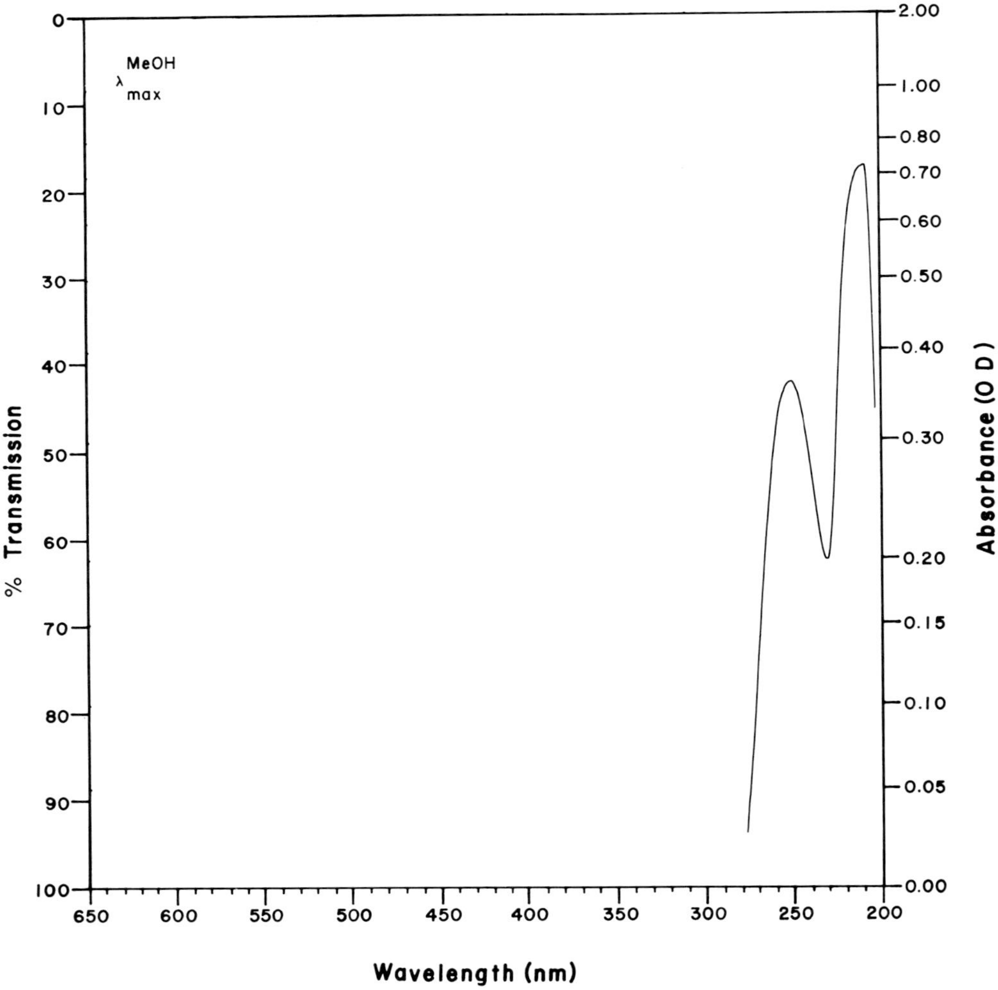

λ_{max}^{MeOH}
% Transmission
0
10
20
30
40
50
60
70
80
90
100
Absorbance (O D)
2.00
1.00
0.80
0.70
0.60
0.50
0.40
0.30
0.20
0.15
0.10
0.05
0.00
650
600
550
500
450
400
350
300
250
200
Wavelength (nm)

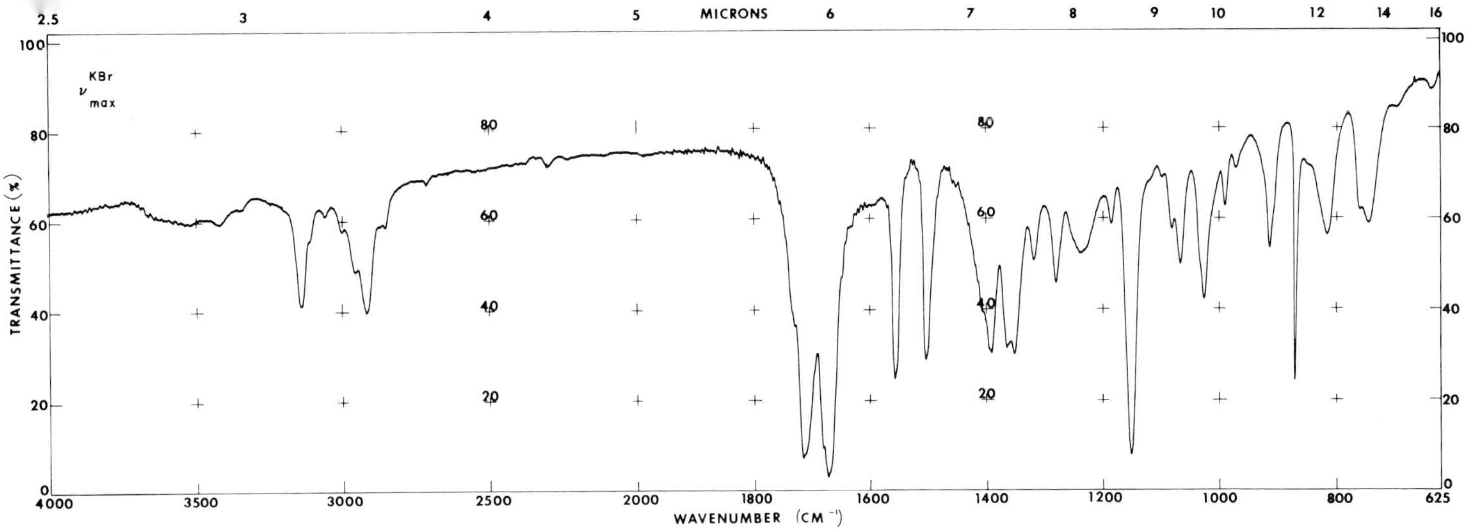
MICRONS
TRANSMITTANCE (%)
KBr
ν max
WAVENUMBER (CM⁻¹)

IPOMEANINE

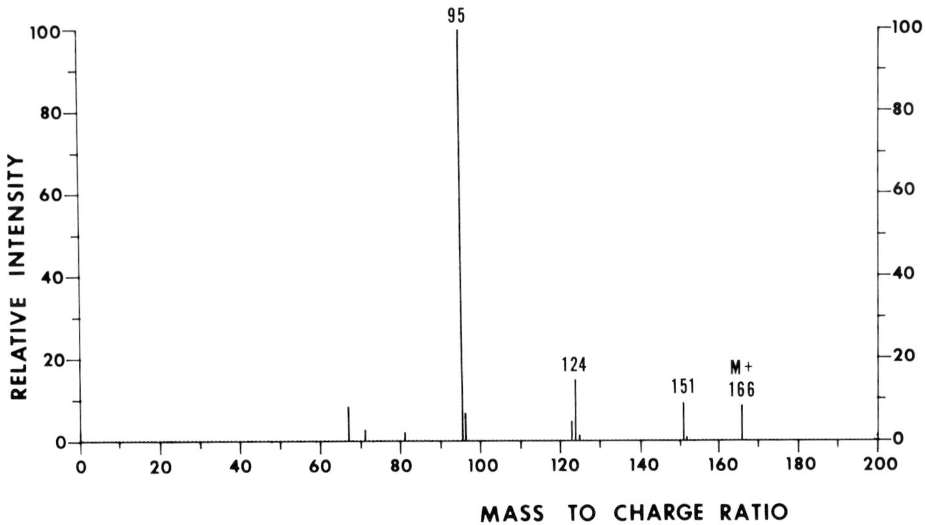
95
124
151
M+
166
RELATIVE INTENSITY
MASS TO CHARGE RATIO

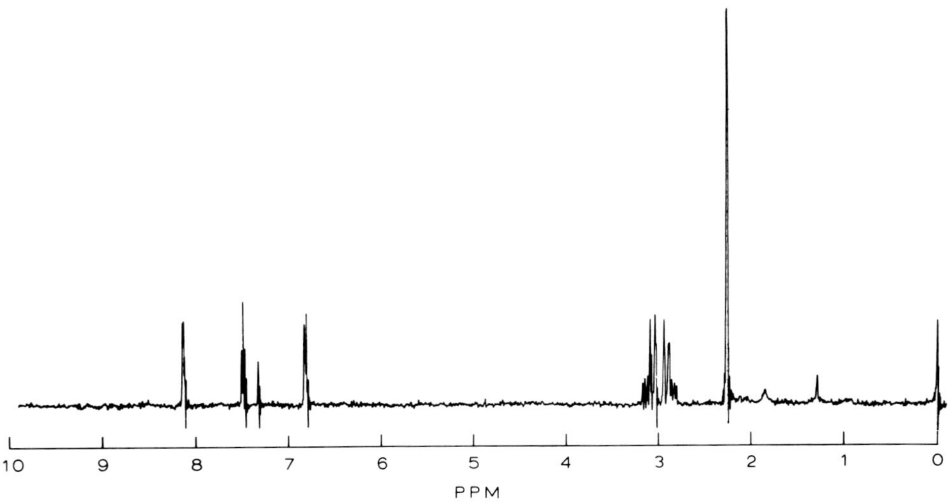

The 100 MHz ^{1}H NMR spectrum of ipomeanine in $CDCl_3$ solution.

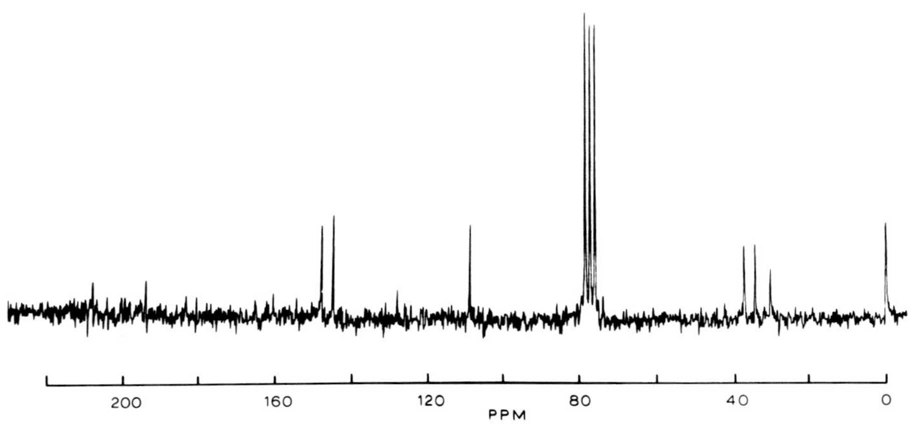

The 25 MHz ^{13}C NMR spectrum of ipomeanine in $CDCl_3$ solution.

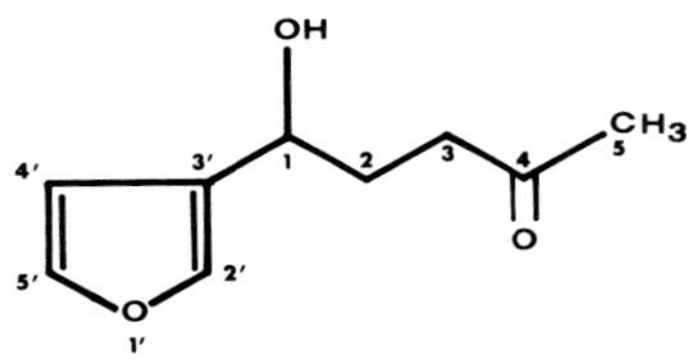

Common name	1-Ipomeanol [1-(3-furyl)-1-hydroxy-4-Pentanone]
Molecular weight	168.0786
Molecular formula	$C_9H_{12}O_3$
General characteristics[2]	$[\alpha]_D^{24}$ +21.9° (*C* = 0.64 in carbon tetrachloride)
UV data[2]	λ_{max}^{MeOH} nm (ϵ): 214(3,160)
Fungal source[7,13]	Produced on *Fusarium solani*-damaged sweet potatoes or as stress metabolite resulting from general damage to sweet potatoes.
Toxicity data[14]	The LD_{50} of 1-ipomeanol was 79 mg/kg (oral), 49 mg/kg (IP), and 34 mg/kg (IV). Primary action on lungs causing edema and congestion; also causes some nephrotoxicity at chronic levels.
TLC data[5,6,14]	*Adsorbent:* silica gel, IB-F (Baker Chem. Co.) *Solvent:* benzene–methanol, 9:1 v/v R_f: not reported *Detection:* spray with Ehrlich's reagent followed by gentle heating.
GLC data[5,6,14]	*Solid support:* Gas Chrom Q, 80/100 mesh *Liquid phase:* 18% UC-W 98 (Applied Science) *Relative retention time:* not reported
^{13}C *NMR data*	C1, 66.1 d; C2, 31.3 t; C3, 39.6 t; C4, 209.0 s; C5, 29.9 q; C2′, 138.8 d; C3′, 128.6 s; C4′, 108.3 d; C5′, 143.1 d
^{1}H *NMR data*	H1, 4.64 (*J* = 6.0); H2, 2.00; H3, 2.56; H5, 2.16; 1-OH, 3.22; H2′, 7.37, H4′, 6.35; H5′, 7.37

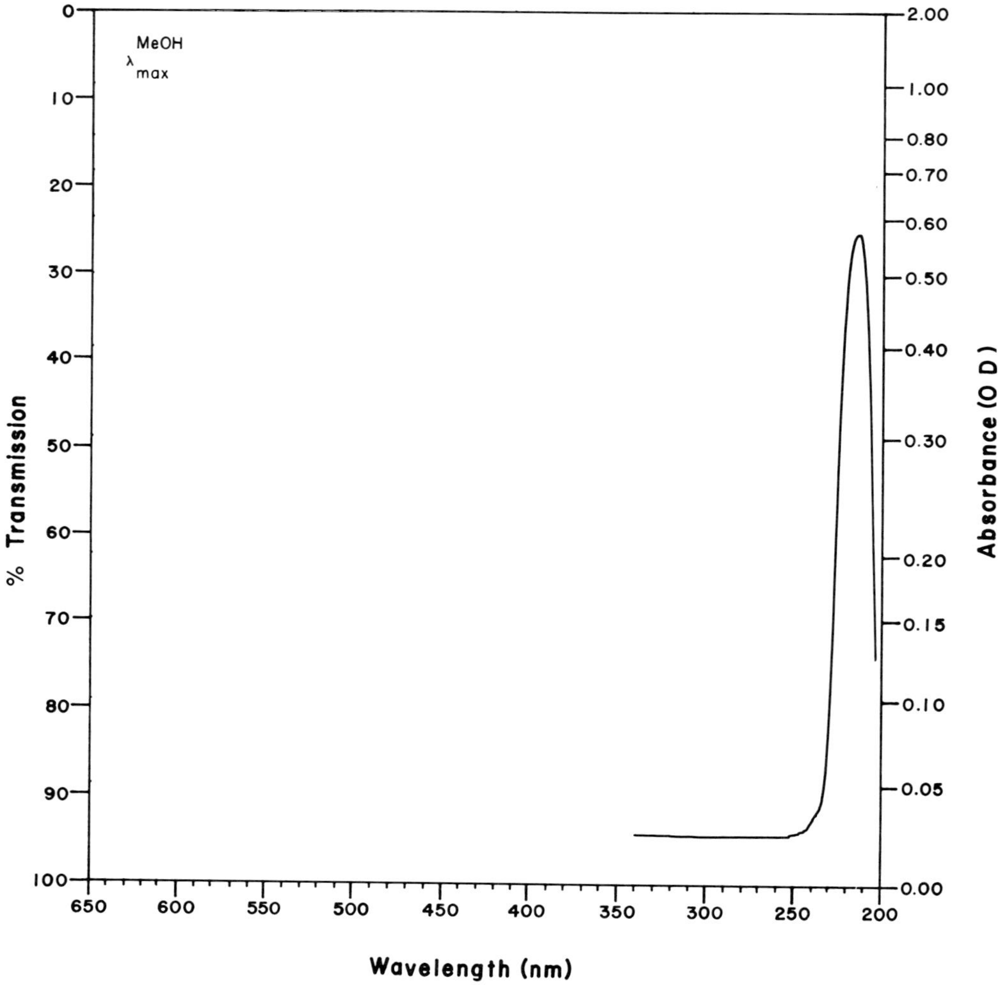

λ max MeOH
% Transmission
0
10
20
30
40
50
60
70
80
90
100
Absorbance (O D)
2.00
1.00
0.80
0.70
0.60
0.50
0.40
0.30
0.20
0.15
0.10
0.05
0.00
650
600
550
500
450
400
350
300
250
200
Wavelength (nm)

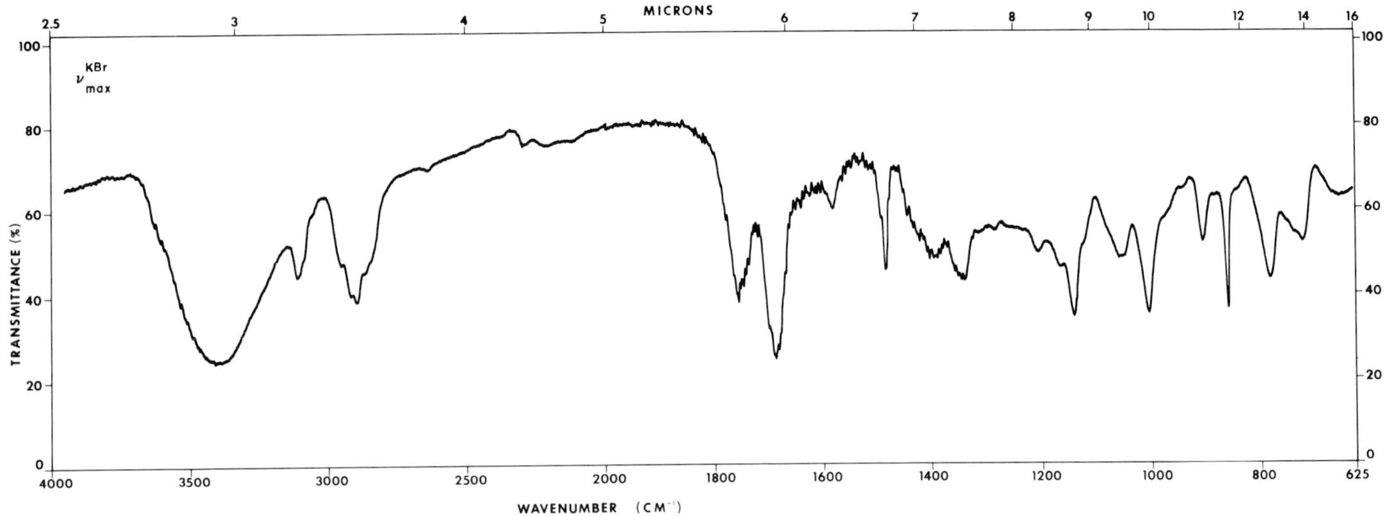
MICRONS
TRANSMITTANCE (%)
ν_{max}^{KBr}
WAVENUMBER (CM⁻¹)

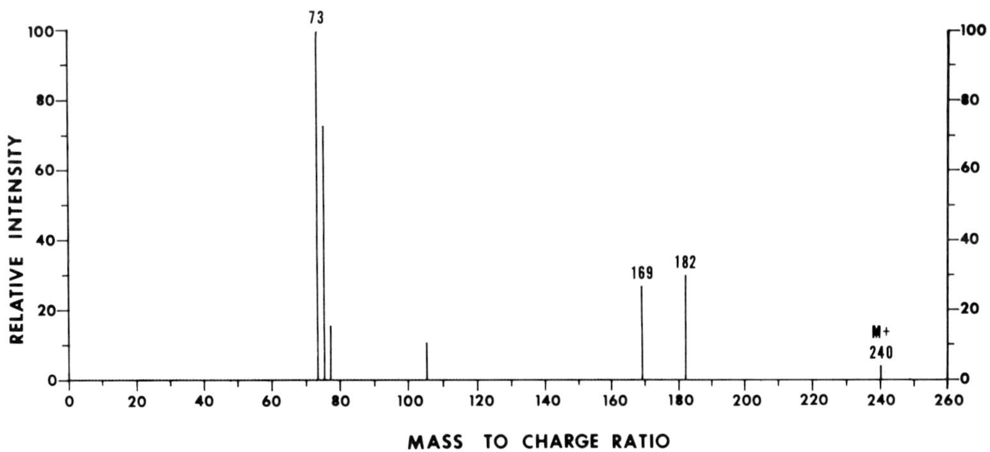
73
169
182
M+
240
RELATIVE INTENSITY
MASS TO CHARGE RATIO

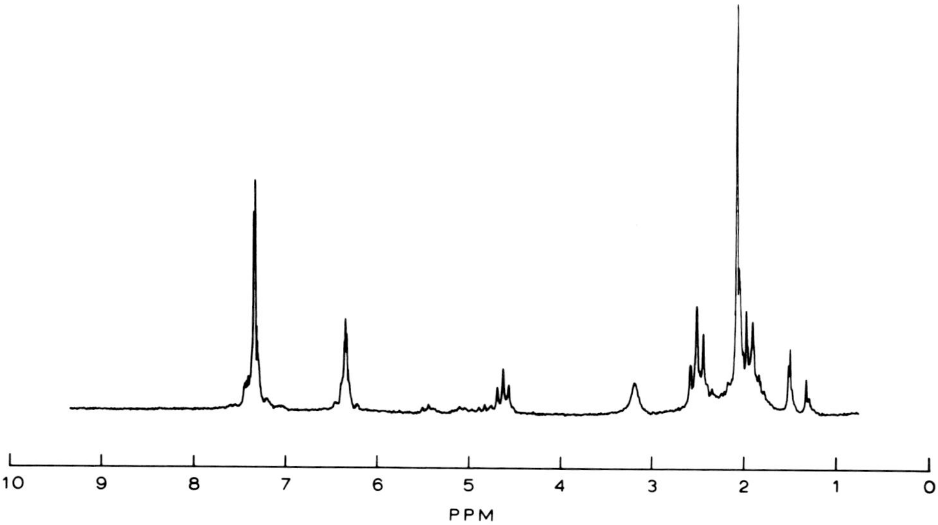

The 100 MHz ^1H NMR spectrum of 1-ipomeanol in $CDCl_3$ solution.

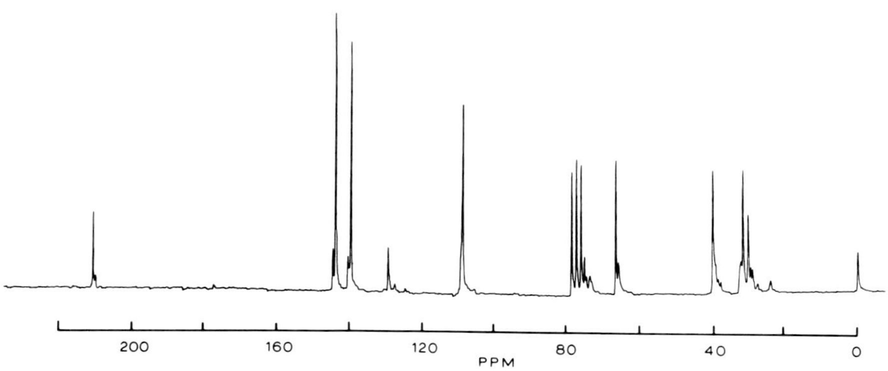

The 25 MHz ^{13}C NMR spectrum of 1-ipomeanol in $CDCl_3$ solution.

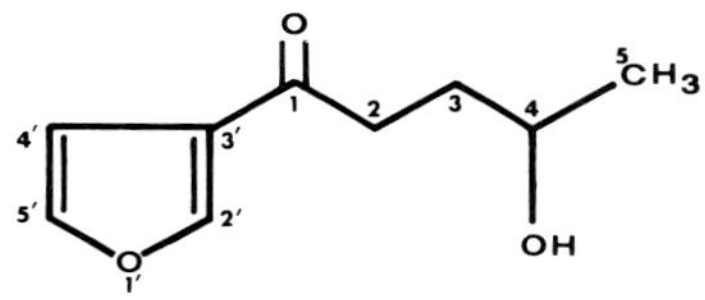

Common name	4-Ipomeanol [1-(3-furyl)-4-Hydroxy-4-pentanone]
Molecular weight	168.0786
Molecular formula	$C_9H_{12}O_3$
General characteristics[2]	$[\alpha]_D^{25}$ +7.86° (C = 1.17 in carbon tetrachloride)
UV data[2]	λ_{max}^{MeOH} nm (ϵ): 211(sh)6,100) and 251(2,970)
Fungal source[13]	Produced on *Fusarium solani-* and *Ceratocystis fimbriata*-infected sweet potatoes; also a stress metabolite in response to general damage to sweet potatoes.
Toxicity data[2,4]	The LD_{50} to mice was 38 mg/kg (oral), 36 mg/kg (IP), and 21 mg/kg (IV). Produces a characteristic pulmonary edema and congestion following a latent period of several hours after dosing. The nephrotoxicity observed with 1,4-ipomeadiol and 1-ipomeanol were not apparent with 4-ipomeanol.
TLC data[5,6,14]	*Adsorbent:* silica gel, IB-F (Baker Chem. Co.) *Solvent:* benzene–methanol, 9:1 v/v R_f: not reported *Detection:* spray with Ehrlich's reagent followed by gentle heating.
GLC data[5,6,14] (as silylated derivative)	*Solid support:* Gas Chrom Q, 80/100 mesh *Liquid phase:* 18% UC-W98 (Applied Science) *Relative retention time:* not reported
Other significant references	Radioisotope studies[3]
^{13}C NMR data	C1, 194.9 s; C2, 36.6 t; C3, 33.0 t; C4, 67.2 d; C5, 23.7 q; C2′, 146.8 d; C3′, 127.3 s; C4′, 108.3 d; C5′, 143.8 d
^1H NMR data	H2, 2.92; H3, 1.88; H4, 3.88; H5, 1.24 (J = 6.0); 4-OH, 2.70; H2′, 8.07; H4′, 6.76; H5′, 7.44

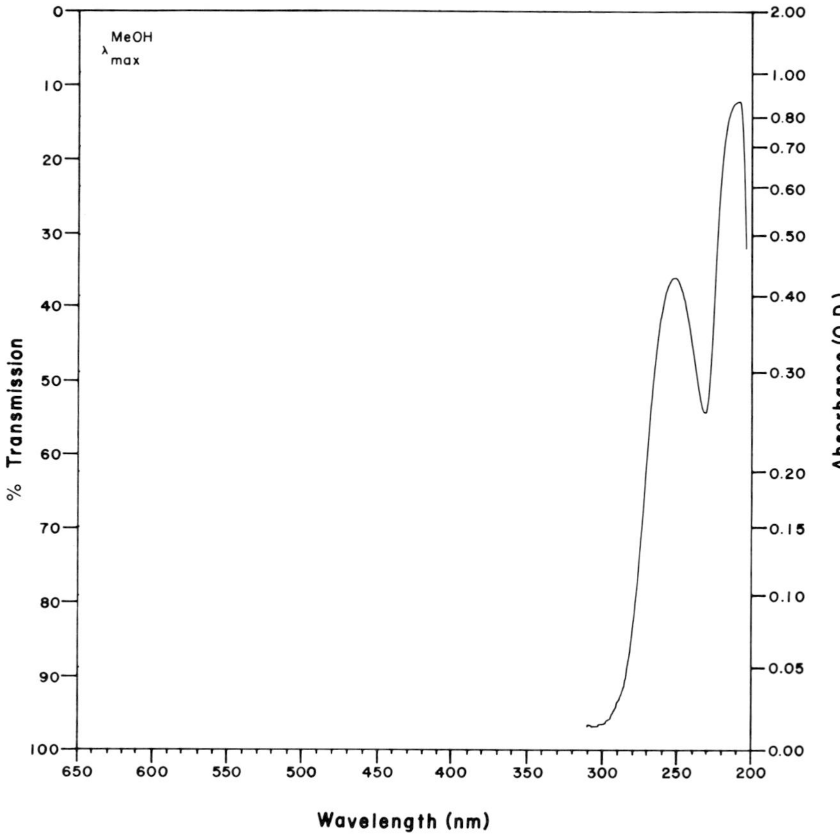

MeOH
λ max
% Transmission
Absorbance (O D)
Wavelength (nm)
0
10
20
30
40
50
60
70
80
90
100
2.00
1.00
0.80
0.70
0.60
0.50
0.40
0.30
0.20
0.15
0.10
0.05
0.00
650
600
550
500
450
400
350
300
250
200

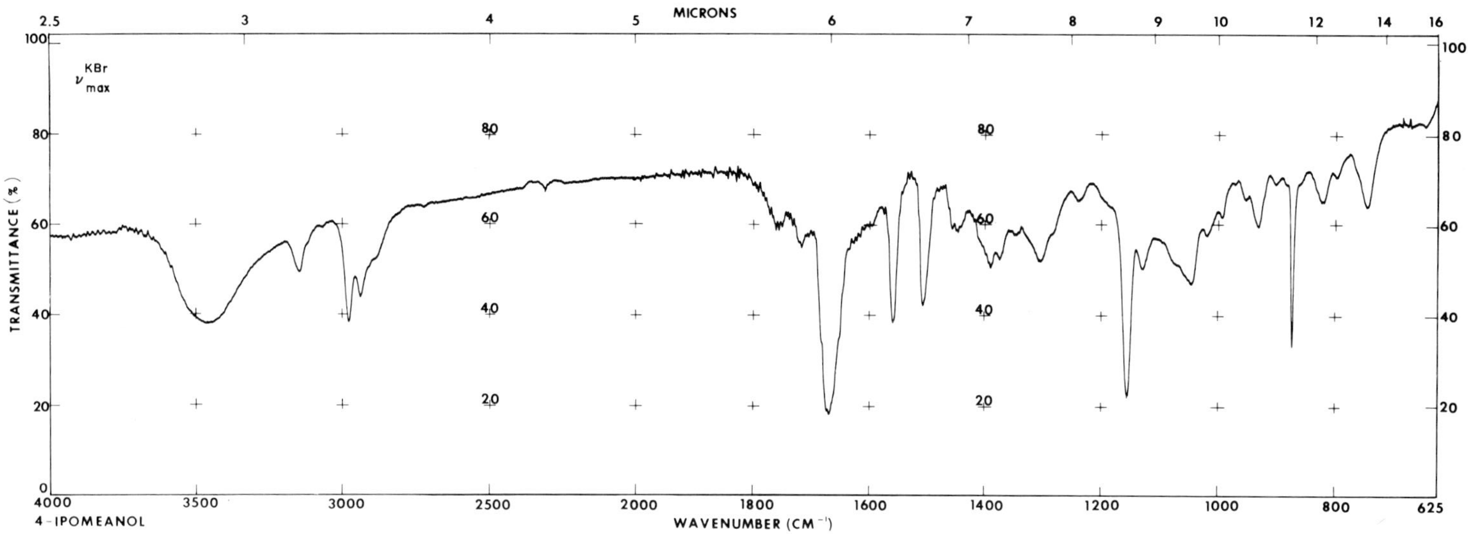
MICRONS
TRANSMITTANCE (%)
ν_{max}^{KBr}
WAVENUMBER (CM⁻¹)
4-IPOMEANOL

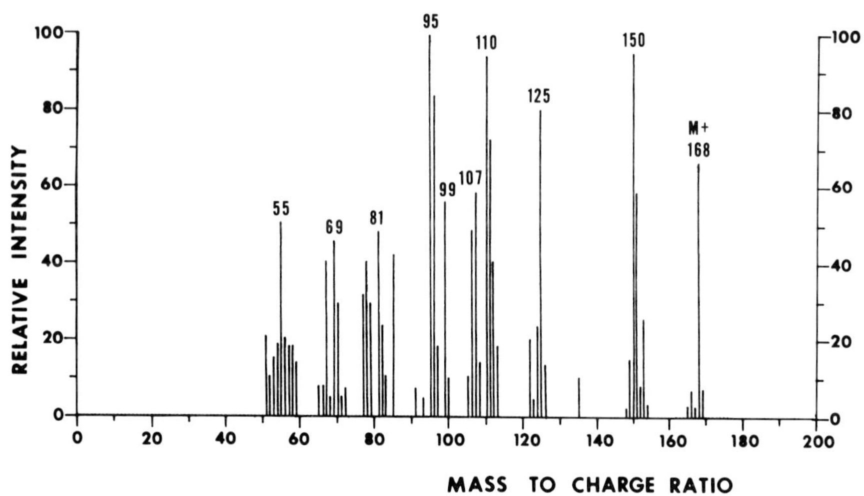
RELATIVE INTENSITY
MASS TO CHARGE RATIO
55
69
81
95
99
107
110
125
150
M+
168

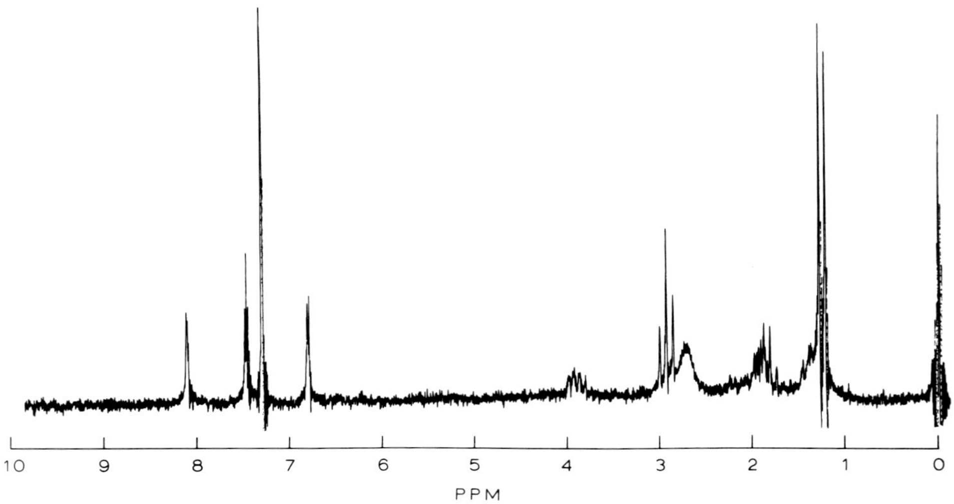

The 100 MHz ^{1}H NMR spectrum of 4-ipomeanol in $CDCl_3$ solution.

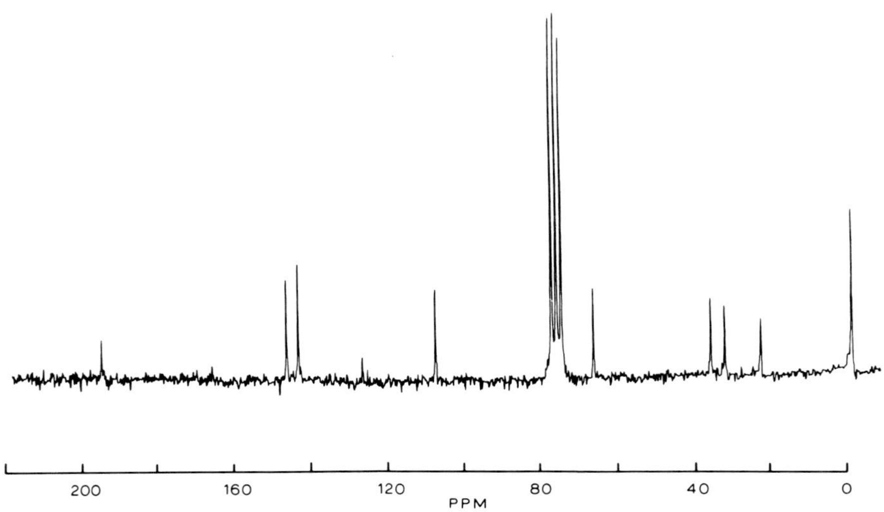

The 25 MHz ^{13}C NMR spectrum of 4-ipomeanol in $CDCl_3$ solution.

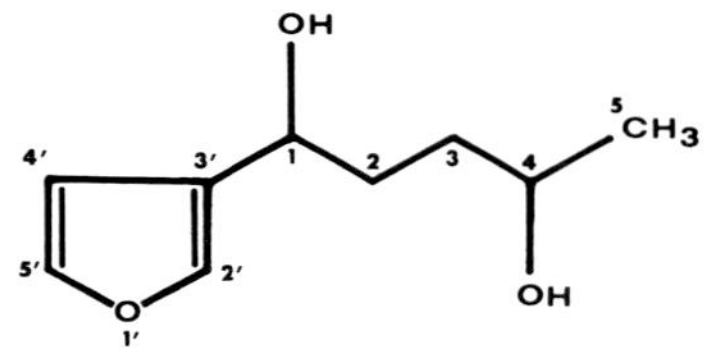

Common name	1,4-Ipomeadiol [1-(3-furyl)-1,4-Pentanediol]
Molecular weight	170.0942
Molecular formula	$C_9H_{14}O_3$
General characteristics[2]	$[\alpha]_D^{25}$ +1.27° (C = 1.21 in carbon tetrachloride)
UV data[2]	λ_{max}^{MeOH} nm (ϵ): 215(3,980)
Fungal source[13]	*Fusarium solani*-infected sweet potatoes; also stress metabolites in sweet potatoes due to general damage of tissue.
Toxicity data[2,4]	The LD_{50} to mice was 104 mg/kg (oral), 67 mg/kg (IP), and 68 mg/kg (IV). Primary site of action is the lungs; however, 1,4-ipomeadiol also shows some nephrotoxicity. In lungs, it produces a characteristic pulmonary edema and congestion following latent period of several hours after dosing. Nonlethal dosages also show nephrotoxicity.
TLC data[5,6,14]	*Adsorbent:* silica gel, IB-F (Baker Chem. Co.) *Solvent:* benzene–methanol, 9:1 v/v R_f: not reported *Detection:* spray with Ehrlich's reagent followed by gentle heating.
GLC data[5,6,14]	*Solid support:* Gas Chrom Q, 80/100 mesh *Liquid phase:* 18% UC-W98 (Applied Science) *Relative retention time:* not reported
*^{13}C NMR data**	C1, 66.8, 67.2 d; C2, 34.8, 35.6 t; C3, 33.8, 34.8 t; C4, 67.8, 68.1 d; C5, 23.5, 23.7 q; C2′, 138.8 d; C3′, 128.9 s; C4′, 108.4 d; C5′, 143.1 d
^1H NMR data	H1, 4.66; H2, 3.34; H3, 1.60; H4, 3.82; H5, 1.20 (J = 7.0); 1-OH, 3.34; 4-OH, 3.34; H2′, 7.37; H4′, 6.37; H5′, 7.37

* Mixture of diastereoisomers.

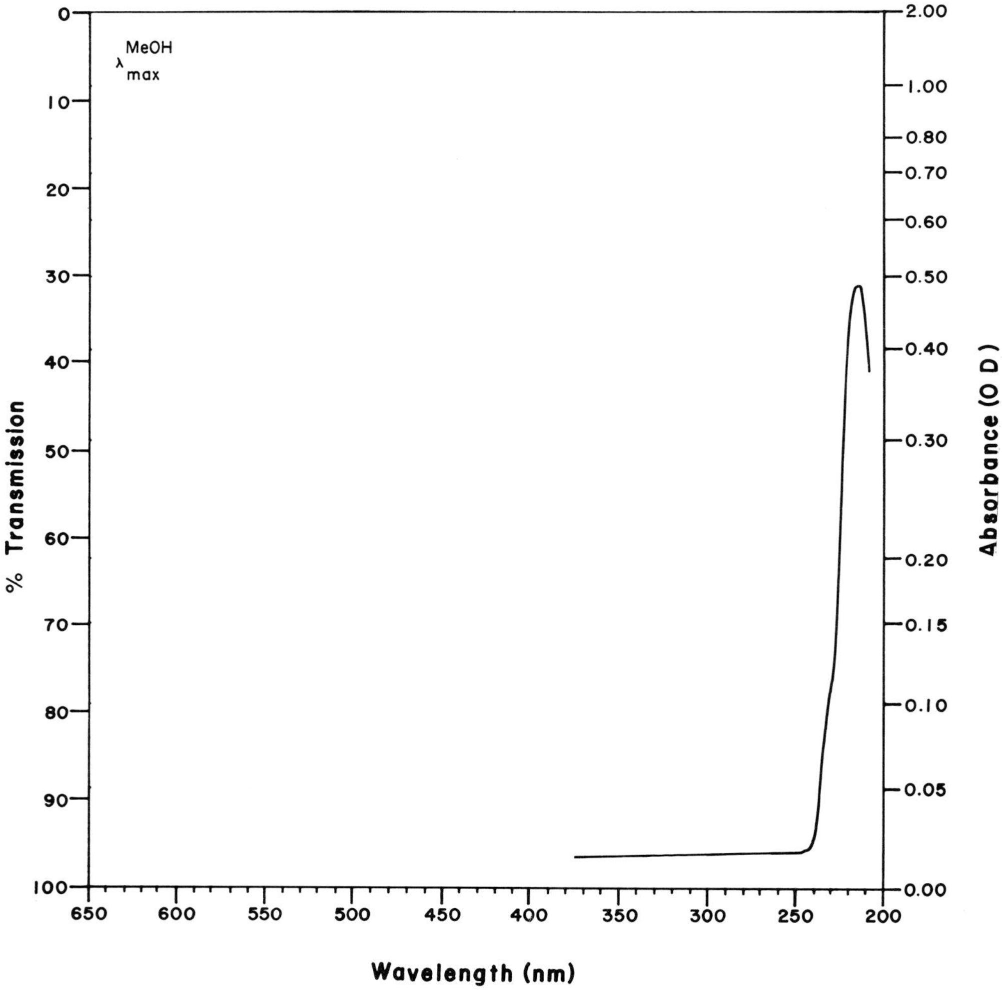

λ_{max}^{MeOH}
% Transmission
Absorbance (O D)
0
10
20
30
40
50
60
70
80
90
100
2.00
1.00
0.80
0.70
0.60
0.50
0.40
0.30
0.20
0.15
0.10
0.05
0.00
650
600
550
500
450
400
350
300
250
200
Wavelength (nm)

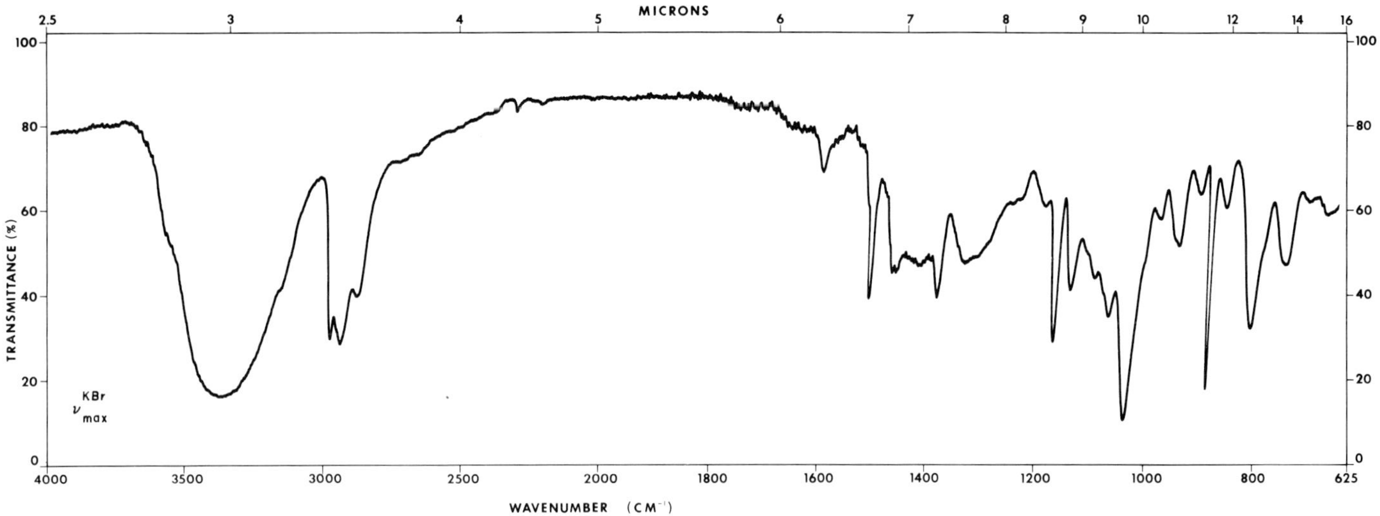
MICRONS
TRANSMITTANCE (%)
KBr
ν_{max}
WAVENUMBER (CM⁻¹)

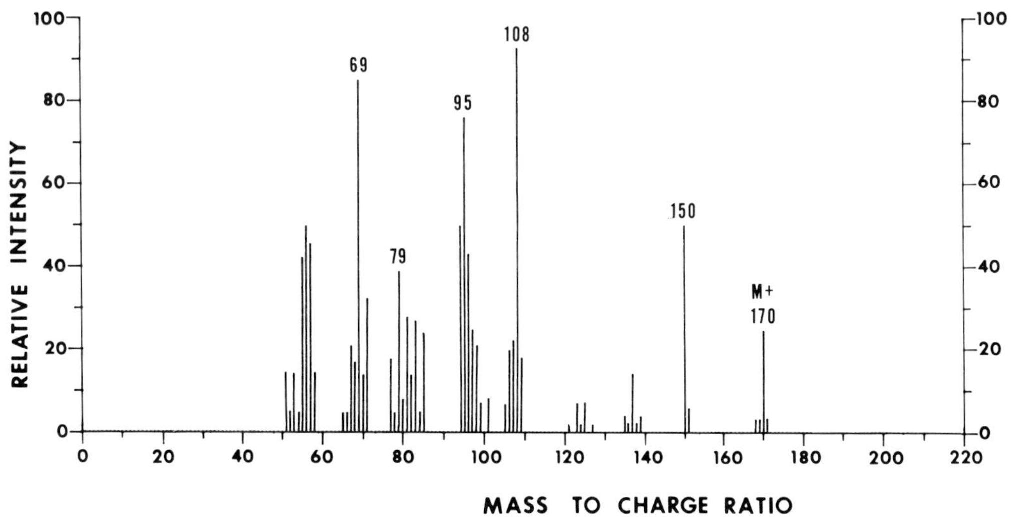
RELATIVE INTENSITY
69
79
95
108
150
M+
170
MASS TO CHARGE RATIO

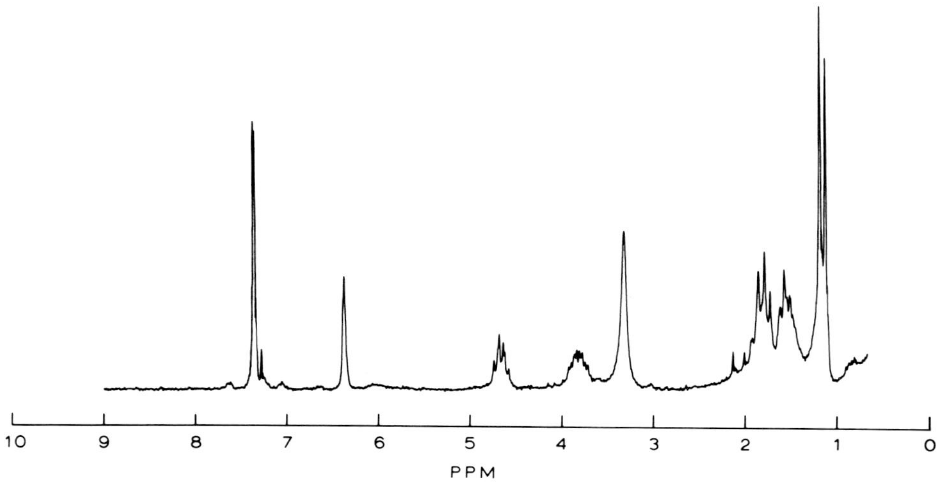

The 100 MHz ^{1}H NMR spectrum of 1,4-ipomeadiol in $CDCl_3$ solution.

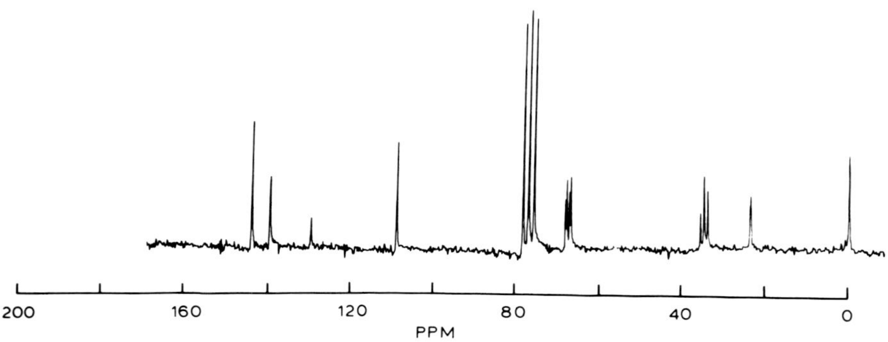

The 25 MHz ^{13}C NMR spectrum of 1,4-ipomeadiol in $CDCl_3$ solution.

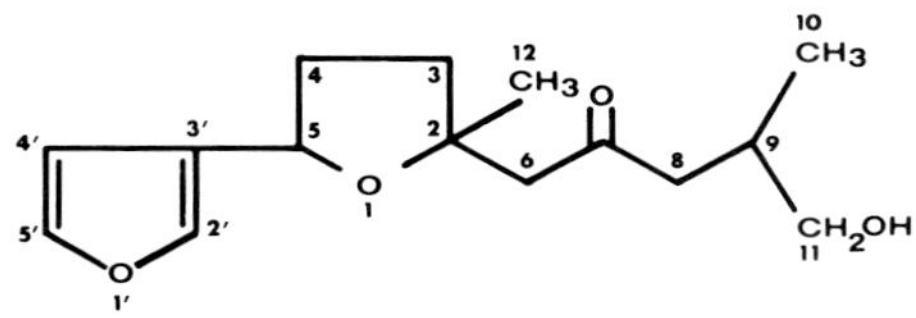

Common name	Ipomeamaronol
Molecular weight	266.1517
Molecular formula	$C_{15}H_{22}O_4$
General characteristics[15]	Colorless oil
UV data[15]	λ_{max}^{MeOH} nm (ϵ): 212(9,100)
IR data[15]	$\nu_{max}^{CHCl_3}$ cm^{-1}: 3400, 1710, 1500, and 875
Fungal source[15]	Mold-damaged sweet potatoes
Toxicity data[15]	Hepatotoxic
Other significant references	Structure[9]
^1H NMR data[9]	H3, NR*; H4, NR; H5, NR; H6, 2.61; H8, NR; H9, NR; H10, 0.87 (J = 6.1); H11, NR; H12, 1.26; H2′, 7.27; H4′, NR; H5′, 7.27

* NR indicates data not reported.

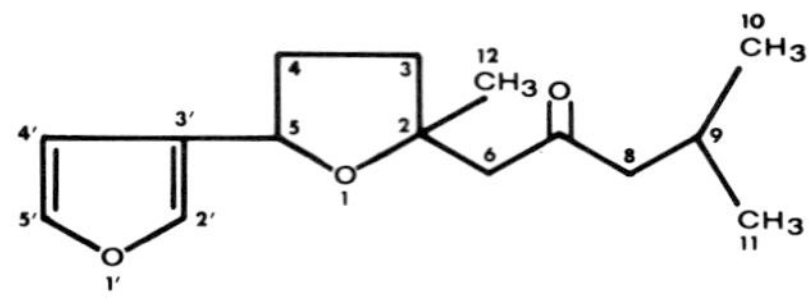

Common name	Ipomeamarone
Molecular weight	250.1568
Molecular formula	$C_{15}H_{22}O_3$
General characteristics[1,6]	Colorless, odorless liquid with a strong, bitter taste $[\alpha]_D^{19.5}$ +32.40° (C = 5.0 in ethanol) Semicarbazone crystals from carbon tetrachloride–petroleum ether, mp 108°–110°C
UV data[1,6]	λ_{max}^{EtOH} nm (ϵ): 211(5,650); semicarbazone 211(12,200), and 234(10,600)
Fungal source[1,6]	Mold-damaged sweet potatoes
Toxicity data[1,6]	Hepatotoxic
TLC data[1,6]	*Adsorbent:* silica gel, IB-F (Baker Chem. Co.) *Solvent:* benzene–methanol, 9:1 v/v R_f: 0.80 *Detection:* spray with Ehrlich's reagent followed by gentle heating; light pink spot changing to dark blue-gray.
GLC data[1,6]	*Solid support:* Chrom W 60/80 mesh *Liquid phase:* 15% Reoplex 400 *Column size:* 8 ft × ¼ in. *Column temp:* 200°C; 80 ml/min *Detector temp:* 300°C *Injector temp:* 300°C *Retention time:* 8 min
Other significant references	Biosynthesis[12]
^1H NMR Data[9]	H3, 1.85; H4, NR*; H5, 4.84 (J = 6.0, 4.2, 1.8); H6, 2.62; H8, 2.29 (J = 6.2); H9, NR; H10, 0.87 (J = 6.0); H11, 0.87 (J = 6.0); H12, 1.29; H2′, 7.30; H4′, 6.29 (J = 1.8); H5′, 7.29 (J = 1.8)

* NR indicates data not reported.

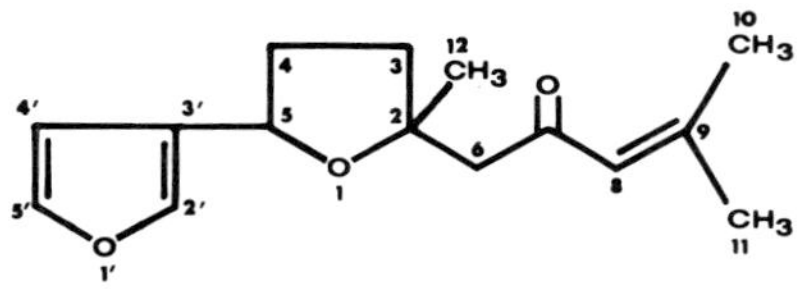

Common name	Dehydroipomeamarone
Molecular weight	248.1412
Molecular formula	$C_{15}H_{20}O_3$
General characteristics[10,11]	Light yellow oil $[\alpha]^{21}_{D}$ +14.7° (C = 0.12 in ethanol)
Toxicity data	Not reported
TLC data[10,11]	*Adsorbent:* Kiesel Gel HF_{245} *Solvent:* *n*-hexane–ethyl acetate, 8:2 v/v R_f: 0.70 *Detection:* Erhlich's spray followed by gentle heating.
Other significant references	Biosynthesis[11]

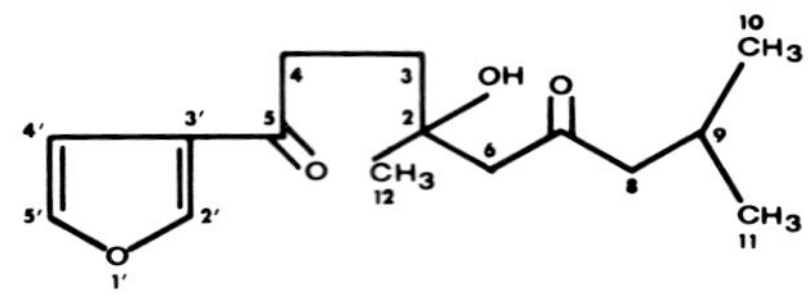

Common name	4-Hydroxymyoporone
Molecular weight	266.1517
Molecular formula	$C_{15}H_{22}O_4$
UV data[8]	λ_{max} nm (ϵ): 251(3,100)
IR data[8]	ν_{max}^{KBr} cm^{-1}: 3480, 3180, 1700, 1560, 1510, 1155, and 875
Fungal source[8]	Mold-damaged sweet potatoes
^{1}*H NMR data*[8]	H3, 1.5–2.4; H4, 2.90; H5, —; H6, 2.55; H8, 1.5–2.4; H9, 1.5–2.4; H10, 0.92; H11, 0.92; H12, 1.22; H2′, 8.08; H4′, 6.77; H5′, 7.17

REFERENCES

1. Akazawa, T., *Arch. Biochem. Biophys.* **90,** 82 (1960).
2. Boyd, M. R., Burka, L. T., Harris, T. M., and Wilson, B. J., *Biochim. Biophys. Acta* **337,** 184 (1974).
3. Boyd, M. R., Burka, L. T., and Wilson, B. J., *Toxicol. Appl. Pharmacol.* **32,** 147 (1975).
4. Boyd, M. R., and Wilson, B. J., *Toxicol. Appl. Pharmacol.* **25,** 459 (1973).
5. Boyd, M. R., and Wilson, B. J., *J. Agric. Food Chem.* **20,** 428 (1972).
6. Boyd, M. R., and Wilson, B. J., *J. Agric. Food Chem.* **19,** 547 (1971).
7. Boyd, M. R., Wilson, B. J., and Harris, T. M., *Nature (London)* **236,** 158 (1972).
8. Burka, L. T., Kuhnert, L., and Wilson, B. J., *Tetrahedron Lett.* p. 4017 (1974).
9. Kato, N., Imaseki, H., Nakashima, N., and Uritani, I., *Tetrahedron Lett.* p. 843 (1971).
10. Oguni, I., and Uritani, I., *Plant Physiol.* **53,** 649 (1974a).
11. Oguni, I., and Uritani, I., *Phytochemistry* **13,** 521 (1974b).
12. Oguni, I., and Uritani, I., *Agric. Biol. Chem.* **35,** 357 (1971).
13. Wilson, B. J., Boyd, M. R., Harris, T. M., and Yang, D. T. C., *Nature (London)* **231,** 52 (1971).
14. Wilson, B. J., and Boyd, M. R., *in* "Mycotoxins" (I. F. H. Purchase, ed.), p. 327. Elsevier, Amsterdam, 1974.
15. Yang, D. T. C., Wilson, B. J., and Harris, T. M., *Phytochemistry* **10,** 1653 (1971).

17

Viridiol Group

In 1966, Grove *et al.*[6] reported on the highly antifungal metabolite viridin from *Gliocladium virens*. Subsequently, Moffatt *et al.*[9] isolated and identified viridiol as a slightly reduced form of viridin. No biological data were given for viridiol. More recently, the demethoxy forms of both metabolites were isolated and identified via X-ray crystallography by two different laboratories.[1,5] One of these studies reported that demethoxyviridiol was toxic to vertebrates and demethoxyviridin was subsequently also found to be toxic to vertebrates.[4,5] These metabolites are modified steroids and are probably derived from a steroidal rather than a diterpenoid biosynthetic pathway.[8]

Viridiol group	Molecular weight	Molecular formula
Demethoxyviridin	322.0841	$C_{19}H_{14}O_5$
Demethoxyviridiol	324.0997	$C_{19}H_{16}O_5$
Viridin	352.0946	$C_{20}H_{16}O_6$
Viridiol	354.1103	$C_{20}H_{18}O_6$

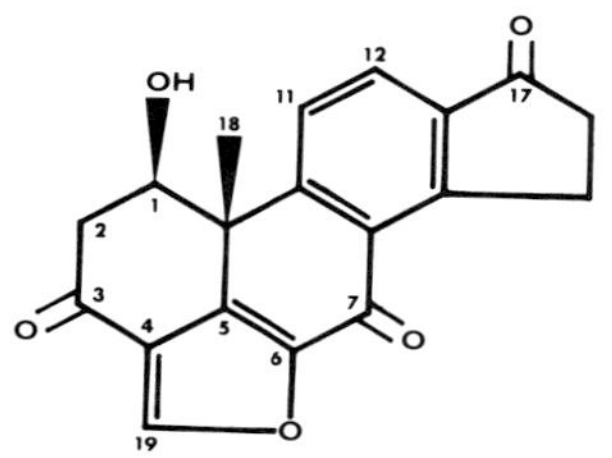

Common name	Demethoxyviridin
Molecular weight	322.0841
Molecular formula	$C_{19}H_{14}O_5$
General characteristics[1]	mp variable depending on solvent of crystallization; two ranges were 145°–160°C (dec.) or 220°–240°C (dec.); crystals from acetone, mp 230°–240°C
UV data[1]	λ_{max} nm (ϵ): 243(27,000) and 303(11,600)
Fungal source[1,3,4]	*Nodulisporium hinnuleum* Smith (ATCC 24911) Unidentified fungus (ACC 3199) *Apiospora camptospora*
Toxicity data[4]	Demethoxyviridin was toxic to day-old cockerels, but the level of toxicity was not determined.
TLC data[4]	*Adsorbent:* silica gel G-HR *Solvent:* toluene–ethyl acetate–formic acid, 5:4:1 v/v/v R_f: 0.57 *Detection:* orange spot after spraying with 50% H_2SO_4 and heat.
Other significant references	X-Ray crystallography[1]
^{13}C NMR data	C1, 70.8 d; C2, 46.5 t; C3, 189.9 s; C4, 122.9 s; C5, 144.3 s; C6, 145.6 s; C7, 172.5 s; C8, 156.9 s; C9, 136.3 s; C10, 41.5 s; C11, 126.2 d; C12, 128.4 d; C13, 156.2 s; C14, 129.6 s; C15, 28.1 t; C16, 35.7 t; C17, 205.4 s; C18, 25.2 q; C19, 149.3 d
^{1}H NMR data	H1, 4.33 (J = 6.0); H2, 2.72 (J = 6.0); H11, 7.83 (J = 8.5); H12, 8.58 (J = 8.5); H15, 2.75; H16, 3.58; H18, 1.58; H19, 8.75; 1-OH, 6.12

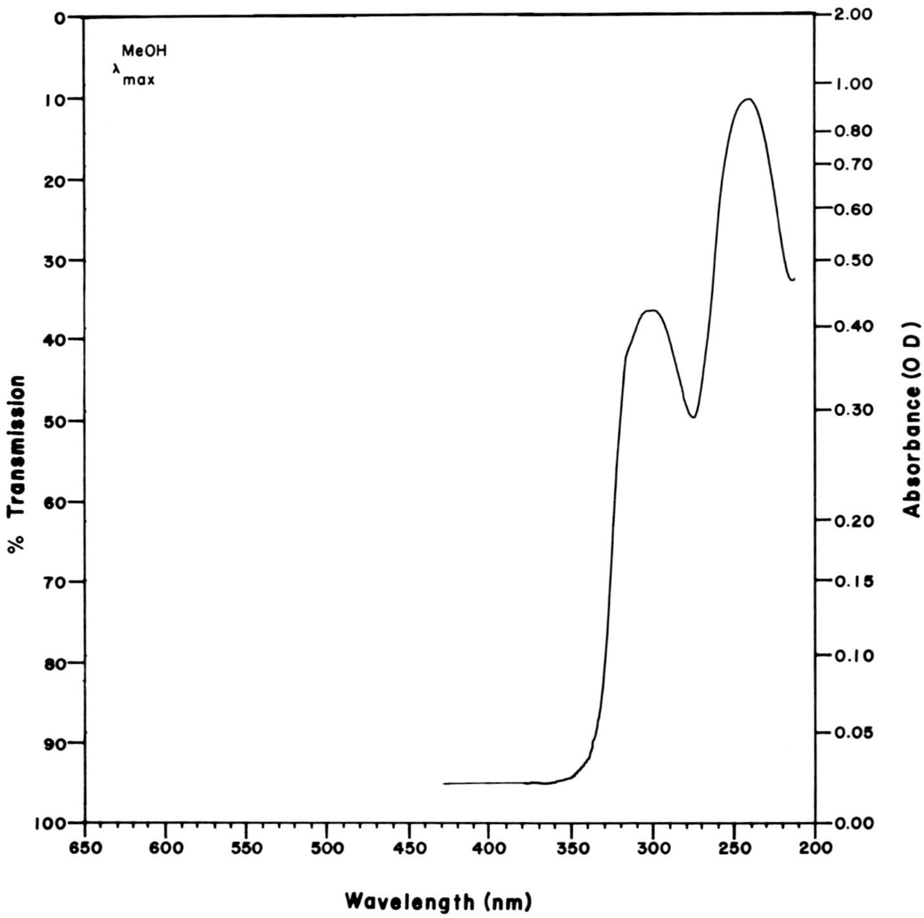
MeOH
λ max
% Transmission
0
10
20
30
40
50
60
70
80
90
100
Absorbance (O D)
2.00
1.00
0.80
0.70
0.60
0.50
0.40
0.30
0.20
0.15
0.10
0.05
0.00
650
600
550
500
450
400
350
300
250
200
Wavelength (nm)

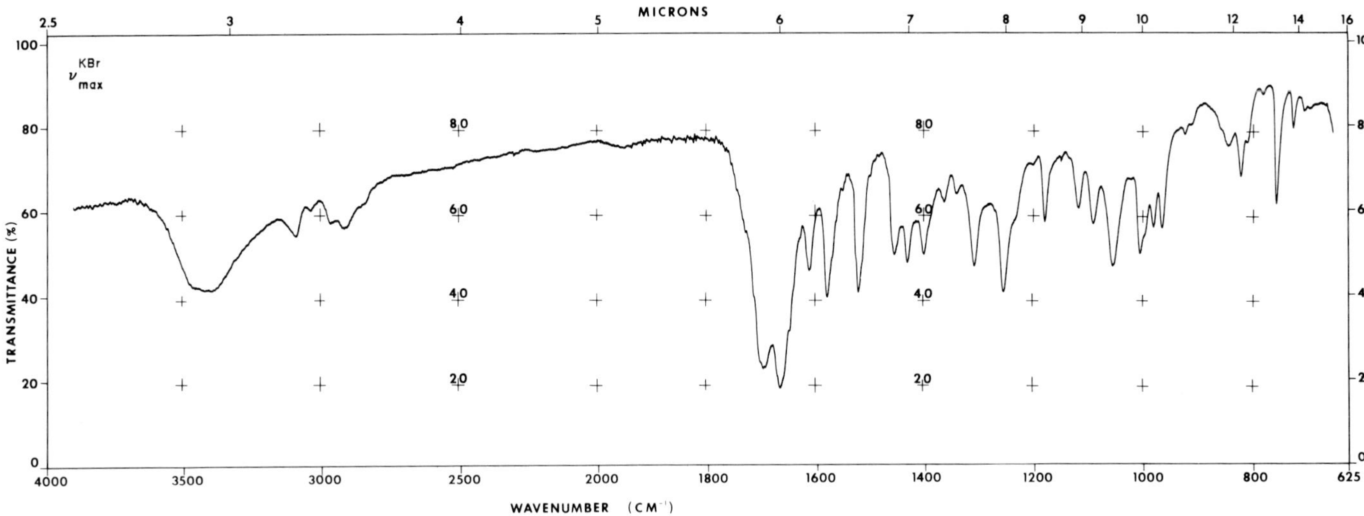
MICRONS
TRANSMITTANCE (%)
KBr
ν_{max}
WAVENUMBER (CM⁻¹)

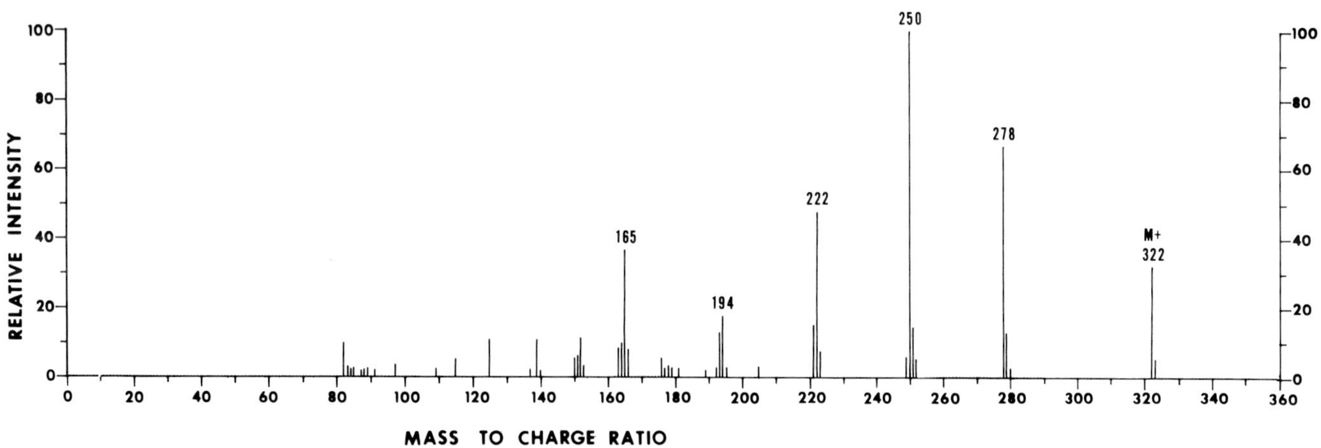
RELATIVE INTENSITY
165
194
222
250
278
M+
322
MASS TO CHARGE RATIO

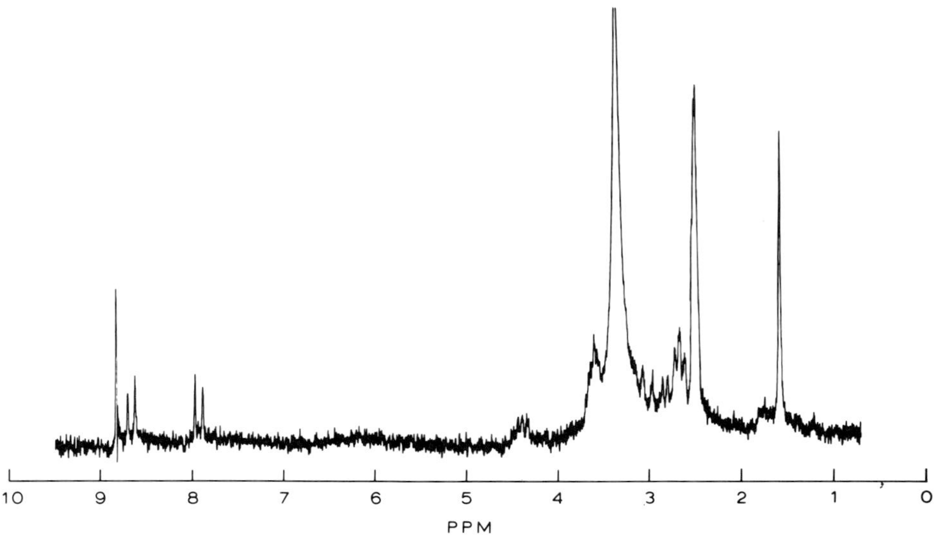

The 100 MHz ^1H NMR spectrum of demethoxyviridin in DMSO solution.

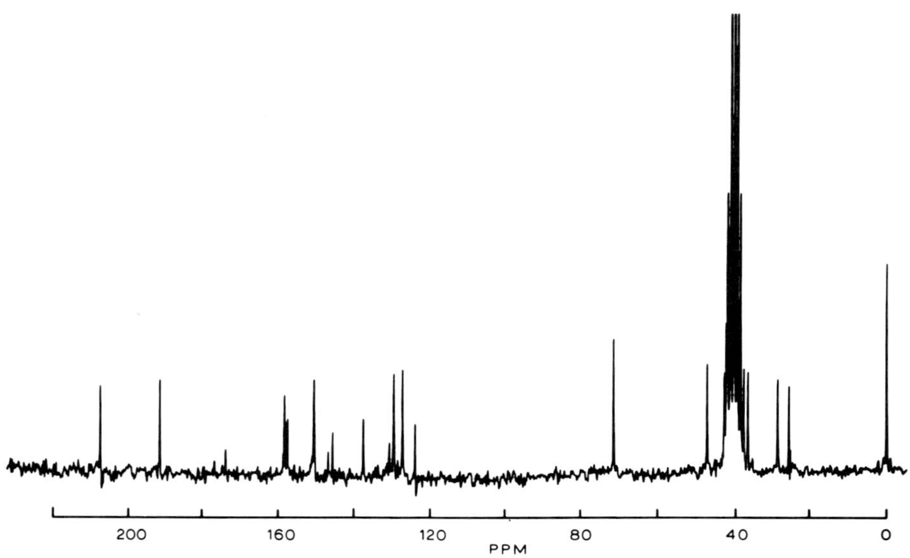

The 25 MHz ^{13}C NMR spectrum of demethoxyviridin in DMSO solution.

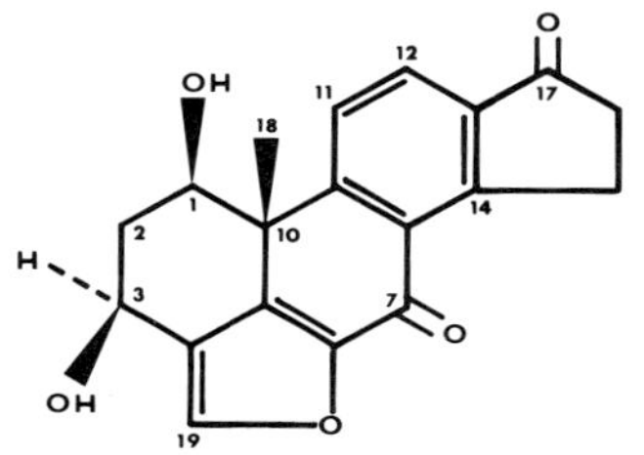

Common name	Demethoxyviridiol
Molecular weight	324.0997
Molecular formula	$C_{19}H_{16}O_5$
General characteristics	Crystals from ethyl acetate, mp 155°–157°C
UV data[1,5]	λ_{max}^{EtOH} nm (ϵ): 251(30,000) and 322(16,000) λ_{max}^{EtOH} nm (ϵ): 249(23,500) and 319(11,400)
Fungal source[1,5]	*Nodulisporium hinnuleum* Smith (ATCC 24911) Unidentified fungus (ACC 3199)
Toxicity data[5]	Demethoxyviridiol had an oral LD_{50} of 4.2 mg/kg in day-old cockerels. It also produced plant-growth regulating and phytotoxic effects in plant assay systems.
TLC data[5]	*Adsorbent:* silica gel G-HR *Solvent:* toluene–ethyl acetate–formic acid, 5:4:1 v/v/v R_f: 0.66 *Detection:* gray-blue spot after spraying with 50% ethanolic H_2SO_4 and heating at 110°C for 5 min.
Other significant references	X-Ray crystallography[1,5]
^{13}C NMR data	C1, 70.3 d; C2, 41.8 t; C3, 59.4 d; C4, 129.6 s; C5, 143.9 s; C6, 143.9 s; C7, 172.2 s; C8, 157.4 s; C9, 135.7 s; C10, 41.8 s; C11, 125.5 d; C12, 127.6 d; C13, 154.4 s; C14, 129.6 s; C15, 27.9 t; C16, 35.6 t; C17, 205.7 s; C18, 27.4 q; C19, 146.0 d
^1H NMR data	H1, 4.00; H2, 2.12; H3, 4.78 (J = 6.0); H11, 7.76 (J = 8.0); H12, 8.56 (J = 8.0); H15, 2.60; H16, 3.55; H18, 1.50; H19, 7.97; OH's, 5.46, 5.70

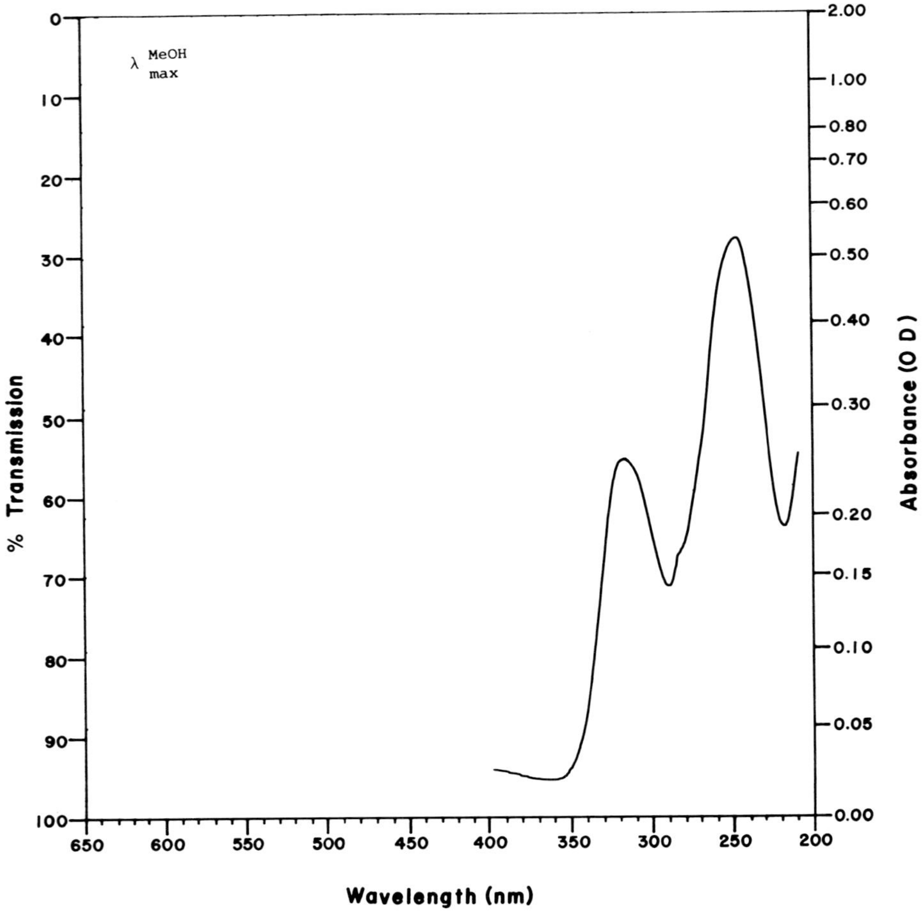
λ MeOH max
% Transmission
0
10
20
30
40
50
60
70
80
90
100
Absorbance (O D)
2.00
1.00
0.80
0.70
0.60
0.50
0.40
0.30
0.20
0.15
0.10
0.05
0.00
650 600 550 500 450 400 350 300 250 200
Wavelength (nm)

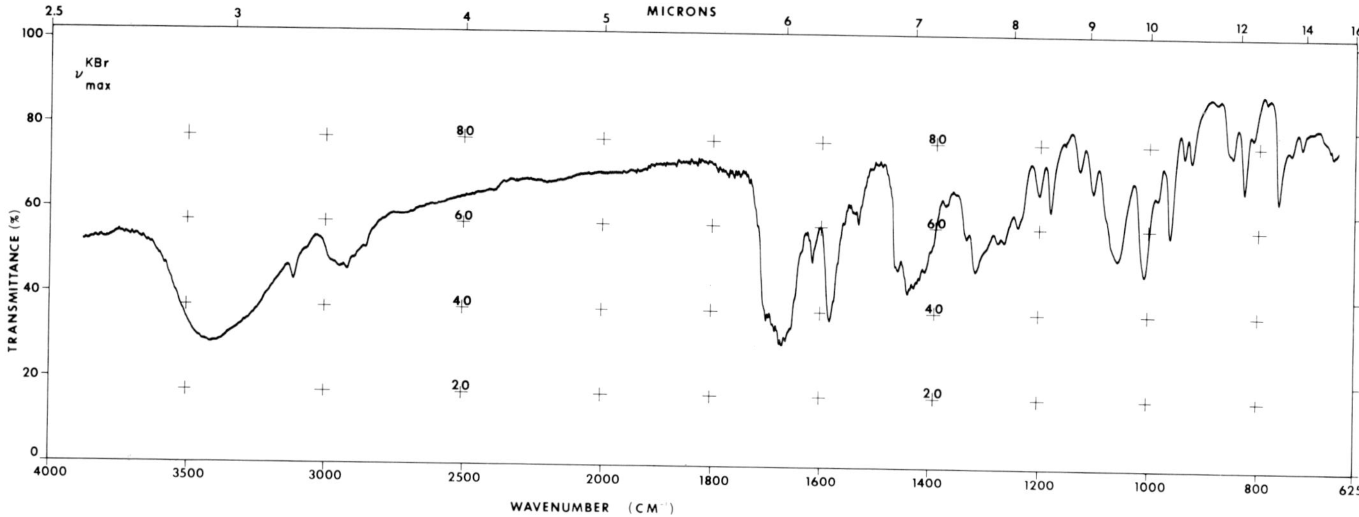
MICRONS
TRANSMITTANCE (%)
KBr
ν max
WAVENUMBER (CM⁻¹)

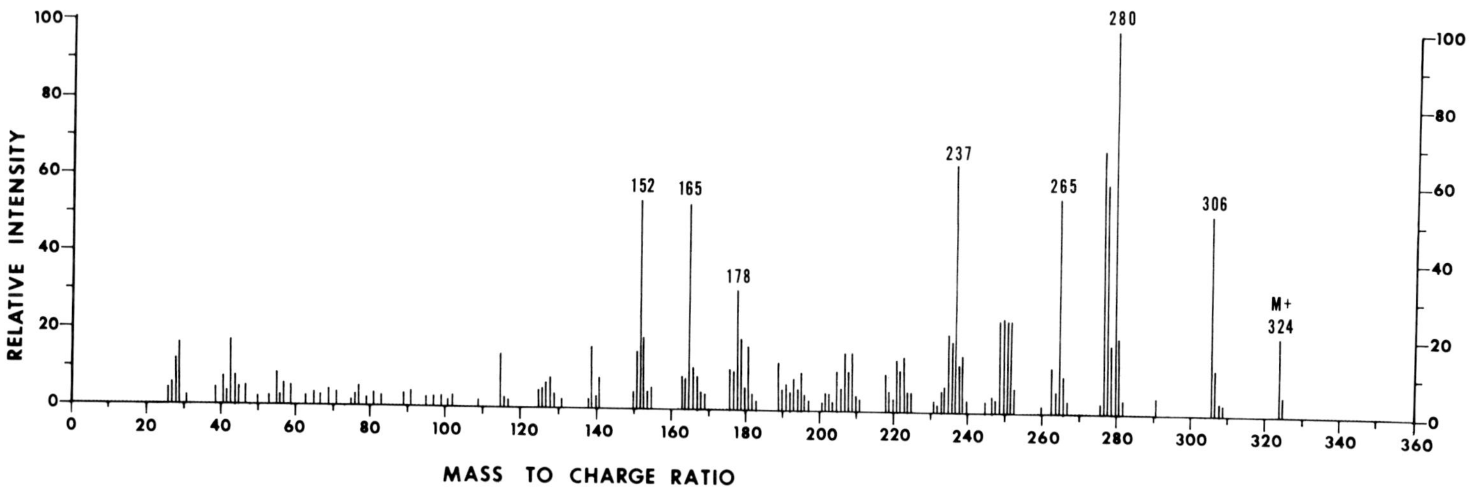
RELATIVE INTENSITY
MASS TO CHARGE RATIO
152
165
178
237
265
280
306
M+
324

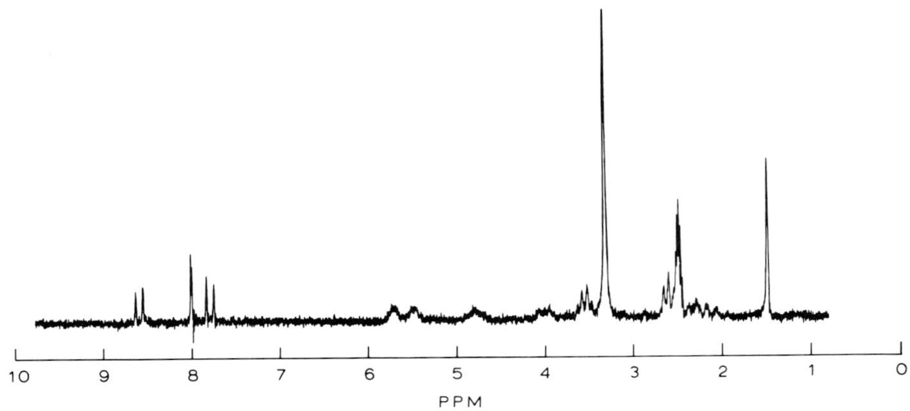

The 100 MHz ^{1}H NMR spectrum of demethoxyviridiol in DMSO solution.

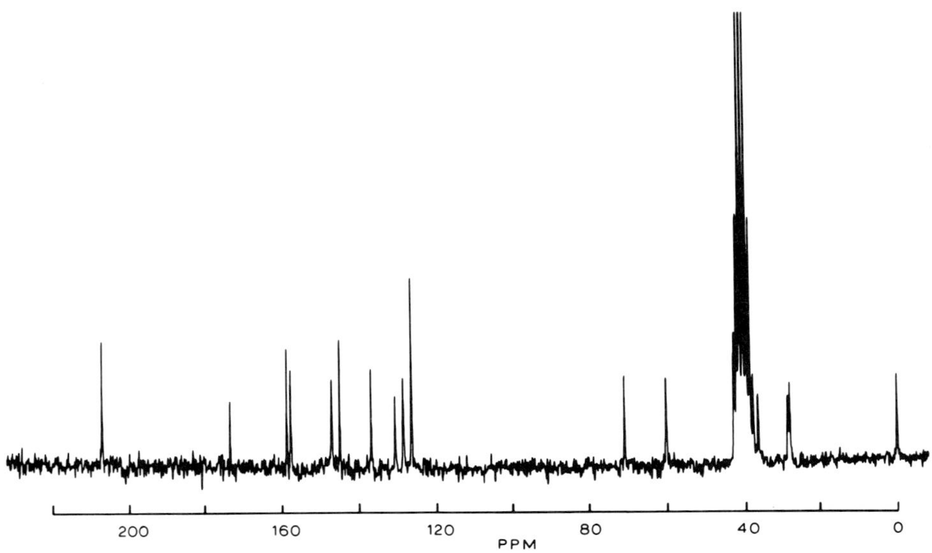

The 25 MHz ^{13}C NMR spectrum of demethoxyviridiol in DMSO solution.

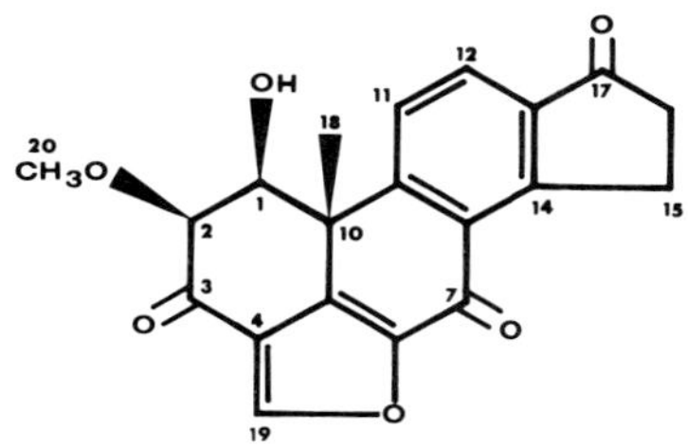

Common name	Viridin
Molecular weight	352.0946
Molecular formula	$C_{20}H_{16}O_6$
General characteristics[6]	Prisms from benzene solution, mp 245°C (dec.); needles from acetone, mp 222°–224°C; plates from methanol, mp 140°C; hemimethanolate, 242°C (dec.) $[\alpha]_D^{19}$ −224° Acetyl derivative: microcrystalline solid, mp 121°–126°C $[\alpha]_D^{17}$ −176°
UV data[6]	λ_{max} nm (ϵ): 242(30,900) and 300(16,500)
IR data	ν_{max}^{Nujol} cm^{-1}: 3390, 3145, 1692, 1675, 1622, 1587, 1532, 1126, 1090, 1070, 1032, 1019, 1005, 983, and 970 ν_{max}^{Nujol}(hemimethanolate) cm^{-1}: 3450, 3105, 1709, 1675, 1622, 1587, 1532, 1120, 1085, 1064, 1015, 1001, 982, and 970
Fungal source[2]	*Gliocladium virens*
Toxicity data[2]	Viridin possesses strong antifungal activity; minimum concentration preventing germination of *Botrytis allii* conidia, 0.005 μg/ml.
Other significant references	X-Ray crystallography[10]; biosynthesis[8]; chemistry[7]; structure and chemistry[6]
^{1}H NMR data[6]	H1, 4.45 (J = 5); H2, 3.95; H11, 8.10 (J = 8); H12, 8.80; H15, 2.90; H16, 3.80; H18, 1.73; H19, 8.45; H20, 3.75

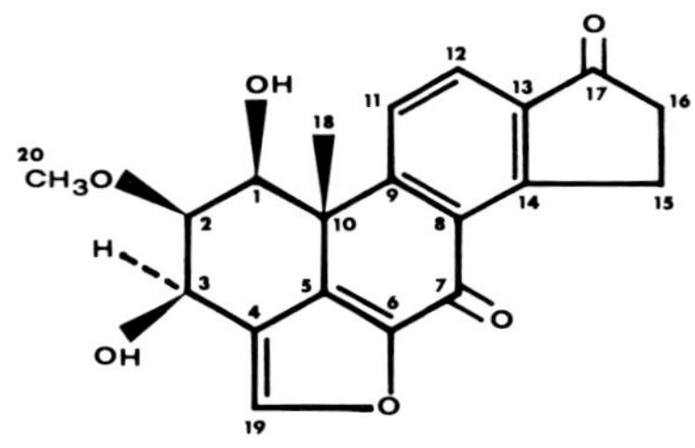

Common name	Viridiol
Molecular weight	354.1103
Molecular formula	$C_{20}H_{18}O_6$
General characteristics[9]	mp 198°–201°C (dec.)
UV data[9]	λ_{max} nm (ϵ): 250(29,500) and 317(11,700)
Fungal source[9]	*Trichoderma viride* Pers. ex Fr. (NRRL 1828)
Toxicity data	Toxicity unknown
TLC data[9]	*Adsorbent:* silica gel G *Solvent:* chloroform–methanol, 9:1 v/v R_f: 0.36 *Detection:* green fluorescence under UV light; red-brown after $AgNO_3$ spray.
^{13}*C NMR data*	C1, 71.7 d; C2, 81.7 d; C3, 61.6 d; C4, 121.9 s; C5, 142.2 s; C6, 142.2 s; C7, 173.1 s; C8, 158.4 s; C9, 136.7 s; C10, 42.2 s; C11, 127.0 d; C12, 127.0 d; C13, 157.6 s; C14, 127.0 s; C15, 28.4 t; C16, 36.4 t; C17, 206.2 s; C18, 30.5 q; C19, 145.3 d; C20, 60.5 q
^{1}*H NMR data*	H1, 4.32; H2, 3.60; H3, 5.12; H11, 7.93 (J = 8.0); H12, 8.25 (J = 8.0); H15, 2.71; H16, 3.64; H18, 1.74; H19, 7.77; H20, 3.75

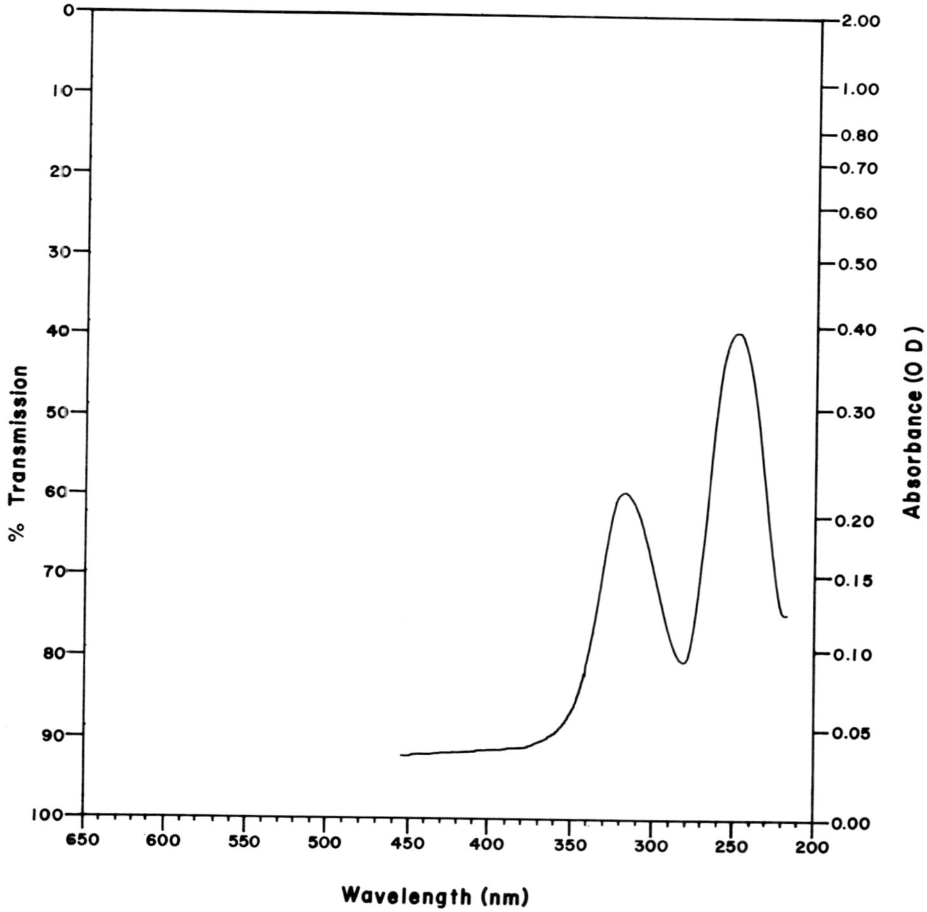
% Transmission
0
10
20
30
40
50
60
70
80
90
100
Absorbance (O D)
2.00
1.00
0.80
0.70
0.60
0.50
0.40
0.30
0.20
0.15
0.10
0.05
0.00
650
600
550
500
450
400
350
300
250
200
Wavelength (nm)

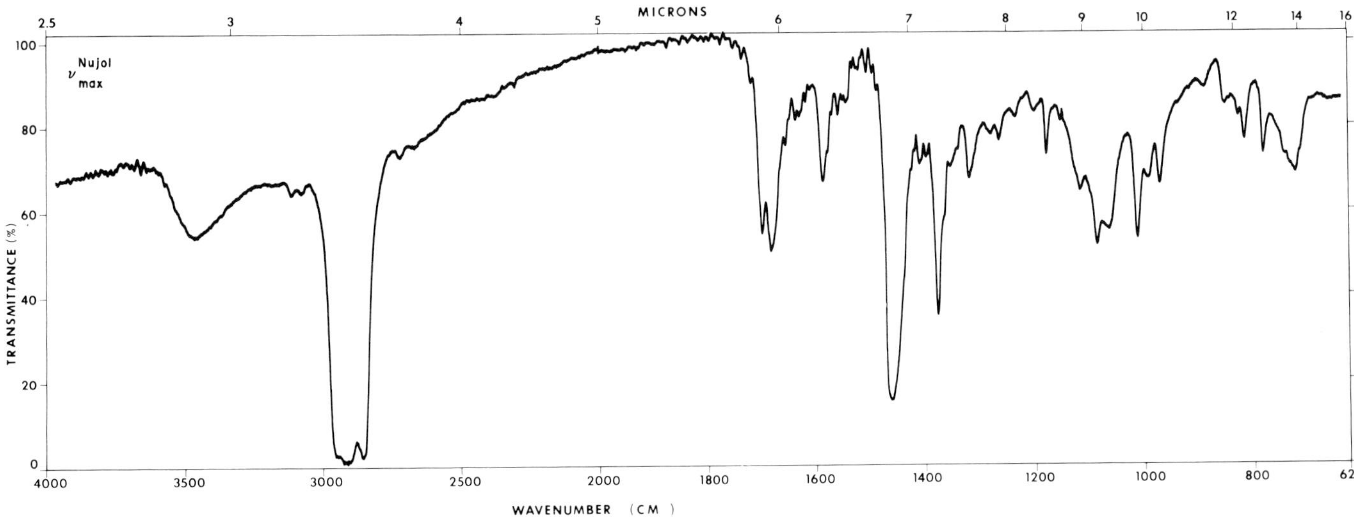
MICRONS
Nujol
ν_{max}
TRANSMITTANCE (%)
WAVENUMBER (CM)

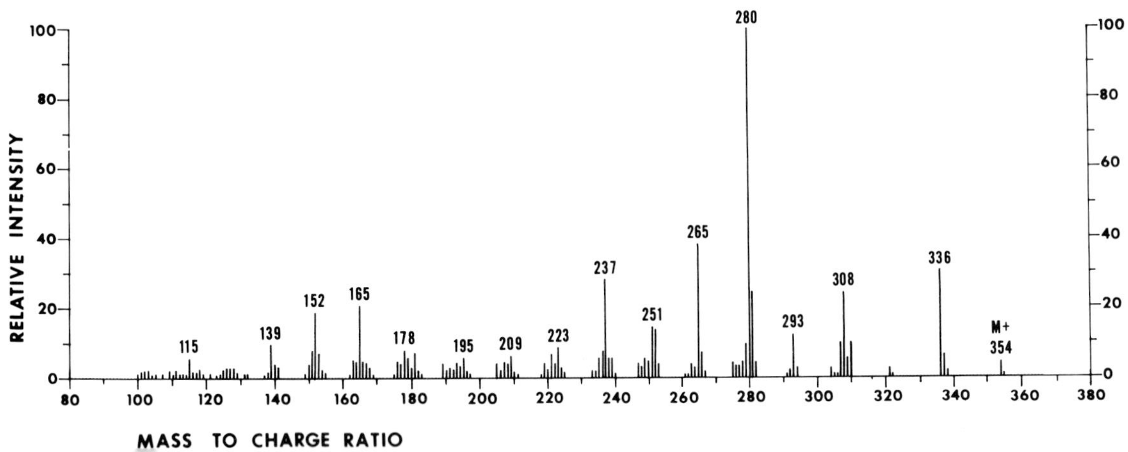
RELATIVE INTENSITY
MASS TO CHARGE RATIO
115
139
152
165
178
195
209
223
237
251
265
280
293
308
336
M+
354

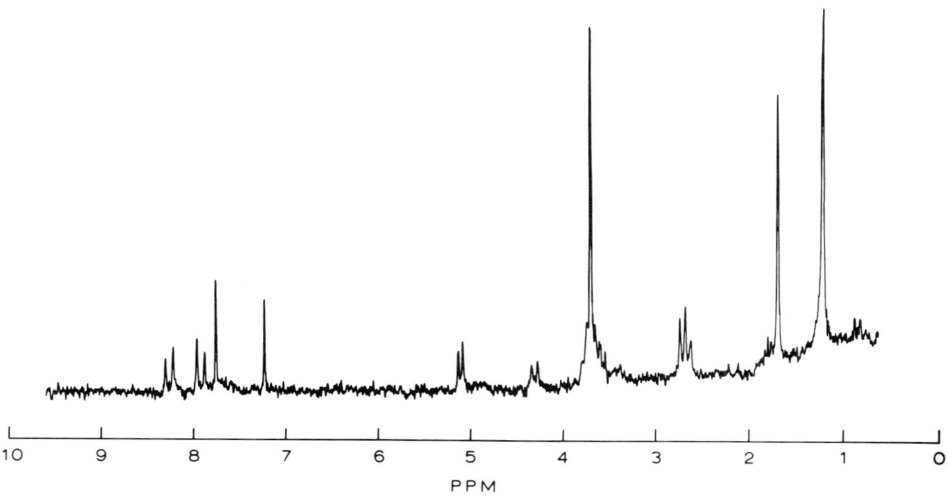

The 100 MHz ^1H NMR spectrum of viridiol in $CDCl_3$ solution.

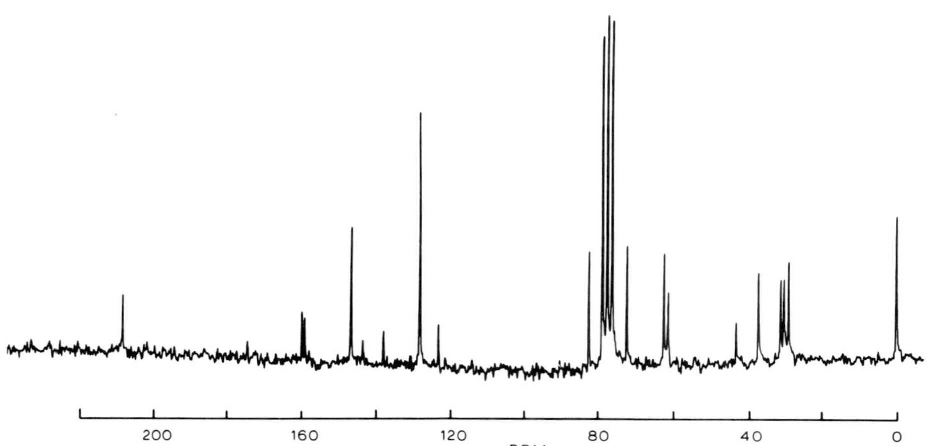

The 25 MHz ^{13}C NMR spectrum of viridiol in $CDCl_3$ solution.

REFERENCES

1. Aldridge, D. C., Turner, W. B., Geddes, A. J., and Sheldrick, B., *J. Chem. Soc., Perkin Trans. 1* p. 943 (1975).
2. Brian, P. W., and McGowan, J. C., *Nature (London)* **156,** 144 (1945).
3. Burrows, B. R., unpublished data.
4. Cole, R. J., unpublished data.
5. Cole, R. J., Kirksey, J. W., Springer, J. P., Clardy, J., Cutler, H. G., and Garren, K. H., *Phytochemistry* **14,** 1429 (1975).
6. Grove, J. F., McCloskey, P., and Moffatt, J. S., *J. Chem. Soc.* p. 743 (1966).
7. Grove, J. F., Moffatt, J. S., and Vischer, E. B., *J. Chem. Soc.* p. 3803 (1965).
8. Grove, J. F., *J. Chem. Soc. C* p. 549 (1969).
9. Moffatt, J. S., Bulock, J. D., and Yuen, T. H., *Chem. Commun.* p. 839 (1969).
10. Neidle, S., Rogers, D., and Hursthouse, M. B., *J. Chem. Soc., Perkin Trans. 2* p. 760 (1972).

18

Aspergillus Toxins

The following fungal metabolites represent a wide diversity in chemical structure and, therefore, cannot be grouped according to chemical similarity. Therefore, they have been placed into four groups, three of which are based on the genus most likely to produce them and one miscellaneous group. These groups are *Aspergillus* toxins, *Penicillium* toxins, *Fusarium* toxins, and miscellaneous toxins. As might be expected, they also produce a wide variety of biological effects. In some cases, only acute toxicity was observed (i.e., aspergillic acid, oosporein, and moniliformin). Many, such as citrinin, possess antibiotic properties in addition to effects on animals and/or higher plants. Slaframine, butenolide, and zearalenone have been implicated strongly in the etiology of naturally occurring toxicoses, whereas others (such as cyclopenin and viridicatin) apparently show little or no acute toxicity to animals. Their chemical size and complexity vary from that of moniliformin, a C_4 cyclobutene-type compound, to that of roseotoxin B, a complex cyclic peptide.

Reviews concerning *Aspergillus* toxins and *Penicillium* toxins have been published.[87]

Aspergillus toxins	Molecular weight	Molecular formula
β-Nitropropionic acid	119.0218	$C_3H_5O_4N$
Kojic acid	142.0266	$C_6H_6O_4$
Terreic acid	154.0266	$C_7H_6O_4$
Terrein	154.0629	$C_8H_{10}O_3$
Fumigatin	168.0422	$C_8H_8O_4$
Spinulosin	184.0371	$C_8H_8O_5$
Maltoryzine	208.0735	$C_{11}H_{12}O_4$
Austdiol	236.0685	$C_{12}H_{12}O_5$
Aspergillic acid	224.1524	$C_{12}H_{20}O_2N_2$
Nigragillin	222.1732	$C_{13}H_{22}ON_2$

(continued)

Aspergillus toxins	Molecular weight	Molecular formula
Austin	500.2046	$C_{27}H_{32}O_9$
Helvolic acid	568.2934	$C_{33}H_{44}O_8$
Fumagillin	458.2304	$C_{26}H_{34}O_7$
Xanthomegnin	574.1111	$C_{30}H_{22}O_{12}$
Viomellein	560.1311	$C_{30}H_{24}O_{11}$
Rubrosulphin	528.1060	$C_{29}H_{20}O_{10}$
Viopurpurin	544.1160	$C_{29}H_{20}O_{11}$

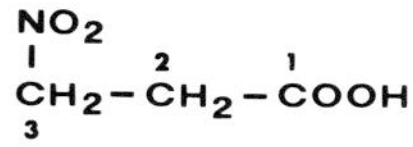

Common name	β-Nitropropionic acid (Hiptagenic acid, Bovinocidin[4,20,59])
Molecular weight	119.0218
Molecular formula	$C_3H_5O_4N$
General characteristics[18]	mp 65°–67°C β-Nitropropionic acid is soluble in water, alcohol, ethyl acetate, acetone, ether, and hot chloroform; fairly soluble in cold chloroform; insoluble in ligroin, petroleum ether, and benzene.
UV data[18]	λ_{max} nm (ϵ): 273(23.7)
Fungal source[17,61,68]	*Aspergillus flavus* *A. oryzae* *A. wentii* *Penicillium atrovenetum* G. Smith
Plant source[87]	*Hiptage mandablota* (benghalensis) Gaertn. (bark component) *Carynocarpus laevigata* *Indigophera spicata* *Viola odorata*
Toxicity data[87]	The LD_{50} of crude β-nitropropionic acid in 25 g mice was approximately 250 mg/kg.In other studies, the LD_{50} of synthetic β-nitropropionic acid (in $NaHCO_3$ solution) was 160–240 mg/kg. Death occurred within 40 min—24 hr after dosing. In mice dosed IV, a dosage of 15–60 mg/kg was not toxic; however, 110–200 mg/kg caused death within 24 hr. Clinical signs were apnea, convulsions, congestion in subcutaneous vessels and lungs, and mottled liver.
Other significant references	Biosynthesis[12,43]; synthesis[18]
^{13}C NMR data	C1, 171.2 s; C2, 30.9 t; C3, 70.1 t
^{1}H NMR data	H2, 2.94 (J = 6.0); H3, 4.64 (J = 6.0)

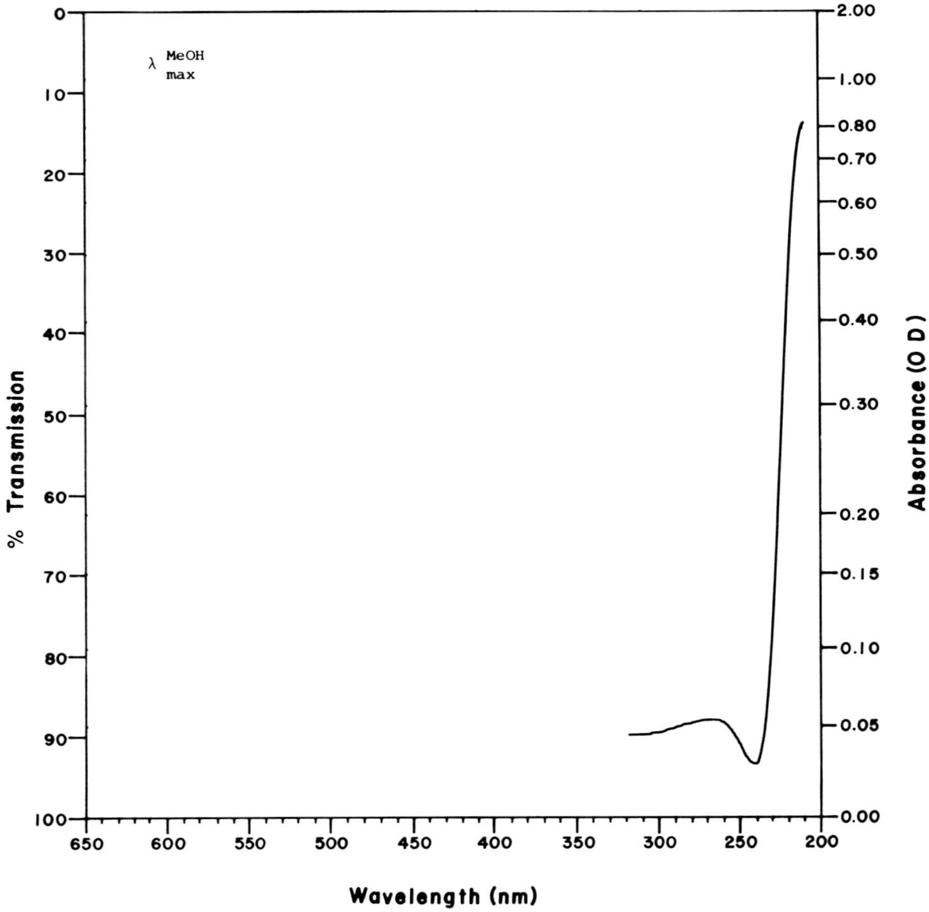

λ MeOH max
% Transmission
0
10
20
30
40
50
60
70
80
90
100
Absorbance (O D)
2.00
1.00
0.80
0.70
0.60
0.50
0.40
0.30
0.20
0.15
0.10
0.05
0.00
650
600
550
500
450
400
350
300
250
200
Wavelength (nm)

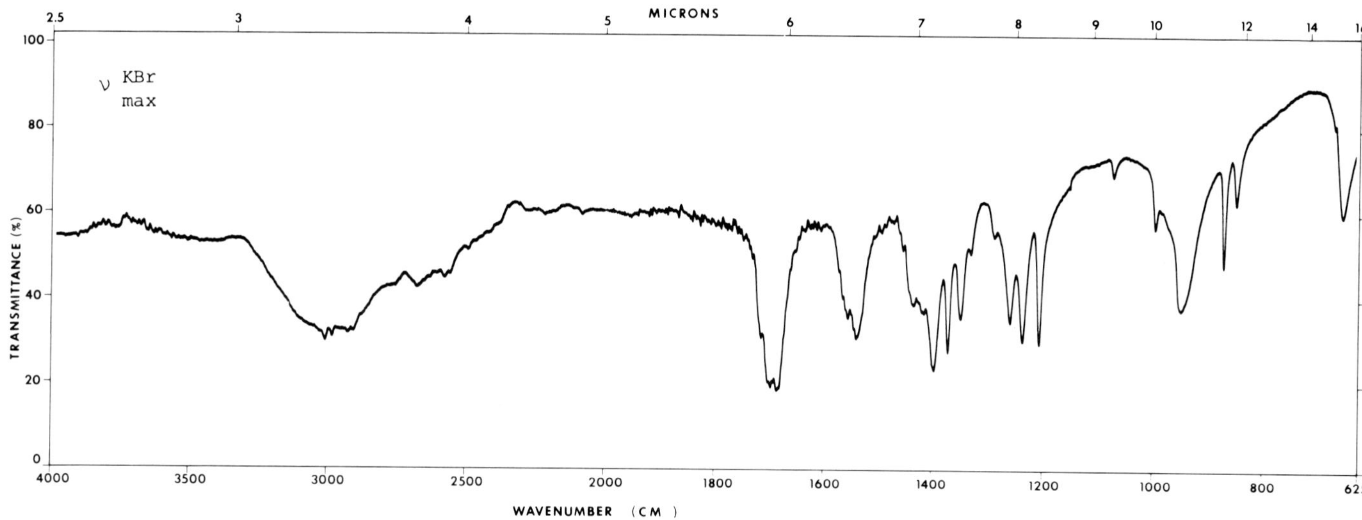
MICRONS
TRANSMITTANCE (%)
WAVENUMBER (CM)
ν KBr max
2.5
3
4
5
6
7
8
9
10
12
14
16
100
80
60
40
20
0
4000
3500
3000
2500
2000
1800
1600
1400
1200
1000
800
625

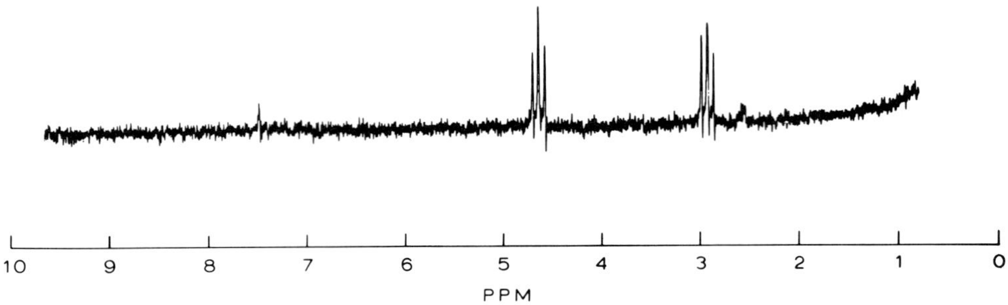

The 100 MHz ^1H NMR spectrum of β-nitropropionic acid in $CDCl_3$/DMSO solution.

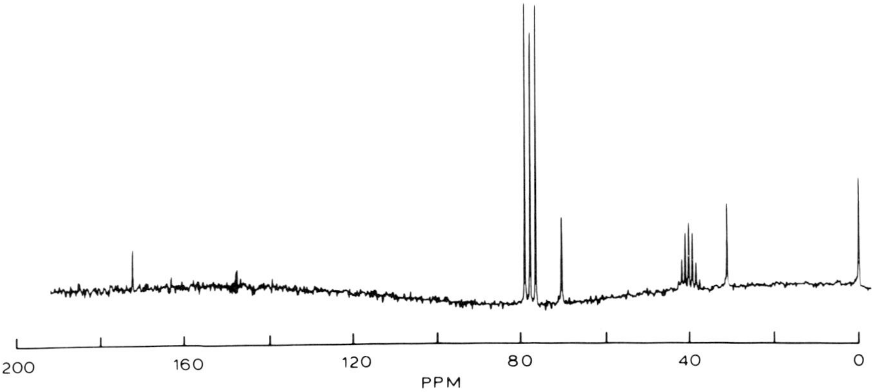

The 25 MHz ^{13}C NMR spectrum of β-nitropropionic acid in $CDCl_3$/DMSO solution.

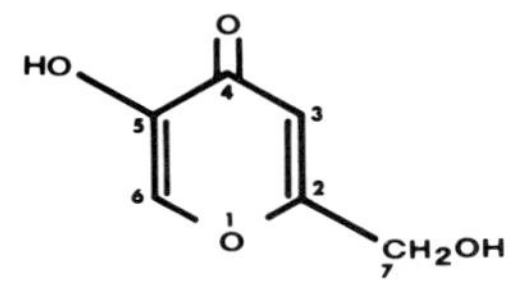

Common name[89]	Kojic acid (2-Hydroxymethyl-5-hydroxy-γ-pyrone)
Molecular weight	142.0266
Molecular formula	$C_6H_6O_4$
General characteristics[89]	White prismatic needles, mp 152°–154°C pK_a 7.90, 8.03 Kojic acid is readily soluble in water, ethanol, methanol, acetone, and ethyl acetate; sparingly soluble in ethyl ether, pyridine, and chloroform; insoluble in benzene and hexane.
UV data[60]	λ_{max}^{HCl} nm (ϵ): 270(7,820) λ_{max}^{HOH} nm (ϵ): 268(8,080) λ_{max}^{NaOH} nm (ϵ): 315(6,560)
Fungal source[10,60,69,87]	*Aspergillus oryzae* *A. luteo-virescens* *A. albus* *A. effusus* Tiraboschi *A. lutescens* *A. alliaceus* *A. awamori* *A. candidus* *A. clavatus* *A. flavus* *A. fumigatus* *A. giganteus* *A. glaucus* *A. gymnosardae* *A. nidulans* *A. parasiticus* *A. tamarii* *A. ustus* *A. wentii* *Penicillium citrinum* *P. daleae* *P. griseofulvum* *P. rubrum* *P. purpurogenum*

Toxicity data[35,38,39,60] Kojic acid possesses antibacterial and antifungal activity. Moderate cardiotoxic and cardiotonic activity have also been reported. The LD_{100} to 12-day-old chick embryos was 16 mg/100 g egg wt. The LD_{50} of Kojic acid in 17 g mice dosed IP was 30 mg/mouse. IV injection in dogs at 150 mg/kg resulted in copious salivation, retching, vomiting, and defecation. A lethal dose was found to be 1 gm/kg.

Other significant references Synthesis[55]; biosynthesis[5]; review[8,10,87]

^{13}C NMR data[51] C2, 167.7 s; C3, 109.6 d; C4, 173.6 s; C5, 145.4 s; C6, 138.9 d; C7, 59.4 t

^{1}H NMR data H3, 6.32; H6, 7.96; H7, 4.29

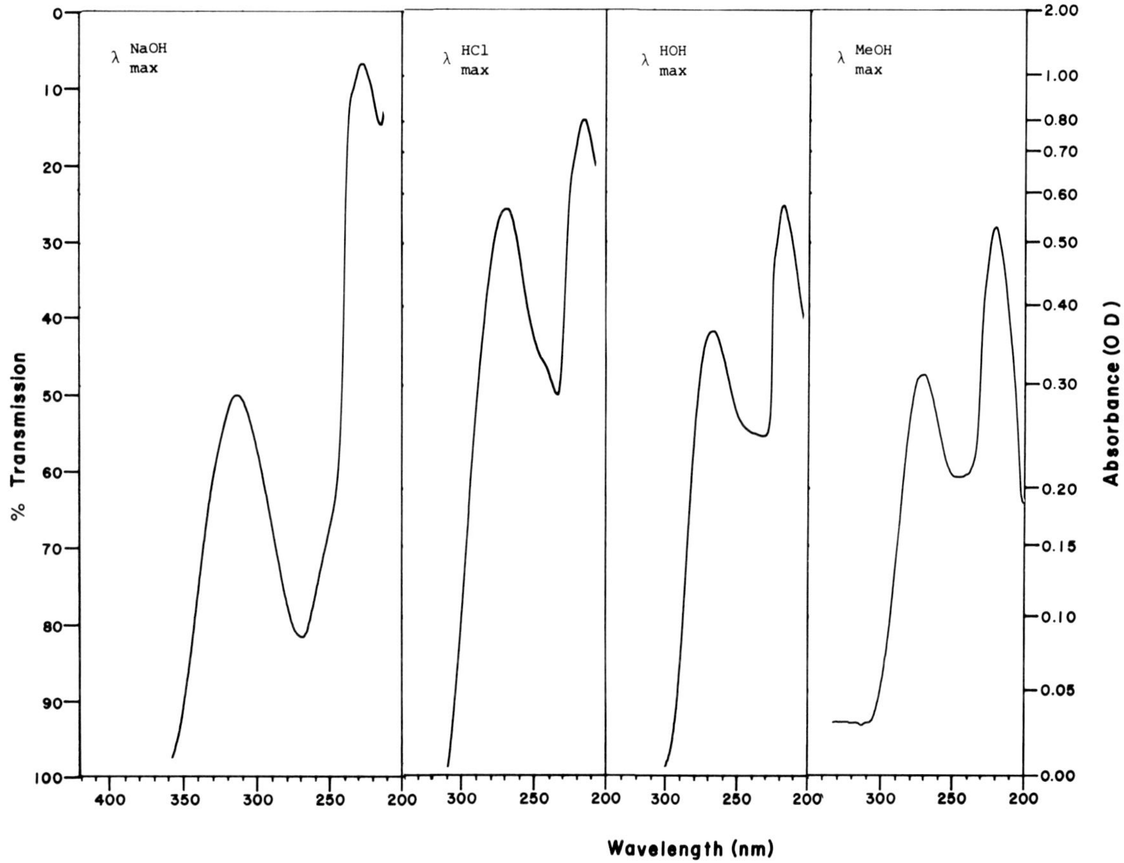
% Transmission
Absorbance (O D)
Wavelength (nm)
λ NaOH max
λ HCl max
λ HOH max
λ MeOH max

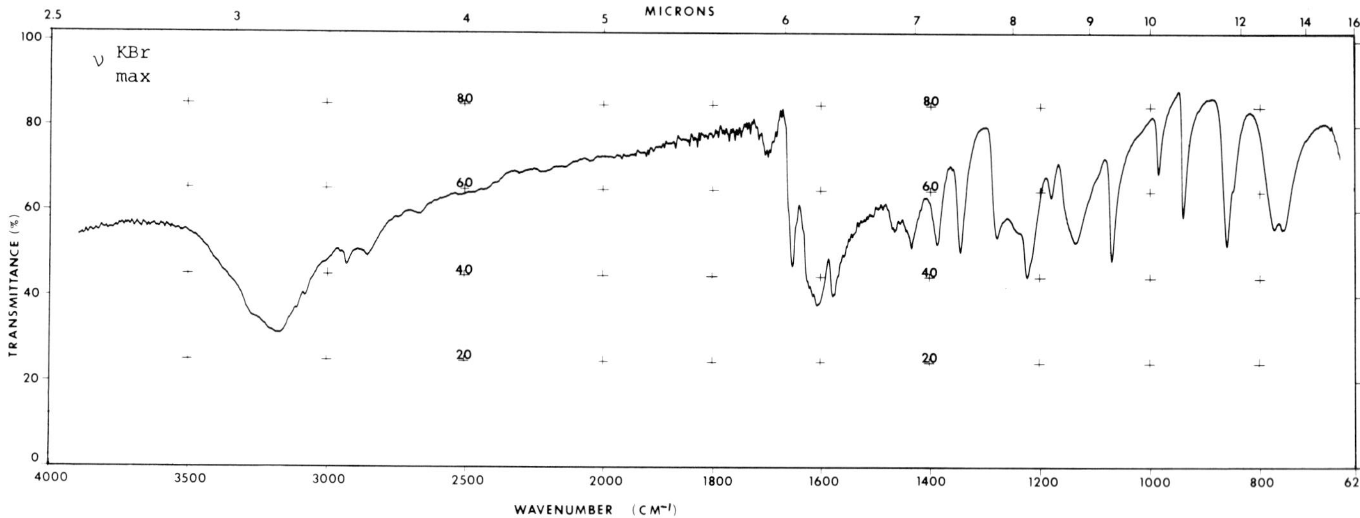
MICRONS
ν KBr max
TRANSMITTANCE (%)
WAVENUMBER (CM⁻¹)

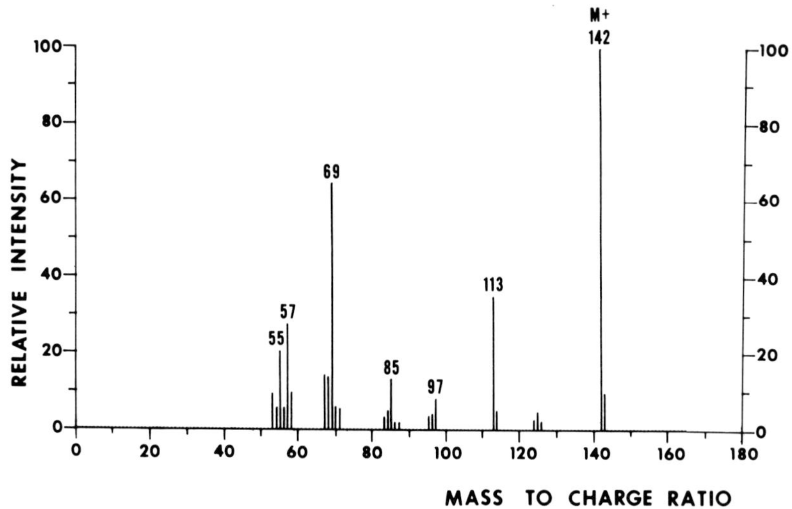
M+ 142
113
97
85
69
57
55
RELATIVE INTENSITY
MASS TO CHARGE RATIO

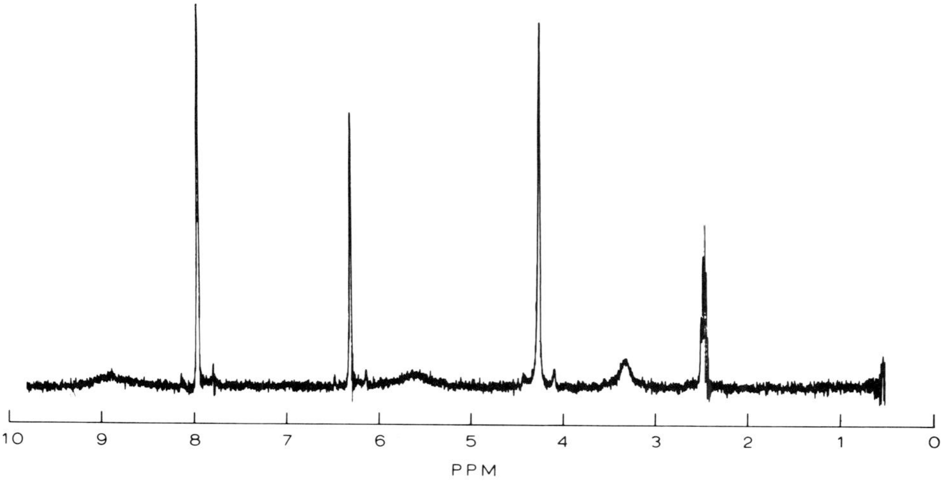

The 100 MHz ^{1}H NMR spectrum of kojic acid in DMSO solution.

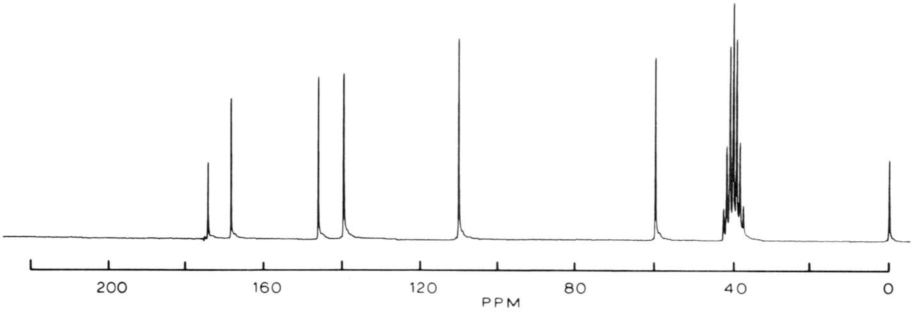

The 25 MHz ^{13}C NMR spectrum of kojic acid in DMSO solution.

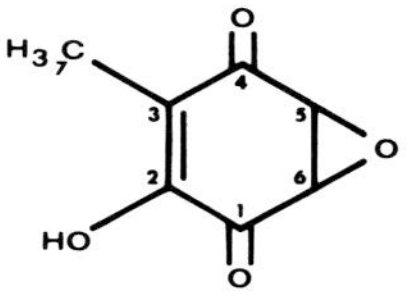

Common name	Terreic acid (5,6-Epoxy-2-hydroxytoluquinone)
Molecular weight	154.0266
Molecular formula	$C_7H_6O_4$
General characteristics[72]	Crystallizes as pale yellow plates from benzene, mp 127°C $[\alpha]_D^{27}$ −12.6° (*C* = 0.86 in 50% aqueous methanol) $[\alpha]_D^{22}$ −16.6° (*C* = 1.0 in chloroform)
UV data[72]	λ_{max}^{EtOH} nm (ϵ): 214(10,700) and 316(7,600)
Fungal source[72]	*Aspergillus terreus* Thom
Toxicity data[49,86]	Terreic acid showed *in vitro* activity against gram-positive and gram-negative bacteria and fungi; *in vivo* tests with mice, however, were unpromising. Terreic acid was reported to be active against the protozoan *Trichomonas vaginalis*. The LD_{50} of terreic acid in mice dosed IV was 71–119 mg/kg.
Other significant references	Synthesis[73]; structure elucidation[72]
^{13}C NMR data	C1, 187.4 s; C2, 151.8 s; C3, 120.4 s; C4, 190.6 s; C5, 53.8 d; C6, 51.5 d; C7, 8.7 q
^{1}H NMR data	H5, 3.85; H6, 3.85; H7, 1.90

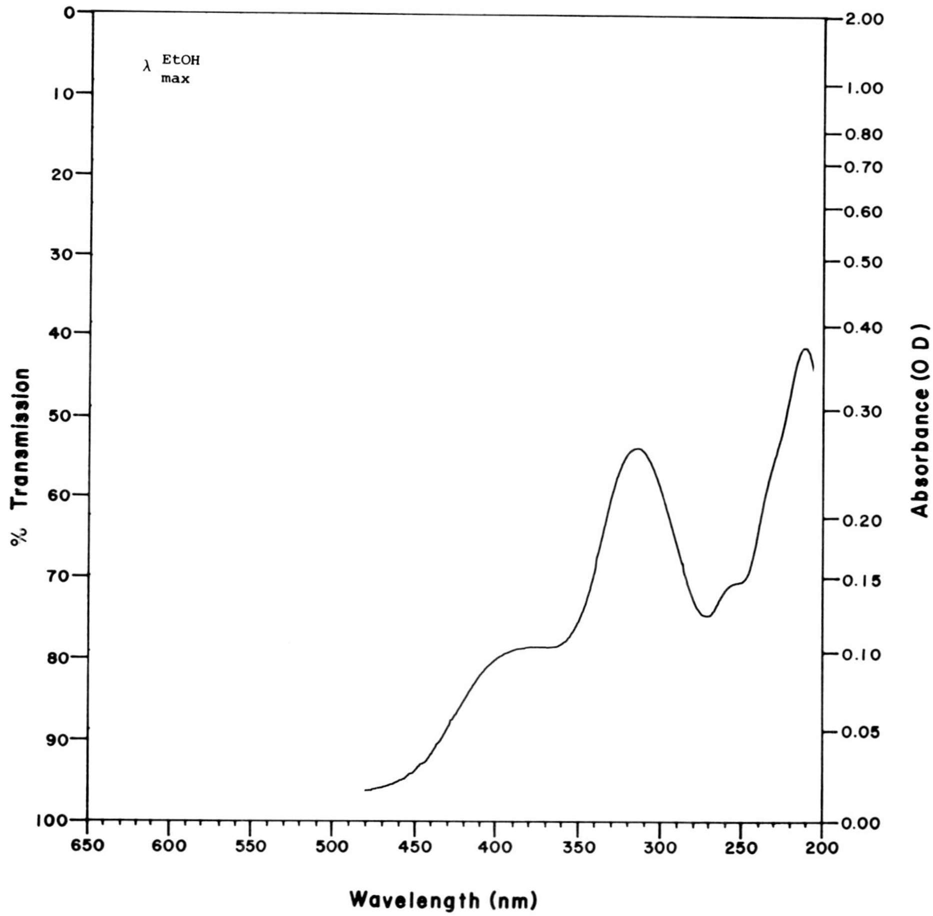
λ EtOH max
% Transmission
0
10
20
30
40
50
60
70
80
90
100
Absorbance (O D)
2.00
1.00
0.80
0.70
0.60
0.50
0.40
0.30
0.20
0.15
0.10
0.05
0.00
650
600
550
500
450
400
350
300
250
200
Wavelength (nm)

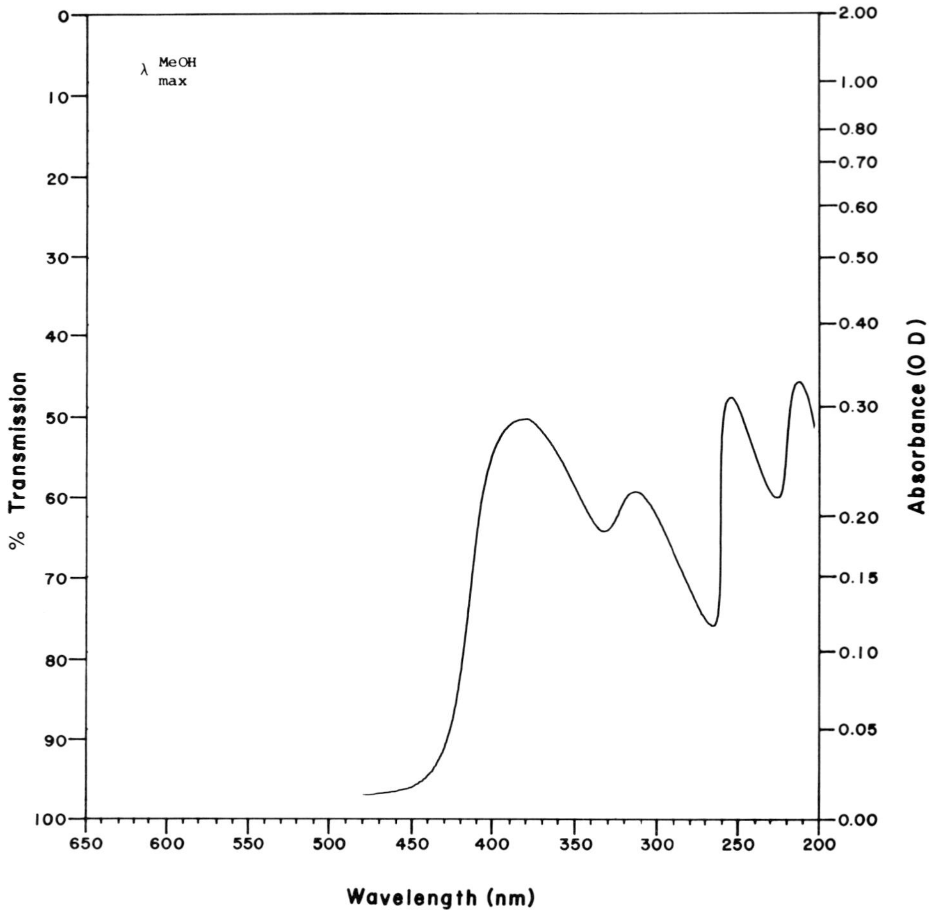

λ MeOH max
% Transmission
0
10
20
30
40
50
60
70
80
90
100
Absorbance (O D)
2.00
1.00
0.80
0.70
0.60
0.50
0.40
0.30
0.20
0.15
0.10
0.05
0.00
650
600
550
500
450
400
350
300
250
200
Wavelength (nm)

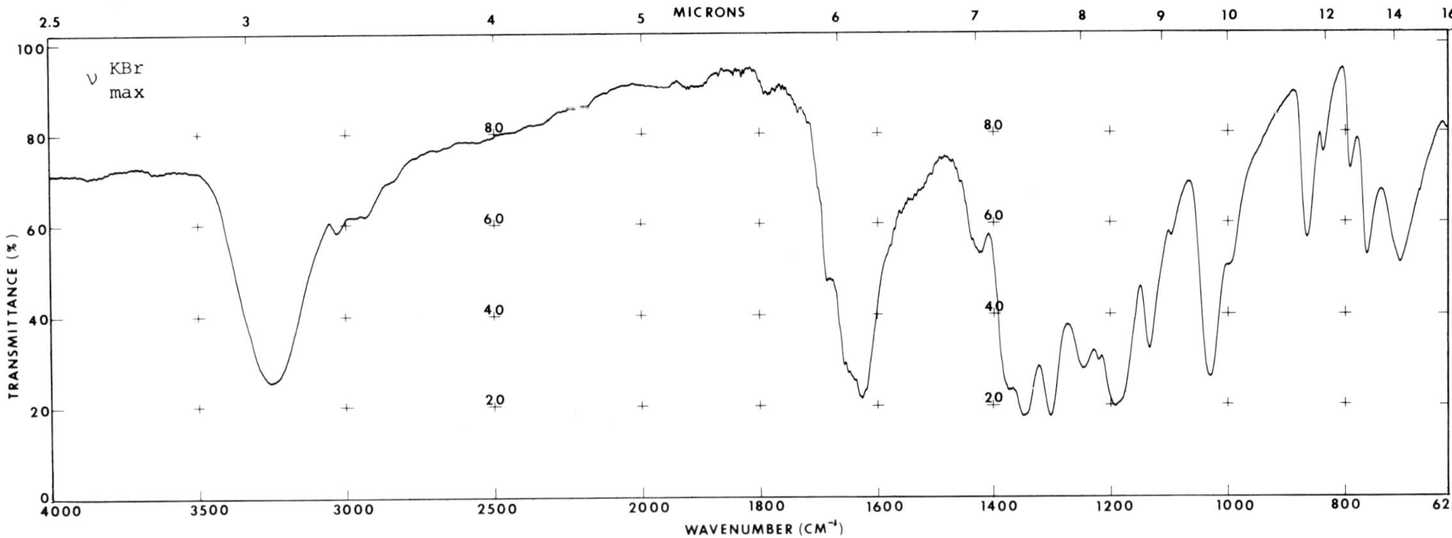
MICRONS
2.5
3
4
5
6
7
8
9
10
12
14
16
ν KBr max
TRANSMITTANCE (%)
100
80
60
40
20
0
4000
3500
3000
2500
2000
1800
1600
1400
1200
1000
800
625
WAVENUMBER (CM^{-1})

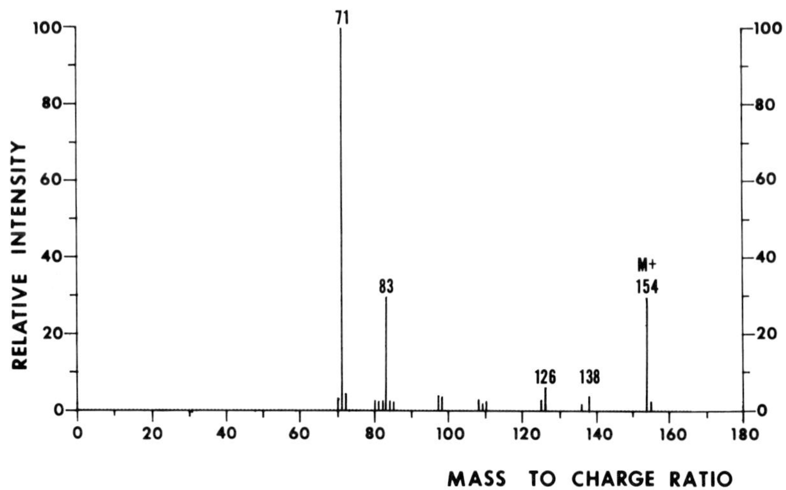
RELATIVE INTENSITY
100
80
60
40
20
0
71
83
126
138
M+
154
0
20
40
60
80
100
120
140
160
180
MASS TO CHARGE RATIO

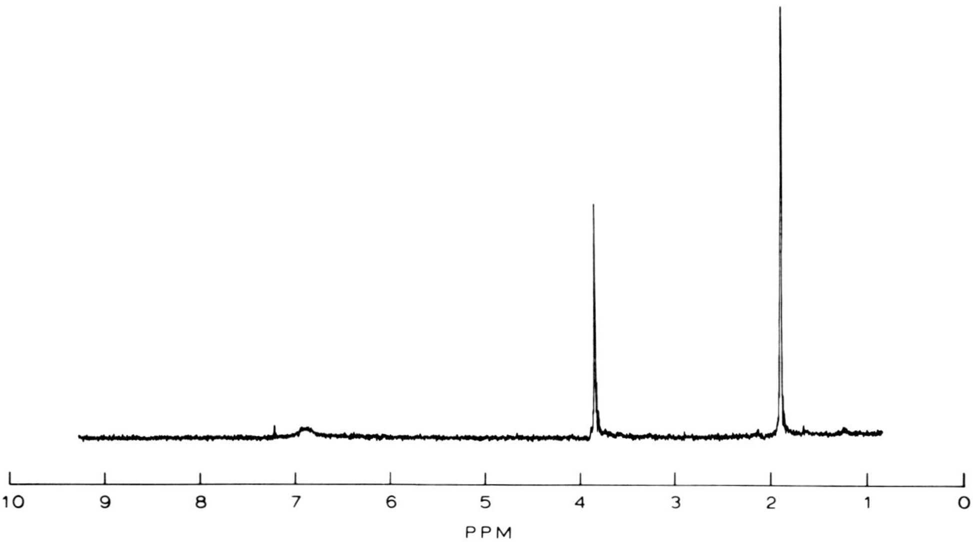

The 100 MHz ^1H NMR spectrum of terreic acid in $CDCl_3$ solution.

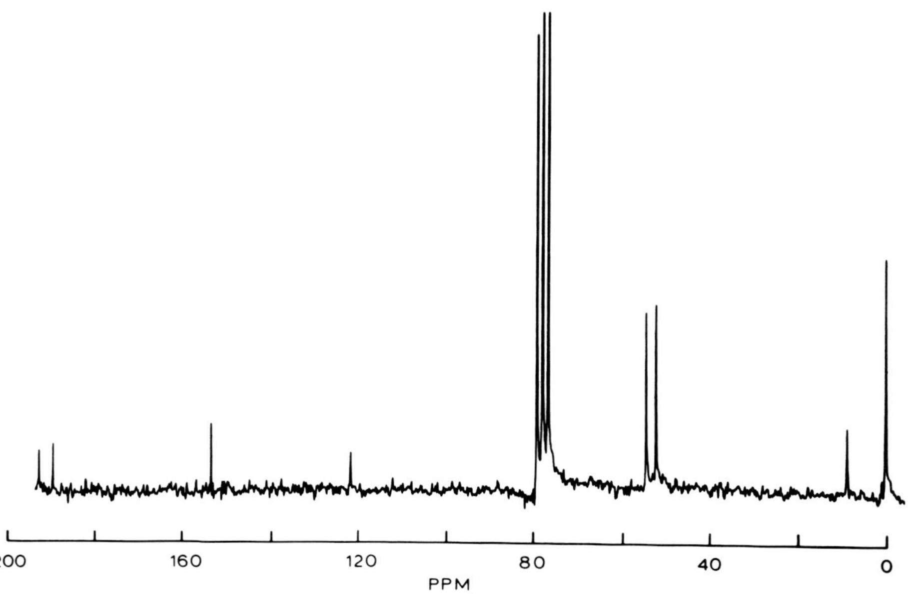

The 25 MHz ^{13}C NMR spectrum of terreic acid in $CDCl_3$ solution.

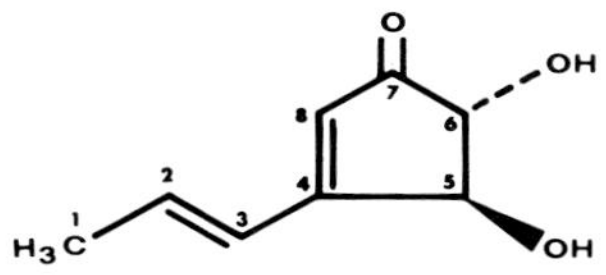

Common name	Terrein (5,6-Dihydroxy-4-[1-propenyl]cyclopent-4-en-7-one)
Molecular weight	154.0629
Molecular formula[40]	$C_8H_{10}O_3$
General characteristics[6,9,28]	Colorless crystals from methylene chloride, mp 121°–122°C $[\alpha]_D^{27}$ +155° (C = 1.0 in water) Colorless needles from ethyl ether or acetone, mp 127°C $[\alpha]_{5461}^{20}$ +185° (C = 1 in aqueous solution) Terrein is soluble in water, ethyl alcohol, and acetone; sparingly soluble in chloroform, ethyl ether, and benzene; insoluble in petroleum ether.
UV data[9,28]	λ_{max}^{EtOH} nm (ϵ): 273(30,000) and 245(125) λ_{max}^{EtOH} nm (ϵ): 275(26,000) and 342(130)
Fungal source[28,40,42,58]	*Aspergillus terreus* (IMI 44339) *A. fischeri* Wehmer *A. stellatus* *Penicillium raistrickii* *Phoma spp.* (NRRL 3188)
Toxicity data	Toxicity unknown
Other significant references	Biosynthesis[11,42]; isolation[24,67]; synthesis[6]
^{13}C NMR data	C1, 18.9 q; C2, 139.1 d; C3, 125.2 d; C4, 168.2 s; C5, 76.3 d; C6, 80.6 d; C7, 203.2 s; C8, 124.5 d
^1H NMR data	H1, 1.95 (J = 7.0); H2, 6.70 (J = 17.0); H3, 6.34 (J = 17.0); H5, 3.92 (J = 2.0); H6, 4.53 (J = 2.0); H8, 5.98; OH, 5.36

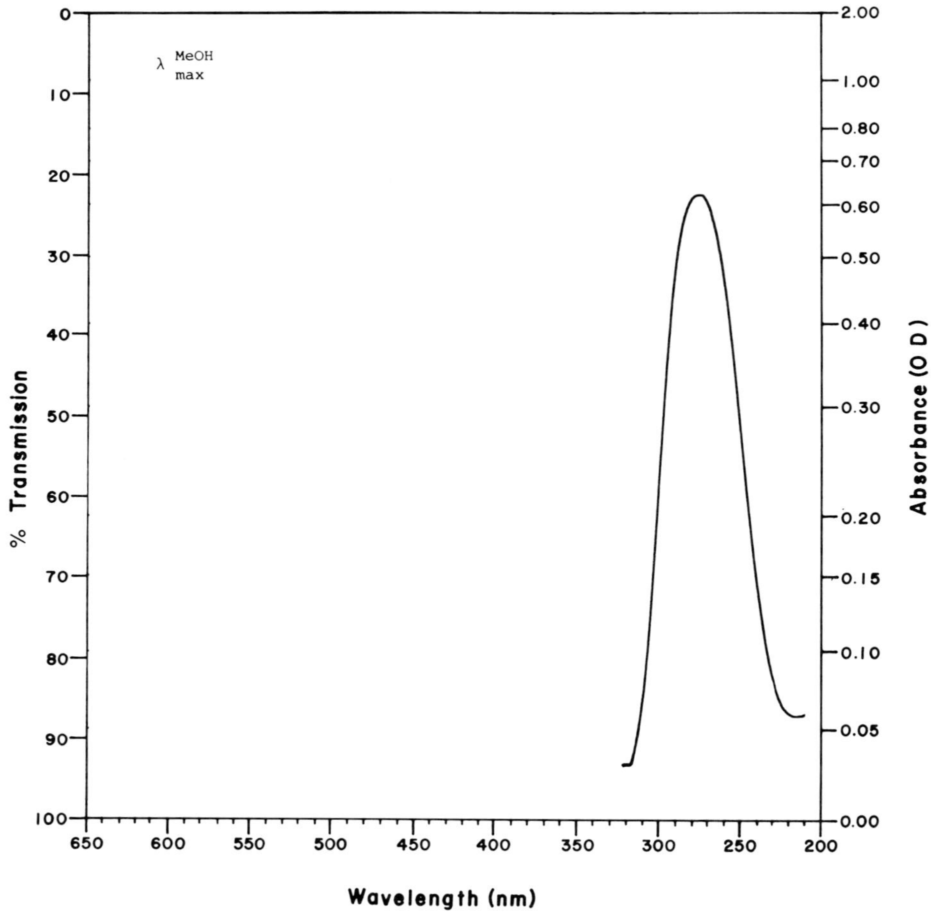

λ MeOH max
% Transmission
0
10
20
30
40
50
60
70
80
90
100
Absorbance (O D)
2.00
1.00
0.80
0.70
0.60
0.50
0.40
0.30
0.20
0.15
0.10
0.05
0.00
650
600
550
500
450
400
350
300
250
200
Wavelength (nm)

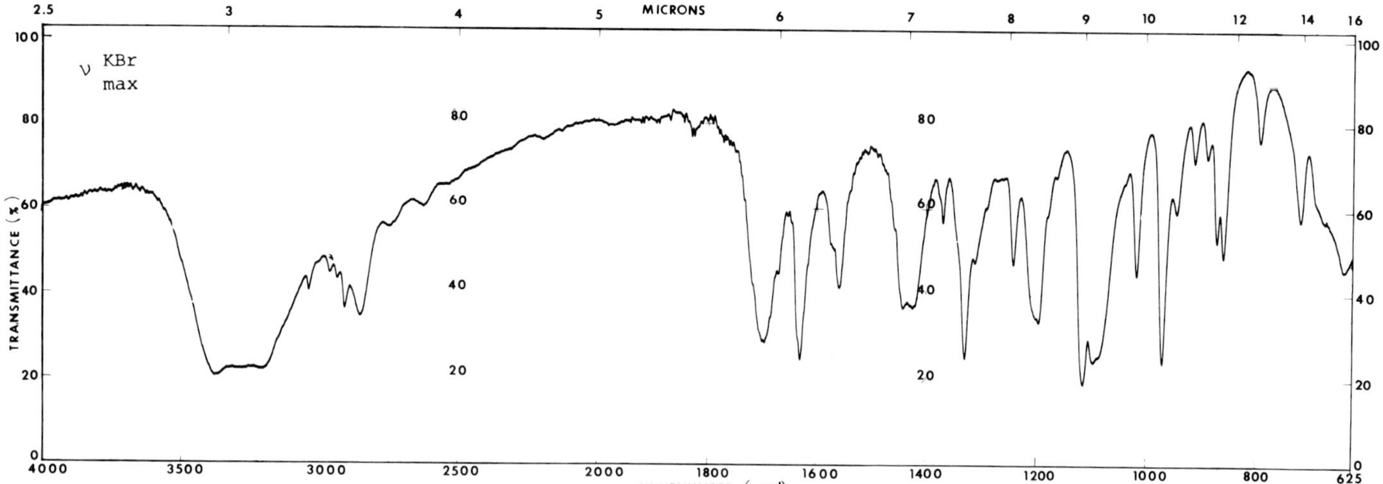
MICRONS
ν KBr max
TRANSMITTANCE (%)
WAVENUMBER (CM⁻¹)

TERREIN

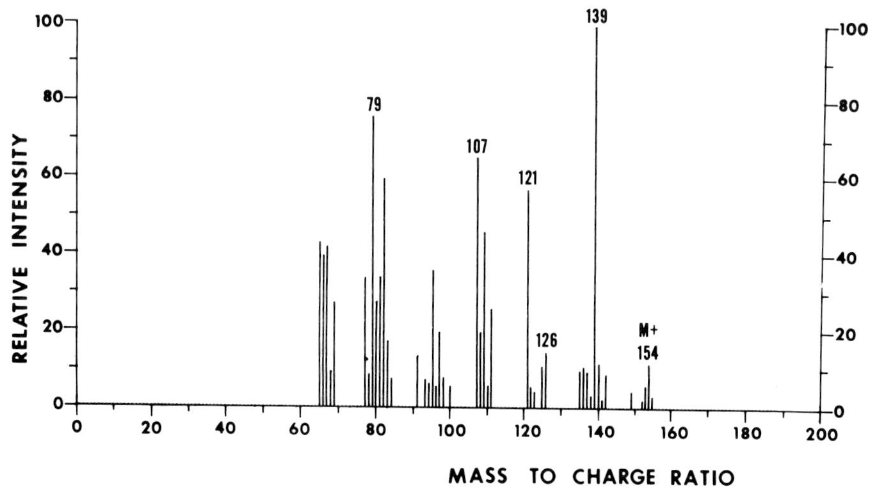
RELATIVE INTENSITY
MASS TO CHARGE RATIO
79
107
121
126
139
M+
154

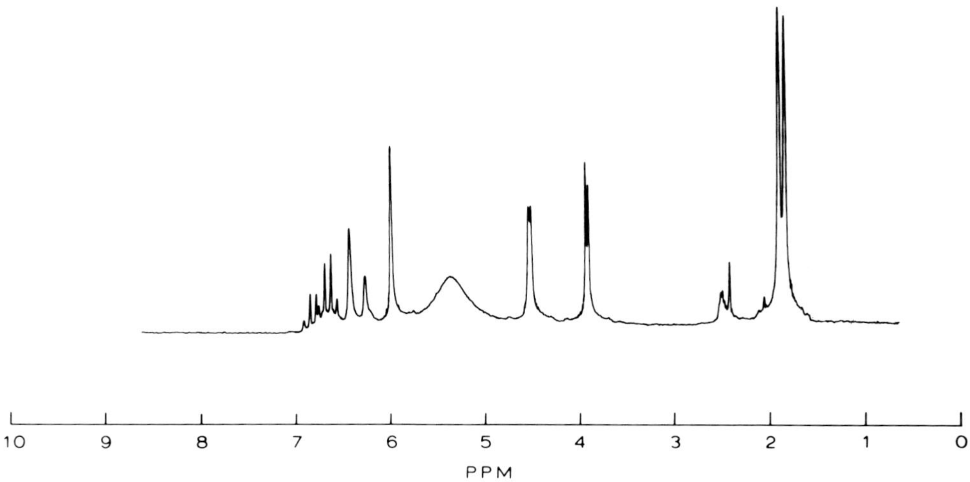

The 100 MHz ^{1}H NMR spectrum of terrein in DMSO solution.

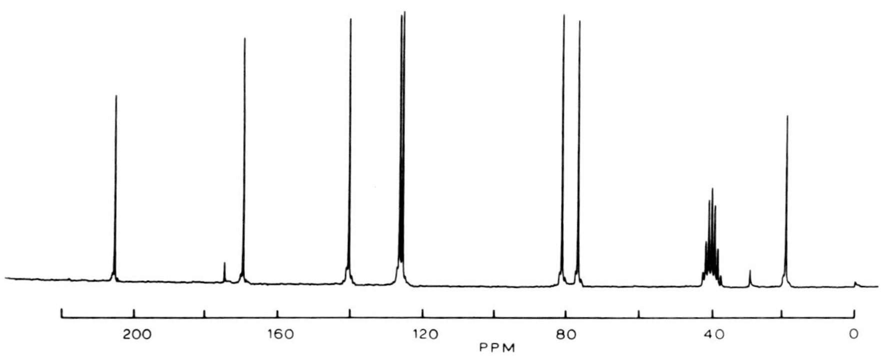

The 25 MHz ^{13}C NMR spectrum of terrein in DMSO solution.

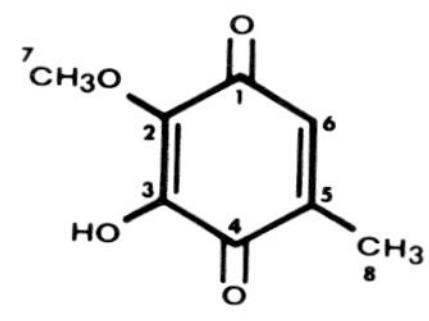

Fumigatin

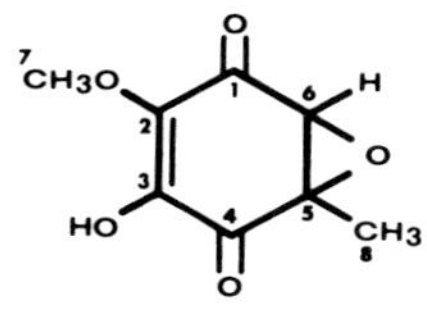

Fumigatin oxide

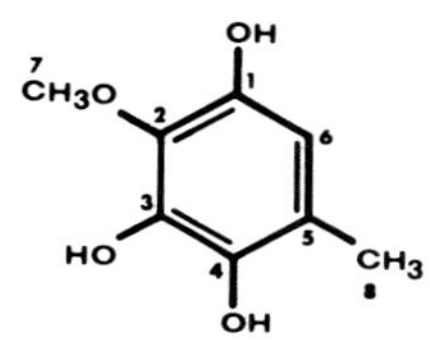

Fumigatin quinol

Common name	Fumigatin (3-Hydroxy-4-methoxy-2,5-toluquinone)
Molecular weight	168.0422 Fumigatin oxide: 184.0371 Fumigatin quinol: 170.0578
Molecular formula	$C_8H_8O_4$ Fumigatin oxide: $C_8H_8O_5$ Fumigatin quinol: $C_8H_{10}O_4$
General characteristics[2]	Fumigatin: maroon-colored needles, mp 116°C. Fumigatin sublimes at high vacuum without decomposition and no change in melting point. Fumigatin monomethyl ether crystallized from petroleum ether as long red needles, mp 59°C. Fumigatin monoacetate crystallized from petroleum ether as canary yellow needles, mp 95°–96°C.
UV data[66]	λ_{max}^{MeOH} nm: 520, 268, and 223 (ϵ_{max} not reported) $\lambda_{max}^{CHCl_3}$ nm: 450 and 285 (ϵ_{max} not reported)

Fungal source[2]	*Aspergillus fumigatus* Fresenius
Toxicity data[87]	Fumigatin has been shown to have strong *in vitro* inhibition against several gram-negative and gram-positive bacteria. Vertebrate toxicity is not known.
TLC data[90]	*Adsorbent:* silica gel slurried with 0.1 *M* oxalic acid instead of water *Solvent:* ethyl acetate–benzene, 2:1 v/v R_f: not reported *Detection:* treatment with ammonia vapor
Other significant references	Biosynthesis[74,90,92]; synthesis[7,73]; chemistry[71,91]
^{13}C *NMR data* (Fumigatin)	C1, 183.3 s; C2, 144.9 s; C3, 141.8 s; C4, 184.6 s; C5, 141.8 s; C6, 131.2 d; C7, 59.9 q; C8, 14.7 q
^{1}H *NMR data* (Fumigatin)	H6, 6.45 (J = 1.9); H7, 3.77; H8, 1.93 (J = 1.9)
^{13}C *NMR data* (Fumigatin oxide)	C1, 183.4 s; C2, 145.8 s; C3, 139.6 s; C4, 184.7 s; C5, 57.9 s; C6, 58.4 d; C7, 59.2 q; C8, 13.7 q
^{1}H *NMR data* (Fumigatin oxide)	H6, 3.82; H7, 3.75; H8, 1.50
^{13}C *NMR data* (Fumigatin quinol)	C1, 138.0 s; C2, 136.1 s; C3, 133.9 s; C4, 142.5 s; C5, 119.6 s; C6, 107.5 d; C7, 60.3 q; C8, 15.7 q
^{1}H *NMR data* (Fumigatin quinol)	H6, 6.16; H7, 3.81, 2.10

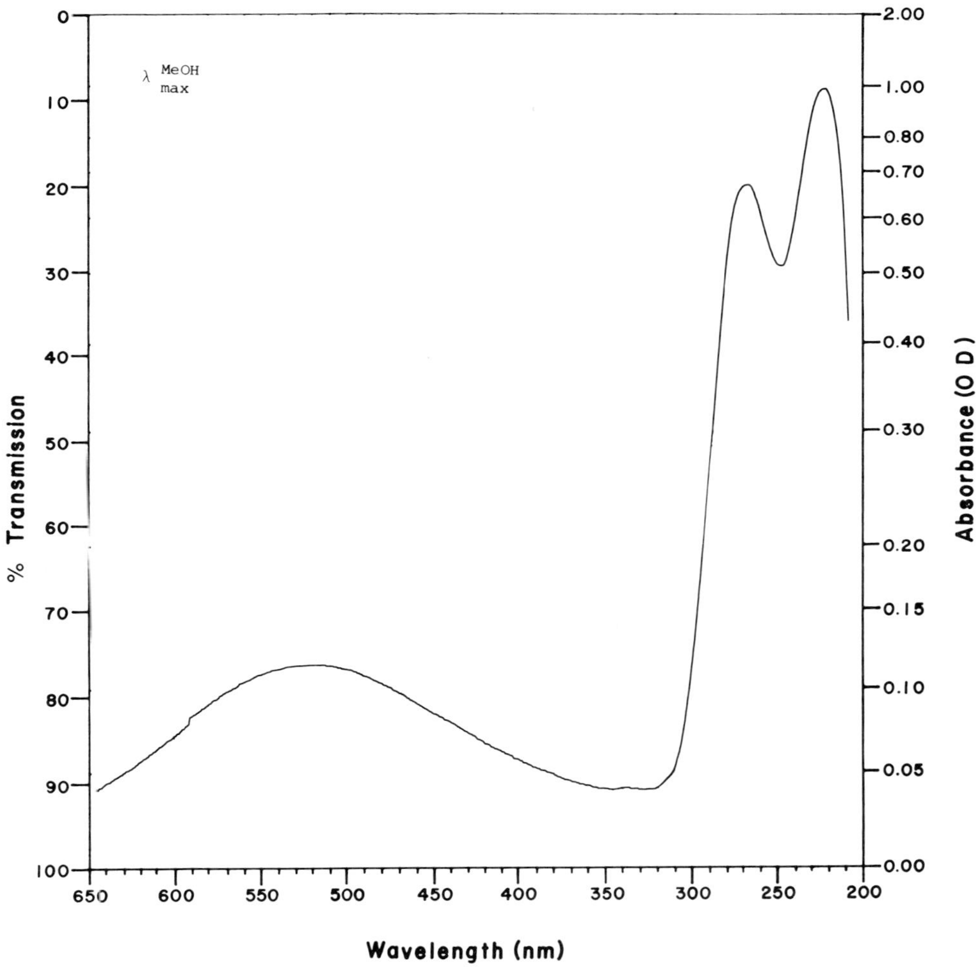
λ_{max}^{MeOH}
% Transmission
Absorbance (O D)
Wavelength (nm)
0
10
20
30
40
50
60
70
80
90
100
2.00
1.00
0.80
0.70
0.60
0.50
0.40
0.30
0.20
0.15
0.10
0.05
0.00
650
600
550
500
450
400
350
300
250
200

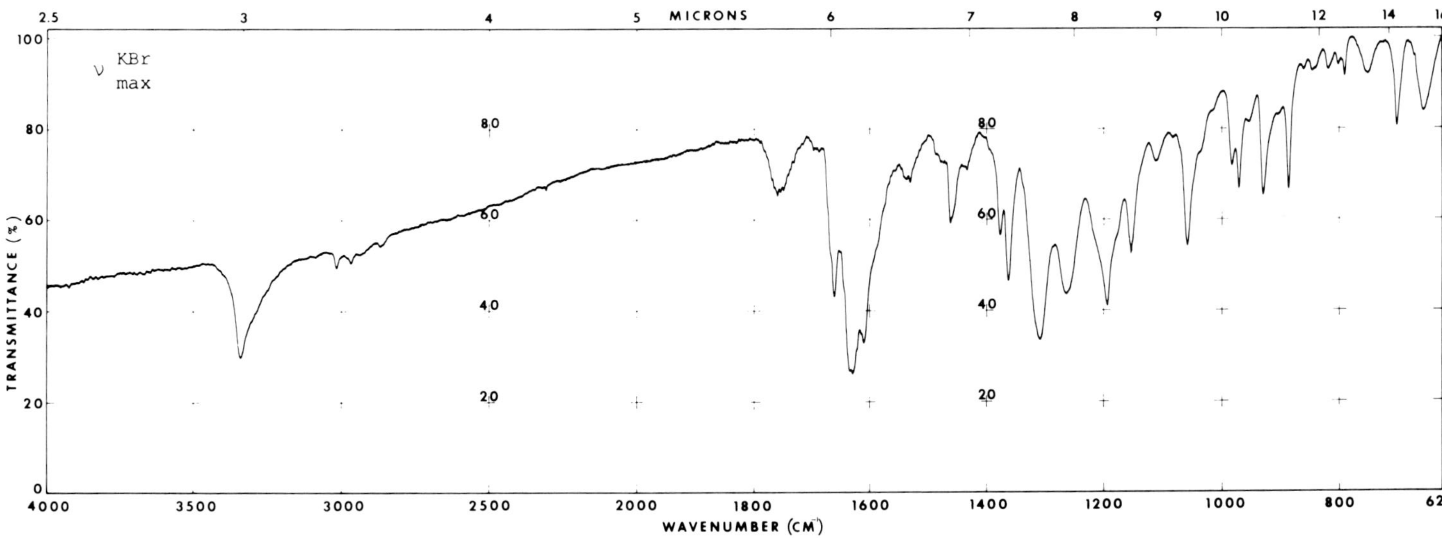
MICRONS
ν_{max}^{KBr}
TRANSMITTANCE (%)
WAVENUMBER (CM⁻¹)
2.5 3 4 5 6 7 8 9 10 12 14 16
4000 3500 3000 2500 2000 1800 1600 1400 1200 1000 800 625
0 20 40 60 80 100

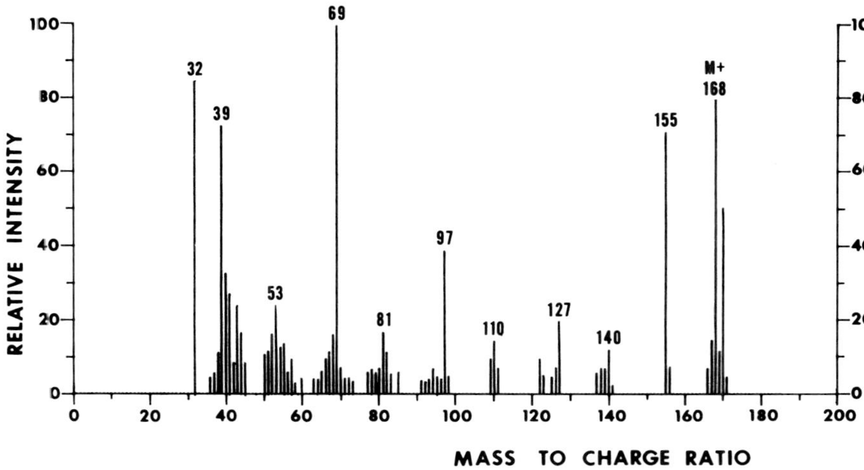

Mass spectrum of fumigatin

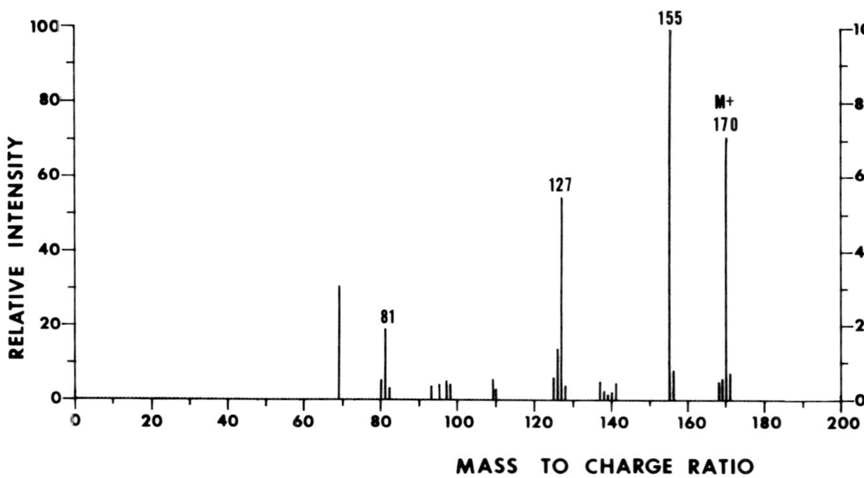

Mass spectrum of fumigatin quinol

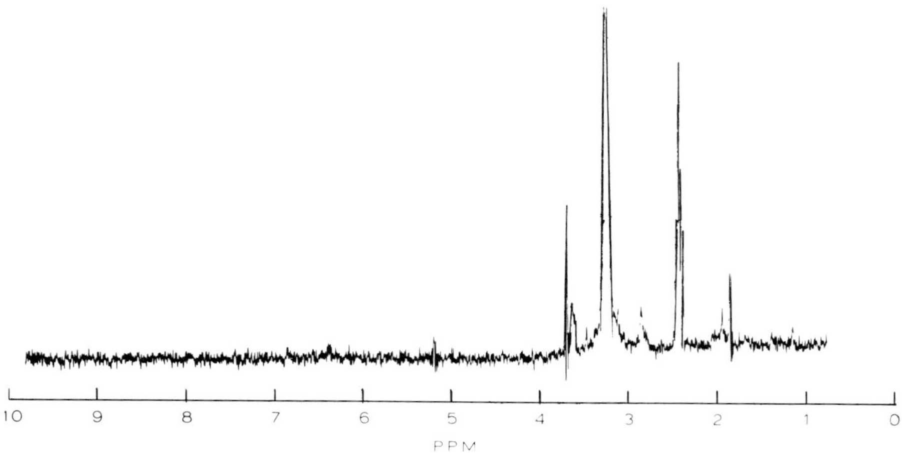

The 100 MHz ^{1}H NMR spectrum of fumigatin in DMSO solution.

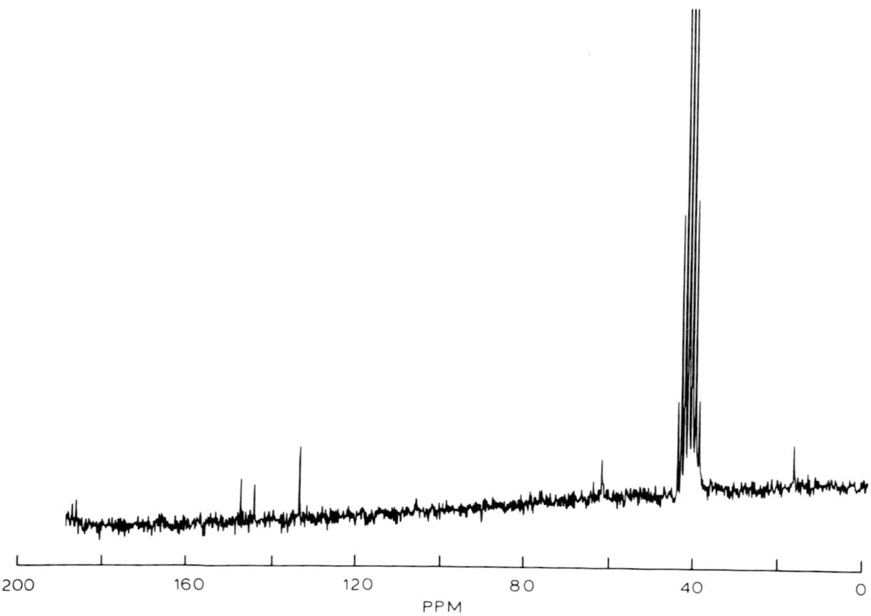

The 25 MHz ^{13}C NMR spectrum of fumigatin in DMSO solution.

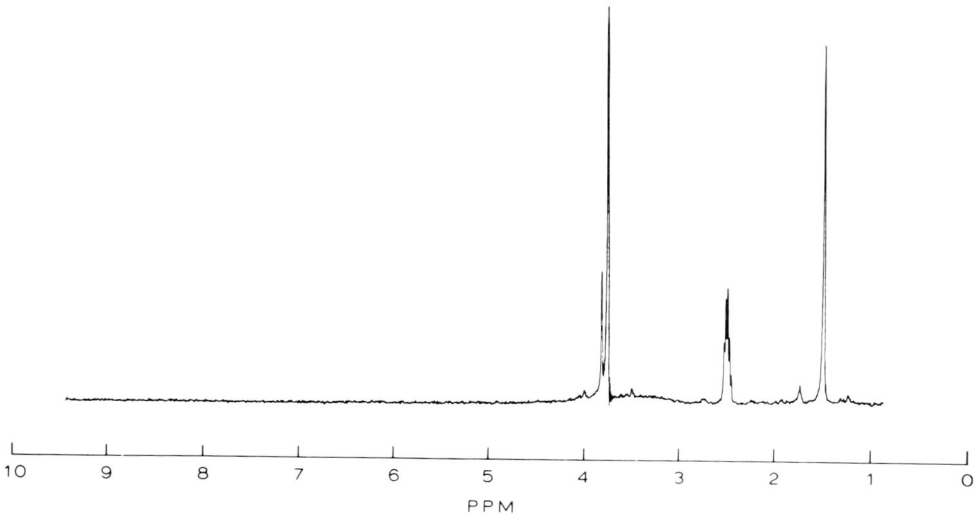

The 100 MHz ^{1}H NMR spectrum of fumigatin oxide in DMSO solution.

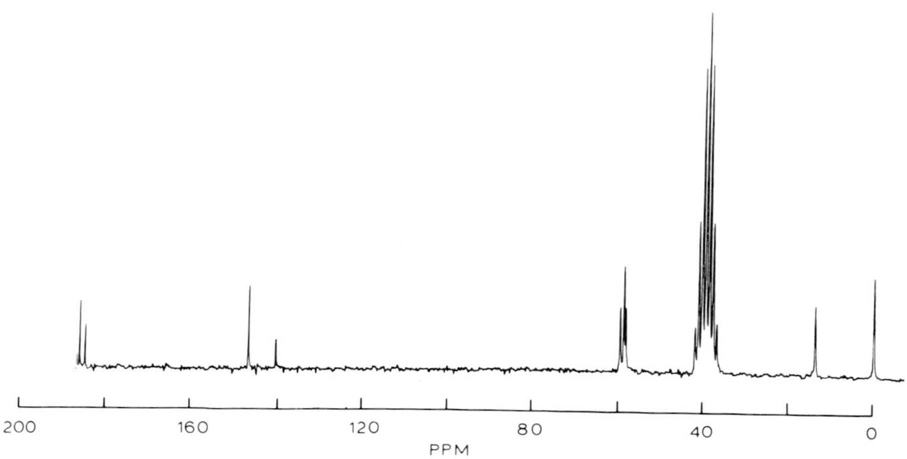

The 25 MHz ^{13}C NMR spectrum of fumigatin oxide in DMSO solution.

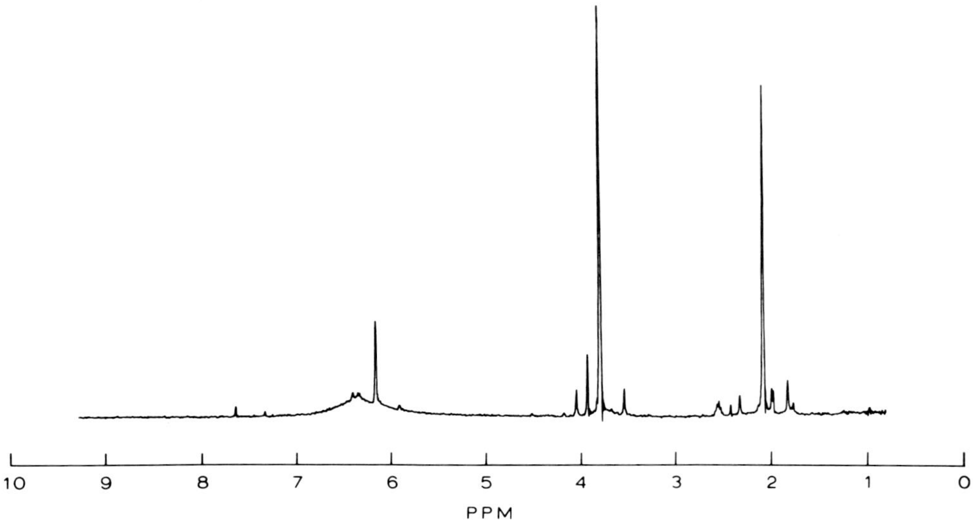

The 100 MHz ^{1}H NMR spectrum of fumigatin quinol in $CDCl_3$/DMSO solution.

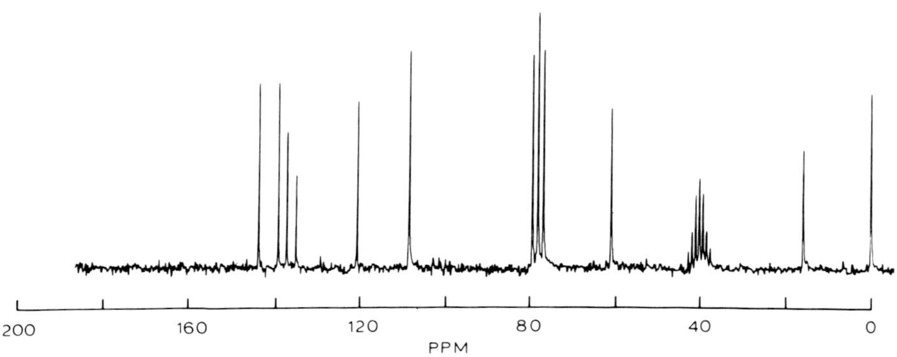

The 25 MHz ^{13}C NMR spectrum of fumigatin quinol in $CDCl_3$/DMSO solution.

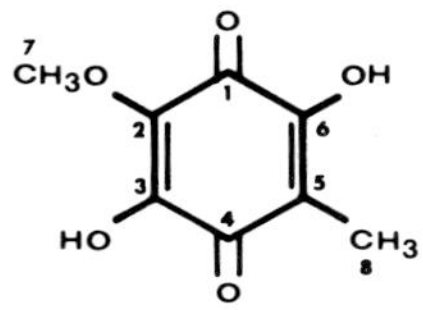

Spinulosin

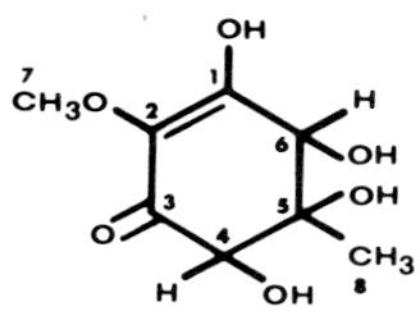

Spinulosin quinol hydrate

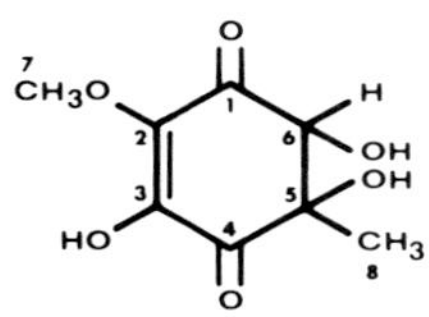

Spinulosin hydrate

Common name[2,13]	Spinulosin (Substance ''Z'') (3,6-Dihydroxy-3-methoxy-1,4-toluquinone)
Molecular weight	184.0371
Molecular formula	$C_8H_8O_5$ Spinulosin quinol hydrate: $C_8H_{12}O_6$, 204.0630, mp 191°C Spinulosin hydrate: $C_8H_{10}O_6$, 202.0474, mp 182°C
General characteristics[13]	Purple-black microcrystals from sublimation at 110°–120°C in high vacuum, mp 201°C
UV data[66]	$\lambda_{max}^{CHCl_3}$ nm: 294 and 460 (ϵ_{max} not reported)
Fungal source[2,13,16]	*Aspergillus fumigatus* Fresenius *Penicillium spinulosum* Thom *P. cinerascens*
Toxicity data[65]	Spinulosin possesses antibacterial activity. Vertebrate toxicity is not known.

Other significant references	Synthesis[2,3,71]
^{13}C NMR data (Spinulosin)	C1, 168.7 s; C2, 135.1 s; C3, 168.7 s; C4, 168.7 s; C5, 111.5 s; C6, 168.7 s; C7, 59.6 q; C8, 7.5 q
^{1}H NMR data (Spinulosin)	H7, 3.73; H8, 1.75
^{13}C NMR data (Spinulosin quinol hydrate)	C1, 176.4 s; C2, 131.9 s; C3, 176.4 s; C4, 75.3, 77.4 d; C5, 78.6, 79.9 d; C6, 75.3, 77.4 d; C7, 59.0 q; C8, 14.5 q
^{1}H NMR data (Spinulosin quinol hydrate)	H7, 3.56; H8, 0.98
^{13}C NMR data (Spinulosin hydrate)	C1, 177.2 s; C2, 146.4 s; C3, 142.4 s; C4, 180.5 s; C5, 77.1 s; C6, 78.1 d; C7, 59.1 q; C8, 20.3 q
^{1}H NMR data (Spinulosin hydrate)	H6, 4.98, H7, 3.80; H8, 1.18

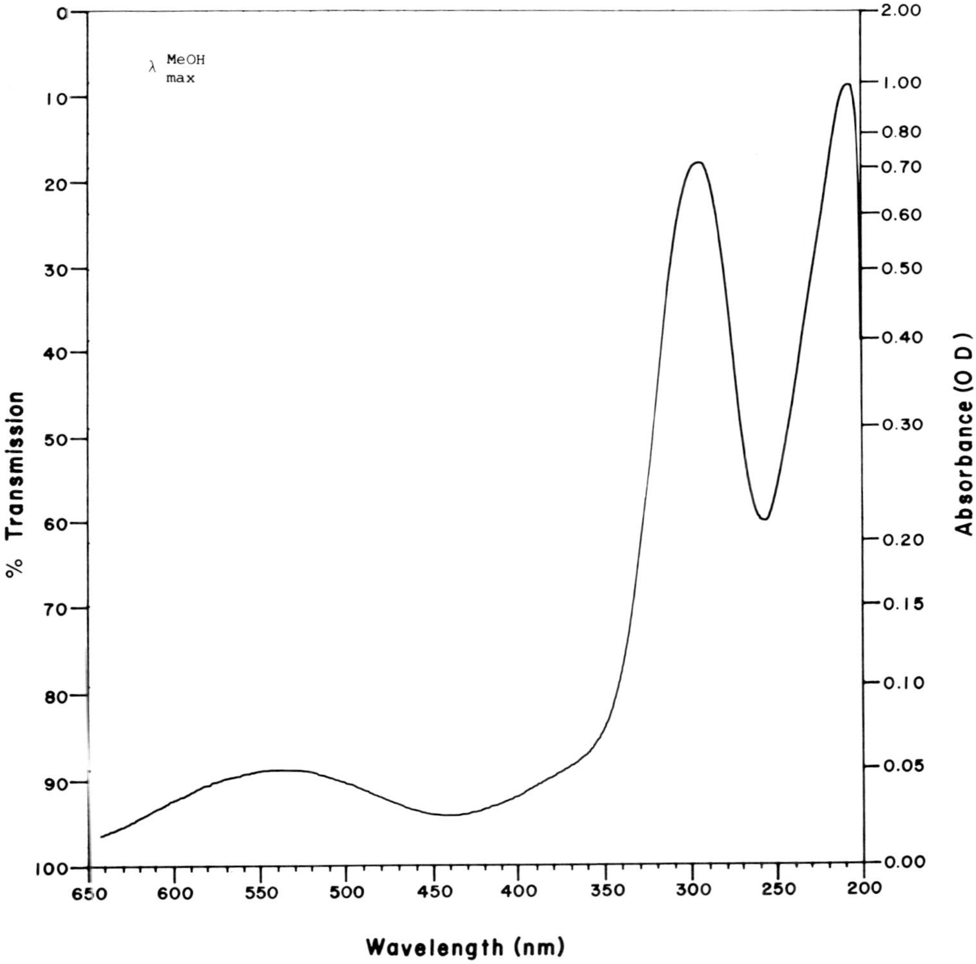
λ MeOH max
% Transmission
Absorbance (O D)
Wavelength (nm)
0
10
20
30
40
50
60
70
80
90
100
2.00
1.00
0.80
0.70
0.60
0.50
0.40
0.30
0.20
0.15
0.10
0.05
0.00
650
600
550
500
450
400
350
300
250
200

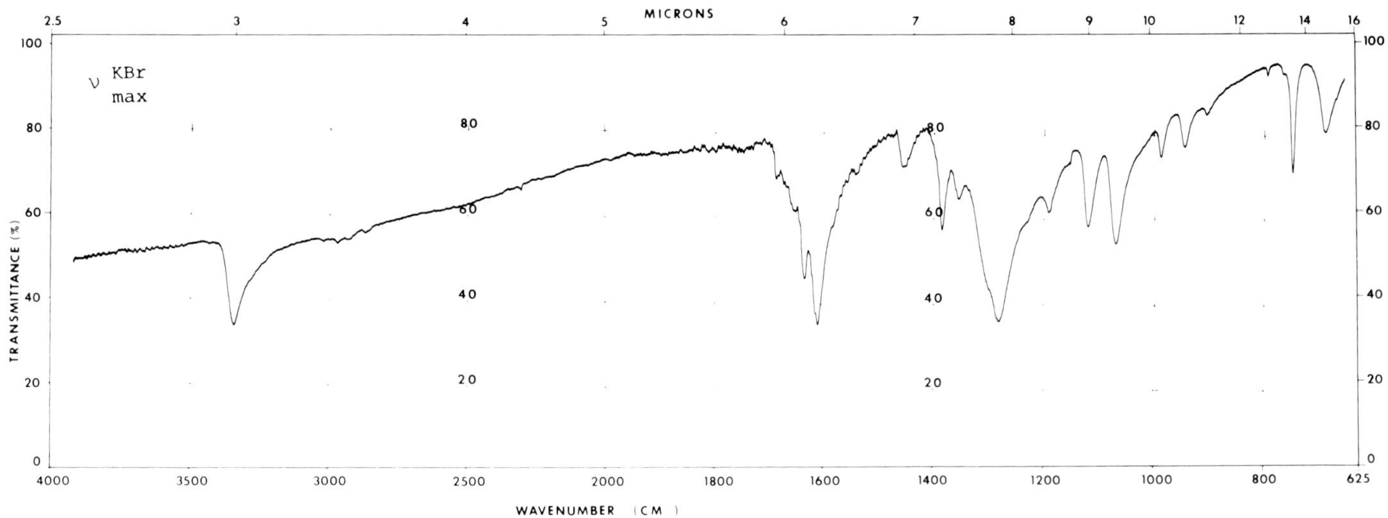
MICRONS
WAVENUMBER (CM)
TRANSMITTANCE (%)
ν KBr max

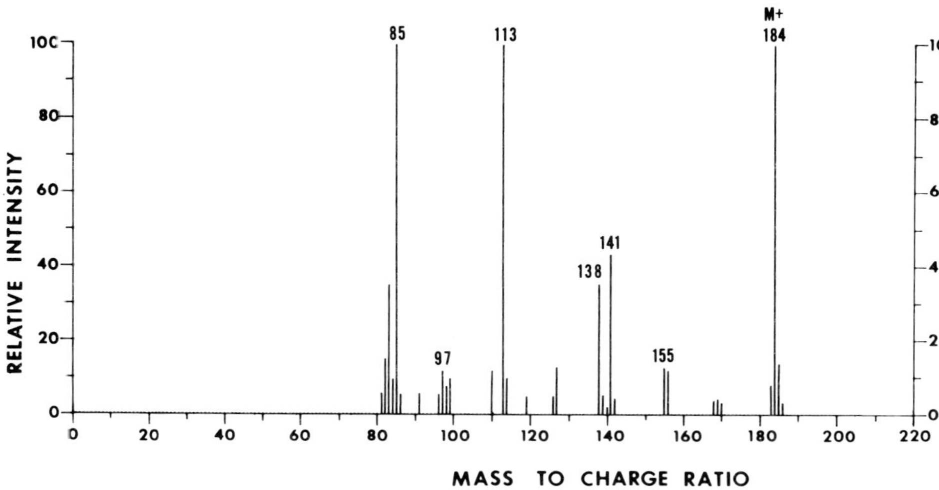

Mass spectrum of spinulosin

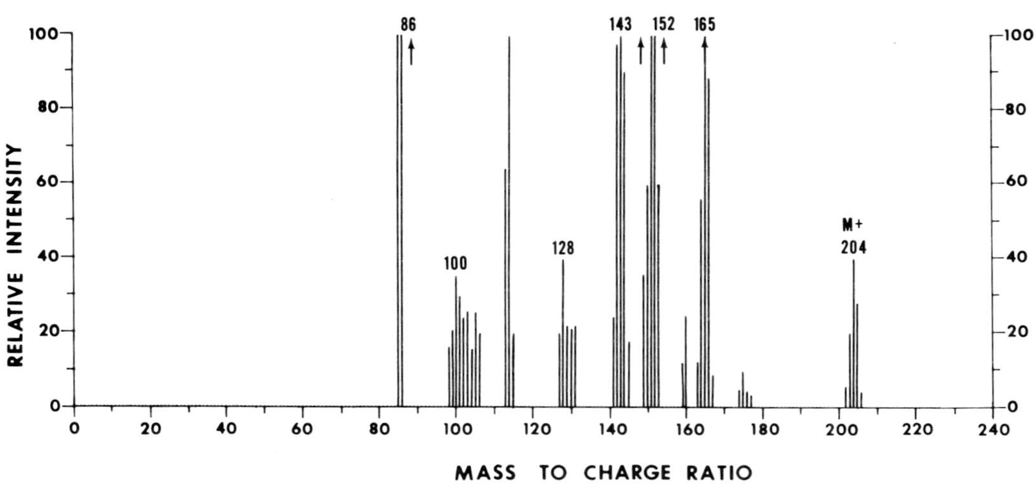

Mass spectrum of spinulosin hydrate

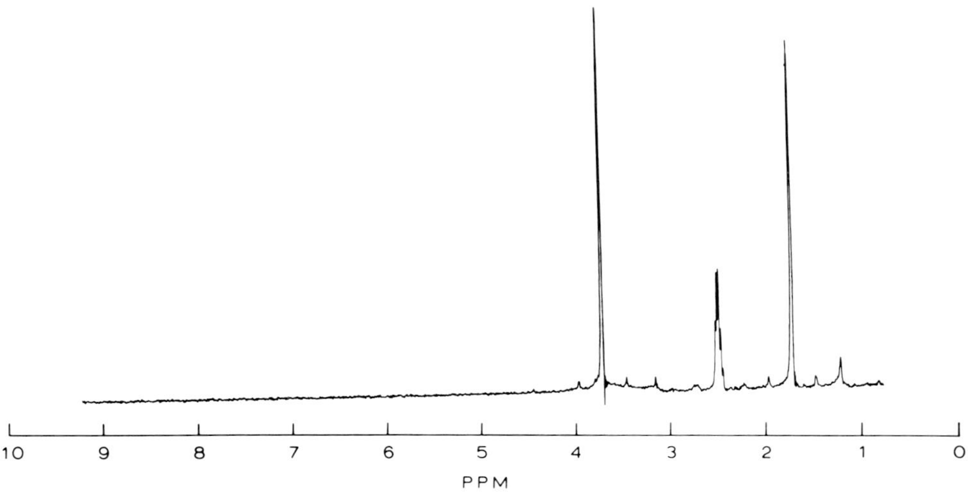

The 100 MHz ^{1}H NMR spectrum of spinulosin in DMSO solution.

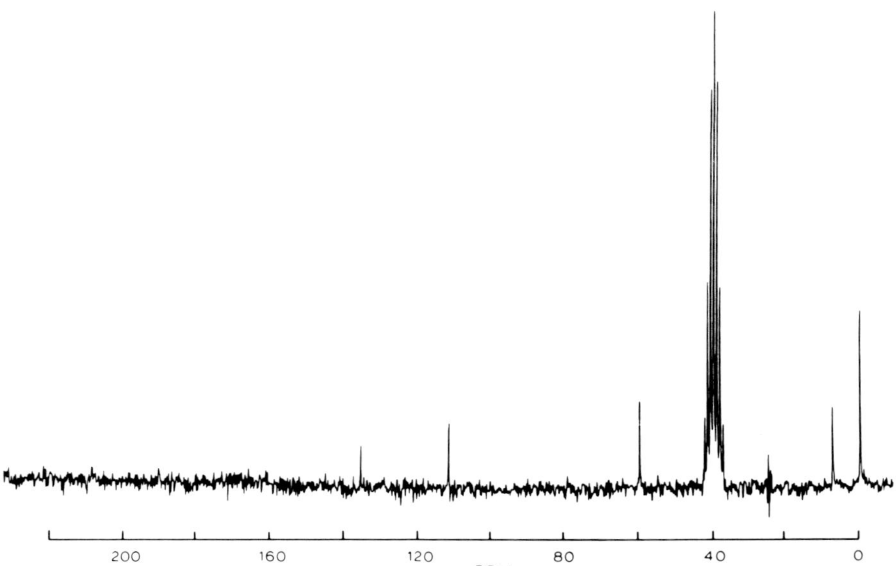

The 25 MHz ^{13}C NMR spectrum of spinulosin in DMSO solution.

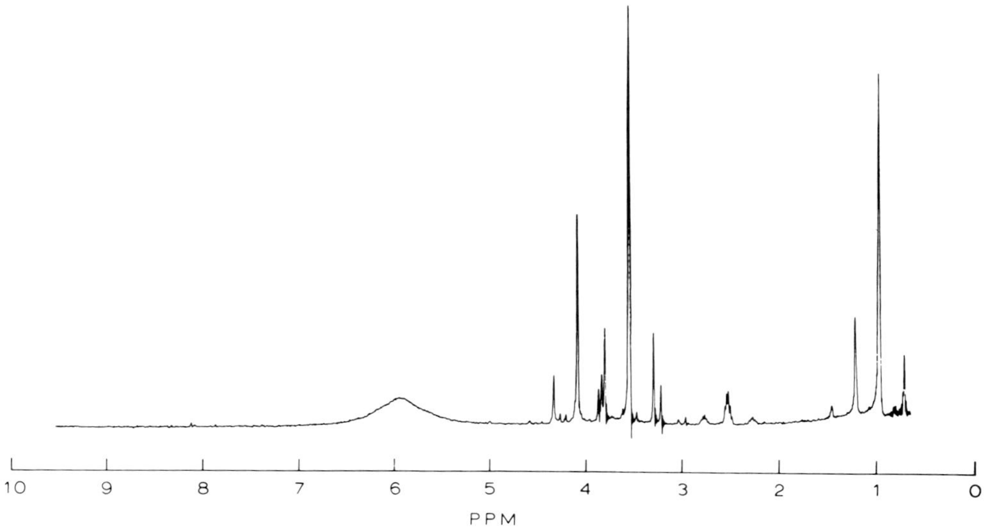

The 100 MHz ^1H NMR spectrum of spinulosin quinol hydrate in $CDCl_3$/DMSO solution.

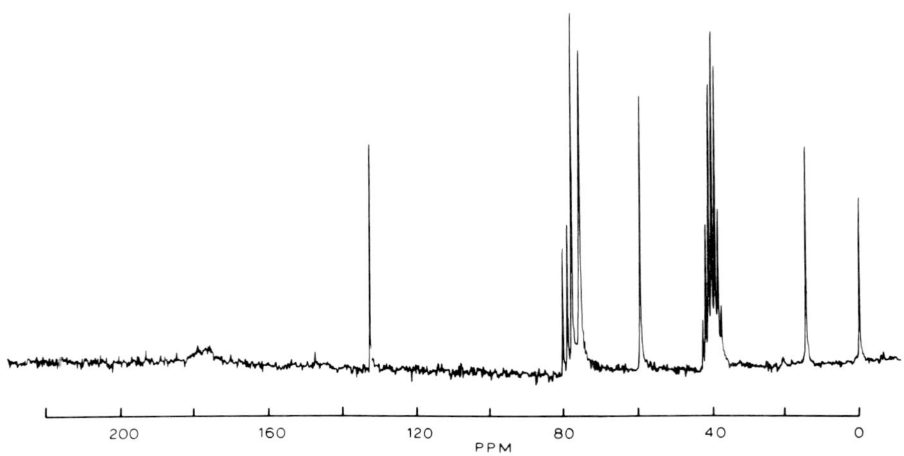

The 25 MHz ^{13}C NMR spectrum of spinulosin quinol hydrate in $CDCl_3$/DMSO solution.

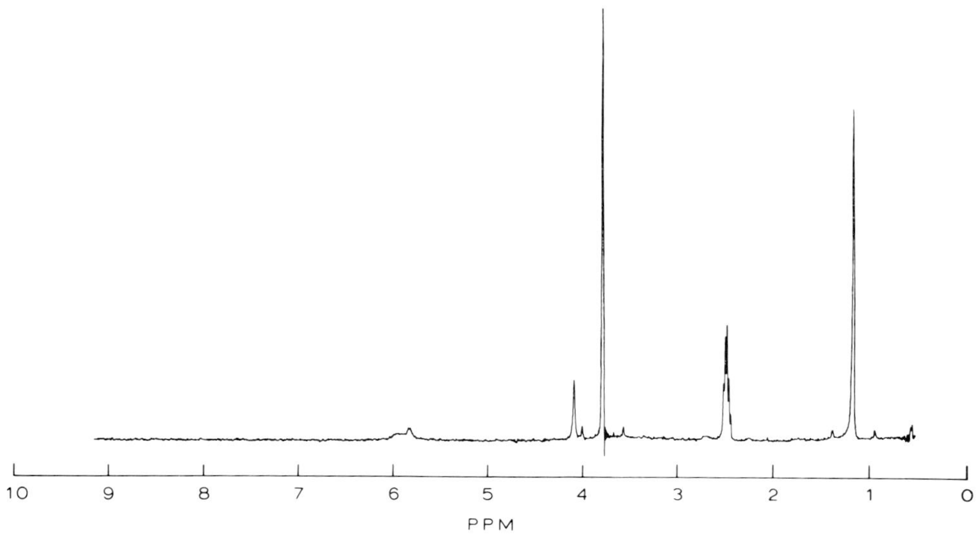

The 100 MHz ^1H NMR spectrum of spinulosin hydrate in DMSO solution.

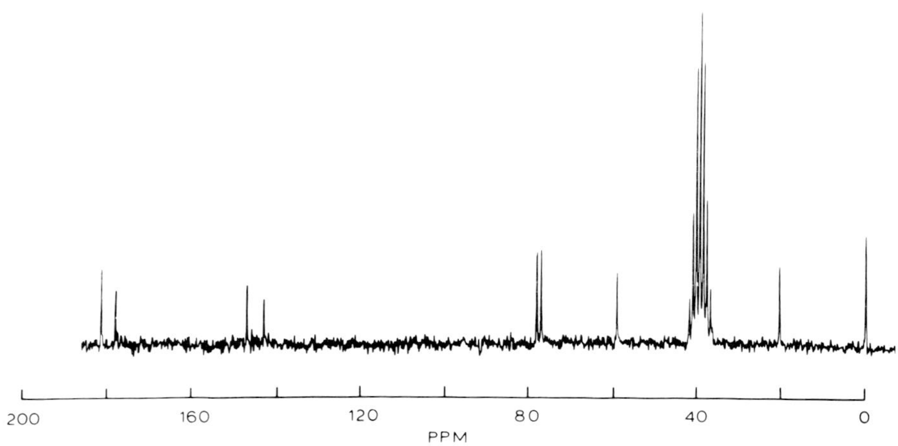

The 25 MHz ^{13}C NMR spectrum of spinulosin hydrate in DMSO solution.

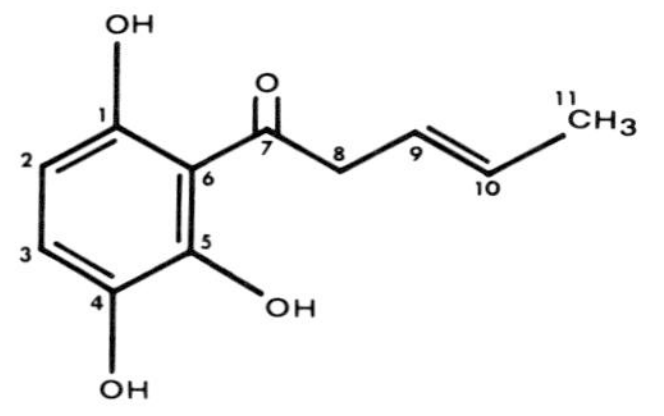

Common name	Maltoryzine [6-(9-Pentenyl)-1,4,5-trihydroxybenzene]
Molecular weight	208.0735
Molecular formula	$C_{11}H_{12}O_4$
General characteristics[44,45]	Pale brownish-yellow needles, mp 68.5°–69°C (dec.) $[\alpha]_D^{25}$ 0° (C = 2.8 in chloroform) Color reactions: brown color with $FeCl_3$; red with diazo compounds; pink with methyl red and decolorized permanganic acid at room temperature. Forms a 2,4-dinitrophenylhydrazone; reddish-brown crystals from water, mp 105.5°–109°C.
UV data[44,45]	λ_{max} nm (ϵ): 220(12,600), 280(1,300), and 320(125)
IR data[44,45]	ν_{max}^{KBr} cm^{-1}: 3300, 1700, 1600, and 1500
Fungal source[44]	*Aspergillus oryzae* var. *microsporus*
Toxicity data[44,45]	The LD_{50} of maltoryzine dosed IP to mice was 3 mg/kg. It caused muscular narcotism in mice and swollen yellow liver. It has been implicated in two cases of feed poisoning in dairy cattle.

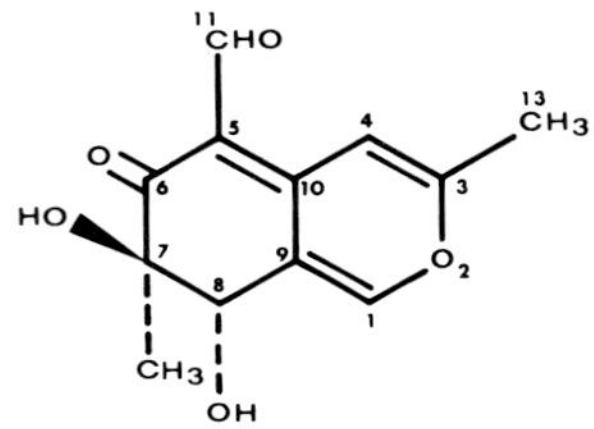

Common name[78]	Austdiol [(7*R*,8*S*)-7,8-Dihydro-7,8-dihydroxy-3,7-dimethyl-6-oxo-6*H*-2-benzopyran-5-carbaldehyde]
Molecular weight	236.0685
Molecular formula	$C_{12}H_{12}O_5$
General characteristics[83]	Yellow needles from chloroform, mp 255°C (dec.) $[\alpha]_D^{26}$ +160.30° (C = 1.25 in pyridine)
UV data[83]	λ_{max}^{EtOH} nm (ϵ): 256(15,100) and 376(24,000)
Fungal source[78]	*Aspergillus ustus* (Bainier) Thom and Church C.S.I.R., 1128; NRRL 5856
Toxicity data[83]	Austdiol is a gastrointestinal toxin. It is reported to cause acute toxicoses in day-old ducklings.
^{13}C *NMR data*	C1, 123.3 d; C3, 166.1 s; C4, 107.4 d; C5, 151.9 s; C6, 200.0 s; C7, 76.3 s; C8, 71.9 d; C9, 109.4 s; C10, 149.2 s; C11, 190.2 d; C12, 19.9 q*; C12, 19.3 q*
^{1}H *NMR data*	H1, 8.00 (J = 2.0); H4, 8.16; H8, 4.72 (J = 2.0); H11, 10.37; H12, 1.90; H13, 1.27

* Assignments may be reversed.

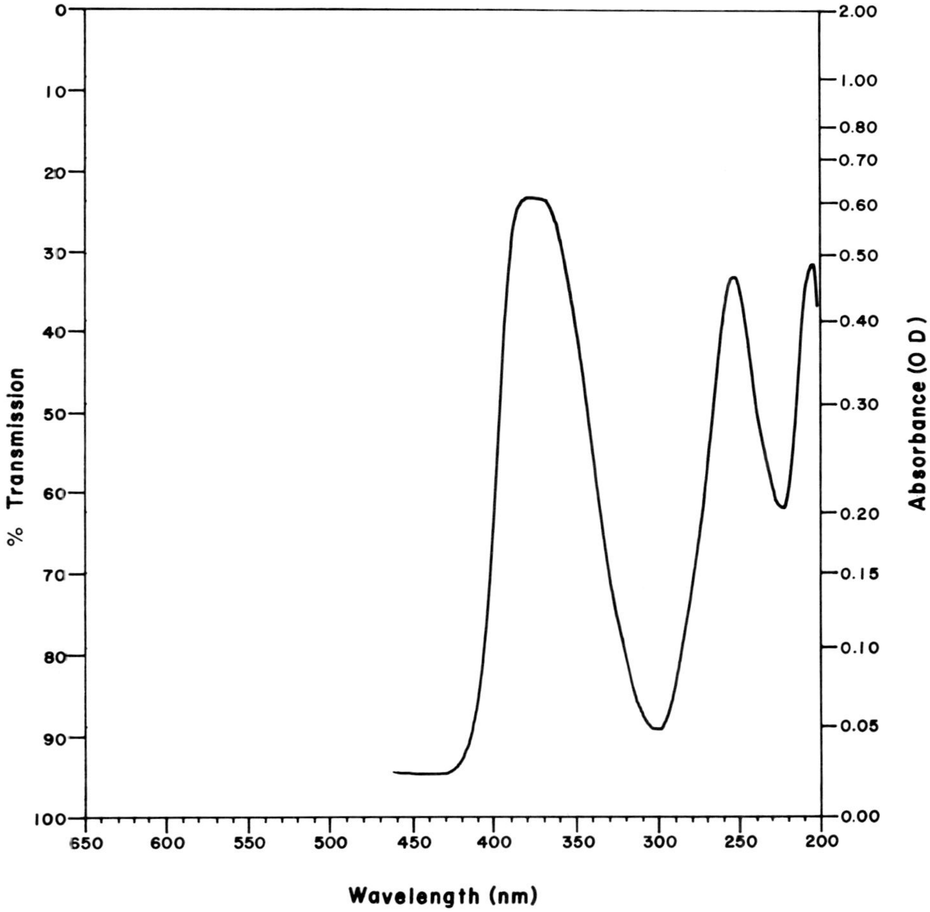
% Transmission
0
10
20
30
40
50
60
70
80
90
100
Absorbance (O D)
2.00
1.00
0.80
0.70
0.60
0.50
0.40
0.30
0.20
0.15
0.10
0.05
0.00
650
600
550
500
450
400
350
300
250
200
Wavelength (nm)

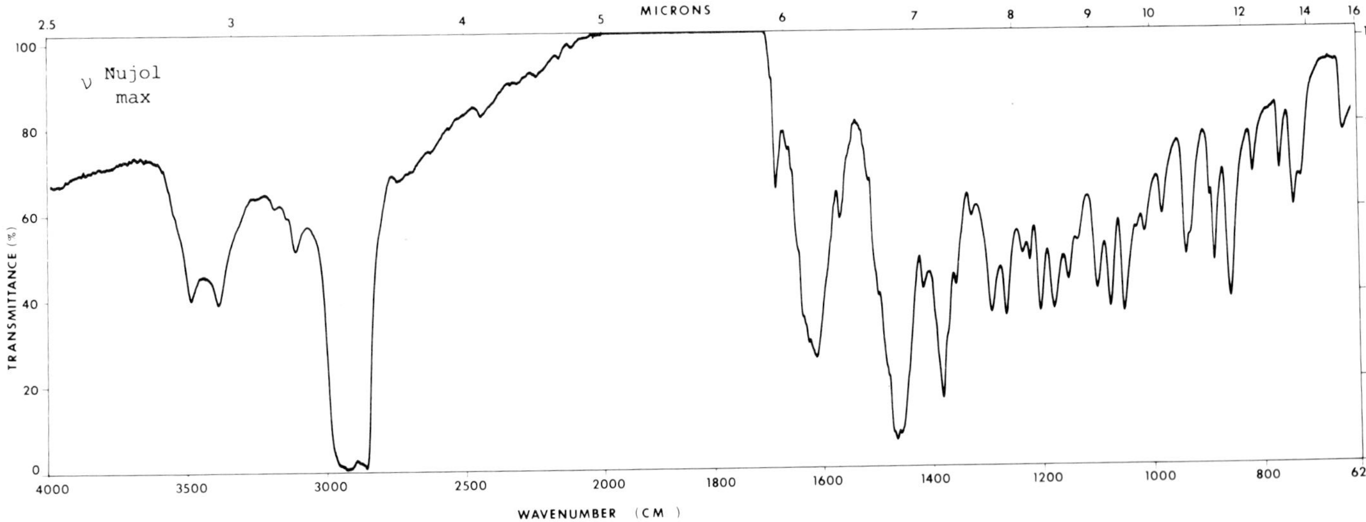
MICRONS
ν Nujol max
TRANSMITTANCE (%)
WAVENUMBER (CM)

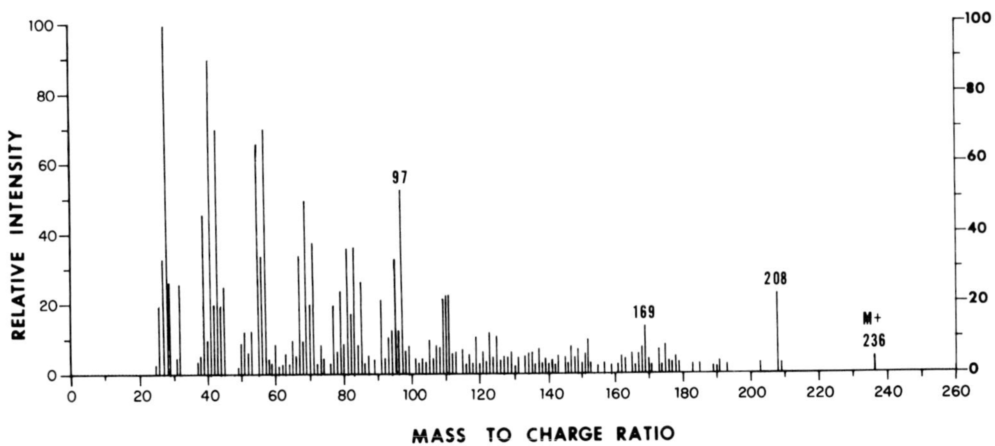
RELATIVE INTENSITY
97
169
208
M+
236
MASS TO CHARGE RATIO

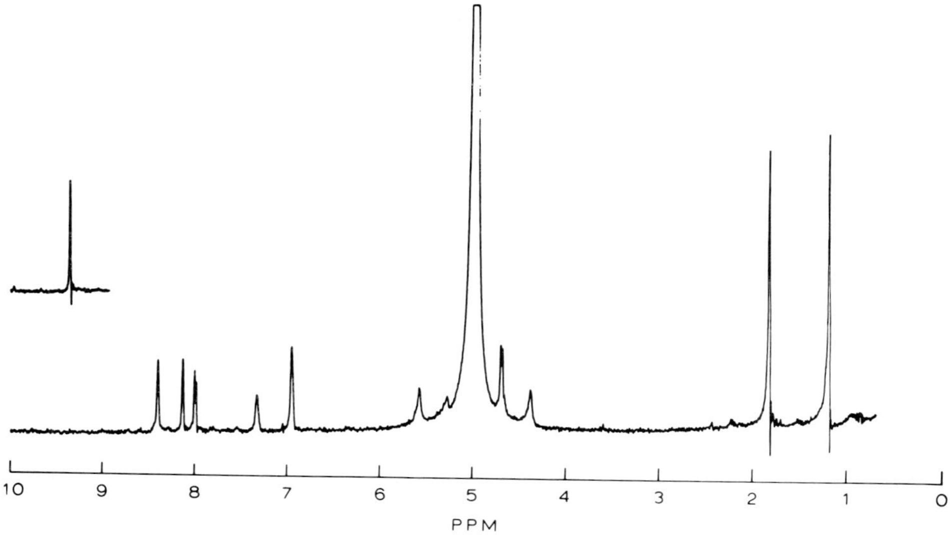

The 100 MHz ^{1}H NMR spectrum of austdiol in pyridine-d_5 solution.

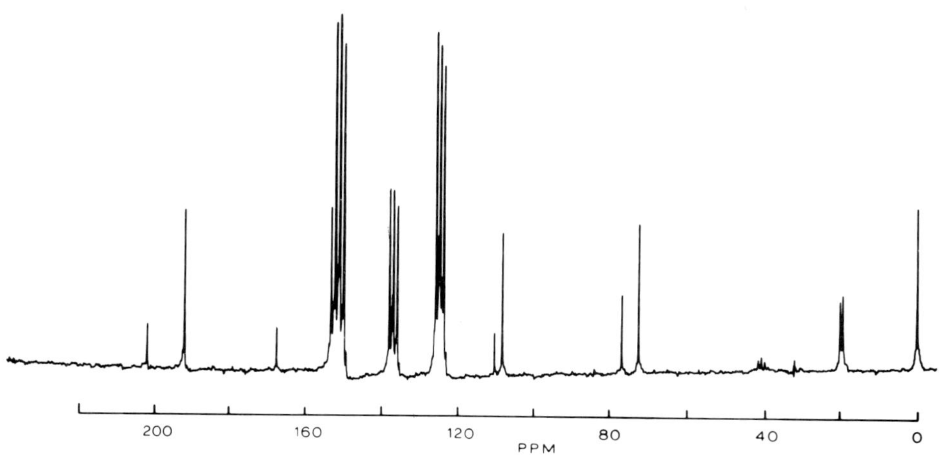

The 25 MHz ^{13}C NMR spectrum of austdiol in pyridine-d_5 solution.

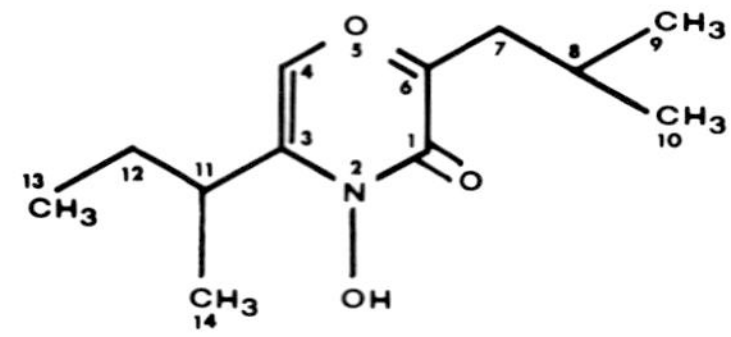

Common name	Aspergillic acid [3-*sec*-Butyl-6-isobutyl-2-hydroxy-2(1*H*)-pyrazinone]
Molecular weight	224.1524
Molecular formula	$C_{12}H_{20}O_2N_2$
General characteristics[31,32]	Yellow rods from methanol, mp 97°–99°C $[\alpha]_D^{18}$ +13.3° (C = 3.9 in ethanol) Soluble in most organic solvents and poorly soluble in water.
UV data[32]	λ_{max}^{EtOH} nm (ϵ): 328(8,500) and 235(10,500)
Fungal source	*Aspergillus flavus* Link
Toxicity data[53,85]	The oral LD_{50} of aspergillic acid was reported to be between 4–5 mg/20 g mouse; LD_{50} (IP) of the sodium salt was between 1–3 mg/20 g mouse. At toxic levels, mice went into convulsions in 5–10 min and died. Aspergillic acid is acutely toxic to mice (LD_{50} 100–150 mg/kg) but shows no chronic toxicity at sublethal dosages. Aspergillic acid shows antibiotic activity *in vitro* to gram-positive and gram-negative bacteria.
Other significant references	Structure elucidation[29,31,32]; synthesis[36,62]; biosynthesis[54]; review[53]
^{13}C NMR data	C1, 22.5 q; C2, 34.1 d; C3, 27.4 t*; C4, 139.5 s; C5, 151.6 s; C6, 151.2 s; C7, 123.6 d; C8, 41.8 t; C9, 27.0 t*; C10, 11.6 q; C11, 17.8 q
^{1}H NMR data	H1, 0.96 (J = 6.4); H2, 3.15; H3, 2.73 (J = 6.2); H7, 7.32; H8, 2.21; H9, 1.79; H10, 0.95 (J = 7.6); H11, 1.35 (J = 6.9)

* Assignments may be reversed.

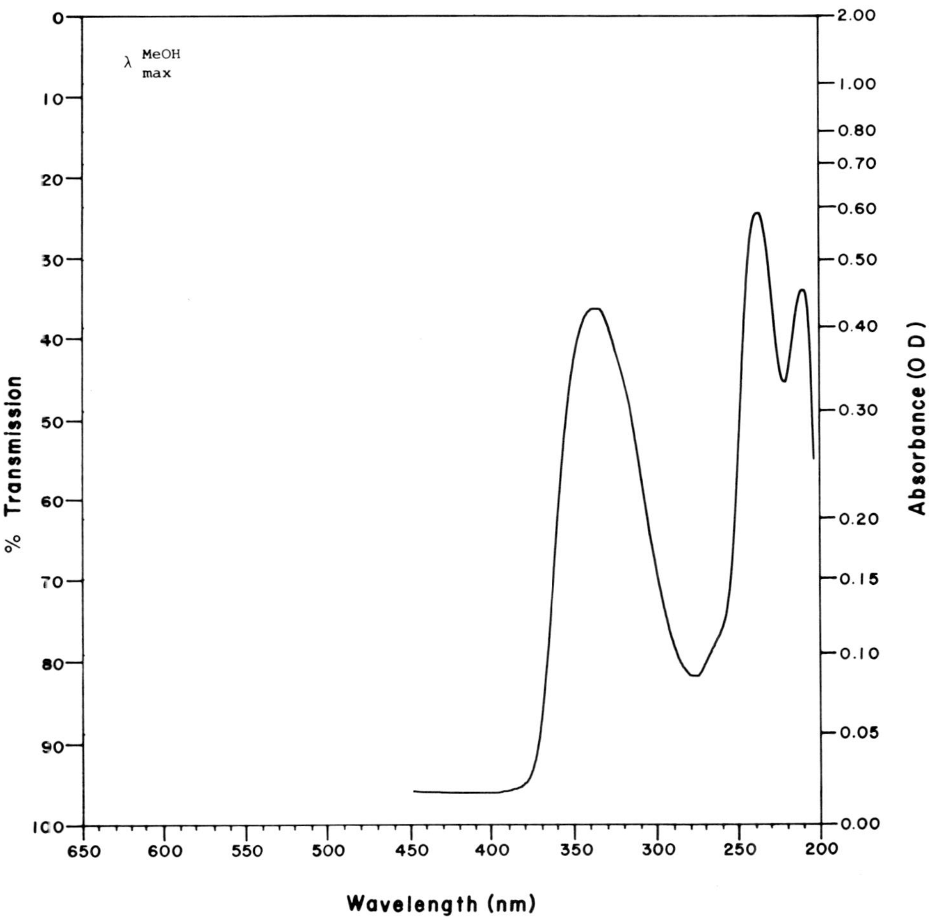
λ MeOH max
% Transmission
Absorbance (O D)
Wavelength (nm)
0
10
20
30
40
50
60
70
80
90
100
650
600
550
500
450
400
350
300
250
200
2.00
1.00
0.80
0.70
0.60
0.50
0.40
0.30
0.20
0.15
0.10
0.05
0.00

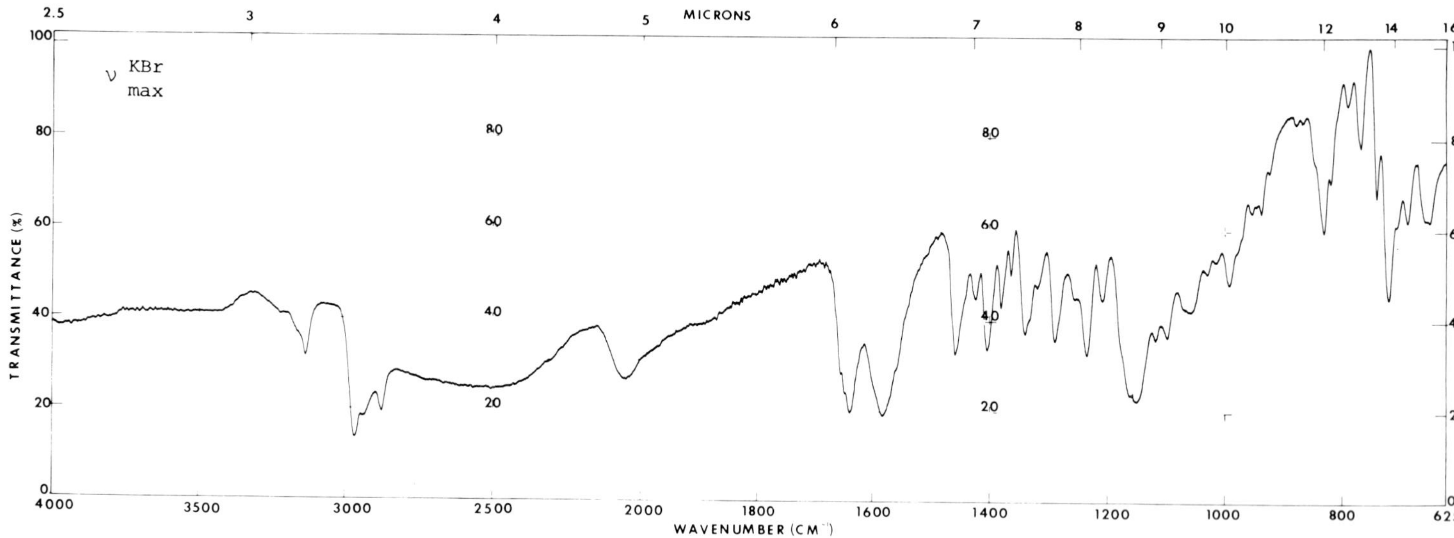
MICRONS
ν KBr max
TRANSMITTANCE (%)
WAVENUMBER (CM⁻¹)

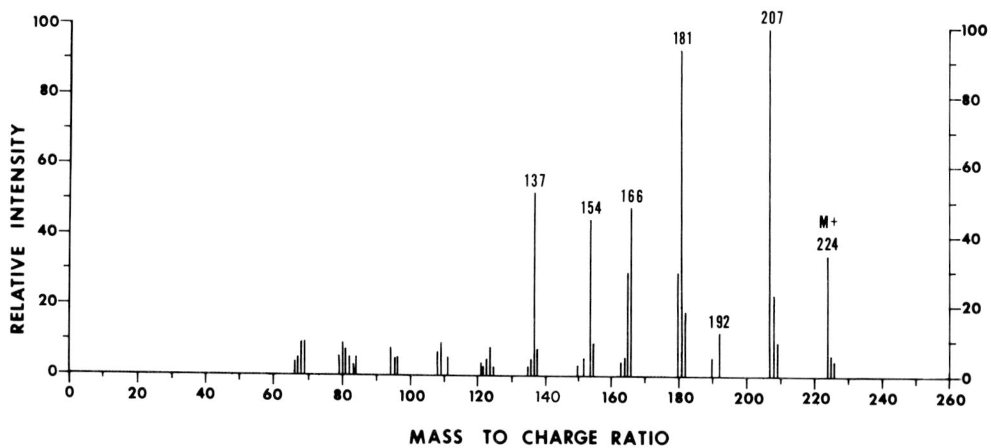
RELATIVE INTENSITY
MASS TO CHARGE RATIO
137
154
166
181
192
207
M+
224

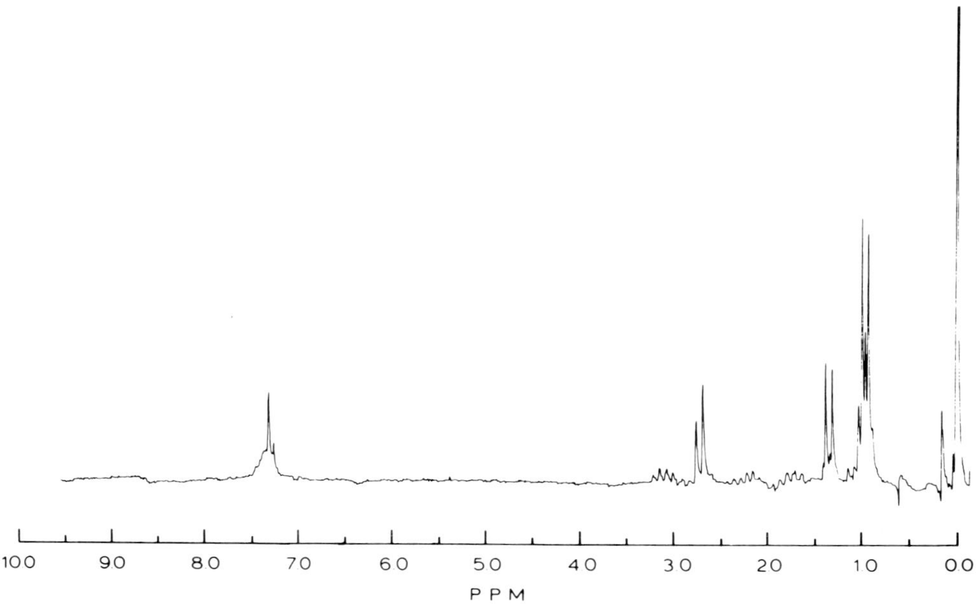

The 100 MHz ^1H NMR spectrum of aspergillic acid in $CDCl_3$ solution.

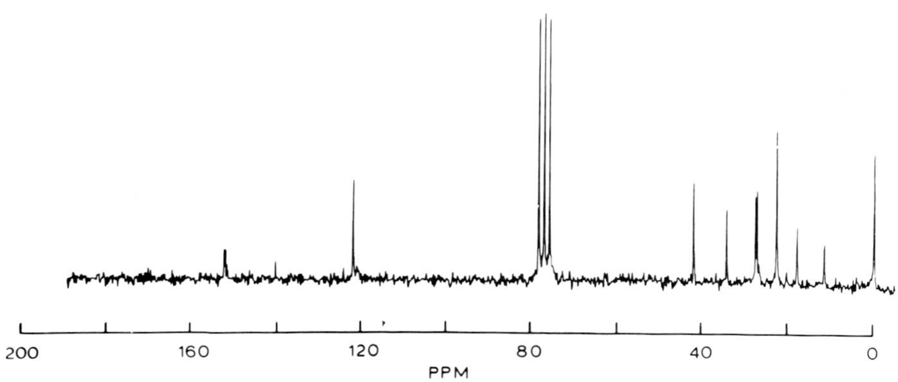

The 25 MHz ^{13}C NMR spectrum of aspergillic acid in $CDCl_3$ solution.

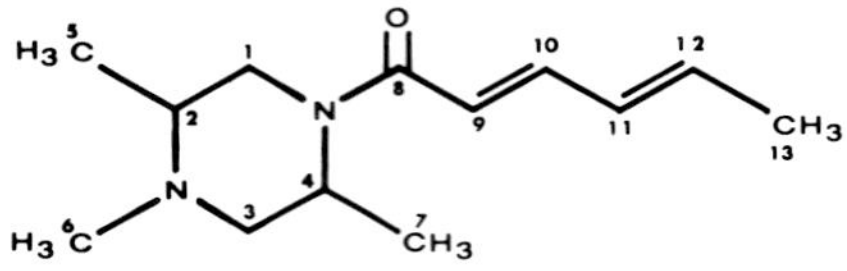

Common name	Nigragillin (*N*-Methyl-*trans*-2,5-dimethyl-*N'*-sorbylpiperazine)
Molecular weight	222.1732
Molecular formula	$C_{13}H_{22}ON_2$
General characteristics[19,25]	mp 129°–132°C $[\alpha]_D^{26}$ +101° (*C* = 1.0 in chloroform)
UV data[19,25]	λ_{max}^{MeOH} nm (ϵ): 262(26,200)
Fungal source[19,25]	*Aspergillus niger* *A. phoenicis*
Toxicity data[25,46]	The LD_{50} of nigragillin dosed orally to day-old cockerels was ≃250 mg/kg. The ED_{50} in silkworms fed nigragillin in their diet was between 20–40 ppm. Visible effects in silkworms such as vomiting were noted down to 10 ppm in the diet. A dosage of 5 μg/g applied topically caused immediate effects such as knockdown and occasionally death in silkworms.
TLC data[25]	*Adsorbent:* silica gel G-HR *Solvent:* toluene–ethyl acetate–formic acid, 5:4:1 v/v/v R_f: 0.19 *Detection:* spraying with ethanolic H_2SO_4 and phosphomolybdic acid and heating at 100°C for 5 min.
^{13}*C NMR data*	C1, 46.8 t; C2, 53.9 d; C3, 51.9 t; C4, 46.8 d; C5, 16.9 q; C6, 42.5 q; C7, 18.4 q; C8, 166.2 s; C9, 117.8 d; C10, 142.7 d; C11, 129.9 d; C12, 136.9 d; C13, 7.4 q

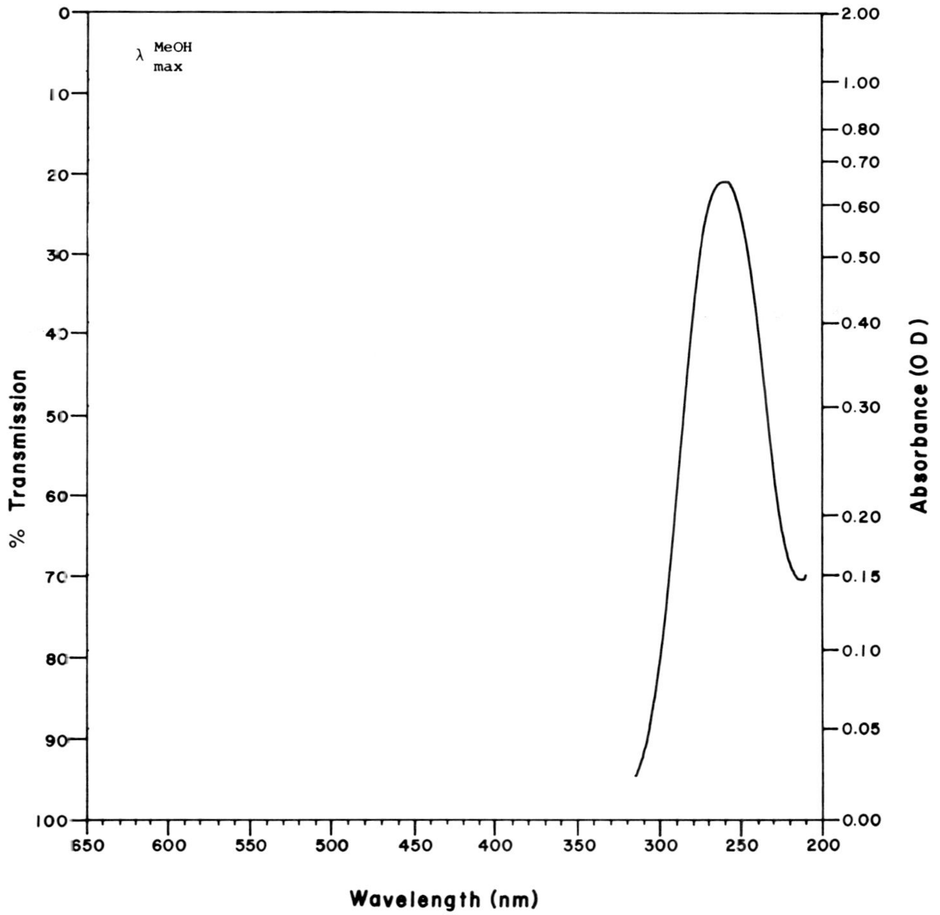
λ MeOH max
% Transmission
Absorbance (O D)
Wavelength (nm)
0
10
20
30
40
50
60
70
80
90
100
2.00
1.00
0.80
0.70
0.60
0.50
0.40
0.30
0.20
0.15
0.10
0.05
0.00
650
600
550
500
450
400
350
300
250
200

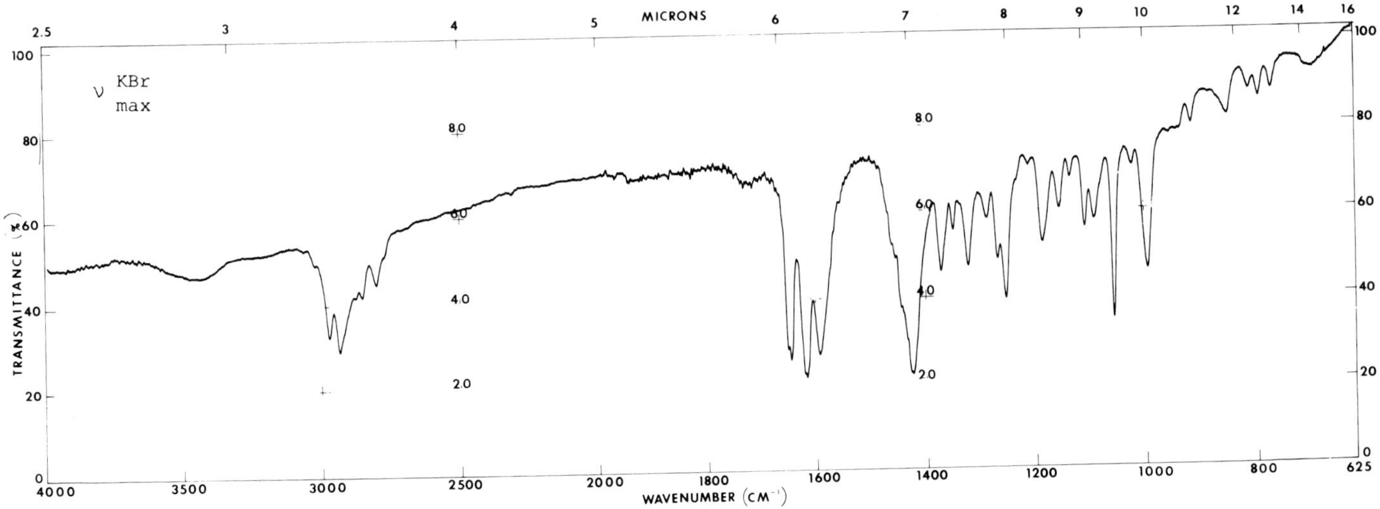

MICRONS
2.5
3
4
5
6
7
8
9
10
12
14
16
ν KBr max
TRANSMITTANCE (%)
100
80
60
40
20
0
8.0
6.0
4.0
2.0
4000
3500
3000
2500
2000
1800
1600
1400
1200
1000
800
625
WAVENUMBER (CM⁻¹)

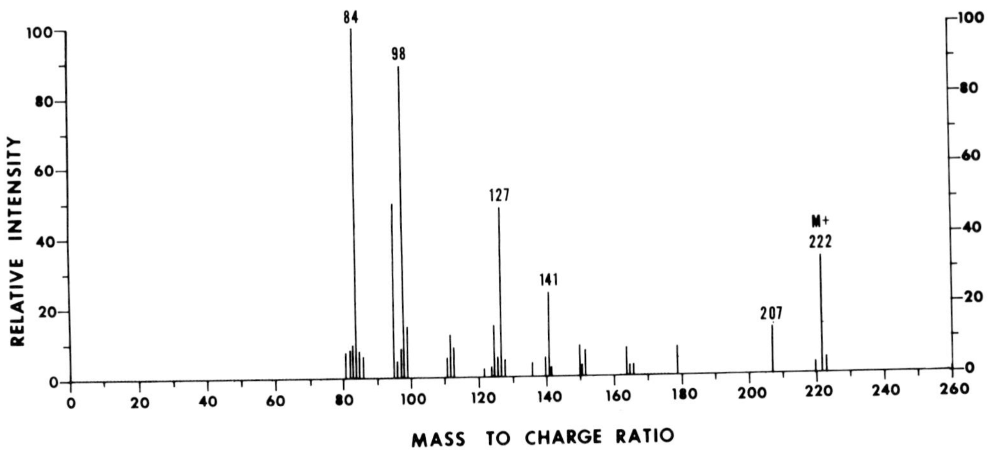

RELATIVE INTENSITY
100
80
60
40
20
0
84
98
127
141
207
M+
222
0
20
40
60
80
100
120
140
160
180
200
220
240
260
MASS TO CHARGE RATIO

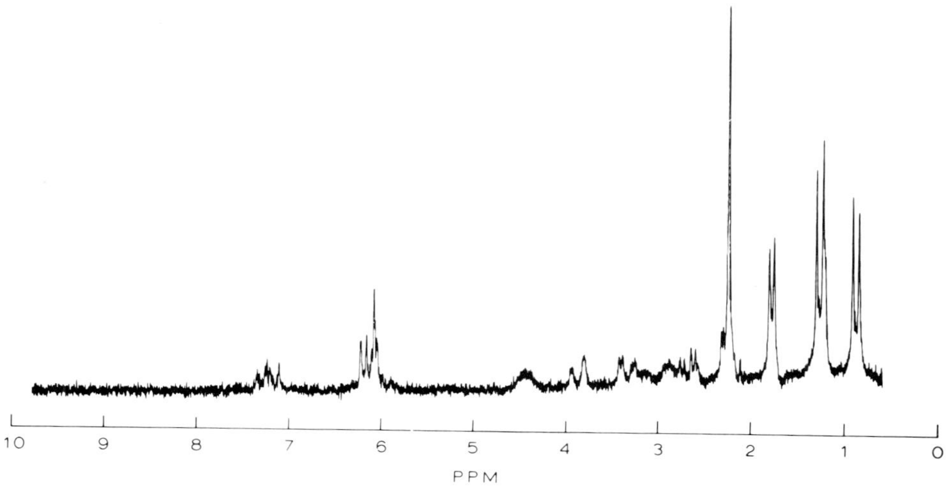

The 100 MHz ^1H NMR spectrum of nigragillin in $CDCl_3$ solution.

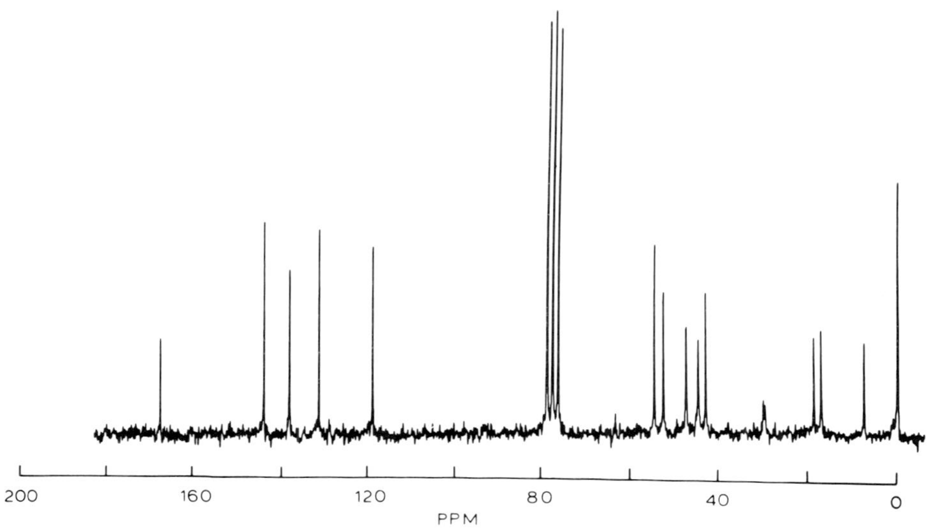

The 25 MHz ^{13}C NMR spectrum of nigragillin in $CDCl_3$ solution.

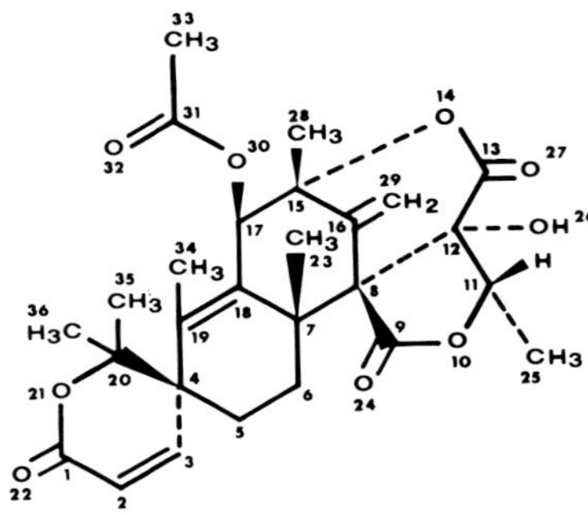

Common name	Austin
Molecular weight[23]	500.2046 The largest detectable ion in the mass spectrum was 442.1627 (M^+ − C_3H_6O) ($C_{24}H_{26}O_8$)
Molecular formula	$C_{27}H_{32}O_9$
General characteristics[23]	Crystals from chloroform–methanol, mp 298°–300°C
UV data[23]	λ_{max}^{EtOH} nm (ϵ): 243(11,900)
Fungal source[23]	*Aspergillus ustus*
Toxicity data[23]	The LD_{50} of austin to day-old cockerels dosed orally was 250–375 mg/kg.
^{13}*C NMR chemical shifts*	C11.5; C15.5; C20.4; C20.7; C22.5; C23.7; C26.0; C26.7; C27.1; C42.2; C46.7; C62.9; C74.8; C78.8; C80.7; C84.2; C85.6; C118.2 t; C120.2 d; C132.7 s; C137.6 s; C143.8 s; C146.6 d; C163.8 s; C168.5 s; C170.1 s; C171.1 s
^{1}*H NMR data*	H2, 6.10 (*J* = 10.0); H3, 6.65 (*J* = 10.0); H11, 4.46 (*J* = 6.8); H17, 6.05; H23, 1.24; H25, 1.34 (*J* = 6.8); 26-OH, 3.64; H28, 1.42; H29a, 5.52 (*J* = 1.5); H29b, 5.74 (*J* = 1.5); H33, 2.06; H34, 1.92; H35, 1.58; H36, 1.66

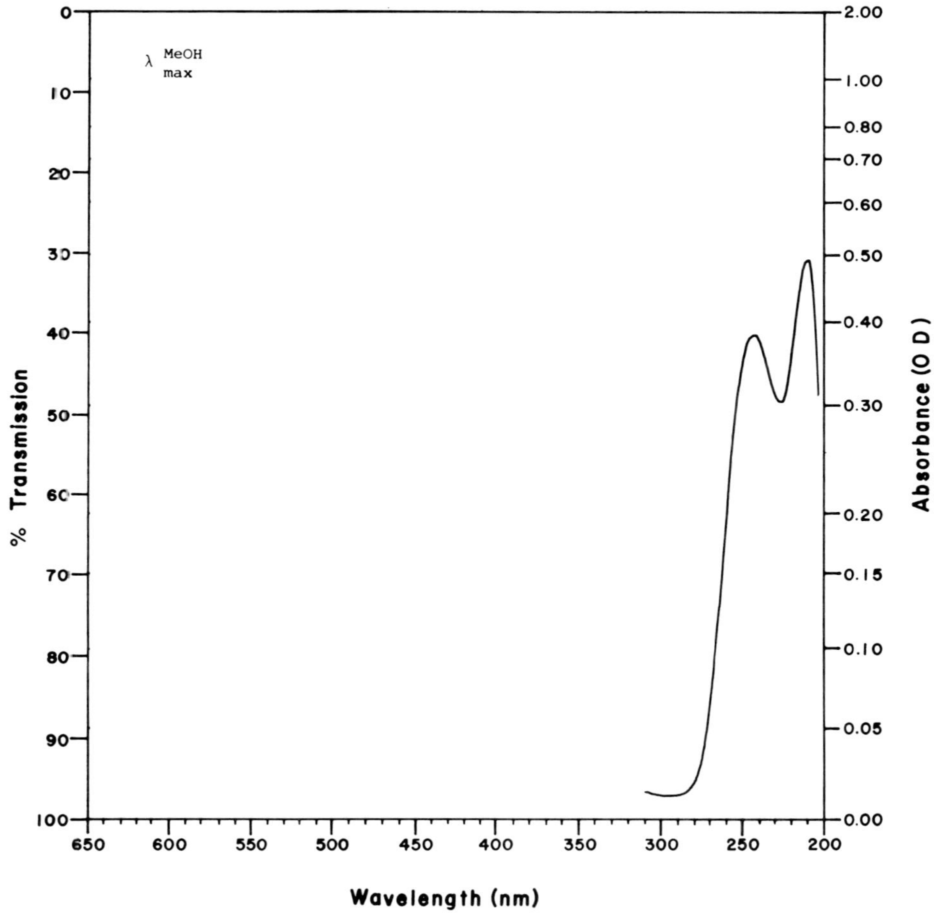

λ_{max}^{MeOH}
% Transmission
Absorbance (O D)
Wavelength (nm)
0
10
20
30
40
50
60
70
80
90
100
2.00
1.00
0.80
0.70
0.60
0.50
0.40
0.30
0.20
0.15
0.10
0.05
0.00
650
600
550
500
450
400
350
300
250
200

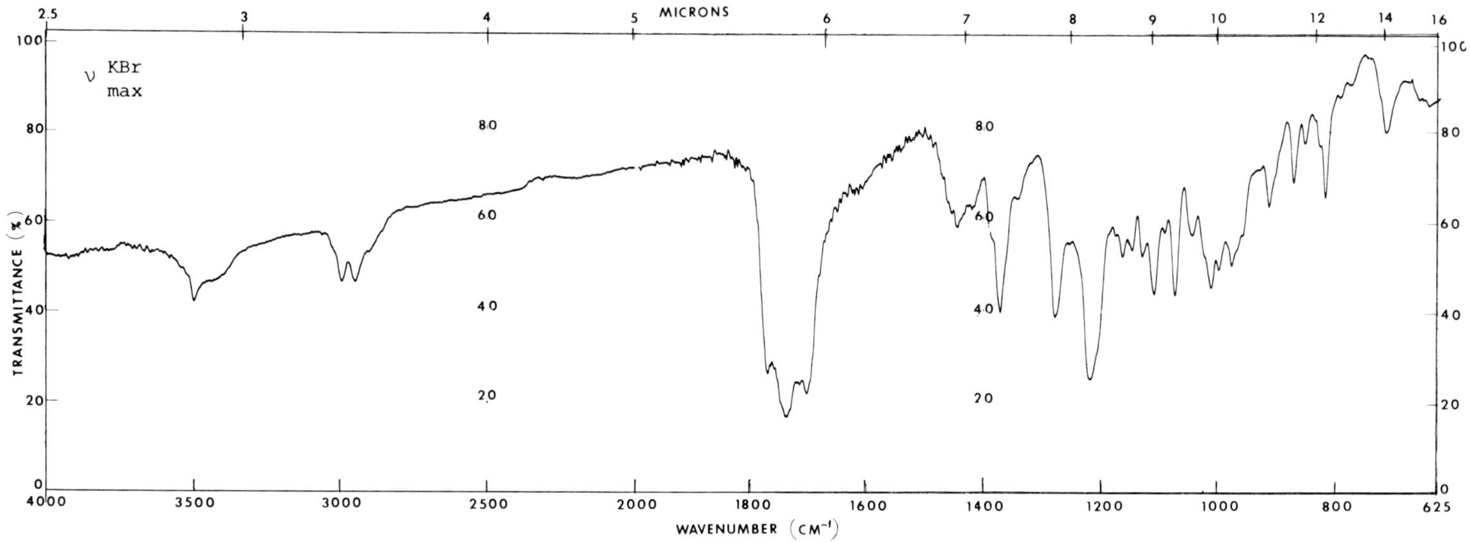
MICRONS
TRANSMITTANCE (%)
ν KBr max
WAVENUMBER (CM⁻¹)

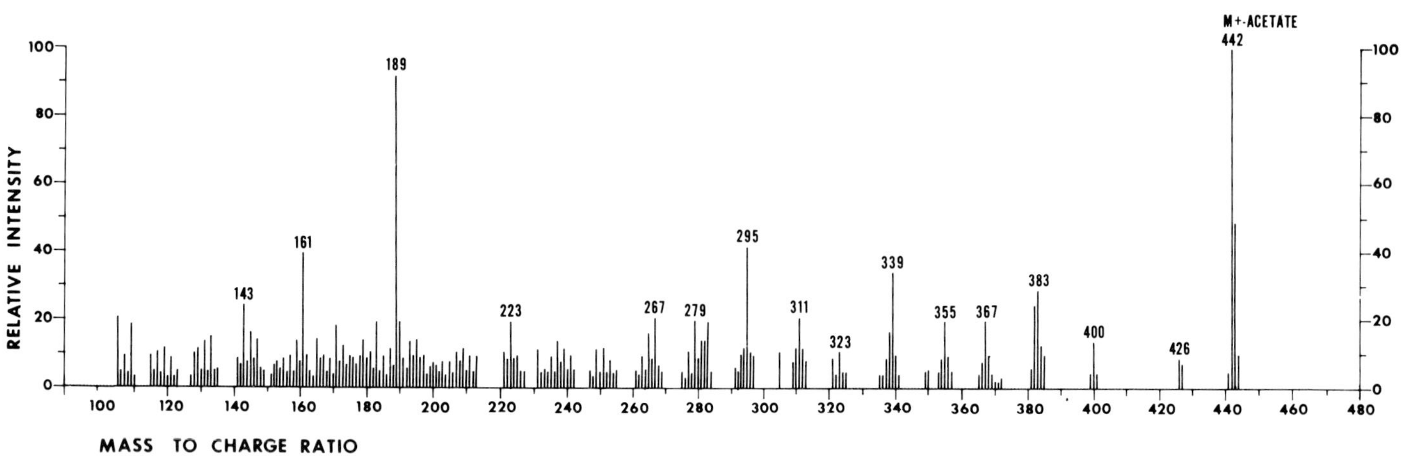
M+-ACETATE
442
189
161
143
223
267
279
295
311
323
339
355
367
383
400
426
RELATIVE INTENSITY
MASS TO CHARGE RATIO

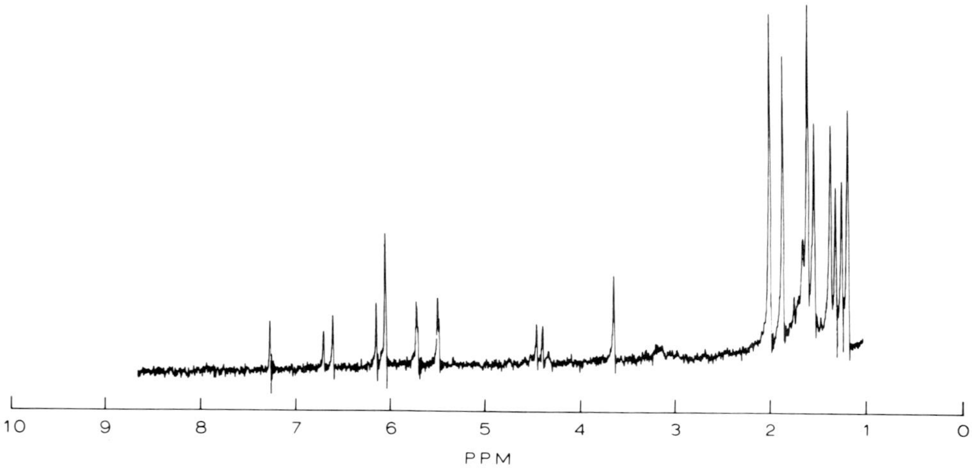

The 100 MHz ^1H NMR spectrum of austin in $CDCl_3$ solution.

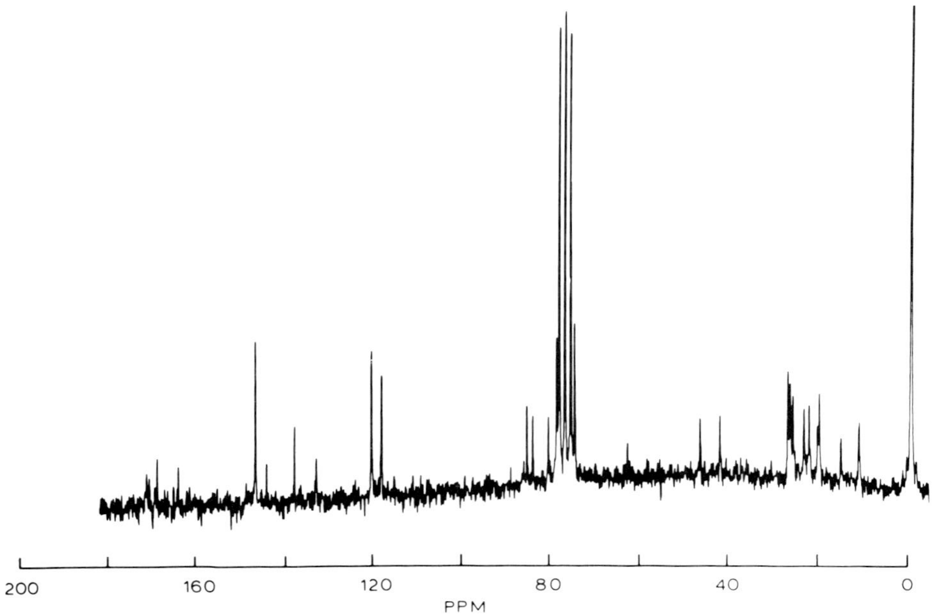

The 25 MHz ^{13}C NMR spectrum of austin in $CDCl_3$ solution.

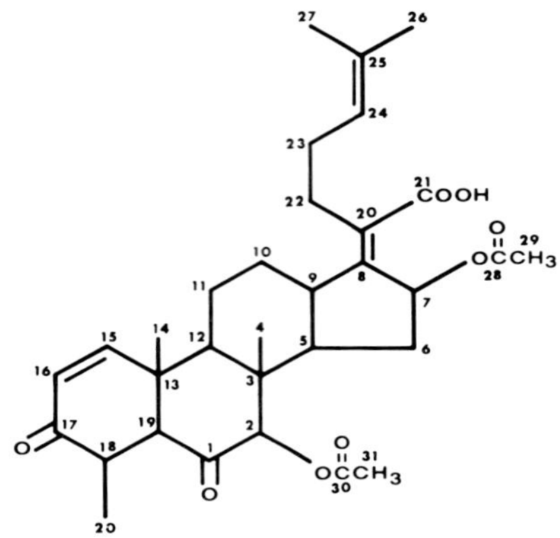

Common name	Helvolic acid (Fumigacin)
Molecular weight[52]	568.2934 The largest detectable ion in the mass spectrum was *m/e* 508 due to $M^+ - 60$
Molecular formula	$C_{33}H_{44}O_8$
General characteristics[1,64,93]	Crystals from aqueous methanol, mp 208°–212°C (dec.). White needles from methanol, mp 188°–189°C, mp 215°C $[\alpha]_D^{25}$ −121° (in chloroform)
UV data[1]	λ_{max}^{EtOH} nm (ϵ): 231(17,400)
Fungal source[22,64,84]	*Aspergillus fumigatus* *Cephalosporium caerulens* *Emericellopsis terricola* van Beyma
Toxicity data[22,84]	Helvolic acid shows antibiotic activity with gram-positive bacteria more sensitive than gram-negative bacteria. Early toxicity reports for helvolic acid were in error due to contamination by gliotoxin. A concentration of 16 mg/20 g mouse showed no definitive effects (route not reported). Other studies reported that the upper limit (IV) in 20 gm mice was 5 mg; 20 mg/mouse was tolerated when administered via gastric route (larger doses were not used).

^{13}C NMR chemical shifts: 13.0, 17.7, 17.9, 18.2, 20.4, 20.6, 23.9, 25.6, 25.9, 27.4, 28.2, 28.5, 30.1, 40.4, 40.7, 41.8, 46.5, 47.2, 49.4, 52.6, 73.4, 73.7, 122.6, 127.6, 130.2, 132.6, 147.4, 156.8, 168.5, 169.8, 200.9, 208.3

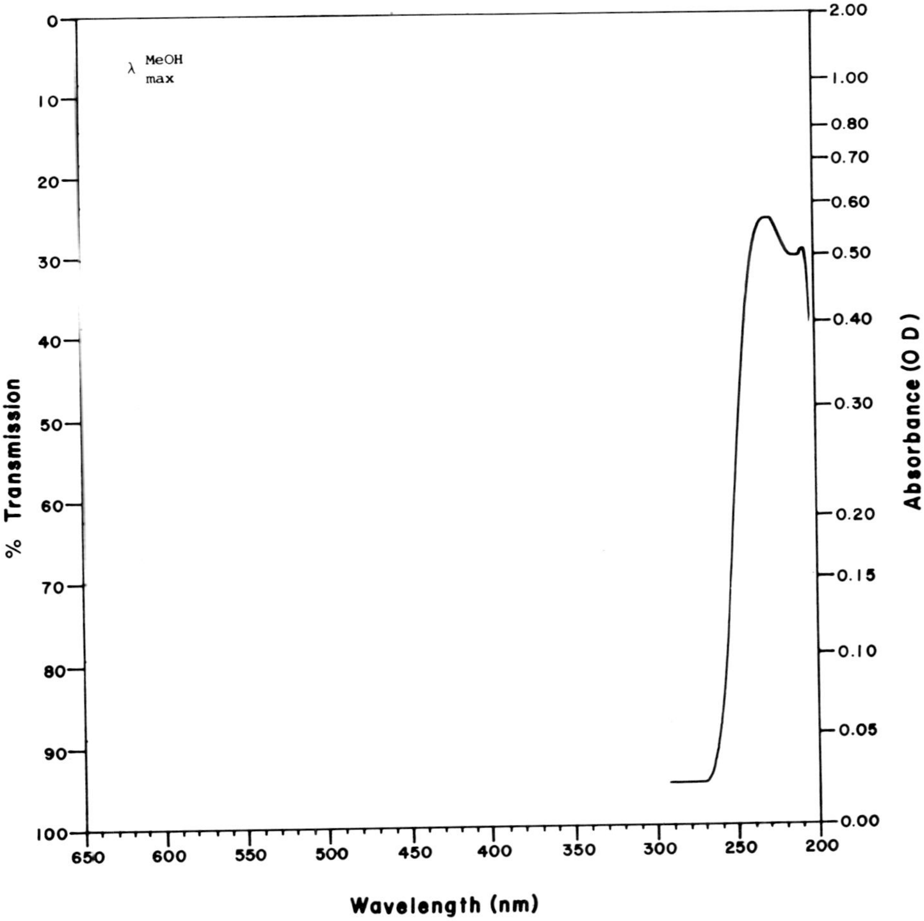

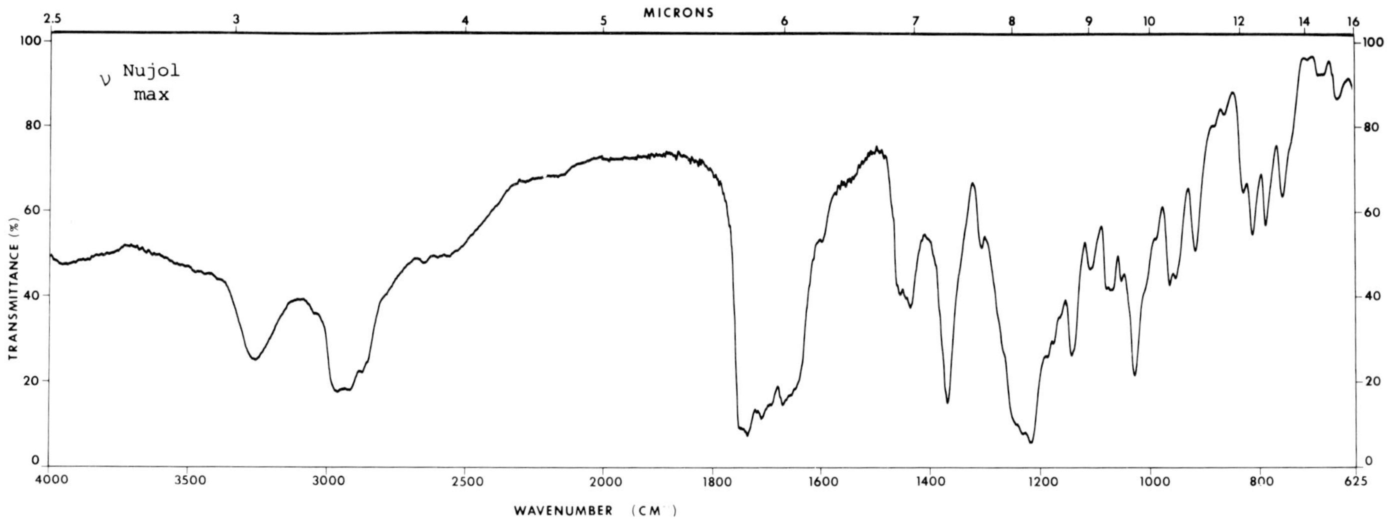
MICRONS
TRANSMITTANCE (%)
ν Nujol max
WAVENUMBER (CM⁻¹)

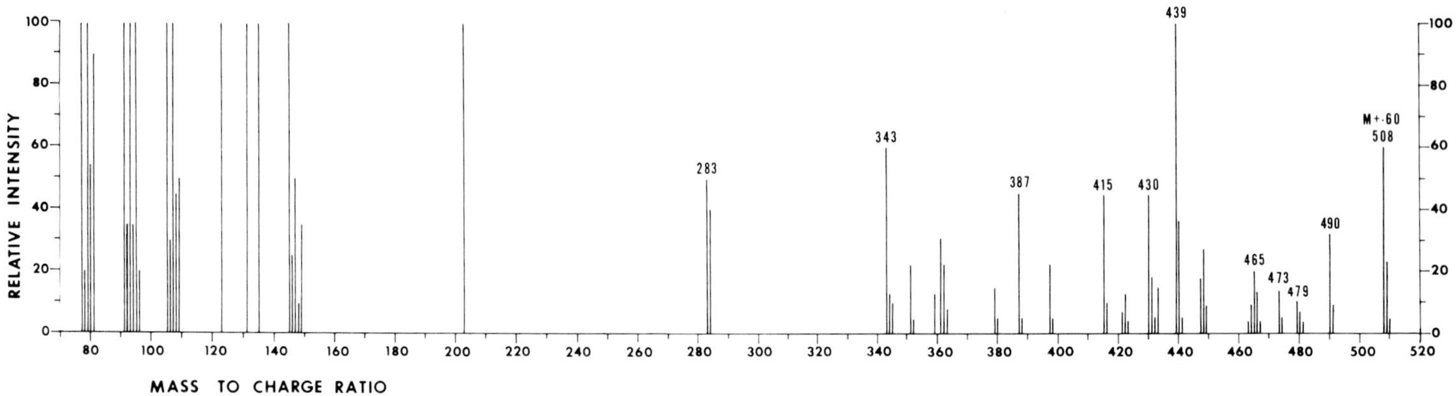
RELATIVE INTENSITY
MASS TO CHARGE RATIO
283
343
387
415
430
439
465
473
479
490
M+·60
508

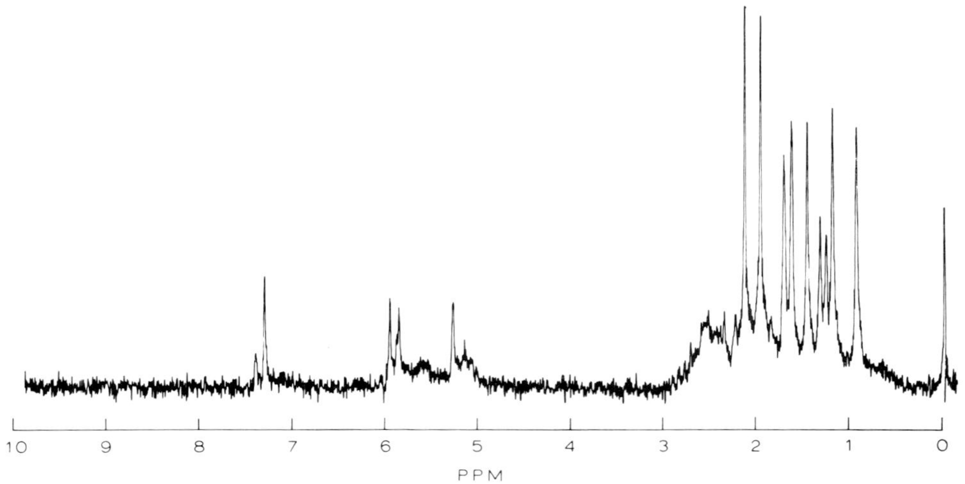

The 100 MHz ^{1}H NMR spectrum of helvolic acid in $CDCl_3$ solution.

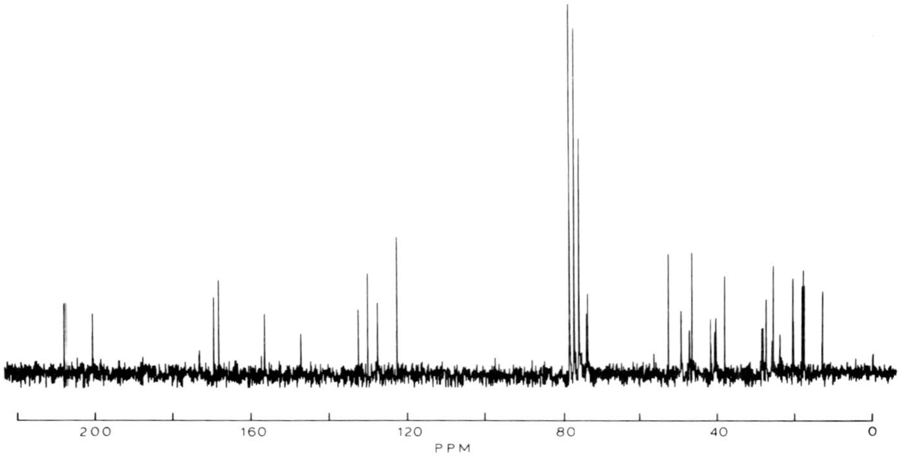

The 25 MHz ^{13}C NMR spectrum of helvolic acid in $CDCl_3$ solution.

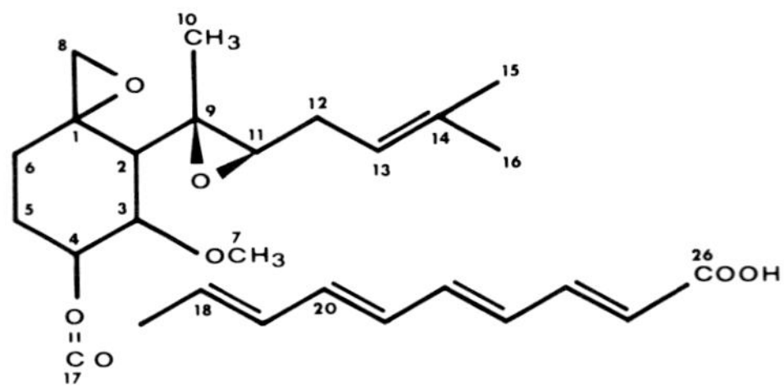

Common name	Fumagillin
Molecular weight	458.2304
Molecular formula	$C_{26}H_{34}O_7$
General characteristics[34,81]	Light-yellow crystals from aqueous methanol, mp 145°–147°C White needlelike crystals from methanol, mp 194°–195°C $[\alpha]_D^{25}$ −26.2° (*C* = 4.5 in 95% ethanol)
UV data[34]	λ_{max}^{EtOH} nm: 239, 304, 322, 336, and 351 (ϵ_{max} not reported)
Fungal source[34]	*Aspergillus fumigatus*
Toxicity data[33,34,37,41,50,57]	Fumagillin possesses antiphage, weak antifungal and antibacterial activity, and strong amebicidal activity against *Entamoeba histolytica*. Fumagillin has been useful in treatment of adult honey bees for infection against the sporozoan parasite, *Nosema apis*. The LD_{50} in mice dosed SC was 800 mg/kg. Orally dosed mice tolerated doses up to 2000 mg/kg.
Other significant references	Chemistry[27,70,79,81,82]; synthesis[26]; structure[80]; X-ray crystallography[56]

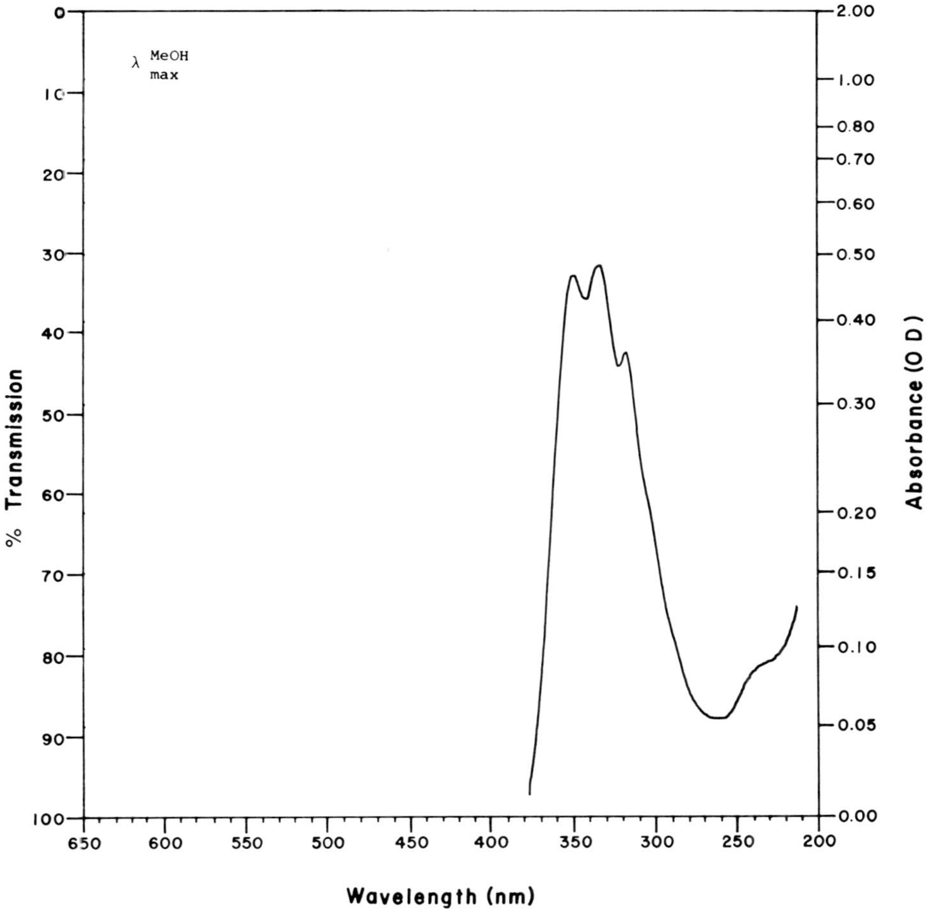

UV-visible spectrum of fumagillin (free).

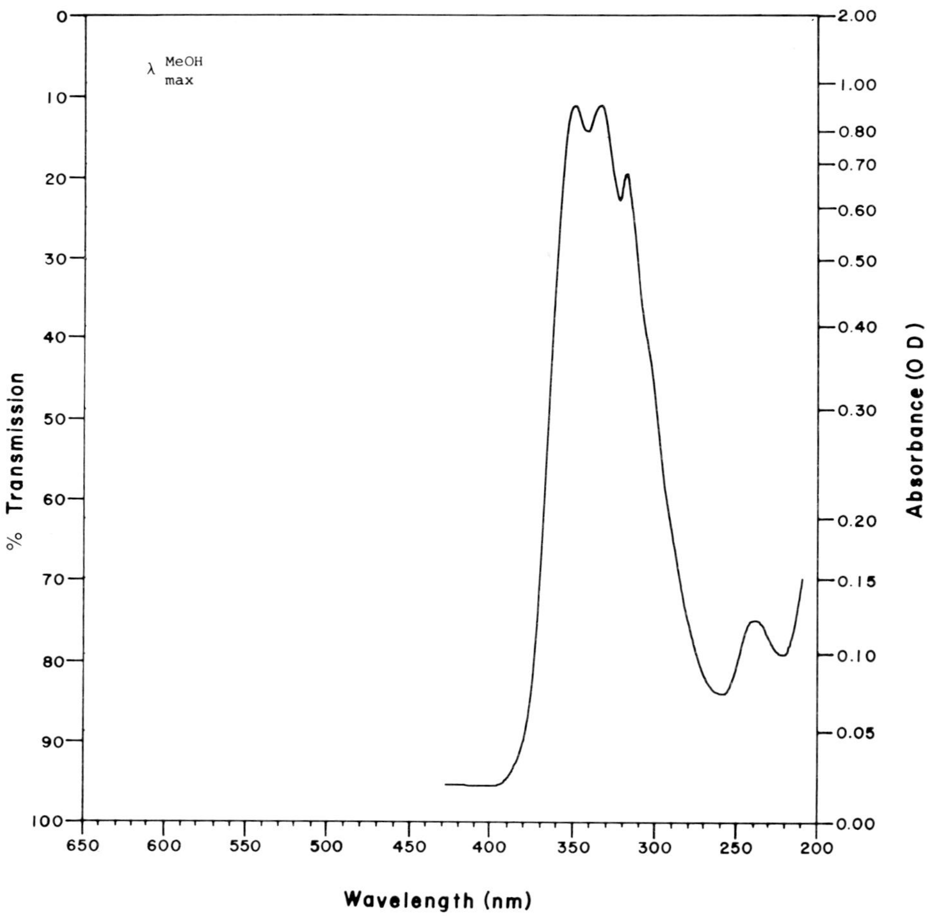

UV-visible spectrum of dicyclohexylamine salt of fumagillin.

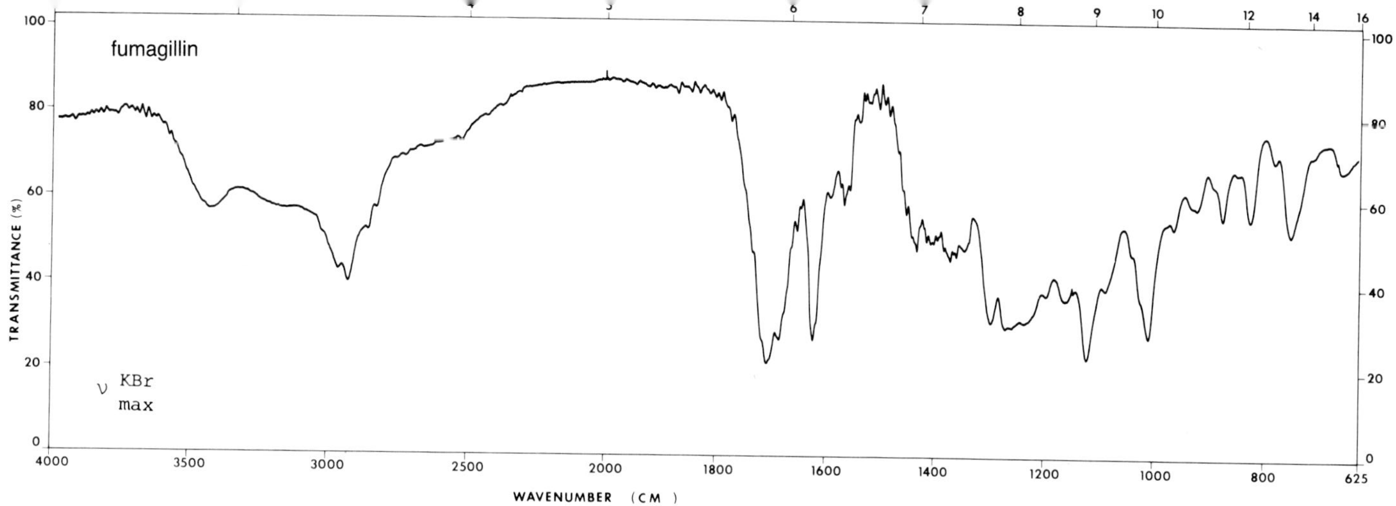
fumagillin
ν KBr max
TRANSMITTANCE (%)
WAVENUMBER (CM)

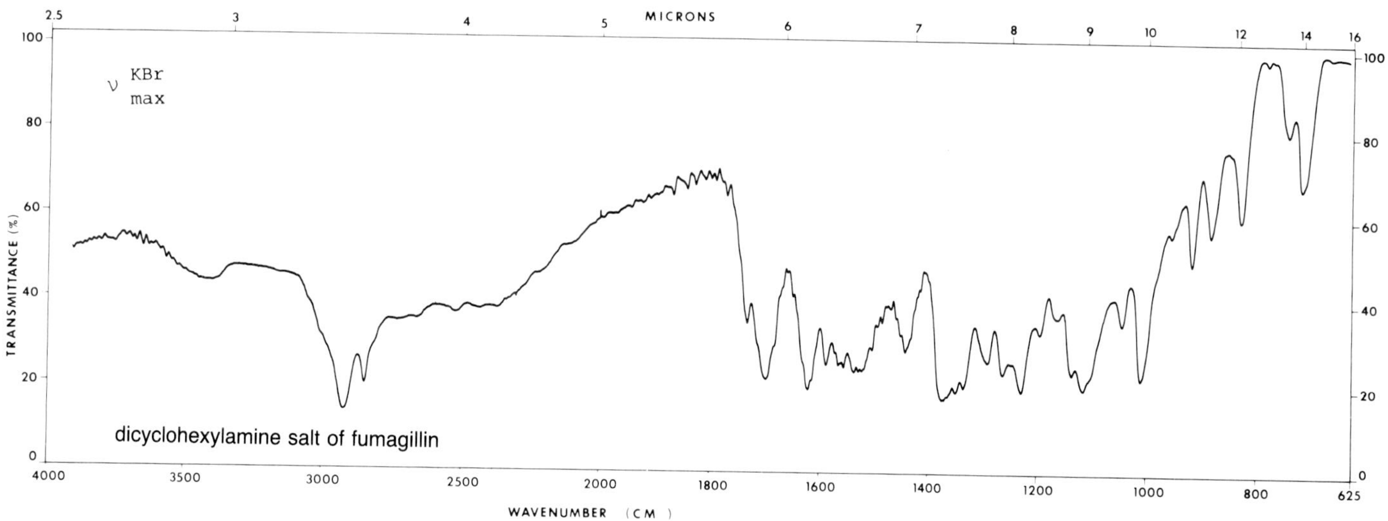
MICRONS
ν KBr max
dicyclohexylamine salt of fumagillin
TRANSMITTANCE (%)
WAVENUMBER (CM)

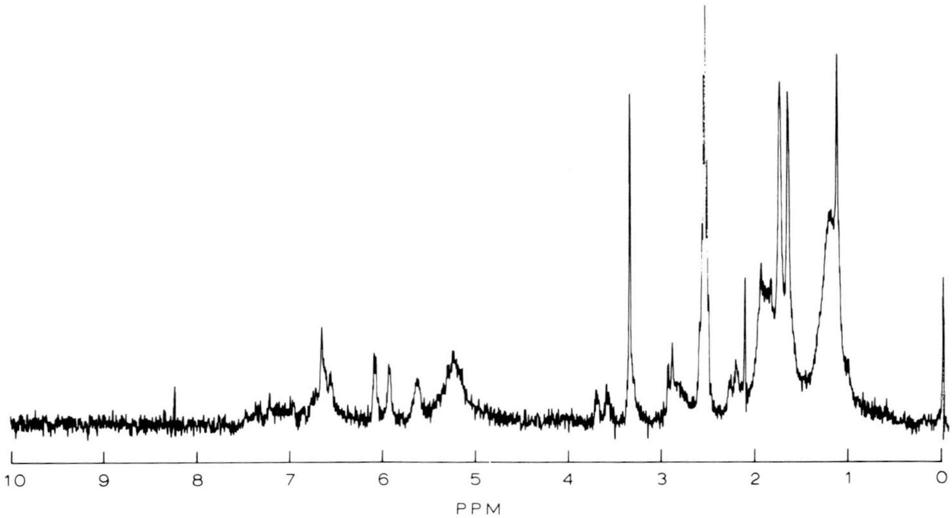

The 100 MHz ^1H NMR spectrum of fumagillin in $CDCl_3$/DMSO solution.

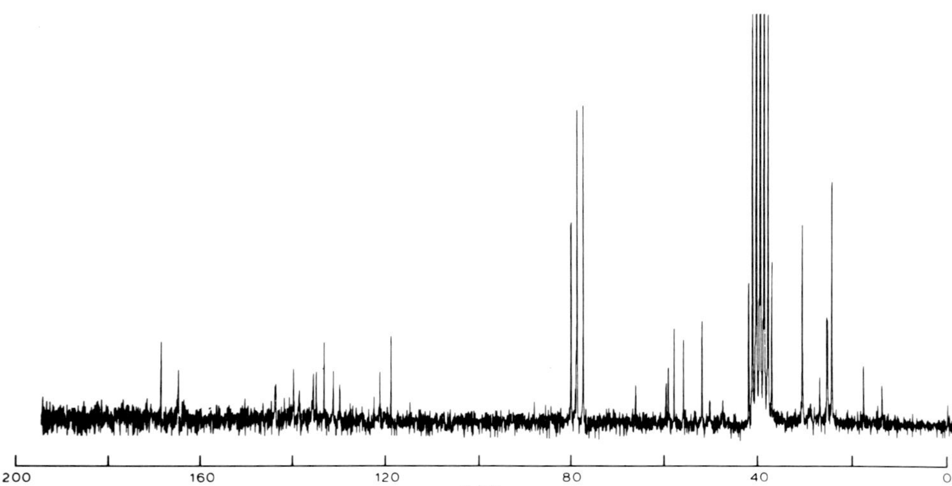

The 25 MHz ^{13}C NMR spectrum of fumagillin in $CDCl_3$/DMSO solution.

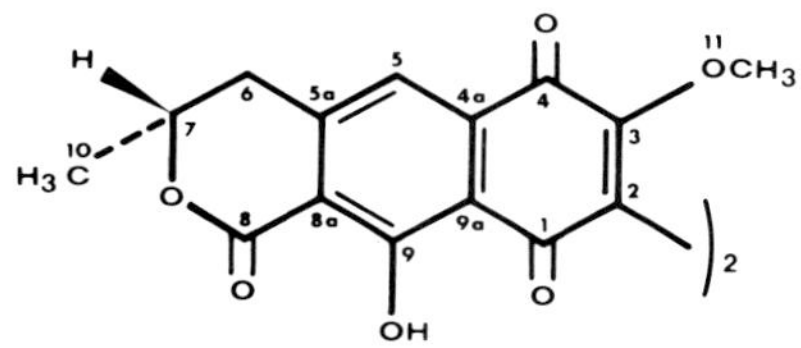

Common name	Xanthomegnin [(−)-3,3′-Bis[2-methoxy-5-hydroxy-7-(2-hydroxypropyl)-8-carboxyl-1,4-naphthoquinone lactone]]
Molecular weight	574.1111
Molecular formula	$C_{30}H_{22}O_{12}$
General characteristics[48]	Orange platelets from chloroform–benzene, mp >264°C (dec.); diacetate crystallized from ethanol–chloroform, mp 227°–228°C. The toxin was insoluble in hydroxylic solvent; slightly soluble in acetone, ethyl acetate, or benzene; moderately soluble in chloroform.
UV data[48]	$\lambda_{max}^{dioxane}$ nm (ϵ): 228(24,500), 289(7,900), and 389(4,500) Xanthomegnin diacetate: $\lambda_{max}^{dioxane}$ nm (ϵ): 262(21,300) and 344(4,700)
IR data[48]	$\nu_{max}^{CHCl_3}$ cm^{-1}: 3389, 1716, 1680, 1622, and 1600
Fungal source[14,30,47,63,77,88]	*Aspergillus ochraceus* *A. melleus* *A. sulphureus* *Penicillium viridicatum* *P. cyclopium* *Trichophyton rubrum* *T. megnini* *T. violaceum* *Microsporium cookei*
Toxicity data[20]	Gross alterations in weanling male Swiss mice fed dietary concentrations of 448 mg/kg xanthomegnin were jaundice, greenish discoloration in the liver. Histologic alterations in the liver were necrotizing cholangitis, periductal edema and pericholangitis, dissiminated focal hepatic necrosis, periductal fibrosis, and hypertrophy and hyperplasia of biliary epithelium.

^{1}*H NMR data*[30]	H6, 7.49; H10, 4.61; H11, 3.04; H12, 1.54; H14, 4.15; 5-OH, 13.11
^{13}*C NMR data*[42a]	C1, 186.1; C2, 122.7; C3, 157.9; C4, 179.7; C4a, 134.6; C5, 116.8; C5a, 148.2; C6, 35.8; C7, 74.4; C8, 161.9; C8a, 117.2; C9, 162.0; C9a, 114.5; C10, 20.4; C11, 61.2

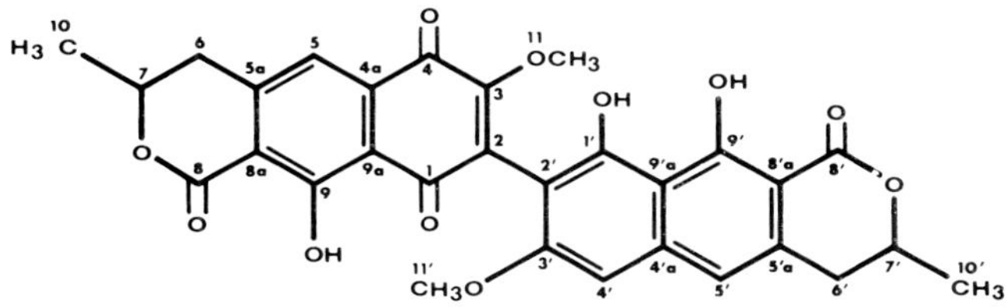

Common name	Viomellein
Molecular weight	560.1311
Molecular formula	$C_{30}H_{24}O_{11}$
General characteristics[75,76]	Reddish-brown plates from chloroform–benzene, mp >275°C (dec.) Triacetate, yellow prisms from benzene–chloroform, mp 194°–198°C
UV data[75,76]	λ_{max}^{MeOH} nm (ϵ): 225(16,800), 264(20,200), and 395(8,200)
IR data[75,76]	ν_{max}^{KBr} cm^{-1}: 3395, 2980, 1730, 1680, and 1590
Fungal source[30,75–77]	*Aspergillus sulphureus* *A. melleus* *A. ochraceus* *Penicillium viridicatum* *P. cyclopium*
Toxicity data[20]	Gross and histological signs in weanling male Swiss mice fed a dietary concentration of 456 mg/kg viomellein were identical to those described for xanthomegnin.
^{13}*C NMR data*[76a]	C1, 188.3; C2, 123.6; C3, 158.2; C4, 180.1; C4a, 134.4; C5, 116.4; C5a, 147.9; C6, 36.3; C7, 74.1; C8, 162.4; C8a, 117.6; C9, 162.8; C9a, 114.8; C10, 20.7; C11, 60.3; C1′, 161.3; C2′, 99.9; C3′, 160.1; C4′, 97.8; C4a′, 140.5; C5′, 116.0; C5a′, 134.0; C6′, 34.6; C7′, 76.5; C8′, 171.2; C8a′, 107.9; C9′, 155.3; C9a′, 105.1; C10′, 20.7; C11′, 55.9
^{1}*H NMR data*[76a]	H5, 7.50; H6, 3.02; H7, 4.63; H10, 1.34; H11, 3.84; H4′, 6.66; H5′, 6.96; H6′, 3.02; H7′, 4.63; H10′, 1.56; H11′, 3.94; H9OH, 13.44; H1′OH, 9.80; H9′OH, 13.88

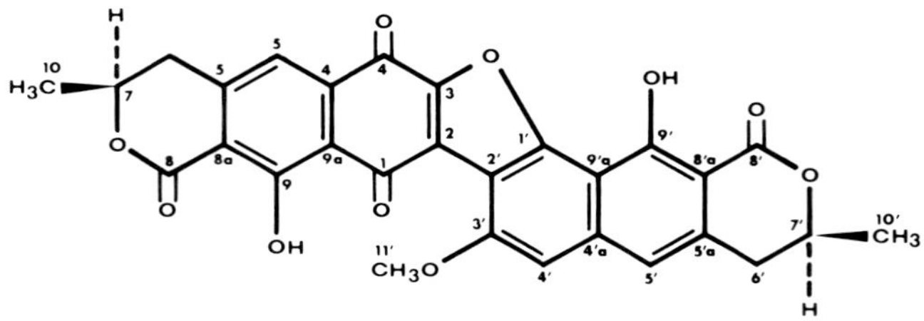

Common name	Rubrosulphin
Molecular weight	528.1060
Molecular formula	$C_{29}H_{20}O_{10}$
General characteristics[30]	Red plates from chloroform–petroleum ether, mp >300°C (dec.) Rubrosulphin diacetate, orange microcrystals, mp 200°–205°C
UV data[30]	$\lambda_{max}^{CHCl_3}$ nm: 280, 357, and 415 (ϵ_{max} not reported)
IR data[30]	ν_{max}^{Nujol} cm^{-1}: 3300(broad), 1725, 1670, 1640, 1610, 1565, 1410, 1254, 1187, 1120, 878, and 827
Fungal source[30,75,76]	*Aspergillus sulphureus* *A. melleus* *Penicillium viridicatum*
Toxicity data	Toxicity not known
^{13}*C NMR data*[76a] (di-*O*-methyl rubrosulphin)	C1, 176.8; C2, 126.2; C3, 150.8; C4, 172.5; C4a, 136.1; C5, 120.4; C5a, 147.1; C6, 36.5; C7, 74.2; C8, 160.1; C8a, 125.6; C9, 162.7; C10, 20.8; C1′, 154.9; C2′, 113.3; C3′, 155.1; C4′, 101.9; C4a′, 138.9; C5′, 119.4; C5a′, 138.6; C6′, 36.5; C7′, 74.2; C8′, 161.7; C8a′, 113.9; C9′, 159.5; C9a′, 111.7; C10′, 20.8; C11′, 55.8; 90CH_3, 63.4; 9′OCH_3, 62.7
^{1}*H NMR data*[76a] (rubrosulphin diacetate)	H5, 7.99; H6, 3.02; H7, 4.68; H10, 1.56; H4′, 6.90; H5′, 7.44; H6′, 3.07; H7′, 4.76; H10′, 1.56; H11′, 4.07; $\underline{CH_3}$CO, 2.51, 2.61

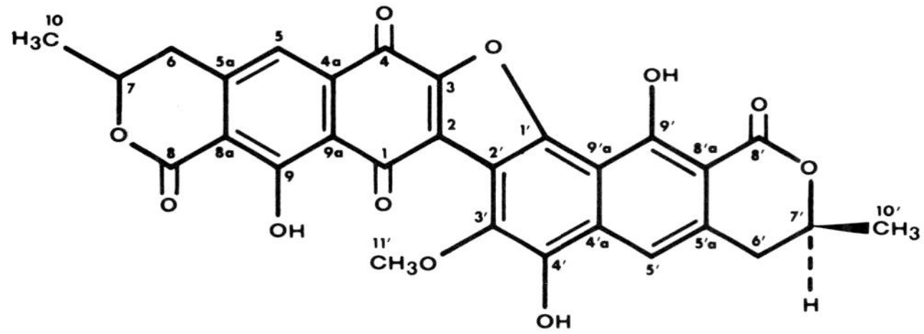

Common name	Viopurpurin
Molecular weight	544.1160
Molecular formula	$C_{29}H_{20}O_{11}$
General characteristics[30]	Purple-black beads from chloroform–petroleum ether, mp >310°C (dec.) Viopurpurin was poorly soluble in organic solvents.
UV data[30]	$\lambda_{max}^{CHCl_3}$ nm (ϵ): 274(37,200), 282(38,700), 377(8,900), and 500(3,000)
IR data[30]	ν_{max}^{Nujol} cm^{-1}: 3200, 1719, 1677, 1635, 1600, 1570, 841, and 820
Fungal source[15,30,75,76]	*Aspergillus sulphureus* *A. melleus* *Penicillium viridicatum* *Trichophyton violaceum*
Toxicity data	Toxicity not known
^{1}H NMR data[30] (viopurpurin triacetate)	H5, 8.02; H6, 3.08; H7, 4.72; H10, 1.54; H5′, 7.58; H6′, 3.16; H7′, 4.72; H10′, 1.54; H11′, 4.02; $\underline{CH_3}CO$, 2.52 (6H), 2.70

REFERENCES

1. Allinger, N. L., and Coke, J. L., *J. Org. Chem.* **26,** 4522 (1961).
2. Anslow, W. K., and Raistrick, H., *Biochem. J.* **32,** 687 (1938).
3. Anslow, W. K., and Raistrick, H., *Biochem. J.* **32,** 803 (1938).
4. Anzai, K., and Suzuki, S., *J. Antibiot.* **13,** 133 (1960).
5. Arnstein, H. R. V., and Bentley, R., *Biochem. J.* **54,** 493, 508, 517 (1953).
6. Auerbach, J., and Weinreb, S. M., *J. Chem. Soc., Chem. Commun.* p. 298 (1974).
7. Baker, W., and Raistrick, H., *J. Chem. Soc.* p. 670 (1941).
8. Barham, H. N., and Smits, B. L., *Trans. Kans. Acad. Sci.* **37,** 91 (1934).
9. Barton, D. H. R., and Miller, E., *J. Chem. Soc.* p. 1028 (1955).
10. Beelik, A., *Adv. Carbohydr. Chem.* **11,** 145 (1944).
11. Birch, A. J., Cassera, A., and Jones, A. R., *Chem. Commun.* p. 167 (1965).
12. Birch, A. J., McLoughlin, B. J., Smith, H., and Winter, J., *Chem. Ind.* (*London*) p. 840 (1960).
13. Birkinshaw, J. H., and Raistrick, H., *Philos. Trans. R. Soc. London, Ser. B* **220,** 245 (1931).
14. Blank, F., Day, W. C., and Just, G., *J. Invest. Dermatol.* **40,** 133 (1963).
15. Blank, F., Ng, A. S., and Just, G., *Can. J. Chem.* **44,** 2873 (1966).
16. Bracken, A., and Raistrick, H., *Biochem. J.* **41,** 569 (1947).
17. Bush, M. T., Goth, A., and Dickinson, H. L., *J. Pharmacol. Exp. Ther.* **84,** 262 (1945).
18. Bush, M. T., Touster, O., and Brockman, J. E., *J. Biol. Chem.* **188,** 685 (1951).
19. Caesau, F., Jansson, K., and Mutschler, E., *Helv. Physiol. Pharmacol. Acta* **44,** 676 (1969).
20. Carlton, W. W., Stack, M. E., and Eppley, R. M., *Toxicol. Appl. Pharmacol.* **38,** 455 (1976).
21. Carter, C. L., and McChesney, W. J., *Nature* (*London*) **164,** 575 (1949).
22. Chain, E., Florey, H. W., Jennings, M. A., and Williams, T. I., *Br. J. Exp. Pathol.* **24,** 108 (1943).
23. Chexal, K. K., Springer, J. P., Clardy, J., Cole, R. J., Kirksey, J. W., Dorner, J. W., Cutler, H. G., and Strawter, B. J., *J. Am. Chem. Soc.* **98,** 6748 (1976).
24. Clutterbuck, P. W., Raistrick, H., and Reuter, F., *Biochem. J.* **31,** 987 (1931).
25. Cole, R. J., unpublished results.
26. Corey, E. J., and Snider, B. B., *J. Am. Chem. Soc.* **94,** 2549 (1972).
27. DiPaolo, J. A., Tarbell, D. S., and Moore, G. E., *Antibiot. Ann.* p. 541 (1958–1959).
28. Dunn, A. W., Entwistle, I. D., and Johnstone, R. A. W., *Phytochemistry* **14,** 2081 (1975).
29. Dunn, G., Newbold, G. T., and Spring, F. S., *Nature* (*London*) **162,** 1779 (1948).
30. Durley, R. C., MacMillan, J., Simpson, T. J., Glen, A. T., and Turner, W. B., *J. Chem. Soc., Perkin Trans. 1* **1,** 163 (1975).
31. Dutcher, J. D., *J. Biol. Chem.* **171,** 321 (1947).
32. Dutcher, J. D., and Wintersteiner, O., *J. Biol. Chem.* **155,** 359 (1944).
33. Eble, T. E., and Garrett, E. R., *J. Am. Pharm. Assoc.* **43,** 536 (1954).
34. Eble, T. E., and Hanson, F. R., *Antibiot. Chemother.* (*Washington, D.C.*) **1,** 54 (1951).
35. Friedmann, T. E., *Science* **80,** 34 (1934).
36. Gallagher, J. J., Newbold, G. T., Sharp, W., and Spring, F. S., *J. Chem. Soc.* p. 4870 (1952).
37. Garrett, E. R., *J. Am. Pharm. Assoc.* **43,** 539 (1954).
38. Giarman, N. J., *J. Pharmacol. Exp. Ther.* **94,** 232 (1948).
39. Giarman, N. J., *J. Pharmacol. Exp. Ther.* **96,** 119 (1949).
40. Grove, J. P., *J. Chem. Soc.* p. 4692 (1954).
41. Hansen, F. R., and Eble, T. E., *J. Bacteriol.* **58,** 527 (1949).
42. Hill, R. A., Carter, R. H., and Staunton, J., *J. Chem. Soc., Chem. Commun.* **10,** 380 (1975).

42a. Höfle, G., and Röser, K., *Chem. Commun.* p. 611 (1978).

43. Hylin, J. W., and Matsumoto, H., *Arch. Biochem. Biophys.* **93,** 542 (1961).
44. Iizuka, H., and Iida, M., *Nature* (*London*) **196,** 681 (1962).
45. Iizuka, H., *in* "Mycotoxins" (I. F. H. Purchase, ed.), p. 405. Elsevier, Amsterdam, 1974.

46. Isogai, A., Horii, T., Suzuki, A., Murakoshi, S., Ikeda, K., Sato, S., and Tamura, S., *Agric. Biol. Chem.* **39,** 739 (1975).
47. Ito, Y., Kawai, K., and Nozawa, Y., *J. Biochem.* (*Tokyo*) **74,** 805 (1973).
48. Just, G., and Day, W. C., *Can. J. Chem.* **41,** 74 (1963).
49. Kaplan, M. A., Hooper, I. R., and Heinemann, B., *Antibiot. Chemother.* (*Washington, D.C.*) **4,** 746 (1954).
50. Katznelson, H., and Jamieson, C. A., *Science* **115,** 70 (1952).
51. Kingsbury, C. A., Cliffton, M., and Looker, J. H., *J. Org. Chem.* **41,** 2777 (1976).
52. Lynch, J. F., Wilson, J. M., Budzikiewicz, H., and Djerassi, C., *Experientia* **19,** 211 (1963).
53. MacDonald, J. C., *Can. J. Biochem.* **51,** 1311 (1973).
54. MacDonald, J. C., *J. Biol. Chem.* **236,** 512 (1961).
55. Maura, K., *Ber. Dtsch. Chem. Ges.* **63,** 25 (1930).
56. McCorkindale, N. J., and Sime, J. G., *Proc. Chem. Soc., London* p. 331 (1961).
57. McCowen, M. C., Callender, M. E., and Lawlis, J. F., Jr., *Science* **113,** 202 (1951).
58. Misawa, M., Nara, T., Nakayama, K., and Kinoshita, S., *Nippon Nogei Kagaku Kaishi* **36,** 699 (1962).
59. Morris, M. P., Pagan, C., and Warmke, H. E., *Science* **119,** 322 (1954).
60. Morton, H. E., Kocholaty, W., Junowicz-Kocholaty, R., and Kelner, A., *J. Bacteriol.* **50,** 579 (1945).
61. Nakamura, S., and Shimoda, C., *J. Agric. Chem. Soc. Jpn.* **28,** 909 (1954).
62. Newbold, G. T., Sharp, W., and Spring, F. S., *J. Chem. Soc.* p. 2679 (1951).
63. Ng, A. S., Just, G., and Blank, F., *Can. J. Chem.* **47,** 1223 (1969).
64. Okuda, S., Iwasaki, N., Tsuda, K., Sano, Y., Hata, T., Udagawa, S., Nakayama, Y., and Yamaguchi, H., *Chem. Pharm. Bull.* **12,** 121 (1964).
65. Oxford, A. E., and Raistrick, H., *Chem. Ind.* (*London*) **61,** 128 (1942).
66. Pettersson, G., *Acta Chem. Scand.* **17,** 1771 (1963).
67. Raistrick, H., and Smith, G., *Biochem. J.* **29,** 606 (1935).
68. Raistrick, H., and Stössl, A., *Biochem. J.* **68,** 647 (1958).
69. Saito, K., *Bot. Mag.* **21,** 240 (1907).
70. Schenck, J. R., Hargie, M. P., and Isarasena, A., *J. Am. Chem. Soc.* **77,** 5606 (1955).
71. Seshadri, T. R., and Venkatansubramanian, G. B., *J. Chem. Soc.* p. 1660 (1959).
72. Sheehan, J. C., Lawson, W. B., and Gaul, R. J., *J. Am. Chem. Soc.* **80,** 5536 (1958).
73. Sheehan, J. C., and Lo, Y. S., *J. Med. Chem.* **17,** 371 (1974).
74. Simonart, P., and Verachtert, H., *Bull. Soc. Chim. Biol.* **49,** 543 (1967).
75. Stack, M. E., Eppley, R. M., Dreifuss, P. A., and Pohland, A. E., *Appl. Environ. Microbiol.* **33,** 351 (1977).
76. Stack, M. E., Eppley, R. M., and Pohland, A. E., *in* "Mycotoxins in Human and Animal Health" (J. V. Rodricks, C. W. Hesseltine, and M. A. Mehlman, eds.), p. 543. Pathotox Publ., Inc., Park Forest South, Illinois, 1977.
76a. Stack, M. E., Mazzola, E. P., and Eppley, R. M., *Tetrahedron Lett.* p. 4989 (1979).
77. Stack, M. E., and Mislivec, P. B., *Appl. Environ. Microbiol.* **36,** 552 (1978).
78. Steyn, P. S., *Tetrahedron* **29,** 107 (1973).
79. Tarbell, D. S., Carman, R. M., Chapman, D. D., Cremer, S. E., Cross, A. D., Huffman, K. R., Kunstmann, M., McCorkindale, N. J., McNally, J. G., Jr., Rosowsky, H., Varino, F. H. L., and West, R. L., *J. Am. Chem. Soc.* **83,** 3096 (1961).
80. Tarbell, D. S., Carman, R. M., Chapman, D. D., Huffman, K. R., and McCorkindale, N. J., *J. Am. Chem. Soc.* **82,** 1005 (1960).
81. Tarbell, D. S., Hoffman, P., Al-Kazimi, H. R., Page, G. A., Ross, J. M., Bogt, H. R., and Wargotz, B., *J. Am. Chem. Soc.* **77,** 5610 (1955).
82. Turner, J. R., and Tarbell, D. S., *Proc. Natl. Acad. Sci. U.S.A.* **48,** 733 (1962).
83. Vleggaar, R., Steyn, P. S., and Nagel, D. W., *J. Chem. Soc., Perkin Trans. 1* p. 45 (1974).
84. Wacksman, S. A., Horning, E. S., and Spencer, E. L., *J. Bacteriol.* **45,** 233 (1943).

85. White, E. C., and Hill, J. H., *J. Bacteriol.* **45,** 433 (1943).
86. Wilkins, W. H., and Harris, G. G. M., *Br. J. Exp. Pathol.* **23,** 166 (1942).
87. Wilson, B. J., *in* "Microbial Toxins" (A. Ciegler, S. Kadis, and S. J. Ajl, eds.), Vol. 6, p. 235. Academic Press, New York, 1971.
88. Wirth, J. G., Beesley, T. E., and Anand, S. R., *Phytochemistry* **4,** 505 (1965).
89. Yabuta, T., *J. Chem. Soc.* **125,** 575 (1924).
90. Yamamoto, Y., Hirai, T., Okada, K., and Saito, K., *Chem. Pharm. Bull.* **22,** 83 (1974).
91. Yamamoto, Y., Nita, K., and Jinbo, A., *Chem. Pharm. Bull.* **15,** 427 (1967).
92. Yamamoto, Y., Nita, K., Oohata, Y., and Furukawa, T., *Chem. Pharm. Bull.* **20,** 931 (1972).
93. Yamazaki, M., personal communication.

19

Penicillium Toxins

A large number of species of the genus *Penicillium* are known to produce toxic metabolites. Although some of these metabolites may be produced by species of other genera, they are grouped under "*Penicillium* Toxins" because they were either isolated initially from a *Penicillium* species or they are known to be produced primarily by *Penicillium* species. These metabolites represent a wide diversity of chemical structures and elicit different biological effects.

A review of toxins produced by *Penicillium* species has been published.[104]

Penicillium toxins	Molecular weight	Molecular formula
Citrinin	250.0841	$C_{13}H_{14}O_5$
Oosporein	306.0375	$C_{14}H_{10}O_8$
Viridicatin	237.0789	$C_{15}H_{11}O_2N$
Verruculotoxin	244.1575	$C_{15}H_{20}ON_2$
Decumbin	280.1674	$C_{16}H_{24}O_4$
Cyclopenin	294.1004	$C_{17}H_{14}O_3N_2$
Cyclopenol	310.0953	$C_{17}H_{14}O_4N_2$
Griseofulvin	352.0713	$C_{17}H_{17}O_6Cl$
Dechlorogriseofulvin	318.1103	$C_{17}H_{18}O_6$
Mycophenolic acid	320.1259	$C_{17}H_{20}O_6$
PR toxin	320.1259	$C_{17}H_{20}O_6$
Eremofortin A	306.1467	$C_{17}H_{22}O_5$
Eremofortin B	248.1412	$C_{15}H_{20}O_3$
Eremofortin C	322.1416	$C_{17}H_{22}O_6$
Citreoviridin	402.2042	$C_{23}H_{30}O_6$
Viridicatumtoxin	565.1947	$C_{30}H_{31}O_{10}N$
Viriditoxin	662.1623	$C_{34}H_{30}O_{14}$

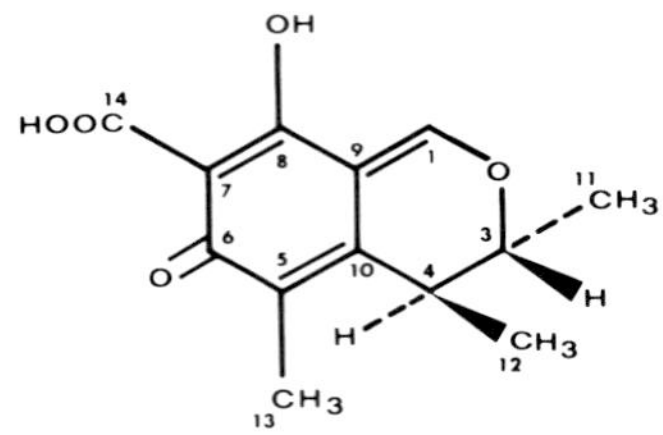

Common name	Citrinin [(3*R*-*trans*)-4,6-Dihydro-8-hydroxy-3,4,5-trimethyl-6-oxo-3*H*-2-benzopyran-7-carboxylic acid]
Molecular weight	250.0841
Molecular formula	$C_{13}H_{14}O_5$
General characteristics[21,86]	Lemon-yellow crystals from absolute ethanol or benzene–cyclohexane, mp 175°C (dec.) $[\alpha]^{23}_{5461}$ −43.1° (*C* = 0.92 in ethanol) Crystals from methanol, mp 178°–179°C Methylate: plates from benzene, mp 139°C (dec.) $[\alpha]^{18}_{D}$ +217.1° (*C* = 0.38 in acetone)
UV data[23,81]	λ^{EtOH}_{max} nm (ϵ): 319(4,700), 253(8,300), and 222(22,300) $\lambda^{3\text{-methyl pentane}}_{max}$ nm (ϵ): 322(8,800) $\lambda^{CHCl_3}_{max}$ nm (ϵ): 332(16,100)
Fungal source[86]	*Penicillium citrinum* Thom *P. implicatum* Biourge *P. lividum* Westling *P. fellutanum* Biourge *P. jenseni* Zaleski *P. citrio-viride* Biourge *P. expansum* *P. notatum* *P. viridicatum* Westling *P. steckii* *Aspergillus terreus* Thom *A. niveus* Blockwitz *A. candidus*
Toxicity data[2,59,98]	Citrinin was originally isolated as an antibiotic but its utility, as such, was negated due to its nephrotoxicity. The LD_{50} of citrinin in rats by SC and IP

administration was 67 mg/kg; in mice it was 35 mg/kg; in guinea pigs dosed SC, it was approximately 37 mg/kg; in rabbits dosed IV, it was 19 mg/kg. Renal toxicity in swine was characterized by lesions and desquamation of renal epithelial cells of the proximal convoluted tubules, dilation of tubules, thickening of basal membranes, and proliferation of cells in the interstitium.

Other significant references

Isolation[50]; structure elucidation[22]; structure and synthesis[25,26,99]; X-ray diffraction[27]; absolute configuration[73]

^{13}C NMR data

C1, 163.8 d; C3, 81.7 d; C4, 34.6 d; C5, 122.7 s; C6, 183.4 s; C7, 100.1 s; C8, 174.2 s; C9, 139.1 s; C10, 107.2 s; C11, 18.2 q; C12, 18.4 q; C13, 9.3 q; C14, 176.9 s

^{1}H NMR data

H1, 8.21; H3, 4.79 (J = 7.0, 1.0); H4, 3.99 (J = 7.0, 1.0); H11, 1.23; H12, 1.35; H13, 2.02; OH's, 15.10, 15.83

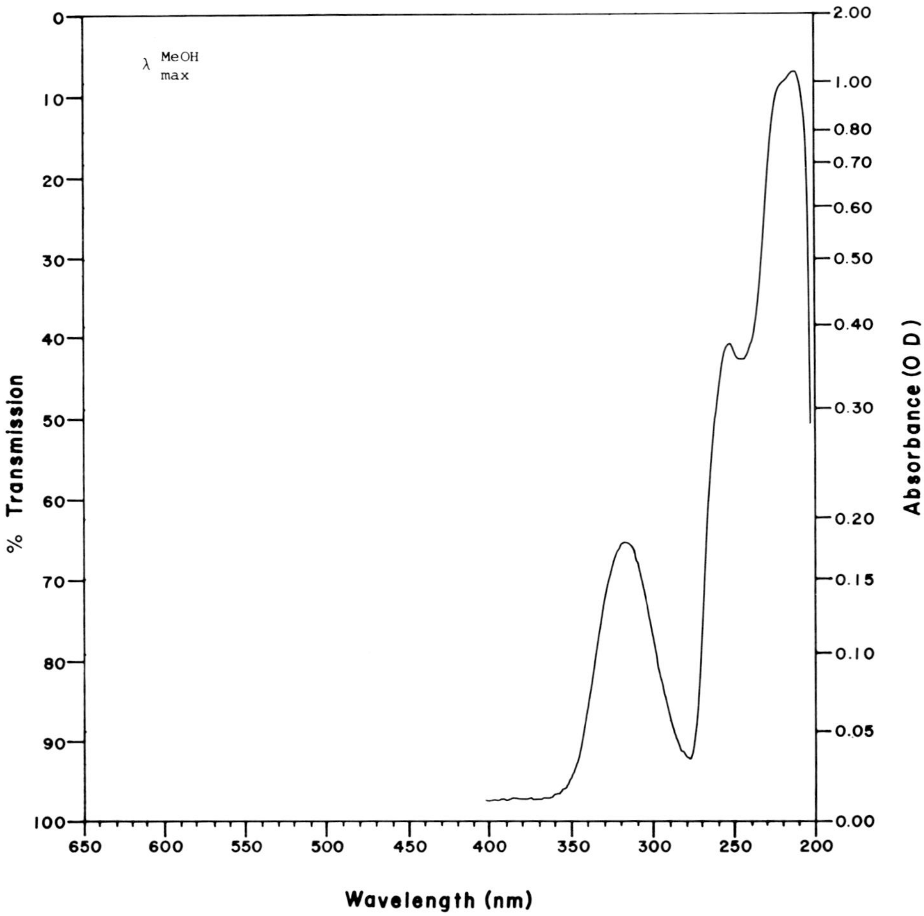

λ MeOH max
% Transmission
0
10
20
30
40
50
60
70
80
90
100
Absorbance (O D)
2.00
1.00
0.80
0.70
0.60
0.50
0.40
0.30
0.20
0.15
0.10
0.05
0.00
650
600
550
500
450
400
350
300
250
200
Wavelength (nm)

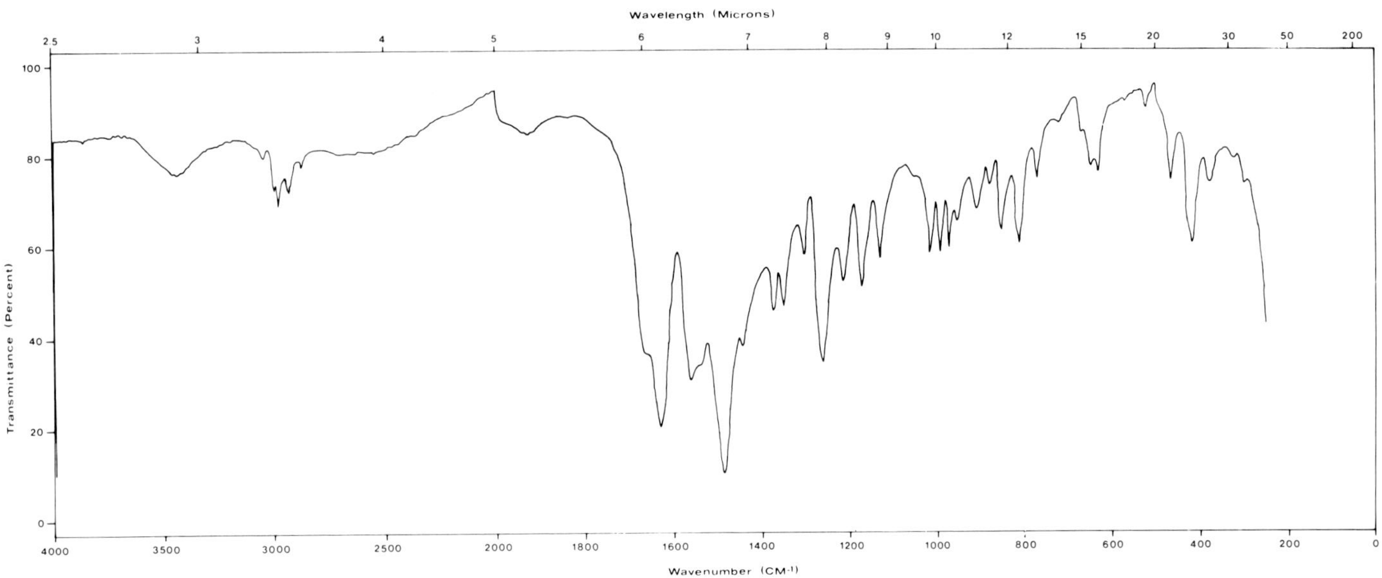
Wavelength (Microns)
2.5
3
4
5
6
7
8
9
10
12
15
20
30
50
200
Transmittance (Percent)
100
80
60
40
20
0
4000
3500
3000
2500
2000
1800
1600
1400
1200
1000
800
600
400
200
0
Wavenumber (CM-1)

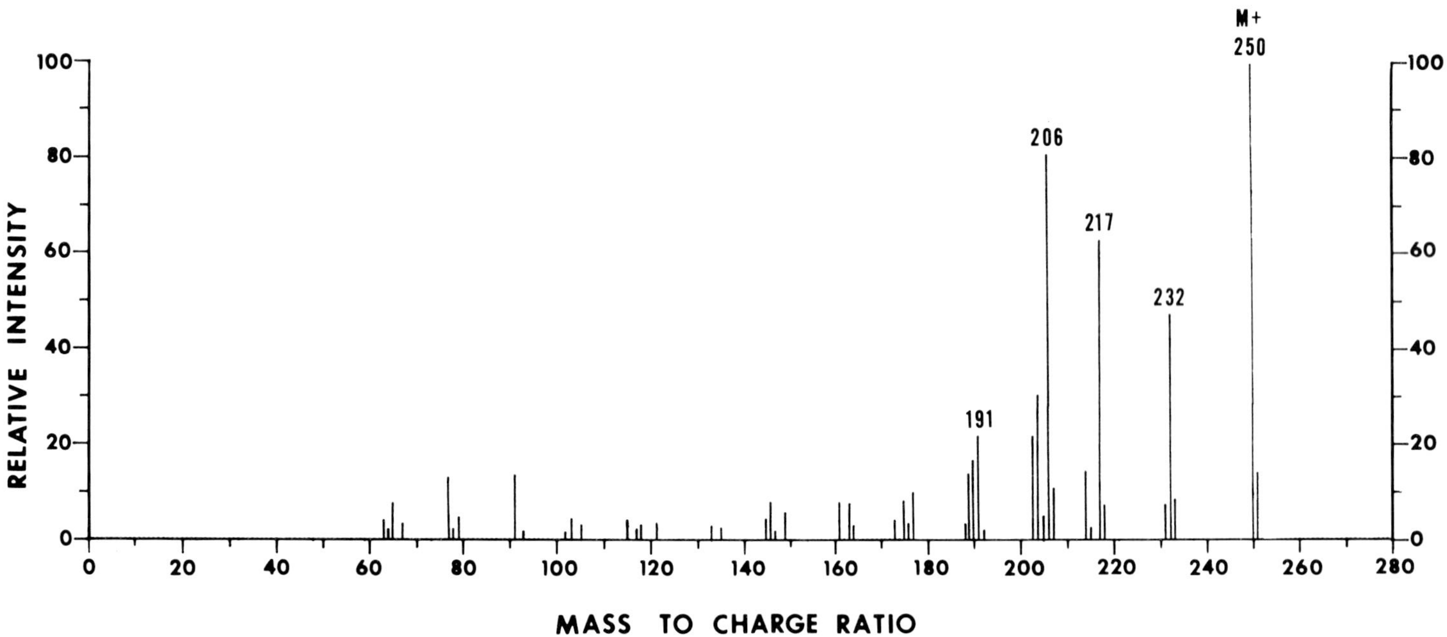
M+
250
206
217
232
191
RELATIVE INTENSITY
MASS TO CHARGE RATIO

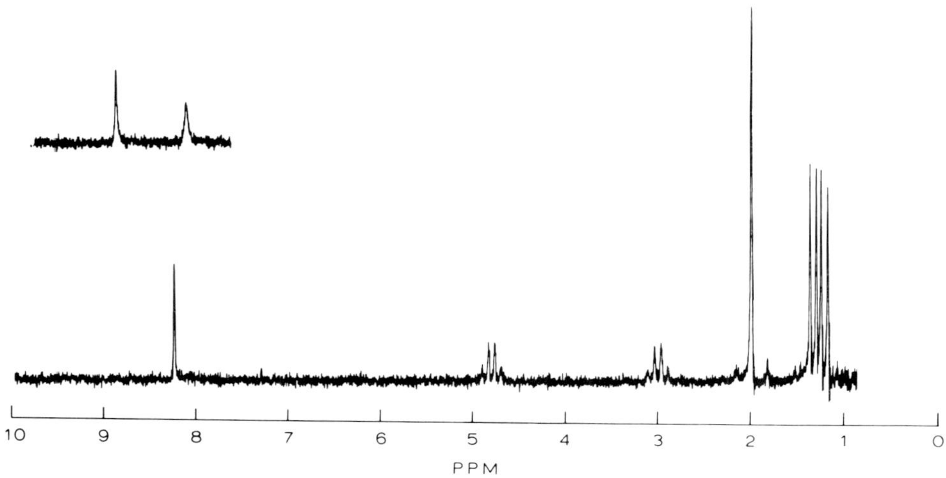

The 100 MHz ^{1}H NMR spectrum of citrinin in $CDCl_3$ solution.

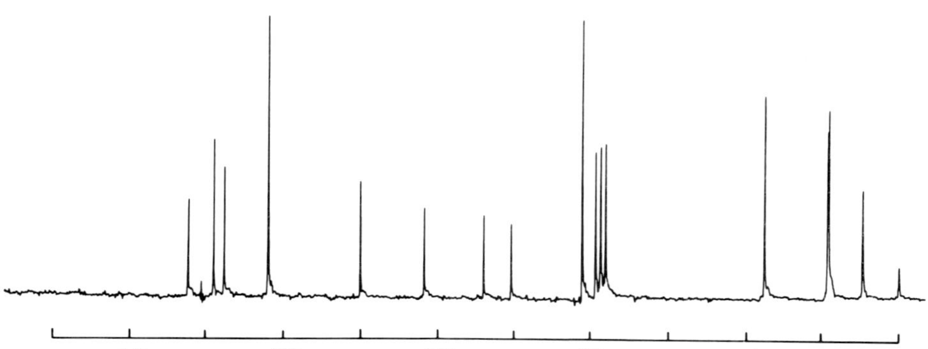

The 25 MHz ^{13}C NMR spectrum of citrinin in $CDCl_3$ solution.

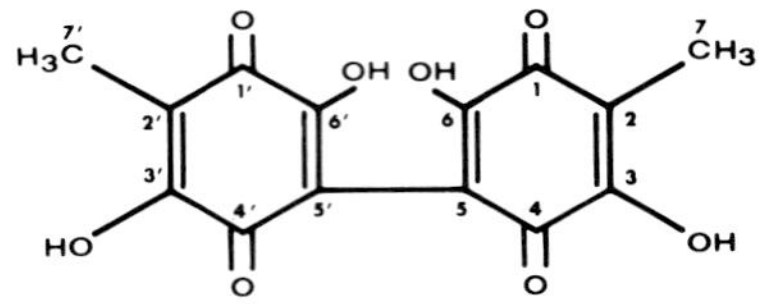

Common name[55,58,63,82,93]	Oosporein (Isoosporin, Chaetomidin) (3,3′,6,6′-Tetrahydroxy-2,2′-dimethyl-5,5′-bi-*p*-benzoquinone)
Molecular weight	306.0375
Molecular formula	$C_{14}H_{10}O_8$
General characteristics[29]	Bronze plates from aqueous methanol, mp 290°–295°C Tetraacetate, yellow needles from methanol, mp 190° C Tetramethyl ether, orange needles from aqueous methanol, mp 123°C
UV data[29]	λ_{max}^{MeOH} nm (ϵ): 208(33,500) and 291(23,000) Tetraacetate: λ_{max}^{EtOH} nm (ϵ): 262(25,700) Tetramethyl ether: λ_{max}^{EtOH} nm (ϵ): 285(25,100) and 394(950)
Fungal source[29,35,37,58,63,97]	*Penicillium phoeniceum* *P. rubrum-* *Oospora coloraus* *Chaetomium aureum* Chivers *C. trilaterale* Chivers *Acremonium spp.* *Beauveria bossiana*
Toxicity data[29,105]	The LD_{50} of oosporein dosed orally to day-old cockerels was 6.12 mg/kg. Oosporein was acutely toxic but no chronic toxicity was observed. Oosporein also produced growth-inhibitory and phytotoxic effects in plants.
TLC data[29]	*Adsorbent:* silica gel G-HR *Solvent:* toluene–ethyl acetate–formic acid, 5:4:1 v/v/v R_f: 0.70 *Detection:* appears as a red spot under visible light

^{13}C *NMR data*	C1, 167.7 s; C2, 112.5 s; C3, 167.7 s; C4, 167.7 s; C5, 107.0 s; C6, 167.7 s; C7, 7.5 q
^{1}H *NMR data*	H7, 1.80

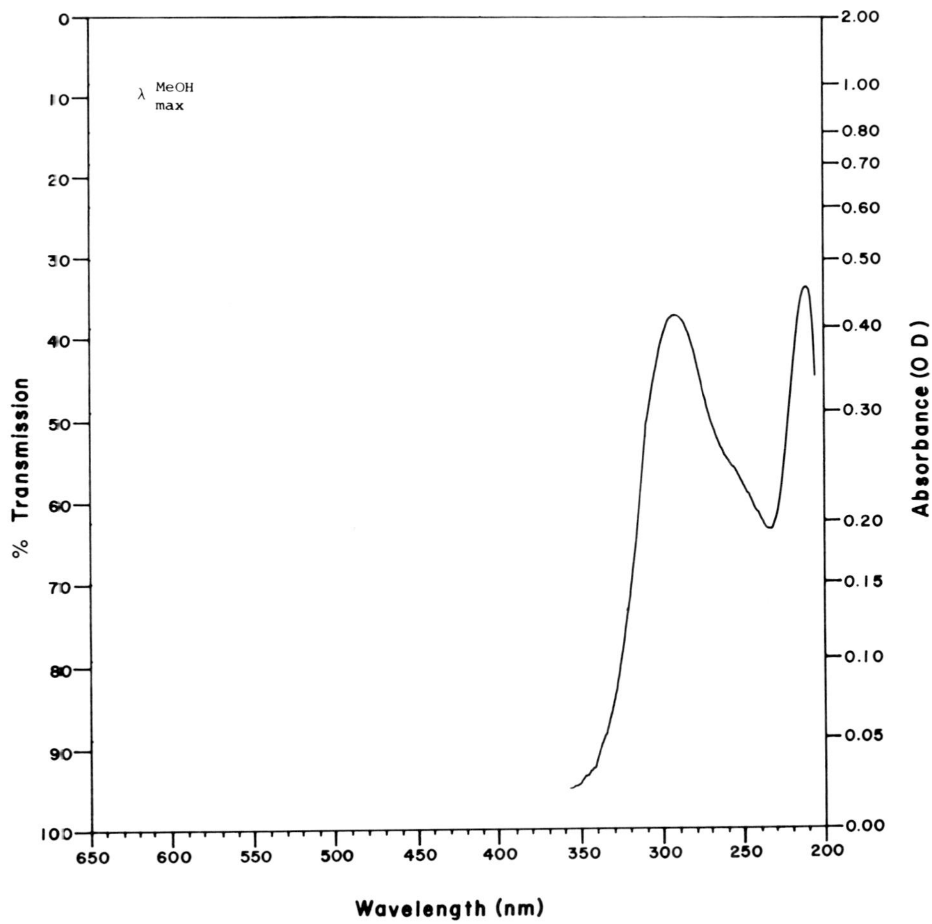

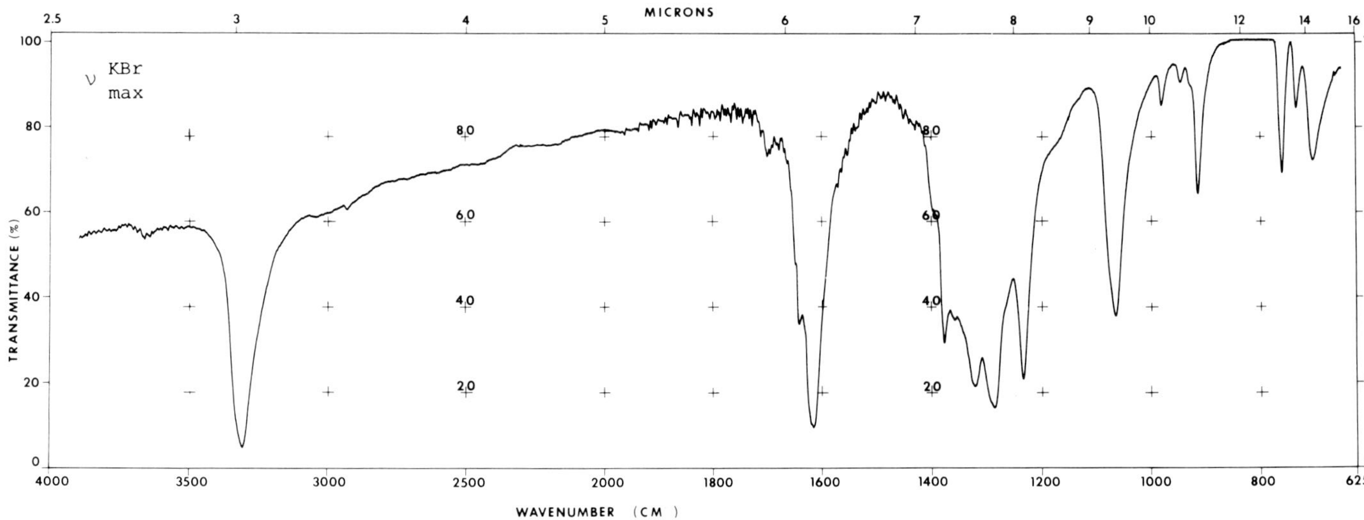
MICRONS
TRANSMITTANCE (%)
ν KBr max
WAVENUMBER (CM)

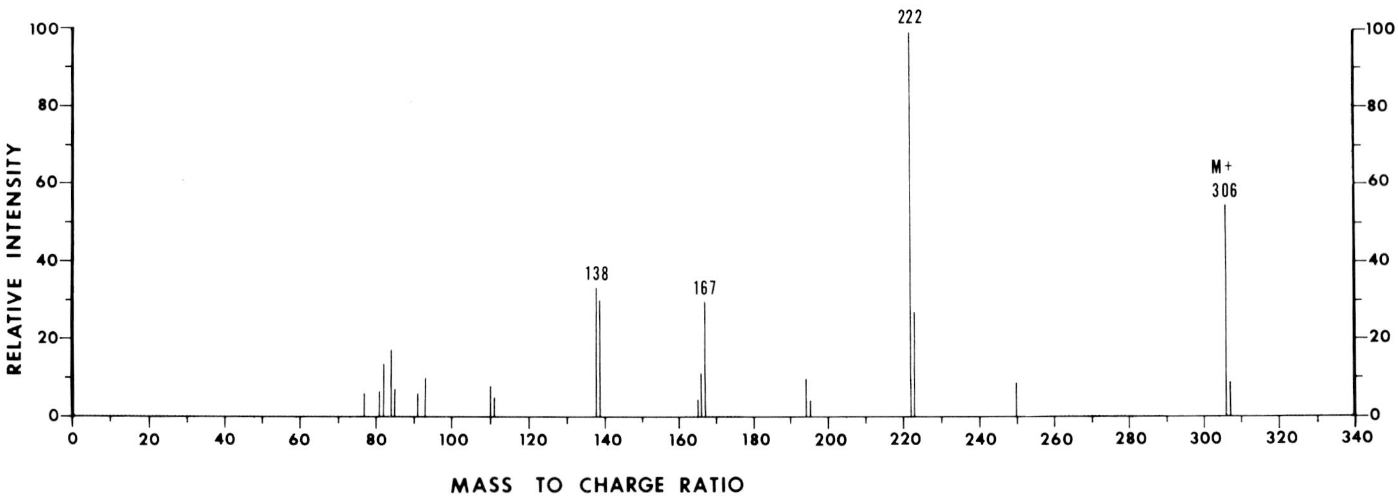
RELATIVE INTENSITY
MASS TO CHARGE RATIO
222
M+
306
138
167

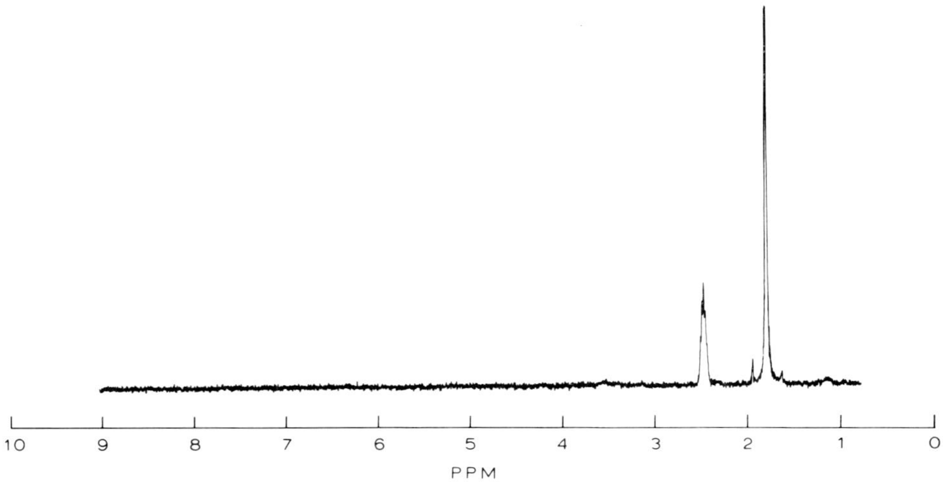

The 100 MHz ^1H NMR spectrum of oosporein in DMSO solution.

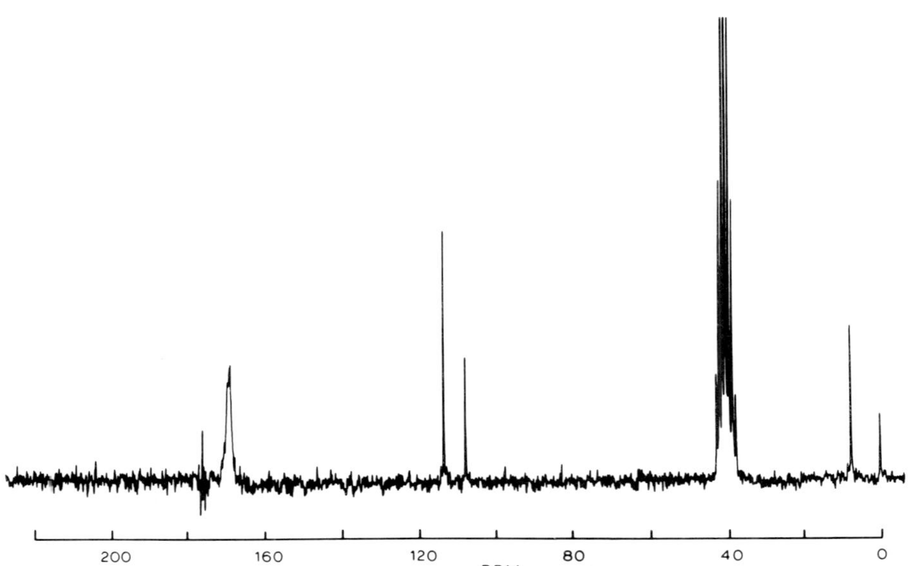

The 25 MHz ^{13}C NMR spectrum of oosporein in DMSO solution.

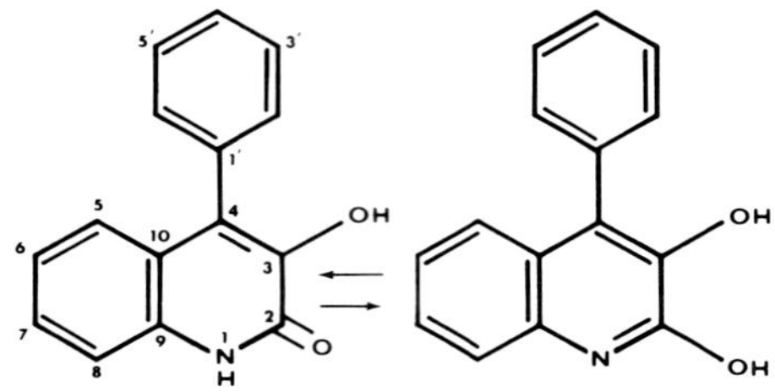

Common name	Viridicatin (2-Keto-3-hydroxy-4-phenylquinoline or 2,3-dihydroxy-4-phenylquinoline (tautomer))
Molecular weight	237.0789
Molecular formula	$C_{15}H_{11}O_2N$
General characteristics[19]	Colorless crystals from ethanol, mp 269°C $[\alpha]_D^{18}$ 0° (C = 1.0 in chloroform) Sparingly soluble in cold organic solvents and in dilute mineral acids; more soluble in boiling concentrated HCl; stable in concentrated mineral acids. Viridicatol: crystals from ethyl acetate, mp 280°C
UV data[34]	λ_{max}^{EtOH} nm: 225, 240(sh), 287, 309, 317, and 330 (ϵ_{max} not reported) Viridicatol: λ_{max}^{MeOH} nm (ϵ): 226(28,200), 284(8,900), 304(sh)(9,100), 316(10,000), and 329(7,500)
Fungal source[34]	*Penicillium viridicatum* Westling
Toxicity data[34]	Antibiotic activity was reported *in vitro* against *Mycobacterium tuberculosis* at a 1:15,000 dilution. It produced no obvious effects against *E. coli*. Vertebrate toxicity is unknown.
Other significant references	Mass spectroscopy[71]
^{13}C NMR data	C2, 157.9 s; C3, 123.7 s; C4, 132.9 s*; C5, 124.0 d; C6, 121.8 d; C7, 126.1 d; C8, 115.0 d; C9, 142.1 s; C10, 120.6 s; C1′, 133.5 s*; C2′,6′, 128.0 d; C3′,5′, 129.5 d; C4′, 127.3 d

* Assignments may be reversed.

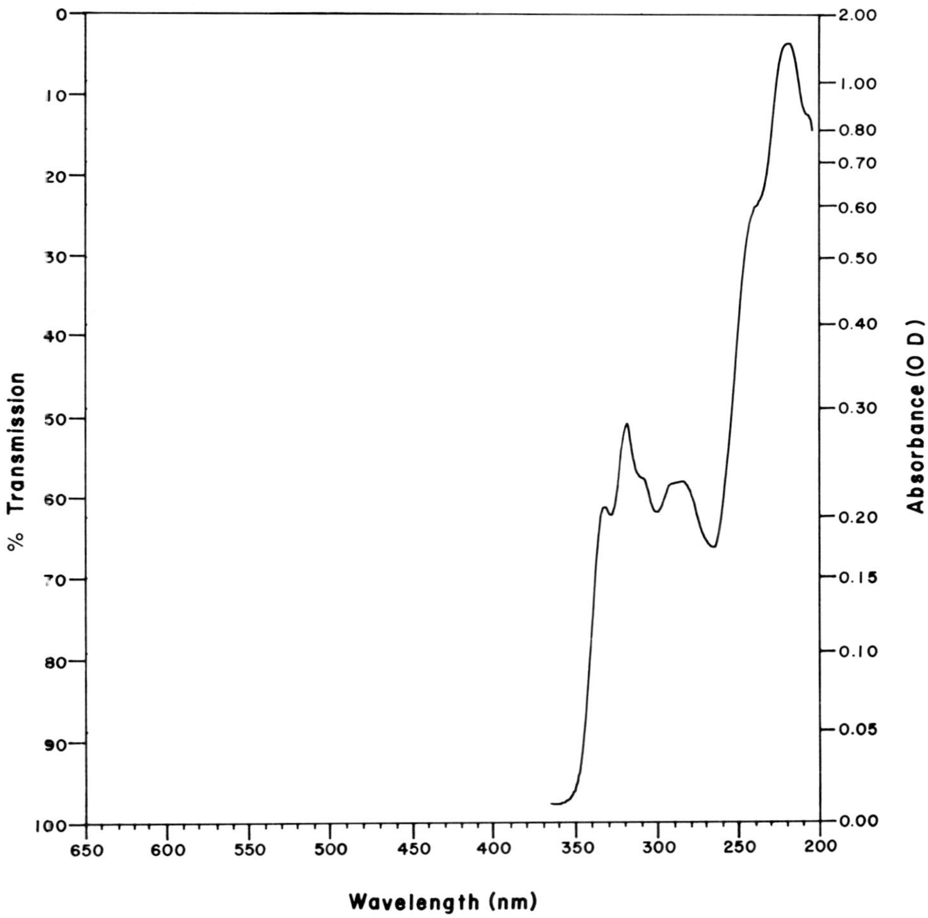
% Transmission
Absorbance (O D)
Wavelength (nm)
0
10
20
30
40
50
60
70
80
90
100
2.00
1.00
0.80
0.70
0.60
0.50
0.40
0.30
0.20
0.15
0.10
0.05
0.00
650
600
550
500
450
400
350
300
250
200

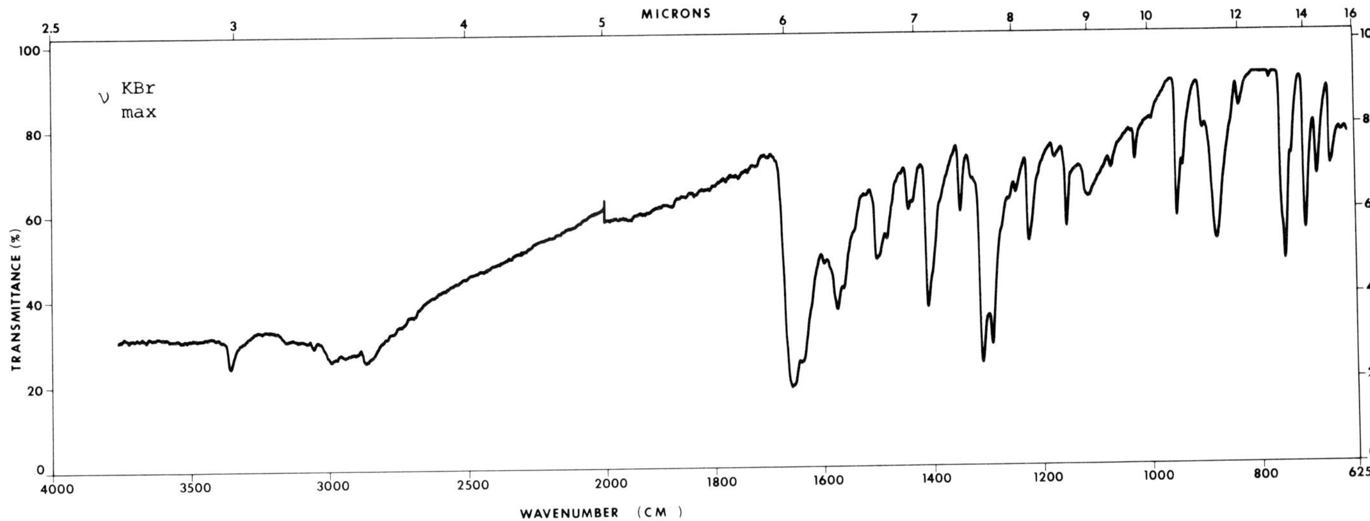
MICRONS
2.5
3
4
5
6
7
8
9
10
12
14
16
TRANSMITTANCE (%)
100
80
60
40
20
0
ν KBr max
4000
3500
3000
2500
2000
1800
1600
1400
1200
1000
800
625
WAVENUMBER (CM)

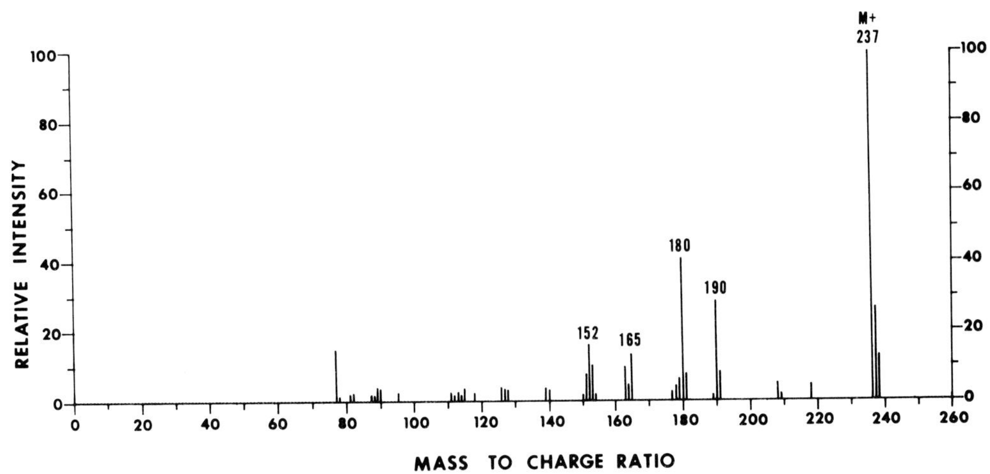
RELATIVE INTENSITY
100
80
60
40
20
0
M+ 237
180
190
152
165
0
20
40
60
80
100
120
140
160
180
200
220
240
260
MASS TO CHARGE RATIO

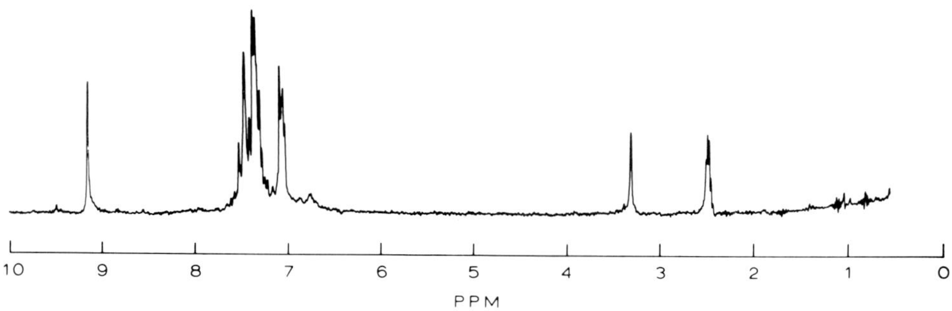

The 100 MHz ^{1}H NMR spectrum of viridicatin in DMSO solution.

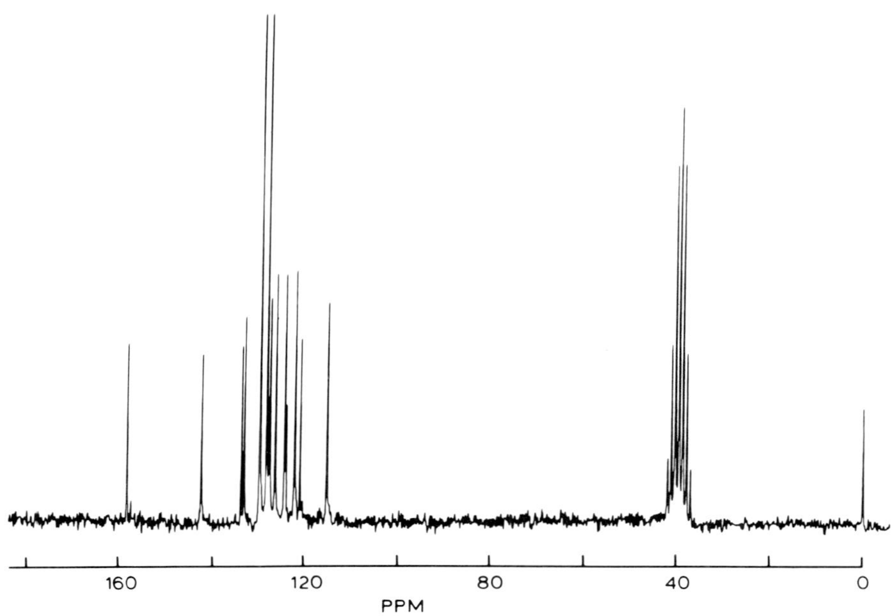

The 25 MHz ^{13}C NMR spectrum of viridicatin in DMSO solution.

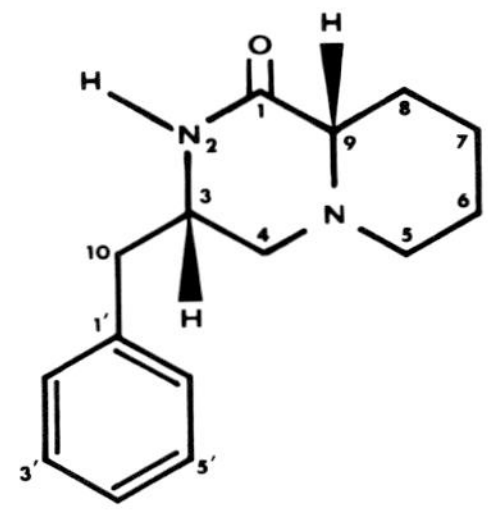

Common name — Verruculotoxin (3-Toluyloctahydro-2*H*-pyrido-[1,2-*a*]pyrazine)

Molecular weight — 244.1575

Molecular formula — $C_{15}H_{20}ON_2$

General characteristics[31,68,69] — Crystals from benzene, mp 152°C
Soluble in chloroform, acetone, ethyl acetate, and ethyl ether; sparingly soluble in toluene, benzene, methanol, and ethanol; and insoluble in hexane and water

UV data[31] — λ_{max}^{MeOH} nm: 240–260 (weak) (ϵ_{max} not reported)

Fungal source[31] — *Penicillium verruculosum* Peyronel

Toxicity data[31,39] — The LD_{50} to day-old cockerels dosed orally was approximately 20 mg/kg or 800 μg per cockerel. An LD_{100} was observed at 25 mg/kg. Acute toxicity was expressed initially by ataxia, prostration, and a complete lack of muscular coordination. Animals appeared to be dead but opened their eyes if disturbed. Later signs of animals receiving a lethal dose were stupor, coma, and death within 4–6 hr after treatment.

Verruculotoxin was shown to act directly on both isolated mammalian and amphibian skeletal muscles to potentiate twitch tension. Twitch tensions were enhanced to 150% of controls.

TLC data[31] — *Adsorbent:* silica gel G-HR
Solvent: chloroform–acetone, 93:7 v/v
R_f: 0.26
Detection: melon-colored spot after ninhydrin spraying and heating at 100°C for 5 min

Other significant references	Structure and synthesis[70]
^{13}C NMR data	C1, 170.7 s; C3, 52.4 d; C4, 55.6 t*; C5, 54.1 t*; C6, 24.6 t*; C7, 25.2 t*; C8, 27.3 t; C9, 65.5 d; C10, 42.2 t; C1′, 137.7 s; C2′,6′, 128.9 d†; C3′,5′, 128.4 d†; C4′, 127.9 d
^{13}C NMR data (verruculotoxin tyrosine analog)	C1, 170.6 s; C3, 52.4 d; C4, 53.6 t*; C5, 55.5 t*; C6, 24.6 t‡; C7, 25.2 t‡; C8, 27.3 t; C9, 65.4 d; C10, 41.1 t; C1′, 128.4 s; C2′,6′, 129.9 d; C3′,5′, 115.3 d; C4′, 155.6 s

* Assignments may be reversed.

† Assignments may be reversed.

‡ Assignments may be reversed.

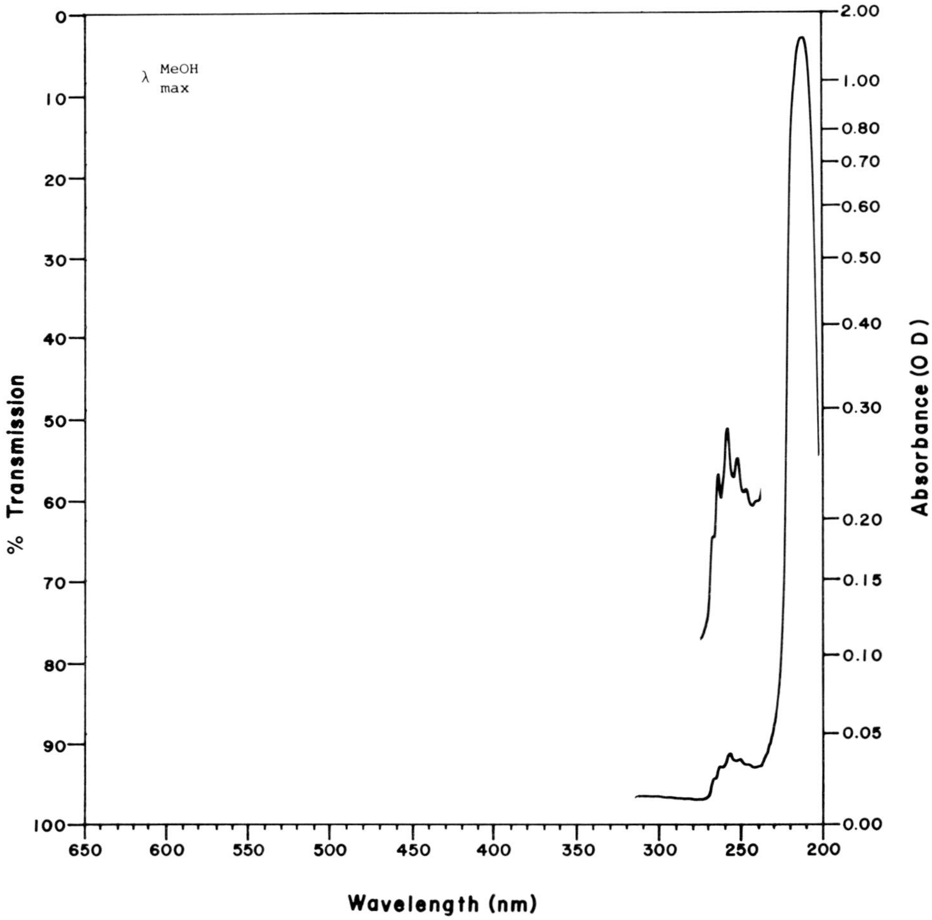
λ MeOH max
% Transmission
0
10
20
30
40
50
60
70
80
90
100
Absorbance (O D)
2.00
1.00
0.80
0.70
0.60
0.50
0.40
0.30
0.20
0.15
0.10
0.05
0.00
650
600
550
500
450
400
350
300
250
200
Wavelength (nm)

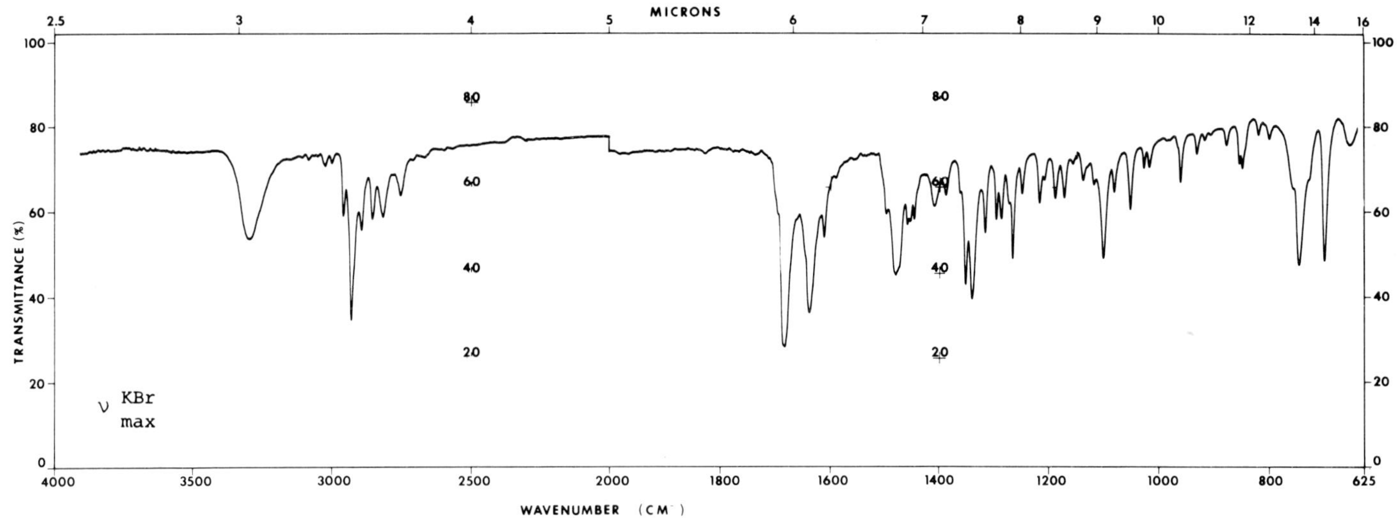
MICRONS
TRANSMITTANCE (%)
WAVENUMBER (CM⁻¹)
ν KBr max

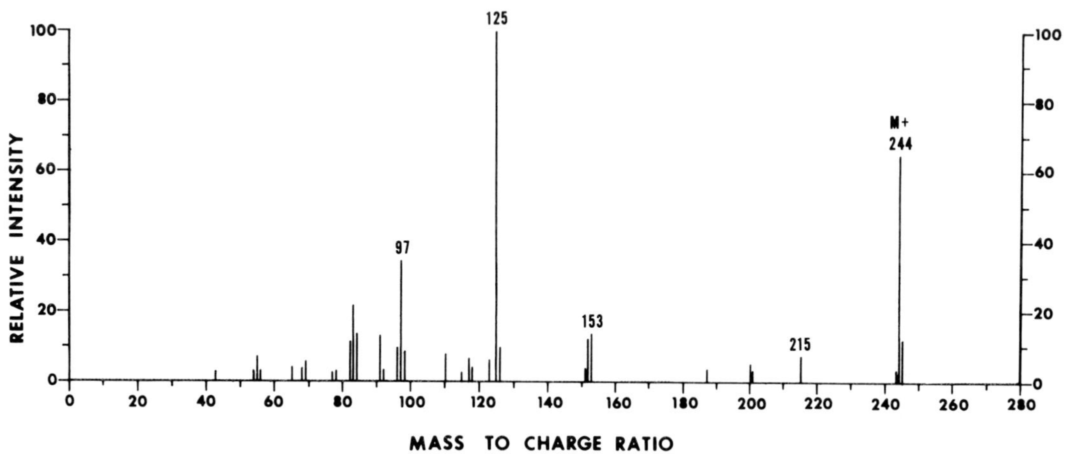
125
97
153
215
M+
244
RELATIVE INTENSITY
MASS TO CHARGE RATIO

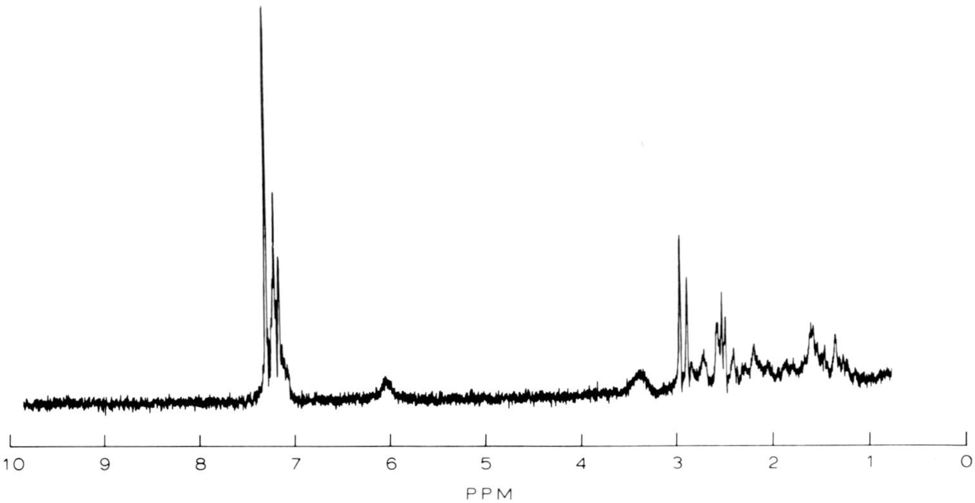

The 100 MHz ^{1}H NMR spectrum of verruculotoxin in $CDCl_3$ solution.

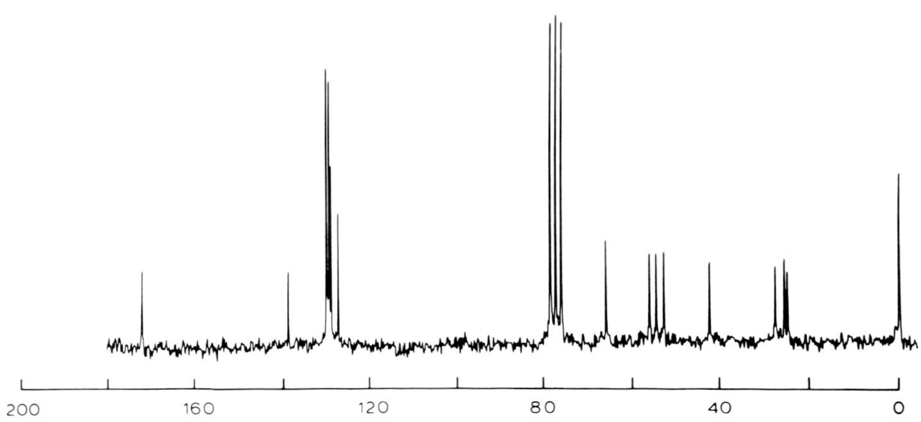

The 25 MHz ^{13}C NMR spectrum of verruculotoxin in $CDCl_3$ solution.

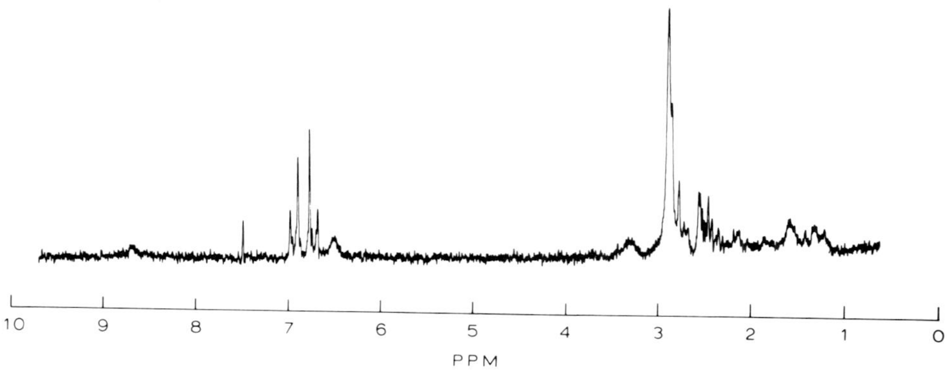

The 100 MHz ^{1}H NMR spectrum of verruculotoxin tyrosine analog in $CDCl_3$/DMSO solution.

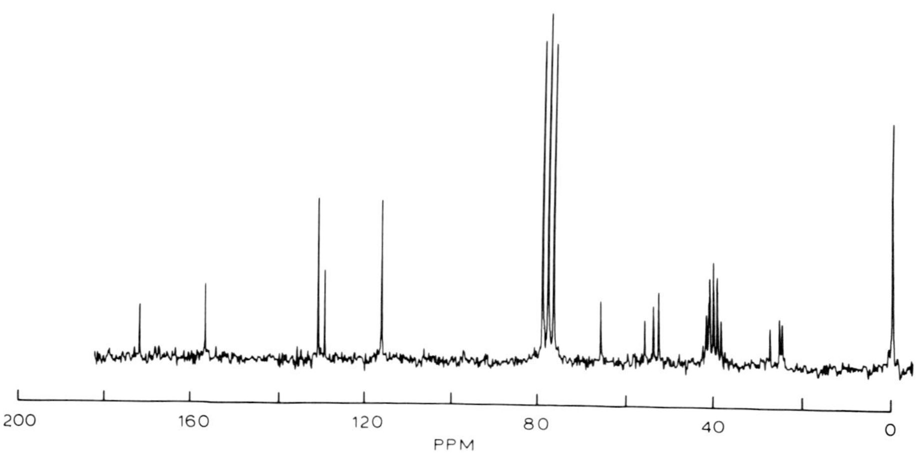

The 25 MHz ^{13}C NMR spectrum of verruculotoxin tyrosine analog in $CDCl_3$/DMSO solution.

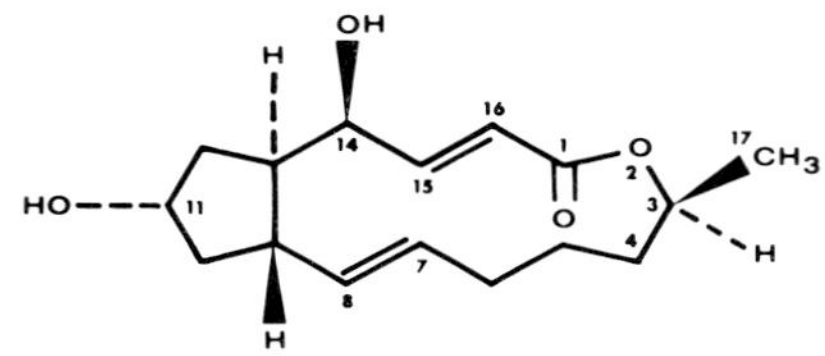

Common name[10,11,13,48,92,94]	Decumbin (Brefeldin A, Ascotoxin, Cyanein)
Molecular weight	280.1674
Molecular formula	$C_{16}H_{24}O_4$
General characteristics[10,11,13,48,92,94]	Crystals from acetone, mp 203°C $[\alpha]_D^{25}$ +91.7° (C = 1.92 in methanol)
UV data[32,48]	λ_{max}^{EtOH} nm (ϵ): 215(11,200)
Fungal source[7,10,13,48,92,94]	*Penicillium brefeldianum* *P. decumbens* Thom *P. cyaneum* *P. simplicissimum* *Ascochyta imperfecta* Peck *Curvularia lunata*
Toxicity data[4,6–10,12,32,92,94]	The LD_{50} of decumbin dosed orally to rats was 275 mg/kg. The toxic signs were a rapid onset of loss of appetite and diarrhea followed by increasing lethargy and cyanosis, labored breathing, stupor, and death in approximately 24 hr. It was also shown to be toxic to goldfish, inhibitory to wheat germination, and antibiotic to *Candida albicans* and other fungi. It inhibited growth of Allium root tips 90% at concentrations of 100 μg/ml and depressed mitotic activity in *Vicia faba* root tips. It strongly inhibited the growth of lucerne and rape seedlings at 20 ppm. The oral LD_{50} in day-old cockerels was 62 mg/kg. Decumbin shows activity against HeLa cells and nematodes.
TLC data[32]	*Adsorbent:* silica gel G-HR *Solvent:* toluene–ethyl acetate–formic acid, 5:4:1 v/v/v R_f: 0.45

Detection: blue spot with phosphomolybdic acid spraying and heating at 100°C for 5 min; H_2SO_4 spraying and heating shows rust-colored spot in normal light; fluorescent rust-colored spot under UV light.

Other significant references

Identification[90]; biosynthesis[33,66]; synthesis[5]

^{13}C NMR data

C1, 165.4 s; C3, 70.4 d; C4, 33.4 t; C5, 26.3 t; C6, 31.4 t; C7, 129.0 d; C8, 136.9 d; C9, 40.8 d; C10, 43.1 t*; C11, 70.7 d†; C12, 43.4 t*; C13, 51.6 d; C14, 74.3 d†; C15, 154.0 d; C16, 116.2 d; C17, 20.6 q

* Assignments may be reversed.
† Assignments may be reversed.

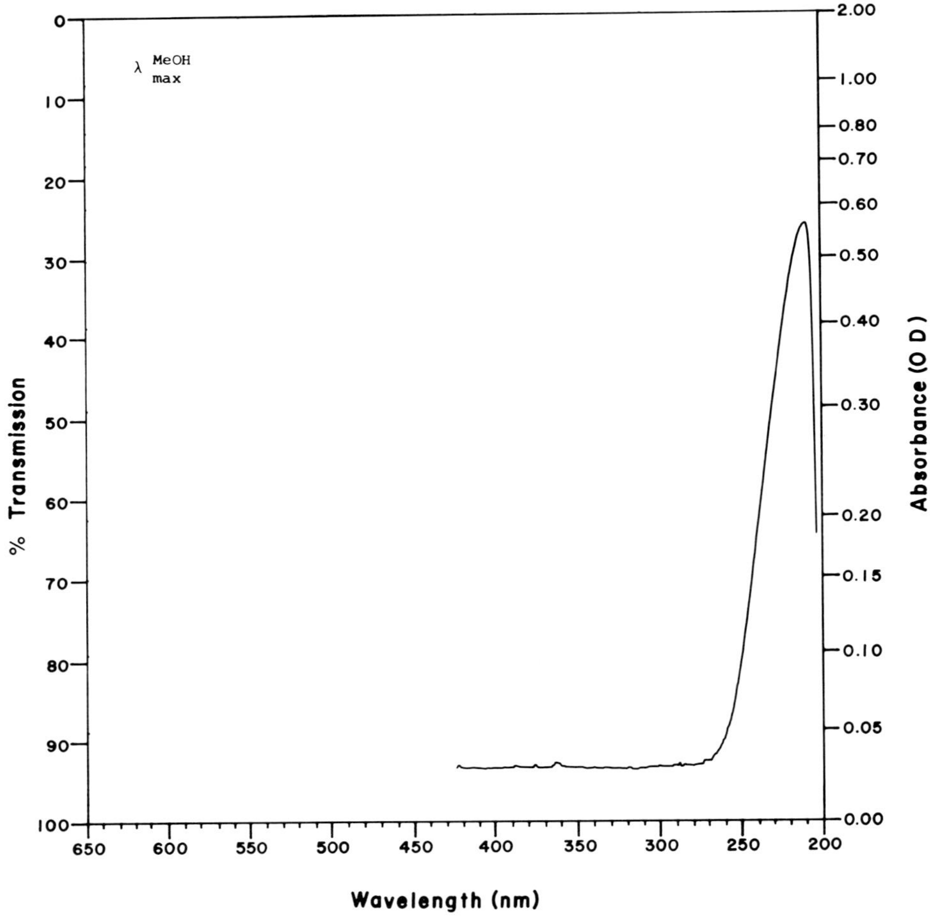

λ MeOH max
% Transmission
0
10
20
30
40
50
60
70
80
90
100
2.00
1.00
0.80
0.70
0.60
0.50
0.40
0.30
0.20
0.15
0.10
0.05
0.00
Absorbance (O D)
650
600
550
500
450
400
350
300
250
200
Wavelength (nm)

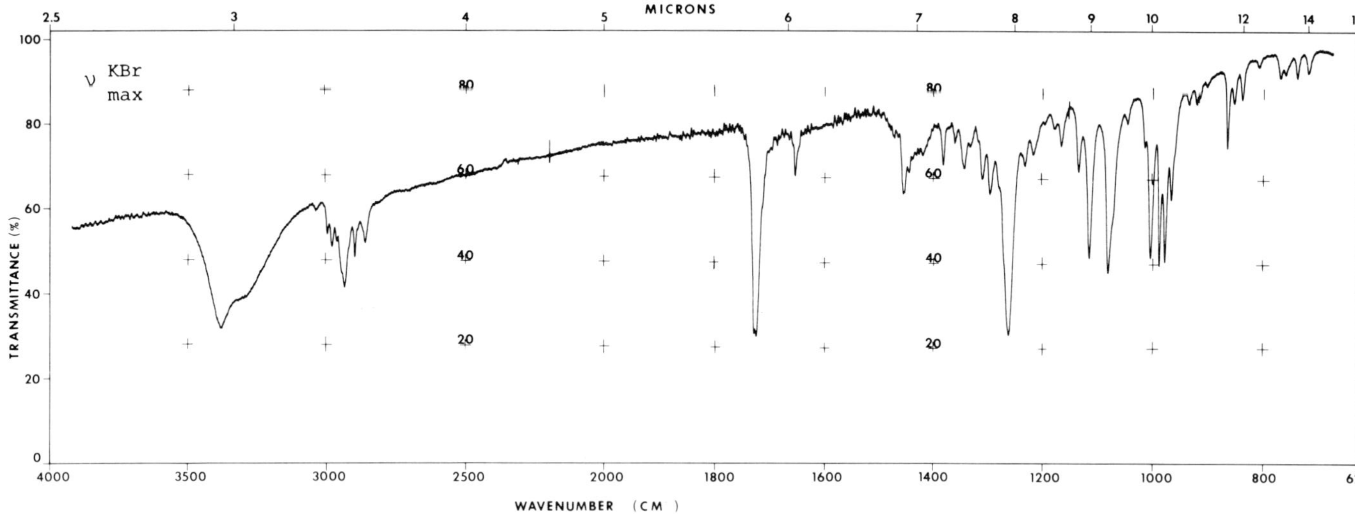
MICRONS
ν KBr max
TRANSMITTANCE (%)
WAVENUMBER (CM)

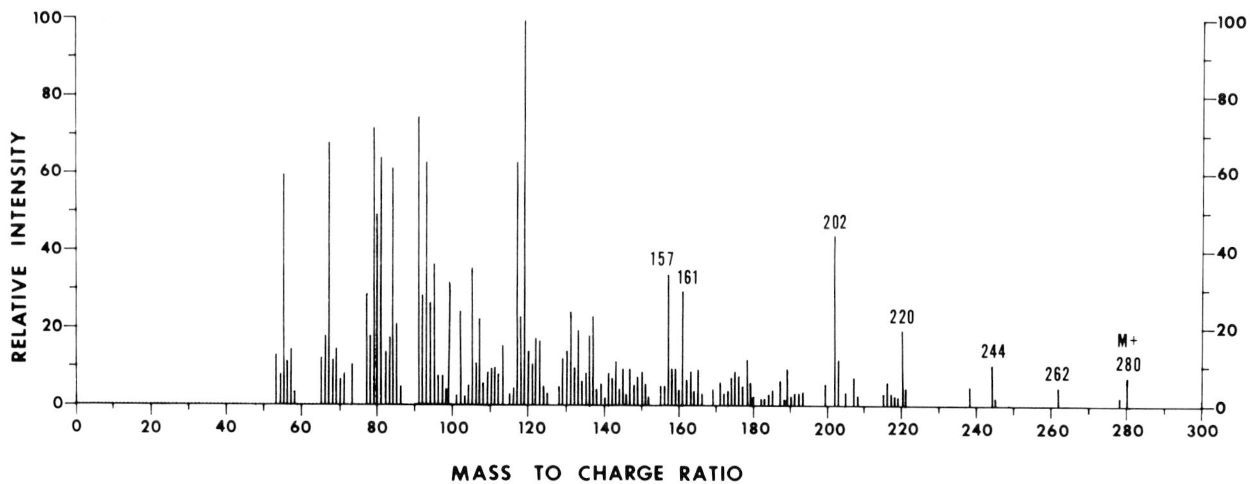
RELATIVE INTENSITY
157
161
202
220
244
262
M+
280
MASS TO CHARGE RATIO

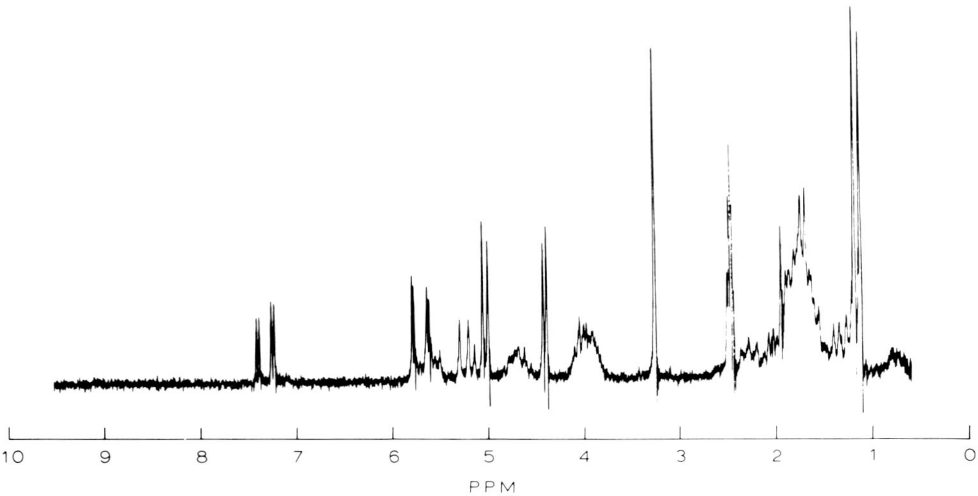

The 100 MHz ^{1}H NMR spectrum of decumbin in DMSO solution.

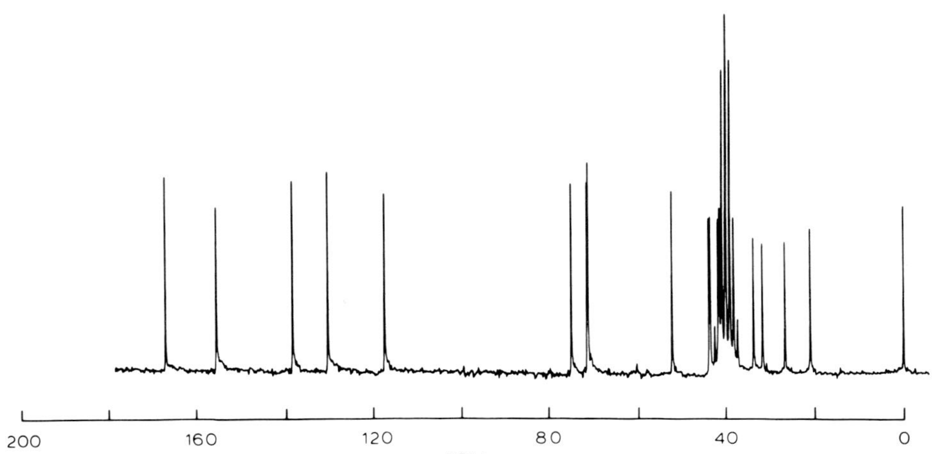

The 25 MHz ^{13}C NMR spectrum of decumbin in DMSO solution.

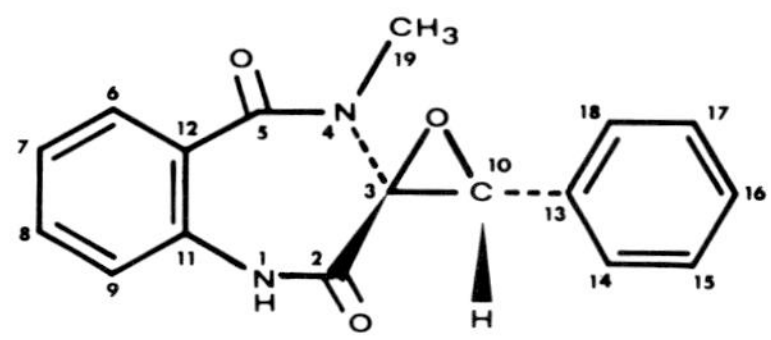

Common name[19]	Cyclopenin (*trans*-3,10-Epoxycyclo[anthranoyl-*N*-methylphenylalanyl])
Molecular weight	294.1004
Molecular formula	$C_{17}H_{14}O_3N_2$
General characteristics[19]	Cyrstals from ethyl acetate, mp 207°C forms hydrate from methanol, mp 110°–120°C (sample contains ½ H_2O of crystallization) $[\alpha]^{20}_{5790}$ −263° (C = 1.0 in ethanol) $[\alpha]^{23}_{5461}$ −301° (C = 1.0 in methanol)
UV data[19]	λ^{MeOH}_{max} nm (ϵ): 211(37,200) and 290(2,060)
Fungal source[19,74]	*Penicillium cyclopium* Westling *P. viridicatum* Westling
Toxicity data[19]	Vertebrate toxicity unknown; it shows slight antibiotic activity against *Micrococcus pyogenes* var. *aureus* and *Escherichia coli.*
Other significant references	Synthesis[70,85]; biosynthesis[41]; mass spectroscopy[71]
^{13}C NMR data	C2, 167.4 s*; C3, 70.2 s; C5, 165.8 s*; C6, 131.3 d†; C7, 125.0 d†; C8, 132.3 d‡; C9, 120.7 d‡; C10, 64.7 d; C11, 130.0 s; C12, 126.7 s; C13, 134.0 s; C14,18, 125.8 d; C15,17, 127.9 d; C16, 128.9 d; C19, 31.4 q

* Assignments may be reversed.
† Assignments may be reversed.
‡ Assignments may be reversed.

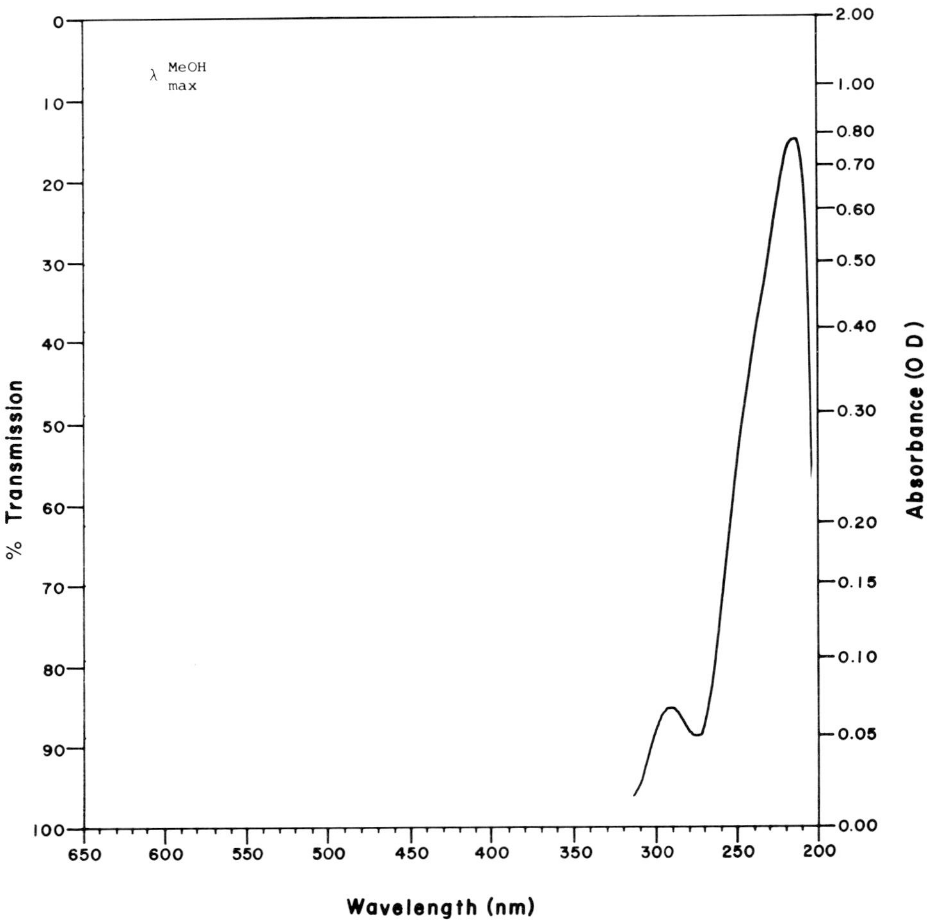

λ MeOH max
% Transmission
Absorbance (O D)
0
10
20
30
40
50
60
70
80
90
100
2.00
1.00
0.80
0.70
0.60
0.50
0.40
0.30
0.20
0.15
0.10
0.05
0.00
650
600
550
500
450
400
350
300
250
200
Wavelength (nm)

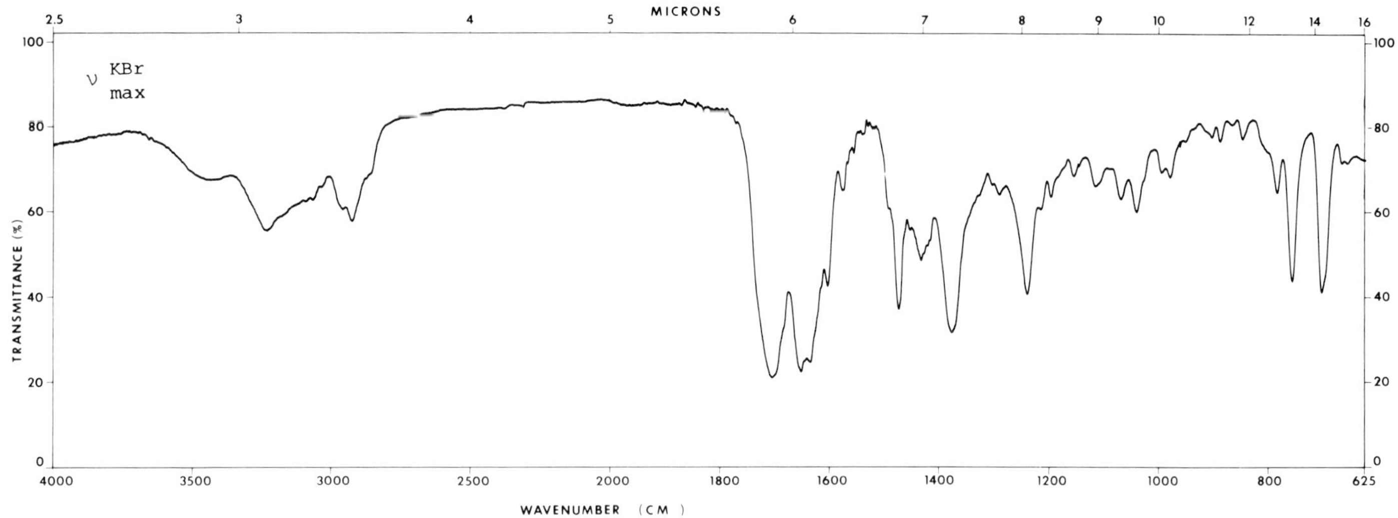
MICRONS
ν KBr max
TRANSMITTANCE (%)
WAVENUMBER (CM)

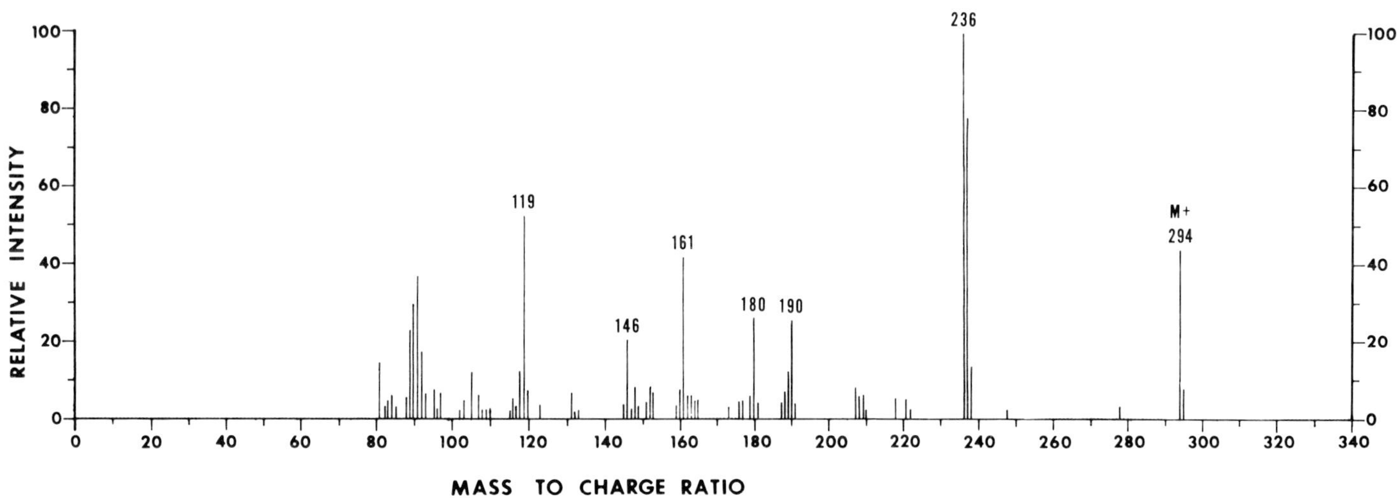
236
119
161
146
180
190
M+
294
RELATIVE INTENSITY
MASS TO CHARGE RATIO

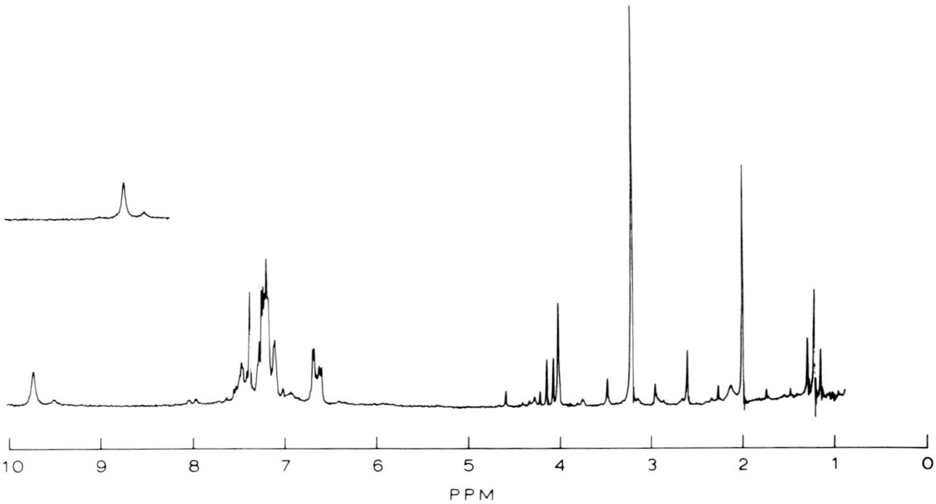

The 100 MHz ^{1}H NMR spectrum of cyclopenin in $CDCl_3$ solution.

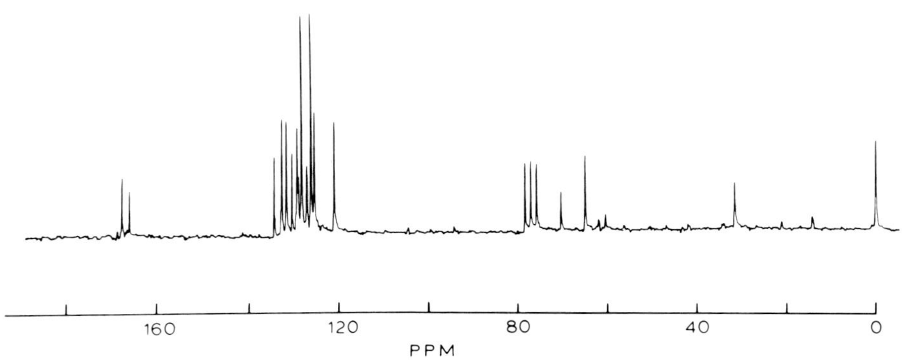

The 25 MHz ^{13}C NMR spectrum of cyclopenin in $CDCl_3$ solution.

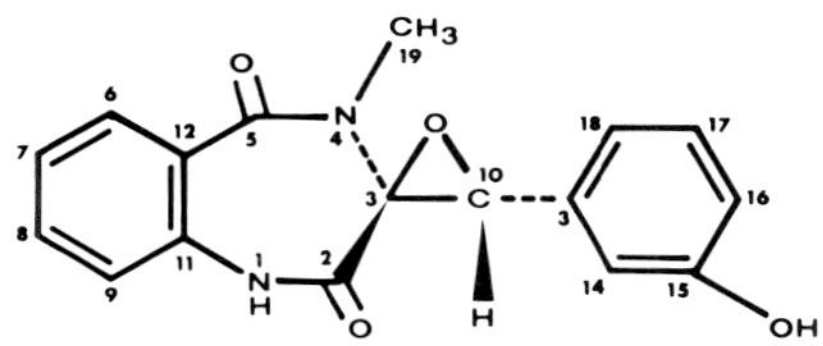

Common name	Cyclopenol (*trans*-3,10-Epoxycyclo[anthranoyl-*N*-methyl-15-hydroxylphenylalanyl])
Molecular weight	310.0953
Molecular formula	$C_{17}H_{14}O_4N_2$
General characteristics[17,74]	Prisms from ethyl acetate–benzene, mp 215°C (dec.) $[\alpha]^{20}_{5461}$ −309° (*C* = 1.3 in methanol) Soluble in ethanol, hot ethyl acetate; sparingly soluble in chloroform; insoluble in carbon tetrachloride, benzene, and ether.
UV data[70]	λ_{max}^{MeOH} nm (ϵ): 285(3,740)
Fungal source[17,74]	*Penicillium cyclopium* Westling (NRRL 3557) *P. viridicatum* Westling
Toxicity data	Toxicity unknown
^{13}C NMR data	C2, 166.9 s; C3, 70.4 s; C5, 166.2 s; C6, 131.2 d*; C7, 124.6 d†; C8, 132.2 d*; C9, 120.9 d†; C10, 64.7 d; C11, 131.9 s; C12, 127.0 s; C13, 135.0 s; C14, 113.1 d‡; C15, 157.0 s; C16, 116.3 d‡; C17, 129.9 d; C18, 117.3 d‡; C19, 31.2 q

* Assignments may be reversed.
† Assignments may be reversed.
‡ Assignments may be reversed.

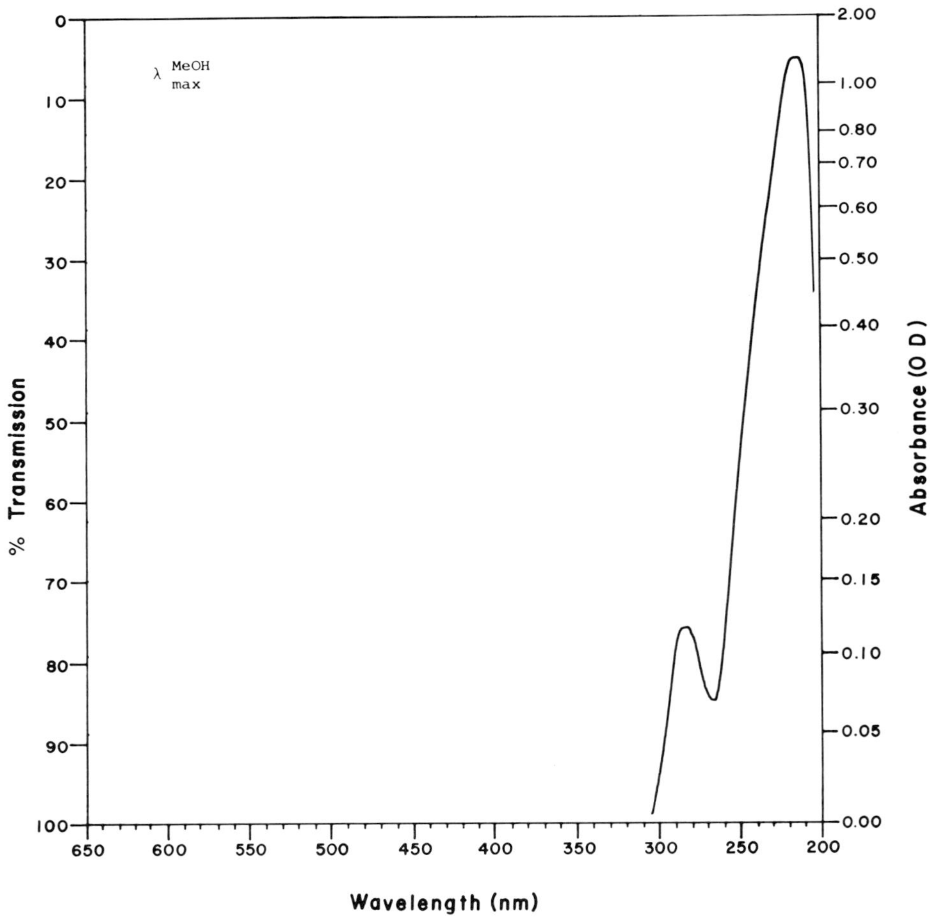
λ MeOH max
% Transmission
0
10
20
30
40
50
60
70
80
90
100
Absorbance (O D)
2.00
1.00
0.80
0.70
0.60
0.50
0.40
0.30
0.20
0.15
0.10
0.05
0.00
650
600
550
500
450
400
350
300
250
200
Wavelength (nm)

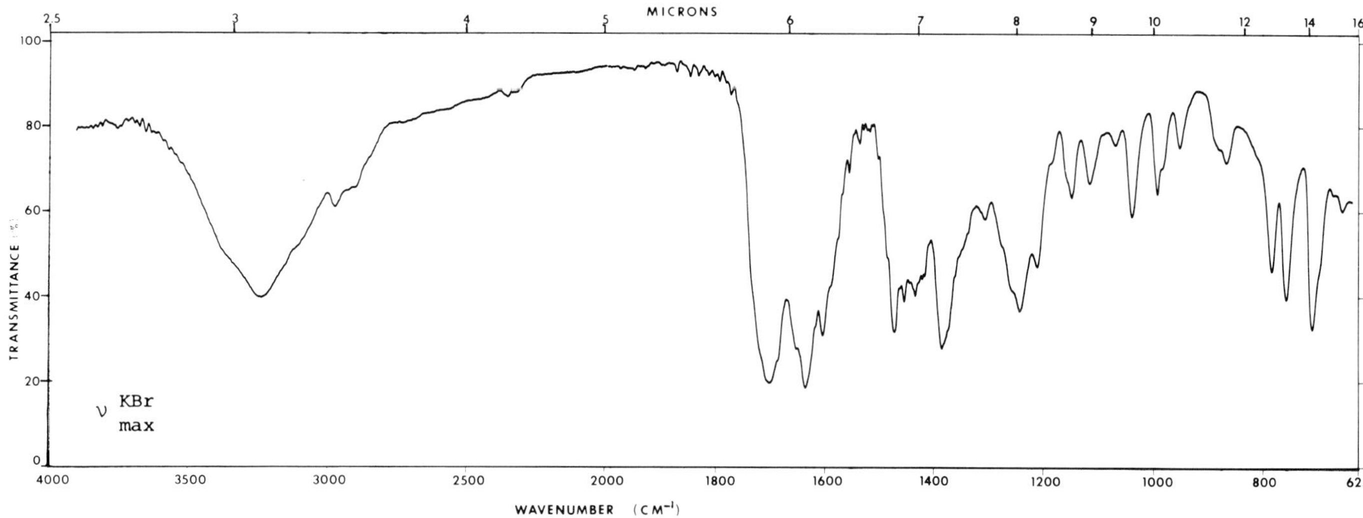
MICRONS
2.5
3
4
5
6
7
8
9
10
12
14
16
100
80
60
40
20
0
TRANSMITTANCE
ν KBr max
4000
3500
3000
2500
2000
1800
1600
1400
1200
1000
800
625
WAVENUMBER (CM⁻¹)

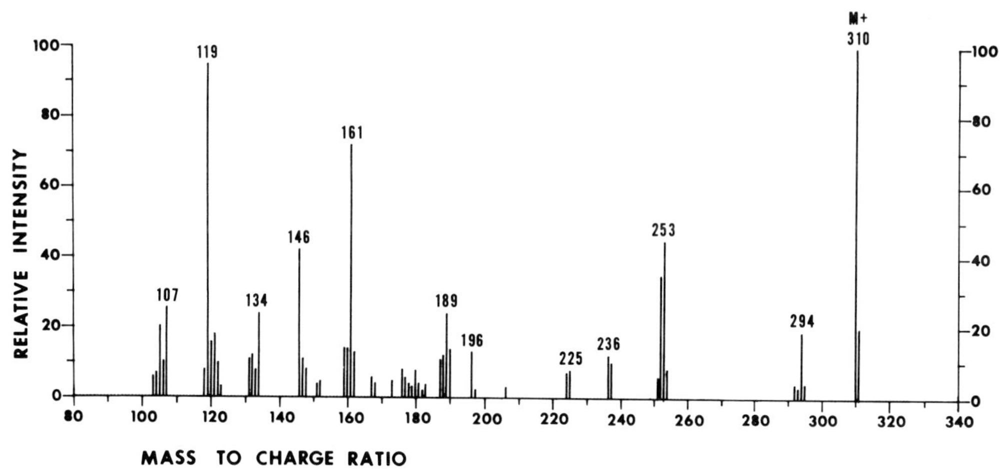
RELATIVE INTENSITY
100
80
60
40
20
0
119
161
146
107
134
189
196
225
236
253
294
M+
310
80
100
120
140
160
180
200
220
240
260
280
300
320
340
MASS TO CHARGE RATIO

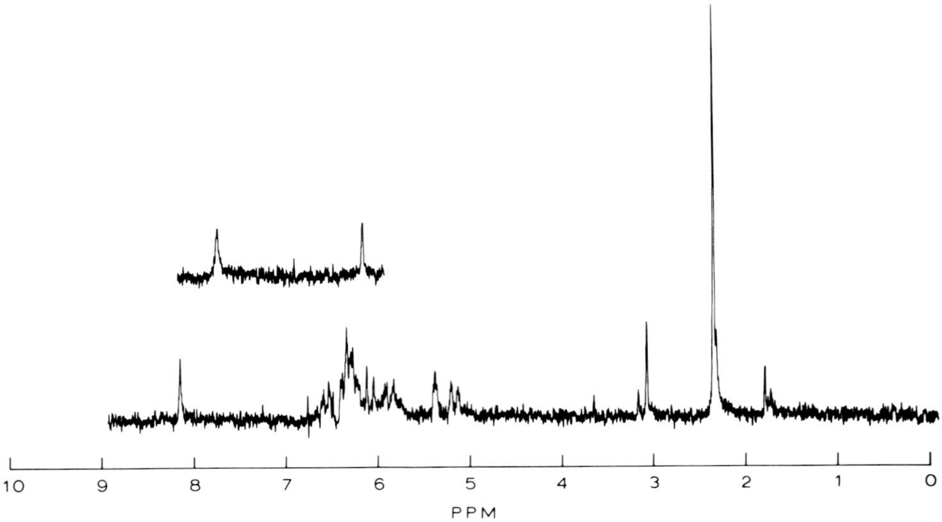

The 100 MHz ^{1}H NMR spectrum of cyclopenol in $CDCl_3$/DMSO solution.

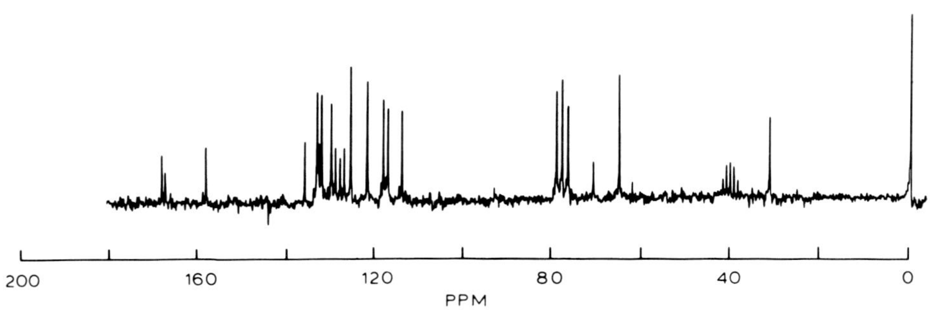

The 25 MHz ^{13}C NMR spectrum of cyclopenol in $CDCl_3$/DMSO solution.

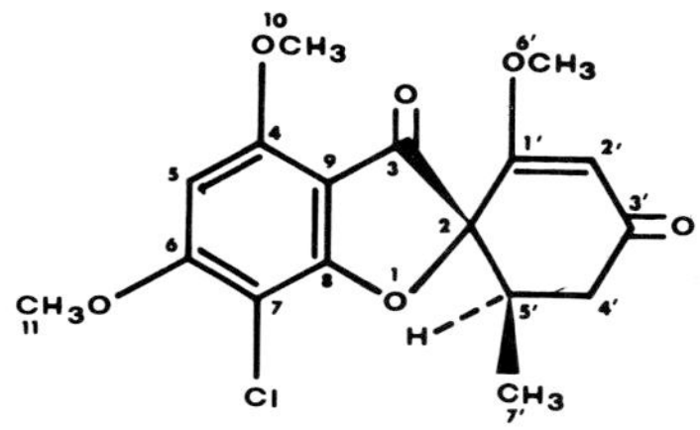

Common name	Griseofulvin ("Curling factor")[7-Chloro-4, 6-dimethoxycoumaran-3-one-2-spiro-1′-(2′-methoxy-6′-methylcyclohex-2′-en-3′-one)]
Molecular weight	352.0713
Molecular formula	$C_{17}H_{17}O_6Cl$
General characteristics[46,72]	mp 220°–221°C $[\alpha]^{21}_{D}$ +337° (C = 1.0 in acetone) $[\alpha]^{19}_{5790}$ +354° $[\alpha]^{19}_{5461}$ +417° (C = 1.0127 in acetone)
UV data[46]	λ^{MeOH}_{max} nm (ϵ): 324(5,200), 291(21,800), 252(12,500), and 236(21,300)
Fungal source[20,72,83,104]	*Penicillium griseofulvum* *P. janczewskii* (*P. nigricans*) *P. patulum* *P. albidum* *P. raciborskii* *P. melinii* *P. urticae* *P. raistrickii* *P. brefeldianum* *P. viridi-cyclopium* *P. brunneo-stoloniferum*
Toxicity data[3,42,57,89,104]	Griseofulvin has been used effectively in the treatment of dermatomycoses in animals and man. The LD_{50} in rats (IV) was 500 mg/kg. At doses of 100 and 200 mg/kg in rats, a marked effect on mitosis was noted in the bone marrow, intestine, and transplanted tumor cells. Mitosis was blocked at metaphase similar to the effects of colchicine in *Allium* root tips. Other

toxic signs in animals included, in skin: angioneurotic edema, erythema, urticaria, vesicular and macular eruptions, photosensitivity, and *Lichen planus* eruptions; other: transient leukopenia, granulocytopenia, punctate basophilia, monocystosis, blurred vision, headaches, disorientation, vertigo, anorexia, nausea, vomiting, diarrhea, thirst, and fatigue. Griseofulvin has also been reported to be teratogenic in rats and cats.

TLC data[30]

Adsorbent: silica gel G-HR
Solvent: chloroform–acetone, 93:7 v/v
R_f: 0.65
Detection: 50% ethanolic H_2SO_4 spraying and heating yields a bright blue fluorescent spot under UV light

GLC data[30]

Solid support: AnaKrom ABS (80 mesh)
Liquid phase: 1% QF-1 and 2% SE-30
Relative retention time (relative to cholestane): 0.81 and 6.31, respectively

Other significant references

Chemistry[43,46,47]; chemical structure[45]; synthesis[36]; X-ray crystallograhy[23]; biosynthesis[49,88]; ^{13}C NMR[60,91]; review[44]

^{13}C NMR data

C2, 90.0 s; C3, 190.0 s; C4, 157.4 s; C5, 90.6 d; C6, 168.5 s; C7, 95.4 s; C8, 164.3 s; C9, 104.1 s; C10, 57.4 q; C11, 56.4 q; C1′, 170.1 s; C2′, 104.4 d; C3′, 195.1 s; C4′, 39.6 t; C5′, 35.7 d; C6′, 56.8 q; C7′, 13.8 q

^{1}H NMR data

H5, 6.51; H10, 4.03; H11, 4.12; H2′, 5.62; H4′, 2.30–2.90; H5′, 2.30–2.90; H6′, 3.70; H7′, 0.88 (J = 7)

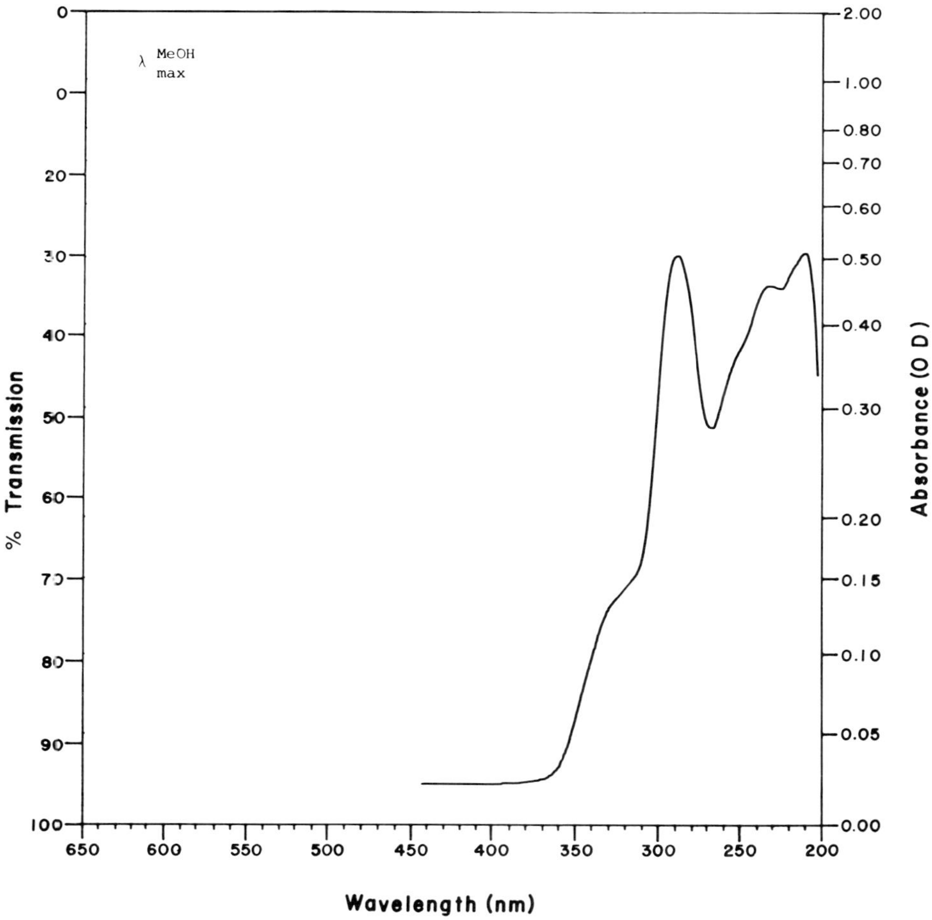

λ MeOH max
% Transmission
Absorbance (O D)
Wavelength (nm)
0
0
20
30
40
50
60
70
80
90
100
2.00
1.00
0.80
0.70
0.60
0.50
0.40
0.30
0.20
0.15
0.10
0.05
0.00
650 600 550 500 450 400 350 300 250 200

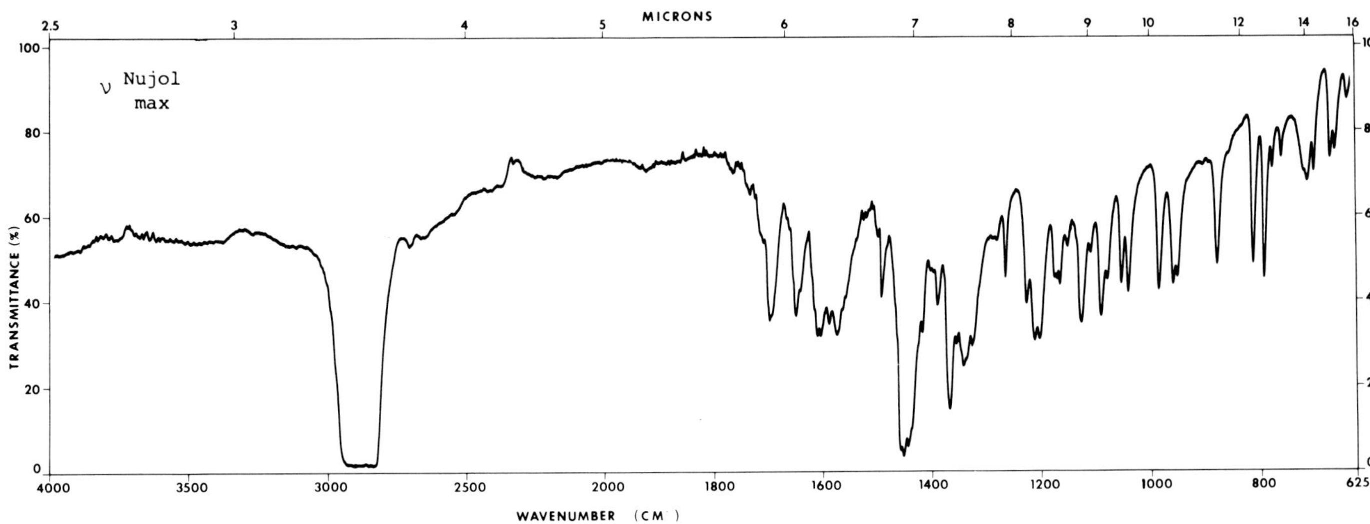
MICRONS
2.5
3
4
5
6
7
8
9
10
12
14
16
100
80
60
40
20
0
ν Nujol max
TRANSMITTANCE (%)
4000
3500
3000
2500
2000
1800
1600
1400
1200
1000
800
625
WAVENUMBER (CM⁻¹)

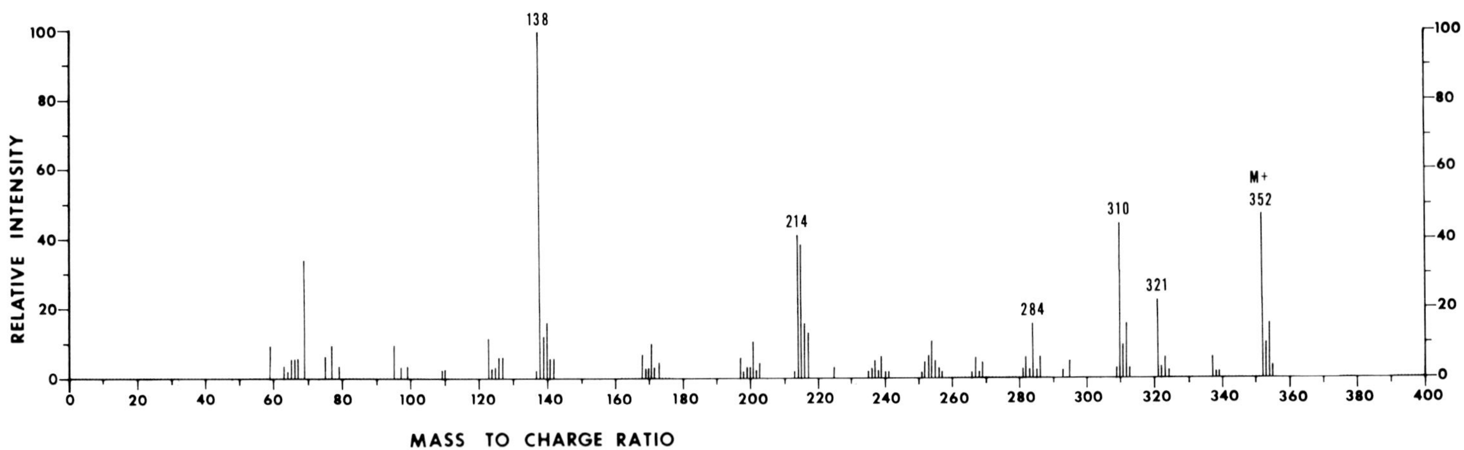
138
214
284
310
321
M+
352
RELATIVE INTENSITY
100
80
60
40
20
0
0
20
40
60
80
100
120
140
160
180
200
220
240
260
280
300
320
340
360
380
400
MASS TO CHARGE RATIO

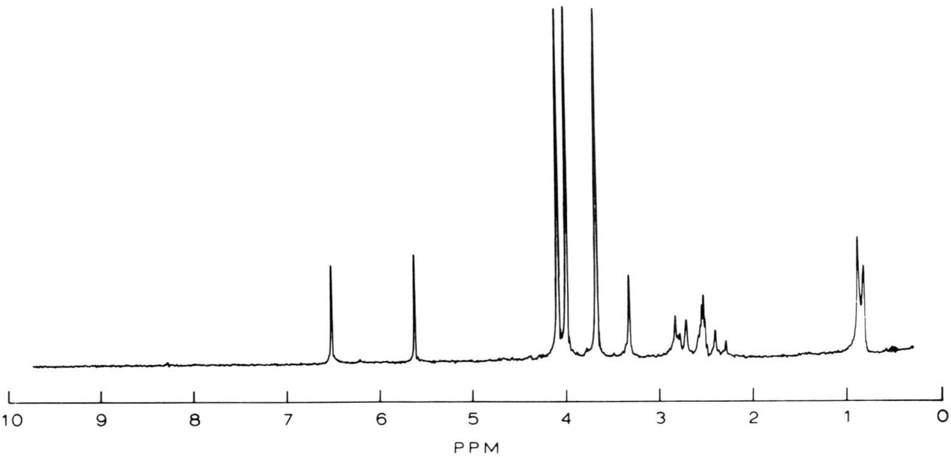

The 100 MHz ^{1}H NMR spectrum of griseofulvin in $CDCl_3$/DMSO solution.

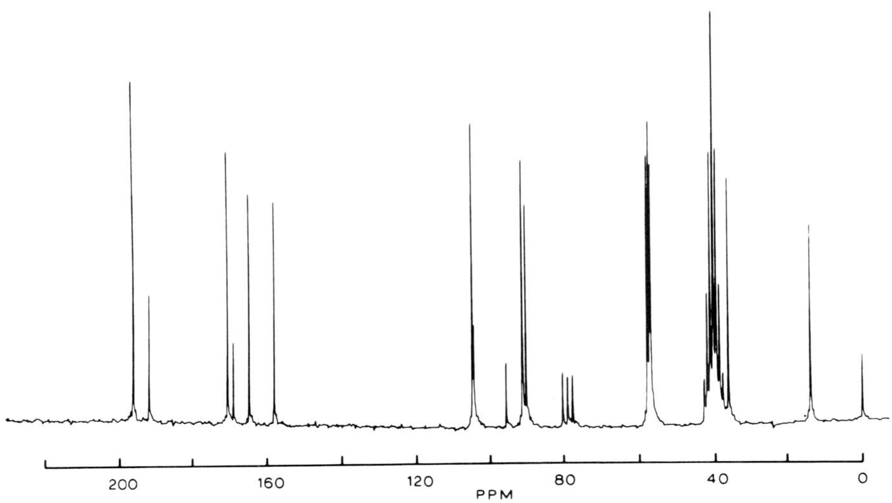

The 25 MHz ^{13}C NMR spectrum of griseofulvin in $CDCl_3$/DMSO solution.

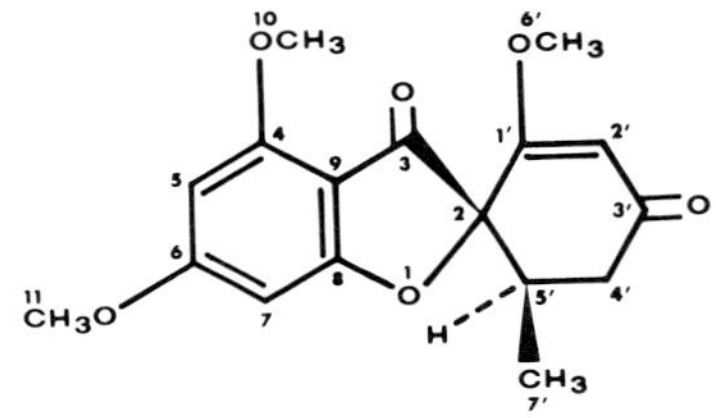

Common name	Dechlorogriseofulvin [4,6-Dimethyoxycoumaran-3-one-2-spiro-1′-(2′-methoxy-6′-methylcyclohex-2′-en-4′-one)]
Molecular weight	318.1103
Molecular formula	$C_{17}H_{18}O_6$
General characteristics[67]	mp 179°–181°C $[\alpha]_D^{19}$ +390° (*C* = 1.0 in acetone)
UV data[30]	λ_{max}^{MeOH} nm (ϵ): 322(4,100), 290(24,500), 253(15,100) and 235(21,400)
Fungal source[30,67]	*Penicillium janczewskii* Zal *P. griseofulvum* Dierckx *P. urticae*
Toxicity data[67]	The biological activity (antifungal) was reported to be a typical griseofulvinlike response in *Botrytis allii* but less active than griseofulvin. Vertebrate toxicity is unknown.
TLC data[30]	*Adsorbent:* silica gel G-HR *Solvent:* chloroform–acetone, 93:7 v/v R_f: 0.65 *Detection:* 50% ethanolic H_2SO_4 spraying and heating yields bright blue fluorescent spot under UV light
GLC data[30]	*Solid support:* AnaKrom ABS (80 mesh) *Liquid phases:* 1% QF-1 and 2% SE-30 *Relative retention time* (relative to cholestane): 0.56 and 4.11, respectively
^{13}*C NMR data*	C2, 89.9 s; C3, 193.0 s; C4, 159.0 s; C5, 88.5 d; C6, 175.9 s; C7, 93.3 d; C8, 170.3 s; C9, 105.6 s; C10, 56.0 q*; C11, 56.0 q*; C1′, 170.1 s; C2′, 104.7 d;

* Assignments may be reversed.

C3′, 197.0 s; C4′, 40.1 t; C5′, 36.6 d; C6′, 56.5 q*; C7′, 14.3 q

^{1}H NMR data H5, 6.20 (J = 2); H7, 6.01 (J = 2); H10, 3.89; H11, 3.90; H2′, 5.52; H4′, 2.36, 3.06; H5′, 2.76; H6′, 3.61; H7′, 0.94

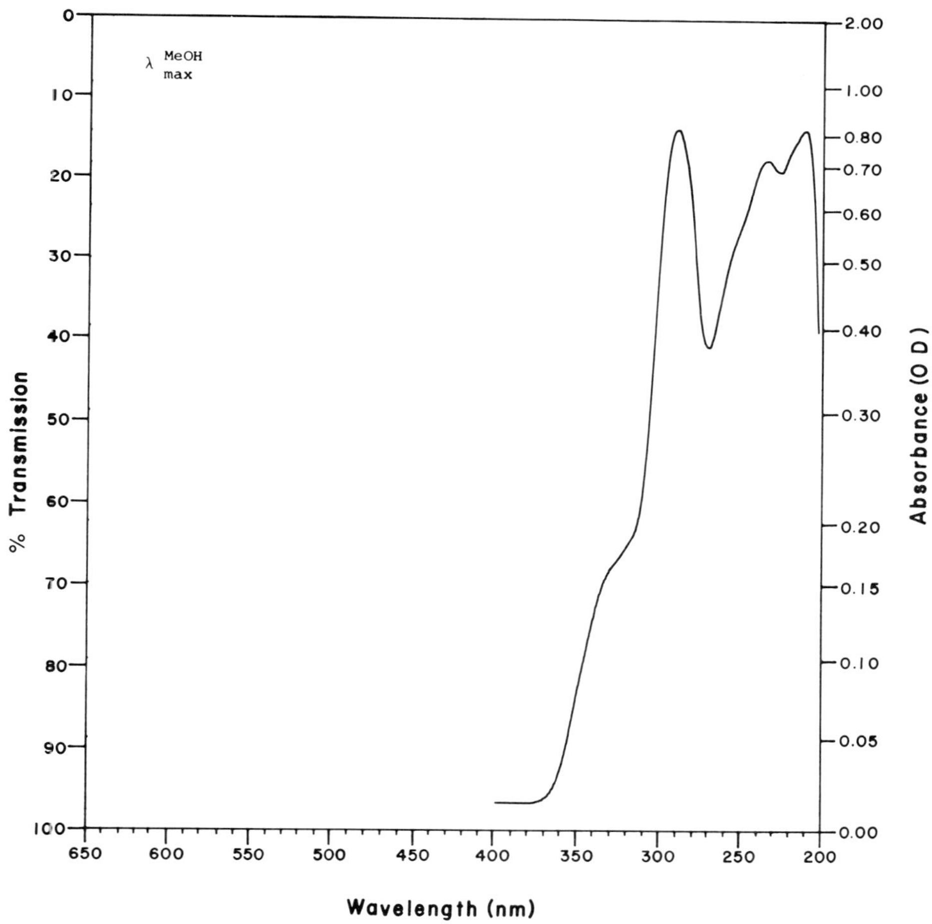

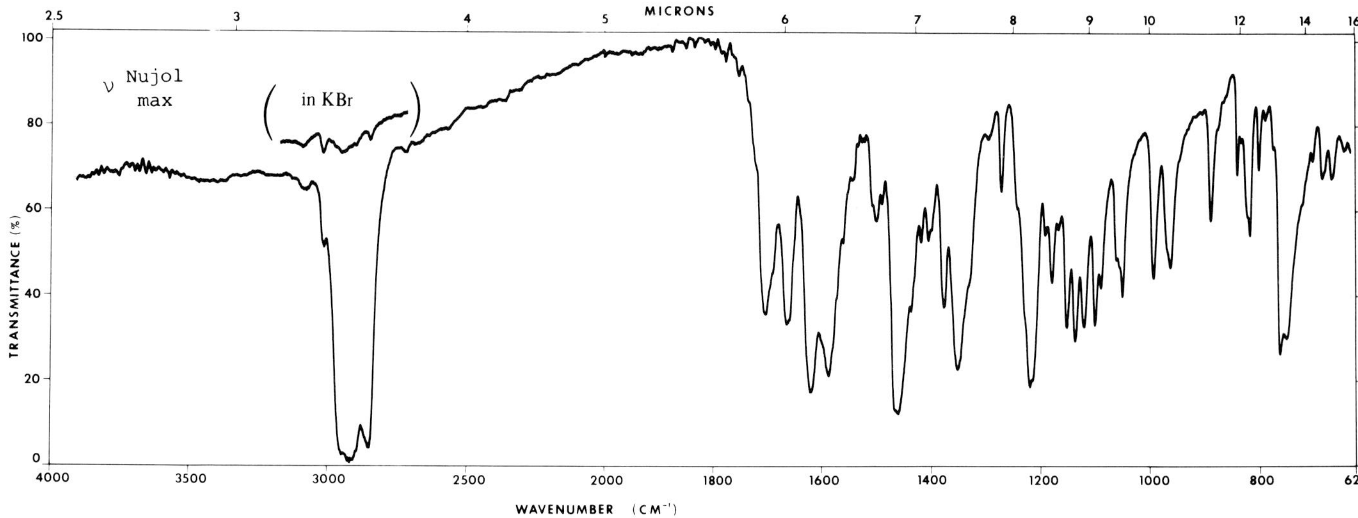
MICRONS
2.5 3 4 5 6 7 8 9 10 12 14 16
TRANSMITTANCE (%)
100 80 60 40 20 0
ν Nujol max
in KBr
4000 3500 3000 2500 2000 1800 1600 1400 1200 1000 800 625
WAVENUMBER (CM⁻¹)

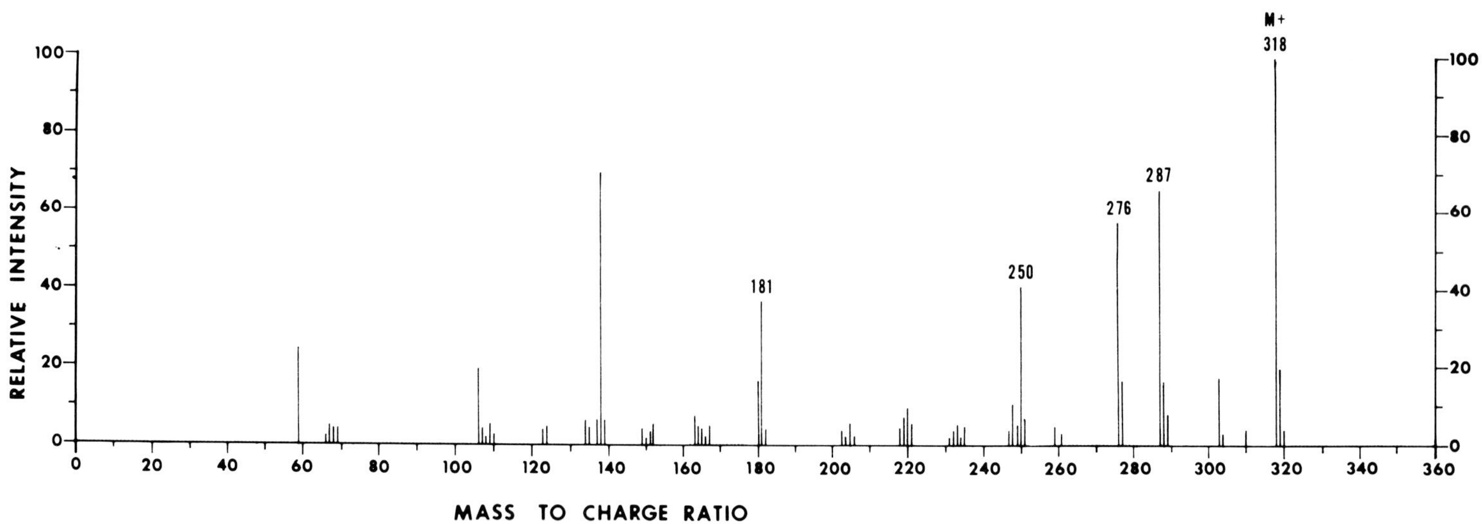
RELATIVE INTENSITY
100 80 60 40 20 0
M+
318
287
276
250
181
0 20 40 60 80 100 120 140 160 180 200 220 240 260 280 300 320 340 360
MASS TO CHARGE RATIO

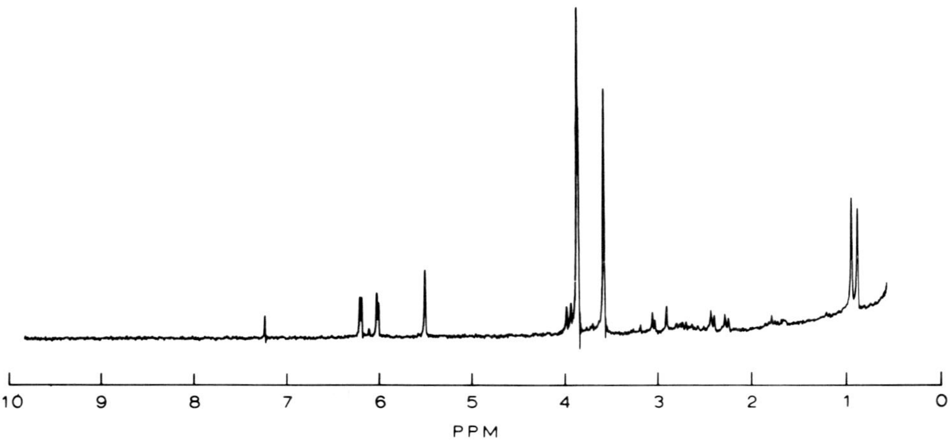

The 100 MHz ^{1}H NMR spectrum of dechlorogriseofulvin in $CDCl_3$ solution.

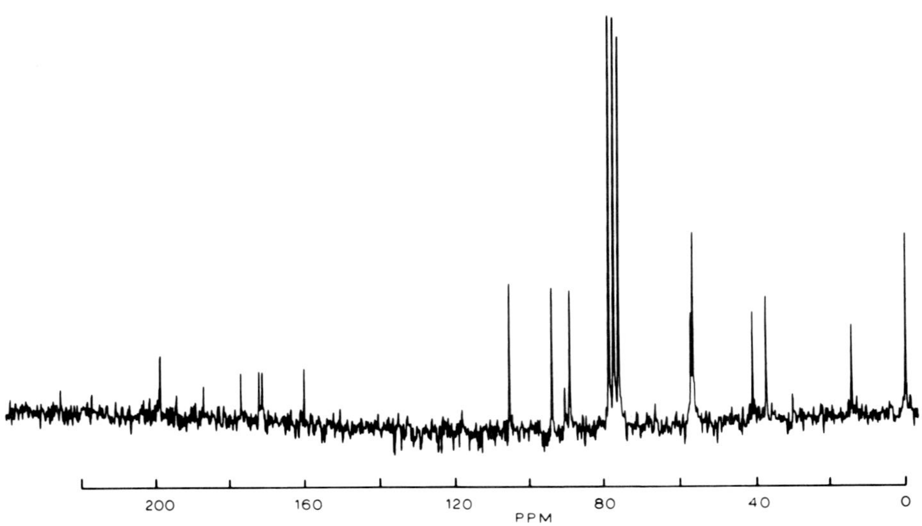

The 25 MHz ^{13}C NMR spectrum of dechlorogriseofulvin in $CDCl_3$ solution.

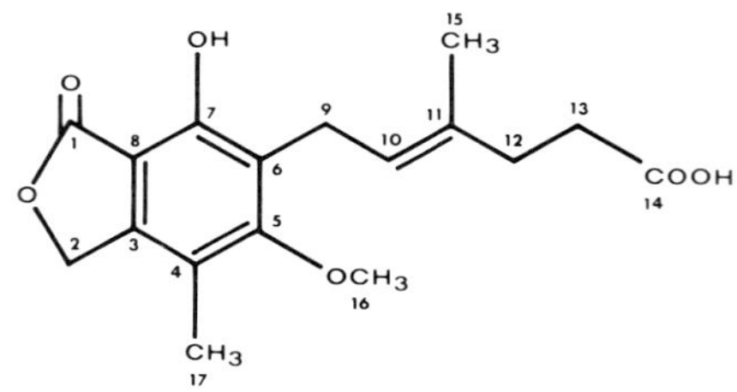

Common name	Mycophenolic acid [6-(4-Hydroxy-6-methoxy-7-methyl-3-oxo-5-phthalanyl)-4-methyl-4-hexenoic acid]
Molecular weight	320.1259
Molecular formula	$C_{17}H_{20}O_6$
General characteristics[28]	Crystals from boiling water or aqueous alcohol, mp 141°C $[\alpha]_D$ 0° Color reactions: blue color with $FeCl_3$ (alcoholic solution); blue-violet with $FeCl_3$ (aqueous solution); heavy precipitate with bromine water; oxidized by permanganate; reduced by palladium and hydrogen. It is a weak acid that is very stable to boiling acids and alkalis; it forms lead, copper, and silver salts.
UV data	λ_{max}^{MeOH} nm: 217, 249, and 304 (ϵ_{max} not reported)
Fungal source[1,40]	*Penicillium brevi-compactum* Dierckx *P. stoloniferum* Thom
Toxicity data[15,24,40,84]	Mycophenolic acid possesses antibiotic activity. It is more effective against gram-positive than against gram-negative bacteria. It has antiviral activity *in vitro;* it was effective *in vivo* against certain solid transplantable neoplasms. The LD_{50} of mycophenolic acid to mice was 2500 mg/kg (PO) and 550 mg/kg (IV). A rhesus monkey given oral doses of 150 mg/kg/day developed abdominal colic, watery and blood-stained feces, and loss of body weight after 2 weeks of dosing. The signs of toxicity subsided when dosing was discontinued. In rats the LD_{50} was 700 mg/kg (PO) and 450 mg/kg (IV).
Other significant references	Identification[16,18]; chemistry[28,65]; biosynthesis[14,40]

^{13}C *NMR data*	11.6 q, 16.1 q, 22.7 t, 32.7 t, 34.3 t, 61.0 q, 70.0 t, 106.3 s, 116.7 s, 122.0 s, 123.0 d, 133.9 s, 144.0 s, 153.7 s, 163.7 s, 172.9 s, 178.9 s
^{1}H *NMR data*	H2, 5.20; H9, 3.39 (J = 7); H10, 5.20 (J = 7); H12, 2.4; H13, 2.4; H15, 1.93; H16, 3.89; H17, 2.17

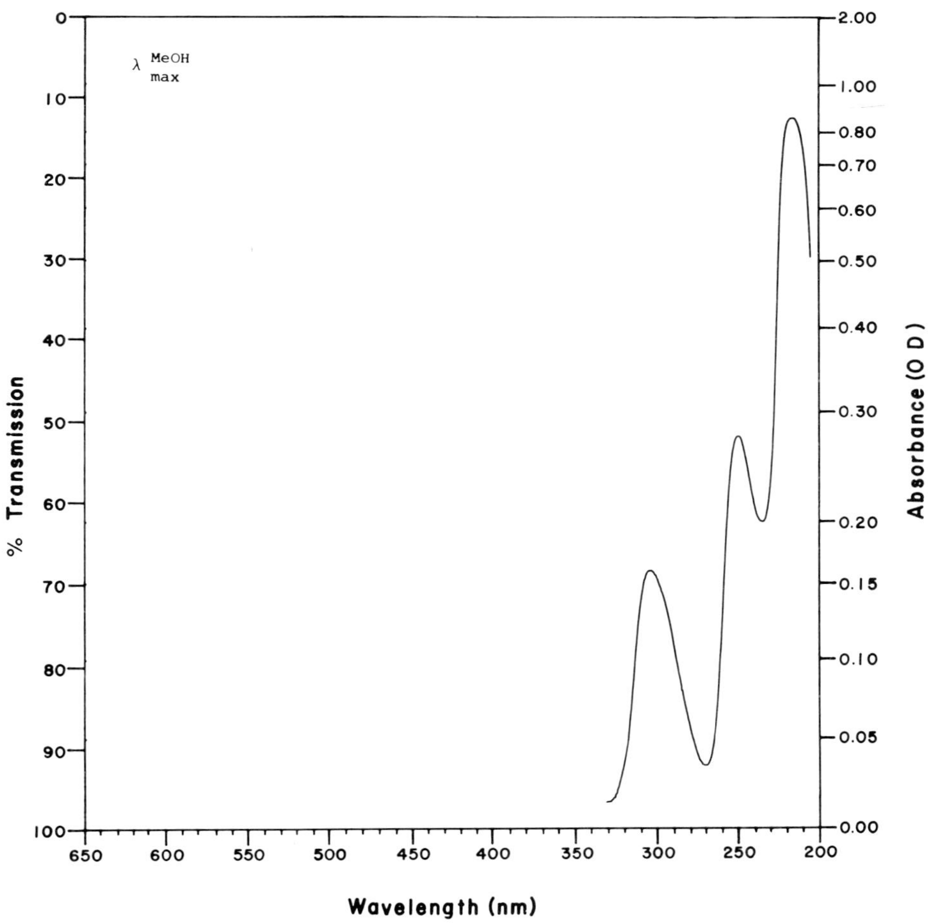

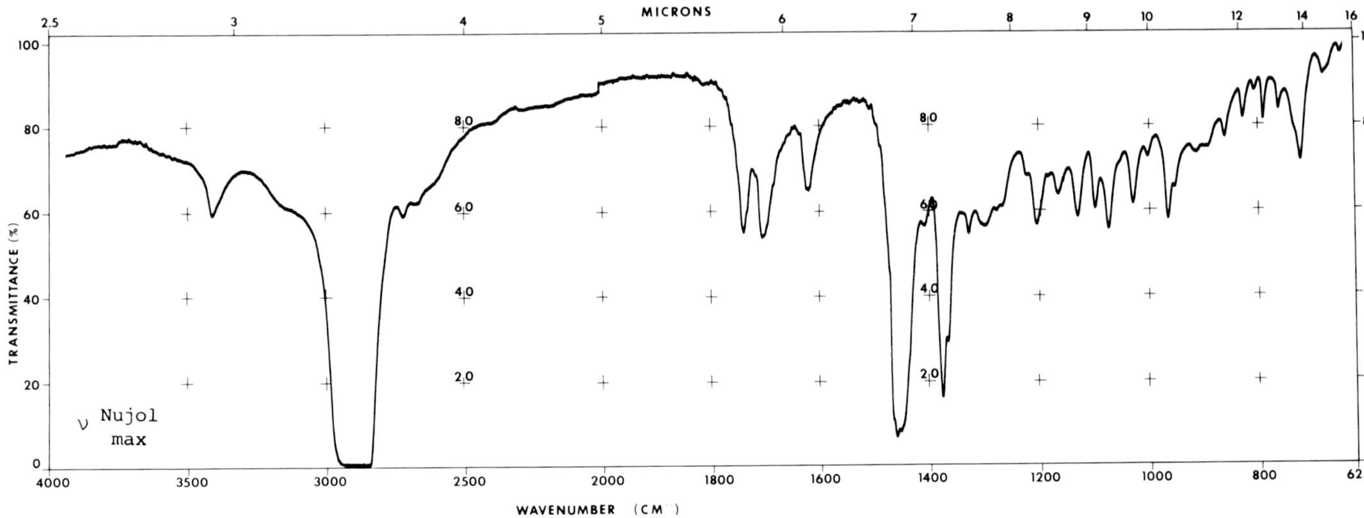
MICRONS
TRANSMITTANCE (%)
ν Nujol
max
WAVENUMBER (CM⁻¹)

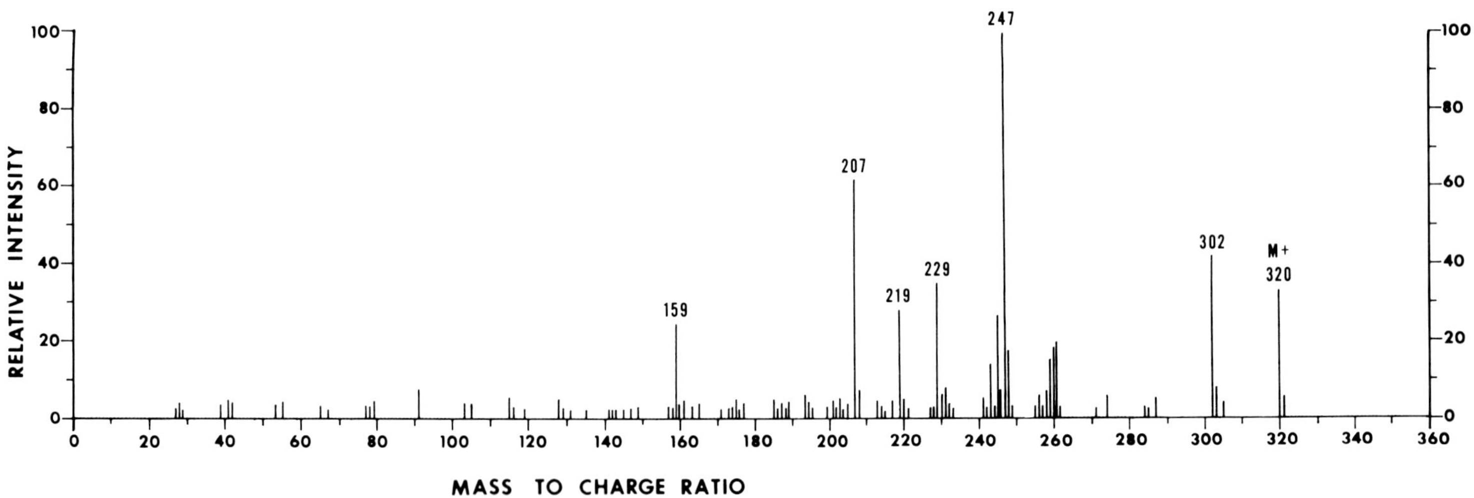
247
207
302
M+
320
229
219
159
RELATIVE INTENSITY
MASS TO CHARGE RATIO

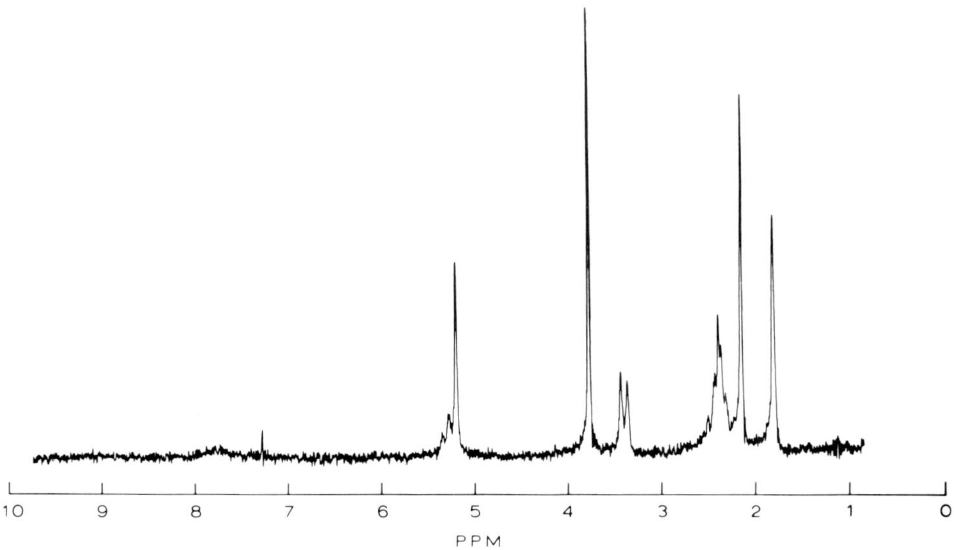

The 100 MHz ¹H NMR spectrum of mycophenolic acid in $CDCl_3$ solution.

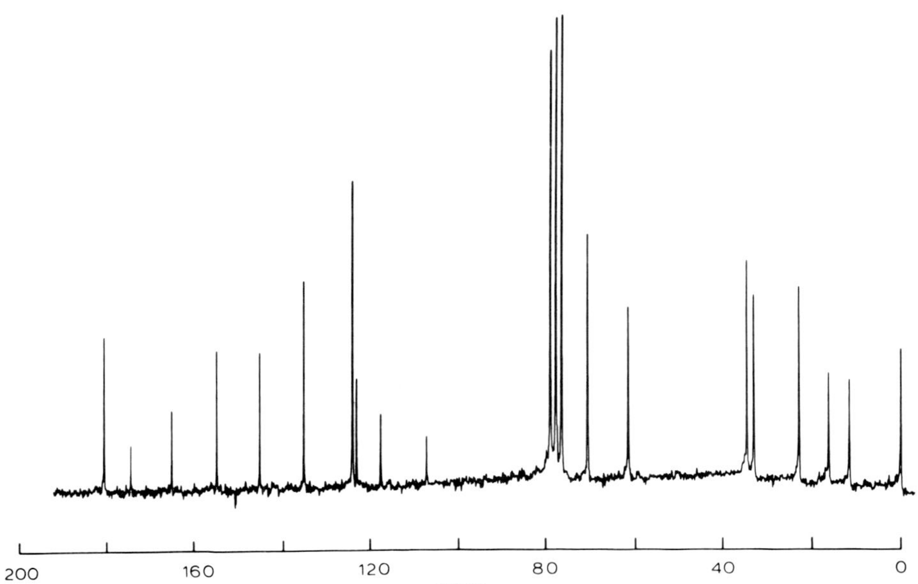

The 25 MHz ¹³C NMR spectrum of mycophenolic acid in $CDCl_3$ solution.

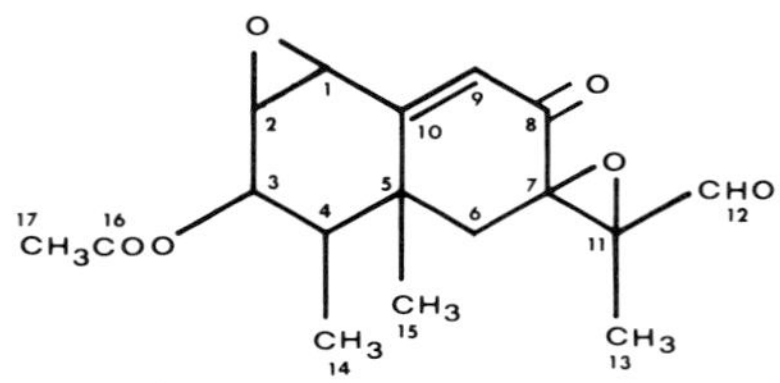

Common name	PR toxin
Molecular weight	320.1259
Molecular formula	$C_{17}H_{20}O_6$
General characteristics[100]	Colorless crystals, mp 155°–157°C $[\alpha]_D^{25}$ +290° (C = 1.34 in chloroform) Soluble in chloroform, carbon tetrachloride, methanol, ethanol, acetone, diethyl ether, and dioxane; insoluble in water, Skellysolve B, HCl, and NaOH. PR–alcohol, mp 152°–154°C
UV data[100]	λ_{max}^{EtOH} nm (ϵ): 249(15,278)
IR data[100]	ν_{max}^{KBr} cm^{-1}: 2945, 1735, 1720, 1680, 1620, 1460, 1435, 1380, 1245, and 1035
Fungal source[102]	*Penicillium roqueforti* (NRRL 849)
Toxicity data[78,102]	LD_{50} in weanling rats (50–60 g): dosed IP was 11 mg/kg; dosed orally was 115 mg/kg administered in 0.1–0.2 ml propylene glycol; LD_{50} in mice dosed IP was 5.8 mg/kg. PR toxin impairs the transcriptional process in liver cells; *in vitro* test results suggest that PR toxin impairs the activity of the RNA polymerase itself. It was shown that PR toxin inhibits both initiation and elongation of the polynucleotide chain.
TLC data[102]	*Adsorbent:* silica gel G *Solvent:* methanol–chloroform, 4:96 v/v R_f: 0.63 *Detection:* dark-blue fluorescent spot under UV light; changing to gray-green fluorescence after brief exposure to UV light and visible as a yellow spot in ordinary light; 50% H_2SO_4 spraying and heating yields yellow spot turning brown.

Other significant references	Review[101]
^{13}C NMR data[83]	C1, 55.9 d*; C2, 55.6 d*; C3, 69.9 d; C4, 42.7 d; C5, 38.0 s; C6, 41.5 t; C7, 67.3 s; C8, 191.6 s; C9, 129.9 d; C10, 164.6 s; C11, 67.3 s; C12, 198.5 d; C13, 21.9 q; C14, 10.1 q; C15, 13.6 q; C16, 170.5 s; C17, 20.7 q
^{1}H NMR data[83]	H1, 3.65 (J = 3.5); H2, 3.96 (J = 5.0, 3.5); H3, 5.16 (J = 5.0); H4, 1.79 (J = 5.0, 6.7); H6, 1.81, 2.16 (J = 14.0); H9, 6.43; H12, 9.75; H13, 1.49; H14, 1.03 (J = 6.7); H15, 1.45 (J = 0.8); H17, 2.16

* Assignments may be reversed.

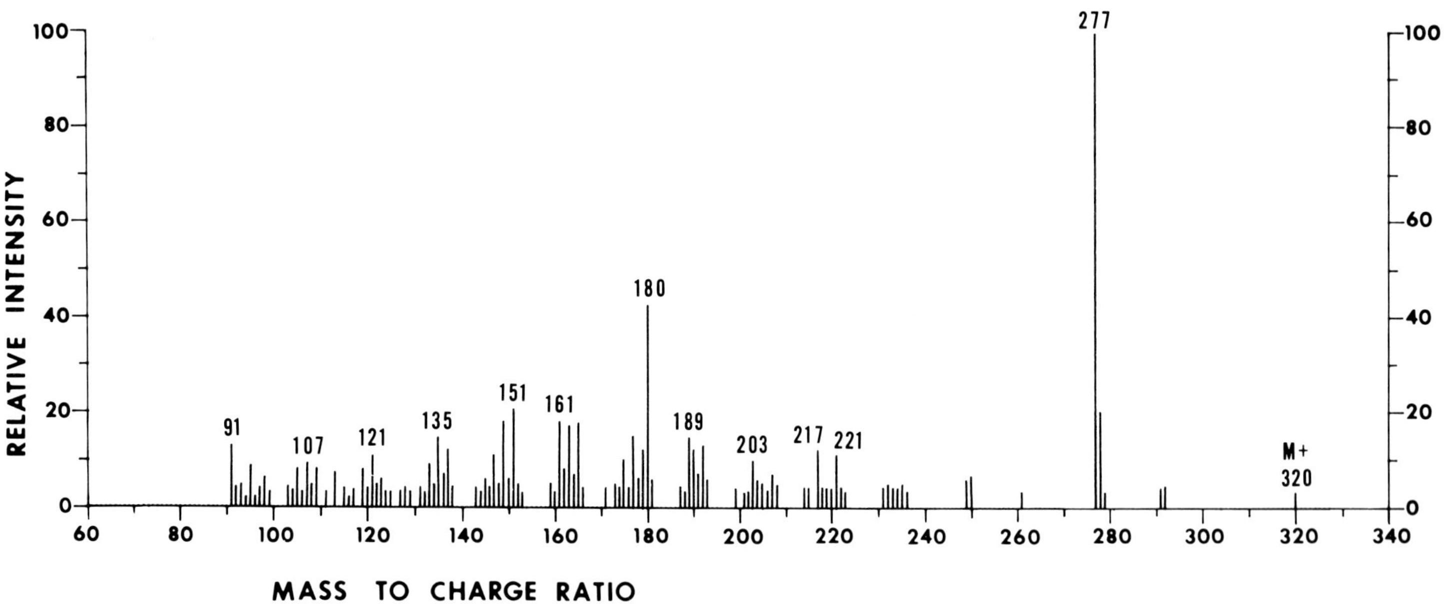

RELATIVE INTENSITY
MASS TO CHARGE RATIO
0
20
40
60
80
100
60
80
100
120
140
160
180
200
220
240
260
280
300
320
340
91
107
121
135
151
161
180
189
203
217
221
277
M+
320

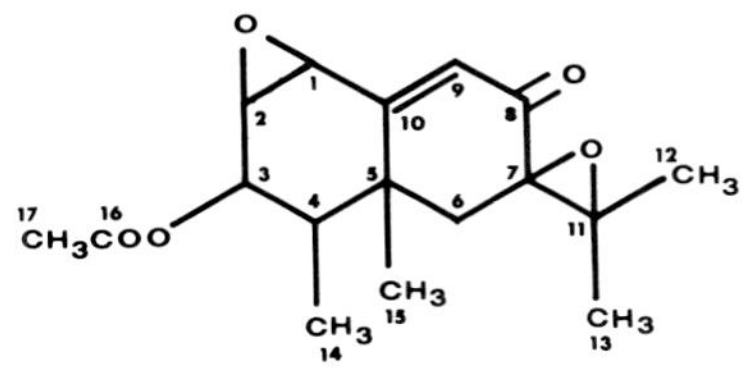

Common name	Eremofortin A
Molecular weight	306.1467 (eremofortin A alcohol, 264.1361)
Molecular formula	$C_{17}H_{22}O_5$ (eremofortin A alcohol, $C_{15}H_{20}O_4$)
General characteristics[75,76]	Crystals, mp 159°–161°C $[\alpha]_D$ +250° Eremofortin A alcohol: Crystals from ethyl ether–isopropyl ether, mp 128°–130°C $[\alpha]_D^{20}$ +143° (C = 1.04 in chloroform)
UV data[75,76]	Eremofortin A alcohol: λ_{max} nm (ϵ): 246(13,200)
IR data[75,76]	Eremofortin A alcohol: ν_{max}^{KBr} cm^{-1}: 3570, 3510, 3380, and 1685
Fungal source[75,76]	*Penicillium roqueforti* (NRRL 849)
Toxicity data	Nontoxic
Other significant references	HPLC[77]

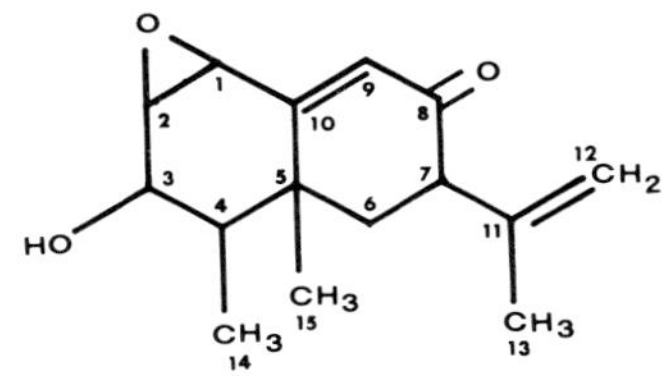

Common name	Eremofortin B
Molecular weight	248.1412
Molecular formula	$C_{15}H_{20}O_3$
General characteristics[76]	Crystals, mp 121°–123°C $[\alpha]_D$ +155°
Fungal source[76]	*Penicillium roqueforti* (NRRL 849)
Toxicity data[76]	Nontoxic
Other significant references	HPLC[77]

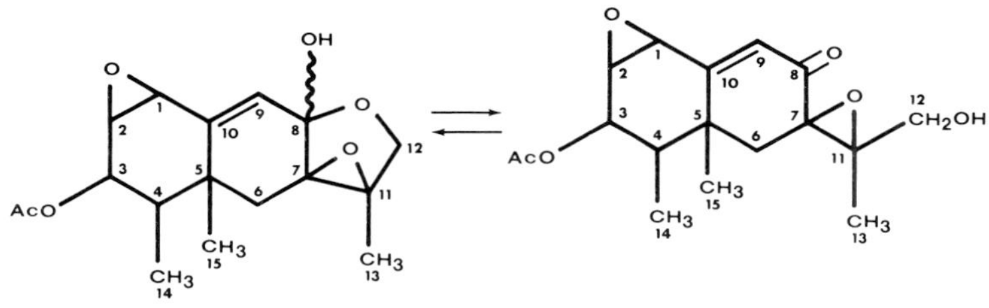

Common name	Eremofortin C
Molecular weight	322.1416
Molecular formula	$C_{17}H_{22}O_6$
General characteristics[75]	Crystals from ethyl ether, mp 122°–126°C Acetate: mp 122°–124°C $[\alpha]_D^{20}$ +161° (C = 1.09 in chloroform) Eremofortin C alcohol crystals from ethyl acetate, mp 170°C
UV data[75]	Acetate: λ_{max} nm (ϵ): 248(16,960)
IR data[75]	ν_{max}^{KBr} cm^{-1}: 3420, 3350, 1685, 1650, and 1620
Fungal source[75]	*Penicillium roqueforti* (NRRL 849)

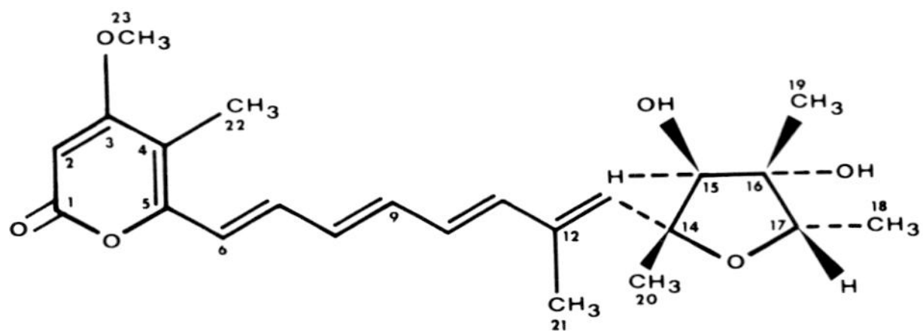

Common name	Citreoviridin
Molecular weight	402.2042
Molecular formula	$C_{23}H_{30}O_6$
General characteristics[87,95]	Yellow crystals from methanol, mp 107°–111°C (methanol solvate) $[\alpha]_D^{18.6}$ −107° (C = 1.0 in chloroform) Citreoviridin is soluble in ethanol, ether, benzene, chloroform, and acetone; insoluble in hexane and water.
UV data[80,87]	λ_{max}^{EtOH} nm (ϵ): 388(48,000), 294(27,000), 286(24,600), 234(10,200), and 204(17,000) λ_{max}^{MeOH} nm (ϵ): 383(44,000), 294(24,800), 285(22,000), 238(10,500), and 206(14,600)
IR data	ν_{max}^{KBr} cm^{-1}: 3500, 1702, 1689, 1654, 1626, 1562, 1531, 1452, 1405, 1249, 1094, 1069, 999, 821, and 811
Fungal source[51–53,64,79,87]	*Penicillium toxicarium* Miyake *P. citreo-viride* Biourge *P. ochrosalmoneum* Udagawa *P. pulvillorum* Turfitt *P. fellutanum*
Toxicity data[95,96]	A dose of 20 mg/kg (SC) to mice was lethal. Characteristic acute signs were ascending paralysis, vomiting, convulsion, and respiratory arrest. Less frequent signs were ataxic posture, spasmodic or hyperkinetic affection of the extremities accompanied by transient stiffness, enforced movements,

and ataxia. Advanced stages were characterized by cardiovascular disturbances, flaccid paralysis, and hypothermia. Dyspnea, gasping, and coma are terminated by respiratory arrest. The LD_{50} in male mice was 11 mg/kg (SC), 7.5 mg/kg (IP), and 20 mg/kg (PO); in female rats the LD_{50} was 3.6 mg/kg (SC).

TLC data[95]

A
Adsorbent: Kieselgel G
Solvent: *n*-hexane–acetone, 1:1 v/v
R_f: 0.45
Detection: bright yellow fluorescent spot under UV light

B
Adsorbent: Kieselgel G
Solvent: ethyl acetate–toluene, 1:1 v/v
R_f: 0.50
Detection: bright yellow fluorescent spot under UV light

C
Adsorbent: silica gel
Solvent: chloroform–methanol–acetone, 45:3:2 v/v/v
R_f: 0.74
Detection: bright yellow fluorescent spot under UV light

Other significant references

Biosynthesis[79]; isolation and structure[87]; densitometric studies[38]

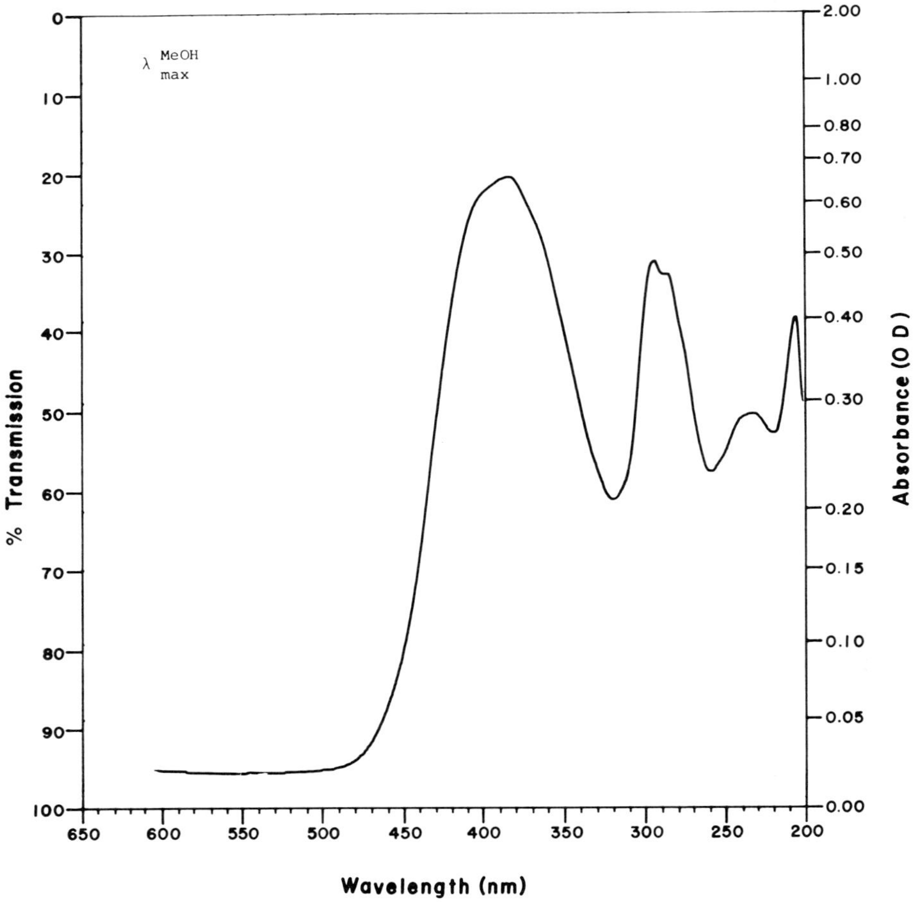

λ MeOH max
% Transmission
0
10
20
30
40
50
60
70
80
90
100
Absorbance (O D)
2.00
1.00
0.80
0.70
0.60
0.50
0.40
0.30
0.20
0.15
0.10
0.05
0.00
650
600
550
500
450
400
350
300
250
200
Wavelength (nm)

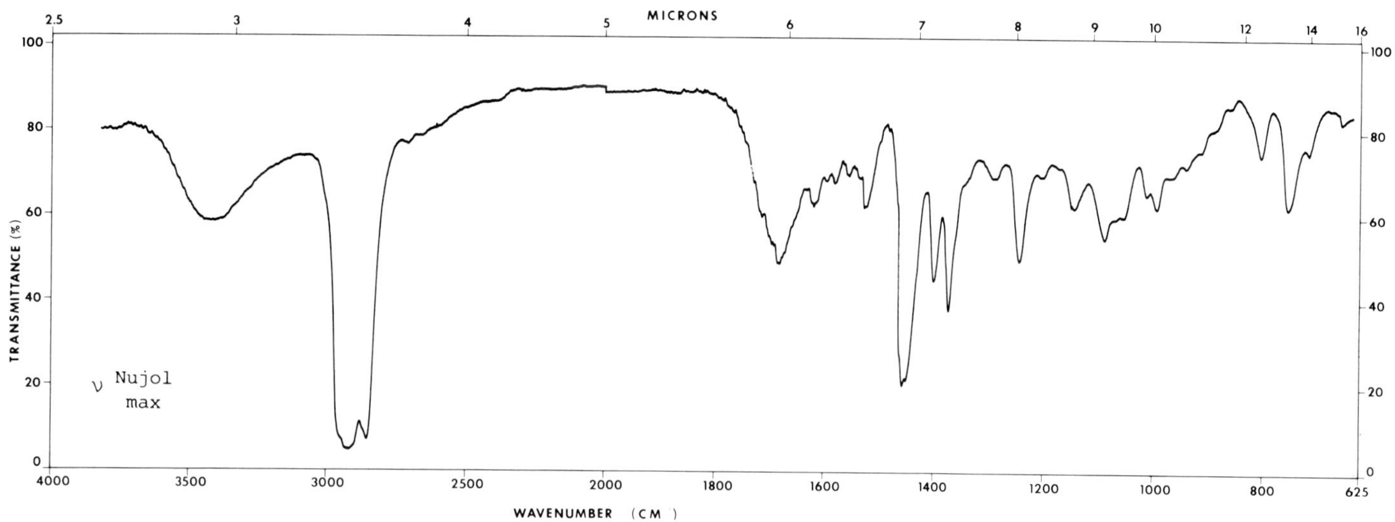
MICRONS
TRANSMITTANCE (%)
WAVENUMBER (CM)
ν Nujol
max

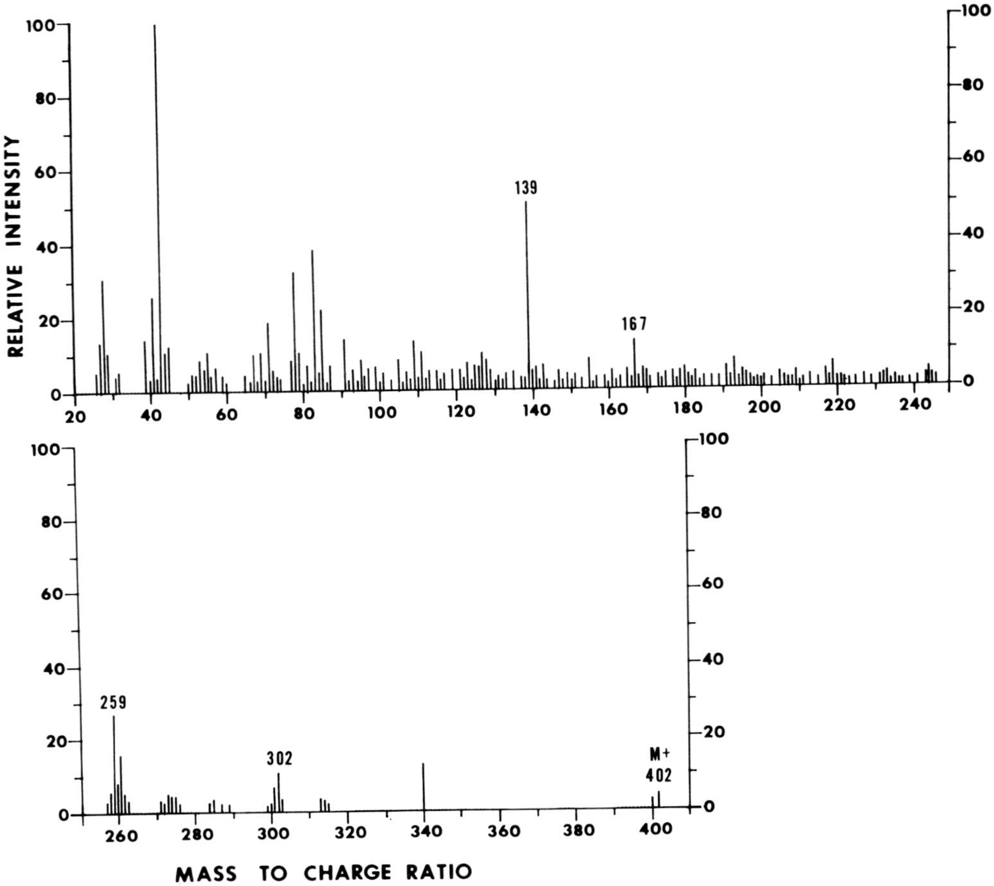
RELATIVE INTENSITY
139
167
259
302
M+
402
MASS TO CHARGE RATIO

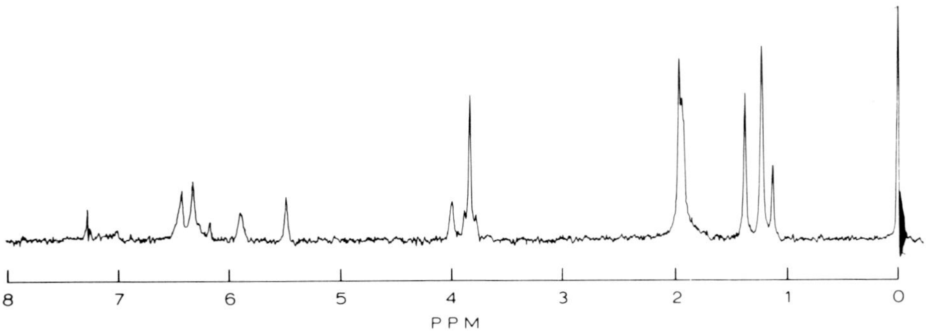

The 100 MHz ^{1}H NMR spectrum of citreoviridin in $CDCl_3$ solution.

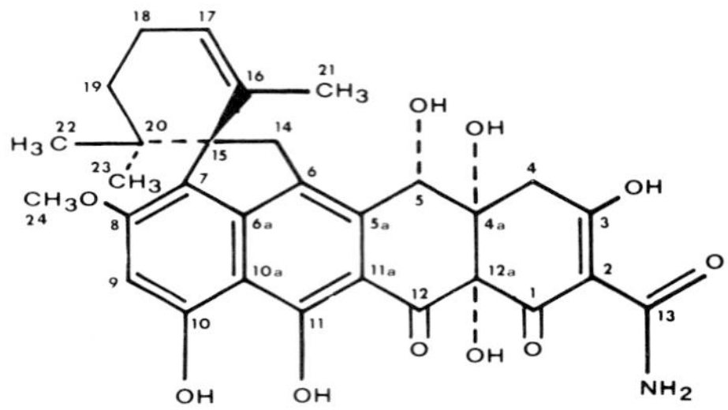

Common name	Viridicatum toxin
Molecular weight	565.1947
Molecular formula	$C_{30}H_{31}O_{10}N$
General characteristics[54]	Bright yellow crystals from chloroform–hexane mp 211°C; from methanol, mp 235°C (dec.) $[\alpha]_D^{23}$ −12° (C = 1.0 in ethanol)
UV data[54]	λ_{max}^{EtOH} nm (ϵ): 237(21,900), 285(32,400), 317(2,760), 331(2,590), 347(1,900), and 424(8,100)
IR data[54]	$\nu_{max}^{CHCl_3}$ cm^{-1}: 3550–3350, 2980, 1725, 1650, 1620, and 1595
Fungal source[54]	*Penicillium viridicatum* Westling (C.S.I.R. 1029)
Toxicity data[54]	The LD_{50} in male Wistar-derived rats (approximately 100 g) dosed in dimethyl sulfoxide via gastric intubation was 122.4 mg/kg. Rats receiving 500 mg/kg died within 24 hr. Histopathological examination showed the presence of hyaline droplet degeneration of the myocardium in all treated animals (50–500 mg/kg). Renal tubular necrosis was present in all animals that died. Hepatocytes exhibited considerable hydropic change with single cells exhibiting necrotic changes; abnormal mitotic figures were common. The spleen was atrophied and there was a paucity of small dark-staining lymphocytes in the splenic nodules.
TLC data[54]	*Adsorbent:* oxalic acid-impregnated silica gel *Solvent:* chloroform–methyl isobutyl ketone, 4:1 v/v R_f: 0.35

	Detection: yellow-brown fluorescent spot under UV light
Other significant references	X-Ray structure[56]
^{13}C *chemical shifts*	20.9, 22.8, 23.9, 25.4, 33.8, 38.4, 40.3, 41.0, 55.4, 59.9, 71.4, 79.9, 99.2, 99.5, 104.7, 105.2, 121.0, 122.3, 123.5, 136.1, 136.6, 146.7, 157.5, 160.2, 165.4, 172.2, 189.8, 192.2, 194.5

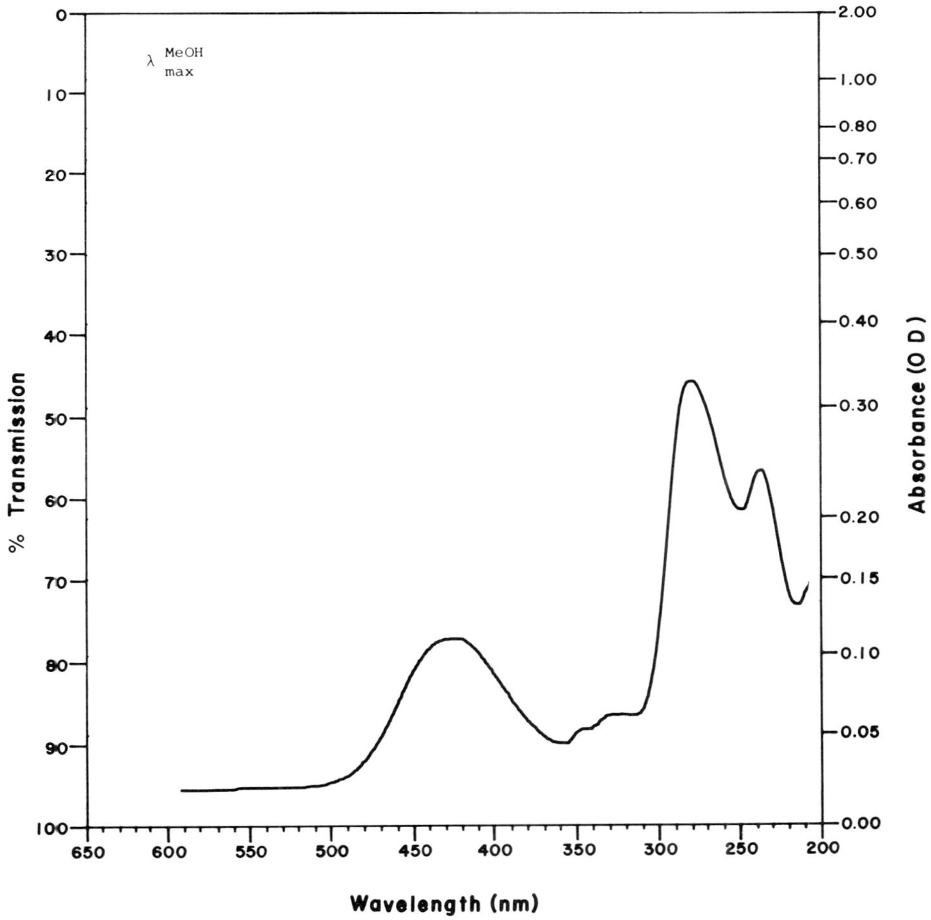

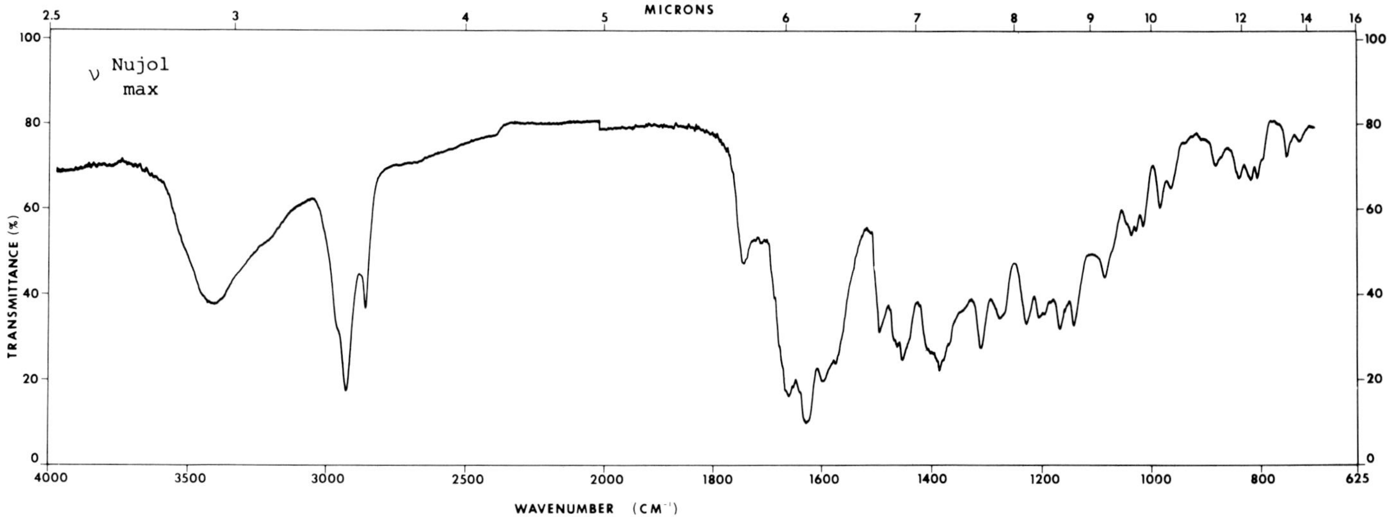
MICRONS
2.5
3
4
5
6
7
8
9
10
12
14
16
TRANSMITTANCE (%)
100
80
60
40
20
0
ν Nujol max
4000
3500
3000
2500
2000
1800
1600
1400
1200
1000
800
625
WAVENUMBER (CM⁻¹)

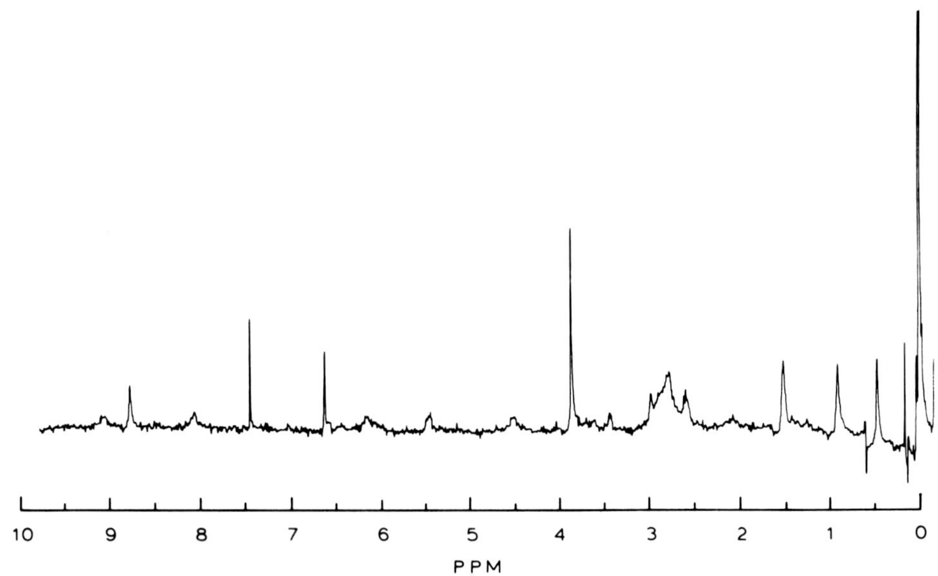
10
9
8
7
6
5
4
3
2
1
0
PPM

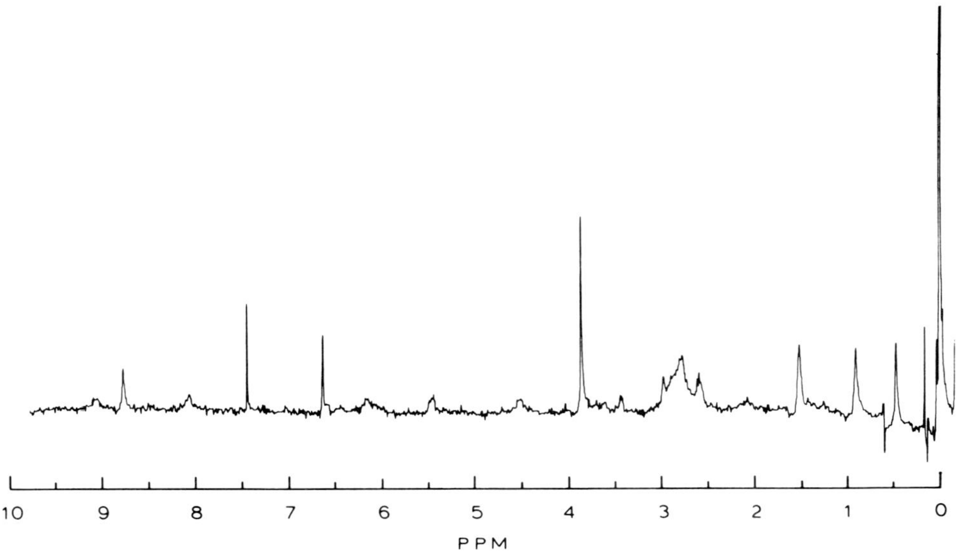

The 100 MHz ^1H NMR spectrum of viridicatum toxin in $CDCl_3$ solution.

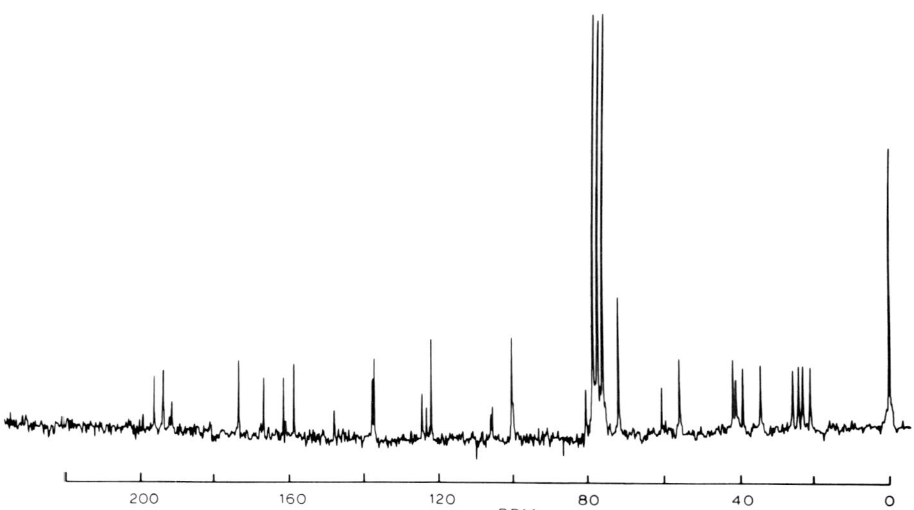

The 25 MHz ^{13}C NMR spectrum of viridicatum toxin in $CDCl_3$ solution.

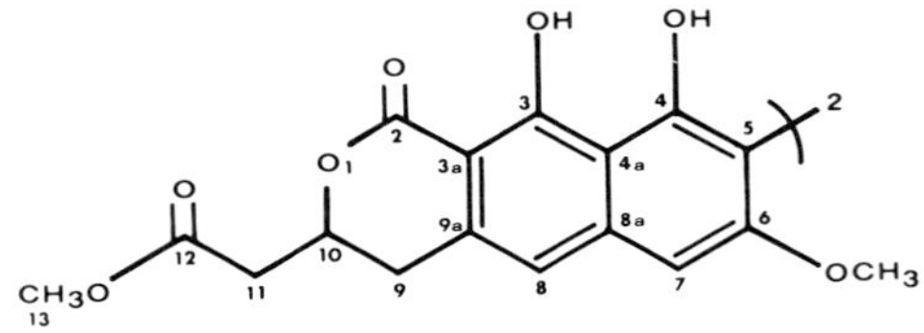

Common name	Viriditoxin
Molecular weight	662.1623
Molecular formula	$C_{34}H_{30}O_{14}$
General characteristics[61,62]	Yellow needle crystals from benzene, mp 242°–245°C (dec.) $[\alpha]_D^{28}$ −202° (in chloroform)
UV data[61]	λ_{max}^{EtOH} nm: 266 and 380 (ϵ_{max} not reported)
Fluorescence emission[103]	(In chloroform solution) excitation maxima: 280 and 385 nm emission bands: 450 and 550 nm
Fungal source[61,62,103]	*Aspergillus viridi-nutans* (Ducker and Thrower) (NRRL 4365) *A. brevipes* (NRRL 576)
Toxicity data[62]	The LD_{50} to 20 g Swiss-Wistar, male mice dosed IP in propylene glycol was 2.8 mg/kg. Viriditoxin showed limited antibiotic activity against *Corynebacterium fascians, Bacillus brevis, B. cereus,* and *Streptococcus faecalis.*
TLC data[61,62]	*Adsorbent:* silica gel G-HR *Solvent:* chloroform–acetonitrile–formic acid, 97:1.5:1.5 v/v/v chloroform–acetone–formic acid, 96:2.5:1.5 v/v/v R_f: 0.20–0.22 solvent system 1 0.18–0.20 in system 2 *Detection:* fluorescent spot under short wavelength UV light
^{13}*C chemical shifts*	33.0 t, 39.5 t, 52.1 q, 56.2 q,d, 98.3 d, 98.6 s, 107.9 s, 110.0 s, 114.1 d, 131.9 s, 139.1 s, 159.1 s, 160.9 s, 163.3 s, 169.5 s, 170.7 s

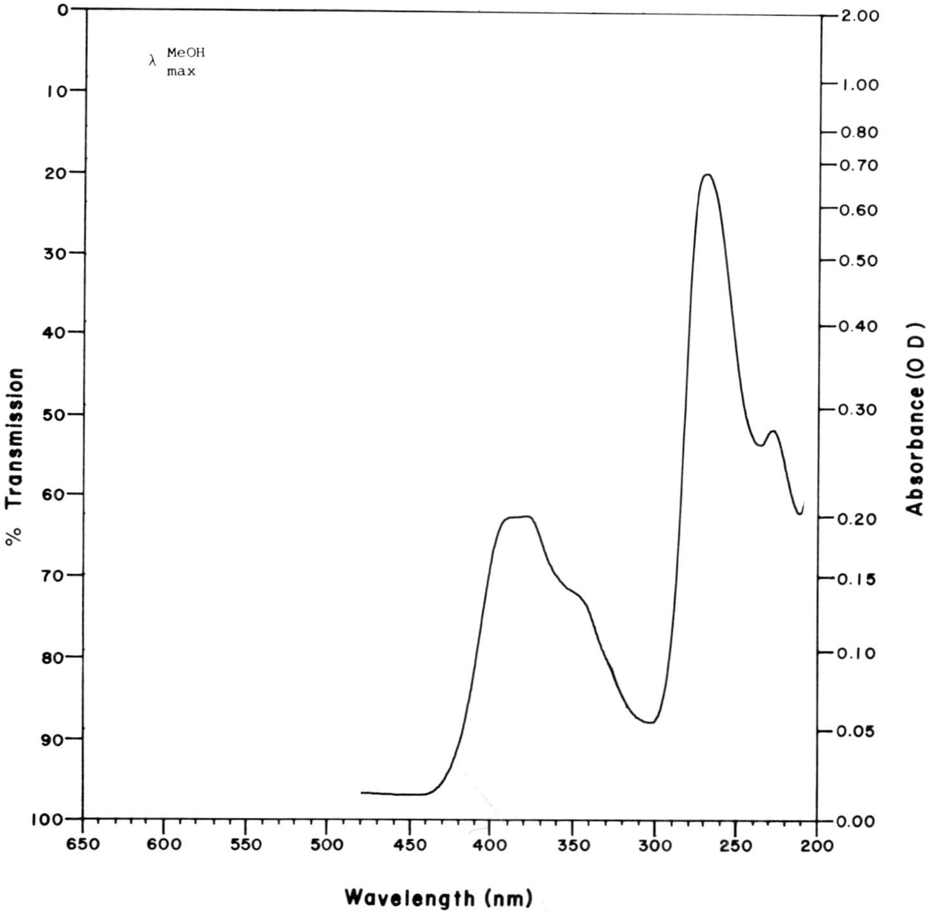

λ MeOH max
% Transmission
0
10
20
30
40
50
60
70
80
90
100
Absorbance (O D)
2.00
1.00
0.80
0.70
0.60
0.50
0.40
0.30
0.20
0.15
0.10
0.05
0.00
650
600
550
500
450
400
350
300
250
200
Wavelength (nm)

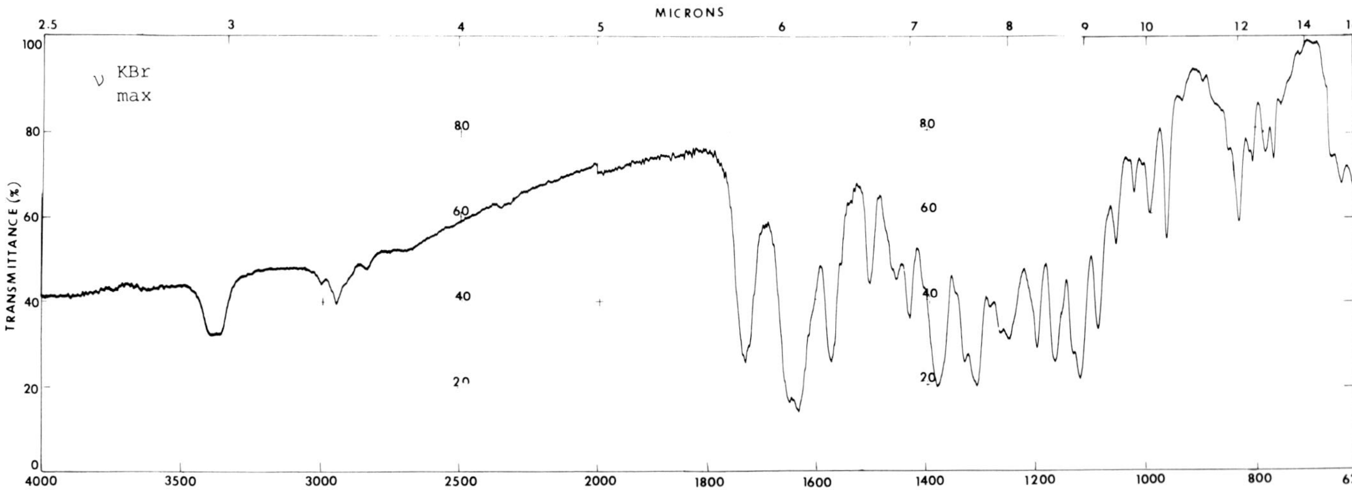

VIRIDITOXIN

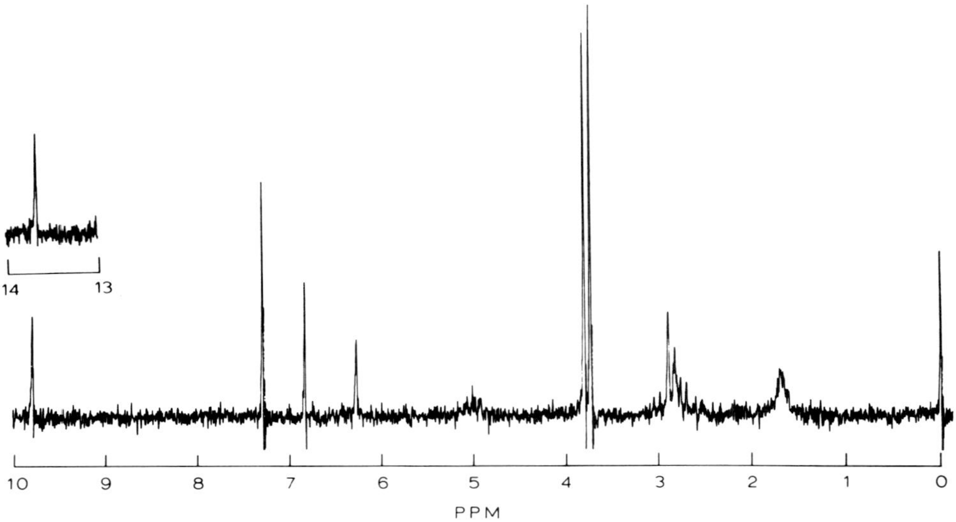

The 100 MHz ¹H NMR spectrum of viriditoxin in $CDCl_3$ solution.

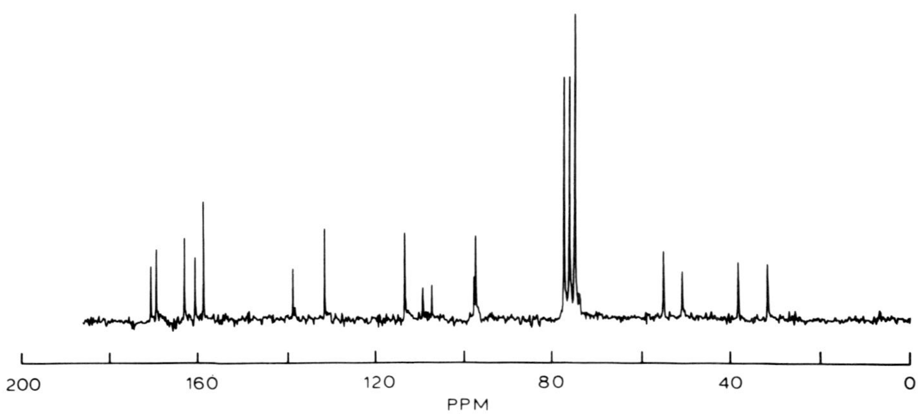

The 25 MHz ¹³C NMR spectrum of viriditoxin in $CDCl_3$ solution.

REFERENCES

1. Alsberg, C. L., and Black, O. F., *U.S., Dep. Agric., Bur. Plant Ind. Bull.* No. 270 (1913).
2. Ambrose, A. M., and DeEds, F., *J. Pharmacol. Exp. Ther.* **88,** 173 (1946).
3. Anderson, D. W., *Ann. Allergy* **23,** 103 (1965).
4. Bacikova, D., Betina, V., and Nemec, P., *Naturwissenschaften* **51,** 445 (1964).
5. Baudouy, R., Crabbe, P., Greene, A. E., Le Drian, C., and Orr, A. F., *Tetrahedron Lett.* p. 2973 (1977).
6. Betina, V., Drobnica, Ľ., Nemec, P., and Zemanová, M., *J. Antibiot.* **17,** 93 (1964).
7. Betina, V., Fuska, J., Kjaer, A., Kutkova, M., Nemec, P., and Shapiro, R. H., *J. Antibiot.* **19,** 115 (1966).
8. Betina, V., Horakova, K., and Barath, Z., *Naturwissenschaften* **49,** 241 (1962).
9. Betina, V., and Murín, A., *Cytologia* **29,** 370 (1964).
10. Betina, V., Nemec, P., Dobias, J., and Barath, Z., *Folia Microbiol. (Prague)* **7,** 353 (1962).
11. Betina, V., Nemec, P., Kovač, Š., Kjaer, A., Shapiro, R. H., *Acta Chem. Scand.* **19,** 16 (1965).
12. Betina, V., and Nemec, P., *Naturwissenschaften* **50,** 696 (1963).
13. Betina, V., Nemec, P., and Strmeňová, H., *Folia Microbiol. (Prague)* **8,** 125 (1963).
14. Birch, A. J., English, R. J., Massy-Westropp, R. A., and Smith, H., *J. Chem. Soc.* p. 369 (1958).
15. Birch, A. J., English, R. J., Massy-Westropp, R. A., and Smith, H., *Proc. Chem. Soc., London* p. 233 (1957).
16. Birkinshaw, J. H., Bracken, A., Morgan, E. N., and Raistrick, H., *Biochem. J.* **43,** 216 (1948).
17. Birkinshaw, J. H., Luckner, M., Mohammed, Y. S., Mothes, K., and Stickings, C. E., *Biochem. J.* **89,** 196 (1963).
18. Birkinshaw, J. H., Raistrick, H., and Ross, D. J., *Biochem. J.* **50,** 630 (1952).
19. Bracken, A., Pocker, A., and Raistrick, H., *Biochem. J.* **57,** 587 (1954).
20. Brian, P. W., Curtis, P. J., and Hemming, H. J., *Trans. Br. Mycol. Soc.* **29,** 1973 (1946).
21. Brown, J. P., Cartwright, N. J., Robertson, A., and Whalley, W. B., *J. Chem. Soc.* p. 859 (1949).
22. Brown, J. P., Cartwright, N. J., Robertson, A., and Whalley, W. B., *Nature (London)* **162,** 72 (1948).
23. Brown, W. A. C., and Sim, G. A., *J. Chem. Soc.* p. 1050 (1963).
24. Carter, S. B., Franklin, T. J., Jones, D. F., Leonard, B. J., Mills, S. D., Turner, R. W., and Turner, W. B., *Nature (London)* **223,** 848 (1969).
25. Cartwright, N. J., Robertson, A., and Whalley, W. B., *J. Chem. Soc.* p. 1563 (1949).
26. Cartwright, N. J., Robertson, A., and Whalley, W. B., *Nature (London)* **163,** 94 (1949).
27. Clarke, S. G., and King, A. J., *Acta Crystallogr.* **4,** 182 (1951).
28. Clutterbuck, P. W., and Raistrick, H., *Biochem. J.* **27,** 654 (1933).
29. Cole, R. J., Kirksey, J. W., Cutler, H. G., and Davis, E. E., *J. Agric. Food Chem.* **22,** 417 (1974).
30. Cole, R. J., Kirksey, J. W., and Holaday, C. E., *Appl. Microbiol.* **19,** 106 (1970).
31. Cole, R. J., Kirksey, J. W., and Morgan-Jones, G., *Toxicol. Appl. Pharmacol.* **31,** 465 (1975).
32. Cole, R. J., unpublished data.
33. Cross, B. E., and Hendley, P., *J. Chem. Soc., Chem. Commun.* p. 124 (1975).
34. Cunningham, K. G., and Freeman, G. G., *Biochem. J.* **53,** 328 (1953).
35. Curtin, T. P., Fitzgerald, G., and Reilly, J., *Biochem. J.* **34,** 1605 (1940).
36. Day, A. C., Nabney, J., and Scott, A. I., *Proc. Chem. Soc., London* p. 284 (1960).
37. Divekar, P. V., Haskins, R. H., and Vining, L. C., *Can. J. Chem.* **37,** 2097 (1959).
38. Engel, G., *J. Chromatogr.* **130,** 293 (1977).
39. Field, D. J., Bowen, J. M., and Cole, R. J., *Toxicol. Appl. Pharmacol.* **46,** 529 (1978).

40. Florey, H. W., Gilliver, K., Jennings, M. A., and Sanders, A. G., *Lancet* **1,** 46 (1946).
41. Framm, J., Nover, L., Azzouny, A. E., Richter, H., Winter, K., Werner, S., and Luckner, M., *Eur. J. Biochem.* **37,** 78 (1973).
42. Gillick, A., and Bulmer, W. S., *Can. Vet. J.* **13,** 244 (1972).
43. Grove, J. F., *Fortschr. Chem. Org. Naturst.* **22,** 203 (1964).
44. Grove, J. F., *in* "Antibiotics" (D. Gottlieb and P. D. Shaw, eds.), Vol. II, p. 123. Springer-Verlag, Berlin and New York, 1967.
45. Grove, J. F., Ismay, D., MacMillan, J., Mulholland, T. P. C., and Rogers, M. A. T., *Chem. Ind. (London)* p. 219 (1951).
46. Grove, J. F., Ismay, D., MacMillan, J., Mulholland, T. P. C., and Rogers, M. A. T., *J. Chem. Soc.* p. 3958 (1952).
47. Grove, J. F., and McGowan, J. C., *Chem. Ind. (London)* p. 647 (1949).
48. Härri, E., Loeffler, W., Sigg, H. P., Stähelin, H., and Tamm, C., *Helv. Chim. Acta* **46,** 1235 (1963).
49. Harris, C. M., Roberson, J. S., and Harris, T. M., *J. Am. Chem. Soc.* **98,** 5380 (1976).
50. Hetherington, A. C., and Raistrick, H., *Philos. Trans. Soc. London, Ser. B* **220,** 269 (1931).
51. Hirata, Y., *J. Chem. Soc. Jpn.* **68,** 74 (1974).
52. Hirata, Y., *J. Chem. Soc. Jpn.* **68,** 63 (1974).
53. Hirata, Y., *J. Chem. Soc. Jpn.* **78,** 104 (1974).
54. Hutchison, R. D., Steyn, P. S., and van Rensburg, S. J., *Toxicol. Appl. Pharmacol.* **24,** 507 (1973).
55. Itahashi, M., Murakami, Y., and Nishikawa, H., *Tokyo J. Agric. Res.* **4,** 281 (1955).
56. Kabuto, C., Silverton, J. V., Akiyama, T., Sankawa, U., Hutchison, R. D., Steyn, P. S., and Vleggaar, R., *J. Chem. Soc., Chem. Commun.* p. 728 (1976).
57. Klein, M. F., and Beall, J. R., *Science* **175,** 1483 (1972).
58. Kogl, F., and Van Wessem, G. C., *Recl. Trav. Chim. Pays-Bas* **63,** 5 (1944).
59. Krogh, P., Hasselager, E., and Friis, P., *Acta Pathol. Microbiol. Scand.* **78,** 401 (1970).
60. Levine, S. G., and Hicks, R. E., *J. Org. Chem.* **40,** 2540 (1975).
61. Lillehoj, E. B., and Ciegler, A., *Can. J. Microbiol.* **18,** 193 (1972).
62. Lillehoj, E. B., and Milburn, M. S., *Appl. Microbiol.* **26,** 202 (1973).
63. Lloyd, G., Robertson, A., Sankey, G. B., and Whalley, W. R., *J. Chem. Soc.* p. 2163 (1955).
64. Locci, R., Merlini, L., Nasini, G., and Locci, J. R., *G. Microbiol.* **13,** 271 (1965).
65. Logan, W. R., and Newbold, G. I., *J. Chem. Soc.* p. 1946 (1957).
66. Mabuni, C. J., Garlaschelli, L., Ellison, R. A., and Hutchinson, C. R., *J. Chem. Soc.* **99,** 7718 (1977).
67. MacMillan, J., *Chem. Ind. (London)* p. 719 (1951).
68. MacMillan, J., *J. Chem. Soc.* p. 1697 (1953).
69. MacMillan, J. G., Springer, J. P., Clardy, J., Cole, R. J., and Kirksey, J. W., *J. Am. Chem. Soc.* **98,** 246 (1976).
70. Martin, P. K., Rapoport, H., Smith, H. W., and Wong, J. L., *J. Org. Chem.* **34,** 1359 (1969).
71. McCamish, M., and White, J. D., *Org. Mass Spectrom.* **4,** 241 (1970).
72. McGowan, J. C., *Trans. Br. Mycol. Soc.* **29,** 188 (1946).
73. Mehta, P. P., and Whalley, W. B., *J. Chem. Soc.* p. 3777 (1963).
74. Mohammed, Y. S., and Luckner, M., *Tetrahedron Lett.* p. 1953 (1963).
75. Moreau, S., Cacan, M., and Lablache-Combier, A., *J. Org. Chem.* **42,** 2632 (1977).
76. Moreau, S., Gaudemer, A., Lablache-Combier, A., and Biguet, J., *Tetrahedron Lett.* p. 833 (1976)
77. Moreau, S., Masset, A., and Biguet, J., *Appl. Environ. Microbiol.* **37,** 1059 (1979).
78. Moule, Y., Jemmali, M., and Rousseau, N., *Chem.-Biol. Interact.* **14,** 207 (1976).
79. Nagel, D. W., Steyn, P. S., and Ferriera, N. P., *Phytochemistry* **11,** 3215 (1972a).
80. Nagel, D. W., Steyn, P. S., and Scott, D. B., *Phytochemistry* **11,** 627 (1972b).

81. Neely, W. C., Ellis, S. P., Davis, N. D., and Diener, U. L., *J. Assoc. Off. Anal. Chem.* **55,** 1122 (1972).
82. Nishikawa, H., *Tohoku J. Agric. Res.* **5,** 281 (1955).
83. Oxford, A. E., Raistrick, H., and Simonart, P., *Biochem. J.* **33,** 240 (1939).
84. Planterose, D. N., *J. Gen. Virol.* **4,** 629 (1969).
85. Richter, H., Winter, K., Elkousy, S., and Luckner, M., *Pharmazie* **29,** 506 (1974).
86. Saito, M., Enomoto, M., and Tatsuno, T., *in* "Microbial Toxins" (A. Ciegler, S. Kadis, and S. J. Ajl, eds.), Vol. 6, p. 358. Academic Press, New York, 1971.
87. Sakabe, N., Goto, T., and Hirata, Y., *Tetrahedron Lett.* p. 1825 (1964).
88. Sato, Y., and Oda, T., *Tetrahedron Lett.* p. 3971 (1976).
89. Scott, F. W., De LaHunta, A., Schultz, R. D., Bistner, S. I., and Riis, R. C., *Teratology* **11,** 79 (1975).
90. Sigg, H. P., *Helv. Chim. Acta* **47,** 1401 (1964).
91. Simpson, T. J., and Holker, J. S. E., *Phytochemistry* **16,** 229 (1977).
92. Singleton, V. L., Bohonos, N., and Ullstrup, A. J., *Nature (London)* **181,** 1072 (1958).
93. Smith, J., and Thomson, R. H., *Tetrahedron* **10,** 148 (1960).
94. Suzuki, Y., Tanaka, H., Aoki, H., and Tamura, T., *Agric. Biol. Chem.* **34,** 395 (1970).
95. Ueno, Y., *in* "Mycotoxins" (I. F. H. Purchase, ed.), p. 283. Elsevier, Amsterdam, 1974.
96. Ueno, Y., and Ueno, I., *Jpn. J. Exp. Med.* **42,** 91 (1972).
97. Vining, L. C., Kelleher, W. J., and Schwarting, A. W., *Can. J. Microbiol.* **8,** 931 (1962).
98. Warren, H. H., Dougherty, G., and Wallis, E., *J. Am. Chem. Soc.* **79,** 3812 (1957).
99. Warren, H. H., Dougherty, G., and Wallis, E. S., *J. Am. Chem. Soc.* **71,** 3422 (1949).
100. Wei, R., Schnoes, H. K., Hart, P. A., and Strong, F. M., *Tetrahedron* **31,** 109 (1975).
101. Wei, R., Schnoes, H. K., Smalley, E. B., Lee, S., Chang, Y., and Strong, F. M., *in* "Animal, Plant and Microbial Toxins" (A. Ohsaka, K. Hayashi, and Y. Sawai, eds.), Vol. 2, p. 137. Plenum, New York, 1976.
102. Wei, R., Still, P. E., Smalley, E. B., Schnoes, H. K., and Strong, F. M., *Appl. Microbiol.* **25,** 111 (1973).
103. Weisleder, D., and Lillehoj, E. B., *Tetrahedron Lett.* **48,** 4705 (1971).
104. Wilson, B. J., *in* "Microbial Toxins" (A. Ciegler, S. Kadis, and S. J. Ajl, eds.), Vol. 6, p. 459. Academic Press, New York, 1971.
105. Wyatt, R., personal communication.

20

Fusarium Toxins

The following three chemically unrelated metabolites are produced by *Fusarium* species that frequently invade and colonize economically important food and feed grains. They, therefore, represent a potential threat to animal health. *Fusarium graminearum,* a frequent contaminant of grain, has been shown to produce all three of these highly toxic metabolites. *Fusarium moniliforme,* a common contaminant of corn, is capable of producing moniliformin and zearalenone in addition to trichothecene mycotoxins.

Butenolide has been implicated in the etiology of a disease of cattle called "fescue foot." Moniliformin is a novel, water-soluble cyclobutenedione that appears to elicit only acute toxicity in experimental animals. Zearalenone shows strong estrogenic and anabolic activity in animals.

Fusarium toxins	Molecular weight	Molecular formula
Moniliformin		
Sodium salt	119.9823	C_4HO_3Na
Potassium salt	136.0945	C_4HO_3K
Free acid	98.0081	$C_4H_2O_3$
Butenolide	141.0425	$C_6H_7O_3N$
(*S*)-Zearalenone	318.1467	$C_{18}H_{22}O_5$
8′-Hydroxyzearalenone	334.1416	$C_{18}H_{22}O_6$

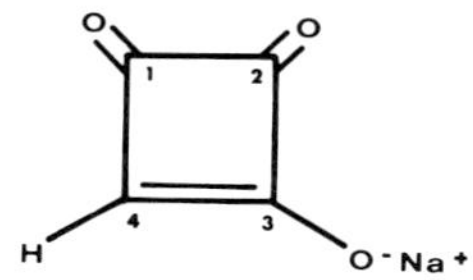

Common name	Moniliformin (3-Hydroxycyclobut-3-ene-1,2-dione)
Molecular weight	119.9823 (sodium salt) 136.0945 (potassium salt) 98.0081 (free acid)
Molecular formula	C_4HO_3Na (sodium salt) C_4HO_3K (potassium salt) $C_4H_2O_3$ (free acid)
General characteristics[6]	Crystals from aqueous methanol, decompose without melting at temperatures up to 350°C (Na or K salt). Crystalline acid, mp 158°C (dec.). Moniliformin occurs naturally as the Na or K salt. pK_a 1.70
UV data[6]	$\lambda_{max}^{H_2O}$ nm (ϵ): 229(19,100) and 260(5,600)
Fungal source[6,12,17,21,22,31]	*Fusarium moniliforme* Sheldon *F. moniliforme* var. *subglutimans* *F. graminearum* *F. fusarioides* *Gibberella fujikuroi* (ATCC 12763)
Toxicity data[4,6,25]	Moniliformin had an LD_{50} of 4.0 mg/kg (160 μg/40 g cockerel) in day-old cockerels dosed orally using sterile distilled water solutions. Death resulted within 45 min after dosing at 40 mg/kg. Gross and histologic lesions in cockerels, dosed at 500 μg and 250 μg/cockerel, that lived more than 2 hr after dosing, were ascites with edema of the mesenteries and small hemorrhages in the proventriculus, gizzard, small and large intestine, and skin. The LD_{50} in mice was 20.9 mg/kg (females) and 29.1 mg/kg (males). Moniliformin also produced plant-growth regulating and phytotoxic effects on plant systems. It was shown that exceedingly low concentrations of moniliformin (<5 μm) selectively inhibited mitochondrial pyruvate and α-ketoglutarate oxidations by 50%.

Other significant references	X-Ray crystallography and synthesis[19]; synthesis[1,2,18]; spectroanalytical parameters[13]; toxicity[12]
^{13}C NMR data	C1, 202.9 s; C2, 207.4 s; C3, 202.9 s; C4, 168.8 d
^{1}H NMR data	H2, 4.80

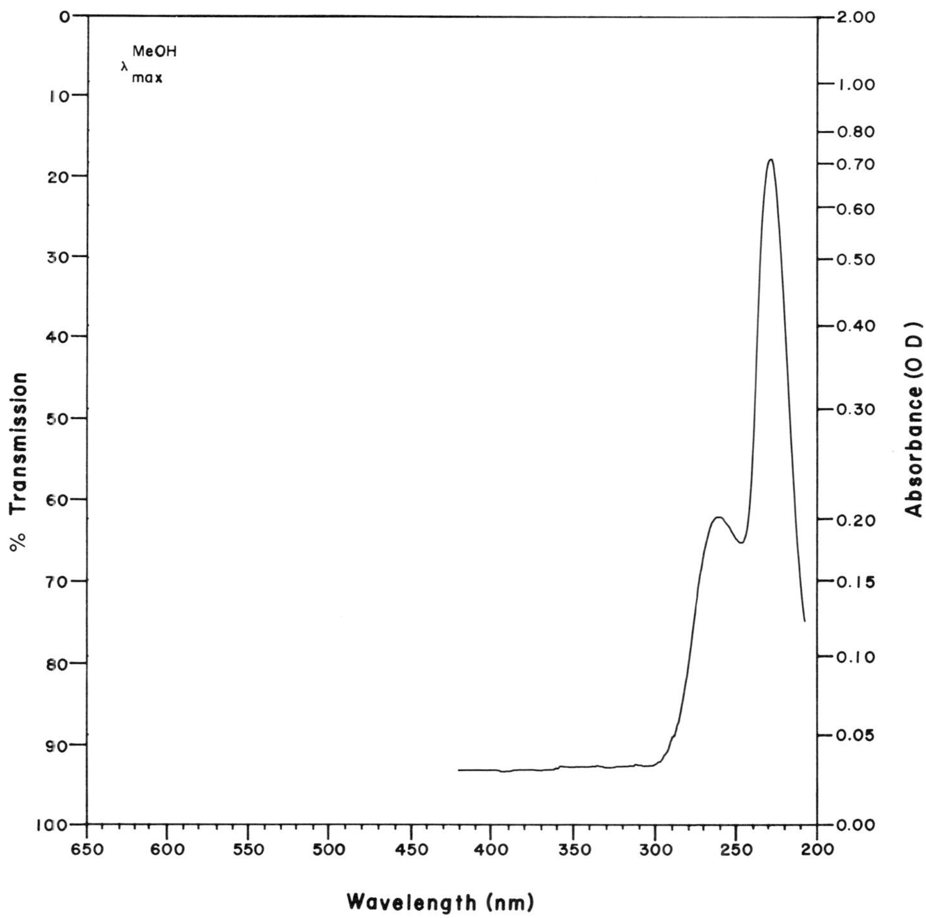

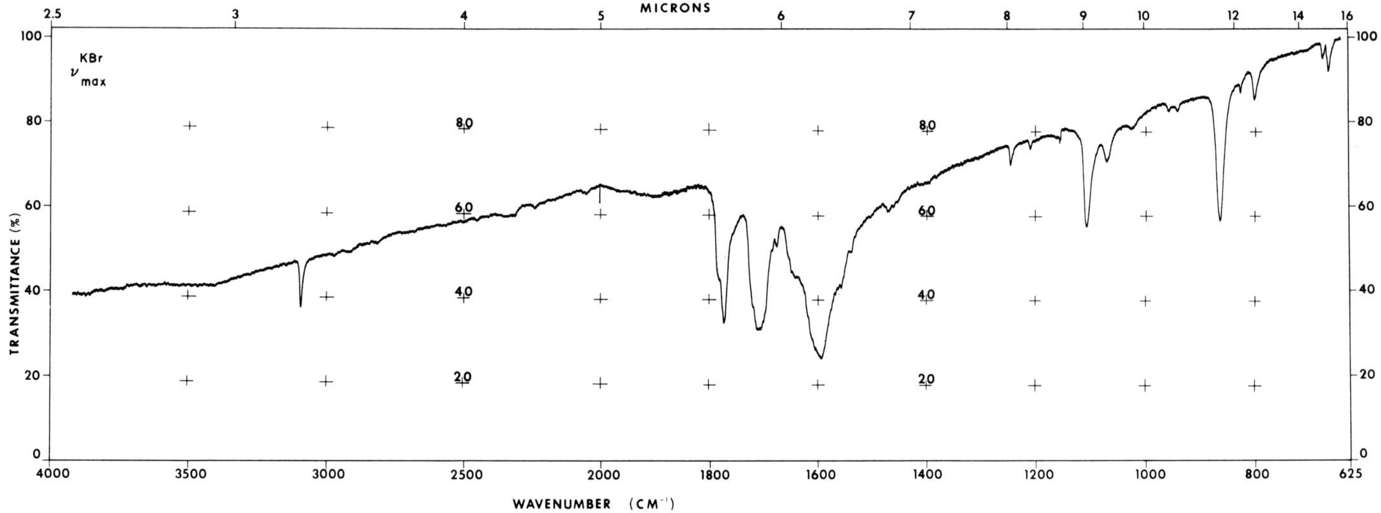
MICRONS
TRANSMITTANCE (%)
KBr
ν_{max}
WAVENUMBER (CM⁻¹)

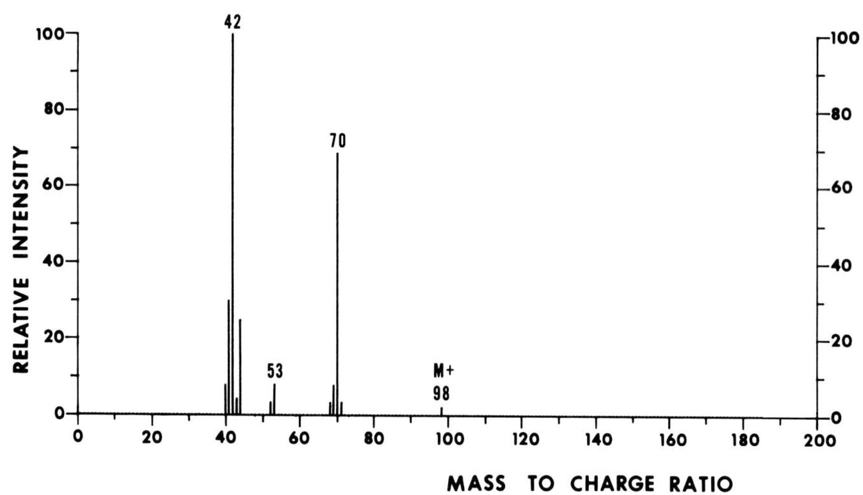
RELATIVE INTENSITY
42
53
70
M+
98
MASS TO CHARGE RATIO

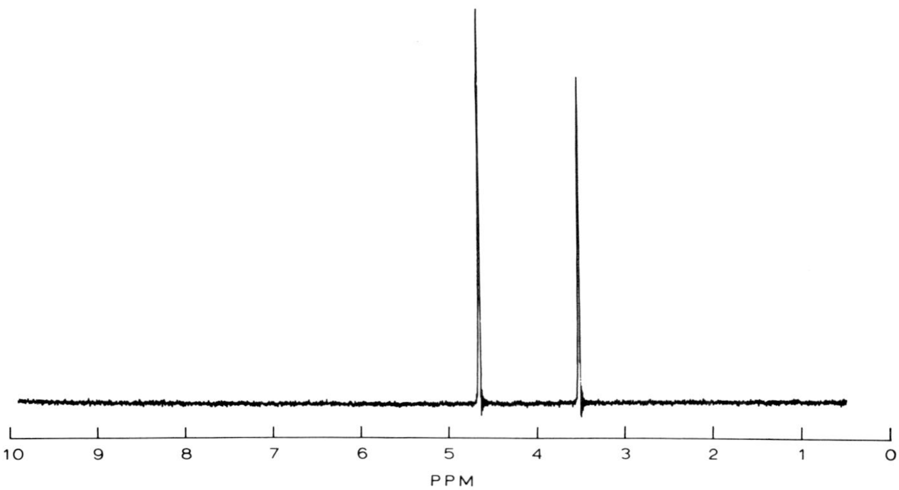

The 100 MHz ^{1}H NMR spectrum of moniliformin in D_2O solution, dioxane as reference.

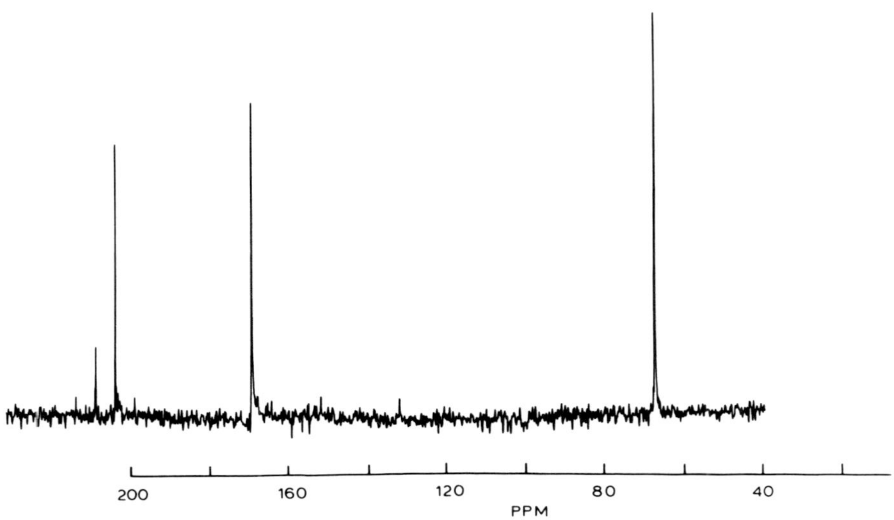

The 25 MHz ^{13}C NMR spectrum of moniliformin in D_2O solution.

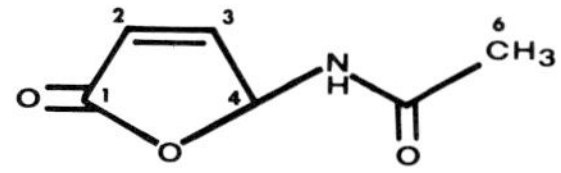

Common name	Butenolide (4-Acetamido-4-hydroxy-2-butenoic acid γ-lactone)
Molecular weight[30,34]	141.0425
Molecular formula	$C_6H_7O_3N$
General characteristics[34]	Crystals from ethyl acetate–cyclohexane, mp 116.5°–118.5°C Crystals from ethyl acetate, mp 115°C $[\alpha]_D$ 0° (C = 1.0 in chloroform, ethanol)
UV data[30]	λ_{max}^{MeOH} nm (ϵ): 202(11,300)
Fungal source[9,28,33]	*Fusarium equiseti* *F. tricinctum* (originally thought to be *F. nivale*) *F. graminearum*
Toxicity data[9,26,32]	Butenolide has been implicated as the causative agent in the gangrenous condition "fescue foot" of cattle that feed on molded tall fescue hay (*Festuca arundinaceae* Schreb.). Synthetic butenolide injected intramuscularly at 3.8 mg/kg into a heifer for 90 days produced dry gangrene at the end of the tail. This is one of the most characteristic signs of cattle grazing tall fescue grass exhibiting signs of "fescue foot." Postmortem examination revealed no other gross abnormalities other than local abscesses which developed at the injection sites.
Other significant references	Chemistry[8]; synthesis[7,30]; identification[34]; review[32]
^{13}C NMR data	C1, 171.0 s*; C2, 123.7 d; C3, 151.9 d; C4, 81.9 d; C5, 170.1 s*; C6, 22.8 q
^{1}H NMR data	H2, 6.15 (J = 5.5); H3, 7.34 (J = 5.5, 1.9); H4, 6.60 (J = 1.9, 9.0); H6, 1.97; NH, 8.47 (J = 9.0)

* Assignments may be reversed.

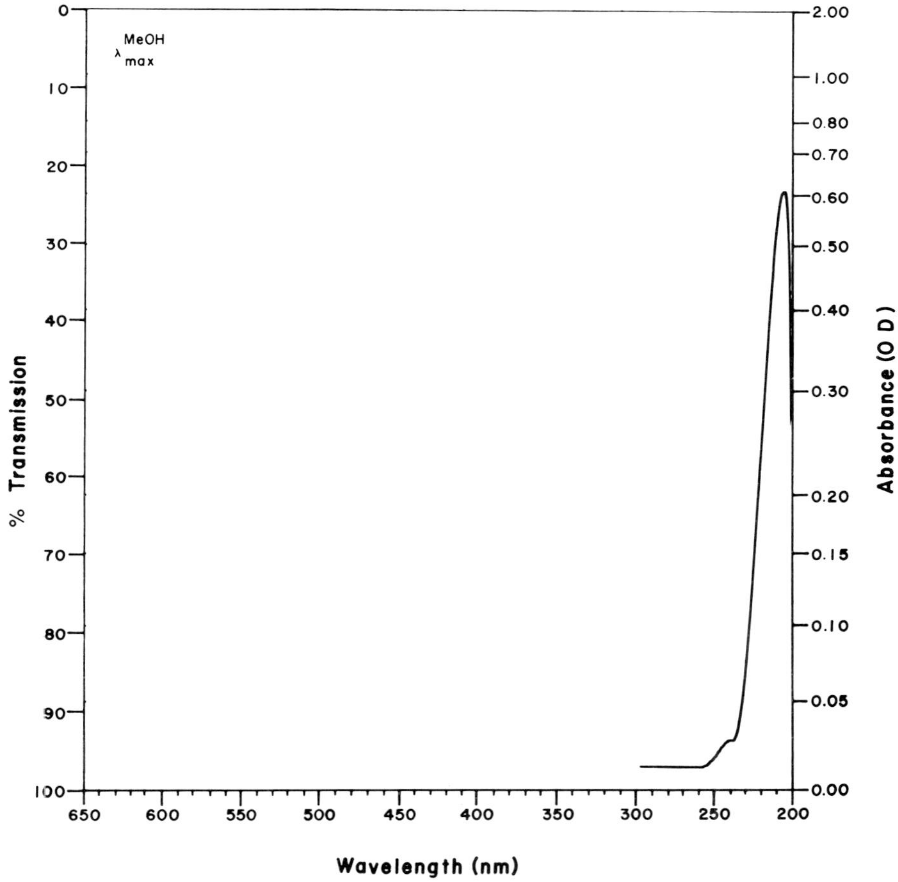

λ_{max}^{MeOH}
% Transmission
Absorbance (O D)
Wavelength (nm)
0
10
20
30
40
50
60
70
80
90
100
2.00
1.00
0.80
0.70
0.60
0.50
0.40
0.30
0.20
0.15
0.10
0.05
0.00
650
600
550
500
450
400
350
300
250
200

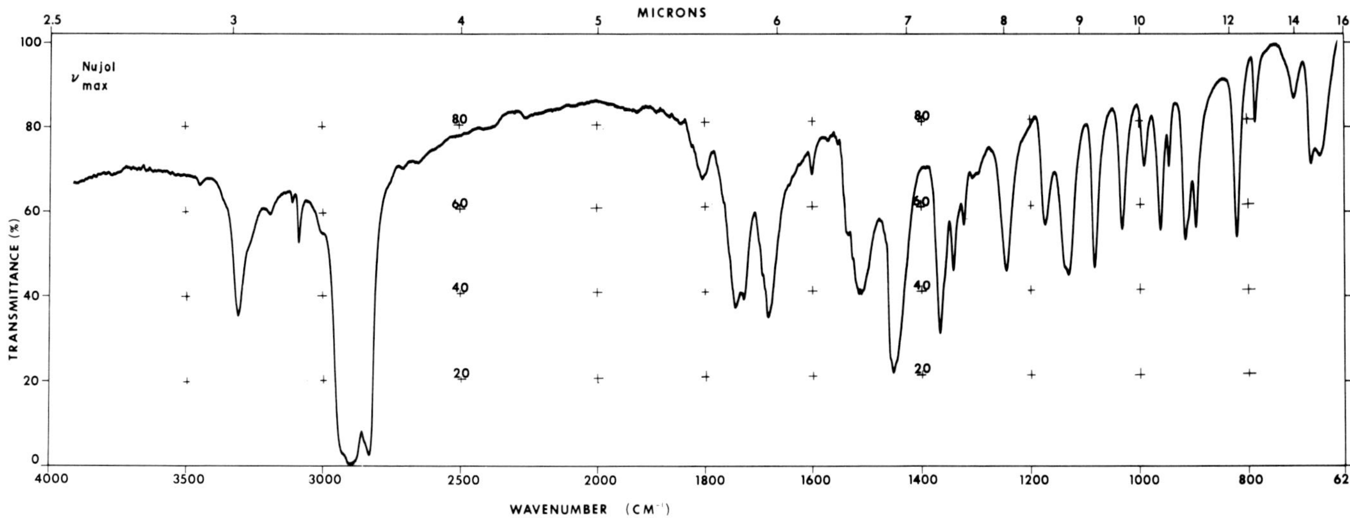
MICRONS
TRANSMITTANCE (%)
Nujol
ν max
WAVENUMBER (CM⁻¹)

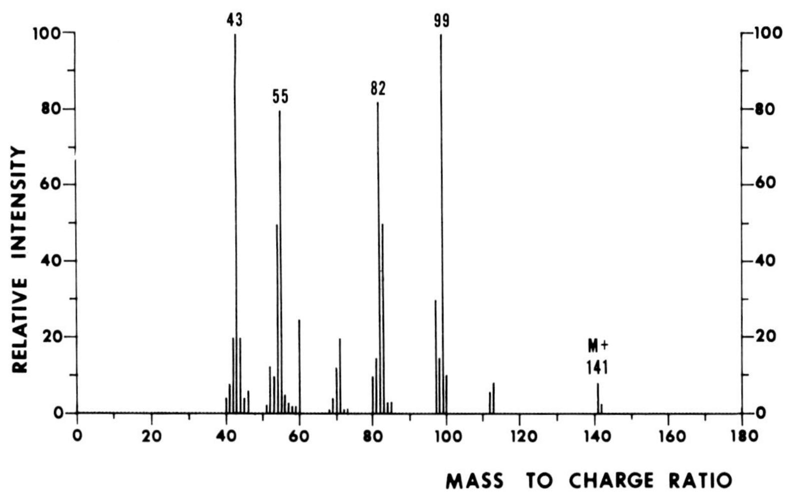
43
55
82
99
M+
141
RELATIVE INTENSITY
MASS TO CHARGE RATIO

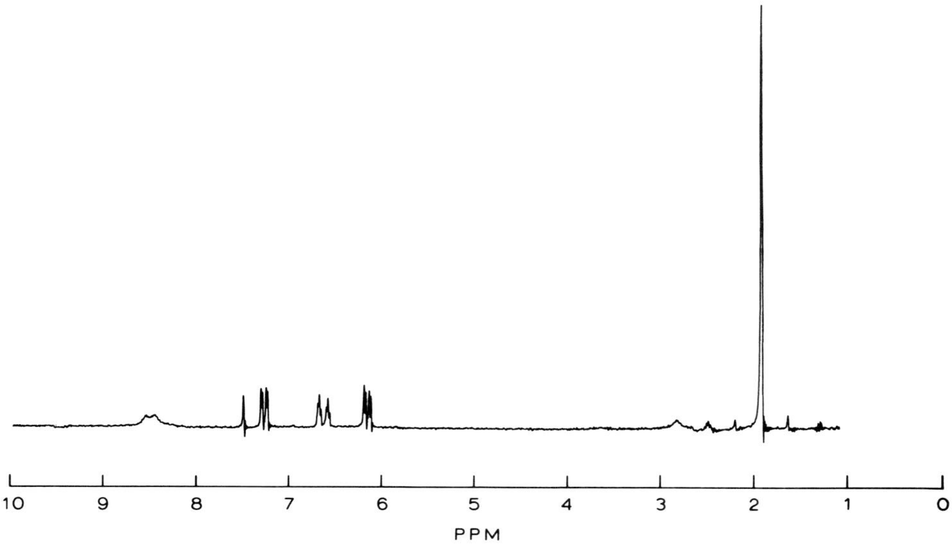

The 100 MHz ^1H NMR spectrum of butenolide in $CDCl_3$/DMSO solution.

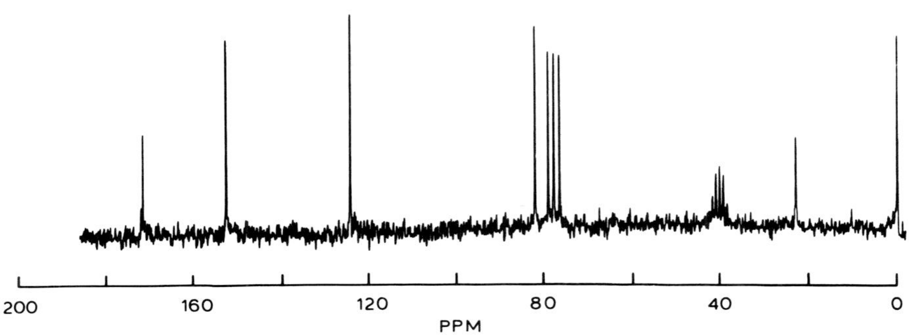

The 25 MHz ^{13}C NMR spectrum of butenolide in $CDCl_3$/DMSO solution.

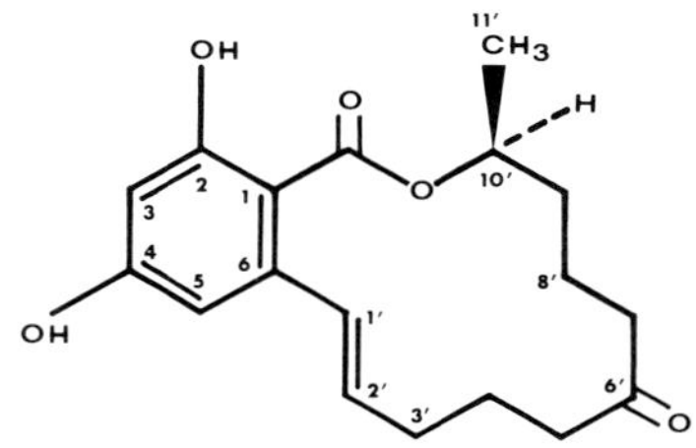

Common name[16]	(*S*)-Zearalenone (F-2)([6-(10-Hydroxy-6-oxo-*trans*-1-undecenyl)-β-resorcyclic acid lactone])
Molecular weight	318.1467
Molecular formula	$C_{18}H_{22}O_5$
General characteristics[27]	White crystals, mp 164°–165°C $[\alpha]^{25}_{546} - 170.5°$ (C = 1.0 in methanol) Soluble in aqueous alkali, ether, benzene, and alcohols; insoluble in water. Dimethyl ether, crystals from acetone–ether, mp 129°–131°C
UV data[27]	λ_{max}^{MeOH} nm (ϵ): 236(29,700), 274(13,909), and 316(6,020)
Fungal source[10,14]	*Fusarium roseum* (*F. graminearum, Gibberella zeae*) *F. tricinctum* *F. lateritium* *F. oxysporum* *F. culmorum* *F. moniliforme* *F. equiseti* *F. gibbosum* *F. avenaceum* *F. nivale* *F. sambucinum* var. *Coeruleum*
Toxicity data[16]	Zearalenone possesses estrogenic and anabolic activity in animals. In the prepuberal gilt, the vulva becomes swollen and edematous. In severe cases, it progresses to vaginal prolapse with the uterus enlarged, edematous, tortuous, and the ovaries shrunken. Pregnant swine abort. Young males may undergo atrophy of the testes and enlargement of the mammary glands.

TLC data[15]	*Adsorbent:* silica gel G *Solvent:* chloroform–ethyl alcohol, 97:3 v/v R_f: 0.50 *Detection:* blue-green fluorescent spot under short wave UV light
Other significant references	Isolation[5,23]; synthesis[24,29]; biosynthesis[20]; reviews[14,16]; structure[27]
^{13}C NMR data	C1, 142.2 d; C2, 164.0 s; C3, 101.1 d; C4, 161.4 s; C5, 107.7 d; C6, 127.0 s; C1′, 132.1 d; C2′, 130.6 d; C3′, 41.8 t; C4′, 21.1 t*; C5′, 33.6 t†; C6′, 209.6 s; C7′, 35.4 t†; C8′, 19.9 t*; C9′, 29.9 t; C10′, 71.9 d; C11′, 19.8 q; C12′, 170.1 s
^{1}H NMR data	H3, 6.24 (J = 2.0); H5, 6.33 (J = 2.0); H1′, 6.92 (J = 15.0); H2′, 4.97 (J = 15.0); H10′, 5.62 (J = 6.5); H11′, 1.23 (J = 6.5); 2OH, 11.92; 4OH, 9.66

* Assignments may be reversed.

† Assignments may be reversed.

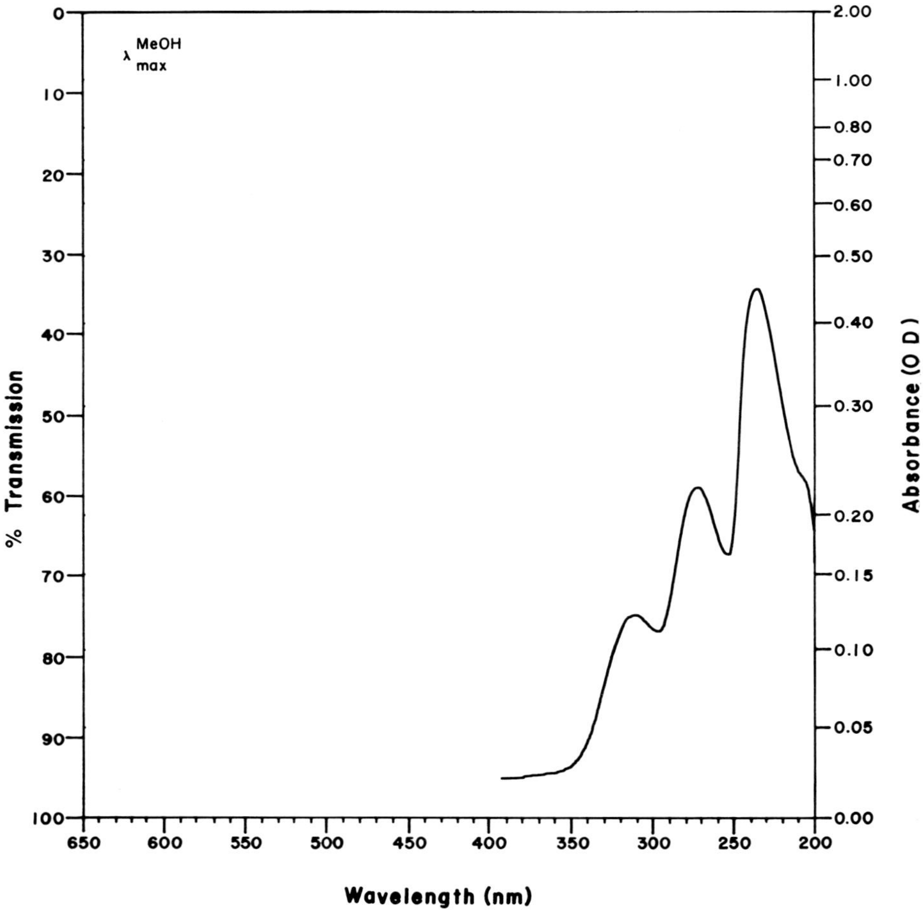

λ MeOH max
% Transmission
0
10
20
30
40
50
60
70
80
90
100
Absorbance (O D)
2.00
1.00
0.80
0.70
0.60
0.50
0.40
0.30
0.20
0.15
0.10
0.05
0.00
650
600
550
500
450
400
350
300
250
200
Wavelength (nm)

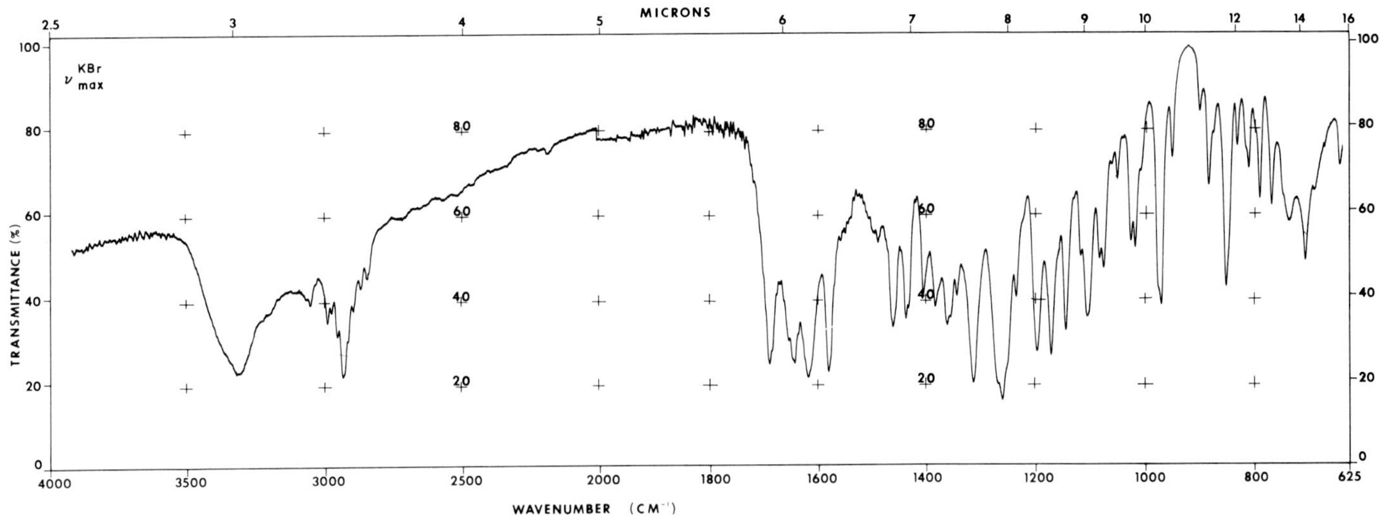
MICRONS
KBr
ν max
TRANSMITTANCE (%)
WAVENUMBER (CM⁻¹)

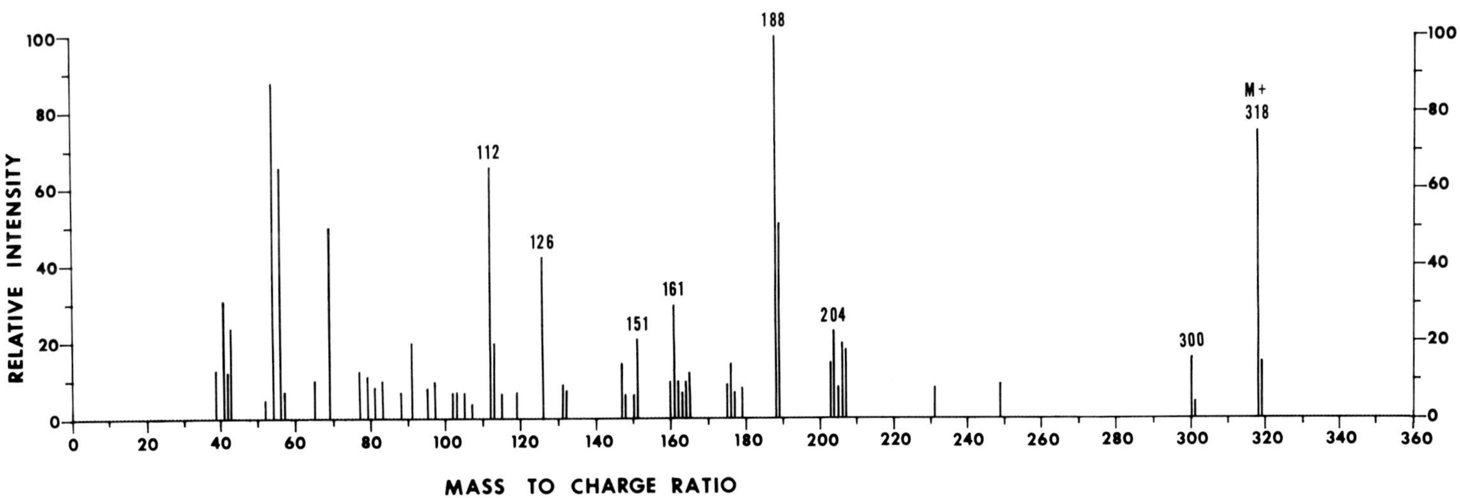
RELATIVE INTENSITY
MASS TO CHARGE RATIO
M+
318
300
204
188
161
151
126
112

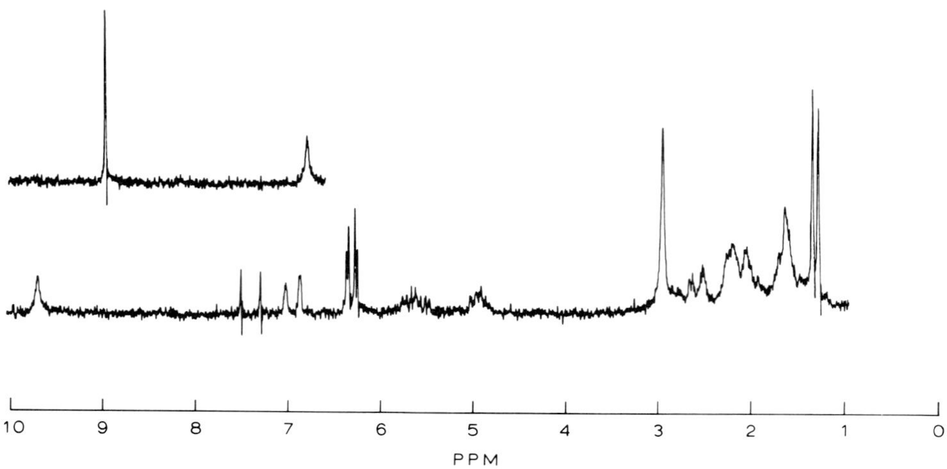

The 100 MHz ^1H NMR spectrum of zearalenone in $CDCl_3$/DMSO solution.

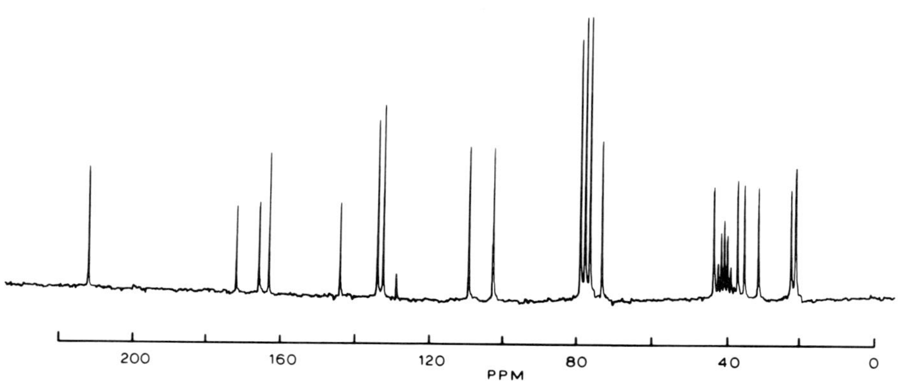

The 25 MHz ^{13}C NMR spectrum of zearalenone in $CDCl_3$/DMSO solution.

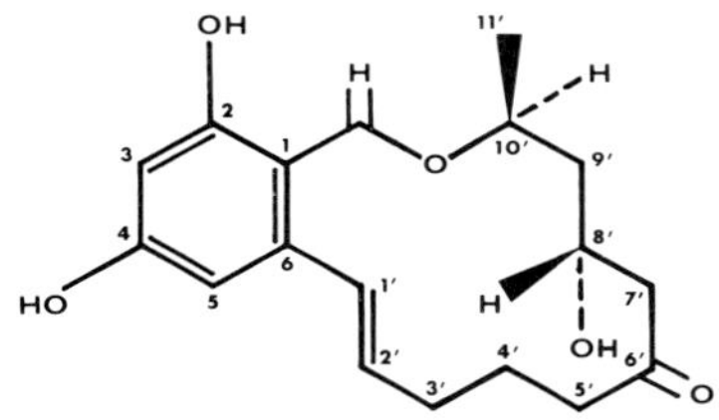

Common name[11]	8′-Hydroxyzearalenone (F-5-3 and F-5-4)
Molecular weight	334.1416
Molecular formula	$C_{18}H_{22}O_6$
General characteristics[3]	Crystals from chloroform–methanol, mp 210°–212°C $[\alpha]_D^{24}$ −149.1° (C = 1.0 in acetone) 8′-Epihydroxyzearalenone: crystals from chloroform–methanol, mp 172°–174°C $[\alpha]_D^{24}$ −53.1° (C = 1.0 in acetone)
UV data[3]	λ_{max}^{EtOH} nm (ϵ): 236(26,900), 274(12,500), and 314(6,000) 8′-Epihydroxyzearalenone: λ_{max} nm (ϵ): 237(28,100), 276(12,500), and 315(5,500)
IR data[3]	ν_{max}^{KBr} cm^{-1}: 3417, 2950, 2920, 1694, 1641, 1611, 1580, 1461, 1440, 1371, 1344, 1311, 1256, 1203, 1169, 1147, 1066, 1013, 959, 858, 820, and 709 8′-Epihydroxyzearalenone: ν_{max}^{KBr} cm^{-1}: 3432–3316, 2927, 1681, 1644, 1611, 1579, 1444, 1387, 1356, 1311, 1256, 1200, 1169, 1106, 1059, 1022, 977, and 849
Fungal source[3]	*Gibberella zeae* (*Fusarium graminearum*)
Toxicity data[3]	They possess estrogenic and anabolic activity similar to that of zearalenone.
^{1}H NMR data[3]	H3, 6.33 (J = 2.5); H5, 6.47 (J = 2.5); H1′, 7.06 (J = 16); H2′, 5.70 (J = 16); H7′, 2.6–3.2; H8′, 4.14; H10′, 5.34; H11′, 1.46; 2-OH, 12.10; 4-OH, 9.12; 8′-OH, 4.14

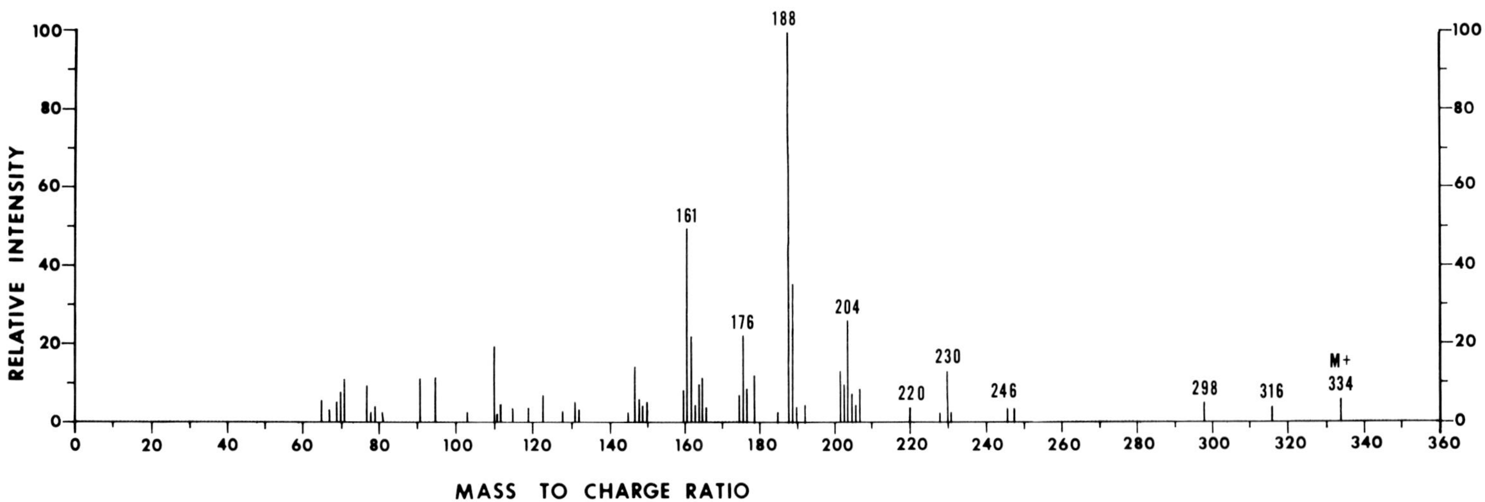

188
161
176
204
220
230
246
298
316
M+
334
RELATIVE INTENSITY
MASS TO CHARGE RATIO

REFERENCES

1. Bellus, D., and Fischer, H., *Adv. Pestic. Sci., Plen. Lect. Symp. Pop. Int. Congr. Pestic. Chem. 4th, 1978,* p. 373 (1979).
2. Bellus, D., Fischer, H., Greuter, H., and Martin, P., *Helv. Chim. Acta* **61,** 1784 (1978).
3. Bolliger, G., and Tamm, C., *Helv. Chim. Acta* **55,** 3030 (1972).
4. Burmeister, H. R., Ciegler, A., and Vesonder, R. F., *Appl. Environ. Microbiol.* **37,** 11 (1979).
5. Christensen, C. M., Nelson, G. H., and Mirocha, C. J., *Appl. Microbiol.* **13,** 653 (1965).
6. Cole, R. J., Kirksey, J. W., Cutler, H. G., Doupnik, B. L., and Peckham, J. C., *Science* **179,** 1324 (1973).
7. Gratz, J. A., Cook, C. E., and Wall, M. E., *Org. Prep. Proced.* **2,** 51 (1970).
8. Grove, M. D., and Weisleder, D., *J. Org. Chem.* **38,** 815 (1973).
9. Grove, M. D., Yates, S. G., Tallent, W. H., Ellis, J. J., Wolff, I. A., Kosuri, N. R., and Nichols, R. E., *J. Agric. Food Chem.* **18,** 734 (1974).
10. Hacking, A., Rosser, W. R., and Dervish, M. T., *Ann. Appl. Biol.* **84,** 7 (1976).
11. Jackson, R. A., Fenton, S. W., Mirocha, C. J., and Davis, G., *J. Agric. Food Chem.* **22,** 1015 (1974).
12. Kriek, N. P. J., Marasas, W. F. O., Steyn, P. S., van Rensburg, S. J., and Steyn, M., *Food Cosmet. Toxicol.* **15,** 579 (1977).
13. Lansden, J. A., Clarkson, R. J., Neely, W. C., Cole, R. J., and Kirksey, J. W., *J. Assoc. Off. Anal. Chem.* **57,** 1392 (1974).
14. Mirocha, C. J., Christensen, C. M., and Nelson, C. H., *in* "Microbial Toxins" (S. Kadis, A. Ciegler, and S. J. Ajl, eds.), Vol. 7, p. 107. Academic Press, New York, 1971.
15. Mirocha, C. J., Schauerhamer, B., and Pathre, S. V., *J. Assoc. Off. Anal. Chem.* **57,** 1104 (1974).
16. Pathre, S. V., and Mirocha, C. J., *Adv. Chem. Ser.* **149,** 178 (1976).
17. Rabie, C. J., Lubben, A., Louw, A. I., Rathbone, E. B., Steyn, P. S., and Vleggaar, R., *J. Agric. Food Chem.* **26,** 375 (1978).
18. Seitz, G., Schmiedel, R., and Mann, K., *Arch. Pharm. (Weinheim, Ger.)* **310,** 991 (1977).
19. Springer, J. P., Clardy, J., Cole, R. J., Kirksey, J. W., Hill, R. K., Carlson, R. M., and Isidor, J. L., *J. Am. Chem. Soc.* **96,** 2267 (1974).
20. Steele, J. A., Lieberman, J. R., and Mirocha, C. J., *Can. J. Microbiol.* **20,** 531 (1974).
21. Steyn, M., Thiel, P. G., and Van Scholkwyk, G. C., *J. Assoc. Off. Anal. Chem.* **61,** 578 (1978).
22. Steyn, P. S., personal communication.
23. Stobb, M., Baldwin, R. S., Tuite, J., Andrews, F. N., and Gillette, K. G., *Nature (London)* **196,** 1318 (1972).
24. Taub, D., Girotra, N. N., Hoffsommer, R. D., Kuo, C. H., Siates, H. L., Weber, S., and Wendler, N. K., *Tetrahedron* **24,** 2443 (1968).
25. Thiel, P. G., *Biochem. Pharmacol.* **27,** 483 (1978).
26. Tookey, H. L., Yates, S. G., Ellis, J. J., Grove, M. D., and Nichols, R. E., *J. Am. Vet. Med. Assoc.* **160,** 1522 (1972).
27. Urry, W. H., Wehrmeister, H. L., Hodge, E. B., and Hidy, P. H., *Tetrahedron Lett.* p. 3109 (1966).
28. Vesonder, R. F., Tjarks, L. W., Ciegler, A., Spencer, G. F., and Wallen, L. L., *Phytochemistry* **16,** 1296 (1977).
29. Vlattas, I., Harrison, I. T., Tokes, L., Fried, J. H., and Cross, A. D., *J. Org. Chem.* **33,** 4176 (1968).
30. White, E. P., *J. Chem. Soc.* p. 346 (1967).
31. White, E. P., personal communication.

32. Yates, S. G., *in* "Microbial Toxins" (S. Kadis, A. Ciegler, and S. J. Ajl, eds.), Vol. 7, p. 191. Academic Press, New York, 1971.
33. Yates, S. G., Tookey, H. L., and Ellis, J. J., *J. Am. Chem. Soc.* **152,** 81A (1966).
34. Yates, S. G., Tookey, H. L., Ellis, J. J., and Burkhardt, H. J., *Tetrahedron Lett.* p. 621 (1967).

21

Miscellaneous Toxins

The following three toxins were grouped under miscellaneous toxins because they show no chemical or common biogenic relationship. Slaframine causes excessive salivation in ruminants, produces prolonged stimulation of pancreatic secretory activity, and selectively stimulates the synthesis of digestive enzymes. Roseotoxin B, a cyclic pentapeptide, is chemically related to the insect toxins called destruxins which are produced by entomophagous fungi.

Miscellaneous toxins	Molecular weight	Molecular formula
Slaframine	198.1368	$C_{10}H_{18}O_2N_2$
Diplodiatoxin	308.1987	$C_{18}H_{28}O_4$
Roseotoxin B	591.3615	$C_{30}H_{49}O_7N_5$

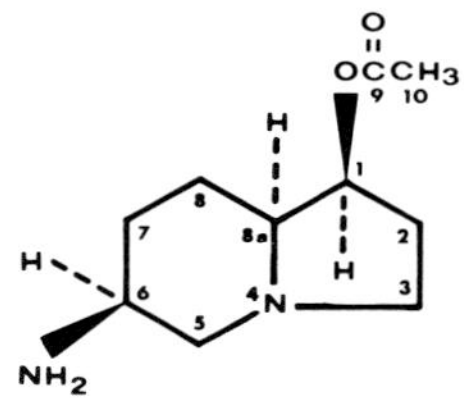

Common name	Slaframine [(1*S*,6*S*,8a*S*)-1-Acetoxy-6-aminooctahydroindolizine]
Molecular weight	198.1368
Molecular formula	$C_{10}H_{18}O_2N_2$
General characteristics[5]	The free base was noncrystalline; dipicrate, mp 183°–184°C; *N*-acetylslaframine; white crystals from hexane–ethanol, mp 143°–146°C; sublimed sample at 100°C produced white needles, mp 143°–146°C. $[\alpha]_D$ 0.0° (*N*-acetylslaframine)
UV data	No significant UV absorption by the free base.
Fungal source[3,12]	*Rhizoctonia leguminicola*
Toxicity data[1–4]	Slaframine causes excessive salivation or slobbering in ruminants feeding on certain fungus-infected legume forage crops. Slaframine has been suggested as a possible research tool for locating acetylcholine receptor sites and as an agent relieving the symptoms of cystic fibrosis. Slaframine produces prolonged stimulation of pancreatic secretory activity and selectively stimulates the synthesis of digestive enzymes. Slaframine must first be activated by the liver, the probable basis for the prolonged production of the active metabolite.
TLC data[3,10]	*Adsorbent:* silica gel *Solvent:* ethyl ether–ethanol, 8:1 v/v R_f: 0.41 (*N*-acetylslaframine) *Detection:* Dragendorff's reagent Paper Chromatography: *Solvent:* butanol–acetic acid–water, 4:1:1 v/v/v R_f: 0.25

	Detection: stains pink following spraying with Dragendorff's reagent
Other significant references	Structure elucidation[5,9,20]; synthesis[6,10]; biosynthesis[11,17]; discovery[7,16]
^{13}C *chemical shifts* (slaframine dipicrate)	18.1, 19.2, 20.6, 23.5, 24.3, 29.2, 44.0, 52.3, 60.5, 66.5, 71.8, 73.4, 168.9

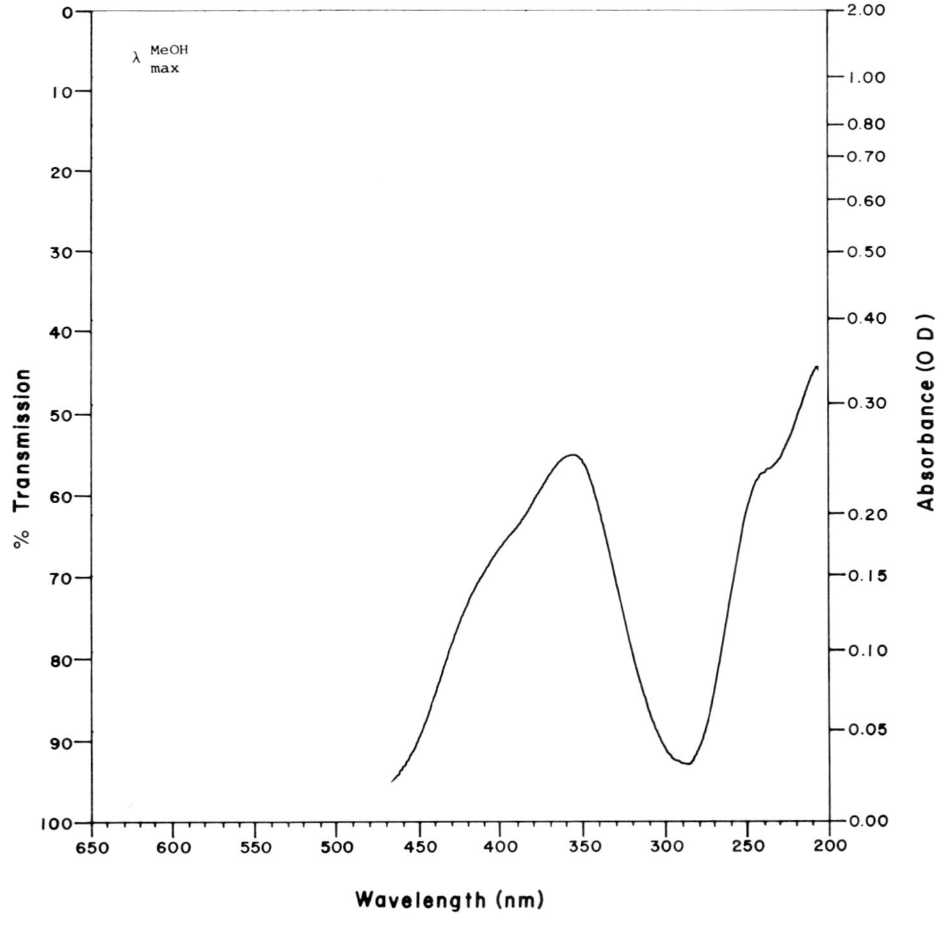

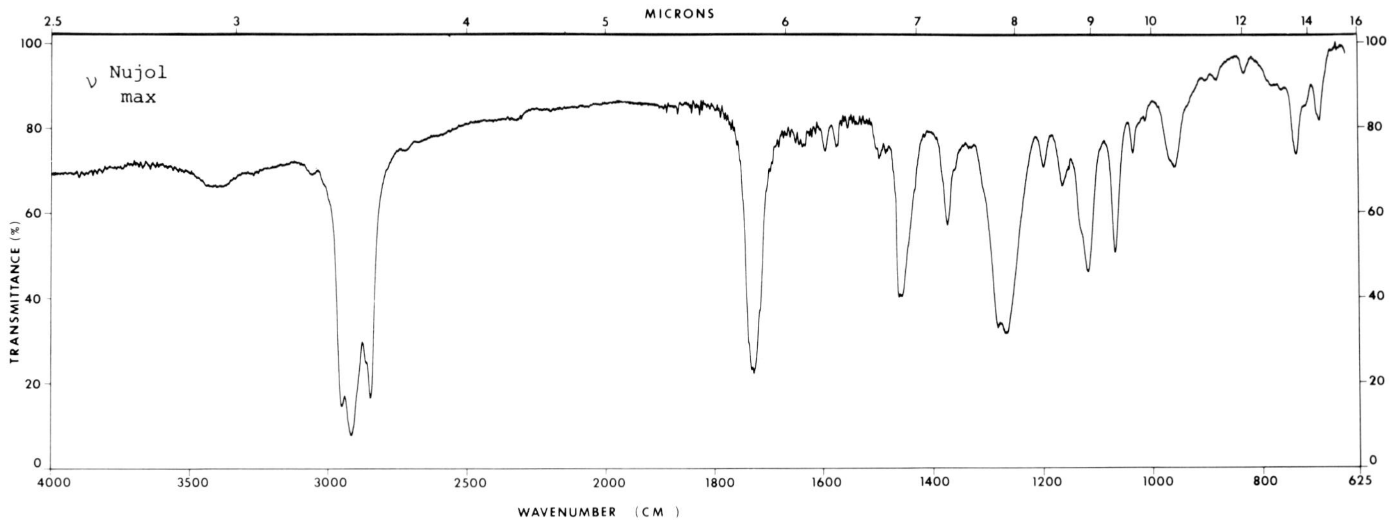
MICRONS
2.5
3
4
5
6
7
8
9
10
12
14
16
100
80
60
40
20
0
TRANSMITTANCE (%)
ν Nujol max
4000
3500
3000
2500
2000
1800
1600
1400
1200
1000
800
625
WAVENUMBER (CM)

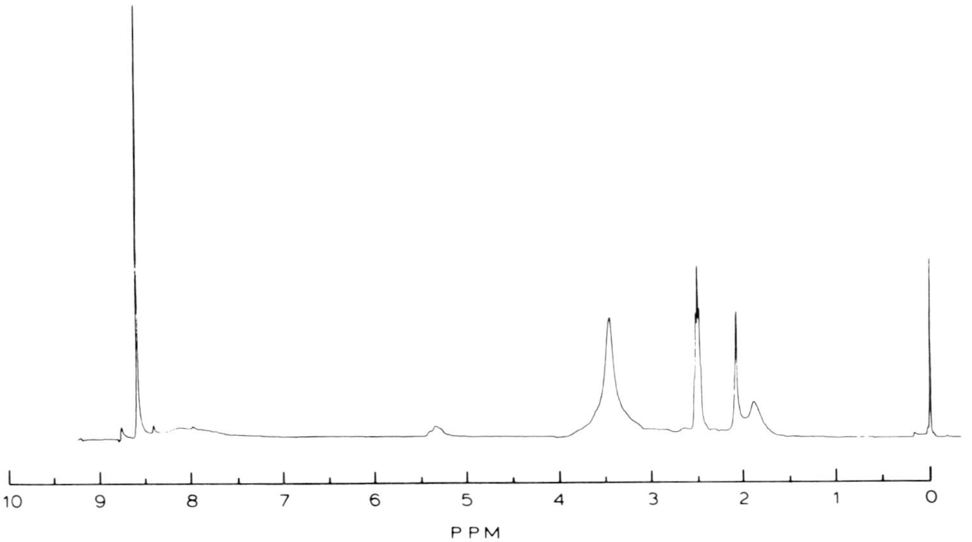

The 100 MHz ^{1}H NMR spectrum of slaframine dipicrate in DMSO solution.

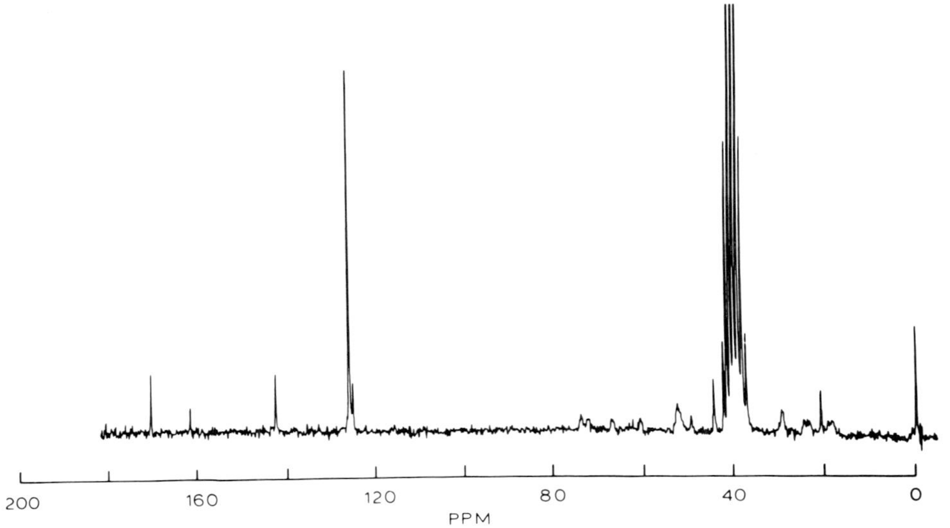

The 25 MHz ^{13}C NMR spectrum of slaframine dipicrate in DMSO solution.

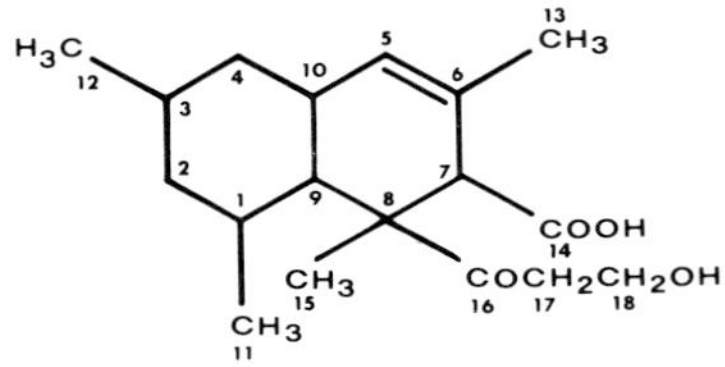

Common name	Diplodiatoxin
Molecular weight	308.1987
Molecular formula	$C_{18}H_{28}O_4$
General characteristics[19]	Crystals from chloroform solution, mp 186°–187°C $[\alpha]_D^{28}$ +101° (C = 4.0 in chloroform)
UV data[19]	λ_{max}^{EtOH} nm (ϵ): 293(50)
CD data[19]	$\Delta\epsilon$ (C = 10.0 in methanol) nm: 315(0), 293(−0.43), 255(0), 224(+11.60), and 205(0)
IR data[19]	$\nu_{max}^{CHCl_3}$ cm^{-1}: 3530, 2960, 2800–2500(broad), 1720, 1700, 1465, 1392, 1078, and 950
Fungal source[19]	*Diplodia maydis* (Berk.) Sacc.
Toxicity data[19]	Details of toxicity not yet reported.
TLC data[19]	*Adsorbent:* Merck precoated silica gel plates *Solvent:* benzene–acetic acid, 4:1 v/v R_f: 0.50 *Detection:* 1% $Ce(SO_4)_2$ in 6N H_2SO_4
^{13}*C NMR data* (diplodiatoxin methyl ester)[16]	C1, 36.2 d; C2, 46.1 t; C3, 44.8 d; C4, 41.9 t; C5, 129.4 d; C6, 126.7 s; C7, 58.4 d; C8, 52.9 s; C9, 33.4 d; C10, 40.4 d; C11, 23.1 q; C12, 22.2 q; C13, 22.6 q; C14, 173.0 s; C15, 16.9 q; C16, 216.1 s; C17, 41.9 t; C18, 58.1 t; OCH_3, 52.1 q
^{1}*H NMR data* (diplodiatoxin methyl ester)[16]	H5, 5.73, 5.76*; H7, 2.68, 2.70*; H9, 2.04 (J = 9.0), 2.09 (J = 9.0)*; H11, 0.55 (J = 6.1), 0.59 (J = 5.9)*; H12, 0.88 (J = 5.7), 0.87 (J = 5.8)*; H13, 1.59, 1.64*; H15, 1.24, 1.24*; H17, 2.86 (J = 5.2), 2.90*; H18, 3.82 (J = 5.2), 3.88*; OCH_3, 3.58

* Data for diplodiatoxin.

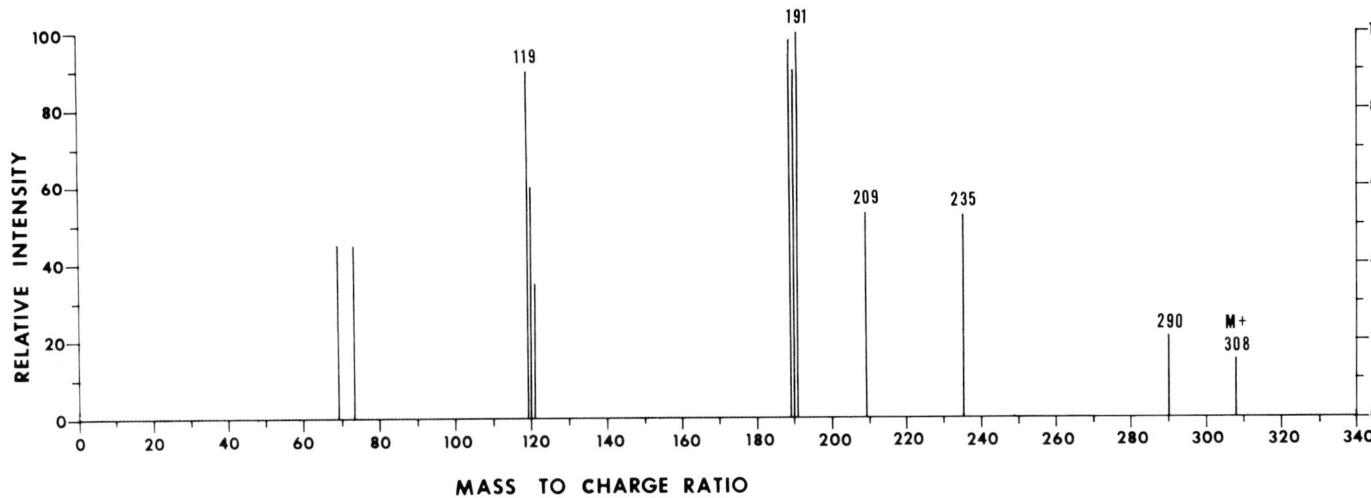
RELATIVE INTENSITY
MASS TO CHARGE RATIO
119
191
209
235
290
M+
308
0
20
40
60
80
100
120
140
160
180
200
220
240
260
280
300
320
340

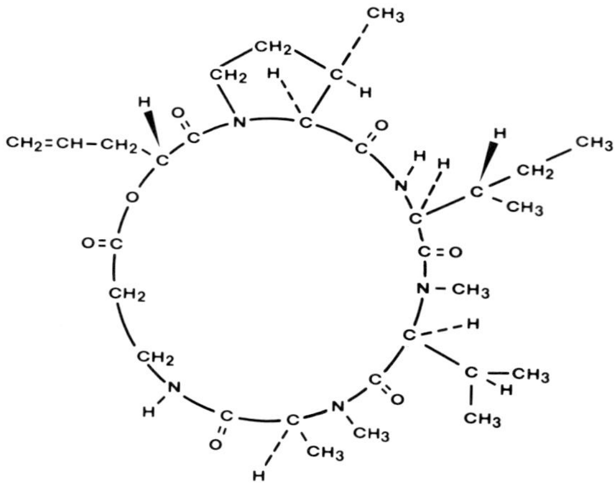

Common name	Roseotoxin B
Molecular weight	591.3615
Molecular formula	$C_{30}H_{49}O_7N_5$
General characteristics[8,18]	Crystals from benzene–hexane, mp 200°–202°C $[\alpha]_D^{25}$ −235° (ethanol) Roseotoxin B was soluble in petroleum ether-chloroform (95:5), carbon tetrachloride, chloroform, ethyl ether, ethyl acetate, and ethanol; sparingly soluble in warm water, and insoluble in petroleum ether.
UV data[8]	λ_{max}^{EtOH} nm: 210 (ϵ_{max} not reported)
Fungal source[13–15]	*Trichothecium roseum* Link
Toxicity data[14,18]	Typical signs in mice given a toxic dose of roseotoxin B were lethargy, tremors, hypermotivity, dyspnea, incoordination, and clonic convulsions. An LD_{100} was reported for a dosage of 166 mg/kg (IP) in mice. The LD_{50} in day-old chickens dosed orally was 12.5 mg/kg.
TLC data[18]	*Adsorbent:* silica gel G-HR *Solvent:* toluene–ethyl acetate–formic acid, 5:4:1 v/v/v R_f: 0.48 *Detection:* iodine vapors
Other significant references	Isolation[13]; chemistry[8]

^{13}C chemical shifts	11.3 q; 15.2 q; 15.3 q; 18.8 q; 19.6 q; 19.9 q; 24.5 t; 27.2 d; 28.0 q; 30.7 q; 30.7 t; 33.2 t; 34.4 t; 35.1 t; 36.1 d; 37.4 d; 45.0 t; 53.3 d; 55.4 d; 57.9 d; 67.1 d; 72.4 d; 119.6 t; 130.6 d; 168.7 s; 169.3 s; 170.5 s; 170.7 s; 173.1 s; 173.3 s
*Probable assignment**	δ-CH_3 (Ile); β-CH_3 (N-MeAla); γ-CH_3 (Ile); γ-CH_3 (N-MeVal); γ-CH_3 (N-MeVal); β-CH_3 (3-MePro); β-CH_2 (HO-Acid); β-CH (N-MeVal); N-CH_3 (N-MeVal); N-CH_3 (N-Me Val); γ-CH_2 (Ile); γ-CH_2 (3-MePro); α-CH_2 (β-Ala); β-CH_2 (β-Ala); β-CH(3-MePro); β-CH (Ile); δ-CH (3-MePro); α-CH (N-MeVal); α-CH (N-MeAla); α-CH (Ile); α-CH (3-MePro); α-CH (HO-Acid); δ-CH_2 (Ho-Acid); γ-CH (HO-Acid); C = O (HO-Acid); C = O (3-MePro); C = O (β-Ala); C = O (N-MeAla); C = O (Ile); C = O (N-MeVal).

* L-isoleucine (Ile), *N*-methylvaline (*N*-MeVal), *N*-methylalanine (N-MeAla), *trans*-3-methylproline (3-MePro), β-alanine (Ala), 2-OH-pentenoic acid (HO-Acid). From Engström *et al.*[8]

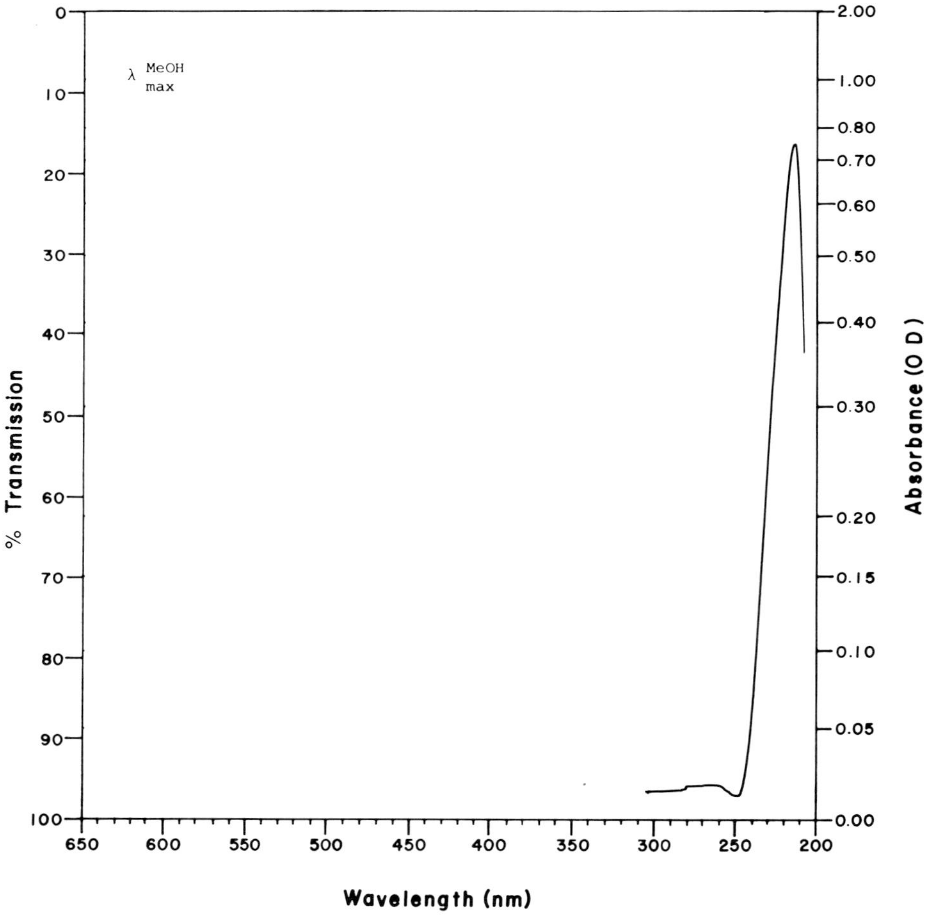
λ MeOH max
% Transmission
Absorbance (O D)
Wavelength (nm)
0
10
20
30
40
50
60
70
80
90
100
2.00
1.00
0.80
0.70
0.60
0.50
0.40
0.30
0.20
0.15
0.10
0.05
0.00
650 600 550 500 450 400 350 300 250 200

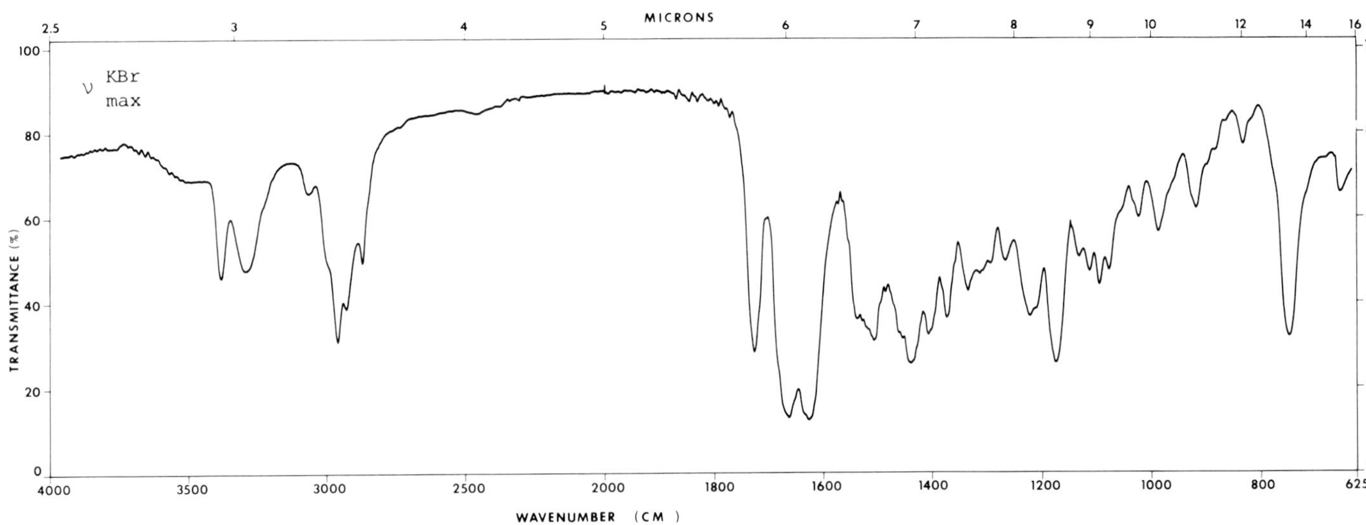
MICRONS
2.5
3
4
5
6
7
8
9
10
12
14
16
ν KBr max
TRANSMITTANCE (%)
100
80
60
40
20
0
4000
3500
3000
2500
2000
1800
1600
1400
1200
1000
800
625
WAVENUMBER (CM)

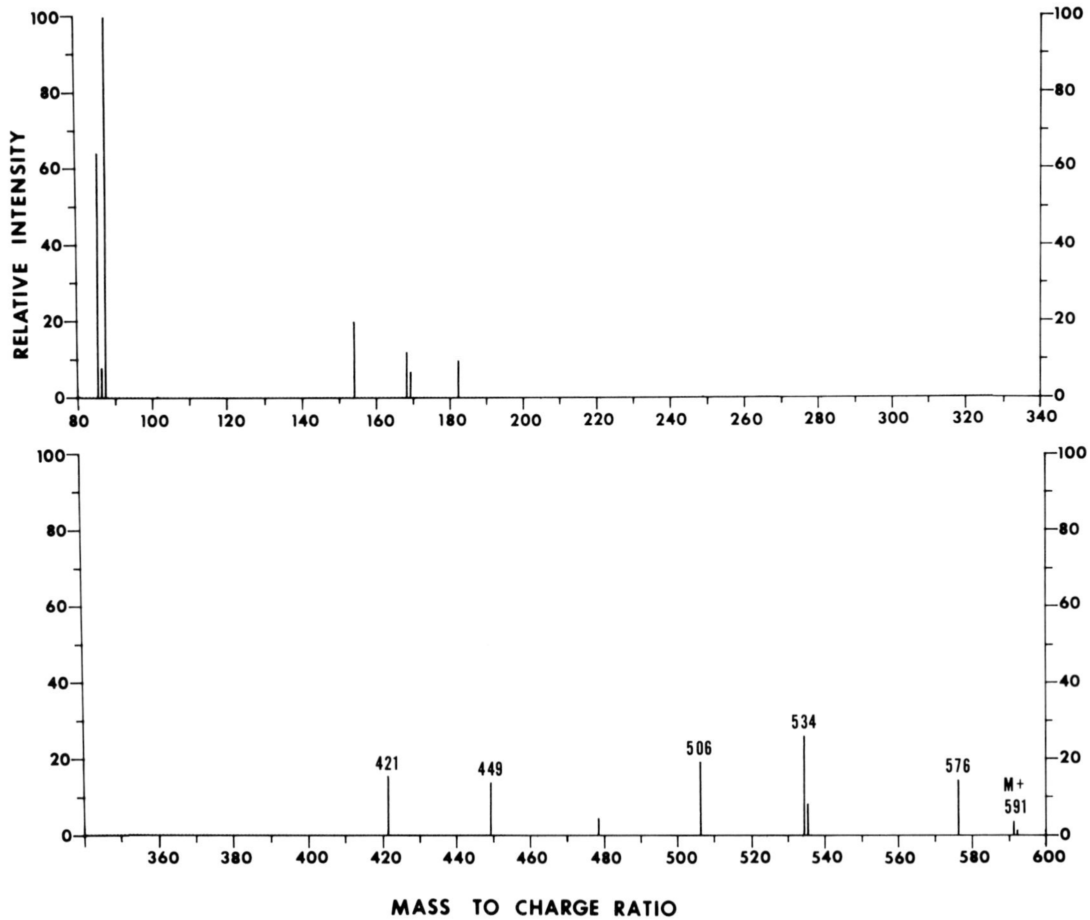
RELATIVE INTENSITY
100
80
60
40
20
0
80
100
120
140
160
180
200
220
240
260
280
300
320
340
421
449
506
534
576
M+
591
360
380
400
420
440
460
480
500
520
540
560
580
600
MASS TO CHARGE RATIO

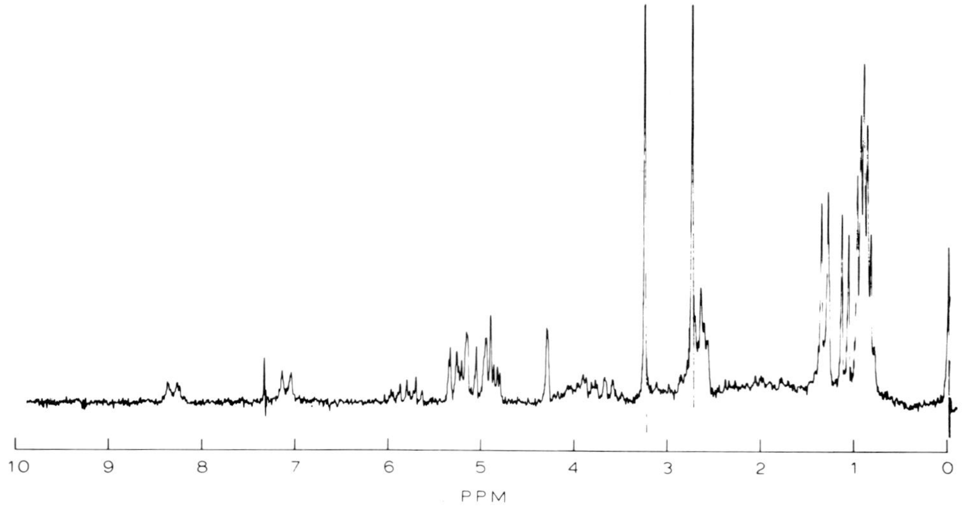

The 100 MHz ^{1}H NMR spectrum of roseotoxin B in $CDCl_3$ solution.

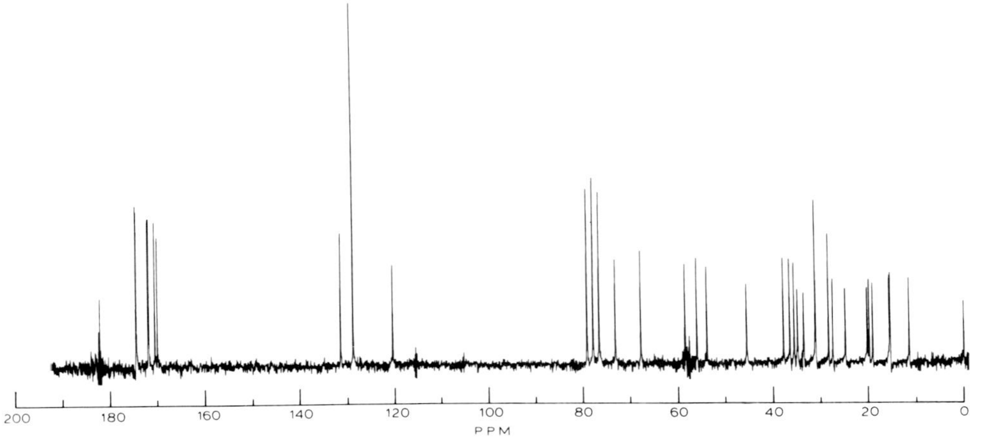

The 25 MHz ^{13}C NMR spectrum of roseotoxin B in $CDCl_3$ solution.

REFERENCES

1. Aust, S. D., *Biochem. Pharmacol.* **18,** 929 (1969).
2. Aust, S. D., *Biochem. Pharmacol.* **19,** 427 (1970).
3. Aust, S. D., and Broquist, H. P., *Nature* (*London*) **205,** 204 (1965).
4. Aust, S. D., Broquist, H. P., and Rinehart, K. L., Jr., *Biotechnol. Bioeng.* **10,** 403 (1968).
5. Aust, S. D., Broquist, H. P., and Rinehart, K. L., Jr., *J. Am. Chem. Soc.* **88,** 2879 (1966).
6. Cartwright, D., Gardiner, R. A., and Rinehart, K. L., *J. Am. Chem. Soc.* **92,** 7615 (1970).
7. Crump, M. H., Smalley, E. B., Henning, J. N., and Nichols, R. E., *J. Am. Vet. Med. Assoc.* **143,** 996 (1963).
8. Engström, G. W., Delance, J. V., Richard, J. L., and Baetz, A. L., *J. Agric. Food Chem.* **23,** 244 (1975).
9. Gardiner, R. A., Rinehart, K. L., Jr., Snyder, J. J., and Broquist, H. P., *J. Am. Chem. Soc.* **90,** 5639 (1968).
10. Gensler, W. J., and Hu, M. W., *J. Org. Chem.* **38,** 3848 (1973).
11. Guengerich, F. P., and Broquist, H. P., *Biochemistry* **12,** 4270 (1973).
12. Rainey, D. P., Smalley, E. B., Crump, M. H., and Strong, F. M., *Nature* (*London*) **205,** 203 (1965).
13. Richard, J. L., Engström, G. W., Pier, A. C., and Tiffany, L. H., *Mycopathol. Mycol. Appl.* **39,** 231 (1969).
14. Richard, J. L., Pier, A. C., and Tiffany, L. H., *Mycopathol. Mycol. Appl.* **40,** 161 (1970).
15. Richard, J. L., Tiffany, L. H., and Pier, A. C., *Mycopathol. Mycol. Appl.* **38,** 313 (1969).
16. Smalley, E. B., Nichols, R. E., Crump, M. H., and Henning, J. N., *Phytopathology* **52,** 753 (1962).
17. Snyder, J. J., Doctoral Thesis, University of Illinois, Urbana (1969).
18. Springer, J. P., Cole, R. J., Dorner, J. W., Cox, R. H., and Richard, J. L., *J. Am. Chem. Soc.* (in preparation).
19. Steyn, P. S., Wessels, P. L., Holzapfel, C. W., Potgieter, D. J. J., and Louw, W. K. A., *Tetrahedron* **28,** 4775 (1972).
20. Whitlock, B. J., Rainey, D. P., Riggs, N. V., and Strong, I. M., *Tetrahedron Lett.* p. 3819 (1966).

Molecular Formula Index

Molecular Weight Index

Compound Index

Microorganism/Plant Source Index